Arid Zone Geomorphology

Arid Zone Geomorphology

Process, Form and Change in Drylands

Third Edition

Edited by

David S. G. Thomas

School of Geography and the Environment
University of Oxford, UK

A John Wiley & Sons, Ltd., Publication

Wiley-Blackwell is an imprint of John Wiley & Sons, formed by the merger of Wiley's global Scientific, Technical and Medical business with Blackwell Publishing.

Registered office: John Wiley & Sons Ltd, The Atrium, Southern Gate, Chichester, West Sussex, PO19 8SQ, UK

Editorial offices: 9600 Garsington Road, Oxford, OX4 2DQ, UK
The Atrium, Southern Gate, Chichester, West Sussex, PO19 8SQ, UK
111 River Street, Hoboken, NJ 07030-5774, USA

For details of our global editorial offices, for customer services and for information about how to apply for permission to reuse the copyright material in this book please see our website at www.wiley.com/wiley-blackwell.

Library of Congress Cataloguing-in-Publication Data

Arid zone geomorphology : process, form and change in drylands / edited by David S. G. Thomas. – 3rd ed.
p. cm.
Includes index.
ISBN 978-0-470-51908-0 (cloth) – ISBN 978-0-470-51909-7 (pbk.)
1. Geomorphology. 2. Arid regions. I. Thomas, David S. G.
GB611.A75 2011
551.41′5–dc22

2010037270

A catalogue record for this book is available from the British Library.

This book is published in the following electronic formats: ePDF 9780470710760; Wiley Online Library 9780470710777; ePub 9780470975695

Set in 10/12pt Times by Aptara Inc., New Delhi, India.

First Impression 2011

For Alice

And in memory of my father
Frederick Thomas

The Geographer who first inspired me
and who passed away when this book was being completed

Contents

List of contributors xvii
Preface to the first edition xix
Preface to the second edition xxi
Preface to the third edition xxiii

I Large-scale controls and variability in drylands **1**

1 Arid environments: their nature and extent 3
David S.G. Thomas

1.1 Geomorphology in arid environments 3
1.2 Arid zone distinctiveness and the quest for explanation 4
1.3 Arid zones: terminology and definitions 5
1.3.1 Terminology 5
1.3.2 Definition 5
1.4 The age of aridity on Earth 7
1.5 The distribution of arid zones 8
1.6 Causes of aridity 9
1.6.1 Atmospheric stability 9
1.6.2 Continentality 9
1.6.3 Topography 9
1.6.4 Cold ocean currents 10
1.7 Climate variability 10
1.8 Dryland ecosystems 10
1.8.1 Arid zone geomorphology 12
1.9 Arid zone geomorphology and people 12
1.10 Organisation of this book 13
References 14

2 Tectonic frameworks 17
Helen Rendell

2.1 Introduction 17
2.2 Tectonic setting of drylands 18
2.3 Uplift and erosion, subsidence and sedimentation 18
2.4 Lengths of record 20
2.5 Existing erosional and depositional records in arid environments 21
2.5.1 Drainage patterns and fluvial systems 21
2.5.2 Playas 22
2.5.3 Desert pavements 22
2.5.4 Aeolian sequences 23

2.6 Selected examples of the geomorphological impact of active tectonics in arid environments 23
2.6.1 Tectonic disruption of fluvial systems 23
2.6.2 Tectonic controls on alluvial sedimentation 23
2.7 Conclusions 24
References 24

3 Climatic frameworks: legacies from the past 27
David S.G. Thomas and Sallie L. Burrough
3.1 Introduction 27
3.2 The significance of arid zone fluctuations in the past 27
3.2.1 Ancient arid zones 27
3.2.2 The development of aridity in the Mesozoic and Cenozoic 28
3.2.3 The Quaternary Period 29
3.2.4 Sedimentary records 29
3.2.5 Marine sediments and palaeoaridity 32
3.2.6 Rock varnish 34
3.2.7 Geomorphological evidence of arid zone change 35
3.2.8 Arid zone contraction 36
3.2.9 Ecological evidence 38
3.3 Dating arid zone fluctuations 39
3.4 Climatic interpretations and issues 39
3.4.1 Aridity during glacial times? 41
3.4.2 Drivers of late glacial tropical aridity 42
3.5 Conclusions 43
References 44

4 Dryland system variability 53
David S. G. Thomas
4.1 A framework for dryland diversity 53
4.2 Geomonotony: how unvarying are the 'flat' drylands of the world? 55
4.3 Within-dryland diversity 57
4.4 Summary issues 58
References 59

5 Extraterrestrial arid surface processes 61
Jonathan Clarke
5.1 Introduction 61
5.2 What does 'aridity' mean beyond Earth? 61
5.3 Why should planetary scientists understand terrestrial arid geomorphology? 62
5.4 What can terrestrial geomorphologists learn from a solar system perspective? 63
5.5 Mars: water-based aridity 64
5.5.1 Overview 64
5.5.2 The history of atmosphere–surface interactions 64
5.5.3 Martian water cycle 66
5.5.4 Surface images 66
5.5.5 The geomorphology of Mars 68
5.5.6 Summary 72
5.6 Titan: methane-based aridity? 72
5.6.1 Methane cycle 73
5.6.2 Surface images 74
5.6.3 Lakes 74
5.6.4 Rock breakdown: process and form 75

5.6.5 Aeolian landforms 75
5.6.6 Fluvial systems 75
5.6.7 Summary 76
5.7 Venus: extreme aridity 76
5.7.1 Surface–atmosphere interaction 76
5.7.2 Surface images 76
5.7.3 Rock breakdown 77
5.7.4 Aeolian landforms 78
5.7.5 Summary 79
5.8 Future Directions 79
References 79

II Surface processes and characteristics 83

6 Weathering systems 85
Heather A. Viles
6.1 Introduction 85
6.2 What makes arid environments unusual in terms of weathering systems? 87
6.3 Theoretical underpinnings of weathering systems research 88
6.4 Current weathering study methods 90
6.5 Linking processes to form in arid weathering systems 92
6.6 Explaining the development of weathering landforms in arid environments 95
6.7 Weathering rates in arid environments 97
6.8 Arid weathering and landscape evolution 97
6.9 Scale and arid weathering systems 98
Acknowledgement 98
References 98
7 Desert soils 101
David L. Dunkerley
7.1 Introduction: the nature and significance of desert soils 101
7.2 Taxonomy of desert soils 103
7.2.1 A note on terminology of near-surface features in desert soils 104
7.3 Some distinctive aspects of desert soil development 104
7.4 Stone-mantled surfaces and desert pavements 105
7.5 Inorganic seals at the soil surface 106
7.5.1 Raindrop properties and raindrop impact seals 106
7.5.2 Factors known to be significant in the formation of raindrop impact seals 109
7.5.3 Depositional seals 109
7.5.4 Effects of seals on infiltration and erosion 110
7.5.5 Biological soil crusts 111
7.5.6 The habitats or niches exploited by microphytic plants in drylands 112
7.5.7 The organisms forming biological soil crusts 113
7.5.8 The classification of biological soil crusts 114
7.5.9 Effects of biological soil crusts on infiltration and overland flow 115
7.5.10 Effects of biological crusts on soil stability and erosion resistance 119
7.5.11 Possible effects of climate change on biological soil crusts 121
7.6 Vesicular soil structures 121
7.6.1 Comparing the infiltrability of biological, raindrop impact and vesicular surfaces 122
7.6.2 Spatial heterogeneity of desert soils 123
7.7 Conclusions 125
References 125

8 Desert crusts and rock coatings 131
David J. Nash

8.1 Introduction 131
8.2 Sodium nitrate deposits 132
8.2.1 General characteristics and distribution 132
8.2.2 Micromorphology, chemistry and mode of formation 133
8.3 Halite crusts 135
8.3.1 General characteristics and distribution 135
8.3.2 Micromorphology and chemistry 136
8.3.3 Mode of formation 137
8.4 Gypsum crusts 137
8.4.1 General characteristics 137
8.4.2 Distribution 139
8.4.3 Micromorphology and chemistry 140
8.4.4 Modes of formation 140
8.5 Calcrete 141
8.5.1 General characteristics 141
8.5.2 Distribution 143
8.5.3 Micromorphology and chemistry 146
8.5.4 Mode of origin 148
8.6 Silcrete 151
8.6.1 General characteristics 151
8.6.2 Distribution 152
8.6.3 Micromorphology and chemistry 153
8.6.4 Mode of formation 156
8.7 Desert rock coatings 158
8.7.1 General controls on formation 158
8.7.2 Rock varnish 159
8.7.3 Silica glazes and iron films 162
8.8 Palaeoenvironmental significance of crusts 163
References 165

9 Pavements and stone mantles 181
Julie E. Laity

9.1 Introduction 181
9.2 Surface types: hamadas and stony surfaces 181
9.2.1 Hamada 181
9.2.2 Stony surfaces: gobi, serir, gibber plains and desert pavements 182
9.3 General theories concerning stony surface formation 185
9.3.1 Deflation 186
9.3.2 Concentration by surface wash and rain splash 186
9.3.3 Upward migration of stones 187
9.3.4 Accretion of aeolian fines 188
9.3.5 Desert pavement formation by aeolian aggradation and development of an accretionary mantle 188
9.4 Stone pavement characteristics 189
9.4.1 Setting 189
9.4.2 Surface clast concentration and characteristics 189
9.5 Processes of pavement formation 190
9.6 Processes of clast size reduction in pavements 192
9.6.1 Pavement soils 192

9.7 Secondary characteristics of pavement surfaces and regional differences in pavement formation 194
9.7.1 Presence of calcium carbonate and carbonate collars 194
9.7.2 Pitting 194
9.7.3 Development of varnish 194
9.7.4 Embedded clasts 194
9.7.5 Clast orientation 195
9.7.6 Clast rubification 195
9.7.7 Development of ventifacted surfaces 195
9.8 Secondary modifications to pavement surfaces 195
9.8.1 Patterns in pavement 195
9.8.2 Animal burrowing, vegetation and stone displacement 195
9.8.3 Regeneration of surfaces by rainfall and runoff events 197
9.8.4 Earthquakes 198
9.8.5 Off-road vehicle disturbance 198
9.8.6 Removal of stones for agriculture 198
9.9 Ecohydrology of pavement surfaces 198
9.9.1 Infiltration in pavements and runoff potential 199
9.9.2 Ecohydrologic relationships and vegetation associations 199
9.10 Relative and absolute dating of geomorphic surfaces based on pavement development 200
9.10.1 Changes in surface characteristics 200
9.10.2 Pavement characteristics and geomorphic surface ages 201
9.10.3 Pavement surfaces as a tool in geomorphic assessment 201
9.11 Conclusions 202
References 204

10 Slope systems 209
John Wainwright and Richard E. Brazier

10.1 Introduction 209
10.1.1 Contexts of slope systems 209
10.2 Badlands 212
10.2.1 Processes and rates of badland evolution 219
10.3 Rock slopes 222
10.3.1 Bare rock or slick-rock slopes 222
10.3.2 Distinctive landforms of rock- and débris-mantled slopes 226
10.4 Conclusion 228
References 229

III The work of water 235

11 Runoff generation, overland flow and erosion on hillslopes 237
John Wainwright and Louise J. Bracken

11.1 Introduction 237
11.2 Infiltration processes 240
11.3 Factors affecting infiltration 241
11.3.1 Controls at the surface–atmosphere interface 241
11.3.2 Subsurface controls 246
11.4 Runoff generation 248
11.4.1 Ponding and surface storage 248
11.4.2 Flow hydraulics 250
11.4.3 Pipes and macropore flow 252
11.4.4 Scales of overland flow 252

11.5 Erosion processes on hillslopes 254
11.5.1 Splash 255
11.5.2 Unconcentrated overland-flow erosion 256
11.5.3 Concentrated overland-flow erosion 256
11.5.4 Patterns and scales of sediment transport 259
11.6 Conclusions 259
References 259

12 Distinctiveness and diversity of arid zone river systems 269
Stephen Tooth and Gerald C. Nanson

12.1 Introduction 269
12.2 Distinctiveness of dryland rivers 270
12.3 Diversity of dryland rivers 273
12.3.1 Higher energy dryland rivers: the Mediterranean region 274
12.3.2 Moderate and lower energy dryland rivers: southern Africa 280
12.3.3 Lower energy dryland rivers: Australia 284
12.4 Reassessing distinctiveness and diversity 289
12.4.1 Downstream flow decreases and localised flood patterns 291
12.4.2 Induration of alluvial sediments 291
12.4.3 Channel–vegetation interactions 293
12.4.4 Fluvial–aeolian interactions 293
12.5 Conclusions 293
References 294

13 Channel form, flows and sediments of endogenous ephemeral rivers in deserts 301
Ian Reid and Lynne E. Frostick

13.1 Introduction 301
13.2 Rainfall and river discharge 302
13.2.1 Storm characteristics 302
13.2.2 Flash flood hydrograph 304
13.2.3 Transmission losses 305
13.2.4 Drainage basin size and water discharge 307
13.3 Ephemeral river channel geometry 309
13.3.1 Channel width 309
13.3.2 Channel bed morphology 309
13.4 Fluvial sediment transport 311
13.4.1 Scour and fill 311
13.4.2 Sediment transport in suspension 314
13.4.3 Sediment transport along the stream bed 317
13.5 Desert river deposits 320
13.5.1 Thin beds 321
13.5.2 Predominance of horizontal lamination in sand beds 322
13.5.3 Mud drapes and mud intraclasts 324
13.6 Conclusions 324
References 327

14 Dryland alluvial fans 333
Adrian Harvey

14.1 Introduction: dryland alluvial fans – an overview 333
14.1.1 Definitions, local occurrence, general morphology 333
14.1.2 Global occurrence and distribution of dryland alluvial fans 334
14.1.3 The role of alluvial fans within dryland fluvial systems 338

14.2 Process and form on dryland alluvial fans 338
14.2.1 Sediment supply, transport and depositional processes 338
14.2.2 Post-depositional modification of dry-region fan surfaces 340
14.2.3 Alluvial fan sediment sequences and spatial variations 342
14.2.4 Alluvial fan morphology and style 345
14.3 Factors controlling alluvial fan dynamics 345
14.3.1 Passive factors: influence on fan morphology 347
14.3.2 Dynamic controls 351
14.4 Alluvial fan dynamics 358
14.4.1 Expressions of fan dynamics 358
14.4.2 Interactions between the dynamic controls: case studies of alluvial fan response to Late Quaternary environmental change 358
14.5 Discussion: significance of dry-region alluvial fans 362
14.5.1 Commonly held myths and outdated concepts 362
14.5.2 Significance to science 362
14.5.3 Significance of dry-region alluvial fans for society 363
Acknowledgements 364
References 364

15 Pans, playas and salt lakes 373
Paul A. Shaw and Rob G. Bryant

15.1 The nature and occurrence of pans, playas and salt lakes 373
15.1.1 Playa and pan terminology 374
15.1.2 General characteristics 374
15.1.3 Origins and development of pans and playas 376
15.2 Pan hydrology and hydrochemistry 379
15.2.1 Inflow and water balance modelling 380
15.2.2 Geochemical processes and mineral precipitation 382
15.2.3 The importance of groundwater: classification of playa and pan types 384
15.2.4 Implications of climate change and human impacts on playa hydrology 385
15.3 Influences of pan hydrology and hydrochemistry on surface morphology 386
15.3.1 Pan topography 386
15.3.2 Surface dynamics: mapping pan surface morphologies using remote sensing 388
15.4 Aeolian processes in pan environments 389
15.4.1 Wind action on the pan surface 391
15.4.2 The emission of fine particles (dust): process and controls 391
15.4.3 Lunette dunes 392
15.4.4 Yardangs 394
15.5 Pans and playas as palaeoenvironmental indicators 394
15.5.1 Identification and dating of pan shorelines 394
15.5.2 Dating and stratigraphy of lunette dunes 394
15.5.3 Stable isotope studies and pan hydrochemical evolution 395
References 395

16 Groundwater controls and processes 403
David J. Nash

16.1 Introduction 403
16.2 Groundwater processes in valley and scarp development 404
16.2.1 Erosion by exfiltrating water: definitions and mechanisms 404
16.2.2 Seepage erosion and valley formation 404
16.2.3 Characteristics of drainage networks developed by groundwater seepage erosion 408
16.2.4 Parameters promoting the operation of groundwater seepage erosion processes 411

16.2.5 Groundwater seepage erosion and environmental change 411
16.2.6 In situ deep-weathering and valley development 412
16.3 Groundwater and pan/playa development 413
16.4 Groundwater and aeolian processes 414
References 418

IV The work of the wind **425**

17 Aeolian landscapes and bedforms 427
David S.G. Thomas

17.1 Introduction 427
17.2 Aeolian bedforms: scales and relationships 427
17.2.1 Scale effects in aeolian bedform development 430
17.3 The global distribution of sand seas 430
17.3.1 Sand sea development 432
17.3.2 Sediment supply in sand seas 432
17.3.3 Sandflow conditions and sand sea development 435
17.3.4 Sand sheets 435
17.4 The global distribution of loess 437
17.4.1 Loess production and distribution 437
17.4.2 Peridesert loess 437
17.5 Dynamic aeolian landscapes in the Quaternary period 439
17.5.1 Dating aeolian landscape change 443
17.6 Conclusions 448
References 448

18 Sediment mobilisation by the wind 455
Giles F. S. Wiggs

18.1 Introduction 455
18.2 The nature of windflow in deserts 456
18.2.1 The turbulent velocity profile 456
18.2.2 Measuring shear velocity (u_*) and wind stress 457
18.2.3 Measuring aerodynamic roughness (z_0) 460
18.2.4 The effect of nonerodible roughness elements on velocity profiles 462
18.3 Sediment in air 464
18.3.1 Grain entrainment 464
18.4 Determining the threshold of grain entrainment 466
18.5 Surface modifications to entrainment thresholds and transport flux 468
18.5.1 Surface crusting 468
18.5.2 Bedslope 469
18.5.3 Moisture content 469
18.6 Modes of sediment transport 471
18.6.1 Suspension 471
18.6.2 Creep 471
18.6.3 Reptation 472
18.6.4 Saltation 472
18.7 Ripples 473
18.7.1 Ballistic ripples 474
18.8 Prediction and measurement of sediment flux 475
18.9 The role of turbulence in aeolian sediment transport 478
18.10 Conclusions 479
References 479

19 Desert dune processes and dynamics 487
Nick Lancaster

19.1 Introduction 487
19.2 Desert dune morphology 487
19.3 Dune types and environments 487
19.3.1 Crescentic dunes 487
19.3.2 Linear dunes 490
19.3.3 Star dunes 494
19.3.4 Parabolic dunes 494
19.3.5 Zibars and sand sheets 494
19.4 Airflow over dunes 494
19.4.1 The stoss or windward slope 495
19.4.2 Lee-side flow 497
19.5 Dune dynamics 497
19.5.1 Erosion and deposition patterns on dunes 497
19.5.2 Long-term dune dynamics 500
19.6 Dune development 502
19.7 Controls of dune morphology 503
19.7.1 Sediment characteristics 503
19.7.2 Wind regimes 503
19.7.3 Sand supply 504
19.7.4 Vegetation 505
19.7.5 Controls of dune size and spacing 505
19.7.6 Dune trends 508
19.8 Dune patterns 509
19.9 Conclusions 511
References 511

20 Desert dust 517
Richard Washington and Giles S. F. Wiggs

20.1 Introduction 517
20.1.1 Dust in a geomorphological context 517
20.1.2 Measuring dust 522
20.1.3 Modelling dust 524
20.1.4 Distribution of dust 525
20.2 Key source areas 526
20.2.1 Bodélé Depression, Chad 526
20.2.2 Saharan Empty Quarter 528
20.2.3 China 530
20.2.4 Southern Africa and Australia 531
20.3 Temporal changes in dust 532
20.3.1 Observational record 532
20.4 Future climate change 532
20.5 Conclusions 532
References 533

21 Wind erosion in drylands 539
Julie E. Laity

21.1 Introduction 539
21.2 The physical setting: conditions for wind erosion 540
21.2.1 Processes of aeolian erosion 540
21.2.2 Yardangs 541
21.2.3 Yardang formative processes 545

21.2.4 Inverted topography 553
21.2.5 Ventifacts 553
21.3 Conclusions 564
References 564

V Living with dryland geomorphology 569

22 The human impact 571
Nick Middleton

22.1 Introduction 571
22.2 Human impacts on soils 571
22.2.1 Terracing and rainwater harvesting 571
22.2.2 Irrigated agriculture 573
22.2.3 Accelerated erosion 574
22.2.4 Grazing 575
22.3 Human impacts on sand dunes 576
22.4 Human impacts on rivers 576
22.4.1 Large dams 576
22.4.2 Urbanisation 577
22.4.3 Changes in vegetation 578
22.5 Cause and effect: the arroyo debate continues 578
22.6 Conclusions 579
References 579

23 Geomorphological hazards in drylands 583
Giles F. S. Wiggs

23.1 Introduction 583
23.2 Aeolian hazards 583
23.2.1 Blowing sand and active dune movement 583
23.2.2 Human disturbance of stable surfaces 585
23.3 The aeolian dust hazard 586
23.4 Agricultural wind erosion 587
23.5 Drainage of inland water bodies 589
23.6 Fluvial hazards 593
23.7 Conclusions 594
References 595

24 Future climate change and arid zone geomorphology 599
Richard Washington and David S. G. Thomas

24.1 Introduction 599
24.2 Climate change projections: basis and uncertainties 599
24.3 Overview of global climate change projections in the context of arid zones 600
24.3.1 Methods of establishing climate change impacts in arid zones 602
24.4 Climate change and dunes 603
24.5 Climate change and dust 605
24.6 Climate change and fluvial systems 607
24.7 Conclusions 607
References 608

Index 611

List of contributors

Dr Louise Bracken, Department of Geography, Durham University, Science Laboratories, South Road, Durham DH1 3LE, UK

Dr Richard Brazier, School of Geography, University of Exeter, The Queen's Drive, Exeter EX4 4QJ, UK

Dr Rob G. Bryant, Department of Geography, University of Sheffield, Sheffield S10 2TN, UK.

Dr Sallie L. Burrough, School of Geography and Environment, Oxford University Centre for the Environment, South Parks Road, Oxford OX1 3QY, UK

Dr Jonathan Clarke, Mars Society Australia, Box 327 Clifton Hill, Victoria 3068, Australia/Australian Centre for Astrobiology, Biological Science Building, University of New South Wales, Kensington, NSW 2052, Australia

Dr David Dunkerley, School of Geography and Environmental Science, Clayton Campus, Monash University, Victoria 3800, Australia

Professor Lynne E. Frostick, Department of Geography, University of Hull, Hull HU6 7RX, UK

Professor Adrian Harvey, School of Environmental Sciences, University of Liverpool, Roxby Building, Chatham Street, Liverpool L69 7ZT, UK

Dr Julie E. Laity, Department of Geography, California State University, Northridge, 18111 Nordhoff Street, Northridge, CA 91330-8249, USA

Professor Nicholas Lancaster, Division of Earth and Ecosystem Sciences, Desert Research Institute, 2215 Raggio Parkway, Reno, NV 89512-1095, USA

Dr Nick Middleton, School of Geography and Environment, Oxford University Centre for the Environment, South Parks Road, Oxford OX1 3QY, UK

Professor Gerald Nanson, School of Earth and Environmental Sciences, University of Wollongong, Wollongong, NSW 2522, Australia

Professor David J. Nash, School of Environment and Technology, University of Brighton, Brighton BN2 4GJ, UK

Professor Ian Reid, Department of Geography, Loughborough University, Loughborough LE11 3TU, UK

Professor Helen Rendell, Department of Geography, Loughborough University, Loughborough LE11 3TU, UK

Professor Paul A. Shaw, Faculty of Science and Agriculture, University of the West Indies, St Augustine, Trinidad and Tobago

Professor David S. G. Thomas, School of Geography and Environment, Oxford University Centre for the Environment, South Parks Road, Oxford OX1 3QY, UK.

Dr Stephen S. Tooth, Institute of Geography and Earth Sciences, Institute of Geography and Earth Sciences, Aberystwyth University, Penglais Campus, Aberystwyth SY23 3DB, UK

Professor Heather A. Viles, School of Geography and Environment, Oxford University Centre for the Environment, South Parks Road, Oxford OX1 3QY, UK

Professor John Wainwright, Department of Geography, University of Sheffield, Sheffield S10 2TN, UK

Dr Richard Washington, School of Geography and Environment, Oxford University Centre for the Environment, South Parks Road, Oxford OX1 3QY, UK

Dr Giles F. S. Wiggs, School of Geography and Environment, Oxford University Centre for the Environment, South Parks Road, Oxford OX1 3QY, UK

Preface to the first edition

Arid environments may not be the most hospitable places on Earth, but the 30 % or more of the global land surface that they cover does support an ever-growing human population and has fascinated travellers, explorers and scientists for centuries. Early geomorphological studies were frequently carried out indirectly, sometimes even unwittingly, by those whose main purpose and motives lay elsewhere: inevitably, but with some notable exceptions, their accounts were descriptive and unscientific. Some would even argue that these traits persisted and dominated desert geomorphological studies well into the second half of this century. Recent years have, however, seen an enhanced rigour in the investigation and explanation of landforms and geomorphological processes in arid lands. New data have been gathered by techniques ranging in scale from the detailed monitoring of processes in the field to remote sensing from space; old theories have been questioned and new ones, based on evidence rather than surmise, have been proposed.

The idea for this volume grew out of these advances and the absence of a recent book which encapsules them (Cooke and Warren's *Geomorphology in Deserts* is 15 years old and Mabbutt's *Desert Landforms* is 11 years old). There have been valuable volumes produced in recent years that deal with specific topics of interest to desert geomorphologists, but none (to my knowledge) that attempts a broader view of arid zone geomorphology. It is hoped that this book fills this gap.

The decision to invite others to contribute chapters was made easily. The geomorphology of arid environments is a huge topic, embracing much of the subject matter of geomorphology as a whole: desert landforms consist of much more than piles of unvegetated sand. Arid and semi-arid environments are very varied, too; involving the expertise of others has therefore inevitably broadened and deepened the basis of the text. While there are inevitably gaps, these have hopefully been kept to a minimum. Many people have provided the help and inspiration needed to turn *Arid Zone Geomorphology* from an idea to a book. Andrew Goudie introduced me to deserts, since which time many people and funding bodies have enabled me to visit them and to conduct research in them: I would particularly like to thank the Shaws in Botswana and Sleaze and Val for showing me Death Valley and other Californian hotspots. During the production of the book the contributors have efficiently met the tasks I have set them, including refereeing other people's chapters; Rod Brown provided additional help in this respect, too, while Chapter 12 also passed through refereeing within the US Geological Survey. The cartographers of many institutions, but especially Paul Coles of the Geography Department, University of Sheffield, produced the diagrams. At the publishers, Iaian Stevenson and Sally Kilmister gave me valuable advice and logistical help. Steve Trudgill inspired me to put a book together in the first place.

Lastly, but most importantly, Liz Thomas not only suffered me during the book's gestation, but helped in a multitude of practical ways and provided a valuable, independent, geomorphological viewpoint. To all of the above, my parents and any I have forgotten to mention, my sincere thanks.

David S.G. Thomas
Sheffield
August 1988

Preface to the second edition

It is almost eight years since the text of the first edition of *Arid Zone Geomorphology* was written, and seven years since the book was first published. Coincidently, on the day of publication, 7 December 1988, the 'Finger of God', pictured on the cover of the first edition, collapsed. So, along with a substantially changed content, this second edition has a new cover.

Since the first edition was produced, much has happened in terms of both geomorphic research in arid environments (or drylands, or deserts: such terms are commonly used interchangeably) and in general and non-scientific interest for such areas. Arid regions are areas of concern, because of population growth, the impacts of desertification and of natural phenomena, particularly droughts. The impending impacts of global warming on these areas and their peoples are also of growing concern. Scientists, including geomorphologists, are responding to the need to know more about the nature and operation of processes in drylands by conducting more research, both fundamental and applied.

This new edition of *Arid Zone Geomorphology* aims to reflect the changes and advances in geomorphological knowledge that have occurred, especially since the publication of the first edition. This has been done in two ways. First, the chapters from the first edition have been updated, in some cases radically. Second, the content of the book has been expanded, with the number of chapters all but doubled and arranged in a new framework of six sections. This has been done to fill gaps in coverage or expand areas of particular interest. In both cases, as with the first edition, experts have been invited to write the chapters of this text rather than one person attempting to summarise and review what amounts to a vast chunk of geomorphology. In the majority of cases the authors of chapters from the first edition have rewritten their own material. In some cases where circumstances have prevented this, new co-authors have conducted the task. For a few themes covered in the first edition, new authors have written material afresh.

In all, the production of this new edition has resulted in 34 researchers from over 25 academic or research institutions making contributions, all involved with research in the fields on which they write. It is this wealth of expertise and the wide-ranging and diverse experience of drylands that it represents that make this book. As editor I am indebted to the cooperation of the contributors for meeting deadlines and to those who have conducted last-minute tasks at my request. The willingness with which writing, updating, changing text, reviewing and other tasks have been taken up is enormously appreciated. The involvement of some new contributors to the second edition has come about through conducting fieldwork in deserts in Africa and Asia with them: Dave Nash and Jo Bullard, whose PhDs I had the privilege to supervise; and Stephen Stokes, Giles Wiggs and Sarah O'Hara. For others, listening to their papers at conferences and meetings in Ahmedabad, India and Hamilton, Canada, and even the UK, or casual conversations over coffee or on fieldtrips, led me to ask them to contribute: David Dunkerley, Gerald Nanson, Jacky Croke, Ed Derbyshire, Helen Rendell, Lillan Berger, Vatche Tchakerian (and, indirectly, Julie Laity) all became victims in this way.

The production of this book has been greatly helped by Kate Schofield, Sam Rewston and Sarah Harmston in the Geography Department office at Sheffield and Paul Coles and Graham Allsopp in Cartographic Services who have produced or updated many of the figures. Iain Stevenson and Katrina Sinclair at the London office of John Wiley & Sons, Ltd have eased production matters. To all those names above, the undergraduates, postgraduates and academics who used the first edition and passed on comments for possible future changes, and especially my wife Lucy, who painstakingly prepared the index and our daughter Mair, who has tolerated the production of this volume since her birth, my sincere thanks.

David S.G. Thomas
Sheffield
January 1996

Preface to the third edition

It was a pleasant surprise when in 2008 the publishers requested that I consider putting together a new edition of *Arid Zone Geomorphology*. It is now thirteen years since the second edition was published and, inevitably, the literature of desert and dryland geomorphology has burgeoned in the intervening period. Research on processes has benefited from many technological advances, no greater than in aeolian geomorphology, where it is now possible to measure airflow and sediment movement in the field over very short (in some cases subsecond) time periods. The benefits of advances in reductionist research have been complemented by developments that allow the bigger picture to be better viewed in space and in time. There are now many options in satellite-based remote sensing, allowing surface conditions and the atmosphere above drylands to be analysed, while ground penetrating radar is permitting the internal structures of some dryland landforms to be viewed. Better reconstructions of past dryland environments, including land surface responses to global climate change, are possible due to a plethora of proxy records being available and enhanced chronometric control of the timings of change are possible due to the advances in luminescence dating. There is therefore much to include in an updated edition.

The format of this edition is much changed from the second, which itself was markedly expanded from the first edition. First is that while some chapters are updates of their equivalents in the second edition, many are new, even if bearing the same or similar titles. The structure of the book is also changed. The introductory chapters in Section 1 are increased from two to five. This has been done to allow a bigger picture to be developed early on prior to the presentation of thematic sections. In previous editions long-term change was not presented until late in the book, whereas now it is integrated through the volume as a whole. This is borne of recognition that to understand landforms and landscapes fully, it is necessary not only to have knowledge of the processes operating today but the inheritance that has occurred from the past. Thus process geomorphology and Quaternary period reconstructions are not artificially divided, as is the common case, into separate research agendas, but instead are integrated as appropriate in individual chapters. Therefore the nature and role of long-term change on drylands, and their former extensions, are presented in Chapter 3 (as opposed to Chapter 26 in the second edition). The diversity of drylands worldwide is also considered in a separate chapter, while arid landscapes on planets other than our own are also considered early on. This reflects the knowledge of drylands that is arising through extraterrestrial research.

Three sections, on surfaces, water and wind, then follow, with a concluding section on human aspects of dryland geomorphology, including the potential impacts of twenty-first century climate change. The regional chapters from the expanded second edition are removed and instead illustrative case studies are included within individual chapters where relevant. This makes for a slimmer book, more akin to the first rather than the second edition.

In 2004 I moved to the University of Oxford after 20 years at Sheffield University, where a significant desert/dryland research group had been established. Oxford now has perhaps the UK's biggest arid land research grouping and this is reflected in the authorship of some of the chapters. We remain in this book truly international in outlook, however, and this is not simply reflected in where the UK-base contributors conduct their desert research – including in the North American arid zone, southern Africa, North Africa, Australia and Arabia – but in internationally renowned researchers from the US, Australia and the West Indies contributing to this volume. I thank them all for their efforts.

A book is not simply down to its contributors, however. In this regard it is key to note that without the efforts of Fiona Woods and Izzy Canning at John Wiley & Sons, Ltd there would not be a third edition. The same can be said for Jan Burke at the School of Geography and Environment at Oxford, whose assistance in the final stages has been immeasurable, I also thank Ailsa Allen who drew many of the figures, Paul Coles at Sheffield whose assistance in tracking down and passing on artwork from the second edition has been gratefully

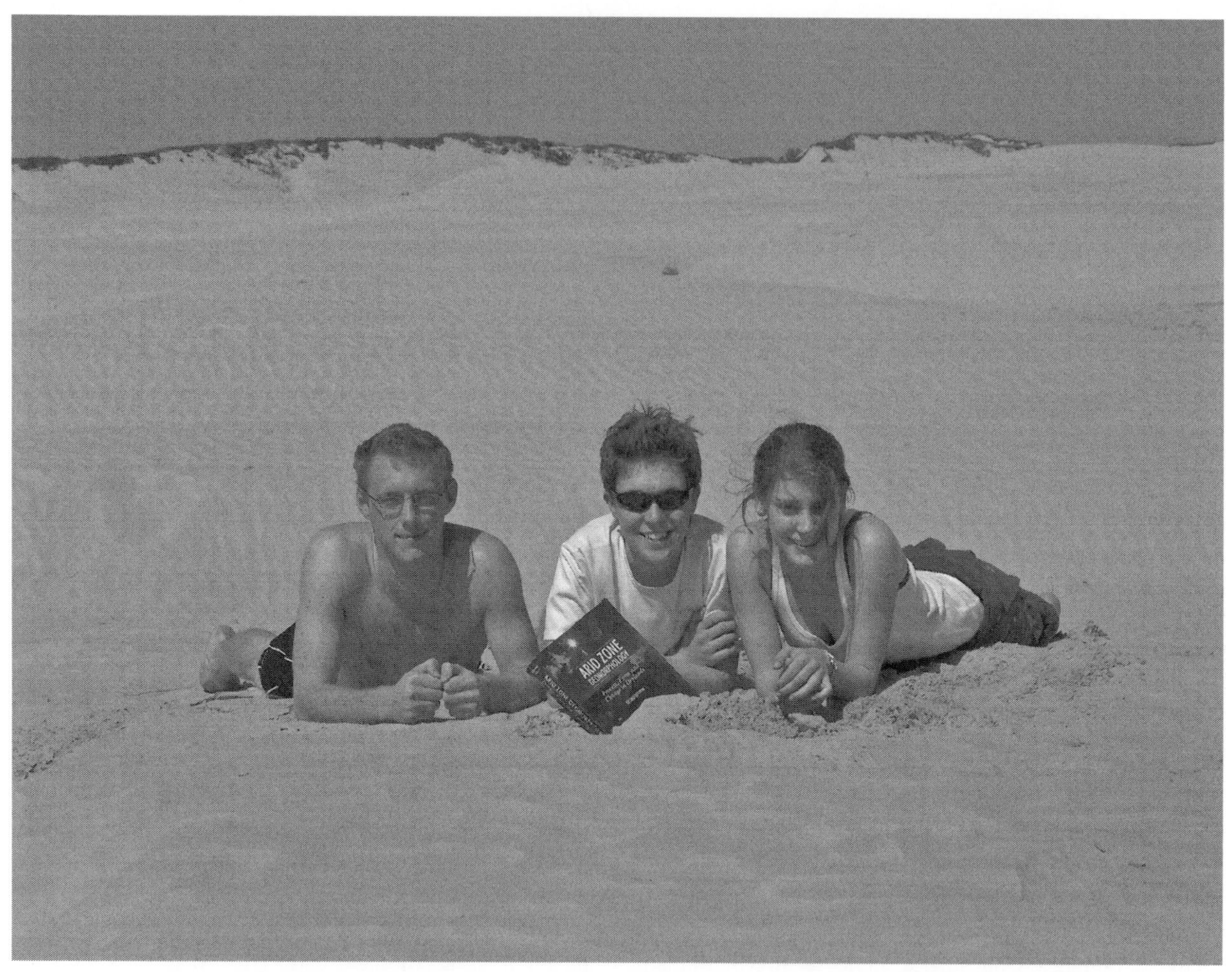

appreciated, and Lucy Heath for considerable help in the dreaded task of preparing the index. There are two further groups to thank. First is the truly inspirational new generation of desert geomorphologists whose doctoral theses I have had the privilege to supervise over the past two decades and whose research appears in some of the chapters. Second are the 300+ undergraduates and Master's students who in fifteen years Giles Wiggs and I (and when at Sheffield, Rob Bryant) have taken on annual dryland geomorphological field trips to Tunisia and the UAE. They have been a good sample of users of *Arid Zone Geomorphology*, and their comments and usage of the book have led to some of the changes in this edition. Thank you all.

David S.G. Thomas
Oxford
March 2010

I

Large-scale controls and variability in drylands

1

Arid environments: their nature and extent

David S.G. Thomas

1.1 Geomorphology in arid environments

Aridity, a deficit of moisture in the environment, is a significant feature of a large part of the Earth's land surface. Aridity is complex, and its environmental manifestations vary from place to place and through time, such that its definition, occurrence and environmental consequences are complex and require careful unravelling. Aridity is also complex and challenging for many life forms, since moisture is such a fundamental requirement for many. Aridity does not simply equate with the concept of deserts, and goes far beyond what are widely regarded as such, so that there is a great diversity within, and between, arid environments. The purpose of this book is to provide explanations for the diversity and nature of arid environments, through an exploration of the land-shaping processes that operate within them.

For much of history and for many human races, arid environments have been areas to avoid, though for those that have been, and continue to be, resourceful and able to adapt, arid regions have proved to be environments that can be effectively and successfully utilised. Lack of surface water, limited foodstuffs and climatic extremes have generally made arid areas unfavourable places for habitation, though for resourceful hunter–gatherer and pastoral–nomadic peoples, living at low population densities, these environments have proved to be places of opportunity. In other contexts, the apparent scarcity of key resources may have driven innovation: it is perhaps no coincidence that early civilisations, in Mesopotamia, in Egypt and in parts of central Asia, developed strategies to cope with aridity, with early agriculture developing, c.4000 years ago, in the Mesopotamian heartland. However, for populations from more temperate and better watered regions, aridity has often proved a significant challenge. Even with the technological advances of the late nineteenth and twentieth centuries that made travel and existence in drylands possible for a greater range of people, arid environments still provide major limitations to the range and extent of human occupations and activities.

European interest in arid environments grew from the late eighteenth century onwards (Heathcote, 1983), usually associated with the quest for natural resources and colonisation, or with attempts at religious conversion. Much of the early 'Western' scientific knowledge concerning such areas came not from specialist scientists but from those whose primary goals were associated with these activities. It has been noted or implied (see, for example, Cooke and Warren, 1973; Cooke, Warren and Goudie, 1993; Goudie, 2002) that early geomorphological research in arid areas was dogged by excessive description, superficiality and secular national terminology. The first characteristic, description, has often been criticised, especially at times when quantification has been a central paradigm in geomorphology. Yet description can be an important prerequisite of rigorous explanation, analysis and deeper investigation. This is no better illustrated than by the work of Dick Grove and Ron Peel (e.g. Grove, 1958, 1969; Peel, 1939), where careful description of land forms and landscapes preceded analysis and the quest for geomorphic explanations of their development and the controls on the processes that shaped them.

In the case of early works, the descriptive component is hardly surprising. For European writers with temperate world origins, desert landscapes must have represented spectacular, bizarre and unusual contrasts to the plant and soil mantled landscapes of many of their homelands. Before early descriptive accounts are totally pilloried, it should also be remembered that geomorphological

Arid Zone Geomorphology: Process, Form and Change in Drylands, Third Edition. Edited by David S. G. Thomas

accounts in the works of early Europeans were usually but by-products of the reasons for their being in deserts in the first place. This also helps to explain the second characteristic, the superficiality of early reports and studies.

The third characteristic attributed above to early works, secularity, arose because national groups tended, until relatively recently, to confine their interests to particular deserts. In Africa, Asia and Australia, early geomorphological investigations were heavily influenced by the distribution of the impacts of European colonialism. Thus Flammand's (1899) account from the Sahara, Passarge's (1904) two volumes on the Kalahari and numerous reports from Australia (e.g. Sturt, 1833; Mitchell, 1837; Spencer, 1896) reflect broader colonial interests of their time. It has been noted, and is now increasingly realised as solutions are sought to dryland environmental problems (e.g. Mortimore, 1989; Thomas and Middleton, 1994; Thomas and Twyman, 2004), that the environmental knowledge vested in indigenous populations was, and remains, considerable and meritorious. Yet this was usually either ignored or unrealised by Europeans entering arid environments for the first time in the nineteenth and through much of the twentieth centuries: such people often saw deserts through eyes more accustomed to their starkly contrasting points of origin, leading to a preoccupation in some cases with the spectacular and unusual landforms they encountered in deserts.

There were, of course, exceptions to these characteristics, even in the nineteenth century. Perhaps most notable were the investigations in the southwestern United States, often with a geomorphological slant, of John Wesley Powell (1875, 1878) and Grove Karl Gilbert (1875, 1877, 1895), the latter regarded by many as the father of modern geomorphology. Their activities were driven by a governmental quest to expand the frontiers of (European) utilisation of North America and their works were essentially early forms of resource appraisal. Some of the early accounts of the geomorphology of the Australian deserts had a similar basis; for example, Thomas Mitchell wrote:

> After summounting the barriers of parched deserts and hostile barbarians, I had at last at length the satisfaction of overlooking from a pyramid of granite a much better country (Mitchell, 1837, p. 275).
>
> We had at last discovered a country ready for the reception of civilised man. (Mitchell, 1837, p. 171)

1.2 Arid zone distinctiveness and the quest for explanation

Early accounts of arid landscapes may, however, be of restricted geomorphological value for a different reason: their focus on unusual and spectacular features was often at the expense of representativeness (but also see Chapter 4, showing that accounts could also focus on the monotony of some dryland regional landscapes). The lack of reliable, systematic, information and data was one reason why theory in arid geomorphology changed rapidly through the first six decades of the twentieth century. As Goudie (1985, p. 122) noted:

> A prime feature of desert geomorphological research over the past century or so has been the rapidity with which ideas have changed, and the dramatic way in which ideas have gone in and out of fashion. This reflects the fact that hypothesis formulation has often preceded detailed and reliable information on form and process, and the fact that different workers have written about different areas where the relative importance of different processes may vary substantially.

Within these changing ideas was a view that arid environments are distinct, even unique, in terms of the operation of geomorphological processes and their resultant landscape outcomes. Early quests for synthesising explanations sought generalisations that were deliberately distinct from those developed for other environments. Davies (1905) produced his cycle of erosion for arid environments based on the belief that fluvial processes in drylands produced distinct outcomes at the landscape scale. This notion of distinctiveness was clearly also present in morphogentic or climatic geomorphology models of explanation (e.g. Birot, 1960; Budel, 1963; Tricart and Cailleux, 1969). While the very terms 'drylands', 'arid zone' and so on clearly imply a climatic delimitation of the extent of these environments, it is debatable whether sweeping models of desert geomorphic explanation are justified, for three reasons. First, notwithstanding that there may be 'a world of difference in the landscapes and geomorphological processes that occur in these different climatic zones' [*within* arid environments] (Goudie, 2002, p. 5) is that drylands themselves are not internally homogeneous; indeed they are markedly diverse climatologically and tectonically (see Chapter 2), which affects seasonality, plant cover, landscape erodibility, sediment types, sediment availability and so on. Second, today's arid regions have been no more immune from the impacts of Quaternary timescale climate changes than any other parts of the Earth's surface, so today's drylands commonly contain landscape and landform expressions inherited from past, different, climatic regimes (Chapter 3). Third, as Parsons and Abrahams (1994, p. 10) succinctly note:

> ... the emphasis of geomorphology has shifted away from morphogenesis within specific areas towards the study of processes *per se*. This shift...in large measure undermines the distinctiveness of desert geomorphology.

Nonetheless, the relative importance of individual processes and the magnitude and frequency of their operation may differ in arid environments compared to other areas. This, together with growing human populations in drylands and the common treatment of them as environmentally distinguishable, is reason enough to pursue arid zone geomorphology in its own right.

The last three decades or so have seen a new rigour enter geomorphological research in arid environments. New techniques have been employed and new methodologies pursued. Landform description for its own sake has largely been eschewed, though it does, of course, still have a valid role in geomorphological research, and has been replaced by studies of process and form, measurement, explanation and application. It is these that this book focuses upon.

1.3 Arid zones: terminology and definitions

Definitions and delimitations of arid environments and deserts abound, varying according to the purpose of the enquiry or the location of the area under consideration. Literary definitions, thoroughly reviewed by Heathcote (1983), commonly employ terms such as 'inhospitable', 'barren', 'useless', 'unvegetated' and 'devoid of water'. Scientific definitions have been based on a number of criteria, including erosion processes (Penck, 1894), drainage patterns (de Martonne and Aufrère, 1927), climatic criteria based on plant growth (Köppen, 1931) and vegetation types (Shantz, 1956). Whatever criteria are used, all schemes involve a consideration of moisture availability, at least indirectly, through moisture balance: the relationship between precipitation and evapotranspiration.

1.3.1 Terminology

'Desert', 'arid zone', 'dryland' and sometimes other terms such as 'thirstland' are all used somewhat interchangeably and imprecisely in both popular and scientific literature. This can lead to confusion regarding differences in levels of moisture availability from place to place. Consider the example of the Kalahari Desert (Thomas and Shaw, 1991) (Figure 1.1), a place that by name is familiar to many people and that at first glance conjures up images of extreme dryness. Yet the Kalahari is widely regarded not as a 'true' desert, and few parts of it today achieve extremes of moisture deficit, particularly when compared with southern Africa's other desert, the Namib. In fact the Kalahari, which spans over ten degrees of latitude from northern South Africa to northern Namibia, embraces environments that range from arid to dry-subhumid and mean annual rainfalls from around 200 mm to over 600 mm (the driest areas of England receive c.500 mm p.a.). Ecosystems range from sparse savanna grassland to subtropical woodland.

Collectively, the Kalahari as described here is undoubtedly 'dryland', given that seasonality of rainfall and potential evaporation give rise to annual moisture deficits. The two uniting factors across the region are structural (the Kalahari occupies an internal structural basin) and sedimentological (the Kalahari is predominantly covered by unconsolidated sands). These sands in fact extend even further north, over a further 15 degrees of latitude, into wet tropical environments. In part the characteristics of the modern Kalahari are a consequence of major climatic and environmental changes in the Late Quaternary period that resulted at times to major expansions of the arid zone (Thomas and Shaw, 2002).

Overall, arid zone and dryland are perhaps more apposite terms than desert to use collectively in the description of moisture deficit regions, and this is reflected in the title and subtitle of this book. However, within the literature, consistency of use is absent, and it is necessary to consider carefully, rather than assume, the environmental and climatic characteristics of areas referred to by any of the terms used in this explanation.

1.3.2 Definition

The origins of direct delimitation of arid environments by consideration of moisture balance date back to 1953 and the growth in concern at the United Nations (specifically UNESCO) with global food production and living in dry regions of the world. The classification scheme,

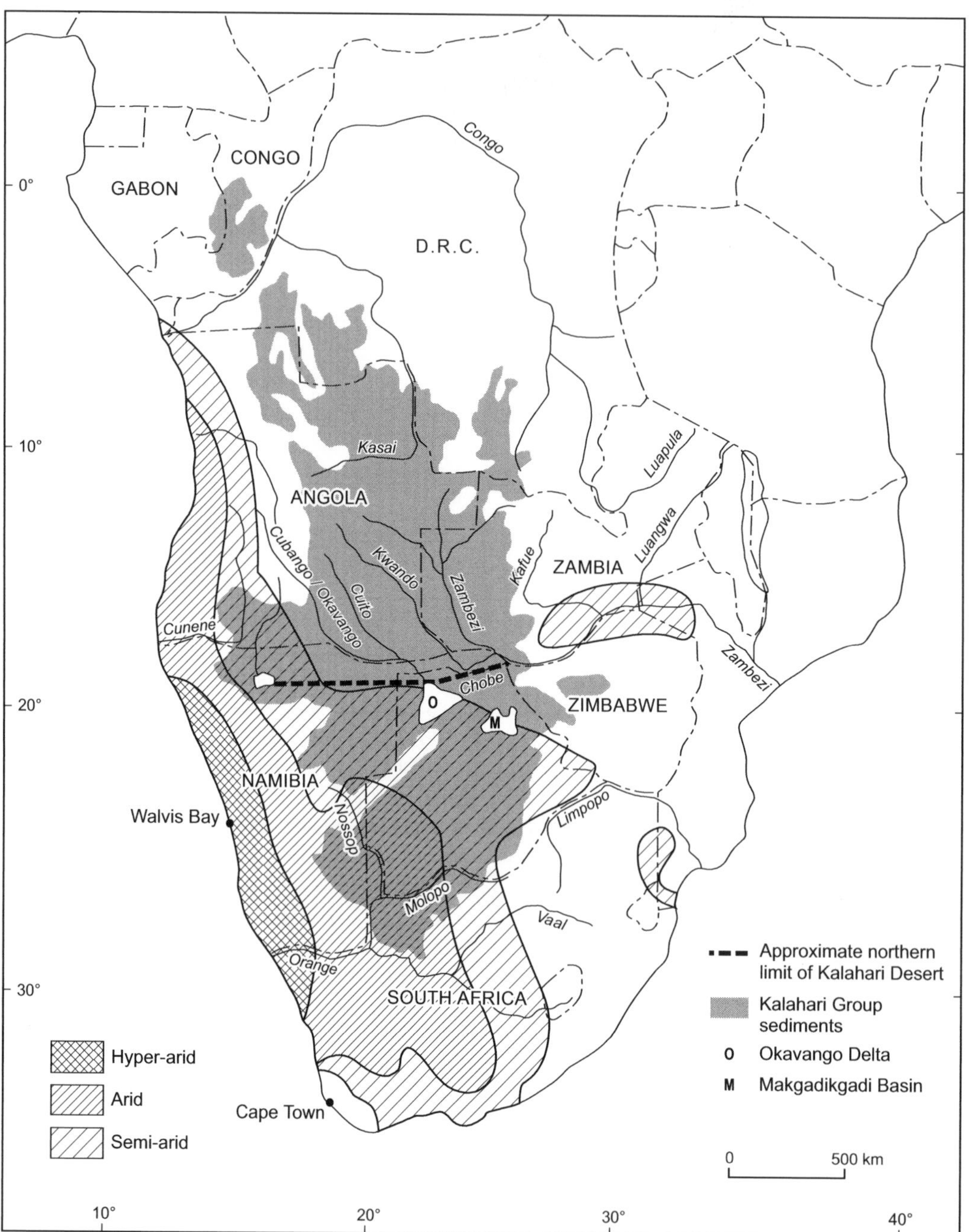

Figure 1.1 The definition of drylands and deserts: the example of the Kalahari. The map shows the distribution of hyper-arid to semi-arid conditions in southern Africa, overlying the distribution of 'Kalahari Group' sediments (dominated by surface sand units). The sediments extend north far beyond today's dryland zone, and in part this is testimony to the impacts of environmental and climatic changes in the Quaternary period. The area referred to as the Kalahari Desert today is broadly coincident with Botswana, the northwest of South Africa and eastern Namibia. Yet this area is predominantly arid to semi-arid, with hyper-arid ('true desert') conditions today restricted to the coastal Namib Desert, with the Kalahari largely being moderately to well vegetated. A further dryland area in southern Africa, the Karoo, is rarely called a desert, yet it is in many respects as worthy of that title as is the Kalahari (map based on data in Thomas and Shaw, 1991).

developed by Peveril Meigs (1953), excluded regions too cold for plant growth (and therefore the polar deserts, notably Antarctica) and utilised Thornthwaite's (1948) indices of moisture availability (*Im*):

$$Im = (100S - 60D)/PET$$

where *PET* is potential evapotranspiration, calculated from meteorological data, and *S* and *D* are, respectively, the moisture surplus and moisture deficit, aggregated on an annual basis from monthly data and taking stored soil moisture into account. Meigs (1953) identified three types of arid environments, delimited by different *Im* index values: semi-arid, arid and hyper-arid. Grove (1977) subsequently attached mean annual precipitation values to the first two categories (200–500 mm and 25–200 mm, respectively), though these are only approximate. Hyper-arid areas with no consecutive months with precipitation have been recorded (Meigs, 1953).

The UN (1977) delimitation of drylands is used as the spatial framework in another desert geomorphology volume (Abrahams and Parsons, 1994). UN (1977) also provided the climatic input to the UNESCO (1979) survey of arid lands utilised in the Cooke, Warren and Goudie (1993) text. The UN (1977) approach defines aridity zones using a *P/PET* index. *PET* is calculated using Penman's formula, which requires a large body of directly measured meteorological data for its calculation and which in practice is not consistently available at the global scale required for dryland delimitation (Hulme and Marsh, 1990). A new assessment of the extent of drylands, based on an aridity index (*AI*), where *AI* = *P/PET* and *PET* is calculated using the simpler Thornthwaite method, has been conducted by Hulme and Marsh (1990) on behalf of UNEP. This new assessment differs from earlier ones by using meteorological data from a fixed time period (a 'time-bounded' study), rather than simply from mean data from the full length of records available, to calculate index values. This is significant given that climate variability can cause mean data to differ depending on the period under consideration (Hulme, 1992). The data used in this new scheme, adopted and utilised in dryland studies such as *The World Atlas of Desertification* (UNEP, 1992), Thomas (1993) and Thomas and Middleton (1994), cover the period 1951–1980 and are based on records from over 2000 meteorological stations worldwide.

The UNEP (1992) classification of drylands also differs from previous estimates by including dry-subhumid areas. This was done because these areas experience many of the climatic characteristics of semi-arid areas (Thomas and Middleton, 1994) and, with UNEP (1992) concerned with desertification, embraces the original application of that term proposed by Aubreville (1949). The delimitation of the different types of dryland environments by *AI* values are dry-subhumid ($AI = 0.50 - <0.65$), semi-arid ($AI = 0.20 - <0.50$), arid ($AI = 0.05 - <0.20$) and hyper-arid ($AI = <0.05$). According to this scheme, these four environments cover about 47 % of the global land area (Figure 1.2). This is significantly more than the areas considered in other schemes (Table 1.1), with this difference principally due to the inclusion of dry-subhumid areas.

In this volume the global *arid zone* is considered to include all elements of this fourfold classification. There are several reasons for this:

1. The divisions between the elements of the classifications are somewhat arbitary. For example, in UN (1977) the boundary between arid and hyper-arid areas was taken as *P/PET* = 0.03, while in UNEP (1992) it is 0.05.

2. In studies prior to UNEP (1992) the climatic data input to the calculation of indices was from nontimebounded and therefore temporally variable data sets.

3. In drylands, annual precipitation frequently varies substantially from year to year, so that in dry-subhumid, semi-arid, arid and hyper-arid areas, the only safe assumption is that any year could be extremely arid (Shantz, 1956).

4. Distinct geomorphic thresholds in terms of processes and landforms have not been identified between the four elements of the scheme.

5. Climatic fluctuations and anthropegenic activities in the twentieth century have caused the expansion of arid surface conditions, especially a decrease in vegetation cover, into some semi-arid environments.

6. Semi-arid areas are often called 'deserts' by their inhabitants.

1.4 The age of aridity on Earth

Desert dune and evaporite sediments preserved in the solid-rock record indicate that aridity has occurred on Earth since Precambrian times (Glennie, 1987), with perhaps the earliest recorded aridity being represented by the c.1800-million-year-old dune sediments in the Hornby Bay Group in the Canadian Northwest Territories (Ross, 1983). The changing configuration of landmasses and

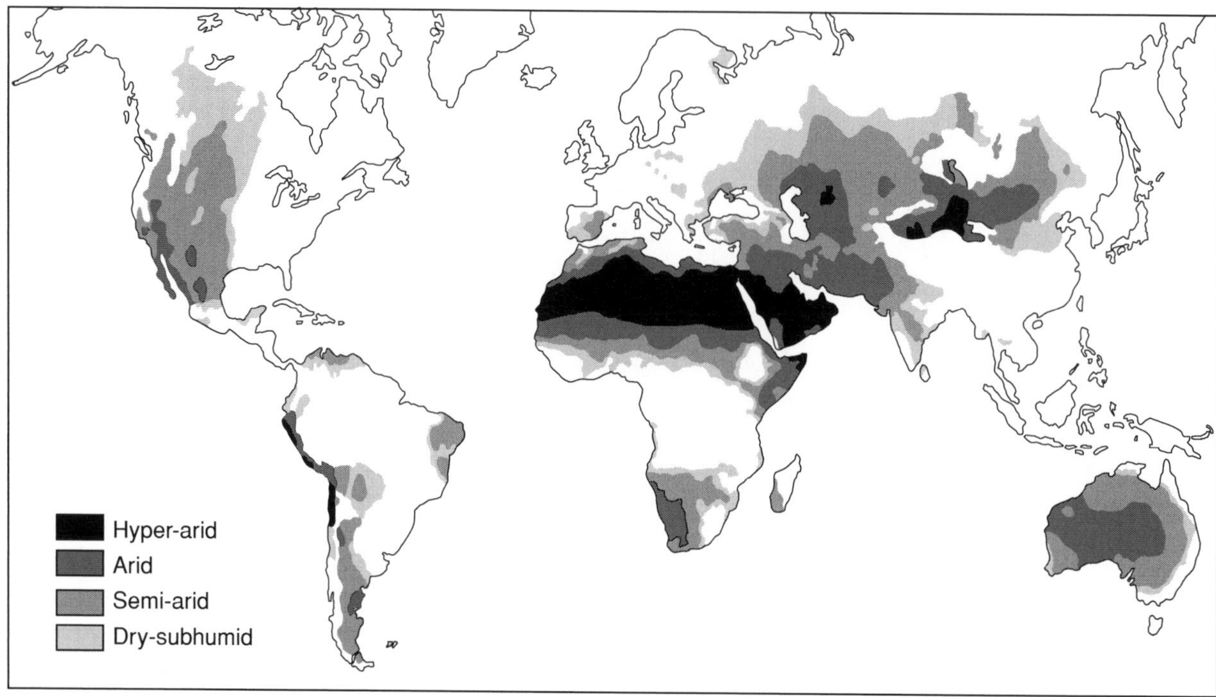

Figure 1.2 The global distribution of drylands.

oceans due to the effects of tectonic plate movements and orogeny, and changes in global climate, have, however, caused the positions and extent of arid zones to change through geologic time.

Sedimentary evidence suggests that the Namib is probably the oldest and most persistent of current arid zones on Earth, dating back 80 million years to the Cretaceous (Ward, Seely and Lancaster, 1983), though such a great antiquity by no means meets with full agreement (see Tankard and Rogers, 1978; Vogel, Rogers and Seely,, 1981). Notwithstanding the effects of subsequent climatic perturbations on the extent and intensity of aridity, many other deserts would seem to date from the Tertiary. Deep sea-core evidence indicates that aeolian material existed off West Africa from 38 million years ago in the Oligocene (Sarnthein, 1978), while other deserts such as those of Australia (Bowler, 1976) and the Atacama (Clark *et al.*, 1967) date from the Miocene.

1.5 The distribution of arid zones

Table 1.2 shows the distribution of arid zones today according to continent. While Africa and Asia each contain almost a third of the global arid zone, inspection of Figure 1.2 clearly shows that Australia is the most arid continent, with approximately 75 % of the land area being arid or semi-arid.

Table 1.1 The extent of the global drylands (expressed as a percentage of the global land area).

Classification	*Dry-subhumid*	*Semi-arid*	*Arid*	*Hyper-arid*	*Total*
Köppen (1931)	—	14.3	12.0	—	26.3
Thornthwaite (1948)	—	15.3	15.3	—	30.6
Meigs (1953)	—	15.8	16.2	4.3	36.3
Shantz (1956)	—	5.2	24.8	4.7	34.7
UN (1977)[a]	—	13.3	13.7	5.8	32.8
UNEP (1992)	9.9	17.7	12.1	7.5	47.2

[a]In Heathcote (1983).

Table 1.2 Distribution of the arid zone by continent (expressed as a percentage of the total global arid zone).

Africa	31.9
Asia	31.7
Australasia	10.8
North America	12.0
South America	8.8
Europe	4.9

Source: UNEP (1992).

Globally, arid areas embrace a range of annual temperature regimes, affected by latitude, altitude and continentality. Cloudsley-Thompson (1969, in Heathcote, 1983) noted that the only common element of temperatures between different arid areas is their range. Meigs (1952) divided arid lands into those that are hot all year round and those with mild, cool and cold winters (Tables 1.3 and 1.4). Variations in temperature affect the seasonal availability of moisture, by influencing evapotranspiration rates and affecting the form of precipitation in relatively high-latitude arid areas. For example, in the arid areas of Canada and central Asia, winter snowfall forms an important component of the annual precipitation budget.

1.6 Causes of aridity

Aridity is characterised by net surface water deficits. It results from climatic, topographic and oceanographic factors that prevent moisture-bearing weather systems reaching an area of the land surface. Four main influences can be identified, which are not mutually exclusive.

Table 1.3 Arid land climates.

	Percentage of arid lands	*Mean temperature (°C)*	
		Coldest month	*Warmest month*
Hot	43	10–30	>30
Mild winter	18	10–20	10–30
Cool winter	15	0–10	10–30
Cold winter	24	<0	10–30

Source: Meigs (1952).

Table 1.4 Examples of arid zones with different climates (classification as in Table 1.3).

Hot	Central Sahara; Arabia; Great Sandy Desert, Australia
Mild	Southern Sahara; Kalahari; Mexican; Simpson, Australia
Cool	Northern Sahara; Turkish Steppes; Atacama; Mojave
Cold	Canadian prairies; Gobi; Turkmenistan; Chinese deserts

1.6.1 *Atmospheric stability*

The major cause of aridity worldwide are the subtropical high-pressure belts: zones of descending, stable air. Tropical and subtropical deserts cover about 20 % of the global land area (Glennie, 1987). In these areas large arid zones are composed of central arid areas surrounded by relatively small, marginal, semi-arid and dry-subhumid belts. Precipitation is very unreliable and largely associated with the seasonal movements of the intertropical convergence zone.

1.6.2 *Continentality*

Distance from the oceans prevents the penetration of rain-bearing winds into the centre of large continents, for example in central Asia. Precipitation and evapotranspiration are both usually lower than in arid areas owing their origins to atmospheric stability, while cold winters are common. In other continents, the failure of dominant easterly trade winds to penetrate to continental interiors (Thompson, 1975), such as in southern Africa, contributes to the continentality effect. Relatively small arid areas are surrounded by an extensive zone of semi-aridity.

1.6.3 *Topography*

Arid areas can occur in the rain shadow of mountain barriers. The Rockies contribute a rain shadow effect in western North America, while in Australia the penetration of easterly trade winds to the interior is further inhibited by the north–south orientation of the Great Divide. Aridity primarily due to atmospheric stability or continentality can therefore also be enhanced by topographic effects.

1.6.4 Cold ocean currents

Cold ocean currents affect the western coastal margins of South America, southern Africa and Australia, giving rise to five west coast subtropical deserts (Meigs, 1966; Lancaster, 1989). These currents reinforce climatic conditions, causing low sea-surface evaporation, high atmospheric humidity, low precipitation (very low rainfall, with precipitation mainly in the form of fog and dew) and a low temperature range. Lack of rainfall in the Namib Desert, western southern Africa, is due both to the impact of the Benguela current on local climates and the failure of easterly rain-bearing winds to penetrate across the continent (Schulze, 1972).

1.7 Climate variability

Interannual variability in precipitation is a marked characteristic of arid regions. Temperate regions may have year-to-year rainfall variability of under 20 %. In the Kalahari it ranges up to 45 % (Thomas and Shaw, 1991) and in the Sahara ranges from 80 to 150 % (Goudie, 2002). Were the areas of individual arid regions to be calculated using annual climate data, they would vary from year to year as precipitation varies. This is reflected environmentally in studies that have monitored, using remote sensing, fluctuations in dryland biomes (e.g. Tucker, Dregne and Newcomb, 1991). Figure 1.3 illustrates interannual rainfall variability and precipitation and temperature trends in example dryland regions during the twentieth century (Hulme, 1996). This analysis shows how 'normal' variability is, particularly in precipitation, and also suggests trends that may be a consequence of global warming impacts. For the Sahel (in itself a very large region within which further subtle spatial variations in variability exist) the long period of rainfall deficit through the 1970s and 1980s may amount effectively to a climate change in terms of its impact on environmental (and social) processes. Interpretation and explanation of the Sahel Great Drought is complex and sometimes controversial (Zeng *et al.*, 1999; Agnew and Chappell, 1999), and may include rainfall changes driven by sea surface temperature changes, the impacts of phenomena such as El Nino and land surface change feedbacks on atmospheric temperature (e.g. Charney, 1975; Zeng *et al.*, 1999).

1.8 Dryland ecosystems

Though this is a book about arid geomorphology, the interplay between plants and the land surface is critical in many geomorphological respects, not least because the means by which plants cope with climatic variability can have marked impacts on the operation of geomorphological processes. The low moisture availability in arid areas has a profound effect on plant growth. As Bloom (1978, p. 314) has noted:

> In the United States, the boundary between humid and semi-arid climates is approximated by the transition westward from medium-height grasses with a continuous turf or sod in the humid regions to short, shallow rooted bunch grasses on otherwise bare ground in semi-arid regions. In arid regions, even the bunch grasses disappear, and the vegetation is, at best, widely spaced shrubs and salt tolerant bushes.

The limited (or absent) vegetation cover is of considerable importance for the operation of geomorphological processes and the development of landforms (Thomas, 1988). The wind can take on the role of geomorphological agent to a degree that cannot occur in other terrestrial environments, except in some coastal locations or places where human activities have interfered with the plant cover. None the less, even limited vegetation can be a very important variable in the operation of arid land geomorphological processes.

Vegetation in arid regions has to cope with the rainfall variability described above, as well as highly seasonal distributions in rainfall and temperature patterns. Different classification schemes exist: a simple scheme based on common plant assemblages associated with decreasing moisture inputs demonstrates issues of plant cover and of plant types (Goddall and Perry, 1979). In Africa and South America, similar schemes exist but with reference to differing forms of savanna vegetation, primarily the changing mix of grass and woody species that comes with moisture availability changes (e.g. Huntley, 1982). Broadly speaking, and unsurprisingly, above-ground biomass tends to increase with increasing available moisture (Figure 1.4).

How plants in arid areas cope with moisture deficits and droughts is particularly important in geomorphological terms. A range of strategies exist (Table 1.5) for coping with moisture deficits and droughts: most obvious is that plant cover is usually less dense in arid areas than in more temperate regions, and an increased plant spacing results in less competition for moisture (Nobel, 1981). In some environments, communities also embrace different rooting depths, allowing water competition to be further reduced. This is in part an explanation for the 'two storey'

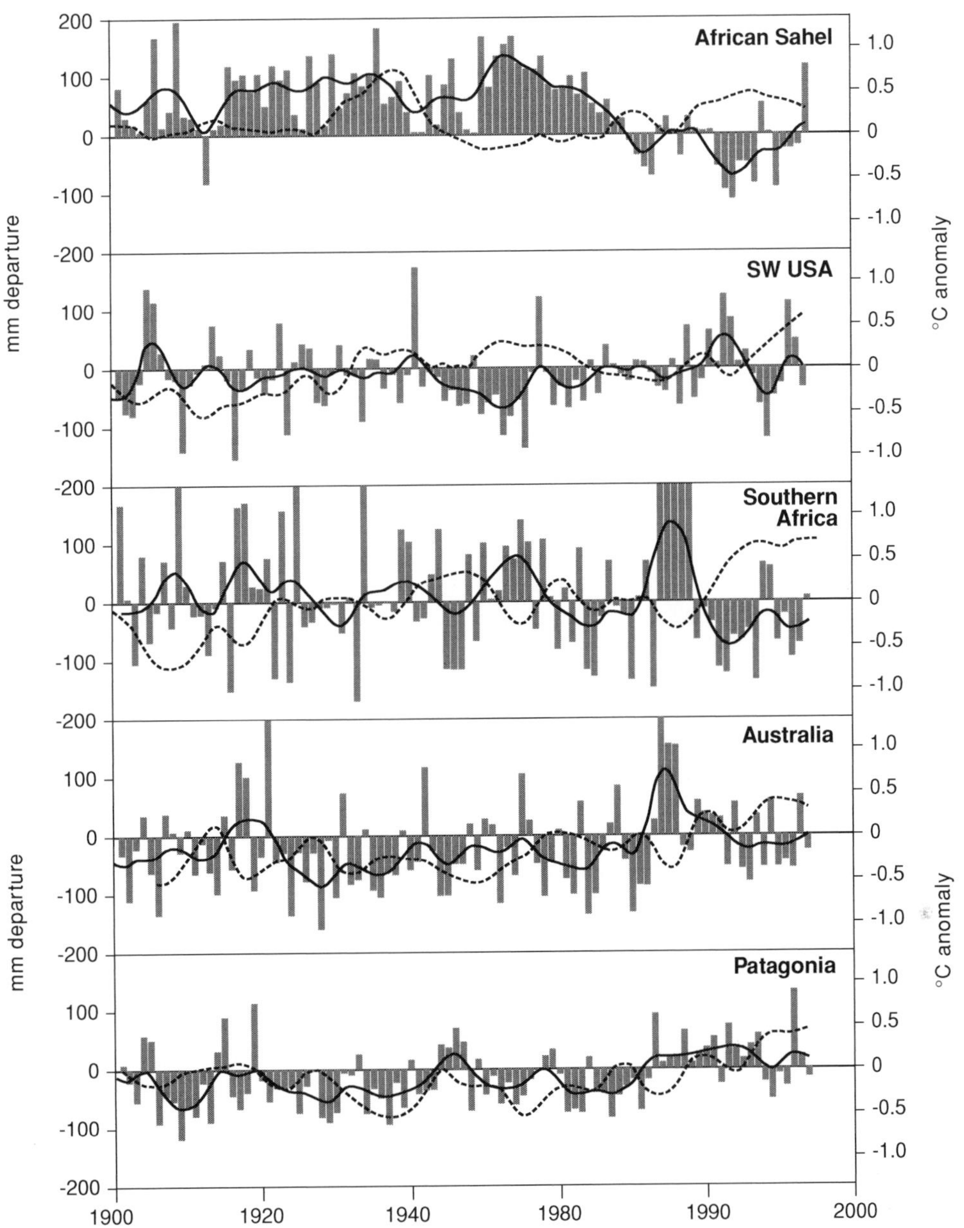

Figure 1.3 Climate variability for selected dryland areas during the twentieth century. Temperature anomalies and precipitation departures are shown as normalised data. Dashed lines are temperature data, histograms are annual rainfall and the solid line is the five-year rainfall running mean. The data show clearly how interannual variability is a 'normal' component of dryland climates, and that longer runs of wetter/drier warmer/cooler conditions can also occur (data adapted from Hulme, 1996).

nature of many savanna plant communities in semi-arid regions.

When droughts occur, plant communities that die back to escape moisture shortages have perhaps the greatest geomorphic significance, and changes in ground cover result. Such plants form part of what are known as nonequilibrium ecosystems (e.g. Ellis and Swift, 1988), where, in contrast to plant communities in more temperate zones, climax plant communities are not achieved over time because of dieback during dry periods. 'Boom and bust' cycles in cover follow temporal and spatial changes in moisture availability, resulting in a patchy mosaic of ground cover. This occurs at the seasonal timescale, with, for example, changes in cover from 80 to 3 % recorded from

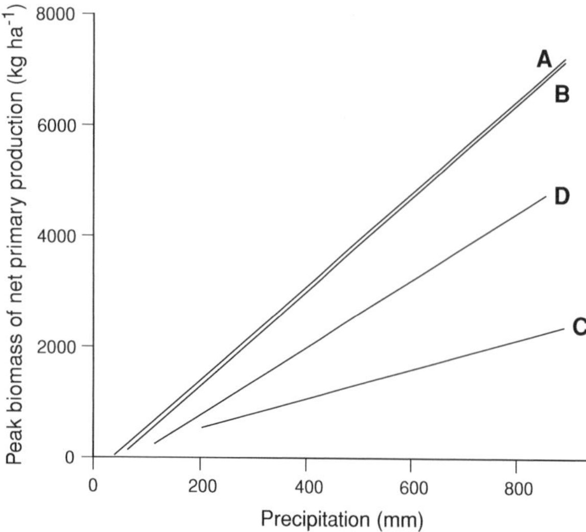

Figure 1.4 Suggested relationships between dryland precipitation and plant activity, as measured by biomass (A, B, C) or primary production (D) (adapted from Bullard, 1997, using data sourced from Desmukh, 1984, Le Houerou and Hoste, 1977, and Sims and Singh, 1978). A is from east and southern Africa, B from Mediterranean Africa, C from the Sahel–Sudan zone and D from the SW USA.

wet to dry seasons in Mtera, Tanzania (Thomas, 1988). During extended dry periods this can result in a marked and prolonged increase in bare ground (Smith *et al.*, 1993; Thomas and Leason, 2005, Figure 1.5), which leads to an expansion in areas that are susceptible to wind and water erosion (e.g. Yeaton, 1988).

1.8.1 Arid zone geomorphology

The role of moisture is often underrated in the assessment of geomorphological activity in arid environments. Surface runoff, whether occurring ephemerally or episodically, is of considerable importance. Even in the driest areas, high-magnitude sheet floods can have significant geomorphic effects, though they have rarely been observed or recorded (but see McGee, 1897, and Rahn, 1967). More important is that even low-intensity rainfall events can generate runoff (Cooke *et al.*, 1982; Goudie, 1985) because of the nature of desert surface conditions. However, the 'spottiness' of desert rainfall events can result in considerable spatial and temporal inequalities in its hydrological and geomorphological effects. The interval between individual rainfall events may also have significant implications for both surface conditions and the operation of specific geomorphic processes.

Many, perhaps all, landforming processes and their morphological expressions are not unique to arid environments. Some processes may operate more favourably or assume a greater relative importance than in other environments, and some landforms may be better developed or better exposed, the latter at least in part due to the limited vegetation cover. On the other hand, for many features, arid conditions may set the possibilities for their development, but their ultimate formation is dependent on suitable materials, lithologies or topographic settings being available.

1.9 Arid zone geomorphology and people

The human population of the global arid zone increased by 63.5 % between 1960 and 1974. By 1979, 15 % (651 million) of the world's population lived in arid lands (Heathcote, 1983). It is now estimated that drylands, as defined in UNEP (1992), support over 2 billion people (UNDP/UNCCD 2010). In some continents, arid areas are central to human occupation. In Africa, for example, 49.5 % of the total population live in arid areas (UNEP, 1992). Arid areas present a wide range of environmental hazards for their occupants, many of which are geomorphological (Table 1.6), which has prompted a strong and growing correlation (Goudie, 1985) between arid zone geomorphology and applied research (see, for example, Cooke *et al.*, 1982). To carry out applied geomorphological research requires a strong underpinning in the salient aspects of geomorphology and allied disciplines. It is to

Table 1.5 Plant coping strategies for arid zone moisture shortages.

	Escape	*Evade*	*Resist*	*Endure*
Plant type	Annuals and ephemerals	Perennials	Perennials	Perennials
Strategy	Dormant/die during dry season; survive as seeds	Tap deep water via extensive roots	Stored water in stems and roots	Strategies for reducing transpiration (inc night photosynthesis)
Example	Grasses, some herbs	Trees	Cacti and other succulents	Shrubs

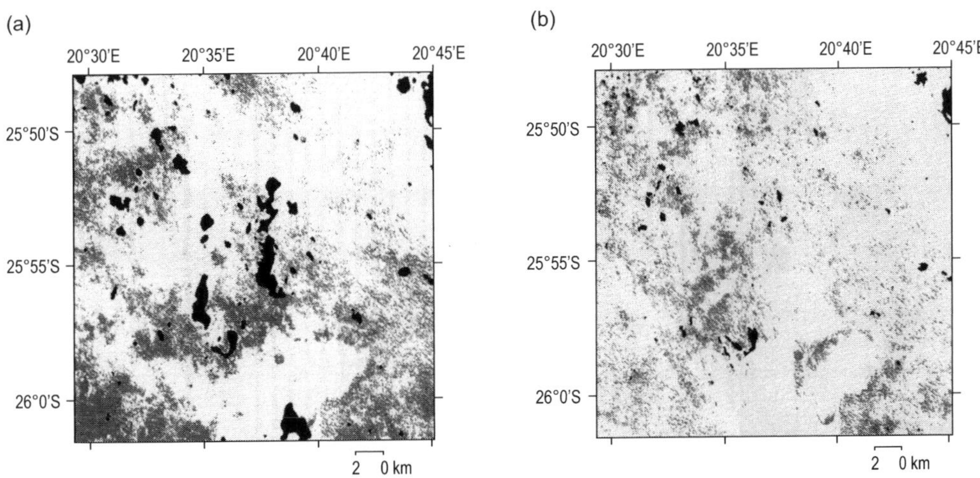

Figure 1.5 Interpreted Landsat TM images from part of the southwest Kalahari, Northern Cape, South Africa. These classified images are of the exact same area but from (a) September 1984 and (b) July 1993. They show how dryland vegetation cover can vary in response to precipitation change. In both images, black areas are river valleys, pan depressions or cloud cover: the remainder is typical Kalahari sand desert. In image (a) classification shows in dark grey areas with less than 14 % vegetation cover on the ground. This image covers the end of a period of several years of deep drought, whereas image (b) is from the dry season of a year in a period of normal (arid, c. 200 mm p.a.) rainfall. The <14 % cover area in 1984 was 10 % of the total scene, reduced to less than 3 % in 1993. The enhanced area, given sandy sediments, in 1984 was subject to increased wind erosion hazard, while the remaining 3 % in the 1993 image is the result of grazing pressure, mainly in association with livestock water points.

these, in the context of the arid zone, that this book is addressed.

1.10 Organisation of this book

There are many ways in which a book covering the geomorphology of arid environments could be organised. In this case, the main chapters have been divided among six sections for convenience, each devoted to a major theme in arid zone geomorphology.

Table 1.6 Some geomorphological hazards.

Water hazards	*Wind hazards*	*Materials hazards*
Flooding following rainfall	Dust storms	Landscales in mountainous areas
Subsidence after water extraction	Dune encroachment	Desiccation contraction of sediments
Gully development	Dune reactivation	Salt weathering
Soil erosion	Soil erosion	

Section One sets the physical framework for considering arid zone geomorphology considering the large-scale controls that shape drylands. The tectonic characteristics and settings of drylands are examined in Chapter 2, with the role of climate change, especially during the Quaternary period of geological time, in leaving environmental legacies investigated in Chapter 3. Chapter 4 addresses the question of how variable drylands are around the world, and on what basis any variability occurs. As environmental systems in extraterrestrial locations, notably Mars, are being explored systematically, many contributions to arid zone geomorphology on Earth are being made. This wider context is pursued in the final chapter of this section. The themes and issues examined in Section One are then utilised as necessary in the thematic sections that then follow, all of which consider both processes and landforms.

Section Two considers land surface processes and characteristics. It is often noted that the relatively vegetation-free surfaces of drylands allow bedrock and material characteristics to exert a greater direct (but not necessarily overriding or all-determining) influence on geomorphic development than in other environments. The six chapters in this section each consider a major aspect of surface conditions and processes. The processes and controls on rock weathering are discussed in Chapter 6, and soil systems, including the role of crusting, in Chapter 7. Chapter 8 considers other forms of crusts – duricrusts – that are a very important part of dryland landscapes. Arid zone surface and penesurface crusts are usually described and distinguished in terms of the dominant or characterising mineral constituent, and may occur within weathering residua and other deposits, soils or bedrock. In Chapter 9, pavements and mantles, which may result from either erosion or deposition, are explored. The remaining two chapters consider processes of surface runoff and flow (Chapter 10) and the development of slope systems in arid regions (Chapter 11).

Sections Three and *Four* examine the arid zone geomorphological process domains of water and wind, respectively. *Section Three* commences with a chapter that considers river systems in arid regions and their distinctiveness, while the processes than shape actual channels, and the channel forms themselves, are dealt with in Chapter 13. The subsequent two chapters in Section Three each deal with a major water-shaped component of arid landscapes: alluvial fans and pans, playas and salt lakes. The final chapter in this section considers how in drylands, so often devoid of surface water, groundwater may influence geomorphological processes.

Section Four contains five chapters, which consider the work of the wind in drylands. Chapter 17 analyses aeolian landscapes, as these provide the context for the range of landforms and processes that wind activity is responsible for. Processes of aeolian transport and sediment movement are then examined in detail in Chapter 18, while Chapter 19 considers the outcomes in terms of sand (focusing on bedforms, particularly dunes) and Chapter 20 considers dust. The final chapter in the section is an examination of wind erosion.

Section Five covers the issue of living with dryland geomorphology in terms of human impacts and natural hazards, and of the potential changes that the global arid zone may experience as a consequence of twenty-first century global warming.

References

Abrahams, A.D. and Parsons, A.J. (eds) (1994) *Geomorphology of Desert Environments*, Chapman Hall, London.

Agnew, C.T. and Chappell, A. (1999) Drought in the Sahel. *GeoJournal*, **48**, 299–311.

Aubreville, A. (1949) *Climats, Forêts et Désertification de l'Afrique tropicale*, Société d'Editions Géographiques Maritimes et Coloniales, Paris.

Birot, P, (1960) *Le Cycle d'Erosion sons les Differents Climats*, University of Brazil, Rio de Janeiro.

Bloom, A.L. (1978) *Geomorphology. A Systematic Analysis of Late Cenzoic Landforms*, Prentice Hall, Englewood Cliffs, NJ.

Bowler, J.M. (1976) Aridity in Australia: age, origins and expression in aeolian landforms and sediments. *Earth Science Reviews*, **12**, 279–310.

Budel, J. (1963) Klima-genetische geomorphologie. *Geographiche Rundschan*, **15**, 269–285.

Bullard, J.E. (1997) Vegetation and dryland geomorphology, in *Arid Zone Geomorphology*, 2nd edn (ed. D.S.G. Thomas), John Wiley & Sons, Ltd, Chichester, pp. 109–131.

Charney, J.G. (1975) Dynamics of deserts band drought in the Sahel. *Quarterly Journal of the Royal Meteorological Society*, **101**, 193–202.

Clark, A H., Meyer, E.S., Mortimer, C. *et al.* (1967) Implications of the isotope ages of ignimbrite flows, southern Atacama desert, Chile. *Nature*, **215**, 723–724.

Cooke, R.U. and Warren, A. (1973) *Geomorphology in Deserts*, Batsford, London.

Cooke, R.U., Warren, A. and Goudie, A.S. (1993) *Desert Geomorphology*, UCL Press, London.

Cooke, R.U., Brunsden, D., Doornkamp, J.C. and Jones, D.K.C. (1982) *Urban Geomorphology in Drylands*, Oxford University Press, Oxford.

Davies, W.M. (1905) The geographical cycle in an arid climate. *Journal of Geology* **38**, 1–27, 136–38.

de Martonne, E. and Aufrère, L. (1927) Map of interior basin drainage. *Geographical Review*, **17**, 414.

Desmukh, I.K. (1984) A common relationship between precipitation and grassland peak biomass for east and southern Africa. *African Journal of Range Ecology*, **22**, 181–186.

Ellis, J.E. and Swift, D.M. (1988) Stability of African pastoral ecosystems: alternative paradigms and implications for development. *Journal of Range Management* **41**, 450–459.

Flammand, G.B.M. (1899) La traversée de l'erg occidental (grands dunes du Sahara oranais). *Annales de Géographie*, **9**, 231–41.

Gilbert, G.K. (1875) Report on the geology of portions of Nevada, Utah, California and Arizona. *Geographical and geological explorations and surveys west of the 100th meridian, Part 1*, Engineers Department, US Army, pp. 21–187.

Gilbert, G.K. (1877) *Report on the Geology of the Henry Mountains*, Government Printing Office, Washington, DC.

Gilbert, G.K. (1895) Lake basins created by wind erosion. *Journal of Geology*, **3**, 47–49.

Glennie, K.W. (1987) Desert sedimentary environments, past and present – a summary. *Sedimentary Geology*, **50**, 135–166.

Goddall, D.W. and Perry, R.W. (eds) (1979) *Arid-Land Ecosystems: Structures, Functioning and Management*, Cambridge University Press, Cambridge.

Goudie, A.S. (1985) Themes in desert geomorphology, in *Themes in Geomorphology* (ed. A. Pitty), Croom Helm, London, pp. 122–140.

Goudie, A.S. (2002) *Great Warm Deserts of the World*, Cambridge University Press, Cambridge.

Grove, A.T. (1958) The ancient ergs of Hausaland, and similar formations on the south side of the Sahara. *Geographical Journal*, **124**, 528–533.

Grove, A.T. (1969) Landforms and climatic change in the Kalahari and Ngamiland. *Geographical Journal*, **135**, 191–212.

Grove, A.T. (1977) The geography of semi-arid lands. *Philosophical Transactions of the Royal Society of London*, Series B, **278**, 457–475.

Heathcote, R.L. (1983) *The Arid Lands: Their Use and Abuse*, Longman, London.

Hulme, M. (1992) Rainfall changes in Africa: 1931-60 to 1961-90. *International Journal of Climatology*, **12**, 685–699.

Hulme, M. (1996) Recent climatic change in the world's drylands. *Geophysical Research Letters*, **23**, 61–64.

Hulme, M. and Marsh, R. (1990) *Global Mean Monthly Humidity Surfaces for 1930–59, 1960–69 and Projected for 2030*, Report to UNEP/GEMS/GRID, Climate Research Unit, University of East Anglia, Norwich.

Huntley, B.J. (1982) Southern African savannas, in *Ecology of Tropical Savannas* (eds B.J. Huntley and B.H. Walker), Springer-Verlag, New York, pp. 101–119.

Köppen, W. (1931) *Die Klimate der Erde*, Berlin.

Lancaster, N. (1989) *The Namib Sand Sea: Dune Forms, Processes and Sediments*, Balkema, Rotterdam.

Le Houerou, H.N. and Hoste, H. (1977) Rangeland production and annual rainfall relations in the Mediterranean Basin and in the African Sahelo-Sudanian zone. *Journal of Range Management*, **30**, 181–189.

McGee, W.J. (1897) Sheetflood erosion. *Bulletin, Geological Society of America*, **78**, 93–98.

Meigs, P. (1952) Arid and semiarid climatic types of the world, in *Proceedings, Eighth General Assembly and Seventeenth International Congress*, International Geographical Union, Washington, pp. 135–138.

Meigs, P. (1953) World distribution of arid and semi-arid homoclimates, in *Arid Zone Hydrology*, UNESCO Arid Zone Research Series 1, pp. 203–209.

Meigs, P. (1966) Geography of coastal deserts. *UNESCO Arid Zones Research*, **28**, 1–40.

Mitchell, T.L. (1837) Account of the recent exploring expedition to the interior of Australia. *Journal of the Royal Geographical Society*, **7**, 271–285.

Mitchell, T.L. (1839) *Three Expeditions into the Interior of Eastern Australia*, Boone, London.

Mortimer, M. (1989) *Adapting to Drought: Farmers, Famine and Desertification in West Africa*, Cambridge University Press, Cambridge.

Nobel, P.S. (1981) Spacing and transpiration of various sized clumps of a desert grass, *Hilaria rigida. Journal of Ecology*, **69**, 735–742.

Parsons, A.J. and Abrahams, A.D. (1994) Geomorphology of desert environments, in *Geomorphology of Desert Environments* (eds A.D. Abrahams and A.J. Parsons), Chapman & Hall, London, pp. 3–12.

Passarge, S. (1904) *Die Kalahari*, Dietrich Riemer, Berlin.

Peel, R.F. (1939) The Gilf Kebir, *Geographical Journal*, **93**, 295–307.

Penck, A. (1894) *Morphologie der Erdoberfläche*, Stuttgart.

Powell, J.W. (1875) *Exploration of the Colorado River of the West*, Washington, DC.

Powell, J.W. (1878) *Report on the Lands of the Arid Region of the United States*, 1962 edn, Harvard University Press, Cambridge, MA.

Rahn, P.H. (1967) Sheetfloods, streamfloods and the formation of pediments. *Annals, Association of American Geographers*, **57**, 593–604.

Ross, G.M. (1983) Bigbear erg: a Proterozoic intermontane eolian sand sea in the Hornby Bay Group, Northwest Territories, Canada, in *Eolian Sediments and Processes* (eds M.E. Brookfield and T.A. Ahlbrandt), Elsevier, Amsterdam, pp. 483–519.

Sarnthein, M. (1978) Neogene sand layers off northwest Africa: composition and source environment, in *Initial Reports of the Deep Sea Drilling Project, 41* (eds Y. Lancelot and E. Seibold), US Government Printing Office, Washington, DC, pp. 939–959.

Schulze, B.R. (1972) South Africa, in *World Survey of Climatology Volume 10, Climates of Africa* (ed. J.F. Griffiths), Elsevier, Amsterdam.

Shantz, H.L. (1956) History and problems of arid lands development, in *The Future of Arid Lands* (ed. G.F. White), American Association for the Advancement of Science, Washington, DC, pp. 3–25.

Sims, P.L. and Singh, J.S. (1978) The structure and function of ten western North American grasslands. III. Net primary production and efficiencies of energy capture and water use. *Journal of Ecology*, **66**, 573–597.

Smith, W.A., Dodd, J.L., Skinner, Q.D. and Rodgers, J.D. (1993). Dynamics of vegetation along and adjacent to an ephemeral channel. *Journal of Range Management*, **46**, 56–64.

Spencer, B. (1896) *Report on the Work of the Horn Scientific Expedition to Central Australia*, Dulan, London.

Sturt, C. (1833) *Two Expeditions into the Interior of Southern Australia during the Years 1828, 1829, 1830 and 1831*, Smith, Elder, London.

Tankard, A.J. and Rogers, J. (1978) Late Cenozoic palaeoenvironments on the west coast of southern Africa. *Journal of Biogeography*, **5**, 319–337.

Thomas, D.S.G. (1988). Arid and semi-arid areas, in *Biogeomorphology* (ed. H.A. Viles), Blackwell, Oxford, pp. 191–221.

Thomas, D.S.G. (1993) Sandstorm in a teacup? Understanding desertification in the 1990s. *Geographical Journal*, **159**, 318–331.

Thomas, D.S.G. and Leason, H.C. (2005) Dunefield activity response to climate variability in the southwest Kalahari. *Geomorphology*, **64**, 117–132

Thomas, D.S.G. and Middleton, N.J. (1994) *Desertification: Exploding the Myth*, John Wiley & Sons, Ltd, Chichester.

Thomas, D.S.G. and Shaw, P.A. (1991) *The Kalahari Environment*, Cambridge University Press, Cambridge.

Thomas, D.S.G. and Shaw, P.A. (2002) Late Quaternary environmental change in central southern Africa: new data, synthesis, issues and prospects. *Quaternary Science Reviews*, **21**, 783–797.

Thomas, D.S.G. and Twyman, C. (2004) Good or bad rangeland? Hybrid knowledge, science and local understandings of vegetation dynamics in the Kalahari. *Land Degradation and Development*, **15**, 215–231.

Thompson, R.D. (1975) The climate of the arid world, University of Reading, Geographical Paper 35.

Thornthwaite, C.W. (1948) An approach towards a rational classification of climate. *Geographical Review*, **38**, 55–94.

Tricart, J. and Cailleux, A. (1969) *Le Modelé des Regions Seches*, Sedes, Paris.

Tucker, C.J., Dregne, H.E. and Newcomb, W.W. (1991) Expansion and contraction of the Sahara Desert from 1980 to 1990. *Science*, **253**, 299.

UN (1977) Status of Desertification in the Arid Regions, Climatic Aridity Index Map and Experimental World Scheme of Aridity and Drought Probability, at a Scale of 1:25,000,000. Explanatory Note, UN Conference on Desertification A/CONF. 74/31, New York.

UNDP/UNCCD (2010). *The forgotten billion: MDG achievement in the drylands.* UNDP Bureau for Development Policy, New York.

UNEP (1992) *World Atlas of Desertification*, Edward Arnold, Sevenoaks.

UNESCO (1979) *Map of the World Distribution of Arid Regions*, MAP Technical Note 4, UNESCO, New York.

Vogel, J.C., Rogers, J. and Seely, M.K. (1981) Summary of SASQUA Congress, May 1981. *South African Journal of Science*, **77**, 435–436.

Ward, J.D., Seely, M.K. and Lancaster, N. (1983) On the antiquity of the Namib. *South African Journal of Science*, **79**, 175–183.

Yeaton, R.I. (1988) Structure and function of the Namib dune grasslands: characteristics of the environmental gradients and species distributions. *Journal of Ecology*, **76**, 744–758.

Zeng, N., Neelin, J.D., Lau, K.-M. and Tucker, C.J. (1999) Enhancement of interdecadal climate variability in the Sahel by vegetation interaction. *Science*, **286**, 1537–1540.

2

Tectonic frameworks

Helen Rendell

2.1 Introduction

The tectonic setting of dryland areas has received comparatively little attention in the literature on arid zone geomorphology. References to tectonics are often limited to discussions of alluvial fans or of pans and playas, where active tectonics may have played some part in the development of these landforms. However, tectonic activity, in its broadest sense, provides the backdrop, in terms of both absolute and relative relief, against which dryland processes operate. Tectonic controls operate on sediment sources and sediment sinks and are, in turn, influenced by both erosion and deposition (Beaumont, Kooi and Willett, 2000). Tectonic settings also influence the timescales over which relative stability or instability have dominated particular areas, and thereby help to determine the continuity or discontinuity of sedimentary records.

The relationship between tectonics and landforms is complex (Summerfield, 1985). Until recently much of tectonic geomorphology has been concerned with using evidence from landforms to infer rates of operation of tectonic processes (Ollier, 1981; Mayer, 1985; Morisawa and Hack, 1985). Conversely, some geomorphological studies have invoked tectonics in order to explain landform evolution (e.g. Currey, 1994b). Scales of investigation are also changing:

> Now that the plate tectonics paradigm has matured, attention has been refocused on mesoscale problems. Among these problems are the more detailed tectonic and geomorphic evolution of plate boundaries and the feedback effects of surface processes, such as denudation and sedimentation, on the primary tectonic processes active at these boundaries (Beaumont, Kooi and Willett, 2000, p. 29).

The history of global climate change during the Quaternary is now well documented as a result of the detailed analysis of ocean sediment cores, spanning the last 2.7 million years (Myr), with ice cores providing a high precision record for the past 0.7 Myr. In common with other environments, dryland landscapes and sediments can be interpreted in the context of these changes (see Chapter 3). Dryland landscapes and sediments can also be used to establish or refine local sequences of climate change. In addition, since landforms may be used to provide insights into recent crustal movements (Burbank and Anderson, 2001), unravelling climatic and tectonic influences remains a key problem (Frostick and Steel, 1993b). Timescales are of critical importance in attempting to resolve the relative dominance of these influences and recent developments in dating both rock surfaces and sediments using cosmogenic nuclides, together with results from fission track thermochronology, have greatly aided resolution (Cockburn *et al.*, 2000).

A major problem is that tectonic influences at the Earth's surface have operated more or less continuously over the last 3 Myr whereas, over the same timescale, climatic conditions have changed in a cyclic manner. The loci of tectonic activity are well specified at the global level and are predominantly, but not exclusively, associated with active plate boundaries. Tectonic activity may appear episodic in the short term (10^1–10^3 yr) (McCalpin, 1993), but is effectively continuous in many areas when viewed over the longer term (10^4–10^7 yr). Rates of tectonic uplift are normally expressed in units of mm/yr or m/1000 yr, even though vertical displacements during individual events may be of the order of 10^2–10^3 mm. Many contemporary dryland areas have only been arid or semi-arid since some time during the Late Tertiary. The major exception may be the Namib Desert, which is thought to

Arid Zone Geomorphology: Process, Form and Change in Drylands, Third Edition. Edited by David S. G. Thomas

date from the Cretaceous (80 million years before present (80 Ma)), but both the Atacama and Australian deserts are considered to date from some point during the Miocene (22–5 Ma). The onset of aridity is recognised to have occurred at a global level c. 3 Ma (Williams, 1994). In addition, all desert areas have experienced climate change during the Late Quaternary, with some changes involving a change to humid or subhumid conditions (Chapter 3). Although Late Quaternary glacial/interglacial cycles have a frequency of 1×10^5 yr, some extremely rapid changes have been identified within these periods, at frequencies of 10^2–10^3 yr (Taylor *et al.*, 1993; Bond *et al.*, 1997, 2001).

The emphasis in this chapter is on the tectonic setting of contemporary dryland areas, but given the issue of timescales mentioned above, problems arise from the fact that the various landforms considered may be out of phase with contemporary climatic conditions. This is a particular problem in the case of erosional landforms such as pediments, for example, which may have evolved over a much longer period of time than that over which arid or semi-arid conditions are thought to have persisted in a particular area (Dohrenwend, 1994).

Arid zones are in many ways ideal places for the study of tectonics (see Burbank and Anderson, 2001). Large structures, fold and fault zones are clearly visible on remotely sensed images of arid zones. Lack of or limited vegetation leads to enhanced visibility of surface expressions of tectonic activity like fault scarps and, in hyper-arid areas such as the Atacama or southern Negev, scarps cutting alluvial fans may remain undegraded for substantial periods of time. Visibility is particularly important given the transient nature of many small features, which may result from seismic disturbance of the ground surface.

Although the scale at which tectonics operate may allow correlations at the macro- or megageomorphological scale of landform evolution, even at a mesoscale, tectonic influences may have considerable significance in arid zones (Beaumont, Kooi and Willett, 2000). In an arid context, for example, fault-related spring lines have a much greater significance than within environments in which water supply is far more plentiful. Also, salt diapirs provide one source of the salts that play a key role in many arid zone rock weathering processes.

This chapter will consider the tectonic setting of contemporary drylands, the controls that tectonics may have in terms of uplift and erosion and subsidence and sedimentation, the issues surrounding the development of a chronological framework for dryland landforms and sediments, the existing record of erosion and sedimentation and, finally, the interaction of active tectonics and contemporary processes.

2.2 Tectonic setting of drylands

Tectonic settings of contemporary dryland areas control the development of landforms and of sedimentary sequences by influencing factors such as sediment supply. It is possible to identify five types of tectonic setting: cratons (shield and platform areas); active continental margins, associated with Cenozoic orogenic belts; older, Phanerozoic, orogenic belts; interorogenic basin and range and inter-cratonic rift zones; and passive continental margins. Examples of each of these are given in Table 2.1.

The currently accepted view of what is 'tectonically stable' has undergone a radical transformation since the late 1960s. However, it is possible to identify particular tectonic settings on the basis of relative stability or instability and the term 'craton' is still used to describe the 'central stable portion of a continent' (Ollier, 1981, p. 75).

The Earth's crust and upper mantle together comprise interlocking lithospheric plates, which move relative to each other. New oceanic crust is continuously created at spreading centres (mid-ocean ridges and rifts), and this increase in crustal material is accommodated by 'loss' of crust by subduction, by folding and thrust faulting, or by movement along transcurrent faults. The most spectacular tectonic activity is associated with active plate margins, and with earthquakes and volcanic activity. Passive (or trailing-edge) continental margins are not without interest, with many of them characterised by the presence of so-called 'great escarpments' (Ollier, 1985a, 1985b). Not all crustal movements are of course compressional, and the development of major rift systems (grabens) and so-called 'basin and range' provinces are products of either mantle-generated or lithosphere-generated rifting (Frostick and Steel, 1993b).

2.3 Uplift and erosion, subsidence and sedimentation

Tectonically driven changes can involve both uplift and subsidence. Whether these changes achieve topographic expression is a function of the degree to which uplift exceeds erosion and to which subsidence exceeds sedimentation. According to Brookfield (1993, p. 16):

> . . . uplift is positive vertical elevation with respect to the geoid (basically mean sea level). This normally refers to net uplift, combining the effects of gross (or total) uplift minus the effects of erosion. Thus: net uplift = gross uplift − erosion,

Table 2.1 Tectonic settings of arid zones.

Contemporary tectonic setting	*Examples*	*Comments*
1. Cratons	Kalahari Great Karoo Australian Desert	Relative stability since the Late Tertiary
2. Active continental margins and Cenozoic orogenic belts	Atacama Sahara Sinai-Negev Arabia-Zagros	Compressional setting, thrust and transcurrent faulting
3. Older orogenic belts	Sahara	Some reactivation of existing fault zones
4. Interorogenic, intercratonic	Sahara (Afar, Ethiopia) Mojave Great Basin Sonora Chihuahua	Extensional tectonic setting, 'pull-apart' basins
5. Passive continental margins	Namib Patagonian	

Note: The physical extent of the Sahara Desert is such that it features in several different categories.

whereas Burbank and Anderson (2001, p. 131) add extra factors to give:

$$\text{Surface uplift} = \text{bedrock uplift} + \text{deposition} - \text{compaction} - \text{erosion.}$$

The interplay between uplift and erosion is more complex and subtle than originally thought (Summerfield, 2000; Montgomery and Brandon, 2002). If erosion keeps pace with gross uplift there is of course no net uplift, but while incision by a major river may keep pace with uplift, as has been the case in the Himalayan region, areal erosion rates within the catchment may not do so, with the consequent development of high relative relief (Burbank and Anderson, 2001). In arid zones, the ability of erosion processes to respond to tectonic displacements is of course affected by the fact that many geomorphic and hydrological processes within these zones are ephemeral in nature.

In an attempt to address the fundamental issue of the comparative rates of uplift and erosion at the global level, Schumm (1963) compared data on maximum rates of uplift (c. 7 mm/yr) and maximum rates of erosion (c. 1 mm/yr) and concluded that uplift greatly exceeds maximum erosion. The problem with this approach, as noted by Vita-Finzi (1986), is that the erosional data are effectively averaged-out areal data whereas the uplift rates may be unrepresentative of uplift at a regional level; that is erosional data maxima are underestimates while uplift data are overestimates. Contemporary data from tectonically active areas show that erosion rates (based on sediment fluxes in river systems) are broadly comparable with current rates of uplift (Table 2.2). Both spatial and temporal scales are important when dealing with erosion rates. Denudation rates appear to be time-dependent, an observation facilitated by the recent developments in cosmogenic and fission-track thermochronology dating. Von Blanckenburg (2006) contrasts the results of denudation rates calculated using sediment gauging data (timescale 10^1–10^2 yr), cosmogenic nuclides (timescale 10^1–10^6 yr)

Table 2.2 Comparison of uplift and denudation rates.

	References
Uplift rates	
10 mm/yr (Karakoram, Pakistan)	Brookfield (1993)
1 mm/yr (Great Himalaya, India)	Brookfield (1993)
2 mm/yr (Makran, Iran)	Vita-Finzi (1986)
2–10 mm/yr (Zagros, Iran)	Vita-Finzi (1986)
Erosion rates	
9 mm/yr (New Zealand)	Hovius , Stark and Allen (1997)
2.5 mm/yr (Central Himalaya)	Galy and France-Lenord (2001)
7 mm/yr (Nanga Parbat)	Shroder and Bishop (2000)

Table 2.3 Estimates of tectonic uplift rates as a function of tectonic setting.

Tectonic setting	*Recent vertical crustal movements*	*Neotectonic warping*
1. Cratons	Less than 1 mm/yr	
2. Active continental margins	Up to 20 mm/yr	Up to 10 mm/yr
3.Older orogenic belts	Up to 5 mm/yr	Up to 1 mm/yr
4. Interorogenic, intercratonic	Up to 10 mm/yr	Up to 5 mm/yr
5. Passive continental margins	Up to 10 mm/yr	Up to 1 mm/yr

Source: Fairbridge (1981).

and fission track thermochronology (timescale 10^6–10^7 yr) for particular regions.

The study of neotectonics has included the documentation of recent vertical crustal movements (RVCMs), which can be determined by instrumentation and measurement, and which are therefore limited to observations made during the last century (Fairbridge, 1981). The order of magnitude of these very recent movements and that of neotectonic measurements determined over longer timescales are summarised in Table 2.3 for the various arid zone tectonic settings identified in Table 2.1. Again there appears to be a scale dependency, with the RVCMs exceeding the neotectonic estimates by up to an order of magnitude.

Depositional sequences may also reflect tectonic influences. Frostick and Steel (1993a, p. 2) identify six ways in which fluvial or lacustrine sedimentation may be affected by tectonic controls:

1. Changes in the overall accommodation space available for filling within the basin.

2. Changes in the direction of tilt and the location and size of the depocentre.

3. Changes in the orientation and character of basin margins and in overall basin size and shape (e.g. by backstepping of marginal faults).

4. Changes in the gradient both at the margins of the basin, through faulting and within the basin (e.g. through tilting of the basin floor).

5. Deflection of the sedimentary systems where tectonically controlled morphological changes act as barriers to sediment transfer both into the basin from the hinterland and between various areas within the basin.

6. Changes in the rate of sediment supply uplift or erosion of the supplying hinterland.

One key characteristic of the supply of water and sediment to arid zone lake basins, and therefore of arid zone fluvial activity, is the coexistence of far-travelled perennial rivers with catchments outside the desert with ephemeral desert streams (Currey, 1994a). The latter have a capacity for transporting sediment during flood events that far exceeds that of similar-size perennial channels. One explanation for this high rate of bedload transport is that armouring of the channel bed has no time to become established in ephemeral bedload streams (Laronne and Reid, 1993). These differences between ephemeral and perennial streams have implications for both the rate and the nature of fluvial and lacustrine sedimentation in arid zone depocentres.

2.4 Lengths of record

From the point of view of process geomorphology, the different tectonic settings are associated with different potentials for sediment generation and sediment supply, and for the disruption of sedimentary sequences in particular climatic contexts. The record of change can be read in both erosional landforms and sequences of sedimentary deposits. The status of different variables changes as a function of the timescales considered. Prior to the development of radiometric dating techniques, the evolutionary state of a particular landscape was assessed with reference to Davisian or Penckian models (Summerfield, 2000; Burbank and Anderson, 2001, Figure 1.2), and this analysis was used to provide the relative age of that landscape (Thornes and Brunsden, 1977). The development of radiometric dating has shifted the burden of proof of antiquity from the landscape itself, but successful dating depends not only on the presence of suitable materials but also on the existence of techniques with appropriate time ranges of application. Until relatively recently most dating in arid zones was based on ^{14}C dating, which can only be used for materials containing organic or inorganic carbon, within the age range 0–45 ka. All ^{14}C dates also require calibration in order to take account of variable rates of production of ^{14}C in the upper atmosphere. With the development of other dating techniques, particularly the use of the cosmogenic nuclides ^{10}Be and ^{26}Al, it has become possible to extend age ranges back beyond that of ^{14}C. In the arid zone context, the main challenges lie in the dating of exposure age of surfaces and of fluvial, lacustrine and aeolian sediments. Taking these in turn, the exposure age of surfaces can be approached via:

1. Dating of varnish coatings (AMS ^{14}C) (Dorn, 1994).
2. Dating the exposure age of rock surfaces using cosmogenic isotopes (Gosse and Phillips, 2001; Cockburn and Summerfield, 2004).
3. Dating of palaeosurfaces – sealed by lava flows (dated using K/Ar or $^{40}Ar/^{39}Ar$) (Dohrenwend, 1994).

The dating of sedimentary deposits (fluvial, lacustrine, aeolian), which is also critical in understanding the role of climate change in the development of dryland landscapes (Chapter 3), may involve the application of the following techniques:

1. ^{14}C dating of organic carbon or inorganic carbonates.
2. Uranium-series dating of inorganic carbonates.
3. ^{36}Cl dating of evaporites (within lacustrine sequences) (e.g. Jannick *et al.*, 1991).
4. Cosmogenic dating of sediment clasts using ^{36}Cl for carbonates and ^{10}Be for quartz or chert (e.g. Matmon *et al.*, 2009).
5. Dating of component sediment grains using luminescence techniques, provided the sediment grains were exposed to light, prior to deposition and burial (e.g. Bristow, Duller and Lancaster, 2007; Sohn *et al.*, 2007), or have been heated to high temperature, e.g. in hearths or by proximity to lava flows (see Box 17.1 in Chapter 17).
6. Conventional K/Ar, $^{40}Ar/^{39}Ar$ or fission-track dating can be used in contexts in which volcanic ashes are intercalated with fluvial or lacustrine sequences.

In addition to the techniques listed above, fission-track thermochronology opens up the possibility of determining long-term rates of uplift and denudation (Gleadow and Brown, 2000).

Access to potentially datable material may be difficult and, depending on the depositional environment, it is often limited to natural cuttings and sections, sometimes revealed by tectonic disruption of sedimentary basins, and to boreholes. Seismic profiling may add considerably to the understanding of sedimentary basin architecture, but such remotely sensed data yield only relative chronological information.

2.5 Existing erosional and depositional records in arid environments

Given the issues associated with the application of the range of dating techniques outlined above to arid zone sequences and landforms, it is difficult to substantiate any assumptions about the lengths of record of erosion or sedimentation in different tectonic settings. Longer records are not necessarily associated with stable tectonic areas; instead tectonic activity may play a key role in ensuring continuity of sediment supply (erosion) and deposition. Data on existing lengths of record are summarised in Table 2.4.

At the global level, pediments tend to be concentrated on granitic terrains. The highest spatial concentration of pediments is reported from the southwest USA, from a basin and range area of interorogenic rifts (Cooke, Warren and Goudie, 1993). The more stable cratonic areas of southern Africa and central Australia also feature pediments, but it is the association with active volcanism that has allowed the development rate of the pediments in the Mojave Desert to be established (Dohrenwend, 1994).

At a more local level, a limited amount of information exists on the slope angles of pediments in Arizona and California in relation to proximity to faults. Data from Cooke, Warren and Goudie (1993, Table 13.1c) indicate that fault-associated pediments have slopes of $2°55' \pm 1°12'$, whereas those not associated with faults had slopes of $2°\ 10' \pm 48'$. Although difference in the means is statistically significant at the 0.01 % level, the problem is that, in physical terms, mean slope angles are very similar and there is significant overlap of the standard deviations.

2.5.1 *Drainage patterns and fluvial systems*

There is considerable evidence of the long-term impact of tectonic activity over time periods of 10^6–10^7 yr on patterns of drainage in contemporary drylands. Frostick and Steel (1993b) discuss the development of patterns of radial drainage in southern Africa in response to doming of the crust at a regional scale. The development of rift systems may also result in the reorganisation of drainage patterns, with some surprising results. In their model for the development of the East African Rift, Frostick and

Table 2.4 Length of record of landforms and depositional sequences in arid zones.

Landform/ depositional sequence	*Length of record and comments*	*References*
Pediments	Mojave desert and Great Basin: Minimum age of relict surfaces 8.9–0.85 Ma Maximum incision rates 47–8 m/Myr Maximum slope retreat rates 365–37 m/Myr	Dohrenwend (1994)
Desert pavements	Cima volcanic field: time required for development 0.2–0.7 Myr	McFadden, Wells and Dohrenwend (1986)
	Surface stability of Negev pavement 1.5–1.8 Ma	Matmon *et al.* (2009)
Lacustrine sequences	Intercalated volcanic ashes dated by $^{40}Ar/^{39}Ar$, used to constrain the age of lacustrine sequences in East African Rift (Lake Turkana) to beyond 3 Ma	Frostick and Reid (1987)
Aeolian sequences – sand seas	Age estimates of sand seas are derived from estimates of sand fluxes and exceed 1 Ma	Cooke. Warren and Goudie (1993)

Reid (1987) note that the backtilting and uplift of footwall blocks essentially diverts drainage away from the main axial depocentre. A similar example is provided by the Red Sea–Gulf of Aden region, where rifting began in the Miocene, in which rivers approaching within 2 km of the west bank of the Red Sea are diverted into the Nile catchment (Frostick and Steel, 1993b).

Tectonic disruption of drainage systems in the southwest USA is invoked by Zimbelman, Williams and Tchakerian (1994) in their argument that the ancestral Mojave River originally continued to flow eastwards to join the Colorado River rather than terminating in the Soda Lake–Silver Lake playas (Brown *et al.*, 1990). Tectonic activity has also played an important role in the Owens River catchment. In a recent study, Orme and Orme (2008, p. 223) conclude that:

> . . . the spillway appears to have been raised relative to Owens Lake since water last overflowed toward Searles Lake. These displacements, presumably linked to regional magmatism and fault activity, have major implications for the behavior of Owens Lake and its linkages southward. Such changes would be superimposed upon those attributable solely to climate change. Conversely, the deformed shoreline record provides new evidence for the later tectonic behavior of the Coso Range and the Owens Valley fault zone within the broader context of the eastern California shear zone (thus) . . ., over the past 23.5 (kyr) at least, Owens Lake has been filling and draining mainly in response to hydroclimatic forcing, but in a structural basin that has been subsiding relative to the rising Coso Range and the Haiwee spillway to the south, and to the Inyo Mountains and Sierra Nevada to the east and west, respectively.

2.5.2 *Playas*

Playa lakes (see Chapter 15) develop in local topographic lows within arid zone drainage systems. Many lakes are set in basins of inland (endoreic) drainage and it is generally recognised that tectonic activity is the fundamental cause of basin closure (Currey, 1994a). The astonishing profusion of playa lakes in the southwest USA, for example, is associated with tectonically active interorogenic basin and range settings (discussed in Chapter 4). Some drainage systems may be complex, with spillways coming into operation during lake high stands (e.g. the Owens River drainage system discussed above).

With the exception of the sequences associated with the East African Rift, which incorporate datable volcanic ashes and extend back to beyond 3×10^6 yr (Frostick and Reid, 1987), and the 2×10^6 yr record of sedimentation in the Owens River–Death Valley system based on ^{36}Cl (Jannick *et al.*, 1991), data on lacustrine/playa sequences still tend to be limited to the age range of ^{14}C of 4.5×10^4 yr (Brown *et al.*, 1990; Currey, 1994b).

2.5.3 *Desert pavements*

The antiquity of desert pavements has recently been demonstrated by Matmon *et al.* (2009) for pavements developed on old alluvial surfaces in the southern Negev. Using cosmogenic ^{10}Be they obtained exposure ages of 1.5–1.8 Ma for chert clasts. The combination of very low slope angles to horizontal surfaces and (contemporary) hyper-aridity provides ideal conditions of surface stability.

These old surfaces have erosion rates estimated to be of the order of 0.25–0.3 m/Myr (0.00025–0.0003 mm/yr).

2.5.4 *Aeolian sequences*

The antiquity of aeolian sequences is sometimes inferred on the basis of the volume of sand incorporated in sand seas or because of the presence of distinct bounding surfaces between different generations of dunes (Lancaster, 1992). Although it is recognised that some sand seas have developed and been reworked, over timescales of the order of 10^6 yr (Cooke, Warren and Goudie, 1993, p. 409), direct chronological information is often lacking. A major exception is in North America for sand ramps in Nevada, where the Bishop ash (K/Ar age 0.74 Ma) is preserved at the base of several of the ramps. The influence of tectonic activity on aeolian activity is likely to be indirect and a function of the timescale considered. Aeolian activity and sedimentation tends to be episodic in nature. In the short term, sediment availability may, for example, be a function of fluvial processes responsible for transporting material to a site for deflation. In the longer term, tectonics may be a major control on erosion and therefore sediment supply. If only short term records are available then it is likely that they will only be amenable to interpretation in terms of climate change rather than tectonism. Luminescence dating of aeolian sands has proved particularly useful in identifying periods of aeolian activity during the Late Quaternary and is discussed at length in Chapters 3 and 17.

2.6 Selected examples of the geomorphological impact of active tectonics in arid environments

2.6.1 *Tectonic disruption of fluvial systems*

The 1980 EI Asnam earthquake in Algeria provides one recent example of tectonic disruption of a fluvial system as a result of surface deformation. Active fold development resulted in the blocking of the Chelif River with the production of a lake some 5 km^2 in extent, and stratigraphical and archaeological evidence is used by King and Vita-Finzi (1981) to argue that local topography is the result of some 30 similar earthquakes over a period of 1.5×10^4 yr.

Large river systems tend to be prone to avulsion, which may or may not be tectonically triggered. Avulsion is when the river relocates itself during a flood event by occupying the lowest point on the floodplain. The history of the Indus River system provides some evidence of depositional sequences responding to crustal deformation. McDougall (1989) hypothesises that a major easterly switch of the Indus course occurred at some point 0.5–0.1 Ma owing to oblique compression in the active right lateral Kalabagh fault zone. Jorgensen *et al.* (1993, p. 310) discuss the interaction of climatic and tectonic influences on the course of the Lower Indus River. They note that:

> ... all of the significant river pattern changes take place at tectonic boundaries ... anomalously steep and gently sloping reaches of the river conform to suspected reaches of uplift and subsidence, and avulsions may mark the points of relative upwarp in contrast to basin subsidence.

2.6.2 *Tectonic controls on alluvial sedimentation*

Although the potential role of tectonics in alluvial fan development clearly involves the topographic setting of the fan, which in many cases may be at a fault-bounded mountain front, the role of active tectonics in fan development is still debated and climatic arguments tend to be dominant (Blair and McPherson, 1994). In their examination of alluvial fan sequences in the Dead Sea rift in Israel, Frostick and Reid (1989) demonstrate that the discrimination between climatically induced and tectonically induced facies changes is extremely difficult. On the basis of field evidence, they argue (p. 537) that:

> ... there is no reason to invoke tectonic destabilization of the river system as a cause of influx of coarse alluvium in the Dead Sea fan sequences ... punctuation of the sedimentary sequence by coarse sediment was achieved by climatic [their emphasis] rather than tectonic factors.

Similarly, Sohn *et al.* (2007) argue that alluvial fan sedimentation in the Black Mountains of southern Death Valley is responding to climatic rather than tectonic forcings. However, Rockwell, Keller and Johnson (1985) argue that active folding and faulting of alluvial fans in the Ventura Valley, California, have produced complex fan morphologies that are quite distinct from those in less active areas along the same mountain front.

2.7 Conclusions

The development of dating techniques, particularly cosmogenic nuclide dating of both rock surfaces and sediment clasts, has provided new data on the rates of uplift and erosion as well as new information on long-term landscape stability. Luminescence dating has also provided new insights into episodic aeolian activity, particularly during the Late Quaternary. Although tectonic settings influence the topography of dryland areas, the critical role played by climate cannot be understated, since climate drives arid zone processes by dictating both the availability of water and the nature of aeolian activity.

References

Beaumont, C., Kooi, H. and Willett, S. (2000) Coupled tectonic-surface process models with applications to rifted margins and collisional orogens, in *Geomorphology and Global Tectonics* (ed. M.A. Summerfield), John Wiley & Sons, Ltd, Chichester, pp. 29–55.

Blair, T.C. and McPherson, J.G. (1994) Alluvial fan processes and forms, in *Geomorphology of Desert Environments* (eds A.D. Abrahams and A.J. Parsons), Chapman & Hall, London, pp. 354–402.

Bond, G., Showers, W., Cheseby, M. *et al.* (1997) A pervasive millennial-scale cycle in North Atlantic holocene and glacial climates. *Science*, **278**, 1257–1266.

Bond, G., Kromer, B., Beer, J., Muscheler, R., Evans, M.N., Showers, W., Hoffmann, S., Lotti-Bond, R., Hajdas, I. and Bonani, G. (2001) Persistent solar influence on North Atlantic climate during the Holocene. *Science*, **294**, 2130–2136.

Bristow, C.S., Duller, G.A.T. and Lancaster, N. (2007) Age and dynamics of linear dunes of the Namib Desert. *Geology*, **35**, 555–558.

Brookfield, M.E. (1993) The interrelations of post collision tectonism and sedimentation in Central Asia, in *Tectonic Controls and Signatures in Sedimentary Successions* (eds L.E. Frostick and R.J. Steel), Special Publication No. 20 of the International Association of Sedimentologists, Blackwell Scientific Publications, Oxford, pp. 13–35.

Brown, W.J., Wells, S.G., Enzel, Y. *et al.* (1990) The Late Quaternary history of pluvial Lake Mojave–Silver Lake and Soda Lake Basins, California, in *At the End of the Mojave: Quaternary Studies in the Eastern Mojave Desert* (eds R.E. Reynolds, S.G. Wells and R.H.I. Brady), Special Publication of the San Bernardino County Museum Association, San Bernardino, CA, pp. 55–72.

Burbank, D.W. and Anderson, R.S. (2001) *Tectonic Geomorphology*, Blackwell Publishing, Oxford.

Cockburn, H.A.P. and Summerfield, M.A. (2004) Geomorphological applications of cosmogenic isotope analysis. *Progress in Physical Geography*, **28**, 1-42.

Cockburn, H.A.P., Brown, R.W., Summerfield, M.A. and Seidl, M.A. (2000) Quantifying passive margin denudation and landscape development using a combined fission-track thermochronology and cosmogenic isotope analysis approach. *Earth and Planetary Science Letters*, **179**, 429–435.

Cooke, R.D., Warren, A. and Goudie, A.S. (1993) *Desert Geomorphology*, UCL Press, London.

Currey, D.R. (1994a) Hemiarid lake basins: hydrographic patterns, in *Geomorphology of Desert Environments* (eds A.D. Abrahams and A.J. Parsons), Chapman & Hall, London, pp. 405–421.

Currey, D.R. (1994b) Hemiarid lake basins: geomorrphic patterns, in *Geomorphology of Desert Environments* (eds A.D. Abrahams and A.J. Parsons), Chapman & Hall, London, pp. 422–444.

Dohrenwend, J.C. (1994) Pediments in arid environments, in *Geomorphology of Desert Environments* (eds A.D. Abrahams and A.J. Parsons), Chapman & Hall, London, pp. 321–353.

Dorn, R.I. (1994) The role of climatic change in alluvial fan development, in *Geomorphology of Desert Environments* (eds A.D. Abrahams and A.J. Parsons), Chapman & Hall, London, pp. 593–615.

Fairbridge, R.W. (1981) The concept of neotectonics: an introduction. *Zeitschrift fur Geomorphologie, Supplementband*, **40**, VII–XII.

Frostick, L.E. and Reid, I. (1987) Tectonic control of desert sediments in rift basins: ancient and modern, in *Desert Sediments: Ancient and Modem* (eds L.E. Frostick and I. Reid), Geological Society of London, Special Publication No. 35, Blackwell Scientific Publications, Oxford, pp. 53–68.

Frostick, L.E. and Reid, I. (1989) Climatic versus tectonic controls of fan sequences: lessons from the Dead Sea, Israel. *Journal of the Geological Society, London*, **146**, 527–538.

Frostick, L.E. and Steel, R.J. (1993a) Tectonics signatures in sedimentary basin fills: an overview, in *Tectonic Controls and Signatures in Sedimentary Successions* (eds L.E. Frostick and R.J. Steel), Special Publication No. 20 of the International Association of Sedimentologists, Blackwell Scientific Publications, Oxford, pp. 1–9.

Frostick, L.E. and Steel, R.J. (1993b) Sedimentation in divergent plate-margin basins, in *Tectonic Controls and Signatures in Sedimentary Successions* (eds L.E. Frostick and R.J. Steel), Special Publication No. 20 of the International Association of Sedimentologists, Blackwell Scientific Publications, Oxford, pp. 111–128.

Galy, A., and France-Lenord, C. (2001) Higher erosion rates in the Himalaya geochemical constraints on riverine fluxes. *Geology*, **29**, 23–26.

Gleadow, A.J.W. and Brown, R.W. (2000) Fission-track thermochronology and the long term denudational response to tectonics, in *Geomorphology and Global Tectonics* (ed. M.A. Summerfield), John Wiley & Sons, Ltd, Chichester, pp. 57–75.

Gosse, J.C. and Phillips, F.M. (2001) Terrestrial in situ cosmogenic nuclides: theory and application. *Quaternary Science Reviews*, **20**, 1475–1560.

Hovius, N., Stark, C.P. and Allen, P.A. (1997) Sediment flux from mountain belt derived by landslide mapping. *Geology*, **25**, 231–234.

Jannick, N.O., Phillips, F.M., Smith, G.I. and Elmore, D. (1991) A 36Cl chronology of lacustrine sedimentation in the Pleistocene Owens River system. *Bulletin of the Geological Society of America*, **103**, 1146–1159.

Jorgensen, D.W., Harvey, M.D., Schumm, S.A. and Flamm, L. (1993) Morphology and dynamics of the Indus River: implications for the Mohen Jo Daro site, in *Himalaya to the Sea: Geology, Geomorphology and the Quaternary* (ed. J.F. Shroder), Routledge, London, pp. 288–326.

King, G.C.P. and Vita-Finzi, C. (1981) Active folding in the Algerian earthquake of 10 October (1980). *Nature*, **292**, 22–26.

Lancaster, N. (1992) Relations between dune generations in the Gran Desierto of Mexico. *Sedimentology*, **39**, 631–644.

Laronne, J.B. and Reid, I. (1993) Very high rates of bedload sediment transport by ephemeral desert rivers. *Nature*, **366**, 148–150.

McCalpin, J.P. (1993) Neotectonics of the northeastern basin and range margin, western USA. *Zeitschrift für Geomorphologie, Supplementband*, **94**, 137–167.

McDougall, J.W. (1989) Tectonically-induced diversion of the Indus River west of the Salt Range, Pakistan. *Palaeogeography, Palaeoclimatology, Palaeoecology*, **71**, 301–307.

McFadden, L.D., Wells, S.G. and Dohrenwend, J.C. (1986) Influence of Quaternary climatic changes on processes of soil development on desert loess deposits of Cima volcanic field. *Catena*, **13**, 361–389.

Matmon, A., Simhai, O., Amit, R. *et al.* (2009) Desert pavement-coated surfaces in extreme deserts present the longest-lived landforms on Earth. *Geological Society of America Bulletin*, **121** (5/6), 688–697.

Mayer, L. (1985) Tectonic geomorphology of the Basin and Range–Colorado Plateau boundary in Arizona, in *Tectonic Geomorphology* (eds M. Morisawa and J.T. Hack), Binghampton Symposia in Geomorphology, International Series No. 15, Allen and Unwin, London, pp. 235–259.

Montgomery, D.R. and Brandon, M.T. (2002) Topographic controls on erosion rates in tectonically active mountain ranges. *Earth and Planetary Science Letters*, **201**, 481–489.

Morisawa, M. and Hack, J.T. (eds) (1985) *Tectonic Geomorphology*, Binghampton Symposia in Geomorphology, International Series No. 15, Allen and Unwin, London.

Ollier, C.D. (1981) *Tectonics and Landform*, Longman, London.

Ollier, C.D. (1985a) Morphotectonics of continental margins with great escarpments, in *Tectonic Geomorphology* (eds M. Morisawa and J.T. Hack), Binghampton Symposia in Geomorphology, International Series No. 15, Allen and Unwin, London, pp. 3–25.

Ollier, C.D. (1985b) The great escarpment of Southern Africa. *Zeitschrift fur Geomorphologie, Supplementband*, **54**, 37–56.

Orme, A.R. and Orme, A.J. (2008) Late Pleistocene shorelines of Owens Lake, California, and their hydroclimatic and tectonic implications, in *Late Cenozoic Drainage History of the Southwestern Great Basin and Lower Colorado River Region: Geologic and Biotic Perspectives* (eds M.C. Reheis, R. Hershler and D.M. Miller), Geological Society of America Special Paper 439, Geological Society of America Inc., Boulder, Colorado, pp. 207–225.

Rockwell, T.K., Keller, E.A. and Johnson, D.L. (1985) Tectonic geomorphology of alluvial fans and mountain fronts near Ventura, Califorrnia, in *Tectonic Geomorphology* (eds M. Morisawa and J.T. Hack), Binghampton Symposia in Geomorphology, International Series No. 15, Allen and Unwin, London, pp. 183–207.

Schumm, S.A. (1963) The disparity between present rates of denudation and orogeny. *US Geological Survey, Professional Paper*, **454H**.

Shroder Jr., J.F. and Bishop, M.P. (2000) Unroofing of the Nanga Parbat Himalaya, in *Tectonics of the Nanga Parbat Syntaxis and the Western Himalaya* (eds M.A. Khan, P.J. Treloar, M.P. Searle and M.Q. Jan), Geological Society of London Special Publication 170, Geological Society of London, London, pp. 163–179.

Sohn, M.F., Mahan, S.A., Knott, J.R. and Bowman, D.D. (2007) Luminescence ages for alluvial-fan deposits in Southern Death Valley: implications for climate-driven sedimentation along a tectonically active mountain front. *Quaternary International*, **166**, pp. 49–60.

Summerfield, M.A. (1985) Plate tectonics and landscape development on the African continent, in *Tectonic Geomorphology* (eds M. Morisawa and J.T. Hack), Binghampton Symposia in Geomorphology, International Series No. 15, Allen and Unwin, London, pp. 27–51.

Summerfield, M.A. (2000) Geomorphology and global tectonics: introduction, in *Geomorphology and Global Tectonics* (ed. M.A. Summerfield), John Wiley & Sons, Ltd, Chichester, pp. 3–12.

Taylor, K.C., Lamorey, G.W., Doyle, G.A. *et al.* (1993) The 'flickering switch' of Late Pleistocene climate change. *Nature*, **361**, 432–436.

Thomes, J.B. and Brunsden, D. (1977) *Geomorphology and Time*, Methuen, London.

Vita-Finzi, C. (1986) *Recent Earth Movements*, Academic Press, London.

von Blanckenburg, F. (2006) The control mechanisms of erosion and weathering at basin scale from cosmogenic nuclides in river sediment. *Earth and Planetary Science Letters*, **242**, 224–239.

Williams, M.A.J. (1994) Cenozoic climatic changes in deserts: a synthesis, in *Geomorphology of Desert Environments* (eds A.D. Abrahams and A.J. Parsons), Chapman & Hall, London, pp. 644–670.

Zimbelman, J.R., Williams, S.H. and Tchakerian, V.P. (1994) Sand transport paths in the Mojave Desert: characterization and dating, Abstracts of a Workshop on Response of Eolian Processes to Global Change, Desert Research Institute, Quaternary Sciences Center, Reno, Nevada, Occasional Paper No.2, pp. 127–131.

3 Climatic frameworks: legacies from the past

David S.G. Thomas and Sallie L. Burrough

3.1 Introduction

In Chapter 1 the broad climatic controls on the distribution of drylands and deserts are presented. Morphological and sedimentary evidence indicate that the world's deserts have experienced significant episodic expansions and contractions during the Quaternary Period, and that these changes are primarily due to global climate changes. The geological record further indicates that the positions of deserts have changed in response to major global tectonic, as well as climatic, developments throughout Earth's history. Therefore, to appreciate fully the development of drylands and the landscapes, landforms and sediments that they possess, it is necessary not only to understand the geomorphological processes that occur within these areas today but also to develop a record of the environmental changes that these regions have experienced in the past.

3.2 The significance of arid zone fluctuations in the past

3.2.1 *Ancient arid zones*

The emphasis in this chapter is on changes in the arid zone during the Quaternary Period of geological time, particularly the Late Quaternary (last ~250 000 years). However, changes over longer geological time periods have also been important in desert evolution. Williams (1994) has noted that as a generalisation, erosional *landscapes* in deserts tend to be significantly older than the depositional *landforms* they possess: in other words, very ancient events have set the scene in which modern deserts have evolved. Sedimentary and tectonic events as far back as the Precambrian have played a role in the nature of some deserts today, such as the Sahara (Williams, 1984).

There is also an economic significance today for understanding the ancient aspects of deserts, as ancient evaporite and aeolian sediments can be important sources of hydrocarbons. The elucidation of the whereabouts and nature of ancient deserts is therefore of paramount economic importance and will be considered briefly. The sporadic existence of deserts has been identified from the rock record as far back as the Proterozoic (Glennie, 1987) (Figure 3.1). Various criteria may be used to credit an ancient sediment with an arid zone origin, including mineralogy, grain-size distribution, bedding structure and the micromorphology of individual particles (Table 3.1). These criteria may allow the depositional context or landform association of sediments to be determined. The utility of these criteria for palaeoenvironmental reconstruction is dependent upon the application of results from studies of today's arid zone phenomena, the so-called *uniformitarian* or *modern analogue* approach.

Studies of ancient desert sediments have led, for example, to the identification of different dune types from bedding structures in the rock record (see, for example, Ahlbrandt and Fryberger, 1982) and to the recognition of complicated depositional systems in ancient sandstones (Crabaugh and Kocurek, 1993). Perhaps most significantly, it has also been recognised that ancient desert systems were as variable and complex as those of today. However, the environmental significance of pre-Devonian aeolian sediments is likely to be rather different from that of modern sand seas, as prior to the colonisation of land surfaces by plants wind activity may have been a more potent geomorphological agent, even in relatively humid environments (Wilson, 1973; Glennie, 1987).

Arid Zone Geomorphology: Process, Form and Change in Drylands, Third Edition. Edited by David S. G. Thomas

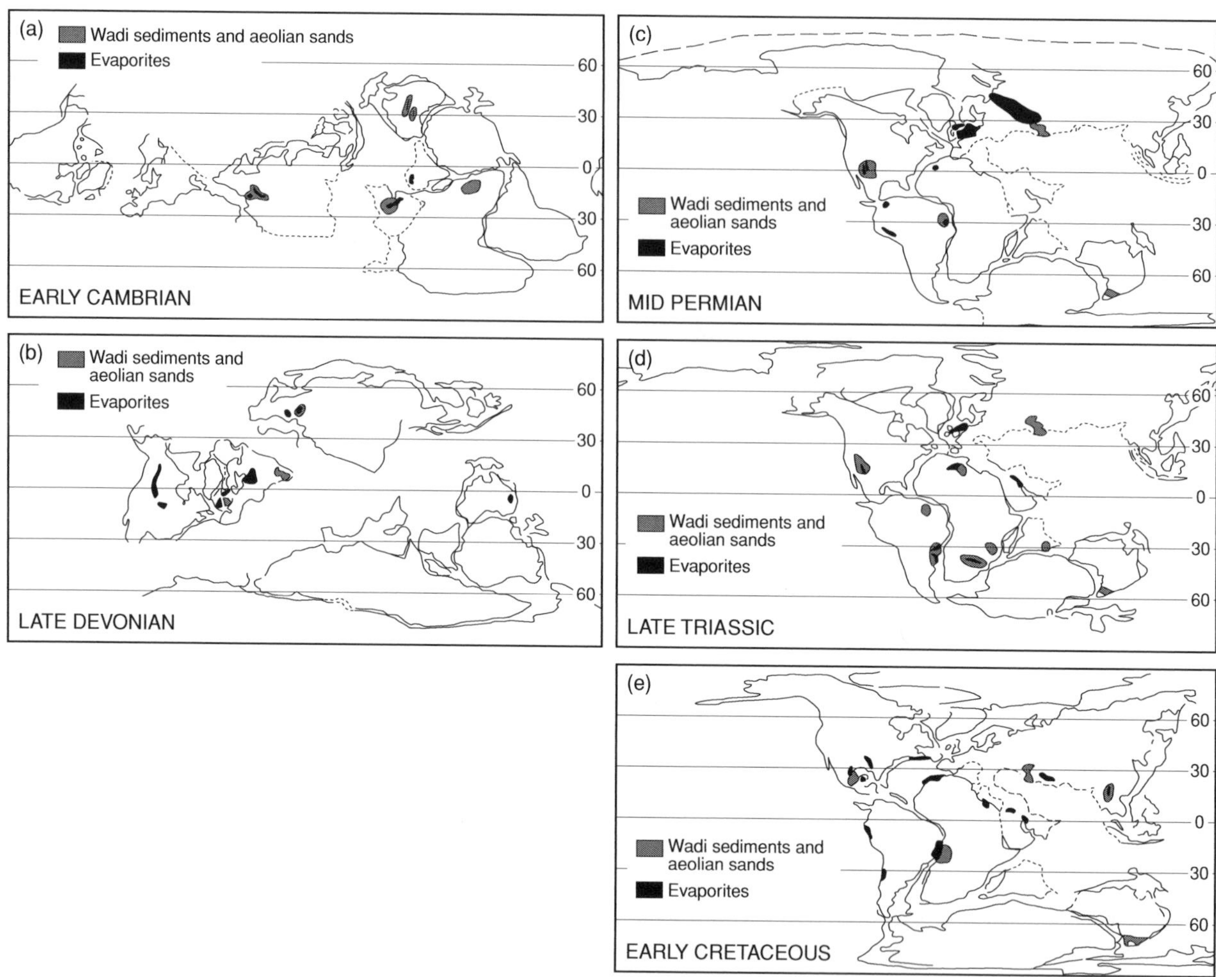

Figure 3.1 Arid zones from the geological record (after Glennie, 1987, Figure 26.1).

3.2.2 The development of aridity in the Mesozoic and Cenozoic

Many of the 'modern' deserts, which have in turn fluctuated in extent in association with Quaternary climate changes, owe their origins to global tectonic changes that commenced in the Mesozoic (from 245 Ma) and continued to develop into the Cenozoic era (from 66 Ma).

The division of Gondwanaland from c.176 Ma provided ultimately for the creation of Antarctica and ice cap growth, the development of atmospheric circulation patterns allowing the growth of subtropical high-pressure belts, and continentality and structural contexts, e.g. in southern Africa and Australia, that ultimately allowed for the development of the Kalahari and Australian deserts (Thomas and Shaw, 1991; Williams, 1994).

The uplift of the Tibetan plateau in the Pliocene at 2–3 Ma had a fundamental effect on atmospheric circulation. This created the opportunities for both global cooling and atmospheric CO_2 drawdown via weathering (Ruddiman and Kutzbach, 1991), and the development of an upper atmospheric easterly jet stream that provided dry subsiding air to eastern Asia, Arabia and the Saharan belt (summarised in Williams *et al.*, 1993). The growth of both Antarctic and northern hemisphere ice caps in this period (Shackleton *et al.*, 1984) also allowed for steeper pole–equator temperature gradients and resultant stronger trade winds, increasing the capacity for aeolian sand transport (Servant *et al.*, 1993). Broadly, it is these events that saw the change in global conditions during the warm Tertiary to the conditions favouring opportunities for glaciations and for aridity in continental interiors and the subtropics.

Table 3.1 Criteria for the identification of arid zone sediments.

Aeolian deposits
(a) Deposits may vary considerably in thickness
(b) Laminae dips range from 0 to 34° (repose angle for sand) unless affected by post-depositional Earth movements. Angles may be reduced by post-depositional compression
(c) Laminae bedding identifiable with structures in modern dunes (e.g. Ahlbrandt and Fryberger, 1982)
(d) Grain sizes range from coarse silt to coarse sand (c. 60–2000 μm, majority in 125–300 μm range
(e) Low silt and clay content (25 %, but see Table 3.2)
(f) Large particles often rounded, smaller are subrounded to subangular (see also Thomas, 1987a)
(g) May be cemented (aeolianite: Gardner, 1983) by haematite or calcium carbonate
(h) May be reddened (Gardner and Pye, 1981)
(i) Distinctive surface micromorphology when viewed using a scanning electron microscope (e.g. Krinsley and Doornkamp, 1973).
Water-lain deposits
(a) Commonly calcite cemented. Locally cemented by gypsum or anhydrite
(b) Conglomerates may be common
(c) Sand fraction may be absent – removed by deflation
(d) Mudflow conglomerates present
(e) Sharp upward decrease in grain size. Indicates rapid water-level fall
(f) Clay pellets, pebbles or flakes common. Due to effects of salt efflorescence (e.g. Bowler, 1986)
(g) Mud cracks common
(h) Often interbedded with aeolian deposits

Source: Modified after Glennie (1987).

3.2.3 The Quaternary Period

Deserts have been described as 'remarkable repositories of [Quaternary] palaeoclimatic information' (Williams *et al.*, 1993, p. 171). For the Quaternary Period, the sedimentological record of arid zone dynamics is often complemented by morphological evidence, in the form of features such as degraded or fossilised sand dunes, inactive rock varnished alluvial fans and palaeolake shorelines. The key, therefore, for identifying past arid zones is the identification of landforms/sediments diagnostic of arid conditions during their formation/deposition in areas that are not arid today. The converse applies for evidence of more humid conditions in drylands during the past: non-arid features found in today's arid zones. In recent decades, the identification of such landforms, often poorly defined or masked by vegetation on the ground, has been greatly enhanced by the availability of high-quality remotely sensed data. However, while in principle this approach to environmental reconstruction sounds seductively simple, in practice this is far from the case, for three main reasons. First, it relies on the assumption that process domains that are distinctively arid can be identified, and that this can be recognised from the resultant landforms and sediments. Second, it cannot be assumed that the climatic controls on these processes can be effectively parameterised to ensure that the arid attribution is correct. Third, it cannot always be assumed that all the features or sediments are not a result of extreme events due to natural short-term climatic variability rather than longer-term climatic change.

While geomorphological evidence is confined to the land surface (though 'drowned' features have been identified in some coastal locations: see Fairbridge, 1964; Sarnthein, 1972; Jennings, 1975), sedimentary data are more wide-ranging. The great advances in Quaternary science that have been forthcoming from the analysis of ocean sedimentary cores (see, for example, discussions in Imbrie and Imbrie, 1979) have also yielded valuable information on fluvial and aeolian inputs into the oceans from arid areas. On land, pollen and other biogenic evidence contained within sediments suitable for preservation can provide key evidence on vegetation community changes, including subtle shifts from arid to semi-arid regimes (Thompson and Anderson, 2000; Scott, 2002).

3.2.4 Sedimentary records

Sedimentary evidence of Quaternary arid zone expansions comes from terrestrial and marine sources. 'Comparative

Table 3.2 Some possible differences between active and ancient aeolian sands.

Active sands	*Ancient sands*
95 % of particles are sand-sized	Higher components of 'fines' due to subsequent inputs or weathering
Bimodal or unimodal size distribution	May be altered by post-depositional inputs/losses
Larger particles more rounded	Rounding may be reduced by post-depositional chemical weathering
Bedding structures are often present	Structure destroyed by burrowing animals and plant root growth
Low carbonate content	Carbonate contents increased through organic inputs

sedimentology' relies on the same uniformitarian principles that are applied in the interpretation of relict landforms, and suffers the same problems. Particularly important are the effects of post-depositional modification (diagenesis), which can alter or masque important diagnostic characteristics.

Aeolian deposits in both marine and terrestrial contexts can be identified by a range of sedimentary characteristics (Table 3.2), though these may be altered significantly after deposition. At Didwana in the Thar Desert, India, for example, a ~20 m high section in aeolian sands, diagnosed by their sedimentary structures and sedimentological characteristics, exposed in a canal cutting indicates that multiple periods of aridity have occurred since 190ka, the basal luminescence date Singhvi *et al.*, (2010). In the United Arab Emirates, quarrying of mega sand dunes has exposed the complex internal structure of dunes of over 60 m thickness. When subject to OSL dating, these reveal the long, and punctuated, records of aeolian accumulation that have occurred in the region during the late Quaternary (Figure 3.2). In other contexts, road cuttings (Figure 3.3) and occasional natural exposures reveal dune internal structures that aid record interpretations. However, in many studies, lack of internal exposures means that sampling for dating and analysis, via drilling (Figure 3.4), is done 'blind', thus limiting interpretation purely to the ages that are obtained by applying luminescence dating (see Chapter 17) and the properties of the sediments themselves. The use of ground penetrating radar may, however, in some contexts assist in overcoming this problem, as it may reveal the structures with dune bodies (Bristow, Lancaster and Duller, 2005).

Interpreting the environmental significance of sand with aeolian attributes is not necessarily simple (e.g. Fitzsimmons, Magee and Amos, 2009). In the Kalahari, the sands found not only in conjunction with relict aeolian landforms but also with those of lacustrine and fluvial origins have many of the attributes of aeolian deposits (Thomas, 1987c). This is because sedimentary characteristics can be inherited from preceding phases of deposition and reworking, or even from parent sediments. The same applies to reddened dune sands, which have sometimes been assumed to indicate aeolian stability and humidity (see Gardner and Pye, 1981) or other aspects of dune landscape history (Bullard and White, 2005). Therefore aeolian attributes in a sediment do not necessarily indicate that aeolian processes were the last mechanism of deposition.

Evaporite deposits are an important component of many arid and semi-arid basin environments (see Chapter 9) and the coastal sebkhas of some arid regions (Glennie, 1987). Lowestein and Hardie (1985) identified a three-stage cycle of ephemeral-playa sedimentation that results in the modification of a salt-crystal structure through dissolution and redeposition and the alternation of mud and salt layers. The sequences they identify in modern playa sediments are also identifiable in the sedimentary record and have assisted in the environmental interpretation of playa deposits preserved in the Quaternary record (see, for example, Hunt and Mabey, 1966; Smith *et al.*, 1983).

Although the minerals within playa salt deposits are somewhat dependent upon the chemistry of inflowing waters (Lowestein and Hardie, 1985), differences in evaporite deposits can yield information about the degree of salinity and hence evaporation rates during past periods of playa sedimentation (Ullman and McLoed, 1986; Risacher and Fritz, 2000). Bromide and chloride concentrations may be particularly valuable in this respect (Allison and Barnes, 1985; Ullman, 1985). Oxygen and carbon isotope ratios from lacustrine sediments may also yield useful information though applications may be more restricted than from ocean sediments (Stuiver, 1970; Gasse *et al.*, 1987; Liu *et al.*, 2009). Lowenstein *et al.* (1994) have analysed the fluid inclusions in halite from a 15 m lake sediment core extracted from the Qaidam basin, western China, for major elements and stable

Figure 3.2 A 60 m high exposure of the internal structure of a sand dune in RAK, UAE. Sampling through the exposed units for OSL dating will reveal the history of accumulation of this ridge during the Late Quaternary. Such an exposure, due to sand quarrying for construction, is unusual: normally dating dune history requires sampling via drilling, when the internal structure is not revealed.

Figure 3.3 Road cutting through a low linear dune in the Strzelecki desert reveals the internal dune structure, including weak palaeosols, which aid paleoenvironmental interpretation of dune history.

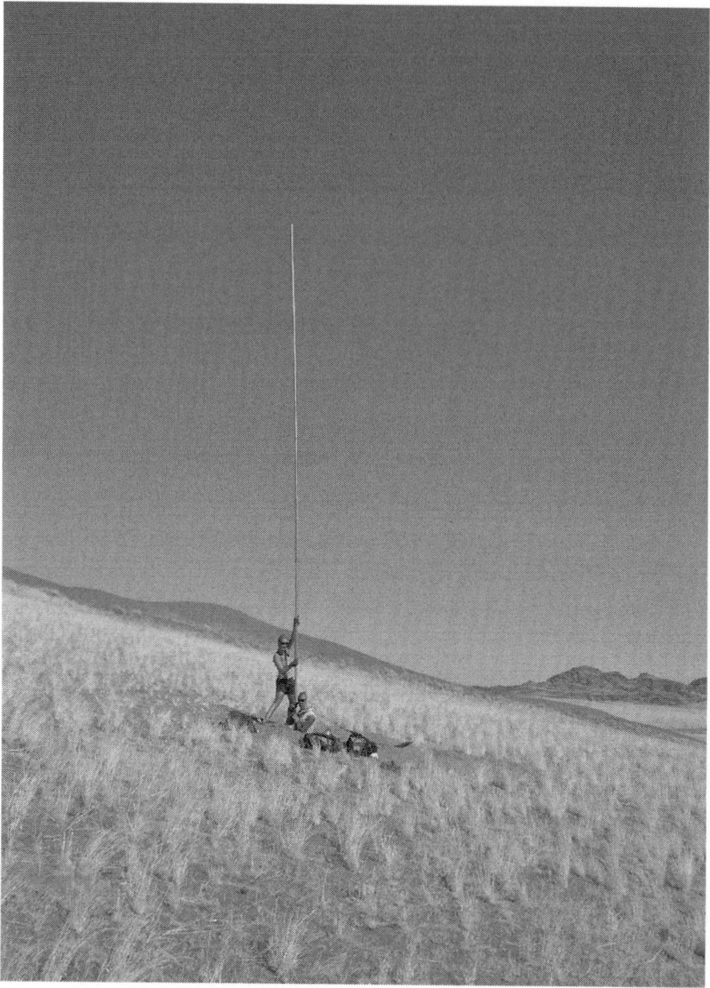

Figure 3.4 Drilling a linear dune in the eastern Namib Desert, Nambia. This approach can yield many samples for OSL dating, but limits analysis of the materials and structures within the dune itself (photo: Abi Stone).

isotopes to produce a uranium series chronology of aridity and atmospheric temperature spanning 50 000 years.

3.2.5 Marine sediments and palaeoaridity

Analyses of terrestrially-derived sediments in ocean cores have contributed greatly to interpreting the timing, intensity and extent of Quaternary arid zone extensions. Much land-derived detritus reaches the oceans through fluvial and aeolian pathways. Down-core changes in the abundance of inputs from these sources provide information on weathering and erosion processes and consequently the hydrological regimes on shore, together with an indication of the intensity and direction of wind systems (see Pokras and Mix, 1985; Stuut *et al.*, 2002; Holz, Stuut and Hentrich, 2004). Analyses of quartz and clay materials from onshore sources have in some cases been supplemented by studies of land-derived pollen (see Parmenter and Folger, 1974; Melia, 1984; Dupont *et al.*, 2000).

Terrestrial inputs are added to the relatively constant supply of oceanic organic carbonates (Bradley, 1999). Therefore, core sections with a relative abundance of carbonates indicate limited terrestrial inputs, with the inverse relationship applying when carbonates are poorly represented. Analysis of particle size (Schroeder, 1985; Stuurt and Lamy, 2004), mineralogy (Windom, 1975; Hashimi and Nair, 1986), quartz grain coloration (Diester-Haass, 1976) and particle surface microrelief (Krinsley, 1978) have been used to determine the transportation pathway and hence palaeoenvironmental significance of the terrigenous components of ocean-core sediments. Recent studies have tended to focus on particle size variations alone (Stuut *et al.*, 2002).

Aeolian dust inputs are represented by a high proportion of quartz silt, which may have red-staining compared with paler fluvial silts (Diester-Haass, 1976). An increased abundance of aeolian silts has been attributed to four main factors (Street, 1981; Stuut *et al.*, 2002): enhanced wind strengths, changes in wind directions, a thickening of the dust transporting layers of the atmosphere and enhanced aridity in the source area (Warren *et al.*, 2007). Enhanced inputs of aeolian dust, and in some cases dune sand, to Atlantic Ocean cores off west Africa (Sarnthein and Diester-Haass, 1977; Kolla, Biscaye and Hanley, 1979; Holz, Stuut and Hentrich, 2004), downwind from the Australian deserts (Thiede, 1979; Hesse and McTainsh, 1999) and in the Arabian Sea (Kolla and Biscaye, 1977) occurred within Quaternary cold periods. The relationship holds for at least the last 600 000 years (Emiliani, 1966; Bowles, 1975), while dust is detectable almost back to 2 million years BP in some core sediments (Parmenter and Folger, 1974).

Comparison of the spatial distribution of Holocene and Late Quaternary aeolian deposits in the east Atlantic cores (Figure 3.5) led Diester-Haass (1976) to propose a southward shift and Kolla, Biscaye and Hanley (1979) the southward expansion of the Sahara during the last glacial maximum. The identification of freshwater diatoms in later Quaternary Atlantic dust as far south as 20°N, interpreted as deflated material from dry lake floors by Parmenter and Folger (1974), also supports the view of Saharan extension at this time.

As well as indicating that some deserts were more extensive at various times during the last glacial cycle, and in some cases centred on the last glacial maximum, the widespread occurrence of aeolian dust in ocean cores also suggests that conditions were windier (Sarnthein *et al.*, 1981; Hesse and McTainsh, 1999; Stuut *et al.*, 2002). The size of aeolian dust particles may not, however, be a good indication of the distance of travel as once supposed, for Schroeder (1985) has shown that the size of mixed-nuclei dust particles responds to water table fluctuations in the source area, not travel distance, though using an end-member model may allow wind- and water-transported silts to be distinguished in ocean cores (Holz, Stuut and Hentrich, 2004). The ratio of high- to low-density dust particles, used by Diester-Haass (1976), may be a better indicator of the transportation ability of the wind and hence atmosphere circulation strengths at the time of deposition.

Fluvial inputs to ocean-core sediments can in some contexts also provide an estimation of the extent of tropical palaeoaridity (e.g. Hemming *et al.*, 1998). Sediments in the offshore Amazon fan, dated to the last glacial period, contain a high proportion of poorly weathered felspathic sand and little kaolinitic clay (Damuth and Fairbridge, 1970). This has been interpreted as an indication of less chemical weathering in source areas due to the spread of aridity to the Brazilian Shield (Damuth and Fairbridge, 1970) or Amazon headwaters (Milliman and Summerhayes, 1975). A similar interpretation has been applied to Late Pleistocene sediments from the Indian continental shelf (Hashimi and Nair, 1986). In other contexts the variations in terrestrial component grain sizes in a core have been used to infer the relative importance of aeolian and fluvial system inputs over time and associated hydrological regime changes (Tjallingii *et al.*, 2008).

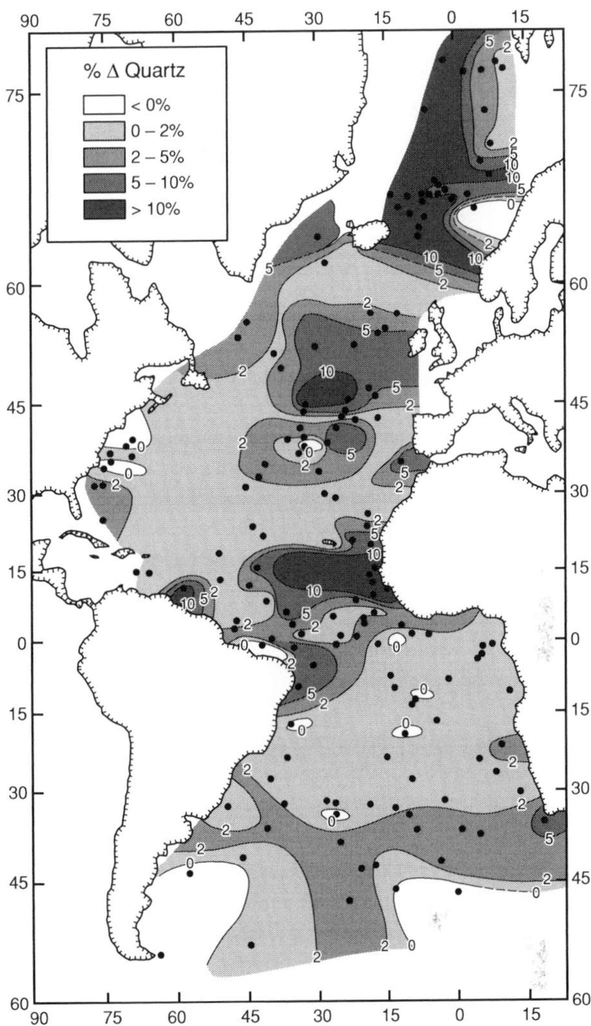

Figure 3.5 Difference in quartz percentage in late glacial and Holocene sections of Atlantic sediment cores. Enhanced late-glacial concentrations in the northern hemisphere tropical zone are interpreted as the product of dust inputs deflated from an expanded Sahara and/or due to stronger circulation systems (after Kolla, Biscay and Hanley, 1979).

Table 3.3 Methods that have been used to estimate the age of rock varnish formation (after Dorn, 1994).

Method	*Theory*	*Precision level* [a]	*Comments/case Examples*
Appearance	Varnish darkens over time	Relative	Controlled by factors other than time
Thickness	As varnish gets older, it grows vertically	Relative	Also controlled by microenvironment
Cover of black surface varnish	Varnish grows laterally away from nucleation centres	Relative	Derbyshire *et al.* (1984)
Other bottom varnish growth	As age increases, undersides of clasts are coated with Fe–clay rock varnish (Mn poor)	Relative	Derbyshire *et al.* (1984)
Trace element trends	Assumes varnish derived from underlying rock.		
	Trace element profiles with depth reflect time	Relative	Bard (1979)
Metal scavenging	Zn, Cu, Ni and other metals increase over time as they are scavenged by Mn–Fe oxides	Relative	Dorn *et al.* (1992b)
Palaeomagnetism	Magnetic field aligned when Fe oxides precipitate	Correlative	Clayton, Verosub and Harrington (1990)
Tephra-chronology	Glass fragments from known volcanic eruptions might be identifiable in rock varnish	Correlative	Harrington (1988)
Varnish geochemical layering	Sequences of chemical (e.g. Mn:Fe, Pb,$\delta^{13}C$) and textural changes correlated from site to site	Correlative	Dorn (1992)
Stratigraphy	Dating material on or under varnish constraints varnish age	Correlative and numerical	Dragovich (1986)
Cation ratio	Mobile cations are leached faster than immobile cations (K+Ca)/Ti	Calibrated	Dorn (1989); Dorn *et al.* (1990)
K–Ar dating	As varnish clays accumulate, they may undergo a diagenesis that refixes K or dates K in Mn oxides	Numerical	Dorn (1989); Vasconcelos *et al.* (1992)
Uranium series	Uranium precipitates with Mn oxides and then decays	Numerical	Knauss and Ku (1980)
Radiocarbon	Accreting varnish can capsulate underlying organic matter	Numerical	Dorn *et al.* (1992a)

[a] *Relative, correlative, calibrate,* and *numerical* are Quaternary dating terms recommended by Colman, Pierce and Birkeland (1987); see text for details.

3.2.6 Rock varnish

In some drylands, especially in the American southwest, valuable palaeoenvironmental information has been gained from rock varnish studies (see also Chapter 6; Dorn, 1990, 2007). The potential for rock varnish palaeoenvironmental contributions is considerable: varnish is widespread in deserts, can possess a long record and there is a growing body of evidence that suggests rock varnish microstratigraphy can provide a detailed environmental record otherwise unavailable from other sources (Dorn, 1994; Liu and Broecker, 2008). However, only varnishes in subaerial situations that represent a continuous deposition are suitable for investigation, while assumptions have to be made about rates of deposition and the correlation and relative dating of microlayer variations to other palaeoclimate records. In some cases it is possible to use absolute age controls (Table 3.3), e.g. by using cosmogenic ^{36}Cl exposure dating or AMS ^{14}C dating of organic material trapped within varnish layers during their deposition (Watchman, 2000; Dorn, 2007).

It is the manganese component of rock varnish, represented in the Fe–Mn ratio that is a particularly useful palaeoenvironmental tool broadly representing changes in alkalinity (Dorn and Oberlander, 1982). In the central Sahara, multilayer rock varnish occurs despite the presently hyper-arid conditions that are not conducive to the biogeochemical processes of Mn enrichment (i.e. low alkalinity and limited fluctuations in Eh and Ph) necessary for varnish formation (Cremaschi, 1996, 2002; Zerboni, 2008). Under these conditions the Mn-rich varnish is a relict of differing conditions in the past and its notable mineralogical differences within the microlayers reflects a shift from humid to arid conditions during the Holocene Period (Zerboni, 2008).

The micromorphology of rock varnish may also yield information on past periods of aeolian activity (Dorn, 1986). When aeolian dust is abundant, the assimilation of particles into the varnish results in a structure of parallel microplatelets, but, with less dust in the atmosphere, accumulation occurs around distinct nuclei (Dorn and Oberlander, 1982). The superimposition of different micromorphologies can therefore result from changes in aeolian activity through time. For example, in the southwestern United States rock-varnish surfaces record periods of dustiness alternating with less dusty episodes, the latter coinciding with high lake-level stages (Dorn, 1986).

The stable isotope content of organic material incorporated in rock varnish is a further palaeoenvironmental data source. Dorn and De Niro (1985) found a significant correlation between ^{13}C content of varnish and moisture in the environment, with Dorn, DeNiro and Ajse (1987) using ^{13}C changes in varnish layers to deduce palaeoclimatic impacts on alluvial fan development.

3.2.7 Geomorphological evidence of arid zone change

3.2.7.1 Arid zone extension

Despite caveats, Fairbridge (1970, p. 99) suggested that 'inherited landforms' are among the most diagnostic evidence of past climates. Table 3.4 lists landforms that have been used to indicate former extensions of arid and semi-arid conditions. Fluvial systems provide some of the most debatable and complex lines of morphological evidence of former aridity, sometimes indicating dramatically oscillating wet–dry climate during the Quaternary (Nanson *et al.*, 2008). The complex controls on tropical watershed fluvial fluxes in the Late Quaternary have recently been reviewed by Thomas (2011). Morphological and sedimentological evidence are strongly coupled, with channel fills and terrace sequences providing the information of past flow conditions. However, as channel flow responds to a range of controls including, for example, climatic and tectonic influences, it can prove difficult to determine palaeoclimatic signals in fluvial sediments and forms (e.g. Frostick and Reid, 1989; Reid, 1994; Tooth, 2005). Arid conditions may favour channel aggradation. Reduced rainfall, often accompanied by a marked seasonality, means diminished stream power and the clogging up of channels with relatively coarse sediment, derived from slopes, because of reduced vegetation cover. Conversely, wetter conditions favour incision and terrace development. The terraces and sediments of the Nile have been interpreted in such a way by Adamson, Gillespie and Williams (1982), and the evidence used to indicate the Late Pleistocene extension of arid conditions. In Amazonia, Andean headwaters also aggraded during the Late Pleistocene arid phase (Baker, 1978), but lowland rivers, including the Amazon itself, incised their courses because sea levels were lower (Tricart, 1975, 1984). One of the biggest problems arrives because irregular flow in arid regimes rarely allows equilibrium forms to develop (Nanson, 1986; Tooth and Nanson, 2000; Tooth and McCarthy, 2006) and therefore diagnostic forms are rare. Flow variability and the ability of rare high magnitude events to conduct significant geomorphic work can also compromise interpretation. Chapters 12 and 13 explore these issues.

Under semi-arid and arid conditions, when vegetation cover is sparse, sediment transported may be more active on slopes than in channels. This may favour the accumulation of thick, lower-slope colluvium aprons. Such deposits have been identified in southeastern Africa and interpreted as evidence of widespread Late Pleistocene semi-arid conditions (Watson, Price Williams and Goudie, 1985; Thomas, 2011).

Alluvial fans may also act as buffers between arid zone slope and fluvial systems, with their sediments providing a valuable record of environmental changes in source areas (see Chapter 14). The proposition that alluvial fans aggrade more favourably under conditions of reduced vegetation cover has been made by authors such as Dorn (1988) and Blair, Clark and Wells (1990). However, the palaeoenvironmental interpretation of fan and slope deposits may be as contentious as that of channel features and sediments (Dorn, 1994), although systematic analysis of sediment changes and inferred flow variations may enhance interpretations (e.g. Harvey and Wells, 1994).

Most apparently diagnostic, in geomorphic terms, of arid zone expansions are sand seas comprising now-stabilised dune systems. Dune inactivity is inferred through a range of criteria, with vegetation cover most widely used, that lead to dunes being designated as relict or fossil forms (cf. Flint and Bond, 1968). The identification and mapping of inactive dune systems has played a large part in theoretical developments regarding the

Table 3.4 Examples of morphological evidence of arid and semi-arid zone extensions.

Landform	*Indicative of*	*Location*	*References*
Slopes			
Pediments in humid environments	Former extension of arid slope systems	Brazil	Bigarella and de Andrade (1965)
Colluvial mantles	Former devegetation due to drier climates	Swaziland	Watson, Price Williams and Goudie (1985)
Dissected alluvial fans	Former high sediment-water ratios due to aridity	Southwestern USA	Lustig (1965)
Alluvial fan construction	Reduction of vegetation in source areas due to aridity	Death Valley Southern Australia	Hunt and Mabey (1966) Williams (1973)
Rivers			
Clogged drainage	Former river blockage by dunes	Niger	Talbot (1980)
Incised channels	Devegetation and lower sea levels	Amazonia	Tricart (1975, 1984)
Aggraded rivers	Lower and more seasonal discharge	Nile	Adamson, Gillespie and Williams. (1982); Williams *et al.* (2010)
Dunes			
Vegetation linear dunes	Former drier climates and past circulation patterns	Australia Southern Sahara Kalahari	Bowler (1976); Fitzsimmons *et al.* (2007) Grove (1958); Lancaster *et al.* (2002) Thomas (1984); Stone and Thomas (2008)
Drowned barchans	Former aridity	Botswana	Cooke (1984)
Vegetated parabolic dunes	Drier conditions in past	Colorado	Muhs (1985); Muhs *et al.* (1986)
Gullied dunes	Former aridity	Sudan	Talbot and Williams (1978)
Lithifield dunes	Former aridity	India	Sperling and Goudie (1975)

expansion of desert conditions since the work of Sarnthein (1978). Such systems are not, however, as simple to interpret in palaeoclimatic terms as was once thought, with a number of issues arising in recent work (e.g. Hesse and Simpson, 2006; Stone and Thomas, 2008; Chase, 2009). The dune form most widely found in the 'relict' state, linear dunes, is perhaps the most stable of all major desert dune types, even when undergoing sediment transport, which both leads to their widespread preservation and inheritance from more arid times past, but which also complicates their clear palaeoclimatic interpretation, especially in situations where dunes lie close to the margins of present day aeolian activity. The palaeoclimatic signal that can be gained from relict dunes, therefore, to some degree depends on their location relative to contemporary environmental conditions (Thomas and Burrough, 2011). Because of their importance, reconstructions based on sand sea evidence are considered in greater depth in Chapter 17.

3.2.8 Arid zone contraction

Various geomorphological sources have yielded information evidencing past periods of greater humidity in present day arid areas (Table 3.5). Studies of closed-basin lake-level fluctuations from shoreline evidence (Figure 3.6) have become significant indicators of environmental changes in many areas of the tropics and subtropics (see reviews by Street and Grove, 1979; Street-Perrott and Roberts, 1983; Street-Perrott, Roberts and Metcalfe, 1985; Currey, 1990; Burrough and Thomas, 2009).

Closed basins are important because their levels adjust to changes in inputs, whereas open lake-basin levels are more likely to remain stable, as input increases can be balanced by changes in overflow. Palaeolake shorelines in the form of terraces or depositional beach ridges occurring within today's arid zone suggest a disequilibrium between these landforms and present day environmental conditions, leading to inferred periods of greater humidity

Table 3.5 Examples of morphological evidence of former more humid conditions in drylands.

Landforms	*Indicative*	*Location*	*References*
Rivers			
Gravel-bed channels remnants left in bas relief by erosion of surrounding soft sediments	Wetter climates/greater channel flow	Oman	Maizels (1987)
Underfit modern channels	Greater flow	Argentina	Baker (1978)
Dry valley networks	Higher rainfall and channel flow or higher groundwater tables and spring sapping	Kalahari, Botswana	Nash, Thomas and Shaw (1994)
Valleys crossed by dunes	Wet-arid climate changes	Southeastern Sahara	Pachur and Kropelin (1987)
Lakes			
Strandlines	Higher former lake levels	Kalahari	Burrough and Thomas (2009)
Slopes and hillsides			
Fossil screes	Colder, possibly moister conditions	Libya	McBurney and Hey (1955)
Cave and speleothem development	Greater local rainfall	Kalahari, Mexico, Somali	Brook, Burney and Cowart (1990)
Alluvial fan aggradation	Increased effective precipitation	California	Harvey and Wells (1994)
Deep-weathering profiles	Wetter climates	Niger	Williams, Abell and Sparkes (1987)

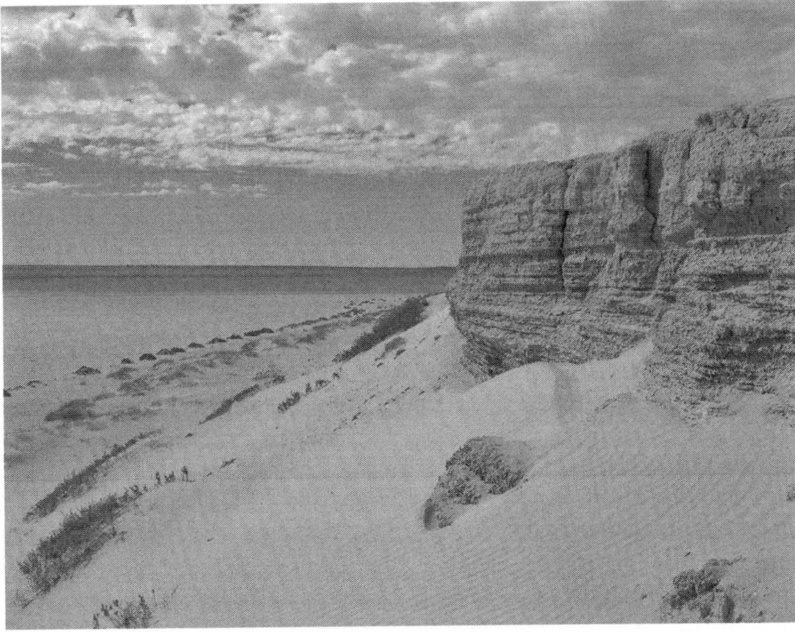

Figure 3.6 Williams Point, Lake Eyre, central Australia. Sediments on the basin margin reveal deposits that have accumulated as shoreline features during periods of more persistant lake occupancy of the basin.

in the past (see, for example, Servant and Servant-Vidary, 1980; Roberts and Wright, 1993; Metcalfe *et al.*, 1994; Drake *et al.*, 2008; Burrough, Thomas and Bailey, 2009).

In the southwestern United States, presently hyper-arid to semi-arid, evidence for more humid conditions during the last Pleistocene is found in the tectonically controlled 'basin and range' physiographic province (Smith and Street-Perrott, 1983). Shoreline and sedimentary features show that many individual basins were once occupied by lakes, some of which were linked together to form huge palaeolakes such as Bonneville (51 700 km^2 in area) and Lahontan (22 442 km^2) (see Chapter 4).

Palaeolake Mega Chad in West Africa (Grove and Warren, 1968) and megalake Makgadikgadi in Botswana (Grey and Cooke, 1977; Burrough, Thomas and Bailey, 2009) were very extensive (350 000 km^2 and 66 000 km^2, respectively) and relatively shallow at their maximum extents. Their limits are defined by relict shorelines that can occur as simple single beach ridges or in groups (beach ridge plains or strand plains) interspersed by swales (narrow, shallow trough-like depressions), such as the Dautsa ridge complex, Lake Ngami (Shaw, 1985). These landforms are highly dependent on the individual hydrological system and can be built in a single event or can exist as composite features accumulating sediment over multiple lake high stands in response to climate reorganisations at both the glacial (Rognon and Williams, 1977; Street, 1981) and sub-Milankovitch scale (Burrough *et al.*, 2007; Thomas *et al.*, 2009). While these geoproxies can be used to identify positive hydrological excursions within a basin, beach ridges are also subject to erosion and migration in response to changing sediment and water supply, so that gaps in the record of beach ridge accretion do not necessarily equate to times of low levels and could instead be a function of differential preservation.

In Australia, groundwater *and* surface-water fluctuations have been identified as important controls of lake-basin development (Bowler, 1986). The presence of lacustrine features and fringing transverse or lunette dunes in lake basins is not necessarily indicative of distinct periods of positive and negative water balance, for seasonal climatic regimes can result in the contemporaneous development of such landforms (Bowler, 1976, 1978). Although fluctuating lake levels produced active fringing lunettes during the last glacial period, the major deflation of dry lakebeds was broadly coincident with the LGM in the arid interior of Australia (Bowler, 1998; Bowler and Price, 1998; Magee and Miller, 1998), implying highly seasonal or arid conditions.

In north and east Africa and the interior of Australia, many lake basins were dry during the LGM, supporting evidence of desert expansion and aridity at that time (Rognon and Williams, 1977). In contrast, the lake basins in the southwestern United States preserve evidence that indicates that enhanced moisture availability characterised these areas during much of the Late Pleistocene due, with varying temporal significance, to increased precipitation and/or ice sheet meltwater inputs. During the early to mid-Holocene lacustrine phases have been documented in north Africa. Together with other offshore marine core evidence that suggests an abrupt change to a wet Sahara, this has given rise to the term the 'African Humid Period' (DeMenocal *et al.*, 2000), believed to be controlled by nonlinear feedback responses to monsoon dynamics, ultimately driven by precession-controlled changes in solar radiation receipt. Evidence that this period of desert contraction extended into Arabia has also been established using lake basin and sedimentary evidence (McClure, 1976; Parker *et al.*, 2006), though there are regional differences in the timings of lake rises and falls (Street, 1981). With improved dating and analysis of proxy records, it has been shown that many lacustrine basins have responded to hydrological changes more frequently and with greater magnitude than previously recognised (e.g. Scholz *et al.*, 2007; Thomas *et al.*, 2009; Burrough, Thomas and Bailey, 2009), their response reflecting not only the basin sensitivity to hydrological change but also the background sensitivity of the environment to abrupt regional or global scale climate shifts.

3.2.9 Ecological evidence

The oxidising conditions of contemporary arid zones results in a paucity of well-preserved organic records so that ecological proxy evidence, commonly utilised in temperate humid environments, is often lacking.

Vegetation change in response to changes in humidity, temperature and CO_2 levels are, however, commonly found in archives of pollen assemblages both outside the desert belt (Kiage and Liu, 2006) and laid down within offshore ocean cores. In addition, novel palaeoenvironmental archives such as speleothems (e.g. Brook, Burney and Cowart, 1990; Holmgren, Carless and Shaw, 1995) and fossilised rock hyrax middens (Scott, Marais and Brook, 2004; Chase *et al.*, 2009) facilitate pollen preservation even under arid conditions, and can be used to provide information on changing environmental conditions and the contraction and expansion of the arid zone during the Quaternary.

In addition to pollen itself, analyses of other, more persistent, vegetation-based biomarkers have provided an additional means of examining past changes in aridity. In particular, aridity is a primary control on the distribution of vegetation using either C3 or C4 photosynthetic pathways. C4 plants are adapted to arid conditions while C3 plants are generally found in more humid climates, so that temporal fluctuations in the proportional abundance of these types can be determined via the analysis of phytolith assemblages (silicified plant cell fossils). In Arabia, the use of phytoliths has helped to establish the abrupt onset of arid conditions at 4100 ca yr BP (Parker *et al.*, 2004). The short life cycle of grasses makes them particularly good indicators of rapid change as they are able to adapt quickly to prevailing environmental conditions. C4/C3 vegetation change is complicated as a proxy for periods of major climate change due to the equally important influence of ambient CO_2 and temperature on the adaptive advantage of the differing photosynthetic pathways (Ficken *et al.*, 2002). In addition to identification of the plant cells themselves and because of the distinct photosynthetic pathways, stable carbon isotope ratios of organic matter in soils, lake sediments, hyrax middens and speleothems can also be used to reflect the isotopic composition of the dominant vegetation assemblage that produced it (e.g. Olago *et al.*, 1999, 2000).

3.3 Dating arid zone fluctuations

Establishing a chronology of arid zone expansions and contractions during the Quaternary Period is subject to the application of relative and absolute dating techniques. In this respect, the advances in isotopic dating methods of the last few decades (see Walker, 2005; Bradley, 1999; and other Quaternary texts for discussions of these methods) have contributed to the overall picture of dynamic environments, to the recognition that some deserts are extremely old (Street, 1981) and to a clearer picture of the geomorphic responses of current desert and peridesert areas to climatic changes. Recent developments and applications of some dating methods, such as the luminescence family of techniques (see Chapter 17 for an explanation of luminescence dating), and the application of radiometric techniques to new data sources, such as hyrax middens (e.g. Chase, 2009), have had major implications for the development of chronologies of climate and environmental changes in drylands (e.g. Singhvi and Porat, 2008). This in part helps to overcome the problems in a dryland context associated with, and relative paucity of materials available for, more generally used methods such as radiocarbon dating (e.g. Deacon, Lancaster and Scott, 1984; Williams, 1985; Thomas and Shaw, 1991; Nanson *et al.*, 1992).

A broad framework for arid zone expansion has been achieved by the placement of aeolian sediments in the long timescales of oceanic cores. Together, the palaeomagnetic and isotopic evidence that they preserve has led to the establishment of the 'master chronology' of Cenozoic temperature and ice volume changes. Within this, major layers of aeolian sediments have been correlated to periods of global cooling extending back 38 million years (Sarnthein and Diester-Haass, 1977; Sarnthein, 1978). This contributed directly to the replacement of the 'glacial equals pluvial' hypothesis with one that equates glaciation with tropical aridity (Street, 1981; Goudie, 1983; Williams, 1985). Like its predecessor, this general framework is not without its problems (see Shaw and Cooke, 1986) and, as more records of past arid zone fluctuations have developed, so more complex patterns of change have been revealed (see recent reviews such as Gasse *et al.*, 2008).

3.4 Climatic interpretations and issues

Resolving the changes in climatic parameters that contributed to fluctuations in the extent of arid zones is often attempted from the terrestrial data sets that are used for environmental reconstructions, and which are summarised in the preceding sections. Most reconstructions focus on proxy data that inevitably point towards precipitation changes. There are limited terrestrial proxies in the arid context that facilitate temperature reconstructions, so studies may resort to data sets, e.g. from marine cores, where isotopic data can be used for temperature reconstruction (e.g. Schefuβ, Schouten and Schneider, 2005). Two examples from major terrestrial data sources can be used to illustrate some of the issues that arise when attempting terrestrial environmental and climatic reconstructions.

In the case of sand dune evidence, a key issue is to try and ascribe the relative roles of temperature and precipitation in leading to dune surface vegetation reduction (erodibility). For dunes dated, for example, to the LGM, precipitation is the likely leading candidate given global temperature reductions. Even if erodibility is increased, sufficient wind energy is required for aeolian activity to occur. The relative role of surface cover change and wind energy change is currently a hotly debated topic (Thomas, Knight and Wiggs, 2005; Chase and Brewer, 2009; Tsoar, 2005; see also Chapter 17) and in such instances other independent proxies, such as records of increased windiness from dust accumulations in offshore marine cores, may be important (Tiedemann, Sarnthein and Shackleton, 1994; Stuut *et al.*, 2002; DeMenocal, 2004).

Table 3.6 Examples of estimates of changes in climatic parameters in the southwestern USA, based on lake-basin studies. The diversity in inferred climatic parameters for each basin illustrates the difficulties of estimating a quantitative value of climatic change from geomorphic data.

Location	*Mean annual temperature (°C)*	*Summer temperature (°C)*	*Annual precipitation (mm)*	*Annual evaporation (mm)*	*Evaporation (%)*	*References*
Lake Lahontan, Nevada			+840		34	Antevs (1952)
Lake Lahontan, Nevada	−5	−5	+270	−410	30	Broecker and Orr (1958)
Lake Lahontan, Nevada	−3	–	+230	−180	16	Mifflin and Wheat (1979)
Lake Estancia, New Mexico	−6.5	−9	+180	−380	34	Leopold (1951)
Lake Estancia, New Mexico	−8	−8	−40	−250	42	Brackenridge (1978)
Lake Estancia, New Mexico	−10.5	−1	−46	−510	45	Galloway (1970)
Spring Valley, Nevada	–	−7	+210	−330	30	Snyder and Langbein (1962)
Spring Valley, Nevada	−8	−8	0	−480	43	Brackenridge (1978)

Lake high stands also have associated difficulties of interpretation because increased rainfall and reduced evaporation can both contribute to a positive hydrological balance (Smith and Street-Perrott, 1983; Bradley, 1985) (see Table 3.6). Various hydrological-balance models have been produced to try and resolve these issues (see Brackenridge, 1978; Street, 1979; Kutzbach, 1980; Bergner, Trauth and Bookhagen, 2003; Duhnforth, Bergner and Trauth, 2006). Additionally, beach ridges and terraces can be constructed and removed at times of stable high lake levels in response to increases or decreases in the rate of sediment supply and wind/wave energy. In all cases, additional biological and geochemical records from the lake floor deposits can potentially help to resolve these issues. Lake basins in drylands are, however, often devoid of surface water for at least part of the year, subjecting basin-floor sediments to aeolian deflation and highly oxidising conditions and resulting in a dearth of organic preservation. This can render a high-precision multiproxy approach, common to temperate and tropical lake investigations highly problematic in arid zones.

In the Makgadikgadi Basin, Botswana, budgetary studies have shown that the highest palaeolake stage cannot be accounted for in local climatological terms alone (Grove, 1969; Shaw, 1985). This has been confirmed by recent studies that have identified links between the palaeolake and increased fluvial inputs from distant, more tropical, sources (Shaw and Thomas, 1988; Burrough and Thomas, 2008; Burrough, Thomas and Bailey, 2009) (Figure 3.7). Where fluvial inputs have clearly made substantial contributions to high lake stages, but the palaeolake tributaries are now dry, as in the case of Lake Bonneville, it is extremely difficult to quantify their inputs in budgetary studies (Smith and Street-Perrott, 1983). This is further complicated in very large basins such as the Lake Eyre basin, megalake Chad and megalake Makgadikgadi, where increases in the surface area of water during lake high stands may have led to water recycling and significantly altered local or regional rainfall patterns that may have contributed to lake systems becoming partially self-sustaining beyond the influence of their initial forcing mechanisms (Coe and Bonan, 1997; Burrough, Thomas and Singarayer, 2009).

A further complication is that investigations in the eastern Kalahari have indicated that some dry valley systems have evolved through the influence of higher groundwater tables and spring sapping, with the magnitude of local humidity increases necessary to account for their development being less than once supposed (Shaw and De Vries, 1988; Nash, Thomas and Shaw, 1994).

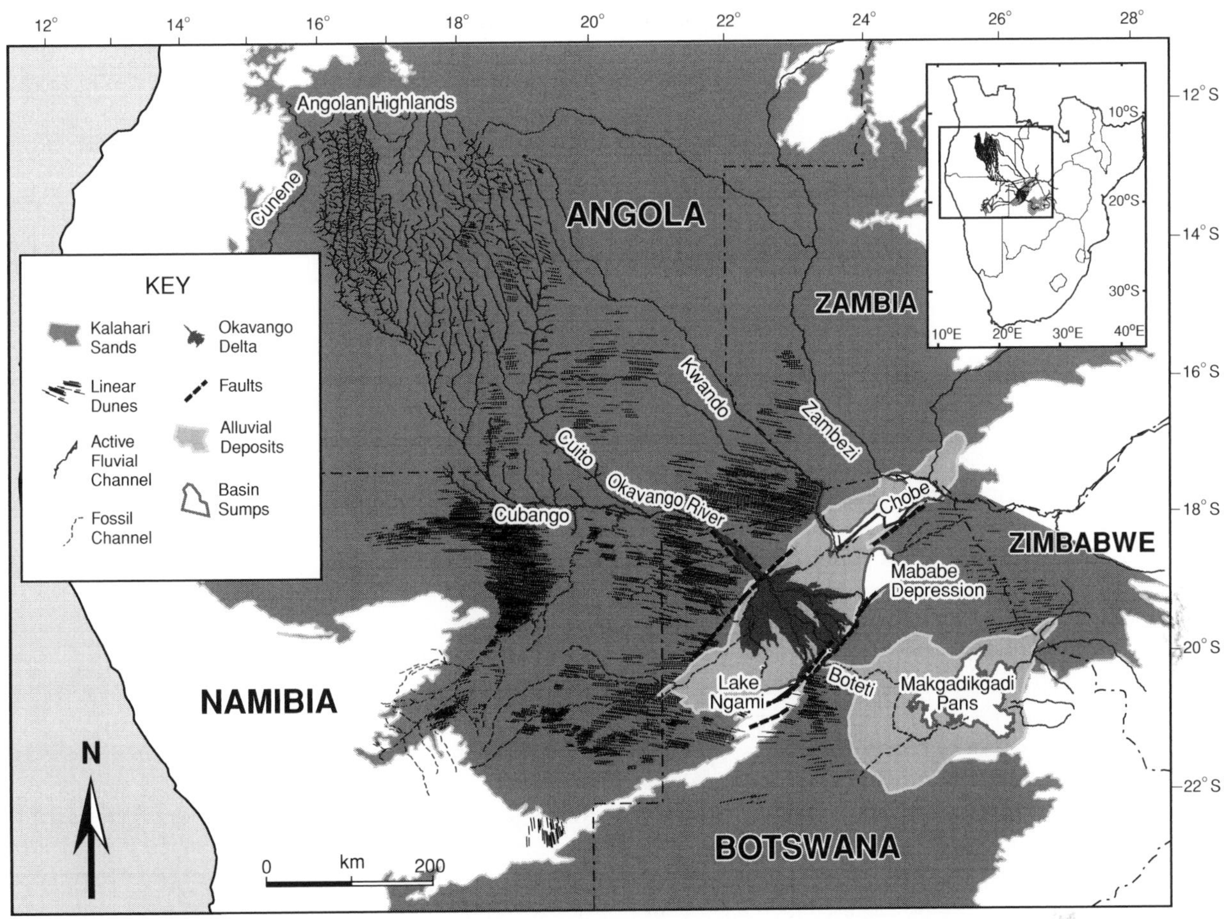

Figure 3.7 Makgadikgadi catchment map from Burrough, Thomas and Bailey (2009). This dryland basin clearly has a catchment that extends to wetter tropical locations, complicating climatic interpretations of lake-level changes in the past.

3.4.1 *Aridity during glacial times?*

As previously noted, shifts in low-latitude aridity are in many cases strongly correlated to variations in orbitally induced solar radiation and its hemispheric distribution. Increasing availability of both palaeoenvironmental evidence and its associated chronologies, however, has also demonstrated that the dynamics of environmental conditions in desert regions can in fact be regionally diverse due in part to feedback mechanisms and the interaction and reorganisation of circulation systems at a number of temporal and spatial scales. In addition, the superposition of abrupt climatic events can sometimes masque broader temporal patterns of aridity. Despite this recognition of variability, there has been a continued affirmation of the paradigm that equates high-latitude glaciations with trends in low-latitude aridity (Sarnthein, 1978; Benson *et al.*, 1997). This makes it appropriate to examine the distribution of expanded deserts, their nature and climatic mechanisms responsible for their development. Because the evidence is clearest, though complex, this will be done with respect to the period of the last glacial (c.80 000 BP to c.12 000 BP with maximum ice extent at c.24–18 000 BP). Global ice expansion had a number of profound effects on oceanic and atmospheric circulation and temperature regimes. Among the most important of these (cf. Nicholson and Flohn, 1980) were the steepening of meridional (pole–equator) temperature gradients; a latitudinal displacement of climatic belts and baroclinic zones including the intertropical convergence zone; and the weakening of the oceanic thermohaline circulation. The southward displacement of the northern hemisphere baraclinic zone and the associated upper westerly jet stream caused an increase in moisture-bearing weather systems reaching parts of the southwestern United States and the Mediterranean region for at least part of the late glacial period.

It has also been suggested that the southward shift of the belt of northern hemisphere subtropical high-pressure cells was accompanied by their expansion and intensification, because of the enhanced Hadley cell circulation caused by steeper temperature gradients (Flohn and Nicholson, 1980). A logical consequence of this would have been the southern extension of the Sahara arid zone as indicated by early studies, which suggested low lake levels in this region (Street and Grove, 1979; Servant and Servant-Vildary, 1980). More recent evidence is less equivocal, with some studies suggesting higher lakes in parts of this region during the last glacial (Scholz *et al.*, 2007) and other studies suggesting enhanced dune activity (Lancaster *et al.*, 2002). This is a further example of a growth in data leading to the emergence of a more complex record of change. Offshore ocean-core records containing dust layers (Sarnthein, 1978; Tjallingii *et al.*, 2008), and in-blown terrestrial phytoliths (Abrantes, 2003), have also yielded important evidence that correlate cold stages and interstadials as times of arid conditions in the southern Sahara and Sahel regions. Other records suggest that even beyond the expanded arid zone, conditions were drier and cooler in parts of tropical Africa (e.g. Livingstone, 1975; Kadomura and Hori, 1990; Barker and Gasse, 2003).

At the last glacial maximum, tropical land areas were about 2–6 °C cooler than today (e.g. Williams *et al.*, 2009). The steeper thermal gradient led to higher wind speeds (see Petit, Briat and Royer, 1981); therefore the Late Pleistocene low-latitude deserts were probably windier as well as colder than their modern counterparts (Hesse and McTainsh, 1999). Over the oceans, this could have increased evaporation, counteracting the effects of reduced sea-surface temperatures on atmospheric moisture levels. It has been suggested that stronger trade winds caused by steeper pressure gradients also led to enhanced upwelling of cold ocean waters in subtropical locations, further enhancing the aridity of coastal deserts such as the Namib and Atacama (Williams *et al.*, 1993).

Other authors suggest several factors, however, that could have mityigated seem to against this. First, the stronger Hadley cell circulation resulted in a greater upwelling of cold water in equatorial oceans (Hays, Imbrie and Shackleton, 1976; Molina-Cruz, 1977; Prell *et al.*, 1980), so that tropical waters were up to 8 °C cooler than today in the main areas of upwelling, potentially enhancing aridity in west coast deserts such as the Namib and Atacama (Williams *et al.*, 1993). Second, the equatorward compression of the atmospheric circulation and the expansion of subtropical highs resulted in the latter becoming even more persistent features than they are today (see Brookfield, 1970; Kolla and Biscaye, 1977). The penetration of northern hemisphere southwest trade winds into west Africa and northwestern India would consequently have been severely restricted (Nicholson and Flohn, 1980), contributing to desert expansion in both areas. Third, lower glacial-age sea levels increased the continentality of areas with broad continental shelves, further limiting the penetration of precipitation. In some areas, e.g. northeastern Australia, this was a major contributory factor in the spread of aridity (see Chappell, 1978). However, a further complication is added, at least in the southern African context, by marine-core studies that suggest through pollen analysis that the Namib Desert may have been wetter during the late glacial due to the equatorward displacement of westerly trade winds (Shi *et al.*, 2001). Whether this enhanced impact of winter westerlies occurred, or even penetrated into the southern African interior (Chase and Meadows, 2007), is hotly debated (Thomas and Burrough, 2011). Similar debates exist for the location of the impact of westerlies on western South America (Ammann *et al.*, 2001).

3.4.2 *Drivers of late glacial tropical aridity*

Late glacial aridity in north Africa and Australia is now well documented and relatively well dated (see Gasse *et al.*, 2008; Williams *et al.*, 2009). In north Africa in particular, the onset of the Holocene brings with it an abrupt step-change towards more humid conditions and the development and expansion of lake bodies and increased fluvial activity (Street and Grove, 1979; Geyh and Thiedig, 2008; Drake *et al.*, 2008). Pokras and Mix (1985) proposed a climatic model that equates north African tropical aridity with phases of high latitude ice growth and maximum humidity with deglaciated conditions at high latitudes. More recently, attention has turned to the role that insolation changes in the tropics may have played in affecting rainfall changes. Thus in Africa a more common hypothesis for humid/arid changes is via the influence of the monsoons, which are in turn governed by precessional variations in summer insolation (Anderson *et al.*, 1988; Kutzbach and Liu, 1997; DeMenocal *et al.*, 2000).

Over shorter timescales, major wet–dry phases at low latitudes either side of the equator in Africa, show a strong coincidence with abrupt warm–cold events in high northern latitudes, presumably related to ITCZ migrations. For example, the younger Dryas cooling event at 11–10 000 BP, recognised in the climate records of the mid latitudes, has now also been identified as causing a period of severe low-latitude aridity, as evidenced from East African lake-level records (Gasse, 2001).

On the Australian continent, environmental change during the last glacial period was regionally variable but can be broadly characterised by the expansion of the arid zone in response to weakened monsoon rains in the northern tropics and reduced moisture carried in on the westerlies in the south. Evidence from dry lake beds suggests that many areas, including the western and southern interior, showed marked aridity between 25 000 BP and 16 000 BP, through to 12 000 BP in the arid core (Bowler and Wasson, 1984). The picture is, however, far from clear-cut and is much more complex than was previously posited. Dune activation occurred in punctuated phases throughout the last glacial, often separated by periods of stability and soil formation (Nanson *et al.*, 1988, 1992; Nanson, Chen and Price, 1992; Fitzsimmons *et al.*, 2007). OSL and radiocarbon-dated regional wetlands (Williams *et al.*, 2001; Williams, Nitschke and Chor, 2006; Williams and Nitschke, 2005) and flood deposits (Haberlah *et al.*, 2010) in the now semi-arid Flinders Range occurred during the period ~35–16 ka, reflecting the complex response of the environment to climatic change.

The severity and extent of India's arid regions have been variable over time and heavily influenced by precipitation from both the southwest monsoon during summer and the penetration of westerly disturbances in the winter (Pant and Kumar, 1997). OSL ages on fossil dunes (e.g. Chawla, Dhir and Singhvi, 1992; Singhvi *et al.*, 1994) and fluvial systems (e.g. Kar *et al.*, 2001) as well as the lacustrine system response (Enzel *et al.*, 1995) have made a significant contribution to the reconstruction of fluctuating environmental conditions through the last glacial. Luminescence dates from relict and active dunes within the Thar Desert have been used to interpret major periods of dune construction at 115–100 kyr, ~75 kyr, ~55 kyr, ~30 kyr and 11–13 kyr (Singhvi and Kar, 2004; Singhvi and Porat, 2008). Dune-building activity in this region has been associated with increased windiness preceding the SW monsoon (Wasson, 1983). Singhvi and Porat argue that when the monsoon is weakened, as is hypothesised for the LGM, winds were weaker and there was minimum aeolian transport. Aeolian activity only resumed with increased intensity of the monsoon and its associated winds at around 14 kyr. At some point shortly after this, increased moisture from an established monsoon decreased the erodibility of sediment, so reducing dune accretion.

Initial evidence from north Africa and Australia (Williams, 1975, 1985; Rognon and Williams, 1977) or from oceanic cores (Sarnthein, 1978) suggested that tropical aridity was the norm during the late glacial. For example, Sarnthein (1978) has written that: 'Today about 10 % of the land area between 30°N and 30°S is covered by active sand deserts . . . [These were] much more widespread 18 000 years ago . . . [characterising] almost 50 % of the land area between 30°N and 30°S, forming two vast belts.'

The analysis of palaeotemperature data by Harrison *et al.* (1984) has indicated that there were considerable departures in the timings and extent of changes in the meridional temperature gradient in the northern and southern hemispheres. Their calculations suggest that, during the late glacial, climatic belts in both hemispheres could have been displaced southwards. This lends support to the views of Tricart (1956), Newell (1973) and Lancaster (1979), who suggested that, over Africa, the meteorological equator (the intertropical convergence zone) may have been displaced south of the equator during the last glacial. As well as latitudinal shifts in the hemispheric circulation systems, longitudinal covariance with ice volume has also been proposed. For example, in southern Africa, present day arid zone expansions are closely linked to expanded westerly circulation associated with intensified blocking anticyclone conditions inhibiting the penetration of moist easterly air from the Indian Ocean to the southern African interior (Tyson, 1986; Thomas and Shaw, 2002).

Increased sophistication of climatic models have allowed predictions of the theoretical configuration of late glacial circulation (Nicholson and Flohn, 1980) to be tested by climate simulations that can include complex interactions of forcings and feedbacks within the environment. The assumed mirroring of conditions in the tropics north and south of the equator (Figure 3.3) was based primarily on the assumption that this would be achieved through the reduction of interhemisphere temperature contrasts at times of maximum ice expansion. In contrast, some model outputs predict a hemispherically asynchronous pattern of aridity in response to orbital controls on insolation, which in turn controls the intensity and penetration of monsoon rain-bearing systems. A growing body of empirical evidence supports this (e.g. Partridge *et al.*, 1997). However, the potential for large-scale reorganisation of circulation systems and differing regional responses to persistent suborbital scale events such as Dansgaard–Oeschger and Heinrich events has also been demonstrated using coupled-ocean atmosphere GCMs (e.g. Thomas *et al.*, 2009).

3.5 Conclusions

There are many issues that remain unresolved in the reconstruction of the extent of past arid zones. Some of these concern the evaluation of morphological and sedimentological evidence. Bowler and Wasson (1984) illustrated how unusual aspects of the relict arid zone landscapes of Australia could be better understood when explanations incorporated an awareness of the environmental conditions preceding a climatic change. This concept of 'geomorphological inheritance' (Bowler and Wasson, 1984) also illustrates that the direction of an environmental change influences the geomorphological response. Knox (1972), for example, demonstrated how a change from aridity to humidity would be represented differently in the geomorphological record when compared to one in the opposite direction.

Other studies have shown that it can be exceedingly difficult to determine the relative contributions of temperature (via evaporation) and precipitation changes to the spread of aridity (Williams, 1985). However, perhaps of overwhelming importance is to recognise the dangers of accepting the simple replacement of a climatic model for the Quaternary, which equated glacial periods with pluvials by an equally simple one that links the growth of ice in high latitudes to the spread of aridity in the tropics.

Better dating resolution and an increase in palaeoenvironmental evidence has since highlighted both the spatial and temporal complexity of environmental responses to climate variations both within and between hemispheres over the last glacial period.

References

Abrantes, F. (2003) A 340,000 year continental climate record from tropical Africa: news from opal phytoliths from the equatorial Atlantic. *Earth and Planetary Science Letters*, **209**, 165–179.

Adamson, D.A., Gillespie, R. and Williams, M.A.J. (1982) Palaeogeography of the Gezira and of the lower Blue and White Nile valleys, in *A Land Between Two Niles* (eds M.A.J. Williams and D.A. Adamson), Balkema, Rotterdam, pp. 165–219.

Ahlbrandt, T.S. and Fryberger, S.G. (1982) Introduction to eolian deposits, in *Sandstone Depositional Environments* (eds R.A. Scholle and D. Spearing), American Association of Petroleum Geologists, Memoirs, vol. **31**, pp. 11–47.

Allison, G.B. and Barnes, C.J. (1985) Estimation of evaporation from the normally 'dry' Lake Frome in South Australia. *Journal of Hydrology*, **78**, 229–242.

Ammann, C., Jenny, B., Kammer, K. and Messerli, B. (2001) Late Quaternary glacier response to humidity changes in the arid Andes of Chile (18–29°S), *Palaeogeography, Palaeoclimatology, Palaeoecology*, **172**, 313–326.

Anderson, P. M., Barnosky, C. W., Bartlein, P. J. *et al.* (1988) Climatic changes of the last 18,000 years: observations and model simulations. *Science*, **241**, 1043–1052.

Antevs, E.A. (1952) Cenozoic climates of the Great Basin. *Geologishe Rundschau*, **40**, 94–108.

Baker, V.R. (1978) Adjustment of fluvial systems to climate and source terrain in tropical and sub-tropical environments, in *Fluvial Sedimentology* (ed. A.D. Miall), Canadian Society of Petroleum Geologists, Memoir, vol. **5**, pp. 211–230.

Bard, J.C. (1979) The development of a palination dating technique utilizing neutron activation and X-ray fluorescence analysis, Unpublished PhD Dissertation, University of California, Berkeley.

Barker, P. and Gasse, F. (2003) New evidence for a reduced water balance in East Africa during the Last Glacial Maximum: implication for model-data comparison. *Quaternary Science Reviews*, **22**, 823–837.

Benson, L., Burdett, J., Lund, S. *et al.* (1997) Nearly synchronous climate change in the Northern Hemisphere during the last glacial termination. *Nature*, **388**, 263–265.

Bergner, A.G.N., Trauth, M.H. and Bookhagen, B. (2003) Paleoprecipitation estimates for the Lake Naivasha basin (Kenya) during the last 175 k.y. using a lake-balance model. *Global and Planetary Change*, **36**, 117–136.

Bigarella, T.J. and de Andrade, G.O. (1965) Contribution to the study of the Brazilian Quaternary, Geological Society of America Special Publication 84, pp. 433–451.

Blair, T.C., Clark, J.S. and Wells, S.C. (1990) Quaternary continental stratigraphy, landscape evolution and application to archaeology: Jarilla piedmont and Tularosa graben floor, White Sand Missile Range, New Mexico. *Bulletin of the Geological Society of America*, **102**, 749–759.

Bowler, J.M. (1976) Aridity in Australia: age, origins and expression in aeolian landforms and sediments. *Earth Science Reviews*, **12**, 279–310.

Bowler, J.M. (1978) Glacial age aeolian events at high and low latitudes: a southern hemisphere perspective, in *Antarctic Glacial History and World Palaeoenvironments* (ed. E.M. van Zinderen Bakker), Balkema, Rotterdam, pp. 149–172.

Bowler, J.M. (1986) Spatial variability and hydrologic evolution of Australian Lake basins: analogue for Pleistocene hydrologic change and evaporite formation. *Palaeogeography, Palaeoclimatology, Palaeoecology*, **54**, 21–41.

Bowler, J.M. (1998) Willandra Lakes revisited: environmental framework for human occupation. *Archaeology in Oceania*, **33**, 120–155.

Bowler, J.M. and Price D.M. (1998) Luminescence dates and stratigraphic analyses at Lake Mungo: review and new perspectives. *Archaeology in Oceania*, **33**, 156–168.

Bowler, J.M. and Wasson, R.J. (1984) Glacial age environments of inland Australia, in *Late Cainozoic Palaeoclimates of the Southern Hemisphere* (ed. J.C. Vogel), Balkema, Rotterdam, pp. 183–208.

Bowles, F.A. (1975) Paleoclimatic significance of quartz/illite variations in cores from the eastern equatorial North Atlantic. *Quaternary Research*, **5**, 225–235.

Brackenridge, G.R. (1978) Evidence for a cold, dry full-glacial climate in the American southwest. *Quaternary Research*, **9**, 22–40.

Bradley, R.S. (1985) *Quaternary Palaeoclimatology*, Allen and Unwin, Boston.

Bradley, R.S. (1999) *Paleoclimatology. Reconstructing Climates of the Quaternary*, 2nd edn, International Geophysics Series, London Academic Press.

Bristow, C.S., Lancaster, N. and Duller, G.A.T. (2005) Combining ground penetrating radar surveys and optical dating to determine dune migration in Namibia. *Journal of the Geomorphological Society*, **162**, 189–196.

Broecker, W.S. and Orr, P.C. (1958) Radiocarbon chronology of Lake Lahontan and Lake Bonneville. *Geological Society of America Bulletin*, **70**, 1009–1032.

Brook, G.A., Burney, D.A. and Cowart, J.B. (1990) Desert paleoenvironmental data from cave speleothems with examples from the Chihuahuan, Somali-Chalbi, and Kalahari deserts. *Palaeogeography, Palaeoclimatology, Palaeoecology*, **76**, 311–329.

Brookfield, M. (1970) Dune trends and wind regime in central Australia. *Zeitschrift für Geomorphologie*, Supplementband **10**, 121–58.

Bullard, J.E. and White, K. (2005) Dust production and the release of iron oxides resulting from the aeolian abrasion of natural dune sands. *Earth Surface Processes and Landforms*, **30** (1), 95–106.

Burrough, S.L. and Thomas, D.S.G. (2008) Late Quaternary lake-level fluctuations in the Mababe Depression: Middle Kalahari palaeolakes and the role of Zambezi inflows. *Quaternary Research*, **69**, 388–403.

Burrough, S.L. and Thomas, D.S.G. (2009) Geomorphological contributions to palaeolimnology on the African continent. *Geomorphology*, **103**, 285–298.

Burrough, S.L., Thomas, D.S.G. and Bailey, R.M. (2009) Mega-Lake in the Kalahari: A Late Pleistocene record of the Palaeolake Makgadikgadi system. *Quaternary Science Reviews*, **28**, 1392–1411.

Burrough, S.L., Thomas, D.S.G. and Singarayer, J.S. (2009) Late Quaternary hydrological dynamics in the Middle Kalahari: Forcing and feedbacks. Earth Science Reviews, **96** (4), 313–326.

Burrough, S.L., Thomas, D.S.G., Shaw, P.A. and Bailey, R.M. (2007) Multiphase Quaternary highstands at Lake Ngami, Kalahari, northern Botswana. *Palaeogeography, Palaeoclimatology, Palaeoecology*, **253**, 280–299.

Chappell, J. (1978) Theories of Upper Quaternary ice ages, in *Climatic Change and Variability: A Southern Perspective* (eds A.B. Pittock *et al.*), Cambridge University Press, Cambridge, pp. 211–225.

Chase, B. (2009) Evaluating the use of dune sediments as a proxy for palaeo-aridity: a southern African case study. *Earth-Science Reviews*, **93**, 31–45.

Chase, B.M. and Brewer, S. (2009) Last Glacial Maximum dune activity in the Kalahari Desert of southern Africa: observations and simulations. *Quaternary Science Reviews*, **28**, 301–307.

Chase, B.M. and Meadows, M.E. (2007) Late Quaternary dynamics of southern Africa's winter rainfall zone. *Earth-Science Reviews*, **84** (3–4), 103–138.

Chase, B.M., Meadows, M.E., Scott, L. *et al.* (2009) A record of rapid Holocene climate change preserved in hyrax middens from southwestern Africa. *Geology*, **37**, 703–706.

Chawla, S., Dhir, R.P. and Singhvi, A.K. (1992) Thermoluminescence chronology of sand profiles in the Thar Desert. *Quaternary Science Reviews*, **11**, 25–32.

Clayton, J.A., Verosub, K.L. and Harrington, C.D. (1990) Magnetic techniques applied to the study of rock varnish. *Geophysical Research Letters*, **17**, 787–790.

Coe, M.T. and Bonan, G.B. (1997) Feedbacks between climate and surface water in northern Africa during the middle Holocene. *Journal of Geophysical Research D: Atmospheres*, **102**, 11087–11101.

Colman, S.M., Pierce, K.L. and Birkeland, P.W. (1987) Suggested terminology for Quaternary dating methods. *Quaternary Research*, **28**, 314–319.

Cooke, H.J. (1984) The evidence from northern Botswana of late Quaternary climatic change, in *Late Cainozoic Palaeoclimates of the Southern Hemisphere* (ed. J.C. Vogel), Balkema, Rotterdam, pp. 265–278.

Crabaugh, M. and Kocurek, G. (1993) Entrada Sandstone: an example of a wet aeolian system, in *The Dynamic and Context of Aeolian Sedimentary System* (ed. K. Pye), Geological Society Special Publication 72, London, pp. 103–126.

Cremaschi, M. (1996) The rock varnish in the Messak Settafet (Fezzan, Libyan Sahara), age, archaeological context, and paleo-environmental implication. *Geoarchaeology – An International Journal*, **11**, 393–421.

Cremaschi, M. (2002) Late Pleistocene and Holocene climatic changes in the central Sahara: the case study of the Southwestern Fezzan Libya, in *Droughts, Food and Culture, Ecological Change and Food Security in Africa's Later Prehistory* (ed. F. Hassan), pp. 65–81.

Currey, D.R. (1990) Quaternary paeolakes in the evolution of semi-desert basins, with special emphasis on Lake Bourneville and the Great Basin, USA. *Palaeogeography, Paleaoclimatology, Palaeoecology*, **76** (3–4), 189–214.

Damuth, J.E. and Fairbridge, R.W. (1970) Arkosic sands of the last glacial stage in the tropical Atlantic off Brazil. *Geological Society of America, Bulletin*, **81**, 189–206.

Deacon, J., Lancaster, N. and Scott, L. (1984) Evidence for late Quaternary climatic change in southern Africa. Summary of the Proceedings of the SASQUA Workshop held in Johannesburg, September (1983), in *Late Cainozoic Palaeoclimate of the Southern Hemisphere* (ed. J.C. Vogel), Balkema, Rotterdam, pp. 391–404.

DeMenocal, P.B. (2004) African climate change and faunal evolution during the Pliocene–Pleistocene. *Earth and Planetary Science Letters*, **220**, 3–24.

DeMenocal, P., Ortiz, J., Adkins, J. *et al.* (2000) Abrupt onset and termination of the African Humid Period: rapid climate responses to gradual insolation forcing. *Quaternary Science Reviews*, **19**, 347–361.

Derbyshire, E., Jijun, L., Perrot, F.A. *et al.* (1984) Quaternary glacial history of the Hunza Valley, Karakoram Mountains, Pakistan, in *The International Karakoram Project*, vol. 2 (ed. K.J. Miller), Cambridge University Press, pp. 456–495.

Diester-Haass, L. (1976) Late Quaternary climatic variations in northwest Africa deduced from east Atlantic sediment cores. *Quaternary Research*, **6**, 299–314.

Dorn, R.I. (1986) Rock varnish as an indicator of aeolian environmental change, in *Aeolian Geomorphology* (ed. W.G. Nickling), The Binghamton Symposia in Gemorphology: International Series 17, Allen and Unwin, Boston, pp. 291–307.

Dorn, R.I. (1988) A rock varnish interpretation of alluvial for development in Death Valley, California. *National Geographic Research*, **4**, 56–73.

Dorn, R.I. (1989) Cation-ratio dating of rock varnish: a geographic perspective. *Progress in Physical Geography*, **13**, 559–596.

Dorn, R.I. (1990) Quaternary alkalinity fluctuations recorded in rock varnish microlamination on western USA volcanics. *Palaeogeography, Palaeoclimatology, Palaeoecology*, **76**, 291–310.

Dorn, R.I. (1992) Palaeoenvironmental signals in rock varnish on petroglyphs. *American Indian Rock Art*, **18**, 1–15.

Dorn, R.I. (1994) Rock varnish as evidence of climatic change, in *Geomorphology of Desert Environments* (eds A.D. Abrahams and A.J. Parsons), Chapman and Hall, London, pp. 539–552.

Dorn, R. I. (2007) Rock varnish, in *Geochemical Sediments and Landscapes* (eds D.J. Nash and S.J. McClaren), Blackwell, London, pp. 246–297.

Dorn, R.I. and De Niro, M.J. (1985) Stable carbon isotope ratios on rock varnish organic matter: a new palaeoenvironmental indicator. *Science*, **227**, 1472–1474.

Dorn, R.I., DeNiro, M.J. and Ajse, H. (1987) Isotopic evidence for climatic influence on allival for development in Death Valley, Gaborone. *Geology*, **15**, 108–110.

Dorn, R.I. and Oberlander, T.M. (1982) Rock varnish. *Progress in Physical Geography*, **6**, 317–367.

Dorn, R.I., Cahill, T.A., Eldred, R.A. *et al.* (1990) Dating rock varnishes by the cation-ratio method with PIXE, ICP, and the electron microprobe. *International Journal of PIXE*, **1**, 157–195.

Dorn, R.I., Clarkson, P.B., Nobbs, M.F. *et al.* (1992a) Radiocarbon dating inclusions of organic matter in rock varnish, with examples from drylands. *Annals of the Association of American Geographers*, **82**, 136–151.

Dorn, R.I., Jull, A.J.T., Donahue, D.J. *et al.* (1992b) Rock varnish on Hualalai and Mauna Kea Volcanoes, Hawaii. *Pacific Science*, **46**, 11–34.

Dragovich, D. (1986) Minimum age of some desert varnish near Broken Hill, New South Wales. *Search*, **17**, 149–151.

Drake, N.A., El-Hawat, A.S., Turner, P. *et al.* (2008) Palaeohydrology of the Fazzan Basin and surrounding regions: the last 7Â million years. *Palaeogeography, Palaeoclimatology, Palaeoecology*, **263**, 131–145.

Duhnforth, M., Bergner, A.G.N. and Trauth, M.H. (2006) Early Holocene water budget of the Nakuru-Elmenteita basin, Central Kenya Rift. *Journal of Paleolimnology*, **36**, 281–294.

Dupont, L.M., Jahns, S., Marret, F. and Ning, S. (2000) Vegetation change in equatorial West Africa: time-slices for the last 150 ka. *Palaeogeography, Palaeoclimatology, Palaeoecology*, **155**, 95–122.

Emiliani, C. (1966) Palaeotemperature analysis of Caribbean cores P6304-8 and P6304-9 and a generalised temperature curve of the past 425,000 years. *Journal of Geology*, **74**, 109–126.

Enzel, Y., Ely, L., Mishra, S. and Baker, V. (1995) High resolution paleohydrological record from NW India: revisiting Lunkaransar Lake, Rajasthan, in South East Asian Conference of International Association of Geomorphologists, p. 40.

Fairbridge, R.W. (1964) African ice-age aridity, in *Problems in Palaeoclimatology* (ed. A.E.M. Nairn), Wiley Interscience, London, pp. 356–363.

Fairbridge, R.W. (1970) World paleoclimatology of the Quaternary. *Revue de Géographie Physique et Géologie Dynamique*, **12**, 97–104.

Ficken, K.J., Wooller, M.J., Swain, D.L. *et al.* (2002) Reconstruction of a subalpine grass-dominated ecosystem, Lake Rutundu, Mount Kenya: a novel multi-proxy approach. *Palaeogeography, Palaeoclimatology, Palaeoecology*, **177**, 137–149.

Fitzsimmons, K.E., Magee, J.W. and Amos, K.J. (2009) Characterisation of aeolian sediments from the Strzelecki and Tirari Deserts, Australia: implications for reconstructing palaeoenvironmental conditions. *Sedimentary Geology*, **218**, 61–73.

Fitzsimmons, K.E., Rhodes, E.J., Magee, J.W. and Barrows, T.T. (2007) The timing of linear dune activity in the Strzelecki and Tirari Deserts, Australia. *Quaternary Science Reviews*, **26**, 2598–2616.

Flint, R.F. and Bond, G. (1968) Pleistocene sand ridges and pans in western Rhodesia. *The Geological Society of America Bulletin*, **79**, 299–314.

Flohn, H. and Nicholson, S. (1980) Climatic fluctuations in the arid belt of the 'Old World' since the Last Glacial Maximum: possible causes and future implications. *Palaeoecology of Africa*, **12**, 3–21.

Frostick, L.E. and Reid, I. (1989) Climate versus tectonic control of fan sequences: lessons from the Dead Sea. *Journal of the Geological Society of London*, **146**, 527–538.

Galloway, R.W. (1970) The full-glacial climate in the southwestern United States. *Annals of the Association of American Geographers*, **60**, 245–256.

Gardner, R.A.M. (1983) Aeolianite, in *Chemical Sediment Sand Geomorphology* (eds A.S. Goudie and K. Pye), Academic Press, London, pp. 265–300.

Gardner, R. and Pye, K. (1981) Nature, origin and palaeoenvironmental significance of red coastal and desert dune sands. *Progress in Physical Geography*, **5**, 514–534.

Gasse, F. (2001) Paleoclimate: hydrological changes in Africa. *Science*, **292**, 2259–2260.

Gasse, F., Fontes, J.C., Plaziat, J.C. *et al.* (1987) Biological remains, geochemistry and stable isotopes for the reconstruction of environmental and hydrological changes in the Holocene lakes from North Sahara. *Palaeogeography, Palaeoclimatology, Palaeoecology*, **60**, 1–46.

Gasse, F., Chalié, F., Vincens, A., Williams, M.A.J. and Williamson, D. (2008) Climatic patterns in equatorial and southern Africa from 30,000 to 10,000 years ago reconstructed from terrestrial and near-shore proxy data. *Quaternary Science Reviews*, **27** (25–26), 2316–2340.

Geyh, M.A. and Thiedig, F. (2008) The Middle Pleistocene Al Mahrúqah Formation in the Murzuq Basin, northern Sahara, Libya evidence for orbitally-forced humid episodes during the last 500,000 years. *Palaeogeography, Palaeoclimatology, Palaeoecology*, **257**, 1–21.

Glennie, K.W. (1987) Desert sedimentary environments, present and past – a summary. *Sedimentary Geology*, **50**, 135–166.

Goudie, A.S. (1983) The arid earth, in *Mega Geomorphology* (eds R.A.M. Gardner and H. Scoging), Oxford University Press, Oxford, pp. 152–171.

Grey, D.R.C. and Cooke, H.J. (1977) Some problems in the Quaternary evolution of the landforms of northern Botswana. *Catena*, **4**, 123–133.

Grove, A.T. (1958) The ancient ergs of Hausaland, and similar formations on the south side of the Sahara. *Geographical Journal*, **124**, 528–533.

Grove, A.T. (1969) Landforms and climatic change in the Kalahari and Ngamiland. *Geographical Journal*, **135**, 191–212.

Grove, A.T. and Warren, A. (1968) Quaternary landforms and climate on the south side of the Sahara. *Geographical Journal*, **134**, 194–208.

Haberlah, D., Williams, M.A.J., Halverson, G., McTainsh, G.H., Hill, S.M., Hrstka, T., Jaime, P., Butcher, A.R. and Glasby, P. (2010) Loess and floods: high-resolution multi-proxy data of Last Glacial Maximum (LGM) slackwater deposition in the Flinders Ranges, semi-arid South Australia. *Quaternary Science Reviews*, **29**, 2673–2693.

Harrington, C.D. (1988) Recognition of components of volcanic ash in rock varnishes and the dating of volcanic ejects plumes. *Geological Society of America Abstracts with Programs*, **20**, 167.

Harrison, S.P., Metcalfe, S.E., Street-Perrott, F.A. *et al.* (1984) A climatic model of the last glacial–interglacial transition based on palaeotemperature and palaeohydrological evidence, in *A Climatic Model of the Last Glacial–Interglacial Transition Based on Palaeotemperature and Palaeohydrological Evidence* (ed. J.C. Vogel), Rotterdam, Balkema, pp. 21–34.

Harvey, A.M. and Wells, S.G. (1994) Late Pleistocene and Holocene changes in hillslope sediment supply to alluival fan systems, Zzyzx California, in *Environmental Change in Drylands: Biogeographical and Geomorphological Perspectives* (eds A.C. Millington and K. Pye), John Wiley & Sons, Ltd, Chichester, pp. 67–84.

Hashimi, N.H. and Nair, R.B. (1986) Climatic aridity over India 11,000 years ago: evidence from feldspar distribution in shelf sediments. *Palaeogeography, Palaeoecology, Palaeoclimatology*, **53**, 309–319.

Hays, J.D., Imbrie, J. and Shackleton, N.J. (1976) Variations in the earth's orbit: pacemaker of the ice ages. *Science*, **194**, 1121–1132.

Hemming, S.R., Biscaye, P.E., Broecker, W.S., Hemming, N.G., Klas, M. and Hajdas, I. (1998) Provenance change coupled with increased clay flux during deglacial times in the western equatorial Atlantic. *Palaeogeography, Palaeoclimatology, Palaeoecology*, **142** (3–4), 217–230.

Hesse, P.P. and McTainsh, G.H. (1999) Last glacial maximum to early Holocene wind strength in the mid-latitudes of the Southern Hemisphere from aeolian dust in the Tasman Sea. *Quaternary Research*, **52**, 343–334.

Hesse, P.P. and Simpson, R.L. (2006) Variable vegetation cover and episodic sand movement on longitudinal desert sand dunes. *Geomorphology*, **81**, 276–291.

Holmgren, K., Carless, W. and Shaw, P.A. (1995) Paleoclimatic significance of the stable isotope composition and petrology of a late Pleistocence stalagmite from Botswana, *Quaternary Research*, **43** (3), 320–328.

Holz, C., Stuut, J.-B.W. and Hentrich, R. (2004) Terrigenous sedimentation processes along the continental margin off NW Africa: implications from grain-size analysis of seabed sediments. *Sedimentology*, **51**, 1145–1154.

Hunt, C.B. and Mabey, D. R. (1966) Stratigraphy and structure of Death Valley, California, US Geological Survey, Professional Paper 494A.

Imbrie, J. and Imbrie, K.P. (1979) *Ice Ages: Solving the Mystery*, Macmillan, London.

Jennings, J.N. (1975) Desert dunes and estuarine fill in the Fitzroy estuary (North West Australia). *Catena*, **2**, 215–262.

Kadomura, H. and Hori, N. (1990) Environmental implications of slope deposits and humid tropical Africa: evidence from southern Cameroon and western Kenya. *Geographical Reports of Tokyo Metropolitan University*, **25**, 213–236.

Kar, A., Singhvi, A.K., Rajaguru, S.N. *et al.* (2001) Reconstruction of the late Quaternary environment of the lower Luni plains, Thar Desert, India. *Journal of Quaternary Science*, **16**, 61–68.

Kiage, L. M. and Liu, K. B. (2006) Late Quaternary paleoenvironmental changes in East Africa: a review of multiproxy evidence from palynology, take sediments, and associated records. *Progress in Physical Geography*, **30** (5), 633–658.

Knauss, K.G. and Ku, T.L. (1980) Desert varnish: potential for age determination via uranium series isotopes. *Journal of Geology*, **88**, 95–100.

Knox, J. C. (1972) Valley alleviation in southwestern Wisconsin. *Annals of the Association of American Geographers*, **62**, 401–410.

Kolla, V. and Biscaye, P.E. (1977) Distribution and origin of quartz in the sediments of the Indian Ocean. *Journal of Sedimentary Petrology*, **47**, 642–649.

Kolla, V., Biscaye, P.E. and Hanley, A.F. (1979) Distribution of quartz in Late Quaternary Atlantic sediments in relation to climate. *Quaternary Research*, **11**, 261–267.

Krinsley, D.H. (1978) The present state and future prospects of environmental discrimination by scanning electron microscopy, in *Scanning Electron Microscopy in the Study of Sediments* (ed. W.B. Whalley), University of East Anglia, Norwich, pp. 169–179.

Krinsley, D.H. and Doornkamp, J. (1973) *Atlas of Quartz Sand Surface Textures*, Cambridge University Press, Cambridge.

Kutzbach, J.E. (1980) Estimate of past climate at paleolake Chad, North Africa, based on a hydrological and energy-balance model. *Quaternary Research*, **14**, 210–223.

Kutzbach, J.E. and Liu, Z. (1997) Response of the African monsoon to orbital forcing and ocean feedbacks in the middle Holocene. *Science*, **278**, 440–443.

Lancaster, N. (1979) Evidence for a widespread late Pleistocene humid period in the Kalahari. *Nature*, **279**, 145–146.

Lancaster, N., Kocurek, G., Singhvi, A. *et al.* (2002) Late Pleistocene and Holocene dune activity and wind regimes in the western Sahara Desert of Mauritania. *Geology*, **30**, 991–994.

Lekach, J. and Schick, A.P. (1982) Suspended sediment in desert floods in small catchments. *Israel Journal of Earth Sciences*, **31**, 144–156.

Leopold, L.B. (1951) Pleistocene climate in New Mexico. *American Journal of Science*, **249**, 152–168.

Li, P-Y and Zhou, L-P (1993) Occurrence and palaeoenvironmental implications of the Late Pleistocene loess along the eastern coast of the Bohai Sea China, in *The Dynamics and Environmental Context of Aeolian Sedimentary Systems* (ed. K. Pye), Geological Society Special Publication 72, London, pp. 293–309.

Liu, T. and Broecker, W.S. (2008) Rock varnish microlamination dating of late Quaternary geomorphic features in the drylands of western USA. *Geomorphology*, **93**, 501–523.

Liu, W., Li, X., Zhang, L., An, Z. and Xu, L. (2009) Evaluation of oxygen isotopes in carbonate as an indicator of lake evolution in arid areas: the modern Qinghai Lake, Qinghai-Tibet Plateau. *Chemical Geology*, **268** (1–2), 126–136.

Livingstone, D.A. (1975) Late Quaternary climatic change in Africa. *Annual Review of Ecology and Systematics*, **6**, 249–280.

Lowestein, T.K. and Hardie, L.A. (1985) Criteria for the recognition of salt-pan evaporites. *Sedimentology*, **32**, 627–644.

Lowenstein, T.K., Spencer, R.J., Wembo, Y. *et al.* (1994) Major-element and stable isotope geochemistry of fluvial inclusions in halite, Qaidam basin, western China: implication for late Pleistocene/Holocene brine evolution and paleoclimates, in *Paleoclimate and Basin Evolution of Playa Systems* (ed. M.R. Rosen), Geological Society of America Special Paper 289, pp. 19–32.

Lustig, L.K. (1965) Clastic sedimentation in Deep Springs Valley, California, US Geological Survey, Professional Paper 352F, pp. 131–192.

McBurney, C.B.M. and R.W. Hey (1955) *Prehistory and Pleistocene Geology in Cyrenian Libya*, Cambridge University Press, Cambridge.

McClure, H.A. (1976) Radiocarbon chronology of Late Quaternary Lakes in the Arabian desert. *Nature*, **263**, 755–756.

Madigan, C.T. (1936) The Australian sand ridge deserts. *Geographical Review*, **26**, 205–227.

Magee, J.W. and Miller, G.H. (1998) Lake Eyre palaeohydrology from 60 ka to the present: beach ridges and glacial maximum aridity. *Palaeogeography, Palaeoclimatology, Palaeoecology*, **144**, 307–329.

Maizels, J.K. (1987) Plio-Pleistocene raised channel systems of the western Sharqiya (Wahiba) Oman, in *Desert Sediments: Ancient and Modern* (eds L.E. Frostick and I. Reid), Geological Society of London Special Publication 35, pp. 31–50.

Melia, M.B. (1984) The distribution and relationship between palynomorphs in aerosals and deep sea sediments off the coast of northwest Africa. *Marine Geology*, **58**, 345–372.

Metcalfe, S.E., Street-Perrott, F.A., O'Hara, S.L. *et al.* (1994) The palaeolimological record of environmental change: examples from the arid frontier of Mesoamerica, in *Environmental Change in Drylands: Biogeographical and Geomorphological Perspective* (eds A.C. Millington and K. Pye), John Wiley & Sons, Ltd, Chichester, pp. 131–145.

Mifflin, M.D. and Wheat, M.M. (1979) Pluvial lakes and estimated pluvial climates of Nevada, Nevada Bureau of Mines and Geology Bulletin 94.

Milliman, J.D. and Summerhayes, C.P. (1975) Quaternary sedimentation on the Amazon continental margin: a model. *Geological Society of America, Bulletin*, **86**, 610–614.

Molina-Cruz, A. (1977) The relation of the southern trade winds to upwelling processes during the last 75,000 years. *Quaternary Research*, **8**, 324–338.

Muhs, D.R. (1985) Age and paleoclimatic significance of Holocene dune sand in northeastern Colorado. *Annals of the Association of American Geographers*, **75**, 566–582.

Muhs, D.R., Stafford, T.W., Cowherd, S.D., Mahan, S.A., Kihl, R., Maat, P.B., Bush, C.A. and Nehring, J. (1996) Origin of the Late Quaternary dune fields of northeastern Colorado. *Geomorphology*, **17** (1–3 Special Issue), 129–149.

Mulcahy, M.J. and Bettenay, E. (1972) Soil and landscape studies in Western Australia. 1. The major drainage division. *Journal of the Geological Society of Australia*, **18**, 349–357.

Nanson, G.C. (1986) Episodes of vertical accretion and catastrophic stripping: a model of disequilibrium flood plain development. *Geological Society of America Bulletin*, **97**, 1467–1475.

Nanson, G.C., Chen, X.Y. and Price, D.M. (1992) Lateral migration, thermoluminescence chronology and colour variation of longitudinal dunes near Birdsville in the Simpson Desert Central Australia. *Earth Surface Processes and Landforms*, **17**, 807–820.

Nanson, G.C., Young, R.W., Price, D.M. and Rust, B.R. (1988) Stratigraphy, sedimentology and late-Quaternary chronology of the Channel Country of western Queensland. *Fluvial Geomorphology of Australia*, 151–175.

Nanson, G.C., Price, D.M., Young, R.W. and Rust, B.R. (1992) Stratigraphy, sedimentology and late Quaternary chronology of the Channel Country of western Queensland, in *Fluvial Geomorphology of Australia* (ed. R.F. Warner), Academic Press, Sydney, pp. 151–175.

Nanson, G.C., Price, D.M., Jones, B.G., Maroulis, J.C., Coleman, M., Bowman, H., Cohen, T.J., Pietsch, T.J. and Larsen, J.R. (2008) Alluvial evidence for major climate and flow regime changes during the Middle and Late Quaternary in eastern central Australia. Geomorphology, **101** (1–2), 109–129.

Nash, D.J., Thomas, D.S.G. and Shaw, P.A. (1994) Timescale, environmental change and dryland valley development, in *Environmental Change in Drylands: Biogeographical and Geomorphological Perspectives* (eds A.C. Millington and K. Pye), John Wiley & Sons, Ltd, Chichester, pp. 25–41.

Newell, R.E. (1973) Climate and the Galapogos Islands. *Nature*, **245**, 91–92.

Nicholson, S.E. and Flohn, H. (1980) African environmental and climatic changes and the general atmospheric circulation in Late Pleistocene and Holocene. *Climatic Change*, **2**, 313–348.

Olago, D.O., Street-Perrott, F.A., Perrott, R.A. *et al.* (1999) Late Quaternary glacial-interglacial cycle of climatic and environmental change on Mount Kenya, Kenya. *Journal of African Earth Sciences*, **29**, 593–618.

Olago, D.O., Street-Perrott, F.A., Perrott, R.A. *et al.* (2000) Long-term temporal characteristics of palaeomonsoon dynamics in equatorial Africa. *Global and Planetary Change*, **26**, 159–171.

Pachur, H.-J. and Kropetin, S. (1987) Wadi Howar: palaeoclimatic evidence from an extinct river system in the southeastern Sahara. *Science*, **237**, 298–300.

Pant, G. B. and Kumar, K. R. (1997) *Climates of South Asia*, John Wiley & Sons, Ltd, Chichester.

Parker, A.G., Eckersley, I., Smith, M.M. *et al.* (2004) Holocene vegetation dynamics in the northeastern Rub' al-Khali desert, Arabian Peninsula: A phytolith, pollen and carbon isotope study. *Journal of Quaternary Science*, **19**, 665–676.

Parker, A.G., Goudie, A.S., Stokes, S., White, K., Hodson, M.J., Manning, M. and Kennet, D. (2006) A record of Holocene climate change from lake geochemical analyses in southeastern Arabia. *Quaternary Research*, **66** (3), 465–476.

Parmenter, C. and Folger, D.W. (1974) Eolian biogenic detritus in deep sea sediments: a possible index of equatorial Ice Age aridity. *Science*, **185**, 695–698.

Partridge, T.C., DeMenocal, P.B., Lorentz, S.A., Paiker, M.J. and Vogel, J.C. (1997) Orbital forcing of climate over South Africa: a 200,000-year rainfall record from the Pretoria saltpan. *Quaternary Science Reviews*, **16** (10), 1125–1133.

Petit, J.R., Briat, M. and Royer, A. (1981) Ice age aerosol content from East Antarctic ice core samples and past wind strength. *Nature*, **293**, 391–394.

Pokras, E.M. and Mix, A.C. (1985) Eolian evidence for spatial variability of late Quaternary climates in tropical Africa. *Quaternary Research*, **24**, 137–149.

Prell, W.L., Hutson, W.H., Williams, D.F. *et al.* (1980) Surface circulation of the Indian Ocean during the last Glacial Maximum, approximately 18,000 BP. *Quaternary Research*, **14**, 309–336.

Reid, I. (1994) River landforms and sediments: evidence of climatic changes, in *Geomorphology of Desert Environments* (eds A.D. Abrahams and A.J. Parsons), Chapman & Hall, London, pp. 571–592.

Risacher, F. and Fritz, B. (2000) Bromine geochemistry of salar de Uyuni and deeper salt crusts, Central Altiplano, Bolivia. *Chemical Geology*, **167**, 373–392.

Roberts, N and Wright, H.E. (1993) Vegetational, lake level and climate history of the Near East and Southwest Asia, in *Global Climates Since the Last Glacial Maximum* (eds H.E. Wright, J.E. Kutcback, T. Webb *et al.*), University of Minnesota Press, Minneopolis, pp. 1994–2017.

Rognon, P. (1987) Late Quaternary climatic reconstruction for the Maghreb (north Africa). *Palaeogeography, Palaeoclimatology, Palaeoecology*, **58**, 11–34.

Rognon, P. and Williams, M.A.J. (1977) Late Quaternary climatic changes in Australia and north Africa: a preliminary interpretation. *Palaeogeography, Palaeoclimatology, Palaeoecology*, **21**, 285–327.

Ruddiman, W.F. and Kutzbach, J.E. (1990) Late Cenozoic plateau uplift and climate change. *Transactions – Royal Society of Edinburgh: Earth Sciences*, **81**, 301–314.

Ruddiman, W.F. and Kutzbach, J.E. (1991) Plateau uplift and climatic change. *Scientific American*, **264**, 42–50.

Sarnthein, M. (1972) Sediments and history of the post-glacial transgression in the Persian Gulf and north-west Gulf of Oman. *Marine Geology*, **12**, 245–266.

Sarnthein, M. (1978) Sand deserts during Glacial Maximum and climatic optimum. *Nature*, **272**, 43–46.

Sarnthein, M. and Diester-Haass, L. (1977) Eolian-sand turbidites. *Journal of Sedimentary Petrology*, **47**, 868–890.

Sarnthein, M., Tetzlaff, G., Koopmann, B. *et al.* (1981) Glacial and interglacial wind regimes over the eastern sub-tropical Atlantic and northwest Africa. *Nature*, **293**, 193–196.

Schefuβ, E., Schouten, S. and Schneider, R.R. (2005) Climatic controls on central African hydrology during the past 20,000 years. *Nature*, **437**, 1003–1006.

Scholz, C.A., Johnson, T.C., Cohen, A.S. *et al.* (2007) East African megadroughts between 135 and 75 thousand years ago and bearing on early-modern human origins. *PNAS*, **104**, 16416–16421.

Schroeder, J.H. (1985) Eolian dust in the coastal desert of the Sudan: aggregate cemented by evaporites. *Journal of African Earth Sciences*, **3**, 370–380.

Scott, L. (2002) Grassland development under glacial and interglacial conditions in southern Africa: review of pollen, phytolith and isotope evidence. *Palaeogeography, Palaeoclimatology, Palaeoecology*, **177**, 47–57.

Scott, L., Marais, E. and Brook, G. A. (2004) Fossil hyrax dung and evidence of late Pleistocene and Holocene vegetation types in the Namib Desert. *Journal of Quaternary Science*, **19**, 829–832.

Servant, M. and Servant-Vidary, S. (1980) L'environment quaternaire du basin du Tchad, in *The Sahara and the Nile* (eds M.A.J. Williams and H. Faure), Balkema, Rotterdam, pp. 133–162.

Servant, M., Maley, J., Turcq, B, Absy, M.L., Brenac, P., Fournier, Mand Ledru, M.P. (1993). Tropical forest changes during the late Quaternary in African and Southern Americas lowlands. Global and Planetary Change, **7**, 25–40.

Shackleton, N.J., Backman, J., Zimmerman, H., Kent, D.V., Hall, M.A., Roberts, D.G., Schnitker, D., Baldauf, J.G., Desprairies, A., Homrighausen, R., Huddlestun, P., Keene, J.B., Kaltenback, A.J., Krumsiek, K.A.O., Morton, A.C., Murray, J.W. and Westberg-Smith, J. (1984) Oxygen isotope calibration of the onset of ice-rafting and history of glaciation in the North Atlantic region. *Nature*, **307** (5952), 620–623.

Shaw, P.A. (1985) Late Quaternary landforms and environmental change in northeast Botswana: the evidence of Lake Ngami and the Mababe Depression. *Transactions, Institute of British Geographers*, NS, **10**, 333–346.

Shaw, P.A. and Cooke, H.J. (1986) Geomorphic evidence for the Late Quaternary palaeoclimate of the Middle Kalahari of Northern Botswana. *Catena*, **13**, 349–359.

Shaw, P.A. and De Vries, J.J. (1988) Duricrusts, ground water and valley development in the Kalahari of southeastern Botswana. *Journal of Arid Environments*, **14**, 245–254.

Shaw, P.A. and Thomas, D.S.G. (1988) Lake Caprivi – a Late Quaternary link between the Zambezi and middle Kalahari drainage systems. *Zeitschrift für Geomorphologie*, NF, **32**, 329–337.

Shi, N., Schneider, R., Beug, H.-J. and Dupont, L.M. (2001) Southeast trade wind variations during the last 135 kyr: evidence from pollen spectra in eastern South Atlantic sediments. *Earth and Planetary Science Letters*, **187**, 311–321

Singer, A. (1984) The paleoclimatic interpretation of clay minerals in sediments – a review. *Earth Science Review*, **21**, 251–293.

Singhvi, A.K. and Kar, A. (2004) The aeolian sedimentation record of the Thar desert. *Proceedings of the Indian Academy of Sciences, Earth and Planetary Sciences*, **113** (3), 371–401.

Singhvi, A.K. and Porat, N. (2008) Impact of luminescence dating on geomorphological and palaeoclimate research in drylands. *Boreas*, **37** (4), 536–558.

Singhvi, A.K., Sharma, Y.P. and Agrawal, D.P. (1982) Thermoluminescence dating of sand dunes in Rajasthan, India. *Nature*, **295**, 313–315.

Singhvi, A.K., Banerjee, D., Rajaguru, S.N., and Kumar, V.S.K. (1994) Luminescence chronology of a fossil dune at Budha Pushkar, Thar Desert: palaeoenvironmental and archaeological implications. *Current Science*, **66**, 770–773.

Singhvi, A.K., Williams, M.A.J., Rajaguru, S.N., Misra, V.N., Chawla, S., Stokes, S., Chauhan, N., Francis, T., Ganjoo, R.K. and Humphreys, G.S. (2010) A $\sim$200 ka record of climatic change and dune activity in the Thar Desert, India. *Quaternary Science Reviews*, **29**, 3095–3105.

Smith, G.I. and Street-Perrott, F.A. (1983) Pluvial lakes of the western United States, in *Late Quaternary Environments in the United States. Volume 1: The Late Pleistocene* (ed. S.C. Porter), University of Minnesota Press, Minneapolis, pp. 190–214.

Smith, G.I., Barczak, V.J., Moulton, G.F. and Liddicoat, T.C. (1983) Core KM-3, on surface-to-bedrock record of Late Cainozoic sedimentation in Searles Valley, California. *US Geological Survey*, Professional Paper 1256.

Smith, H.T.U. (1963) Eolian geomorphology, wind direction and climatic change in North Africa, US Air Force, Cambridge Research Laboratories, Report AF19 (628).

Snyder, C.T. and Langbein, W.B. (1962) The Pleistocene Lake in Spring Valley, Nevada, and its climatic implications. *Journal of Geophysical Research*, **67**, 2385–2394.

Sombroek, A.G., Mbuvi, J.P. and Okwaro, H.W. (1976) Soils of the semi-arid savanna zone of Northeastern Kenya, Kenya Soil Survey, Paper M2.

Sperling, C.H.B. and Goudie, A.S. (1975) The miliolite of western India: a discussion of the aeolian and marine hypotheses. *Sedimentary Geology*, **13**, 71–75.

Stone, A.E.C. and Thomas, D.S.G. (2008) Linear dune accumulation chronologies from the southwest Kalahari, Namibia: challenges of reconstructing late Quaternary palaeoenvironments from aeolian landforms. *Quaternary Science Reviews*, **27**, 1667–1681.

Street, F.A. (1979) Late Quaternary precipitation estimates for the Ziway-Shala Basin, southern Ethiopia. *Palaeoecology of Africa*, **11**, 135–143.

Street, F.A. (1981) Tropical palaeoenvironments. *Progress in Physical Geography*, **5**, 157–185.

Street, F.A. and Grove, A.T. (1979) Global maps of lake-level fluctuations since 20,000 yr BP. *Quaternary Research*, **12**, 83–118.

Street-Perrott, F.A. and Roberts, N. (1983) Fluctuations in closed-basin lakes as an indicator of past atmospheric circulation patterns, in *Variations in the Global Water Budget* (eds F.A. Street-Perrott, M. Beran and R. Ratcliffe), Reidel, Dordrecht, pp. 331–345.

Street-Perrott, F.A., Roberts, N. and Metcalfe, S. (1985) Geomorphic implications of Late Quaternary hydrological and climatic change in the northern hemisphere tropics, in *Environmental Change and Tropical Geomorphology* (eds T. Douglas and T. Spencer), George Allen and Unwin, London, pp. 165–183.

Stuiver, M. (1970) Oxygen and carbon isotope ratios of freshwater carbonates as climatic indicators. *Journal of Geophysical Research*, **75**, 5247–5257.

Stuut, J.-B.W. and Lamy, F. (2004) Climate variability at the southern boundaries of the Namib (southwestern Africa) and Atacama (northern Chile) coastal deserts during the last 120,000 yr. *Quaternary Research*, **62**, 301–309.

Stuut, J.W., Prins, M.A., Schneider, R.R. *et al.* (2002) A 300-kyr record of aridity and wind strength in southwestern Africa: inferences from grain-size distributions of sediments on Walvis Ridge, SE Atlantic. *Marine Geology*, **180**, 221–233.

Talbot, M.R. (1980) Environmental responses to climatic change in the West African Sahel over the past 20,000 years, in *The Sahara and the Nile* (eds M.A.J. Williams and H. Faure), Balkema, Rotterdam, pp. 37–62.

Talbot, M.R. and Williams, M.A.J. (1978) Erosion of fixed sand dunes in the Sahel, central Niger. *Earth Surface Processes*, **3**, 107–113.

Thiede, J. (1979) Wind regime over the Late Quaternary southwest Pacific Ocean. *Geology*, **7**, 259–262.

Thomas, D.S.G. (1984) Ancient ergs of the former arid zones of Zimbabwe, Zambia and Angola. *Transactions, Institute of British Geographers*, NS, **9**, 75–88.

Thomas, D.S.G. (1987a) The roundness of aeolian quartz sand grains. *Sedimentary Geology*, **52**, 149–153.

Thomas, D.S.G. (1987b) Research strategies and methods for Quaternary science: the case of Southern Africa, School of Geography, Oxford, Research Paper Series 39.

Thomas, D.S.G. (1987c) Discrimination of depositional environments, using sedimentary characteristics, in the Mega Kalahari, central southern Africa, in *Desert Sediments: Ancient and Modern* (eds L. Frostock and I. Reid), Geological Society of London Special Publication 35, Blackwell, Oxford, pp. 293–306.

Thomas, D.S.G and Burrough, S.L. (2011) Interpreting geoproxies of late Quaternary climate change in African drylands: implications for understanding environmental and early human behaviour. *Quaternary International*. In press.

Thomas, D.S.G., Knight, M. and Wiggs, G.F.S. (2005) Remobilization of southern African desert dune systems by twenty-first century global warming. *Nature*, **435**, 1218–1221.

Thomas, D.S.G. and Shaw, P.A. (1991) *The Kalahari Environment*, Cambridge University Press, Cambridge.

Thomas, D.S.G. and Shaw, P.A. (2002) Late Quaternary environmental change in central southern Africa: new data, synthesis, issues and prospects. *Quaternary Science Reviews*, **21**, 783–797.

Thomas, D.S.G., Bailey, R., Shaw, P.A. *et al.* (2009) Late Quaternary highstands at Lake Chilwa, Malawi: frequency, timing and possible forcing mechanisms in the last 44 ka. *Quaternary Science Reviews*, **28**, 526–539.

Thomas, M.F. (2003) Late Quaternary sediment fluxes from tropical watersheds. *Sedimentary Geology*, **162**, 62–81.

Thompson, R.S. and Anderson, K.H. (2000) Biomes of western North America at 18,000, 6000 and 0 14C yr BP reconstructed from pollen and packrat midden data. *Journal of Biogeography*, **27**, 555–584.

Tiedemann, R., Sarnthein, M. and Shackleton, N.J. (1994) Astronomic timescale for the Pliocene Atlantic d18O and dust flux records from Ocean Drilling Program Site 659. *Paleoceanography*, **9**, 619–638.

Tjallingii, R., Claussen, M., Stuut, J.-B.W. *et al.* (2008) Coherent high- and low-latitude control of the northwest African hydrological balance. *Nature, Geoscience*, **1**, 670–675.

Tooth, S. (2005) Splay formation along the lower reaches of ephemeral rivers on the northern plains of arid central Australia. *Journal of Sedimentary Research*, **75** (4), 636–649.

Tooth, S. and McCarthy, T.S. (2006) Wetlands in drylands: key geomorphological and sedimentological characteristics, with emphasis on examples from southern Africa. *Progress in Physical Geography*, **31**, 3–41.

Tooth, S. and Nanson, G.C. (2000) Equilibrium and nonequilibrium conditions in dryland rivers. *Physical Geography*, **21**, 183–211.

Tricart, J. (1956) Tentative de correlation des périodes pluviales africaines et des périodes glaciaires. *Comptes Rendus Sommaires de la Société Géologique de France*, **9–10**, 164–167.

Tricart, J. (1975) Influence des oscillations climatiques recentes sur le modèle en Amazonie orientale (Région de Santarem) d'après les images de radar latéral. *Zeitschrift für Geomorphologie*, NF, **19**, 140–163.

Tricart, J. (1984) Evidence of Upper Pleistocene dry climates in northern South America, in *Environmental Change and Tropical Geomorphology* (eds I. Douglas and T. Spencer), Allen and Unwin, London, pp. 197–217.

Tsoar, H. (2005) Sand dunes mobility and stability in relation to climate. *Physica A: Statistical Mechanics and Its Applications*, **357**, 50–56.

Tyson, P.D. (1986) *Climatic Change and Variability in Southern Africa*, Oxford University Press, Cape Town.

Ullman, W.J. (1985) Evaporation rates from a salt pan: estimates from chemical profiles in near surface groundwaters. *Journal of Hydrology*, **79**, 365–373.

Ullman, W.J. and McLoed, L.C. (1986) The Late Quaternary salinity record of Lake Frome, South Australia: evidence from Na^+ in stratigraphically preserved gypsum. *Palaeogeography, Palaeoclimatology, Palaeoecology*, **54**, 153–169.

Vasconcelos, P.M., Becker, T.A., Renne, P.R. and Brimball, G.H. (1992) Age and duration of weathering by 40K–40Ar snd 40Ar/39Ar analysis of potassium–manganese oxides. *Science*, **258**, 451–455.

Walker, M. (2005) *Quaternary Dating Methods*, John Wiley & Sons, Ltd, Chichester,. 286 pp.

Warren, A., Chappell, A., Todd, M.C. *et al.* (2007) Dust-raising in the dustiest place on earth. *Geomorphology*, **92**, 25–37.

Wasson, R.J. (1983) Geomorphology, Late Quaternary stratigraphy and palaeoclimatology of the Thar dunefield. *Zeitschrift fur Geomorphologie, Supplementband*, **45**, 117–151.

Watchman, A. (2000) A review of the history of dating rock varnishes. *Earth Science Reviews*, **49**, 261–277.

Watson, A., Price Williams, D. and Goudie, A.S. (1985) The palaeoenvironmental interpretation of colluvial sediments and palaeosols in Southern Africa. *Palaeogeography, Palaeoclimatology, Palaeoecology*, **45**, 255–249.

Williams, G.E. (1973) Late Quaternary piedmont sedimentation, soil formation and palaeoclimates in arid South Australia. *Zeitschrift für Geomorphologie*, NF, **17**, 102–125.

Williams, M.A.J. (1975) Late Pleistocene tropical aridity synchronous in both hemispheres? *Nature*, **253**, 617–618.

Williams, M.A.J. (1984) Geology, in *Key Environments. Sahara Desert* (ed. J.L. Cloudsley-Thompson), Pergamon Press, Oxford, pp. 31–39.

Williams, M.A.J. (1985) Pleistocene aridity in tropical Africa, Australia and Asia, in *Environmental Change and Tropical Geomorphology* (eds I. Douglas and T. Spencer), George Allen and Unwin, London, pp. 219–233.

Williams, M.A.J. (1994) Cenozoic climate changes in deserts: a synthesis, in *Geomorphology of Desert Environments* (eds A.D. Abrahams and A.J. Parsons), Chapman & Hall, London, pp. 644–670.

Williams, M.A.J., Abell, P.I. and Sparks, B.W. (1987) Quaternary landforms, sediments, depositional environments and gastropod isotope ratios at Adrar Bous, Ténéré Desert of Niger, in *Desert Sediments Ancient and Modern* (eds L. Frostick and I. Reid), Geographical Society of London Special Publication 35, pp. 105–125.

Williams, M. and Nitschke, N. (2005) Influence of wind-blown dust on landscape evolution in the Flinders Ranges, South Australia. *South Australian Geographical Journal*, **104**, 26–37.

Williams, M., Nitschke, N. and Chor, C. (2006) Complex geomorphic response to late pleistocene climatic changes in the arid flinders ranges of South Australia. *Geomorphologie: Relief, Processus, Environnement*, **4**, 249–258.

Williams, M.A.J., Dunkerley, D.L., DeDeckter, P. *et al.* (1993) *Quaternary Environments*, Edward Arnold, London.

Williams, M., Prescott, J.R., Chappell, J. *et al.* (2001) The enigma of a late Pleistocene wetland in the Flinders Ranges, South Australia. *Quaternary International*, **82–85**, 129–144.

Williams, M., Cook, E., Van Der Kaars, S. *et al.* (2009) Glacial and deglacial climatic patterns in Australia and surrounding regions from 35 000 to 10 000 years ago reconstructed from terrestrial and near-shore proxy data. *Quaternary Science Reviews*, **28**, 2398–2419.

Williams, M.A.J., Williams, F.M., Duller, G.A.T., Munro, R.N., El Tom, O.A.M., Barrows, T.T., Macklin, M., Woodward, J., Talbot, M.R., Haberlah, D. and Fluin, J. (2010) Late Quaternary floods and droughts in the Nile Valley, Sudan: new evidence from optically stimulated luminescence and AMS radiocarbon dating. *Quaternary Science Reviews*, **29** (9–10), 1116–1137.

Wilson, I.G. (1973), Ergs. *Sedimentary Geology*, **10**, 77–106.

Windom, H.L. (1975) Eolian contributions to marine sediments. *Journal of Sedimentary Petrology*, **45**, 520–529.

Zerboni, A. (2008) Holocene rock varnish on the Messak plateau (Libyan Sahara): chronology of weathering processes. *Geomorphology*, **102** (3–4), 640–651.

4 Dryland system variability

David S. G. Thomas

This short chapter explores the geomorphological variability within drylands. We have already noted in Chapter 1 that specific landforming processes are not unique to arid environments. Some processes are, however, more readily facilitated, or occur on a greater scale, in drylands than elsewhere. For example, beyond coastal contexts and those of poorly managed agricultural land, aeolian processes have their greatest landform expressions in drylands than in any other environment. Similarly, salt-weathering processes have their greatest propensity to operate in certain dryland situations, because of moisture deficits.

For many landforms, arid conditions may set the possibilities for development, but their ultimate formation is dependent on suitable materials, lithologies or topographic settings being available. As Abrahams and Parsons (1994) noted, the assumption that existed of a 'distinct' desert geomorphology was based on a belief that 'similarity of climate throughout desert areas outweigh differences that may arise from other influences' (pp. 9–10). Arid regions occupy a variety of structural and tectonic settings (Murphy, 1968), discussed in Chapter 2, which range from a tectonically active montane setting to stable continental cratons (Heathcote, 1983; Thomas, 1988). Drylands of course embrace a range of climatic contexts, with differences in 'effective moisture' between the 'zones' along the hyper-arid–dry-subhumid continuum (Chapter 1), as well as differences in the nature and duration of seasons (affecting precipitation receipt) and in temperature (affecting moisture loss). A third layer that contributes to difference and diversity is environmental and climatic history, especially during the Late Quaternary Period (Chapter 3), which has left a legacy of palimpsest landforms indicative of different geomorphic regimes in the past. The climatic and structural variations present within and between arid regions, together with the impact of climatic changes during the Quaternary Period, are therefore important variables contributing to the present appearance of arid zone landscapes.

4.1 A framework for dryland diversity

Clements *et al.* (1957) and Mabbutt (1976) provided early attempts to classify the landforms present in some arid areas (Table 4.1). Notwithstanding the limitations of such spatially extensive generalisations, the data they present serves to illustrate some important issues, e.g. the importance of erosional and water-worked features in arid areas with high and frequent changes in relative relief, such as in much of the southwestern United States, and the greater (but not necessarily predominant) importance of the wind as a geomorphological agent in regions with extensive areas of more limited relative relief, such as in the Australian arid zone. Even the Sahara, the ultimate 'sandy desert' in the eyes of many nonspecialists, is not especially sandy, with less than a third of its area being sand seas (including dunefields). The Sahara is actually dominated by mountainous zones, with sand seas sitting either in intermontaine depressions (see Chapter 17) or in areas to the south of the Sahara where, in the Sahelian zone, they represent legacies from more extensive aridity during periods of the Late Quaternary.

There is, in fact, no simple classification of dryland environments, nor really is there a need for one, save as a basis for seeking explanations for understanding the drivers of processes, as well as the controls on their occurrence and resultant environmental expressions. However, illustrating the geographical diversity and distribution of different desert/dryland 'types' does provide a useful

Arid Zone Geomorphology: Process, Form and Change in Drylands, Third Edition. Edited by David S. G. Thomas

Table 4.1 Landscape components in different arid regions (expressed as a percentage of area[a])

SW USA	*Sahara*	*Libya*	*Arabia*	*Australia*[b]	
Mountains	38.1	43	39	47	16
Low-angle bedrock surfaces	0.7	10	6	1	14
Alluvial fans	31.4	1	1	4	
River plains	1.2	1	3	1	13
Dry watercourses	3.6	1	1	1	
Badlands	2.6	2	8	1	–
Playas	1.1	1	1	1	1
Sand seas	0.6	28	22	26	38
Desert flats[c]	20.5	10	18	16	18
Recent volcanic deposits	0.2	3	1	2	–

[a]Percentages given are only approximate, with the degree of accuracy differing between areas.
[b]From Mabbutt (1976). Categories used by Mabbutt do not necessarily coincide with those used in other areas: included for comparison only. The remaining data are from a study by Clements *et al.* (1957) for the US Army.
[c]Undifferentiated: includes areas bordering playas.

context for the systematic analysis of geomorphological processes and landforms that occur in the chapters that form the heart of this book. Chapter 1 has already explained the different climatic and oceanographic mechanisms that led to dryland development, and in Chapter 2 (Table 2.1) specific drylands are linked to specific tectonic contexts. However, neither of these approaches explains the resultant landscapes, the geomorphological outcomes. Mabbutt (1969) provided a summary of physiographic contexts in deserts (also reproduced in Goudie, 2002), which can be adapted to provide the following overview:

Upland drylands. Geology controls relief and bedrock is widely exposed, resulting in weathering and often erosion – controlled landscapes. Depending on the degree of relief and on latitudinal setting, temperature may be as important a control as moisture deficit on plant cover and on the weathering processes that occur.

Piedmont drylands. Transition zones between upland and basin settings. Sediment transport may be dominant but depositional (e.g. alluvial fan) and erosional (e.g. pediment) landscapes both occur.

Stony deserts. These often occur at the foot of pediments or in basin locations where either outwash or deflation brings gravel material to a dominant surface location. Stony deserts can also occur in structural plateau locations.

Fluviolacustrine landscapes. River systems are not an obvious feature of drylands, but can form significant landscape components, either through channel systems inherited from wetter past climates or as systems originating from neighbouring temperate or tropical regions or in dryland uplands. Some channel systems reach the oceans but in a number of important dryland regions internal drainage leads to lacustrine or salt flat sump regions.

Drylands dominated by aeolian deposits. Aeolian landscapes occur in regions beyond the limit of fluvial activity, but may derive much of their sediment from floodplains, lake basins and other water-derived contexts. To this framework provided by Mabbutt (1969), which applies to sand seas, can be added loess landscapes, which tend to blanket plateau top or plateau edge locations.

Depending on the scale of analysis, individual deserts or drylands may embrace one or more of the broad physiographic contexts described above, while structural frameworks, among other factors, may result in the transitions between different zones occurring over large or small distances. This is well illustrated, for example, by the relative lack of mesoscale landscape variations over large distances in the Kalahari, Saudi Arabian and Australian arid zones, with their stable, cratonic settings (Rendell, Chapter 2 in this volume) and the relatively low relief over hundreds of kilometres in these drylands (e.g. Thomas and Shaw, 1991). By way of contrast, the intercratonic/interorogenic contexts of the deserts of the western USA and eastern Sahara (in Ethiopia) result in much greater transitions in relief, and accordingly greater diversity, and over shorter distances, between physiographic zones.

Goudie (2002) illustrates a further complication in desert diversity. He notes that desert regions that at first can appear remarkably similar can often display, on deeper analysis, markedly contrasting characteristics, as with the Atacama of South America and Namib of South Africa:

> At first sight they might be expected to be very similar, [with] narrow coastal fringes in a similar latitude with cold currents offshore and with mountains bounding them on their eastern sides . . . they are both hyper-arid and foggy. However there the similarity ends. The Atacama is more profoundly arid, . . . its

fogs more localized, it has less extensive dune fields, and is dominated by nitrates rather than by gypsum (Goudie, 2002, pp. 17–18).

These differences are principally due to the contrasting tectonic settings of these two deserts.

The previous discussion has provided a framework that allows explanations to be sought for the diversity that occurs between different desert and dryland areas. The following sections illustrate the nature of drylands that display considerable, and those that at first glance may display little, internal diversity. The purpose is not to provide an account of the entire world's drylands: this is best gained by reference to the detailed studies that prevail (e.g. Thomas and Shaw, 1991, on the Kalahari and Lancaster, 1989, on the Namib) or summary texts such as Goudie (2002).

4.2 Geomonotony: how unvarying are the 'flat' drylands of the world?

Many of the apparently flat dryland landscapes are dominated by depositional (often aeolian) landscapes, but certainly not all, as 'flat' erosional landscapes in the form of pediments and stony plains are also important dryland elements. It is stable cratonic settings (Chapter 2) that most obviously create the framework for 'flat' (depositional or erosional) landscapes to develop. Such landscapes can initially appear unvarying but, as the example below indicates, marked geomorphological complexity emerges on deeper analysis.

Livingstone (1857) described the semi-arid to arid Kalahari Desert of central southern Africa as 'remarkably flat' (p. 47), while in 1912 its southern region was described as 'dreary and depressing' because of its monotonous and repetitive landscape of vegetated 'sand hills' (Hodson, 1912). Similarly 'unimpressed' interpretations of flat (or repetitive) dryland landscapes by early European travellers can be found for parts of western Australia, the North American mid-west and so on. While this observation seemingly contradicts the previously observed tendency for early landform accounts to focus on the spectacular and unusual (see Chapter 1), it might also provide a contributory explanation as to why the unusual was often greeted with such enthusiasm.

The 'geomonotony'[1] of the apparently flat and endless swathes of terrain that some desert and dryland regions present on the ground actually belies considerable variability and complexity. This has most obviously been revealed not from the ground but from the air. To this end, works such as *A study of global sand seas* (McKee, 1979), which utilised early Landsat satellite imagery to map the complexity and diversity of the dune systems within individual sand seas, have done much to lay down a basis for subsequent explanations of diversity, based on analysis of meteorological station data and field measurement.

McKee (1979) was not, however, the first time that the generation of a bigger picture revealed within-desert complexity. Alexander Du Toit used aerial survey as the basis for his systematic *Kalahari Reconnaisance* (Du Toit, 1926) from which a better understanding of the topography and hydrological connections of the region emerged, allowing a scheme for dramatic river diversion and irrigation in the region (Schwarz, 1920) to be quashed. From a purely geomorphological perspective, the use of aerial survey, via air photograph analysis, has played a significant role in underpinning analysis of the complexity of 'flat' drylands, which is no better illustrated than by Grove (1958) for the Sahel and Grove (1969) for the Kalahari.

Away from its uplifted margins, the Kalahari displays no more than c.100 m relative relief over lateral distances of up to 1000 km, save for isolated inselbergs. Largely a sand-filled interior cratonic basin with a mean elevation of 1000 m a.s.l. (Thomas and Shaw, 1991), Grove's (1969) analysis paved the way for systematic investigation of its actual considerable geomorphological diversity (Figure 4.1) that is not obviously apparent on the ground. Though much of the Kalahari is covered by sands that have been shaped by the wind into dunefields (now primarily vegetated) that show regional variations in orientations and dune type, the widespread occurrence of alluvial and lacustrine deposits was also highlighted, along with the distribution of largely dysfunctional drainage systems and the local role of tectonic movements in disrupting both drainage lines and locally displacing dune patterns.

Grove's work provided the basis of systematic geomorphological analyses of component parts of the Kalahari landscape (e.g. dunes: Lancaster, 1981; Thomas, 1984; pans: Lancaster, 1978; lake basins with associated palaeoshorelines: Cooke, 1979; Shaw, 1984; valleys: Nash, Thomas and Shaw, 1994). Beyond a landform-based analysis, complexity has also been revealed with the sands that themselves characterise the Kalahari, in terms of their source regions and provenance as defined by heavy mineral (e.g. Baillieul, 1975) and wider statistical (Thomas, 1987) analysis. Within the depositional sediments of the Kalahari are also significant suites

[1] With acknowledgement to Dr Frank Eckardt, remote sensor and geomorphologist, who has used this term to describe the experience many people get when driving through parts of the Kalahari Desert.

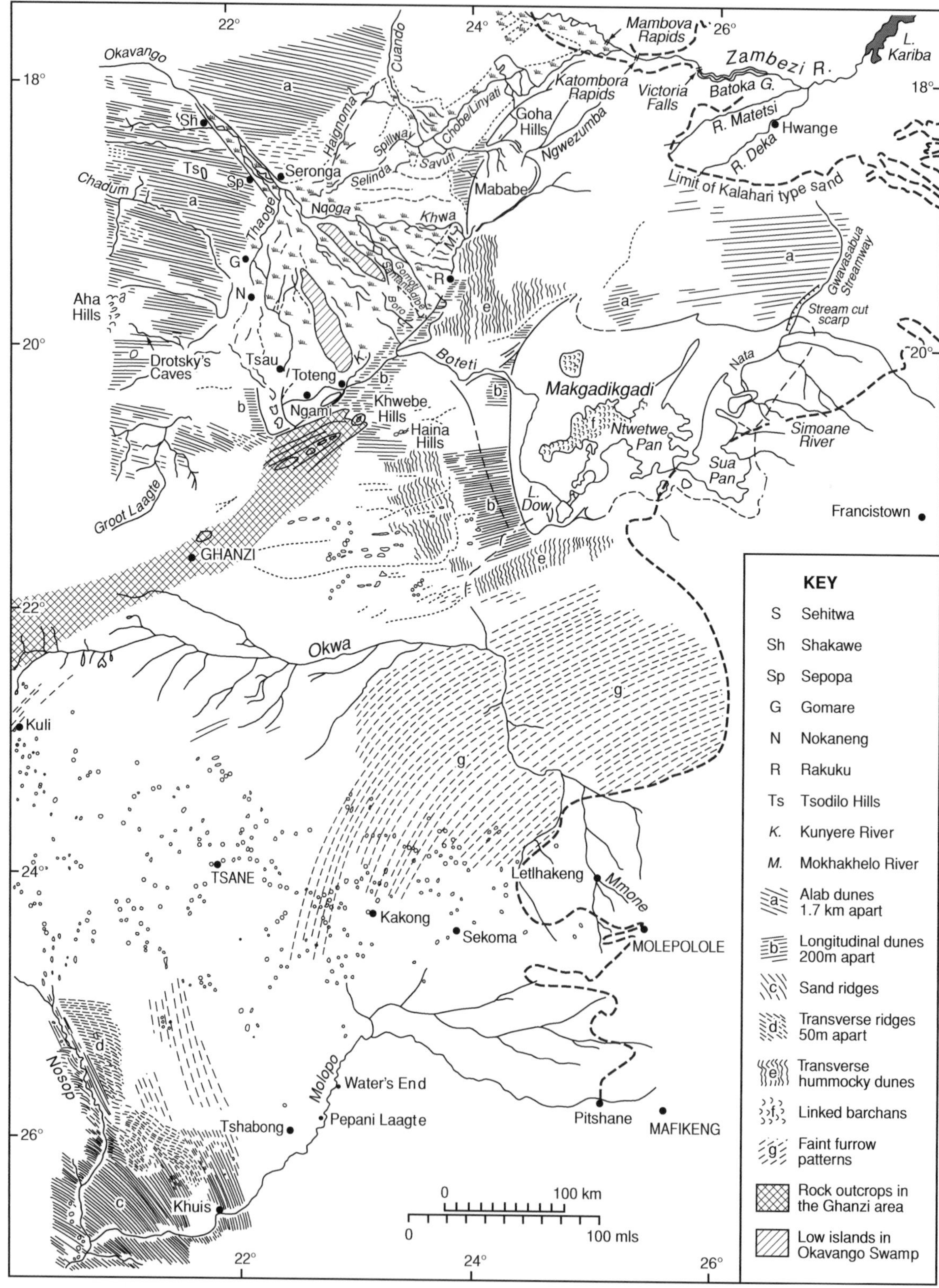

Figure 4.1 Landform diversity in the Kalahari Desert. Originally mapped from air photographs by Grove (1969), subsequent analysis in the field has demonstrated how Quaternary climate change has played a significant role in shaping the depositional landforms that dominate the interior of southern Africa.

of duricrusts (silcretes and calcretes, and hybrid forms predominate), which locally can contribute markedly to variability in surface geomorphological expressions, notably in the context of river valley systems (Nash, Shaw and Thomas, 1994).

Overall, geomorphic variability within the Kalahari is accountable in terms of three primary factors in order of ascending significance. First are local tectonic influences (e.g. Shaw and Thomas, 1993; McFarlane and Eckardt, 2007) that include, via a subtle subsiding graben on the northern edge of the Kalahari proper, the context for the development of the Okavango Delta, which is in reality a very-low gradient alluvial fan. Second are variations in sediment availability and supply, impacting upon the opportunities for depositional landform development. Third, and perhaps most important for the resultant overall landscape, is the role of climatic changes during the Late Quaternary Period. Grove (1969) prophetically suggested the importance of the latter, but it has only been with the advent and application of suitable geochronometric techniques (see Chapter 3) that the complexity of landscape and geomorphic evolution in the region is being recognised (e.g. Burrough, Thomas and Bailey, 2009; Stone and Thomas, 2008).

4.3 Within-dryland diversity

The extensive Prairie and Great Plains regions of North America are, like the Kalahari described above, regions that today are arid to semi-arid, flat and extensive. These 'Greater American Deserts' also require careful geomorphological interpretation in order to reveal their inherent, but sometimes subtle, complexity. Also, like the Kalahari, they owe many of their present characteristics to the influences of Late Quaternary climatic changes. The remaining drylands of North America (Figure 4.2) have markedly contrasting physiographic settings to the plains, being located in intraorogenic basin settings (see Table 2.1), where tectonics have set the scene for more obvious geomorphological variability than occurs in more stable locations.

This is well illustrated by the Great Basin Desert. This covers almost a third of the North American arid zone (over 400 000 km^2) and has an altitudinal range from over 4300 m a.s.l. in western Nevada to 86 m below sea level in Death Valley. Within this region, climatic conditions range from hyper-arid to semi-arid, with rain shadow effects created by the mountains themselves locally enhancing aridity to extreme levels (e.g. in Death Valley, California). The Rocky Mountains mark the eastern margin of the Great Basin and the Sierra Nevada and Cascade Range the western edge, but this is a simplistic representation of

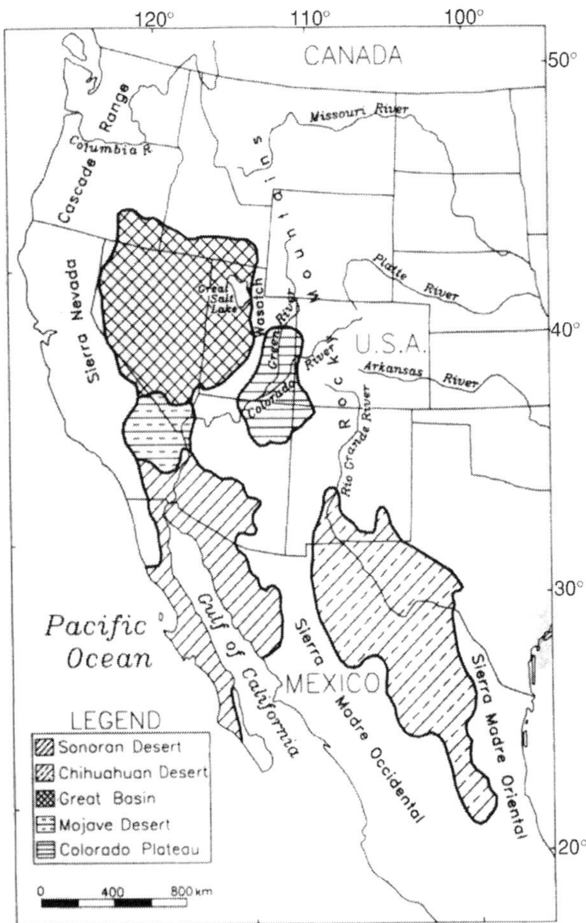

Figure 4.2 Drylands of the USA.

the setting, for in between lie over 150 separate basins that themselves are separated by 160 mountain belts that trend north–south (Dohrenwend, 1987; Grayson, 1993; Goudie, 2002).

This setting gives rise to marked, and obvious, geomorphological diversity within the total system. Some elements of the totality are, however, similar in context to the otherwise contrasting Kalahari: fans, sand seas and the remnants of playa lakes inherited from wetter climatic periods are all present. Because of the basin-and-range setting, however, spatial variability is marked and contrasts within the landscape arise over shorter distances that in drylands with more flat terrain. Erosional landscape components, on mountain slopes and on pediments, predominate over depositional settings, which are confined to basin floors or the alluvial fans of the interface zone. Thus sand seas, where sediment derived from slope and lake basin systems has been reworked by the wind, are smaller and more localised (e.g. within Death Valley), palaeolakes are more numerous – over one hundred of

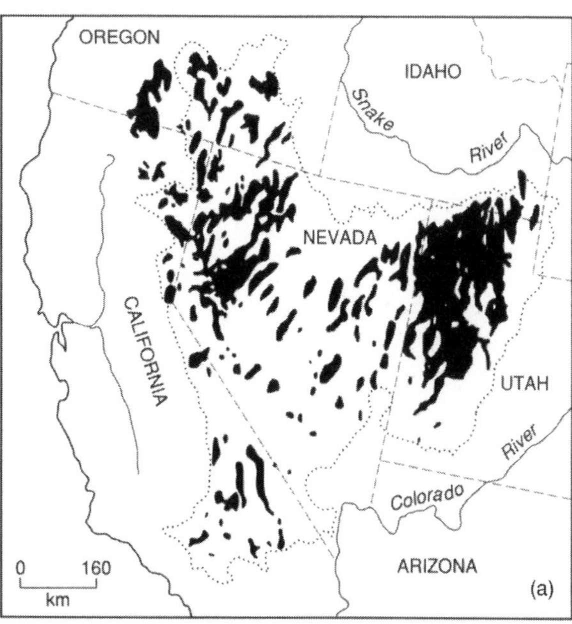

Figure 4.3 Distribution of lake basins, at their maximum extent, in the Great Basin.

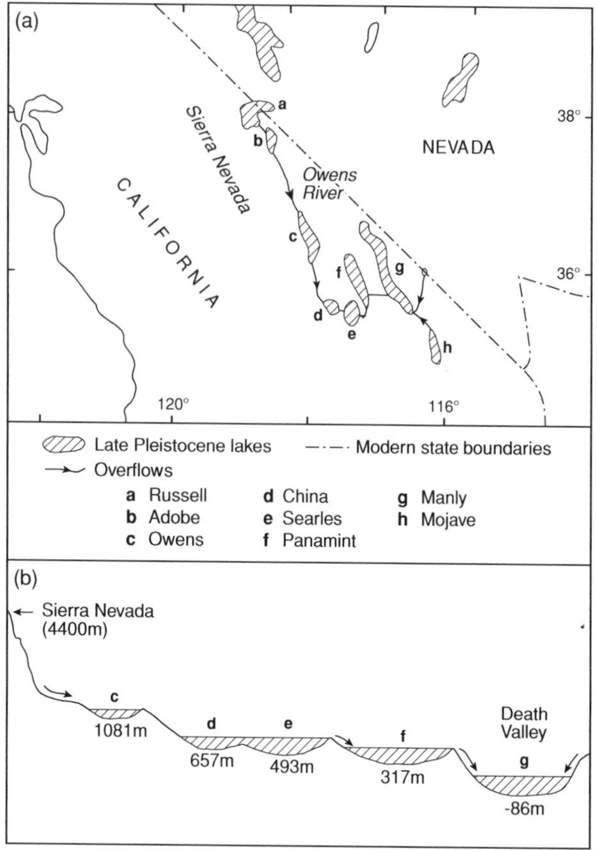

Figure 4.4 (a) Death Valley, California, has a mean annual precipitation of only 41 mm. Note evaporite deposits in the foreground and alluvial fan in the background. (b) and (c) During the Late Pleistocene the valley was occupied by Lake Manly, which at its greatest was 183 m deep and 1600 km in area. This was one of a chain of lakes, linked by overspill channels, supplied by moisture from the Sierra Nevada via the Owens River. (b) and (c) These are respectively based on diagrams in Flint (1971) and Smith and Street-Perrott (1983).

the individual basins possessed water bodies during the Late Quaternary – and fans are common at the mountain front–basin interface.

The large number of lake basins does not mean lakes were small (Figure 4.3). Sediments in and shorelines around basins have provided important data for reconstructing regional environmental change in the Late Quaternary through the direct contribution of increased precipitation, versus the role, for example, of glacial meltwater, in feeding high lakes (e.g. Li, Lowenstein and Blackburn, 1996). During wetter phases, individual lakes coalesced where topography facilitated this or lost their closed status and flowed one into another (Figure 4.4). The largest, Lake Bonneville, covered almost 52 000 km^2 and attained a maximum depth of 370 m (Grayson, 1993). Its sump is today's Great Salt Lake.

The active tectonics that created the basin-and-range terrain have continued to impact markedly and obviously on the operation of geomorphological processes in the Great Basin. Alluvial fans (see Chapter 13) are a very significant landscape feature of this desert region. Not only are fans commonly trenched due to persistent uplift, but their very presence is at least in part due to the highly effective weathering and erosional environment that uplift creates in the mountain zones. In many contexts, erosive pediments are a significant feature of the zone between mountain fronts and basins (Dohrenwend, 1994).

4.4 Summary issues

Although the explanation above has focused on two desert regions to exemplify the broad characteristics of the extreme ends of dryland landscapes, other case studies could have illustrated the points under consideration equally well. Nor do these examples illustrate the full diversity that occurs with the global arid zone. Indeed, it could be argued that each dryland region is distinctive in its own right: different interplays between erosional and depositional processes, and the balance between tectonic, sediment and climatic controls (past and present), are the broad determinants of the geomorphological characteristics of the world's dryland regions.

Close analysis, in the field or via remote sensing, shows that even the great flat desert expanses in shield contexts possess marked internal geomorphological variability. Ultimately, it is the scale of analysis that determines how varying or unvarying different dryland regions are. The Kalahari may be a large extent of sand, covering up to 2.5 million km^2 when its northern most, equator-ward extensions (that are not today arid) are included in assessments, but within that total area, particularly when evidence of past climatic events and conditions, and local tectonic activity are taken into account, significant geomorphological and sedimentological variability emerges. This is both in terms of the operation of geomorphological processes in the past and in terms of the sensitivity of different areas to specific processes today.

Since the 1970s, satellite imagery has contributed markedly to the recognition of within- and between-dryland geomorphic variability. Couple to this the advances that have occurred in the capacity to unravel the role of past environmental change and tectonic activity, and the growth in application of robust and contextualised (rather than generalised) geomorphological processes studies (Thomas and Wiggs, 2008), and a better picture emerges: not only of the diversity of dryland regions but in the explanations for that diversity. No longer is it legitimate to overgeneralise about the geomorphology of arid regions, because the capacity exists to base interpretation upon an ever-growing recognition of the complexity of arid environmental systems and the occurrence and interplay of the geomorphological processes that operate, and have operated, within them.

References

Abrahams, A.D. and Parsons, A.J. (1994) *Geomorphology of Desert Environments*, Chapman & Hall, London.

Baillieul, T.A. (1975) A reconnaissance survey of the cover sands of the republic of Botswana. *Journal of Sedimentary Petrology*, **45**, 494–503.

Burrough, S.L., Thomas, D.S.G. and Bailey, R. (2009) Mega-lake in the Kalahari. A 250 kyr record of Palaeolake Makgadikgadi. *Quaternary Science Reviews*, **28**, 1392–1411.

Clements, T., Merriam, R.H., Eymann, J.L. *et al.* (1957). A study of desert surface conditions, US Army Environmental Protection Research Division Technical Report EP-53, Natick, MA.

Cooke, H.J. (1979) The origin of the Makgadikgadi pans. *Botswana Notes and Records*, **11**, 37–42.

Dohrenwend, J.C. (1987) The basin and range, in *Geomorphic Systems of North America* (ed. W. Graf), Geological Society of America, Centennnial Special Volume **2**, Boulder, Colorado.

Dohrenwend, J.C. (1994) Pediments in arid environments, in *Geomorphology of Desert Environments* (eds A.D. Abrahams and A.J. Parsons), Chapman & Hall, London, pp. 321–353.

Du Toit, A.L. (1926) *Report of the Kalahari Reconnaissance*, Department of Irrigation, Pretoria.

Flint, R.F. (1971) *Glacial and Quaternary Geology*, Wiley Interscience, New York.

Goudie, A.S. (2002) *Great Warm Deserts of the World*, Blackwell, Oxford.

Grayson, D.K. (1993) *The Desert's Past: A Natural Prehistory of the Great Basin*, Smithsonian Institution Press, Washington, DC.

Grove, A.T. (1958) The ancient ergs of Hausaland, and similar formations on the south side of the Sahar. *Geographical Journal*, **124**, 528–533.

Grove, A.T. (1969) Landforms and climatic change in the Kalahari and Ngamiland. *Geographical Journal*, **135**, 191–212.

Heathcote, R.L. (1983) *The Arid Lands: Their Use and Abuse*, Longman, London.

Hodson, A.W. (1912) *Trekking the Great Thirst. Travel and Sport in the Kalahari*, Fisher Unwin, London.

Lancaster, I.N. (1978) The pans of the southern Kalahari, Botswana. *Geographical Journal*, **144**, 80–98.

Lancaster, N. (1981) Palaeoenvironmental implications of fixed dune systems in southern Africa. *Palaeogeogarphy, Palaeoclimatology, Palaeoecology*, **33**, 327–346.

Lancaster, N. (1989) *The Namib Sand Sea: Dune Forms, Processes and Sediments*, Balkema, Rotterdam.

Li, J., Lowenstein, T.K. and Blackburn, I.R. (1996) Response of evaporate mineralogy to inflow water sources and climate during the past 100 k.y. in Death Valley, California. *Bulletin, Geological Society of America*, **109**, 1361–1371.

Livingstone (1857) *Missionary Travels and Researches in South Africa*, Ward Lock, London.

Mabbutt, J.A. (1969) Landforms of arid Australia, in *Arid Lands of Australia* (eds R.O. Slayter and R.A. Perry), ANU Press, Canberra, pp. 11–32.

Mabbutt, J.A. (1976) Physiographic setting as an indication of inherent resistance to desertification. *WGDAL*, **1976**, 189–197.

McFarlane, M.J. and Eckardt, F.D. (2007) Palaeodune morphology associated with the Gumare fault of the Okavango graben in the Botswana/Namibia borderland: a new model of tectonic influence. *South African Journal of Geology*, **110**, 535–542.

McKee, E.D. (1979) Introduction to a study of global sand seas, US Geological Survey, Professional Paper 1052, pp. 1–19.

Murphy, R.E. (1968) Landform regions of the world. *Annals, Association of American Geographers*, **58**: Map Supplement 9.

Nash, D.J., Shaw, P.A. and Thomas, D.S.G. (1994) Timescales, environmental change and dryland valley development, in *Environmental Change in Drylands* (eds A.C. Millington and K. Pye), John Wiley & Sons, Ltd, Chichester, pp. 25–41.

Nash, D.J., Thomas, D.S.G. and Shaw, P.A. (1994) Duricrust development and valley evolution: process-landform links in the Kalahari. *Earth Surface Processes and Landforms*, **19**, 299–317.

Schwarz, E.H.L. (1920) *The Kalahari, or Thirstland Redemption*, Maskew Miller, Cape Town, South Africa.

Shaw, P.A. (1984) An historical note on outflows of the Okavango system. *Botswana Notes and Records*, **16**, 127–130.

Shaw, P.A. and Thomas D.S.G. (1993) Geomorphology, sedimentation and tectonics in the Kalahari Rift. *Israel Journal of Earth Sciences*, **41**, 87–94.

Smith, G.I. and Street-Perrott, F.A. (1983) Pluvial lakes of the western United States, in *Late Quaternary Environments in the United States. Volume 1: The Late Pleistocene* (ed. S.C. Porter), University of Minnesota Press, Minneapolis, pp. 190–214.

Stone, A.E.C. and Thomas D.S.G. (2008) Linear dune accumulation chronologies from the southwest Kalahari, Namibia: challenges of reconstructing Late Quaternary palaeoenvironments from aeolian landforms. *Quaternary Science Reviews*, **27**, 1667–1681.

Thomas, D.S.G. (1984) Ancient ergs of the former arid zone of Zimbabwe, Zambia and Angola. *Transactions, Institute of British Geographers*, **NS9**, 75–88.

Thomas, D.S.G. (1987) Discrimination of depositional environments, using sedimentary characteristics, in the mega Kalahari, central southern Africa, in *Desert Sediments, Ancient and Modern* (eds L.E. Frostick and I. Reid), Geological Society of London Special Publication 35, Blackwell Scientific Publlications, Oxford, pp. 293–306.

Thomas, D.S.G. (1988) Discrimination of depositional environments, using sedimentological characteristics, in the Mega Kalahari, central southern Africa, in *Desert Sediments, Ancient and Modern* (eds L. Frostick and I. Reid), Geological Society Special Publication 35, Blackwell, Oxford.

Thomas, D.S.G. and Shaw, P.A. (1991) *The Kalahari Environment*, Cambridge University Press, Cambridge.

Thomas, D.S.G. and Wiggs, G.F.S. (2008) Aeolian system responses to global change: challenges of scale, process and temporal integration. *Earth Surface Processes and Landforms*, **33**, 1396–1418.

5

Extraterrestrial arid surface processes

Jonathan Clarke

5.1 Introduction

Planetary geomorphology is one of the emerging frontiers of arid geomorphological research (Tooth, 2009). In earlier editions of this work (Wells and Zimbelman, 1989, 1997) extraterrestrial aridity could only be discussed with reference to Mars, although the nature and implications of extraterrestrial aridity had long exercised the imaginations of some writers (e.g. Frank Herbert's 1965 novel *Dune*). More recent data allow the question to be considered over a much wider range of solar system bodies than just Earth and Mars (Figure 5.1). This chapter provides a review of the nature of 'arid geomorphology' in a solar system context, what it might mean in environments quite different from those of Earth, why planetary scientists need to be cognisant with terrestrial geomorphology and terrestrial geomorphologists familiar with the surfaces of other bodies in the solar system. It will then briefly review the landscapes of Mars, Titan and Venus and the features and processes relevant to each that can be considered comparable to terrestrial arid landforms.

5.2 What does 'aridity' mean beyond Earth?

Even the most cursory glance at images from space exploration missions reveals scenes that are strongly reminiscent of terrestrial arid landscapes, be they dry gullies on Mars or dune fields on Titan. These features occur on bodies whose surface temperatures, atmospheric pressures, gravity and composition are very different from that of Earth, which raises the question as to what 'aridity' might even mean on these bodies.

On Earth 'aridity' is defined by moisture deficits mostly strongly evidenced in vegetation distribution (Chapter 1). It is the presence or absence of vegetation that determines the nature of many of the features seen as characteristic of arid environments, such as wind erosion and deposition, or the formation of arroyos and braided stream deposits. Vegetation is therefore widely used as a proxy for evaporation in the absence of direct climatic data. However, this criterion cannot be used where it is absent, not only on other planets but for 90 % of Earth's history prior to the evolution of terrestrial vegetation (Davis and Gibling, 2010). On Earth areas of dunes will form wherever there is sufficient sand supply and vegetation cover is minimal, even in high rainfall. Much the same applies for the formation of arroyos and braided stream deposits.

Other proxies must therefore be used for moisture deficits, the most useful of which is the accumulation of salts in soils and in the terminal parts of drainage systems. Another proxy is the evidence for the role of ephemeral moisture and salt growth in the breakdown of rock surfaces, although on Earth these are not exclusive to arid environments. This allows aridity in the terrestrial sense to be applied to landscapes on Mars where water moisture is at least ephemerally present and may have been more widespread in the past.

However, the absence of moisture is not in itself a sufficient criterion. While the surfaces of the Moon or Mercury, for example, may be lacking in moisture, they are also lacking in atmosphere (though perhaps not traces of water in other states; see Clark, 2009, and Slade, Butler and Muhleman, 1992), and the application of a climate-defined term such as aridity in such an environment stretches it past breaking point. What are we to make of bodies such as Titan, which, despite average

Arid Zone Geomorphology: Process, Form and Change in Drylands, Third Edition. Edited by David S. G. Thomas

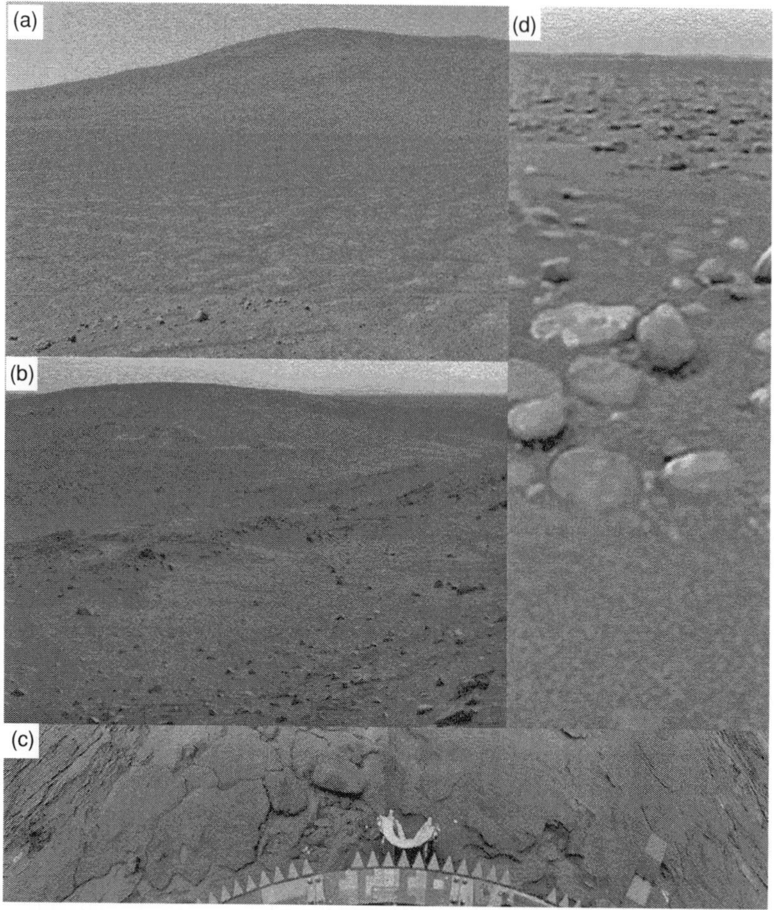

Figure 5.1 Arid landscapes of four worlds: (a) Earth (Atacama desert); (b) Mars (Columbia Hills, Gusev Crater); (c) Venus (Phoebe Regio); (d) Titan (Xanadu Regio). Image (b) NASA; (c), original *Venera* data reprocessed by Don Mitchell (used by permission); (d), NASA/ESA.

surface temperatures well below the range of liquid water, have a wide range of landforms that on Earth would be considered typical of humid (apparently permanent lakes and streams, Mitri *et al.*, 2007) and arid (large-scale longitudinal dune fields, Lorenz *et al.*, 2006)?

To allow for these observations it is proposed that on a planetary scale 'arid geomorphology' should be expanded to include those bodies with a solid surface and with an atmosphere sufficiently dense to allow any liquid phase to condense and play a role analogous to water in some way. These definitions both exclude the gas giants such as Jupiter and airless bodies like the Moon. This definition includes Venus and Titan as bodies that are partly to completely arid with very different condensable atmospheric liquids (sulfides on Venus and methane on Titan). At present it also excludes bodies such as Triton, where the atmosphere modifies the surface (Greeley, 1994) but is too attenuated to allow the condensation of a liquid phase.

5.3 Why should planetary scientists understand terrestrial arid geomorphology?

Our ability to study the surfaces of other bodies in the solar system is still severely limited in many respects. Planetary geomorphologists must rely heavily on terrestrial analogues for the features they observe, with all the limitations this entails. This requires a high level of familiarity with terrestrial surface processes across a wide range of environments and regions. There is an unfortunate reliance on a comparatively small number of analogues from the southwest of the United States; this should be expanded

Figure 5.2 Tracks in the sand: desert exploration in our solar system. (a) Tracks in the Namib Sand Sea: University of Oxford fieldwork, 2007 (photo: David Thomas). (b) Tracks of NASA's *Opportunity* Rover, 2004, Meridiani Planum, Mars (photo: NASA).

to include the diverse analogue landscapes that can be found elsewhere in the deserts of the world, both the well-studied hot deserts and the somewhat neglected cold deserts.

5.4 What can terrestrial geomorphologists learn from a solar system perspective?

Based on comments to the author by terrestrial geomorphologists, some find planetary geomorphology superficial and simplistic. This is partly due to the immense limitation of the discipline compared to terrestrial geomorphology, being restricted largely to images from spacecraft in orbital or on flyby trajectories, and with very limited ground truth. Data are collected by instruments designed to meet engineering constraints, not optimised for collecting geomorphic data. Similarly, landing sites are necessarily prioritised according to spacecraft safety, not geomorphic interest. Because of these restrictions planetary geomorphology is somewhat comparable to the earliest phases of the exploration of terrestrial deserts (see Chapter 1), where the literature is dominated by first impressions (sometimes quite erroneous). The heyday of planetary geomorphology is yet to come with the development of high-resolution images, long-range remotely operated vehicles and, eventually, crewed missions returning to the Moon and then to Mars and beyond, and building on the foundation of terrestrial desert exploration (Figure 5.2).

Even in this early stage there are many lessons that terrestrial geomorphologists can profitably learn from studying planetary surfaces. Not the least of these is the limits to climatic geomorphology as presently understood, evidenced by the similarity of rocky surfaces on Earth, Mars,

Venus and Titan, despite the radically different conditions and compositions on the surface of each, or the similarity in dune patterns on Earth and Titan. Furthermore, planetary landscapes provide natural laboratories that can test a range of hypotheses, including the role of ice complexity in landscape processes, the significance of moisture in extremely low abundances, and the degree to which vegetation influences the expression of geomorphic processes. Ultimately there is the possibility that, through understanding palaeoclimatic records across the solar system, e.g. in the polar deposits of Mars, this will assist in clarifying the role of solar variability in climate change. Lastly, an appreciation of extraterrestrial geomorphology can encourage research into specific terrestrial analogues and spur study of previous underresearched aspects (Clarke, 2008).

5.5 Mars: water-based aridity

5.5.1 *Overview*

Mars is the third largest rocky body in the solar system. It has a diameter that is 53 % of Earth's, a mass that is 10.7 % of Earth's and the surface gravity is consequently 38 % of the terrestrial value. The atmosphere is predominantly (95.7 %) CO_2 with minor (2.7 %) N_2. The surface pressure is 6.36 millibars at the mean radius, varying from 4.0 to 8.7 millibars depending on season. Pressure varies between 11.6 and 0.3 millibars over an altitude range of −8.2 km below the Mars areoid and +21.2 km above Mars areoid. The mean surface temperature is about −46 °C with highs of 20 °C at the ground surface at midday on the equator and lows of −130 °C during the polar winters. The atmosphere is active with global and regional dust storms, local dust devils and with clouds, snow and frost composed of water and carbon dioxide ices.

The surface of Mars (Figure 5.3) is the most studied of any solar system body after the Earth and Moon. The data sources and literature are therefore very extensive. However, interpretations of some surface features are controversial; see, for example, the discussion of recent gully forms (Malin *et al.*, 2006; McEwan *et al.*, 2007; Heldmann *et al.*, 2009). This brief section will therefore only summarise some of the more salient aridity-related features.

5.5.2 *The history of atmosphere–surface interactions*

Mars is uniquely a two-volatile planet with both CO_2 and water condensing from its atmosphere on to the surface (Kargel, 2004). Frozen CO_2 passes back into the gaseous phase without going through a liquid phase but the pressure and temperatures are high enough for ephemeral moisture films and adsorbed water in many locations on Mars (Möhlmann, 2004; Richardson and Mischna, 2005). Evaporation rates are very high, creating a largely dry surface, although shallow ice extends over much of the surface in the mid and high latitudes (Mitrofanov *et al.*, 2003). In these respects Mars resembles terrestrial cold deserts.

Mars appears to undergo orbitally forced climate changes analogous to those of Earth (Head *et al.*, 2003). Large changes in obliquity from 15 to 35 degrees occur at intervals of ~40 ka. This, in conjunction with the orbital eccentricity and changes to areocentric longitude at perihelion, may lead to 'glacial' periods of ~1.75 Ma duration and 'interglacial' periods of 0.65 Ma. Polar ice is most stable at periods of low obliquity and equatorial ice during periods of high obliquity (Head *et al.*, 2005). Water is therefore shuttled between the poles and the equator, being lost by sublimation from one and deposited at the other as snow. At present Mars is in an intermediate state, with ice stable at the poles but having being lost through sublimation at the equator, except where protected by thick regolith.

Over a much longer time frame Mars appears to have evolved from earlier epochs with much more abundant liquid water, including lakes and seas, through a transitional phase to the present cold and dry regime. These different epochs of water activity may have led to characteristic mineralogical signatures in the martian surface (Bibring *et al.*, 2006). The earliest epoch is dominated by phyllosilicates formed during water-rich conditions and has been termed the 'Phyllocian' era. Sulfates formed in a second era (the 'Theiikian'), which saw increased aridity. The 'Siderikian' era of the last ~3.5 Ga is dominated by anhydrous ferric oxides in a slow, dry weathering environment. Kargel (2004) gives a good summary of the ideas associated with the evolution of the surface environment with respect to the availability of water and whether this evolution was a simple linear process or one of greater complexity involving longer-term cycles of wetter and drier epochs or episodic water release events driven by volcanism or impacts. These eras can be compared with the more widely used martian timescale based on crater densities (Cattermole, 2001):

- Amazonian (younger limits 0.70–1.80 Ga; older limits 2.3–3.55 Ga).
- Hesperian (younger limits 1.80–3.10 Ga; older limits 3.70–3.80 Ga).

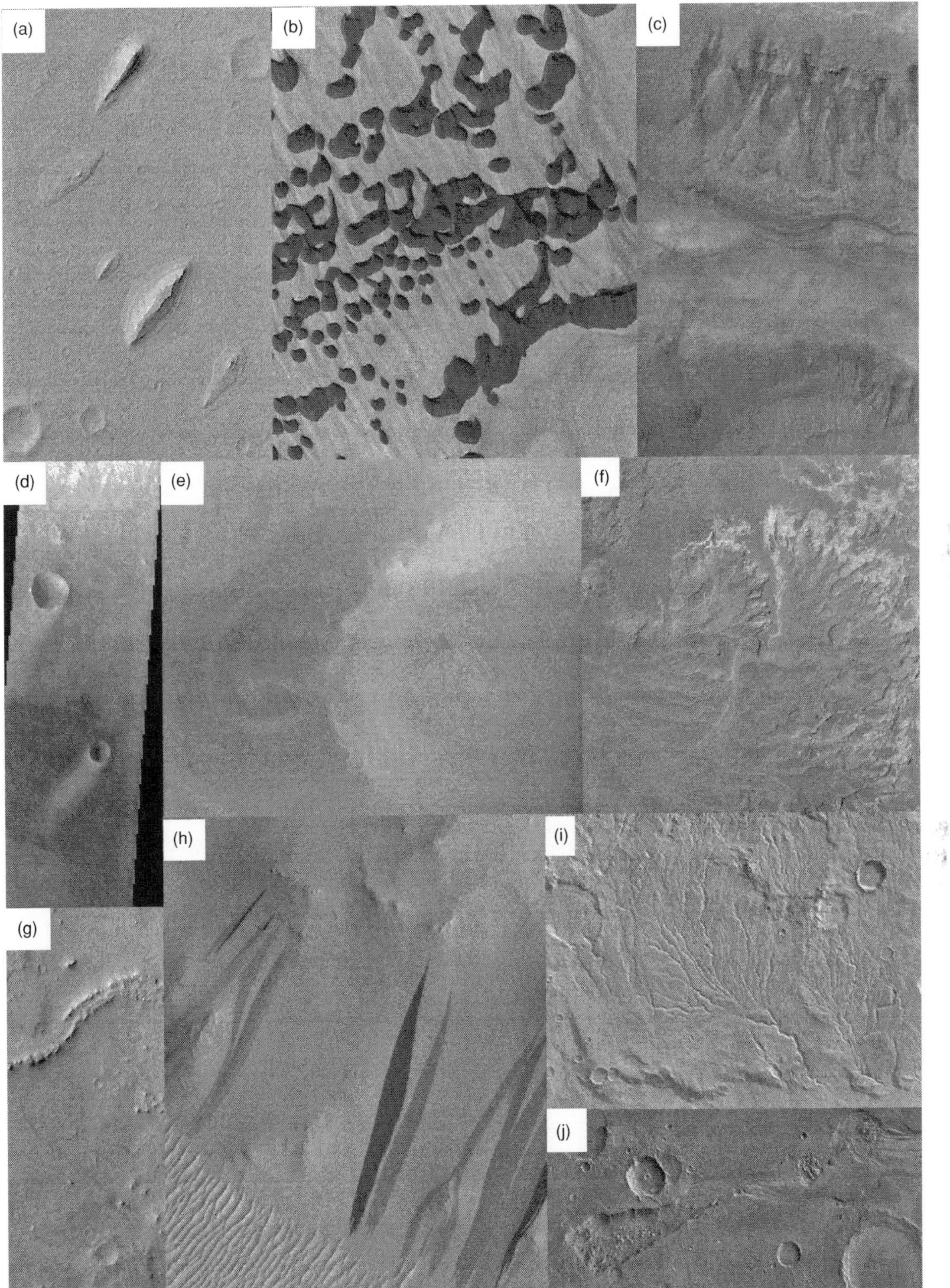

Figure 5.3 Arid land forms of Mars seen in orbital imagery: (a) yardangs (MOC image); (b) Barchan dunes of dark basaltic sand in the north polar erg of Mars (MOC image); (c) gullies formed by episodic water flows from two levels of crater wall (MOC image); (d) bright wind streaks downwind of craters (THEMIS image); (e) regional summer dust storm blowing off the north polar cap of Mars (MOC image); (f) delta in Eberswalde crater (MOC image); (g) inverted channel in Miyamoto crater (MOC image); (h) darker (younger) and lighter (older) slope streaks (MOC image); (i) dendritic drainage system, Warrego Vallis (*Viking* orbiter image); (j) Outflow channel (Ravi Vallis) (*Viking* orbiter image).

- Noachian (younger limits 3.50–3.85 Ga; older limit >4.60 Ga).

The alteration eras correspond loosely to the cratering eras – Noachian with the Phyllocian, Hesperian with the Theiikian and Amazonian with the Siderikian.

5.5.3 Martian water cycle

The expression of the water cycle on Mars has therefore varied greatly through the planet's history, reflecting the planetary scale geophysical evolution and in turn driving the extant geomorphic processes.

The presence of a thick, greenhouse atmosphere in the Noachian appears to have led to a hydrological cycle similar to that of Earth. Precipitation (not necessarily rainfall) led to runoff and the formation of dendritic valley networks like Warrego Vallis (Ansan and Mangold, 2006) that discharged ultimately into a northern ocean (Carr and Head, 2003). This active, Earth-like hydrologic cycle may have been accompanied by at least local deep weathering with low-temperature clay–carbonate alternation (Wray *et al.*, 2009). The martian atmosphere may have been lost with the global magmatic after the cessation of the core dynamo (Kargel, 2004; Cattermole, 2001).

The Hesperian marks a transition from the active water cycle of the Noachian to the largely inactive cycle of the Amazonian. Water, released by magmatic or impact events formed outflow channels, but was refrozen into the upper martian crust and regolith in what was essentially a one-way process. Atmospheric water was also deposited at the poles forming ice deposits, though probably not the ice deposits seen today.

The Amazonian appears to have been characterised by a water cycle consisting of the solid and vapour phases only, and the transfer of that between latitudes where sublimation is dominant and its trapping where condensation dominates (Head *et al.*, 2003). The water may be deposited as surface ice deposits and subsequently buried.

On Earth an active water cycle involving all three phases has been operating for its entire geologic history. As a result the whole hydrosphere has been cycled perhaps 1000 times (Kargel, 2004). On Mars the cycling has been far less since it ended sometime between 3 and 4 Ga.

5.5.4 Surface images

Many thousands of images have been returned from six locations on Mars (Figure 5.4), four of which are essentially point sites and two are multikilometre traverses.

The *Viking 1* spacecraft was the first to operate successfully on Mars. It landed on the western slopes of the Chryse Planitia basin (Mutch *et al.*, 1976a) and revealed a rough surface with possible bedrock surfaces, numerous rocks of 20 cm size, scattered blocks and boulders up to 1 m by 3 m in size, and small aeolian dunes. Many of the rocks are pitted through differential erosion or weathering or because of their original texture. The exhaust jets of the landing engines locally exposed a polygonally cracked resistive horizon just beneath the surface. This was incorrectly termed 'duricrust' (Mutch *et al.*, 1976a) when in reality it more closely resembles the lightly cohesive or weakly indurated surface crusts of terrestrial arid environments (Thomas, Clarke and Pain, 2005). Over time, minor rearrangement by the wind of fine-grained surface material was noted and there was a small slump on the face of a dune. This showed the presence of weak induration on that dune (Jones *et al.*, 1979).

Viking 2 landed on the northern plains of Mars at Utopia Planitia on the ejecta blanket of Mie crater (Mutch *et al.*, 1976b). A flat landscape supported large numbers of blocks, some more than 20 cm across, scattered with a much higher density than at the *Viking 1* site. Rock surfaces were heavily pitted. Actual lithological variability appeared lower than at the previous site and there were no small dunes or bedrock outcrops. The surface was composed of a weakly indurated crust broken up by large polygons ~8 m across, suggesting subsurface ice. No ice was encountered during the Lander's excavation programme (Jones *et al.*, 1979) but data from hydrogen mapping and studies of ice distribution (Mitrofanov *et al.*, 2003) shows that there was probably an ice table within a metre of the surface. Ripples of wind-deposited material occurred in the polygonal troughs. Over the operational period of the mission small changes were observed (Jones *et al.*, 1979), consisting of minor aeolian reworking and the build-up and sublimation of water ice frosts.

The *Pathfinder* Lander touched down on Ares Vallis and soon after deployed a small rover, *The Sojourner*, which allowed data to be collected over a much wider area. The landing site was where a large outflow channel debouched on to the southern part of Chryse Planitia (Ward *et al.*, 1999). The surface relief was complex, with probable water-eroded bedrock hills and crater rims in the distance, rounded cobbles and boulders, some possibly imbricate, scattered across an irregular sedimentary surface that showed possible small channels, and a surface deposit of small aeolian dunes. Rock surfaces showed numerous ventifacts and there were signs of relatively recent stripping by wind of significant thicknesses of soil. Boulders ranged from 5–20 cm up to ~2 m in size (Smith

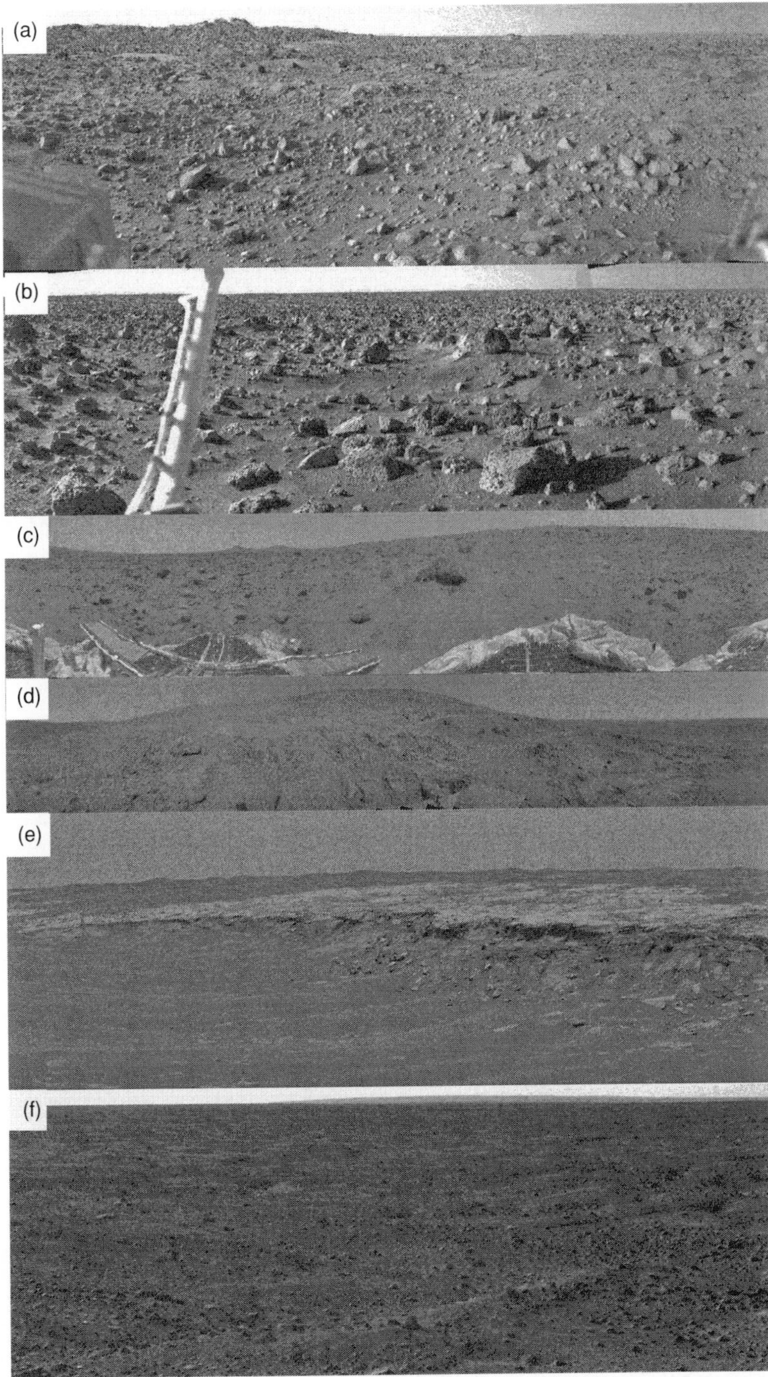

Figure 5.4 Mars landing sites: (a) impact-gardened basaltic plain with aeolian drifts, Chryse Planita (*Viking 1* panorama); (b) deflated ejecta blanket from Mie crater, Utopia Planita (*Viking 2* panorama); (c) outwash deposit of Ares Vallis with possible boulder imbrication and potentially water scoured hills (*Pathfinder* panorama); (d) bedrock outcrops, Columbia Hills, Gusev Crater (*Spirit* image); (e) fractured evaporitic bedrock overlain by dunes of dark basaltic sand, Erebus Crater, Meridiani Planum (*Opportunity* image); (f) polygonal terrain of Vasitas Borealis (*Phoenix* image). All photos NASA.

et al., 1997). The spacecraft did not remain operational for long enough to detect surface changes.

The *Spirit* Rover (Crumpler *et al.*, 2005; Haskin *et al.*, 2005) was deployed after landing in Gusev crater. From its landing site on the basaltic crater floor covered mostly by very small loose rocks and scattered wind ripples, the Rover drove across to the Columbia Hills, composed of bedrock predating the crater fill. *Spirit* then climbed Husband Hill and descended past some very dark basaltic dunes to the volcanic feature known as Home Plate. At the time of writing *Spirit* had travelled more than 7 km but was badly bogged, a not unusual experience for a desert traveller. Noteworthy images have been obtained of tafoni-style weathering, insolation fracturing and well-developed ventifacts that illustrate a range of processes operating on the surface that are also common in terrestrial arid environments. *Spirit* has also observed dust devils that have removed dust from the surface, as have short-lived wind gusts (Greeley *et al.*, 2006). Heightened dust deposition during and after regional and global dust storms has also been observed.

The *Opportunity* Rover landed on almost the opposite side of the planet at Meridiani Planum (Soderblom *et al.*, 2004; Sullivan *et al.*, 2005). The landscape here is very different, comprising an extremely flat plain of horizontal evaporitic sediments punctuated by variably eroded impact craters. The surface is mantled by a lag of ironstone concretions eroding out of the sedimentary bedrock and veneered by small dunes or large ripples of basaltic sand. At the time of writing *Opportunity* has covered more than 12 km in its traverse from its landing site, visiting successively larger craters that expose progressively deeper parts of the stratigraphy. In the course of its journey the Rover has been twice bogged, once seriously and once briefly, but was successfully extricated on both occasions. Noteworthy images reminiscent of terrestrial arid environments have included dedo-like projections (Thomas, Clarke and Pain, 2005) attributable to wind erosion, surficial soil crusts, probably from salt cementation, and a range of wind ripples and small dunes. As with *Spirit*, dust devils and wind gusts have removed dust from the surface (and the Rover), and heightened dust deposition during and after regional and global dust storms has been measured. Winter frosts have been imaged on the Rover.

Phoenix, in the most recent mission to land on Mars, touched down on the martian arctic plains at 68°N (Smith *et al.*, 2009). It revealed a very flat landscape dominated by patterned ground with polygonal cracking, mounding and stone sorting (Mellon *et al.*, 2009), similar to those of terrestrial polar dry deserts (Levy, Head and Marchant, 2009). The efflux from the landing rocket engines removed the top layers of regolith, exposing a water ice table at a few centimetres depth. The polygonal patterned ground occurred at two scales, 20–25 m large polygons and small 2–3 m polygons. The properties of the soil were largely unexpected. Rather than being loose and acidic, it proved to be exceptionally sticky and alkaline, containing traces of perchlorate salts, which significantly lowered the eutectic temperature of the water and may control regolith–atmosphere water exchange (Smith *et al.*, 2009). A number of active processes were observed over the course of the mission, despite its relatively short duration. These included the partial sublimation of water ice exposed by the rocket efflux and in excavations, frost deposition as winter approached, dust storms and water ice snow.

5.5.5 The geomorphology of Mars

Martian surface features are diverse (Figures 5.5 and 5.6) and are only summarised below with respect to the most significant geomorphological elements related to aridity. Other processes – cratering, tectonics and volcanism – are also highly important but are not considered here.

5.5.5.1 Lakes

Mars presently has no bodies of standing water, frozen or unfrozen. In the past, especially in the Noachian and Hesperian eras when the hydrological system appears to have been more active, there were lakes (Irwin *et al.*, 2005) and possibly a northern sea (Carr and Head, 2003). The *Opportunity* Rover has imaged a diverse suite of sulfate evaporites associated with ancient lake deposits on Meridiani Planum, including lake bed, lake margin and lake derived dune facies (Grotzinger *et al.*, 2005). The complex facies mosaic of these sediments suggests deposition in groundwater-fed discharge complexes, of which the Australian boinkas may provide a terrestrial analogue (Baldridge *et al.*, 2010).

5.5.5.2 Soils

The complex processes of the martian surface, especially the potential role of ephemeral moisture (Thomas, Clarke and Pain, 2005), can allow the formation of differentiated surface regolith with some affinities to soils of both hot and cold deserts. Soil-like features have been observed at all landing sites to date, including resistant layers (*Viking 1* and *Viking 2*, Mutch *et al.*, 1976a, 1976b), highly variable mechanical properties of fine-grained surface materials from unconsolidated to high compressive to strong resistant (*Pathfinder*, Ward *et al.*, 1999), enrichment in salts

Figure 5.5 Arid landforms of Mars seen in surface imagery: (a) rock disaggregation, possibly from insolation or salt weathering, Gusev crater; (b) tessellated megaripples, Endurance Crater, Meridiani Planum; (c) thermal contraction polygons, Vasitas Borealis; (d) dry particulate flow in sand, Husband Hill, Gusev Crater; (e) multiple aeolian facets on rock in Columbia Hills, Gusev Crater; (f) incipient honeycomb weathering, Columbia Hills, Gusev Crater.

related to moisture (*Opportunity*, Soderblom *et al.*, 2004) including those of sulfur, chlorine and bromine (*Spirit*, Haskin *et al.*, 2005; Crumpler *et al.*, 2005), carbonate deposition (*Phoenix*, Boynton *et al.*, 2009) and ice segregations (*Phoenix*, Smith *et al.*, 2009). Further studies of soil profiles and processes on Mars await future surface missions able to trench or core deeper into the soil and perform detailed analyses at various depths.

5.5.5.3 *Rock breakdown processes*

The long-term evolution of surface–hydrosphere interactions (Bibring *et al.*, 2006) has led to a secular change in rock breakdown processes on Mars (Chevrier and Mathe, 2007). The earliest epochs are characterised by chemical rock breakdown similar to that experienced in humid terrestrial environments. As water became less common the environment was subject to more sulfate-dominated chemical and physical weathering. In the most recent time physical processes assisted by moisture-driven processes dominate. Surface images of ongoing processes indicate active rock breakdown by aeolian-driven grains (e.g. Bridges *et al.*, 1999), exfoliation and disaggregation (Thomas, Clarke and Pain, 2005), with abundant evidence for a role by surface salts (Jagoutz, 2006). The *Phoenix* Lander imaged recent fracturing on the polar zone (Mellon *et al.*, 2009) and freezing is probably an important component to rock fragmentation even in the martian tropics. However, much still remains to be understood about the rock fragmentation processes on Mars, e.g. what process generated the large quantities of dark basaltic sands (e.g. at the *Opportunity* site, Soderblom *et al.*, 2004) that form many of the large dune fields, such as the circumpolar erg (Warner and Farmer, 2008).

5.5.5.4 *Aeolian landforms*

Wind is the most active process currently shaping the martian surface. Erosion features are visible over a wide range of scales, including deflation surfaces, leaving pebbly lags, ventifacts (Bridges *et al.*, 1999, 2007) and yardangs (Zimbelman and Griffin, 2010). Wind is also responsible for the stripping of dust mantles (Bridges *et al.*, 2007). Large-scale denudation through the Amazonian has been controlled largely by the wind (Zimbelman and Griffin,

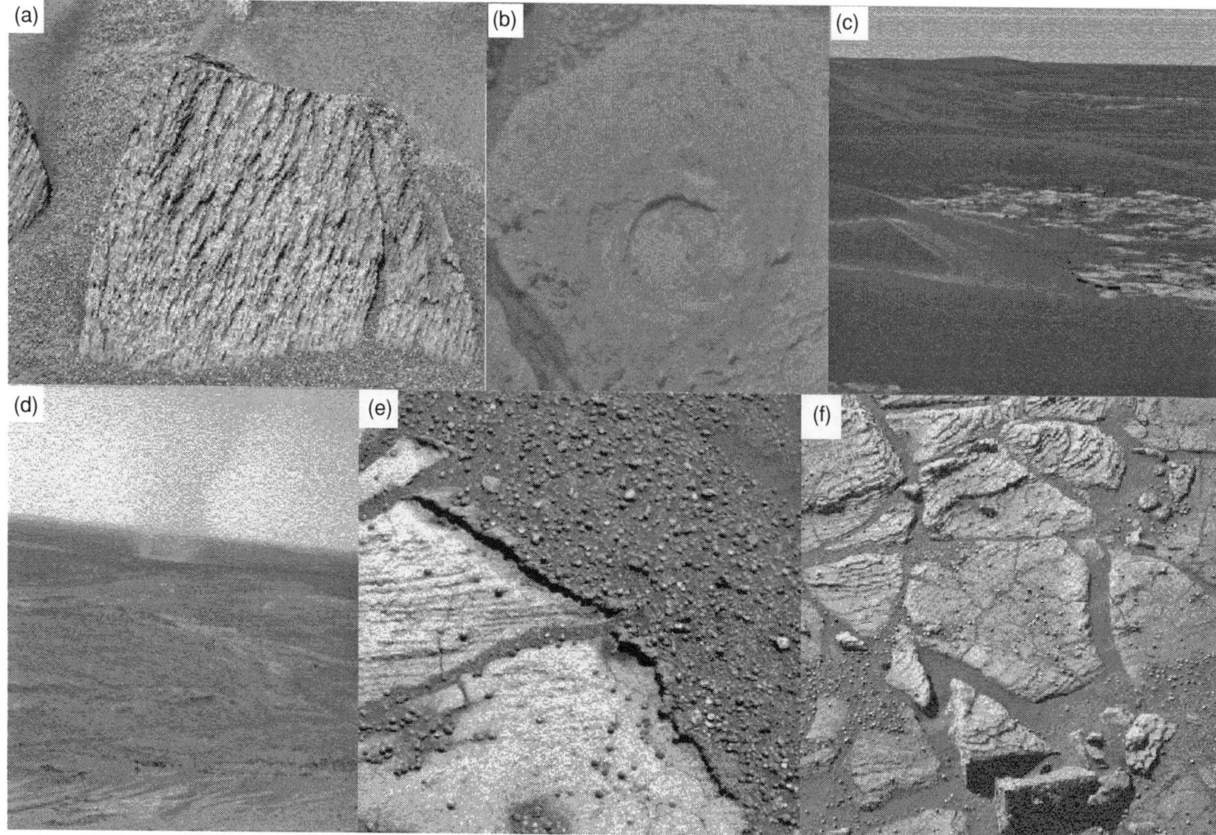

Figure 5.6 More arid landforms of Mars seen in surface imagery: (a) wind-eroded grooves on boulders near top of Husband Hill, Gusev Crater; (b) dedos formed by erosion of boulder at Fram Crater, Meridiani Planum, (c) dark basaltic sand dunes over fractured sedimentary bedrock, Meridiani Planum; (d) dust devil on the floor of Gusev Crater; (e) surface crusts, Meridiani Planum; (f) dirt cracking (cf. Ollier, 1965), Meridiani Planum.

2010), and among other things has led to relief inversion of older fluvial features (Pain, Clarke and Thomas, 2007).

Aeolian deposits are diverse and occur at all scales. Dust deposition occurs over large areas of the planet (Bridges *et al.*, 2007), analogous to terrestrial loess and parna (Greeley and Williams, 1994). Grain size of these deposits is much smaller than those that occur on Earth, typically being 2–40 micrometres. Tails of aeolian deposits are evident at almost all landing sites and small dunes are common in Lander images (Mutch *et al.*, 1976a; Ward *et al.*, 1999; Greeley *et al.*, 2006; Sullivan *et al.*, 2005). Many are low albedo and probably made up of basaltic sand, but light coloured dunes are also present. Some of these may be coated by light coloured dust but others are plausibly composed of lighter-coloured grains. Ice is a likely component of polar dunes (Feldman *et al.*, 2008). Wind streaks form downwind of obstacles such as craters and are of two types, light and dark. Some dark coloured streaks (e.g. Geissler *et al.*, 2008) are where wind has deposited dark basaltic sand or removed the bright coloured surface dust. Light coloured streaks are where light coloured material, sand or dust has been deposited. Dunes are common and diverse in size and setting, and range from scattered occurrences to those in craters or downwind of obstacles through to large ergs (Warner and Farmer, 2008). Intermediate scaled features, termed transverse aeolian ridges, are especially common (Bourke, Bullard and Barnouin-Jha, 2004). The most common dune morphologies are barchans, barchanoids and domes, but transverse and rectilinear dunes are also common, the latter in craters subject to interfering wind directions. Longitudinal dunes are rare. Dune migration has not yet been detected although erosion of smaller dunes has been noted (Bourke, Edgett and Cantor, 2008).

5.5.5.5 Slope streaks

Slope streaks are among the most enigmatic features actively forming on the surface of Mars. First recognised in *Viking* orbiter images, slope streaks were widely recognised in the high-resolution images collected by the Mars orbital camera (MOC) onboard the *Mars Global Surveyor* spacecraft (Sullivan *et al.*, 2001). Slope streaks occur in equatorial regions with thick dust mantles. They have an elongated fan-shaped plan and apparently originate at a point and extend downslope. Their terminations are often wedge-shaped and sometimes branching or braided. Slope streaks cross small topographic obstacles but flow round larger ones. When newly formed they are much darker than the surrounding material. They progressively lighten or fade, in some cases even becoming brighter than their surrounds. Slope streaks at MOC resolution (~3 m) show no textural difference from the surrounding terrain and appear neither raised or depressed in contrast to it. Evidence suggests that the global rate of formation has increased over an observation period of seven years (Schorghofer *et al.*, 2007).

The most widely accepted explanation is that slope streaks are dry flows, formed by mass flow of dust which exposes the underlying darker bedrock. These dark streaks are progressively lightened by deposition of new layers of dust (Sullivan *et al.*, 2001). This explanation appears to have been borne out by high-resolution (~25 cm) images from the HiRISE camera onboard the *Mars Global Surveyor* spacecraft, which showed apparent textural changes inside a streak as opposed to outside and the inside surface was slightly depressed with a distinct scarp compared to outside (Phillips, Burr and Beyer, 2007). Nevertheless, alternative explanations for at least some slope streaks also appear compelling, e.g. Head *et al.*'s (2007) report of slope streaks in Antarctic dry valleys which are morphologically very similar and show similar behaviour. The Antarctic slope streaks form by near-surface melting-derived saline water travelling downslope along the top of the ice table, wicking and dampening the surface to cause the streak. These streaks are also initially dark and then brighten over time. Slight differences in surface texture were also reported. Salt efflorescence on the surface of the streak can cause brightening over the surround terrain, a feature reported from some martian streaks (Schorghofer *et al.*, 2007) and difficult to explain by removal of surface dust by dry flow. Without ground truthing from a surface mission it will be difficult to distinguish between these and other hypotheses, but multiple formative processes for what is superficially the same morphology cannot be discounted.

5.5.5.6 *Glacial and periglacial processes*

The polar caps of Mars are the most extensive deposits of exposed ice on the surface. Radar soundings from orbiting spacecraft (Phillips *et al.*, 2008) have revealed well-layered internal stratigraphy. Evidence of glacier-like flow has not, however, been forthcoming to date. The same applies to smaller surface ice accumulations in subpolar craters, e.g. Louth Crater (Brown *et al.*, 2008). However, ice accumulations at lower latitudes that are buried by surface mantles do show signs of glacial flow.

Extant ice deposits (Neukum *et al.*, 2004) showing signs of glacial flow have been found on the high mountains of Tharsis (Head and Marchant, 2003). The temperatures in these environments indicate that flow would resemble those of terrestrial cold-based mountain glaciers. Elsewhere on Mars, landforms apparently shaped by glacial flow (Holt *et al.*, 2008) are commonplace. The ice is buried too deep to be detected by neutron spectrometers but has shown up in orbiting ground-penetrating radar profiling. Diverse landforms are present including potential terminal and ground moraines and apparent ice surface features indicative of viscous flow.

Martian surface landforms show many similarities with those of terrestrial cold deserts. As previously noted, the landscape of the *Phoenix* landing site was characterised by sorted stone patterns, polygonal terrain and a shallow ice table (Mellon *et al.*, 2009; Levy, Head and Marchant, 2009). Polygons were also observed at the *Viking 2* landing site (Mutch *et al.*, 1976b; Jones *et al.*, 1979). Some of the most striking evidence for periglacial processes was documented by Balme and Gallagher (2009) from the equatorial regions. They reported a diverse suit of features including pitted mound and cone structures (resembling pingos) and scarps with downslope deposits, dendritic channels, gullies, blocky debris and hummocky terminal deposits (resembling retrogressive thaw slumps in thermokarst environments). They postulate an evolutionary sequence of morphologies beginning with subsidence of polygonised surfaces and ending with scarp-bounded basins fed by interpolygon channels. This apparently active breakdown of inferred subsurface permafrost in the martian tropics may be related to the instability of equatorial ice deposits deposited under different climatic conditions, as postulated by Head *et al.* (2003, 2005).

5.5.5.7 *Fluvial systems*

The surface of Mars is characterised by many different channel morphologies. They include dendritic channels showing high stream order, fretted terrain valleys commonly inferred to be due to sapping (Harrison and Grimm, 2005), outflow channels associated with catastrophic events (Baker, 2001) and inverted channels (Pain, Clarke and Thomas, 2007). In addition there are the extremely young gullies found on crater walls, and occasionally other steep slopes (Heldmann *et al.*, 2005; Kossacki and Markiewicz, 2004) whose formation appears ongoing although their genesis remains controversial.

Dendritic channels such as Warrego Vallis are of great antiquity; they are found only on the oldest terrains and are heavily cratered. Their formation is most likely due to epochs early in martian history when precipitation-driven runoff was more common (Harrison and Grimm, 2005). The depositional termini of these channels are sometimes marked by crater-filling fans and deltas, most spectacularly that of Holden North East Crater (now known as Eberswalde; Bhattacharya *et al.*, 2005). The meandering channel forms with channel cut-offs and scroll bars of the delta indicate multiple sustained flows similar to that of terrestrial deltas (Bhattacharya *et al.*, 2005). Landscape lowering has resulted in some of these channels being exhumed and inverted, forming sinuous and often dendritic ridges on the martian landscape (Pain, Clarke and Thomas, 2007).

The dendritic channels are often modified by the fretted and outflow valleys, due to a shift to drier, colder climates where water flows occurred after singular events, such as large-scale impacts or volcanism and melting permafrost (Baker, 2001). In most cases, these features can be traced back to source areas, where collapse and removal of presumably ice-rich subsurface materials formed chaotic terrain, with evidence for volcanism occurring in the last few Myr (Neukum *et al.*, 2004); these events, although rare, are likely to continue into the future. The scale of some of the catastrophic outflow events was colossal, dwarfing their terrestrial equivalents by two to three orders of magnitude (Baker, 2001).

Like slope streaks, gullies are an ongoing process on the surface of Mars, whose origin is highly controversial (Malin *et al.*, 2006; McEwan *et al.*, 2007; Heldmann *et al.*, 2009). As with terrestrial gully forms, it is likely that these features are polygenic, formed by several processes, some unique to Mars, such as dry mass flows and CO_2 gasification. Of those that show sinuous channels and lobate distributaries, many, although probably not all, are related to fluvial or hyper-concentrated flows. (Heldmann *et al.*, 2005, 2009). The mostly likely source of water is the melting of shallowly buried snow or ice (Kossacki and Markiewicz, 2004), given that the run-out distances of gully-forming flows is close to that modelled for pure water.

Lastly, it should recognised that there are many relatively small mound and outflow features on the surface of Mars whose origins are enigmatic but which may have been formed through the discharge of artesian waters as springs to form mounded deposits (Bourke *et al.*, 2007). These deposits may form a continuum with both cold desert features such as open system pingos and hydrothermal systems. Like terrestrial desert springs, martian spring deposits are of considerable potential interest to astrobiologists as targets for sample return missions (Allen and Oehler, 2008).

5.5.6 *Summary*

The surface of Mars is deceptively Earth-like in many respects. Familiar landforms abound with analogues in terrestrial cold and hot deserts. These should not blind the geomorphologists to the fact that Mars is none the less not Earth and the expression of familiar processes may show subtle and perhaps profound differences. In addition Mars will have its own unique set of processes, especially those related to the condensation and volatilisation of CO_2, that may mask, overprint or mimic more familiar terrestrial processes. Lastly, the massive changes in geomorphic style during to the large-scale evolution of the planet have resulted in a unique landscape recording immense climatic change.

5.6 Titan: methane-based aridity?

Titan is the largest satellite of Saturn and the second largest in the solar system. Titan is larger than the planet Mercury and its diameter is 74 % of Mars. The internal structure is inferred to consist of a sepentinite core overlain by an icy mantle and crust (Fortes *et al.*, 2007). Titan's density is therefore low, resulting in a surface gravity only 14 % of Earth's. Despite this, Titan has a very dense atmosphere, with a surface pressure of 1.46 bars. The atmosphere is very cold, with an average temperature of −180 °C. The atmosphere is composed of 98.4 % nitrogen and 1.6 % methane and contains few clouds but a thick haze layer at altitudes above 40 km (Tomasko *et al.*, 2005). The surface temperatures on Titan are such that methane–ethane can condense and fall as rain, acting in a manner analogous to water on Earth (Lunine and Atreya, 2008).

Since 2004 the *Cassini* spacecraft in orbit round Saturn has achieved multiple encounters with Titan. Infrared imaging and spectroscopy have provided information on the composition of the surface (e.g. Soderblom *et al.*, 2007a) and radar on the morphology (e.g. Elachi *et al.*, 2006). Some notable geomorphic features are shown in Figure 5.7. However, only part of the surface of Titan has been imaged in any detail and many unexpected features may well emerge.

Fly-by data on the surface of Titan has been supplemented by a single Lander. The *Huygens* probe was released by *Cassini* during its first encounter with Titan and

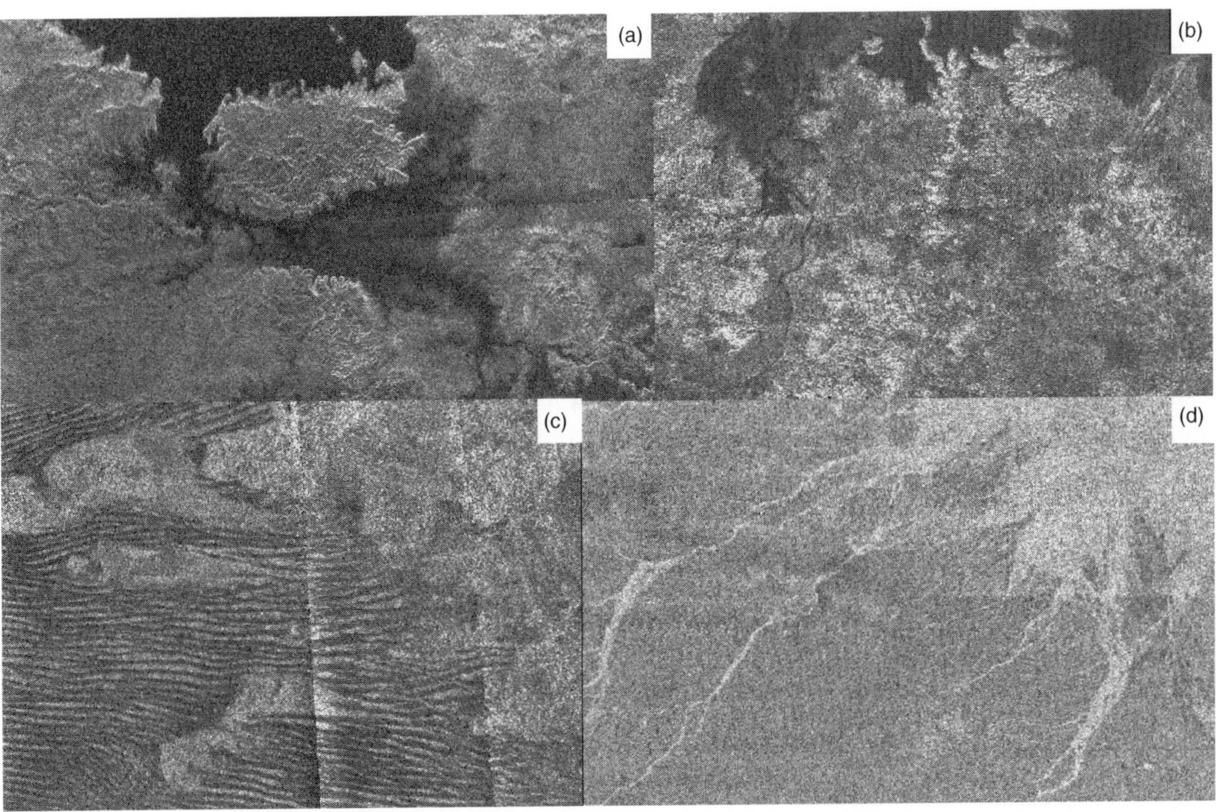

Figure 5.7 Titan as seen by *Cassini* during encounters: (a) probable methane/ethane lakes (dark) surrounded by uplands (bright) on Titan; note ria-like indented shoreline; *Cassini* radar image, width ~300 km; (b) radar-dark channels (possibly filled by flowing methane/ethane) with delta features along shore of probable hydrocarbon lake; *Cassini* radar image, width ~200 km; (c) radar-dark longitudinal dunes adjacent to radar-bright highlands; *Cassini* radar image, image width ~400 km; (d) radar-bright channel sediments and outwash plain; *Cassini* radar image, width ~350 km. Images NASA/ESA/University of Arizona.

provided detailed in situ data on atmospheric properties and surface composition (Lebreton *et al.*, 2005). It also imaged the surface (or ranging over) of several kilometres as it descended through the atmosphere. The final images returned by the *Huygens* probe were of the surface itself.

5.6.1 Methane cycle

Methane on Titan plays an analogous role to water on Earth (Lunine and Atreya, 2008). Methane is evaporated from long-lasting and deep (probably tens of metres or more deep) lakes in the north polar region and condenses to form clouds, which in turn precipitate methane rain to the surface. The relative humidity of methane at the tropics is high (45 %). This would be enough to trigger convective rainfall on Earth but does not do so on Titan to the same extent because Titan receives ~1 % of the solar energy that Earth does. The methane cycle is therefore much less active that Earth's water cycle. Despite this, the *Huygens* Lander encountered methane drizzle during its descent (Tomasko *et al.*, 2005) and the surface it landed on appeared to be saturated with liquid methane, even though visually dry (Zarnecki *et al.*, 2005). Transient events in the surface images that are consistent with methane raindrops were observed (Karkoschka *et al.*, 2007). Rainfall generates runoff and methane rivers (Soderblom *et al.*, 2007a; Jaumann *et al.*, 2008; Lunine and Lorenz, 2009), which flow into methane lakes in the north polar region, evaporate or recharge 'methanifers'.

The methane cycle is not completely closed as irreversible ultraviolet-driven photochemical reactions in the atmosphere form complex hydrocarbons and nitriles. These make up the atmospheric haze of Titan and are removed by methane rainfall (Lunine and Lorenz, 2009; Lunine and Atreya, 2008).

Seasonal variations are expected over the 29.5 year orbital period of Saturn. Those observed to date include the frequency of clouds. Modelling suggests that the shorelines of shallower lakes shift hundreds of metres to tens of kilometres in a year (Hayes *et al.*, 2008; Mitri *et al.*, 2007). Over longer timescales, the hemispheric dichotomy between lake-filled basins in the north polar region and almost completely liquid-free basins in the south invite explanations involving orbitally forced climatic cycles (Aharonson *et al.*, 2009) with surface liquids accumulating first at one pole and then the other.

5.6.2 Surface images

The *Huygens* probe landed on the margins of Xanadu Regio close to a boundary between light highland areas and dark lowland areas. Descent and surface images by the probe (Soderblom *et al.*, 2007a) showed that it landed in a lowland area that appears swept by episodic floods (Figure 5.8). The landing site itself is about ~40 % covered (Karkoschka *et al.*, 2007) by rounded pebbles of icy material on a darker and finer grained substrate. The pebbles are 13–16 cm in diameter (Tomasko *et al.*, 2005). Rock abundance for those >5 cm varies, in the foreground (within 80 cm of the Lander), rocks are more common than in the middle distance (80–160 cm). Rock abundance increases again further away. This variation may indicate a possible stream bed. Elongated dark sediment trails behind some of the larger rocks are aligned with the possible stream bed and is consistent with the orientation of channel-like features seen further away in the descent images (Tomasko *et al.*, 2005). The mechanical properties of the surface are consistent with sediment saturated with a liquid, probably methane (Zarnecki *et al.*, 2005). Methane rain was encountered during *Huygens*' descent (Tomasko *et al.*, 2005); the heat of the Lander also appears to have evaporated methane from the surface. Whether this relatively 'wet' environment (in the sense of the presence of liquid methane) is the norm, or the exception, remains to be determined.

5.6.3 Lakes

Lakes of methane or similar hydrocarbons such as ethane have long been proposed as present on Titan's surface, based on the extant surface temperature and pressure and the presence of hydrocarbon gases in the atmosphere (Lunine, Stevenson and Yung, 1983). Initial observations by *Cassini* indicated that any lakes covered only a tiny fraction of the surface (Mitri *et al.*, 2007). Radar images of the north polar region in 2006 uncovered more than 75 liquid methane-filled lakes at latitudes above 70° N (Stofan *et al.*, 2007) plus a large number of apparently dry lake bed depressions. Lakes are almost absent from southern circumpolar latitudes (Aharonson *et al.*, 2009).

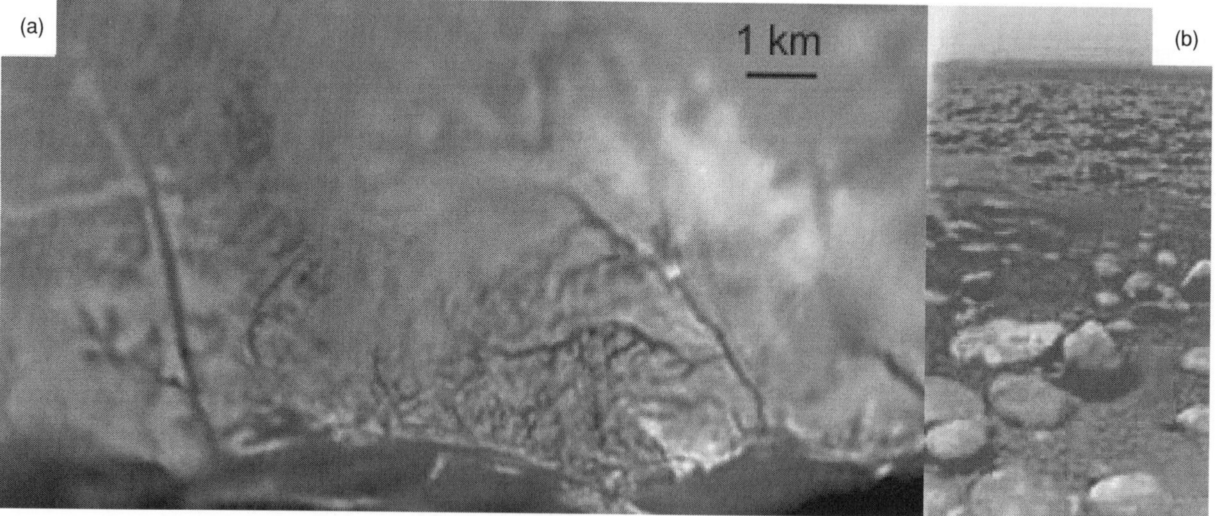

Figure 5.8 Titan as seen by *Huygens* during its descent to the surface of Titan: (a) two types of channels formed by flowing methane/ethane, composite image from an altitude of ~40 km, varying resolution due to atmospheric haze and different altitude of composite components; channels cut through optical bright terrain and filled with optically dark material and terminate in an optically dark plain; stubby channels (left) possibly cut by sapping methane/ethane; contrast with dendritic channels, possibly formed by runoff of methane/ethane rain; (b) ground view from *Huygens* Lander showing optically bright cm-scale pebbles on optically dark depositional plain; note tails of fine-grained sediment to pebbles in distance and low relief ridge on skyline. Images NASA/ESA.

The lakes appear to be bounded by bedrock with bedrock islands and complex shorelines. How the lake depressions formed is not known, but none are obviously associated with impact craters. Bourgeois *et al.* (2008) suggested that dissolution of the substrate by liquid methane was one possible process, analogous to karst developed on calcretes beneath playa lakes of Namibia. Further similarities with terrestrial playa lakes were noted by Lorenz, Jackson and Hayes (2010).

Dendritic drainage is visible at depths of at least 10 m below the surface of lakes owing to the transparency of methane liquids at the relevant radar wavelengths (Lunine and Lorenz, 2009). This indicates alternating lake levels and, together with the rias, indicates periods of low lake level with fluvial incision that are followed by high lake levels and drowning of the fluvial features.

5.6.4 Rock breakdown: process and form

Prior to the *Cassini* mission, Lorenz and Lunine (1996) postulated that one of the most effective means of rock breakdown on Titan was by differential solubility. Titan's crust was considered to be composed largely of water, ammonia and water–ammonia ices with solid hydrocarbons. These materials have different solubilities with respect to methane-rich liquids. This proposal may have been borne out by the apparent morphological similarity of Titan's lakes to solution-generated lakes in arid environments on Earth (Bourgeois *et al.*, 2008), a suggestion that requires further investigation. Dissolution of ices by liquid methane also implies subsequent deposition through evaporation of the methane. This is likely to leave behind a residue of solid hydrocarbons and crystallisation of ices. These deposits are potentially analogous to terrestrial salts. The ability of such deposits to cause rock breakdown is a subject for further research.

Lorenz and Lunine (1996) also dismissed diurnal temperature variations as a likely mechanism for rock breakdown. This position is probably still valid because of the low diurnal temperature range on Titan.

5.6.5 Aeolian landforms

Likewise, Lorenz and Lunine (1996) dismissed aeolian processes as a likely cause of rock breakdown because of the modelled low wind velocity. However, the discovery of large linear dune fields was one of the first major discoveries of the *Cassini* mission (Elachi *et al.*, 2006). Although no ventifacts are visible in the *Huygens* descent images, the presence of dunes and the implications of abundant sand and at least episodically strong winds implies that, at least locally, wind may be a major process in rock disintegration.

The third *Cassini* flyby of Titan revealed the presence of dark, linear features. Their distribution and configuration indicated that they are large, longitudinal dunes similar to those in many terrestrial deserts such as in Australia and Namibia (Rubin and Hesp, 2009). The dunes are radar dark, occur in lowland areas along the equator, are oriented east–west and appear to have been formed by a predominantly westerly wind. Like terrestrial longitudinal dunes, those on Titan show tuning fork junctions, diverge around some topographic highs and terminate against others. Where terminated by bedrock rises they pick up again in the downwind direction (Lunine and Lorenz, 2009). The dunes appear to be composed of organic-rich materials (Soderblom *et al.*, 2007b). The origin of the sand-sized particles is presently unknown; the sand may have been eroded from the highlands, formed by aggregation of airborne hydrocarbon particles (Lunine and Lorenz, 2009), from some as yet unknown process, or a combination of all of these.

5.6.6 Fluvial systems

Lorenz and Lunine (1996) raised the possibility of fluvial erosional features on the surface of Titan in the pre-*Cassini* era. Modelling studies (Burr *et al.*, 2006) showed that liquid methane is potentially even more effective at transporting sediments under titanian conditions than water is on Earth. Thus it is not surprising that a range of fluvial landforms are visible on Titan. In the region of the *Huygens* landing site two distinct types of channels are visible (Soderblom *et al.*, 2007a): deeply incised dendritic drainage networks up to fourth order and short, stubby low-order drainages that follow apparent structural features. The dendritic, high-order channels are attributed to runoff and the stubby, low-order channels to discharge from methanifers. Valley slopes of up to 30 degrees occur on the sides of both channels, indicating a competent substrate. River channels elsewhere on Titan are typically hundreds of kilometres in length and in some cases are more than a thousand kilometres long (Jaumann *et al.*, 2008). Channel morphology is consistent with rare (decadal or century scale intervals) but intense (20–50 mm per hour) rainfall events, analogous to those that occur in terrestrial desert regions. The drizzle observed during *Huygens*' descent (Tomasko *et al.*, 2005) is not sufficient to generate runoff events, although it appears sufficient to moisten the surface (Tomasko *et al.*, 2005). Lorenz (2005) likened these catastrophic events to the fictitious 'methane monsoon' describe by Arthur Clarke in his novel *Imperial Earth* (1976) which is partly set on Titan.

Radar-bright delta-shaped features are apparent where some channels debouche on to the lowlands; these are interpreted as fan deposits (Lorenz *et al.*, 2008). Meandering channel forms are also present (Lorenz *et al.*, 2008) and the morphology of estuaries where some channels enter lakes is strongly reminiscent of rias.

5.6.7 Summary

Titan remains enigmatic. With an active methane cycle, large lakes, drizzle and apparently wet surfaces in ostensibly dry areas, the surface in some respects resembles that of a terrestrial humid landscape. Conversely, the dry lakebeds stained by dissolved residues from the evaporation of liquid methane, evidence for major fluctuations in lake levels, large dune fields, rare but intense rainfall events and highly variable river flow, are all strongly reminiscent of terrestrial aridity. Further complexity is introduced by the slow pace of Titan's seasons – a full circuit of the sun takes almost 30 years and observations of the surface have to date comprised but a fifth of that period. Combined with the lower insolation, the rate of environmental variability of Titan may be much lower than of Earth.

5.7 Venus: extreme aridity

Venus has an equatorial diameter of 94.9 % and a mass of 81.5 % that of Earth's. The surface of Venus is obscured by high-altitude clouds of sulfuric acid droplets. The atmosphere is composed almost completely of CO_2 and the average surface pressure is ~60 bars, equivalent to an ocean depth on Earth of ~600 m. As a result of the dense CO_2 atmosphere the greenhouse effect is extreme and surface temperatures average ~400 °C. There is considerable topographic variation, with pressures and temperatures of 41 bars and 374 °C in the highlands and 96 bars and 465 °C typical of the lowlands. This range means that impact, volcanic, aeolian and weathering processes may operate very differently in the highlands and lowlands (Greeley, 1994). Conversely, diurnal variations from the slow, retrograde rotation of the planet are minimal because of the efficient equalization of latitudinal temperature differences by atmospheric circulation.

Much of the surface of Venus remains enigmatic, despite more than 20 missions to the planet and 13 probes that successfully reached the surface, four of which returned images from their landing sites. Variable-resolution global radar coverage of the cloud-obscured surface is available from the *Pioneer 12*, *Venera 15* and *16*, and *Magellan* missions, and from terrestrial radar astronomical studies the best resolution is 125 m. It is these images that have provided our understanding of the larger-scale geomorphology of Venus (Figure 5.9).

5.7.1 Surface–atmosphere interaction

The corrosive atmosphere and high temperatures and pressures suggest that chemical weathering may be significant on Venus (McGill *et al.*, 1983; Nozette and Lewis, 1982). Arvidson *et al.* (1992) demonstrated that lava flows of different ages in Sedna Planita showed systematic changes in radar backscattered characteristics that were greater than could be explained by aeolian action alone. They hypothesised that the flow modification was due in part to weathering.

A wide range of trace volatiles are known to occur or are likely to occur in the atmosphere of Venus (Brackett, Fegley and Arvidson, 1995). Condensation of these phases as metal sulfides and halides at higher altitudes, where temperatures are almost 100 degrees cooler than the lowlands, are hypothesised to produce weathering analogous to terrestrial salt weathering. Therefore it may be significant that topographic features on Venus more than 3 km above datum are coated with a material of low emissivity and high dielectric constant. This is consistent with the condensing of heavy metal sulfides from the atmosphere (Schaefer and Fegley, 2004). The most likely candidates are galena (lead sulfide) and/or bismuthite (bismuth sulfide). Tellurium compounds have also been suggested. Some of these condensed phases may well undergo melting, leading to the possibility not only of analogues to salt weathering on Earth and Mars but also of the effects of moisture.

Unfortunately all four landers that have imaged the surface touched down at too low an elevation to observe this phenomenon directly. However, if previous atmospheric temperatures were lower, it is possible that these processes may have operated at some of the landing sites and rock breakdown features caused by condensed atmospheric materials might still be preserved.

5.7.2 Surface images

Black and white images of the surface of Venus were obtained by *Venera 9* and *10* (Florenskii *et al.*, 1983) while higher resolution colour images were obtained by *Venera 13* and *14* (Moroz, 1983). These show that most of the

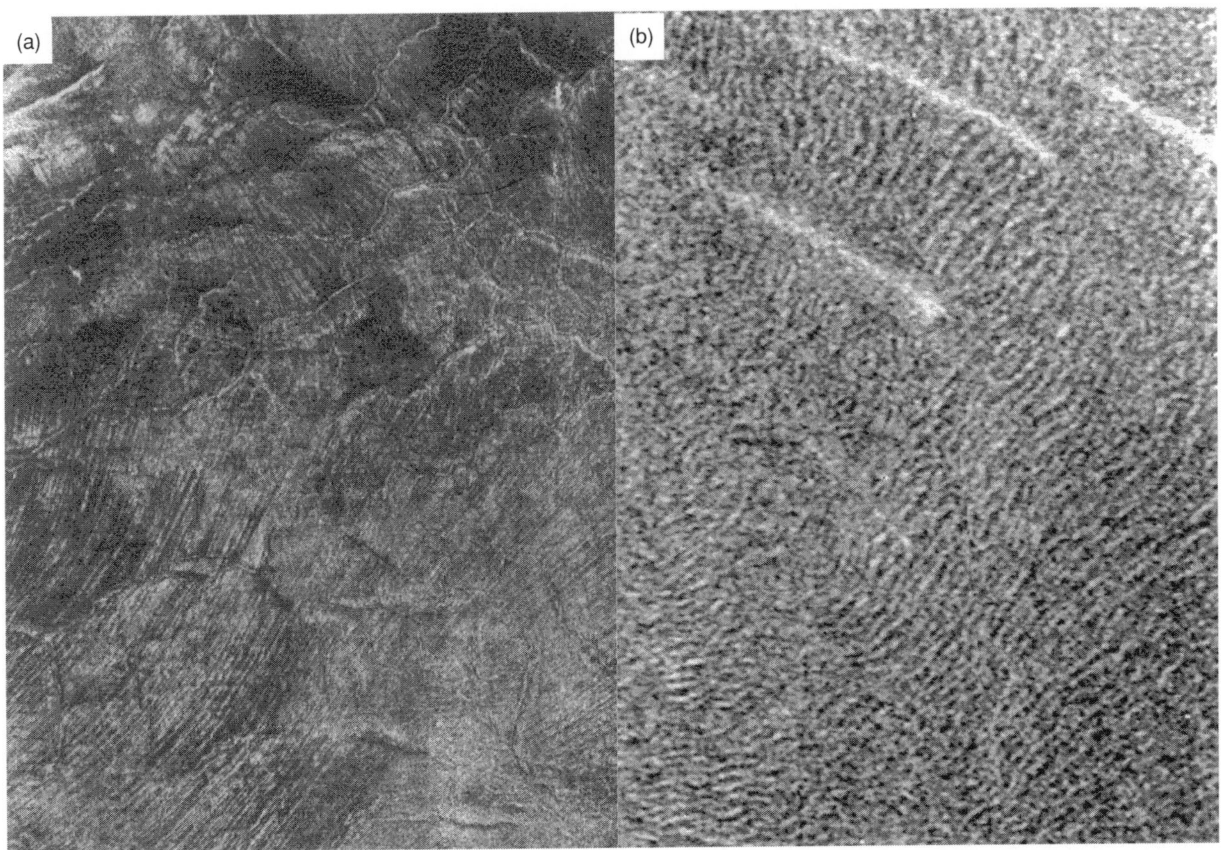

Figure 5.9 Aeolian features on Venus captured by *Magellan* radar imagery: (a) possible yardangs centred at 9 N, 60.7 E; image width ~95 km; (b) wind streaks and transverse dunes, part of the Fortuna-Meshkenet dune field; image width ~37 km. NASA images courtesy of R. Greeley.

surface consists of rocky material, with loose regolith in the minority (Figure 5.10). The surface of Venus as revealed by the various missions displays surface properties as variable as on Earth or Mars, which can be summarised on a mission-bymission basis.

Venera 9 landed on the eastern side of the highlands of Rhea Mons and recorded the first images by any spacecraft from the surface of another planet. The image revealed scattered slightly flattened to equidimensional, angular to subangular rocks. Possible layering, picked out by differential weathering, was seen in several rocks. The rocks were resting on or were partly buried by a poorly sorted granular regolith.

The landing site for *Venera 10* is south of *Venera 9* on the southwestern side of Theia Mons. The surface is composed of large flat slabs of rock with wavy fracture patterns separated by patches of very dark granular regolith. Textures similar to terrestrial sandy corrosion and cellular weathering can be seen (Florenskii, Basilevskii and Pronin, 1977).

Landings sites for *Venera 13* and *14* probes were the eastern flanks of the highlands of Phoebe Regio, with *Venera 14* to the east of *Venera 13*. The *Venera 13* image resembled that of *Venera 10*, except the fine regolith was only slightly darker than the bedrock but more abundant. The higher resolution images returned by these missions showed that the loose regolith was composed of very poorly sorted angular fragments.

Venera 14 imaged a flat plain almost completely covered by slabby bedrock. Fractures are linear to scalloped, and loose regolith is minor. Possible dirt cracking (*senso*, Ollier, 1965) is visible. Areas in the distance with more extensive loose regolith are visible.

5.7.3 Rock breakdown

Radar data (Greeley, 1994) suggest that only a quarter of the surface is composed of loose material. This is consistent with the *Venera* images and suggests that rock

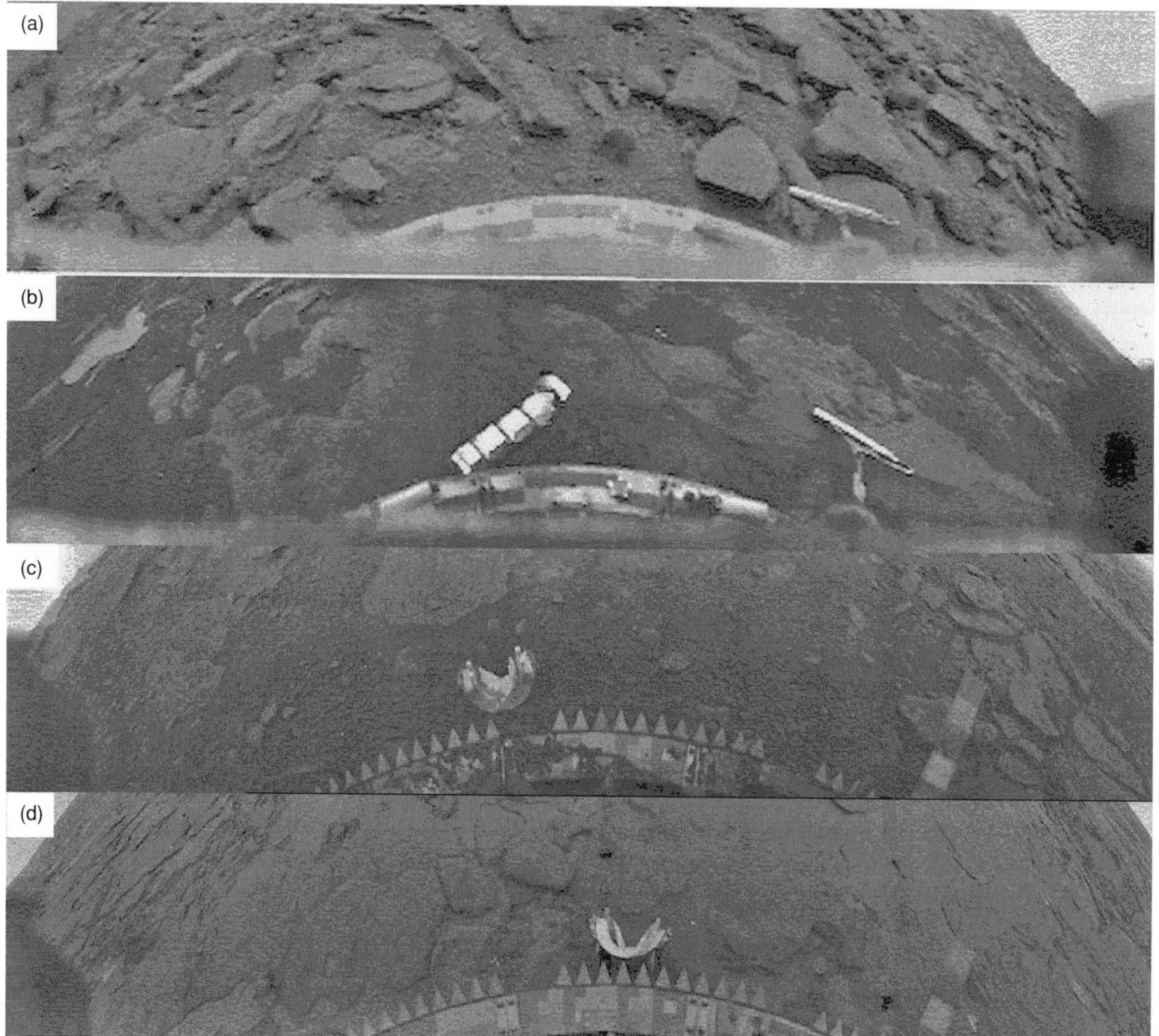

Figure 5.10 The surface of Venus: (a) *Venera 9* – angular layered rocks and coarse interstitial regolith; (b) *Venera 10* – slabby rocky outcrops and patches of dark fine-grained regolith; (c) *Venera 13* – slabby rocky outcrops and patches of dark, fine to medium-grained regolith; higher relief area and possible valley is visible near horizon in top left-hand corner; possible dirt cracking is visible on the left-hand side, just above the colour chart protruding from the Lander; (d) *Venera 14* – smooth plain with slabs of fractured rock and minimal fine regolith. All images are 180-degree panoramas using a semi-circular imaging scanner. Original data reprocessed by Don P. Mitchell, and used by permission.

breakdown may not be particularly effective on the surface. Angular rocks in images shows that some physical weathering is, however, an ongoing process although the nature of these processes is not known. The paucity of loose materials suggests that rock breakdown rates are slow or countered by some form of cementation. Exfoliation is unlikely, given the minimal diurnal change in temperature. One possibility is that chemical weathering may assist mechanical breakdown to the rock. Particular interesting are textures reminiscent of terrestrial sandy corrosion and cellular weathering at the *Venera 10* (Florenskii, Basilevskii and Pronin, 1977). If correct this may indicate a weathering role for traces of an ephemeral liquid on at least parts of the surface of Venus.

5.7.4 Aeolian landforms

Measured surface wind velocities on Venus are only 1–2 m/s, but this is sufficient to move sand particles.

Loose material from radar data and *Venera* images appear to be only a minor component of the surface, but scattered dune fields have been observed and streaks of bright material downwind of craters are not uncommon (1992, 1997; Greeley, Bender and Thomas, 1994). Two notable dune fields, Aglaonice and Meshkenet, are located near Aglaonice Crater and Meshkenet Tessera, respectively. It is possible that radar look-angle effects may result in the underrepresentation of dunes in the radar images (Weitz *et al.*, 1994).

Possible yardangs have been reported from near Mead Crater (Greeley *et al.*, 1992, 1997; Greeley, Bender and Thomas, 1994). Clear signs of aeolian erosion, such as ventifacts or microyardangs are absent in the surface images (Trego, 1992). It is likely that many more aeolian features occur at scales too small to be resolved by the *Magellan* radar; data on these await more surface images, higher resolution radar surveys or descent images from future missions.

5.7.5 Summary

With at most only a minor role played by traces of liquid phases, Venus is probably the most arid planet in the solar system.

5.8 Future Directions

Extraterrestrial research opportunities for arid geomorphologists are extensive. Mars, with ongoing surface and orbital missions and a succession of future missions being planned, offers the most possibilities. The data archive from previous missions is immense and barely examined. At present there are no plans to return to Venus, although missions are being studied. Nevertheless, scope exists for further research on the *Magellan* data set and the *Venera* surface images. With Titan it is unlikely we will see further images from the surface for many decades hence. However, radar mapping by the *Cassini* spacecraft is ongoing and much work still needs to be done.

Two fruitful avenues of terrestrial research are the experimental modelling of planetary surface conditions and the search for terrestrial analogues. Modelling of the surface conditions on Mars, Titan or Venus allows processes to be studied and hypotheses tested. I suggest that studies of rock or ice breakdown under the surface conditions of these bodies and the interaction of liquids other than water with surface materials may be particularly promising.

Analogue research allows the geomorphologists to study terrestrial counterparts of extraterrestrial processes. It is particularly helpful in relation to Mars, where the literature is extensive but biased towards North American examples. Much more work needs to be done examining geomorphic analogues for martian features from other continents. In particular, I would emphasise the value of studying cold desert processes, often sadly neglected in almost all reviews of both arid and periglacial geomorphology. However, analogue research is not limited to martian landforms but also to those of Titan, where there are excellent terrestrial counterparts to many features, especially the longitudinal dunes. Such analogue research not only provides insights into extraterrestrial processes but stimulates new perspectives on terrestrial geomorphology.

References

Aharonson, O., Hayes, A.G., Lunine, J.I. *et al.* (2009) *Nature Geoscience*, **2**, 851–854.

Allen, C.C. and Oehler, D.Z. (2008) A case for ancient springs in Arabia Terra, Mars. *Astrobiology*, **8**(6) DOI: 10.1089/ast.2008.0239.

Ansan, V. and Mangold, N. (2006) New observations of Warrego Valles, Mars: evidence for precipitation and surface runoff. *Planetary and Space Science*, **54**, 219–242.

Arvidson, R.E., Greeley, R., Malin, M.C. *et al.* (1992) Surface modification of Venus as inferred from *Magellan* observations of plains. *Journal of Geophysical Research*, **97**, 13,303–13,317.

Baker, V.R. (2001) Water and the martian landscape. *Nature*, **412**, 228–236.

Baldridge, A., Hook, S.J., Bridges, N.T. *et al.* (2010) *Phylosilicate and Sulfate Layering in Interplaya Dunes; Analogs for Mars Intercrater Deposits*, Abstracts 41st Lunar and Planetary Science Conference, Abstract #2268.

Balme, M.R. and Gallagher, C. (2009) An equatorial periglacial landscape on Mars. *Earth and Planetary Science Letters*, **285**, 1–15.

Bhattacharya, J.P., Tobias H.D., Payenberg, T.H.D. *et al.* (2005) Dynamic river channels suggest a long-lived Noachian crater lake on Mars. *Geophysical Research Letters*, **32** (L10201). DOI:10.1029/2005GL022747.

Bibring, J.-P., Langevin, Y., Mustard, J.F., et al. and the OMEGA Team (2006). Global mineralogical and aqueous Mars history derived from OMEGA/Mars Express data. *Science*, **312**, 400–404.

Bourgeois, O., Lopez, T., Le Mouélic, S. *et al.* (2008) A surface dissolution/precipitation model for the development of Lakes on Titan, based on an arid terrestrial analogue: the pans and calcretes of Etosha (Namibia), Abstracts of the 39th Lunar and Planetary Science Conference, Abstract 1733.

Bourke, M.C., Bullard, J.E. and Barnouin-Jha, O.S. (2004) Aeolian sediment transport pathways and aerodynamics at troughs on Mars. *Journal of Geophysical Research*, **109** (E07005), DOI: 10.1029/2003JE002155.

Bourke, M.C., Edgett, K.S. and Cantor, B.A. (2008) Recent aeolian dune change on Mars. *Geomorphology*, **94**, 247–255.

Bourke, M.C., Clarke, J., Manga, M. *et al.* (2007) Spring mounds and channels at Dalhousie, Central Australia, Abstracts of the 37th Lunar and Planetary Science Conference, Abstract 2174.

Boynton, W.V., Ming, D.W., Kounaves, S.P. *et al.* (2009) Evidence for calcium carbonate at the Mars *Phoenix* landing site. *Science*, **325**, 61–64.

Brackett, R.A., Fegley Jr, B. and Arvidson, R.E. (1995) Volatile transport on Venus and implications for surface geochemistry and geology. *Journal of Geophysical Research*, **100**, 1,553–1,563.

Bridges, N.T., Greeley, R., Haldemann, A.F.C. *et al.* (1999) Ventifacts at the Pathfinder landing site. *Journal of Geophysical Research*, **104**, 8595–8615.

Bridges, N.T., Geissler, P.E., McEwen, A.S. *et al.* (2007) Windy Mars: a dynamic planet as seen by the HiRISE camera. *Geophysical Research Letters*, **34**, L23205, DOI:10.1029/2007GL031445.

Brown, A.J., Byrne, S., Tornabene, L.L. and Roush, T. (2008) Louth crater: evolution of a layered water ice mound. *Icarus*, **196**, 433–445.

Burr, D.M., Emery, J.P., Lorenz, R.D. *et al.* (2006) Sediment transport by liquid surficial flow: application to Titan. *Icarus*, **181**, 235–242.

Carr, M.H. and Head III, J.W. (2003) Oceans on Mars: an assessment of the observational evidence and possible fate. *Journal of Geophysical Research*, **108**, DOI:10.1029/2002JE001963.

Cattermole, P. (2001) *Mars: The Mystery Unfolds*, Oxford University Press, 181 pp.

Chevrier, V. and Mathe, P.E. (2007) Mineralogy and evolution of the surface of Mars: a review. *Planetary and Space Science*, **55**, 289–314.

Clark, R.N. (2009) Detection of adsorbed water and hydroxyl on the Moon. *Science*, **326**, 562–564.

Clarke, J.D.A. (2008) Extraterrestrial regolith, Chapter 13 in *Regolith Textbook* (eds K.A. Scott and C.P. Pain), CSIRO Publishing, Perth, pp. 377–407.

Crumpler, L.S., Squyres, S.W., Arvidson, R.E. *et al.* (2005). Mars Exploration Rover geologic traverse by the *Spirit* Rover in the Plains of Gusev Crater, Mars. *Geology*, **33**, 809–812.

Davis, N.S. and Gibling, M.R. (2010) Cambrian to Devonian evolution of alluvial systems: the sedimentological impact of the earliest land plants. *Earth Science Reviews*, **98**, 171–200.

Elachi, C., Wall, S., Janssen, M. *et al.* (2006) Titan Radar Mapper observations from Cassini's T3 fly-by. *Nature*, **441**, 709–713.

Feldman, W.C., Bourke, M.C., Elphic, R.C. *et al.* (2008) Hydrogen content of sand dunes within Olympia undae. *Icarus*, **196**, 422–432.

Florenskii, K.P., Basilevskii, A.T. and Pronin, A.A. (1977) The first panoramas of the Venusian surface – geological-morphological analysis of pictures, in *Space Research XVII* (ed. M.J. Rycroft), Pergamon Press, Oxford and New York, pp. 645–649.

Florenskii, K.P., Bazilevskii, A.T., Burba, G.A. *et al.* (1983) Panorama of the *Venera* 9 and 10 landing sites, in *Venus* (eds D.M. Hunten, L. Colin, T.M. Donahue and V.I. Moroz), The University of Arizona Press, Tucson, pp. 137–153.

Fortes, A.D., Grindrod, P.M., Trickett, S.K. and Vocaldo, L. (2007) Ammonium sulfate on Titan: possible origin and role in cryovolcanism. *Icarus*, **188**, 139–153.

Geissler, P.E., Johnson, J.R., Sullivan, R. *et al.* (2008) First in situ investigation of a dark wind streak on Mars. *Journal of Geophysical Research*, **113** (E12S31), DOI:10.1029/2008JE003102.

Greeley, R. (1994) *Planetary Landscapes*, 2nd edn, Chapman and Hall, New York, 286 pp.

Greeley, R., Bender, K. and Thomas, P.E. (1994) Wind-related features and processes on Venus: summary of *Magellan* results. *Icarus*, **115**, 399–420.

Greeley, R. and Williams, S.H. (1994) Dust deposits on Mars; the 'Parna' analog. *Icarus*, **110**, 165–177.

Greeley, R., Arvidson, R.E., Elachi, C. *et al.* (1992) Aeolian features on Venus: preliminary *Magellan* results. *Journal of Geophysical Research*, **97**, 13,319–13,345.

Greeley, R., Bender, K.C., Saunders, R.S. *et al.* (1997) Aeolian processes and features on Venus, in *Venus II* (eds S.W. Bougher, D.M. Hunten and R.J. Phillips) The University of Arizona Press, Tucson, pp. 547–589.

Greeley, R., Arvidson, R.E., Barlett, P.W. *et al.* (2006) Gusev crater: wind-related features and processes observed by the Mars Exploration Rover *Spirit*. *Journal of Geophysical Research*, **111** (E02S09), DOI:10.1029/2005JE002491.

Grotzinger, J.P., Arvidson, R.E., Bell III, J.F. *et al.* (2005). Stratigraphy and sedimentology of a dry to wet eolian depositional system, Burns formation, Meridiani Planum, Mars. *Earth and Planetary Science Letters*, **240**, 11–72.

Harrison, K.P. and Grimm, R.E. (2005) Groundwater-controlled valley networks and the decline of surface runoff on early Mars. *Journal of Geophysical Research*, **110** (E12S16), DOI:10.1029/2005JE002455.

Haskin, L.A., Wang, A., Jolliff, B.L. *et al.* (2005) Water alteration of rocks and soils on Mars at the *Spirit* Rover site in Gusev crater. *Nature*, **436**, 66–69.

Hayes, A., Aharonson, O., Callahan, P. *et al.* (2008) Hydrocarbon lakes on Titan: distribution and interaction with a porous regolith. *Geophysical Research Letters*, **35**, L09204, DOI:10.1029/2008GL033409.

Head, J.W. and Marchant, D.R. (2003) Cold-based mountain glaciers on Mars: Western Arsia Mons. *Geology*, **31**, 641–644.

Head, J., Mustard, J.F., Kreslavsky, K.A. *et al.* (2003) Recent ice ages on Mars. *Nature*, **426**, 797–802.

Head, J.W., Neukum, G., Jaumann, R., *et al.* and The HRSC Co-Investigator Team (2005) Tropical to mid-latitude snow and ice accumulation, flow and glaciation on Mars. *Nature*, **434**, 346–351.

Head, J.W., Marchant, D.R., Dickson, J.L. *et al.* (2007) Slope streaks in the Antarctic dry valleys: characteristics, candidate formation mechanisms, and implications for slope streak formation in the martian environment, Abstracts of the 38th Lunar and Planetary Science Conference, Abstract 1935.

Heldmann, J.L., Toon, O.B., Pollard, W.H. *et al.* (2005) Formation of Martian gullies by the action of liquid water flowing under current Martian environmental conditions. *Journal of Geophysical Research*, **110** (E05004), DOI:10.1029/2004JE002261.

Heldmann, J.L., Conley, C., Brown, A.J. *et al.* (2009) Possible liquid water origin for Atacama Desert mudflow and recent gully deposits on Mars. *Icarus*, DOI: 10.1016/j.icarus.2009.09.013.

Holt, J.W., Safaeinili, A., Plaut, J.F. *et al.* (2008) Radar sounding evidence for buried glaciers in the southern mid-latitudes of Mars. *Science*, **322**, 1235, DOI: 10.1126/science.1164246.

Irwin III, R.P., Howard, A.D., Craddock, R.A. and Moore, J.M. (2005) An intense terminal epoch of widespread fluvial activity on early Mars: 2. Increased runoff and paleolake development. *Journal of Geophysical Research*, **110** (E12S15), DOI:10.1029/2005JE002460.

Jagoutz, E. (2006) Salt-induced rock fragmentation on Mars: the role of salt in the weathering of martian rocks. *Advances in Space Research*, **38**, 696–700.

Jaumann, R., Brown, R.H., Stephan, K. *et al.* (2008) Fluvial erosion and post-erosional processes on Titan. *Icarus*, **197**, 526–538.

Jones, K.L., Arvidson, R.E., Guiness, E.A. *et al.* (1979) One Mars year: *Viking*: Lander imaging operations. *Science*, **204**, 799–806.

Kargel, J.S. (2004) *Mars – A Warmer, Wetter Planet*, Springer Praxis Books, New York.

Karkoschka, E., Tomasko, M.G., Doose, L.R. *et al.* (2007) DISR imaging and the geometry of the descent of the *Huygens* probe within Titan's atmosphere. *Planetary and Space Science*, **55**, 1896–1935.

Kossacki, K.J. and Markiewicz, W.J. (2004) Seasonal melting of surface water ice condensing in martian gullies. *Icarus*, **171**, 272–283.

Lebreton, J.-P., Witasse, O., Sollazzo, C. *et al.* (2005) An overview of the descent and landing of the *Huygens* probe on Titan. *Nature*, **438**, 758–764.

Levy, J.S., Head, J.W. and Marchant, D.R. (2009) *Geophysical Research Letters*, **36** (L21203), DOI:10.1029/2009GL040634.

Lorenz, R.D. (2005) Titan's methane monsoon: evidence of catastrophic hydrology from Cassini RADAR. *Bulletin American Astronomical Society*, **37**, 53.07.

Lorenz, R.D., Jackson, B. and Hayes, A. (2010) Racetrack and Bonnie Claire: southwestern US playa lakes as analogs for Ontario Lacus, Titan. *Planetary and Space Science*, **58**, 724–731.

Lorenz, R.D. and Lunine, J.I. (1996) Erosion on Titan: past and present. *Icarus*, **122**, 79–91.

Lorenz, R.D., Wall, S., Radebaugh, J. *et al.* (2006) The sand seas of Titan: Cassini RADAR observations of longitudinal dunes. *Science*, **312**, 724–727.

Lorenz R.D., Lopes, R.M., Paganelli, F., *et al.* and The Cassini RADAR Team (2008) Fluvial channels on Titan: initial Cassini RADAR observations. *Planetary and Space Science*, **56**, 1132–1144.

Lunine, J.I. and Atreya, S.K. (2008) The methane cycle on Titan. *Nature Geoscience*, **1**, 159–164.

Lunine, J.I. and Lorenz, R.D. (2009) Rivers, lakes, dunes, and rain: crustal processes in Titan's methane cycle. *Annual Review of Earth and Planetary Sciences*, **37**, 299–320.

Lunine, J.I., Stevenson, D.J. and Yung, Y.L. (1983) Ethane ocean on Titan. *Science*, **222**, 229–1230.

McEwen, A.S., Hansen, C.J., Delamere, W.A. *et al.* (2007) A closer look at water-related geologic activity on Mars. *Science*, **317**, 1706–1709.

McGill, G.E., Warner, J.L., Malin, M. *et al.* (1983) Topography, surface properties, and tectonic evolution, in *Venus* (eds D.M. Hunten, L. Colin, T.M. Donahue and V.I. Moroz), The University of Arizona Press, Tucson, pp. 69–130.

Malin, M.C., Edgett, K.S., Posiolova, L.V. *et al.* (2006) Present-day impact cratering rate and contemporary gully activity on Mars. *Science*, **314**, 1573–1577.

Mitri, G., Showman, A.P., Lunine, J.I. and Lorenz, R.D. (2007) Hydrocarbon lakes on Titan. *Icarus*, **186**, 385–394.

Mitrofanov, I.G., Zuber, M.T., Litvak, M.L. *et al.* (2003) CO_2 snow depth and subsurface water–ice abundance in the Northern Hemisphere of Mars. *Science*, **300**, 2081–2084.

Mellon, M.T., Malin, M.C., Arvidson, R.E. *et al.* (2009) The periglacial landscape at the *Phoenix* landing site. *Journal of Geophysical Research*, **114** (E00E06), DOI:10.1029/2009JE003418.

Möhlmann, D.T.R. (2004) Water in the upper martian surface at mid- and low-latitudes: presence, state, and consequences. *Icarus*, **168**, 318–323.

Moroz, V.I. (1983) Summary of preliminary results of the *Venera 13* and *14* missions, in *Venus* (eds D.M. Hunten, L. Colin, T.M. Donahue and V.I. Moroz), The University of Arizona Press, Tucson, pp. 45–68.

Mutch, T.A., Binder, A.B., Huck, F.O. *et al.* (1976a) The surface of Mars: the view from the *Viking 1* Lander. *Science*, **193**, 791–801.

Mutch, T.A., Grenander, S.U., Jones, K.L. *et al.* (1976b) The surface of Mars: the view from the *Viking 2* Lander. *Science*, **194**, 1277–1283.

Neukum, G., Jaumann, R., Hoffmann, H., *et al.* and The HRSC Co-Investigator Team (2004) Recent and episodic volcanic and glacial activity on Mars revealed by the high resolution stereo camera. *Nature*, **432**, 971–979.

Nozette, S. and Lewis, J.S. (1982) Venus – chemical weathering of igneous rocks and buffering of atmospheric composition. *Science*, **216**, 181–183.

Ollier, C.D. (1965) Dirt cracking – a type of insolation weathering. *Australian Journal of Science*, **27**, 236–237.

Pain, C.F., Clarke, J.D.A. and Thomas, M. (2007) Inversion of relief on Mars. *Icarus*, **190**, 478–491.

Phillips, C.B., Burr, D.M. and Beyer, R.A. (2007) Mass movement within a slope streak on Mars. *Geophysical Research Letters*, **34**, L21202, DOI:10.1029/2007GL031577.

Phillips, R.J., Zuber, M.T., Smrekar, S.E. *et al.* (2008) Mars north polar deposits: stratigraphy, age, and geodynamical response. *Science*, **320** (1182). DOI: 10.1126/science.1157546.

Richardson, M.I. and Mischna, M.A. (2005) Long-term evolution of transient liquid water on Mars. *Journal of Geophysical Research*, **110** (E03003), DOI: 10.1029/2004JE002367.

Rubin, D.M. and Hesp, P.A. (2009) Multiple origins of linear dunes on Earth and Titan. *Nature Geoscience*, **2**, 653–658.

Schaefer, L. and Fegley, B.J. (2004) Heavy metal frost on Venus. *Icarus*, **168**, 215–219.

Schorghofer, N., Aharonson, O., Gerstell, M.F. and Tatsumi, L. (2007) Three decades of slope streak activity on Mars. *Icarus*, **191**, 132–140.

Slade, M.A., Butler, B.J. and Muhleman, D.O. (1992) Mercury radar imaging: evidence for polar ice. *Science*, **258**, 635–640.

Smith, P.H., Bell III, J.F., Bridges, N.T. *et al.* (1997) Results from the Mars *Pathfinder* camera. *Science*, **278**, 1758–1764.

Smith, P.H., Tamppari, L.K., Arvidson, R.E. *et al.* (2009) H_2O at the *Phoenix* landing site. *Science*, **326**, 58–61.

Soderblom, L.A., Anderson, R.C., Arvidson, R.E. *et al.* (2004) Soils of Eagle crater and Meridiani Planum at the opportunity Rover landing site. *Science*, **306**, 1723–1726.

Soderblom, L.A., Kirk, R.L., Lunine, J.I. *et al.* (2007a). Correlations between Cassini VIMS spectra and RADAR SAR images: implications for Titan's surface composition and the character of the *Huygens* probe landing site. *Planetary and Space Science*, **55** (13), 2025–2036. DOI:10.1016/j.pss.2007.04.014.

Soderblom, L.A., Tomasko, M.G., Archinal, B.A. *et al.* (2007b) Topography and geomorphology of the *Huygens* landing site on Titan. *Planetary and Space Science*, **55** (13), 2015–2024. DOI: 10.1016/j.pss.2007.04.015.

Stofan, E.R., Elachi, C., Lunine, J.I. *et al.* (2007) The lakes of Titan. *Nature*, **445**, 61–64, DOI: 10.1038/nature05438.

Sullivan, R., Thomas, P., Veverka, J. *et al.* (2001) Mass movement slope streaks imaged by the Mars orbiter camera. *Journal of Geophysical Research*, **106**, 23,607–23,633.

Sullivan, R., Banfield, D., Bell III, J.F. *et al.* (2005) Aeolian processes at the Mars Exploration Rover Meridiani Planum landing site. *Nature*, **436**, 58–61.

Thomas, M., Clarke, J.D.A. and Pain, C.F. (2005) Weathering, erosion and landscape processes on Mars identified from recent Rover imagery, and possible earth analogues. *Australian Journal of Earth Sciences*, **52**, 365–378.

Tomasko, M.G., Archinal, B., Becker, T. *et al.* (2005) Rain, winds and haze during the *Huygens* probe's descent to Titan's surface. *Nature*, **438**, 765–778.

Tooth, S. (2009) Arid geomorphology: emerging research themes and new frontiers. *Progress in Physical Geography*, **33**, 251–287.

Trego, K.D. (1992) Yardang identification in Magellan imagery of Venus. *Earth, Moon, and Planets*, **58**, 289–290.

Ward, A.W., Gaddis, L.R., Kirk, R.L. *et al.* (1999) General geology and geomorphology of the Mars Pathfinder landing site. *Journal of Geophysical Research*, **104**, 8555–8571.

Warner, N.H. and Farmer, J.D. (2008) Importance of aeolian processes in the origin of the north polar chasmata, Mars. *Icarus*, **196**, 368–384.

Weitz, C.M., Plaut J.J., Greeley, R. and Saunders, R.S. (1994) Dunes and microdunes on Venus: Why were so few found in the *Magellan* data? *Icarus*, **112**, 282–295.

Wells, G.L. and Zimbelman, B.J. (1989) Extraterrestrial arid surface processes, in *Arid Zone Geormorphology*, 1st edn (ed. D.S.G. Thomas), Bellhaven Press, London, pp. 335–358.

Wells, G.L. and Zimbelman, B.J. (1997) Extraterrestrial arid surface processes, in *Arid Zone Geormorphology* (ed. D.S.G. Thomas), John Wiley & Sons, Ltd, Chichester, pp. 659–690.

Wray, J.J., Murchie, S.L., Squyres, S.W. *et al.* (2009) Diverse aqueous environments on ancient Mars revealed in the southern highlands. *Geology*, **37**, 1043–1046.

Zarnecki, J.C., Leese, M.R., Hathi, B. *et al.* (2005) A soft solid surface on Titan as revealed by the *Huygens* Surface Science Package. *Nature*, **438**, 792–795.

Zimbelman, J.R. and Griffin, L.J. (2010) HiRISE images of yardangs and sinuous ridges in the lower member of the Medusae Fossae Formation, Mars. *Icarus*, **205**, 198–210.

II

Surface processes and characteristics

6 Weathering systems

Heather A. Viles

6.1 Introduction

Weathering is a very neglected part of the geomorphic system in all environments. While in comparison with other Earth surface processes weathering is often slow and hard to discern, it is undoubtedly a key influence on landscape development in two main ways, i.e. through its production of individual (often small-scale) landforms and through its contribution to overall denudation rates. The 'holy grail' for weathering geomorphologists is to understand these two influences and how they might be linked to one another. Arid environments provide an extreme case of the interactions between weathering and geomorphology, and are thus of particular interest. It is noticeable that most scientists who study weathering in arid environments also do so in a range of other environments; thus there is much cross-fertilisation of ideas based on comparisons of weathering systems in different climatic zones. While weathering systems (which can be defined as the two-way, complex and emerging interactions between weathering processes and the materials on which they operate) in arid environments do not perhaps have the immediate visual appeal of aeolian systems (especially as manifested through extensive dunefields) or fluvial systems (such as networks of wadis), they are at the heart of some exceptionally beautiful (and often highly valued) landscapes (Figure 6.1). The intensely dissected and finely sculpted sandstone outcrops of Utah in the USA, the Wadi Rum area in Jordan and the Acacus Massif in Libya, for example, all demonstrate the distinctive and often enigmatic role of weathering in arid environments.

Within arid environments, there is general agreement that weathering processes are involved in the production of a range of intriguing small- and medium-scale landforms, and also contribute to the evolution of landscapes more generally through the generation of sediment and shaping of relief. However, despite much recent research progress, three major challenges remain for scientists studying desert weathering systems: i.e. to explain the genesis of some enigmatic desert weathering landforms, to ascertain the tempo of weathering processes in dryland environments and to elucidate the contribution that weathering makes to overall landscape evolution within drylands. Looking at the first of these challenges, Figure 6.2 illustrates some of the unusual and complex weathering features characteristic of arid environments. There are still no general theories for the development of many of these weathering features such as alveoli and tafoni, despite a wealth of testable hypotheses and a wide range of empirical studies from many dryland areas. It is still unclear whether any weathering features are found uniquely in arid environments or, indeed, to what extent they are purely weathering features or influenced by other geomorphic processes.

The second challenge for scientists, measuring rates of weathering, has been the subject of debate since the earliest days of desert geomorphology. Workers such as Peltier in the 1950s believed that desert environments were characterised by extremely low rates of weathering, while from the 1970s and 1980s onwards, scientists such as Andrew Goudie pioneered the view that some desert areas are weathering 'hot spots' with very high rates (Goudie and Watson, 1984). Problems of quantifying weathering rates meaningfully over short and long timespans, coupled with the diversity of desert conditions, means that this debate still rages. Difficulties in addressing the first two challenges means that we are still a long way from overcoming the third challenge of understanding the overall contribution of weathering to landscape evolution in

Arid Zone Geomorphology: Process, Form and Change in Drylands, Third Edition. Edited by David S. G. Thomas

Figure 6.1 Arid landscapes dominated by weathering can be attractive and are often heritage sites as shown by these sandstone outcrops forming part of the World Heritage Site in the Acacus Massif, Libya.

Figure 6.2 Examples of complex weathering features from arid environments: (a) alveoli covering a granitic rock face in the Swakop Valley, Namibia; (b) tor-like sandstone outcrops near Keetmanshoop, Namibia.

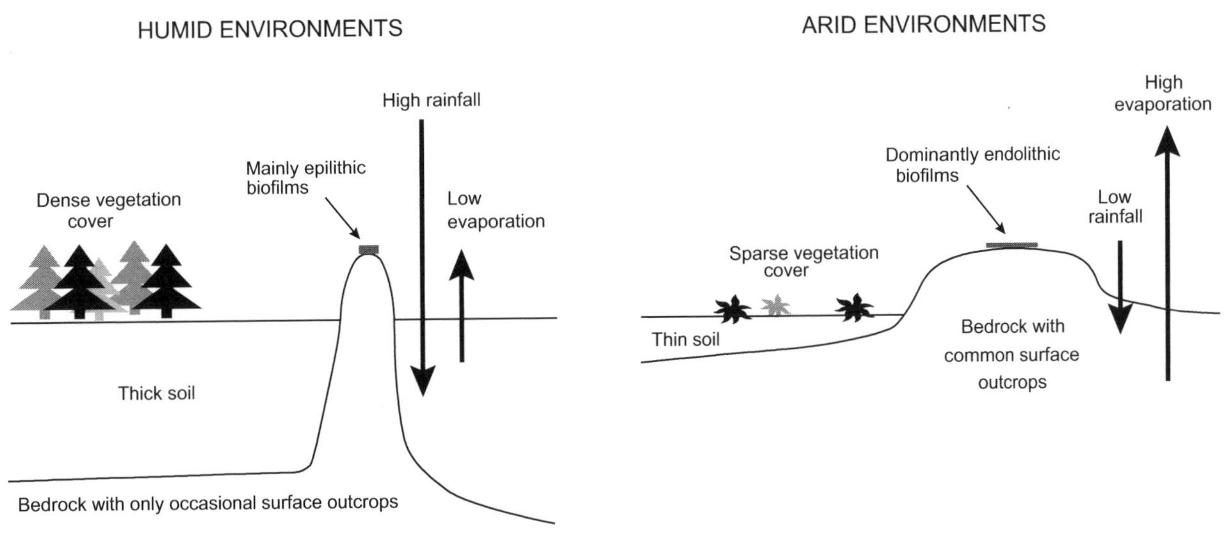

Leaching and runoff of weathering products	Weathering products often build up in situ
Precipitation an important control of weathering rate	Weathering rate heavily influenced by variability in precipitation and evaporation over time and space
Vegetation and biofilms usually have indirect influence on weathering	Vegetation and biofilms usually have direct influence on weathering
Weathering system characterised by equilibrium conditions	Weathering system characterised by non-equilibrium conditions

Figure 6.3 A conceptual diagram summarising the major elements of arid versus humid weathering systems.

drylands. This chapter aims to explore recent progress in these three areas, after an introduction to the key facets of arid environments as relevant to weathering systems, some theoretical underpinnings to weathering research and a review of current study methods, which are allowing us unprecedented insights into weathering systems in arid and other environments.

6.2 What makes arid environments unusual in terms of weathering systems?

Arid environments possess three very important characteristics, which have a dominant influence on their weathering systems, i.e. extreme environmental conditions (Chapters 1), nonequilibrium conditions (or variation over time, Chapter 2) and a diversity of environmental conditions (or variation over space, Chapter 3 and 4). Arid environments are extreme environments in the sense that they are highly moisture-limited. This lack of moisture is matched by a lack of vegetation and soils in comparison with many other terrestrial environments, which has a clear influence on weathering systems (Figure 6.3). For one thing, the lack of soils makes bare rock surfaces more common and thus small-scale weathering landforms (which develop predominantly on bare rock surfaces) more frequent. While the lack of moisture and soils is generally coupled with sparse vegetation, the vegetation that does survive under these conditions often has a more clear-cut influence on weathering than it does in other environments. For example, plants exploit crevices in rocks in order to find moisture, shelter and anchorage in moisture-limited and soil-poor environments. In so doing, they can cause weathering through root growth pressures and chemical effects of associated endophytic bacteria (Puente, Li and Bashan, 2009). Of much more widespread importance is the relative importance of rock surface biofilms (mixed communities of lower plants and microorganisms) to weathering. Such biofilms, because of the harsh growth conditions in arid environments, often have very intimate connections with the underlying rock and, as a result, can play a large role in both chemical and physical weathering (Viles, 2008).

The extreme nature of arid environments has further, more complex dimensions that affect weathering. For example, while moisture is highly limited in dryland environments, evaporation and capillary rise are as important as precipitation and runoff (especially within hot arid environments). This means that rocks and minerals at the Earth's surface in arid environments are exposed to very complex moisture movements in comparison, for example, with those in humid tropical environments where downward leaching of weathered material dominates. In arid environments there is more commonly a two-way system of water movements within soils and rock outcrops, i.e. 'up and down' or 'in and out' – to put it more simply, but rather crudely. Furthermore, moisture inputs to weathering systems in arid environments do not only take the form of rainfall, and thus quantifying moisture availability can be much more complex than simply monitoring annual rainfall amounts. For example, within coastal subtropical deserts, such as the Atacama Desert in Peru and Chile, South America, and the Namib Desert in Namibia, Africa, fog can be a very important source of moisture to many rock surfaces. In many desert environments condensation and dew may also be important, while widely within arid environments groundwater contributes to the moisture available for weathering. Quantifying these multiple sources of moisture and their transformations and movements over time and space can be highly difficult, especially given the sparse meteorological data available on the ground in many arid environments. Within cold arid environments, the presence of water in the form of ice, and the possibility for transformations of water from solid to liquid or gas phases, makes things even more complicated.

The extreme nature of arid environments is also characterised by nonequilibrium conditions in climate, vegetation and geomorphology. In terms of climate, nonequilibrium conditions are visible in the high interannual and interdecadal variability within arid zone climates (Viles and Goudie, 2003). For vegetation, nonequilibrium conditions have been recognised by ecologists, who have identified bi- or multiple-stable states in vegetation systems as a response to changing climatic conditions. For example, a range of studies have shown that arid and semi-arid vegetation systems can oscillate between patchy (often patterned) and bare conditions depending on rainfall. Such bi-stable or multiple-stability conditions can also be identified in arid environment geomorphic systems, which can, for example, flip between aeolian and fluvial process regimes (Bullard and Livingstone, 2002). These nonequilibrium conditions will have a strong effect on weathering systems. For example, runs of dry and wet years will influence the dominant weathering processes and their rate of operation at any one arid site. Furthermore, extrapolations of short-term monitoring studies of weathering processes to the longer term might be quite misleading if they were undertaken, for example, during extreme El Nino conditions.

Arid environments also possess huge spatial diversity of environmental conditions within and between areas. Thus, some deserts are much drier than others, while within any one desert area some individual sites may be much drier than others. For example, across the Namib Desert from west (coastal) to east (inland) annual rainfall amounts rise from under 20 mm per year along the coast at Swakopmund to almost 50 mm per year at Ganab (over 100 km inland), while fog receipt drops in reverse from around 35 mm a year at the coast to around zero 100 km inland (Viles, 2005). Variability does not only relate to climatic conditions, but also includes differences in environmental history, tectonic setting and geology. Some deserts are of much greater antiquity than others, some have experienced quite complex swings between arid and wetter environmental conditions and so on. Tectonically there are, for example, huge differences between the Atacama and Namib deserts, which, in broad climatic terms, are otherwise quite similar. Most deserts exhibit a broad range of rock types – with, for example, the Namib Desert having exposures of basalt (in the north), marble and granites (in the central areas) as well as a range of schists and other rocks. Such spatial variability makes it very difficult to make any meaningful generalisations about the nature, tempo and geomorphological importance of weathering systems within the arid zone. High spatial variability in many deserts can make extrapolations from one place to another only tens of km away difficult (Viles and Goudie, 2000).

6.3 Theoretical underpinnings of weathering systems research

Despite the hostility of the desert environment, a large number of studies have been carried out on weathering in areas as diverse as the southwestern USA, the Sahara, the Namib, the Atacama and the Negev ever since the earliest explorers visited. For many years weathering research, in arid and other environments, fit broadly speaking into the CLORPT framework, which identified the key controls on weathering (and soil development) as climate, organisms, relief, parent material and time (e.g. Brunsden, 1979). Quantification of those factors was thought to

Table 6.1 Factors controlling arid weathering systems at different scales.

Name	*Spatial scale*	*Temporal scale*	*Weathering system*	*Major controlling factors*	*Major weathering effects*
Microscale	mm to cm	Days to years	Cobble-sized clast, small weathering feature on rock surface (e.g. rills, micropits)	Pore water characteristics, microclimate, individual plant root, individual lichen, biofilm, mineralogy, grain size, porosity	Granular disintegration, pitting, microflaking, cracking, dissolution
Small scale	Metres	Decades to hundreds of years	Large boulder, medium-sized weathering feature on rock outcrop (e.g. tafoni)	Groundwater and internal moisture regimes, topoclimate, patterns of climatic variability such as ENSO, individual shrubs and vegetation patches, variability in biofilms, lithology, variations in hardness and porosity	Blistering, flaking, exfoliation, cracking, case hardening
Medium scale	Kilometres	Millennia	Wadi-side slopes, large rock outcrops, desert pavements, closed depressions	Groundwater histories, regional climate and climate change, ecosystems, neotectonics, stratigraphy	Cracking, rockfall, crust formation (calcrete, gypcrete), weathering- and transport-limited situations
Large scale	100s of km	Millions of years	Massifs, gravel plains, large inland basins	Long-term hydroclimatic changes, biomes and biome shifts, tectonics	Surface lowering, sediment production, histories of weathering- and transport-limited situations

enable the nature and rate of weathering to be ascertained and process-form links to be understood. However, in reality weathering has been found to be far more variable over time and space than the CLORPT framework suggested. Some workers have proposed other similar but more complex frameworks, such as Pope, Dorn and Dixon (1995), who identified a much wider range of controlling factors affecting the geographical variation in weathering. Such factors can be viewed as operating in some hierarchical fashion. Other workers suggest that nonlinear dynamical systems (NLDSs) or nonequilibrium perspectives might provide a better framework and enable better understanding of weathering processes and their contributions to geomorphology (Viles, 2001).

One key idea underpinning today's weathering systems research, which can be derived from both hierarchical and NLDS approaches, is that scale matters. A preoccupation with the importance of scale is nothing new in geomorphology, as Schumm and Lichty's landmark paper on scale and causality in 1965 testifies. However, weathering scientists have been slow to take up the scale challenge, and it is only in recent years that technology has been developed that has allowed us to start proper investigation of weathering systems at different scales. The factors that control weathering systems vary depending upon the scale of analysis, where scale implies both the length of time considered and the area of concern. Table 6.1 illustrates some key factors thought to exert control over weathering systems in arid environments at different scales. In many cases hypothesised links between factors and weathering systems have not been properly investigated and further research is urgently required. Another way of looking at the statement 'scale matters' in terms of arid environment weathering systems is in terms of the different contributions of weathering to geomorphology at different scales. We will revisit this idea at the end of the chapter.

6.4 Current weathering study methods

A revolution has occurred in the methods used to study weathering in recent years, contributing to a flowering of research in arid and other environments and a growing ability to answer some key questions. Four types of method are now available, i.e. monitoring, nondestructive testing, experimentation (including modelling) and dating, which throw light on weathering systems when used individually or together. Many of the methods described below have not yet been fully trialled in hot arid environments, but have proved to be highly effective in other extreme environments such as the Antarctic and the built environment. Monitoring methods are now available, thanks to the automation, ruggedisation and miniaturisation of datalogging technology, which can monitor both weathering processes and the environmental conditions that control weathering over long periods of time at high temporal resolution. For example, time lapse photography can be used to monitor the detachment of grains or flakes of rock while iButtons and other equipment can monitor the fluctuation of temperature and relative humidity at the rock surface (Figure 6.4(a)). A large number of studies have now collected an impressive range of microclimatic data from weathering systems in hot and cold arid environments (e.g. Hall, 1997, 1998; Viles, 2005; Hall and Andre, 2006; Sumner, Hedding and Meiklejohn, 2007; McKay, Molaro and Marinova, 2009). Scientists have been slower to implement time-lapse videography and other high-tech methods to monitor the weathering response to such

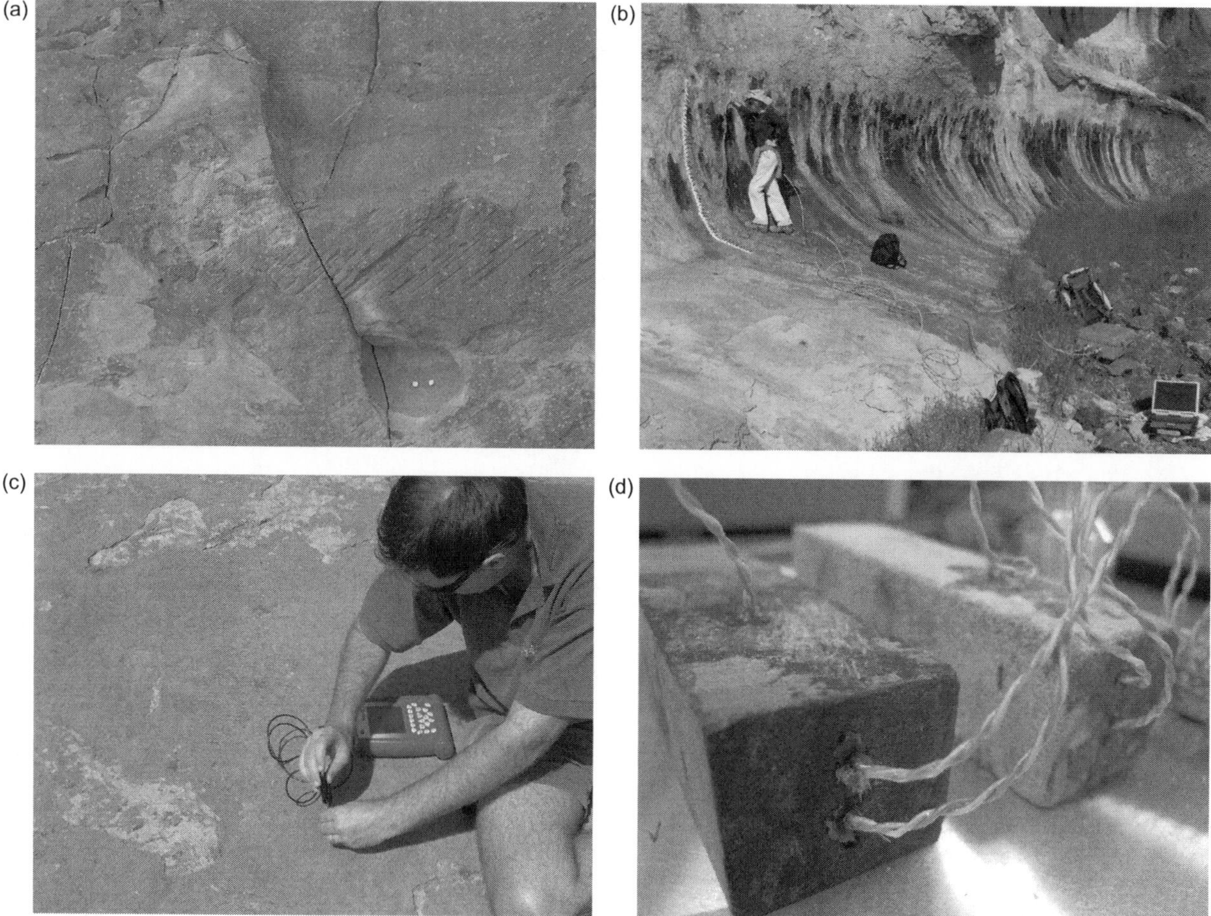

Figure 6.4 Techniques for studying weathering in arid environments: (a) two iButtons in a c. 10 cm wide tafone on sandstone in the Messak Settafet, Libya, deployed to monitor rock surface temperature and relative humidity regimes; (b) two-dimensional resistivity survey of subsurface moisture regimes in a cavernous weathering feature, Drakensberg Mountains, South Africa (image courtesy of Lisa Mol); (c) the Equotip hardness tester; (d) monitored blocks within an arid environment weathering simulation set up in a programmable environmental cabinet.

microclimatic fluctuations in arid environments, probably because of cost and logistical constraints, but these have huge future possibilities.

Monitoring of surface change can now be done using a family of methods that build on the established and still valuable techniques of surface profiling and the microerosion meter (MEM), which capture data on rock surface topography and topographic change in a spatially limited, but quick, fashion (e.g. Stephenson *et al.*, 2004). Laser scanning, optical scanning and photogrammetric techniques can now be used to collect high-resolution topographical data sets. From these, geomorphometric measurements (e.g. roughness measurements) can be taken to quantify topography at a range of scales. Furthermore, these techniques can be used to monitor surface change as a result of weathering – with before and after data collected at very high resolution allowing detailed comparisons of surface topography. Most examples of their use in weathering research so far come from laboratory-based experimental studies (Birginie and Rivas, 2005; Bourke *et al.*, 2008). However, increased portability of such systems means that they can now be used in the field even in remote and challenging environments, while casting techniques can also be used to capture topographic detail in the field for subsequent scanning in the laboratory (Ehlmann *et al.*, 2008).

A wide range of nondestructive testing methods is now available, which can be used to survey conditions on and within weathering rock surfaces in the field in order to investigate weathering. For example, a range of resistivity-based techniques are available to monitor surface or subsurface moisture regimes within porous materials, which help in understanding some of the key factors affecting weathering at the small scale (Figure 6.4(b)). While such techniques have mainly been used in mountain and built environments, they have huge potential for helping to understand weathering in desert environments (Sass, 2005; Sass and Viles, 2006; Mol and Viles, 2010). Although weathering occurs at or near the surface of a rock or mineral it is often influenced by deep-seated movements of water and thus knowledge of conditions and moisture movements under the surface is of huge importance. Geophysical techniques such as ground penetrating radar (GPR) have also recently been applied to understanding the development of flaking and undoubtedly have wider applications to weathering in arid environments (Denis *et al.*, 2009). Several nondestructive techniques are also available to monitor rock hardness in the field, which can be used to evaluate weathering, based on the principle that weathering will (in general) reduce the hardness of rock surfaces. The Schmidt Hammer, Duroscope and Equotip devices have all been used to investigate rock weathering in arid and other environments (Figure 6.4(c)). These devices are (relatively) cheap, easily portable and quick to use; thus they are ideal for collecting large data sets under challenging field conditions (Goudie, 2006; Aoki and Matsukura, 2007).

Experimental studies, while not in themselves new, have dramatically increased in number and sophistication in recent years as a result of the availability of better equipment to control and monitor environmental conditions. Environmental cabinets enable researchers to control temperature and relative humidities and cycle them, as well as use lamps to provide radiant heating. The basic approach of many weathering experiments of relevance to arid environments has been to subject samples (usually cut blocks of stone) to a weathering regime (often involving salts, moisture and temperature fluctuations) over a period of time (Figure 6.4(d)). Assessment of weathering has usually been made by making comparative observations of shape, size, weight, surface topography, strength and so on, before and after (and sometimes during) the experiment (see, for example, Goudie, 1986; Robinson and Williams, 2001; Goudie, Wright and Viles, 2002; Elliott, 2008). Smith *et al.* (2005) write perceptively about the problems of many such weathering experiments. Flaws within experimental designs used and limitations of available simulation chambers mean that it is very difficult to link results from experiments to real-world conditions. Increased realism has recently been brought into experiments of arid weathering systems through the use of pre-treated blocks to simulate an inherited 'weathering history' (Warke, 2007). Thus, progress is being made in bridging the gap between the field and the laboratory.

Weathering experiments that are highly limited in spatial and temporal scale also raise problems. Thus, we cannot expect too much of them in terms of contributing to questions about weathering systems over large spatial scales and long time scales. Furthermore, most arid weathering experiments have been of intermediate to high complexity – and perhaps the time is now right for more simple experiments, linked to the development of computer-based models, to allow the development of a better understanding of what is really happening in weathering. An example is the use of environmental scanning electron microscopy (ESEM) to simulate the behaviour of salt solutions and small samples of salt-affected stone under very simple regimes (Doehne, Carson and Pasini, 2005). The fact that the ESEM allows us to view the changes as they occur means that we can, at last, watch weathering happen! On the other hand, weathering experiments can also usefully become more complex and more realistic, and here the use of field experiments is very attractive. Modifications to the natural environment (such as putting out test blocks whose starting properties are well characterised) allow the creation of a field experiment, but natural conditions free us from having to simulate the environment during the experiment. Such field experiments, which have a long history in arid weathering studies (e.g. Goudie and Watson, 1984), now benefit hugely from the improvements in microclimatic monitoring and so on, mentioned above, which now allow a really good assessment of conditions during the experiment.

Computer-based modelling allows virtual experimentation and is a very powerful tool for areas such as weathering systems research where processes of interest are slow and hard to monitor. However, weathering geomorphologists have been relatively slow to take up such techniques beyond conceptual modelling (usually in the form of diagrams illustrating potential weathering systems evolutionary pathways). Some exceptions are the work of Huinink, Pel and Kopinga (2004) and Turkington and Phillips ((2004)), both of whom have addressed cavernous weathering, and Moores, Pelletier and Smith (2008), who have focused on solar radiation, moisture and cracking. Modelling has the huge advantage that it can address questions at any temporal and spatial scale. The growth of microclimatic data collection in recent years can help the development and testing of models of arid weathering systems.

The laboratory and field-based methods discussed above provide a largely small-scale and short-term understanding of arid weathering. In distinction, dating techniques allow us to determine the age of weathering-related phenomena and thus can enlarge our view of the rates of weathering. Cosmogenic dating techniques in particular have proved to be very helpful in answering questions about long-term rates of weathering and have been widely used in arid environments. Cosmogenic dating works by measuring the amount of cosmogenic isotopes produced by cosmic ray bombardment of minerals. The longer a mineral has been exposed at the Earth's surface the more cosmic rays it has received and thus the more cosmogenic isotopes are found. Commonly used cosmogenic isotopes for weathering and denudation studies are ^{10}Be and ^{26}Al, which are produced by cosmic ray bombardment of silicate minerals. Studies have found extremely low rates of denudation over million year timespans in arid environments, such as in Namibia, where Van der Wateren and Dunai (2001) found rates of 0.1–1 m/Ma over the past 5 Myr and Bierman and Caffee (2001) found rates of 1–16 m/Ma over the Pleistocene. It is also worth mentioning here that, as well as using dating techniques to quantify rates of weathering, the development of weathering-related features such as weathering rinds has been used as a relative dating technique and ^{14}C dating of desert varnish has also been used to investigate weathering and palaeoenvironmental histories (Dorn, 1998; Zerboni, 2008).

6.5 Linking processes to form in arid weathering systems

At the dawn of the twentieth century it was largely envisaged that physical weathering processes dominated in arid environments. Indeed, the picture was even simpler than this, with an orthodox view emerging that heating and cooling as a result of changing receipt of solar radiation diurnally and over shorter timespans was the major (sometimes the only) process operating. This process is often called insolation weathering. The perceived dominance of insolation weathering (while easy to understand given the lack of techniques to probe more deeply) was based on flimsy evidence and a degree of circular reasoning. Extremes of temperature are frequently observed in arid environments, as is cracking and flaking and other weathering features that could be caused by temperature fluctuations. A plausible link between the two could thus be made. Early experimental work by Griggs queried the efficacy of insolation weathering and indicated that it was much more effective in the presence of moisture (Griggs, 1936). More recently there has been a resurgence of interest in insolation weathering and its role in cracking boulders on desert pavements (McFadden *et al.*, 2005). Similar thermal cracking can occur in arid environments as a result of fires (Dorn, 2003). As well as cracking, insolation weathering has been linked to exfoliation of the outer layers of boulders and the granular disintegration of rocks comprised of minerals with different thermal properties.

A century of research has now confirmed that a much larger array of weathering processes operates in arid environments, as depicted in Table 6.2. The prevalence of extreme fluctuations of temperature and relative humidity in many desert areas means that physical weathering processes are preeminent. However, the presence of salts in groundwater and their concentration by evaporation of rainfall and other moisture sources means that salt weathering has now come to be seen as a dominant process in many deserts. Indeed, salt weathering has now become the new orthodoxy. Salt weathering operates as salts exert pressure on pore walls within the near-surface zone of a rock in three main ways – through repeated crystallisation and dissolution, through hydration and through expansion on heating (see Goudie and Viles, 1997, for a general overview of salt weathering). Cooke and Sperling (1985) carried out some simple, elegant experiments that

Table 6.2 Weathering processes known to operate in arid environments.

Physical	*Biological*	*Chemical*
Insolation weathering	Biophysical	Hydrolysis
Fire and thermal shock	Root wedging	Dissolution (of limestone)
Salt weathering	Lichen thalli growth pressures	Iron oxidation
Crystallisation	Lichen thalli wetting and drying	Hydration
Hydration	Biochemical	
Thermal expansion	Biodissolution	
	Chelation	

demonstrated the relative importance of salt crystallisation versus hydration under controlled conditions. Since their experiments, a range of other workers have tried to isolate and investigate different facets of the salt weathering role of a variety of salts on different rocks (e.g. Yu and Oguchi, 2009). While salt weathering in arid environments is basically a physical weathering process it does, of course, rely on chemical transformations (the dissolution of salts in water and their recrystallisation) and requires the presence of water. Some salts are also capable of hydration (a chemical process whereby water molecules are taken up into their crystalline structure), which has also been shown to be an effective agent of salt weathering. Salt weathering has been linked to flaking, blistering and granular disintegration of salt-affected rock surfaces (Figure 6.5(a)).

Other important weathering processes thought to operate in arid environments are biochemical and biophysical weathering by rock surface biofilm communities. A rich microflora inhabits many desert areas, even where higher vegetation growth is prohibited by the harsh conditions. Lichens, algae, fungi and bacteria can exist on very little water, and can in some cases extract moisture from humid air without the need for any sort of liquid water. They can also often tolerate quite saline environments, and many microorganisms have evolved effective chemical sunscreens to enable them to survive in conditions where they receive very high solar radiation. A range of microorganisms and lichens can also extract nutrients directly from the rock surface and are thus highly able to survive in nutrient-limited arid environments. Many of these physiological adaptations mean that these biofilm communities are highly effective agents of weathering. In some cases they can bore their way into the rock surface (up to a few millimetres in depth) through a combination of chemical and physical means, in order to seek shade, nutrients and water. This process of boring causes weathering directly and also weakens the rock surface so that it can be exploited by other agents of weathering. Figures 6.5(b) and (c) depict examples of biological weathering by rock surface biofilms in arid environments. Biofilms and lichens can cause weathering in a number of ways. Paradise (1997) illustrates, through a detailed microscope study of lichens from the genus *Xanthoparmelia* on Red Mountain, Arizona, that lichens can produce both biophysical weathering (towards the centre of the thallus) and biochemical weathering (towards the edges). Biological weathering has been linked to exfoliation of thin flakes and granular disintegration of rocks where biofilms occur, as well as the case hardening of sandstones in deserts (Viles and Goudie, 2004).

Chemical weathering processes, despite the large-scale lack of moisture, can play an important role in arid environments, but one that often leads to the accumulation and consolidation of minerals and rocks rather than their breakdown and removal. A suite of chemical weathering processes has been noted from a range of environments, including the dissolution of soluble minerals, iron oxidation, hydrolysis of silicate minerals and a number of other chemical transformations. Where even small amounts of moisture are present any susceptible minerals will be affected by such processes, often producing very small weathering features diagnostic of chemical weathering (e.g. microscale karren features as shown in Figure 6.5(d)). In desert environments, the high rates of evaporation mean that water is often only transient and thus the products of chemical weathering tend to be deposited within or on the affected soils and rocks, rather than being washed away as dissolved or suspended load. However, Pope, Dorn and Dixon (1995) reanalysed data on chemical denudation across climatic regions (drawn from solute loads of rivers) and found no clear relationship with climate – arid and semi-arid areas do not appear to have distinctly lower rates of chemical denudation and so chemical weathering may be more important than other workers have assumed. Chemical weathering

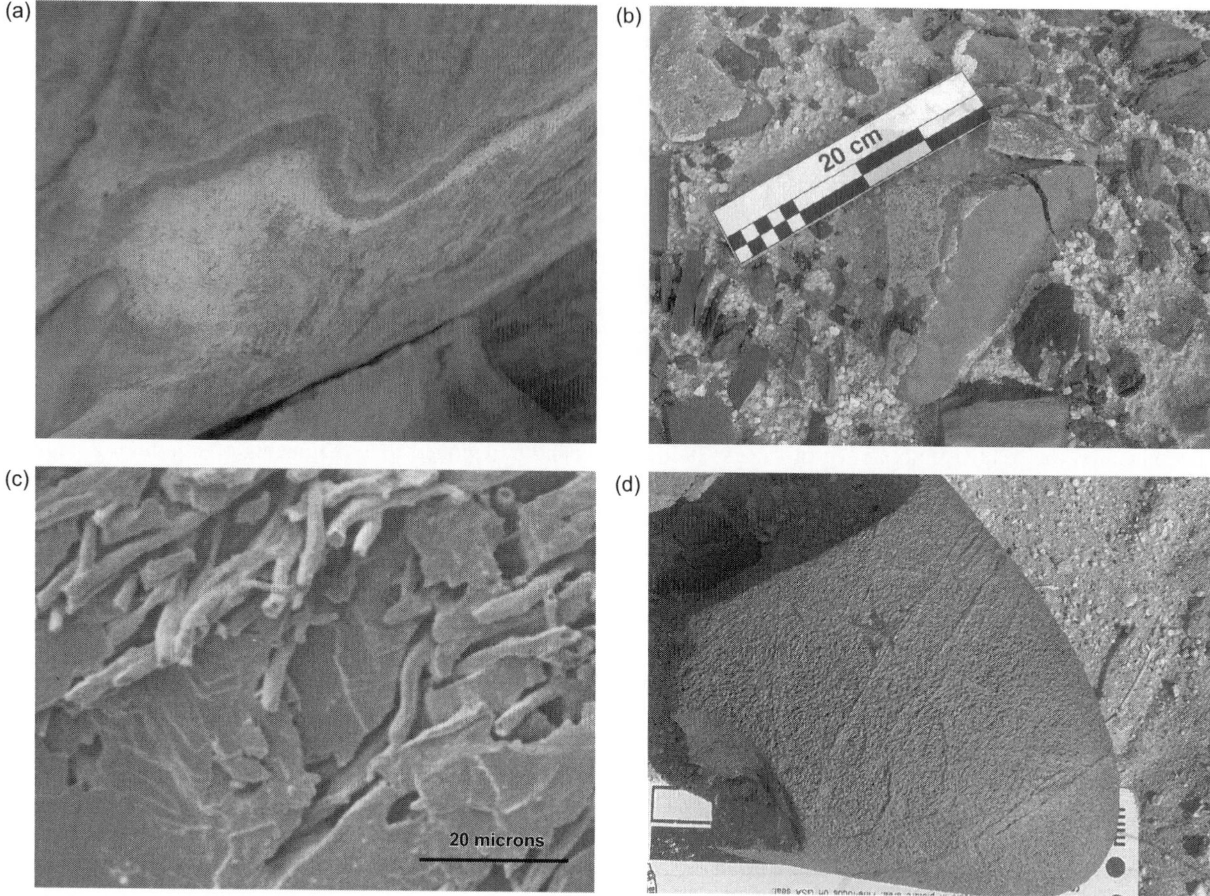

Figure 6.5 Evidence for arid weathering processes: (a) salt crystallization on the underside of a tafone roof, Messak Settafet, Libya; (b) lichens growing on the leeward side of pebbles in the Namib Desert; (c) SEM micrograph of a cross-section through the lower parts of a lichen thallus growing on marble in the Namib Desert showing the production of boreholes by fungal hyphae; (d) microsolutional features on basalt pebble, northern Namib Desert (image courtesy of Dr Mary Bourke).

in arid environments has been associated with small-scale karren on soluble rocks, case hardening and the production of calcrete, gypcrete and other indurated layers in soils, as well as the growth of tufa and speleothem deposits. However, in several places there is plausible evidence that such indicators of chemical weathering date back to periods when wetter conditions prevailed, as they are now starting to become deteriorated through the action of subsequent weathering processes, as found on limestone outcrops in southern Tunisia by Smith, Warke and Moses (2000). Thus, large-scale data (from solute loads) and small-scale observations of weathering processes and features can give different pictures of the importance of chemical weathering in today's arid environments.

Much of our knowledge of weathering processes in arid environments comes from observations in the field, coupled with laboratory analyses of samples collected in the field. This approach might be called 'abductive weathering science', where one deduces the processes that operated from a detailed analysis of the material left behind. Laboratory experiments have also been used to corroborate such field evidence and test whether or not particular processes are effective under certain conditions and what kinds of deterioration they cause. Monitoring of microenvironmental conditions in the field has also helped to test whether specific processes could be operating, but as yet there has been no real monitoring of individual weathering processes under field conditions. The small-scale nature of the weathering processes discussed above has, with currently available technologies, precluded this. For example, while we can use time-lapse photography to watch flakes become detached from rock surfaces in the field we cannot, as yet, monitor salt crystallisation and dissolution events repeatedly within cracks and pores to prove that such a process is causing the flaking.

6.6 Explaining the development of weathering landforms in arid environments

A weathering landform can be defined as a landform (usually relatively small in size, ranging from centimetres to metres in dimensions) produced entirely or dominantly by weathering processes. A range of weathering landforms has been identified from arid environments, including tafoni, alveoli, exfoliated boulders, weathering pits and rills. Most of these features are also found in other environments, although they are often more highly developed and commonly found in drylands. Named weathering landforms are ideal types; in reality rock surfaces display an often baffling relief, which can be hard to describe and categorise into discrete features. An atlas of commonly observed weathering features, which includes many observed in arid environments, can be found in Bourke and Viles (2007).

Cavernous weathering features are one of the most obvious weathering landforms found in many desert areas, occurring in both hot and cold deserts and on a range of rock types (granites, metamorphic rocks, sandstones and limestone). Cavernous weathering is an umbrella term used to describe small (alveoli) and large (tafoni) weathering hollows found, usually, on vertical rock surfaces. Alveoli tend to occur in clusters and networks of alveoli can fill entire rock faces, while tafoni can occur as individual features or in small groups. The factors controlling cavernous weathering are still debated, although salt weathering, thermal regime and the influence of wind have been often invoked as key agents (Rodriguez-Navarro and Sebastian, 1999; Strini, Gugliemin and Hall, 2008). While many detailed studies have been made of cavernous weathering in a range of arid environments, often coming up with particular explanations for the individual study site, there has not yet been a generalised explanation. However, self-organisation has been proposed as a general condition basic to cavernous weathering, despite the range of individual processes involved in different locations. Turkington and Phillips ((2004)), for example, propose that the development of cavernous weathering on sandstone outcrops in the Valley of Fire State Park, Nevada, is a self-organisational response to dynamical instability within the weathering system. McBride and Picard (2000) illustrate how subsurface water flow within tuff at Crystal Peak, Utah, experiences self-organisation and precipitates calcite at the surface in regular patterns. The calcite then causes salt weathering and leads to the development of tafoni.

How can we explain the development of weathering landforms in arid environments that may have taken decades to millennia to form? As Schumm (1991) has noted, explaining the development of landforms of all types is beset by many problems, including convergence (different processes can produce the same landform outcome), divergence (the same process can produce very different landform outcomes) and multiplicity (multiple processes acting together to cause a landform outcome). There are also problems of environmental change and scale. Measurements, or inferences, of processes acting today, for example, may be irrelevant in explaining a landform that was largely produced by processes acting in former times under different environmental conditions. The complexity of many arid weathering regimes and the resultant palimpsest of active and inherited weathering features makes process–form links complex even at the small scale (Smith, Warke and Moses, 2000). Short-term and small-scale information on processes derived from one part of a landform may be inadequate to explain larger features, which may be influenced by a range of different processes acting across them. These issues have proved to be serious stumbling blocks for scientists working on weathering systems in arid environments, although progress is being made (see Box 6.1).

Box 6.1 Understanding cavernous weathering in the Atacama Desert, Chile

Pilot studies of cavernous weathering features ranging in dimensions from a few cm to several metres in the northern Atacama Desert illustrate the challenges of linking processes to landforms. Tafoni and alveoli are widely distributed here and are often found together in complex arrangements (Figure 6.6(a)). Two sites were studied, one at Punta Pateche at around sea level on the coast just south of Iquique and one in the Lluta Valley at around 2500 m a.s.l. and c. 60 km inland from Arica. A range of igneous rocks were found at both sites. Measurements of depth and diameter of 12 tafoni on one large boulder at Punta Pateche showed a range in sizes from about 5 cm depth and diameter to 30 cm depth and diameter. At Lluta Valley 10 tafoni on individual boulders on a hillside were measured and found to have diameters and depths ranging from 50 to 150 cm.

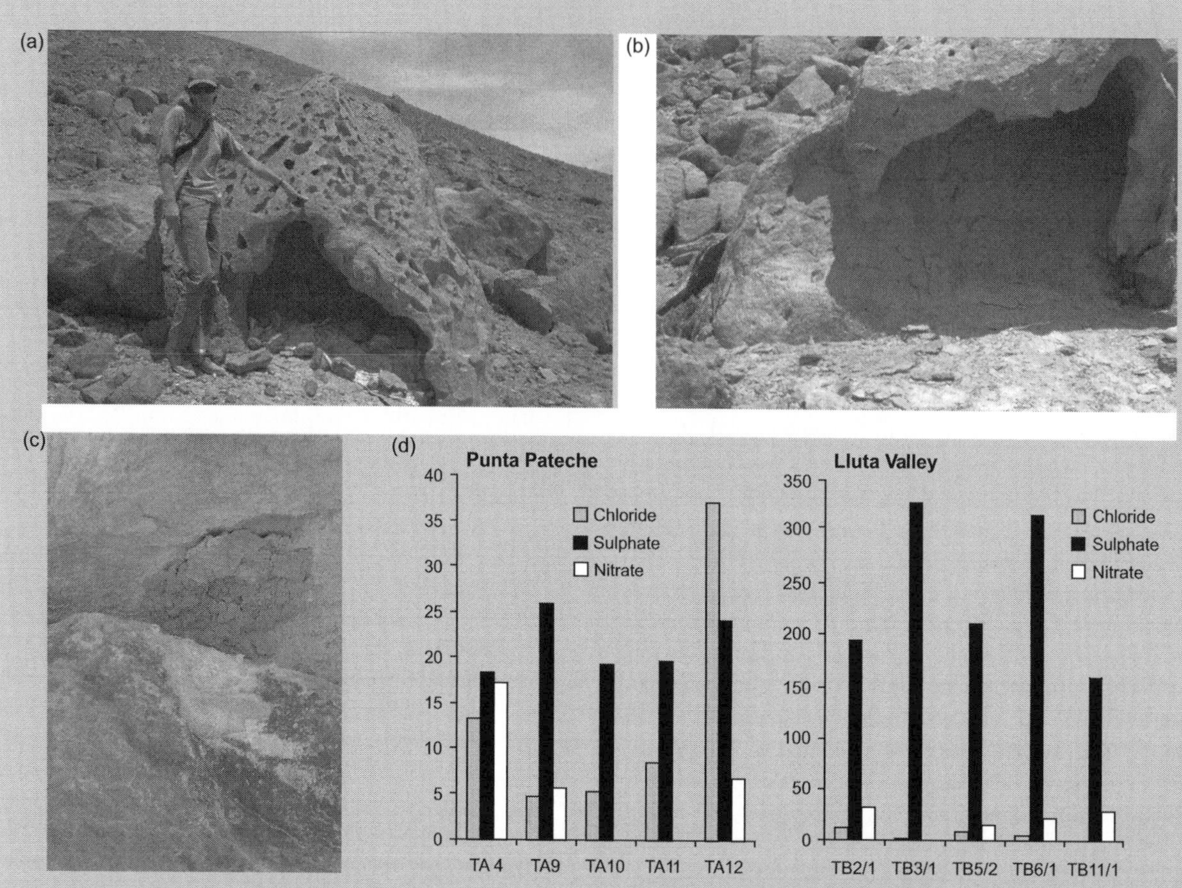

Figure 6.6 Cavernous weathering in the Atacama Desert, Chile: (a) boulder with one large tafone and many alveoli, Lluta Valley, Chile; (b) large flakes on the floor of tafone having been weathered from the upper surface; (c) crystalline salt deposits from inside tafoni; (d) graphs showing concentrations of chloride, nitrate and sulfate from tafoni at Punta Pateche and Lluta Valley measured in microgrammes per gram.

At both sites, especially in the Lluta Valley, tafoni development appeared to be highly active, as shown in Figure 6.6(b), with many newly detached large flakes on the floor of the tafoni. The remarkable feature at both sites was the presence of hard, thick framboidal crystalline salt deposits within the tafoni, and especially within flaking material at the sides and back of the tafoni. These deposits looked similar to many cave minerals found in subterranean environments, and were clearly different to the powdery salt efflorescences often found in tafoni (see Figure 6.6(c)). Basic geochemical analyses of these salt deposits are summarized in Figure 6.6(d). As can be seen, the site at Lluta Valley contains very high levels of sulfate and higher levels of nitrate than Punta Pateche, with similar levels of chloride at both sites.

These simple observations indicate that in these hyper-arid, but foggy, conditions salt weathering is highly likely to be a key cause of the present-day development of these tafoni. Differences in the factors controlling tafoni development are evidenced between the two sites. The field evidence also indicates that these tafoni are actively and rapidly developing. Further monitoring studies need to be carried out to provide a better understanding of how these tafoni are developing today and the key weathering processes involved, in comparison with those from other arid environments. However, concepts such as self-organisation need to be invoked in order to explain the bigger questions of the genesis and development of the entire system of cavernous weathering features in these areas.

6.7 Weathering rates in arid environments

In 1950 Louis Peltier produced a conceptual graph illustrating the nature and rate of weathering in different environments around the world (Peltier, 1950). This identified arid environments as not only environments dominated by physical weathering but also as being places where weathering was extremely slow. Such a view of desert landscapes as being fossilised because of slow weathering rates was indeed another geomorphological orthodoxy during the first half of the twentieth century. The supposedly slow rates of weathering were deduced from the lack of moisture and thus the limited array of potential weathering processes. Furthermore, slow weathering rates were inferred from the thin soils present (as higher rates of weathering usually lead to the accumulation of thicker soil layers built up from the debris and residue produced by weathering). The more recent focus on salt weathering as the key agent of weathering in arid environments has been coupled with a growing belief that, in fact, many desert areas are characterised by rather rapid rates of weathering.

Despite the wide array of techniques used to study weathering in the field and laboratory, it is still very hard to quantify weathering rates in any general way. The measurements that do exist tend to be highly place- and time-specific. For example, Viles and Goudie (2004) monitored weathering rates at a range of different locations in and around a coastal salt pan in Namibia using test blocks (some of which were cut from local rocks) and found high rates in some rock types in the most saline parts. Because some sensitive rock tablets disintegrated during the two year exposure period while others showed only minimal weight loss (and in some cases weight gain), it proved very difficult to find a common metric for rock breakdown rates. Thus, it was impossible to go beyond saying that weathering was fast for some rock types in some locations and slow in others. However, one clear finding was that there were indeed hot spots of weathering activity within salty, near-coastal environments, lending support to the view that in some parts of arid environments weathering rates can be rapid. Rather conflicting views of the rates of weathering in deserts are obtained from cosmogenic dating, which estimates denudation rates over millions of years. Over these timespans weathering in the Namib and other deserts has been found to be slow relative to other environments. These findings confirm the importance of scale in answering questions about arid zone weathering.

6.8 Arid weathering and landscape evolution

How weathering contributes to overall landscape evolution within desert environments remains a subject of debate. While it is clear that progress is being made in linking weathering to the development of individual small-scale landforms, the bigger picture is much more complex and blurred. Several studies have illustrated, often through laboratory experimentation, that weathering is capable of producing fine-grained sediments such as desert loess (Smith, Wright and Whalley, 2002). Salt weathering 'hotspots' thus might be important sources of silt-sized sediment or dust, as evidenced along the coastal fringes of the Namib Desert by Viles and Goudie (2007). Studies have also indicated that weathering contributes to rock fall and the evolution of rock slope profiles within arid areas, as, for example, found by Goudie *et al.* (2002) and Migon *et al.* (2005) in the Wadi Rum area of Jordan. In cliffed slopes in jointed rocks, salt weathering and other processes become concentrated along joints and fractures, contributing to the destabilisation of blocks and so encouraging rock fall. Once the fallen boulders are concentrated along the foot of the cliff, in the absence of effective fluvial erosion, the prevailing weathering rate will determine whether or not they remain in situ protecting the foot of the cliff from undercutting. Very different slope profiles will result depending on the rates of weathering in joints as compared with those on boulders.

Large closed depressions in arid environments may also owe much of their development to weathering, as recorded by Aref, El-Khoriby and Hamdan (2002) for the Qattara depression in Egypt. Here, differences in groundwater salinity between east and west portions of the basin control the activity of salt weathering. Within the western part of the basin, highly saline groundwater leads to very active salt weathering, causing denudation and the production of fine-grained sediment, which are then blown away to form nearby lunette dunes. This process has been going on since the start of the Quaternary, although the initial excavation of the depression started much earlier and was caused by denudation under wetter environmental conditions.

A big debate has emerged in the geomorphological literature in recent years over the role of groundwater sapping in the formation of amphitheatre-headed canyons in many arid areas, which exemplifies the complexity of linking weathering processes to the development of larger-scale features. Amphitheatre-headed valleys have steep headwalls and show little evidence that precipitation-fed overland flow has been involved in their creation. Several workers have proposed that they develop from groundwater sapping of these headwalls. Groundwater sapping involves the action of springwater in weathering and eroding the base of a rock scarp, encouraging rock fall and collapse and gradual retreat of the scarp slope. Hoke *et al.* (2004) invoke such processes in the formation of large canyons (quebradas) in the Chilean Atacama Desert. However, Lamb *et al.* (2008) rule out such a groundwater origin for Box Canyon, Idaho, illustrating (among other lines of evidence) that there are no signs of enhanced weathering on the headwall. Better data collection on the nature and rates of groundwater sapping processes to be found in arid areas would help test hypotheses over the formation of amphitheatre-headed canyons.

The concepts of weathering-limited and transport-limited situations are highly important to any assessment of the influence of weathering on arid environmental landscape development. Looking, for example, at a near-vertical rock slope, if transport at the base is the limiting factor then weathering will produce debris at a faster rate than it is removed, producing a protective buttress at the foot of the slope. Over time, such transport-limited conditions will lead to slope retreat. The influence of weathering on slope formation under these circumstances will be changeable, as a result of the declining slope angle. If, however, the same slope is subjected to weathering-limiting conditions, the debris will be removed as soon as it is created through weathering, thus allowing parallel treatment of slopes and a more constant contribution of weathering to slope development and landscape evolution (as slope angles and weathering regimes do not change over time). Thus, the nature of the longer-term influence of weathering on landscape development is highly influenced by the nature and rate of removal of material at the base of slopes (through fluvial and aeolian processes). Weathering scientists have been slow to engage in these bigger questions of arid landscape development, but have much to offer in terms of data to calibrate models.

6.9 Scale and arid weathering systems

As Table 6.1 identifies, weathering in arid environments has importance for geomorphology at many different scales, but the nature of the link between weathering and geomorphology changes drastically according to the scale under consideration. Furthermore, the factors that need to be invoked to explain the nature and rate of weathering at different scales also vary hugely. The consequences are that how we understand weathering depends very greatly on the scale of interest, and it is very difficult to apply data from one scale to another. While so far weathering scientists have been extremely good at understanding what is happening at the microscale (often based on laboratory experimentation), these findings have not yet been successfully translated to the small scale (i.e. the scale of most weathering landforms), the medium scale (for which quantification of weathering rates are clearly important) or the large scale (i.e. the landscape evolution scale).

In future more integrated research is needed in order to grapple more fully with the varying importance of weathering across different temporal and spatial scales. Such integration should involve more closely interrelated laboratory, field, experimental and dating-based studies. Furthermore, in order to establish more clearly the role of weathering in arid zone geomorphology as a whole, weathering scientists should get involved in larger-scale geomorphological and palaeoenvironmental research.

Acknowledgement

To Andrew Goudie for almost 25 years of sharing his arid weathering experience with me.

References

Aoki, H. and Matsukura, Y. (2007) A new technique for non-destructive field measurement of rock-surface strength: an application of the Equotip hardness tester to weathering studies. *Earth Surface Processes and Landforms*, **32**, 1759–1769.

Aref, M.A.M., El-Khoriby, E. and Hamdan, M.A. (2002) The role of salt weathering in the origin of the Qattara depression, Western Desert, Egypt. *Geomorphology*, **45**, 181–195.

Bierman, P.R. and Caffee, M. (2001) Slow rates of rock surface erosion and sediment production across the Namib Desert and escarpment, southern Africa. *American Journal of Science*, **301**, 326–358.

Birginie, J.M. and Rivas, T. (2005) Use of a laser camera scanner to highlight the surface degradation of stone samples subjected to artificial weathering. *Building and Environment*, **40**, 1011–1020.

Bourke, M.C. and Viles, H.A. (2007) *A Photographic Atlas of Rock Breakdown Features in Geomorphic Environments*, Planetary Science Institute, Tucson.

Bourke, M.C., Viles, H., Nicoli, J. *et al.* (2008) Innovative applications of laser scanning and rapid prototype printing to rock breakdown experiments. *Earth Surface Processes and Landforms*, **33**, 1614–1621.

Brunsden,, D. (1979) Weathering, in *Process in Geomorphology* (eds C. Embleton and J.B. Thornes), Edward Arnold, London, pp. 73–129.

Bullard, J.E. and Livingstone, I. (2002) Interactions between aeolian and fluvial systems in dryland environments. *Area*, **34**, 8–16.

Cooke, R.U. and Sperling, C.H.B. (1985) Laboratory simulation of rock weathering by salt crystallisation and hydration processes in hot arid environments. *Earth Surface Processes and Landforms*, **10**, 541–555.

Denis, A., Huneau, F., Hoerle, S. and Salomon, A. (2009) GPR data processing for fractures and flakes detection in sandstone. *Journal of Applied Geophysics*, **68**, 282–288.

Doehne, E., Carson, D. and Pasini, A. (2005) Combined ESEM and CAT scan: the process of salt weathering. *Microscopy and Microanalysis*, **11**, 416–417.

Dorn, R.I. (1998) *Rock Coatings*, Elsevier, Amsterdam.

Dorn, R.I. (2003) Boulder weathering and erosion associated with a wildfire, Sierra Ancha Mountains, Arizona. *Geomorphology*, **55**, 155–171.

Ehlmann, B.E., Viles, H.A. and Bourke, M.C. (2008) Quantitative morphologic analysis of boulder shape and surface texture to infer environmental history: a case study of rock breakdown at the Ephrata fan, Channeled Scabland, Washington. *Journal of Geophysical Research*, **113**, F020212, DOI:10.1029/2007JF000872.

Elliott, C. (2008) Influence of temperature and moisture availability on physical rock weathering along the Victoria Land coast, Antarctica. *Antarctic Science*, **20**, 61–67.

Goudie, A.S. (1986) Laboratory simulation of 'the wick effect' in salt weathering of rock. *Earth Surface Processes and Landforms*, **11**, 275–285.

Goudie, A.S. (2006) The Schmidt hammer in geomorphological research. *Progress in Physical Geography*, **30**, 703–718.

Goudie, A.S. and Viles, H.A. (1997) *Salt Weathering Hazards*, John Wiley & Sons, Ltd, Chichester.

Goudie, A.S. and Watson, A. (1984) Rock block monitoring of rapid salt weathering in southern Tunisia. *Earth Surface Processes and Landforms*, **9**, 95–98.

Goudie, A.S., Wright, E. and Viles, H.A. (2002) The roles of salt (sodium nitrate) and fog in weathering: a laboratory simulation of conditions in the northern Atacama Desert, Chile. *Catena*, **48**, 255–266.

Goudie, A.S., Mgon, P., Allison, R.J. and Rosser, N. (2002) Sandstone geomorphology of the Al Quwayra area of southern Jordan. *Zeitschrift fur Geomorphologie*, **46**, 365–390.

Griggs, D.T. (1936) The factor of fatigue in rock exfoliation. *Journal of Geology*, **44**, 783–796.

Hall, K. (1997) Rock temperatures and implications for cold region weathering 1: new data from Viking Valley, Alexander Island, Antarctica. *Permafrost and Periglacial Processes*, **8**, 69–90.

Hall, K. (1998) Rock temperatures and implications for cold region weathering 2: new data from Rothera, Adelaide Island, Antarctica. *Permafrost and Periglacial Processes*, **9**, 47–55.

Hall, K. and Andre, M.-F. (2006) Temperature observations in Antarctic tafoni: Implications for weathering, biological colonisation and tafoni formation. *Antarctic Science*, **18**, 377–384.

Hoke, G.D., Isacks, B.L., Jordan, T.E. and Yu, J.S. (2004) Groundwater-sapping origin for the giant quebradas of northern Chile. *Geology*, **32**, 605–608.

Huinink, H.P., Pel, L. and Kopinga, K. (2004) Simulating the growth of tafoni. *Earth Surface. Processes and Landforms*, **29**, 1225–1233.

Lamb, M.P., Dietrich, W.E., Aciego, S.M. *et al.* (2008) Formation of Box Canyon, Idaho, by megaflood: implications for seepage erosion on Earth and Mars. *Science*, **320**, 1067–1070.

McBride, E.F. and Picard, M.D. (2000) Origin and development of tafoni in Tunnel Spring Tuff, Crystal Peak, Utah, USA. *Earth Surface Processes and Landforms*, **25**, 869–879.

McFadden, L.D., Eppes, M.C., Gillespie, A.R. and Hallet, B. (2005) Physical weathering in arid landscape due to diurnal variation in the direction of solar heating. *GSA Bulletin*, **117**, 161–173.

McKay, C.P., Molaro, J.L. and Marinova, M.M. (2009) High-frequency rock temperature data from hyper-arid desert environments in the Atacama and Antarctic dry valleys and implications for weathering. *Geomorphology*, **110**, 182–187.

Migon, P., Goudie, A.S., Allison, R. and Rosser, N. (2005) The origin and evolution of footslope ramps in the sandstone desert environment of south west Jordan. *Journal of Arid Environments*, **60**, 303–320.

Mol, L., and Viles, H.A. (2010) Geoelectric investigations into sandstone moisture regimes: implications for rock weathering and the deterioration of San Rock Art in the Golden Gate Reserve, South Africa. *Geomorphology*, 280–287, DOI: 10.1016/j.geomorph.2010.01.008.

Moores, J.E., Pelletier, J.D. and Smith, P.H. (2008) Crack propagation by differential insolation on desert surface clasts. *Geomorphology*, **102**, 472–481.

Paradise, T.R. (1997) Disparate sandstone weathering beneath lichens. Red Mountain, Arizona. *Geografiska Annaler*, **70A**, 177–184.

Peltier, L. (1950) The geographic cycle in periglacial regions as it is related to climatic geomorphology. *Annals, Association of American Geographers*, **40**, 214–236.

Pope, G.A., Dorn, R.I. and Dixon, J.C. (1995) A new conceptual model for understanding geographical variations in weathering. *Annals, Association of American Geographers*, **85**, 38–64.

Puente, M.E., Li, C.Y. and Bashan, Y. (2009) Rock-degrading endophytic bacteria in cacti. *Environmental and Experimental Botany*, **66**, 389–401.

Robinson, D.A. and Williams, R.B.G. (2001) Experimental weathering of sandstone by combinations of salts. *Earth Surface Processes and Landforms*, **25**, 1309–1315.

Rodriguez-Navarro, C. and Sebastian, E. (1999) Origins of honeycomb weathering: the role of salts and wind. *GSA Bulletin*, **111**, 1250–1255.

Sass, O. (2005) Rock moisture measurements: techniques, results and implications for weathering. *Earth Surface Processes and Landforms*, **30**, 347–359.

Sass, O. and Viles, H.A. (2006) How wet are these walls? Testing a novel technique for measuring moisture in ruined walls. *Journal of Cultural Heritage*, **7**, 257–263.

Schumm, S.A. (1991) *To Interpret the Earth*, Cambridge University Press, Cambridge.

Schumm, S.A. and Lichty, R.W. (1965) Time, space and causality in geomorphology. *American Journal of Science*, **263**, 110–119.

Smith, B.J., Warke, P.A. and Moses, C.A. (2000) Limestone weathering in contemporary arid environments: a case study from southern Tunisia. *Earth Surface Processes and Landforms*, **25**, 1343–1354.

Smith, B.J., Wright, J.S. and Whalley, W.B. (2002) Sources of non-glacial loess-sized quartz silt and the origins of 'desert loess'. *Earth-Science Reviews*, **1242**, 1–26.

Smith, B.J., Warke, P.A., McGreevy, J.P. and Kane, H.L. (2005) Salt-weathering simulations under hot desert conditions: agents of enlightenment of perpetuators of preconceptions? *Geomorphology*, **67**, 211–227.

Stephenson, W.J., Taylor, A.T., Hemmingsen, M.A. *et al.* (2004) Short-term microscale topographical changes of coastal bedrock on shore platforms. *Earth Surface Processes and Landforms*, **29**, 1663–1673.

Strini, A., Gugliemin, M. and Hall, K. (2008) Tafoni development in a cryotic environment: an example from Northern Victoria Land, Antarctica. *Earth Surface Processes and Landforms*, **33**, 1502–1519.

Sumner, P.D., Hedding, D.W. and Meiklejohn, R.I. (2007) Rock surface temperatures in southern Namibia and implications for thermally-driven physical weathering. *Zeitschrift fur Geomorphologie*, **NF 5**, 133–147.

Turkington, A.V. and Phillips, J.D. (2004) Cavernous weathering, dynamical instability and self-organization. *Earth Surface Processes and Landforms*, **29**, 665–675.

Van Der Wateren, F.M. and Dunai, T.J. (2001) Late Neogene passive margin denudation history – cosmogenic isotope measurements from the central Namib Desert. *Global and Planetary Change*, **30**, 271–307.

Viles, H.A. (2001) Scale issues in weathering studies. *Geomorphology*, **41**, 63–72.

Viles, H.A. (2005) Microclimate and weathering in the central Namib Desert, Namibia. *Geomorphology*, **67**, 189–209.

Viles, H.A. (2008) Understanding dryland landscape dynamics: Do biological crusts hold the key? *Geography Compass*, **2/3**, 899–919.

Viles, H.A. and Goudie, A.S. (2000) Weathering, geomorphology and climatic variability in the central Namib Desert, in *Linking Climate Change to Land Surface Change* (eds S.J. McLaren and D.R. Kniveton), Kluwer, Dordrecht, The Netherlands, pp. 65–82.

Viles, H.A. and Goudie, A.S. (2003) Interannual, decadal and multidecadal scale climatic variability and geomorphology. *Earth-Science Reviews*, **61**, 105–131.

Viles, H.A. and Goudie, A.S. (2004) Biofilms and case hardening on sandstones from Al-Quwayra, Jordan. *Earth Surface Processes and Landforms*, **29**, 1473–1485.

Viles, H.A. and Goudie, A.S. (2007) Rapid salt weathering in the coastal Namib Desert: implications for landscape development. *Geomorphology*, **85**, 49–62.

Warke, P.A. (2007) Complex weathering in drylands: implications of 'stress' history for rock debris breakdown and sediment release. *Geomorphology*, **85**, 30–48.

Yu, S. and Oguchi C.T. (2009) Complex relationships between salt type and rock properties in a durability experiment of multiple salt-rock treatments. *Earth Surface Processes and Landforms*, **34**, 2096–2110.

Zerboni, A. (2008) Holocene rock varnish on the Messak plateau (Libyan Sahara): chronology of weathering processes. *Geomorphology*, **102**, 640–651.

7 Desert soils

David L. Dunkerley

7.1 Introduction: the nature and significance of desert soils

In contrast to wetter areas, where soil mantles of pedogenically altered decomposed bedrock cover most or all of the landscape, the surfaces of deserts may be only patchily covered with soil. The surface over many desert uplands is formed of outcropping bedrock, and in lower-lying areas there may be a cover of aeolian or fluvial sediments so little modified that it hardly amounts to 'soil' by any common definition (e.g. many fall into the *entisol* order of the US system of taxonomy). In semi-arid areas, the effects of pedogenesis become more apparent, and a richer array of soil types has been described from these environments. Factors that contribute to climatic gradients in soil properties include various climatic factors such as rainfall and temperature, and also increasing biomass and larger accessions of organic detritus in humid environments. Rainfall in drylands tends to decline with distance from the ocean. Mean rates of rainfall decline are often around 1 mm/km, but the trend is really exponential in form and regressions show that the rainfall declines by 50% over distances of about 400 km (Makarieva and Gorshkov, 2007; Sheil and Murdiyarso, 2009). In addition to driving a decrease in plant cover, this pattern of diminishing rainfall results in less intense leaching of dryland soils, and consequently quite steep regional gradients in dryland soil properties occur. In areas of extensive forest, rainfall decline with distance inland is much less evident owing to the biotic recycling of precipitation, and biomass and soil properties vary less steeply.

In the overview presented in this chapter, we will not concentrate on the nutrient status or cropping potential of desert soils. Rather, our goal is to review those aspects that relate to the geomorphic evolution of the materials and surfaces upon which desert soil development takes place, and on the particular properties of desert soils that make their hydrologic response and erosional behaviour distinctive. In interesting ways, the development of desert soils is intimately connected with landscape development and with the history of environmental change in the arid and semi-arid regions, especially through the Quaternary Period (see Chapter 3). Despite the lack of protective vegetation, some of the features of desert soils may contribute to markedly lower rates of erosion than are seen in wetter climates. Desert stone pavements in the hyper-arid Negev, for example, have been shown to armour the surface to an extent that yields exceptionally stable and long-lived landforms (Matmon *et al.*, 2009).

As a background to the detailed study of desert soils, we can note that the importance of these soils extends beyond the margins of the global drylands, and extends their significance even to global environments. Dust is often entrained from deserts and carried downwind to wetter environments where vegetation taps dust, or rainout carries it to the land surface. The result is soils that contain exotic silts and clays, and whose depth and texture have been modified by the accessions of desert dust. Grazing pressure and other forms of land use in the dryland source areas accelerate the deflation of dusts. The dusts have additional effects so that, for instance, if the dust settles on snowfields, it reduces the albedo and promotes melting. Painter *et al.* (2007) showed that the duration of snow cover in the seasonally blanketed San Juan Mountains of Colorado was reduced by up to ~1 month by the enhanced absorption of solar radiation caused by desert dust. They speculated that the shortened snow season may have been established in the 1800s following the introduction of settlement and pastoralism in the western USA, with its associated soil disturbance. Such changes in snowmelt

Arid Zone Geomorphology: Process, Form and Change in Drylands, Third Edition. Edited by David S. G. Thomas

clearly have the potential to alter the volume and timing of water resources derived from the snowfields, and, moreover, such effects may be prone to further strengthening if global and regional climate changes cause drying and increased emission of desert dusts. Additional wider significance for desert soils comes from the suggestion that they may constitute a significant sink for methane, a strong greenhouse gas (Striegl *et al.*, 1992; McLain and Martens, 2005). In a similar way, the carbonates that accumulate at depth in many desert soils have a significant place in global carbon cycling and storage (Hirmas, Amrhein and Graham, 2010). Finally, it is not possible to consider the wider climatic significance of desert soils without mentioning the effects of the iron fertilisation of marine phytoplankton that results when the dust settles to the sea surface (e.g. Jickells *et al.*, 2005; Aumont, Bopp and Schulz, 2008). Phytoplankton growth then removes carbon dioxide, another important greenhouse gas, from the atmosphere. Moreover, the radiative forcing caused by atmospheric loadings of dust (Yoshioka *et al.*, 2007) promotes a feedback link between periods of dry, dusty climates and rainfall suppression. The long-distance transport of desert dust may even be implicated in the transport of microorganisms, for example, from northern Africa across the Atlantic Ocean to the Americas (Griffin, 2007).

These few examples show that while they may have limited fertility and little pedologic development compared with agricultural soils, desert soils are by no means unimportant. It is appropriate therefore to consider that desert soils have both autochthonous significance (influencing water partitioning, overland flow and erosion within the desert itself) and allochthonous significance (the offsite and global effects of dust on soils, albedos, the oceans and global climate). Desert soils then are not only affected by climatic changes but are also drivers of global environmental change. However, in the remainder of this chapter, our main focus will be on contemporary soils and the features developed in and on them that are important to dryland hydrology, geomorphology and ecology. Though deserts may superficially appear to be vast and immune from degradation, the truth is quite the opposite. In particular, desert soils exhibit features that are in fact quite fragile and that can easily be damaged by people, stock or vehicles (e.g. Adams *et al.*, 1982). Informed land stewardship therefore also benefits from a knowledge of these soils.

There are marked differences in the hydrologic role of soils in drylands and in the humid zone. In many humid areas, soil infiltrability is high when judged against the common rain rates, and the occurrence of Hortonian overland flow is uncommon except in rare events with sustained high rain rates or where soils have been compacted, and it may be absent altogether. Near-surface soil horizons are characteristically porous and permeable, owing to the abundant organic matter and the overturning of the soil produced by the population of soil fauna that is supported by the organic matter. Soil depth is a second characteristic of enormous significance, since it determines the capacity for water storage within the soil column. Even in the humid zone, rainfall is intermittent, and water storage in the soil is vital to the maintenance of plant cover and of baseflow in perennial streams. Variations in soil depth can therefore result in wide fluctuations in soil moisture availability within a single climatic environment and can magnify the soil moisture variability caused by climatic gradients in rainfall (e.g. Hamerlynck, McAuliffe and Smith, 2000). Even if soils are highly permeable, if they are shallow and have little capacity to store water, plant moisture stress arises relatively soon after rain, and prolonged drought can result in the mortality of plants. Under the same conditions of rainfall and soil permeability, plants in deeper soils survive. The available pore spaces in a shallow soil can fill relatively quickly during rain and consequently shallow soils may partition more rain into saturation overland flow, therefore providing a functionally significant soil moisture store that is smaller than might be anticipated from climatological information (such as the mean annual rainfall). In extreme cases, this can amount to a form of 'pedologic aridity' that we will shortly see is very common in drylands.

In drylands, soil infiltrability is often only moderate judged against rain rates and locally may be very low. We will examine some of the causes of this later, but low levels of organic matter, less abundant biopores produced by burrowing organisms and various kinds of surface seals and crusts all contribute. Hortonian overland flow is consequently more common in some drylands. However, as we shall see, it is primarily the surface properties of dryland soils that are the key determinant of their hydrologic behaviour and not the bulk properties of the deeper soil column. A large proportion of the surface of dryland soils, often 70–80%, may be devoid of vegetation cover, owing primarily to the scarcity of water and other resources, and is exposed to wind, rain, frost and strong solar radiation. Allelopathy, the inhibition of the growth of neighbouring plants by the production of inhibitory chemicals, is also involved in the wide spacing of vascular plants (e.g. Halligan, 1973).

The lack of cover creates three circumstances of enormous importance to the hydrology of dryland soils:

1. It leaves the soil surface exposed to the energetic impacts of raindrops, as well as to the erosional and depositional effects of overland flow, and to wind, ground

ice and intense solar radiation and heating. Various consequences of exposure to beating rain have been described, but a critical one is the physical breakdown of soil aggregates caused by the repeated raindrop impacts and the production of a 'seal' or 'crust' at the surface, which may be of very low permeability. This will be discussed in detail later.

2. The open 'interspaces', the gaps between the canopies of vascular plants, provide an environment where there is sufficient light and moisture for other specialised organisms such as algae and lichens. Some of these can draw moisture from the air, or derive it from dew in sufficient amounts, and can build extensive communities that are collectively known as biological soil crusts (BSCs). These biological communities may also occur beneath plant canopies that allow sufficient light and moisture to reach the soil surface.

Consequently, dryland soil surfaces can have properties quite different from the deeper subsurface where raindrop energy has no effect, and where light does not penetrate. Over vast areas of the drylands, it is the uppermost few millimetres of the soil that determine infiltrability and resistance to erosion (Patrick, 2002). As noted earlier, it is not uncommon for the infiltrability of dryland interspaces to be quite low. Thus, even in rain events of moderate rain rate, water is partitioned into overland flow and little may infiltrate. This perpetuates the dry conditions that exclude vascular plants from most of the landscape, and is an example of pedologic aridity. On the other hand, the surface flow can move downslope where, if it can be absorbed as run-on water at a more permeable site, it can supplement the scant rainfall and allow more plant growth than would otherwise be possible. This leads to a third characteristic of many dryland landscapes, that is intimately connected with soil properties.

3. Soil hydraulic properties and soil water balance can vary greatly over short distances (metres to tens of metres) and soil properties, spatially, are in fact often very patchy. The soils present a mosaic pattern of areas having contrasting properties, some areas exhibiting high infiltrability while adjacent patches show much lower infiltrability. There can be considerable contrasts in soil properties between subcanopy sites, which are shaded and receive inputs of organic material, and more open interspace sites. The production of overland flow on one patch or set of patches can drive quite important transfers of the key ecosystem resource, water, to other patches downslope where it can be absorbed. In other words, considerations of scale become critical when seeking to understand desert soils and their moisture regime. It is worth noting that other resources may move with flows of water, including organic litter particles, seeds and nutrients. The contrasting locations in the landscape, either generating or absorbing overland flow, create an important patch structure that can greatly modify the hillslope or larger-scale responses of the landscape to rain events. Because of this structure, overland flow may only move small distances across the landscape before being reabsorbed.

In later sections, we will consider the first two factors in greater depth, beginning with the key effects of raindrop impact on bare soils and then turning to the nature of the communities of nonvascular plants forming biological soil crusts. First, we begin our examination of desert soils with a short examination of their classification and nomenclature.

7.2 Taxonomy of desert soils

Desert soils are not widely documented. Those classification schemes for soils developed primarily to support food and fibre production in wetter areas reflect the importance of drainage, nutrient status, horizon characteristics and other aspects that relate to the capacity of the soil to support functions such as cultivation, irrigation and crop growth. They do not in general lend themselves well to the classification of desert soils. Desert soils may show very little horizon differentiation, a key taxonomic tool in humid areas, but often show features such as accumulations of soluble salts, which are rare in soils used for agriculture. Furthermore, many of the most important features from the hydrologic and geomorphic perspectives, such as desert stone mantles, surface seals due to beating rain, microbiotic crusts and vesicularity in near-surface layers, are developed with little dependence on pedologic characteristics, such as the nutrient status, mineralogy or horizonation of the deeper regolith. Rather, they occur in soils of diverse depth and mineralogy. Thus, though of extreme importance in landscape behaviour, because of their effect on rainfall partitioning and the production of overland flow, these features too are of relatively little use taxonomically, and instead depend on factors including the distance from the nearest vascular plant, position in the landscape and the evolutionary history of the site.

Table 7.1 lists the major soil groups recognised in the US classification system. A fuller examination of the properties of these soil groups (e.g. see Nettleton and Peterson, 1983; US Department of Agriculture, 1988; Watson, 1992) will confirm that many geomorphically significant

Table 7.1 The major soil orders of the arid and semi-arid regions (modified after Dregne, 1976).

Soil order	*Primary characteristics*	*Total global land area occupied (km²)*	*Percentage of global land area occupied*
Aridisols	Plant growth restricted by dryness and/or salinity all year	16.6×10^6	11.3
Alfisols	Moderate base saturation; an argillic horizon; some plant-available water seasonally	3.1×10^6	2.1
Entisols	Almost no horizon differentiation; little-altered sedimentary materials	19.2×10^6	13.1
Mollisols	Thick, dark, base- and organically-rich epipedon	5.5×10^6	3.7
Vertisols	Deep, cracking clay soils, with shrink–swell features common	1.9×10^6	1.3

features that are mentioned later in this chapter are not included in the classification rules, and we thus sidestep soil taxonomy, turning to the important issues of soil function in the landscape.

7.2.1 A note on terminology of near-surface features in desert soils

Throughout this chapter, the term 'soil seal' is used to refer to inorganic structures at and near the soil surface, since the key hydrologic effect of these structures is partially to seal the soil against water entry (and the reciprocal escape of air from the soil pore spaces). The term 'soil crust' is reserved for reference to organic or biological soil crusts (BSCs), since these often form quite distinct surface layers that can be separated from the underlying soil in small flakes or sheets. This usage is not universal in the published literature, but it is helpful for clarity of expression to adopt distinct terminology. Some authors refer to raindrop impact seals as 'seals' only when they are wet and adopt the name 'crust' for the same features when they have become air dry.

7.3 Some distinctive aspects of desert soil development

Desert soils develop in a wide range of parent materials, including the extensive fluviatile deposits of the piedmont slopes flanking upland areas, as well as fluvial and aeolian materials of diverse origins and ages. Averaged over regional scales, erosion rates in many desert areas are low, and the intensity of leaching is low owing to aridity. On balance, many desert soils are consequently typified by *accumulations* of materials such as salts and windblown dusts entrained from playa lakes and other erodible surfaces upwind, since these materials accumulate faster than they are eroded, especially during intervals of dry and dusty climate. Local weathering products and organic matter are also components of desert soils (Marion *et al.*, 2008). As noted in the Introduction, particular surfaces within deserts do lose dusts to the global atmospheric transport system, and dust fallout and washout in rain are known from widespread and remote locations, and contribute to sedimentation over large areas of the oceans (see Chapter 18). In the desert of western New South Wales, Australia, it has been estimated that at the end of the last glacial arid phase, regional landscapes may have been blanketed by significant depths of windblown materials from the Australian interior, with present-day A-horizons more recently emplaced by slopewash processes (see Chartres, 1982, 1983). Rates of dust accession to NSW soils decrease eastwards (downwind, see Cattle, McTainsh and Elias, 2009). Tiller, Smith and Merry (1987) cite aeolian dust accumulation rates from sites in Australia of up to 32 t/km^2 a (equivalent to more than 15 mm/ka). Rates of loess accumulation from sites in China also cited by these authors are up to 70 mm/ka. Windblown materials can be incorporated into desert soils, providing allogenic minerals, salts and much more abundant clay than could be produced in situ by ordinary weathering processes. Thus, argillic horizons in desert soils are often attributable to dust incorporation. In contrast, many humid zone soils are dominated by autogenic clays weathered in situ from the parent materials, though small accessions of dust are involved in their pedogenesis.

Because of the slow overall pace of landscape change, desert soil-forming processes involve events over long time periods. Materials such as calcium and bicarbonate ions, delivered in dilute solution in rainwater, progressively accumulate in the soil. Subsoils, often alkaline because of accumulated carbonates, may slowly develop carbonate enrichment to the point where fully indurated *petrocalcic* horizons evolve (see Chapter 8). The timescales over which this occurs can be studied using several dating methods. Continuing accretion in carbonate horizons can result in their upper surface progressively approaching the ground surface, over substantial fractions of Quaternary time. These time-related developments have been studied with a view to using the depth and thickness of carbonate layers to estimate the age of desert surfaces within which such accumulations occur (e.g. Marion, 1989). The voluminous carbon storage in soil carbonates of the drylands is a significant component of the global carbon cycle (Kraimer, Monger and Steiner, 2005), and this provides yet another example of the way in which desert soils exert an influence that extends well beyond the drylands themselves.

The detailed history of most desert soils remains unknown, and is likely to be complex. However, the present-day morphology of such soils has been studied in greater detail. We will turn now to consider some of the key morphologic features found in many desert soils that are important in setting their place in the hydrologic and geomorphic functioning of the landscape. The features present often constitute a mixture of ages, some being rather young and some quite old. Desert soils are truly features that are *polygenetic*, forming a palimpsest of past events.

7.4 Stone-mantled surfaces and desert pavements

Many desert soils carry a surface veneer of stones, often coated with a desert varnish containing at least some allochthonous (foreign) components (Figure 7.1). The stone veneer often overlies a stone-free subsoil. Dregne (1976, p. 42) argued that stone mantles were usually the result of the removal of fines by wind or water, leaving the gravel as a *lag* deposit. It appears, however, that the stones can be concentrated at the surface by other means. A mechanism now widely envisaged is that windblown dusts, settling on to a desert surface, are washed down between surface stones, perhaps passing into the regolith along desiccation cracks (see Chapter 19). Weathered rock debris is thus kept exposed and continually rides to the top of the accumulating soil materials, being itself too large to be washed into soil crevices. Thus, far from signifying wind erosion, desert stone mantles may reflect quite the opposite, a local *accumulation* of significant amounts of material; the soils may thus deepen with time in a process that has been termed 'cumulic pedogensis' (McFadden *et al.*, 1998; Gustavson and Holliday, 1999; Ugolini *et al.*, 2008). Soil development through time as stone pavements and

Figure 7.1 Desert pavement capping soils in arid northern South Australia: (a) pavement of well-sorted stones with significant exposure of silts; (b) pavement of poorly sorted stones resulting in very high areal coverage and little exposed fine soil material; (c) view across arid desert pavement to residual hills in the distance. Scattered patches of grass can be seen, marking locations where the pavement is less complete, and where water infiltration is possible.

vesicular horizons develop includes progressive changes to infiltrability and surface water partitioning (Young *et al.*, 2004). The development and significance of stone mantles are considered more fully in Chapter 9, and we will not consider them further here.

7.5 Inorganic seals at the soil surface

Another very widespread feature of desert soils is the inorganic seal. Seals are formed at and near the surface of susceptible soils by several mechanisms related to the behaviour of raindrops, surface ponding of water or overland flow. In detail, there is a great variety in the characteristics of seals, related to the texture and composition of the soil materials involved, the rainfall environment, the amount of plant, litter and stone cover, the landuse and many other factors. For an example of a detailed classification system, see Valentin and Bresson (1992). Here, we will divide seals into two classes, which accommodate most forms:

1. ***Raindrop impact seals*** (often called structural seals) that result from the mechanical work done by raindrops in rearranging and packing near-surface soil materials.
2. ***Depositional seals*** that result when transported fine particles are laid down in layers on the soil surface, either in ephemeral surface ponding or simply as overland flow that is absorbed or comes to rest.

The complexity of seals can be envisaged by considering intermediate classes resulting from deposition occurring while the last stages of rain were still falling, for example, or with variations in soil texture, especially the amounts of silt and clay. Dryland surfaces often contain a mosaic of inorganic seals over quite small distances, with raindrop impact seals on the freely draining, higher parts of the microtopography and depositional seals along the threads of flow that follow the lower areas (Figure 7.2).

7.5.1 Raindrop properties and raindrop impact seals

Given the importance of raindrops in driving the breakdown of soil aggregates and the structural rearrangement of the breakdown products at the soil surface, we need to consider briefly the properties of these drops.

Small raindrops are held into nearly spherical shapes by the surface tension of water. Surface tension results in a force on the water within the drop that causes the internal pressure to be positive. The pressure is only slightly higher than atmospheric for large raindrops, but for very small droplets the internal pressure can be 50% higher than atmospheric pressure. Large droplets are deformed to a flattened shape as they fall through the air and strike the soil at a terminal velocity of about 30 km/h. The kinetic energy of the falling drop is then dissipated in various ways, including splashing if there is water ponded on the surface and the breakdown of susceptible soil aggregates. When the soil surface is wet, water is thrown out laterally from the point of impact, and the lateral jets, which can have speeds 3–10 times greater than the fall speed of the raindrops (e.g. Ghadiri and Payne, 1981), can exert considerable tearing forces on any protruding soil particles, the disruption caused being a function of factors including the mechanical strength (especially tensile strength) of the soil particle and the depth of any water ponded on the soil surface (Hartley and Alonso, 1991). Splash and surface shear are amplified by shallow surface ponding (Ferreira and Singer, 1985), but shear forces decline to low levels once the ponding reaches a depth equivalent to about three diameters of the incident raindrops (Hartley and Alonso, 1991). We will see later that algal filaments and other organic structures that permeate the uppermost soil layers can greatly increase the cohesiveness of soil materials, so that they are resistant to disruption by splashing rain.

A simple calculation suggests the enormous numbers of drop collisions that occur on exposed soil surfaces during rain. If we consider a rain event delivering 10 mm and suppose for simplicity that all of the raindrops are spheres uniformly 1.5 mm in diameter, then over each square metre during the storm there are about 5.65×10^6 drop impacts. Every point on the soil surface would be struck multiple times. In the case of a freshly tilled or cultivated soil, these impacts gradually reduce the surface roughness, flattening and rounding the surface form (e.g. Jester and Klik, 2005). In drylands, bare soil and sediment surfaces take on a subdued surface microtopography arising from raindrop impact, generally into shallow overland flows. The surface typically is not completely smooth, but rather has very shallow depressions, which act as splash traps for coarse sand and granules, and coarse organic fragments. A distinctive feature that enables these sealed surfaces to be distinguished from those supporting colonies of crust microorganisms (discussed later) is that they crack upon drying, but this leaves the surface flat, without any tendency for curling or disruption of the surface. The effect of drop impacts on the soil can be especially great in convective rain events, when rapid updrafts of heated air sweep smaller drops away from the surface, allowing only

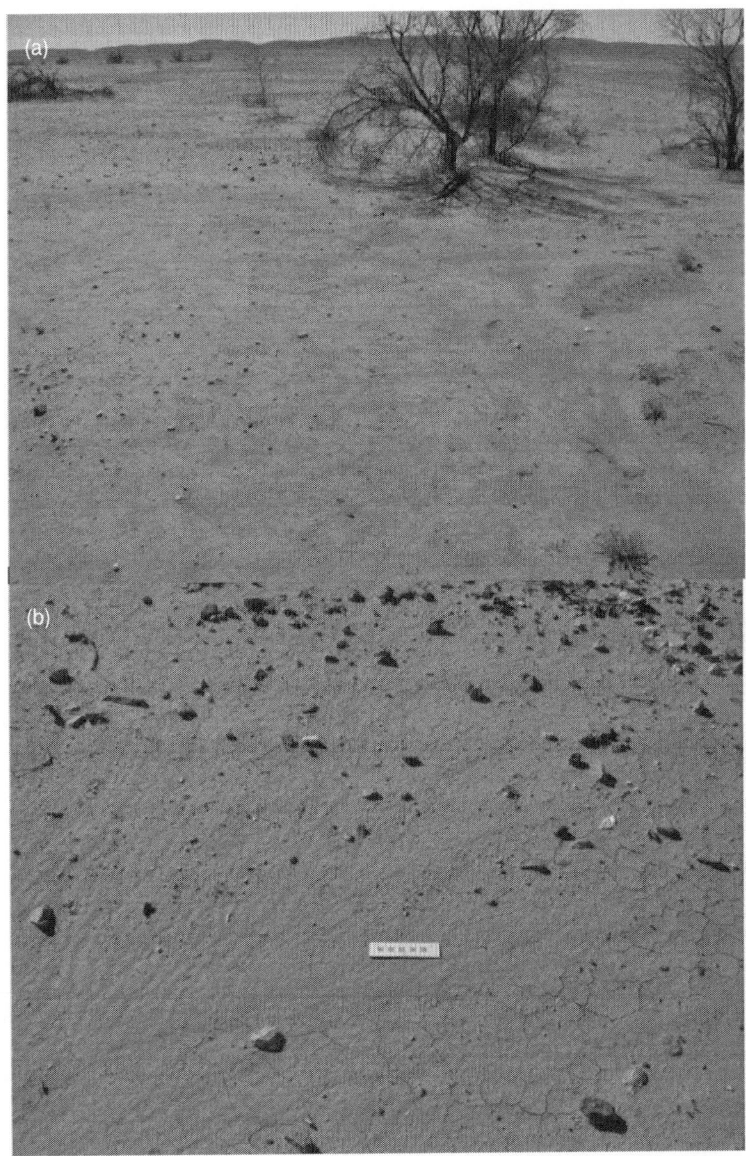

Figure 7.2 Inorganic soil surface seals, western NSW, Australia: (a) view across a low-gradient landscape where the soil surface is generally covered with inorganic seal and scattered stones; (b) close-up of part of the foreground from A, showing depositional seals along nonincised flow paths followed by overland flow.

larger and heavier drops to strike the ground. Additionally, high rain rates (intensities) can arise in convective storms, so that the bombardment of the soil can be particularly intense as a result of the combined effects of large drops and high rain rates. Raindrop impacts are a major driver of interrill soil erosion, but our concern here is with their role in soil seal development. As remarked above, the kinetic energy of falling raindrops is dissipated in various kinds of deformation of the soil surface. In the hypothetical 1 h storm, the power dissipated by the rain is about 0.14 W/m^2. However, if we take an intense event of 100 mm/h, the power being dissipated rises to about 1.4 W/m^2. These are not strikingly high values, but the drops work on soil material that is subject at the same time to other forces, such as those arising from the hydration and swelling of clays, the escape of trapped air and so on. Moreover, the soil surface is exposed to repeated rain events and some of these are likely to contain short bursts of very intense rain. The effects of intense bursts are not well known, but it is possible that even though they may last for just minutes, they are responsible for much of the breakdown of soil aggregates.

A feature commonly reported from soils subjected to prolonged raindrop impact is a layer up to a few mm thick, just below the soil surface, where porosity is reduced in comparison with the deeper layers and where correspondingly the bulk density is higher. It has been suggested that this layer results from the 'inwashing' of fine particles from the surface, the fine particles lodging in void spaces and obstructing them. Physical compaction might also be involved (Moss, 1991). This has been termed the 'washed-in layer' or the 'compacted layer'. Raindrop impact at the soil surface is considered to be the process that releases fine particles by the breakdown of soil aggregates. Experimental studies, including the analysis of density using X-ray imaging, have shown that bulk density in sealed soils typically exhibits a maximum at the soil surface, declines steeply through the underlying 3–5 mm of soil and then remains at a constant value, unless other processes intervene (e.g. Bresson, Moran and Assouline, 2004). The surface bulk density can be 0.1–0.3 g/cm^3 (roughly 25–30%) greater than that of the bulk soil below a depth of 5 mm. Again using simulated rain and X-ray density analysis, Fohrer *et al.* (1999) reported peak bulk densities of about 1.84 g/cm^3 at a depth of about 1 mm in sealed soils. The bulk soil below, unaffected by raindrop compaction effects, had a bulk density about 1.45 g/cm^3, nearly 0.4 g/cm^3 lower than the densest part of the seal. Others have reported little evidence of significant washed-in fines following experimentation with simulated rain and microscopic examination of the resulting soils. For example, Tarchitzky *et al.* (1984) suggested that early obstruction of near-surface pores as soil aggregates were broken down would of itself reduce the likelihood of further inwashing. It seems likely that soil and eroded sediment textural properties determine whether or not inwashing is important. In particular, the relationship between pore sizes among the soil matrix grains and the proportions of silt and clay in materials delivered with infiltrating water is likely to determine the probability that the clogging of near-surface pores would occur.

If the near-surface soil materials are compacted during raindrop impact seal development, then it would be anticipated that the proportion of void space would decline. This has been confirmed experimentally in tests on soils sealed under simulated rain at 65 mm/h, using mercury injection porosimetry to assess voids in the range 0.005–100 μm diameter (Vázquez *et al.*, 2008). Results showed that after 260 mm of rain had been applied, the total pore space in initially unsealed soil (315.6 cm^3/kg) declined by nearly 9% to 288.0 cm^3/kg. Perhaps more significantly, the distribution of pore sizes also changed, with a 32% decline in the space occupied by large pores (50–100 μm diameter). In this size, the total pore space declined from 18.5 cm^3/kg to 12.6 cm^3/kg. Large pores contribute to the permeability of soils and a decline in their abundance following compaction under raindrop impact explains some of the effect of soil seals in reducing soil water uptake rates, which is quantified shortly.

A second layer has also been described, lying above the washed-in layer. This is a surficial layer of silt particles, 0.1–0.5 mm thick, which McIntyre (1958a, 1958b) reported to be lacking in void spaces. He termed this the 'skin seal' and showed that a 0.1 mm seal reduced infiltration by as much as a 1 mm washed-in layer. The low permeability of the skin seal would restrict the reciprocal escape of air from the soil pore spaces that must occur if water is to enter them. Consequently, infiltrating water can result in compression of the air trapped within the pore spaces, and once the soil is damp and deformable, the air pressure deforms the pore spaces until they are roughly spherical, since this minimises the pressure on the walls of the void. The spherical pore spaces created by air pressure within the soil are termed 'vesicles'. In this way a vesicular horizon, containing abundant closely packed voids, can form beneath the surface, where the surface has low permeability to air and water. These features are considered in more detail later.

It is thus clear that raindrop impact can result in several kinds of layers close to the soil surface that may exhibit lack of void spaces, low permeability, high bulk density or spherical void spaces that are not interconnected, and which therefore do not contribute to permeability. Vesicles remain air-filled even when the soil is saturated, because the soil suction force retains water in the soil matrix. Thus, every potential infiltration pathway that contains a vesicle is excluded from the infiltration process. A square metre of the soil surface that is underlain by vesicles covering 75% of the area therefore actually only presents 0.25 m^2 through which infiltration can take place. With reference to raindrop impact seals, the particular layers that are formed will vary with the rate of surface erosion, the texture and chemical composition of the soil materials, the energy delivered by the rain and so on. Much of what is known in relation to these effects has been derived from experimentation using artificial soil mixtures and artificial rain, and cannot be taken to represent exactly the processes occurring in nature. Moreover, soil seals are of particular significance in agricultural soils of the humid zone. Following ploughing, these soils can be left bare and exposed to striking rain just as dryland soils are, and much of what has been learned about seal behaviour relates to agricultural soils, not dryland soils. For example, the wetting rate is known to influence aggregate breakdown in tilled soils (see Fan *et al.*, 2008), but this effect has not been as well explored for dryland soils

(Ben-Hur and Lado, 2008). Consequently, our understanding of dryland seal formation is not as yet complete, but we have considerable evidence of the effects of seals once they have formed. A range of views about the relative importance of compaction, inwashing, pore blockage, structural rearrangement of clay particles and other mechanisms that may be involved in seal formation was reviewed by Assouline (2004). One problematic issue is distinguishing between structures produced *during* rain from those that result *after* rain, as clays settle to the surface and the soil dries out. In laboratory experimental work, the soil trays are often allowed to dry prior to sampling the seal and it is not clear how the seal properties change through drying. The lack of correspondence of experimental conditions with those in the field needs to be borne in mind at all times. For example, Tanaka, Yokoi and Kyuma (1992) made a study of soil surface sealing using simulated rain at an intensity of nearly 270 mm/h, but it is not clear how their findings relate to field conditions in drylands, where much of the rain arrives at rates of <5 mm/h (Dunkerley, 2008a).

7.5.2 Factors known to be significant in the formation of raindrop impact seals

Field and laboratory studies have shown that seal formation and seal properties can be affected by a range of local factors. Though there is much to be resolved in terms of fine detail, the factors fall into two categories:

1. Properties of the rainfall.

2. Properties of the soil materials.

Seal formation involves the structural rearrangement of soil materials (such as aggregate breakdown, reorganisation of particles and pore spaces, inwashing of fines, plugging of pore spaces and so on), all of which represent work done by the energy of the raindrops. Therefore, the kinetic energy of the rain has been identified as a parameter likely to be of importance. However, it is difficult to know how best to describe a complex phenomenon such as rainfall, and other parameters including rain rate (intensity), total depth of rain and total kinetic energy delivered have been explored. The complex details are not necessary here, but it can be noted that for unsealed experimental soils exposed to rain, lower kinetic energy of incident rain seems to be associated with less densely compacted and more permeable seals. However, in drylands, once a seal is formed, the behaviour of subsequent rain would appear to have less relevance. As seals develop, it may be that successively higher levels of kinetic energy or rain rate in storms incrementally increase the particle packing density, perhaps over a period of years. Much would depend upon the rate at which the surface was being lowered by erosion, since in the case of such a surface, the seal would need to be renewed as the erosion progressed.

The soil materials involved in seal formation actually comprise a physicochemical system. The interaction of soil particles with water is partly governed by physical factors such as drop or interparticle collisions and partly by electrochemical processes such as clay hydration, clay dispersion, slaking and osmosis. In these latter processes, both the chemistry of the soil and of the rainwater are significant. The soil characteristics that have been considered as potentially significant include texture, organic matter content, sodicity and clay mineralogy. Again it is not fully established just how particular soil characteristics affect seal formation. Clay particles, for example, disperse in water when water molecules move between clay platelets and separate them to the extent that the Van der Waal's forces between the platelets are no longer sufficiently strong to hold the soil aggregates together. Dispersed clays are then able to be rearranged structurally or to move into the deeper soil pore spaces where pore clogging may result. Owing to their crystalline microstructures, smectitic soils (containing smectite or montmorillonite clays) tend to be readily dispersible and are susceptible to seal formation, while kaolinitic soils are much less dispersible and are less prone to sealing (Lado and Ben-Hur, 2004). Soils with abundant clays may exhibit aggregate stability that is too high to permit seal development, while soils with <10% clay may release too few particles for widespread pore clogging. Thus, soils with perhaps 20–30% clay-size particles may be the most prone to seal formation (Singer and Shainberg, 2004).

7.5.3 Depositional seals

Depositional seals are laid down where water slows and transported silts and clays settle to the soil surface forming thin, dense, bedded sediment layers. Valentin and Bresson (1992) distinguished depositional seals of two kinds, deposited either from flowing or still water, but noted that in fact most depositional seals involve components of both. The thickness of depositional seals can grow to many millimetres if there are successive episodes of deposition. Curling and cracking of these seals is commonly observed as they dry, because the uppermost layers, exposed to sun and wind, dry and shrink while the soil beneath is still moist and deformable. Often the cracking upper layers separate from the material beneath along a surface of low cohesive strength, such as a relatively coarse or sandy layer. It has been shown that the mineralogy-dependent flocculation of clays can result in depositional seals having varying densities and hydraulic conductivities (Lado, Ben-Hur and Shainberg, 2007), and given that electrochemical forces are involved, the composition of the transporting water, including its salinity, are certain to affect the properties of depositional seals.

7.5.4 Effects of seals on infiltration and erosion

An example of the effect of soil seals on infiltrability was provided by Ries and Hirt (2008) from studies of abandoned fields in the Spanish drylands. Using cylinder infiltrometer tests, they reported a final infiltrability of 4.6 mm/h for crusted silty soils, compared with 10.2 mm/h for uncrusted soils. Thus, the presence of a seal reduced the infiltrability in this area by more than 50%. For small plots under simulated rain (40 mm/h) Ries and Hirt (2008) also demonstrated runoff ratios on sealed soils of up to 80%, and averaging 61.3%, more than double the value on vegetated plots. Similarly low infiltrability values were reported from sealed silty soils at an arid site in Jordan, where the mean infiltrability was just 3.2 mm/h (Al-Qinna and Abu-Awwad, 1998). A final example can be drawn from work done in Israel (Carmi and Berliner, 2008) in which micropermeameter measurements were collected on sealed and nonsealed (tilled) soils. The sealed soils exhibited a mean infiltrability of 5.05 mm/h, while for nonsealed soils the value was about 20 mm/h. Thus, the soil seal in this study reduced the infiltrability by about 75%. Other data on infiltrability of various kinds of soil seals from west Africa were presented by Casenave and Valentin (1992). Experimental studies have confirmed that the conductivity of soil seals can be sufficiently low that the soil beneath conducts only unsaturated flow (e.g. Morin, Benyamini and Michaeli, 1981). In this situation, the hydraulic conductivity that is manifested by the soil below the crust is dramatically lower than the saturated hydraulic conductivity, since strong soil suction forces arise during unsaturated conditions, as well as surface tension effects at the air–water interface in partially filled void spaces.

It is clear from studies such as those just mentioned that seals act to throttle infiltration to much lower rates than would occur in the absence of the seal. Therefore, it follows that a larger fraction of the rain falling on a sealed surface must be partitioned into surface ponding or overland flow. The prospect of increased overland flow brings with it the risk of increased erosion and transport of soil materials, and this is a second major area of significance associated with soil surface seals. There is a great diversity in the methods and approaches used to study these phenomena, and we will consider only some examples of what has been learned.

The effects of seals on water erosion in drylands remain only partially understood. Singer and Shainberg (2004) argued that if there is abundant erodible sediment and erosion is transport limited (i.e. there is insufficient capacity in the overland flow to remove all the available material), then sealing by promoting more runoff may increase surface erosion. On the other hand, they argued, if erosion is detachment-limited (i.e. the surface is somewhat erosion resistant and provides a smaller load of particles than could be carried away by the overland flow), then seal development is only likely to increase the extent of detachment limitation, and hence reduce the erosion rate. In the field, conditions are more complex than these analyses suggest. It has to be remembered that there may be runon water arriving from upslope, perhaps with an associated sediment load, so that erosion over a sealed surface may depend upon its landscape context. Moreover, the soil surface in the field may have a variable cover of organic litter and stones, and the interaction of these materials with rain and overland flow cannot readily be isolated from those of seals on the exposed soil. Such issues were explored at a field site in New Mexico under simulated rainfall (Neave and Rayburg, 2007). The soils were mechanically broken up prior to the start of experiments, and it was found that penetration resistance increased progressively as seals developed through successive applications of rain, separated by time for soil drying. Soil surfaces protected from raindrop impact by mesh screens nevertheless showed similar behaviour, perhaps because of the occurrence of depositional seals. However, the results showed complex patterns of runoff coefficients and sediment yields as a function of other surface properties such as stone and organic litter cover. The highest sediment yields were recorded by Neave and Rayburg (2007) from partially sealed, partially stone covered plots, rather than from bare, sealed ones. This may reflect the formation of concentrated flows passing downslope between the stones and forming more efficient transport pathways for sediment. A related argument has been raised in relation to surface roughness and microrelief. Where sealed soils are flat, overland flow may occur as a relatively uniform sheetflow, with shallow depth and low flow speed. This would allow greater time for water absorption. On the other hand, a site with an undulating surface would carry overland flow in faster-moving flow threads, tracing out the lower-lying pathways across the surface. These faster flows would drain water more quickly from the soil surface, and hence tend to increase the runoff ratio and decrease the total water depth infiltrated (Carmi and Berliner, 2008). Considerations of this kind make it clear that, in real drylands, the effects of seals cannot readily be isolated and that, instead, it is more informative to seek a comprehensive understanding of the multiple controls on infiltration, runoff production and sediment transport processes. These multiple drivers of infiltration, runoff and erosion include of course the influence of biota such as termites (e.g. Mando, Stroosnijder and Brussaard, 1996). Where soils are cultivated, such as some of the Spanish drylands (e.g. Ries and Hirt, 2008) there are differences in seal properties, infiltrability and erosion rates among fields of different ages and at local scale between ridges and furrows within the fields. Thus, the runoff and erosional response of any sizeable area is the aggregate of many disparate local and microscale effects of soil surface features. As Ries and Hirt (2008) showed, enhanced runoff from sealed areas can result in gully development further downslope, so that again we can see that the consequences of seal formation cannot be established solely from core or small plot tests on the sealed surfaces themselves. Rather, their position and role in the larger landscape must be considered. Finally, of course, we can expect differing responses between small and large rain events, and between single, isolated rain events and multiple, closely spaced events. This introduces great complexity in the temporal response to seal formation, in parallel with the spatial complexity.

A formalised study of the effect of rain event arrival on soil seals (Fohrer *et al.*, 1999) compared the effects of a single event of 60 mm with the same total rain depth delivered in five separate events, each of 12 mm, separated in time. In the 'multiple events' treatment, the infiltration rate declined much more rapidly with time than during the single event. Consequently, even relatively short, low rain rate events on sealed soils could yield overland flow if they followed a prior wetting event. The effects in a particular case would depend on the extent of soil drying achieved between each event, with the possibility of surface cracking, for example, modifying the processes in later events. Carmi and Berliner (2008) reported similar findings from a study of field plots containing sealed soils at an arid site in Israel. Their results showed that the correlation between runoff coefficients and rainfall rate was higher for events closely spaced in time and weaker for widely spaced events, reflecting soil moisture levels and, in particular, suction gradients within the soil.

Studies of particular processes on sealed soil surfaces can be used to explore the means by which sealing affects soil dislodgment. For example, Slattery and Bryan (1992) studied seal development and erosion in a laboratory flume using simulated rain. They found that splash detachment increased with time during rain as initial soil aggregates were broken down by raindrop impact, thus providing a supply of splashable particles. As the surface seal developed, however, and the disaggregated particles were partly washed away and partly packed into the surface seal, the rate of soil splash declined with time to a stable, lower value. Similar findings have been reported by others (e.g. Roth and Helming, 1992).

Seals are believed to reduce wind erosion, because of the stable packing of particles into the seal, which reduces the flux of abrasive grains passing over the soil surface. This is a function of soil texture, however. On sandy soils in Niger, it was shown that despite the formation of a sieving (structural) seal, in which some fines are moving downwards among the framework grains, leaving a friable sandy surface, the eroded flux of fine sediment particles was not supply-limited. Moreover, as in the case of runon arriving from upslope, there can be sources of abrasive sand particles upwind from areas that are sealed. In such a situation, the saltating grains can progressively cause a loss of structural integrity in soil seals, with a consequent increase in the flux of eroded particles (e.g. Hupy, 2004). As with the case of raindrop impact on exposed soils quantified earlier, the number of impacts of saltating grains with the soil surface is likely to be very large indeed, on the order of $10^3/\text{cm}^2$ s in a typical cloud of moving sand (Rice, Willetts and McEwan, 1996). Only moderate wind conditions are needed for such progressive impact damage to occur (Rice, McEwan and Mullins, 1999). Most studies of wind erosion on inorganic seals are short-term experimental investigations made in tunnels with fan-driven airflow. In drylands, whether a seal survives drought conditions and sand-blasting effects presumably depends upon the upwind sand supply and upon the duration of dry and windy conditions.

Having considered something of the properties and significance of inorganic soil seals, we now turn to the second major class of dryland soil surface features, the biological soil crusts.

7.5.5 *Biological soil crusts*

Where the dryland surface is neither eroding nor aggrading too quickly, the uppermost parts of the soil commonly form the habitat for hardy varieties of nonvascular plant and other organisms, including algae, lichens (the symbiosis of an alga and a fungus) and bryophytes (mosses, liverworts and so on). These terricolous (soil-dwelling) organisms can form extensive coverings on the soil surface, commonly referred to as 'biological soil crusts' (BSCs). Communities of organisms can also colonise rock materials in drylands (Figure 7.3) (Friedmann, Lipkin and

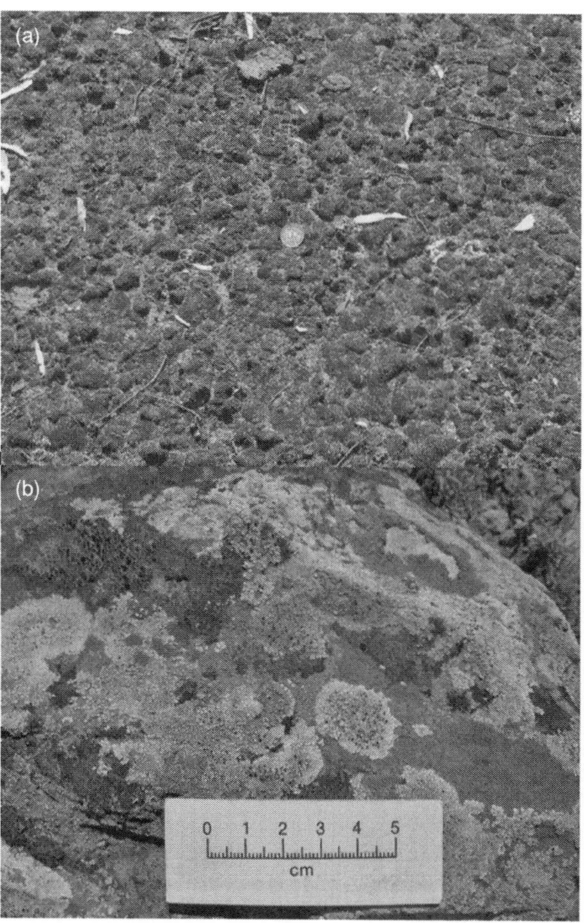

Figure 7.3 Biological crusts, arid western NSW, Australia: (a) rugose crust of lichens and bryophytes on soil, where coin used for scale is 23 mm in diameter; (b) lichen crust on calcareous bedrock outcrop.

Ocampo-Paus, 1967). Through a set of mechanisms somewhat different from those just discussed in relation to inorganic seals, BSCs affect infiltration, overland flow and erosion. We will now consider these biological crusts and their place in dryland ecohydrological and erosional processes. Biological soil crusts have considerable significance in the sustainable management of drylands that are used for pastoralism (e.g. Harper and Marble, 1988; West, 1990; Bowker *et al.*, 2008) but the impact of grazing pressure on BSCs is not covered here. In addition to the surface and near-surface crust organisms, microbial communities, including for instance heterotrophic bacteria together with bacterivorous nematodes, rotifers and so on, occur with depth in dryland soils, sometimes preferentially in the sheltered ‘resource islands’ located beneath and near the canopies of vascular plants but elsewhere distributed more uniformly over the landscape (e.g. Herman *et al.*, 1995; Berkeley, Thomas and Dougill, 2005; Goberna *et al.*, 2007; Housman *et al.*, 2007). The structure of nutrient pools, the distribution of organic matter and other effects arising from these organisms may have their own influence on soil hydraulic properties. The rhizosphere (the zone of microbial activity associated with the root systems of plants) forms another important component of dryland soils that has significance for nutrient pools and soil processes. Some discussion of the microorganisms of the rhizosphere (as well as BSCs) can be found in Bhatnagar and Bhatnagar (2005).

The identification and taxonomy of biological crust organisms is a specialised pursuit. Reference should be made to works such as Rosentreter, Bowker and Belnap (2007) on the biological soil crusts of the western USA, which contains a helpful glossary and collection of URLs for internet sources of additional information. This and other reports can be downloaded from www.soilcrust.org. For the Australian drylands, Eldridge and Tozer (1997) is a well-illustrated guidebook that includes dichotomous keys to aid the identification of lichens, mosses and liverworts. Works on the biology of key organism groups include the extensive review of soil algae by Metting (1981), the overview of crusts and their ecological roles by Evans and Johansen (1999) and the major treatment of the biology and ecology of the cyanobacteria by Whitton and Potts (2000). Belnap and Lange (2001) provided a comprehensive treatment of BSCs, with global coverage.

7.5.6 The habitats or niches exploited by microphytic plants in drylands

Organisms exploit a range of habitats in drylands, growing on or within soil or rock, and also beneath small stones (Figure 7.4). In the most hyper-arid drylands, conditions are too severe for autotrophic photosynthetic organisms to survive. Lichens and other crust organisms are largely poikilohydric (lacking the ability to control desiccation) and are thus vulnerable to weather and climate conditions, especially if hot conditions occur while the organisms are moist. For example, the central Atacama Desert in Chile lacks any photosynthetic plants or cyanobacteria, and indeed bacteria are only found with any abundance within the soils at depths greater than about 20 cm (Moser, 2008). However, beyond such intensely dry locations, microphytic plants are found to be very widespread in locations where sufficient light reaches the ground. In the hypolithic niche, algae and cyanobacteria are often found growing beneath translucent quartz pebbles, as evidenced by patches of green coloration on the undersides of stones. The intensity of solar radiation is greatly attenuated in this

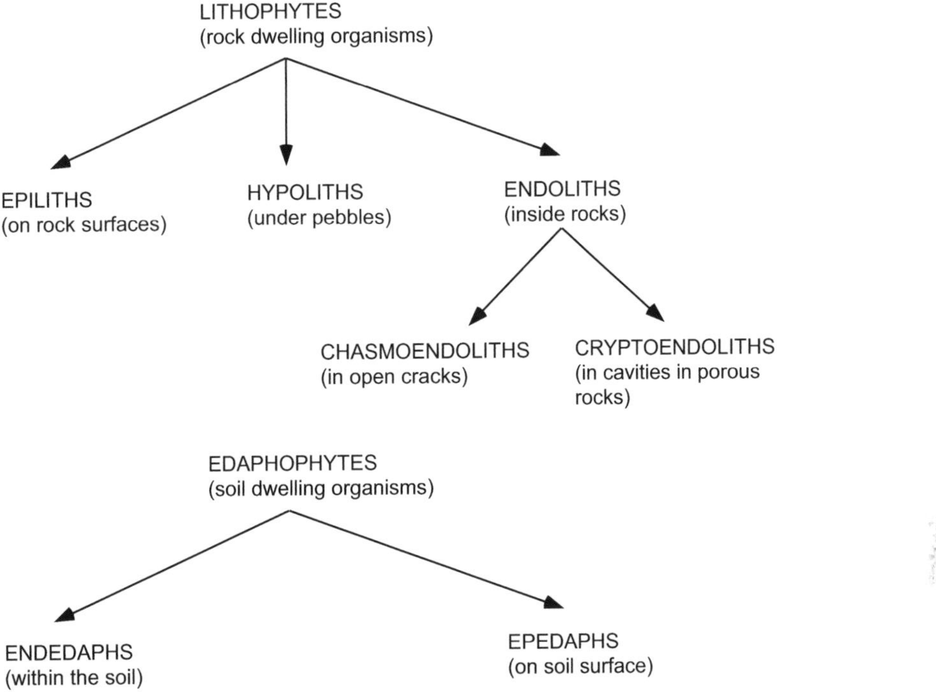

Figure 7.4 Classification of rock and soil niches inhabited by desert microorganisms.

environment, and the temperature can be up to 8 °C cooler than the surrounding soil during the day (Azua-Bustos, 2008). On soil and rock surfaces, desiccated microphytic plants can survive extreme temperatures. At several hyper-arid BSC sites in Chinese deserts, all having mean annual rainfalls <25 mm, soil temperatures range from −24 to 54 °C (Pointing *et al.*, 2007). Many crust organisms are able to source at least a part of their moisture requirements from very minor rain events, or from fog, dew or atmospheric moisture (De Vries and Watling, 2008). In the western Negev, it has been estimated that dew and fog may support 3.2–9.4% of the total annual time for which BSCs on dunes are sufficiently wet to be able to photosynthesise (Kidron, Herrnstadt and Barzilay, 2002). Owing to their capacity to exploit even minor sources of moisture, crust organisms are able to survive in harsh conditions where vascular plants cannot and BSCs are often found in the open interspaces between dryland shrubs and grasses. It is common experience that many of these organisms can begin photosynthesis within minutes of water becoming available, green coloration from chlorophyll rapidly becoming visible. Despite their ability to inhabit very harsh environments, the diversity of organisms inhabiting soil and rock environments in the drylands is still more strongly dependent on the availability of liquid water within each niche than on temperature or meteorological rainfall (Warren-Rhodes *et al.*, 2006; Pointing *et al.*, 2007). Warren-Rhodes *et al.* (2006) studied the abundance of hypolithic cyanobacteria along a rainfall gradient in the Atacama Desert in Chile. By counting a minimum of 1000 stones at each sample location, they found that along a transect where mean annual rainfall declined from 21 mm/a to about 2.5 mm/a, the fraction of quartz pebbles colonised on the sides or base declined from 27.6 to <0.1%. A major decline occurred for annual rainfalls less than 5 mm/a or where periods of around 10 years with no rain were experienced. With increasing hyper-aridity, the size of stone that was colonised also increased. It is interesting to note from the findings of Warren-Rhodes *et al.* (2006) that at the driest sample site in the Atacama the hypolithic organisms had on average only 75 ± 15 hours per year during which there was sufficient water and light for photosynthesis to proceed. The organisms therefore spend most time in a dormant state.

7.5.7 The organisms forming biological soil crusts

A great diversity of organisms can be found in BSCs, including lichens, bryophytes (mosses, liverworts), algae,

Table 7.2 The major groups of organisms found in biological soil crusts (after Evans and Johansen, 1999).

Organism group	*Species that are commonly found in BSCs*
Cyanobacteria	*Microcoleus vaginatus*
	Nostoc commune
	Schizothrix calcicola
Lichens	*Collema tenax*
	Fulgensia desertorum
	Psora decipiens
	Catapyrenium squamulosum
Mosses	*Tortula ruralis*
	Pterygoneurum ovatum
	Bryum spp.

cyanobacteria and fungi (e.g. yeasts, moulds, mycorrhizal fungi). A review of BSCs and their ecosystem roles (Evans and Johansen, 1999) concluded that a relatively small group of cosmopolitan species commonly dominate, with many other taxa being localised or quite rare. The common dominants are listed in Table 7.2, following Evans and Johansen (1999).

In the upper few mm of many soils, the mineral particles are permeated by a mesh of biological filaments of various kinds, produced by these organisms (Figures 7.5 and 7.6). These include the polysaccharide sheaths that enclose the multicellular cyanobacterial filaments, the rhizoids of mosses and fungal hyphae associated with lichens. Cyanobacterial sheaths release secretions that include glucose, mannitol, arabinose and galactose (Zhang, 2005), which bind adjacent soil particles into aggregates that include the sheath as a kind of structural element. Given that the cyanobacterial filaments extend from their sheaths when the soil is wet and grow towards the soil surface (Belnap, Prasse and Harper, 2001), the filamentous binding elements can permeate the uppermost soil, producing a crust that can with care be separated from the deeper soil without breaking. The filaments are entangled with the soil particles, and this adds to the aggregation caused by the exudates already referred to. The binding effect of the algal filaments can remain for some time even if the organisms are dead.

Some soil crust organisms can fix atmospheric nitrogen, converting atmospheric N_2 to ammonia (NH_4^+), and release much of this into the surrounding soil, together with significant amounts of carbon. Estimates of the flux of nitrogen to the soil vary widely, but values in the range 1–10 kg/ha a are common, and it has been suggested that the tolerance of heat and water stress exhibited by crusts, and their additions to the nutrient pools when rain returns, may contribute significantly to the recovery of soil fertility following drought (Aranibar *et al.*, 2003).

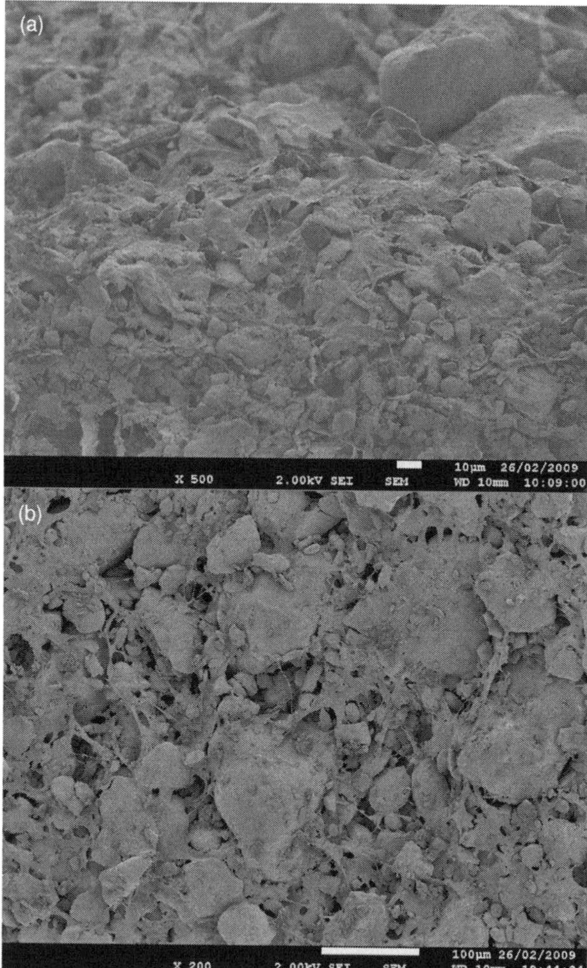

Figure 7.5 Scanning electron micrographs of soil particles stabilised by the exudates of soil algae: (a) part of a soil surface showing extensive drapes; (b) section of the uppermost mm of the soil showing particles linked by bridges and necks. Note the differing scale bars on the two images. Soils from western NSW, Australia. Images by the Monash Centre for Electron Microscopy.

7.5.8 The classification of biological soil crusts

Owing to their wide taxonomic diversity and range of structural forms, a system for classifying BSCs is very necessary. The classification of Belnap (2006) is widely cited, though it offers no distinction among different forms of very common cyanobacterial crust, for example. Moreover, the classification is entirely based on visual assessment of crust morphology. The classification is not

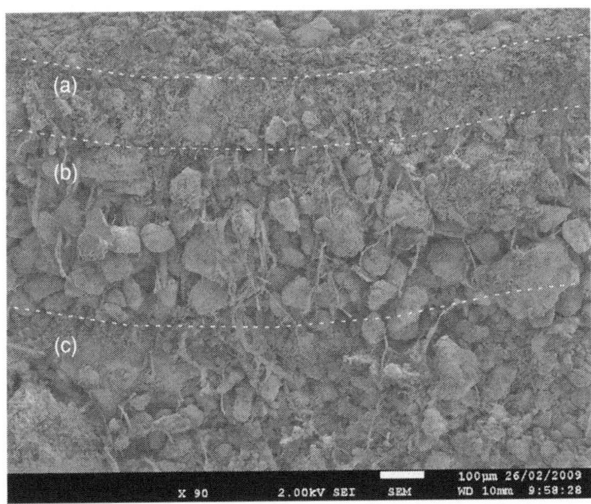

Figure 7.6 Cross-section of the uppermost part of a soil from western NSW Australia, colonised by cyanobacteria. The large particles in zone (b) are silts and are enmeshed by algal filaments. The soil surface (a) is enriched in fine silts and clays and the deepest zone (c) shows a less extensive mesh of biological filaments and poorer sorting of the mineral particles. Image by the Monash Centre for Electron Microscopy.

numerically or quantitatively based. Further, the classification is simply a morphologic one, since there is no evidence that the different classes are internally coherent and functionally significant in terms of differing infiltration or other process outcomes. The Belnap classification describes four classes of crust:

Smooth crusts (Figure 7.7). These crusts, dominated by cyanobacteria colonising the uppermost mm or so of the soil, occur where the soil is not deformed by freezing during winter. The surface tends to be quite smooth, with a low roughness drag on overland flow. However, these crusts commonly exhibit polygonal fracture systems caused by repeated shrinkage after rain. The polygonally cracked crust is typically made up of plates that are on average six-sided, which is consistent with this morphology being a space-filling one caused by shrinkage and swelling. Because the surface dries out and shrinks more rapidly than the slightly deeper materials, the polygonal plates curl at the edges to form a pedicled structure (i.e. separated from the soil except for a relatively small support that remains attached), and in this state the surface roughness is greater.

Rugose crusts (Figure 7.8). These contain more lichens and bryophytes than the smooth crusts, and typify locations with slightly better moisture availability. The surface roughness is higher than for smooth crusts and may exhibit many small depressions where water can pond. The frictional drag on overland flow would be considerably greater than for smooth crusts. Rugose lichen and bryophyte crusts are often quite leathery, and able to deform under a load without tearing or rupturing.

Pinnacled crusts. These are found in drylands where soil freezing occurs, and the deformation of the surface results in very marked surface microrelief, measured in cm. Overland flow between the prominent peaks and ridges of the crust surface may contribute to the development of the marked surface features (Belnap, 2006). Pinnacled crusts are quite fragile, and are readily damaged by crushing when exposed to pedestrian traffic or livestock grazing. Figure 7.9 shows a pinnacled crust in Utah, USA.

Rolling crusts. These are similar to pinnacled crusts, but greater crust biomass (related to better moisture conditions) limits the amount of deformation during soil freezing, yielding a surface with less dramatic microrelief.

7.5.9 *Effects of biological soil crusts on infiltration and overland flow*

There has been considerable research interest in the effects of BSCs on surface water balance via infiltration and overland flow. The crusts are often darker than the mineral soil surface where crusts are not present, and this can result in soil temperatures being up to 14 °C warmer than uncrusted sites nearby (Belnap, Prasse and Harper, 2001). This may drive greater evaporative losses from the soil. Additionally, some crusts are mildly hydrophobic (Yair, 2001) and in addition it is suspected that the additional extracellular materials added to the void spaces between mineral particles may reduce permeability and force more rain to be partitioned into surface ponding and overland flow. This effect may be exacerbated by the swelling of the crust organisms when wet. Cyanobacterial sheaths can very rapidly absorb water equivalent to 12 times their dry weight and expand in volume by 10 times (Belnap and Gardner, 1993; Yair, 2001). Crusts comprised principally of cyanobacteria often form quite smooth surfaces, and this would facilitate overland flow.

In contrast, crusts dominated by lichens and bryophytes are often quite rough (rugose), containing ridges and depressions of cm amplitude and many voids where water can be held (Figure 7.10). These would certainly increase the capacity of the surface to pond and detain water and reduce the speed of overland flow once the roughness elements were overtopped in large rain events.

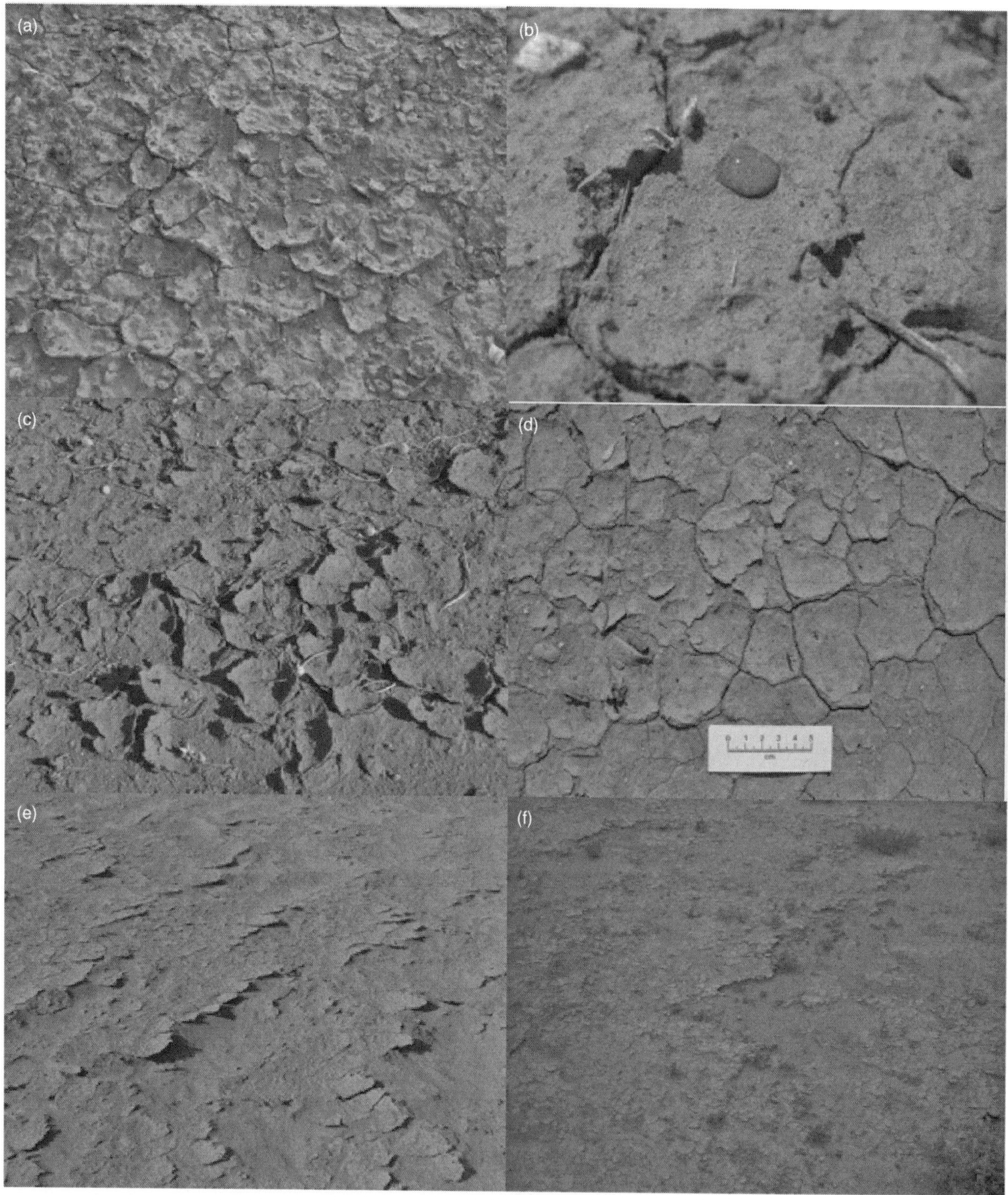

Figure 7.7 Examples of smooth, cyanobacterial biological soil crusts from western NSW, Australia: (a) characteristic pedicled (concavely upward curled) cyanobacterial plates; (b) closeup of a water drop on one of the plates in A showing a tendency to water repellency; (c) cyanobacterial plates in drought, showing a tendency to curl and separate from the underlying soil; (d) soil surface with less strongly developed BSC of cyanobacteria and a greater proportion of inorganic seals; (e), (f) scarplets in cyanobacterial crusts arising from wind abrasion during drought, with underlying silts exposed to deflation.

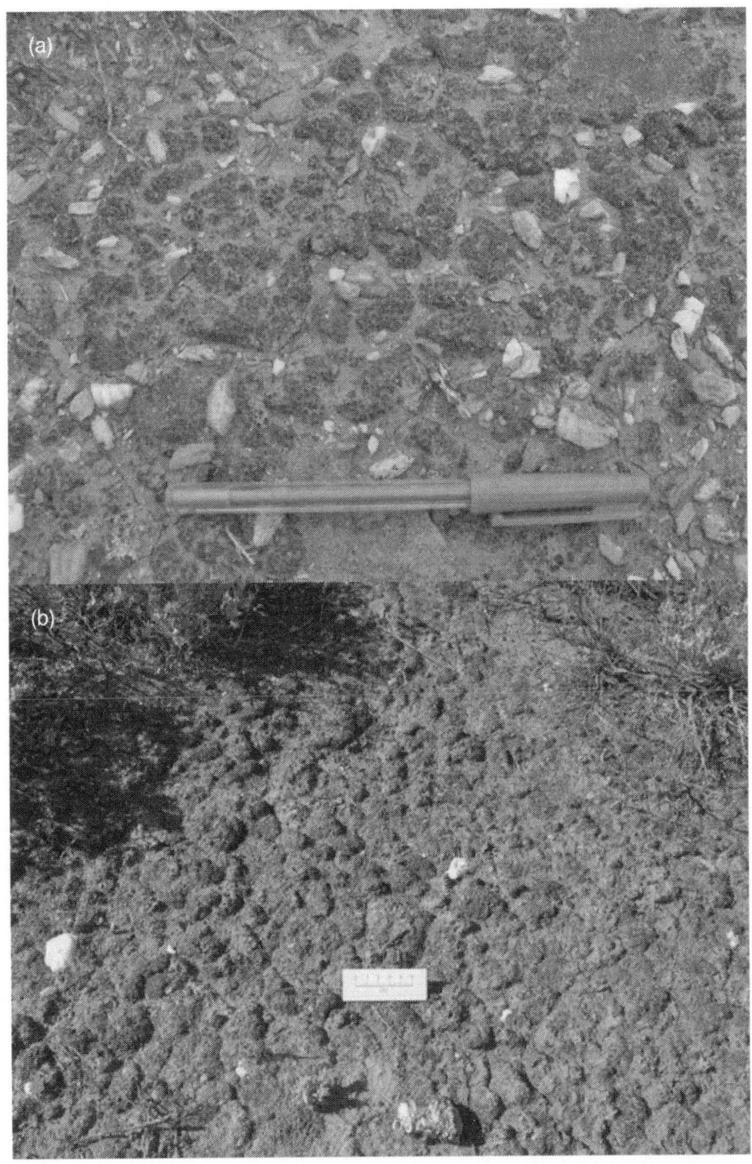

Figure 7.8 Rugose crusts, western NSW, Australia: (a) surface comprised of BSC, exposed silts and quartz pebbles, in dry condition; (b) surface largely covered by bryophyte crust. Note the marked surface roughness and considerable depression storage capacity.

The role of BSCs is further complicated because the soil fertility benefits that they confer may foster more abundant soil invertebrates able to contribute to soil bioporosity. Unsurprisingly, given the variation in soil properties and BSC types, the literature reveals no clear consensus about the effects of crusts on surface water balance or infiltration (Eldridge and Greene, 1994). It is possible that the effects of crusts are indeed different in rain events of varying size. Mild water repellency of some crusts may reduce the soil uptake of water from small events, but the invigoration of crust organisms and consequent swelling, together with roughening of the surface, may promote water retention in larger events. The swelling of polysaccharide sheaths during wetting evidently blocks pore spaces within the crust and may contribute to the partitioning of water into surface ponding or overland flow (Kidron, Yaalon and Vonshak, 1999). The surface roughness, lower flow speeds and greater surface detention and depression storage capacity all seem likely to contribute to this. Moreover, it is important to recall the patchy nature of many dryland surfaces, referred to at the start of this chapter. Even if overland flow is promoted in some rain events

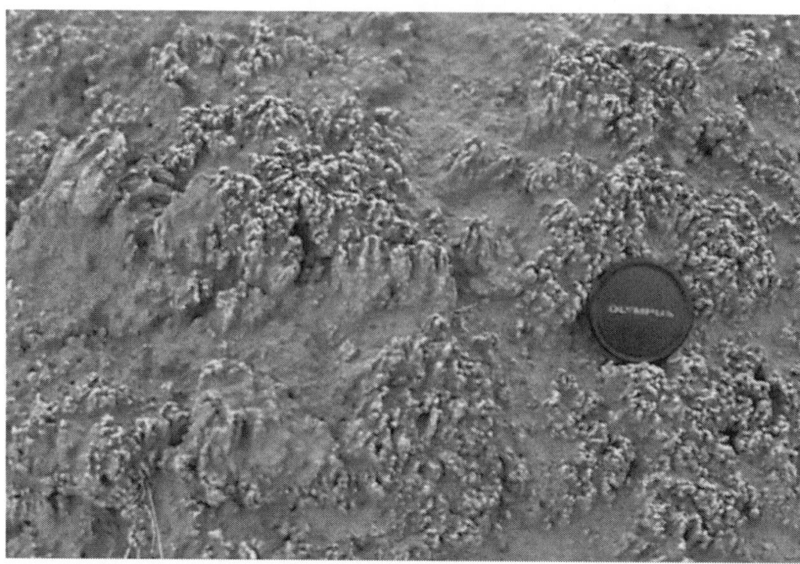

Figure 7.9 Pinnacled crust, Utah USA. The lens cap is 49 mm in diameter. Soil surfaces such as the one illustrated have very marked relief, but are quite fragile and become readily degraded under pressure of people, vehicles or grazing stock.

on some crusts, the runoff water may well pass downslope to other locations where infitrability is higher, and there be absorbed as ecologically important runon water. In other words, the importance and role of BSCs cannot necessarily be evaluated simply by local analyses of water balance on the crusted patches themselves. Thus the role of BSCs (just as for surface seals) cannot be evaluated without undertanding their landscape context and the co-variation of other soil properties with landscape position. For example, in mosaic vegetation communities, the distribution of BSCs is linked to the distribution of vascular plants, and other features relevant to infiltrability, such as surface stone and litter cover, may also be linked spatially. The work of Issa *et al.* (1999) on patterned vegetation in Niger provides an example. They showed that there was systematic variation in the nature and extent of crusting across the repeating pattern of contour-aligned vegetation patches and intergroves, with maximal crust development immediately upslope of grove margins. At such locations, water ponds briefly during and after rain and sediments are laid down. They typically exhibit multiple surface features, including cyanobacterial colonisation of depositional, inorganic seals. More work needs to be done on the overall, hillslope-scale role of these crusts. Issa *et al.* (1999) maintained that the zone of maximal BSC development just above the grove boundary constituted a relatively moist environment and would foster progressive upslope colonisation by vascular plants. However, the infiltrability of the soils in such locations is typically at a minimum within a patterned plant community, so that overland flow is partitioned into the adjoining grove. Thus, in fact, conditions appear to mitigate against the colonisation of these sites by plants. The important point here is that the crust is of pivotal importance within the patterned plant community, both for the offsite impacts of the overland flow that is generated, as well as for the local effects on soil moisture immediately below the crust. Other field studies reinforce the need for context always to be considered when analysing the nature and role of crusts. Ram and Yair (2007) examined crusts along a rainfall gradient from 86 mm to 160 mm annual rainfall at a site near the Egypt–Israel border. They found that the amount of soil moisture available to support vascular plants was actually less at the higher rainfall sites than at the more arid ones, owing to the abstraction of water on to thicker biological crusts there. At the drier sites, there was less abstraction and overland runoff could also arise and locally be concentrated downslope to create relatively moist microsites of advantage to vascular plants. These results are consistent with those of Grishkan, Zaady and Nevo (2006), who also studied soil crust organisms along a rainfall gradient in Israel. They found that there was only a weak influence of rainfall on the spatial variation in biological crust properties, which instead were largely governed by microenvironmental characteristics such as levels of soil moisture and organic matter.

Studies of the behaviour and environments of BSCs on north- and south-facing dune flanks in the Hallamish dune field, Israel, further illustrate the effects of microclimate and crust context. North-facing study plots carrying

Figure 7.10 Closer views of bryophyte crusts, dominated by *Riccia limbata*, western NSW, Australia. Note the extensive coverage of the soil surface in both (a) and (b), providing protection from striking raindrops: (a) in active state following rain; (b) appearance following a long rainless period.

a moss-dominated crust yielded on average 3.2 times more runoff than south-facing plots dominated by cyanobacteria (Kidron, Barzilay and Sachs, 2000). Despite this, the crust biomass was greater on the north-facing slopes, the explanation apparently being that the cooler aspect allowed soil wetness to persist for much longer following rain, so permitting greater time for crust growth. Significant effects of slope aspect were also found in a study of lichen crusts in the Tabernas Desert in Spain, where morning sun on east-facing slopes shortened the time for which dew was present and able to support photosynthesis (Del Prado and Sancho, 2007).

Experimental studies have often involved the 'scalping' or careful removal of crusts, with infiltration tests then made on the exposed substrate (e.g. Graetz and Tongway, 1986; Williams, Dobrowolski and West, 1995; Eldridge, Zaady and Shachak, 2000). Results have suggested that BSCs may result in considerable reductions in infiltration during ponding, but it is not clear whether such results primarily reflect the effects of the crust removal and soil disturbance and not of the crust itself. Presumably, in the absence of BSCs, soil surfaces might develop inorganic seals, and then the infiltrability would again be modified.

7.5.10 Effects of biological crusts on soil stability and erosion resistance

The situation is somewhat better understood in relation to local effects of crusts on erosional processes. In essence, many crusts provide structures that link and bind soil particles into a more coherent structural framework able to offer greater resistance to dislodgment by raindrop splash, overland flow and wind. Indeed, the surface roughness of mosses is known to promote the trapping of aeolian dust, increasing accessions of exogenous materials to some dryland soils (Danin and Ganor, 1991). As with soils generally, the increased contribution of organic matter is likely also to produce a better aggregated and more stable soil structure. For example, laboratory splash cup experiments with soils inoculated with cyanobacteria or autoclaved to prevent their growth suggested that splash losses were reduced by >90% on both fine and coarse textured substrates by cyanobacterial colonies (Hill, Nagarkar and Jayawardena, 2002). This result cannot be extrapolated directly to field conditions, of course, owing to the lack of strict correspondence of conditions (lack of overland flow and so on). In a multiyear study in semi-arid badlands in southeastern Spain, it was found that bare sites eroded at moderate rates but that slopes with lichen crusts were essentially stable (Lázaro *et al.*, 2008). This does not establish that the lichens directly reduce erosion, since these organisms only become established on stable sites (Eldridge and Greene, 1994), so that the pattern of crust types might be affected by spatially varying erosion rates, as well as partly contributing to that pattern.

Studies have often employed simulated rain of a single intensity in order to explore the effects of BSCs on erosion. For example, Kinnell, Chartres and Watson (1990) exposed soil monoliths to overland flow and the impact of simulated rain at 64 mm/h. They found that sediment concentrations on uncrusted (but sealed) soils were 3–4 times higher, at 2–4 g/L, than from samples having BSCs (approximately 0.2–0.8 g/L). Results on splash loss of soil monoliths under simulated rain at 45 mm/h showed that soil loss declined exponentially as the fraction of the soil surface covered by BSC increased (Eldridge and Greene, 1994). In tests lasting 20 minutes, soil loss fell from >300 g/m^2 with no BSC cover, to <50 g/m^2 at 75% crust cover and finally to <20 g/m^2 with 100% cover. The explanations for this behaviour in terms of the processes active on the surface were considered to involve the effects of the crusts on surface roughness and on aggregate stability. Additional surface roughness may have resulted in ephemeral ponding of water at some locations that was sufficiently deep for the mechanical effects of raindrop impact on the surface to be cushioned. The structural binding effect of the crust organisms is also likely to diminish the breakdown of soil aggregates into smaller components that can be more readily moved by airsplash.

BSCs also modify the nature and rate of wind erosion processes on dryland soils (Williams *et al.*, 1995). Using a portable wind tunnel at field sites in Utah, Belnap and Gillette (1997) showed that intact moss and lichen crusts, with considerable surface roughness, were more than 500 times as resistant to wind erosion than was bare sand and more than 60 times the resistance of flat, cyanobacterial crusts. Additional data of this kind were presented by Belnap and Gillette (1998), who showed that stones lying on the soil and inorganic seals also conferred wind erosion resistance. In both of these studies, the enhanced resistance to wind erosion arising from biological crusts was shown to be diminished quite readily by surface trampling or disturbance. At dryland sites in southeastern Australia, Eldridge and Leys (2003) used a portable wind tunnel to explore the wind resistance of crusted soils that had been mechanically disturbed by hand (using raking and rotary cultivation). They showed that as the cover fraction of BSC diminished, the flux of eroded soil increased logarithmically. Very similar trends were shown for both loamy and sandy soil materials, though the fluxes of wind-eroded soil were greater for the sandy soils (Figure 7.11).

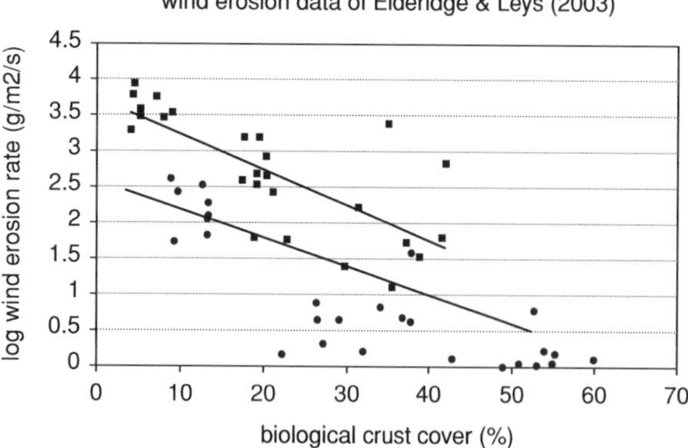

Figure 7.11 Plots of data from Eldridge and Leys (2003) showing the way in which the wind erosion rate of loamy and sandy soils declines with increasing stabilisation of the soil surface by biological soil crusts. The soils had been artificially disturbed by raking prior to the experiments, in order to simulate the impacts of grazing stock.

The various experiments with BSCs and wind erosion cannot fully represent natural conditions owing to their very small spatial and temporal scales. The very short-term wind erosion experiments need to be supplemented by longer-term studies of the cover of biological crusts in drylands exposed to the range of stresses from people, grazing stock, wind and occasional rain events.

Damage to biological crusts by vortex wind systems (willy willies or dust devils), which are common over many drylands in summer (e.g. Oke, Dunkerley and Tapper, 2007; Oke, Tapper and Dunkerley, 2007), have not been explored. Often, wind systems of this kind carry sand particles that can be very abrasive. In a similar way, winds arriving with a considerable upslope fetch (not represented in small wind tunnel trials) can be loaded with erosive sand grains. In the face of such conditions, BSCs, especially the thinner cyanobacterial crusts, which may be curled at the edges and so allow the entry of wind and eroding particles, can readily be fragmented and carried away (Figure 7.7). In collisions with saltating sand grains, the mechanical strength characteristics of the crusts are especially significant. Measurements of relevant properties, such as compressive and tensile strength, have been made in the laboratory. For example, Wang, Zhou and Zheng (2006) measured the tensile strength of a crust from Shapotou in China, and reported rupture strengths of 30–42.8 kPa. These values are comparable to those reported for natural aggregates from soils of southern Australia (Dexter and Chan, 1991), suggesting that the additional strength arising from the microorganisms in the crusted sand can be similar to that arising from the cohesion generated in soils by silts and clays. The elastic behaviour that some BSCs exhibit allows surface deformation to reduce the impact stress caused by saltating grains, but nevertheless sustained impacts progressively cause a loss of structural integrity (McKenna Neuman, Maxwell and Boulton, 1996; McKenna Neuman and Maxwell, 2002). In the drylands of western NSW, Australia, the rainfall climate is strongly modulated by multiyear quasi-cycles of above and below average rainfall, related to the ENSO phenomenon. In long and severe droughts connected with this climatic system, abrasion of cyanobacterial crusts has been observed during several extended droughts, becoming progressively more complete through the course of several months. Clearly aligned, wind abrasion and scour features result on exposed surfaces and wind scour lifts away the crust, which fragments further during transportation across the soil surface. As in the case of wind abrasion of inorganic seals discussed earlier, critical factors in the wind erosion of BSCs include the location and availability of upwind sources of entrainable sand particles and the strength and persistence of sand-transporting winds, and not merely the mechanical strength properties of the crusts themselves.

7.5.11 Possible effects of climate change on biological soil crusts

Changes in temperature, atmospheric gas composition and rainfall events and annual rain amount all have the potential to affect BSCs. Impacts that reduce the cover or integrity of these crusts have the potential to reduce the soil surface stabilisation achieved in the drylands. Belnap, Phillips and Smith (2007) reported that the concentrations of the pigments that protect Mojave Desert cyanobacteria from damage by the intense ultraviolet radiation in sunlight were correlated with rainfall amounts. They reasoned that any decline in rainfall in coming decades has the potential to diminish BSC cover owing to increasing stress on the organisms, and perhaps mortality. It is then possible to envisage effects of increasing dust fluxes on the various aspects of the global climate mentioned at the start of this chapter.

7.6 Vesicular soil structures

It has been noted earlier in this chapter that the effects of seals and crusts on infiltration, overland flow and erosion can only be evaluated in isolation in a somewhat artificial manner, since in the field multiple parameters affect these processes. Many dryland locations exhibit compound surfaces, where for instance a biological crust has developed on a stabilised inorganic seal. Elsewhere, a patchwork of seals and crusts may exist within quite small areas. Finally, the effects of vascular plants may extend into the interspaces where seals and crusts occur, modifying the properties of the soil materials through additional provision of organic matter, the effects of root systems extending beyond the limits of the plant canopy and so on (e.g. Dunkerley, 2000). The role of surface stone mantles in modifying infiltration and overland flow, as well as wind erosion, have also been widely reported. Often vesicular horizons are associated with surface stone cover, or with seals and crusts, such as the 'coarse pavement crust' of Valentin and Bresson (1992) (see Table 7.2). The vesicles form when the reciprocal escape of air is hindered during the entry of water into soils during rain or runon events. Positive air pressure within the pore spaces of damp and deformable soils results in pores of up to a few mm diameter, which are roughly spherical (Figure 7.12). There vesicles remain air filled, even when the soil is saturated, and reduce the volume of soil matrix available to conduct infiltrating

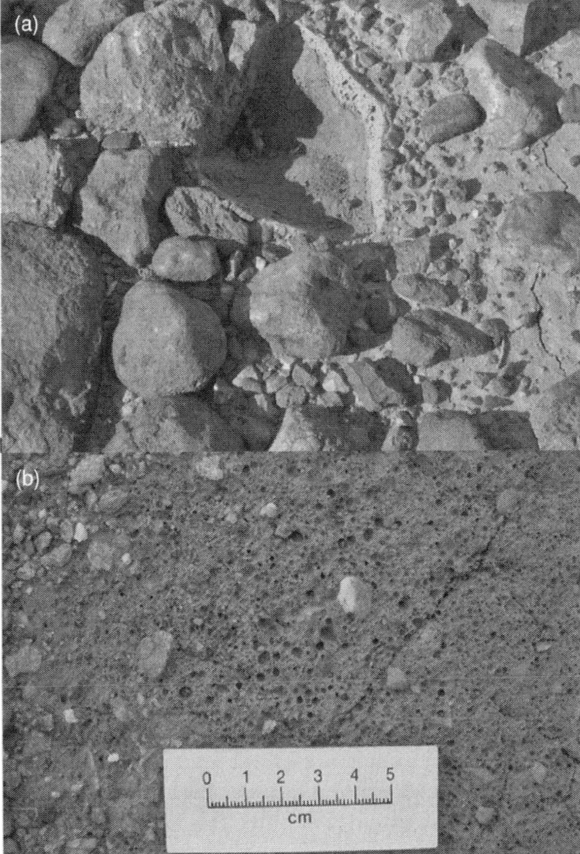

Figure 7.12 Vesicular layers in desert soils: (a) northern South Australia, where a large stone has been removed from this stony tableland soil to reveal the peripheral vesicles developed around the margins of the stone; (b) field of vesicles lying beneath a fine gravel pavement, western NSW. The veneer of stones has been carefully scraped away in order to reveal the vesicular horizon beneath. Some desiccation cracks traverse the image.

water downwards through the soil. Given that vesicular layers signal limited air escape, they also indicate quite low soil infiltrabilities. Given that dense cyanobacterial crusts can cause vesicle formation in the underlying soil, the recognition of the hybrid entity (BSC plus associated vesicles) makes more functional sense than a treatment of the two features separately.

It is clear that the presence of vesicular horizons immediately below the sealed or crusted soil surface or under pavement stones provides a throttle on infiltration. However, as mentioned previously, the hydrological effects of a surface feature such as this, or of seals or crusts, cannot be fully appreciated solely from local, onsite effects. The partitioning of rainfall among infiltration, ponding or overland flow of course depends upon the incident rain

Table 7.3 Infiltration rates measured in unsealed, sealed, crusted and vesicular soils, Fowlers Gap, western NSW, Australia (unpublished data of the author).

Soil surface type	*Mean infiltration rate (mm/h)*	*Standard deviation of infiltration test results (mm/h)*	*Number of tests*
Unsealed and uncrusted soil	30.4	6.8	3
Biological crust in grazing exclosure	11.9	4.3	5
Inorganic seal in lower intergrove, banded shrubland	6.2	1.2	5
Vesicular horizon beneath fine stone mantle	0.8	n/a	1

rate and the rate of delivery of runon water from upslope. Two implications arise from this. The first is that the effect of a vesicular layer will almost certainly vary between stratiform rainfall, with relatively low rain rates and convective rainfall, with higher rain rates and rain kinetic energy. Therefore, the overall effect will be dependent upon the kind of rainfall climate experienced at any particular site being considered. It is also probable that in many areas this behaviour will result in the effects of a vesicular layer varying seasonally. For example, for the Mojave Desert, Hamerlynck, McAuliffe and Smith (2000) suggested that vesicular horizons would have their greatest effect in summer, by throttling the infiltration from high rain rate convective storms, and be less significant in more prolonged, lower intensity winter rains arising, for example, from slow-moving frontal systems. The second implication is that the effect of a vesicular layer on the hydrologic response will probably depend upon its location within the landscape. Upslope locations may receive smaller volumes of runon water and downslope locations a larger amount. Thus, the aggregate rate of delivery of water to the soil surface will vary with topographic position and the fraction that can be taken in through soils carrying a vesicular layer will vary similarly.

7.6.1 Comparing the infiltrability of biological, raindrop impact and vesicular surfaces

We have seen that there is uncertainty attached to the effects of various soil surface characteristics on infiltrability. Part of this undoubtedly relates to actual variability in the field and between different kinds of surfaces (stable or eroding, runoff or runon locations, surfaces having sandy or finer-textured soils and so on) in the various drylands. It may be useful to compare the three kinds of features discussed in this chapter (inorganic seals, biological crusts and vesicular horizons) within the one small dryland area. To do this, Table 7.3 presents results from the Fowlers Gap Arid Zone Research Station in the drylands of far western New South Wales, Australia. There are extensive soil materials in this area that exhibit quite comparable bulk properties, because the material is derived from a blanket of Late Quaternary aeolian dusts set down over this landscape. The dust materials were sourced from major endorheic drainage systems of the continental interior. In order to assess semi-natural surface properties, the infiltrability of biological crusts (including lichen crusts and bryophyte crusts) were measured by cylinder infiltrometry within a grazing exclosure protected from hard-hoofed animals for more than 30 years. Using identical apparatus, infiltrability was determined from intergrove soils carrying raindrop impact seals, from soils under a pavement of fine stones with an extensive vesicular horizon and from soils not affected by seals, crusts or vesicularity from grove locations where soil shrink–swell phenomena mix the soil and maintain a low bulk density and a friable texture.

The unsealed and uncrusted soils exhibited a mean infiltrability of 30.4 mm/h. The soils containing a vesicular horizon exhibited the lowest infiltrability (approx. 0.8 mm/h), or only 2.6% of the unsealed rate. Biological crust sites exhibited a mean of 11.9 mm/h, or nearly 40% of the unsealed rate. Finally, sealed intergrove soils showed a lower mean infiltrability of 6.2 mm/h, or about 20% of the unsealed rate. We can therefore put these soils into their sequence of increasing infiltrability: vesicular soils < sealed soils < biological crusts < unsealed and uncrusted soils. Finally, we can compare these rates with the rain rates recorded at this location. The median rate for rain events defined using a 6 hour minimum interevent time (Dunkerley, 2008b) is 2.0 mm/h. The maximum event rain rate from a 6 year rainfall record was 49.7 mm/h, but only 5.3% of events have a rain rate exceeding 10.0 mm/h. Therefore, vesicular soils can be expected to generate surface ponding and overland flow in most rain events, while

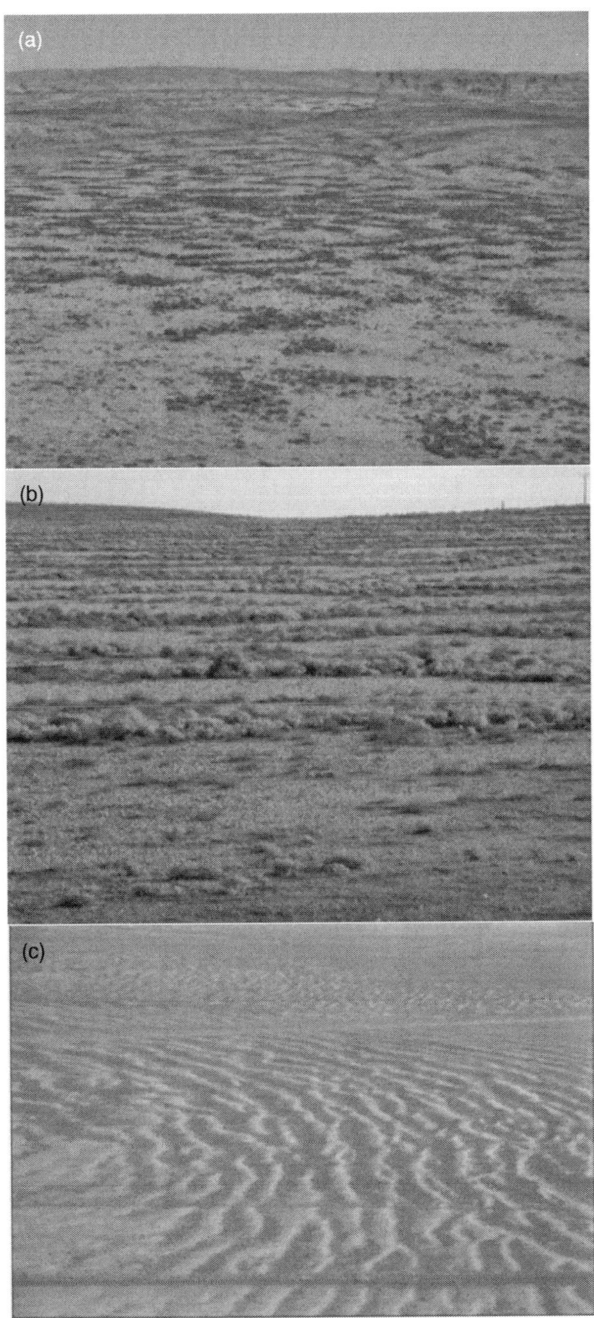

Figure 7.13 Vegetation and soil mosaics from arid western NSW, Australia: (a) an irregular, patchy mosaic of chenopod shrubs on undulating terrain in the northern Barrier Ranges, where intergrove soils in this landscape are heavily mantled with stones, but the groves carry very few; (b) a strikingly regular mosaic in chenopod shrubs, where, in this instance, the intergrove soils are very impermeable owing to a surface covering of densely packed quartz pebbles, while the grove soils are bare, friable and much more permeable owing to biological activity; (c) aerial view of a contour-aligned mosaic in mixed grassland and shrubland, on flat terrain, showing a road cross in the upper part of the image. The darker bands are friable and permeable soils densely vegetated with tussock grasses. The pale intergroves carry biological soil crusts, veneers of surface stones, and are underlain by vesicular horizons; thus they are very impermeable and so dense, hard and dry that growth of vascular plants is totally absent.

the unsealed and uncrusted soils would do so only very rarely. On the basis of the cylinder infiltrometer tests, sealed and biologically crusted soils would be expected to generate surface ponding and overland flow quite rarely in this environment, perhaps a few times per year. However, it remains possible that the conductivity of seals tested when dry is higher than would be measured during active raindrop impact. Moreover, runon water from impermeable patches of soil (e.g. with vesicular horizons or significant stone cover) would add to the flux of incident water at some sites, so allowing runoff to become more spatially integrated than might be suggested by point-based infiltrability data. Infiltrability procedures employing simultaneous simulated rainfall and runon flow have been developed (e.g. see Lei *et al.*, 2006) and these may be helpful on tilled soils. However, in drylands, the surface seal is generally fully developed, so that ponded tests should yield results reasonably indicative of seal properties. A consideration raised by these data is that it may be fruitful to consider the relationship between soil surface infiltrabilities, under various seals and crusts, and the relevant local rainfall event properties. This may prove more informative than attempts to resolve the question of whether particular seal or crust types increase or decrease water uptake. Such a comparison emphasises the fact that the presence of a seal or crust may have little effect on water uptake in some events but a significant effect in others. Thus, it is not solely the properties of the seal or crust that are of significance, but how those seals or crusts behave during real rainfall events of varying magnitude within the dryland concerned.

7.6.2 Spatial heterogeneity of desert soils

A very distinctive characteristic of many desert soils is that their properties, at least those of their near-surface horizons, vary in dramatic ways over quite small distances (Figure 7.13). The kinds of changes involved mean that, while taxonomically the soils might be similar, in terms of their function in the landscape they most certainly are not. In many areas, it has been found that key nutrients (notably nitrogen) are concentrated beneath and around woody shrubs, to form 'resource islands' (Gutiérrez *et al.*, 1993; Cross and Schlesinger, 1999; Ewing *et al.*, 2007). Very often the spatial variability of desert soils can be regarded as involving repeated patterns of two distinct regolith forms, forming what is termed a *binary mosaic* or *two-phase mosaic*. One component of the mosaic consists of soils supporting relatively dense vascular plant cover (grasses, shrubs, small trees) while the other is relatively devoid of plants. For example, Tongway and Ludwig (1990) have described binary mosaics in semi-arid mulga woodland (dominant tree species *Acacia aneura*) where groves of trees alternate downslope with treeless intergroves. In association with this pattern are changes in the abundance of soil fauna, notably termites, which preferentially inhabit the wooded subareas (e.g. see Whitford, Ludwig and Noble, 1992). Thus, the mosaic actually involves a spatial patterning in biological activity, nutrient cycling, organic matter decomposition and related aspects of soil development.

Spatial mosaics of soil like this have been reported widely from arid and semi-arid areas (Dunkerley and Brown, 1995). Their functioning is important to the continued existence of the plant communities that grow in these spatial mosaics. The key to this lies in the hydrologic response of the soil surfaces. The intergroves have bare, sealed surfaces often underlain by vesicular horizons. They are quite impermeable and absorb very little of the rain that falls on them. Rather, the rainwater is efficiently shed as runoff, and this trickles downslope to infiltrate in the more protected, more organically rich and more porous soil of the vegetated phase of the mosaic. Little or no water passes through a typical grove on to the next intergrove downslope. Thus, with a 50/50 mosaic pattern, the groves receive water equivalent to almost twice the climatological rainfall, while the intergroves are about twice as arid as the rainfall would suggest. It has been argued recently that binary mosaics may be an ancient feature of the desert landscape, developing from changes triggered by the transition from last-glacial to Holocene climatic environments (Dunkerley and Brown, 1995). For the genetic taxonomy of soils, this behaviour is problematic, because it results in soils having quite different leaching characteristics, salinity and moisture contents within metres of each other, and developed under a uniform external climatic environment. For example, Montaña *et al.* (1988), in a study of patchy vegetation in the Mexican Chihuahuan Desert, found mean organic matter levels of 2.6% within vegetation patches but only 0.8% in the intervening bare spaces. These differences occurred over spatial scales of 10–100 m. In western Australia, Mabbutt and Fanning (1987) reported that, beneath mulga bands 10–20 m in width, a siliceous hardpan was typically located more deeply within the soil beneath groves than beneath intergroves. In fact, they described the depression of the hardpan as forming a 'trench' beneath the groves, which might act as a trap for water infiltrating there. Apart from this, the main differences in the grove and intergrove soils were restricted to the upper few cm.

The hydrologic efficiency of the runoff/runon process in two-phase mosaics was demonstrated by Tongway and Ludwig (1990) using artificial rainfall. In their mulga intergroves, runoff began after only 7 minutes of rain at 29 mm/h, equivalent to 3.4 mm rain depth. In contrast, mulga grove soils showed no runoff from this rain application. Some of the other clear differences between the mulga grove soils and those of the intergrove runoff areas are summarised in Table 7.4. It is evident that the contrasts are especially marked in the top few cm of the soil.

Other forms of spatial heterogeneity have been reported from desert sites. For example, Rostagno, del Valle and

Table 7.4 Summary of grove and intergrove soil properties, semi-arid mulga woodland, NSW, Australia (after Tongway and Ludwig, 1990).

	Electrical conductivity (μS/cm)		*Exchangeable calcium (meq/100 g)*		*Cation exchange capacity (meq/100 g)*		*Organic carbon (%)*	
Depth (cm)	*Intergrove*	*Grove*	*Intergrove*	*Grove*	*Intergrove*	*Grove*	*Intergrove*	*Grove*
0–1	22.0	57.9	2.69	4.78	7.38	10.50	0.71	1.97
1–3	16.5	36.3	2.46	3.99	6.63	9.37	0.43	0.95
3–5	22.1	29.4	3.68	5.05	8.40	9.21	0.40	0.71
25–50	44.2	17.9	5.05	3.56	9.21	7.87	0.23	0.30

Videla (1991) described a Patagonian landscape of shrub mounds and mound interspaces. Beneath shrub mounds, total nitrogen, organic carbon, electrical conductivity, exchangeable sodium percentage, exchangeable cation levels and infiltration rates were significantly higher than in interspace soils. Rostagno *et al.* inferred that erosion dominates in interspace areas while accumulation promotes soil evolution within the shrub mounds. Similar mounded topography, with mounds 3–5 m long and 1–3 m wide, and a relief of about 6.4 cm, was reported from arid acacia shrubland in Western Australia by Mott and McComb (1974). Once again the mound soils had higher organic carbon and total nitrogen levels. Moisture characteristics for the two soils were very similar, but a hardpan was located typically 10 cm deeper beneath mounds, so that soils there were overall about 16 cm deeper than those of the mound interspace. The greater soil depth thus provided a greater reservoir of soil moisture to support the mound vegetation.

7.7 Conclusions

We still have much to learn about desert soils. In many ways, they differ from the soils of wetter areas both in origin and development, and in their present-day role in landscape hydrology and geomorphology. Given the many possible forms of anthropogenic climate change presently underway and the mounting pressure from a burgeoning world population, it is important that more attention be paid to the understanding and management of dryland soils. Some have already envisaged elevated levels of soil loss under enhanced greenhouse effect environmental scenarios (e.g. see Dregne, 1990). Changing temperature, rainfall and other climatic conditions in coming decades have the potential to cause changes in some of the surface properties discussed in this chapter, which may ripple through many linked environmental processes, driving significant change in the world's drylands. To understand the kinds of changes that might arise and to support the development of management policy and practice, we need to develop a deeper knowledge of the vulnerability or resilience of the distinctive dryland features considered here.

References

Adams, J.A., Stolzy, L.H., Endo, A.S. *et al.* (1982) Desert soil compaction reduces annual plant cover. *California Agriculture*, September–October 1982, 6–7.

Al-Qinna, M.I. and Abu-Awwad, A.M. (1998) Infiltration rate measurements in arid soils with surface crust. *Irrigation Science*, **18**, 83–89.

Aranibar, J.N., Anderson, I.C., Ringrose S. and Macko, S.A. (2003) Importance of nitrogen fixation in soil crusts of southern African arid ecosystems: acetylene reduction and stable isotope studies. *Journal of Arid Environments*, **54**, 345–358.

Assouline, S. (2004) Rainfall-induced soil surface sealing: a critical review of observations, conceptual models, and solutions. *Vadose Zone Journal*, **3**, 570–591.

Aumont, O., Bopp, L. and Schulz, M. (2008) What does temporal variability in Aeolian dust deposition contribute to sea-surface iron and chlorophyll distributions? *Geophysical Research Letters*, **35**, DOI: 10.1029/2007GL031131.

Azua-Bustos, A. (2008) A first glance at the microenvironmental conditions allowing the colonization of quartzes by hypolithic microorganisms on the Atacama Desert, Chile. *Astrobiology*, **8**, 427.

Belnap, J. (2006) The potential roles of biological soil crusts in dryland hydrologic cycles. *Hydrological Processes*, **20**, 3159–3178.

Belnap, J. and Gardner, J.S. (1993) Soil microstructure in soils of the Colorado Plateau: the role of the cyanobacterium *Microcoleus vaginatus*. *Great Basin Naturalist*, **53**, 40–47.

Belnap, J. and Gillette, D.A. (1997) Disturbance of biological soil crusts: impacts on potential wind erodibility of sandy desert soils in southeastern Utah. *Land Degradation and Development*, **8**, 355–362.

Belnap, J. and Gillette, D.A. (1998) Vulnerability of desert biological soil crusts to wind erosion: the influences of crust development, soil texture, and disturbance. *Journal of Arid Environments*, **39**, 133–142.

Belnap, J. and Lange, O.L. (eds) (2001) *Biological Soil Crusts: Structure, Function, and Management*, Ecological Studies 150, Springer, Berlin, 503 pp.

Belnap, J., Phillips, S.L. and Smith, S.D. (2007) Dynamics of cover, UV-protective pigments, and quantum yield in biological soil crust communities of an undisturbed Mojave Desert shrubland. *Flora*, **202**, 674–686.

Belnap, J., Prasse, R. and Harper, K.T. (2001) Influence of biological soil crusts on soil environments and vascular plants, Chapter 21, pp. 281–300, in *Biological Soil Crusts: Structure, Function, and Management* (eds J. Belnap and O.L. Lange), Springer, Berlin, 503 pp.

Ben-Hur, M. and Lado, M. (2008) Effects of soil wetting conditions on seal formation, runoff, and soil loss in arid and semi-arid soils – a review. *Australian Journal of Soil Research*, **46**, 191–202.

Berkeley, A., Thomas, A.D. and Dougill, A.J. (2005) Cyanobacterial soil crusts and woody shrub canopies in Kalahari rangelands. *African Journal of Ecology*, **43**, 137–145.

Bhatnagar, A. and Bhatnagar, M. (2005) Microbial diversity in desert ecosystems. *Current Science*, **89**, 91–100.

Bowker, M.A., Miller, M.E., Belnap, J. *et al.* (2008) Prioritizing conservation efforts through the use of biological soil crusts as ecosystem function indicators in an arid region. *Conservation Biology*, **22**, 1533–1543.

Bresson, L.L., Moran, C.J. and Assouline, S. (2004) Use of bulk density profiles from X-radiography to examine structural crust models. *Soil Science Society of America Journal*, **68**, 1169–1176.

Carmi, G. and Berliner, P. (2008) The effect of soil crust on the generation of runoff on small plots in an arid environment. *Catena*, **74**, 37–42.

Casenave, A. and Valentin, C. (1992) A runoff capability classification system based on surface features criteria in semi-arid areas of West Africa. *Journal of Hydrology*, **130**, 231–249.

Cattle, S.R., McTainsh, G. and Elias, S. (2009) Aeolian dust deposition rates, particle-sizes and contributions to soils along a transect in semi-arid New South Wales, Australia. *Sedimentology*, **56**, 765–783.

Chartres, C.J. (1982) The pedogenesis of desert loam soils in the Barrier Range, western New South Wales. I. Soil parent materials. *Australian Journal of Soil Research*, **20**, 269–281.

Chartres, C.J. (1983) The pedogenesis of desert loam soils in the Barrier Range, western New South Wales. II. Weathering and soil formation. *Australian Journal of Soil Research*, **21**, 1–13.

Cross, A.F. and Schlesinger, W.H. (1999) Plant regulation of soil nutrient distribution in the northern Chihuahuan Desert. *Plant Ecology*, **145**, 11–25.

Danin, A., and Ganor, E. (1991) Trapping of airborne dust by mosses in the Negev Desert, Israel. *Earth Surface Processes and Landforms*, **16**, 153–162.

Del Prado, R. and Sancho, L.G. (2007) Dew as a key factor for the distribution pattern of the lichen species *Teloschistes lacunosus* in the Tabernas Desert (Spain). *Flora*, **202**, 417–428.

De Vries, M.C. and Watling, J.R. (2008) Differences in the utilization of water vapour and free water in two contrasting foliose lichens from semi-arid southern Australia. *Austral Ecology*, **33**, 975–985.

Dexter, A.R. and Chan, K.Y. (1991) Soil mechanical properties as influenced by exchangeable cations. *Journal of Soil Science*, **42**, 219–226.

Dregne, H.E. (1976) *Soils of Arid Regions*, Elsevier, Amsterdam, 237 pp.

Dregne, H.E. (1990) Impact of climate warming on arid region soils, in *Soils on a Warmer Earth* (eds H.W. Scharpenseel, M. Schommaker and A. Ayoub), Elsevier, Amsterdam, pp. 177–184.

Dunkerley, D.L. (2000) Hydrologic effects of dryland shrubs: defining the spatial extent of modified soil water uptake rates at an Australian desert site. *Journal of Arid Environments*, **45**, 159–172.

Dunkerley, D.L. (2008a) Rain event properties in nature and in rainfall simulation experiments: a comparative review with recommendations for increasingly systematic study and reporting. *Hydrological Processes*, **22**, 4415–4435.

Dunkerley, D.L. (2008b) Identifying individual rain events from pluviograph records: a review with analysis of data from an Australian dryland site. *Hydrological Processes*, **22**, 5024–5036.

Dunkerley, D.L. and Brown, K.J. (1995) Runoff and runon areas in patterned chenopod shrubland, arid western New South Wales, Australia: characteristics and origin. *Journal of Arid Environments*, **30**, 41–55.

Eldridge, D.J. and Greene, R.S.B. (1994) Microbiotic soil crusts: a review of their roles in soil and ecological processes in the rangelands of Australia. *Australian Journal of Soil Research*, **32**, 389–415.

Eldridge, D.J. and Leys, J.F. (2003) Exploring some relationships between biological soil crusts, soil aggregation and wind erosion. *Journal of Arid Environments*, **53**, 457–466.

Eldridge, D., and Tozer, M.E. (1997) A practical guide to soil lichens and bryophytes of Australia's dry country, Published by the Department of Land and Water Conservation, Government of New South Wales, Sydney, Australia, 80 pp.

Eldridge, D.J., Zaady, E. and Shachak, M. (2000) Infiltration through three contrasting biological soil crusts in patterned landscapes in the Negev, Israel. *Catena*, **40**, 323–336.

Evans, R.D. and Johansen, J.R. (1999) Microbiotic crusts and ecosystem processes. *Critical Reviews in Plant Sciences*, **18**, 183–225.

Ewing, S.A., Southard, R.J., Macalady, J.L. *et al.* (2007) Soil microbial fingerprints, carbon, and nitrogen in a Mojave Desert creosote-bush ecosystem. *Soil Science Society of America Journal*, **71**, 469–475.

Fan, Y., Lei, T., Shainberg, I. and Cai, Q. (2008) Wetting rate and rain depth effects on crust strength and micromorphology. *Soil Science Society of America Journal*, **72**, 1604–1610.

Ferreira, A.G. and Singer, M.J. (1985) Energy dissipation for water drop impact into shallow pools. *Soil Science Society of America Journal*, **49**, 1537–1542.

Fohrer, N., Berkenhagen, J., Hecker, J.-M. and Rudolph, A. (1999) Changing soil and surface conditions during rainfall single rainstorm/subsequent rainstorms. *Catena*, **37**, 355–375.

Friedmann, I., Lipkin, Y. and Ocampo-Paus, R. (1967) Desert algae of the Negev (Israel). *Phycologia*, **6**, 185–195.

Ghadiri, H. and Payne, D. (1981) Raindrop impact stress. *Journal of Soil Science*, **32**, 41–49.

Goberna, M., Pascual, J.A., García, C. and Sánchez, J. (2007) Do plant clumps constitute microbial hotspots in semiarid Mediterranean patchy landscapes? *Soil Biology and Biochemistry*, **39**, 1047–1054.

Graetz, R.D. and Tongway, D.J. (1986) Influence of grazing management on vegetation, soil structure, nutrient distribution and the infiltration of applied rainfall in a semi-arid chenopod shrubland. *Australian Journal of Ecology*, **11**, 347–360.

Griffin, D.W. (2007) Atmospheric movement of microorganisms in clouds of desert dust and implications for human health. *Clinical Microbiology Reviews*, **20**, 459–477.

Grishkan, I., Zaady, E. and Nevo, E. (2006) Soil crust microfungi along a southward rainfall gradient in desert ecosystems. *European Journal of Soil Biology*, **42**, 33–42.

Gustavson, T.C. and Holliday, V.T. (1999) Eolian sedimentation and soil development on a semiarid to subhumid grassland, Tertiary Ogallala and Quaternary Blackwater Draw formations, Texas and New Mexico high plains. *Journal of Sedimentary Research*, **69**, 622–634.

Gutiérrez, J.R., Mserve, P.L. Contreras, L.C. *et al.* (1993) Spatial distribution of soil nutrients and ephemeral plants underneath and outside the canopy of Porlieria chilensis shrubs (Zygophyllaceae) in arid coastal Chile. *Oecologia*, **95**, 347–352.

Halligan, J.P. (1973) Bare areas associated with shrub stands in grassland: the case of *Artemisia californica*. *BioScience*, **23**, 429–432.

Hamerlynck, E.O., McAuliffe, J.R. and Smith, S.D. (2000) Effects of surface and sub-surface soil horizons on the seasonal performance of *Larrea tridentata* (creosotebush). *Functional Ecology*, **14**, 596–606.

Harper, K.T. and Marble, J.R. (1988) A role for nonvascular plants in management of arid and semiarid rangelands, in *Application of Plant Sciences to Rangeland Management and Inventory* (ed. P.T. Tueller), Martinus Nijhoff, Amsterdam, pp. 135–169.

Hartley, D.M. and Alonso, C.V. (1991) Numerical study of the maximum boundary shear stress induced by raindrop impact. *Water Resources Research*, **27**, 1819–1826.

Herman, R.P., Provencio, K.R., Herrera-Matos, J. and Torrez, R.J. (1995) Resources islands predict the distribution of heterotrophic bacteria in Chihuahuan Desert soils. *Applied and Environmental Microbiology*, **61**, 1816–1821.

Hill, R.D., Nagarkar, S. and Jayawardena, A.W. (2002) Cyanobacterial crust and soil particle detachment: a rain chamber experiment. *Hydrological Processes*, **16**, 2989–2994.

Hirmas, D.R., Amrhein, C. and Graham, R.C. (2010) Spatial and process-based modeling of soil inorganic carbon storage in an arid piedmont. *Geoderma*, **154**, 486–494.

Housman, D.C., Yeager, C.M., Darby, B.J. *et al.* (2007) Heterogeneity of soil nutrients and subsurface biota in a dryland ecosystem. *Soil Biology and Biochemistry*, **39**, 2138–2149.

Hupy, J.P. (2004) Influence of vegetation cover and crust type on wind-blown sediment in a semi-arid climate. *Journal of Arid Environments*, **58**, 167–179.

Issa, O.M., Trichet, J., Défarge, C. *et al.* (1999) Morphology and microstructure of microbiotic soil crusts on a tiger bush sequence (Niger, Sahel). *Catena*, **37**, 175–196.

Jester, W. and Klik, A. (2005) Soil surface roughness measurement – methods, applicability, and surface representation. *Catena*, **64**, 174–192.

Jickells, T.D., An, Z.S., Andersen, K.K. *et al.* (2005) Global iron connections between desert dust, ocean biogeochemistry, and climate. *Science*, **308**, 67–71.

Kidron, G.J., Barzilay, E. and Sachs, E. (2000) Microclimate control upon sand microbiotic crusts, western Negev Desert, Israel. *Geomorphology*, **36**, 1–18.

Kidron, G.J., Herrnstadt, I. and Barzilay, E. (2002) The role of dew as a moisture source for sand microbiotic crusts in the Negev Desert, Israel. *Journal of Arid Environments*, **52**, 517–533.

Kidron, G.J., Yaalon, D.H. and Vonshak, A. (1999) Two causes for runoff initiation on microbiotic crusts: hydrophobicity and pore clogging. *Soil Science*, **164**, 18–27.

Kinnell, P.I.A., Chartres, C.J. and Watson, C.L. (1990) The effect of fire on the soil in a degraded semi-arid woodland. II. Susceptibility of the soil to erosion by shallow rain-impacted flow. *Australian Journal of Soil Research*, **28**, 755–777.

Kraimer, R.A., Monger, H.C. and Steiner, R.L. (2005) Mineralogical distinctions of carbonates in desert soils. *Soil Science Society of America Journal*, **69**, 1773–1781.

Lado, M. and Ben-Hur, M. (2004) Soil mineralogy effects on seal formation, runoff and soil loss. *Applied Clay Science*, **24**, 209–224.

Lado M., Ben-Hur M. and Shainberg, I. (2007) Clay mineralogy, ionic composition, and pH effects on hydraulic properties of depositional seals. *Soil Science Society of America Journal*, **71**, 314–321.

Lázaro, R., Cantón, Y., Solé-Benet, A. *et al.* (2008) The influence of competition between lichen colonization and erosion on the evolution of soil surfaces in the Tabernas badlands (SE Spain) and its landscape effects. *Geomorphology*, **102**, 252–266.

Lei, T., Pan, Y., Liu, H. *et al.* (2006) A runoff-on-ponding method and models for the transient infiltration capability process of sloped soil surface under rainfall and erosion impacts. *Journal of Hydrology*, **319**, 216–226.

Mabbutt, J.A. and Fanning, P. (1987) Vegetation banding in arid Western Australia. *Journal of Arid Environments*, **12**, 41–59.

McFadden, L.D., McDonald, E.V., Wells, S.G. *et al.* (1998) The vesicular layer and carbonate collars of desert soils and pavements: formation, age and relation to climate change. *Geomorphology*, **24**, 101–145.

McIntyre, D.S. (1958a) Permeability measurements of soil crusts formed by raindrop impact. *Soil Science*, **85**, 185–189.

McIntyre, D.S. (1958b). Soil splash and the formation of surface crusts by raindrop impact. *Soil Science*, **85**, 261–266.

McKenna Neuman, C. and Maxwell C. (2002) Temporal aspects of the abrasion of microphytic crusts under grain impact. *Earth Surface Processes and Landforms*, **27**, 891–908.

McKenna Neuman, C., Maxwell, C. and Boulton, J.W. (1996) Wind transport of sand surfaces crusted with photoautotrophic microorganisms. *Catena*, **27**, 229–247.

McLain, J.E.T. and Martens, D.A. (2005) Studies of methane fluxes reveal that desert soils can mitigate global climate change, in US Department of Agriculture, Forest Service: Proceedings RMRS-P-36, pp. 496–499.

Makarieva, A.M. and Gorshkov, V.G. (2007) Biotic pump of atmospheric moisture as driver of the hydrological cycle on land. *Hydrology and Earth System Sciences*, **11**, 1013–1033.

Mando, A., Stroosnijder, L. and Brussaard, L. (1996) Effects of termites on infiltration into crusted soil. *Geoderma*, **74**, 107–113.

Marion, G.M. (1989) Correlation between long-term pedogenic $CaCO_3$ formation rate and modern precipitation in deserts of the American southwest. *Quaternary Research*, **32**, 291–295.

Marion, G.M., Verburg, P.S.J., Stevenson, B. and Arnone J.A. (2008) Soluble element distributions in a Mojave Desert soil. *Soil Science Society of America Journal*, **72**, 1815–1823.

Matmon, A., Simhai, O., Amit, R. *et al.* (2009) Desert pavement-coated surfaces in extreme deserts present the longest-lived landforms on Earth. *Geological Society of America Bulletin*, **121**, 688–697.

Metting, B. (1981) The systematics and ecology of soil algae. *The Botanical Review*, **47**, 195–312.

Montaña, C., Ezcurra, E., Carrillo, A. and Delhoume, J.P. (1988) The decomposition of litter in grasslands of northern Mexico: a comparison between arid and non-arid environments. *Journal of Arid Environments*, **14**, 55–60.

Morin, J., Benyamini, Y. and Michaeli, A. (1981) The effect of raindrop impact on the dynamics of soil surface crusting and water movement in the profile. *Journal of Hydrology*, **52**, 321–335.

Moser, D. (2008) Substantial bacterial diversity in surface soil from the Atacama Desert. *Astrobiology*, **8**, 429.

Moss, A.J. (1991) Rain-impact soil crust. I. Formation on a granite-derived soil. *Australian Journal of Soil Research*, **29**, 271–289.

Mott, J.J. and McComb, A.J. (1974) Patterns in annual vegetation and soil microrelief in an arid region of Western Australia. *Journal of Ecology*, **62**, 115–126.

Neave, M. and Rayburg, S. (2007) A field investigation into the effects of progressive rainfall-induced soil seal and crust development on runoff and erosion rates: the impact of surface cover. *Geomorphology*, **87**, 378–390.

Nettleton, W.D. and Peterson, F.F. (1983) Aridisols, in *Pedogenesis and Soil Taxonomy. II. The Soil Orders* (eds L.P. Wilding, N.E. Smeck and G.F. Hall), Elsevier, Amsterdam, pp. 165–215.

Oke, A.M.C., Dunkerley, D.L. and Tapper, N.J. (2007) Willy-willies in the Australian landscape: sediment transport characteristics. *Journal of Arid Environments*, **71**, 216–228.

Oke, A.M.C., Tapper, N.J. and Dunkerley, D.L. (2007) Willy-willies in the Australian landscape: the role of key meteorological variables and surface conditions in defining frequency and spatial characteristics. *Journal of Arid Environments*, **71**, 201–215.

Painter, T.H., Barrett, A.P., Landry, C.C. *et al.* (2007) Impact of disturbed desert soils on duration of mountain snow cover. *Geophysical Research Letters*, **34**. DOI: 10.1029/2007GL030284.

Patrick, E. (2002) Researching crusting soils: themes, trends, recent developments and implications for managing soil and water resources in dry areas. *Progress in Physical Geography*, **26**, 442–461.

Pointing, S.B., Warren-Rhodes, K.A., Lacap, D.C. *et al.* (2007) Hypolithic community shifts occur as a result of liquid water availability along environmental gradients in China's hot and cold hyperarid deserts. *Environmental Microbiology*, **9**, 414–424.

Ram, A. and Yair, A. (2007) Negative and positive effects of topsoil biological crusts on water availability along a rainfall gradient in a sandy arid area. *Catena*, **70**, 437–442.

Rice, M.A., McEwan, I.K. and Mullins, C.E. (1999) A conceptual model of wind erosion of soil surfaces by saltating particles. *Earth Surface Processes and Landforms*, **24**, 383–392.

Rice, M.A., Willetts, B.B. and McEwan, I.K. (1996) Wind erosion of crusted soil sediments. *Earth Surface Processes and Landforms*, **21**, 279–293.

Ries, J.B. and Hirt, U. (2008) Permanence of soil surface crusts on abandoned farmland in the Central Ebro Basin/Spain. *Catena*, **72**, 282–296.

Rosentreter, R., Bowker, M. and Belnap, J. (2007) *A Field Guide to Biological Soil Crusts of Western U.S. Drylands*, United States Government Printing Office, Denver, Colorado, 103 pp.

Rostagno, C.M., del Valle, H.F. and Videla, L. (1991) The influence of shrubs on some chemical and physical properties of an aridic soil in north-eastern Patagonia, Argentina. *Journal of Arid Environments*, **20**, 179–188.

Roth, C.H. and Helming, K. (1992) Dynamics of surface sealing, runoff formation and interrill soil loss as related to rainfall intensity, microrelief and slope. *Zeitschrift für Pflanzenernährung und Bodenkunde*, **155**, 209–216.

Sheil, D. and Murdiyarso, D. (2009) How forests attract rain: an examination of a new hypothesis. *BioScience*, **59**, 341–347.

Singer, M.J. and Shainberg, I. (2004) Mineral soil surface crusts and wind and water erosion. *Earth Surface Processes and Landforms*, **29**, 1065–1075.

Slattery, M.C. and Bryan, R.B. (1992) Laboratory experiments on surface seal development and its effect on interrill erosion processes. *Journal of Soil Science*, **43**, 517–529.

Striegl, R.G., McConnaughey, T.A., Thorstenson, D.C., Weeks, E.P. and Woodward, J.C. (1992) Consumption of atmospheric methane by desert soils. *Nature*, **357**, 145–147.

Tanaka, U., Yokoi, Y. and Kyuma, K. (1992) Morphological characteristics of soil surface crusts formed under simulated rainfall. *Soil Science and Plant Nutrition*, **38**, 655–664.

Tarchitzky, J., Banin, A., Morin, J. and Chen, Y. (1984) Nature, formation and effects of soil crusts formed by water drop impact. *Geoderma*, **33**, 135–155.

Tiller, K.G., Smith, L.H. and Merry, R.H. (1987) Accession of atmospheric dust east of Adelaide, South Australia, and the implications for pedogenesis. *Australian Journal of Soil Research*, **25**, 43–54.

Tongway, D.J. and Ludwig, J.A. (1990) Vegetation and soil patterning in semi-arid mulga lands of eastern Australia. *Australian Journal of Ecology*, **15**, 23–34.

Ugolini, F.C., Hillier, S., Certini, G. and Wilson, M.J. (2008) The contribution of Aeolian material to an aridisol from southern Jordan as revealed by mineralogical analysis. *Journal of Arid Environments*, **72**, 1431–1447.

US Department of Agriculture, Soil Conservation Service (1988) *Soil Taxonomy. A Basic System of Soil Classification for*

Making and Interpreting Soil Surveys, Krieger Publishing Co., Malabar, Florida, 754 pp.

Valentin, C. and Bresson, L.-M. (1992) Morphology, genesis and classification of crusts in loamy and sandy soils. *Geoderma*, **55**, 225–245.

Vázquez, E.V., Ferreiro, J.P., Miranda, J.G.V. and González, A.P. (2008) Multifractal analysis of pore size distributions as affected by simulated rainfall. *Vadose Zone Journal*, **7**, 500–511.

Wang, Z.-T., Zhou, Y.-H. and Zheng, X.-J. (2006) Tensile test of natural microbiotic crust. *Catena*, **67**, 139–143.

Warren-Rhodes, K.A., Rhodes, K.L., Pointing, S.B. *et al.* (2006) Hypolithic cyanobacteria, dry limit of photosynthesis, and microbial ecology in the hyperarid Atacama Desert. *Microbial Ecology*, **52**, 389–398.

Watson, A. (1992) Desert soils, in *Weathering, Soils and Paleosols* (eds I.P. Martini and W. Chesworth), Developments in Earth Surface Processes 2, Elsevier, Amsterdam, pp. 225–260, 618 pp.

West, N.E. (1990) Structure and function of microphytic soil crusts in wildland ecosystems of arid to semi-arid regions. *Advances in Ecological Research*, **20**, 179–223.

Whitford, W.G., Ludwig, J.A. and Noble, J.C. (1992) The importance of subterranean termites in semi-arid ecosystems in south-eastern Australia. *Journal of Arid Environments*, **22**, 87–91.

Whitton, B.A. and Potts, M. (eds) (2000) The Ecology of Cyanobacteria, Kluwer, Dordrecht, 669 pp.

Williams, J.D., Dobrowolski, J.P. and West, N.E. (1995) Microphytic crust influence on interrill erosion and infiltration capacity. *Transactions of the American Society of Agricultural Engineers*, **38**, 139–146.

Williams, J.D., Dobrowolski, J.P., West, N.E. and Gillette, D.A. (1995) Microphytic crust influence on wind erosion. *Transactions of the American Society of Agricultural Engineers*, **38**, 131–137.

Yair, A. (2001) Effects of biological soil crusts on water redistribution in the Negev Desert, Israel: a case study in longitudinal dunes, Chapter 22, pp. 303–314, in *Biological Soil Crusts: Structure, Function, and Management* (eds J. Belnap and O.L. Lange), Springer, Berlin, 503 pp.

Yoshioka, M., Mahowald, N.M., Conley, A.J. *et al.* (2007) Impact of desert dust radiative forcing on Sahel precipitation: relative importance of dust compared to sea surface temperature variations, vegetation changes, and greenhouse gas warming. *Journal of Climate*, **20**, 1445–1467.

Young, M.H., McDonald, E.V., Caldwell, T.G. *et al.* (2004) Hydraulic properties of a desert soil chronosequence in the Mojave Desert, USA. *Vadose Zone Journal*, **3**, 956–963.

Zhang, Y. (2005) The microstructure and formation of biological soil crusts in their early developmental stage. *Chinese Science Bulletin*, **50**, 117–121.

8

Desert crusts and rock coatings

David J. Nash

8.1 Introduction

Even in the driest regions of the world, the occurrence of crusts at or near the land surface (termed *duricrusts* when partially or fully indurated) testifies to the mobilisation and precipitation of minerals in the presence of water. The most widespread varieties of desert crust are *calcrete* (cemented by calcium carbonate), *silcrete* (silica-cemented), *gypcrete* (gypsum-cemented) and halite crust (or *salcrete*). A number of other minerals form less widespread hardened crusts, including *ferricrete* and *dolocrete*, and a wide range of evaporite crusts composed of minerals such as thenardite (Na_2SO_4), mirabilite ($Na_2SO_4{\cdot}10H_2O$), glauberite ($Na_2Ca(SO_4)_2$) and epsomite ($MgSO_4{\cdot}7H_2O$) also exist. Even less common evaporites such as nitratite (or nitratine, $NaNO_3$), natron ($Na_2CO_3{\cdot}10H_2O$) and trona ($NaH(CO_3){\cdot}2H_2O$) may form crusts in desert basins in volcanic regions such as the East African Rift Valley and parts of the Atacama Desert of Chile.

Desert crusts have a variable geomorphological influence. Some are ephemeral features and have little long-lasting impact upon the landscape. For example, halite crusts are prone to rapid dissolution by water and may only survive for a few months or years after formation. Nevertheless, these and crusts such as gypcrete can have a significant effect when they encase and immobilise sand dunes (Watson, 1985a, 1985b; Chen, Bowler and Magee, 1991a, 1991b). Other crusts, such as calcrete and silcrete, are persistent features in the landscape once indurated (e.g. Pain and Ollier, 1995). Calcrete crusts tens of metres thick are not uncommon and can mantle extensive areas of desert terrain, protecting the underlying materials from subaerial weathering and erosion (Figure 8.1). By reducing denudation rates, these crusts play an important role in the accumulation and preservation of thick sedimentary sequences (see Nash and McLaren, 2007).

The majority of desert crusts are the product of specific arid zone hydrological or pedological processes. As such, the occurrence of crusts as relict surficial deposits, or in the geological record, may provide valuable palaeoclimatic information. However, the processes and precise environmental controls involved in crust formation are not always well understood. Furthermore, there are several structural forms of most of the main types of desert crust. Some crusts closely resemble features that form under markedly different environmental conditions – for example, some silcretes may be the products of weathering in the humid tropics while others form in predominantly semi-arid regions. Hence, it is essential that the macro- and micromorphological characteristics and chemical properties of desert crusts, plus the processes involved in their formation, are fully appreciated before they are employed as evidence in palaeoenvironmental reconstructions.

This chapter provides an overview of the main types of desert duricrust, beginning with those that are found only in hyper-arid regions (sodium nitrate deposits) through to silcrete, which may form in wet-dry subtropical regions (Figure 8.2). It also describes the major types of rock coating, before considering the palaeoenvironmental significance of duricrusts and rock coatings. Throughout the chapter, the varieties of duricrust are treated separately. Forms such as silcrete, calcrete and gypcrete, however, are only the end members of a spectrum of duricrust types. Hybrid or *intergrade* forms can occur when, for example, diagenetic alteration of a duricrust results from water percolation. A range of minerals may also be precipitated within pores during the latest stages of cementation. Thus it is possible to find not only silcrete and calcrete

Arid Zone Geomorphology: Process, Form and Change in Drylands, Third Edition. Edited by David S. G. Thomas

Figure 8.1 Duricrusts as erosional controls: (a) calcrete caprocks preserving palaeosurface remnants near Nkob, central Morocco; (b) pedogenic silcrete sheet capping a mesa near Tibooburra, New South Wales, Australia.

in a landscape but also siliceous calcrete (or *sil-calcrete*) and calcareous silcrete (or *cal-silcrete*), depending upon the proportions of silica and calcium carbonate present (Nash and Shaw, 1998). Forms such as calcretised ferricrete (*cal-ferricrete*) or iron-cemented calcrete (*ferro-calcrete*) and silcrete (*ferro-silcrete*) have also been identified in some environments (Nash, Shaw and Thomas, 1994; Lee and Gilkes, 2005; Ramakrishnan and Tiwari, 2006). Finally, it should be noted that not all of the types of crust described are found exclusively in warm desert environments; some can persist when the climate becomes wetter. Similar crusts may also form in humid climates or in regions where, despite low temperatures, high rates of evaporation promote aridity. As such, desert crusts often have distributions that defy classification according to simple climatic parameters.

8.2 Sodium nitrate deposits

8.2.1 General characteristics and distribution

Sodium nitrate (nitratite or nitratine, $NaNO_3$) is highly soluble compared with other salts – at 35 °C its solubility in water is 49.6 %, whereas that of NaCl is 26.6 % and of calcium sulfate 0.21 % (Goudie and Heslop, 2007) – and therefore only persists in extremely dry environments. The one location where deposits are found in great abundance and thickness is the hyper-arid Atacama Desert of Chile, where high-grade *salitre* deposits were mined commercially as fertiliser from the 1820s until after World War I (Figure 8.3). Sodium nitrate deposits have also been described in various cave (Hill and Forti, 1997) and open sites (e.g. Johannesson and Gibson, 1962;

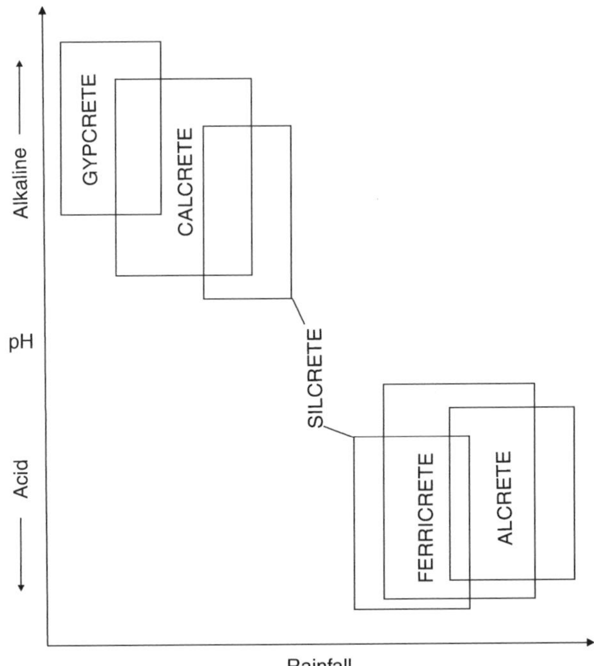

Figure 8.2 Schematic representation of the relationship between the formation of selected duricrusts and prevailing environmental conditions (after Summerfield, 1991). Note that in many instances local factors such as topography and drainage may outweigh broad climatic controls, particularly in the case of nonpedogenic crusts.

Claridge and Campbell, 1968; Ericksen, Hosterman and St Amand, 1988; Böhlke, Ericksen and Revesz, 1997; Graham *et al.*, 2008). Most is known about sodium nitrate deposits from the Atacama; for further information about similar crusts elsewhere, see the recent review by Goudie and Heslop (2007).

High-grade sodium nitrate deposits in the Atacama occur along a 700 by 30 km strip inland of the Coastal Range between 19°30′S and 26°S (Ericksen, 1981). Most deposits are found below 2000 m a.s.l. but some may extend up to 4000 m a.s.l. They reach greatest thicknesses on the lower slopes of hills and piedmont plains but also occur in a range of topographic settings from hilltops to the centres of valleys. Many *salars* (see Chapter 15) in the Atacama contain thick nitrate crusts, including the Pampa Blanca, Pampa Lina, Salar del Carmen and Salar de Lagunas. Nitrate deposits occur on and within a wide range of rock and sediment types, and there appears to be little lithological control on their different structural types and mineral assemblages (Goudie and Heslop, 2007).

The high solubility of sodium nitrate in water means that large volumes of the salt can be precipitated within rock pores, where it can cause severe rock disintegration. Nitratite is also hygroscopic (it absorbs atmospheric moisture) and deliquescent (it dissolves in that moisture). When relative humidity exceeds a critical level (73.8 % at 25 °C for pure sodium nitrate; see Tereschenko and Malyutin, 1985), water condenses on and dissolves mineral surfaces; when the relative humidity falls back below that level, sodium nitrate will recrystallise. Ericksen (1981, 1983) recognised two main types of nitrate deposit: regolith cement (or *alluvial caliche*, not to be confused with the term *caliche* used to describe calcrete in some regions) and impregnated bedrock (or *bedrock caliche*). A typical alluvial caliche profile contains several zones. At the top is a 10–30 cm thick *chuca* zone, a powdery to poorly cemented leached surface layer containing clastic fragments, some gypsum and anhydrite, and salts such as bloedite, humberstonite and thenardite. Beneath this is a moderate to well cemented 0.5–2.0 m thick *costra* zone. Below this is a well cemented 1.0–3.0 m thick *caliche* zone, containing layers of white nitrate-rich deposits referred to as *caliche blanco*. The caliche zone grades downwards into weakly cemented (*conjelo*) or uncemented (*coba*) regolith. Bedrock caliche consists of layers of nitrate minerals that have forced open rock fissures during crystallisation. These may be several tens of cm thick and, in advanced cases of replacement, may so disrupt the bedrock that it consists of minor rock fragments surrounded by dominant *caliche blanco*.

8.2.2 Micromorphology, chemistry and mode of formation

With the exception of some near-surface veins in bedrock (Ericksen and Mrose, 1972) and some *caliche blanco*, the matrix of sodium nitrate deposits rarely consists of pure nitratite. Ericksen, Hosterman and St Amand (1988), for example, report deposits from the Atacama that contain anhydrite, borax, darapskite, glauberite, gypsum, halite, tincalconite and trona in addition to nitratite. The most important of these impurities is halite, which may exceed the percentage of nitrate present (Penrose, 1910).

Relatively few accounts of the micromorphology of sodium nitrate sequences have been published. Pueyo, Chong and Vega (1998) describe a range of microscale features in deposits from the Antofagasta region of Chile. Here, sodium nitrate deposits contain silt and other detrital material cemented by a mix of salt crystals of different compositions. Layers of relatively pure sulfate minerals (bloedite and humberstonite) may be interspersed with veins of purer nitrate (Ericksen and Mrose, 1972). Small chalcedony nodules may also be present. Due to the high solubility of sodium nitrate, deposits commonly contain

Figure 8.3 Abandoned sodium nitrate workings in the Atacama Desert: (a) west of Oficina Salitirera Victoria; (b) at Oficina Salitirera Humberstone, Tarapacá region, northern Chile.

evidence of multiple episodes of precipitation, dissolution (in the form of solution cavities), repeated fracturing and displacive growth. Hard, spheroidal cakes of anhydrite, called *losa*, may also be present, produced at the surface of the deposit by prolonged slow leaching (Ericksen, 1981).

Goudie and Heslop (2007) provide an overview of the theories that have been put forward to account for the development of sodium nitrate deposits in Chile. An early view was that nitrate crusts were remnants of ancient seaweed deposits left stranded by falling sea levels (Forbes, 1861). This was soon dismissed (Newton, 1869) on the grounds that there were no marine deposits at the altitudes where nitrate sequences occurred. A related theory was that the nitrates were remnants of guano deposits (Gautier, 1894), but, again, the altitudes at which caliche occurs makes this unlikely. A third proposal by Mueller (1968) was that weathering in the higher rainfall zones of the Andes had released solutes that accumulated within closed basins at lower altitudes and ultimately formed chloride and sulfate deposits. The slopes around these basins were fed by capillary concentration, and this led to the zones of nitrate accumulation.

The most widely accepted theories for explaining the abundance of nitrate accumulations in the Atacama centre on the role of atmospheric salt deposition. Claridge and Campbell (1968) were the first to propose that Chilean nitrate deposits had built up as a result of the accumulation of oxidised atmospheric nitrogen, together with compounds of iodine, sulfur and chlorine derived by evaporation from the Pacific. Ericksen (1981) added that biological activity, bedrock weathering and volcanic emissions from the Andes may have played a role, with volcanic sources also highlighted by other authors (Searl and Rankin, 1993; Chong, 1994; Searl, 1994; Oyarzun and Oyarzun, 2007). Ericksen (1981) suggested that atmospherically derived deposits were leached, enriched in their most soluble components and redistributed by rainwater to accumulate on lower hillsides, at breaks in slopes and in salars. Evidence suggests that the Atacama may have been arid since the Early Eocene and hyper-arid since the Middle Miocene or Late Pliocene (Clarke, 2006). Such extended arid conditions, associated with an almost complete absence of biological activity and leaching, would have permitted large amounts of nitrate to build up even from modest rates of inputs. Stable N, O and S isotope studies in the Atacama, Antarctica and Mojave (Böhlke, Ericksen and Revesz, 1997; Michalski *et al.*, 2002) support the hypothesis that nitrate-rich deposits represent long-term accumulations of atmospheric deposition, with a minor contribution by microbial fixing and oxidation of reduced N compounds in areas with slightly higher rainfall (Böhlke and Michalski, 2002).

Coastal fogs (locally known as *camanchaca*) have also been proposed as a nitrate source in Chile (Ericksen, 1981). Ridge sites close to the Pacific experience fog on up to 189 days per year (Cereceda and Schemenauer, 1991), with fog moisture fluxes of about 8.5 L/m/day at the coast, declining to 1.1 L/m/day 12 km inland (Cereceda *et al.*, 2002). Analyses of coastal fog moisture (e.g. Eckardt and Schemenauer, 1998) estimate NO_3 concentrations at around 1.6 mg/L. The relative purity of fog water and the limited amounts that are deposited inland suggest that fog can only be a minor contributor to nitrate accumulation. Indeed, S and Sr isotopic studies of Atacama aerosols and sediments (Rech, Quade and Hart, 2003) indicate that the spatial distribution of high-grade nitrate deposits corresponds to areas that receive the lowest fluxes of ocean- or *salar*-derived salts, which may dilute atmospheric nitrate fallout.

8.3 Halite crusts

8.3.1 General characteristics and distribution

Halite (NaCl) deposits most commonly develop in sabkhas (Chapter 15) or in the basins of ephemeral lakes that are subject to periodic evaporation to dryness (Eugster and Kelts, 1983; Lowenstein and Hardie, 1985). Here, hard crusts of almost pure halite form horizontal beds on the surface. These are often white, though the presence of microorganisms such as the flagellate *Dunaliella salina* can give a pink colour. Periodic influxes of rainwater or runoff often result in the dissolution and subsequent reprecipitation of the deposits. In a lacustrine or lagoonal setting (Figure 8.4), dissolution may increase brine salinity, thus preventing further dissolution of underlying evaporites and allowing thick salt sequences to accumulate (Morris and Dickey, 1957; Busson and Perthuisot, 1977). The largest of the terrestrial evaporite deposits, which include interbedded gypsum and halite, may cover thousands of square kilometres and reach thicknesses of hundreds of metres (e.g. Salar de Uyuni, Bolivia; see Risacher and Fritz, 2000). In the geological record, preserved halite deposits may be difficult to distinguish from marine evaporites formed by the evaporation of blocked near-coastal or epicontinental seas.

In addition to lacustrine halite accumulations, powdery halite efflorescences and encrustations are common in many arid and hyper-arid environments (e.g. Eswaran, Stoops and Abtahi, 1980; Pye, 1980; Basyoni and Mousa, 2009). These deposits are usually dissolved by even small amounts of rainfall, but reform during subsequent dry conditions. There are few examples of halite crusts that have accreted beneath the land surface, either close to the water table or in the soil zone, in the same way as many calcretes and gypsum crusts. Pedogenic halite crusts, which have a strong structural resemblance to columnar gypsum crusts, have been described in the Namib Desert (Kaiser and Neumaier, 1932; Watson, 1983a, 1983b, 1985a) and the Chilean Tacna Desert (Mortensen, 1927). Subsurface, phreatic crusts have been reported from coastal sabkhas along the Arabian Gulf (Shearman, 1963; Patterson and Kinsman, 1978) and in lagoonal environments in Western Australia (Arakel, 1980).

Halite crust formation associated with lacustrine and phreatic evaporative processes is usually restricted to areas where mean annual rainfall is less than about 200 mm (Stankevich, Imameev and Garanin, 1983; Goudie and Cooke, 1984). In contrast, pedogenic halite crusts in the Namib and Atacama Deserts occur only where annual rainfall is less than about 25 mm. Since the sources of salts forming pedogenic halite crusts are often atmospheric, the underlying materials can be highly variable. Though pedogenic halite crusts are found in association with gypsum crusts, usually there are no distinct stratigraphic associations. Lagoonal, lacustrine and sabkha halite crusts may be interbedded with gypsum (Phleger, 1969) or anhydrite. Other characteristic facies, as well as structural and textural features, are well documented (e.g. Shearman, 1966; Arakel, 1980; Warren and Kendall, 1985; Schreiber, 1986).

Figure 8.4 A halite crust on the floor of Umm as Samim playa in Oman: (a) halite polygons on the playa bed; (b) well-developed halite crystals in the upper layers of the salt crust; (c) halite efflorescences emerging from pressure ridges between halite polygons.

Halite crusts do not have a lasting effect on the landscape owing to their susceptibility to dissolution. However, halite is important in desert weathering (see Chapter 6) and pedogenesis (see Chapter 7). Saline crusts may also impact upon aeolian processes by consolidating the sand on dune slipfaces (Nickling and Ecclestone, 1981; Nickling, 1984). Halite 'salt scalds' in soils are often associated with rising saline groundwater tables which, in some cases, are a function of land management practices such as removal of vegetation or overirrigation (Chivas, 2007). Understanding the factors that control the formation of halite crusts can also be of economic significance, as many petroleum reserves are associated with evaporite sequences.

8.3.2 Micromorphology and chemistry

Several distinct micromorphologies are displayed by halite deposited in lacustrine and sabkha evaporite sequences (see Warren, 2006, for a recent overview). Halite precipitated in shallow-water bodies usually grows as upward-pointing chevron-, cube- or cornet-shaped crystals. Cloudy crystals – some containing poikilitic inclusions – are normally primary precipitates, while clear crystals are interpreted as secondary diagenetic precipitates or syndepositional void-fills (Warren, 2006). Saccharoidal textures form during rapid, sporadic halite precipitation, and occur as bridge cements that consolidate the host grains in thin salcretes. Halite oolites (halolites) and pisoids (halopisoids) have also been described in the shallow Tuz Gölü saline lake of Central Anatolia, Turkey (Tekin *et al.*, 2006). The halolites are spherical features ranging from 0.7 to 2.0 cm in diameter, composed of a nucleus of coarse-grained halite crystals surrounded by halite-dominated concentric laminae. They form in the swash zone where halite crystals roll around in halite-rich mud. Granular crystal textures predominate in phreatic crusts (Arakel, 1980). Hopper crystals with negative-crystal brine inclusions are also common (Arakel, 1980; Lowenstein and Hardie, 1985). Little information is available on the micromorphology of pedogenic halite crusts. However, there is evidence that crystal habits differ depending on the depositional environment. Surface efflorescences of halite can be made up of fibrous crystals, while cubic or trigonal pyramids crystallise in porous media (Eswaran, Stoops and Abtahi, 1980).

The mineralogical and chemical characteristics of halite crusts are closely related to the environment of deposition (see Chapter 15). In lacustrine settings, the initial chemical composition of the evaporating brine controls the sequence of minerals that precipitate from it by a process termed fractional crystallisation (Hardie and Eugster, 1970; Eugster and Hardie, 1978). Evaporite minerals associated with primary halite and gypsum deposits such as celestite ($SrSO_4$) (Evans and Shearman, 1964; Magee, 1991) and polyhalite ($K_2Ca_2Mg(SO_4)\cdot 2H_2O$) (Holster, 1966) are formed by diagenesis. Pedogenic halite crusts may contain significant quantities of accessory minerals, particularly sulfates and, less commonly, nitrates. For example, XRD and SEM analyses of saline soils from the Las Vegas Basin, USA, identified the presence of halite, bloedite, eugsterite, gypsum, hexahydrite, mirabilite, sepiolite thenardite, vivianite and possibly kainite (Buck *et al.*, 2006).

8.3.3 *Mode of formation*

The requirements for the development of a halite crust in a lacustrine setting are a source of solutes, their transport to an accumulation basin and their evaporation. Hardie, Smoot and Eugster (1978) and Smoot and Lowenstein (1991) identify a number of subenvironments where this combination of factors occurs. In perennial saline lakes (e.g. Lake Chad, the Dead Sea and the Caspian Sea), evaporite minerals including halite accumulate as cumulus crystals that initially precipitate at the water surface and sink to the lake floor. Evaporite crusts may also be precipitated directly on shallow lake beds. Logan (1987), for example, describes upward-pointing chevron crystals of halite forming units up to 5 m thick in Lake McLeod, Western Australia. Halite deposits may exhibit ripple or cross-bedded structures if subaqueously reworked. In playa environments (Figure 8.4(a) and (b)), salts accumulate during successive cycles of flooding and evaporation (Lowenstein and Hardie, 1985). Subaqueous salt hoppers and rafts, together with inwashed clastic sediments, may bury and preserve earlier layered salts. Displacive and diagenetic salts, including gypsum and halite, may also precipitate within pore spaces. Ultimately, a polygonal surface crust may form, with halite efflorescences emerging from the pressure ridges between polygons (Figure 8.4(c)). Many playas have only a thin surface halite crust, which redissolves upon flooding and reprecipitates by evaporation after flood-borne clastic deposition. By this mechanism the halite crust is preserved as the uppermost horizon, with clastic sediment progressively accumulating 'beneath' the halite layer (Chivas, 2007). The saline mudflat areas surrounding perennial saline lakes and playas are associated with the development of halite efflorescences and intrasediment (displacive) evaporite minerals. In some systems, these may be zoned, with the most soluble minerals towards the lowest portion of the mudflat (e.g. Saline Valley, California; see Hardie, 1968). Dry mudflats, where the groundwater table is deeper, are associated with minor efflorescences and intrasediment salts related to fluctuating water tables (Smoot and Lowenstein, 1991).

The formation of some terrestrial halite crusts is a pedogenic process. In the Namib (Watson, 1983b) and Atacama Deserts (Ericksen, 1981), halite crusts are illuvial accretions formed by similar processes to pedogenic calcretes and gypcretes. Salt deposited at the surface as dust or in fog moisture in the Namibian and Chilean Deserts is leached into the solus where it recrystallises when the soil moisture evaporates. Provided the soil's moisture storage capacity is not exceeded, the salts accumulate as an illuvial horizon. Amit and Yaalon (1996) describe cubic halite accumulations within mature reg soils in the Negev Desert, formed by crystallisation from supersaturated soil solutions at the depth of maximum water penetration. Highly soluble salts are less likely to persist because even infrequent saturation of the solus will cause flushing. The different solubilities, and therefore vertical mobilities (Yaalon, 1964), of the common crust-forming minerals can result in the formation of two-tiered crusts if rainfall is sufficient to mobilise less soluble salts but not to flush the more soluble. The less soluble salt will accumulate in an illuvial horizon above the more soluble one. Calcretes above gypsum crusts are found in North Africa (e.g. Horta, 1980) and gypsum crusts above halite crusts in the central Namib Desert.

8.4 Gypsum crusts

8.4.1 *General characteristics*

Gypsum crusts (often referred to as gypcretes) (Figure 8.5) have been defined as 'accumulations at or within 10 m of the land surface from 0.10 to 5.0 m thick containing more than 15 % by weight gypsum ... and at least 5.0 % by weight more gypsum than the underlying bedrock' (Watson, 1985a, p. 855). Gypcretes have received less attention than other duricrusts. In part, this reflects a more

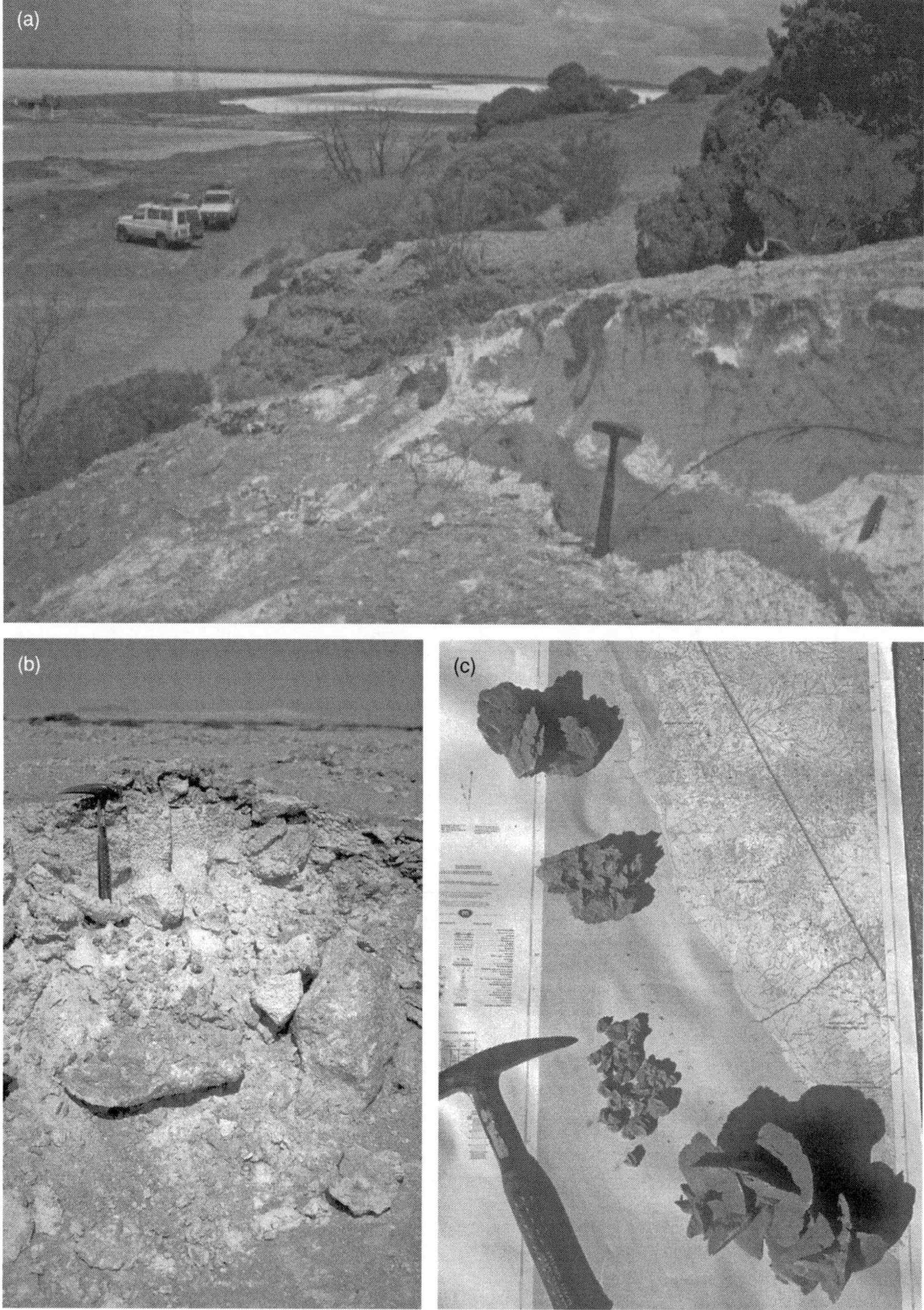

Figure 8.5 Examples of gypsum crusts. (a) An island of Archaean bedrock draped by gypcrete in the halite-encrusted floor of Lake Lefroy, near Kambalda, Western Australia. The white outcrop in the foreground is uncemented very fine-grained gypsum (locally called *kopi*), that has been deflated from the playa floor and recrystallised under the influence of percolating rainwater. (b) Powdery pedogenic gypsum crust from the Tumas River in the Namib Desert, Namibia. This crust, unusually, is enriched in uranium. (c) Desert rose crystals sampled from sabkha sediments between Terrace Bay and Agate Beach along the Skeleton Coast, Namibia. Photograph (a) courtesy of Allan Chivas; photographs (b) and (c) courtesy of Frank Eckardt.

restricted geographical distribution which, like sodium nitrate crusts, results from the greater susceptibility of gypsum to dissolution by rainwater. This limits the crusts to some of the Earth's most arid regions. In addition, the less widespread occurrence of sources of gypsum ($CaSO_4 \cdot 2H_2O$), compared to calcium carbonate or silica, hinders crust formation. Nonetheless, gypcretes have been reported from virtually every continent of the world (Watson, 1983a).

Three main forms of gypsum crust have been identified: (a) horizontally bedded crusts; (b) subsurface crusts, sometimes known as *croute de nappe*, composed either of large, lenticular crystals (between 1 mm and 0.50 m in diameter; desert rose crusts) or mesocrystalline material (crystal diameters from 50 μm to 1.0 mm); and (c) surface crusts (Figure 8.5(a) and (b)), composed mainly of alabastrine gypsum (crystallites less than 50 μm in diameter), occurring as columnar crusts, powdery deposits or superficial cobbles (Watson, 1979, 1983a, 1985a). There is no typical profile. Desert rose crusts (Figure 8.5(c)) can reach thicknesses of up to 5 m (e.g. Kulke, 1974) and range in colour from white or grey to green or red, depending on the host material. Columnar surface crusts are usually 1–2 m thick and white or grey in colour, with roughly hexagonal columns, 0.25–0.75 m in diameter, extending through the full thickness. The mesocrystalline, columnar and cobble forms of gypsum crust are probably genetically related; morphological and chemical differences result from diagenesis and, possibly, degradation of pedogenic crusts. However, just as some calcretes are nonpedogenic in origin, the bedded and desert rose forms of gypsum crust are also genetically distinct. This has profound implications for the significance of these desert crusts as palaeoenvironmental indicators (Watson, 1988).

8.4.2 Distribution

Gypsum crusts are found in all the Earth's warm deserts, though their extent varies. Factors such as climate, geology, topography and hydrology are critical to their formation and preservation. Most gypsum crusts are found in areas where mean annual rainfall is less than 250 mm (Watson, 1985a). In North Africa, for example, there appears to be a transition from calcretes to gypsum crusts as mean annual rainfall drops below this amount (Pervinquière, 1903). Rarely, the upper rainfall limit reaches 300 mm; this is the case in Iraq (Tucker, 1978) and Rajasthan (Srivastava, 1969). Here, high mean monthly temperatures, especially during the wetter months, promote high rates of evaporation and a net monthly soil moisture deficit is maintained throughout much of the year. As with calcrete distribution, there also appears to be an annual rainfall threshold below which gypsum crusts are less common. In the central Namib Desert, for example, there is a gradual transition from gypsum as rainfall drops below about 25 mm (Watson, 1983b, 1985a); more soluble halite crusts form preferentially under such hyperarid conditions (Hartley and May, 1998).

The main areas of widespread gypsum crusts are in North Africa, particularly central Algeria and Tunisia (e.g. Coque, 1955a; Watson, 1979, 1985a, 1988; Drake, Eckardt and White, 2004). Gypcretes are documented in the Namib Desert (e.g. Martin, 1963; Scholz, 1972; Watson, 1981, 1985a, 1988; Eckardt *et al.*, 2001), Australia (e.g. Arakel and McConchie, 1982; Warren, 1982; Jacobson, Arakel and Chen, 1988; Chen, Bowler and Magee, 1991a, 1991b; Chivas *et al.*, 1991; Magee, 1991; Milnes, Thiry and Wright, 1991; Chen, 1997) and central Asia (e.g. Tolchel'nikov, 1962; Evstifeev, 1980). They are found throughout the Middle East from Egypt (e.g. Ali and West, 1983; Aref, 2003), Israel (e.g. Dan *et al.*, 1972, 1982; Amit and Gerson, 1986), Jordan (Turner and Makhlouf, 2005) and Syria (Eswaran and Zi-Tong, 1991) to Iraq (Tucker, 1978), Iran (Gabriel, 1964), Saudi Arabia (Al Juaidi, Millington and McLaren, 2003) and Kuwait (El Sayed, 1993). Occurrences in the southwest USA (e.g. Reheis, 1987; Harden *et al.*, 1991; Buck and van Hoesen, 2002) and South America (Risacher, 1978; Hartley and May, 1998; Rech, Quade and Hart, 2003) are more sporadic. Gypcretes are also reported from Antarctica, where strong winds promote high rates of evaporation resulting in evaporite crystallisation (Gibson, 1962; Lyon, 1978).

Gypsum crusts often form an erosion-resistant horizon at the land surface. Perhaps because of their greater susceptibility to dissolution, however, gypcretes rarely create the mesa-and-butte landscapes associated with more durable crusts such as silcrete and calcrete. Nevertheless, in southern Tunisia gypsum crusts mantle several generations of pediment slopes (glacis), apparently protecting relict surfaces from erosion (Coque, 1955b, 1962). In many regions, gypsum crusts are found in and around large hydrological basins (e.g. the chotts of Tunisia and Algeria, salinas of South America, salt lakes of Australia and playas in the central Namib). While some crusts are lacustrine evaporites (Warren, 1982) and others are phreatic precipitates (Kulke, 1974; Risacher, 1978), many occur on hill crests and steep slopes beyond the phreatic zone (Figure 8.5(a)). These are pedogenic crusts, which blanket the landscape. Such crusts may develop directly on unweathered bedrock such as granite, basalt, marble, limestone and clay (Watson, 1985a, 1988) or on unconsolidated sediments such as colluvium, alluvium and dune sand (Watson, 1979).

8.4.3 Micromorphology and chemistry

The solubility of gypsum in pure water is about 2.0 g $CaSO_4$/Land varies very little with temperature (Hill, 1937). However, in the presence of sodium chloride, gypsum solubility is raised to 8.0 g $CaSO_4$/L at 35 °C when the solvent contains 100 g NaCl/L (Zen, 1965). The presence of sodium ions appears to influence the crystalline habit of gypsum precipitates (Masson, 1955; Edinger, 1973), such that the micromorphology of gypsum crusts is often more genetically diagnostic than that of many calcretes and silcretes. Horizontally bedded gypsum crusts are structurally distinct from the other forms of accumulation, and are characterised by size-grading of packed microcrystalline gypsum in strata up to about 0.10 m thick. Within individual beds there is a gradation from lenticular crystals (0.50 mm in diameter) at the base to alabastrine crystallites (less than 50 μm in diameter) at the top. Subsurface desert rose crusts are distinct not only because of the large, lenticular crystals they contain but also because these crystals incorporate clastic host material. Poikilitic inclusions are evident even in relict crusts that have experienced pronounced diagenesis (Watson, 1988). Phreatic gypsum accumulations from Australia (Arakel and McConchie, 1982) are characteristically lensoid or wedge shaped, and comprise discoidal or acicular gypsum crystals with a massive, tabular or clotted micromorphology. Mesocrystalline subsurface crusts are pedogenic accretions that contain fibrous gypsum, a rare but highly significant crystalline texture. As with pedogenic calcretes, it is possible that this habit indicates displacive crystallisation within the host material. The various forms of surface crust – columnar, powdery and cobble – are characterised by alabastrine textures. These are largely diagenetic, resulting from the dissolution of lenticular crystals and reprecipitation of alabastrine material, a process similar to sparmicritisation in calcretes (Kahle, 1977). Most surface crusts appear to be exhumed mesocrystalline crusts (Watson, 1985a, 1988), the cobble form representing a stage in the degradation of columnar crusts to powdery residua by dissolution and leaching. Exhumation of the illuvial mesocrystalline crusts is often brought about by aeolian erosion of the unconsolidated overburden.

The three forms of gypsum crust show considerable chemical and mineralogical diversity, primarily due to variations in the degree of induration. However, horizontally bedded gypsum crusts generally contain between 50 and 80 % gypsum, desert rose crusts between 50 and 70 % gypsum and the mesocrystalline subsurface form and surface forms up to 90 % gypsum. The other main constituents are quartz grains and variable amounts of calcium carbonate, with limited iron minerals inherited from the host sediment (Watson, 1983a, 1985a). Clay minerals reported in gypcretes from Wyoming, USA, include kaolinite, smectite and mica, ranging in abundance from 10 to 30 %, with minor traces of palygorskite (Reheis, 1987). Authigenic palygorskite is, in fact, more prevalent in some pedogenic gypsum crusts than in calcretes (Yaalon and Wieder, 1976; Reheis, 1987). This may be related to the development of two-tiered calcrete-gypsum crusts as a result of differential leaching of soluble minerals (Drever and Smith, 1978; Watson, 1985a).

Small quantities of other elements may also be present. Sodium is an important constituent of many African gypcretes and is suggested to reflect the role of groundwater in its formation (Watson, 1983a). In addition to significant differences in the content of gypsum and clastic components, sodium levels are highest in lacustrine evaporite crusts, followed by phreatic gypsum crusts, and are lowest in the pedogenic forms, with magnesium levels showing a similar trend. Evolution from the illuvial mesocrystalline form through columnar surface crusts to the cobble form results in a decrease in Na and Sr concentrations. This is probably related to the release of ions coprecipitated with the lenticular gypsum (Kushnir, 1980, 1981) when it is dissolved and reprecipitated as alabastrine material. The lack of crystal growth following nucleation limits further coprecipitation.

8.4.4 Modes of formation

Gypsum crust formation can occur via pedogenic and nonpedogenic pathways. Pedogenic gypsum accumulation has been attributed to either capillary rise or illuviation, with the latter model most widely embraced (Watson, 1979, 1983a, 1985a; Chen, 1997; Eckardt *et al.*, 2001). The upward movement and precipitation of salts from gypsum-saturated soilwater or groundwater by capillary rise would rapidly plug desert soil profiles and prevent further evaporation, and, significantly, requires large quantities of water to generate even a thin crust (Watson, 1979, 1983a, 1985a). Under the illuvial model, gypsum deposited at the surface is dissolved by rainwater and leached into the solus, where it precipitates during subsequent desiccation. Where gypsum accumulates deeper within a soil, a past wetter climate may be inferred (Reheis, 1987). Displacive accretion is achieved mainly by uplift of the overlying host material, a process similar to that envisaged in the formation of some desert pavements (see Chapter 9). The evolution of the surface forms of gypsum crust is contingent upon the exhumation of illuvial crusts. Though some crusts exhibit replacive features, evidence of displacive crystallisation is more common.

Table 8.1 Genesis and post-depositional changes of gypsum sediments (Chen, 1997).

Mode of genesis	*Example*
Primary crystallisation	
Surface brine	*Coastal zone.* Large (>2 mm) compact selenite and laminated columnar/prismatic crystals, commonly occurring with halite and carbonates
	Salt lakes. Relatively large (>0.5 mm) prismatic crystals, laminated and commonly interbedded with mud
Groundwater	Large (commonly >5 mm) crystals, near watertable, including sand and mud. Mostly shortened in *c* axis, commonly interlocked forming aggregates (e.g. desert rose)
Transportation and redeposition	
By water	Fractured and corroded crystals or crystal fragments in lacustrine and alluvial sediments
By wind	Gypsum sands in dunes along playa lake margins, with crystals relatively well sorted and with near-horizontal orientation
Post-deposition alteration	
Below capillary fringe	Overgrowth and dissolution
Above capillary fringe	Dissolution, leaching and recrystallisation, forming very fine crystals (<0.1 mm)

Chen (1997) conceptualises the processes involved in pedogenic gypsum crust formation as a series of genetic and diagenetic steps (Table 8.1). Prior to accumulating within a soil profile, gypsum may be precipitated from a saturated solution such as a surface brine or groundwater, most commonly via evaporation. Gypsum crystals are then exhumed, eroded, transported and redeposited at the ultimate site of gypcrete formation (by either the wind or water), before being incorporated into the soil profile. Once within the profile, the gypsum may be subject to diagenetic alteration either above or below the zone of capillary rise to generate the pedogenic gypcrete profile.

The main sources of gypsum for pedogenic gypcrete formation are windblown sand, dust or aerosols (Watson, 1979, 1985a; Dan *et al.*, 1982; Amit and Gerson, 1986; Bao, Thiemens and Heine, 2001). Gypsum may be deflated from playa surfaces (Coque, 1955a, 1955b, 1962; Watson, 1985a) or reworked hydromorphic gypsum crusts (Reheis, 1987; Chen, Bowler and Magee, 1991a, 1991b). Fog, sea spray, biogenic sulfur and marine evaporite deposits have been suggested as sulfur sources for gypsiferous crusts in Tunisia, Australia, the Atacama and Namib Desert (Martin, 1963; Carlisle *et al.*, 1978; Ericksen, 1981; Watson, 1985a; Chivas *et al.*, 1991; Day, 1993; Eckardt and Spiro, 1999; Rech, Quade and Hart, 2003; Drake, Eckardt and White, 2004). However, Eckardt and Schemenauer (1998) have demonstrated that the ionic content of Namib fog is too low to act as a major sulfur source.

Nonpedogenic models of gypcrete formation have been put forward to describe the deposition of gypsum in lacustrine environments or close to the groundwater table. For example, Jacobson, Arakel and Chen (1988) suggested that the formation of extensive gypsum deposits beneath central Australian palaeolakes was driven by interactions between vadose and phreatic groundwaters and infiltrating meteoric waters. Watson (1985a, 1988) attributed the formation of lacustrine gypcretes in the Namib Desert and Tunisia to two mechanisms. Bedded, lacustrine evaporites probably precipitated in shallow-water environments. The presence of size-graded strata indicates that the water body evaporated to dryness because late in the evaporative cycle there was minimal ionic migration to the growth faces on alabastrine crystallites. In contrast, phreatic, desert rose crusts probably accreted as gypsum precipitates from evaporating groundwater (Castens-Seidell and Hardie, 1984) where the water table was within 1 or 2 m of the land surface. Precipitation as a result of dilution by less saline meteoric water (Pouget, 1968) is unlikely because it does not allow for the prolonged hydrochemical stability that is required for the growth of large crystals.

8.5 Calcrete

8.5.1 General characteristics

Calcrete is a term coined by Lamplugh (1902, 1907) that is now used to describe 'a near surface accumulation of predominantly calcium carbonate, which occurs in a variety of forms from powdery to nodular, laminar and massive. It results from the cementation and displacive and replacive introduction of calcium carbonate into soil profiles, sediments and bedrock, in areas where vadose and shallow

Table 8.2 A classification of calcrete types (Carlisle, 1983).

Calcrete classification	*Incorporated calcrete types*
Pedogenic calcrete	Caliche; Kunkar; Nari
Nonpedogenic superficial calcrete	Laminar crusts; case hardening
Nonpedogenic gravitational zone calcrete	Gravitational zone calcrete
Nonpedogenic groundwater calcrete	Valley calcrete; channel calcrete; deltaic calcrete; alluvial fan calcrete
Nonpedogenic detrital and reconstituted calcrete	Recemented transported calcrete; brecciated and recemented calcretes

phreatic groundwaters are saturated with respect to calcium carbonate' (Wright, 2007, p. 10). This definition is modified from Wright and Tucker (1991) and builds upon upon earlier definitions by Reeves (1976), Watts (1980) and Goudie (1973, 1983). Calcrete is referred to as *caliche* in the North American literature, but should not be confused with the sodium nitrate deposit of the same name. Calcretes have been classified in numerous ways, but the most fundamental distinction is between varieties that develop within soil profiles, in the vadose zone, and those that develop around the water table-capillary fringe or below (Table 8.2). Pedogenic calcretes owe their origin to largely vertical illuvial–eluvial processes, whereby calcium carbonate is redistributed within the soil profile; nonpedogenic calcretes form typically as a result of carbonate precipitation from groundwater. Recent overviews of calcretes are provided by Alonso-Zarza (2003), Wright (2007) and Dixon and McLaren (2009).

Calcretes are found both within sedimentary successions, where they indicate either a depositional hiatus or the post-sedimentary introduction of calcium carbonate into a profile, and at landform surfaces (Figure 8.1(a)), where they reflect subaerial weathering and pedogenesis after the creation of that land surface (Candy *et al.*, 2003; Candy and Black, 2009). They are generally white, cream or grey in colour, though pink mottling and banding is common, and occur in a variety of forms (Table 8.3). In pedogenic calcretes (Figures 8.6 and 8.7(a) and (b)), many of these forms fall within an evolutionary continuum (Gile, Peterson and Grossman, 1966; Netterberg, 1969, 1980; Reeves, 1970; Goudie, 1983), which has been used as a tool for comparing the relative morphogenetic development, or stage, of profiles (Bachman and Machette, 1977; Machette, 1985; see Figure 8.8 and Table 8.4). Calcified soils and chalky or powder calcretes (Stage I) may develop into nodular calcretes (Stage II) as calcium carbonate concretions grow within the host sediment or soil (Wieder and Yaalon, 1974; Magaritz, Kaufman and Yaalon, 1981) (Figure 8.7(a)). These concretions eventually coalesce to form honeycomb calcrete (Stage III), with the voids being filled with residual, unconsolidated host material. As the surface horizons become plugged, a hardpan calcrete (Stages IV and V) may evolve (Gile, Peterson and Grossman, 1966; Yaalon and Singer, 1974). Laminar calcretes consisting of finely banded carbonate often cap hardpans (James, 1972; Klappa, 1979a; Arakel, 1982; Warren, 1983; Verrecchia *et al.*, 1995) and may also encase nodules and boulders (Goudie, 1983). Brecciated calcrete (sometimes referred to as boulder calcrete; see Netterberg, 1969, 1978) may be the final step (Stage VI) in the evolution of a profile and represents the onset of solutional degradation of a hardpan. The whole sequence may be buried by upper soil horizons containing carbonate pisoliths (coated clasts), unless these more friable horizons have been eroded (Candy and Black, 2009). Many profiles also contain indurated root mats, indicating the importance of biogenic precipitation of carbonate in some situations (e.g. Klappa, 1980).

While this sequence is an idealised evolutionary model, mature pedogenic calcrete profiles (Figure 8.6) often consist of a laminar zone developed upon a platy or massive hardpan overlying a zone of nodules and chalky carbonate (Arakel, 1982; Goudie, 1983). The lower parts of the sequence may be truncated if the calcrete has developed within a veneer of sediment situated directly upon bedrock. Well-developed profiles are usually between 1.0 and 5.0 m thick. Hardpans are most commonly in the range 0.3–0.5 m thick (Goudie, 1984). Laminar calcrete zones are rarely more than 0.25 m thick (Goudie, 1983), but some forms can reach 2.0 m and exhibit a domal form (Wright, 2007).

Care needs to be taken when attempting to describe nonpedogenic calcretes using the Bachman and Machette (1977) scheme (Table 8.4). Although pedogenic and nonpedogenic calcretes share many macromorphological similarities (Wright, 2007), nonpedogenic varieties do not normally exhibit the well-organised profile structure shown in Figure 8.6, and may not go through the same stages of development (e.g. Arakel and McConchie, 1982; Arakel, 1986, 1991; Jacobson, Arakel and Chen, 1988; Khadkikar *et al.*, 1998; Tandon and Andrews, 2001; Nash and McLaren, 2003). Groundwater and channel calcretes also appear to develop much more rapidly than pedogenic varieties (e.g. Nash and Smith, 1998), so a densely cemented groundwater calcrete may represent carbonate accumulation over a much shorter timescale than a pedogenic hardpan of equivalent thickness and induration. Many calcrete profiles are polygenetic

Table 8.3 Morphological varieties of calcrete (Wright, 2007).

Calcrete type	*Characteristics*
Calcareous soil	Very weakly cemented or uncemented soil with small carbonate accumulations as grain coatings, patches of powdery carbonate including needle-fibre calcite, carbonate-filled fractures and small nodules
Calcified soil	Friable to firmly cemented soil with scattered nodules; 10–50 % carbonate
Chalky or powder	Fine loose powder of carbonate as a continuous body with little or no nodule development, consisting of micrite or microspar, with etched silicate grains, peloids and root and fungal-related microfeatures
Pedotubule or rhizocretionary	All or nearly all of the secondary carbonate occurs as root encrustations or calcifications, or around burrows, having a predominantly vertical structure
Nodular or glaebular	Soft-to-highly indurated concretions of carbonate, or carbonate cemented host material; the nodules can range in shape from spherical to elongate and typically consist of micrite or less commonly microsparite
Honeycomb	Partial coalescence of nodules with softer internodular areas produces a honeycomb-like effect
Mottled	Equivalent of nodular features where the host sediment is carbonate-dominated. It consists of irregular mottles of typically micritic carbonate that has cemented and replaced the original host grains
Hardpan or massive	Consists of a sheet-like indurated layer, which typically has a sharp top and a gradational base into chalky or nodular calcrete. It may reach thicknesses of >1 m in pedogenic calcretes and many metres in groundwater forms, and may be blocky or prismatic
Platy	Consists of cm-thick plates, up to tens of cm in diameter, commonly found above hardpan or chalky layers. Plates may be tabular or wavy in form and may exhibit crude lamination. Some are fragmented calcified root mats
Laminar	Consists of millimetre-scale sheets of laminated light or dark coloured carbonate. They commonly occur capping hardpan layers but can also occur within chalky layers or in the host sediment or soil. Most are only a few cm thick but some forms reach 2 m
Stringer	Closely related to laminar calcretes but not always well laminated. These are sheets of subvertical to subhorizontal carbonate, usually only a few cm thick, that penetrate into carbonate-rich hosts and are related to root mats
Pisolitic	Consists of mm-to-cm sized coated grains in layers typically only a few cm thick but up to metres thick at the bases of slopes. Pisolith laminae are generally micritic. Commonly occur above laminar calcretes
Breccia or conglomeratic	Disrupted hardpans or other forms, with fracturing due to mechanical processes and roots and tree heave

and incorporate complex pedogenic and nonpedogenic carbonate signatures (Alonso-Zarza, 2003). Furthermore, the superposition of numerous calcrete horizons can create complex composite sequences (Watts, 1980; Kaemmerer and Revel, 1991; Nash and Smith, 1998).

8.5.2 Distribution

Calcretes are the most widespread variety of desert duricrust. Soils containing calcic or petrocalcic horizons are estimated to cover some 20 million km^2 or 13 % of the Earth's land surface (Yaalon, 1988). This figure does not take into account the large areas over which nonpedogenic calcretes outcrop or subcrop. Groundwater calcrete, for example, covers many tens or even hundreds of thousands of square kilometres of Australia (Wright, 2007) and southern Africa. Calcic soil horizons develop where there is an annual net moisture deficit, such that carbonate precipitated during a drier seaon is not leached away in the following wetter season. Most contemporary pedogenic calcretes form in areas with warm to hot temperatures (mean annual temperature 16–20 °C) and low, seasonal rainfall. Goudie (1983) suggested that a mean annual precipitation in the range 400–600 mm was typical, while Royer (1999) concluded from a data set containing 1481 studies that carbonate-bearing soils correlate with a mean annual precipitation of <760 mm. Annual precipitation alone, however, does not provide a precise distributional control. The seasonality of rainfall and average temperature during the wetter months determine the broad pattern of evapotranspiration. This, more than any other factor,

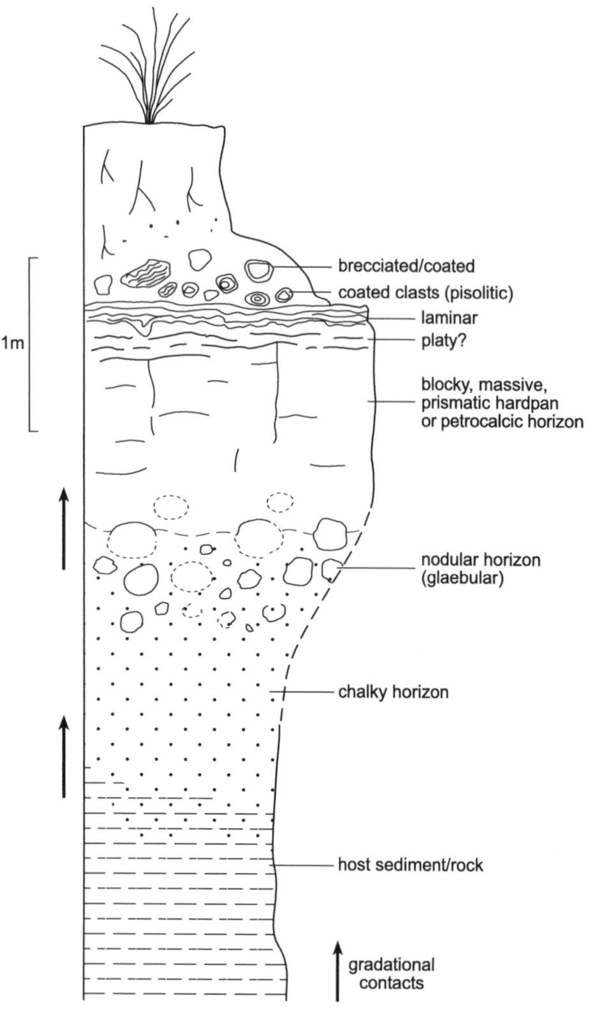

Figure 8.6 Idealised pedogenic calcrete profile showing a range of macroforms (from Wright, 2007).

limits the formation and preservation of most desert carbonate crusts. In southwest Australia, calcretes are best developed where mean annual rainfall reaches 800 mm but mean annual evaporation is 1900 mm (Semeniuk and Searle, 1985).

Such environmental boundaries apply mainly to pedogenic calcretes. Nonpedogenic crusts such as groundwater calcrete may accrete under more arid conditions and rhizocretionary calcretes in more humid environments (Mann and Horwitz, 1979; Semeniuk and Searle, 1985), such that the maximum mean annual rainfall under which calcretes can develop may be as high as 1000 mm (Mack and James, 1994). It has also been demonstrated experimentally that the freezing of calcium carbonate-saturated water leads to calcite precipitation (Ek and Pissart, 1965). Relict calcretes have been recorded from Pleistocene cold arid periods in various regions (e.g. Vogt and Corte, 1996) and might be similar to carbonate crusts currently forming in the Arctic (e.g. Bunting and Christensen, 1980; Lauriol and Clarke, 1999). Candy (2002) has described rhizogenic calcretes from Oxygen Isotope Stage 12 deposits in Norfolk, UK, probably formed during a period of climatic amelioration. Local factors may be of greater importance than macroclimatic variables in some environments (Wright, 2007). For example, Strong, Giles and Wright (1992) describe a 4 m thick Holocene calcrete from North Yorkshire, UK, that is suggested to have formed as a result of evapotranspiration promoted by the rapid free drainage of a host Pleistocene glacial gravel.

Pedogenic calcretes require long periods of time within a soil profile to reach maturity. As a result, the majority are found on surfaces that are, or were, geomorphologically stable. In southeast Spain, Candy and Black (2009) have identified that calcretes formed during interglacials/interstadials (when climates were semi-arid and a garrigue vegetation cover was present) are abundant compared to calcretes that date to glacial/stadial episodes (when the climate was more arid and slopes less stable). Most pedogenic calcretes mantle undulating or gently sloping terrain, with desert alluvial fans (e.g. Lattman, 1973; Wright and Alonso-Zarza, 1990; Stokes, Nash and Harvey, 2007), river terraces (e.g. Khadkikar *et al.*, 1998; Khadkikar, Chamyal, Ramish, 2000; Candy *et al.*, 2003; Candy, Black and Sellwood, 2004a) and pediments (e.g. Hüser, 1976; Van Arsdale, 1982; Dhir *et al.*, 2004) often calcretised. The presence of a well-developed calcrete can influence landscape development since it creates an impervious layer and may promote the erosion of overlying soil horizons. In some landscapes, pedogenic and groundwater calcrete layers act as caprocks (Maizels, 1990; Kaemmerer and Revel, 1991; Nash and Smith, 1998), while the erosion of noncemented sediments adjacent to valley calcretes may lead to the creation of inverted relief (Reeves, 1983; Mann and Horwitz, 1979). The presence of channel calcretes within fluvial networks has been shown to have an impact upon both morphology and long profile development and may limit the depth to which scour can occur (Nash and Smith, 2003; McLaren, 2004; Nanson *et al.*, 2005).

Extensive plateau calcretes are widespread in North America (Bretz and Horberg, 1949; Brown, 1956; Reeves, 1970; Machette, 1985), North Africa (Moseley, 1965; Abdel Jaoued, 1987) and the Middle East (Chapman, 1974). In southern Africa, exposures of calcrete occur throughout the Kalahari, with greatest thicknesses around palaeolakes and palaeodrainage features (Mabbutt, 1955; Netterburg, 1980; Nash, Shaw and Thomas, 1994; Blümel and Eitel, 1994; Nash and McLaren, 2003). There are

Figure 8.7 Examples of pedogenic and nonpedogenic calcrete profiles: (a) powdery to nodular pedogenic calcrete exposed near New Hanehai, central Botswana; (b) pedogenic and groundwater calcretes developed in the upper and lowermost units of Quaternary alluvial fan gravels, Tabernas Basin, southeast Spain; (c) valley calcrete exposed within a river terrace in the floor of the Okwa Valley, central Botswana; (d) section through a channel calcrete that forms a waterfall in a tributary to the Rambla de Tabernas, southeast Spain.

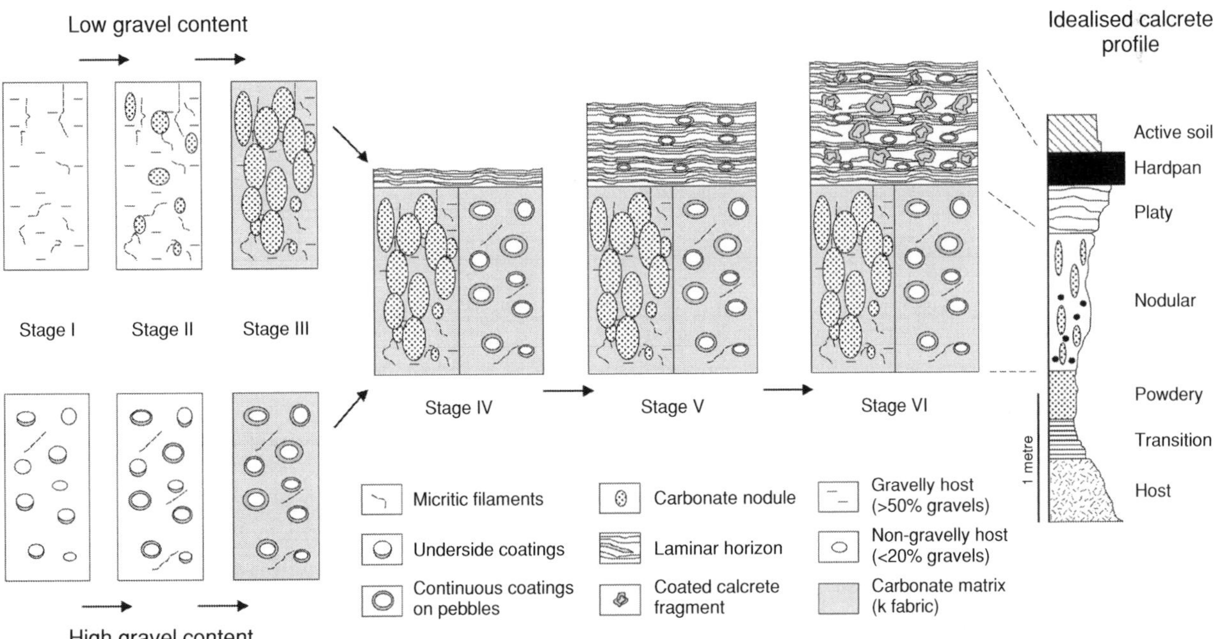

Figure 8.8 Stages of development of pedogenic calcretes developed in gravel-poor and gravel-rich sediments (based on Machette, 1985, and Alonso-Zarza, 2003).

Table 8.4 Stages in the morphogenetic sequence of carbonate deposition during calcrete formation (Bachman and Machette, 1977).

Stage	*Diagnostic carbonate morphology*
I	Filaments or faint carbonate coatings, including thin discontinuous coatings on the underside of pebbles
II	Firm carbonate nodules few to common but isolated from one another. The matrix in between nodules may include friable interstitial carbonate accumulations. Continuous pebble coatings present
III	Coalesced nodules in disseminated carbonate matrix
IV	Platy, massive indurated matrix, with relict nodules visible in places. The profile may be completely plugged with weak incipient laminar carbonate coatings on upper surfaces. Case hardening is common on vertical exposures
V	Platy to tabular, dense and firmly cemented. Well-developed laminar layer on upper surfaces. Scattered incipient pisoliths may be present in the laminar zone. Case hardening common
VI	Massive, multilaminar and brecciated profile, with pisoliths common. Case hardening common

comparatively few descriptions from Asia, though calcretes can be common in both pedogenic contexts (e.g. Tandon and Narayan, 1981; Dhir *et al.*, 2004, 2010; Khadkikar, 2005; Durand *et al.*, 2007) and in association with alluvial deposits (e.g. Khadkikar *et al.*, 1998; Khadkikar, Chamyal and Ramesh, 2000). In Australia, many calcretes have been interpreted as nonpedogenic crusts (Mann and Horwitz, 1979; Arakel, 1986, 1991; Jacobson, Arakel and Chen, 1988), but pedogenic varieties are also extensive (Warren, 1983; Semeniuk and Searle, 1985; McQueen, Hill and Foster, 1999).

The development of both pedogenic and nonpedogenic calcretes often bears little relationship to the materials upon which they accrete. Calcretes can occur on most sediments and rock types, both fresh or weathered, ranging from undifferentiated metamorphics (Durand *et al.*, 2007), granites (Scholz, 1972) and volcanics (Hay and Reeder, 1978) to gypsum bedrock (Lattman and Lauffenburger, 1974), alluvium (Khadkikar *et al.*, 1998), dune sand (Warren, 1983; Dhir *et al.*, 2004, 2010) and loess (Reeves, 1970; Gustavson and Holliday, 1999). It has been suggested that pedogenic calcretes form preferentially on basic rocks (Lattman, 1973; Wells and Schultz, 1979). However, this presupposes that the source of carbonate was from bedrock as opposed to surface inputs (Capo and Chadwick, 1999; Chiquet *et al.*, 1999). Arakel (1986) suggested that rock type is important in determining the distribution of pedogenic and groundwater calcretes in central Australia. There are numerous examples of calcretes formed by in situ alteration of limestone (Blank and Tynes, 1965; El Aref, Abdel Wahab and Ahmed, 1985) or chalk (Yaalon and Singer, 1974). Clearly, an immediate supply of calcium carbonate is conducive to calcrete formation but hydrological or atmospheric inputs from distant sources can be predominant.

8.5.3 Micromorphology and chemistry

Calcretes contain a variety of microtextures (Figure 8.9), which provide insights into the way the crust developed. Calcium carbonate in the majority of calcretes is in the form of micrite – aggregates of calcite crystallites less than about 100 μm in diameter. Micrite may occur in a variety of forms (Figure 8.9(a)), including 2–3 μm microcrystalline interflorescences, rounded calcite grains, calcans, neocalcans, neocalcitons surrounding pore spaces, concretions and nodules or glaebules (Sehgal and Stoops, 1972; Rabenhorst, Wilding and Girdner, 1984; Drees and Wilding, 1987; West *et al.*, 1988). Nodules exist in orthic, disorthic and allothic forms; orthic nodules appear to be similar to the surrounding matrix and developed in situ, disorthic nodules exhibit evidence of displacement and allothic nodules are relict forms inherited from a pre-existing soil (Wieder and Yaalon, 1974). The presence of micrite can indicate that precipitation was driven by rapid evaporation. However, micrite formation is not always syngenetic with calcretisation, since sparry calcite may alter to micrite through dissolution and reprecipitation (Kahle, 1977). Other common calcite crystal textures include microspar (Figure 8.9(b)), flower spar and both random and tangential fibres or needles (James, 1972), which may crystallise from carbonate-saturated water (Knox, 1977). The presence of well-developed rhombic microspar and sparry calcite crystals may indicate that carbonate precipitation occurred under relatively moist microconditions (cf. Nash and Smith, 2003). Needle fibre calcite and calcified filaments often occupy void spaces within powder, nodular, platy and hardpan calcrete types (Phillips and Self, 1987; Phillips, Milnes and Foster, 1987). Calcified filaments include biogenic structures (Figure 8.9(c)) such as root hairs, algal filaments

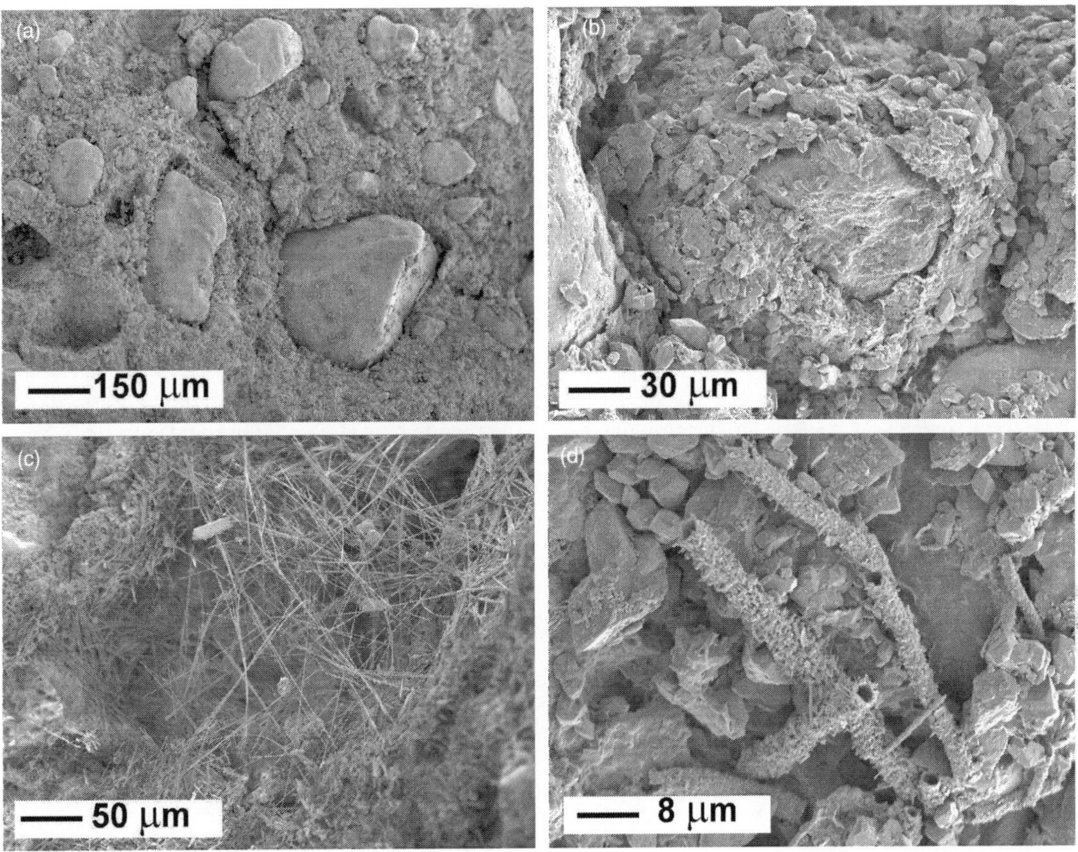

Figure 8.9 Scanning electron microscope images of calcrete cements from valley calcretes exposed in the Okwa and Hanehai valleys in the central Kalahari, Botswana: (a) α-fabric micritic cement with quartz grains and shell fragments; (b) quartz grain with microspar grain-coating cement; (c) hollow fungal filaments encrusted with micrite and small bladed and acicular calcite crystals; (d) needle-fibre and calcified fungal filaments (plus possible fruiting spores) lining a void with needle-fibres crossing the void.

and fungal hyphae (Klappa, 1978, 1979a, 1979b; Verrecchia and Verrecchia, 1994). Various concentric carbonate forms may also be present, including pellets, ooids and pisoliths (Bachman and Machette, 1977; Hay and Wiggins, 1980). Rhombic calcite crystals within calcrete pisoliths may accrete by the displacement of host material (Folk, 1971; Chafetz and Butler, 1980). Other nodular structures are usually micritic, as are the laminae which characterise many calcretes.

Two end members of calcrete microtexture can be identified (Wright, 1990). Alpha fabrics, which correspond to the k-fabrics of Gile, Peterson and Grossman (1966), consist of skeletal grains within a micritic to sparitic crystalline groundmass. They are often densely cemented and may contain mineral grains or clasts 'floating' within the carbonate matrix. Considerable care is needed when using crystal size within alpha-fabrics as a palaeoclimatic indicator or as a means to distinguish groundwater from pedogenic calcretes, since recrystallisation of micrite may have occurred (Wright, 2007). This can lead to the development of microspar and even sparry cements over time and can only be identified with certainty through the use of cathodoluminescence petrography (Budd, Pack and Fogel, 2002). Beta-fabrics, in contrast, exhibit abundant biogenic features, most notably the calcified remains of fungi and roots. Fungi can produce carbonate textures including alveolar septal fabrics and needle-fibre calcite (Verrecchia and Verrecchia, 1994; Borsato *et al.*, 2000) (Figure 8.9(d)). Calcification associated with plant roots also generates a range of diagnostic textures (Klappa, 1980; Warren, 1983; Semeniuk and Searle, 1985; Alonso-Zarza, 1999). Some calcrete laminae have been recognised to be of organic origin (Klappa, 1979a, 1979b; Wright, 1989; Verrecchia *et al.*, 1991) and one crystal texture consisting of millimetre-long calcite prisms has been attributed to a hypothetical colonial bacterium, *Microcodium* (e.g. Esteban, 1974; Klappa, 1978), or to the calcification of the cortical cells of roots (Kosir, 2004).

Goudie (1973), from a sample of 300 bulk chemical analyses, showed that calcretes on average comprised 79.28 % calcium carbonate (42.62 % CaO), 12.30 % silica, 3.05 % MgO, 2.03 % Fe_2O_3 and 2.12 % Al_2O_3. These figures, however, mask considerable chemical variability. This is in part due to the diverse ways in which calcretes can form, but is also related to the increase in carbonate content as a calcrete profile develops. Hutton and Dixon (1981) have shown from studies in South Australia that pedogenic hardpan calcretes are the most calcareous and powder calcretes the least, and that CaO levels typically decline down-profile with a parallel rise in the proportion of MgO (see also McQueen, Hill and Foster, 1999). In contrast, CaO contents of thick groundwater calcretes from southeast Spain have been shown to be remarkably homogenous (Nash and Smith, 1998).

Much of the recent work on pedogenic calcrete chemistry has focused upon understanding the range of factors that influence C and O stable isotope compositions. C and O isotopes have been used, for example, to determine the palaeotemperature and range of C_3 versus C_4 vegetation types present during calcrete formation and the partial pressure of atmospheric CO_2 during diagenesis (e.g. Andrews *et al.*, 1998; Cerling, 1999; Deutz *et al.*, 2001; Dhir *et al.*, 2010). However, as Kelly, Black and Rowan (2000) and Deutz, Montanez and Monger (2002) discuss, considerable care in interpretation is needed. The chemical signatures of pedogenic calcretes are invariably time-averaged as a result of calcite dissolution, recrystallisation and crystal overgrowth during profile development. This severely limits the temporal resolution of O and C stable isotope interpretations; conclusions should only be drawn where a calcrete is isotopically heterogenous (Wright, 2007). Strontium isotopes have been used successfully to identify the sources of carbonates within individual calcrete profiles. For example, Chiquet *et al.* (1999) used $^{87}Sr/^{86}Sr$ ratios to demonstrate that local weathered granite contributed at most 33 % (and as little as 3 %) of the Sr within pedogenic calcretes in central Spain, with atmospheric inputs of Ca and Sr from dust dominating. Similar results have been produced in New Mexico, where $^{87}Sr/^{86}Sr$ analyses suggest that atmospheric contributions to soil carbonate comprise at least 94 % and more likely 98 % of the total (Capo and Chadwick, 1999).

The carbonate mineralogy of calcretes is dominated by low-Mg calcite (Wright and Tucker, 1991) with variable amounts of dolomite present. In Australia, dolomite abundances are highest within the fine carbonate powders at the base of pedogenic calcrete profiles and decrease upwards with increasing induration (Hutton and Dixon, 1981). If abundances of diagenetic dolomite are particularly high, a dolocrete, as opposed to a calcrete, may be produced. The diagenetic processes leading to dolocrete formation can be determined by both the mineralogy of the host materials (Hay and Reeder, 1978; Hay and Wiggins, 1980; Hutton and Dixon, 1981) and the introduction of foreign ions (El Aref, Abdel Wahab and Ahmed, 1985). Significant proportions of high-Mg calcite have been reported in calcretes from Australia (Hutton and Dixon, 1981; McQueen, Hill and Foster, 1999) and the Kalahari (Watts, 1980). The latter study showed that calcrete profiles exhibit significant variations in the percentage of high-Mg calcite and dolomite. These variations in turn appear to be related to the occurrence of authigenic silica and silicates, notably the clay minerals palygorskite and sepiolite. Aragonite may also occur in some calcretes (Watts, 1980; Milnes and Hutton, 1983).

Quartz is the most important non-carbonate mineral in most calcretes, both in bulk samples and within the clay-size fraction. Silica may be present in other forms, including opal and chalcedony, as a result of mineral replacement or emplacement during diagenesis (e.g. Brown, 1956; Reeves, 1970; Arakel *et al.*, 1989; Nash and McLaren, 2003). Minor noncarbonate minerals include glauconite, grossularite, gypsum, haematite, magnetite, muscovite, rutile, tourmaline and zircon (Aristarain, 1971; Goudie, 1973). The other main components of calcretes are the clays palygorskite, sepiolite, illite, kaolinite, montmorillonite and chlorite (e.g. Aristarain, 1971; Bachman and Machette, 1977; Hay and Wiggins, 1980; Watts, 1980; Milnes and Hutton, 1983). In some calcretes, clays may be detrital (Beattie and Haldane, 1958; Shadfan and Dixon, 1984) or related to waterlogging (Hodge, Turchenek and Oades, 1984). In many instances, however, clay minerals are authigenic, produced by chemical interactions between high- and low-Mg calcite, silica and dolomite during diagenesis. There is strong evidence, for example, that sepiolite and palygorskite (attapulgite) are authigenic under alkaline conditions (Hassouba and Shaw, 1980; Watts, 1980; Mackenzie, Wilson and Mashhady, 1984; Singer, 1984). The types of authigenic clay mineral within a calcrete may also reflect the provenance of cementing agents. Relatively high levels of palygorskite and sepiolite in calcretes from the Murray Basin of South Australia are suggested to indicate that calcrete formation (and clay neoformation) occurred in an environment with abundant available Mg (Hutton and Dixon, 1981).

8.5.4 Mode of origin

The carbonate within a calcrete can come from a variety of sources (Figure 8.10). These include solid and dissolved carbonate introduced into the profile from

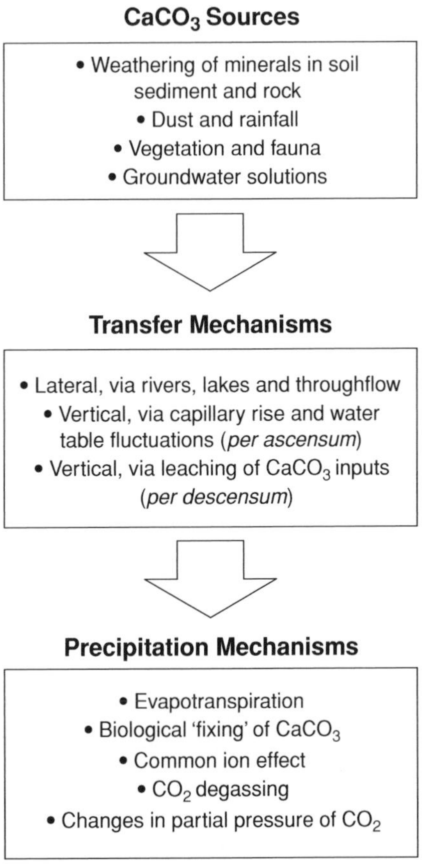

$CaCO_3$ Sources

- Weathering of minerals in soil sediment and rock
- Dust and rainfall
- Vegetation and fauna
- Groundwater solutions

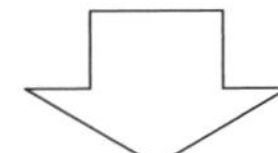

Transfer Mechanisms

- Lateral, via rivers, lakes and throughflow
- Vertical, via capillary rise and water table fluctuations (*per ascensum*)
- Vertical, via leaching of $CaCO_3$ inputs (*per descensum*)

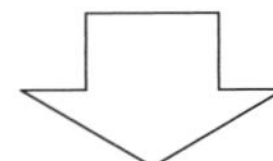

Precipitation Mechanisms

- Evapotranspiration
- Biological 'fixing' of $CaCO_3$
- Common ion effect
- CO_2 degassing
- Changes in partial pressure of CO_2

Figure 8.10 Sources, transfer mechanisms and factors triggering carbonate precipitation during calcrete development (after Goudie, 1984).

above (atmospheric dust, rainfall, volcanic ash, sea-spray, surface runoff, carbonate-rich plant materials and shells) or below (groundwater and weathered bedrock; see Goudie, 1983; Cailleau, Braissant and Verrecchia, 2004; Garvie, 2004). The contribution of specific sources varies spatially and with calcrete type. Atmospheric dust and rainfall inputs are of importance for pedogenic carbonate build-up (Capo and Chadwick, 1999; Chiquet *et al.*, 1999; Monger and Gallegos, 2000) and subsurface sources critical for groundwater, valley and channel calcrete development (e.g. Arakel and McConchie, 1982; Arakel, 1986, 1991; Nash and McLaren, 2003; Nash and Smith, 2003).

A variety of mechanisms to transfer carbonate to the site of calcrete formation has been described. These include vertical transfers within the soil or sediment profile driven by percolation, capillary rise and watertable fluctuations (the *per ascensum* and *per descensum* models of Goudie, 1983), and lateral transfers associated with the movement of carbonate-rich surface or groundwater, for example, within lacustrine or channel-margin settings. These mechanisms are not mutually exclusive, as carbonate may be moved laterally into the vicinity of the site of calcrete formation and then concentrated into particular horizons by vertical water movements. The importance of these transfer mechanisms for the formation of the various calcrete types is discussed below.

Calcium carbonate dissolved in water may be precipitated by a variety of processes including evaporation, evapotranspiration, an increase in pH to above pH 9.0 (Goudie, 1983), a decrease in the partial pressure of CO_2 within the soil or sediment profile (Schlesinger, 1985), a loss of CO_2 by degassing (e.g. as temperature increases; Barnes, 1965), freezing (Ek and Pissart, 1965) and the common ion effect (Wigley, 1973). Organic agency is important in the formation of some calcretes, with carbonate fixation driven by the life processes of organisms including algae (Kahle, 1977), fungi (Knox, 1977; Verrecchia, 1990; Verrecchia, Dumont and Rolko, 1990; Verrecchia, Dumont and Verrecchia, 1993), bacteria (Lattman and Lauffenburger, 1974; Schmittner and Giresse, 1999; Loisy, Verrecchia and Dufour, 1999; Braissant *et al.*, 2003) and cyanobacteria (Verrecchia *et al.*, 1995). Calcrete laminae, in particular, have been suggested to be of organic origin (Klappa, 1979a, 1979b; Wright, 1989; Verrecchia *et al.*, 1991). The formation of some calcretes is related to calcite precipitation around plant roots (Klappa, 1979a, 1979b, 1980; Warren, 1983; Semeniuk and Searle, 1985; Alonso-Zarza, 1999), but it is uncertain if this is a widespread phenomenon. Termites may also fix calcite in soils, as can earthworms and slugs (Canti, 1998; Monger and Gallegos, 2000).

Pedogenic calcrete formation can take place either through a simple, progressive process or a more dynamic one if carbonate build-up is interrupted by periods of erosion (Wright, 2007). In the progressive case, calcrete formation in a carbonate-poor host proceeds via the downward translocation, precipitation and accumulation of carbonate over time. If the host is carbonate-rich, this process may involve more complex phases of dissolution and reprecipitation. The progressive illuviation model forms the basis for the time-dependent stages of calcrete development shown in Table 8.4 and Figure 8.8. By the time Stage IV of this sequence has been reached, a profile is effectively plugged by a petrocalcic hardpan horizon. The subsequent ponding of water above the hardpan may promote the development of a laminar calcrete. Profile formation may also take place by the accumulation of calcified root mats, with rhizogenic calcretes up to 2 m thick formed in this way (Wright *et al.*, 1995). If the host is carbonate-rich and well indurated, calcretisation may instead take place along a progressively downward migrating alteration front. Eventually, the profile may break up, with features such as brecciated cobbles and bedrock and pseudo-anticlines formed (e.g. Price, 1925; Jennings and Sweeting, 1961; N. L. Watts, 1977, 1978, 1980). This sequence may be complicated if, for example, landscape erosion occurs, as this may cause reworking and a lowering of the carbonate profile by leaching and translocation.

There are debates over why brecciation occurs (Klappa, 1979c, 1980; Braithwaite, 1983), but it is probably a product of root growth (Klappa, 1980) or progressive displacive crystallisation during carbonate accretion (N. L. Watts, 1978). Displacive crystallisation occurs when the calcium carbonate content exceeds the host material's volumetric porosity (Gardner, 1972) and is common in noncarbonate hosts as calcite is unable to form adhesive bonds with noncarbonate materials (Chadwick and Nettleton, 1990). Chemical replacement rather than physical displacement of host material has been advocated to explain the purity of some pedogenic calcretes. Hubert (1978), for example, held that palaeosol calcretes in Connecticut show signs of 50–95 % replacement; McFarlane (1975) suggested that a fabric of floating garnet grains in some Kenyan calcretes represented the residuum of a replaced host material. Many calcretes exhibit signs of replacement of clay (Hay and Reeder, 1978), feldspars and quartz (Burgess, 1961; Chapman, 1974), but the latter may be diagenetic rather than syngenetic features. Displacive, replacive and passive void-filling can all take place (Yaalon and Singer, 1974; Watts, 1980; Nash, Shaw and Thomas, 1994), but their relative importance varies according to host lithology and environmental conditions.

Less research has been undertaken into the origins of nonpedogenic calcretes, despite the fact that the thickest calcrete accumulations in the world are associated with carbonate precipitation in the capillary fringe or phreatic zone. Nonpedogenic calcretes are frequently referred to as 'groundwater calcretes', a term often considered synonymous with phreatic, channel, valley or alluvial fan calcrete. These are, however, distinct calcrete types, and while there is a link between them, it is questionable whether such terms should be used interchangeably. The distinction between phreatic and groundwater calcretes is also unclear in the literature. Groundwater calcretes (*sensu stricto*) range from thin layers of nodules to large bodies of massive carbonate (Wright, 2007) and form by precipitation in the zone of capillary rise directly above the watertable (although Jacobson, Arakel and Chen, 1988, suggest that precipitation can also occur below this level). Phreatic calcretes, in contrast, are those in which cementation has occurred at or below the watertable (Arakel, 1986; Wright and Tucker, 1991). In both types, carbonate precipitation may be triggered by mechanisms including evaporation, evapotranspiration and degassing. Phreatophytic rhizocretions and calcified root mats produced by deep-rooting plants have been described in groundwater calcretes (Semeniuk and Meagher, 1981), suggesting that organic agency may be a factor in formation. Some groundwater calcretes are of economic significance as they are enriched in ions such as strontium (Kulke, 1974) or, more unusually, uranium minerals such as carnotite (Carlisle *et al.*, 1978; Arakel and McConchie, 1982; Briot, 1983).

Calcretes associated with drainage systems may form ribbon-like bodies extending many hundreds of kilometres in length. Some of the best developed examples in Australia are over 10 km in width and many metres in thickness (cf. Arakel and McConchie, 1982; Carlisle, 1983; Arakel, 1986, 1991; Morgan, 1993). Nash and McLaren (2003) distinguish between valley calcretes, which cement alluvium within broad, shallow, drainage courses but do not necessarily occupy the full valley width (e.g. Carlisle *et al.*, 1978; Mann and Horwitz, 1979; Carlisle, 1983; Reeves, 1983; Arakel, 1986, 1991; Jacobson, Arakel and Chen, 1988; Arakel *et al.*, 1989), and channel calcretes, which cement alluvium within confined impermeable bedrock channels or exhumed palaeochannels and may occupy the full channel cross-section (e.g. Maizels, 1990; Rakshit and Sundaram, 1998; Nash and Smith, 2003; McLaren, 2004). Valley calcretes in the Kalahari (Figure 8.7(c)) comprise sand-sized sediments bound by massive micritic cements and are suggested to have formed by relatively rapid calcium carbonate precipitation in a near-surface setting and close to the watertable, driven by evaporation or evapotranspiration (Nash and McLaren, 2003). In contrast, channel calcretes from southeast Spain (Figure 8.7(d)) are cemented by micrite and pore-filling sparite, with an increasing percentage of euhedral sparite crystals towards the base of the profile (Nash and Smith, 2003). These cements developed in conjunction with a fluctuating watertable, with the downward increase in crystal size caused by the greater duration of wetting in basal zones. Channel calcretes in southeast Spain occur most commonly at sites of gradient change where tributaries feed into main fluvial trunk valleys, so carbonate deposition may have been triggered by the common ion effect at sites of subsurface water mixing (Nash and Smith, 1998, 2003).

Calcretes preserved within alluvial fans form sheet-like bodies (Mack, Cole and Trevino, 2000) and reach total thicknesses of over 200 m (Maizels, 1987). Carbonate deposition may occur by the common ion effect, especially where alluvial fans debouch on to playa surfaces and fan groundwaters are forced to mix with more saline lake waters (Colson and Cojan, 1996). The distribution of calcrete across a fan surface reflects variations in particle size and the residence time of the solum in the zone of carbonate accumulation (Wright, 2007), itself affected by changing patterns of stable and unstable areas of the fan surface (Wright and Alonso-Zarza, 1990). Mack, Cole and Trevino (2000) describe calcretes from the Palomas Basin in northern Chihuahua, which are preferentially developed at the distal margins of alluvial fans where playa-fan sediments become finer and the watertable is at a shallower depth. Stokes, Nash and Harvey (2007) have analysed an Early–Middle Quaternary alluvial fan in southeast Spain that is encased and 'fossilised' by calcrete. In contrast to the example from Chihuahua, near-surface proximal fan calcretes contain increased quantities of interstitial sparite and microsparite cement, attributed to funnelling of groundwater through the fan head area, with the proportion of sparry cements also increasing with depth.

Pedogenic calcrete can be distinguished from nonpedogenic varieties by the presence of calcified filaments, needle-fibre calcite and fungal microfossils in the former (Vaniman, Chipera and Bish, 1994; Pimentel, Wright and Azevedo, 1996). As noted above, phreatophytic rhizocretions and root mats can also occur within nonpedogenic crusts. Wright (2007) suggests that pedogenic calcretes tend to be better developed within finer-grained sediments whereas groundwater calcretes are more prevalent in permeable, coarse deposits. Many valley, channel and alluvial fan calcretes tend to exhibit more coarsely crystalline cements (Nash and Smith, 1998, 2003; Mack, Cole and Trevino, 2000), although Wright and Tucker (1991) note the dominance of micritic cements in the majority of groundwater calcretes. Calcretes of mixed genesis can occur (Alonso-Zarza, 2003), with nonpedogenic forms developing pedogenic features as hydrological conditions change (Mann and Horwitz, 1979; Arakel, 1982, 1986). Calcretes of differing origins may form within the same sediment profile. Nash and Smith (1998), for example, describe alluvial fan sediments from southeast Spain that contain thick groundwater calcretes in their lowermost units and well-developed pedogenic calcretes on their upper surfaces (Figure 8.7(b)). Clear chemical distinctions are often masked by diagenesis, although Manze and Brunnacker (1977) suggested that $^{12}C/^{13}C$ and $^{16}O/^{18}O$ ratios can distinguish the two types. At present, there is little information on stable isotopes from nonpedogenic calcretes (Manze and Brunnacker, 1977; Jacobson, Arakel and Chen, 1988), but with further research it may be possible to recognise phreatic and pedogenic origins.

8.6 Silcrete

8.6.1 *General characteristics*

Silcrete has been described as a 'brittle, intensely indurated rock composed mainly of quartz clasts cemented by a matrix which may be a well-crystallised quartz, or amorphous (opaline) silica' (Langford-Smith, 1978, p. 3). It is a product of the cementation or replacement of surficial materials such as rock, sediment, saprolite or soil by various forms of secondary silica (Milnes and Thiry, 1992). Silicification occurs in near-surface environments and is a low-temperature physicochemical process, distinguishing silcrete from the various silica-cemented sedimentary rocks (Summerfield, 1979, 1981, 1983a, 1983b; Milnes and Thiry, 1992). Many silcretes also exhibit a porphyroclastic texture in thin section, as distinct from orthoquartzites where the texture is more even grained (Hutton, Twidale and Milnes, 1978). See Nash and Ullyott (2007) for a recent review.

Well-developed silcrete horizons are between 0.5 and 3 m thick, though >7 m thicknesses have been documented in the Kalahari (Nash, Thomas and Shaw, 1994) and >15 m beyond the arid zone in the Paris basin (Thiry and Simon-Coinçon, 1996). Silcrete may also occur as lenses within laterite (Wright, 1963; Alley, 1977; Langford-Smith and Watts, 1978) and calcrete profiles (Summerfield, 1982; Nash, Shaw and Thomas, 1994). Silcrete varies in appearance from opaline- or cryptocrystalline silica-cemented material ('Albertinia'-type silcrete; Frankel, 1952, Smale, 1973) to conglomeratic varieties (called 'puddingstone' in the UK; Summerfield and Goudie, 1980). Most commonly, silcrete consists of silica-cemented quartz grains, a variety sometimes termed 'terrazzo silcrete' (Smale, 1973; Soegaard and Eriksson, 1985), or a quartzitic material with a subconchoidal fracture through the grains and matrix (Summerfield, 1983a). Silcrete may also occur as a late-stage patina on rock outcrops (Hutton *et al.*, 1972; McFarlane, Borger and Twidale, 1992). A variety of terms has been used in the description of silcrete profiles, including massive, columnar, bulbous, nodular, glaebular and mammilated, reflecting the numerous surface morphologies of many exposures. The colour of silcrete is also highly variable: grey, white, buff, brown, red and green varieties have all been observed (Nash and Ullyott, 2007).

A number of classification schemes have been developed to distinguish the various silcrete types (e.g. Goudie, 1973; Smale, 1973; Wopfner, 1978). Summerfield (1983a, 1983b) suggested a classification based upon micromorphology (Table 8.5), which has been widely used and

Table 8.5 Micromorphological classification of silcrete (Summerfield, 1983a, 1983b).

Fabric type	*Fabric characteristics*
GS-fabric	*Grain-supported fabric* – skeletal grains (i.e. grains >30 μm diameter) constitute a self-supporting framework. Subdivided by cement type: (a) optically continuous quartz overgrowths (b) chalcedonic overgrowths (c) microquartz/cryptocrystalline/opaline silica in-fill
F-fabric	*Floating fabric* – skeletal grains comprise >5 % but float in matrix, grain solution or fretting common. Subtypes: (a) massive (glaebules absent) (b) glaebular (glaebules present)
M-fabric	*Matrix fabric* – skeletal grains comprise <5 %. Subtypes: (a) massive (glaebules absent) (b) glaebular (glaebules present)
C-fabric	*Conglomeratic fabric* – skeletal grains include fractured bedrock, gravel or duricrust fragments > 4 mm

adapted to describe other duricrusts. Milnes and Thiry (1992) divided silcretes into two groups, also on the basis of micromorphology, pedogenic (or complex) and groundwater (or simple), with Thiry (1999) adding a third category of silicification associated with evaporites. The classification shown in Figure 8.11 builds upon this scheme, subdividing silcretes into pedogenic and 'non-pedogenic' varieties, with the latter grouped on the basis of geomorphological context into groundwater, drainage-line and pan/lacustrine types.

8.6.2 Distribution

Silcretes are most widespread in Australia and in the Cape coastal zone and Kalahari basin of southern Africa, and have been documented in western Europe, North Africa, the Arabian Gulf and, to a lesser extent, in North and South America (see Table 4.6 in Nash and Ullyott, 2007, for an extensive reference list). Silcretes are exposed in a variety of geomorphological settings, of which the most important are as caprocks on residual hills or escarpments (Figure 8.12). Exposures of silcrete in low-lying parts of the landscape are rare, although outcrops have been described in, or adjacent to, valley floors and ephemeral lakes (Summerfield, 1982; Taylor and Ruxton, 1987; Thiry, Ayrault and Grisoni, 1988; Nash, Shaw and Thomas, 1994; Nash, Thomas and Shaw, 1994; Nash, McLaren and Webb, 2004; Shaw and Nash, 1998). Silcretes also outcrop as layers or lenses partway up slopes (Mountain, 1952; Young, 1978; Thiry, Ayrault and Grisoni, 1988; Milnes, Thiry and Wright, 1991), representing exhumed horizons that formed at earlier stages of landscape evolution.

There is a lack of consensus over the precise controls upon silcrete distribution. In Australia, many silcretes are exposed in areas that are less humid than where laterites are found but not as arid as those where calcretes predominate (Young, 1978; Mann and Horwitz, 1979; Milnes

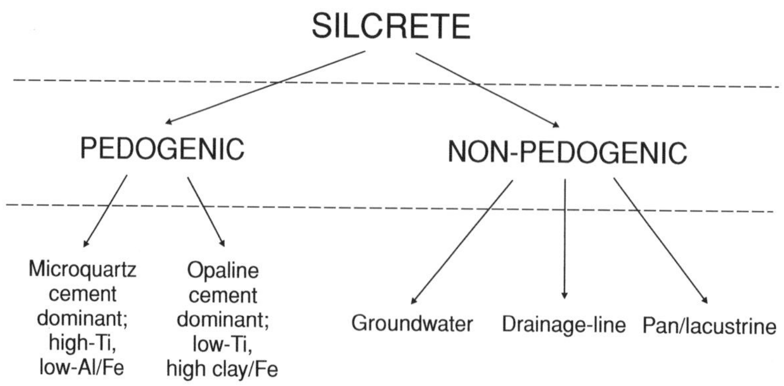

Figure 8.11 Geomorphological classification of silcrete (Nash and Ullyott, 2007).

Figure 8.12 Examples of silcrete profiles: (a) in situ pedogenic silcrete profile developed within deeply weathered bedrock, near Inniskillin, Western Cape, South Africa; (b) massive drainage-line silcrete at Samedupe Drift in the floor of the ephemeral Boteti River, Botswana; (c) superposed groundwater silcrete lenses developed within Fontainebleau Sand, Bonnevault Quarry, Paris basin, France (with Medard Thiry for scale).

and Thiry, 1992). This prompted the suggestion that Australian silcretes originated under semi-arid or markedly seasonal, wet and dry climates. However, even authors that accepted this view had differing opinions as to whether the crusts were the result of weathering (Brückner, 1966; McGowran, Rutland and Twidale, 1977), hydrological processes (Smale, 1973, 1978) or pedogenesis (Hutton, Twidale and Milnes, 1978; Callen, 1983). It was also argued that many silcretes were relict horizons from deep chemical weathering profiles which formed under humid or seasonally wet, tropical conditions (Woolnough, 1930; S. H. Watts, 1977, 1978a, 1978b; Butt, 1985) and were preserved under subsequent aridity (Twidale, 1983). In contrast, silcretes in the Kalahari are found in association with calcretes (Summerfield, 1982; Shaw and De Vries, 1988; Nash, Shaw and Thomas, 1994), and there are numerous sites where silica has replaced calcrete cements (Nash and Shaw, 1998; Nash, McLaren and Webb, 2004). While the presence of silcrete as lenses within lateritic (Wright, 1963; Alley, 1977; Langford-Smith and Watts, 1978) or calcrete profiles (Summerfield, 1982) may suggest formation under tropical or semi-arid conditions respectively, such occurrences do not necessarily prove simultaneous development (Smale, 1973).

8.6.3 Micromorphology and chemistry

Silcretes are chemically simple (Table 8.6), by definition comprising >85 % (and more often >95 %) silica with minor amounts of titanium, iron and aluminium oxides and resistate trace elements (Summerfield, 1983a). Titanium is usually present as anatase and may be disseminated throughout the matrix or concentrated within geopetal structures. TiO_2 may be enriched to >1 % (Summerfield, 1979, 1983d; Thiry and Millot, 1987) and exceed 20 % in silcrete 'skins' (Hutton *et al.*, 1972). TiO_2 content may be related to the environment of silicification but is also

Table 8.6 Bulk chemistry of silcretes in Australia and southern Africa (analyses by X-ray fluorescence).

Region		SiO_2	TiO_2	Al_2O_3	Fe_2O_3	*MnO*	*MgO*	*CaO*	Na_2O	K_2O	P_2O_5	SO_3	*LOI*[f]
Inland Australia[a]	Mean	94.81	2.09	0.50	0.76	0.01	0.10	0.18	0.07	0.08	0.05	no data	1.22
$n = 72$	SD[e]	*5.06*	*4.06*	*0.77*	*1.50*	*0.01*	*0.09*	*0.20*	*0.16*	*0.09*	*0.08*	*no data*	*1.18*
Eastern Australia[b]	Mean	97.58	0.42	0.30	0.79	0.02	0.10	0.02	0.02	0.03	0.02	0.004	0.47
$n = 63$	SD	*1.76*	*0.36*	*0.37*	*1.28*	*0.05*	*0.14*	*0.02*	*0.02*	*0.06*	*0.02*	*0.001*	*0.44*
Kalahari[c]	Mean	91.63	0.12	1.69	0.86	0.01	0.98	0.85	0.36	0.95	0.01	no data	2.80
$n = 48$	SD	*4.04*	*0.06*	*0.98*	*0.54*	*0.00*	*0.66*	*1.51*	*0.35*	*0.86*	*0.00*	*no data*	*1.51*
Cape Coastal[d]	Mean	95.04	1.79	0.61	1.28	0.01	0.28	0.13	no data	0.05	0.04	no data	0.99
$n = 66$	SD	*2.54*	*0.58*	*0.46*	*1.70*	*0.00*	*0.20*	*0.37*	*no data*	*0.12*	*0.03*	*no data*	*0.67*

[a]Calculated from analyses in Hutton *et al.* (1972), Senior and Senior (1972), S.H. Watts (1977), Callender (1978), Senior (1978), Wopfner (1978), O'Neill (1984), Collins (1985), Thiry and Milnes (1991), Van Dijk and Beckmann (1978), Tait (1998) and Webb and Golding (1998).
[b]Calculated from analyses in Taylor and Smith (1975), O'Neill (1984), Collins (1985) and Webb and Golding (1998).
[c]Calculated from analyses in Summerfield (1982), Nash and Shaw (1998), Nash, Thomas and Shaw (1994) and Nash, McLaren and Webb (2004).
[d]Calculated from analyses in Frankel and Kent (1938), Bosazza (1939), Frankel (1952) and Summerfield (1983d).
[e]Standard deviation.
[f]Loss on ignition.

influenced by the host material mineralogy and the ratio of matrix to detrital grains (e.g. Nash, Thomas and Shaw, 1994; Webb and Golding, 1998). Alumina content usually reflects the presence of authigenic, illuviated or inherited clays. Haematite and goethite may be retained during silicification (Meyer and Pena dos Reis, 1985; Bustillo and Bustillo, 1993; Armenteros, Bustillo and Blanco, 1995; Ballesteros, Talegón and Hernández, 1997). Glauconite has been reported in some Kalahari silcretes and may indicate suboxic, partially reducing groundwater conditions during silicification (Nash, McLaren and Webb, 2004).

The examination of silcretes in thin section and under scanning electron microscope can provide important insights into their diagenetic history. At this scale, silcretes can be seen to comprise varying proportions of detrital minerals, silica cements and void spaces (which may be partially or completely filled with secondary silica or other minerals). The nature of these components reflects not only the diagenetic processes operating during formation, but may, in part, be inherited from the host sediment, regolith or bedrock.

Four main forms of silcrete can be identified at a microscale (Table 8.5): those that have quartzitic grain-supported (GS-) or conglomeratic (C-) fabrics (where clasts of >4 mm diameter are present), those where skeletal grains comprise >5 % of the silcrete but are dispersed to create a fabric of floating clasts (F-fabric silcretes) and matrix-dominated varieties with <5 % skeletal grains (M-fabric silcretes). These fabric types are not mutually exclusive. For example, F-fabrics may grade into GS- or M-fabrics within individual profiles or even single samples (Ullyott *et al.*, 2004). All four fabrics occur across the range of silcrete types. However, GS-fabrics are most common in arid zone silcretes. M- and F-fabric silcretes are suggested to form by pedogenesis within deeply weathered profiles (S. H. Watts, 1977, 1978a; Summerfield, 1979, 1983a) or via the replacement of pre-existing calcrete matrix materials (Nash, Thomas and Shaw, 1994; Nash, McLaren and Webb, 2004), the fabrics developing by silicification of the former matrix or by displacive crystallisation (Butt, 1985).

Silcretes may be cemented by opal, chalcedony (a microcrystalline fibrous form of silica comprising intergrowths of quartz and the silica polymorph moganite; see Heaney, 1995), cryptocrystalline silica or quartz. The matrix of F- and M-fabric silcretes usually consists of microquartz, cryptocrystalline or opaline silica and may include silt- or clay-sized detrital quartz, anatase, iron oxides and clay minerals (Frankel and Kent, 1938; Smale, 1973). GS-fabric silcretes are commonly cemented by microquartz or cryptocrystalline silica and may exhibit syntaxial quartz overgrowth cements. These indicate that silica precipitation occurred slowly (Thiry and Millot, 1987) and that intergranular impurities or coatings were absent (Heald and Larese, 1974). More ordered silica polymorphs typically occur towards the top of a profile, particularly within pedogenic silcretes. Thiry and Millot (1987) identified a sequence of silicification starting with amorphous opal and progressing through chalcedony to microquartz and megaquartz. The variety of silica within the matrix depends not only upon the polymorph initially precipitated but also the diagenetic history of the material. Silica polymorphs may transform by dissolution and recrystallisation into other silica species in a stepwise manner

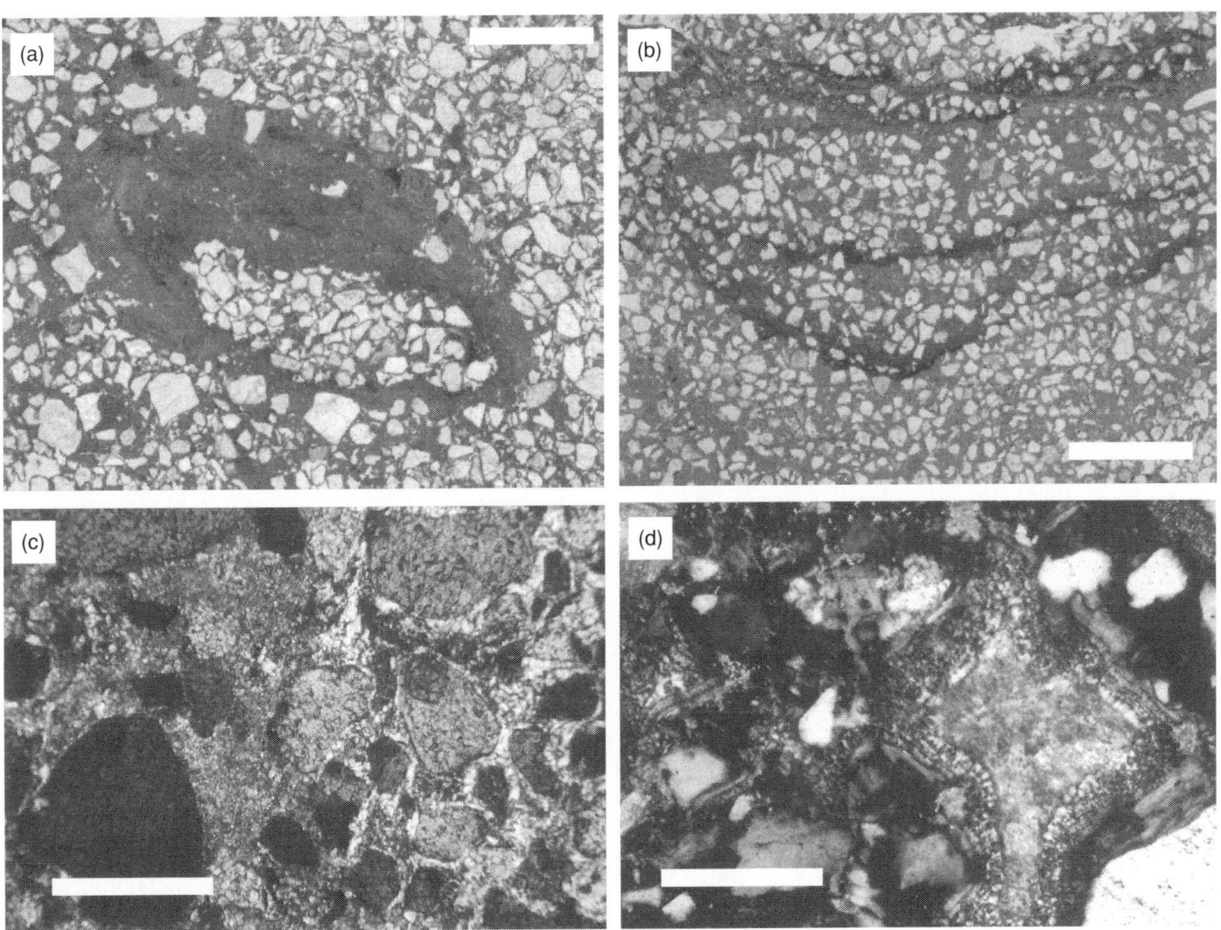

Figure 8.13 Photomicrographs of pedogenic and groundwater silcretes: (a) grain supported to floating fabric glaebular pedogenic silcrete from Stuart Creek, South Australia, consisting of quartz grains surrounded by a microquartz and opal matrix (plain polarised light; scale bar 2 mm); (b) grain supported to floating fabric pedogenic silcrete from Stuart Creek, South Australia, consisting of quartz grains surrounded by a microquartz and opal matrix and exhibiting anatase-rich geopetal laminations (plain polarised light; scale bar 2 mm); (c) partially silicified nonpedogenic calcrete from a depth of 4.0 m beneath the floor of Kang Pan, near Kang, Botswana, showing patchy replacement of carbonate cement (left of view) by chalcedonic and cryptocrystalline silica (right of view), with a sharp boundary between carbonate and silica cements (cross-polarised light; scale bar 0.5 mm); (d) complex floating fabric pan/lacustrine silcrete from Sua Pan, Botswana, consisting of quartz grains cemented by chalcedonic and cryptocrystalline silica, which is partially replaced by micritic calcite. The sample also contains voids which are lined by opal and chalcedony and infilled with late-stage calcite (cross-polarised light; scale bar 0.5 mm). Micrographs (a) and (b) both courtesy of John Webb.

over time (Dove and Rimstidt, 1994). Amorphous opal-A transforms to near-amorphous opal-CT, better-ordered opal-CT, cryptocrystalline quartz or chalcedony and finally microcrystalline quartz (see Nash and Hopkinson, 2004). The extent of paragenesis is determined by processes including silica complexation, adsorption by clay minerals and the neoformation of clays and other silicates (Williams, Parks and Crerar, 1985), and may be accompanied by textural changes.

A range of geopetal structures have been described. The most common are illuvial colloform features (Figure 8.13(a) and (b)), cusp-like structures containing layers of silica and TiO_2, or silica and iron or manganese oxides (Frankel and Kent, 1938; Terry and Evans, 1994; Ballesteros, Talegón and Hernández, 1997; Thiry *et al.*, 2006). Laminated and unlaminated conical or cap-like structures have been identified on top of sediment clasts (Callen, 1983; Van Der Graaff, 1983; Thiry and Milnes, 1991). Glaebules (concretionary or nodular structures; see Brewer, 1964) are documented within F- and M-fabric silcretes of presumed pedogenic origin (Summerfield, 1983a, 1983d). They can be distinguished due to darker anatase or iron oxide staining, and have either a concentrically zoned or undifferentiated internal fabric. Glaebules are often used as diagnostic indicators of pedogenesis. However, some glaebular structures within Kalahari silcretes formed at or near the watertable (Shaw and Nash, 1998) or were inherited by silica replacement of pedogenic calcretes (Nash, Thomas and Shaw, 1994), so caution is needed. It is also important to ascertain whether glaebules formed in situ during silicification or had a detrital origin (Hill, Eggleton and Taylor, 2003).

Void spaces within silcretes may be filled by a variety of minerals (Figure 8.13(c) and (d)). The most common sequence is for massive or laminated opal to line void margins, followed by layers of chalcedony, microquartz and megaquartz (Summerfield, 1982; Thiry and Milnes, 1991; Rodas *et al.*, 1994). The increasing organisation of silica minerals towards the centres of voids may be accompanied by fractionation of silica isotopes, with Basile-Doelsch, Meunier and Parron (2005) reporting δ^{30}Si values as low as -5.7 ‰ for megaquartz cements within groundwater silcretes. The nature of the host material may also influence silica mineralogy (e.g. S. H. Watts, 1978a, 1978b; Summerfield, 1982; Wopfner, 1983; Webb and Golding, 1998), with opal prevalent in clay-rich substrates and microquartz common in more porous sediments or carbonates (Callen, 1983; Thiry and Ben Brahim, 1990; Benbow *et al.*, 1995). Void fills of other minerals such as calcite (Summerfield, 1982, 1983c; Nash and Shaw, 1998), iron oxides or zeolites (Terry and Evans, 1994) are also reported.

8.6.4 Mode of formation

There is considerable debate concerning the mode of formation of silcrete. This is not helped by the fact that, with the exception of biogenic silcretes in Botswana (Shaw, Cooke and Perry, 1990), duripans in North America (Flach *et al.*, 1969; Chadwick, Hendricks and Nettleton, 1989; Dubroeucq and Thiry, 1994) and dorbanks in South Africa (Ellis and Schloms, 1982), most are relict features. There are a number of potential silica sources. Chemical weathering of silicate minerals is the most important, although replacement of quartz by carbonates may be significant in some environments. Surface solution of quartz dust may provide a source in deserts (Summerfield, 1983a), together with biological inputs from silica-rich plants and microorganisms. Silica may be transported to the site of silcrete accretion by two mechanisms. Quartz dust, plant phytoliths, sponge spicules and diatoms may be blown considerable distances by the wind. All other transfers rely upon the movement of silica in solution, usually in the form of undissociated monosilicic acid (either as the monomer H_4SiO_4 or the dimer $H_6Si_2O_7$; see Dove and Rimstidt, 1994). Waterborne transfer mechanisms can be divided into lateral or vertical movements, although both can occur in combination. Lateral transfers are of importance for the development of nonpedogenic silcretes (Stephens, 1971; Hutton *et al.*, 1972; Hutton, Twidale, Milnes, 1978). Silica transfer in river water, for example, is a key control upon the formation of drainage-line silcretes (Shaw and Nash, 1998), while pan/lacustrine silcrete development is dependent upon the movement of dissolved silica into topographic lows. Subsurface lateral water transfers are vital for groundwater silcrete formation and may provide a silica source for other silcrete types. Vertical transfer mechanisms involve movements of silica-rich solutions supplied from surface waters, leaching of overlying sediments (S. H. Watts, 1978a, 1978b; Marker, McFarlane and Wormald, 2002) or aeolian or biological sources (Goudie, 1973). Dissolved silica may be washed down through a soil or sediment or drawn upwards from the water table. In pedogenic silcretes, downward-moving silica may be supplied by dissolution during profile development, with silica mobilisation and precipitation promoted by wetting and drying respectively (Thiry, 1999).

Silica precipitation in near-surface environments is controlled by the concentration of silica in solution and the duration of wetting/drying cycles (Knauth, 1994). If a solution is supersaturated with respect to amorphous silica, silica monomers may polymerise and aggregate to form colloids, which may in turn precipitate as opal-A (Williams, Parks and Crerar, 1985). Quartz may nucleate from dilute solutions providing there are no inhibiting minerals or ions present (chlorite, illite, haematite, Fe or Mg), the solutions are slow moving and there is a template such as a quartz grain to initiate deposition (e.g. Heald and Larese, 1974; Ollier, 1978; Summerfield, 1982). Crystal size and order are controlled by the speed of precipitation, itself influenced by the host material permeability and evaporation rate.

Silica precipitation is also influenced by pH, Eh, evaporation, the presence of other elements and minerals in solution, biological life processes and, to a minor extent in near-surface settings, temperature and pressure. Silica is only weakly soluble in neutral pH water (10 ppm at 25 °C for quartz; see Siever, 1972) with solubility increasing significantly only at above pH 9.0 (Dove and Rimstidt, 1994). The solubility of amorphous silica reaches 800 ppm at pH 10.6 (Alexander, Heston and Iler, 1954); such alkaline conditions are not unusual in saline lakes (see Chapter 15). Following the dissolution of silicates at pH 9.5 to 10.5, desiccation can lower the pH to about 7.0, precipitating chert (Peterson and Van Der Borch, 1965;

Wheeler and Textoris, 1978) or hydrous silicates such as magadiite ($NaSi_7O_{13}(OH_3)\cdot 3H_2O$), which convert to chert or opaline silica gel when leached by dilute water (Renaut, Tiercelin and Owen, 1986). However, this does not imply that all silcretes are of lacustrine origin as silica gels also form in terrestrial settings (Kaiser, 1926; Elouard and Millot, 1959).

Environmental Eh has a complex effect upon silica solubility. Quartz placed in solutions containing ferrous iron, for example, becomes more soluble once exposed to oxidising conditions (Morris and Fletcher, 1987). Evaporation is significant in the development of drainage-line (Shaw and Nash, 1998), pan/lacustrine (Stephens, 1971; Wopfner, 1978, 1983; Milnes, Thiry and Wright, 1991) and shallow groundwater silcretes (e.g. Lamplugh, 1907; Waugh, 1970; Mountain, 1952) and the upper parts of pedogenic silcrete profiles (Van Der Graaf, 1983; Thiry and Millot, 1987; Webb and Golding, 1998). Silica solubility may be reduced if adsorption on to iron or aluminium oxides occurs. The presence of NaCl and oxides and ions including Fe^{3+}, UO_2^{+}, Mg^{2+}, Ca^{2+}, Na^{2+} and F^{-} may enhance solubility (Dove and Rimstidt, 1994). Where significant quantities of salts are present in solution, silica colloids may aggregate to form hydrated opaline silica gels (Iler, 1979). The mixing of silica-rich groundwater with downward-percolating water rich in NaCl may precipitate silica (Frankel and Kent, 1938; Frankel, 1952; Smale, 1973); under high salinity, chalcedony may precipitate directly (Heaney, 1993). Biogenic processes may be of local significance for silica fixation (Shaw, Cooke and Perry, 1990; McCarthy and Ellery, 1995).

Pedogenic silcretes typically develop on stable basement or basin-marginal areas over time periods of $>10^6$ years (Thiry, 1978; Callen, 1983; Milnes and Thiry, 1992). Some of the best developed profiles are in Australia and along the south coast of South Africa (Figure 8.12(a)). It is unlikely that many pedogenic silcretes formed under present-day arid climates; most are instead relict features and date from the Palaeogene or Early Neogene (Nash and Ullyott, 2007). Thiry (1999) identifies two types of pedogenic silcrete: those with abundant microquartz and Ti-enrichment that lack clay minerals and iron oxides, and more opaline 'duripans' in which clay and iron are retained. Formation occurs due to cycles of downward flushing of silica-bearing water followed by silica precipitation, possibly under seasonal climates with high rates of evaporation (Webb and Golding, 1998). This produces geopetal colloform structures along water pathways, together with glaebules and rootlets, and may be associated with evidence of bioturbation (Summerfield, 1983d; Thiry *et al.*, 2006; Terry and Evans, 1994; Lee and Gilkes, 2005). Pedogenic silcrete profiles are usually complex. A 'typical' profile contains two sections (Figure 8.14): an upper part, which exhibits a columnar structure with well-ordered silica cements, and a lower part, which is weakly cemented by poorly ordered silica (Milnes and Thiry, 1992). Semi-continuous or lenticular pedogenic silcretes, pedogenic silcrete 'skins', discrete nodules and veins have also been described.

Groundwater silcretes occur as discontinuous lenses or sheets in a range of weathered and unweathered materials (e.g. Senior and Senior, 1972; Wopfner, 1983; Meyer and Pena dos Reis, 1985; Thiry, Ayrault and Grisoni, 1988; Lee and Gilkes, 2005). They develop near the ground surface but have been documented at depths of up to 100 m (Thiry, 1999; Basile-Doelsch, Meunier and Parron, 2005), sometimes underlying or overlying pedogenic silcretes. Groundwater silcretes exhibit simple fabrics with good preservation of host structures. The main genetic models arise from work in the Paris basin, where silcretes occur in arenaceous formations and lacustrine limestones (Figure 8.12(c)). Formation in arenaceous bedrock is thought to have taken place around zones of groundwater outflow within valleys, with dissolved silica supplied by the chemical breakdown of clay minerals in the host sediment (Figure 8.15). Silicification occurred during various phases of Plio-Quaternary landscape incision to create a series of superposed silcrete lenses. Development is thought to have been relatively rapid, with each lens taking 30 000 years to form (Thiry, Ayrault and Grisoni, 1988). Similar processes have been proposed for groundwater silcrete formation in Australia (Thiry and Milnes, 1991; Simon-Coinçon *et al.*, 1996; Lee and Gilkes, 2005). Carbonate-hosted groundwater silcretes may have derived their silica from overlying sands and soils, with silicification proceeding by infilling of karst voids and carbonate replacement (Ribet and Thiry, 1990; Thiry and Ben Brahim, 1997).

Drainage-line silcretes are closely related to groundwater types but develop *within* alluvial fills in current or former fluvial systems as opposed to within bedrock *marginal* to valley systems (Nash and Ullyott, 2007). Silcretes develop at sites that are subject to seasonal wetting/ drying or in zones of watertable fluctuation (Taylor and Ruxton, 1987; McCarthy and Ellery, 1995). One of the best examples is the Mirackina Conglomerate in South Australia, which formed as a result of silica-rich groundwater moving through channel alluvium within a 200 km long palaeodrainage system (Barnes and Pitt, 1976; McNally and Wilson, 1995). Few genetic models for drainage-line silicification exist. Silcretes exposed in the Boteti River, Botswana (Figure 8.12(b)), are suggested to have formed by the accumulation of clastic and phytolithic silica from floodwater and dissolved silica in groundwater, with silica precipitation induced by

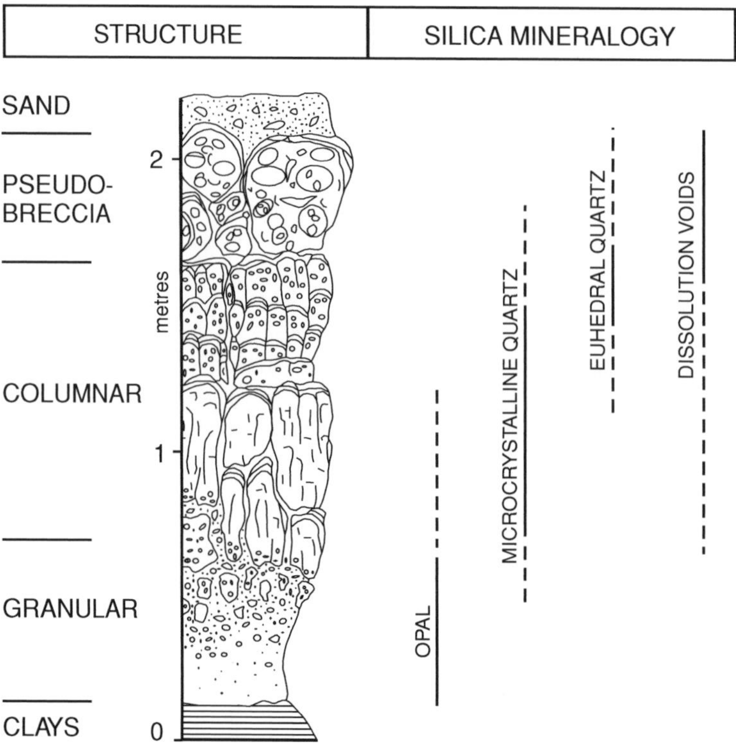

Figure 8.14 Schematic representation of a 'typical' pedogenic silcrete profile, showing silcrete structures and the vertical arrangement of silica cements (after Milnes and Thiry, 1992; Thiry, 1999).

evapotranspiration from seasonal pools (Shaw and Nash, 1998). Massive silcrete layers within the channel alluvium are suggested to have developed in response to salinity shifts associated with movements of the wetting front during flood events.

Pan/lacustrine silcretes develop within, or adjacent to, ephemeral lakes, pans or playas (Goudie, 1973). Phases of silica mobility and precipitation are driven by changes in pH and salt concentration during cycles of flooding and evaporation (Summerfield, 1982). The zone of maximum mixing and precipitation occurs around the lake margin and immediately above the water table (Thiry, 1999), which may explain why silicification is often linked to lake regression (Ambrose and Flint, 1981; Bustillo and Bustillo, 1993, 2000; Armenteros, Bustillo and Blanco, 1995). Biological fixing of silica may also occur. Sua Pan in Botswana contains sheet-like silcretes developed from desiccated colonies of the silica-fixing cyanobacteria *Chloriflexus* (Shaw, Cooke and Perry, 1990). Pan/lacustrine silcretes may contain a variety of minerals in addition to silica, depending upon the chemistry of the cemented lake sediments. Glauconite-illite occurs within green pan silcretes in the Kalahari (Summerfield, 1982). Silica replacement is commonplace in such environments and may lead to the development of intergrade cal-silcrete and sil-calcrete crusts (Nash and Shaw, 1998).

8.7 Desert rock coatings

8.7.1 General controls on formation

Thin rock coatings occur on almost all exposed rock surfaces found in deserts. Over a dozen varieties have been identified (Table 8.7), some of which exist in different types. Many coatings interdigitate, producing complex microstratigraphic sequences (Dorn, 2007). Some, such as iron films and rock varnish, appear similar but have contrasting origins and hence palaeoenvironmental significance. The most common coatings found in deserts are rock varnish, silica glaze and iron films. Carbonate skins and salt crusts are also important in some localities. For further information see Dorn (1998, 2007, 2009).

For any rock coating to form, a bare and stable rock surface is required. In some cases, coatings may develop

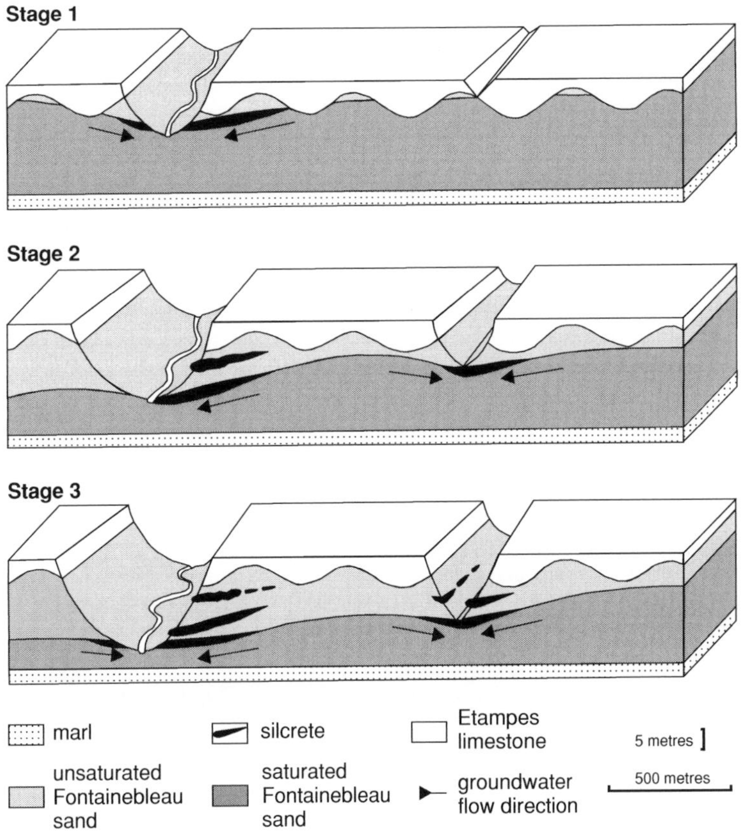

Figure 8.15 Model of groundwater silcrete development in the Paris basin (after Thiry, Ayrault and Grisoni, 1988) showing the development of superposed silcrete lenses within bedrock adjacent to a progressively incising valley system.

in subsoil environments or fissures and are 'exhumed' during soil erosion or rock spalling; these coatings are not considered here. During the initial stages of development, faster-growing coatings will tend to dominate over slower-growing types (Dorn, 2009). In many environments, this means that lithobiontic coatings (e.g. those associated with lichen and fungal growth; see Souza-Egipsy *et al.*, 2004) will out-compete the formation of most natural inorganic coatings. Many fast-growing lithobionts are capable of weathering rock surfaces (Viles, 1995), including pre-existing rock coatings. Assuming that a stable rock surface is available, a mechanism is required to transport the rock coating constituent minerals to the rock face. In many cases, constituent availability may directly determine the type of coating that develops (Dorn, 1998). For example, Fe- and Mn-rich water flowing over a rock face may result in the formation of a heavy metal skin rather than rock varnish (Dorn, 2009), since a supply of clay minerals (often absent from such waters) is essential to the development of a varnish (Potter and Rossman, 1977).

8.7.2 Rock varnish

Rock varnish (often called 'desert varnish') ranges in thickness from between 2.0 μm and 1.0 mm (Krumbein and Jens, 1981; Dorn and Oberlander, 1982; Dorn, 1991). It is usually orange, grey, brown or black in colour (Figure 8.16) and has a lustrous appearance. Most varnishes comprise alternating layers of light and dark material ranging in thickness from nanometres up to 20 μm, the darker laminae containing a higher proportion of MnO_2 (Perry and Adams, 1978; Dorn, 1984; Dragovich, 1988; Krinsley, Dorn and Tovey, 1995). The composition of individual layers is linked to environmental conditions, since Mn fixation is dependent on pH; alkaline conditions restrict fixation and orange-coloured varnish develops (Dorn and Oberlander, 1982). Varnishes exhibit other microtextures including dendritic growth patterns (Billy and Cailleux, 1968) and lamellate and botryoidal structures (Dorn and Oberlander, 1982). A variety of factors can affect the textures present, including the rate of dust supply during formation, the clay mineralogy, the relative

Table 8.7 Major categories of rock coating (Dorn, 2007).

Coating	*Description*
Carbonate skin	Coating composed primarily of carbonate, usually $CaCO_3$, but sometimes combined with magnesium
Case hardening agents	Addition of cementing agent to rock matrix material; the agent may be manganese, sulfate, carbonate, silica, iron, oxalate, organisms or anthropogenic
Dust film	Light powder of clay- and silt-sized particles attached to rough surfaces and in rock fractures
Heavy metal skins	Coatings of iron, manganese, copper, zinc, nickel, mercury, lead and other heavy metals on rocks in natural and human-altered settings
Iron film	Composed primarily of iron oxides or oxyhydroxides
Lithobiontic coatings	Organic remains form the rock coating, e.g. lichens, moss, fungi, cyanobacteria, algae.
Nitrate crust	Potassium and calcium nitrate coatings on rocks, often in caves and rock shelters in limestone areas
Oxalate crust	Mostly calcium oxalate and silica with variable concentrations of magnesium, aluminium, potassium, phosphorous, sulfur, barium and manganese. Often found forming near or with lichens. Usually dark in colour, but can be as light as ivory
Phosphate skin	Various phosphate minerals (e.g. iron phosphates or apatite) that are mixed with clays and sometimes manganese; can be derived from decomposition of bird excrement
Pigment	Human-manufactured material placed on rock surfaces by people
Rock varnish	Clay minerals, Mn and Fe oxides, and minor and trace elements; colour ranges from orange to black in colour produced by variable concentrations of different manganese and iron oxides
Salt crust	The precipitation of chlorides on rock surfaces
Silica glaze	Usually clear white to orange shiny luster, but can be darker in appearance, composed primarily of amorphous silica and aluminium, but often with iron
Sulfate crust	Composed of the superposition of sulfates (e.g. barite, gypsum) on rocks; not gypsum crusts, which are sedimentary deposits

Figure 8.16 A 10 to 100 μm-thick rock varnish (darker debris slope surface on upper half of image) masks the appearance of the host igneous and metamorphic rocks in a road cut between Death Valley, California, and Las Vegas, Nevada, USA. Photograph courtesy of Ronald Dorn.

abundance of Mn, the presence of epilithic organisms, the morphology of the underlying rock, aeolian abrasion and the proximity of the varnish to soil (Dorn, 1986, 2007).

Rock varnish comprises a mixture of about two-thirds clay minerals (commonly mixed-layer illite-montmorillonite) cemented to the host rock by typically one-quarter manganese and iron oxyhydroxides (birnessite and haematite, respectively), the remaining constituents comprising minor and trace elements (Potter and Rossman, 1979; Dorn and Oberlander, 1982). Varnish Mn : Fe ratios vary from less than 1 : 1 to over 50 : 1, in contrast with Mn : Fe ratios of about 1 : 60 in the Earth's crust (Dorn, 2007). In general, blacker varnishes contain more manganese (typically 20 % MnO_2 by weight) while iron predominates in orange varnish (10 % FeO_2 by weight and less than 3 % MnO_2; see Potter and Rossman, 1977; Perry and Adams, 1978; Elvidge and Moore, 1979). In addition to Si, Al, Mn and Fe, major elements found in most varnishes include K, Na and Ti. The proportion of other elements varies considerably. Ba and Sr are the most abundant trace elements, but Cu, Ni, Zr, Pb, V, Co, La, Y, B, Cr, Sc and Yb, in order of decreasing concentration, are found in all varnishes (Engel and Sharp, 1958; Lakin *et al.*, 1963). Trace heavy metals generally covary with Mn abundance, and sometimes with Fe, due to the scavenging properties of the Mn and Fe oxyhydroxides (Thiagarajan and Lee, 2004; Tebo *et al.*, 2004).

Rock varnish may be of geomorphic importance in protecting rocks from weathering (Merrill, 1898). However, Engel and Sharp (1958) found that varnish is readily destroyed by mechanical flaking and dissolution under wet conditions. Resistance to removal is related to chemistry, clay mineral content and the roughness of the underlying substrate, with Si-rich varnish on originally rough substrate having the highest resistance and some Fe-rich varnishes on smooth rock being sufficiently soft to rub off with a fingertip (Oberlander, 1994). Dorn and Oberlander (1982) reported hardnesses approaching 6.5 Moh and Allen (1978) held that in the Sonoran Desert it is destroyed only by sandblasting.

Any coherent explanation for the origin of rock varnish must be tied to, and explained by, the processes that lead to (a) the enhancement of Mn concentrations on surfaces and (b) the fixation of clay minerals by Mn and Fe oxyhydroxides to such surfaces (von Humboldt, 1812; Lucas, 1905; Engels and Sharp, 1958). Theories of varnish development can be divided into two groups, according to whether the coating formed from an internal or external source (Drake, Heydeman and White, 1993). Much early work suggested that varnishes developed as moisture was drawn out of rocks, precipitating the minerals that were dissolved within (Merrill, 1898; Linck, 1901; Blake, 1905). However, this process of 'sweating' would be expected to form a leached and weathered zone beneath the varnish. Occasionally this is present (Allen, 1978; Glasby *et al.*, 1981) but the coating is technically then a weathering rind.

The consensus today is that the majority of the constituents that make up a rock varnish are derived from allochthonous sources (Allen, 1978; Perry and Adams, 1978; Elvidge and Moore, 1979). Such sources help explain the presence of trace elements that are absent from the host rock (Lakin *et al.*, 1963; Knauss and Ku, 1980). Four conceptual models have been proposed to explain the build-up of a varnish over time. The first invokes abiotic surficial chemical weathering to increase Mn : Fe ratios (Linck, 1901; Engel and Sharp, 1958; Marshall, 1962; Whalley, 1984; Smith and Whalley, 1988, 1989). Under this model, small pH/Eh fluctuations towards more acid conditions dissolve Mn but not Fe (Krauskopf, 1957). The released Mn is then fixed in clays following the evaporation of surface water or a further change in pH. Dorn (1989, 2007, 2009) presents a detailed critique of this model, identifying a number of features of varnishes that are inconsistent with a purely abiotic origin. First, in addition to occurring in deserts, varnishes are found in environments that are too wet and acidic to oxidise Mn (Dorn, 1998). Second, varnishes found in wetter climates are typically higher in Mn than those in arid, more alkaline, settings (Dorn, 1990), a result that would not be predicted by the high pH requirements of abiotic oxidation. Third, Mn-rich rock varnish is not common in some environments, such as coastal fog deserts, where repeated pH/Eh fluctuations might be expected to enhance Mn levels. Fourth, given that dust deposition and pH/Eh fluctuations occur at least annually in all but the most arid environments, varnishes should accrete many orders of magnitude faster than is seen in nature (Liu and Broecker, 2000, 2007). Overall, while abiotic processes may be involved in varnish formation (Bao, Thiemens and Heine, 2001), including clay cementation (Potter and Rossman, 1977; Dorn, 1998) and trace element enhancement during wetting (Thiagarajan and Lee, 2004), these processes alone would not generate a rock varnish.

A second conceptual model proposes that microorganic processes generate and bind the constituents that produce rock varnish; studies in North Africa (Drake, Heydeman and White, 1993), North America (e.g. Dorn and Oberlander, 1981, 1982; Palmer *et al.*, 1985; Nagy *et al.*, 1991), South America (Jones, 1991), Australia (Staley *et al.*, 1983; Dorn and Dragovich, 1990) and the eastern Mediterranean (Krumbein and Jens, 1981; Hungate *et al.*, 1987) all support such an origin. A wide range of lithobionts (or their remains) have been identified in, on or under rock varnish, including lichen, fungi, bacteria, cyanobacteria, pollen, peptides, refractory organic fragments, fatty acid methyl esters and amino acids from gram-positive chemoorganotrophic bateria (Dorn, 2009). Laudermilk (1931) and Krumbein and Jens (1981), for example, suggested that lichen oxidise and concentrate Mn, with algae and fungi also implicated (Bauman, 1976; Borns *et al.*, 1980; Krumbein and Jens, 1981; Grote and Krumbein, 1992). However, microcolonial fungi have been shown to either coexist with or actively dissolve rock varnish (Borns *et al.*, 1980; Staley *et al.*, 1983; Dragovich, 1993). Fungal respiration is also inhibited by the presence of montmorillonite (Stortsky and Rem, 1967), one of the main clay minerals in rock varnish. This may suggest instead that Mn fixation is accomplished by bacteria (Dorn and Oberlander, 1981, 1982). Dorn and Dragovich (1990) successfully cultured nine Mn-oxidising bacteria, and the dendritic patterns in some varnishes are similar to those produced during Fe and Mn oxidation by *Bacillus cereus* (Billy and Cailleux, 1968). Rock varnishes are 'home' to vast numbers of bacteria, with analyses of samples collected from desert pavement cobbles in eastern California revealing the presence of 10^8 cells of bacteria per gram dry weight of varnish (Kuhlman *et al.*, 2005, 2006). There are, however, two major limitations of the biogenic model. First, the association of lithobionts with varnishes is insufficient to explain Mn enhancement or clay cementation, as there is no unequivocal process by which these organisms can produce rock varnish. Second, if all the organisms observed in or on rock varnish were actively involved in varnish genesis, rates of formation should be several orders of magnitude higher than observed (Dorn, 2007). Where rates of varnish accumulation have been measured, typically only one to twenty bacterial diameter equivalents accumulate each thousand years (Liu and Broecker, 2000).

The most recent model to explain varnish formation is silica binding (Perry *et al.*, 2006). Under this model, silica is dissolved from dust and other minerals, then gels, condenses and hardens, binding clay minerals, organics and other detrital components to a rock surface. Silica binding cannot explain varnish formation alone as the process contains no mechanism for enhancing Mn levels. Dorn (2007, 2009) highlights other flaws, including the inability of the process to explain the dominance of clay minerals in varnishes and the rapid speed with which silica precipitation proceeds compared with known rates of varnish accretion. As the model relies on high temperatures to 'bake' the silica coating, it cannot explain the occurrence of subsurface and cold climate varnishes (e.g. Whalley, 1984; Whalley *et al.*, 1990; Dorn and Dragovich, 1990; Dorn *et al.*, 1992).

In the absence of one wholly convincing hypothesis for varnish formation, Dorn (2007) has suggested a fourth model in which abiotic and biotic processes act in tandem. Under this polygenetic model, slow-growing Mn- and Fe-fixing bacteria colonise dust layers on rock surfaces. The decay of Mn- and Fe-encrusted bacterial casts under weakly acidic conditions allows Mn and Fe to be mobilised from the cell walls, move a few nanometres and abiotically cement clay minerals. High-resolution transmission electron microscope analyses by Dorn (1998) and Krinsley (1998) have shown that varnishes are a product of the nanometre-scale combination of clay minerals and the slow accumulation and decay of Mn- and Fe-encrusted bacterial casts. This model explains the abundance of Mn and Fe within varnishes and, when combined with other abiotic mechanisms (Thiagarajan and Lee, 2004), may also account for the enhanced levels of many trace elements found in varnishes.

8.7.3 Silica glazes and iron films

Silica glaze ranges from micrometres up to a millimetre in thickness, and comprises mostly amorphous silica (in the form of opal) or other more crystalline silica polymorphs such as moganite (Perry *et al.*, 2006; Dorn, 2009). It varies in colour from charcoal black to ivory and can be almost transparent to completely opaque and dull to highly reflective. Dorn (1998) recognises six categories of glaze. Type I is a homogeneous deposit of amorphous silica. Type II includes a proportion of detrital minerals, cemented by amorphous silica. Type III is cemented dominantly by silica but incorporates 5–40 oxide weight % FeO and 5–30 % Al_2O_3. This variety often has a brown appearance due to the increased iron content and has been documented in a variety of environments (e.g. Watchman, 1992; Curtiss, Adams and Ghiorso, 1985; Mottershead and Pye, 1994). Types IV and V each contain around 50 % amorphous silica, with aluminium the main other component in the former type and iron in the latter. Type VI glazes, dominated by Al_2O_3, are relatively rare.

Silica glaze can form rapidly, with Curtiss, Adams and Ghiorso (1985) describing glazes developed on historic lava flows in Hawaii. It can slow rock weathering processes and, if the glaze has penetrated sufficiently, harden weathering rinds (Dorn, 2009). Sources of silica are ubiquitous in deserts and include weathered bedrock, dust, rainfall, groundwater and biological remains such as diatom tests and plant phytoliths. Silica is also dissolved readily in most weathering environments. Once transported in solution to the site of glaze formation, dissolved silica most probably precipitates on rock surfaces as a gel or directly as amorphous silica under the influence of evaporation or possibly organic complexation (cf. Krauskopf, 1957; Watchman, 1992; Perry *et al.*, 2006). Full details of the processes driving silica mobilisation and precipitation are given in the section concerning silcrete (Section 8.6.4 above).

A range of mechanisms are required to explain the formation of the various types of silica glaze. Type I and II glazes probably developed through the direct precipitation of silica gels. The other varieties require additional processes to mobilise and precipitate aluminium and iron. Dorn (2009) suggests that the presence of abundant aluminium within Type III, IV and VI glazes could be due to the incorporation of soluble aluminium silicate complexes ($(Al(OSi(OH)_3)^{2+})$ released during the weathering of phyllosilicate minerals. Al–Si complexes can be mobilised by gentle wetting events and ultimately bond to an existing silica glaze (Zorin *et al.*, 1992). The presence of enhanced iron contents in Type III and V glazes is more easily explained by the strong adherence of Fe released by weathering to silica through Fe–O–Si bonds (Dove and Rimstidt, 1994).

Iron films are widespread in desert and other environments and occur in three varieties (Dorn, 1998). Type I iron films consist of homogeneous iron with few other constituents. Type II iron films include additional Si and Al, but iron remains the dominant component. In Type III films, iron constitutes less than a third by weight of the coating but is present in sufficient quantities to give a strong red/orange colour. The mobilisation and fixation of iron may involve abiotic processes, with the oxidation of Fe^{2+} to Fe^{3+} occurring rapidly above pH 5. The formation of chemical Fe–O–Si bonds and interactions with interstratified clays may be key to the development of Type II and III films respectively (Dorn, 2009).

8.8 Palaeoenvironmental significance of crusts

Desert crusts and rock coatings are of value as palaeoenvironmental indicators only if their mode of origin, age and the environmental conditions during their formation are correctly interpreted. Ideally, environments where crusts are forming today should be used as analogues; this is feasible for calcrete, gypsum and halite crusts and some rock coatings, but not for silcrete, where the majority of documented examples are relict features.

Sodium nitrate, gypsum and halite crusts can be valuable palaeoclimatic indicators because they attest to extreme aridity. However, their susceptibility to dissolution limits their preservation at the land surface to regions where aridity has persisted. This can provide valuable information on the ages of deserts if the crusts can be dated (Dan *et al.*, 1982; Reheis, 1987; Watson, 1988). One line of evidence for the early initiation of the Atacama Desert, for example, is the existence of a gypsum crust preserved beneath an ignimbrite dated to around 9.5 million years (Hartley and May, 1998). The high solubility of gypsum implies that aridity has prevailed since that time. As both phreatic and pedogenic forms of halite and gypsum crust usually accrete within a host material, assessing the age of incorporated artefacts and other datable substances may not be representative of the true ages (Watson, 1988). Furthermore, gypsum and halite crusts form in several different ways, so environmental interpretations must proceed carefully. Isotopic studies of gypsum crusts have provided insights into their origins (Carlisle *et al.*, 1978; Sofer, 1978; Drake, Eckardt and White, 2004) and the sources of their constituent materials (Chivas *et al.*, 1991; Eckardt and Spiro, 1999; Rech, Quade and Hart, 2003). Most reported occurrences of halite and gypsum crusts in the stratigraphic record have been interpreted as lacustrine, lagoonal or sabkha evaporites (Schreiber, 1986). In part, this reflects a lack of clear criteria for identifying pedogenic evaporite facies.

Valuable information can be obtained by coring modern playa floors to recover sequences of interbedded evaporites and clastic sediments (Smith, 1979; Smith and Bischoff, 1997). Microfossil, particle-size and mineralogical assessments, plus analyses of the chemistry and isotopic signatures of salts, pore water and fluid inclusions, are all routinely used to elucidate lacustrine palaeoenvironments (e.g. Spencer, Eugster and Jones, 1985; Teller and Last, 1990). Fluid inclusions within halite crystals can yield homogenisation temperatures for the time when halite precipitation took place (Roberts and Spencer, 1995; Lowenstein *et al.*, 1999; Lowenstein and Brennan, 2001). Direct temperature estimates for evaporite sequences may also be provided by amino acid palaeothermometry, which produces diagenetic temperature information for broad depositional phases (Kaufman, 2003).

The presence of a pedogenic calcrete in the landscape or geological record is an indicator that semi-arid or arid conditions prevailed at the time of its formation. Calcretes have been identified in rocks from every continent and dating from the Proterozoic to the present (see Alonso-Zarza, 2003; Wright, 2007). Careful identification of the type of calcrete is needed before any palaeoenvironmental inferences are drawn; some calcretes have been interpreted as altered reef carbonates or cold climate carbonate accumulations. Stable carbon and oxygen isotopic analyses of primary carbonates within pedogenic calcretes can be used in the reconstruction of palaeoclimates, past vegetation and CO_2 concentrations (Cerling, 1999; Alonso-Zarza, 2003). Both $\delta^{13}C$ and $\delta^{18}O$ levels are dependent upon the depth beneath the soil surface at which the samples are obtained, decreasing rapidly to become almost constant at 10–50 cm depth (Quade, Cerling and Bowman, 1989). The oxygen isotope composition of a calcrete is directly related to that of the meteoric water from which it formed. In arid zones (with <250 mm annual rainfall), values of $\delta^{18}O$ lower than −5 ‰ do not occur and areas receiving less than 350 mm have $\delta^{18}O$ values greater than −2 ‰ (Talma and Netterberg, 1983). The $\delta^{13}C$ values of soil carbonates at depths below 30 cm depend on the isotopic composition of the soil CO_2, itself controlled by the amount of atmospheric CO_2 that penetrates the air-soil interface, the density of vegetation cover and the proportions of plants present that use the C_3, C_4 and CAM photosynthetic pathways (Cerling, 1999). Low $\delta^{13}C$ values are normally taken to indicate the dominance of C_3 plants while heavier values suggest greater proportions of C_4 and CAM plant communities.

The reliable dating of calcretes is essential if they are to be used as an effective palaeoenvironmental indicator. The earliest age estimates for calcrete were based on relative dating. However, the possibility of radiocarbon dating calcretes (Williams and Polach, 1971; Magaritz, Kaufman and Yaalon, 1981) led to their widespread use in palaeoenvironmental studies. Great care is needed to ensure that only one generation of carbonate is represented in the dated sample (Rust, Schmidt and Dietz, 1984) and the appropriate correction factors are applied to take into account natural variations in stable isotope levels (Salomons and Mook, 1976; Salomons, Goudie and Mook, 1978). More recent studies have used U/Th dating to establish the age of carbonates. This technique is prone to the same sampling issues as radiocarbon dating (Ku *et al.*, 1979; Schlesinger, 1985), but many problems can be overcome through microsampling of carbonate phases. For example, a suite of Middle Pleistocene to Holocene pedogenic calcretes in the Sorbas basin, southeast Spain, have been dated using this approach (Kelly, Black and Rowan, 2000; Candy, Black and Sellwood, 2004a, 2004b, 2005).

The use of silcrete as a palaeoenvironmental indicator has been the subject of considerable debate (see Nash and Ullyott, 2007). However, recent advances in our understanding of silcrete genesis provide a starting point for reevaluating palaeoenvironmental signals. Pedogenic silcrete is now thought to form mainly under climates with alternating wet/humid and dry/arid seasons or periods (e.g. Thiry, 1981; Callen, 1983; Meyer and Pena dos Reis, 1985; Curlík and Forgác, 1996; Webb and Golding, 1998), although the pH range under which silicification occurs remains disputed (Summerfield, 1979, 1983a; Thiry and Milnes, 1991; Terry and Evans, 1994). Groundwater, drainage-line and pan/lacustrine silcrete formation is strongly influenced by palaeotopography, which controls both the watertable position and groundwater flow; silicification under both high pH (Summerfield, 1982; Meyer and Pena dos Reis, 1985; Nash and Shaw, 1998) and low pH (Wopfner, 1978, 1983; Meyer and Pena dos Reis, 1985; Taylor and Ruxton, 1987; Thiry, Ayrault and Grisoni, 1988; Thiry and Milnes, 1991) conditions have been suggested. Recent analyses of an Australian groundwater silcrete by Alexandre *et al.* (2004) also suggest that silicification may have occurred under cooler and wetter conditions than at present.

Siliceous duricrusts are difficult to date, although the development of techniques for dating diagenetic events (McNaughton, Rasmussen and Fletcher, 1999) and K–Mn oxides in weathered profiles (Vasconcelos, 1999; Vasconcelos and Conroy, 2003) offers scope for optimism. Radtke and Brückner (1991) used electron spin resonance to estimate the ages of Australian silcretes; however, their use of bulk samples rendered the derived dates almost meaningless. Most studies use relative dating approaches, where the ages of silcretes are determined from their stratigraphic position. This concept has been employed widely in Australia where the presence of plant fossils and the relationship between silcretes and basalts of known age have been used to age-constrain outcrops (Webb and Golding, 1998; Taylor and Eggleton, 2001). Most pedogenic silcretes date from the Palaeogene or Early Neogene (Thiry, 1999), although some have been attributed to the Mesozoic (Langford-Smith, 1978; Wopfner, 1978; Ballesteros, Talegón and Hernández, 1997), Late Neogene and Quaternary (Dubroeucq and Thiry, 1994; Curlík and Forgác, 1996). Australian groundwater silcretes have mostly been ascribed a mid-Tertiary age (Stephens, 1971; Alley, 1973; Young, 1978; Ambrose and Flint, 1981).

Rock varnish may coat surfaces within a few decades (Engel and Sharp, 1958; Dorn and Meek, 1995), but is more likely to take hundreds or thousands of years to form (Elvidge, 1982; Dorn and DeNiro, 1985). Hence, some varnishes may provide extremely long palaeoenvironmental records (see also Chapter 3). A variety of varnish properties has been used to infer past environments, including analyses of micromorphological changes (Dorn, 1986), stable carbon isotope ratios (Dorn and DeNiro, 1985), organic carbon ratios (Dorn *et al.*, 2000), trace element geochemistry (Fleisher *et al.*, 1999), ^{17}O excesses in sulfates (Bao, Thiemens and Heine, 2001) and interlayering with other rock coatings (Dragovich, 1986). Contrary to popular opinion, rock varnish does not darken with age – Dorn (1998) identifies eleven factors other than time that lead to darkening. Attempts have been made to date rock varnish using the $^{230}Th/^{234}U$ technique corroborated with $^{231}Pa/^{235}U$ (Knauss and Ku, 1980), as well as cation-ratio ($K^{+} + Ca^{2+}/Ti^{4+}$) and radiocarbon methods (Dorn *et al.*, 1986, 1989). These techniques, and particularly the cation-ratio method, have been the subject of debate concerning their accuracy, comparability and reliability (Bierman and Gillespie, 1991, 1992a, b, 1994, 1995; Dorn, 1992, 1995; Cahill, 1992; Harrington and Whitney, 1995). Perhaps the greatest potential application for rock varnish is through the analysis of varnish microlaminations (Dorn, 2007). The basic assumption of the technique is that varnish microstratigraphy is influenced by regional climatic variations; black Mn-rich layers record wet intervals, while orange and yellow Mn-poor layers indicate drier periods (Dorn, 1990; Jones, 1991; Liu and Dorn, 1996; Liu, 2003). The climatic signals recorded in varnish have been proven to be regionally contemporaneous (Liu and Dorn, 1996; Liu *et al.*, 2000), such that varnish microstratigraphy now has the potential to be used as a dating tool in a similar way to tephrachronology.

References

Abdel Jaoued, S. (1987) Calcretes and dolocretes of Upper Palaeocene–Miocene (Bou-Loufa Formation) in southern Tunisia. *Bulletin de la Société Géologique de France*, **3**, 777–781.

Al Juaidi, F.A., Millington, A. and McLaren, S. (2003) Evaluating image fusion techniques for mapping geomorphological features on the eastern edge of the Arabian Shield (Central Saudi Arabia). *Geographical Journal*, **169**, 117–131.

Alexander, G.B., Heston, W.M. and Iler, R.K. (1954) The solubility of amorphous silica in water. *Journal of Physical Chemistry*, **58**, 453–455.

Alexandre, A., Meunier, J.D., Llorens, E. *et al.* (2004) Methodological improvements for investigating silcrete formation: petrography, FT-IR and oxygen isotope ratio of silcrete quartz cement, Lake Eyre Basin (Australia). *Chemical Geology*, **211**, 261–274.

Ali, Y.A. and West, I. (1983) Relationships of modern gypsum nodules in sabkhas of loess to composition of brines and sediments in northern Egypt. *Journal of Sedimentary Petrology*, **53**, 1151–1168.

Allen, C.C. (1978) Desert varnish of the Sonoran Desert – optical and electron probe analysis. *Journal of Geology*, **86**, 743–752.

Alley, N.F. (1973) Landscape development of the mid-north of South Australia. *Transactions of the Royal Society of South Australia*, **97**, 1–17.

Alley, N.F. (1977) Age and origin of laterite and silcrete duricrusts and their relationship to episodic tectonism in the mid-north of South Australia. *Journal of the Geological Society of Australia*, **24**, 107–116.

Alonso-Zarza, A.M. (1999) Initial stages of laminar calcrete formation by roots: examples from the Neogene of central Spain. *Sedimentary Geology*, **126**, 177–191.

Alonso-Zarza, A.M. (2003) Palaeoenvironmental significance of palustrine carbonates and calcretes in the geological record. *Earth-Science Reviews*, **60**, 261–298.

Ambrose, G.J. and Flint, R.B. (1981) A regressive Miocene lake system and silicified strandlines in northern South Australia; implications for regional stratigraphy and silcrete genesis. *Journal of the Geological Society of Australia*, **28**, 81–94.

Amit, R. and Gerson, R. (1986) The evolution of Holocene reg (gravelly) soils in deserts – an example from the Dead Sea region. *Catena*, **13**, 59–79.

Amit, R. and Yaalon, D.H. (1996) The micromorphology of gypsum and halite in reg soils – the Negev Desert, Israel. *Earth Surface Processes and Landforms*, **21**, 1127–1143.

Andrews, J.E., Singhvi, A.K., Kailath, A.J. *et al.* (1998) Do stable isotope data from calcrete record Late Pleistocene monsoonal climate variation in the Thar Desert of India? *Quaternary Research*, **50**, 240–251.

Arakel, A.V. (1980) Genesis and diagenesis of Holocene evaporitic sediments in Hutt and Leeman Lagoons, Western Australia. *Journal of Sedimentary Petrology*, **50**, 1305–1326.

Arakel, A.V. (1982) Genesis of calcrete in Quaternary soil profiles, Hutt and Leeman Lagoons, Western Australia. *Journal of Sedimentary Petrology*, **52**, 109–125.

Arakel, A.V. (1986) Evolution of calcrete in palaeodrainages of the lake Napperby area, central Australia. *Palaeogeography, Palaeoclimatology, Palaeoecology*, **54**, 283–303.

Arakel, A.V. (1991) Evolution of Quaternary duricrusts in Karinga Creek drainage system, central Australian groundwater discharge zone. *Australian Journal of Earth Sciences*, **38**, 333–347.

Arakel, A.V. and McConchie, D. (1982) Classification and genesis of calcrete and gypsum lithofacies in palaeodrainage systems of inland Australia and their relationship of carnotite mineralization. *Journal of Sedimentary Petrology*, **52**, 1149–1170.

Arakel A.V., Jacobson G., Salehi M. and Hill C.M. (1989) Silicification of calcrete in palaeodrainage basins of the Australian arid zone. *Australian Journal of Earth Sciences*, **36**, 73–89.

Aref, M.A.M. (2003) Classification and depositional environments of Quaternary pedogenic gypsum crusts (gypcrete) from east of the Fayum Depression, Egypt. *Sedimentary Geology*, **155**, 87–108.

Aristarain, L.F. (1971) On the definition of caliche deposits. *Zeitschrift für Geomorphologie*, **NF 15**, 274–289.

Armenteros, I., Bustillo, M.A. and Blanco, J.A. (1995) Pedogenic and groundwater processes in a closed Miocene basin (northern Spain). *Sedimentary Geology*, **99**, 17–36.

Bachman, G.O. and Machette, M.N. (1977) *Calcic soils and calcretes in the southwestern United States*, US Geological Survey, Open File Report 77-794.

Ballesteros, E.M., Talegón, J.G. and Hernández, M.A. (1997) Palaeoweathering profiles developed on the Iberian Hercynian basement and their relationship to the oldest Tertiary surface in central and western Spain, in *Palaeosurfaces: Recognition, Reconstruction and Palaeoenvironmental Interpretation* (ed. M. Widdowson), Geological Society Special Publication 120, London, pp. 175–185.

Bao, H.M., Thiemens, M.H. and Heine, K. (2001) Oxygen-17 excesses of the Central Namib gypcretes: spatial distribution. *Earth and Planetary Science Letters*, **192**, 125–135.

Barnes, I. (1965) Geochemistry of Birch Creek, Inyo County, California. *Geochemica et Cosmochimica Acta*, **29**, 85–112.

Barnes, L.C. and Pitt, G.M. (1976) The Mirackina Conglomerate. *Quarterly Geological Notes of the Geological Survey of South Australia*, **59**, 2–6.

Basile-Doelsch, I., Meunier, J.D. and Parron, C. (2005) Another continental pool in the terrestrial silicon cycle. *Nature*, **433**, 399–402.

Basyoni, M.H. and Mousa, B.A. (2009) Sediment characteristics, brine chemistry and evolution of Murayr Sabkha, Arabian (Persian) Gulf, Saudi Arabia. *Arabian Journal for Science and Engineering*, **34**, 95–123.

Bauman, A.J. (1976) Desert varnish and marine ferromanganese oxide nodules: congeneric phenomena. *Nature*, **259**, 387–388.

Beattie, J.A. and Haldane, A.D. (1958) The occurrence of palygorksite and barytes in certain parna soils of the Murrumbidgee Region, New South Wales. *Australian Journal of Science*, **20**, 274–275.

Benbow, M.C., Callen, R.A., Bourman, R.P. and Alley, N.F. (1995) Deep weathering, ferricrete and silcrete, in *Bulletin 54: The Geology of South Australia, Volume 2: The Phanerozoic* (eds J.F. Drexel and W.V. Preiss), Geological Survey of South Australia, pp. 201–207.

Bierman, P.R. and Gillespie, A.R. (1991) Accuracy of rock varnish chemical analyses – implications for cation-ratio dating. *Geology*, **19**, 196–199.

Bierman, P.R. and Gillespie, A.R. (1992a) Accuracy of rock varnish chemical analyses – implications for cation-ratio dating – reply. *Geology*, **20**, 470.

Bierman, P.R. and Gillespie, A.R. (b) Accuracy of rock varnish chemical analyses – implications for cation-ratio dating – reply. *Geology*, **20**, 471–472.

Bierman, P.R. and Gillespie, A.R. (1994) Evidence suggesting that methods of rock-varnish cation-ratio dating are neither comparable nor consistently reliable. *Quaternary Research*, **41**, 82–90.

Bierman, P.R. and Gillespie, A.R. (1995) Evidence suggesting that methods of rock-varnish cation-ratio dating are neither comparable nor consistently reliable - reply. *Quaternary Research*, **43**, 274–276.

Billy, C. and Cailleux, A. (1968) Depots dendritiques d'oxydes de fer and manganese par action bacterienne. *Comptes Rendus Hebdomadaires des Séances de l'Académie des Sciences*, **266D**, 1643–1645.

Blake, W.P. (1905) Superficial blackening and discoloration of rocks especially in desert regions. *Transaction of the American Institute of Mining Engineers*, **35**, 371–375.

Blank, H.R. and Tynes, E.W. (1965) Formation of caliche in situ. *Bulletin of the Geological Society of America*, **76**, 1387–1391.

Blümel, W.D. and Eitel, B. (1994) Tertiäre Deckschichten und Kalkkrusten in Namibia: Entstehung und geomorphologische Bedeutung. *Zeitschrift für Geomorphologie*, **38**, 385–403.

Böhlke, J.K., Ericksen, G.E. and Revesz, K. (1997) Stable isotope evidence for an atmospheric origin of desert nitrate deposits in northern Chile and Southern California, USA. *Chemical Geology*, **136**, 135–152.

Böhlke, J. and Michalski, G. (2002) Atmospheric and microbial components of desert nitrate deposits indicated by variations in δ18O, δ17O, δ16O isotope ratios. *EOS Transactions of the AGU*, **83** (47), Fall Meeting Supplement of Abstracts.

Borns, D.J., Adams, J.B., Curtiss, B. *et al.* (1980) The role of micro-organisms in the formation of desert varnish and other rock coatings: SEM study. *Geological Society of America, Abstracts with Programs*, **12**, 390.

Borsato, A., Frisia, S., Jones, B. and Van Der Borg, K. (2000) Calcite moonmilk: crystal morphology and environment of formation in caves in the Italian Alps. *Journal of Sedimentary Research*, **70**, 1171–1182.

Bosazza, V.L. (1939) The silcrete and clays of the Riversdale–Mossel Bay area. *Fulcrum, Johannesburg*, **32**, 17–29.

Braissant, O., Guillaume, C., Dupraz, C. and Verrecchia, E.P. (2003) Bacterially induced mineralization of calcium carbonate in terrestrial environments: the role of exopolysaccharides and amino acids. *Journal of Sedimentary Research*, **73**, 485–490.

Braithwaite, C.J.R. (1983) Calcrete and other soils in Quaternary limestones: structures, processees and applications. *Journal of the Geological Society of London*, **140**, 351–363.

Bretz, J.H. and Horberg, L. (1949) Caliche in south-eastern New Mexico. *Journal of Geology*, **57**, 491–511.

Brewer, R.C. (1964) *Fabric and Mineral Analysis of Soils*, Robert E. Kreiger, New York.

Briot, P. (1983) L'environnement hydrogéochimique du calcrète uranifère de Yeelirrie (Australie Occidentale). *Mineralium Deposita*, **18**, 191–206.

Brown, C.N. (1956) The origin of caliche in the north-eastern llana Estacado, Texas. *Journal of Geology*, **64**, 1–15.

Brückner, W.D. (1966) Origin of silcretes in Central Australia. *Nature*, **209**, 496–497.

Buck, B.J. and van Hoesen, J.G. (2002) Snowball morphology and SEM analysis of pedogenic gypsum, southern New Mexico, USA. *Journal of Arid Environments*, **51**, 469–487.

Buck, B.J., Wolff, K., Merkler, D.J. and McMillan, N.J. (2006) Salt mineralogy of Las Vegas Wash, Nevada: morphology and subsurface evaporation. *Soil Science Society of America Journal*, **70**, 1639–1651.

Budd, D.A., Pack, S.M. and Fogel, M.L. (2002) The destruction of paleoclimatic isotopic signals in Pleistocene carbonate soil nodules of Western Australia. *Paleogeography, Palaeoclimatology, Palaeoecology*, **188**, 249–273.

Bunting, B.T. and Christensen, L. (1980) Micromorphology of calcareous crusts from the Canadian High Arctic. *Geol. Foren. Stockholm Fordland*, **100**, 361–367.

Burgess, I.C. (1961) Fossil soils in the Upper Old Red Sandstone of South Ayrshire. *Transactions of the Geological Society of Glasgow*, **24**, 138–153.

Busson, G. and Perthuisot, J.P. (1977) Intérêt de la Sabkha el Melah (Sud tunisien) opour l'interprétation des séries évaporitiques anciennes. *Sedimentary Geology*, **19**, 139–164.

Bustillo, M.A. and Bustillo, M. (1993) Rhythmic lacustrine sequences with silcretes from the Madrid Basin, Spain: geochemical trends. *Chemical Geology*, **107**, 229–232.

Bustillo, M.A. and Bustillo, M. (2000) Miocene silcretes in argillaceous playa deposits, Madrid Basin, Spain: petrological and geochemical features. *Sedimentology*, **47**, 1023–1037.

Butt, C.R.M. (1985) Granite weathering and silcrete formation on the Yilgarn Block, Western Australia. *Australian Journal of Earth Sciences*, **32**, 415–432.

Cahill, T.A. (1992) Accuracy of rock varnish chemical analyses – implications for cation-ratio dating – comment. *Geology*, **20**, 469.

Cailleau, G., Braissant, O. and Verrecchia, E. P. (2004) Biomineralization in plants as a long term carbon sink. *Naturwissenschaften*, **91**, 191–194.

Callen, R.A. (1983) Late Tertiary 'grey billy' and the age and origins of surficial silicifications (silcrete) in South Australia. *Journal of the Geological Society of Australia*, **30**, 393–410.

Callender, J.H. (1978) A study of the silcretes near Marulan and Milton, New South Wales, in *Silcrete in Australia* (ed. T. Smith), University of New England Press, Armidale, pp. 209–221.

Candy, I. (2002) Formation of a rhizogenic calcrete during a glacial stage (Oxygen Isotope Stage 12): its palaeoenvironmental stratigraphic significance. *Proceedings of the Geologists' Association*, **113**, 259–270.

Candy, I. and Black, S. (2009) The timing of Quaternary calcrete development in semi-arid southeast Spain: investigating the role of climate on calcrete genesis. *Sedimentary Geology*, **218**, 6–15.

Candy, I., Black, S. and Sellwood, B.W. (2004a) Complex response of a dryland river system to Late Quaternary climate change: implications for interpreting the climatic record of fluvial sequences. *Quaternary Science Reviews*, **23**, 2513–2523.

Candy, I., Black, S. and Sellwood, B.W. (2004b) Quantifying timescales of pedogenic calcrete formation using U-series disequilibria. *Sedimentary Geology*, **170**, 177–187.

Candy, I., Black, S. and Sellwood, B.W. (2005) U-series isochron dating of immature and mature calcretes as a basis for constructing Quaternary landform chronologies; examples from the Sorbas basin, southeast Spain. *Quaternary Research*, **64**, 100–111.

Candy, I., Black, S., Sellwood, B.W. and Rowan, J.S. (2003) Calcrete profile development in Quaternary alluvial sequences, Southeast Spain: implications for using calcretes as a basis for landform chronologies. *Earth Surface Processes and Landforms*, **28**, 169–185.

Canti, M. (1998) Origin of calcium carbonate granules found in buried soils and Quaternary deposits. *Boreas*, **27**, 275–288.

Capo, R.C. and Chadwick, O.A. (1999) Sources of strontium and calcium in desert soil and calcrete. *Earth and Planetary Science Letters*, **170**, 61–72.

Carlisle, D. (1983) Concentration of uranium and vanadium in calcretes and gypcretes, in *Residual Deposits* (ed. R.C.L. Wilson), Geological Society of London Special Publication 11, pp. 185–195.

Carlisle, D., Merifield, P.M., Orme, A.R. and Kolker, O. (1978) The distributuon of calcretes and gypcretes in Southwestern United States and their uranium favorability. Based on a Study of Deposits in Western Australia and South West Africa (Namibia), University of California, Los Angeles, Open File Report 76-002-E.

Castens-Seidell, B. and Hardie, L.A. (1984) Anatomy of a modern sabkha in a rift valley setting, northwest Gulf of California, Baja California, Mexico. *Bulletin of the American Association of Petroleum Geologists*, **68**, 460.

Cereceda, P. and Schemenauer, R.S. (1991) The occurrence of fog in Chile. *Journal of Applied Meteorology*, **30**, 1097–1105.

Cereceda, P., Osses, P., Larrain, H., *et al.* (2002) Advective, orographic and radiation fog in the Tarapacá region, Chile. *Atmospheric Research*, **64**, 261–271.

Cerling, T.E. (1999) Stable isotopes in palaeosol carbonates, in *Palaeoweathering, Palaeosurfaces and Related Continental Deposits* (eds M. Thiry and R. Simon-Coinçon), International Association of Sedimentologists Special Publication 27, pp. 43–60.

Chadwick, O.A., Hendricks, D.M. and Nettleton, W.D. (1989) Silicification of Holocene soils in Northern Monitor Valley, Nevada. *Soil Science Society of America Journal*, **53**, 158–164.

Chadwick, O.A. and Nettleton, W.D. (1990) Micromorphological evidence of adhesive and cohesive forces in soil cementation. *Developments in Soil Science*, **19**, 207–212.

Chafetz, H.S. and Butler, J.C. (1980) Petrology of recent caliche pisolites, spherulites, and speleothem deposits from central Texas. *Sedimentology*, **27**, 497–518.

Chapman, R.W. (1974) Calcareous duricrust in Al-Hasa, Saudi Arabia. *Bulletin of the Geological Society of America*, **85**, 119–130.

Chen, X.Y. (1997) Pedogenic gypcrete formation in arid central Australia. *Geoderma*, **77**, 39–61.

Chen, X.Y., Bowler, J.M. and Magee, J.W. (1991a) Gypsum ground: a new occurrence of gypsum sediment in playas of central Australia. *Sedimentary Geology*, **72**, 79–95.

Chen, X.Y., Bowler, J.M. and Magee, J.W. (1991b) Aeolian landscapes in central Australia – gypsiferous and quartz dune environments from Lake Amadeus. *Sedimentology*, **38**, 519–538.

Chiquet, A., Michard, A., Nahon, D. and Hamelin, B. (1999) Atmospheric input vs in situ weathering in the genesis of calcretes: an Sr isotope study at Galvez (Central Spain). *Geochimica et Cosmochimica Acta*, **63**, 311–323.

Chivas, A.R. (2007) Terrestrial evaporites, in *Geochemical Sediments and Landscapes* (eds D.J. Nash and S.J. McLaren), Blackwell, Oxford, pp. 330–364.

Chivas, A.R., Andrew, A.S., Lyons, W.B. *et al.* (1991) Isotopic constraints on the origin of salts in Australian playas. 1. Sulphur. *Palaeogeography, Palaeoclimatology, Palaeoecology*, **84**, 309–332.

Chong G.D. (1994) The nitrate deposits of Chile, in *Tectonics of the Southern Central Andes* (eds K.-J. Reutter, E. Scheuber and P.J. Wigger), *Springer*, Berlin, pp. 303–316.

Claridge, G.G.C. and Campbell, I.B. (1968) Origin of nitrate deposits. *Nature*, **217**, 428–430.

Clarke, J.D.A. (2006) Antiquity of aridity in the Chilean Atacama Desert. *Geomorphology*, **73**, 101–114.

Collins, N. (1985) *Sub-basaltic silcrete at Morrisons, central Victoria*, Unpublished Honours Thesis, Department of Earth Sciences, University of Melbourne, Australia.

Colson, J. and Cojan, I. (1996) Groundwater dolocretes in a lake-marginal environment: an alternative model for dolocrete formation in continental settings (Danian of the Provence Basin, France). *Sedimentology*, **43**, 175–188.

Coque, R. (1955a) Les croûtes gypseuses du Sud tunisien. *Bulletin de la Société des Sciences naturelles de Tunisie*, **8**, 217–236.

Coque, R. (1955b) Morphologie et croûte dans le Sud tunisien. *Annales de Geographie*, **64**, 359–370.

Coque, R. (1962) *La Tunisie Présaharienne*, Armand Colin, Paris.

Curlík, J. and Forgác, J. (1996) Mineral forms and silica diagenesis in weathering silcretes of volcanic rocks in Slovakia. *Geologica Carpathica*, **47**, 107–118.

Curtiss, B., Adams, J.B. and Ghiorso, M.S. (1985) Origin, development and chemistry of silica-aluminium rock coatings from the semi-arid regions of the island of Hawaii. *Geochimica et Cosmochimica Acta*, **49**, 49–56.

Dan, J., Yaalon, D.H., Koymudjisky, H and Raz, Z. (1972) The soil association map of Israel (1:1,000,000). *Israel Journal of Earth Sciences*, **21**, 29–49.

Dan, J., Yaalon, D.H., Moshe, R. and Nissim, S. (1982) Evolution of reg soils in southern Israel and Sinai. *Geoderma*, **28**, 173–202.

Day, J.A. (1993) The major ion chemistry of some southern African saline systems. *Hydrobiologia*, **267**, 37–59.

Deutz, P., Montanez, I.P. and Monger, H.C. (2002) Morphology and stable and radiogenic isotope composition of pedogenic carbonates in Late Quaternary relict soils, New Mexico, USA: an integrated record of pedogenic overprinting. *Journal of Sedimentary Research*, **72**, 809–822.

Deutz, P., Montanez, I.P., Monger, H.C. and Morrison, J. (2001) Morphology and isotopic heterogeneity of Late Quaternary pedogenic carbonates: implications for paleosol carbonates as paleoenvironmental proxies. *Palaeogeography, Palaeoclimatology, Palaeoecology*, **166**, 293–317.

Dhir, R.P., Tandon, S.K., Sareen, B.K., *et al.* (2004) Calcretes in the Thar desert: genesis, chronology and palaeoenvironment. *Proceedings of the Indian Academy of Sciences – Earth and Planetary Sciences*, **113**, 473–515.

Dhir, R.P., Singhvi, A.K., Andrews, J.E. *et al.* (2010) Multiple episodes of aggradation and calcrete formation in Late Quaternary aeolian sands, Central Thar Desert, Rajasthan, India. *Journal of Asian Earth Sciences*, **37**, 10–16.

Dixon, J.C. and McLaren, S.J. (2009) Duricrusts, in *Geomorphology of Desert Environments* (eds A.J. Parsons and A.D. Abrahams), Springer, pp. 123–151.

Dorn, R.I. (1984) Cause and implications of rock varnish microchemical laminations. *Nature*, **310**, 767–770.

Dorn, R.I. (1986) Rock varnish as an indicator of aeolian environmental change, in *Aeolian Geomorphology* (ed. W.G. Nickling), Allen and Unwin, Winchester, MA, pp. 291–307.

Dorn, R.I. (1989) A comment on 'A note on the characteristics and possible origins of desert varnished form southeast Morocco' by Drs Smith and Whalley. *Earth Surface Processes and Landforms*, **14**, 167–170.

Dorn, R.I. (1990) Quaternary alkalinity fluctuations recorded in rock varnish microlaminations on western USA volcanics. *Palaeogeography, Palaeoclimatology, Palaeoecology*, **76**, 291–310.

Dorn, R.I. (1991) Rock varnish. *American Scientist*, **79**, 542–553.

Dorn, R.I. (1992) Accuracy of rock varnish chemical analyses – implications for cation-ratio dating – comment. *Geology*, **20**, 470.

Dorn, R.I. (1995) Evidence suggesting that methods of rock-varnish cation-ratio dating are neither comparable nor consistently reliable – comment. *Quaternary Research*, **43**, 272–273.

Dorn, R.I. (1998) *Rock Coatings*, Elsevier, Amsterdam.

Dorn, R.I. (2007) Rock varnish, in *Geochemical Sediments and Landscapes* (eds D.J. Nash and S.J. McLaren), Blackwell, Oxford, pp. 246–297.

Dorn, R.I. (2009) Desert rock coatings, in *Geomorphology of Desert Environments*, 2nd edn (eds A.J. Parsons and A.D. Abrahams), Springer, Berlin, pp. 153–186.

Dorn, R.I. and DeNiro, M.J. (1985) Stable carbon isotope ratios of rock varnish organic matter: a new paleoenvironmental indicator. *Science*, **227**, 1472–1474.

Dorn, R.I. and Dragovich, D. (1990) Interpretation of rock varnish in Australia: case studies from the arid zone. *Australian Geographer*, **21**, 18–31.

Dorn, R.I. and Meek, N. (1995) Rapid formation of rock varnish and other rock coatings on slag deposits near Fontana, California. *Earth Surface Processes and Landforms*, **20**, 547–560.

Dorn, R.I. and Oberlander, T.M. (1981) Microbial origin of desert varnish. *Science*, **213**, 1245–1247.

Dorn, R.I. and Oberlander, T.M. (1982) Rock varnish. *Progress in Physical Geography*, **6**, 317–367.

Dorn, R.I., Bamforth, D.B., Chaill, T.A. *et al.* (1986) Cation-ratio and accelerator radiocarbon dating of rock varnish on Mojave artifacts and landforms. *Science*, **231**, 830–833.

Dorn, R.I., Jull, A.J.T., Donahue, D.J. *et al.* (1989) Accelerator mass spectrometry radiocarbon dating of rock varnish. *Bulletin of the Geological Society of America*, **101**, 1363–1372.

Dorn, R.I., Krinsley, D.H., Liu, T.Z. *et al.* (1992) Manganese-rich rock varnish does occur in Antarctica. *Chemical Geology*, **99**, 289–298.

Dorn, R.I., Stasack, E., Stasack, D. and Clarkson, P. (2000) Through the looking glass: analyzing petroglyphs and geoglyphs with different perspectives. *American Indian Rock Art*, **27**, 77–96.

Dove, P.M. and Rimstidt, J.D. (1994) Silica–water interactions, in *Silica: Physical Behaviour, Geochemistry and Materials Applications* (eds P.J. Heaney, C.T. Prewitt and G.V. Gibbs), Reviews in Mineralogy 29, Mineralogical Society of America, Washington, pp. 259–308.

Dragovich, D. (1986) Minimum age of some desert varnish near Broken Hill, New South Wales. *Search*, **17**, 149–151.

Dragovich, D. (1988) A preliminary electron microprobe study of microchemical laminations in desert varnish in western New South Wales. *Earth Surface Processes and Landforms*, **13**, 259–270.

Dragovich, D. (1993) Distribution and chemical composition of microcolonial fungi and rock coatings from arid Australia. *Physical Geography*, **14**, 323–341.

Drake, N.A., Eckardt, F.D. and White, K.H. (2004) Sources of sulphur in gypsiferous sediments and crusts, and pathways of gypsum redistribution in southern Tunisia. *Earth Surface Processes and Landforms*, **29**, 1459–1473.

Drake, N.A., Heydeman, M.T. and White, K.H. (1993) Distribution and formation of rock varnish in southern Tunisia. *Earth Surface Processes and Landforms*, **18**, 31–41.

Drees, L.R. and Wilding, L.P. (1987) Micromorphic record and interpretation of carbonate forms in the Rolling Plains of Texas. *Geoderma*, **40**, 157–175.

Drever, J.I. and Smith, C.L. (1978) Cyclic wetting and drying of the soil zone as an influence on the chemistry of ground-water in arid terrains. *American Journal of Science*, **278**, 1448–1454.

Dubroeucq, D. and Thiry, M. (1994) *Indurations siliceuses dans des sols volcaniques. Comparison avec des silcrètes anciens.* Transactions of the 15th World Congress of Soil Sciences, Acapulco, Mexico, 10–16 July 1994, 6a, pp. 445–459.

Durand, N., Gunnell, Y., Curmi, P. and Ahmad, S.M. (2007) Pedogenic carbonates on Precambrian silicate rocks in South India: origin and paleoclimatic significance. *Quaternary International*, **162**, 35–49.

Eckardt, F.D. and Schemenauer, R.S. (1998) Fog water chemistry in the Namib Desert, Namibia. *Atmospheric Environment*, **32**, 2595–2599.

Eckardt, F.D. and Spiro, B. (1999) The origin of sulphur in gypsum and dissolved sulphate in the Central Namib Desert, Namibia. *Sedimentary Geology*, **123**, 255–273.

Eckardt, F.D., Drake, N.A., Goudie, A.S. *et al.* (2001) The role of playas in the formation of pedogenic gypsum crusts of the central Namib Desert. *Earth Surface Processes and Landforms*, **26**, 1177–1193.

Edinger, S.R. (1973) The growth of gypsum. An investigation of the factors which affect the size and growth rates of the habit faces of gypsum. *Journal of Crystal Growth*, **18**, 217–224.

Ek, C. and Pissart, A. (1965) Dépôt de carbonate de calcium par congélation et teneur en bicarbonate des eaux résiduelles. *Comptes Rendus Hebdomadaires des Séances de l'Académie des Sciences*, **260**, 929–932.

El Aref, M., Abdel Wahab, S. and Ahmed, S. (1985) Surficial calcareous crust of caliche type along the Red Sea coast, Egypt. *Geologische Rundschau*, **74**, 155–163.

Ellis, F. and Schloms, B.H.A. (1982) A note on the Dorbanks (duripans) of South Africa. *Palaeoecology of Africa*, **15**, 149–157.

Elouard, P. and Millot, G. (1959) Observations sur les silifications du Lutétien en Mauritanie et dans la vallée du Sénégal. *Bulletin du Service de la Carte Géologique d'Alsace et de Lorraine*, **12**, 15–19.

El Sayed, M.I. (1993) Gypcrete of Kuwait – field investigation, petrography and genesis. *Journal of Arid Environments*, **25**, 199–209.

Elvidge, C.D. (1982) Re-examination of the rate of desert varnish formation reported south of Barstow, California. *Earth Surface Processes and Landforms*, **7**, 345–348.

Elvidge, C.D. and Moore, C.B. (1979) A model for desert varnish formation. *Geological Society of America, Abstracts with Programs*, **11**, 271.

Engel, C.G. and Sharp, R. (1958) Chemical data on desert varnish. *Bulletin of the Geological Society of America*, **69**, 487–518.

Ericksen, G.E. (1981) Geology and origin of the Chilean nitrate deposits. *USGS Professional Paper*, **1188**, 1–37.

Ericksen, G.E. (1983) The Chilean nitrate deposits. *American Scientist*, **71**, 366–374.

Ericksen, G.E., Hosterman, J.W. and St. Amand, P. (1988) Chemistry, mineralogy, and origin of the Clay-Hill nitrate deposits, Amagosa River valley, Death Valley region, California, U.S.A. *Chemical Geology*, **67**, 85–102.

Ericksen, G.E. and Mrose, M.E. (1972) High purity veins of soda-niter, $NaNO_3$, and associated saline minerals in the Chilean nitrate deposits. *USGS Professional Paper*, **800-B**, B43-B49.

Esteban, M. (1974) Caliche textures and 'Microcodium'. *Bollettina della Società Geologie Italiana*, **92** (Suppl.), 105–125.

Eswaran, H., Stoops, G. and Abtahi, A. (1980) SEM morphologies of halite. *Journal of Microscopy*, **120**, 343–352.

Eswaran, H. and Zi-Tong, G. (1991) Properties, genesis, classification and distribution of soils with gypsum, in *Occurrence, Characteristics and Genesis of Carbonate, Gypsum and Silica Accumulation in Soils* (ed. W.D. Nettleton), Soil

Science Society of America, Special Publication 26, pp. 89–119.

Eugster, H.P. and Hardie, L.A. (1978) Saline lakes, in *Lakes: Chemistry, Geology, Physics* (ed. A. Lerman), Springer, New York, pp. 237–293.

Eugster, H.P. and Kelts, K. (1983) Lacustrine chemical sediments, in *Chemical Sediments and Geomorphology* (eds A.S. Goudie and K. Pye), Academic Press, London, pp. 321–368.

Evans, G. and Shearman, D.J. (1964) Recent celestite from the sediments of the Trucial Coast of the Persian Gulf. *Nature*, **202**, 385–386.

Evstifeev, Y.G. (1980) Extra-arid Gobi soils. *Problems of Desert Development*, (2), 17–26.

Flach, K.W., Nettleton, W.D., Gile, L.H. and Cady, J.G. (1969) Pedocementation: induration by silica, carbonates, and sesquioxides in the Quaternary. *Soil Science*, **107**, 442–453.

Fleisher, M., Liu, T., Broecker, W. and Moore, W. (1999) A clue regarding the origin of rock varnish. *Geophysical Research Letters*, **26**, 103–106.

Folk, R.L. (1971) Caliche nodule composed of calcite rhombs, in *Carbonate Cements* (ed. O.P. Bricker), John Hopkins University Press, Baltimore, MD, pp. 167–168.

Forbes, D. (1861) On the geology of Bolivia and southern Peru. *Quarterly Journal of the Geological Society of London*, **17**, 7–62.

Frankel, J.J. (1952) Silcrete near Albertina, Cape Province. *South African Journal of Science*, **49**, 173–182.

Frankel, J.J. and Kent, L.E. (1938) Grahamstown surface quartzites (silcretes). *Transactions of the Geological Society of South Africa*, **15**, 1–42.

Gabriel, A. (1964) Zum Problem des Formenschatzes in extremariden Raumen. *Mitteilungen der Õsterreichischen Geographischen Gesellschaft*, **106**, 3–15.

Gardner, L.R. (1972) Origin of the Mormon Mesa caliche, Clark County, Nevada. *Bulletin of the Geological Society of America*, **83**, 143–155.

Garvie, L.A.J. (2004) Decay-induced biomineralization of the saguaro cactus (*Carnegiea gigantean*). *American Mineralogist*, **88**, 1879–1888.

Gautier, A. (1894) Sur un gisement de phosphates de chaux et d'alumine contenant des espèces rares ou nouvelles et sur la genèse des phosphates et nitres naturels. *Annales de Mines*, ser. **9** (5), 5–53.

Gibson, G.W. (1962) Geological investigations of southern Victoria Land, Antarctica. Part 8: Evaporite salts in the Victoria Valley region. *New Zealand Journal of Geology and Geophysics*, **5**, 361–374.

Gile, L.H., Peterson, F.F. and Grossman, R.B. (1966) Morphological and genetic sequences of carbonate accumulation in desert soils. *Soil Science*, **101**, 347–360.

Glasby, G.P., McPherson, J.G., Kohn, B.P. *et al.* (1981) Desert varnish in Southern Victoria Lane, Antarctica. *New Zealand Journal of Geology and Geophysics*, **24**, 389–397.

Goudie, A.S. (1973) *Duricrusts in Tropical and Subtropical Landscapes*, Clarendon Press, Oxford.

Goudie, A.S. (1983) Calcrete, in *Chemical Sediments and Geomorphology* (eds A.S. Goudie and K. Pye), Academic Press, London, pp. 93–131.

Goudie, A.S. (1984) Duricrusts and landforms, in *Geomorphology and Soils* (eds K.S. Richards, R.R. Arnett and S. Ellis), Allen and Unwin, London, pp. 37–57.

Goudie, A.S. and Cooke, R.U. (1984) Salt efflorescences and saline lakes: a distributional analysis. *Geoforum*, **15**, 563–585.

Goudie, A.S. and Heslop, E. (2007) Sodium nitrate deposits and efflorescences, in *Geochemical Sediments and Landscapes* (eds D.J. Nash and S.J. McLaren), Blackwell, Oxford, pp. 391–408.

Graham, R.C., Hirmas, D.R., Word, Y.A. and Amrhein, C. (2008) Large near-surface nitrate pools in soils capped by desert pavement in the Mojave Desert, California. *Geology*, **36**, 259–262.

Grote, G. and Krumbein, W.E. (1992) Microbial precipitation of manganese by bacteria and fungi from desert rock and rock varnish. *Geomicrobiology Journal*, **10**, 49–57.

Gustavson, T.C. and Holliday, V.T. (1999) Eolian sedimentation and soil development on a semi-arid to subhumid grassland, Tertiary Ogallala and Quaternary Blackwater Draw Formations, Texas and New Mexico High Plains. *Journal of Sedimentary Research*, **69**, 622–634.

Harden, J.W., Taylor, E.M., Reheis, M.C. and McFadden, L.D. (1991) Calcic, gypsic and siliceous soil chronosequences in arid and semi-arid environments, in *Occurrence, Characteristics and Genesis of Carbonate, Gypsum and Silica Accumulation in Soils* (eds W.D. Nettleton), Soil Science Society of America, Special Publication 26, pp. 1–16.

Hardie, L.A. (1968) The origin of the recent non-marine evaporite deposits of Saline Valley, Inyo County, California. *Geochimica et Cosmochimica Acta*, **32**, 1279–1301.

Hardie, L.A. and Eugster, H.P. (1970) The evolution of closed-basin brines. *Mineralogical Society of America Special Publication*, **3**, 273–290.

Hardie, L.A., Smoot, J.P. and Eugster, H.P. (1978) Saline lakes and their deposits: a sedimentological approach, in *Modern and Ancient Lake Sediments* (eds A. Matter and M.E. Tucker), International Association of Sedimentologists, Special Publication 2, Blackwell, Oxford, pp. 7–42.

Harrington, C.D. and Whitney, J.W. (1995) Evidence suggesting that methods of rock-varnish cation-ratio dating are neither comparable nor consistently reliable – comment. *Quaternary Research*, **43**, 268–271.

Hartley, A.J. and May, G. (1998) Miocene gypcretes from the Calama Basin, northern Chile. *Sedimentology*, **45**, 351–364.

Hassouba, H. and Shaw, H.F. (1980) The occurrence of palygorskite in Quaternary sediments of the coastal plain of south-west Egypt. *Clay Minerals*, **15**, 77–83.

Hay, R.L. and Reeder, R.J. (1978) Calcretes of Olduvai Gorge and Ndolanya Beds of northern Tanzania. *Sedimentology*, **25**, 649–673.

Hay, R.L. and Wiggins, B. (1980) Pellets, ooids, sepiolite and silica in three calcretes of the south-western United States. *Sedimentology*, **27**, 559–576.

Heald, M.T. and Larese, R.E. (1974) Influence of coatings on quartz cementation. *Journal of Sedimentary Petrology*, **44**, 1269–1274.

Heaney, P.J. (1993) A proposed mechanism for the growth of chalcedony. *Contributions to Mineralogy and Petrology*, **115**, 66–74.

Heaney, P.J. (1995) Moganite as an indicator for vanished evaporites: a testament reborn? *Journal of Sedimentary Research*, **A65**, 633–638.

Hill, A.E. (1937) Transition temperature of gypsum to anhydrite. *Journal of the American Chemical Society*, **59**, 2242–2244.

Hill, S.M., Eggleton, R.A. and Taylor, G. (2003) Neotectonic disruption of silicified palaeovalley systems in an intraplate, cratonic landscape: regolith and landscape evolution of the Mulculca range-front, Broken Hill Domain, New South Wales. *Australian Journal of Earth Sciences*, **50**, 691–707.

Hill, C.A. and Forti, P. (1997) *Cave Minerals of the World*, 2nd edn, National Speleological Society, Huntsville, Alabama.

Hodge, T., Turchenek, L.W. and Oades, J.M. (1984) Occurrence of palygorskite in ground-water rendzinas (Petrocalcic calciaquolls) in south-east South Australia, in *Palygorskite-Sepiolite: Occurrences, Genesis and Uses* (eds A. Singer and E. Galan), Developments in Sedimentology 37, Elsevier, Amsterdam, pp. 199–210.

Holster, W.T. (1966) Diagnetic polyhalite in recent salt from Baja California. *American Mineralogist*, **51**, 99–109.

Horta, J.C. de O.S. (1980) Calcrete, gypcrete and soil classification in Algeria. *Engineering Geology*, **15**, 15–52.

Hubert, J.F. (1978) Paleosol caliches in the New Haven Arkose, Newark group, Connecticut. *Palaeogeography, Palaeoclimatology, Palaeoecology*, **24**, 151–168.

Hungate, B., Danin, A., Pellerin, N.B. *et al.* (1987) Characterisation of manganese-oxidising (MnII–MnIV) bacteria from Negev Desert rock varnish – implications in desert varnish formation. *Canadian Journal of Microbiology*, **33**, 939–943.

Hüser, K. (1976) Kalkrusten in Namib-Randbereich des mittleren Südwestafrika. *Mitteilungen der Basler Afrika Bibliographien*, **15**, 51–81.

Hutton, J.T. and Dixon, J.C. (1981) The chemistry and mineralogy of some South Australian calcretes and associated soft carbonates and their dolomitization. *Journal of the Geological Society of Australia*, **28**, 71–79.

Hutton, J.T., Twidale, C.R. and Milnes, A.R. (1978) Characteristics and origin of some Australian silcretes, in *Silcrete in Australia* (ed. T. Langford-Smith), Department of Geography, University of New England, Armidale, NSW, pp. 19–39.

Hutton, J.T., Twidale, C.R., Milnes, A.R. and Rosser, H. (1972) Composition and genesis of silcretes and silcrete skins from the Beda valley, southern Arcoona plateau, South Australia. *Journal of the Geological Society of Australia*, **19**, 31–39.

Iler, R. (1979) *The Chemistry of Silica: Solubility, Polymerization, Colloid and Surface Properties and Biochemistry*, John Wiley & Sons, Ltd, New York.

Jacobson G., Arakel A.V. and Chen, Y.J. (1988) The central Australian groundwater discharge zone – evolution of associated calcrete and gypcrete deposits. *Australian Journal of Earth Sciences*, **35**, 549–565.

James, N.P. (1972) Holocene and Pleistocene calcareous crust (caliche) profiles: criteria for subaerial exposure. *Journal of Sedimentary Petrology*, **42**, 817–836.

Jennings, J.N. and Sweeting, M.M. (1961) Caliche pseudo-anticlines in the Fitzroy Basin, Western Australia. *American Journal of Science*, **259**, 635–639.

Johannesson, J.K. and Gibson, G.W. (1962) Nitrate and iodate in Antarctica salt deposits. *Nature*, **194**, 567–568.

Jones, C.E. (1991) Characteristics and origin of rock varnish from the hyperarid coastal deserts of northern Peru. *Quaternary Research*, **35**, 116–129.

Kaemmerer, M. and Revel, J.C. (1991) Calcium carbonate accumulation in deep strata and calcrete in Quaternary alluvial formations of Morocco. *Geoderma*, **48**, 43–57.

Kahle, C.F. (1977) Origin of subaerial Holocene calcareous crust: role of algae, fungi and sparmicritisation. *Sedimentology*, **24**, 413–435.

Kaiser, E. (1926) *Die Diamentenwüste Südwest-Afrikas*, vol. **2**, Dietrich Reimer, Berlin.

Kaiser, E. and Neumaier, F. (1932) Sand-Steinsalz-Kristellskellete aus der Namib Südwestafrikas. *Zentralblatt für Mineralogie, Geologie und Paläontologie*, **6A**, 177–188.

Kaufman, D.S. (2003) Amino acid paleothermometry of Quaternary ostracodes from the Bonneville Basin, Utah. *Quaternary Science Reviews*, **22**, 899–914.

Kelly, M., Black, S. and Rowan, J.S. (2000) A calcrete-based U/Th chronology for landform evolution in the Sorbas basin, southeast Spain. *Quaternary Science Reviews*, **19**, 995–1010.

Khadkikar, A.S. (2005) Elemental composition of calcites in late Quaternary pedogenic calcretes from Gujarat, western India. *Journal of Asian Earth Sciences*, **25**, 893–902.

Khadkikar, A.S., Chamyal, L.S. and Ramesh, R. (2000) The character and genesis of calcrete in Late Quaternary alluvial deposits, Gujarat, western India, and its bearing on the interpretation of ancient climates. *Palaeogeography Palaeoclimatology Palaeoecology*, **162**, 239–261.

Khadkikar, A.S., Merh, S.S., Malik, J.N. and Chamyal, L.S. (1998) Calcretes in semi-arid alluvial systems: formative pathways and sinks. *Sedimentary Geology*, **116**, 251–260.

Klappa, C.F. (1978) Biolithogenesis of Microcodium. *Sedimentology*, **25**, 489–522.

Klappa, C.F. (1979a) Lichen stromatolites: criterion for subaerial exposure and a mechanism for the formation of laminar calcretes (caliche). *Journal of Sedimentary Petrology*, **49**, 387–400.

Klappa, C.F. (1979b) Calcified filaments in Quaternary calcretes: organo-mineral interactions in the subaerial vadose environment. *Journal of Sedminary Petrology*, **49**, 955–968.

Klappa, C.F. (1979c) Comment on 'Displacive calcite: evidence from recent and ancient calcretes'. *Geology*, **7**, 420–421.

Klappa, C.F. (1980) Brecciation textures and tepee structures in Quaternary calcrete (caliche) profiles from eastern Spain:

the plant factor in their formation. *Geological Journal*, **15**, 81–89.

Knauss, K.G. and Ku, T.-L. (1980) Desert varnish: potential for age dating via uranium-series isotopes. *Journal of Geology*, **88**, 95–100.

Knauth, L.P. (1994) Petrogenesis of chert, in *Silica: Physical Behaviour, Geochemistry and Materials Applications* (eds P.J. Heaney, C.T. Prewitt and G.V. Gibbs), Reviews in Mineralogy 29, Mineralogical Society of America, Washington, pp. 233–258.

Knox, G.J. (1977) Caliche profile formation, Saldanha Bay (South Africa). *Sedimentology*, **24**, 657–674.

Kosir, A. (2004) Microcodium revisited: root calcification products of terrestrial plants on carbonate-rich substrates. *Journal of Sedimentary Research*, **74**, 845–857.

Krauskopf, K.B. (1957) Separation of manganese from iron in sedimentary processes. *Geochimica et Cosmochimica Acta*, **12**, 61–84.

Krinsley, D. (1998) Models of rock varnish formation constrained by high resolution transmission electron microscopy. *Sedimentology*, **45**, 711–725.

Krinsley, D., Dorn, R.I., and Tovey, N.K. (1995) Nanometrescale layering in rock-varnish – implications for genesis and paleoenvironmental interpretation. *Journal of Geology*, **103**, 106–113.

Krumbein, W.E. and Jens, K. (1981) Biogenic rock varnishes of the Negev Desert (Israel): an ecological study of iron and manganese transformation by cyanobacteria and fungi. *Oecologia*, **50**, 25–38.

Ku, T.-L., Bull, W.B., Freeman, S.T. and Knauss, K.G. (1979) Th230–U234 dating of pedogenic carbonates in gravelly desert soils of Vidal Valley, southeastern California. *Bulletin of the Geological Society of America*, **90**, 1063–1073.

Kuhlman, K.R., Allenbach, L.B., Ball, C.L. *et al.* (2005) Enumeration, isolation, and characterization of ultraviolet (UV-C) resistant bacteria from rock varnish in the Whipple Mountains, California. *Icarus*, **174**, 585–595.

Kuhlman, K.R., Fusco, W.G., Duc, M.T.L. *et al.* (2006) Diversity of microorganisms within rock varnish in the Whipple Mountains, California. *Applied and Environmental Microbiology*, **72**, 1708–1715.

Kulke, H. (1974) Zur Geologie und Mineralogie der Kalkunder Gipskrusten Algeriens. *Geologische Rundschau*, **63**, 970–998.

Kushnir, J. (1980) The coprecipitation of strontium, magnesium, sodium, potassium and chloride ions with gypsum. An experimental study. *Geochemica et Cosmochimica Acta*, **44**, 1471–1482.

Kushnir, J. (1981) Formation and early diagenesis of varved evaporite sediments in a coastal hypersaline pool. *Journal of Sedimentary Petrology*, **51**, 1193–1023.

Lakin, H.W., Hunt, C.B., Davidson, D.F. and Oda, U. (1963) *Variations in minor-element content of desert varnish*, United States Geological Survey, Professional Paper 475B, pp. 28–31.

Lamplugh, G.W. (1902) Calcrete. *Geological Magazine*, **9**, 575.

Lamplugh, G.W. (1907) The geology of the Zambezi Basin around the Batoka Gorge (Rhodesia). *Quarterly Journal of the Geological Society of London*, **63**, 162–216.

Langford-Smith, T. (1978) A select review of silcrete research in Australia, in *Silcrete in Australia* (ed. T. Langford-Smith), Department of Geography, University of New England, Armidale, NSW, pp. 1–11.

Langford-Smith, T. and Watts, S.H. (1978) The significance of coexisting siliceous and ferruginous weathering products at select Australian localities, in *Silcrete in Australia* (eds T. Langford-Smith), Department of Geography, University of New England, Armidale, NSW, pp. 143–165.

Lattman, L.H. (1973) Calcium carbonate cementation of alluvial fans in southern Nevada. *Bulletin of the Geological Society of America*, **84**, 3013–3028.

Lattman, L.H. and Lauffenburger, S.K. (1974) Proposed role of gypsum in the formation of caliche. *Zeitschrift für Geomorphologie, NF, Supplementband*, **20**, 140–149.

Laudermilk, J.D. (1931) On the origin of desert varnish. *American Journal of Science*, ser. 5, **21**, 51–66.

Lauriol, B. and Clarke, J. (1999) Fissure calcretes in the Arctic: a paleohydrologic indicator. *Applied Geochemistry*, **14**, 775–785.

Lee, S.Y. and Gilkes, R.J. (2005) Groundwater geochemistry and composition of hardpans in southwestern Australian regolith. *Geoderma*, **126**, 59–84.

Linck, G. (1901) Über die dunkeln Rinden der Gesteine der Wüste. *Jenaische Zeitschrift für Naturwissenschaft*, **35**, 329–336.

Liu, T. (2003) Blind testing of rock varnish microstratigraphy as a chronometric indicator: results on Late Quaternary lava flows in the Mojave Desert, California. *Geomorphology*, **53**, 209–234.

Liu, T. and Broecker, W.S. (2000) How fast does rock varnish grow? *Geology*, **28**, 183–186.

Liu, T. and Broecker, W.S. (2007) Holocene rock varnish microstratigraphy and its chronometric application in drylands of western USA. *Geomorphology*, **84**, 1–21.

Liu, T. and Dorn, R.I. (1996) Understanding spatial variability in environmental changes in drylands with rock varnish microlaminations. *Annals of the Association of American Geographers*, **86**, 187–212.

Liu, T., Broecker, W.S.., Bell, J.W. and Mandeville, C. (2000) Terminal Pleistocene wet event recorded in rock varnish from the Las Vegas Valley, southern Nevada. *Palaeogeography, Palaeoclimatology, Palaeoecology*, **161**, 423–433.

Logan, B.W. (1987) The Macleod evaporite basin, Western Australia. *American Association of Petroleum Geologists*, Memoir 44.

Loisy, C., Verrecchia, E.P. and Dufour, P. (1999) Microbial origin for pedogenic micrite associated with a carbonate paleosol (Champagne, France). *Sedimentary Geology*, **126**, 193–204.

Lowenstein, T. and Brennan, S.T. (2001) Fluid inclusions in paleolimnological studies of chemical sediments, in *Tracking*

Environmental Change Using Lake Sediments, Vol. 2: Physical and Chemical Models (eds W.M. Last and J.P. Smol), Kluwer, Dordrecht, pp. 189–216.

Lowenstein, T.K. and Hardie, L.A. (1985) Criteria for the recognition of salt-pan evaporites. *Sedimentology*, **32**, 627–6244.

Lowenstein, T.K., Li, J., Brown, C. *et al.* (1999) 200 k.y. paleoclimate record from Death Valley salt core. *Geology*, **27**, 3–6.

Lucas, A. (1905) *The Blackened Rocks of the Nile Cataracts and of the Egyptian Deserts*, National Printing Department, Cairo.

Lyon, G.L. (1978) The stable isotope geochemistry of gypsum, Miers Valley, Antarctica. *Bulletin of the New Zealand Department of Scientific and Industrial Research*, **220**, 97–103.

Mabbutt J.A. (1955) Erosion surfaces in little Namaqualand and the ages of surface deposits in the south-western Kalahari. *Transactions of the Geological Society of South Africa*, **58**, 13–29.

McCarthy, T.S. and Ellery, W.N. (1995) Sedimentation on the distal reaches of the Okavango Fan, Botswana, and its bearing on calcrete and silcrete (ganister) formation. *Journal of Sedimentary Research*, **A65**, 77–90.

McFarlane, M.J. (1975) A calcrete from the Namanga–Bissel area of Kenya. *Kenyan Geographer*, **1**, 31–43.

McFarlane, M.J., Borger, H. and Twidale, C.R. (1992) Towards an understanding of the origin of titanium skins on silcrete in the Beda Valley, South Australia, in *Mineralogical and Geochemical Records of Paleoweathering* (eds J.-M. Schmitt and Q. Gall), Ecole des Mines de Paris, Mém. Sciences de la Terre 18, pp. 39–46.

McGowran, B., Rutland, R.W.R. and Twidale, C.R. (1977) Discussion: age and origin of laterite and silcrete duricrusts and their relationship to episodic tectonism in the mid-north of South Australia. *Journal of the Geological Society of Australia*, **24**, 421–422.

Machette, M.N. (1985) Calcic soils of the south-western United States. *Geological Society of America Special Paper*, **203**, 1–21.

Mack, G.H. and James, W.C. (1994) Palaeoclimate and global distribution of paleosols. *Journal of Geology*, **102**, 360–336.

Mack, G.H., Cole, D.R. and Trevino, L. (2000) The distribution and discrmination of shallow, authigenic carbonates in the Pliocene–Pleistocene Palomas Basin, southern Rio Grande Rift. *Bulletin of the Geological Society of America*, **112**, 643–656.

Mackenzie, R.C., Wilson, M.J. and Mashhady, A.S. (1984) Origin of palygorksite in some soils of the Arabian peninsula, in *Palygorskite-Sepiolite: Occurrences, Genesis and Uses* (eds A. Singer and E. Galan), Developments in Sedimentology 37, Elsevier, Amsterdam, pp. 177–186.

McLaren, S. (2004) Characteristics, evolution and distribution of Quaternary channel calcretes. *Earth Surface Processes and Landforms*, **29**, 1487–1507.

McNally, G.H. and Wilson, I.RE. (1995) Silcretes of the Mirackina Palaeochannel, Arckaringa, South Australia. *Australian Geological Survey Organisation Journal of Australian Geology and Geophysics*, **16**, 295–301.

McNaughton, N.J., Rasmussen, B. and Fletcher, I.R. (1999) SHRIMP uranium–lead dating of diagenetic xenotime in siliciclastic sedimentary rocks. *Science*, **285**, 78–80.

McQueen, K.G., Hill, S.M. and Foster, K.A. (1999) The nature and distribution of regolith carbonate accumulations in southeastern Australia and their potential as a sampling medium in geochemical exploration. *Journal of Geochemical Exploration*, **67**, 67–82.

Magaritz, M., Kaufman, A. and Yaalon, D.H. (1981) Calcium carbonate nodules in soils: 18O/16O and 12C/13C ratios and 14C contents. *Geoderma*, **25**, 157–172.

Magee, J.W. (1991) Late Quaternary lacustrine, groundwater, aeolian and pedogenic gypsum in the Prungle Lakes, southeastern Australia. *Palaeogeography, Palaoclimatology, Palaeoecology*, **84**, 3–42.

Maizels, J. (1987) Plio-Pleistocene raised channel systems of the western Sharqiya (Wahiba), Oman, in *Desert Sediments: Ancient and Modern* (eds L.E. Frostick and I. Reid), Geological Society of London Special Publication 35, pp. 31–50.

Maizels, J. (1990) Raised channel systems as indicators of palaeohydrologic change: a case study from Oman. *Palaeogeography, Palaeoclimatology, Palaeoecology*, **76**, 241–277.

Mann, A.W. and Horwitz, R.C. (1979) Groundwater calcrete deposits in Australia: some observations from Western Australia. *Journal of the Geological Society of Australia*, **26**, 293–303.

Manze, U. and Brunnacker, K. (1977) Über das Verhalten des Sauerstoff- under Kohlenstroof-Isotope in Kalkkursten und Kalktuffen des mediterranen Raumes und der Sahara. *Zeitschrift für Geomorphologie*, NF, **21**, 343–353.

Marker, M.E., McFarlane, M.J. and Wormald, R.J. (2002) A laterite profile near Albertinia, Southern Cape, South Africa: its significance in the evolution of the African surface. *South African Journal of Geology*, **105**, 67–74.

Marshall, R.R. (1962) Natural radioactivity and the origin of desert varnish. *Transactions of the American Geophysical Union*, **43**, 446–447.

Martin, H. (1963) A suggested theory for the origin and a brief description of some gypsum deposits of South-West Africa. *Transactions of the Geological Society of South Africa*, **66**, 345–351.

Masson, P.H. (1955) An occurrence of gypsum in south-west Texas. *Journal of Sedimentary Petrology*, **25**, 72–77.

Merrill, G.P. (1898) Desert varnish. *Bulletin of the United States Geological Survey*, **150**, 389–391.

Meyer, R. and Pena dos Reis, R.B. (1985) Palaeosols and alunite silcretes in continental Cenozoic of Western Portugal. *Journal of Sedimentary Petrology*, **55**, 76–85.

Michalski G., Savarino, J., Böhlke J.K. and Thiemens, M. (2002) Determination of the Total Oxygen Isotopic composition of nitrate and the calibration of a δ17O nitrate reference material. *Analytical Chemistry*, **74**, 4989–4993.

Milnes, A.R. and Hutton, J.T. (1983) Calcretes in Australia, in *Soils: An Australian Viewpoint*, Academic Press, London, pp. 119–162.

Milnes, A.R. and Thiry, M. (1992) Silcretes, in *Weathering, Soils and Paleosols* (eds I.P. Martini and W. Chesworth), Developments in Earth Surface Processes 2, Elsevier, Amsterdam, pp. 349–377.

Milnes, A.R., Thiry, M. and Wright, M.J. (1991) Silica accumulations in saprolite and soils in South Australia, in *Occurrence, Characteristics and Genesis of Carbonate, Gypsum and Silica Accumulation in Soils* (ed. W.D. Nettleton), Soil Science Society of America, Special Publication 26, pp. 121–149.

Monger, H.C. and Gallegos, R.A. (2000) Biotic and abiotic processes and rates of pedogenic carbonate accumulation in the southwestern United States – relationship to atmospheric CO_2 sequestration, in *Global Climate Change and Pedogenic Carbonates* (eds R. Lal, J.M. Kimble, H. Eswaran and B.A. Stewart), CRC Press/Lewis Publishers, Boca Raton, pp. 273–289.

Morgan, K.H. (1993) Development, sedimentation and economic potential of palaeoriver systems of the Yilgarn Craton of Western Australia. *Sedimentary Geology*, **85**, 637–656.

Morris, R.C. and Dickey, P.A. (1957) Modern evaporite deposition in Peru. *Bulletin of the American Association of Petroleum Geologists*, **41**, 2467–2474.

Morris, B.A. and Fletcher, I.A. (1987) Increased solubility of quartz following ferrous–ferric iron reactions. *Nature*, **330**, 558–561.

Mortensen, H. (1927) Der Formenschatz der nor-chilenischen Wüste, ein Beitrag zum Gesetz der Wüstenbildung. *Abhaldlungen der Gesellschaft der Wissenschalften zu Göttingen*, NF 12.

Moseley, F. (1965) Plateau calcrete, calcreted gravels, cemented dunes and related deposits of the Maallegh–Bomba region of Libya. *Zeitschrift für Geomorphologie*, **NF 9**, 167–185.

Mottershead, D.N. and Pye, K. (1994) Tafoni on coastal slopes, South Devon, U.K. *Earth Surface Processes and Landforms*, **19**, 543–563.

Mountain, E.D. (1952) The origin of silcrete. *South African Journal of Science*, **48**, 201–204.

Mueller, G. (1968) Genetic histories of nitrate deposits from Antarctica and Chile. *Nature*, **219**, 1131–1134.

Nagy, B., Nagy, L.A., Rigali, M.J., *et al.* (1991) Rock varnish in the Sonoran Desert – microbiologically mediated accumulation of manganiferous sediments. *Sedimentology*, **79**, 542–553.

Nanson, G.C., Jones, B.G., Price, D.M. and Pietsch, T.J. (2005) Rivers turned to rock: Late Quaternary alluvial induration influencing the behaviour and morphology of an anabranching river in the Australian monsoon tropics. *Geomorphology*, **70**, 398–420.

Nash, D.J. and Hopkinson, L. (2004) A reconnaissance laser Raman and Fourier transform infrared survey of silcretes from the Kalahari Desert, Botswana. *Earth Surface Processes and Landforms*, **29**, 1541–1558.

Nash, D.J. and McLaren, S.J. (2003) Kalahari valley calcretes: their nature, origin and environmental significance. *Quaternary International*, **111**, 3–22.

Nash, D.J. and McLaren, S.J. (2007) Introduction: geochemical sediments in landscapes, in *Geochemical Sediments and Landscapes* (eds D.J. Nash and S.J. McLaren), Blackwell, Oxford, pp. 1–9.

Nash, D.J. McLaren, S.J. and Webb, J.A. (2004) Petrology, geochemistry and environmental significance of silcrete–calcrete intergrade duricrusts at Kang Pan and Tswaane, central Kalahari, Botswana. *Earth Surface Processes and Landforms*, **29**, 1559–1586.

Nash, D.J. and Shaw, P.A. (1998) Silica and carbonate relationships in silcrete–calcrete intergrade duricrusts from the Kalahari Desert of Botswana and Namibia. *Journal of African Earth Sciences*, **27**, 11–25.

Nash, D.J., Shaw, P.A. and Thomas, D.S.G. (1994) Duricrust development and valley evolution – process-landform links in the Kalahari. *Earth Surface Processes and Landforms*, **19**, 299–317.

Nash, D.J. and Smith, R.F. (1998) Multiple calcrete profiles in the Tabernas Basin, SE Spain: their origins and geomorphic implications. *Earth Surface Processes and Landforms*, **23**, 1009–1029.

Nash, D.J. and Smith, R.F. (2003) Properties and development of channel calcretes in a mountain catchment, Tabernas Basin, southeast Spain. *Geomorphology*, **50**, 227–250.

Nash, D.J., Thomas, D.S.G. and Shaw, P.A. (1994) Siliceous duricrusts as palaeoclimatic indicators: evidence from the Kalahari Desert of Botswana. *Palaeogeography, Palaeoclimatology, Palaeoecology*, **112**, 279–295.

Nash, D.J. and Ullyott, J.S. (2007) Silcrete, in *Geochemical Sediments and Landscapes* (eds D.J. Nash and S.J. McLaren), Blackwell, Oxford, pp. 95–143.

Netterberg, F. (1969) Ages of calcrete in southern Africa. *South African Archaeological Bulletin*, **24**, 88–92.

Netterberg, F. (1978) Dating and correlation of calcretes and other pedocretes. *Transactions of the Geological Society of South Africa*, **81**, 379–391.

Netterberg, F. (1980) Geology of southern African calcretes. I. Terminology, description, macrofeatures and classification. *Transactions of the Geological Society of South Africa*, **83**, 255–283.

Newton, W. (1869) The origin of nitrate in Chili. *Geological Magazine* NS, Decade **4** (3), 339–342.

Nickling, W.G. (1984) The stabilizing effect of bonding agents on the entrainment of sediment by wind. *Sedimentology*, **31**, 111–117.

Nickling, W.G. and Ecclestone, M. (1981) The effects of soluble salts on the threshold shear velocity of fine sand. *Sedimentology*, **28**, 505–510.

Oberlander, T.M. (1994) Rock varnish in deserts, in *Geomorphology of Desert Environments* (eds A.D. Abrahams and A.J. Parsons), Chapman and Hall, London, pp. 107–119.

Ollier, C.D. (1978) Silcrete and weathering, in *Silcrete in Australia* (ed. T. Langford-Smith), Department of Geography, University of New England, Armidale, NSW, pp. 13–17.

O'Neill, G. (1984) *Geochemistry of silcrete in the Sunbury Area*, Unpublished Honours Thesis, Department of Earth Sciences, University of Melbourne, Australia.

Oyarzun, J. and Oyarzun, R. (2007) Massive volcanism in the altiplano-puna volcanic plateau and formation of the huge Atacama Desert nitrate deposits: a case for thermal and electric fixation of atmospheric nitrogen. *International Geology Review*, **49**, 962–968.

Pain, C.F. and Ollier, C.D. (1995) Inversion of relief – a component of landscape evolution. *Geomorphology*, **12**, 151–165.

Palmer, F.E., Staley, J.T., Murray, R.G.E. and Counsell, T. (1985) Identification of manganese oxidising bacteria from desert varnish. *Geomicrobiology Journal*, **4**, 343–360.

Patterson, R.J. and Kinsman, D.J.J. (1978) Marine and continental groundwater sources in a Persian Gulf coastal sabkha. *American Association of Petroleum Geologists, Studies in Geology*, **4**, 381–397.

Penrose, R.A.F. (1910) The nitrate deposits of Chile. *Journal of Geology*, **18**, 1–32.

Perry, R.S. and Adams, J.B. (1978) Desert varnish: evidence for cyclic deposition of manganese. *Nature*, **276**, 489–491.

Perry, R.S., Lynne, B.Y., Sephton, M.A. *et al.* (2006) Baking black opal in the desert sun: the importance of silica in desert varnish. *Geology*, **34**, 737–740.

Pervinquière, L. (1903) *Étude Géologique de la Tunisie central*, De Rudeval, Paris.

Peterson, M.N.A. and Van Der Borch, C.C. (1965) Chert: modern inorganic deposition in a carbonate-precipitating locality. *Science*, **149**, 1501–1503.

Phillips, S.E., Milnes, A.R. and Foster, R.C. (1987) Calcified filaments: an example of biological influences in the formation of calcrete in South Australia. *Australian Journal of Soil Research*, **25**, 405–428.

Phillips, S.E. and Self, P.G. (1987) Morphology, crystallography and origin of needle-fibre calcite in Quaternary pedogenic calcretes of South Australia. *Australian Journal of Soil Research*, **25**, 429–444.

Phleger, F.B. (1969) A modern evaporite deposit in Mexico. *Bulletin of the American Association of Petroleum Geologists*, **53**, 824–8300.

Pimentel, N.L., Wright, V.P. and Azevedo, T.M. (1996) Distinguishing early groundwater alteration effects from pedogenesis in ancient alluvial basins: examples from the Palaeogene of southern Portugal. *Sedimentary Geology*, **105**, 1–10.

Potter, R.M. and Rossman, G.R. (1977) Desert varnish: the importance of clay minerals. *Science*, **196**, 1446–1448.

Potter, R.M. and Rossman, G.R. (1979) The manganese and iron oxide mineralogy of desert varnish. *Chemical Geology*, **25**, 79–94.

Pouget, M. (1968) Contribution à l'étude des croûtes et encroûtements gypseux de nappe dans le sud tunisien. *Cahiers ORSTOM*, Série Pédologie, **6**, 309–365.

Price, W.A. (1925) Caliche and pseudo-anticlines. *Bulletin of the American Association of Petroleum Geologists*, **9**, 1009–1017.

Pueyo J.J., Chong, G. and Vega, M. (1998) Mineralogia y evolucion de las salmueras madres en el yacimento de nitratos Pedro de Valdivia, Antofagasta, Chile. *Revista Geologica de Chile*, **25**, 3–15.

Pye, K. (1980) Beach salcrete and eolian sand transport: evidence from North Queensland. *Journal of Sedimentary Petrology*, **50**, 257–261.

Quade, J., Cerling, T.E. and Bowman, J.R. (1989) Systematic variations in the carbon and oxygen isotopic composition of pedogenic carbonate along elevation transects in the southern Great Basin, United States. *Geological Society of America Bulletin*, **101**, 464–475.

Rabenhorst, M.C., Wilding, L.P. and Girdner, C.L. (1984) Airborne dust in the Edwards Plateau region of Texas. *Soil Science Society of America Journal*, **48**, 621–627.

Radtke, U. and Brückner, H. (1991) Investigation on age and genesis of silcretes in Queensland (Australia) – preliminary results. *Earth Surface Processes and Landforms*, **16**, 547–554.

Rakshit, P. and Sundaram, R.M. (1998) Calcrete and gypsum crusts of the Thar Desert, Rajasthan, their geomorphic locales and use as palaeoclimatic indicators. *Journal of the Geological Society of India*, **51**, 249–255.

Ramakrishnan, D. and Tiwari, K.C. (2006) Calcretized ferricretes around the Jaisalmer area, Thar Desert, India: their chemistry, mineralogy, micromorphology and genesis. *Turkish Journal of Earth Sciences*, **15**, 211–223.

Rech, J.A., Quade, J. and Hart, W.S. (2003) Isotopic evidence for the source of Ca and S in soil gypsum, anhydrite and calcite in the Atacama Desert, Chile. *Geochimica et Cosmochimica Acta*, **67**, 576–586.

Reeves, C.C. (1970) Origin, classification, and geological history of caliche on the southern High Plains, Texas and eastern New Mexico. *Journal of Geology*, **78**, 352–362.

Reeves, C.C. (1976) *Caliche: Origin, Classification, Morphology and Uses*, Estacado Books, Lubbock, TX.

Reeves, C.C. (1983) Pliocene channel calcrete and suspenparallel drainage in West Texas and New Mexico, in *Residual Deposits* (ed. R.C.L. Wilson), Geological Society of London Special Publication 11, pp. 179–183.

Reheis, M.C. (1987) Gypsic soils on the Kane Alluvial Fans. Big Horn County, Wyoming, Bulletin of the United States Geological Survey, 1590C.

Renaut, R.W., Tiercelin, J.J. and Owen, R.B. (1986) Mineral pricipatation and diagenesis in the sediments of the Lake Bogoria basin, Kenya Rift Valley, in *Sedimentation in the African Rifts* (eds L.E. Frostick, R.W. Renaut, I. Reid and J.J. Tiercelin), Blackwell Scientific, Oxford, pp. 159–175.

Ribet, I. and Thiry, M. (1990) Quartz growth in limestone: example from water-table silicification in the Paris Basin. *Chemical Geology*, **84**, 316–319.

Risacher, F. (1978) Genèse d'une croûte de gypse dans un basin de l'Altiplano bolivien. *Cahiers ORSTOM*, série Géologie, **10**, 91–100.

Risacher, F. and Fritz, B. (2000) Bromine geochemistry of Salar de Uyuni and deeper salt crusts, central Altiplano, Bolivia. *Chemical Geology*, **167**, 373–392.

Roberts, S.M. and Spencer, R.J. (1995) Paleotemperatures preserved in fluid inclusions in halite. *Geochimica et Cosmochimica Acta*, **59**, 3929–3942.

Rodas, M., Luque, F.J., Mas, R. and Garzon, M.G. (1994) Calcretes, palycretes and silcretes in the Paleogene detrital sediments of the Duero and Tajo Basins, Central Spain. *Clay Minerals*, **29**, 273–285.

Royer, D.L. (1999) Depth to pedogenic carbonate horizons as a paleoprecipitation indicator? *Geology*, **27**, 1123–1126.

Rust, U., Schmidt, H. and Dietz, K. (1984) Palaeoenvironments of the present day arid south western Africa 30,000–5000 BP: results and problems. *Palaeoecology of Africa*, **16**, 109–148.

Salomons, W., Goudie, A. and Mook, W.G. (1978) Isotopic compostion of calcrete deposits from Europe, Africa and India. *Earth Surface Processes*, **3**, 43–57.

Salomons, W. and Mook, W.G. (1976) Isotope geochemistry of carbonate dissolution and reprecipitation in soils. *Soil Science*, **122**, 15–24.

Schlesinger, W.H. (1985) The formation of caliche in soils of the Mojave Desert, California. *Geochimica et Cosmochimica Acta*, **49**, 57–66.

Schmittner, K.-E. and Giresse, P. (1999) Micro-environmental controls on biomineralization: superficial processes of apatite and calcite precipitation in Quaternary soils, Roussillon, France. *Sedimentology*, **46**, 463–476.

Scholz, H. (1972) The soils of the central Namib Desert with special consideration of the soils in the vicinity of Gobabeb. *Madoqua*, **2**, 33–51.

Schreiber, B.C. (1986) Arid shorelines and evaporites, in *Sedimentary Environments and Facies* (ed. H.G. Reading), Blackwell Scientific, Oxford, pp. 189–228.

Searl, A. (1994) Discussion of a petrographic study of the Chilean nitrates. *Geological Magazine*, **131**, 849–852.

Searl, A. and Rankin, S. (1993) A preliminary petrographic study of the Chilean nitrates. *Geological Magazine*, **130**, 319–333.

Sehgal, J.L. and Stoops, G. (1972) Pedogenic calcite accumulations in arid and semi-aid regions of the Indo–Gangetic alluvial plain of erstwhile Punjab. *Geoderma*, **8**, 59–72.

Semeniuk, V. and Meagher, T.D. (1981) Calcrete in Quaternary coastal dunes in south Western Australia: a capillary-rise phenomenon associated with plants. *Journal of Sedimentary Petrology*, **51**, 47–68.

Semeniuk, V. and Searle, D.J. (1985) Distribution of calcrete in Holocene coastal sands in relationship to climate, southwestern Australia. *Journal of Sedimentary Petrology*, **55**, 86–95.

Senior, B.R. (1978) Silcrete and chemically weathered sediments in southwest Queensland, in *Silcrete in Australia* (ed. T. Langford-Smith), Department of Geography, University of New England, Armidale, NSW, pp. 41–50.

Senior, B.R. and Senior, D.A. (1972) Silcrete in southwest Queensland. *Bulletin of the Bureau of Mineral Resources, Geology and Geophysics of Australia*, **125**, 23–28.

Shadfan, H. and Dixon, J.B. (1984) Occurrence of palygorskite in the soils and rocks of the Jordan Valley, in *Palygorskite-Sepiolite: Occurrences, Genesis and Uses* (eds A. Singer and E. Galan), Developments in Sedimentology 37, Elsevier, Amsterdam, pp. 187–198.

Shaw, P.A., Cooke, H.J. and Perry, C.C. (1990) Microbialitic silcretes in highly alkaline environments: some observations from Sua Pan, Botswana. *South African Journal of Geology*, **93**, 803–808.

Shaw, P.A. and De Vries, J.J. (1988) Duricrust, groundwater and valley development in the Kalahari of south-east Botswana. *Journal of Arid Environments*, **14**, 245–254.

Shaw, P.A. and Nash, D.J. (1998) Dual mechanisms for the formation of fluvial silcretes in the distal reaches of the Okavango Delta Fan, Botswana. *Earth Surface Processes and Landforms*, **23**, 705–714.

Shearman, D.J. (1963) Recent anhydrite, gypsum, dolomite, and halite from the coastal flats of the Arabian shore of the Persian Gulf. *Proceedings of the Geological Society of London*, **1607**, 63–64.

Shearman, D.J. (1966) Origins of marine evaporites by diagenesis. *Transactions of the Institution of Mining and Metallurgy*, **75B**, 208–215.

Siever, R. (1972) Silicon: solubilities of compounds which control concentrations in natural waters, in *Handbook of Geochemistry* (ed. K.H. Wedepohl), Springer-Verlag, Berlin, pp. 14-H-1–14-H-7.

Simon-Coinçon, R., Thiry, M. and Quesnel, F. (2000) Siderolithic palaeolandscapes and palaeoenvironments in the northern Massif Central (France). *Comptes Rendus de l'Academie des Sciences Series II*, **330**, 693–700.

Simon-Coinçon, R., Milnes, A.R., Thiry, M. and Wright, M.J. (1996) Evolution of landscapes in northern South Australia in relation to the distribution and formation of silcretes. *Journal of the Geological Society, London*, **153**, 467–480.

Singer, A. (1984) Pedogenic palygorskite in the arid environment, in *Palygorskite-Sepiolite: Occurrences, Genesis and Uses* (eds A. Singer and E. Galan), Developments in Sedimentology 37, Elsevier, Amsterdam, pp. 169–175.

Smale, D. (1973) Silcretes and associated silica diagenesis in southern Africa and Australia. *Journal of Sedimentary Petrology*, **43**, 1077–1089.

Smale, D. (1978) Silcretes and associated silica diagenesis in southern Africa and Australia, in *Silcrete in Australia* (ed. T. Langford-Smith), Department of Geography, University of New England, Armidale, NSW, pp. 261–279.

Smith, G.I. (1979) Subsurface stratigraphy and geochemistry of Late Quaternary evaporites, Searles Lake, California, United States Geological Survey Professional Paper 1043.

Smith, G.I. and Bischoff, J.L. (1997) An 800,000-year paleoclimate record from core OL-92, Owens Lake, southeast California, Geological Society of America, Special Paper 317.

Smith, B.J. and Whalley, W.B. (1988) A note on the characteristics and possible origins of desert varnish from southeast

Morocco. *Earth Surface Processes and Landforms*, **13**, 251–258.

Smith, B.J. and Whalley, W.B. (1989) A note on the characteristics and possible origins of desert varnish from southeast Morocco. A reply to comments by R.I. Dorn. *Earth Surface Processes and Landforms*, **14**, 171–172.

Smoot, J.P. and Lowenstein, T.K. (1991) Depositional environments of non-marine evaporites, in *Evaporites, Petroleum and Mineral Resources* (ed. J.L. Melvin), Elsevier, Amsterdam, pp. 189–347.

Soegaard, K. and Eriksson, K.A. (1985) Holocene silcretes from NW Australia: implications for textural modification of pre-existing sediments. *Geological Society of America, Abstracts with Programs*, **17**, 722.

Sofer, Z. (1978) Isotopic composition of hydration water in gypsum. *Geochimica et Cosmochimica Acta*, **42**, 1141–1149.

Souza-Egipsy, V., Wierzchos, J., Snacho, C. *et al.* (2004) Role of biological soil crust cover in bioweathering and protection of sandstones in a semi-arid landscape (Torrollones de Gabarda, Huesca, Spain). *Earth Surface Processes and Landforms*, **29**, 1651–1666.

Spencer, R.J., Eugster, H.P. and Jones, B.F. (1985) Geochemistry of Great Salt Lake, Utah II: Pleistocene–Holocene evolution. *Geochimica et Cosmochimica Acta*, **49**, 739–747.

Srivastava, K.K. (1969) Gypsum. *Bulletin of the Geological Survey of India*, **31A**.

Staley, J.T., Jackson, M.J., Palmer, F.E. *et al.* (1983) Desert varnish coatings and microcolonial fungi on rocks of the Gibson and Great Victorian Deserts, Australia. *BMR Journal of Geology and Geophysics*, **8**, 83–87.

Stankevich, E.F., Imameev, A.N. and Garanin, I.V. (1983) Formation of salt lake deposits in the arid regions of Middle Asia and Kazakhstan. *Problems of Desert Development*, **1983** (5), 19–24.

Stephens, C.G. (1971) Laterite and silcrete in Australia: a study of the genetic relationships of laterite and silcrete and their companion materials, and their collective significance in the formation of the weathered mantle, soils, relief and drainage of the Australian continent. *Geoderma*, **5**, 5–52.

Stokes, M., Nash, D.J. and Harvey, A.M. (2007) Calcrete 'fossilisation' of alluvial fans in SE Spain: the roles of groundwater, pedogenic processes and fan dynamics in calcrete development. *Geomorphology*, **85**, 63–84.

Stortsky, G. and Rem, L.T. (1967) Influence of clay minerals on microorganisms. IV: Montmorillonite and kaolinite on fungi. *Canadian Journal of Microbiology*, **13**, 1535–1550.

Strong, G.E., Giles, J.R.A. and Wright, V.P. (1992) A Holocene calcrete from North Yorkshire, England: implications for interpreting palaeoclimates using calcretes. *Sedimentology*, **39**, 333–347.

Summerfield, M.A. (1979) Origin and palaeoenvironmental interpretation of sarsens. *Nature*, **281**, 137–139.

Summerfield, M.A. (1981) *The nature and occurrence of silcrete*, Southern Cape Province, South Africa, Oxford University School of Geography, Research Papers 28.

Summerfield, M.A. (1982) Distribution, nature and genesis of silcrete in arid and semi-arid southern Africa. *Catena, Supplement*, **1**, 37–65.

Summerfield, M.A. (1983a) Silcrete, in *Chemical Sediments and Geomorphology* (eds A.S. Goudie and K. Pye), Academic Press, London, pp. 59–91.

Summerfield, M.A. (1983b) Petrology and diagenesis of silcrete from the Kalahari Basin and Cape coastal zone, southern Africa. *Journal of Sedimentary Petrology*, **53**, 895–909.

Summerfield, M.A. (1983c) Silcrete as a palaeoclimatic indicator: evidence from southern Africa. *Palaeogeography, Palaeoclimatology, Palaeoecology*, **41**, 65–79.

Summerfield, M.A. (1983d) Geochemistry of weathering profile silcretes, southern Cape Province, South Africa, in *Residual Deposits* (ed. R.C.L. Wilson), Geological Society of London Special Publication 11, pp. 167–178.

Summerfield, M.A. (1991) *Global Geomorphology*, Longman, Harlow.

Summerfield, M.A. and Goudie, A.S. (1980) The sarsens of southern England: their palaeoenvironmental interpretation with reference to other silcretes, in *The Shaping of Southern England* (eds D.K.C. Jones), Academic Press, London, pp. 71–100.

Tait, M. (1998) *Geology and landscape evolution of the Mt Wood Hills area, near Tibooburra, northwestern New South Wales*, Unpublished Honours Thesis, Department of Earth Sciences, La Trobe University, Australia.

Talma, A.S. and Netterberg, F. (1983) Stable isotope abundances in calcretes, in *Residual Deposits* (ed. R.C.L. Wilson), Geological Society of London Special Publication 11, pp. 221–233.

Tandon, S.K. and Andrews, J.E. (2001) Lithofacies associations and stable isotopes of palustrine and calcrete carbonates: examples from an Indian Maastrichtian regolith. *Sedimentology*, **48**, 339–355.

Tandon, S.K. and Narayan, D. (1981) Calcrete conglomerate, case-hardened conglomerate and cornstone: a comparative account of pedogenic and non-pedogenic carbonates from the continental Siwalik Group, Punjab, India. *Sedimentology*, **28**, 353–367.

Taylor, G. and Eggleton, R.A. (2001) *Regolith Geology and Geomorphology*, John Wiley & Sons, Ltd, Chichester.

Taylor, G. and Ruxton, B.P. (1987) A duricrust catena in southeast Australia. *Zeitschrift für Geomorphologie*, **31**, 385–410.

Taylor, G. and Smith, I.E. (1975) The genesis of sub-basaltic silcretes from the Monaro, New South Wales. *Journal of the Geological Society of Australia*, **22**, 377–385.

Tebo, B.M., Bargar, J.R., Clement, B.G. *et al.* (2004) Biogenic manganese oxides: properties and mechanisms of formation. *Annual Review Earth and Planetary Science*, **32**, 287–328.

Tekin, E., Ayyildiz, T., Gündoğan, I. and Orti, F. (2006) Modern halolites (halite oolites) in the Tuz Gölü, Turkey. *Sedimentary Geology*, **195**, 101–112.

Teller, J.T. and Last, W.M. (1990) Palaeohydrological indicators in playas and salt lakes, with examples from Canada,

Australia, and Africa. *Palaeogeography, Palaeoclimatology, Palaeoecology*, **76**, 215–240.

Tereschenko, O. and Malyutin, S. (1985) Hygroscopicity of sodium nitrate (Chile saltpetre). *Journal of Applied Chemistry (USSR)*, **85**, 810–813.

Terry, D.O. and Evans, J.E. (1994) Pedogenesis and paleoclimatic implications of the Chamberlain Pass Formation, Basal White River Group, Badlands of South Dakota. *Palaeogeography, Palaeoclimatology, Palaeoecology*, **110**, 197–215.

Thiagarajan, N. and Lee, C.A. (2004) Trace-element evidence for the origin of desert varnish by direct aqueous atmospheric deposition. *Earth and Planetary Science Letters*, **224**, 131–141.

Thiry, M. (1978) Silicification des sédiments sablo-argileux de l'Yprésien du sud-est du bassin de Paris. Genèse et évolution des dalles quartzitiques et silcrètes. *Bulletin du BRGM (deuxième serie)*, **1**, 19–46.

Thiry, M. (1981) Sédimentation continentale et altérations associées: calcitisations, ferruginisations et silicifications. *Les Argiles Plastiques de Sparnacien du Bassin de Paris. Sci. Géol., Mém.*, **64**.

Thiry, M. (1999) Diversity of continental silicification features: examples from the Cenozoic deposits in the Paris Basin and neighbouring basement, in *Palaeoweathering, Palaeosurfaces and Related Continental Deposits* (eds M. Thiry and R. Simon-Coinçon), International Association of Sedimentologists, Special Publication 27, Blackwell Science, Oxford, pp. 87–127.

Thiry, M., Ayrault, M.B. and Grisoni, J. (1988) Groundwater silicification and leaching in sands: example of the Fontainebleau Sand (Oligocene) in the Paris Basin. *Bulletin of the Geological Society of America*, **100**, 1283–1290.

Thiry, M. and Ben Brahim, M. (1990) Silicifications pédogénétiques dans les dépôts hamadiens du piémont de boudenib (Maroc). *Geodinamica Acta*, **4**, 237–251.

Thiry, M. and Ben Brahim, M. (1997) Ground-water silicifications in the calcareous facies of the Tertiary piedmont deposits of the Atlas Mountain (Hamada du Guir, Morocco). *Geodinamica Acta*, **10**, 12–29.

Thiry, M. and Millot, G. (1987) Mineralogical forms of silica and their sequence of formation in silcretes. *Journal of Sedimentary Petrology*, **57**, 343–352.

Thiry, M. and Milnes, A.R. (1991) Pedogenic and groundwater silcretes at Stuart Creek opal field, South Australia. *Journal of Sedimentary Petrology*, **61**, 111–127.

Thiry, M. and Simon-Coinçon, R. (1996) Tertiary palaeoweatherings and silcretes in the southern Paris Basin. *Catena*, **26**, 1–26.

Thiry, M., Milnes, A.R., Rayot, V. and Simon-Coinçon, R. (2006) Interpretation of palaeoweathering features and successive silicifications in the Tertiary regolith of inland Australia. *Journal of the Geological Society, London*, **163**, 723–736.

Tolchel'nikov, Y.S. (1962) Calcium sulphate and carbonate neoformations in sandy desert soils. *Soviet Soil Science*, **1962**, 643–650.

Tucker, M.E. (1978) Gypsum crusts (gypcrete) and patterned ground from northern Iraq. *Zeitschrift für Geomorphologie*, **NF 22**, 89–100.

Turner, B.R. and Makhlouf, I. (2005) Quaternary sandstones, northeast Jordan: age, depositional environments and climatic implications. *Palaeogeography, Palaeoclimatology, Palaeoecology*, **229**, 230–250.

Twidale, C.R. (1983) Australian laterites and silcretes: ages and significance. *Revue de Geologie Dynamique et de Geographie Physique*, **24**, 35–45.

Ullyott, J.S., Nash, D.J., Whiteman, C.A. and Mortimore, R. (2004) Distribution, petrology and mode of development of silcretes (sarsens and puddingstones) on the eastern South Downs, UK. *Earth Surface Processes and Landforms*, **29**, 1509–1539.

Van Arsdale, R. (1982) Influence of calcrete on the geometry of arroyas near Buckeye, Arizone. *Geological Soceity of America, Bulletin*, **93**, 20–26.

Van Der Graaff, W.J.E. (1983) Silcrete in Western Australia: geomorphological settings, textures, structures, and their possible genetic implications, in *Residual Deposits* (ed. R.C.L. Wilson), Geological Society of London Special Publication 11, pp. 159–166.

Van Dijk, D.C. and Beckmann, G.G. (1978) The Yuleba Hardpan, and its relationship to soil-geomorphic history, in the Yuleba–Tara region, Southeast Queensland, in *Silcrete in Australia* (eds T. Langford-Smith), University of New England Press, Armidale, pp. 73–91.

Vaniman, D.T., Chipera, S.J. and Bish, D.L. (1994) Pedogenesis of siliceous calcretes at Yucca Mountain, Nevada. *Geoderma*, **63**, 1–17.

Vasconcelos, P.M. (1999) K–Ar and 40Ar/39Ar geochronology of weathering processes. *Annual Review of Earth and Planetary Sciences*, **27**, 183–229.

Vasconcelos, P.M. and Conroy, M. (2003) Geochronology of weathering and landscape evolution, Dugald River valley, NW Queensland, Australia. *Geochimica et Cosmochimica Acta*, **67**, 2913–2930.

Verrecchia, E.P. (1990) Litho-diagenetic implications of the calcium oxalate–carbonate biogeochemical cycle in semiarid calcretes, Nazareth, Israel. *Geomicrobiology Journal*, **8**, 87–99.

Verrecchia, E.P., Dumont, J.-L. and Rolko, K.E. (1990) Do fungi building limestones exist in semi-arid regions? *Naturwissenschaften*, **77**, 584–586.

Verrecchia, E.P., Dumont, J.-L. and Verrecchia, K.E. (1993) Role of calcium oxalate biomineralisation by fungi in the formation of calcretes: a case study from Nazareth, Israel. *Journal of Sedimentary Petrology*, **63**, 1000–1006.

Verrecchia, E.P. and Verrecchia, K.E. (1994) Needle-fiber calcite: a critical review and a proposed classification. *Journal of Sedimentary Research*, **A64**, 650–664.

Verrecchia, E.P., Ribier, J., Patillon, M. and Rolko, K.E. (1991) Stromatolitic origin for desert laminar crusts. *Naturwissenschaften*, **77**, 505–507.

Verrecchia, E.P., Freytet, P., Verrecchia, K.E. and Dumont, J.L. (1995) Spherulites in calcrete laminar crusts: biogenic

$CaCO_3$ precipitation as a major contributor to crust formation. *Journal of Sedimentary Research*, **65**, 690–700.

Viles, H.A. (1995) Ecological perspectives on rock surface weathering: towards a conceptual model. *Geomorphology*, **13**, 21–35.

Vogt, T. and Corte, A.E. (1996) Secondary precipitates in Pleistocene and present cryogenic environments (Mendoza Precordillera, Argentina, Transbailalia, Siberia, and Seymour Island, Antarctica). *Sedimentology*, **43**, 53–64.

von Humboldt, A. (1812) *Personal Narrative of Travels to the Equinoctial Regions of America during the Years 1799–1804* (translated and edited by T. Ross in 1907), George Bell & Sons, London.

Warren, J.K. (1982) The hydrological setting, occurrence and significance of gypsum in Late Quaternary salt lakes in South Australia. *Sedimentology*, **29**, 609–637.

Warren, J.K. (1983) Pedogenic calcrete as it occurs in Quaternary calcareous dunes in coastal South Australia. *Journal of Sedimentary Petrology*, **53**, 787–796.

Warren, J.K. (2006) *Evaporites: Sediments, Resources and Hydrocarbons*, Springer, Berlin.

Warren, J.K. and Kendall, C.G.St.C. (1985) Comparison of sequences formed in marine sabkhas (subaerial) and salina (subaqueous) settings – modern and ancient. *Bulletin of the American Association of Petroleum Geologists*, **69**, 1013–1023.

Watchman, A. (1992) Composition, formation and age of some Australian silica skins. *Australian Aboriginal Studies*, **1992** (1), 61–66.

Watson, A. (1979) Gypsum crusts in deserts. *Journal of Arid Environments*, **2**, 3–20.

Watson, A. (1981) Vegetation polygons in the central Namib Desert near Gobabeb. *Madoqua*, **2**, 315–325.

Watson, A. (1983a) Gypsum crusts, in *Chemical Sediments and Geomorphology* (eds A.S. Goudie and K. Pye), Academic Press, London, pp. 133–161.

Watson, A. (1983b) Evaporite sedimentation in non marine environments, in *Chemical Sediments and Geomorphology* (eds A.S. Goudie and K. Pye), Academic Press, London, pp. 163–185.

Watson, A. (1985a) Structure, chemistry and origins of gypsum crusts in southern Tunisia and the central Namib Desert. *Sedimentology*, **32**, 855–875.

Watson, A. (1985b) The control of wind blown sand and moving dunes: a review of methods of sand control, with observations from Saudi Arabia. *Quarterly Journal of Engineering Geology*, **18**, 237–252.

Watson, A. (1988) Desert gypsum crusts as palaeoenvironmental indicators: a micropetrographic study of crusts from southern Tunisia and the central Namib Desert. *Journal of Arid Environments*, **15**, 19–42.

Watts, N.L. (1977) Pseudo-anticlines and other structures in some calcretes of Botswana and South Africa. *Earth Surface Processes*, **2**, 63–74.

Watts, S.H. (1977) Major element geochemistry of silcrete from a portion of inland Australia. *Geochemica et Cosmochimica Acta*, **41**, 1164–1167.

Watts, N.L. (1978) Displacive calcite: evidence from recent and ancient calcretes. *Geology*, **6**, 699–703.

Watts, S.H. (1978a) A petrographic study of silcrete from inland Australia. *Journal of Sedimentary Petrology*, **48**, 987–994.

Watts, S.H. (1978b) The nature and occurrence of silcrete in the Tibooburra area of northwestern New South Wales, in *Silcrete in Australia* (eds T. Langford-Smith), Department of Geography, University of New England, Armidale, NSW, pp. 167–185.

Watts, N.L. (1980) Quaternary pedogenic calcretes from the Kalahari (southern Africa): mineralogy, genesis and diagenesis. *Sedimentology*, **27**, 661–686.

Waugh, B. (1970) Petrology, provenance and silica diagnosis of the Penrith Sandstone (lower Permian) of northwest England. *Journal of Sedimentary Petrology*, **40**, 1226–1240.

Webb, J.A. and Golding, S.D. (1998) Geochemical mass-balance and oxygen-isotope constraints on silcrete formation and its palaeoclimatic implications in southern Australia. *Journal of Sedimentary Research*, **68**, 981–993.

Wells, S.G. and Schultz, J.D. (1979) Some factors influencing Quaternary calcrete formation and distribution in alluvial fill of arid basins. *Geological Society of America, Abstracts with Programs*, **11**, 538.

West, L.T., Drees, L.R., Wilding, L.P. and Rabenhorst, M.C. (1988) Differentiation of lithogenic and pedogenic carbonate forms in Texas. *Geoderma*, **43**, 271–287.

Whalley, W.B. (1984) High altitude rock processes, in *The International Karakoram Project*, vol. **1** (ed. K.J. Miller), Cambridge University Press, Cambridge, pp. 365–373.

Whalley, W.B., Gellatly, A.F., Gordon, J.E. and Hansom, J.D. (1990) Ferromanganese rock varnish in North Norway: a subglacial origin. *Earth Surface Processes and Landforms*, **15**, 265–275.

Wheeler, W.H. and Textoris, D.A. (1978) Triassic limestone and chert of playa origin in North Carolina. *Journal of Sedimentary Petrology*, **48**, 765–776.

Wieder, M. and Yaalon, D.H. (1974) Effect of matrix composition on carbonate nodule crystallisation. *Geoderma*, **11**, 95–121.

Wigley, T.M.L. (1973) Chemical evolution of the system calcite–gypsum–water. *Canadian Journal of Earth Sciences*, **10**, 306–315.

Williams, L.A., Parks, G.A. and Crerar, D.A. (1985) Silica diagenesis, I. Solubility controls. *Journal of Sedimentary Petrology*, **55**, 301–311.

Williams, G.E. and Polach, H.A. (1971) Radiocarbon dating of arid-zone calcareous palaeosols. *Bulletin of the Geological Society of America*, **83**, 3069–3086.

Woolnough, W.G. (1930) The influence of climate and topography in the formation and distribution of products of weathering. *Geological Magazine*, **67**, 123–132.

Wopfner, H. (1978) Silcretes of northern South Australia and adjacent regions, in *Silcrete in Australia* (ed. T. Langford-Smith), Department of Geography, University of New England, Armidale, NSW, pp. 93–141.

Wopfner, H. (1983) Environment of silcrete formation: a comparison of examples from Australia and the Cologne

Embayment, West Germany, in *Residual Deposits* (ed. R.C.L. Wilson), Geological Society of London Special Publication 11, pp. 151–157.

Wright, R.L. (1963) Deep weathering and erosion surfaces in the Daly River basin, Northern Territory. *Journal of the Geological Society of Australia*, **10**, 151–163.

Wright, V.P. (1989) Terrestrial stromatolites and laminar calcretes: a review. *Sedimentary Geology*, **65**, 1–13.

Wright, V.P. (1990) A micromorphological classification of fossil and recent calcic and petrocalcic microstructures, in *Soil Micromorphology: A Basic and Applied Science* (ed. L.A. Douglas), Developments in Soil Science 19, Elsevier, Amsterdam, pp. 401–407.

Wright, V.P. (2007) Calcrete, in *Geochemical Sediments and Landscapes* (eds D.J. Nash and S.J. McLaren), Blackwell, Oxford, pp. 10–45.

Wright, V.P. and Alonso-Zarza, A. (1990) Pedostratigraphic models for alluvial fan deposits: a tool for interpreting ancient sequences. *Journal of the Geological Society of London*, **147**, 8–10.

Wright, V.P. and Tucker, M.E. (1991) Calcretes: an introduction, in *Calcretes* (eds V.P. Wright and M.E. Tucker), International Association of Sedimentologists Reprint Series Volume 2, Blackwell Scientific Publications, Oxford, pp. 1–22.

Wright, V.P., Platt, N.H., Marriott, S.B. and Beck, V.H. (1995) A classification of rhizogenic (root-formed) calcretes, with examples from the Upper Jurassic–Lower Cretaceous of Spain and Upper Cretaceous of southern France. *Sedimentary Geology*, **100**, 143–158.

Yaalon, D.H. (1964) Downward movement and distribution of ions in soil profiles with limited wetting, in *Experimental Pedology* (eds E.G. Hallsworth and D.V. Crawford), Butterworth, London, pp. 159–164.

Yaalon, D.H. (1988) Calcic horizon and calcrete in aridic soils and paleosols: progress in the last twenty two years, Soil Science Society of America Agronomy Abstracts.

Yaalon, D.H. and Singer, S. (1974) Vertical variation in strength and porosity of calcrete (nari) on chalk, Shefala, Israel and interpretation of its origin. *Journal of Sedimentary Petrology*, **44**, 1016–1023.

Yaalon, D.H. and Wieder, M. (1976) Pedogenic palygorskite in some arid brown (calciorthid) soils in Israel. *Clay Minerals*, **11**, 73–80.

Young, R.W. (1978) Silcrete in a humid landscape: the Shoalhaven Valley and adjacent coastal plans of southern New South Wales, in *Silcrete in Australia* (ed. T. Langford-Smith), Department of Geography, University of New England, Armidale, NSW, pp. 195–207.

Zen, E.-A. (1965) Solubility measurements in the system $CaSO_4$–NaCl–H_2O at 35°, 50° and 70° and one atmosphere pressure. *Journal of Petrology*, **6**, 124–164.

Zorin, Z.M., Churaev, N., Esipova, N. *et al.* (1992) Influence of cationic surfactant on the surface charge of silica and the stability of acqueous wetting films. *Journal of Colloid and Interface Science*, **152**, 170–182.

9

Pavements and stone mantles

Julie E. Laity

9.1 Introduction

In arid and semi-arid environments, soils commonly contain rock fragments, whose size and cover affect a range of geomorphic processes, including wind and water erosion, infiltration, evaporation, runoff generation and soil formation processes. A mosaic of soils and geomorphic surfaces develops, ranging in age from those only recently formed to relict features from the Pleistocene or even earlier. This chapter examines the nature of two widespread covers: stony mantles and rocky or bouldery surfaces (hamadas).

Mantles of stony material have been given many names according to their geographic location, including stone pavement (desert pavement), reg, serir, gobi and gibber plain. The degree to which these terms are synonymous is unclear, as a systematic global comparison of the different named surfaces has not been conducted. Considerable research efforts have been devoted to understanding the nature and formation of stone pavements. This interest is merited, as they are estimated to be the most extensive of desert surfaces, covering more than 50 % of the land (Evenari, 1985). By contrast, hamadas have received much less attention and these features are poorly defined.

Although hamadas and stone pavements are discussed separately in this chapter, their distinction is often blurred. As Mabbutt (1977) points out, hamadas may become smoothed over time as fragments weather and interstitial soil develops through aeolian and fluvial inputs, or wash and animal movements redistribute material. Therefore, hamada slopes may be transitional to pavements, both spatially (an upper hamada slope and lower pavement region) and temporally (hamadas are weathered to form pavements). In some cases, no distinction between the two types of surfaces is made in the literature. Twidale (1994) considers the gibber plain to be equivalent to a reg or hamada and Sharon (1962) uses the term hamada in the context of a serir.

The formation of desert pavements has been intensively investigated over many decades, particularly in the arid southwestern United States and Israel, where modern models of pavement formation were largely developed. This interest shows little sign of waning. The expansion of research to new regions, such as the UAE, Patagonia and Australia, has provided additional data, which show that stony mantle formation, while sharing many processes in common, also varies spatially, reflecting differing parent materials, weathering processes and aeolian environments.

9.2 Surface types: hamadas and stony surfaces

9.2.1 Hamada

Hamada (*hammada*) is an Arabic word denoting difficult bouldery terrain that is 'unfruitful' to cross. In Australia, the term *stony tableland* has also been employed (Mabbutt, 1977). Although hamada surfaces are extensive, their development and geomorphic significance have received little study. Additionally, they are not always clearly defined and there is some contradiction in the use of terminology. Sharon (1962, p. 130) described hamada in the Negev and Sinai Deserts as being 'covered by quite closely backed angular rock fragments forming a desert pavement', and later noted that the surface should be classified as serir (p. 132), although 'hamada' followed local Israeli and Arab usage. Zohary (1945, p. 14) used hamada in the context of 'gravelly desert plains', more similar to

Arid Zone Geomorphology: Process, Form and Change in Drylands, Third Edition. Edited by David S. G. Thomas

Figure 9.1 Outcrop or rock hamada with residual in situ origin, Israel. Note the angularity of the rocks that make up this surface. A coin (28 mm in diameter) shows the scale of the features. (Photograph courtesy of Antony Orme).

the serir surfaces of other authors, whereas Mabbut (1977) considered hamada to be formed of residual rocks or boulders. The latter definition appears to be the most common usage and the way in which the term will be employed in this chapter.

There are two basic types of hamada surfaces. The first is an outcrop or rock hamada: throughout much of the Sahara, hamada has the connotation of a structural tableland, generally in association with flat-bedded rocks or horizontal weathered crusts (Figure 9.1).

Structural hamada surfaces may be reinforced by the deposition of duricrusts, including silcretes or calcretes. Most rock hamada have a residual in situ origin; i.e. they have not been transported. They are often mantled by angular and unworked rocks. However, the surface may be considerably modified by aeolian erosion. Hobbs (1917) describes the extensive Libyan Hamada as having a blinding polish and being 'ruffled' by sand abrasion, such that the surface appears like waves on the ocean.

The second type is boulder hamada (Figure 9.2). Some boulders have a residual origin, as shown by their lack of sorting, angularity and affinity with the rock below (Mabbutt, 1977). Others, often rounded in form, have been deposited by transport from local sources. Secondary processes may modify the boulders, changing their texture or form. Weathering breaks down the rocks, creating either rougher or smoother surfaces. Rounding may occur in place (Mabbutt, 1977) or the materials may become more angular as planar facets develop by aeolian abrasion (Figure 9.2). Czajka (1972) describes wedge-shaped ventifacts in a boulder hamada on the Argentine Puna Plateau at altitudes of 1500–5000 m.

Fürst (1965) noted that there is sometimes a spatial continuum between stony surface forms, with rock hamada in tablelands, boulder hamada downslope and reg at lower altitudes. Materials are stripped from higher altitudes and moved to lower elevations.

It is likely that many hamada are of considerable antiquity. These largely planar surfaces are protected from dissection and channel development by their rocky character and moderate gradients, which tend to disperse sediment-laden water. Like pavements, they are best developed and maintained when protected from active fluvial erosion. Dark coatings of rock varnish that cover exposed hamada document the age of the surface and shed light on climatic transitions from wet environments to progressively more arid conditions (Cremaschi, 1996; Zerboni, 2008). In some areas, the varnish-coated surfaces are partially stripped by wind erosion (Cremaschi, 1996).

9.2.2 *Stony surfaces: gobi, serir, gibber plains and desert pavements*

Despite the extent and significance of stony mantles in global deserts, regional differences have not been systematically compared. The literature suggests that they vary in a number of aspects, including clast size, shape, degree

Figure 9.2 Boulder hamada to the east of the Sierra Nevada, California. The boulders were transported short distances from nearby mountains and then isolated from further fluvial activity by faulting along the southern margin of Owens Lake. The boulders have become more angular over time from abrasion. The hamada, mantled by loose sand, is transitional to a developing pavement.

of rounding and texture, development of desert varnish, substrate, age and the degree to which the particles interlock. Furthermore, many gravel mantles, although differing in name, share enough characteristics that they are essentially of the same type. Thus, a 'pavement' in one region may be a 'serir' in another. Conversely, there may be significant differences in form or developmental history that are not yet fully understood. This section will summarise what is known about these different named surfaces. Following this initial regional examination, the most widely studied of these types, the desert pavement, will be examined in more detail.

9.2.2.1 Gobi

The term 'gobi' is Mongolian, referring to a region of flat gravel pavement (Wang *et al.*, 2006) that forms the basis of the Gobi Desert. In Mongolia, 14 % of the land is desert, with surfaces extensively covered in gravel (Yang *et al.*, 2004). 'Gobi' is also used in arid parts of China (Li *et al.*, 2006), where vast expanses of unvegetated gravel plains occur (Figure 9.3). In northern China, gobi covers about 44 % of the desert (Yang *et al.*, 2004).

Gravel-covered regions are further subdivided into gobi of accumulation and gobi of denudation (Yang *et al.*, 2004), but these terms are not clearly defined. 'Gobi of accumulation' appears to refer to the extensive piedmont plains that cover the inland Quaternary basin surfaces of China and form a 'gobi zone', which, in cross-section, is filled with a very thick sequence of gravel beds (Mengxiong, 1994). Much of the rain and melting snow and ice from the mountains disappear into these gravel sequences. In piedmont areas, removed from the presence of sand, the gravel cover may reach 100 % (Figure 9.3). In recent deposits, the surface clasts retain a rounded form and soil is absent.

Many Chinese dune fields are surrounded by gobi surfaces. One general theory for pavement formation is that it represents a deflationary lag. Although this idea has been largely discredited for accretionary desert pavements in southwestern North America and elsewhere (see the discussion later in this chapter), deflation may play a role where both gravel and sand are present, including some gobi.

To date, there appears to be little research on the nature of gobi in central Asia. My own brief travels suggest at least three types of gravel surfaces: (1) fresh, relatively unmodified gravel surfaces with rounded fluvial clasts; (2) gobi in sandy environments (potentially 'deflationary gobi', but not yet properly documented); and (3) well-knit, varnished pavements, with smoothed surfaces and an underlying stone-free and silty substrate, similar to accretionary stone pavements of southwestern North America. The unmodified gobi occur where active storms and flooding bring gravel from the mountains on to the piedmont surfaces: the gravels and cobbles are well rounded and largely unvarnished, and lack the silt-rich subsurface horizons common to desert pavements of more stable

Figure 9.3 Extensive piedmont gravel plains cover inland Quaternary basins in China. As shown here, they commonly have ~100 % gravel cover, extend over vast expanses and lack vegetation. This surface is largely unmodified, with rounded fluvial clasts.

environments (Figure 9.3). Pavements with angular interknit particles, dark varnish and a stone-free silty substrate appear to be uncommon, but are observed in areas protected from fluvial erosion (Figure 9.4).

Gobi may develop secondary characteristics. The clasts can be coated on their undersides by calcium carbonate and on their upper surfaces by varnish. Ventifacts may form, with abrasion features such as polished faces or pits (Li *et al.*, 2006). At high altitudes in periglacial arid regimes, freeze-thaw polygons occur.

The percentage of gravel coverage necessary to constitute a gobi is not clearly defined. Gobi surfaces on a clifftop above the Mogao Grottoes, China, have clast cover values ranging from a low of 9 % near the Mingsha Mountains to a high of 66 %. For the wind regime of this region, a value of 65 % is thought to eliminate erosion of sand in the substrate (Wang *et al.*, 2006). The gravel protects the underlying sediments by increasing the aerodynamic roughness of the surface boundary layer (McKenna Neuman, 1998). The greater the local wind velocity, the larger the percentage of gravel cover is needed to prevent erosion. In wind tunnel experiments where gravels were initially veneered by a 50 mm layer of sand, the air, at speeds from 2 to 30 m/s, was allowed to blow until no additional sand was removed ('equilibrium gravel coverage') (Wang *et al.*, 2006). The equilibrium coverage increased with wind speed (e.g. 25–35 % at 8 m/s and 60–75 % at 18 m/s).

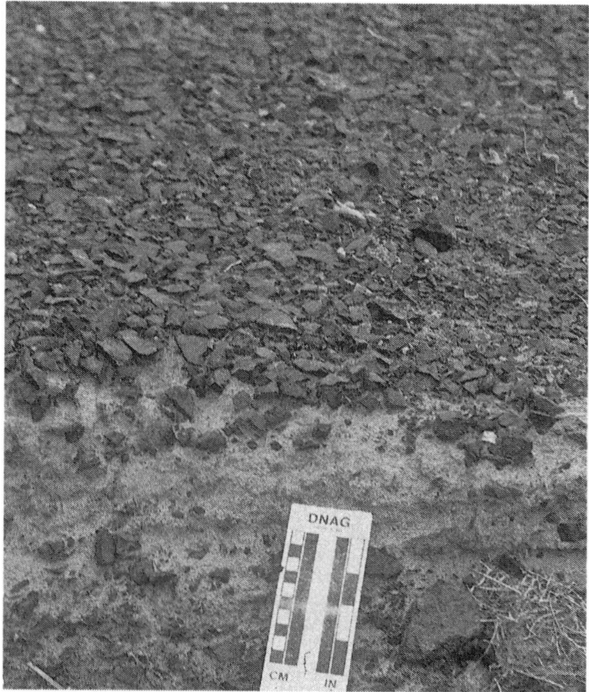

Figure 9.4 This pavement in western China differs considerably from the gravel plains shown in Figure 9.3. A monolayer of angular, heavily varnished pebbles sits on top of a thick (25+ cm) layer of wind-derived silt. It strongly resembles the pavement from the Mojave Desert shown later in Figure 9.9.

9.2.2.2 *Reg or serir*

In North Africa and the Middle East, the terms 'reg' or 'serir' are used to describe extensive sheets of gravel with little or no vegetation (Evenari, Yaalon and Gutterman, 1974). *Reg* is an Arabic term meaning 'becoming smaller', and surfaces of finer gravel produce trafficable desert pavements in Algeria, Israel and areas of the Sahara. *Serir* replaces reg in the central Sahara (Libya and Egypt) (Mabbutt, 1977). Saharan regs cover vast expanses of low-angle surfaces in the desert, extending for immense distances from residual mountain ranges (Symmons and Hemming, 1968). The Tibesti Serir, for example, is an alluvial plain of low gradient (0.1 to 0.2%) and little relief (less than 50 cm) (Fürst, 1965). The rounding of some stones in regs suggests either transport over a considerable distance or alluvial stones of a residual nature, derived from beds rich in pebbles (Peel, 1968).

Desert soils that form on stable, well-drained, sandy to coarse gravelly alluvial parent material are referred to as reg soils (Amit and Gerson, 1986; Amit and Yaalon, 1996). In hyper-arid parts of the Negev Desert, these soils show characteristics similar to desert pavements in North America, with high amounts of soluble salts in the soil (mainly halite and gypsum), a vesicular A-horizon and desert pavement at the surface. Salt plays a major pedogenic role, with salt-shattered gravel in the B- and C-horizons and gypsic and sometimes petrosalic horizons developed (Dan *et al.*, 1982; Amit and Gerson, 1986; Amit and Yaalon, 1996).

9.2.2.3 *Gibber plains or stony mantles or stony desert*

Extensive stony plains in Australia are termed gibber plains, stony mantles or stony desert. The term stony tableland soil has been applied to fine-textured, saline soils with a high gypsum content and an abundance of stones in the upper part of the profile and on the surface (Jessup, 1960).

Gibber plains vary in character according to local sources of bedrock, with limestone gibbers, for example, adjacent to the northern Flinders Ranges (Twidale, 1994) and silcrete gibbers mantling a deflated surface in the Sturt Stony Desert (Thomas, Clarke and Pain, 2005). The silcrete fragments are derived from the cappings of dissected, silcrete-capped plateaus (Twidale, 1994).

Multiple processes may be involved in gibber plain formation. According to Twidale (1994), vertical sorting and churning by the wetting and drying of clays in the soil (gilgai action) concentrate stones at the surface. Thomas, Clarke and Pain (2005) believe that both deflation and heave processes play a role in the formation of the Sturt Stony Desert.

9.2.2.4 *Desert or stone pavements*

The terms desert or stone pavement are used in southwestern North America, particularly in the Mojave and Sonoran Deserts, where such surfaces are common and well developed. This nomenclature is becoming more widely applied, however, and has recently been used to describe stony surfaces in other areas of the world, including Chile and Argentina (Bailey *et al.*, 2007; Sauer, Schellmann and Stahr, 2007), Israel (Matmon *et al.*, 2009), the UAE and Oman (Al-Farraj and Harvey, 2000) and Egypt (Adelsberger and Smith, 2009). Pavement surfaces have also been reported in southern Africa (Francis *et al.*, 2007). Desert pavements have received detailed and systematic study and, relative to other surfaces described earlier, are well defined. They are discussed in more detail later in this chapter.

9.3 General theories concerning stony surface formation

Four main theories have been proposed to explain the concentration of surface stones common to gobi, reg, gibber plains or pavement surfaces: (1) deflation; (2) concentration by rain beat and surface wash; (3) upward migration of stones by heave processes; and (4) upward displacement of clasts by aeolian aggradation of fine materials beneath the pavement surface. Of these, the most oft-cited introductory textbook explanation relates pavement formation to the deflation of fines from the surface. Research over the past few decades, however, has shown that pavement formation is highly complex and cannot be explained by simple deflationary processes. Indeed, some of the best-studied pavements, most notably those of the American southwest, appear to have evolved in conjunction with long-term aeolian deposition rather than erosion.

Although a truly global perspective on stony surface formation is still lacking, research to date suggests that several different processes may act in concert, with aeolian aggradation, for example, as the principal process, but surface wash and subsurface heave contributing to long-term development. Additionally, numerous secondary processes – chemical, physical and biological – influence the ultimate surface and subsurface expression of the pavement. The following sections summarise the four major theories of pavement formation. Following this introduction, the best studied of these, the aeolian aggradation model, is examined in more detail.

9.3.1 *Deflation*

According to the deflationary model, pavements are considered to be lags or veneers caused by the wind erosion of fines, a process that lowers the surface until the stones are sufficiently close to act as a desert armour (Blake, 1904; Cooke, 1970; and others). This mechanism was once widely cited to explain all types of desert pavement formation (see, for example, Symmons and Hemming, 1968). Today, however, it is generally recognised that wind erosion is largely ineffective once the stone covering reaches between about 50 % (Peel, 1968) and 65 % (Wang *et al.*, 2006), with the actual value at which a lag bed stabilises dependent on stone spacing, angularity and wind velocity (McKenna Neuman, 1998; Wang *et al.*, 2006). As the stone cover increases, the stones take up an increasing portion of total fluid stress relative to that which is acting on the interstitial fines (McKenna Neuman, 1998). As a result, with increasing stone coverage, the surface becomes stable with respect to wind erosion.

Deflation is not thought to account for the development of very well-knit pavements (clasts in close physical contact, with little intergravel space visible), in which there is a substantial thickness of silty soil beneath the surface clasts and where clast coverage may be 65 % or greater. Although a well-knit pavement may not develop from deflation, the stones do act to protect the surface from wind erosion (Wang *et al.*, 2006). If the stones are removed and the subsurface disturbed, minor lowering of the surface by deflation occurs (Symmons and Hemming, 1968). In general, however, there is little deflation from stony surfaces unless the pavement and subsurface soils are heavily disturbed, as during military operations (Cooke, 1970).

Although deflation continues to be cited as a mechanism of stony surface formation (e.g. Thomas, Clarke and Pain, 2005; Bailey *et al.*, 2007), there are as yet no detailed studies of the process. Deflationary pavements appear best developed where aeolian sands and gravels occur together (McFadden *et al.*, 1992; Li *et al.*, 2006; Wang *et al.*, 2006; Al-Farraj, 2008). The removal of sand may help in the concentration of gravel to form initial lag deposits. In places, the role of wind erosion is made evident by strongly abraded surfaces with numerous ventifacts. Atacama Desert pavements in exposed coastal zones and in the high Andes contain ventifacts and may form largely by deflation (Cooke, 1970). Thomas, Clarke and Pain (2005) infer that the ventifacted basaltic plains of the Martian landing sites (MER, Mars Pathfinder and Viking) represent deflated surfaces with remnant lag deposits.

Not all pavements that appear at first glance to be deflationary (covered by sand, adjacent to dunes, and surfaced by ventifacts) are formed by this mechanism. Incipient pavements in southern Owens Valley (Figure 9.5(a) and (b)) with ventifacts and a veneer of sand, have a soil characteristic of accretionary pavements, with a silt-rich substrate that is largely impermeable to water infiltration.

9.3.2 *Concentration by surface wash and rain splash*

The concentration of stones by surface wash has also been proposed as a mechanism of pavement development. Runoff removes superficial fines and re-sorts the coarse residue (Cooke, 1970). However, like the deflation theory, the removal of fines by wash does not account for the development of a stone-free subsurface zone.

Rain splash may also contribute to pavement formation. The impact of high intensity raindrops helps form patchy, discontinuous pavement on semi-arid hillslopes where the erosion of fines is concentrated in areas between plants. Soil detachment, depletion and lowering gradually increase the percentage of cover of coarse lag material (Parsons, Abrahams and Simanton, 1992).

Surface wash may cause a slight lowering of pavement surfaces, expose buried clasts and redistribute pebbles for short distances. In several experiments where plots were cleared of coarse particles, wash processes helped rapidly to reestablish pavements (Sharon, 1962; Cooke, 1970; Wainwright, Parsons and Abrahams, 1999). Sharon (1962) cleared a plot leaving only 5–10 % of embedded clasts. Within five years, new stones, largely 2–6 cm in diameter, covered 60–80 % of the surface. Initially, wind erosion removed some of the loose material found at the surface. Subsequent to this, water erosion dominated, washing away soil both above and between previously embedded stones, and lowering the surface on average 15–20 mm. The rate of water erosion was high at first, and then diminished as stone coverage increased. Water flow also redistributes clasts over short distances (Figure 9.6), concentrating them into localised patchy pavements that may represent an initial substage of accretionary mantle development (Williams and Zimbelman, 1994). Furthermore, creep processes may reduce some of the original depositional relief of a surface (Denny, 1965, 1967).

Understanding the role of raindrop impact and wash processes on pavement formation and regeneration is important in determining potential recovery rates for surfaces disturbed by construction or off-road vehicle use (Figure 9.6). Where the surface material is mixed into the soil (rather than removed), experimental results suggest that raindrop processes help a pavement to recover quickly (within a year) following disturbance (Wainwright, Parsons and Abrahams, 1999).

Figure 9.5 (a) This surface shows the transition from a hamada to a pavement south of Owens Lake, California. Many weathered and wind-abraded boulders remain amid surface pebbles whose clast cover ranges from ~50 to 70 %. (b) Salts blown in from the exposed saline surface of Owens Lake are actively weathering the rocks shown in Figure 9.5(a). The granular disintegration of granites (upper two large rocks) and the angular splitting of fine-grained rocks (darker material) result in smaller clasts, which over time may form an interlocking pavement.

9.3.3 Upward migration of stones

The absence of stones beneath a desert pavement has been accounted for by an upward migration of gravel through a clay-rich B-horizon via alternating shrinking and swelling associated with wetting and drying and/or freezing and thawing (Springer, 1958; Jessup, 1960; Cooke, 1970; Mabbutt, 1977). Soils that exhibit shrink–swell tendencies contain expansive clays and swell and heave on wetting and shrink and crack on drying. Over many cycles, the stones are displaced upwards to the surface. This process is believed to be significant in the stony tablelands of Australia (Jessup, 1960; Twidale, 1994; Thomas, Clarke and Pain, 2005). Laboratory experiments indicate that although some upward movement of particles is possible, partial burial may occur also by displacement (Cooke, 1970).

The upward migration of gravel is potentially less effective in other geographic regions. In the Mojave Desert, for example, rainfall infiltration is limited by the arid

Figure 9.6 A pavement scraped clear during construction, which illustrates (1) the role of wash processes in the recovery of disturbed pavements through the redistribution of particles over short distances and (2) ped boundaries in the A-horizon. The A-horizon is commonly characterised at the surface by a small-scale polygonal pattern of cracks, which form a three-dimensional ped system of vertical columns. Fine sands, silt and solutes are transported to depth through the polygonal cracks. Location: Father Crowley overlook, Panamint Valley, California.

conditions and extensive freezing does not occur. Furthermore, continuous upward migration over time should produce stones with different ages at the surface, and cosmogenic dating and varnish studies indicate that pavement stones are of a similar age (McFadden, Wells and Jercinovich, 1987; Wells *et al.*, 1995; Liu and Broecker, 2008).

The saturation and desiccation of desert soils causes expansion and contraction features, such as polygonal structures in the A-horizon and the development of gilgai in Australia. This process is most effective where the clay content is high.

9.3.4 Accretion of aeolian fines

Whereas models that emphasise deflation or lowering by surface wash imply that pavements are largely zones of erosion, the accretionary mantle model considers them to be zones of deposition. This concept, first outlined in detailed papers in the 1980s, represents a significant departure from earlier studies. Based on studies of pavement formation on volcanic fields in the Mojave Desert, Wells *et al.* (1985) and McFadden, Wells and Jercinovich (1987) proposed a model of upward surface growth, which has been subsequently applied to other landforms with pavements, including alluvial fans, beach ridges and fluvial terraces. In their model, the stones originally present on a landform are maintained at the surface as the landscape evolves. The surface rises and smoothes owing to the development of cumulate soils beneath the clasts in response to the incorporation of aeolian silts and clays. In Israel, Amit and Gerson (1986) also detailed how gravelly alluvial terrains evolve as dust and salts are introduced into the Dead Sea region. In recent years, new research has shown that this model may be extended to areas beyond western North America, including the Libyan Plateau of Egypt (Adelsberger and Smith, 2009). Figure 9.4 shows pavement development by this process in western China. The accretionary mantle model of pavement formation is the basis for most modern research on pavements. Additional details are provided in the following sections.

9.3.5 Desert pavement formation by aeolian aggradation and development of an accretionary mantle

Desert pavements are highly complex systems. Their development can be understood by considering key factors that affect soils in desert regions, including vegetation type and density, faunal activity, the nature and rate of rock weathering, climate (precipitation and temperature), hydrology (surface runoff and infiltration of water), aeolian inputs (of dust, fine sand and salts) and time. As time passes and the climate changes, the morphological, physical, chemical, mineralogical and biological properties of the pavement alter. It is believed that the characteristics of a mature pavement, including smooth surfaces, low relief, lack of vegetation and closely knit particles, require an extended period, typically thousands of years, to develop.

9.4 Stone pavement characteristics

9.4.1 Setting

Physical factors that affect the development of pavements include surface stability, slope and elevation. Pavements develop where surfaces are stable, i.e. are not subject to significant water erosion (Peel, 1968; Cooke, 1970; Ugolini *et al.*, 2008; Adelsberger and Smith, 2009; Matmon *et al.*, 2009). Associated with the stability of the surface are low slope angles. Pavements in Jordan, for example, have developed on a 1 % slope (Ugolini *et al.*, 2008). Although the relationship is not yet well studied, it appears that elevation limits pavement formation, as a result of altitudinally related increases in rainfall and disruptive vegetation cover (Quade, 2001; Marchetti and Cerling, 2005).

Pavements are built on two different types of surfaces: (1) those of an alluvial character and (2) those of a residual character (Cooke, 1970). The first is characterized by gravel of mixed composition and distant origin, including pavements formed on beach ridges, alluvial fans and stream terraces. The second incorporates clasts from residual deposits (e.g. volcanic pyroclastic deposits) or bedrock surfaces (e.g. basalt flows).

9.4.2 Surface clast concentration and characteristics

On surfaces where stones are highly concentrated, the term stone or desert pavement is used, particularly in North America. The pavement usually consists of closely packed gravel in a surface layer a few centimetres thick (one- to two-particle-thick layer) that rests on, or is embedded in, a soil that may contain dispersed stones, or may be relatively free of stones.

With respect to percent clast cover, desert pavements have been described as distinctive features in which at least 65 % of the surface is covered in stones (Wood, Graham and Wells, 2002, 2005). This is not a hard and fast rule, however, and may vary with location. For example, in Israel, Amit *et al.* (1996) described 'weak' gravel covers of 10–20 %, 'moderately developed' desert pavements covering ~50 % of the surface and a 'well-developed' cover of >85 %. Clast coverage varies with slope gradient, with increases in slope leading to size sorting and variations in clast coverage (Poesen *et al.*, 1998). For the Mojave Desert, Quade (2001) found that pebble densities were greatest at lower altitudes (>90 %) and decreased systematically with altitude at a rate of ~3 % per 100 m. At pebble densities <60 %, characteristics common to a mature pavement are lost, such as surface smoothness and the interlocking of clasts.

The gravel layer may consist of angular to rounded clasts, with the form dependent on the origin of the particles, their age and degree of weathering. Young pavements on beach ridges may have rounded particles (Figure 9.7), whereas those on alluvial fans (particularly near sources of salts such as playas) often have more angular clasts. Cooke (1970) divided the clasts into *primary particles*, which are similar in size, form or lithology to the original surface, and *secondary particles*, derived from the primary particles by disintegration (splitting or granular disintegration) or added later as human artefacts (Casana, Herrmann and Qandil, 2009). The presence of secondary Neolithic or Palaeolithic artefacts on pavement surfaces provides evidence of their stability (Peel, 1968).

Splitting of particles creates smaller fragments, tends to increase their angularity and increases surface particle density. The degree of splitting is a function of time: in this respect, older surfaces often show greater breakdown (Cooke, 1970). However, splitting is also related to the concentration of aerosolic salts that are deposited on surfaces and, when alluvial fans are adjacent to saline playas, may be greater at the distal than proximal ends. Recently split fragments may lack the varnish that characterises older clasts. Where granitic boulders are present, granular disintegration produces coarse grus at the surface (Figure 9.5(b)). In this case, some fragments, which were originally very angular in form, may actually show some rounding with time (Al-Farraj, 2008).

The physical characteristics of pavements are often more variable than they first appear, even across a single surface of an alluvial fan. A spatially complex pattern exists, wherein pavement units differ in the percent clast cover, degree of packing, particle size, angularity, degree of varnish cover, sorting and the degree to which the particles are embedded (Wood, Graham and Wells, 2002). These differing characteristics can be used to create desert pavement mosaic map units (Wood, Graham and Wells, 2002), with boundaries that range from very sharp to gradual.

Figure 9.7 A tightly interknit pavement surface in which the clasts are formed of rounded pebbles rather than angular fragments. There is little secondary modification to the surface of this Holocene beach ridge. Location: Death Valley, California.

9.5 Processes of pavement formation

The establishment of a stone pavement requires a geomorphologically stable area where the processes of stone sorting, surface creep and clast disintegration are dominant over concentrated wash and fluvial processes. Mature stone pavement surfaces are characterised by low relief, as lower areas are gradually infilled, and the entire surface rises by infiltration of aeolian material. On alluvial fan surfaces, for example, the original bar and swale microtopography is reduced as a smooth vegetation-free surface develops (Matmon *et al.*, 2009). In most cases, the surface clast size becomes progressively smaller as rocks weather (Figure 9.8). Salt-rich aeolian fines accumulate in clast fractures, and wetting and drying result in volumetric changes related to salt crystal growth and/or shrinking and swelling of clay. Over time, clasts fracture and are vertically and laterally displaced. Additional aeolian fines, as well as weathered granules (grus), are deposited between the fragments, further enhancing their separation.

The formation of smooth pavement surfaces is related to the entrapment and infiltration of dust. Landscapes that initially have surface obstructions, such as plants or rocks, trap dust, beginning a period of pedogenesis, which results in an upwardly thickening aeolian mantle (Blank, Young and Lugaski, 1996). Rocky surfaces trap several tens of times more dust than adjacent pebble-free surfaces. In Patagonia, for example, fine aeolian particles have been deposited into the interstitial spaces in Holocene and Pleistocene gravelly beach deposits to produce a single massive soil structure beneath the desert pavement (Sauer, Schellmann and Stahr, 2007). Entrapment rates increase as the rock fragments become smaller, more flattened and elongated, and the total cover density increases (Goossens, 2006). Pavement stones trap dust brought in by the wind from surrounding plains. The fine dust, often rich in salt, accelerates mechanical weathering of the surface rocks and accumulates within the pore spaces between clasts. As the aeolian layer grows in thickness, clay neogenesis proceeds until the soil has the strength and shrink–swell potential to displace stones upwards (Blank, Young and Lugaski, 1996). The surface gravels continue to trap dust and thicken the B-horizon, and the deposition of aeolian fines creates an accretionary mantle, in which the clasts undergo syndepositional lifting with the surface (Figures 9.8 and 9.9) (Wells *et al.*, 1985; Wells, McFadden and Dohrenwend, 1987).

The accretionary mantle model of pavement formation implies that many pavements are 'born at the land surface', with pavement gravels exposed at the surface rather than having been deeply buried in the underlying soil and excavated by wind or water erosion (McFadden, Wells and Jercinovich, 1987). This theory has been supported by a number of studies on the age of surface clasts. Cosmogenic ray surface exposure dating in the Cima volcanic field (Wells *et al.*, 1995) showed that the ages of pavement clasts were similar to those of their source bedrock. Varnish microlamination (VML) dating also supports the

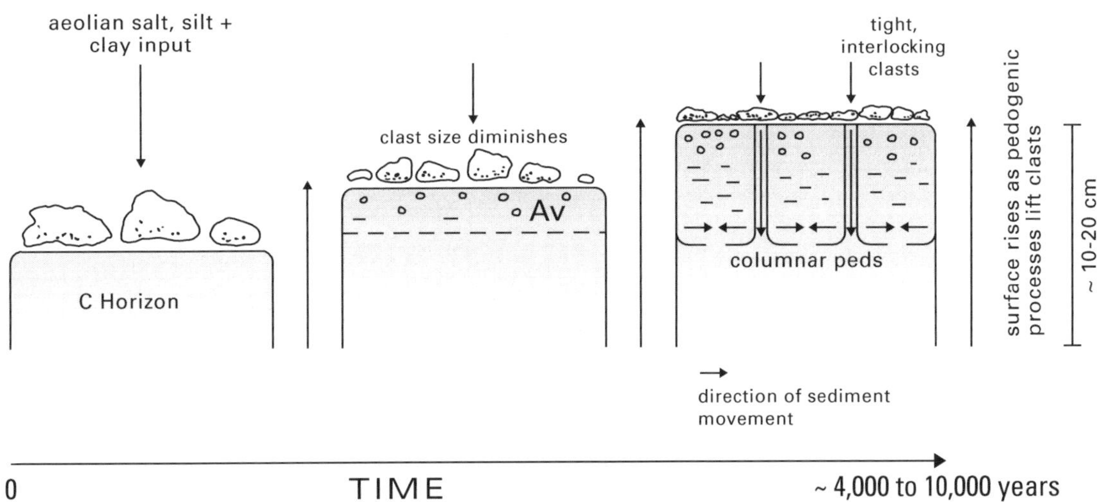

Figure 9.8 A simplified conceptual diagram of soil pavement formation according to the aeolian aggradation model. Over time, the surface rises owing to the input of fine aeolian sediment. A thin monolayer of clasts is preserved at the surface, with the rocks becoming smaller and more tightly interlocked over time as they weather. A vesicular (Av) horizon forms under the clasts. As aerosolic fines accumulate, distinct columnar shapes develop. The aeolian fines and solutes may be transported downwards through the polygonal cracks and the dust moved horizontally along platy boundaries to ped interiors. The establishment of a mature desert pavement may take 4000–10 000 years or longer. After Anderson, Wells and Graham, 2002, and Young *et al.*, 2004.

Figure 9.9 A heavily varnished pavement (similar to that shown in Figure 9.4) located in the east-central Mojave Desert, California, near the Cowhole Mountains. The gravel sits above 30+ cm of silty material that is largely free of stones.

hypothesis that pavement clasts have been continuously exposed since their formation (Liu and Broecker, 2008).

9.6 Processes of clast size reduction in pavements

The comminution of boulders and gravel over time produces smaller particles that eventually interlock to form a tightly knit pavement. The reduction in clast size results from two principle processes – splitting and granular disintegration. Splitting reduces rock size and tends to produce angular fragments that form the basic framework of the clast monolayer. Granular disintegration provides even smaller particles, which commonly infill the spaces between the split fragments, leading to a tightly packed veneer (Figure 9.5(b)). In a very mature pavement, fracturing of clasts near the surface may be absent, suggesting that there is a point in time after which there is no further reduction in clast size (Amit, Gerson and Yaalon, 1993; Matmon *et al.*, 2009).

The production of secondary particles by splitting is common in pavement development. The principal processes invoked are insolation, frost and salt weathering. Insolation weathering refers to the disintegration of rocks by volumetric changes associated with seasonal or diurnal changes in temperature. In the American southwest, many boulders have cracks with a north–south (N–S) orientation, suggesting they may be genetically related to thermal stresses caused by the differential heating and cooling of the rock associated with the daily movement of the sun across the sky (McFadden *et al.*, 2005). A pronounced N–S orientation of vertical cracks was also recorded for dark chert cobbles on the Libyan Plateau in central Egypt (Adelsberger and Smith, 2009). Solar-controlled breakdown of chert may be a function of the dark colour or other physical properties of the rock, which aid in forming initial cracks that may be further propagated by other processes, such as salt weathering. Crack growth may also be enhanced by hydration, and crack tips with minimal solar exposure and greater moisture retention will tend to propagate (Moores, Pelletier and Smith, 2008).

Frost weathering results from the volumetric expansion of ice in confined spaces. Some deserts experience subfreezing conditions in the winter, the extent of which may have been even greater during past climate regimes. It is possible, therefore, that this process may be active in some deserts or have produced some of the split fragments on older pavement surfaces.

The role of salts in desert pavement formation is well documented. They play an important role in shattering surface clasts, making them both smaller and changing their initial shape. The weathering processes that fracture clasts are sensitive to climatically induced variations in the flux of aeolian fines. Aerosolic salts are a significant component of clast fractures and are found in the soils beneath many stone pavements. Their abundance is indicated by high soil electroconductivities and the presence of secondary gypsum. Weathering is conspicuous near the margins of saline playas (Figure 9.5(b)) and along the late Pleistocene shorelines of pluvial lakes. Medium- to coarse-grained plutonic and metamorphic rocks are particularly susceptible to this process (McFadden, Wells and Jercinovich, 1987).

The degree of salt weathering varies spatially, both within a given region (e.g. closer to or farther from a playa) and from region to region. Moreover, salt can affect rocks both at the surface and within the soil. In the Negev Desert, Amit, Gerson and Yaalon (1993) documented gravel shattering by salts within reg soils. Not all pavement surfaces are shattered by salt. In Jordan, relatively young pavements of low soil salinity lacked salt-shattered clasts, although salts increased clast porosity, allowing calcite to infill voids (Ugolini *et al.*, 2008). Although salt weathering is often associated with an increase in angularity of the particles, the reverse may occur. In South Australia, silcrete boulders, which have been subject to many pre-weathering processes involving brecciation and recementation, break up along previous fracture lines rather than modern stress patterns. The break-up produces small, very angular and uniformly sized clasts that are not subject to further diminution. The highly saline Lake Eyre environment chemically weathers the clast edges, decreasing their angularity over time (Al-Farraj, 2008).

9.6.1 Pavement soils

Most stone pavements are underlain by soils (see also Chapter 7), many of which share common characteristics. Over time, the thickness and qualities of the pavement soils change. However, even young soils show some degree of evolution.

The fine-grained pavement substrate is derived largely from the windblown suspended load, and is composed of clays, silts and sandy silts (McFadden, Wells and Jercinovich, 1987; Reheis *et al.*, 1995; Sauer, Schellmann and Stahr, 2007). It is generally free of coarse particles. The A-horizon is commonly characterised at the surface by a small-scale polygonal pattern of cracks, which form a three-dimensional ped framework of vertical cracks (Symmons and Hemming, 1968; Wood, Graham and Wells, 2005; Meadows, Young and McDonald, 2008) (Figures 9.6 and 9.8). Although soil columns are widespread beneath pavements, they are not always found

(Wood, Graham and Wells, 2005). Between major periods of silt deposition, a lower aeolian flux rate permits development of soils in the deposits. Soil formation proceeds beneath the surface concurrently with pavement development. Over time, soil depth increases, horizons become more differentiated and the B-horizon becomes more argillic and red in colour. $CaCO_3$ accumulates in the soil and soluble salts are precipitated.

Pavement soils are defined by several horizons. At the surface is the gravelly desert pavement and intergravel crust (Amit and Gerson, 1986). Beneath this is an ochric A-horizon with vesicular pores (Av-horizon, where v stands for vesicular), which is the key pedogenic feature associated with desert pavement (Figure 9.8). The vesicles, spherical voids that are generally a few mm in diameter, produce a honeycomb-like appearance in the soil (Figure 9.10). The Av-horizon does not conform to established criteria for horizon nomenclature and furthermore exhibits complex properties, with some similarities to both B- and C-horizons (Anderson, Wells and Graham, 2002). However, the designation of vesicular horizons as Av in pavement studies is now a widespread convention.

The Av-horizon contains significant quantities of silt, clay and other constituents such as calcium carbonate, gypsum or other soluble salts (McFadden *et al.*, 1998). It differs from the parent material in both mineralogy and grain size and is the horizon most similar in composition to atmospheric dust (Reheis *et al.*, 1995). The disconnected spherical or ovoid vesicles, with walls strengthened by calcium carbonate cementation, probably result from soil air that cannot escape freely after a rainfall event (Evenari, Yaalon and Gutterman, 1974; McFadden *et al.*, 1998). The expansion of air may be favoured by heating of the soil after wetting events, and the zone of most recent vesicle formation is in the uppermost few millimetres or centimetres of the Av-horizon (McFadden *et al.*, 1998). The vesicles are better developed beneath the clasts than in the interclast spaces, because the stones act to trap the heated and expanding air (Evenari, Yaalon and Gutterman, 1974; Ugolini *et al.*, 2008). Valentin (1994) found that vesicular soils are also better developed beneath embedded clasts than free-floating stones, but Brown and Dunkerley (1996) found a weakly negative relationship between the abundance of vesicles and the average depth to which a stone is embedded. The occurrence of a subpavement vesicular soil is widely reported in deserts, including Israel (Evenari, Yaalon and Gutterman, 1974), Niger (Valentin, 1994), Egypt (Adelsberger and Smith, 2009), Australia and the southwestern United States. As the Av-horizon thickens, older vesicles are buried and the increasing weight of overlying material deforms and flattens them, potentially favouring the development of a fine, platy structure that lacks vesicularity (McFadden *et al.*, 1998).

Figure 9.10 The occurrence of a subpavement vesicular horizon (Av-horizon) is reported from many deserts. The spherical or ovoid vesicles are not connected and their walls are commonly strengthened by calcium carbonate cementation.

In some, but not all, regions, vesicular horizons show a pronounced columnar structure, caused by the alternating shrinking and swelling of clays in the increasingly clay-enriched Av-horizon, and a platy soil structure (McFadden, Wells and Jercinovich, 1987; Anderson, Wells and Graham, 2002; Young *et al.*, 2004). Fine sand, silt and solutes are transported below the surface through the polygonal cracks by rain splash, surface wash and translocation, and the walls of the peds are coated with loose silt. The dust may be transported horizontally along platy boundaries to ped interiors (Figure 9.8) (Anderson, Wells and Graham, 2002).

Soil pH of the Av-horizon commonly ranges from 7.0 to 8.9. Electroconductivity (EC) values are usually low, indicating the leaching of salts. However, seasonal efflorescence may occur as a result of capillary updraw (McFadden *et al.*, 1998). The soil carbonate content ranges considerably – from <0.1 % to 30 %. High values are associated with soils developed in limestone-rich parent materials. Carbonate may be disseminated within the ped interior or may occur as segregated coatings on the bottom of peds. The interior of the ped may be more strongly effervescent than its top or sides (McFadden *et al.*, 1998). For example, ped interiors at the Cima Volcanic Field contain about 12 % $CaCO_3$, whereas the sediments adhering to the sides have only 2 % $CaCO_3$ (Anderson, Wells and Graham, 2002).

Underlying the Av-horizon is a cambic (Bw) or argillic (Bt) horizon enriched in silt and clay and commonly containing pedogenic $CaCO_3$ (Bk-horizon) and soluble salts (By- or Bz-horizon) (Reheis *et al.*, 1995). These deeper horizons represent dilutions of the original parent material by dust and may contain products of in situ weathering. The B-horizon is virtually gravel-free. The C-horizon has a higher stone content than either the A- or B-horizons (Cooke, 1970). It may contain introduced fines and salts (chlorides, carbonates, gypsum) and gravel shattered by mechanical weathering (Amit and Gerson, 1986).

Variations in soil chemistry are related to ecohydrologic conditions. Nitrate levels in soils beneath pavements of the Mojave Desert may be one to two orders of magnitude higher than found in nearby soils that lack pavement (Graham *et al.*, 2008). Nitrate accumulates in soils via atmospheric deposition under very arid conditions, where infiltration and leaching are minimal. Graham *et al.* (2008) found that the highest concentrations of nitrate beneath pavements are within 0.1–0.6 m in depth. The depth distribution trends of chloride and nitrate are similar, reflecting minimal leaching in the upper half-metre of the soil (Figure 9.11). These trends are consistent throughout widely separated locations within the Mojave Desert. The high salinity of pavement soils and the lack of available water limits root access to the stored nitrate. Cooke (1970) notes that although pavement soils in the Atacama and Mojave Deserts are broadly similar in character, greater aridity causes the Atacama soils to have calcium carbonate at higher levels in the profile.

9.7 Secondary characteristics of pavement surfaces and regional differences in pavement formation

Pavements have many secondary characteristics that account for local or regional differences. These include changes to the surface texture or colour, coatings of carbonate, clast rubification, pitting, wind abrasion features or varnish, as well as variations in clast axis orientation and the degree to which they are embedded in the surface. In some regions, human artefacts are also present on the pavement surface.

9.7.1 Presence of calcium carbonate and carbonate collars

Many pavement clasts show the development of a distinctive calcium carbonate 'ring' at their interface with the soil surface or a continuous coating within clast fractures. The rings are often referred to as carbonate collars (Sena *et al.*, 1994). They occur as a coating of carbonate that forms as a belt, typically orientated parallel to the ground surface, but truncated both at the ground line and at a relatively shallow depth below it (McFadden *et al.*, 1998). The rings may be displaced or tilted with respect to the ground surface (Wood, Graham and Wells, 2002). In the Providence Mountains, California, the number of tilted clasts, measured as the degree of concordance with the surface of both varnish lines and carbonate collars, has been shown to increase with age (Sena *et al.*, 1994). In extreme cases, coarse pavement particles may be cemented together by calcium carbonate, as described by Cooke (1970) for the Atacama Desert.

9.7.2 Pitting

Some surface clasts show pitting, but this characteristic appears to be less universal than others. Pitting is rarely recorded for pavements of the American Southwest. In the Negev Desert, Israel, on the oldest desert pavement yet recorded ($\geq$2.5 Ma), only slight to moderate pitting was recorded (Matmon *et al.*, 2009). However, pitting is extensive on exposed limestone gravel in reg soils in the Dead Sea region of Israel, and the degree varies greatly according to the age of the surface, with pitting much more marked on older pavements. Pitting depth varies, with Amit and Gerson (1986) recording a range from 0.1–0.5 cm (low values) to 1.5–2.5 cm (very high values).

9.7.3 Development of varnish

Desert varnish (Chapter 8) on pavement clasts is common (Figures 9.4 and 9.9). Heavily varnished clasts are often found on very old pavements (Matmon *et al.*, 2009) and, in general, the progressive darkening of varnish is a useful measure of the age or maturity of a surface (Al-Farraj and Harvey, 2000). However, this relationship is not universally applicable, as in some areas varnish development shows complex patterns related to weathering (Al-Farraj and Harvey, 2000), with varnish inhibited in areas subject to intense salt shattering or wind abrasion. The lithology of the clasts may also play a role, with varnish absent on the feldspars of granitic detritus, but very dark in tone on pavements composed of metamorphic and volcanic clasts (Christenson and Purcell, 1985).

9.7.4 Embedded clasts

Clasts may either lie loosely on the surface or be embedded into the A-horizon. There has been little comparative study of these characteristics. The degree to which clasts are embedded appears to vary considerably. Sharon (1962) refers to well-embedded clasts comprising 5–10 % of all clasts within his study area in Israel. Wood, Graham and Wells (2005) note that within a single area of desert pavement, different mosaics form, within which some clasts lie loosely on the surface and others are more deeply embedded.

9.7.5 Clast orientation

There are relatively few studies of the alignment or fabric of clasts. Adelsberger and Smith (2009) found no relationship between pavement clast orientation and other characteristics, such as slope, indicating little effect from gravity-driven creep or overland flow. Whereas some pavements showed a small degree of orientation, others contained clasts whose orientation appeared to be completely random. However, Abrahams *et al.* (1990) observed that pavement clasts on slopes had their long axes aligned parallel to the downslope direction.

9.7.6 Clast rubification

Clast rubification refers to the development of an orange coating on the undersides (ventral sides) of stones in a desert pavement. The undersides of clasts are potentially more stable than the upper sides, being protected from surface weathering and sand blasting, and Helms, McGill and Rockwell (2003) found that the degree of ventral varnish was more strongly correlated with age than either dorsal varnish or soil properties. Both the percentage of clasts rubified (McFadden, Wells and Jercinovich, 1987) and the degree of reddening of a pavement clast's underside are considered to be indicative of the relative age of the pavement (Helms, McGill and Rockwell, 2003; Valentine and Harrington, 2006), although it is likely that such features vary considerably according to location. In southern Nevada, Valentine and Harrington (2006) noted that pavement clasts on a young (75–80 ka) volcanic cone showed much less rubification than on an older (ca. 1 Ma) cone. On the Cima volcanic field, California, only 10–20 % of clast undersides on flows younger than 0.4 Ma were reddened, whereas all clasts showed rubification on flows older than 0.4 Ma. The degree of reddening, as determined by hue, also increased on the older flows (McFadden, Wells and Jercinovich, 1987). For Holocene surfaces in southeastern California, Helms, McGill and Rockwell (2003) noted that the ventral sides of clasts reddened, brightened and darkened systematically with age. The differences between Pleistocene surfaces were subtler. Rubification may develop rapidly and was observed on surfaces whose ages were inferred to be less than 500 years old (Helms, McGill and Rockwell, 2003).

9.7.7 Development of ventifacted surfaces

On pavements that are subject to blowing sand, ventifacts may form (Peel, 1968; Cooke, 1970; Li *et al.*, 2006). There have been few studies of this process, and it is infrequently recorded. One aspect of wind abrasion is the removal of desert varnish, which may complicate studies of the relative ages of surfaces.

9.8 Secondary modifications to pavement surfaces

9.8.1 Patterns in pavement

Several authors have noted the development of patterns in pavement. These may be related either to frost heave under earlier climatic conditions or to the effects of swelling and shrinking clays associated with wetting and drying (Peel, 1968). In modern high-altitude gobi on the Qinghai-Tibet Plateau, freeze–thaw polygons are present (Li *et al.*, 2006).

9.8.2 Animal burrowing, vegetation and stone displacement

Several biological processes disrupt the formation of pavements. These include animal burrowing, plant growth and cryptobiotic crust formation. The disruption of pavement by cryptobiotic crusts has received little attention, but Quade (2001) noted this effect on pavement in the Mojave Desert at altitudes between 1300 and 1500 m.

Vegetation growth affects pavement development and, in turn, is affected by the presence of pavements. As altitude increases in deserts, rainfall and vegetation cover also rise. For the Mojave Desert, Quade (2001) determined three divisions of pavement formation related to altitude, climate change and potentially disruptive vegetation cover: (1) surfaces at elevations >1900 m that lack pavement owing to vegetation cover in both glacial and interglacial periods; (2) surfaces between ~1900 and 400 m that show strongly developed pavement, but may undergo pavement disruption and loss owing to vegetation growth during glacial periods; and (3) surfaces below

~400 m that undergo pavement formation during both interglacial and glacial phases. In Utah, Marchetti and Cerling (2005) observed pavements present on a fluvial terrace at 1780–1710 m, but not on a nearby terrace at an elevation of ~2050 m. Other studies in the western United States, however, suggest that pavement formation is not simply 'reset' during glaciations, as pavements on volcanic cones of different ages (75–80 ka age Lathrop Wells cone and ~ 1 Ma Red Cone) at altitudes of 800–900 m show clear differences according to their age, implying that the surface properties were not strongly disturbed or obliterated by vegetation growth. Further work will be necessary to clarify the impact of past variations in climate and altitudinal changes in vegetation growth on pavement formation.

Animals modify pavements or contribute to their formation by laterally displacing stones at the surface and by burrowing in the subsurface. Haff and Werner (1996) cleared small patches of stones and documented pavement recovery by repeat photography. The cross-surface movement of stones by animals is one means by which topographic highs are lowered and depressions filled, contributing to the nearly flat surface characteristic of mature pavements.

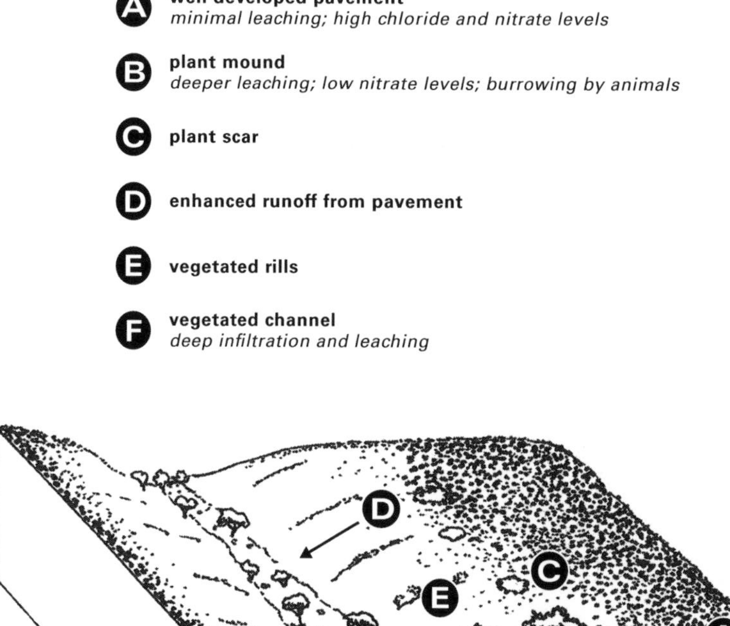

Figure 9.11 A schematic diagram of ecohydrologic conditions in a Mojave Desert pavement. On well-developed pavements (A) rainfall infiltrates to shallow depths and leaching is minimal, causing the accumulation of high concentrations of chloride and nitrate. Vegetation is largely absent from these smooth surfaces owing to the saline nature of the soil and its low infiltration capacity. Some vegetation may be found near the margins of pavement (B). The shrubs trap aeolian fines and burrowing animals displace nonvarnished or carbonate-coated clasts upward, causing the mounds to be lighter in colour than the surrounding pavement. Macropore channels from shrub roots and burrows promote deeper leaching depths than found beneath the pavement. Plant scars (C) on barren pavements are widespread in the American southwest, indicating that vegetation and animal life was once more widespread. The low infiltration capacity of the pavement enhances runoff, channelling water through rill systems (E) to vegetated ephemeral washes (F) that dissect the pavements. Deep infiltration into the channels leaches salts, providing a favourable environment for plants.

A
Well developed pavement.
B
Plant mound with small animal burrows.
C
Plant scar.
F
Vegetated channel fed by pavement runoff.

Figure 9.12 Photographs of features corresponding to (A), (B), (C) and (F) in Figure 9.11.

The degree of animal activity and burrowing appears to be higher on pavements with more vegetation and exposed bare soil (Wood, Graham and Wells, 2005). In turn, burrowing and bioturbation help to produce a mosaic of different surface characteristics within a desert pavement. In general, animals prefer to dig in friable soils underneath large plants and therefore pavements tend to lack burrows (Figure 9.12). On a 580 000-year-old basalt flow in the Cima Volcanic field, six distinctive surface mosaic types were identified, ranging from three classified as desert pavement (surface clast coverage >65 %) to three that appeared as bare ground (<65 % clast coverage). One of the bare ground surface mosaics (GB2) consisted of a 58 % cover of subangular carbonate-encrusted medium to coarse gravel fragments that appeared to result from excavations by burrowing mammals (Wood, Graham and Wells, 2005). Burrows were absent on the pavement mosaics.

The presence of plant scars on barren pavements in arid parts of the American southwest suggests that vegetation and animal life were once more widespread (McAuliffe and McDonald, 2006). Bioturbation by burrowing animals produced plant scar mounds and depressions that are readily visible on aerial imagery (Figure 9.11 and 9.12(c)).

Gravity-driven processes, often aided by animal movement, affect the formation of pavement on slopes. In addition, hydraulic action influences the alignment or fabric of clasts, which may be oriented with their long axes parallel to the downslope direction (Abrahams *et al.*, 1990).

9.8.3 Regeneration of surfaces by rainfall and runoff events

Pavements whose surfaces have been disturbed can recover by rainfall and runoff events (Figure 9.6). Wainwright, Parsons and Abrahams (1999) conducted a study at Walnut Gulch, Arizona, in which surface stones were mixed with the fine substrate to investigate the effect of raindrop erosion processes on stone pavement regeneration. Significant accumulations of coarse particles were observed within five 5-minute artificial rainfall events. A disturbed plot with an initial stone cover of 41 % increased to a 71 % cover following the experiment – close to the 77 % cover of the pre-disturbance pavement. About 10 runoff events are annually recorded at the site of the study, suggesting that pavements would be able to regenerate on an annual cycle following disturbance.

9.8.4 Earthquakes

In seismically active areas, earthquakes displace surface clasts, thereby influencing pavement development. Haff (2005) noted a number of disturbances to desert pavements generated by the October 1999 Hector mine earthquake in the Mojave Desert (with a moment magnitude, M_W, of 7.1). These included zones of wholesale gravel displacement, rotated and displaced cobbles, and moat formations around loosened boulders. Displaced clasts tended to move downslope. The effects of seismic shaking are similar to animal displacement, rainbeat and wash processes.

9.8.5 Off-road vehicle disturbance

Owing to their trafficable nature, pavements are commonly traversed by off-road vehicles. Despite their strong appearance, however, the pavements are easily disrupted by traffic. Tracks from World War II military exercises (1942–1944) are still discernable in the Mojave Desert, suggesting that the time for full recovery may be decades (Belnap and Warren, 2002). Off-road vehicles produce both compression and shear forces, and penetration resistance is significantly higher inside than outside the tracks. The recovery of soils in hot deserts is slow, as clay mineral expansion during wetting and drying cycles, freeze–thaw heaving and biological activity are limited. Vehicle activity affects the distribution of clast sizes, pushing the desert pavement downwards into the soil and allowing the rut to infill with smaller particles (Belnap and Warren, 2002).

Desert pavements have been studied as part of military characterizations of trafficable surfaces. In the Sonoran Desert, vehicles were tested over an endurance course in which the degree of desert pavement development was qualitatively ranked from 'none' to 'strongly developed' (Bacon *et al.*, 2008).

9.8.6 Removal of stones for agriculture

Desert pavement surfaces often occur in regions that are too arid to permit widespread agriculture. Nonetheless, where recent high rates of population growth have put pressure on available land, stones have been, and continue to be, removed to permit cultivation. In Jordan, access to groundwater has allowed increasingly intensive agriculture and caused extensive clearance of pavements to provide new land. Following irrigation, there are chemical and physical changes to the soil, including development of a surface crust (Allison *et al.*, 2000).

There is also documentation of pavement stones being cleared for agriculture in the past. The ancient settlers of the Negev removed pavement stones more than a thousand years ago (Sharon, 1962). The renewal of the pavement since that time is evidence of the dynamic nature of the surface.

9.9 Ecohydrology of pavement surfaces

In most deserts, rainfall infiltrates only to shallow depths and soil moisture values are low. As a consequence, there is little leaching of soils and dissolved materials are deposited in the upper horizons. Spatially, there is considerable variability in infiltration, as water seeps into porous surfaces, including sandy soils, soils around plants, and stream beds or other alluvial surfaces, but runs off from less permeable covers such as pavements (Cornet, Delhoume and Montaña, 1988; Abrahams and Parsons, 1991; Dunkerley, 2002; Belnap, 2003).

Ecohydrologic studies examine the relationship between plant cover, soils and surface characteristics. In deserts, the striking differences in soil properties that commonly occur over short distances may create a mosaic landscape. A binary or two-phase system of soils often develops, with one component supporting a relatively dense vascular plant cover (patches, groves or bands) and the other component consisting primarily of unvegetated ground. The bare ground is relatively impermeable and sheds water downslope to vegetated bands. In the plant patches, the soils are soft and friable, with high infiltration capacities. As a result of the differences in soil properties, the vegetated zones may receive water amounts close to twice the climatological rainfall (Cornet, Delhoume and Montaña, 1988), whereas the bare patches are more arid than rainfall totals suggest. The result is a highly heterogeneous landscape (Cornet, Delhoume and Montaña, 1988; Seghieri and Galle, 1999; Dunkerley and Brown, 1997). The patchiness of the vegetation may help to conserve scarce water and nutrients within part of the landscape. Areas of desert pavement in the American southwest appear to be part of this type of ecohydrologic system.

The two critical mechanisms of water transport are overland flow and infiltration. Infiltration affects the development of vegetation on the pavement, whereas overland flow transfers water to plants adjacent to the pavement. In addition, water flow redistributes sediment, organic litter and seeds. Many desert pavements are largely devoid of vegetation owing to the absence of soil moisture (Musick, 1975) caused by the reduced infiltration capacity of the soil and salt accumulation at shallow depths. Where plants occur, they are smaller in stature and lower in density than on nearby nonpavement areas (McAuliffe and McDonald, 2006). The low infiltration capacity of the pavement enhances runoff, channelling water to adjacent wash zones or dune areas where plants grow vigorously (Figures 9.11 and 9.12). The redistribution of runoff and retardation of plant growth is a function of evolutionary changes to the hydrologic properties of the pavement, discussed below.

9.9.1 Infiltration in pavements and runoff potential

As pavements and soils evolve, their hydrologic properties change. In the primary stages of soil evolution, porosity (30–40 %) and permeability (equivalent to 60–80 mm/h of precipitation) are high, and runoff is negligible (Amit and Gerson, 1986). These properties diminish with time. The introduction of fines and salts into the soil profile retards the infiltration of water (Sena *et al.*, 1994; Young *et al.*, 2004). As a result, stone pavement surfaces with mature soils, embedded fragments and a well-knit surface texture tend to have very low infiltration rates, with little absorption of rain (Wells *et al.*, 1985; Amit and Gerson, 1986; Valentin, 1994; Wood, Graham and Wells, 2005). The infiltration rates decrease with increasing clast cover (Abrahams and Parsons, 1991) and change with soil age (Young *et al.*, 2004). Thus, runoff is relatively high from well-developed pavement soils.

Sheet flow from pavement funnels water into nearby drainage courses (Symmons and Hemming, 1968). In the Eastern Desert of Egypt, desert pavement had the lowest infiltration rates of the surface types studied by Foody, Ghoneim and Arnell (2004) (0.7 mm/h), whereas unconsolidated wadi deposits had the highest (140.1 mm/h) .The depth of infiltration is marked by the location of the wetting front. Following 52 mm of precipitation in the Sonoran Desert, the wetting front in the central area of a well-developed varnished pavement was 139 mm, increasing to 271 mm just outside the pavement margin (McAuliffe and McDonald, 2006). The runoff from the pavements creates a favourable soil moisture environment in the adjacent channels.

Drainage density on pavements may increase as a function of decreasing infiltration capacity over time. On Cima volcanic field pavements, drainage density (km/km^2) increases with age: a 0.14 my flow has a density of 0.6, whereas a 0.99 my old flow has a density of 14.9 (Wells *et al.*, 1985). Runoff increases as the soil becomes plugged with carbonates and clay and distinct soil horizons develop (Amit and Gerson, 1986), eventually eroding away pavements. At Cima, fluvial stripping has reduced the mantle thickness of the older flow surfaces (Wells *et al.*, 1985) and in the Nahal Zéelim sequence of alluvial fans in Israel gullying is widespread (Amit and Gerson, 1986).

Where pavements are extensive, their low permeability gives rise to rapid rates of runoff and the potential for flash flooding. The spatial distribution of different surface classes can be assessed using satellite data and the potential for flooding determined by hydrological modelling (Foody, Ghoneim and Arnell, 2004).

9.9.2 Ecohydrologic relationships and vegetation associations

There are two basic plant associations with accretionary pavements. The first is vegetated ephemeral washes that lie adjacent to and receive runoff from the pavements and the second is 'islands' of desert scrub development within the pavement itself.

On dissected alluvial fans (Chapter 14), pavements often form elongate interfluves. The relatively impermeable pavements shed flow to adjacent ephemeral washes (Wood, Graham and Wells, 2005; McAuliffe and McDonald, 2006). In the channels, water infiltrates more deeply, leaching salts (Wood, Graham and Wells, 2005) and providing a favourable environment for plants (Figures 9.11 and 9.12).

On the surface of the pavement itself, vegetation is largely lacking owing to the saline nature of the soil and its limited infiltration capacity. However, the presence of plant scar mounds and depressions indicates that long-lived perennial plants were more common in the past (McAuliffe and McDonald, 2006). In the Sonoran Desert of Arizona, modern vegetation is largely limited to stream channels, but prominent, light-coloured plant scar mounds (up to 25 cm in elevation and 200–600 cm in diameter) are widespread on old (probably Pleistocene age), varnished pavements of alluvial fans. The light surface colour is a result of the lack of varnish on surface clasts, coatings of pedogenic carbonate on noncalcareous clasts and fragments of indurated pedogenic carbonate. The clasts are smaller than those of the surrounding pavement. Other than ephemeral plants and very young creosote bushes (*Larrea tridentata*), vegetation is absent from the mounds. The mounds are thought to represent a temporal stage that follows the disappearance of large perennial plants, below which burrow systems and bioturbation mounds were once common. Burrowing rodents originally brought the small clasts to the surface. Following the death of the plant, deflation removed fine sediment, leaving behind a concentrated lag of small clasts.

Why did creosote bushes once grow on pavements that are largely vegetation-free today? McAuliffe and McDonald (2006) hypothesise that the more effective Late Pleistocene precipitation not only provided more moisture to the plants but also leached the soils to greater depths, reducing the salinity of the substrate. Today, under conditions of contemporary drought, the mortality of modern creosote bush on pavement is greater in the central pavement area than peripheral areas, owing to the more xeric soil moisture environment.

In North America, the percent shrub cover seems to be strongly correlated with the percent clast cover, decreasing as pavement cover increases. Within a given pavement, vegetation is most likely in areas with the highest amount of bare soil exposed. In areas of contemporary pavement where shrubs cluster, the hydrologic character of the soil differs considerably from nonshrub surfaces. Aeolian sand is trapped by the shrubs and later incorporated into the soil horizons, the organic fraction increases and macropore channels from shrub roots promote deeper infiltration of water. Leaching beneath shrubs removes soluble salts to below the 50-cm depth (Wood, Graham and Wells, 2005).

9.10 Relative and absolute dating of geomorphic surfaces based on pavement development

Desert pavements are commonly examined in order to understand other geomorphic processes. These include the age of surfaces, rates and processes of erosion and deposition, the development of alluvial fans and faulting and tectonism.

Pavements and their underlying soils have been widely used as a relative age dating tool owing to the pronounced changes that characterise their development over time (Al-Farraj and Harvey, 2000; Al-Farraj, 2008). The differentiation of surfaces into age groups based on pavement development indices has been used to correlate geomorphic surfaces within a region. Cooke (1970), for example, noted that the best-developed pavements in Chile occurred on the oldest and highest fluvial terraces. Pavement characteristics have been used to map a number of Quaternary surficial deposits, including alluvial fans (Shlemon, 1978; Christenson and Purcell, 1985; Al-Farraj and Harvey, 2000), stream terraces, shorelines and beach ridges (Sauer, Schellmann and Stahr, 2007; Al-Farraj, 2008), and lava flows (Wells *et al.*, 1985; McFadden, Wells and Jercinovich, 1987; Williams and Zimbelman, 1994). Changes in the chemical composition of varnish and the presence of organic carbon in varnish coating gravel of pavements have been employed to estimate the age of the underlying materials on which the pavement formed (McFadden, Wells and Jercinovich, 1987; Dorn, 1988; Liu and Broecker, 2008).

9.10.1 Changes in surface characteristics

The relative dating of surfaces and the correlation of geomorphic surfaces by pavement development is possible because pavement characteristics tend to change in a systematic way within any given region as the surface matures and becomes older. Common surface modifications include a reduction in clast sizes, better particle sorting, increased angularity and an increase in the surface area occupied by interlocking smooth pavement. In Israel, Amit and Gerson (1986) studied pavement evolution by examining 15 terraces on a vast sublacustrine delta formed by the Holocene drop of Lake Lisan, the precursor of the Dead Sea. Similar to the findings of Cooke (1970), pavement maturity, as determined by percent cover of the terrace surface, sorting, shape of fragments, length of fragments, degree of pitting and percent cover, was greatest on the oldest terrace and least on the youngest. In Oman, a similar progression was noted on fans and wadi terraces of different ages (Al-Farraj and Harvey, 2000). With time, the topography is modified and smoothed, with a systematic decrease in bar-and-swale microtopography and a rounding of gully and terrace edges on older surfaces (Pelletier, Cline and DeLong, 2007).

Soil development proceeds concurrently with pavement evolution and involves increases in soil thickness, fines content (silt plus clay), redness of the B-horizon and carbonate accumulation, as well as progressive horizon development (Al-Farraj and Harvey, 2000; Young *et al.*, 2004). For early pavement formation in many areas, vesicular A-horizons constitute the initial soil horizon, weakly developed colour B-horizons are present below the vesicular A-horizon of Middle to Early Holocene soils and a weak to moderately strong and usually nongravelly argillic horizon is present below the vesicular A-horizon of Late Pleistocene and older fan soils.

In Israel, the terminal soil thickness was reached early in pavement development, with soils of several hundred to about one thousand years old being the same thickness as those 14 000 years in age (Amit and Gerson, 1986). By contrast, other locations, including Oman (Al-Farraj and Harvey, 2000) and the Providence Mountains, California (Sena *et al.*, 1994), showed an increase in soil thickness over time: these locations are of much greater age, however, and have probably experienced several cycles of dust deposition.

9.10.2 Pavement characteristics and geomorphic surface ages

Although it is known that pavement changes systematically with time, it is a more difficult step to assign ages to surfaces based on these assumptions. In general, fan surfaces of Early Holocene age are thought to exhibit partial pavement development and Late Pleistocene surfaces are characterised by smooth, well-armored, tightly knit and heavily varnished pavements (Christenson and Purcell, 1985; Bull, 1991; Al-Farraj and Harvey, 2000), indicative of long-term surface stability. Amit and Gerson (1986) suggest that at least 100 000 years is required for the development of a mature smooth pavement surface consisting of fully shattered small clasts. There are exceptions, however, with some Early to Middle Holocene alluvial surfaces exhibiting well-developed pavements (Quade, 2001) and VML dates indicating that pavements in California can form in less than 10 000 years (Liu and Broecker, 2008). Clearly, the development of the pavement and its underlying soil are influenced by dust and salt deposition rates, which not only vary with time but also are higher near source areas, such as playas. Thus, it is likely that rates of pavement development vary spatially (Pelletier, Cline and DeLong, 2007). In addition, Wood, Graham and Wells (2002) caution that pavements often demonstrate a mosaic of disjunct fabrics and textures within short distances, although they may appear monotonous, flat and uniform at first glance. As such, caution must be used when applying rock weathering of pavement as a relative age dating technique.

Furthermore, while pavement development characteristics may be useful in assigning relative ages to young pavements or separating young from old pavements, very old pavements have often attained an equilibrium form and show little in the way of distinguishable changes from place to place. Varnish microlamination dating by Liu and Broecker (2008) confirm that once a pavement forms, it may survive for 75 000 years or longer without significant disturbance. On the eastern Libyan Plateau, the presence of Middle and Upper Palaeolithic artefacts suggests that the surface has seen 100 000 years of stability. Clast parameters, such as size and orientation, show no distinguishable changes over space and so do not allow pavements of different ages to be distinguished (Adelsberger and Smith, 2009).

The very great age of pavements in areas of the Sahara–Arabian Deserts is further supported by the research of Matmon *et al.* (2009), who concluded that abandoned alluvial surfaces of the Paran Plains in the southern Negev Desert, Israel, represent one of the longest-lived landforms on Earth. Heavily varnished chert clasts cover more than 95 % of the surface, which is thought to have not been disturbed since emplacement. The pavement and associated reg soils began to form at the latest 1.5–2.0 Ma ago. The surface has remained exceptionally stable, owing to long-term hyper-aridity that limits surface runoff, vegetation and bioturbidity. Indeed, the erosion rates are the lowest that have been documented to date on Earth.

9.10.3 Pavement surfaces as a tool in geomorphic assessment

Given the paucity of numerical dates in arid environments, there is often an emphasis on relative age dating and the use of soil development to understand geomorphic development and to establish a paleoseismic fault chronology. Where dated surfaces are available, a chronosequence of soils may be established on alluvial fans and the displacement of these soils used to study active tectonism.

A determination of surface ages is critical in studies of the key controlling factors in alluvial fan development, such as climate, climate change and tectonism. In characterising the age of alluvial fan surfaces, a variety of morphostratigraphic and weathering characteristics are used, including the relative development of desert pavement. Key characteristics include the degree of disintegration of surface clasts, the preservation of bar and swale morphology (changes in microtopography), the percent coverage of the surface by pavement, the degree of soil maturity and the development of rock varnish (Christenson and Purcell, 1985; Amit *et al.*, 1996; Spelz *et al.*, 2008).

Changes in soil properties and pavement development with time are used as a basis for tectonic studies, including estimations of fault scarp ages and slip rates. Pavements and their associated soils provide a valuable resource owing to the difficulty of applying radiometric methods in extremely arid environments. In desert areas of the American southwest and the Middle East (Amit *et al.*, 1996), many faults displace alluvial fan surfaces. Indeed, alluvial fans provide the single most important geomorphic marker for defining slip rates in North America (Spelz *et al.*, 2008).

9.11 Conclusions

Contemporary desert landscapes reflect the integration of geologic conditions and geomorphic processes operating over long time frames. They thus reflect a succession of former events and environments. Stony surfaces are widespread and show a range in forms associated with geologic conditions, age and process. A continuum of rocky or gravelly surfaces develops from hamada through desert pavement. There is potentially a range of processes responsible for their formation. Cooke (1970) suggested that pavements may illustrate the concept of equifinality, with similar forms produced in different areas by a variety of processes operating on different initial conditions. It is likely that a single set of processes cannot explain all occurrences of gravel-covered surfaces.

In sandy soils, deflation may play a role. Deflated surfaces are recorded from gobi (China), gibber plains (Australia) and possibly the Martian landing sites. Unlike accretionary pavements, however, there has been little detailed study of the pavement of deflation. The presence of sand in the groundmass between separated clasts is not sufficient to claim that the pavement has formed by wind erosion.

Where sand is largely absent, pavements appear to develop by a variety of geomorphic processes that result in stone sorting, surface creep and stone displacement, clast disintegration and syndepositional lifting of surface clasts by the accumulation of salt- and carbonate-rich aeolian fines (Dohrenwend, 1987). These processes operate at different levels of relative importance as the pavements evolve. Aeolian processes have significantly influenced pavement soil development. The silt, clay, carbonates and soluble salts accumulated in desert soils are attributable largely to the incorporation of windblown dust rather than to chemical weathering of soil parent materials. Although Quaternary aeolian activity has been episodic, the rate of incorporation of aeolian fines has been slow enough to permit cumulate soil development and to cause uplift of pre-existing soil pavements. Varnish data from alluvial fans in the southwestern United States suggest that varnish development has occurred on stones exposed continuously since abandonment of the fan surface (no stones reflect the emergence of clasts during the Holocene into Late Pleistocene pavements, as would develop if there was upward migration of gravel). This suggests that desert pavements (1) are born at the land surface and (2) remain at the land surface via aeolian deposition and simultaneous development of cumulate soils beneath the pavements. Pavements appear to require some degree of stability for their full development, occurring beyond the limits of modern drainage or standing above active channels through dissection or tectonic processes. Formation times are related to rates of salt and dust accumulation, a function of the distance from source areas (particularly playas) and the prevailing wind direction (Pelletier, Cline and DeLong, 2007). Leaching rates also influence soil formation. Thus, stone pavement development is strongly linked to long-term climate change.

Desert pavements play an important ecohydrologic role in deserts. Mature pavements form a strongly interlocking carapace that sheds water rapidly. Plants are largely absent from pavements owing to salts within the soil and to rapid runoff from the impermeable surface. However, vegetation in adjacent wadis or dune areas benefits from the enhanced water supply provided by the runoff. If pavement surfaces are extensive, they tend to increase the flood potential of a region.

Stony surfaces remain an important focus of geologic, geomorphic, hydrologic and archaeologic studies in arid lands. To date, detailed studies of pavements have largely focused on the American southwest and Israel. The recent expansion of research into new regions has increased our understanding of the diversity of pavement types and processes, but a more systematic approach to pavement and hamada studies is required in order to understand the similarities and dissimilarities of stony surface development across the desert realm.

Box 9.1 Pavements of the Mojave Desert: the Cima and Pisgah Volcanic Fields

The lava flows of the Mojave Desert, California, have provided the basis for some of the most detailed research on the relationship between aeolian deposition, pedogenesis and pavement formation (Dohrenwend *et al.*, 1984; Wells *et al.*, 1985, 1995; McFadden, Wells and Jercinovich, 1987; Anderson, Wells and Graham, 2002; Helms, McGill and Rockwell, 2003; Williams and Zimbelman, 1994; Wood, Graham and Wells, 2002; Graham *et al.*, 2008; Meadows, Young and McDonald, 2008). The best studied of these volcanic deposits is the Cima Volcanic Field in the east-central Mojave Desert, which consists of around 40 cinder cones and 60 associated basalt flows,

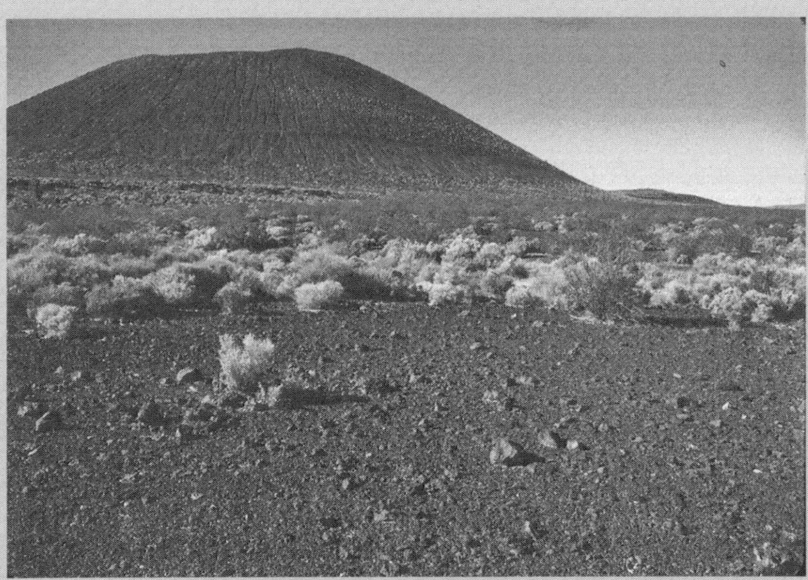

Figure 9.13 The pavements of the Cima Volcanic Field, Mojave Desert, California, are very well studied and provide an excellent example of the type of inflationary surfaces common in the southwestern United States. The pavement sits above a great thickness (from 100 to 300 cm) of aeolian-derived fine sand, silt and clay. Surface-exposure dates show that the pavement clasts are the same age as the underlying basalt flows. In the background is one of the 40 cinder cones in the area and, in the foreground, the smooth appearance of a well-developed pavement.

formed from the Miocene to Holocene (Figure 9.13). Not only are pavements extensive on these flows, but also dating has provided age control, allowing the study of pavement evolution over time.

Cima represents an excellent example of the type of inflationary desert pavements commonly observed in the southwestern United States. The cumulic aeolian epipedon (a soil horizon that forms at or near the surface, at the boundary between the soil and air) steadily accumulates in thickness as a function of dust deposition rates. The accretionary mantle at Cima is very thick and relatively clast free. It consists of a 1- to 3-m-thick accumulation of aeolian fine sand, silt and clay (Wells *et al.*, 1985).

One of the most important aspects of the lava flow studies at Cima is that the surfaces are dateable. K–Ar dating of the underlying bedrock on which pavements formed provides an important age control. In situ cosmogenic ^{3}He surface-exposure dates indicate that the pavements are isochronous with the underlying basalt flows (Wells *et al.*, 1995), suggesting that the pavement clasts were never buried, but rose upward on a vertically accreting aeolian mantle and were maintained as the underlying soil evolved (Wells *et al.*, 1985; McFadden, Wells and Dohrenwend, 1986; Anderson, Wells and Graham, 2002).

By examining flows of different ages, Wells *et al.* (1985) and McFadden, Wells and Jercinovich (1987) were able to demonstrate that the morphology of flow surfaces changes markedly over time in concert with pavement development. Relative to more recent flows, which show numerous surface features, older flows lack constructional forms, have very low relief and possess extensive stone pavements. The pavement forms as locally derived basaltic rubble, consisting of mechanically weathered blocks of flow rock derived from constructional highs, moves into topographic lows by mass wasting and other slope processes. In turn, aeolian fines accumulate in the topographic lows, owing to local reductions in near-surface wind velocity. Basalt flows, with high initial surface roughness and permeability, are very effective traps for salts and aeolian materials derived from nearby playas and distal piedmont areas (Wells *et al.*, 1985). Accumulation appears not to be continuous, but episodic, with occasional rapid influxes of material. The most active dust incorporation occurs during periods of playa expansion and vegetation reduction (McFadden *et al.*, 1998). Over time, the surface roughness of the flow reduces and the trapping effect is dampened. The extremely high permeability of flows inhibits fluvial erosion of the trapped silt.

In contrast to the mature pavements of the Cima volcanic field, young well-developed stone mosaics on the Pisgah basalt flow lack underlying salt and dust. Williams and Zimbelman (1994) propose that sheetfloods play a dominant role in the incipient growth of small pavement areas (about 1 m^2). Aeolian processes do not appear to affect these early stages of development. Such pavements may represent an initial substage of accretionary mantle development, wherein the mosaic is formed prior to soil development, rather than contemporaneously with mantle accretion.

References

Abrahams, A.D. and Parsons, A.J. (1991) Relation between infiltration and stone cover on a semiarid hillslope, southern Arizona. *Journal of Hydrology*, **122**, 49–59.

Abrahams, A.D., Soltyka, N., Parsons, A.J. and Hirsch, P.J. (1990) Fabric analysis of a desert debris slope: Bell Mountain, California. *Journal of Geology*, **98**, 264–272.

Adelsberger, K.A. and Smith, J.R. (2009) Desert pavement development and landscape stability on the Eastern Libyan Plateau, Egypt. *Geomorphology*, **107**, 178–194.

Al-Farraj, A. (2008) Desert pavement development on the lake shorelines of Lake Eyre (South), South Australia. *Geomorphology*, **100**, 154–163.

Al-Farraj, A. and Harvey, A.M. (2000) Desert pavement characteristics on wadi terrace and alluvial fan surfaces: Wadi Al-Bih, U.A.E. and Oman. *Geomorphology*, **35**, 279–297.

Allison, R.J., Grove, J.R., Higgitt, D.L. *et al.* (2000) Geomorphology of the eastern Badia basalt plateau, Jordan. *The Geographical Journal*, **166**, 352–370.

Amit, R. and Gerson, R. (1986) The evolution of Holocene reg (gravelly) soils in deserts – an example from the Dead Sea region. *Catena*, **13**, 59–79.

Amit, R., Gerson, R. and Yaalon, D.H. (1993) Stages and rate of the gravel shattering process by salts in desert Reg soils. *Geoderma*, **57**, 295–324.

Amit, R. and Yaalon, D.H. (1996) The micromorphology of gypsum and halite in reg soils – the Negev Desert, Israel. *Earth Surface Processes and Landforms*, **21**, 1127–1143.

Amit, R., Harrison, J.B.J., Enzel, Y. and Porat, N. (1996) Soils as a tool for estimating ages of Quaternary fault scarps in a hyperarid environment – the southern Arava valley, the Dead Sea Rift, Israel. *Catena*, **28**, 21–45.

Anderson, K., Wells, S. and Graham, R. (2002) Pedogenesis of vesicular horizons, Cima Volcanic Field, Mojave Desert, California. *Soil Science Society of America Journal*, **66**, 878–887.

Bacon, S.N., McDonald, E.V., Baker, S.E. *et al.* (2008) Desert terrain characterization of landforms and surface materials within vehicle test courses at U.S. Army Yuma Proving Ground, USA. *Journal of Terramechanics*, **45**, 167–183.

Bailey, J.E., Self, S., Wooller, L.K. and Mouginis-Mark, P.J. (2007) Discrimination of fluvial and eolian features on large ignimbrite sheets around La Pacana Caldera, Chile, using Landsat and SRTM-derived DEM. *Remote Sensing of Environment*, **108**, 24–41.

Belnap, J. (2003) The world at your feet: desert biological soil crusts. *Frontiers in Ecology and the Environment*, **1**, 181–189.

Belnap, J. and Warren, S.D. (2002) Patton's tracks in the Mojave Desert, USA: an ecological legacy. *Arid Land Research and Management*, **16**, 245–258.

Blake, W.P. (1904) Origin of pebble-covered plains in desert regions. *Transactions of the American Institute of Mining Engineers*, **34**, 161–162.

Blank, R.R., Young, J.A. and Lugaski, T. (1996) Pedogenesis on talus slopes, the Buckskin Range, Nevada, USA. *Geoderma*, **71**, 121–142.

Brown, K.J. and Dunkerley, D.L. (1996) The influence of hillslope gradient, regolith texture, stone size and stone position on the presence of a vesicular layer and related aspects of hillslope hydrologic processes: a case study from the Australian arid zone. *Catena*, **26**, 71–84.

Bull, W.B. (1991) *Geomorphic Responses to Climate Change*, Oxford University Press, New York.

Casana, J., Herrmann, J.T. and Qandil, H.S. (2009) Settlement history in the eastern Rub al-Khali: Preliminary Report of the Dubai Desert Survey (2006–2007). *Arabian Archaeology and Epigraphy*, **20**, 30–45.

Christenson, G.E. and Purcell, C. (1985) Correlation and age of Quaternary alluvial-fan sequences, Basin and Range province, southwestern United States. *Geological Society of America Special Paper*, **203**, 115–122.

Cooke, R.U. (1970) Stone pavements in deserts. *Association of American Geographers*, **60**, 560–577.

Cornet, A.P., Delhoume, J.P. and Montaña, C. (1988) Dynamics of striped vegetation patterns and water balance in the Chihuahuan Desert, in *Diversity and Pattern in Plant Communities* (eds H.J. During, M.J.A. Werger and H.J. Willems), SPB Academic Publishing, The Hague, pp. 221–231.

Cremaschi, M. (1996) The rock varnish in the Messak Settafet (Fezzan, Libyan Sahara), age, archaeological context, and paleo-environmental implication. *Geoarchaeology: An International Journal*, **11**, 393–421.

Czajka, W. (1972) Windschliffe als Landschaftmerkmal. *Zeitschrift für Geomorphologie*, **16**, 27–53.

Dan, J., Yaalon, D.H., Moshe, R. and Nissim, S. (1982) Evolution of reg soils in southern Israel and Sinai. *Geoderma*, **28**, 173–202.

Denny, C.S. (1965) Alluvial fans in the Death Valley region California and Nevada, US Geological Survey Professional Paper 466.

Denny, C.S. (1967) Fans and pediments. *American Journal of Science*, **265**, 81–105.

Dohrenwend, J.C. (1987) Basin and range, in *Geomorphic Systems of North America* (ed. W.L. Graf), Centennial Special, vol. 2, Geological Society of America, Boulder, CO, pp. 303–342.

Dohrenwend, J.C., McFadden, L.D., Turrin, B.D. and Wells, S.G. (1984) K–Ar dating of the Cima Volcanic fFeld, eastern Mojave Desert, California: Late Cenozoic volcanic history and landscape evolution. *Geology*, **12**, 163–167.

Dorn, R.I. (1988) A rock varnish interpretation of alluvial fan development in Death Valley, Callifornia. *National Geographic Research*, **4**, 56–73.

Dunkerley, D.L. (2002) Infiltration rates and soil moisture in a groved mulga community near Alice Springs, arid central Australia: evidence for complex internal rainwater redistribution in a runoff–runon landscape. *Journal of Arid Environments*, **51**, 199–219.

Dunkerley, D.L. and Brown, K.J. (1997) Desert soils, in *Arid Zone Geomorphology: Process, Form and Change in Drylands*, 2nd edn (ed. D.S.G. Thomas), John Wiley & Sons, Ltd, Chichester, pp. 55–68.

Evenari, M. (1985) The desert environment, in *Hot Deserts and Arid Shrublands 12A* (eds M. Evenari, I. Noy-Meir and D.W. Goodall), Elsevier Science Publishing Company, New York, pp. 1–22.

Evenari, M., Yaalon, D.H. and Gutterman, Y. (1974) Note on soils with vesicular structure in deserts. *Zeitschrift Für Geomorphologie*, N.F. **18**, 162–172.

Foody, G.M., Ghoneim, E.M. and Arnell, N.W. (2004) Predicting locations sensitive to flash flooding in an arid environment. *Journal of Hydrology*, **292**, 48–58.

Francis, M.L., Fey, M.V., Prinsloo, H.P. *et al.* (2007) Soils of Namaqualand: compensations for aridity. *Journal of Arid Environments*, **70**, 588–603.

Fürst, M. (1965) Hammada – Serir-Erg. Eine morphogenetische Analyse des nordöstlichen Fezzan (Libyen). *Zeitschrift für Geomorphologie*, **9**, 385–421.

Goossens, D. (2006) Field experiments of aeolian dust accumulation on rock fragment substrata. *Sedimentology*, **42**, 391–402.

Graham, R.C., Hirmas, D.R., Wood, Y.A. and Amrhein, C. (2008) Large near-surface nitrate pools in soils capped by desert pavement in the Mojave Desert, California. *Geology*, **36**, 259–262.

Haff, P.K. (2005) Response of desert pavement to seismic shaking, Hector Mine earthquake, California, 1999. *Journal of Geophysical Research*, **110**, F02006, DOI: 10.1029/2003JF000054.

Haff, P.K. and Werner, B.T. (1996) Dynamical processes on desert pavements and the healing of surficial disturbances. *Quaternary Research*, **45**, 38–46.

Helms, J.G., McGill, S.F. and Rockwell, T.K. (2003) Calibrated, Late Quaternary age indices using clast rubification and soil development on alluvial surfaces in Pilot Knob Valley, Mojave Desert, southeastern California. *Quaternary Research*, **60**, 377–393.

Hobbs, W.H. (1917) The erosional and degradational processes of deserts, with especial reference to the origin of desert depressions. *Annals of the Association of American Geographers*, **7**, 25–60.

Jessup, R.W. (1960) The Stony Tableland soils of the Australian arid zone and their evolutionary history. *Journal of Soil Science*, **11**, 188–196.

Li, X. Yi, C., Chen, F. *et al.* (2006) Formation of proglacial dunes in front of the Puruogangri Icefield in the central Qinghai-Tibet Plateau: implications for reconstructing paleoenvironmental changes since the Lateglacial. *Quaternary International*, **154–155**, 122–127.

Liu, T. and Broecker, W.S. (2008) Rock varnish microlamination dating of late Quaternary geomorphic features in the drylands of western USA. *Geomorphology*, **93**, 501–523.

Mabbutt, J.A. (1977) *Desert Landforms*, MIT Press, Cambridge, MA.

McAuliffe, J.R. and McDonald, E.V. (2006) Holocene environmental change and vegetation contraction in the Sonoran Desert. *Quaternary Research*, **65**, 204–215.

McFadden, L.D., Wells, S.G. and Dohrenwend, J.C. (1986) Influences of Quaternary climatic changes on processes of soil development on desert loess deposits of the Cima Volcanic Field, California. *Catena*, **13**, 361–389.

McFadden, L.D., Wells, S.G. and Jercinovich, M.J. (1987) Influences of eolian and pedogenic processes on the origin and evolution of desert pavements. *Geology*, **15**, 504–508.

McFadden, L.D., Wells, S.G., Brown, W.J. and Enzel, Y. (1992) Soil genesis on beach ridges of pluvial Lake Mojave: implications for Holocene lacustrine and eolian events in the Mojave Desert, southern California. *Catena*, **19**, 77–97.

McFadden, L.D., McDonald, E.V., Wells, S.G. *et al.* (1998) The vesicular layer and carbonate collars of desert soils and pavements: formation, age and relation to climate change. *Geomorphology*, **24**, 101–145.

McFadden, L.D., Eppes, M.C., Gillespie, A.R. and Hallet, B. (2005) Physical weathering in arid landscapes due to diurnal variation in the direction of solar heating. *Geological Society of America Bulletin*, **117**, 161–173.

McKenna Neuman, C. (1998) Particle transport and adjustments of the boundary layer over rough surfaces with an unrestricted, upwind supply of sediment. *Geomorphology*, **25**, 1–17.

Marchetti, D.W. and Cerling, T.E. (2005) Cosmogenic 3He exposure ages of Pleistocene debris flows and desert pavements in Capitol Reef National Park, Utah. *Geomorphology*, **67**, 423–435.

Matmon, A., Simhai, O., Amit, R. *et al.* (2009) Desert pavement-coated surfaces in extreme deserts present the longest-lived landforms on Earth. *Geological Society of America Bulletin*, **121**, 688–697.

Meadows, D.G., Young, M.H. and McDonald, E.V. (2008) Influence of relative surface age on hydraulic properties and infiltration on soils associated with desert pavements. *Catena*, **72**, 169–178.

Mengxiong, C. (1994) Characteristics of inland Quaternary basins in Northwest China with reference to their hydrological significance. *Engineering Geology*, **37**, 61–65.

Moores, J.E., Pelletier, J.D. and Smith, P.H. (2008) Crack propagation by differential insolation on desert surface clasts. *Geomorphology*, **102**, 472–481.

Musick, H.B. (1975) Barrenness of desert pavement in Yuma County, Arizona. *Journal of the Arizona Academy of Sciences*, **10**, 24–28.

Parsons, A.J., Abrahams, A.D. and Simanton, J.R. (1992) Microtopography and soil-surface materials on semi-arid piedmont hillslopes, southern Arizona. *Journal of Arid Environments*, **22**, 107–115.

Peel, R.F. (1968) Comment on 'Wind-stable stone-mantles in the southern Sahara'. *Geographical Journal*, **134**, 463–465.

Pelletier, J.D., Cline, M. and DeLong, S.B. (2007) Desert pavement dynamics: numerical modeling and field-based calibration. *Earth Surface Processes and Landforms*, **32**, 1913–1927.

Poesen, J.W., van Wesemael, B., Bunte, K. and Benet, A.S. (1998) Variation of rock fragment cover and size along semiarid hillslopes: a case-study from southeast Spain. *Geomorphology*, **23**, 323–335.

Quade, J. (2001) Desert pavements and associated rock varnish in the Mojave Desert: How old can they be? *Geology*, **29**, 855–858.

Reheis, M., Goodmacher, J.C., Harden, J.W. *et al.* (1995) Quaternary soils and dust deposition in southern Nevada and California. *Geological Society of America Bulletin*, **107**, 1003–1022.

Sauer, D., Schellmann, G. and Stahr, K. (2007) A soil chronosequence in the semi-arid environment of Patagonia (Argentina). *Catena*, **71**, 382–393.

Seghieri, J. and Galle, S. (1999) Run-on contribution to a Sahelian two-phase mosaic system: soil water regime and vegetation life cycles. *Acta Oecologica*, **20**, 209–217.

Sena, G., Connell, S., Wells, S. and Anderson, K. (1994) Investigation of surficial processes active on fan pavement surfaces using tilted carbonate collars, Providence Mountains, California, in *Geological Investigations of An Active Margin* (eds S.F. McGill and T.M. Ross), Cordilleran Section Guidebook, 27th Annual Meeting, Geological Society of America, Boulder, CO, pp. 210–213.

Sharon, D. (1962) On the nature of hamadas in Israel. *Zeitschrift für Geomorphologie*, **6**, 129–147.

Shlemon, R.J. (1978) Quaternary soil-geomorphic relationships, southeastern Mojave Desert, California and Arizona, in *Quaternary Soils* (ed. W.C. Mahaney), Geo Books, Norwich, pp. 187–207.

Spelz, R.M., Fletcher, J.M., Owen, L.A. and Caffee, M.W. (2008) Quaternary alluvial-fan development, climate and morphologic dating of fault scarps in Laguna Salada, Baja California, Mexico. *Geomorphology*, **102**, 578–594.

Springer, M.E. (1958) Desert pavement and vesicular layer of some desert soils in the desert of the Lahontan Basin, Nevada. *Soil Science Society America Proceedings*, **22**, 63–66.

Symmons, P.M. and Hemming, C.F. (1968) A note on wind-stable stone-mantles in the southern Sahara. *Geographical Journal*, **34**, 60–64.

Thomas, M., Clarke, J.D.A. and Pain, C.F. (2005) Weathering, erosion and landscape processes on Mars identified from recent rover imagery, and possible Earth analogues. *Australian Journal of Earth Sciences*, **52**, 365–378.

Twidale, C.R. (1994) Desert landform evolution: with special reference to the Australian experience. *Cuaternario y Geomorfología*, **8**, 3–31.

Ugolini, F.C., Hillier, S., Certini, G. and Wilson, M.J. (2008) The contribution of aeolian material to an Aridisol from southern Jordan as revealed by mineralogical analysis. *Journal of Arid Environments*, **72**, 1431–1447.

Valentin, C. (1994) Surface sealing as affected by various rock fragment covers in West Africa. *Catena*, **23**, 87–97.

Valentine, G.A. and Harrington, C.D. (2006) Clast size controls and longevity of Pleistocene desert pavements at Lathrop Wells and Red Cone volcanoes, southern Nevada. *Geology*, **34**, 533–536.

Wainwright, J., Parsons, A.J. and Abrahams, A.D. (1999) Field and computer simulation experiments on the formation of desert pavement. *Earth Surface Processes and Landforms*, **24**, 1025–1037.

Wang, W., Dong, Z., Wang, T. and Zhang, G. (2006) The equilibrium gravel coverage of the deflated gobi above the Mogao Grottoes of Dunhuang, China. *Environmental Geology*, **50**, 1077–1083.

Wells, S.G., McFadden, L.D. and Dohrenwend, J.C. (1987) Influence of late Quaternary climatic changes on geomorphic and pedogenic processes on a desert piedmont, eastern Mojave Desert, California. *Quaternary Research*, **27**, 130–146.

Wells, S.G., Dohrenwend, J.C., McFadden, L.D. *et al.* (1985) Late Cenozoic landscape evolution on lava flow surfaces of the Cima volcanic field, Mojave Desert, California. *Geological Society of America Bulletin*, **96**, 1518–1529.

Wells, S.G., McFadden, L.D., Poths, J. and Olinger, C.T. (1995) Cosmogenic 3He surface-exposure dating of stone pavements: Implications for landscape evolution in deserts. *Geology*, **23**, 613–616.

Williams, S.H. and Zimbelman, J.R. (1994) Desert pavement evolution: an example of the role of sheetflood. *The Journal of Geology*, **102**, 243–248.

Wood, Y.A., Graham, R.C. and Wells, S.G. (2002) Surface mosaic map unit development for a desert pavement surface. *Journal of Arid Environments*, **52**, 305–317.

Wood, Y.A., Graham, R.C. and Wells, S.G. (2005) Surface control of desert pavement pedologic process and landscape function, Cima Volcanic Field, Mojave Desert, California. *Catena*, **59**, 205–230.

Yang, X., Rost, K.T., Lehmkuhl, F. *et al.* (2004) The evolution of dry lands in northern China and in the Republic of Mongolia since the Last Glacial Maximum. *Quaternary International*, 118–119, 69–85.

Young, M.H., McDonald, E.V., Caldwell, T.G. *et al.* (2004) Hydraulic properties of a desert soil chronosequence in the Mojave Desert, USA. *Vadose Zone Journal*, **3**, 956–963.

Zerboni, A. (2008) Holocene rock varnish on the Messak plateau (Libyan Sahara) Chronology of weathering processes. *Geomorphology*, **102**, 640–651.

Zohary, M. (1945) Outline of the vegetation in Wadi Araba. *The Journal of Ecology*, **32**, 204–213.

10
Slope systems

John Wainwright and Richard E. Brazier

10.1 Introduction

Outwith the major sand seas, hillslopes are ubiquitous in the arid zone. Even in some areas dominated by dunes, surface stabilization and evolution are affected by slope processes to a certain extent (e.g. Kidron, 1999; Kidron and Yair, 2001; Parsons *et al*., 2003). The understanding of slope systems is therefore fundamental to geomorphology in general and to arid-zone geomorphology in particular. While there have been a number of key texts relating to slopes over the last 40 years (e.g. Carson and Kirkby, 1972; Young, 1972; Parsons, 1988; Selby, 1993), most have been out of print for a long time, and more recent overviews from the tectonic perspective of landform evolution give little explicit reference to them (e.g. Burbank and Anderson, 2001). This omission is problematic, as the exact nature and rates of slope evolution have significant feedbacks on landscape evolution as a whole.

Slope systems need to be evaluated, as do other geomorphic systems, in terms of linkages between process and form. One issue in terms of slope form is the very wide range of potential slope shapes and configurations that can occur. Dalrymple, Blong and Conacher (1968) produced a useful descriptive approach based on nine types of slope unit (Figure 10.1). Their qualitative model allows any slope to be described in terms of these nine types, and the different types may be repeated in any order. Only plateau areas – type 1 – need to be present, and as Parsons (1988) points out, the flexibility of the model is also its limitation in terms of distinguishing any particular slope. It also suffers from being static and two-dimensional, thereby missing any important characteristics relating to catchment shape, and the feedbacks, for example, between drainage density, catchment response and channel (type 9) behaviour. One advantage of the approach, though, is that it does require consideration to be given in terms of linkages and boundary conditions between the different elements in a landscape.

More recent approaches have exploited digital elevation model (DEM) data to analyse the three-dimensional form of slopes and their landscape position in more detail (e.g. Pike, 2000), although they too suffer from an inability to account for landscape dynamics. The latter require a characterization of the land-forming processes and their spatiotemporal variability. The following chapter considered these dynamics in terms of runoff generation and overland-flow erosion processes, and it is useful to develop these dynamics further in terms of the patterns in which they occur on slopes, and are replaced or complemented by other slope-forming and modifying processes such as creep, mass movement and weathering (Dietrich and Dunne, 1978; Selby, 1993) (Chapter 6). A general process model for soil-covered and rock slopes is shown in Figure 10.2, showing some of the complexities of understanding slope systems and their evolution. This chapter will attempt to unravel some of these complexities in relation to specific types of slope system common in drylands.

10.1.1 Contexts of slope systems

The last two decades have seen a renaissance of what might be considered to be geological approaches to understanding landform evolution (e.g. Burbank and Anderson, 2001), in part linked into the development of Earth System Science (Wainwright, 2009a, and see Chapter 1). At the largest scale, the tectonic setting provides a fundamental control on the location of some arid areas, and the relative stability of intracratonic surfaces when compared to active continental margins or intermediate zones

Arid Zone Geomorphology: Process, Form and Change in Drylands, Third Edition. Edited by David S. G. Thomas

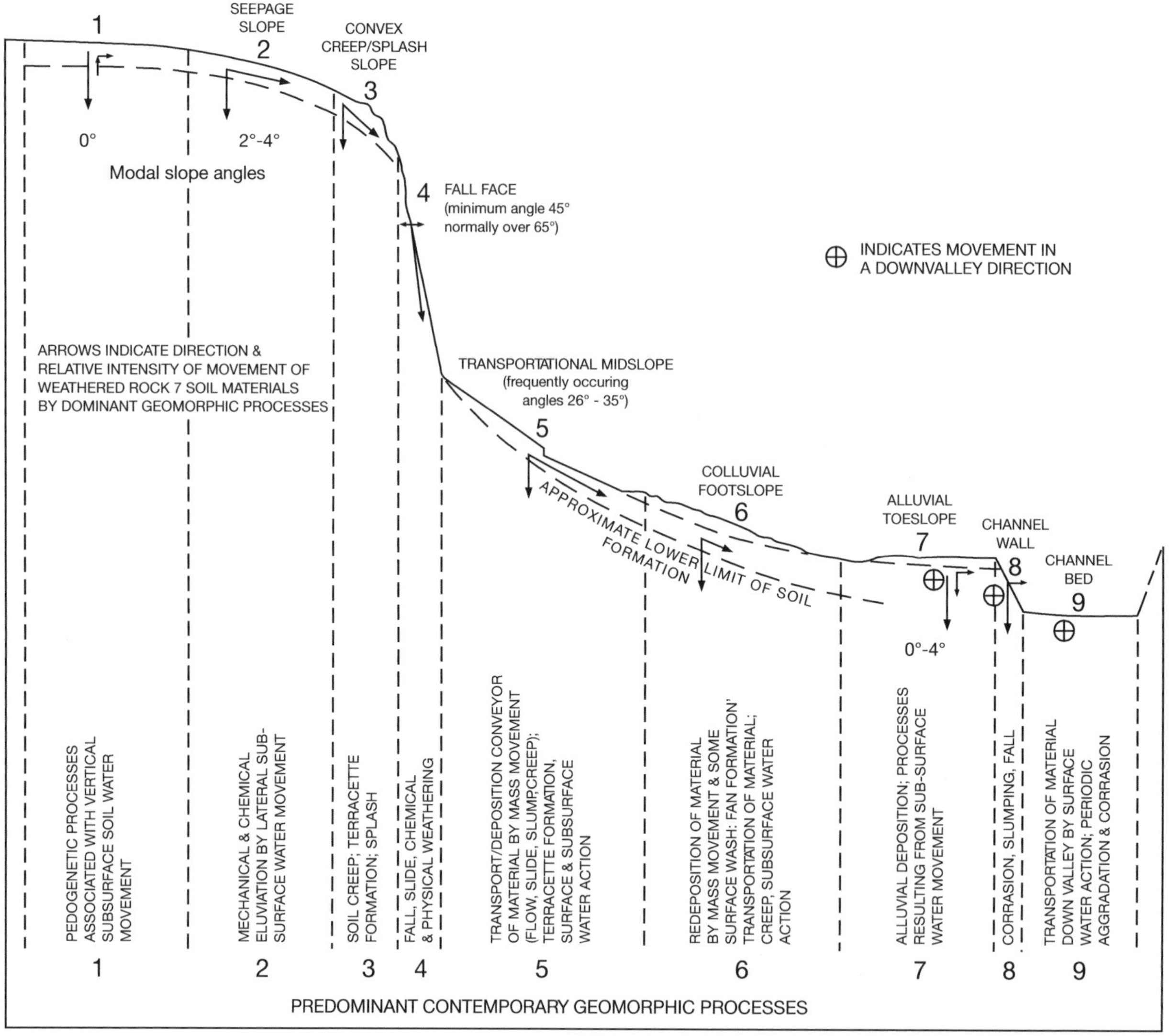

Figure 10.1 The nine-unit hillslope form model of Dalrymple, Blong and Conacher (1968) as modified by Parsons (1988) showing the different slope units, angles and dominant process mechanisms.

provides a further fundamental control in terms of vertical crust movements (Chapters 1 and 2). In the simplest sense, greater vertical movements generate steeper slopes and thus faster erosion rates, but they also cause landscapes to cross thresholds, so that other processes such as mass movements can become important. More recent studies by Willett (1999) have demonstrated that because of the feedbacks between erosion and compensatory isostatic uplift, the relative position of an arid zone to a mountain belt is important in controlling large scale erosion rates and landscape form. Thus, the location of drylands relative to orogenic zones and the position of the latter in relation to ocean currents and atmospheric circulation can lead to significant changes in the patterns and rates of landscape evolution. The tectonic structure of a landscape will control variability in evolution from the hillslope scale, e.g. as a result of the distribution of joints on rock faces, to regional scale as a result of faults and patterns of unloading affecting the material strength of bedrock.

At smaller scales, a geological framework is important because of the controls exerted by lithology on weathering and erosion processes. Of course, there is a further link here to tectonics, but also suggests that the past history of a continent can cause it to evolve in significantly different ways. Lithological variability is also significant over scales from kilometres to micrometres, with changes in rock type, bedding and particle arrangement all being potentially important sources of variation in material strength and thus ability to respond to the forces imposed by various slope-forming processes.

(a) soil-covered slope-erosion system

Soil
Mixing, weathering
Saprolite
Weathering
Bedrock
RIDGE
Creep, weathering
Creep, weathering
Soil
Mixing, weathering
Saprolite
Weathering
Bedrock
SIDE SLOPE
Splash, wash
Hollow
Dissolved load transport
Dissolved load discharge
Translational slide or flow
Suspended load transport
Suspended load discharge
Bed material attrition
Rill and gully
Sediment in tributary channels
Debris torrent, high stream discharge
Debris fan
Debris fan erosion
Sediment on main valley floor
Bed and bank erosion
Active sediment in main channel
Bedload transport
Bedload discharge
Transfer processes
Storage systems
Outputs

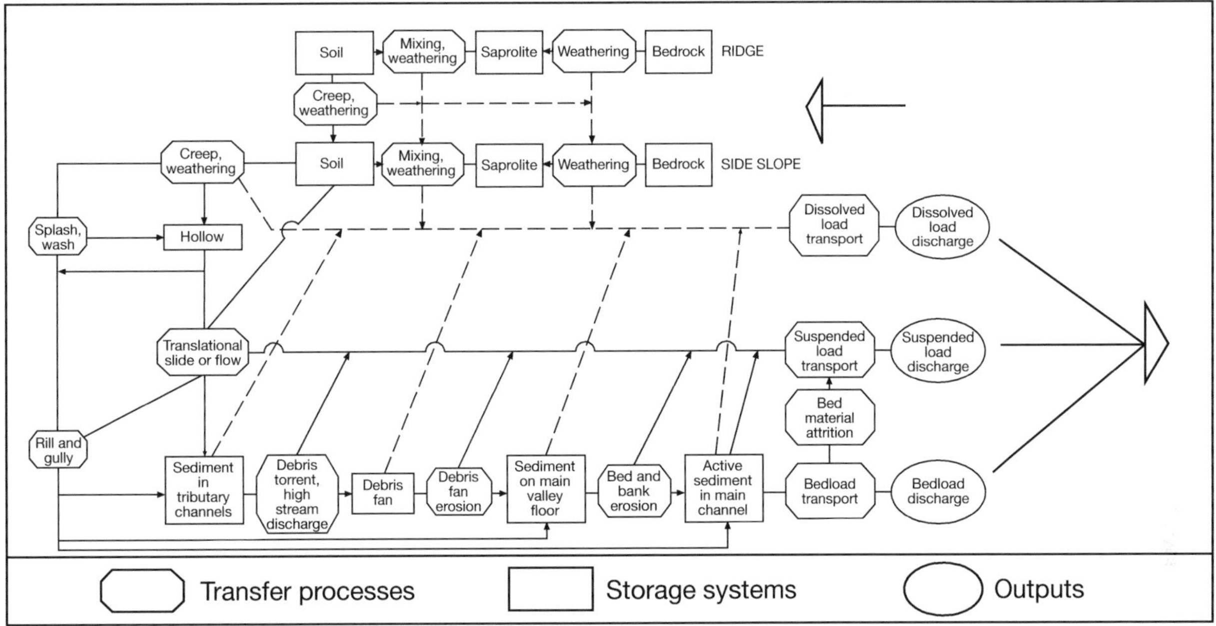

(b) rock-slope-erosion system

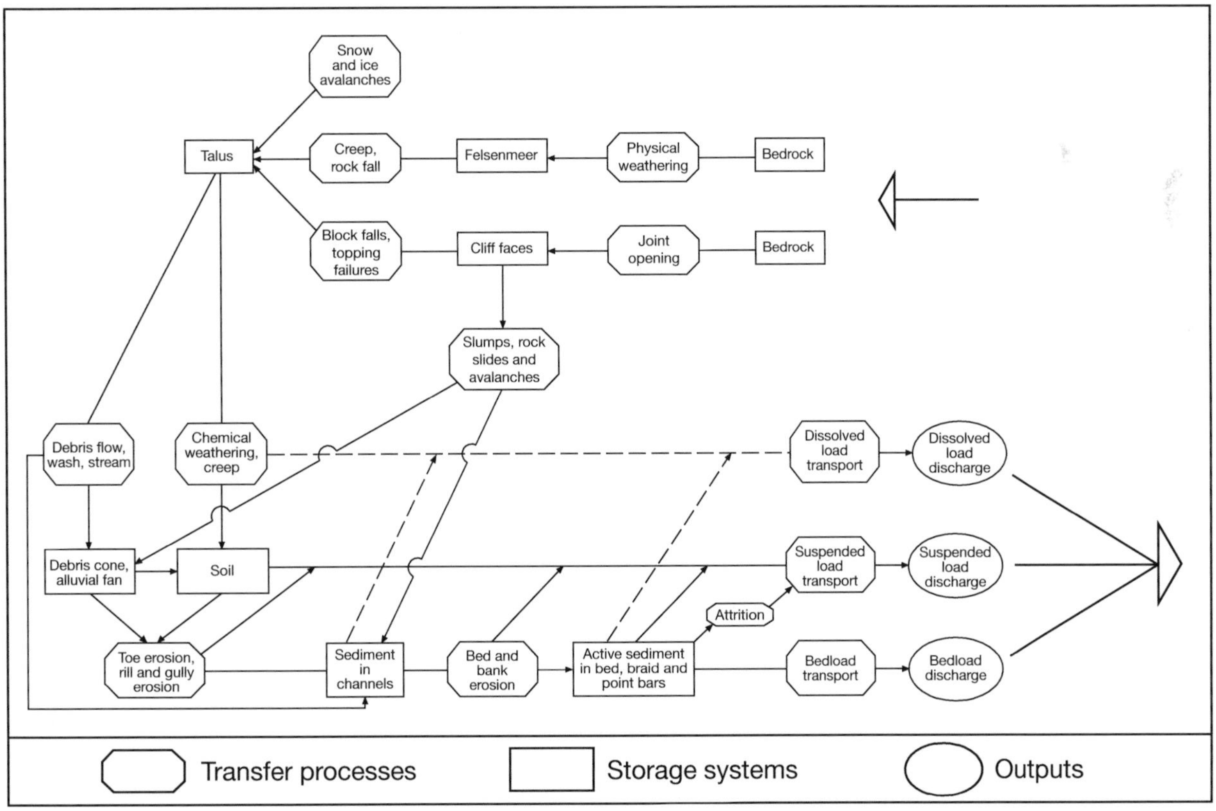

Snow and ice avalanches
Talus
Creep, rock fall
Felsenmeer
Physical weathering
Bedrock
Block falls, topping failures
Cliff faces
Joint opening
Bedrock
Slumps, rock slides and avalanches
Debris flow, wash, stream
Chemical weathering, creep
Dissolved load transport
Dissolved load discharge
Debris cone, alluvial fan
Soil
Suspended load transport
Suspended load discharge
Attrition
Toe erosion, rill and gully erosion
Sediment in channels
Bed and bank erosion
Active sediment in bed, braid and point bars
Bedload transport
Bedload discharge
Transfer processes
Storage systems
Outputs

Figure 10.2 A process-based approach to understanding slope systems: the store and transfer model of Dietrich and Dunne (1978) as modified by Selby (1993). The slope–erosion system for (a) soil-covered landscapes and (b) rock slopes.

A first-order control of these processes is due to climate and climate variability. The previous chapter discussed climate controls on the type and nature of runoff and erosion processes in drylands. Since the work of Langbein and Schumm (1958), semi-arid regions have been seen to be the dominant region where such processes are important, and more recent studies have shown that *ceteris paribus* this importance holds (e.g. Kosmas *et al.*, 1997). In drier conditions, less erosion seems to be a result of less cumulative available energy in storm events. More humid conditions seem to be limited by more complete vegetation cover. However, the ecogeomorphology, in particular the distribution and pattern of vegetation, is the other major first-order control of these processes in drylands. It is not possible to consider climate and vegetation separately, as they covary, certainly at spatial scales of more than a few tens of kilometres (Wainwright, 2009b). The nature and type of runoff and erosion processes are very closely controlled by vegetation quantity and distribution.

In general terms, at short timescales the evolution of slopes can be considered in relation to the effectiveness of climatically driven processes as modified by the vegetation. The specific behaviour of a particular location will be controlled by lithological variability and especially the relative weathering and erosion rates at a point, producing a first-order response in terms of whether the slopes are regolith- or rock-dominated. These behaviours are further controlled by boundary conditions of the responses of neighbouring slope units. At longer timescales, geological structure and context affect slope evolution and these controls are further modified by climatic changes. Inasmuch as a number of drylands had significantly different climates through the Pleistocene and earlier (see Chapters 3 and 4), it is also important to consider the inheritance of past landscape-forming processes on slope forms. These different controls are now considered in the context of two types of slope system – badlands and rock slopes – which are considered as stereotypical of drylands.

10.2 Badlands

Badlands are heavily dissected, strongly gullied landscapes, often with very sparse vegetation cover. The name apparently derives from early French colonists in northern America encountering terrain that was bad, or difficult, to cross. A number of North American badlands have become almost type localities – notably those in Dinosaur Provincial Park, Alberta, Canada (e.g. Bryan, Campbell and Yair, 1986) and the Henry Mountains (e.g. Gilbert, 1877; Howard, 2009). However, badlands are also common in Mediterranean environments (De Ploey, 1974, 1989; Wainwright and Thornes, 2003; Imeson, Kwaad and Verstraten, 1982) and in other drylands in Africa (e.g. Boardman *et al.*, 2003; Eriksson, Reuterswald and Christiansson, 2003; Feoli, Gallizia Vuerich and Woldu, 2002; Achten *et al.*, 2008), Chile (Maerker *et al.*, 2008), India (Joshi, Tambe and Dhawade, 2009) and China (Liu *et al.*, 1985). Badland topography has been considered to be a 'model landscape', but detailed investigations generally demonstrate complex process-form relationships, in some cases over small spatial scales (Figure 10.3, but note that the spatial scales in the figure are caricatures to a certain extent). Valley-bottom and hillslope gullies, extensive rilling, piping at a range of scales and mass movements all accompany unconcentrated overland-flow and splash erosion as important landscape-forming processes in badlands.

Most badlands are to be found on unconsolidated or weakly consolidated bedrock, particularly marls and recent alluvium, which can erode rapidly. Caution is required, though, in interpreting them as areas of ongoing areas of high erosion, as several studies have suggested that some badlands – or at least large areas within them – can be stable over millennia (Wise, Thornes and Gilman, 1982, but cf. Wainwright, 1994; Howard, 1997; Díaz-Hernández and Juliá, 2006). Evidence from badlands at the humid end of the dryland spectrum (or even in fully humid locations: Lam, 1977; Segerstrom, 1950; Hicks, Gomez and Trustrum, 2000; Parkner *et al.*, 2006) suggests that the lack of vegetation (e.g. on former industrial sites in Wales: Bridges and Harding, 1971; Haigh, 1979; and in the US: Schumm, 1956; and in unconsolidated tephras following volcanic eruptions in Japan: Lin and Oguichi, 1985) or the disturbance of vegetation (e.g. in subhumid locations in the French Prealps: Ballais, 1996; and following clearance phases on Easter Island: Mieth and Bork, 2005) are important in providing triggering conditions for badland formation. This interpretation is consistent with the vegetation–erosion interactions considered conceptually by Thornes (1985). There has been some suggestion that biological crusts and lichen (see following chapter) can provide stabilization mechanisms on some badland slopes (Alexander *et al.*, 1994; Loppi, Boscagli and De Dominicis, 2004). Once incision has occurred and slopes have become steep, it is less likely that vegetation will be able to recolonise. For example, Bochet, García-Fayos and Poesen (2009) considered plant covers and species richness on slopes of different aspects in eastern Spain (mean annual precipitation 373 mm) and found that north-facing slopes above 63° were unable to support vegetation. This threshold decreased to 50 and 46° for east- and west-facing slopes and was as low as 41° for south-facing slopes. These different thresholds can lead to strong asymmetries in badlands (Figure 10.4; see also Butler and Goetz, 1986, for a similar example in the

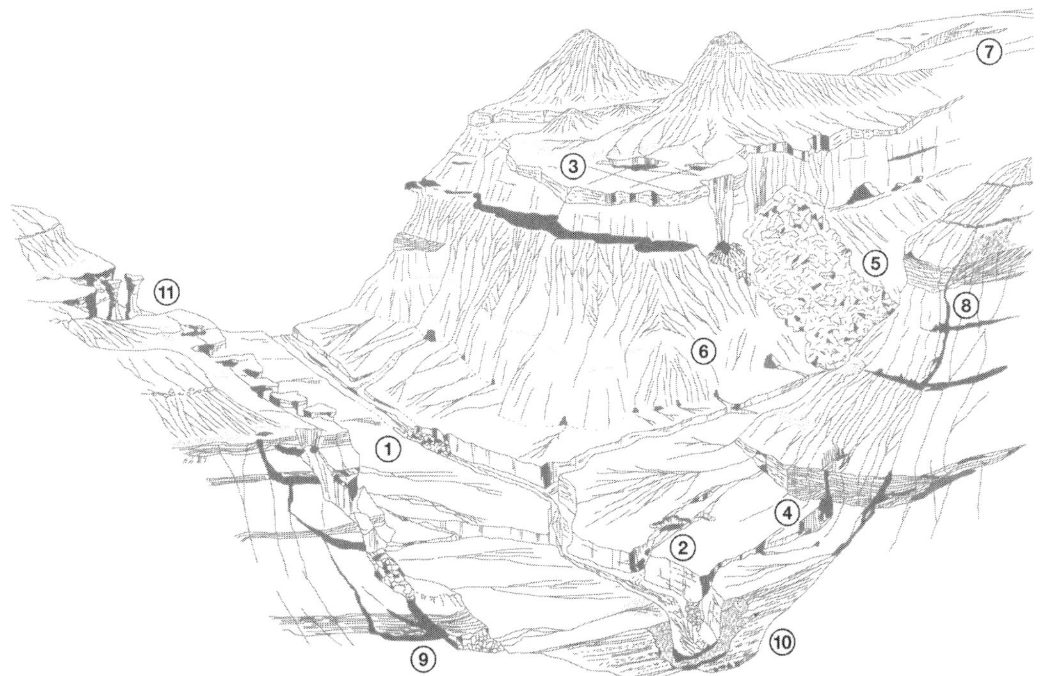

Figure 10.3 Composite image of a badland landscape showing the potential range and variability of landscape units and processes (cf. Figure 10.1) (Campbell, 1997): 1, headcuttting by valley-bottom gully in alluvial fill with accompanying slump and collapse by basal sapping; 2, pipe-induced collapse initiating discontinuous gully; 3, surface drainage into sinks and pipe shafts controlled by structural discontinuities; 4, valley-side gullies fed by convergent rill flow; 5, percoline-controlled large-scale piping generating slope collapse and triggering mass movements; 6, near-surface micropiping producing honeycomb erosion of slope surfaces by rill collapse; 7, gully heads expanding into undissected surface; 8, early stage of meso-scale pipe development; 9, mature meso-scale piping initiating slope collapse, large gullies and tributary formation; 10, multiple cut-and-fill gully deposits in bedrock-floored valley; and 11, hoodoos formed by dissection of resistant caprock.

Figure 10.4 The Tabernas Badlands in Almería, Spain, showing asymmetry of slope form relating to aspect and vegetation differences.

badlands of North Dakota). The results of Cerdà and García-Fayos (1997) suggest that slope-related differences may relate more to different erosional behaviour, rather than differences in runoff production. Regüés, Guàrdia and Gallart (2000) demonstrate an opposite asymmetry related to aspect, where the north-facing slopes in Pyrenean badlands have lower vegetation cover and higher erosion rates as a result of frost action in the winter. Once significant erosion episodes have taken place, the seedbank will be severely depleted or absent (Guàrdia and Recatalá, 1992; García-Fayos, García-Ventoso and Cerdà, 2000; Guàrdia, Gallart and Ninot, 2000), so the ability of plants to disperse seeds over long distances was identified by Bochet, García-Fayos and Poesen (2009) as a primary trait of colonising species. They also suggested that plants producing seeds with mucilaginous coatings that are able to bind to biological crusts also have a significant advantage in the colonization process. Some grass species in the Mediterranean seem to have adapted specially to colonising in these types of condition (Danin, 2004; Guàrdia, Raventós and Caswell, 2000). García-Fayos, García-Ventoso and Cerdà (2000) noted that the extreme badland environment – even in relatively temperate conditions in the Pyrenean foothills – leads to high rates of seedling mortality even once germination takes place. Pintado *et al.* (2005) point to higher lichen covers on north-facing slopes in the Tabernas badlands in southern Spain. This difference may be an important differential stabilisation mechanism here (e.g. Alexander *et al.* 1994). Once colonisation takes place and runoff and erosion are reduced, plants that reproduce vegetatively may take over (Bochet, García-Fayos and Poesen, 2009), although Guerrero-Campo, Palacio and Montserrat-Martí (2008) also found that plants with vegetative reproduction were best able to sustain themselves on actively eroding slopes.

As well as vegetation, Torri, Calzolari and Rodolfi (2000) identified lithology, structure and bedrock, climate and human activities as significant controls on badland formation and behaviour. Lithological controls may relate to regional variation – not least the localization of unconsolidated bedrocks – as well as to smaller-scale variability. Cerdà (2002) notes that badlands near Alacant in Spain may be forming as a result of diapiric behaviour of the clays in the bedrock, leading to incision of the main river channel and thus gullying and badland development. Salt diapirs near to the *terres noires* badlands studied by Wainwright (1996a) in southern France (Figure 10.5) may have a similar effect. At Cerdà's study site, the behaviour of surfaces varies spatially as a function of whether marl, clay or sandy bedrocks are at the surface. The former two surfaces undergo shrink–swell changes as a result of wetting and drying cycles, producing heavily cracked conditions, especially in the dry summer months (Figure 10.6). As a result, ponding and runoff occur more rapidly, and runoff coefficients are higher in autumn, when the surfaces are less cracked. The marls produced most runoff, followed by the sands, due to the presence of a surface crust (see the

Figure 10.5 Localized badlands near to Propiac, Drôme, France, showing a small runoff event in 1993 with well-developed rill flows (Wainwright, 1996b). Note the control of lithology – the alluvial deposits in the foreground are used for vine cultivation but also support trees.

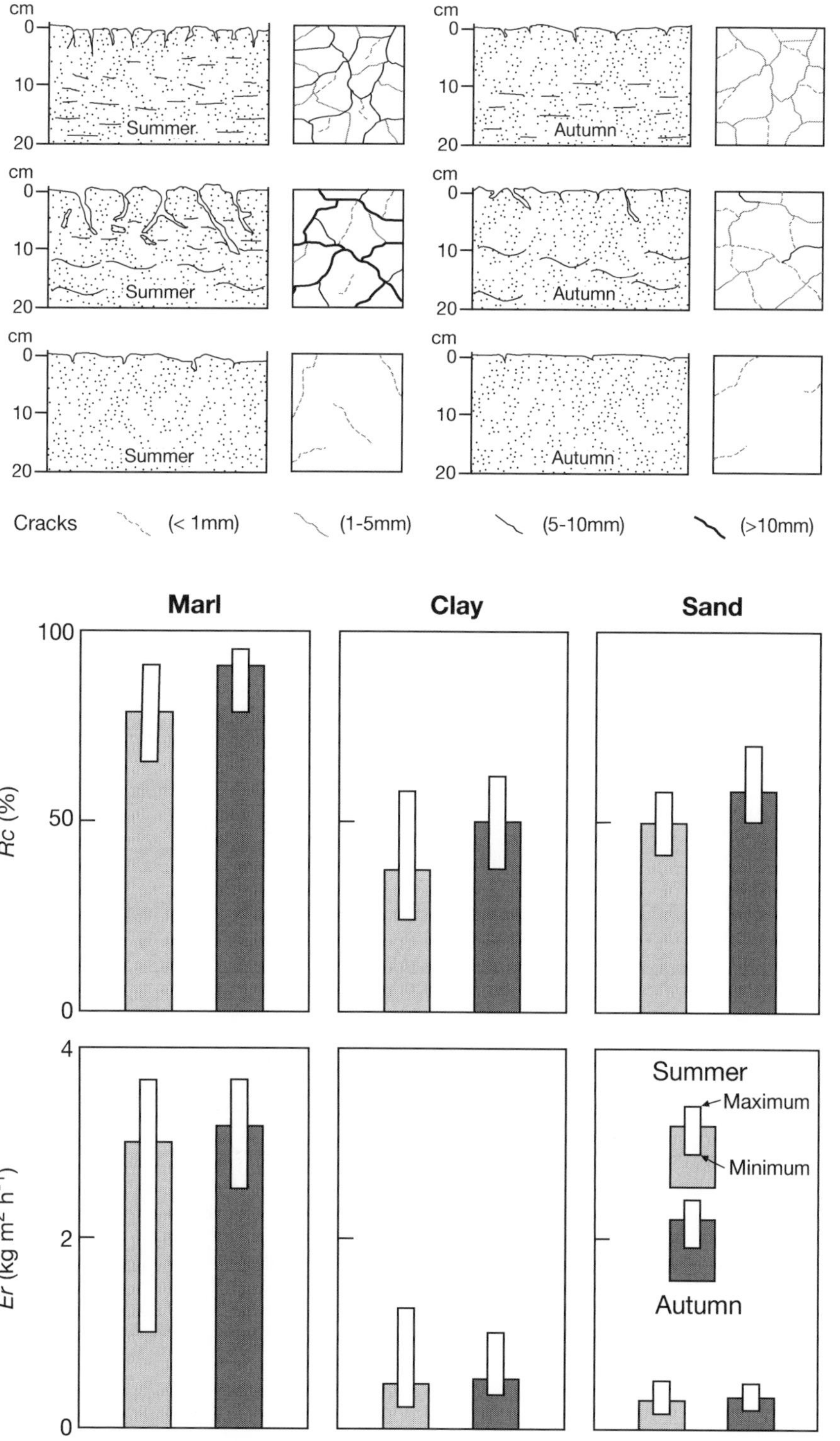

Figure 10.6 Surface and soil-profile morphological changes as a result of different patterns of shrink–swell behaviour on marl, clay and sand bedrocks in badlands near to Alacant, Spain, and resulting lithological and seasonal differences in infiltration and erosion response (Cerdà, 2002).

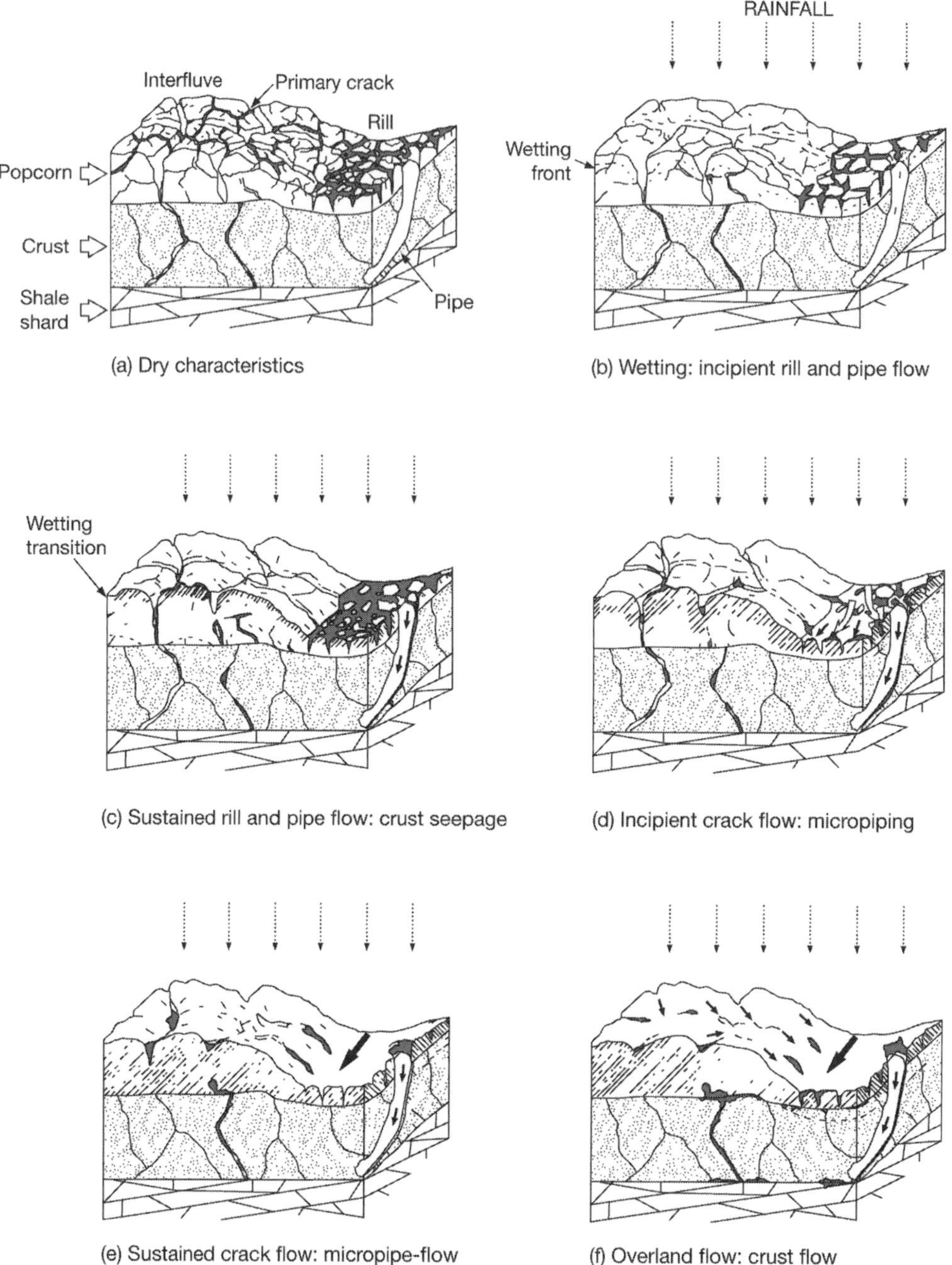

Figure 10.7 Successive stages of surface wetting and surface change and their interactions with runoff development on shale badlands (from work on the Dinosaur Badlands, Alberta, Canada by Hodges and Bryan, 1982).

previous chapter). Conversely, the sands were less eroded than the clays, in part because of the crust, in part due to density differences and in part due to lower aggregate stabilities in the clay.

Interbedding of different lithologies can produce significantly different runoff and erosion responses over very short spatial scales, leading to the development of complex slope forms in badlands (Figure 10.7). Campbell (1997) produced a generic model of slope response, where sand-dominated bedrock surfaces pass from splash- to interrill- and then rill-dominated surfaces, often within a space of a few tens of centimetres to metres. Bentonitic mudstones, on the other hand, are dominated by piping and discontinuous rills (often formed by pipe collapse where they are not fed from upslope rills) with localised mudflows. Shrink–swell processes lead to the development of a typical 'popcorn' surface. Interbedding may lead to oscillations between these types before leading to a basal pediment, perhaps with small fans at the break in slope and distributary channels building up a surface of very thinly bedded units. At Dinosaur Provincial Park in Alberta, Kasanin-Grubin and Bryan (2007) evaluated differences between rills on mudrock and sandstone bedrocks. They found that while rill systems on the latter remained fairly static over a two-year period, those on mudrock varied significantly, both in terms of morphology and the extent to which a 'popcorn' layer was present. Structural controls such as the locations of tectonic joints and bedding planes have been demonstrated as being significant in controlling the position and magnitude of soil pipes by both Torri and Bryan (1997) and Farifteh and Soeters (1999) (Figure 10.8).

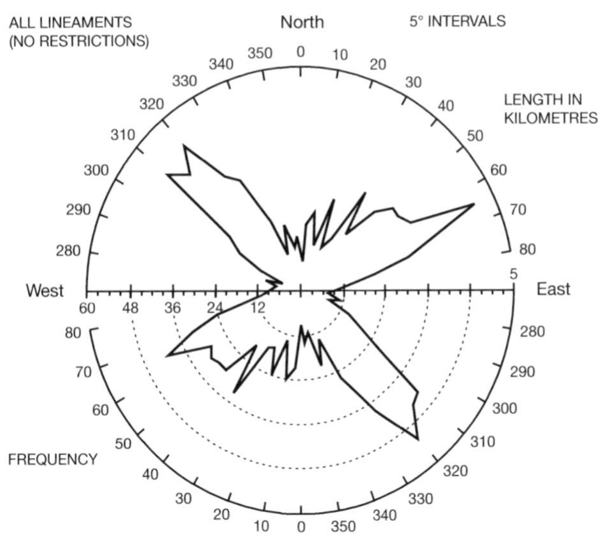

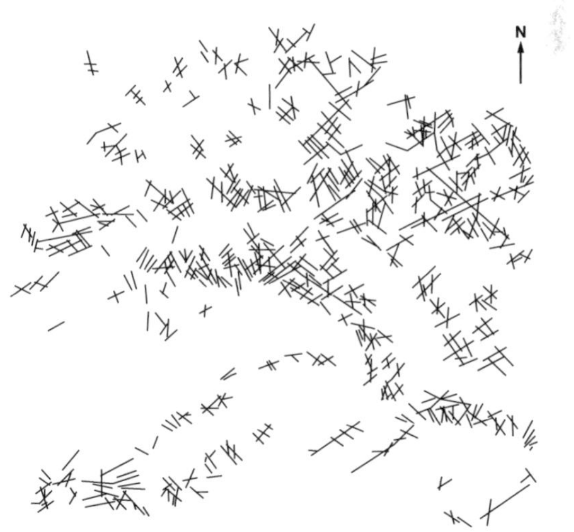

Figure 10.8 Rose diagram and spatial pattern of lineations passing through soil pipes in badlands in Basilicata, Italy, demonstrating the occurrence of preferential spatial patterns related to structural controls at the landscape scale (after Farifteh and Soeters, 1999).

These same controls are also important in controlling the style and extent of mass movements. In areas where the bedrock is thinly bedded and dipping, shallow translational slides can be significant (see the discussion in Howard, 2009), and the orientation of the dip relative to gullies can lead to distinct asymmetries in morphology, as bedrock dipping with the side slope is much more likely to fail than bedrock dipping into the side slope. That gullies may pick lines of weakness following dipping bedrock means this pattern is not uncommon. Farifteh and Soeters (2006) suggested that the size and spacing of joints and faults in basilicata was responsible for the patterns of biancane (small, isolated) rather than calanchi (large, gullied) badland systems. Griffiths *et al.* (2005) noted that landslides in badlands in the Aguas basin in southern Spain were more likely to occur on specific lithological boundaries (e.g. 39 % of all landslides where multiple lithologies were present occurred where the lower unit was a calcareous mudstone). Elsewhere in their study area, major landslides were found in areas of anomalously rapid uplift.

Climate has been mentioned above in the ways it controls aspect-related differences in badland evolution through both insolation and frost action, as moderated through the action of vegetation. The ability of different types of vegetation to be sustained in different climatic settings also has a significant impact on the development of badlands in drylands. Climatic variability over longer timescales can have a significant impact on badland formation. Bryan, Campbell and Yair (1986) suggested that in the Dinosaur Provinical Park in Alberta, badland evolution has followed broad phases of incision and filling relating to relatively wetter and drier phases through the Holocene (see also Evans, Campbell and Lemmen, 2004, and similar arguments for the evolution of southern French badlands in Descroix and Gautier, 2002), although Evans (2000) has demonstrated that this landscape is also strongly controlled by inherited features from past glacial and periglacial processes. Initial formation of the South Dakota badlands seems to have been triggered by gullying as a response to base-level change (Mather, Stokes and Griffiths, 2002; Howard, 2009), and base-level change has also been implicated in Spanish gully systems (Griffiths *et al.*, 2005), so that large-scale climate variability can also be seen to be important. However, gully formation and evolution is a complex process that may or may not indicate climatic control (Cooke and Reeves, 1976).

On much shorter timescales, badlands typically respond rapidly to the extreme events that characterise drylands (see Chapter 11). Wainwright (1996a) used rainfall-simulation experiments to demonstrate that the high rates of runoff production on *Terres Noires* badlands in southern France during high-intensity rainfall correspond to rapid decreases in the surface friction, leading to high velocity flows. These conditions lead to concentrated flows for much of the slope length, and thus high erosion rates. In rainfall-simulation experiments on similar badlands, Oostwoud Wijdenes and Ergenzinger (2003) generated miniature débris flows. A subsequent modelling study (Wainwright, 1996b) suggested that the small area of badlands (when combined with relatively small agricultural areas) were responsible for disproportionate amounts of runoff and erosion entering the channel system in the catastrophic floods at Vaison-la-Romaine in 1992. The low friction and steep slopes also meant that the badlands typically produced runoff much earlier in the storm than surrounding areas.

Likewise, the monitoring results of Cantón *et al.* (2001) suggested that the 5 % of storm events in which more than 80 mm of rain fell produced around 35 % of all the sediment exported in a small catchment in the Tabernas badlands over a two-year period. Francke *et al.* (2008) found that badlands in northern Spain similarly produced very high sediment yields in intense late summer storms, but had a much more moderate response in lower intensity storms in the autumn, although García-Ruiz *et al.* (2008) found less variability in subhumid badlands in the same region. Torri *et al.* (1999) explained the effects of high rainfall intensities as being due to rapid mechanical crusting, producing a rapid decrease in infiltration rates (Chapter 11), together with the same sort of decrease in surface roughness observed by Wainwright (1996a). For the soils of Torri *et al.*, the dominant effect on infiltration and thus runoff occurred once a cumulative rainfall energy of about 0.5 kJ/m^2 had been reached. These studies all point to important variations relative to when a particular storm occurred. Kasanin-Grubin and Bryan (2007) took this idea further in demonstrating that the seasonal dynamics of rills in Alberta, as discussed above, relate to changes in surface conditions that are a function of the duration and intensity of rainfall input. Short wetting cycles lead to 'popcorn'-type surfaces on smectite-rich mudstones, while longer wetting cycles will produce a breakdown of aggregates and thus much smoother surfaces (Figure 10.9). Different patterns of storms in different years could

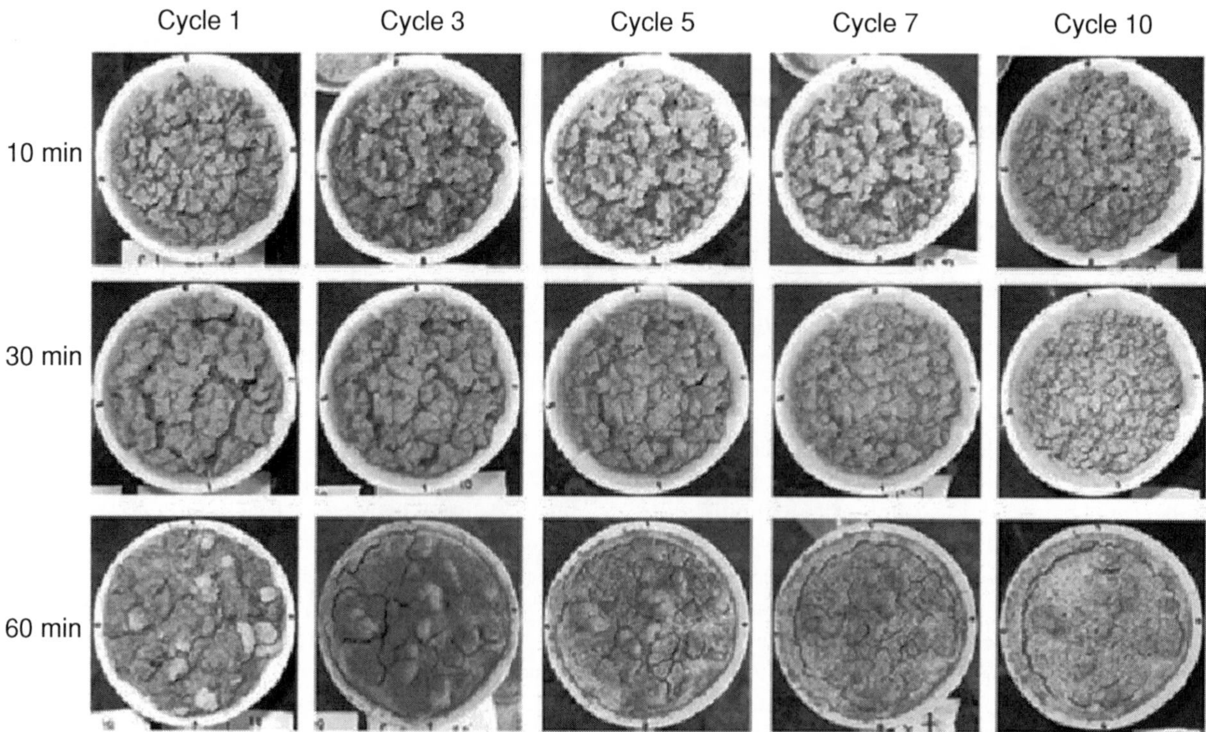

Figure 10.9 Experimental weathering experiment showing the evolution of a smectite-rich mudrock from the Dinosaur Provincial Park, Alberta, during different cycles of application of simulated rainfall. The rainfall was at an intensity of 45 mm/h and the figure shows outcomes of repeated cycles of 10-, 30- and 60-minute durations (from Kasanin-Grubin and Bryan, 2007).

thus explain variability of response in subsequent events of the same magnitude.

A final major control on badland formation and evolution is that of human activity. Gullying as a result of vegetation clearance has been suggested as a triggering mechanism for the subsequent spread of erosion at times ranging from prehistoric to the present, over a wide geographical range (Italy: Torri *et al.*, 1999; Greece: King and Sturdy, 1994; France: Ballais, 1996; Spain: Castro *et al.*, 1998; South Africa: Boardman *et al.*, 2003; Easter Island: Mieth and Bork, 2005; Tanzania: Eriksson, Reuterswald and Christiansson, 2003). As noted above, gullying can have multiple causes, and in some areas more detailed landscape-history reconstructions have shown the initial phase of erosion may not have been anthropic, even if it was subsequently exacerbated by human action (e.g. in North Dakota: Gonzalez, 2001). There has also been much emphasis on badland reclamation in recent years in Mediterranean Europe, largely as a consequence of agricultural subsidies, e.g. using mechanical levelling (Phillips, 1998a; Clarke and Rendell, 2000; Capolongo *et al.*, 2008) or various forms of green engineering (Rey, 2009). Such approaches have not often been successful, even at fully reestablishing native vegetation to stabilise slopes at the humid end of the spectrum (Vallauri, Aronson and Barbero, 2002), and have often led to significant increases in surface-erosion rates (Piccarreta *et al.*, 2006; Robinson and Phillips, 2001), piping (Romero Díaz *et al.*, 2007) and mass movements (Linares *et al.*, 2002). In several cases, opposition to this reclamation has been on the basis that badlands form a significant part of the aesthetic and cultural landscape (Phillips, 1998b, 1998c, and the discussion in Wainwright and Thornes, 2003).

10.2.1 Processes and rates of badland evolution

Because of the high perceived rates of badland evolution, there has been much attention on quantification of these rates. Comparison of plot- and catchment-based rates of overland flow is problematic because they are highly scale-dependent in a nonlinear way (Parsons *et al.*, 2004, 2006). Point measurements of surface lowering using erosion pins may be less problematic in this respect, and show rates ranging from 0.7 mm/a in caprocks to 77 mm/a in slope-foot settings (Table 10.1). However, caution must still be applied when interpreting these data, as measurements may tend to be focused where there is measurable erosion, leading to bias, and some surface change may be due to surface expansion and contraction, especially in smectite-rich mudstone bedrocks. Several of the studies in the table also show areas of net deposition in badlands, especially on micropediments and in the development of alluvial valley bottoms (Figure 10.3). Gullying has been noted above to be an important process in the initiation of badlands, but is also a significant source for sediment production (Kirkby and Bull, 2000; Nogueras *et al.*, 2000). The review of Poesen *et al.* (2003) suggests gully erosion may make up anything between 10 and 94 % of total catchment erosion, although none of their figures are explicitly derived from badlands, so it is hard to be more specific here. As noted experimentally by Kosov *et al.* (1978, cited in Sidorchuk, 1999), during the first 5 % of the lifetime of a gully, it will typically have developed >90 % of its length, 85 % of its depth, 60 % of its area and 35 % of its volume (Figure 10.10). After about 15 % of its lifetime, the length has stopped extending (usually by running out of a catchment area sufficient to provide enough concentrated flow: Faulkner, 1974; Thornes, 1985; Kemp née Marchington, 1990), the depth is nearly 90 %, the area 85 % and the volume about 55 % of the final amount. These nonlinear patterns of evolution may go a long way in explaining the apparent long-term stability of some gullied badland systems, as noted above. They are also consistent with the observation that even if the Zin badlands in the Negev have at times eroded more rapidly, the mean rate of ground lowering is equivalent to 0.75 mm/a over a period of 75 ka (Yair, Goldberg and Brimer, 1982).

Piping is extensively found on clay-rich bedrocks (e.g. Farifteh and Soeters, 1999; Torri and Bryan, 1997; but cf. Howard, 2009). Pipes are hydraulically efficient and erode rapidly, both by mechanical erosion and dissolution (hence the alternative name of 'pseudokarst': Halliday, 2007). Direct measurements on badland pipe flows are sparse, but in the Chinese loess belt, Zhu, Luk and Cai (2002) showed that for 65 % of storm events, pipe flows preceded the main runoff peak, and even for short storms, pipe flows could peak up to 30 minutes before the main subbasin response. They found a wide range of sediment concentrations in the pipe flows, with a maximum of 893.2 g/L. While some pipes are relatively ephemeral features and may form, be destroyed and reform on a storm-to-storm or seasonal basis – especially beneath 'popcorn'-type surfaces – larger, structural pipes may be much more permanent features of badlands. All sizes of pipe are liable to eventual collapse as progressive erosion leads to enlargement and thus instability of the pipe roof. Pipe collapse is thus implicated in rill and gully initiation. This process is likely to be cyclical, as the new rills and gullies will lead to a drop in the perched watertable during a storm event and thus the formation of new pipes following lower lines of weakness in the bedrock. Pipe development along vertical joints may also be significant in producing concentrated water flows to the head of steep side slopes,

Table 10.1 Surface-lowering rates on badlands as estimated using erosion pins.

Site	*Mean (or range) surface lowering (mm)*	*Standard deviation (mm)*	n	*Reference*
Central Karoo, South Africa				
Interfluves	5.6	2.0	10	Keay-Bright and Boardman (2009)
Channels	2.6	5.7	10	
Sidewalls	16.7	9.6	7	
Footslopes	4.7	4.4	10	
Tuscany, Italy				
Biancane	60			Della Seta *et al.* (2009)
'Embryonic'	77			
Slope foot	74			
Pediment	2.8			
Rill	26			
Calanchi	55			
Interrill without caprock	0.7			
Caprock ridge				
Caprock				
South Tuscany, Italy				
Calanchi	20			Ciccacci *et al.* (2008)
Biancane	15			
Basilicata, Italy				
Biancane	9.1	2.2	10	Clarke and Rendell (2006)
Massa Abate	11.3	4.6	32	
Pozzo Varisana	14.7	5.0	72	
Serra Pizzuta	18.7	4.6	15	
Massa Soldano				
Basilicata, Italy				
Calanchi	9.7	1.7	20	Clarke and Rendell (2006)
Serra Pizzuta	7.2	3.6	15	
Mesola della Zazzara	9.5	2.2	14	
Massa Soldano				
Basilicata, Italy				
Calanchi	5.3–13.6			Alexander (1982)
Biancane	22.8–39.7			
Drôme, France	2–14		36	Wainwright (1996a)
Hautes Alpes, France	10–20			Bufalo, Oliveros and Quélennec (1989)
Vallcebre, Pyrenees, Spain	9			Solé *et al.* (1992)
Ebro basin, Spain	5–9			Benito, Gutiérrez and Sancho (1992)
Ebro basin, Spain	1		377	Desir and Marín (2007)
Tabernas, Almería, Spain	12			Lázaro *et al.* (2008)
Ugijar, Granada, Spain	6–27			Scoging (1982)
Dinosaur Provincial Park, Alberta, Canada	3.4	2.2		Campbell (1982)
Badlands National Monument, South Dakota, USA	22.7	5.0		Schumm (1962)

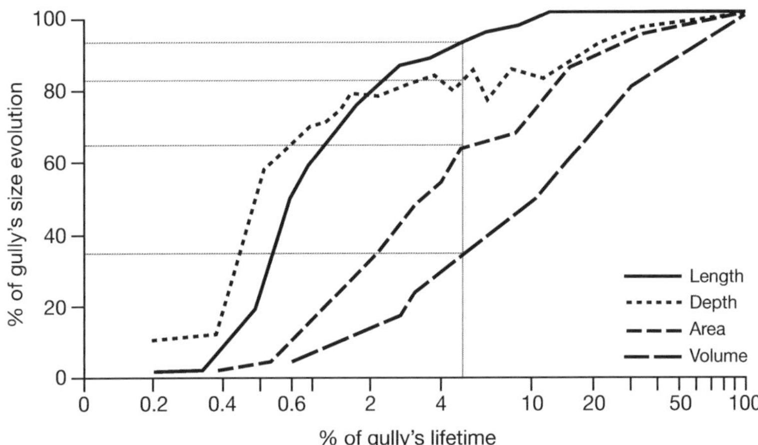

Figure 10.10 Gully evolution based on the experimental data of Kosov *et al.* (1978) cited in Sidorchuk (1999), showing the relative rates of evolution of the different morphological characteristics relative to the lifetime of the gully. Note the different extents at 5 % of the gully lifetime.

thus reducing their stability and leading to the triggering of large mass movements. At a smaller scale, Collison (1996) demonstrated that this mechanism can also be important in the collapse and retreat of gully headwalls.

As well as the small translational mass movements noted above, badlands can produce large-scale landslides as well. The best-studied case is that of Super Sauze at Barcelonette in the *Terres Noires* of the French Alps (Maquaire *et al.*, 2003). This style of landslide is typically triggered by extended periods of rain (Flageollet *et al.*, 1999) and so is typically restricted to the more humid end of the badland spectrum. Initially triggered as a series of large block falls and translational slides in the 1960s, it has continued to develop as a series of earthflows (Flageollet, Malet and Maquaire, 2000; Maquaire *et al.*, 2003). The chaotic surfaces produced by the block falls have been smoothed by the flows and subsequent rainfall events have superimposed a new dendritic gully drainage on the surface. Flageollet *et al.* (1999) demonstrated the importance of older gully systems in producing discontinuities that promoted the initial instability and Maquaire *et al.* (2003) showed that the material strength may take up to 150 years to recover. This evidence suggests a cyclicity in dominant processes in this setting, which is not inconsistent with that interpreted for the much drier Henry Mountains badlands by Howard (1997). Mass movements have been noted on large scales in badlands in Alberta (De Lugt and Campbell, 1992), Italy (Piccarreta *et al.*, 2006; Ciccacci *et al.*, 2008) and Spain (Griffiths *et al.*, 2005). As well as contributing to large-scale evolution, mudflows have also been demonstrated in extreme events on a within-slope scale (Oostwoud Wijdenes and Ergenzinger, 1998; Ciccacci *et al.*, 2008; Godfrey, Everitt and Martín Duque, 2008).

While not a dominant process in most badlands, Godfrey (1997) has demonstrated the local importance of wind erosion in the Mancos Shale badlands in Utah. Near ridge crests, he found that local acceleration was able to provide sufficient lift force to entrain surface crusts. This 'vacuuming' of the crust could further develop microcirques up to 1 m tall and 3 m wide along ridges, and in some exposed cases cliffs of up to 10 m tall.

In summary, despite their superficial simplicity, slope systems in badlands evolve according to a complex set of interrelated processes over a range of timescales. Comparison of Figures 10.1 and 10.3 suggest that the slope forms that evolve lead to various interactions in terms of boundary conditions set up between the dominant slopes and processes operating on different slope units. For this reason, a holistic approach, including modelling (Howard, 1997), is important for understanding badland evolution. As noted in Chapter 11, there is an increasing focus on ideas of connectivity for understanding the large-scale behaviour and evolution of hydrologic and geomorphic systems, and badlands are no different in this respect. Faulkner (2008) has developed these ideas most fully, differentiating between meso-scale, closed-system and macro-scale, in open-system explanations. In the former, connectivity first increases as erosional links between parts of the system increase, and then decreases again as slope angles decrease and depositional processes become more important. However, as noted above, connectivity between landscape elements (e.g. through pipe development and destruction from crust weathering) may be more dynamic over shorter timescales as well (see also Faulkner, Alexander and Zukowskyj, 2008, and Della Seta *et al.*, 2009, for further examples in this context). Similarly, the Super Sauze earthflow example suggests cyclicity in the development of chaotic, disconnected surfaces, upon which connected drainage patterns redevelop (Maquaire *et al.*, 2003). Discontinuity of flows in drylands has long been recognised as a way of producing disconnected systems and cycles of erosion and deposition (e.g. Schumm, 1973). Such cycles have been demonstrated on timescales from annual to decadal on badland pediments and over several centuries in the main channel system in the Mancos Shale badlands by Godfrey, Everitt and Martín Duque (2008). The macro-scale explanation uses variations in climate and tectonics to evaluate how these internal processes might be affected over different timescales. This approach thus ties into the discussion above about tectonic and climatic impacts on badlands, but also other elements of landscape interconnectivity such as river capture (e.g. Mather, Stokes and Griffiths, 2002; Griffiths *et al.*, 2005).

10.3 Rock slopes

Rock slopes occur where the geological substrate is exposed via processes such as water and wind erosion (of soils) and mass movement (of parent material). The relative speed of erosion relative to weathering processes in drylands (Chapter 6) results in these conditions occurring frequently. In arid zones, such geology is often sedimentary in nature, overlying (or underlying) older and deeper crystalline bedrock, as in the slick-rock, sandstone slopes of the southwestern United States (Figure 10.11).

Figure 10.11 Slick rock sandstone of the Entrada series in South western USA.

Rock slopes are found in the shield-platform deserts of the world (see Mabbutt's 1977 classification), across Africa, Arabia, Australia and India. Their form is characterised by retreating cuestas, mesas and buttes, with rate of formation being controlled by the geological form of the dominant bedrock in interaction with the rate of surface weathering and mass movement processes. Herein, rock slopes are categorised as (1) gravity-controlled, where slopes are steep and slope failure is typified by rockfalls leading to predominantly bare profiles or (2) rock (débris)-mantled slopes that are lower in gradient and are typically controlled by both mass movement and erosional processes. In many dryland locations, these different slope units occur as a continuum of slope evolution with steep, bare rock slopes occupying the upper section of a hillslope, near the divide or free face of a cuesta and rock or debris-mantled slopes occupying a mid-slope location, where debris has accumulated. Lower slope sections are often overlain by soils, which are not dealt with here (see Chapters 6 and 11).

Rock slopes are characterised by a range of landforms including (1) bare rock or slick-rock slopes, (2) inselbergs/bornhardts and (3) substrate ramps, all of which tend to form beneath the harder caprocks or cuestas mentioned above. Numerous authors argue that rates of formation of these landforms may vary between similar strata, due to the control exerted by the overlying caprocks, which either protect the rock slopes from weathering or enhance the processes of weathering that lead to distinctive landform development.

10.3.1 Bare rock or slick-rock slopes

Massively bedded or unfaulted bedrock is common in arid zones and can form into very durable slopes that weather at slow rates and are characterised by their clean faces of varying steepness. The strength of such slopes has been estimated using the Rock Mass Strength classification (Selby, 1982), which illustrates that rocks with lower mass strengths weather to lower gradients compared to those with high mass strengths. Clearly, in sedimentary exposures, variability of mass strength may be evident. Such variability is thought to control the rate of weathering and can lead to distinct changes in the gradients of bare rock slopes across sedimentary boundaries or bedding planes in the underlying bedrock.

10.3.1.1 Inselbergs/bornhardts

Highly durable rocks found in arid zones such as granites tend to weather via exfoliation, whereby layers (akin to the

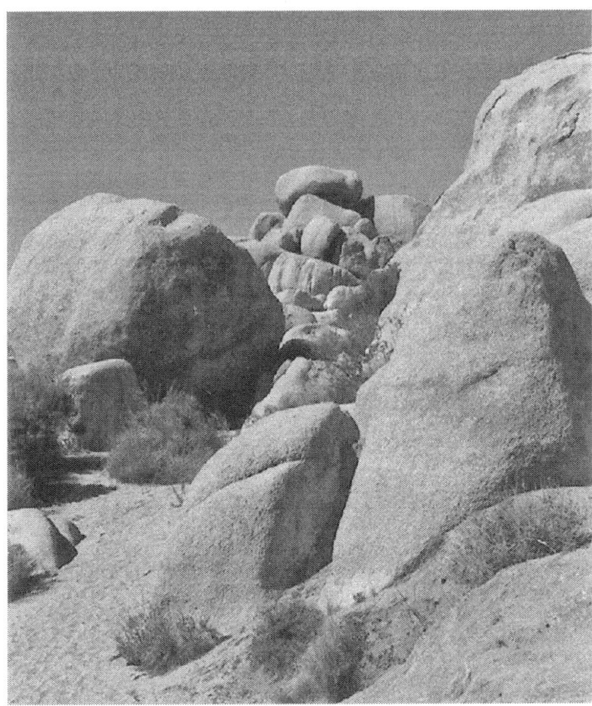

Figure 10.12 Granite inselbergs.

skin of an onion) peel away from the bedrock slope to reveal fresh surfaces (see Figure 10.12). This process, often termed 'sheeting', leads to domed formations, with convex slope forms and very abrupt changes in gradient with the surrounding landscape. Numerous hypotheses exist to describe the process of sheeting and how it leads to the formation of inselbergs. These hypotheses are based on either (1) rock-formation processes – planes are formed as the batholiths intrude into overlying material or crust material is compressed to form concentric and parallel faults within the granite – or (2) post-formation processes – stress releases after formation and secondary shearing or faulting within the massive bedrock form concentric layers of weakness. It is likely that these processes are not mutually exclusive and can operate alongside each other depending on the characteristics of the bedrock. For example, granite batholiths that exhibit stress-release sheeting (from internal stresses within the bedrock) may also be recovering from the removal of overlying material that has been eroded, thus unloading the slopes and allowing them to expand due to lower compression from the (now absent) overlying material.

The bornhardts of East Africa (named after the German geologist Willhelm Bornhardt) show classic dome-shaped forms, within massively bedded igneous rocks. However, similar forms are also evident in fine-grained sandstones and other sedimentary rocks, which have not been subject to the same formation mechanisms and must therefore be formed by different processes. The formation of Ayers Rock in Australia is a resistant sandstone that has survived weathering processes that have lowered the surrounding landscape to leave the domes clearly elevated from the plains. Thus, the key characteristics of such dome-like formations in arid zones are perhaps that they are all, regardless of the processes that formed them, harder rocks than those found in the local landscape and have thus been able to resist weathering since their formation.

10.3.1.2 Substrate ramps

Substrate ramps are formed in erodible strata at the base of caprocks and are often characterised by a distinct change in gradient from the near-vertical caprock to the 30–40° ramp, where the caprock is still present (Figure 10.13). Upslope or headward extension of these features is either due to fluvial processes of basal sapping or erosion, which undermine caprock and lead to failure and collapse or the effect of gravity on the vertical caprock (perhaps in combination with freeze and thaw action), which will also promote collapse. Such collapse may result in the extension of a debris-covered substrate ramp, covered by large amounts of talus or talus ledges (see below). Substrate ramps may be formed in highly erodible material, such as limestones or sandstones, which are prone to erosion by sheet wash and dissolution and may therefore become gullied or highly dissected by overland flow. Figure 10.14 illustrates the various forms of rock slopes and shows how substrate ramps often intersect with the more durable caprocks well below the margin between the two strata.

10.3.1.3 Rock-slope erosion features

The process of sheeting, which exfoliates layers of rock to form dome-like rock slopes, is described above. In parallel to sheeting, it is also common for bare rock slopes to be weathered via fluvial processes, which sculpt slopes via dissolution of the bedrock, a process that can be accelerated with slightly acidic precipitation in contact with calcareous, sedimentary bedrock.

In addition, wind action leads to the erosion of bare rock slopes, in areas such as the sandstone Colorado plateau, southwestern USA. The work of Loope *et al.* (2008) discusses the importance of the wind-sculpting of bare rock slopes, leading to the formation of a variety of features, which hitherto may have been interpreted as fluvial landforms.

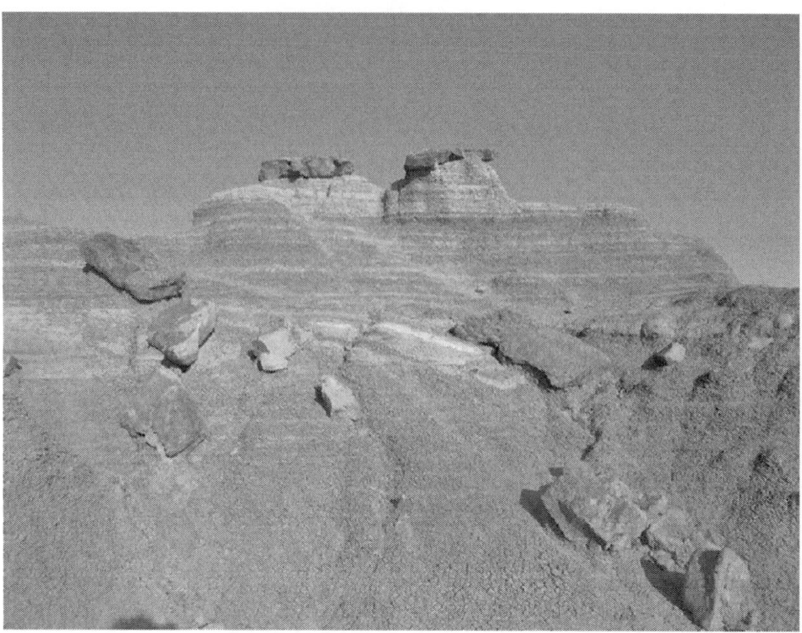

Figure 10.13 Desert caprocks overlying more erodible sandstone substrates.

10.3.1.4 *Aeolian scour pits*

Such features may range in size from a few to many metres in diameter and depth and tend to be formed by circulating vortices of wind, which entrain individual sand particles and deposit them as sand dunes further downslope or within the scour pit base. Loope *et al.* (2008) suggest that fluvial action cannot be responsible for the formation of these features as sufficient catchment areas are not present to provide enough energy for water erosion processes to entrain sediment. These features occur in bedrock that is highly exposed to the dominant wind direction and where the wind blows consistently throughout the year.

10.3.1.5 *Treads and risers/aeolian ripples*

In similar localities to the scour pits described above, rock slopes are dominated by small-scale ripple features that cover many of the Navajo sandstone slick-rock slopes in the study area (Loope, Rowe and Joeckel, 2001). These features are in part controlled by the binding action of microbial layers in the surface rock, rather than the strata or bedding of the rock, in interaction with the consistent abrasion by the wind. Figure 10.15 shows the characteristic form of these aeolian erosional features and the extent to which they cover entire faces of exposed bedrock slopes. It is argued by Loope *et al.* (2008) that such ripples are

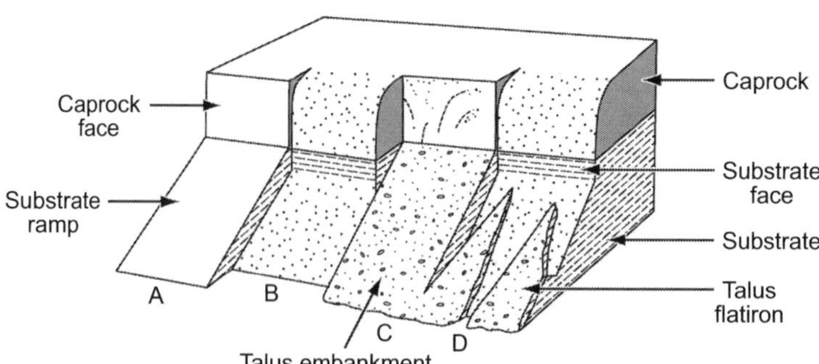

Figure 10.14 Model of compound scarp development: A, generalized form of compound scarp composed of caprock face and substrate ramp; B, erosion of caprock brow and substrate ramp, with downward cliff expansion into substrate; C, failure of substrate face and collapse of unsupported caprock, producing talus embankment on substrate ramp; D, erosion of caprock brow, talus embankment and substrate ramp; erosion of talus embankment produces talus flatirons; downward cliff extension into substrate approaches threshold for next collapse (after Oberlander, 1997).

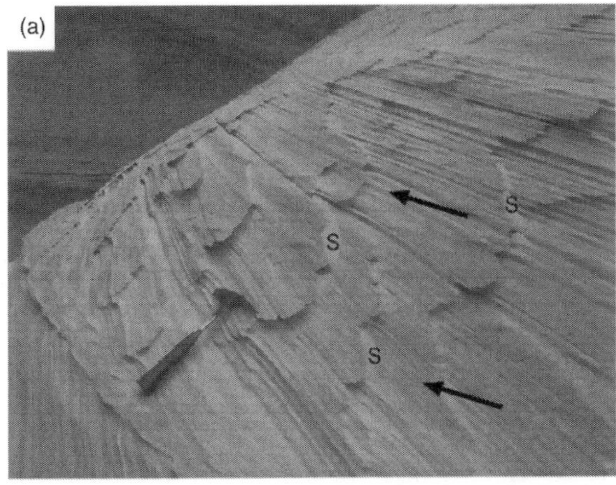

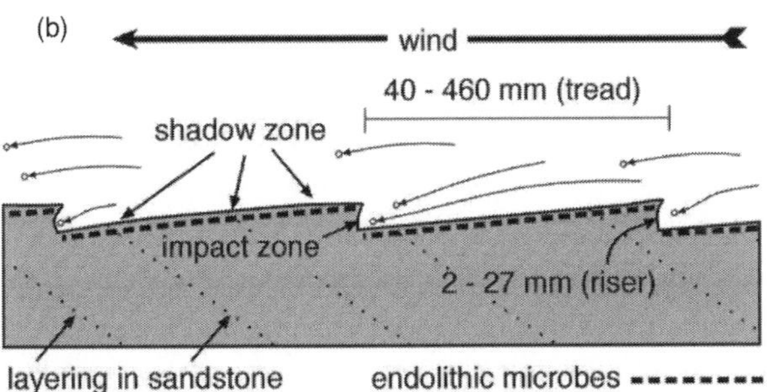

Figure 10.15 Characteristics of treads and risers cut into Navajo sandstone: A, upwind-facing, overhanging risers (up to 27 mm high) on sloping sandstone surface – note loose sand (S) beneath overhangs; B, schematic showing treads, risers and microbial colonies migrating to the left (downwind). Differential erosion is controlled by binding action of microbes, not by rock structure (after Loope *et al.*, 2008).

analogous with erosional bedforms in windblown sand (Chapter 18), and as such exhibit a ripple index (ratio of wavelength to height) of 14/41, which is comparable to the range observed by Bagnold (1954).

10.3.1.6 Rates of hillslope processes in rock slopes

Bare rock slopes tend to evolve as a function of the geological characteristics of the strata and, as such, they are often termed as weathering or detachment limited. The supply of material to these slopes from overlying caprock is periodic and significantly lower than the potential transport rate. Weathering rates will depend upon the composition of the bedrock – friable sandstones, with large pore spaces, being more susceptible to weathering than finer-grained bedrock, for example. However, there are also biological controls on bare rock weathering rates in arid zones, as has been illustrated by the reviews of Cooke, Warren and Goudie (1993) and more recently by Chen, Blume and Beyer (2000). Such authors illustrate that biological controls may accelerate weathering rates (e.g. Robinson and Williams, 2000) or that they may protect the bare rock surface from weathering processes (Kurtz and Netoff, 2001). Souza-Egipsy *et al.* (2004) demonstrate that areas of rock that are covered by biological crusts of fungi and cyanobacteria, which are more resistant to water erosion than neighbouring areas of bare rock (Figure 10.16). There is an analogy here with the protection afforded by desert varnish to bedrock that is otherwise susceptible to tafoni formation (Mellor, Short and Kirkby, 1997) (see Chapters 6 and 8).

10.3.1.7 Off-loading

The block-by-block wasting of a rock slope is clearly an episodic process that may be controlled to some extent by the material properties of the bedrock – high mass strength rocks being more resistant to gravitational forces than low mass strength rocks. However, the actual rates of retreat are more likely to be determined by the rate of basal sapping, where this is the dominant weathering agent, which is dependent upon the propensity of the underlying bedrock to weathering via solution and the subsequent collapse of the overlying strata. Where rocks are susceptible to exfoliation, then rates of retreat may be controlled by either the rates of expansion or pressure release, as described above. Some desert environments exhibit low rates of off-loading under current climatic conditions and are therefore perceived to be in equilibrium (but see Bracken and Wainwright 2006); any small rock falls that do occur do not alter slope form greatly and simply contribute to the rock-mantled slopes beneath (discussed below).

10.3.1.8 Basal sapping

Rates of slope retreat caused by basal sapping are controlled by the rates of groundwater supply to the margin between the substrate and overlying caprock, in conjunction with the mass strength of the overlying strata. Basal sapping may occur where groundwater exfiltrates at a springline and thus a relatively constant supply of water may undermine the overlying strata. The role of groundwater or spring sapping has been observed in various arid zone landscapes (Oberlander, 1997; Ahnert, 1960; Laity and Malin, 1985) and is thought to be significant particularly where permeable strata overly impermeable strata, such as shales. As arid zones are ephemeral environments, in terms of water supply, the more diffuse seepage of water at various points within a rock slope are perhaps less well understood. Sapping due to such 'seepage erosion'

Figure 10.16 Biological crusts of fungi and cyanobacteria (black) affording protection to a sandstone rock slope (after Souza-Egipsy *et al.*, 2004).

is therefore likely to be an important process, but will be highly episodic and may contribute to slower rates of slope retreat. Seepage erosion will also serve to undermine the overlying strata via processes of solution and fluvial erosion, which is often evidenced by solution features, small scour pits or flutes within the rock slope immediately downslope of the seepage point.

10.3.1.9 Evolution through time

Rates of slope-profile evolution are slow, when compared with soil-mantled hillslopes. Episodes of erosion, which move significant quantities of material, through rock fall, for example, lead to maximum rates of slope retreat of ca. 0.1 m/a (Oberlander, 1997). Such data tend to be based on the inference of stratigraphic surveys (e.g. Sancho *et al.*, 1988) and as such are highly variable estimates of the long-term evolution of rock slopes.

10.3.2 Distinctive landforms of rock- and débris-mantled slopes

Unlike their bare rock counterparts, mantled slopes may be well protected from contemporary weathering processes and thus may exhibit significantly different forms and rates of evolution. Such slopes are mantled by a range of materials, which exhibit different particle sizes and often different parent materials as they may be sourced from locations upslope at significant distances from their current location. In arid zones these mantled slopes lie between the previously discussed rock slopes and the badlands or soil-mantled slopes discussed above. The form of rock-mantled hillslopes is greatly dependent upon the parent material that supplies the slope, although herein slopes are only considered that lie on gradients less than the angle of repose for this material. As mentioned above, mantled slopes may occur as part of the continuum of hillslope that is dominated by bare rock slopes, or cuestas, in its upper reaches, through to well-developed, soil-mantled slopes in the floodplains or at the toe-slope locations. A typical hillslope form may therefore be steep and convex at the divide, grading into rectilinear and then concave at the toe of the slope. As gradient changes (decreases) in a downslope direction, it is common to see downslope fining of the mantle, with coarser particles left in situ and finer particles transported downslope due to fluvial erosion processes. A detailed summary of the downslope fining that is common on arid zone hillslopes is given in Parsons, Abrahams and Howard (2009). Examples of distinctive landforms on mantled slopes include: (1) talus (or scree) embankments, (2) talus flatirons, (3) talus ledges and (4) stone pavements. These features are discussed below in terms of their typical form in arid zones, the dominant processes that control these forms and the rates of slope evolution that occur.

10.3.2.1 Talus embankments

Talus embankments are formed when overlying caprock is undermined and collapses onto a substrate ramp, covering the underlying, weaker substrate and often protecting it from further weathering. Such talus-mantled slopes may continue to evolve if supply of material from the caprock is continuous (though episodic), but will often stabilise if supply of material is limited. Layers of talus may be up to 5 m in thickness (Schmidt, 2009) and form part of a continuum of sediment movement from the headwalls of the cuesta, or scarp face, to the lower gradient slopes below. Talus embankments are thus supplied with material due to gravity-controlled processes such as off-loading and will remain active until this supply is truncated when the steeper faces above become stabilised.

10.3.2.2 *Talus flatirons*

Talus flatirons are formed when existing talus embankments become disconnected from the steeper slopes above and are subsequently dissected by fluvial processes, such as gullying or mass movement of material, either as rockfall or as rock slides, leaving areas of talus still mantling the substrate ramp (Figure 10.14). These features may therefore evolve back to 'clean' substrate ramps, unless there is a continuation of supply of talus from the caprock, in which case more composite, layered flatirons may form where numerous episodes of talus supply to a substrate ramp occur. If there is no continued supply of material, the mode of formation is described as the noncyclic model (Schmidt, 1996), as the flatiron will develop as a supply-limited feature. If there is continued supply of material, or fluctuations in supply, due to extrinsic factors such as changing climates, then the climatocyclic model of talus flatiron evolution will take place. In this model, changing climate leads to alterations in the processes that dominate the hillslope. In very dry periods, gullying and rock collapse within the flatiron surface is common and the flatiron will dissect, whereas during wetter periods, unconcentrated erosion processes will dominate (Schmidt, 2009).

10.3.2.3 *Talus ledges*

Talus ledges form when underlying substrate ramps are characterised by nonlinear profiles, such that topographic hollows or plateaus accumulate talus due to breaks of slope that slow the downslope movement of material supplied from the caprock. Ledges may stabilise or may continue to supply rock slopes below during episodic mass movement or gullying and deposits may be reworked if slopes become active, particularly in wetter periods.

10.3.2.4 *Stone pavements and crusts*

Stone pavements, discussed in Chapter 21 and also considered in a soils context in Chapter 7, are common features in arid zones, characterised by an armouring layer of clasts with a matrix of fine particles that overlies poorly developed soils. Such surfaces are common as desert soils and are often very stony and subject to high rates of overland flow and erosion, which are two causal mechanisms for stone-pavement formation (Parsons, Abrahams and Simanton, 1991). The effect of stone pavements on the hydrology and geomorphology of mantled hillslopes is twofold. As stone cover increases, the soil becomes more protected from what are typically low-frequency but high-magnitude (and high-intensity) rainfall events. Compaction of the soil by raindrop impact is thus reduced and infiltration rates remain relatively high. In this scenario, overland flow and subsequent erosion of the slope is decreased. In a second scenario, which may in fact occur in close proximity to the latter, the stones, which are impermeable, generate more overland flow, which leads to increasing erosion downslope and a significant loss of the fine material that comprises the matrix of the stone pavement (Abrahams and Parsons, 1991a). It is this second scenario that leads to downslope fining on arid zone hillslopes, as finer particles tend to travel greater distances than coarser counterparts, leaving coarser fractions in situ in the upper slopes and depositing finer particles downslope when a significant change in slope gradient occurs (Parsons *et al.*, 2006).

Where mantled slopes are dominated by stony soils with a fine matrix of soil, then mechanical and biological surface crusts may also form that influence overland flow and subsequent slope evolution due to fluvial processes. Mechanical crusts may be caused by the compaction effect of raindrop impact and will lead to a thin (1–2 mm) surface crust, which reduces infiltration and promotes overland flow (Abrahams and Parsons, 1991b). Biological crusts form where lichens or mosses are present in the fine matrix of the hillslopes, leading to the growth of a surface crust that is also thought to reduce infiltration rates (see the detailed discussion of mechanical and biological crusts in Chapter 7 and 11). Belnap (2006) reviews the research conducted on biological soil crusts from a wide range of arid landscapes and concludes that they have no consistent control on infiltration rates. However, Belnap also concludes that, in all cases, biological soil crusts reduce the amount of erosion on mantled hillslopes, due to the reduction in available fine sediment for removal.

10.3.2.5 *Rates of hillslope processes in rock- and débris-mantled slopes*

Rock-mantled slopes tend to evolve as a function of the transport capacity of the processes (mass movements, gullying and overland flow), which remove the thin mantle of materials overlying the bare rock surface. Experiments have been designed to monitor the rates of movement of particles due to overland flow over such hillslopes in order to determine the rates of hillslope evolution through time. Early work in the southwestern USA by Abrahams *et al.* (1984) observed the rates of movement of a range of particle sizes >8 mm and showed that hillslope gradient exerted a strong control over the travel distance of particles (see also Abrahams, Parsons and Hirsh, 1985), which travelled further with longer overland flow distances. This work has been developed in the last 20 years through related experiments. Parsons, Abrahams and Luk (1991) and Parsons, Abrahams and Simanton (1991) show that there is a selective erosion of finer particles from debris-covered hillslopes, leaving coarser particles in situ and leading to the development of desert pavements, overlying the débris mantle. Furthermore, Abrahams and Parsons (1991b) show that overland flow from débris-mantled hillslopes may subsequently be controlled by stone pavement cover as well as stone size, which itself increases with hillslope gradient. Finally, Parsons *et al.* (2004) postulate that sediment loss from debris-mantled desert hillslopes is strongly controlled by the travel distance of individual particles, with coarser particles travelling very short distances and fine particles travelling similar distances to overland flow, on an event basis. Such work illustrates that low-gradient, mantled hillslopes are dominated by hydrological processes and will evolve episodically, when sufficiently intense rainfall events occur to entrain particles via overland flow. Parsons *et al.* (2006) provided empirical support for this interpretation for slopes at Walnut Gulch in Arizona.

Rates of talus flatiron development have been studied by Gutiérrez *et al.* (1998) and Gutiérrez Elorza and Sesé Martinez (2001), who studied the rates of slope evolution under flatirons in the Ebro and Almazán Basins in Northern Spain. They concluded that scarp retreat rates were of the order of 1–10 mm/a and corresponded well with changes in climate over the last 30,000 years. Thus, it seems that rates of talus-slope evolution are often controlled by changing climates, which control the all-important supply of moisture to the arid zone. Schmidt (2009) summarises the important characteristics of talus-flatiron evolution in five stages: (1) when moisture supply is high, the slope will develop a concave profile, as supply of material is plentiful. Such flatirons may then be starved of supply if aridity increases or they may continue to develop, particularly if: (2) highly resistant caprocks are evident to maintain supply of material that will armour the slope surface during periods of dissection (drier periods). Some talus flatiron slopes become highly protected as calcium carbonates are washed into the matrix, forming a well-cemented talus covering. (3) The thickness ratio of harder caprock strata to softer substrate is critical, as thin overlying caprock is insufficient to supply the flatiron surface and thus the flatiron will become disconnected from the source material. Conversely, very thick caprock strata may oversupply the substrate slope, which does not allow the development of a concave hillslope profile. (4) If numerous strata of varying resistance to weathering are present, then the talus flatiron may not develop a fully concave profile and the talus ledges described above may dominate. (5) Finally, truncation of the toe slope by fluvial action will lead to steepening of the lower profile, increasing the removal of talus material from the flatiron base. This model can be seen as a specific example of the ways in which different elements of the slope system control each other by the interaction of material supply and process rates, producing boundary conditions across different slope units (Figure 10.1).

In certain locations, such as the semi-arid southwestern USA (Schmidt and Meitz, 1996) and Northern Spain (Gutiérrez *et al.*, 1998, 2001), it is clear that talus slopes are currently evolving at significant rates, particularly during periods of enhanced rainfall. During drier periods, or in drier locations, flatirons tend to become disconnected from the adjacent supply of caprock material and will dissect, until bare rock substrates are exposed.

10.4 Conclusion

In one respect, badlands and rock slopes can be considered to be very similar in that they are essentially bedrock surfaces in many drylands. However, the lack of a significant thickness of regolith *sensu stricto* in the case of badlands is outweighed by the typically unconsolidated nature of the bedrock, so they tend to behave at the extreme end of what might be expected for soil-covered landscapes. Thus, key controls for distinguishing between the behaviour of these stereotypical drylands must be sought elsewhere. Such controls can be seen to be extrinsic, in the sense of lithological, tectonic and climatic variability, and thus operate on timescales from decadal to millions of years. Explaining the patterns occurring within these slope systems then requires an understanding of the interactions between processes of weathering (Chapter 6), runoff and sediment transport (Chapter 11).

The interactions between these processes lead to the evolution of spatial variability in slope form, and this variability then feeds back to the ways in which different slope elements interact with each other. In revisiting the process-form dualism, geomorphologists need to develop models of these interactions that can account for their complexity and the ways in which they evolve over multiple timescales. As noted also in Chapter 11, there is an emerging set of ideas about landscape connectivity, which are likely to form the basis of better models of landscape evolution that are underpinned by a strong process understanding.

References

Abrahams, A.D. and Parsons, A.J. (1991a) Relation between sediment yield and gradient on debris-covered hillslopes, Walnut Gulch, Arizona. *Geological Society of America Bulletin*, **103**, 1109–1113.

Abrahams, A.D. and Parsons, A.J. (1991b) Relation between infiltration and stone cover on a semiarid hillslope, southern Arizona. *Journal of Hydrology*, **122**, 49–59.

Abrahams, A.D., Parsons, A.J. and Hirsch, P.J. (1985) Hillslope gradient – particle size relations: evidence for the formation of debris slopes by hydraulic processes in the Mojave Desert. *Journal of Geology*, **93**, 347–357.

Abrahams, A.D., Parsons, A.J., Cooke, R.U. and Reeves, R.W. (1984) Stone movement on hillslopes in the Mojave Desert, California: a 16-year record. *Earth Surface Processes and Landforms*, **9**, 365–370.

Achten, W.M.J., Dondeyne, S., Mugogo, S. *et al.* (2008) Gully erosion in South Eastern Tanzania: Spatial distribution and topographic thresholds. *Zeitschrift für Geomorphologie*, **52**, 225–235.

Ahnert, F. (1960) The influence of Pleistocene climates upon the morphology of cuesta scarpes on the Colorado Plateau. *Annals of the Association of American Geographers*, **50**, 139–156.

Alexander, D. (1982) Difference between 'calanchi' and 'biancane' badlands in Italy, in *Badland Geomorphology and Piping* (eds R.B. Bryan and A. Yair), GeoBooks, Norwich, pp. 71–85.

Alexander, R.W., Harvey, A.M., Calvo, A. *et al.* (1994) Natural stabilization mechanisms on badland slopes – Tabernas, Almería, Spain, in *Effects of Environmental Change on Drylands: Biogeographical and Geomorphological Perspectives* (eds A.C. Millington and K. Pye), John Wiley & Sons, Ltd, Chichester, pp. 85–111.

Bagnold, R.A. (1954) *The Physics of Blown Sand and Desert Dunes*, Methuen, London.

Ballais, J.-L. (1996) L'âge du modelé de roubines dans les Préalpes du Sud: l'exemple de la region de Digne. *Géomorphologie: Relief, Processus, Environnement*, (4), 61–68.

Belnap, J. (2006) The potential roles of biological soil crusts in dryland hydrologic cycles. *Hydrological Processes*, **20**, 3159–3178.

Benito, G., Gutiérrez, M. and Sancho, C. (1992) Erosion rate in badland areas of the Central Ebro basin (NE-Spain). *Catena*, **19**, 269–286.

Boardman, J., Parsons, A.J., Holland, R. *et al.* (2003) Development of badlands and gullies in the Sneeuberg, Great Karoo, South Africa. *Catena*, **50**, 165–184.

Bochet, E., García-Fayos, P. and Poesen, J. (2009) Topographic thresholds for plant colonization on semi-arid slopes. *Earth Surface Processes and Landforms*, **34**, 1758–1771.

Bracken, L.J. and Wainwright, J. (2006) Geomorphological equilibrium: myth and metaphor? *Transactions of the Institute of British Geographers, NS*, **31**, 167–178.

Bridges, E.M. and Harding, D.M. (1971) Micro-erosion processes and factors affecting slope development in the Lower Swansea Valley. *Institute of British Geographers Special Publication*, **3**, 65–79.

Bryan, R.B., Campbell, I.A. and Yair, A. (1986) Postglacial geomorphic development of the Dinosaur Provincial Park badlands, Alberta. *Canadian Journal of Earth Science*, **24**, 135–146.

Bufalo, M., Oliveros, C. and Quélennec, R.E. (1989) L'érosion des Terres Noires dans la région du Buëch (Hautes-Alpes). Contribution à l'étude des processus érosifs sur le bassin versant représentatif (BVRE) de Saint-Genis. *La Houille Blanch*, **1989** (3/4), 193–195.

Burbank, D.W. and Anderson, R.S. (2001) *Tectonic Geomorphology*, Blackwell, Oxford.

Butler, J. and Goetz, H. (1986) Vegetation and soil-landscape relationships in the North Dakota badlands. *American Midland Naturalist*, **116**, 378–386.

Campbell, I.A. (1982) Surface morphology and rates of change during a ten-year period in the Alberta badlands, in *Badland Geomorphology and Piping* (eds R.B. Bryan and A. Yair), GeoBooks, Norwich, pp. 221–237.

Campbell, I.A. (1997) Badlands and badland gullies, in *Arid Zone Geomorphology*, 2nd edn (ed. D.S.G. Thomas), John Wiley and Sons, Ltd, Chichester, pp. 261–291.

Cantón, Y., Domingo, F., Solé-Benet, A. and Puidefábregas, J. (2001) Hydrological and erosional response of a badlands system in semiarid SE Spain. *Journal of Hydrology*, **252**, 65–84.

Capolongo, D., Pennetta, L., Piccarreta, M. *et al.* (2008) Spatial and temporal variations in soil erosion and deposition due to land-levelling in a semi-arid area of Basilicata (Southern Italy). *Earth Surface Processes and Landforms*, **33**, 364–379.

Carson, M.A. and Kirkby, M.J. (1972) *Hillslope Form and Process*, Cambridge University Press, Cambridge.

Castro, P.V., Chapman, R.W., Gili, S. et al. (eds) (1998) Aguas Project. Paleoclimatic reconstruction and the dynamics of human settlement and land-use in the area of the Middle Aguas (Almería) in the South-East of the Iberian Península, EUR 18036, European Commission, Brussels.

Cerdà, A. (2002) The effect of season and parent material on water erosion on highly eroded soils in eastern Spain. *Journal of Arid Environments*, **52**, 319–337.

Cerdà, A. and García-Fayos, P. (1997) The influence of slope angle on sediment, water and seed losses on badland slopes. *Geomorphology*, **18**, 77–90.

Chen, J., Blume, H.P. and Beyer, L. (2000) Weathering of rocks induced by lichen colonization – a review. *Catena*, **39**, 121–146.

Ciccacci, S., Galiano, M., Roma, M.A. and Salvatore, M.C. (2008) Morphological analysis and erosion rate evaluation in badlands of Radicofani area (Southern Tuscany – Italy). *Catena*, **74**, 87–97.

Clarke, M.L. and Rendell, H.M. (2000) The impact of the farming practice of remodelling hillslope topography on badland morphology and soil erosion processes. *Catena*, **40**, 229–250.

Clarke, M.L. and Rendell, H.M. (2006) Process-form relationships in Southern Italian badlands: erosion rates and implications for landform evolution. *Earth Surface Processes and Landforms*, **31**, 15–29.

Collison, A.C. (1996) Unsaturated strength and preferential flow as controls on gully head development, in *Advances in Hillslope Processes* (eds M.G. Anderson and S.M. Brooks), John Wiley and Sons, Ltd, Chichester, pp. 753–770.

Cooke, R.U. and Reeves, R.W. (1976) *Arroyos and Environmental Change in the American Southwest*, Clarendon Press, Oxford.

Cooke, R.U., Warren, A. and Goudie, A.S. (1993) *Desert Geomorphology*, UCL Press, London.

Dalrymple, J.B., Blong, R.J. and Conacher, A.J. (1968) A hypothetical nine-unit land-surface model. *Zeitschrift für Geomorphologie*, **12**, 60–76.

Danin, A (2004) Arundo (Gramineae) in the Mediterranean reconsidered. *Willdenowia*, **34**, 361–369.

Della Seta, M., Del Monte, M., Fredi, P. and Lupia Palmieri, E. (2009) Space–time variability of denudation rates at the catchment and hillslope scales on the Tyrrhenian side of Central Italy. *Geomorphology*, **107**, 171–177.

De Lugt, J. and Campbell, I.A. (1992) Mass movements in the badlands of Dinosaur Provincial Park, Alberta, Canada. *Catena Supplement*, **23**, 75–100, Catena, Cremlingen.

De Ploey, J. (1974) Mechanical properties of hillslopes and their relation to gullying in central semi-arid Tunisia. *Zeitschrift für Geomorphologie Supplementband*, **21**, 177–190.

De Ploey, J. (1989) *Soil-Erosion Map of Western Europe*, Catena, Cremlingen.

Descroix, L. and Gautier, E. (2002) Water erosion in the southern French Alps: climatic and human mechanisms. *Catena*, **50**, 53–85.

Desir, G. and Marín, C. (2007) Factors controlling the erosion rates in a semi-arid zone (Bardenas Reales, N.E. Spain). *Catena*, **71**, 31–40.

Díaz-Hernández, J.L. and Juliá, R. (2006) Geochronological position of badlands and geomorphological patterns in the Guadix-Baza basin (SE Spain). *Quaternary Research*, **65**, 467–477.

Dietrich, W.E. and Dunne, T. (1978) Sediment budget for a small catchment in mountainous terrain. *Zeitscrift für Geomorphologie, Supplementband*, **29**, 191–206.

Eriksson, M., Reuterswald, K. and Christiansson, C. (2003) Changes in the fluvial system of the Kondoa Irangi Hills, central Tanzania, since 1960. *Hydrological Processes*, **17**, 3271–3285.

Evans, D.J.A. (2000) Quaternary geology and geomorphology of the Dinosaur Provincial Park area and surrounding plains, Alberta, Canada: the identification of former glacial lobes, drainage diversions and meltwater flood tracks. *Quaternary Science Reviews*, **19**, 931–958.

Evans, D.J.A., Campbell, I.A. and Lemmen, D.S. (2004) Holocene alluvial chronology of One Tree Creek, southern Alberta, Canada. *Geografisker Annaler*, **86A**, 117–130.

Farifteh, J. and Soeters, R. (1999) Factors underlying piping in the Basilicata region, southern Italy. *Geomorphology*, **26**, 239–251.

Farifteh, J. and Soeters, R. (2006) Origin of biancane and calanchi in East Aliano, southern Italy. *Geomorphology*, **77**, 142–152.

Faulkner, H. (1974) An allometric growth model for competitive gullies. *Zeitschrift für Geomorphologie Supplementband*, **21**, 76–87.

Faulkner, H. (2008) Connectivity as a crucial determinant of badland morphology and evolution. *Geomorphology*, **100**, 91–103.

Faulkner, H, Alexander, R. and Zukowskyj, P. (2008) Slope-channel coupling between pipes, gullies and tributary channels in the Mocatán catchment badlands, Southeast Spain. *Earth Surface Processes and Landforms*, **33**, 1242–1260.

Feoli, E., Gallizia Vuerich, L. and Woldu, Z. (2002) Processes of environmental degradation and opportunities for rehabilitation in Adwa, Northern Ethiopia. *Landscape Ecology*, **17**, 315–325.

Flageollet, J.-C., Malet, J.-P. and Maquaire, O. (2000) The 3D structure of the Super-Sauze earthflow: a first stage towards modelling its behaviour. *Physics and Chemistry of the Earth, Part B: Hydrology, Oceans and Atmosphere*, **25**, 785–791.

Flageollet, J.-C., Maquaire, O., Martin, B. and Weber, D. (1999) Landslides and climatic conditions in the Barcelonnette and Vars basins (Southern French Alps, France). *Geomorphology*, **30**, 65–78.

Francke, T., López-Tarazón, J.A., Vericat, D. *et al.* (2008) Flood-based analysis of high-magnitude sediment transport using a non-parametric method. *Earth Surface Processes and Landforms*, **33**, 2064–2077.

García-Fayos, P., García-Ventoso, B. and Cerdà, A. (2000) Limitations to plant establishment on eroded slopes in southeastern Spain. *Journal of Vegetation Science*, **11**, 77–86.

García-Ruiz, J.M., Regüés, D., Alvera, B. *et al.* (2008) Flood generation and sediment transport in experimental catchments affected by land use changes in the central Pyrenees. *Journal of Hydrology.* **356**, 245–260.

Gilbert, G.K. (1877) Report on the geology of the Henry Mountains, US Geographical and Geological Survey of the Rocky Mountain Region, Washington, DC.

Godfrey, A.E. (1997) Wind erosion of Mancos Shale badland ridges by sudden drops in pressure. *Earth Surface Processes and Landforms*, **22**, 345–352.

Godfrey, A.E., Everitt, B.J. and Martín Duque, J.F. (2008) Episodic sediment delivery and landscape connectivity in the Mancos Schale badlands and Fremont River system, Utah, USA. *Geomorphology*, **102**, 242–251.

Gonzalez, M.A. (2001) Recent formation of arroyos in the Little Missouri Badlands of southwestern North Dakota. *Geomorphology*, **38**, 63–84.

Griffiths, J.S., Hart, A.B., Mather, A.E. and Stokes, M. (2005) Assessment of some spatial and temporal issues in landslide initiation within the Río Aguas Catchment, South-East Spain. *Landslides*, **2**, 183–192.

Guàrdia, R., Gallart, F. and Ninot, J.M. (2000) Soil seed bank and seedling dynamics in badlands of the upper Llobregat Basin (Southeastern Pyrenees). *Catena*, **40**, 189–202.

Guàrdia, R., Raventós, J. and Caswell, H. (2000) Spatial growth and population dynamics of a perennial tussock grass (*Achnatherum calamagrostis*) in a badland area. *Journal of Ecology*, **88**, 950–963.

Guàrdia, R. and Recatalá, T.M. (1992) La reserve de semillas en una cuenca de badlands. *Pirineos*, **140**, 29–36.

Guerrero-Campo, J., Palacio, S. and Montserrat-Martí, G. (2008) Plant traits enabling survival in Mediterranean badlands in northeastern Spain suffering from soil erosion. *Journal of Vegetation Science*, **19**, 457–464.

Gutiérrez, M., Sancho, C., Arauzo, T. and Pena, J.L. (1998) Scarp retreat rates in semi-arid environments from talus flatirons (Ebro basin, North East Spain). *Geomorphology*, **25**, 11–21.

Gutiérrez Elorza, M. and Sesé Martinez, V.H. (2001) Multiple talus flatirons, variations of scarp retreat rates and the evolution of slopes in Almazán basin (semi-arid central Spain). *Geomorphology*, **38**, 19–29.

Haigh, M.J. (1979) Ground retreat and slope evolution on plateau-type colliery spoil mounds at Blaenavon, Gwent. *Transactions of the Institute of British Geographers*, **NS 4**, 321–328.

Halliday, W.R. (2007) Pseudokarst in the 21st century. *Journal of Cave and Karst Studies*, **69**, 103–113.

Hicks, D.M., Gomez, B. and Trustrum, N.A. (2000) Erosion thresholds and suspended sediment yields, Waipoa River basin, New Zealand. *Water Resources Research*, **36**, 1129–1142.

Howard, A.D. (1997) Badland morphology and evolution: interpretation using a simulation model. *Earth Surface Processes and Landforms*, **22**, 211–227.

Howard, A.D. (2009) Badlands and gullying, in *Geomorphology of Desert Environments*, 2nd edn (eds Parsons, A.J. and Abrahams, A.D.), Springer, Berlin, pp. 265–299.

Imeson, A.C., Kwaad, F.J.P.M. and Verstraten, J.M. (1982) The relationship of soil physical and chemical properties to the development of badlands in Morocco, in *Badland Geomorphology and Piping* (eds R.B. Bryan and A Yair), GeoBooks, Norwich, pp. 47–70.

Joshi, V., Tambe, D. and Dhawade, G. (2009) Geomorphometry and fractal dimension of a riverine badland in Maharashtra. *Journal of the Geological Society of India*, **73**, 355–370.

Kasanin-Grubin, M. and Bryan, R. (2007) Lithological properties and weathering response on badland hillslopes. *Catena*, **70**, 68–78.

Keay-Bright, J. and Boardman, J. (2009) Evidence from field-based studies of rates of soil erosion on degraded land in the central Karoo, South Africa. *Geomorphology*, **103**, 455–465.

Kemp née Marchington, A.C. (1990) Towards a dynamic model of gully growth, in *Erosion, Transport and Deposition Processes (Proceedings of the Jerusalem Workshop, March–April 1987)* (eds D.E. Walling, A. Yair and S. Bercowicz), IAHS Publication 189, Wallingford, pp. 121–134.

Kidron, G.J. (1999) Differential water distribution over dune slopes as affected by slope position and microbiotic crust, Negev Desert, Israel. *Hydrological Processes*, **13**, 1665–1682.

Kidron, G.J. and Yair, A. (2001) Runoff-induced sediment yield over dune slopes in the Negev Desert. 1: Quantity and variability. *Earth Surface Processes and Landforms*, **26**, 461–474.

King, G.P.C. and Sturdy, D. (1994) Tectonics, in *Understanding the Natural and Anthropogenic Causes of Soil Degradation and Desertification in the Mediterranean Basin, Volume 1: Land Degradation in Epirus* (ed. S.E. van der Leeuw), Final Report on Contract EV5V-CT91-0021, EU, Brussels, pp. 13–45.

Kirkby, M.J. and Bull, L.J. (2000) Some factors controlling gully growth in fine-grained sediments: a model applied to south-east Spain. *Catena*, **40**, 127–146.

Kosmas, C.S., Danalatos, N., Cammeraat, L.H. *et al.* (1997) The effect of land use on runoff and soil erosion rates under Mediterranean conditions. *Catena*, **29**, 45–59.

Kurtz, H.D. and Netoff, D.I. (2001) Stabilization of friable sandstone surfaces in a desiccating, wind-abraded environment of south-central Utah by rock surface microorganisms. *Journal of Arid Environments*, **48**, 89–100.

Laity, J.E. and Malin, M.C. (1985) Sapping processes and the development of theatre-headed valley networks in the Colorado plateau. *Bulletin of the Geological Society of America*, **96**, 203–217.

Lam, K.C. (1977) Patterns and rates of slopewash on the badlands of Hong Kong. *Earth Surface Processes*, **2**, 319–332.

Langbein, W.B. and Schumm, S.A. (1958) Yield of sediment in relation to mean annual precipitation. *Transactions of the American Geophysical Union*, **39**, 1076–1084.

Lázaro, R., Cantón, Y., Solé-Benet, A. *et al.* (2008) The influence of competition between lichen colonization and erosion on the evolution of soil surfaces in the Tabernas badlands (SE Spain) and its landscape effects. *Geomorphology*, **102**, 252–266.

Lin, Z. and Oguchi, T. (2004) Drainage density, slope angle, and relative basin position in Japanese bare lands from high-resolution DEMs. *Geomorphology*, **63**, 159–173.

Linares, R., Rosell, J., Pallí, L. and Roqué, C. (2002) Afforestation by slope terracing accelerates erosion. A case study in the Barranco de Barcedana (Conca de Tremp, N.E. Spain). *Environmental Geology*, **42**, 11–18.

Liu, T., An, Z., Yuan, B. and Han, J. (1985) The loess-palaeosol sequence in China and climatic history. *Episodes*, **8**, 21–28.

Loope, D.B., Rowe, C.M. and Joeckel, R.M. (2001) Annual monsoon rains recorded by Jurassic dunes. *Nature*, **412**, 64–66.

Loope, D.B., Seiler, W.M., Mason, J.A. and Chan, M.A. (2008) Wind scour of Navajo Sandstone at the Wave (Central Colorado Plateau, USA). *Journal of Geology*, **116**, 173–183.

Loppi, S., Boscagli, A. and De Dominicis, V. (2004) Ecology of soil lichens from Pliocene clay badlands of central Italy in relation to geomorphology and vascular vegetation. *Catena*, **55**, 1–15.

Mabbutt, J.A. (1977) *Desert Landforms*, Australian National University Press, Canberra.

Maerker, M., Castro, C.P., Pelacani, S. and Baeurle, M.V.S. (2008) Assessment of soil degradation susceptibility in the Chacabuto Province of central Chile using a morphometry based response units approach. *Geografia Fisica e Dinamica Quaternaria*, **31**, 47–53.

Maquaire, O., Malet, J.-P., Remaître, A. *et al.* (2003) Instability conditions of marly hillslopes: towards landsliding or gullying? The case of the Barcelonnette Basin, South East France. *Engineering Geology*, **70**, 109–130.

Mather, A.E., Stokes, M. and Griffiths, J.S. (2002) Quaternary landscape evolution: a framework for understanding contemporary erosion, southeastern Spain. *Land Degradation and Development*, **13**, 89–109.

Mellor, A., Short, J. and Kirby, S.J. (1997) Tafoni in the El Chorro area, Andalucia, southern Spain. *Earth Surface Processes and Landforms*, **22**, 817–833.

Mieth, A. and Bork, H.-R. (2005) History, origin and extent of soil erosion on Easter Island (Rapa Nui). *Catena*, **63**, 244–260.

Nogueras, P., Burjachs, F., Gallart, F. and Puidefábregas, J. (2000) Recent gully erosion in the El Cautivo badlands (Tabernas, S.E. Spain). *Catena*, **40**, 203–215.

Oberlander, T.M. (1997) Slope and pediment systems, in *Arid Zone Geomorphology*, 2nd edn (ed. D.S.G. Thomas), John Wiley & Sons, Ltd, Chichester, pp. 135–163.

Oostwoud Wijdenes, D.J. and Ergenzinger, P. (1998) Erosion and sediment transport on steep marly hillslopes, Draix, Haute-Provence, France: an experimental study. *Catena*, **33**, 179–200.

Oostwoud Wijdenes, D.J. and Ergenzinger, P. (2003) Erosion and sediment transport on steep marly hillslopes, Draix, Haute-Provence, France: an experimental field study. *Catena*, **33**, 179–200.

Parkner, T., Page, M.J., Marutani, T. and Trustrum, N.A. (2006) Development and controlling factors of gullies and gully complexes, East Coast, New Zealand. *Earth Surface Processes and Landforms*, **31**, 187–199.

Parsons, A.J. (1988) *Hillslope Form*, Routledge, London.

Parsons, A.J., Abrahams, A.D. and Howard, A.D. (2009) Rock mantled slopes, in *Geomorphology of Desert Environments*, 2nd edn (eds A.J. Parsons and A.D. Abrahams), Springer, Berlin, pp. 233–263.

Parsons, A.J., Abrahams, A.D. and Luk, S.-H. (1991) Size characteristics of sediment in interrill overland-flow on a semiarid hillslope, southern Arizona. *Earth Surface Processes and Landforms*, **16**, 143–152.

Parsons, A.J., Abrahams, A.D. and Simanton, J.R. (1991) Microtopography and soil surface materials on semi-arid piedmont slopes, southern Arizona. *Journal of Arid Environments*, **22**, 107–115.

Parsons, A.J., Wainwright, J., Schlesinger, W.H. and Abrahams, A.D. (2003) The role of overland flow in sediment and nitrogen budgets of mesquite dunefields, southern New Mexico. *Journal of Arid Environments*, **53**, 61–71.

Parsons, A.J., Wainwright, J., Powell, D.M. *et al.* (2004) A conceptual model for understanding and predicting erosion by water. *Earth Surface Processes and Landforms*, **29**, 1293–1302.

Parsons, A.J., Brazier, R.E., Wainwright, J. and Powell, D.M. (2006) Scale relationships in hillslope runoff and erosion. *Earth Surface Processes and Landforms*, **31**, 1384–1393.

Phillips, C.P. (1998a) Reclaiming the erosion susceptible landscape of the Italian badlands for arable cultivation. *Land Degradation and Development*, **9**, 331–346.

Phillips, C.P. (1998b) The badlands of Italy: a vanishing landscape? *Applied Geography*, **18**, 243–257.

Phillips, C.P. (1998c) The Crete Senesi, Tuscany: a vanishing landscape? *Landscape and Urban Planning*, **41**, 19–26.

Piccarreta, M., Capolongo, D., Boenzi, F. and Bentivengam, M. (2006) Implications of decadal changes in precipitation and land use policy to soil erosion in Basilicata, Italy. *Catena*, **65**, 138–151.

Pike, R.J. (2000) Geomorphometry – diversity in quantitative surface analysis. *Progress in Physical Geography*, **24**, 1–20.

Pintado, A., Sancho, L.G., Green, T.G.A. *et al.* (2005) Functional ecology of the biological soil crust in semiarid SE Spain: sun and shade populations of *Diploschistes diacapsis* (Ach.) Lumbsch. *The Lichenologist*, **37**, 425–432.

Poesen, J., Nachtergaele, J., Verstraeten, G. and Valentin, C. (2003) Gully erosion and environmental change: importance and research needs. *Catena*, **50**, 91–133.

Regüés, D., Guàrdia, R. and Gallart, F. (2000) Geomorphic agents versus vegetation spreading as causes of badland occurrence in a Mediterranean subhumid mountainous area. *Catena*, **40**, 173–187.

Rey, F. (2009) A strategy for fine sediment retention with bioengineering works in eroded marly catchments in a mountainous Mediterranean climate (southern Alps, France). *Land Degradation and Development*, **20**, 210–216.

Robinson, D.A. and Phillips, C.P. (2001) Crust development in relation to vegetation and agricultural practice on erosion susceptible, dispersive clay soils from central and southern Italy. *Soil & Tillage Research*, **60**, 1–9.

Robinson, D.A. and Williams, R.B.G. (2000) Accelerated weathering of a sandstone in the High Atlas Mountains of Morocco by an epilithic lichen. *Zeitschrift für Geomorphologie*, **44**, 513–528.

Romero Díaz, A., Marín Sanleandro, P., Sánchez Soriano, A. *et al.* (2007) The causes of piping in a set of abandoned agricultural terraces in southeast Spain. *Catena*, **69**, 282–293.

Sancho, C., Gutierrez, M., Pena, J.L. and Burillo, F. (1988) A quantitative approach to scarp retreat starting from triangular slope facets, central Ebro basin, Spain. *Catena Supplement*, **13**, 139–146.

Schmidt, K.-H. (1996) Talus and pediment flatirons – indicators of climatic change on scarp slopes on the Colorado plateau, USA. *Zeitschrift für Geomorpholgie Supplementband*, **103**, 135–158.

Schmidt, K.-H. (2009) Hillslopes as evidence of climate change, in *Geomorphology of Desert Environments*, 2nd edn (eds A.J. Parsons and A.D.Abrahams), Springer, Berlin, pp. 675–694.

Schmidt, K.-H. and Meitz, P. (1996) Cuesta scarp forms and processes in different altitudinal belts of the Colorado plateau as indicators of climate change, in *Advances in Hillslope Processes* (eds M.G. Anderson and S.M. Brooks), John Wiley & Sons, Ltd, Chichester, pp. 1079–1097.

Schumm, S.A. (1956) Evolution of drainage systems and slopes at Perth Amboy, New Jersey. *Bulletin of the Geological Society of America*, **73**, 719–724.

Schumm, S.A. (1962) Erosion on miniature pediments in Badlands National Monument, South Dakota. *Bulletin of the Geological Society of America*, **73**, 719–724.

Schumm, S.A. (1973) Fluvial geomorphology, in *Proceedings of the 4th Annual Geomorphology Symposium Series, Binghamton*, Allen and Unwin, London, pp. 299–311.

Scoging, H.M. (1982) Spatial variations in infiltration, runoff and erosion on hillslopes, in *Badland Geomorphology and Piping* (eds R.B. Bryan and A. Yair), GeoBooks, Norwich, pp. 89–112.

Segerstrom, K. (1950) *Erosion studies at Parícutin, State of Michoacán, Mexico*, US Geological Survey Bulletin 965-A, Washington DC.

Selby, M.J. (1982) Rock mass strength and the form of some inselbergs in the central Namib Desert. *Earth Surface Processes and Landforms*, **7**, 489–497.

Selby, M.J. (1993) *Hillslope Materials and Processes*, 2nd edn, Oxford University Press, Oxford.

Sidorchuk, A. (1999) Dynamic and static models of gully erosion. *Catena*, **37**, 401–414.

Solé, A., Josa, R., Pardini, G. *et al.* (1992) How mudrock and soil physical properties influence badland formation at Vallcebre (Pre-Pyrenees, NE Spain). *Catena*, **19**, 287–300.

Souza-Egipsy, V., Wierzchos, J., Sancho, C. *et al.* (2004) Role of biological soil crust cover in bioweathering and protection of sandstones in a semi-arid landscape (Torrollones de Gabarda, Huesca, Spain). *Earth Surface Processes and Landforms*, **29**, 1651–1661.

Thornes, J.B. (1985) The ecology of erosion. *Geography*, **70**, 222–235.

Torri, D. and Bryan, R.B. (1997) Micropiping processes and biancana evolution in southeast Tuscany. *Geomorphology*, **20**, 219–235.

Torri, D., Calzolari, C. and Rodolfi, G. (2000) Badlands in changing environments: an introduction. *Catena*, **40**, 119–125.

Torri, D., Regüés, D., Pellegrini, S. and Bazzoffi, P. (1999) Within-storm soil surface dynamics and erosive effects of rainstorms. *Catena*, **38**, 131–150.

Vallauri, D.R., Aronson, J. and Barbero, M. (2002) An analysis of forest restoration 120 years after reforestation on badlands in the southwestern Alps. *Restoration Ecology*, **10**, 16–26.

Wainwright, J. (1994) Anthropogenic factors in the degradation of semi-arid regions: a prehistoric case study in Southern France, in *Effects of Environmental Change on Drylands* (eds A.C. Millington and K. Pye), John Wiley & Sons, Ltd, Chichester, pp. 285–304.

Wainwright, J. (1996a) A comparison of the infiltration, runoff and erosion characteristics of two contrasting 'badland' areas in S. France. *Zeitschrift für Geomorphologie Supplementband*, **106**, 183–198.

Wainwright, J. (1996b) Hillslope response to extreme storm events: the example of the Vaison-la-Romaine event, in *Advances in Hillslope Processes* (eds M.G. Anderson and S.M. Brooks), John Wiley & Sons, Ltd, Chichester, pp. 997–1026.

Wainwright, J. (2009a) Earth-system science, in *Blackwell Companion to Environmental Geography* (eds N. Castree, D. Liverman, B. Rhoads and D. Demerritt), Blackwell, Oxford, pp. 145–167.

Wainwright, J. (2009b) Desert ecogeomorphology, in *Geomorphology of Desert Environments*, 2nd edn (eds A.J. Parsons and A.D. Abrahams), Springer, Berlin, pp. 21–66.

Wainwright, J. and Thornes, J.B. (2003) *Environmental Issues in the Mediterranean: Processes and Perspectives from the Past and Present*, Routledge, London.

Willett, S.D. (1999) Orogeny and orography: the effects of erosion on the structure of mountain belts. *Journal of Geophysical Research*, **104**, 28957–28982.

Wise, S.M., Thornes, J.B. and Gilman, A. (1982) How old are the badlands? A case study from southeast Spain, in *Badland Geomorphology and Piping* (eds R.B. Bryan and A. Yair), GeoBooks, Norwich, pp. 259–277.

Yair, A., Goldberg, P. and Brimer, B. (1982) Long term denudation rates in the Zin-Havarim badlands, northern Negev, Israel, in *Badland Geomorphology and Piping* (eds R.B. Bryan and A. Yair), GeoBooks, Norwich, pp. 279–291.

Young, A. (1972) *Slopes*, Oliver and Boyd, Edinburgh.

Zhu, T.X., Luk, S.-H. and Cai, Q.G. (2002) Tunnel erosion and sediment production in the hilly loess region, North China. *Journal of Hydrology*, **257**, 78–90.

III

The work of water

11

Runoff generation, overland flow and erosion on hillslopes

John Wainwright and Louise J. Bracken

11.1 Introduction

It may appear paradoxical to consider the role of water flows in arid regions, yet they are one of the most important landscape-forming processes in many drylands. As noted in Chapter 1, there are a number of reasons why aridity occurs. In the tropical and subtropical zones, low annual rainfalls are a function of extended dry periods and a relatively short rainy season. For example, in the US southwest, over half the annual rainfall typically falls between July and September (Osborn and Renard, 1969; Nielson, 1986; Wainwright, 2005); in the southern Mediterranean, precipitation is concentrated over a few winter months, while the northern shores tend to experience peaks in autumn and again in spring (Wainwright and Thornes, 2003, Bracken, Cox and Shannon, 2008). Rainfall in these environments is predominantly convective, producing large pulses of rain in a matter of minutes or hours. High rainfall intensities lead to the crossing of process thresholds and corresponding high energies produce the potential for significant amounts of erosion. The position of these thresholds is affected also by the nature of dry spells. Cold ocean current deserts tend to have significant amounts of their precipitation in the form of fog, which can contribute to the formation of biological crusts (Belnap, 2006). The extent of periods of aridity leads to distinctively patchy vegetation characteristics (Wainwright, 2009a), which often provide extensive bare surface areas that enhance runoff production. These periods and patches also enhance aeolian activity (see Section 4), which can itself feed back to thresholds of runoff production, e.g. by the formation of mechanical crusts (e.g. Valentin, 1993).

Runoff is generated by three basic mechanisms. In the first mechanism, precipitation arriving at the surface exceeds the capacity of the surface to absorb the precipitation (Figure 11.1). The rate at which the surface can absorb the precipitation is called the infiltration rate. The infiltration rate creates the threshold for runoff generation in one of two ways. First, the rainfall intensity can exceed the instantaneous infiltration rate of an unsaturated soil. This mechanism produces infiltration excess or Hortonian overland flow (named after the Robert Horton, who lived from 1875 to 1945 and made many significant advances to our understanding of hydrology and erosion in the first half of the twentieth century). Second, saturation of the soil can produce saturation overland flow (sometimes called Dunne overland flow after the geomorphologist/hydrologist Thomas Dunne) as precipitation arrives on an already saturated surface. In reality, the distinction between these two mechanisms is somewhat artificial, as even saturated soils will enable infiltration at a low rate (the general exception being in soils with high contents of swelling clays) and so they can still be considered to be producing runoff by infiltration excess. These conditions may be relatively common in cases where storms follow one another on successive days (Lange *et al.*, 2003).

The second basic mechanism of runoff generation is by exfiltration, or return flow. Exfiltration occurs when saturated soils receive lateral flows from upslope, which causes them to exceed their capacity for soil-moisture storage. Where slope angles get steeper in the downslope direction, or where planform concavities occur, convergence of subsurface flows can occur, producing concentrations of moisture that can lead to exfiltration. An

Arid Zone Geomorphology: Process, Form and Change in Drylands, Third Edition. Edited by David S. G. Thomas

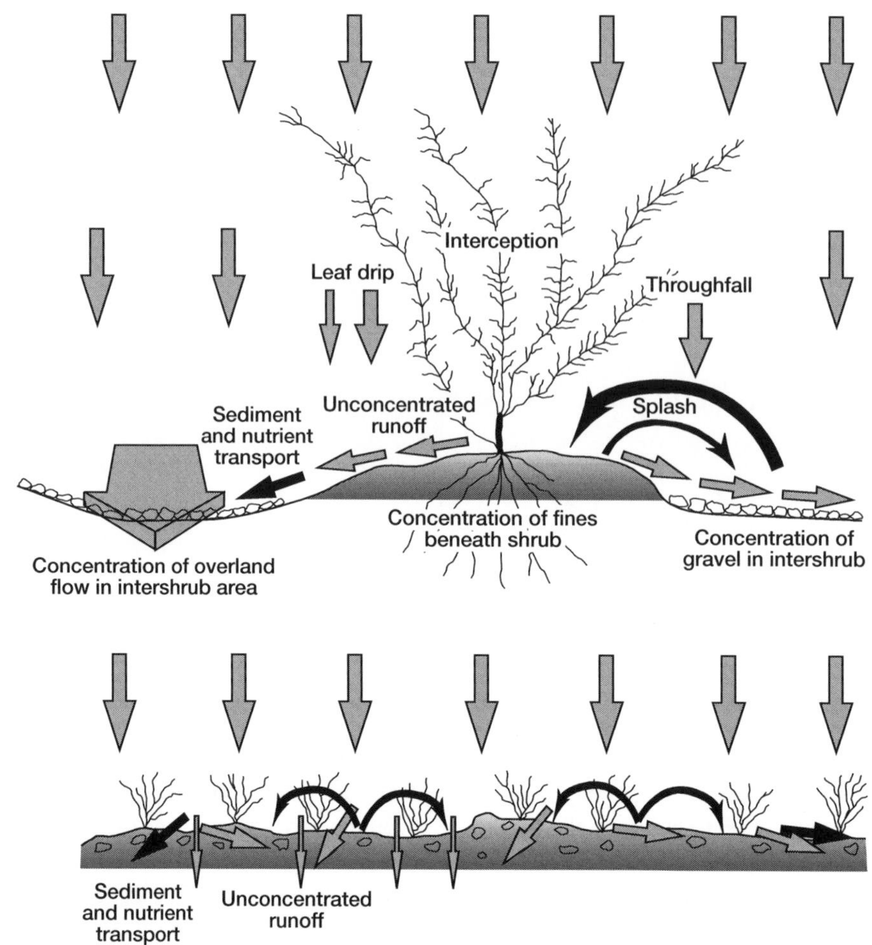

Figure 11.1 Conceptual structure of rainfall–runoff processes (after Wainwright, Parsons and Abrahams, 2000).

often overlooked case in which such a mechanism could occur in drylands is where soils are thin and sitting on relatively impermeable bedrock, so that the local soil-moisture-storage capacity is low. Dryland soils tend to have intrinsically slow rates of formation (Chapter 7) or are thin because of removal by erosion processes.

In the third mechanism, the formation of subsurface pipes can deliver runoff rapidly from slopes to channels. In certain drylands, pipes are a significant method of runoff production. In all three cases, an understanding of the hydraulic characteristics of the soil, in particular the infiltration rate, is fundamental for understanding rates of runoff production. However, rates of infiltration and incident precipitation can be modified by a range of other factors, notably interception, stemflow and leafdrip, emphasising further the feedbacks from vegetation type and cover. Between storm events, soil moisture can change significantly and rapidly, especially in the hot arid regions, affecting initial infiltration rates for the next event and thus changing thresholds for all three types of runoff-producing mechanism. Vegetation can be a further factor in this feedback. Key runoff-producing areas tend to be found on steep slopes, either abandoned after agriculture or with sparse vegetation, and are composed of runoff-promoting soils such as marls (Bull *et al.*, 2000). These areas do not necessarily relate to the channel network and the mosaic pattern they form is key to producing floods in ephemeral channels. These conceptual models of runoff are in contrast to the variable source area (VSA) developed for humid regions in the 1960s (Betson, 1964; Hewlett and Hibbert, 1967; Dunne and Black, 1970), which proposed that saturated areas produce most of the storm runoff as the water table rises to the soil surface over an expanding area as rainfall continues. Saturation overland flow initially spreads up low-order tributaries, then up unchannelled swales and gentle footslopes of hillsides (Dunne, Moore and Taylor, 1975). The position and expansion of variable areas contributing runoff is related to geology,

Figure 11.2 Overland flows showing the discontinuous nature of flow and the presence of flow threads: (a) following a storm event at Jornada LTER site, New Mexico; (b) during a rainfall-simulation event on a desert grassland at Walnut Gulch, Arizona (see Parsons *et al.*, 1997); (c) concentrated rill flows at Jornada LTER site, New Mexico; and (d) flow threads and rills developed on an initially smooth experimental slope at Long Ashton, Bristol, in experiments investigating the movement of archaeological materials on slopes.

topography, soils, rainfall characteristics and vegetation (Dunne and Black, 1970; Dunne, Moore and Taylor, 1975).

Recently, the different conceptual models of runoff production have been the focus of critique, with calls for a new theory of runoff generation (e.g. McDonnell, 2003; Ambroise, 2004; Bracken and Croke, 2007). Hydrological connectivity is one possible concept of runoff generation and flood production that could provide a way forward, but currently there is much confusion in the literature about how the term is used and how it relates to existing research. There has been considerable research on aspects of hydrological 'connectivity', although not always referred to as such, including runoff generation at the patch, hillslope and catchment scales, which makes it difficult to draw work together into a single theoretical model of runoff connectivity (e.g. Fitzjohn, Ternan and Williams, 1998; Cammeraat and Imeson, 1999; Ludwig, Wiens and Tongway, 2000). A consistent definition of the term, however, remains difficult to discern from published studies in hydrology and geomorphology.

Once formed, runoff on dryland slopes may be highly discontinuous because of spatial and temporal variability in precipitation and because of spatial variability in surface properties. Reinfiltration of runoff is usually called runon on hillslopes, although once in concentrated flows in rills or gullies, the channel-based term of transmission loss is more generally used. Again, this distinction is more one of terminology rather than reality, and there is essentially a continuum between the two. Similarly, the hydraulics of overland flows was characterised in the early literature as either sheetflow (Horton, 1945) or rill flow. Since Emmett (1970) challenged the realism of unconcentrated flows as occurring as continuous 'sheets' and numerous studies since (e.g. Abrahams, Parsons and Luk, 1989; Huang, 1990; Baird, Thornes and Watts, 1992; Dunkerley, 2004; Parsons and Wainwright, 2006) have emphasised the lateral variability of such flows into distinct threads of faster and slower flows (Figure 11.2), it is disappointing that the use of the term still persists. While there may be a continuum in form between flow threads and rills (and ultimately gullies), there may still be a useful threshold in terms of process (Bull and Kirkby, 1997; Kirkby and Bracken, 2009). In particular, as flow concentrates so that it becomes sufficiently competent to produce sediment detachment and thus the formation of rills (and gullies), flow depths and erosion rates typically increase by an order of magnitude, modifying the nature of feedbacks to the infiltration and runoff processes. The principal characteristic of overland flows on hillslopes is the high relative roughness of the surface compared to the flow depth. This roughness significantly affects the pathways and rates of flow transfers and the scientific and methodological basis for understanding these processes is still relatively young. Recent attempts to combine all of the relevant information have been within a connectivity framework.

11.2 Infiltration processes

Infiltration is the critical threshold for understanding runoff generation and thus flooding and related geomorphic processes. This importance is reflected by the amount of research carried out to evaluate and predict infiltration rates under specific conditions. Yet, in many cases, the definition of this threshold is far from simple and many applications result in unhelpful calibrations of models so that any understanding achieved through measurement is undone. The difficulties in making such predictions lie in part in the use of simplified conceptual models, which are problematic in a range of conditions that pertain in drylands. The standard model for infiltration starts off with an exposition of Darcy's 'law':

$$q = -K_s \frac{dH}{dz} \tag{11.1}$$

where q is the rate of flow [L/T], K_s is a constant usually called the saturated hydraulic conductivity [L/T] and dH/dz [L/L] is the pressure gradient. The negative sign relates to the convention of measuring the pressure gradient negative downwards. It is informative to recognise that the law is not a strict one scientifically, but was based on empirical observations (Henri Darcy was an engineer responsible for the water supply in the French city of Dijon, who was interested in characterising the rates of flow through beds of sand used to filter the water). It is based on a number of assumptions, specifically homogeneity of material, saturated flow, steady-state conditions and relatively low rates of viscous flow. Indeed, when these conditions are met, the equation can be derived from the Navier–Stokes equations, which describe fluid motions more completely (Neuman, 1977; Hassanizadeh, 1986). Standard methods exist for determining K_s as a function of particle-size data, based on laboratory testing of homogeneous materials (e.g. Campbell, 1985), but most arid region soils are rather heterogeneous in nature.

Most dryland soils are rarely saturated, but conditions of unsaturated infiltration were addressed in the early twentieth century by Buckingham (1907) and more conceptually by Richards (1931), who coupled Darcy's equation to a one-dimensional continuity equation. The Richards equation exists in several forms, the most straightforward of which is

$$\frac{\partial \theta}{\partial t} = \frac{\partial}{\partial z}\left(K(\theta)\left[1 + \frac{\partial \psi(\theta)}{\partial z}\right]\right) \tag{11.2}$$

where θ is the volumetric soil-moisture content [L^3/L^3], $K(\theta)$ is the hydraulic conductivity [L/T], i.e. the hydraulic conductivity as a function of the moisture content of the soil (and equal to K_s when the soil is saturated), and $\psi(\theta)$ is the soil suction as a function of the moisture content [L]. At saturation, $K(\theta) = K_s$ by definition, and the soil suction is zero, so that Equations (11.1) and (11.2) can be seen to be equivalent at this point. Soil suction is a highly nonlinear and hysteretic function of soil moisture, which is one reason why the Richards equation is extremely difficult to solve in practice (see Baird, 2003). Simplified approximations $K(\theta)$ and $\psi(\theta)$, usually ignoring the hysteresis, have been developed from laboratory testing (e.g. Brooks and Corey, 1964; Van Genuchten, 1980). Where soils are layered, the values of the parameters can be estimated for each layer, but this approach does not overcome lateral heterogeneity, which is another important characteristic of dryland soils.

The steady-state assumption of the Darcy–Richards approach is also rarely met in dryland conditions, not least because of the high temporal variability of rainfall when it does occur. Not only will the rainfall input be changing dramatically but also will the pressure head. Unsteady flows may also develop from feedbacks within the system. De Rooij (2000) reviewed the literature on infiltration that occurs as discrete fingers rather than as a uniform wetting front and suggested that unsteady infiltration is likely to occur when there is acceleration of the flow through the profile. In dryland soils, a higher flow rate just below the surface is likely to occur in soils that have high stone covers (especially at times when vesicular horizons have developed below the stones), mechanical or microbial crusts, or water-repellent layers developed from burning (see below). Furthermore, the temperature of the surface will also have a significant effect on producing such feedbacks. It has been noted that

$$K_s = k\frac{\rho g}{\eta} \qquad (11.3)$$

where k is the effective permeability of the soil, ρ is the density of the fluid (kg/m^3), g is acceleration due to gravity (m/s^2) and η is the viscosity of the fluid (m^2/s) (Richards, 1931; Swartzendruber, 1962), which means that K_s will be 50 % higher at 20 °C than 5 °C, and so will be typically accelerating when relatively cold convective rainfall hits the warm ground and starts to infiltrate. These changes are dominated by the temperature control on the fluid viscosity. In reality, the fluid is not pure water but a changing mix of water and solutes. An increase in solute content as the precipitation starts to interact with the surface and subsurface soil is likely to counter the amount of acceleration noted above, although the extent to which this will occur is a function of the soil and precipitation chemistry, and has received little investigation (but see Levy, Goldstein and Mamedov, 2005).

The assumption that relatively low rates of viscous flow are maintained again works well for homogeneous, fine soils, but can start to break down in heterogeneous soils and soils where macropores or cracks occur. In the former case, the pressure gradient becomes ill-defined (Beven, 2001), whereas in the latter, flow is too rapid. Disagreement exists in the literature as to whether the breakdown is due to the onset of turbulent flows or whether it occurs in transitional 'nonlinear' flows (Hassanizadeh and Gray, 1987).

It should be clear from the discussion above that standard techniques for the assessment of hydraulic conductivity do *not necessarily* give an indication of the infiltration rate, either because of conceptual limitations or because of the specific nature of dryland soils. For these reasons, it is preferable *not* to use the terminology of $K(\theta)$ or K_s to refer to infiltration characteristics; instead they will be called unsaturated and saturated (or final) infiltrations rates in the discussion below. Measurement of infiltration also has an implicit scale. It is not applicable at the scale of individual pores, and begins to break down conceptually (not least because of the difficulty of identifying the necessary pressure head driving the process) at larger scales (e.g. Reggiani *et al.*, 1999; Beven, 2001). It is thus necessary to identify the range of factors that affect infiltration rates in drylands before returning to the discussion of practical methods of estimating infiltration at different spatial and temporal scales.

11.3 Factors affecting infiltration

11.3.1 Controls at the surface–atmosphere interface

Many dryland surfaces are very stony (Chapter 7). Cooke, Warren and Goudie (1993) put forward a range of mechanisms to explain stone concentrations at or near the surface, including deflation of fines by wind, removal of fines by splash and water erosion, and temperature or wetting/drying cycles. It is likely that all of these processes are active to a greater or lesser extent at a specific location. The predominance of physical rather than chemical weathering (Chapter 6; see Wainwright, 2009b) also helps to explain the initial presence of many stone fragments, which persist because of the low and discontinuous presence of water in the profile.

On pavement surfaces, Musick (1975) measured infiltration rates of more than an order of magnitude lower than on adjacent unpaved surfaces in the Sonoran Desert at Yuma in Arizona. At Walnut Gulch, Arizona, each one percent increase in pavement cover led to a measured decrease of final infiltration rate of 0.36 to 0.84 mm/h^{-1} (Abrahams and Parsons, 1991a; Parsons *et al.*, 1997). Higher percentage pavement covers also produced much more rapid changes in unsaturated infiltration, so that the final infiltration rate was reached much more rapidly.

Poesen (1992) found that the effect of stones in general could be of the same order of magnitude as noted in the above studies. However, he noted that it is important to distinguish between stones that are embedded in the surface, which are likely to reduce infiltration as suggested above, and stones that are lying on the surface, which seem to *increase* the amount of infiltration (Bunte and Poesen, 1994). The latter effect seems to be due to the prevention of seals developing around the stones (Figure 11.3). The spacing between stones is also important, not least because of feedbacks with flow hydraulics and small-scale runoff–runon linkages (see below; Kumar Mandal *et al.*, 2005).

Bare surfaces may also exhibit extensive amounts of crust development. Mechanical crusts occur typically on sandy soils. Valentin and Bresson (see Valentin and Bresson, 1992; Casenave and Valentin, 1992; Valentin, 1993) distinguish between structural, erosional and depositional crusts (Figure 11.4). Structural crusts can be formed by a sieving mechanism, in which loose, coarser sands overlay a well-packed finer sand, silt and clay horizon by the repeated action of raindrop impacts. Impacts from saltating aeolian sand would have a similar effect. Pavements are seen as the extreme end-member of this type of crust.

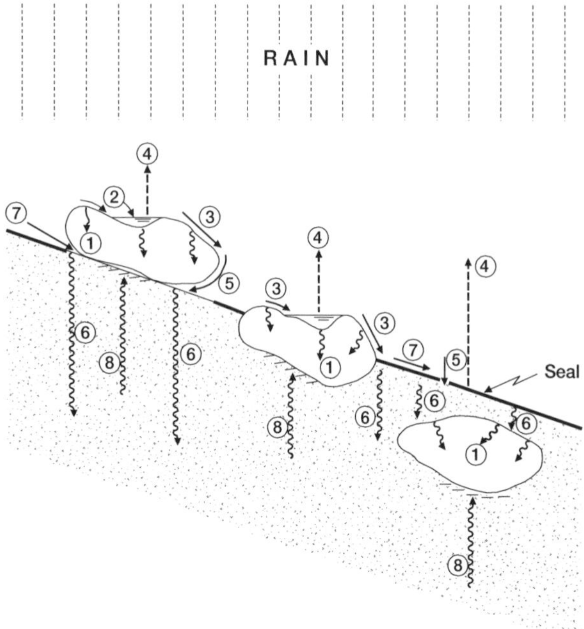

Figure 11.3 Hydrological processes occurring on and in soils with rock fragments at or near the surface: 1, water absorption; 2, interception and depression storage; 3, rockflow; 4, evaporation; 5, infiltration; 6, percolation; 7, overland flow; 8, capillary rise (from Poesen and Lavee, 1994).

Valentin and Bresson's second type of structural crust occurs by slaking, and is perhaps better considered as a surface seal (see below). Erosional and depositional crusts are the result of selective entrainment and deposition of soils, resulting in the formation of smooth surfaces where surface pores are infilled with silt- and clay-sized particles. Depositional crusts can also be in the form of saline deposition, which are most closely associated with playas (Chapter 15). Graef and Stahr (2000) show that crusts are almost ubiquitous in Niger but vary as a complex function of soil, slope and land-use conditions, so that very different crust types can be found over spatial scales of a few square metres.

Valentin (1993) gives an example of an erosional crust in Burkino Faso, where the final infiltration rate increased from 1.6 to 6 mm/h when the crust was destroyed by tillage. However, over an entire rainfall season, the difference was less marked (285 mm compared to 241 mm total runoff), which is likely to be due to the reformation of the crust after the initial storm events. Puigdefábregas *et al.* (1999) noted the presence of sieving crusts based on soil micromorphological evidence at the Rambla Honda in southern Spain, but noted that the infiltration rates associated were much higher than those measured in West Africa. Thus, it is important to recognize that crusts seem to modify the absolute rates of infiltration based on other soil characteristics, rather than providing the principal control. Martinez-Mena, Castillo and Albaladejo (2002) demonstrated that crusts also formed rapidly on marly soils in southern Spain, so that they are by no means just limited to sandy soils.

There is an increasing body of work which suggests that biological processes are also highly significant in crust formation. Belnap (2006) provides an excellent review of biological crusts and their effects in drylands. Biological crusts may be formed by cyanobacteria and lichens, but green algae, microfungi, bacteria and bryophytes may also live in the top few millimetres to centimetres of the soil. They bind the soil together both directly, e.g. in the form of filaments that can be seen with the naked eye if a section of crust is held to the light, and chemically, by the production of polysaccharides secreted on the outer surface of the organisms. Belnap notes that many previous studies on the effects of biological crusts have been inconclusive because they fail to account for the different types of crust and confounding variables that are modified in standard experimental treatments. She aims to overcome these limitations by defining four types of crust characteristic of different environments (Figure 11.5). Typically, as conditions go from hyper-arid to semi-arid then cool and cool-cold semi-arid, cyanobacteria in the crust decrease and lichens increase, leading to smooth, rugose, pinnacled and rolling crusts, respectively. However, cyanobacteria will be dominant in very sandy or saline soils, or in locations where there is a high proportion of swelling clays; lichens favour carbonate-, gypsum- or silt-rich soils, leading to crusts of the different types in other climate régimes. Pinnacled crusts have the biggest effect on promoting infiltration by producing hollows in which ponded water infiltrates without running off. Rolling crusts may increase infiltration to a lesser extent, while the smooth and rugose crusts tend to decrease infiltration by blocking the pores at the surface. Belnap *et al.* (2005) note that biological crusts may make up as much as 70 % of the surface of dryland soils. They also suggest that behaviour of biological crusts is likely to be dynamic with relation to the size, duration and temporal spacing of rainfall events. After extended dry periods, the first wetting may kill 30 to 50 % of the bacterial biomass, so that subsequent events will be less affected until regrowth occurs. Regrowth can be rapid during a monsoon season, for example, but depends on the exact timing and magnitude of subsequent rainfall pulses (Cable and Huxman, 2004). For this reason, infiltration and surface-roughness characteristics may be expected to be highly dynamic at different times in the year. Biological crusts may lead to other spatial feedbacks in that they are typically orientated in an east–northeast direction to

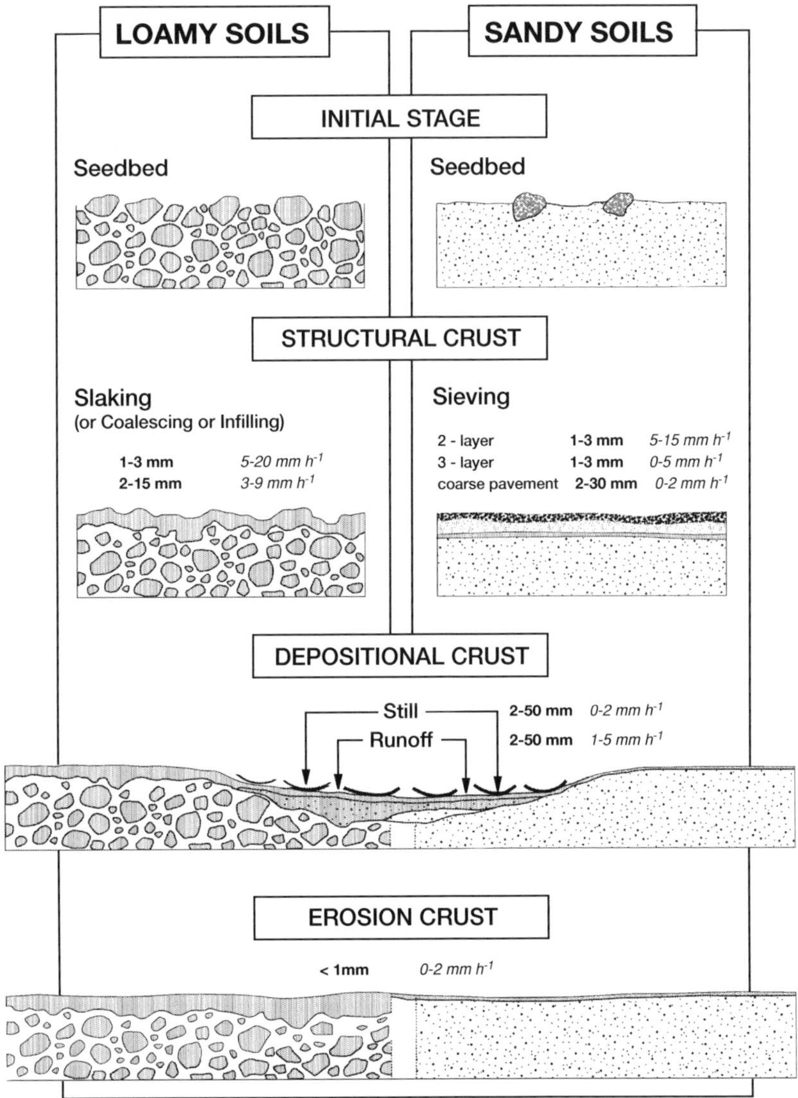

Figure 11.4 Development of crust types and characteristics based on observations in West Africa by Valentin and Bresson (1992).

maximize photosynthesis, thus affecting flow pathways on low-angled slopes in semi-arid regions. Kidron (1999) also found a distinct difference between runoff on north-facing slopes of longitudinal dunes in the Negev compared to the south-facing slopes.

The latter typically had half the bacterial biomass and produced 3.2 times less runoff. Maestre *et al.* (2002) also found biological crusts in the interspace between *Stipa tenacissima* grass clumps in southeast Spain and suggested that they were important controls, increasing runoff in these locations and making it available downslope for the grasses, in a similar way as has been suggested for semi-arid sites in the US and elsewhere (Belnap, 2006). Hyper-arid areas can often produce significant biological crusts, sustained by the interception of fog and dew as a moisture source (Kidron, 2005; Lange *et al.*, 2007; Shanyengana *et al.*, 2002; Warren-Rhodes *et al.*, 2007).

Sealing of the soil surface can occur by slaking of aggregates of fine particles following wetting, so that clays in particular collect in surface pores, producing highly impermeable surfaces (Figure 11.6). Where swelling clays are present, the surface may become essentially totally impermeable (Sposito, 1972). The extent to which surfaces are vulnerable to slaking can be measured by evaluating the aggregate stability, e.g. by observing the number of drops of water required for the aggregate to start to

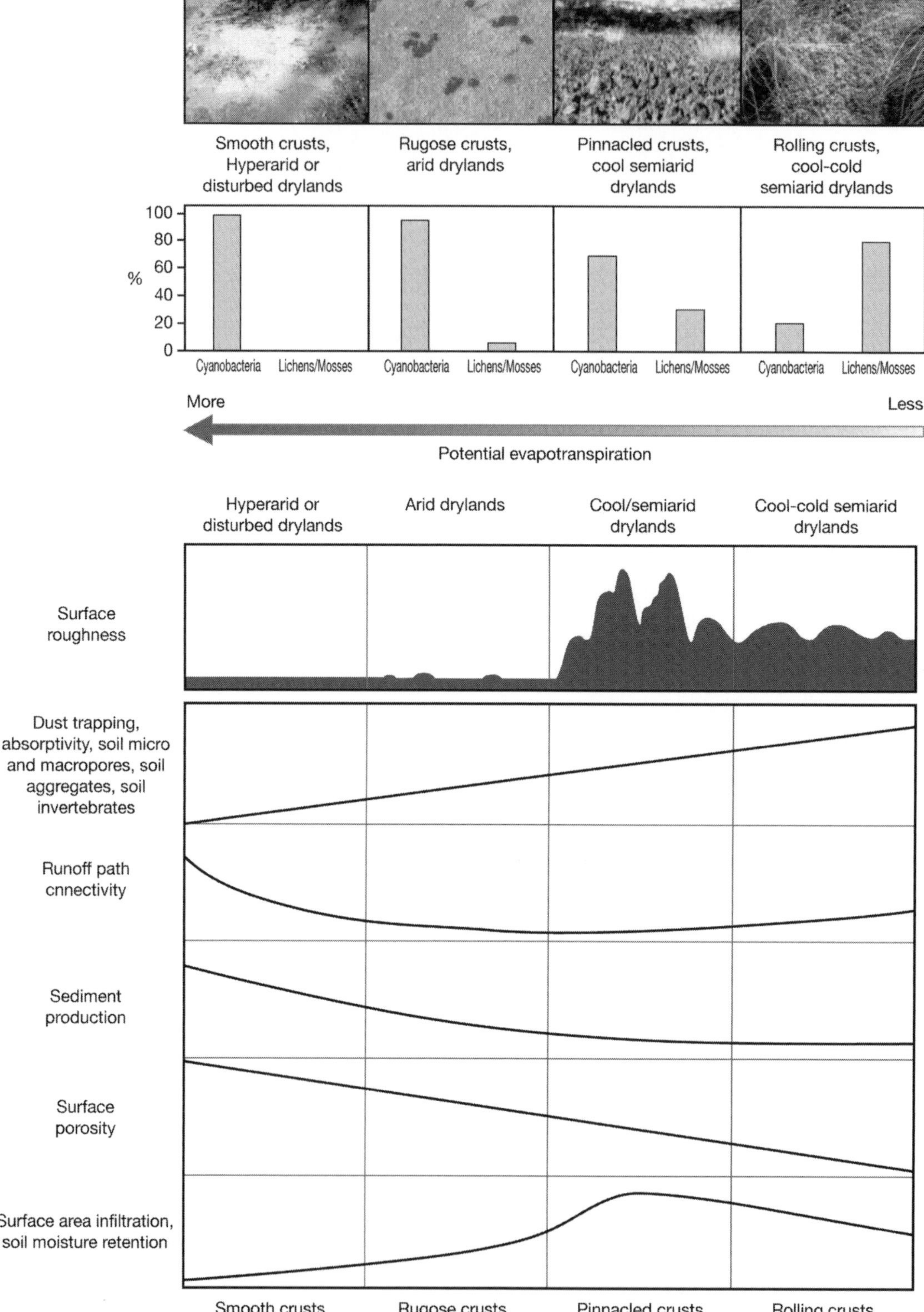

Figure 11.5 Classification of biological crust types and their hydrologic and geomorphic effects (Belnap, 2006).

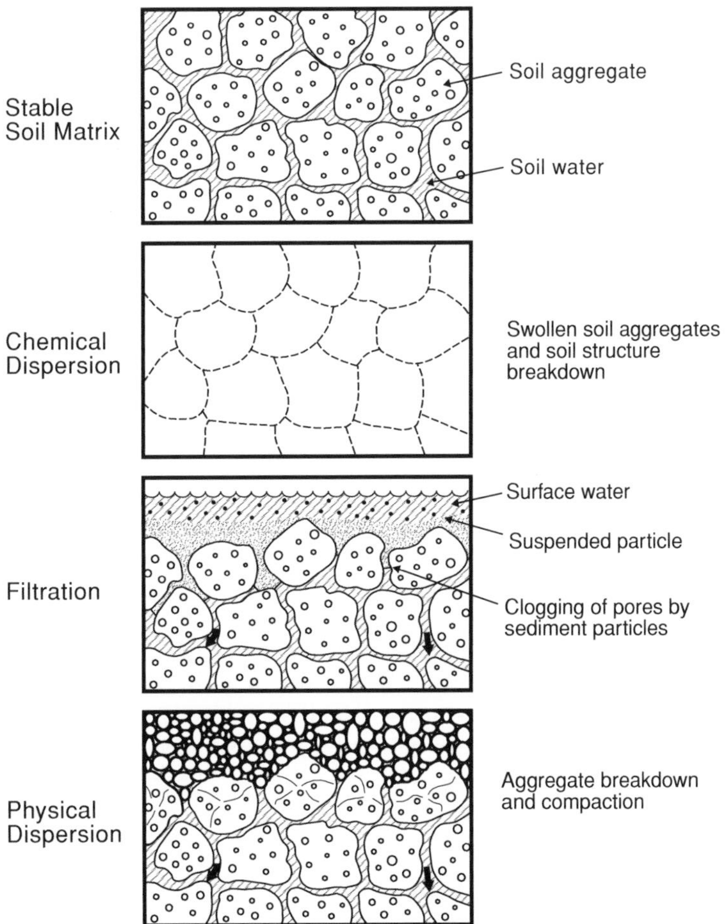

Figure 11.6 A schematic representation of the different mechanisms of seal development (Römkens, Prasad and Whisler, 1990).

break up. Le Bissonnais (1996) has reviewed the various mechanisms of aggregate break-up – slaking, breakdown by differential swelling, mechanical breakdown by raindrop impact and physicochemical dispersion (Mualem, Assouline and Rohdenburg, 1990; Römkens, Prasad and Whisler, 1990) – and suggested that as well as the different effects of soil size and chemistry, the rate of wetting is also important. Thus, larger storm events are likely to increase the rate of aggregate breakdown and thus produce more sealed surfaces. Abu-Awwad (1997) noted that crusts on silty, calcareous soils with low organic matter components in Jordan tended to form after any wetting of the soil, so that the impact of wetting rate will depend on the initial soil conditions. However, Luk and Cai (1990) demonstrated using micromorphological analyses that seal formation could be dynamic within a single event, so that the effects on infiltration could both increase and decrease. Mualem, Assouline and Rohdenburg (1990) suggest that most seals are complex and may be formed by a variety of these mechanisms, so that the controls on their formation and destruction during a single event is also likely to be highly variable. Aggregate stability is reduced in dryland soils because of the relative paucity of organic matter, and the spatial pattern of organic matter – e.g. clustered beneath shrubs or grass clumps – is likely to provide a feedback in terms of infiltration and vegetation patterns (Dunne, Zhang and Aubry, 1991; Boix-Fayos *et al.*, 1998). There is also a feedback between sediment contained in the infiltrating water and crust formation. Abu-Sharar and Salameh (1995) measured reductions of 80–85 % in the infiltration rate when the water contained as little as 0.07 % fine clay, as compared to infiltration from clear water. Thus, as a storm event continues and more fines are moved in the flow, there is likely to be a dynamic feedback with runoff, infiltration and erosion rates.

The effects of vegetation in drylands are frequently overlooked, yet have a significant impact (Wainwright, 2009a). Vegetation can intercept rainfall, reducing the amount of water available at the ground surface for infiltration. Abrahams, Parsons and Wainwright (2003) measured the maximum canopy storage in the desert shrub creosotebush, which was found to be 5 mm of rain, and a similar amount was found for juniper (Owens, Lyons and Alejandro, 2006). In small events, this effect might be significant, so that over a rainfall season in New Mexico, Martinez-Meza and Whitford (1996) estimated about 44 % of rainfall was intercepted by creosotebush, about 45 % for tarbush and 37 % for mesquite. Dunne, Zhang and Aubry (1991) note that the interception of rainfall also increases infiltration by the reduction of rainfall energy, which in turn reduces the likelihood of crust formation. Even for sparse shrub canopies, Wainwright, Parsons and Abrahams (1999) demonstrated a 30 % reduction in energy in this way. Stemflow generated by plants can, on the other hand, increase the point source of precipitation at the ground surface by concentrating it from the larger canopy area. Proportions of 5–10 % of precipitation have been commonly reported for stemflow (Martinez-Meza and Whitford, 1996; Owens, Lyons and Alejandro, 2006), although values as high as 42 % have been recorded for acacia and eucalyptus species in Australia (Pressland, 1973; Nulsen *et al.*, 1986). Abrahams, Parsons and Wainwright (2003) note that it is important to look at shrub understoreys to evaluate the impact of stemflow.

Where no understorey was present, they found that the high point source can lead to some shrubs producing small amounts of runoff in a storm event before surrounding bare surfaces start to pond. However, when a grass understorey was present, all of the stemflow was usually found to infiltrate – with an important feedback to water availability to maintain plant growth. Spatial patterns that emerge as a result of the feedbacks between increased infiltration under and near vegetation are recognized as an important characteristic of the ecogeomorphology of drylands in many parts of the world (Cerdà, 1997; Dunkerley, 2002; Janeau, Mauchamp and Tarin, 1999; Gutierrez and Hernandez, 1996; Pariente, 2002; Wainwright, Parsons and Abrahams, 2002; Ludwig *et al.*, 2005; Van Der Kamp, Hayashi, and Gallen, 1967; Smit and Rethman, 2000).

Disturbance of the surface through digging by small mammals has been noted to have different effects on infiltration. Moorhead, Fisher and Whitford (1988) and Mun and Whitford (1990) have suggested that such digging increases porosity and thus infiltration in the Chihuahuan Desert, although Neave and Abrahams (2001) were unable to demonstrate a significant different in water yield between areas with digging and those without. Larger mammals can also have an impact on infiltration, most notably by compaction, although Hiernaux *et al.* (1999) noted in Niger that compaction was only significant at high stocking densities. At moderate rates of animal presence, infiltration rates were observed to increase slightly due to the effect of the animals on breaking up existing crusts. There are also likely to be increases in such conditions due to enhanced aggregate stability in the presence of higher levels of organic matter.

Fire is an important process in many drylands (Naveh, 1975; Wainwright and Thornes, 2003; Snyman, 2003). The effect of burning is dependent on temperature conditions reached during a fire, which is a function of fuel availability as well as temperature, wind and other local conditions. Giovannini, Lucchesi and Giachetti (1988) found experimentally that sand content changed little in burnt soils for temperatures up to 170 °C, but between 220 and 460 °C there was a rapid rise so that sand became the dominant particle size due to aggregation and fusing of particles, but higher temperatures produced little further effects. Loss of organic matter, and thus aggregate stability is important, especially at temperatures above 170 °C (Giovannini, Lucchesi and Giachetti, 1990). Water repellency can be developed as organic matter is vapourized at or below the surface (as a function of the temperature reached and its transmission through the soil profile; see Neary, Ryan and DeBano, 2008), although Cerdà and Doerr (2007) have suggested that calcic soils may reduce the likelihood of water repellency occurring. Certain vegetation types – such as pines in the Mediterranean and creosotebush in the US – are more likely to release water-repellent chemicals into the soil. Thus, there is likely to be a significant spatial variability of the effects of fire, with some locations having reduced infiltration and others increased (Imeson *et al.*, 1992; Lavee *et al.*, 1995).

Water repellency can also develop in many dryland soils as a result of the breakdown of phenolic, turpene and related compounds from plants at ambient temperatures. There have been suggestions that these substances commonly evolve in dryland plants as a way of competing with neighbours (but see the discussion in Fitter and Hay, 1987) or as a means of reducing herbivory (Hyder *et al.*, 2002). Beyond the reduction of infiltration rates that water repellency produces, Ritsema and Dekker (2000) have demonstrated that in sandy soils it can lead to the development of significant amounts of preferential – sometimes known as 'fingered' – flow, which is a further reason why the Darcy–Richards approximation is problematic in such conditions.

11.3.2 Subsurface controls

A major subsurface control on infiltration is due to the effects of roots, which produce distinct flow pathways

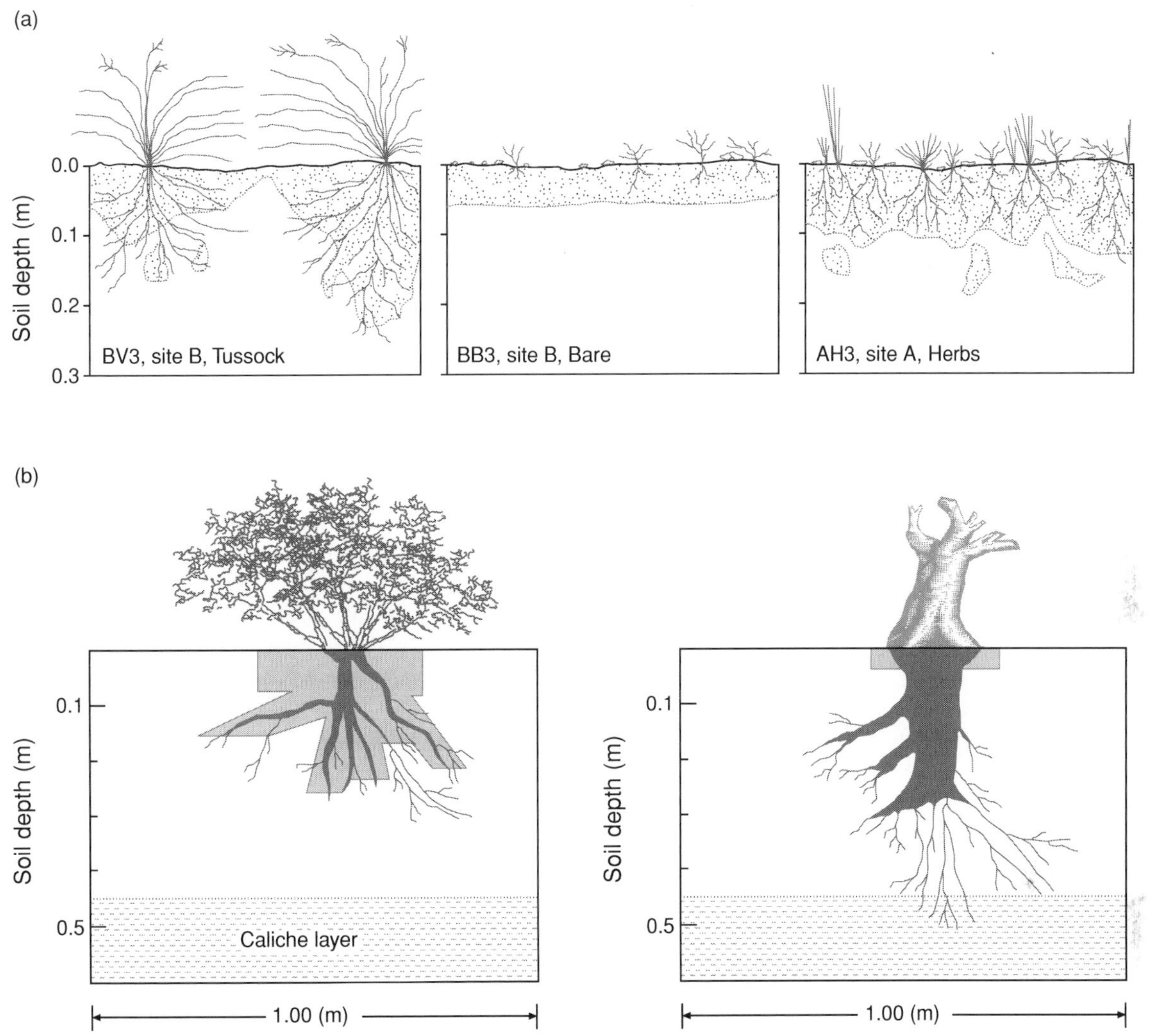

Figure 11.7 The use of dye-tracing studies to demonstrate the effects of plant roots on subsurface flows in storm events: (a) comparison of tussock grass (*Stipa tenecissima*), bare and small shrubs in Alacant and Murcia, Spain (Cerdà, 1997) and (b) comparison of creosotebush (*Larrea tridentata*) and honey mesquite (*Prosopis glandulosa*) in New Mexico (Martinez-Meza and Whitford, 1996).

as they grow and extend, and form macropores as the plant dies (Dunne, Zhang and Aubry, 1991; Devitt and Smith, 2002; Leffler *et al.*, 2005). Root decay seems to be relatively rapid in drylands where breakdown is aided by termite activity (Whitford, Stinnett and Anderson, 1988), and such breakdown typically produces organic matter, which increases aggregate stability. Dye-tracing experiments have been used in a number of dryland environments to demonstrate the close linkage between subsurface flow pathways and plant roots (Cerdà, 1997; Martinez-Meza and Whitford, 1996; Wang *et al.*, 2007; Ryel *et al.*, 2003, 2004) (Figure 11.7). Archer, Quinton and Hess (2002) suggested that the difference in root structure between shrubs and grasses in Spain produced a higher apparent hydraulic conductivity in soils beneath the former.

Because saturation is rarely reached or maintained in drylands, lateral throughflow is typically rare. Similarly, vertical percolation is low and in many cases leads to the development of petrocalcic or other subsurface horizons (Chapter 7). The difficulty plant roots have in penetrating these horizons – most likely because of feedbacks with soil-water chemistry – is often a significant feedback on the ability of desert plants to maintain themselves (Cunningham and Burk, 1973; McAuliffe, 1994; Escoto-Rodríguez and Bullock, 2002; Gibbens and Lenz,

2001). Other developments of soil horizonisation, either at the surface (Abrahams and Parsons, 1991a; Young *et al.*, 2004) or in the subsurface (Hamerlynck, McAuliffe and Smith, 2000; Hamerlynck *et al.*, 2002), have been noted as having significant effects on reducing infiltration rates.

As throughflow and deep percolation are rare, mechanisms of spatial variability in water flow are largely surface-driven, as a result of runoff–runon processes (Wainwright and Parsons, 2002; Rockström, Jansson and Barron, 1998; and see the further discussion below) or as a result of the pattern of vegetation, or the interaction of the two. Vegetation draws moisture out of the soil by transpiration, and there are increasing numbers of observations of upwards vertical movement of moisture by a process known as hydraulic lift (Richards and Caldwell, 1987; Yoder and Nowak, 1999). This process occurs when there are relatively saturated conditions at depth, allowing plant roots to take up water and move it upwards internally until transpiration stops (e.g. at night-time), at which point the water may flow back out of root pores into surrounding drier soils because of the hydraulic gradient.

Soil-moisture variability in space and time is important in controlling infiltration and runoff generation by the control it exerts on the hydraulic gradient at the start of a rainfall event. Turnbull, Wainwright and Brazier (2010) demonstrated that both grass and shrub vegetation in the Chihuahuan Desert caused the more rapid drying of the soil than on bare surfaces (Figure 11.8) – suggesting the greater importance of transpiration compared to evaporation – and the evolution of related spatial structure in the soil moisture (with the further feedbacks noted above). As storm events may occur more rapidly than the decline to 'background' dry conditions as seen in this example and in Ceballos *et al.* (2002), the differential effect and the development of connectivity of flows (see below) as affected by vegetation in this way may be critical in understanding variability in the behaviour of the system during the rainfall season.

Subsurface animal activity may also be an important control on infiltration. Areas with termites in unvegetated areas in New Mexico were found to have significantly higher infiltration rates (88 $\pm$ 6 mm/h) than areas without termites (51 $\pm$ 7 mm/h) by Elkins *et al.* (1986). Cammeraat *et al.* (2002) also suggested that ant nests in Spain contributed to changes in infiltration by physical modification of the soil, but that chemical changes caused the behaviour to vary according to moisture conditions (and, thus, the time of year). The breakdown of plant materials by the ants produces conditions of water repellency in the soil, so that, for initially dry conditions, areas with ant nests will have relatively low infiltration rates, and initially wet conditions or low intensity storms allowing wetting to occur will produce higher infiltration rates. The presence of burrows produces significant macropores in the soil, and these burrows may often be co-located with vegetation as animals attempt to avoid predation, which is another reason why infiltration rates under plants may be higher than elsewhere.

Macropores may also form where the soil surface becomes cracked. In drylands, this phenomenon most commonly occurs as a result of desiccation, providing a rapid link from the surface to subsurface flows. It is commonly found in badlands, especially in those dominated by very fine clays, particularly montmorillonite (Bryan, Imeson and Campbell, 1984), and where they can develop into large-scale pipes (Torri and Bryan, 1997; Farifteh and Soeters, 1999). Govers (1991) has suggested that the size and spacing of desiccation cracks depends on landscape location, with those on slopes being larger and closer together than those on plateaus and aspect-related differences in insolation increasing the extent of cracking.

A further subsurface effect is that of air entrapment. Baird (1997) suggested that the entrapment of air in the profile can cause pores to be blocked to the passage of infiltrating water, citing experimental evidence from Constantz, Herkelrath and Murphy (1988) that this effect could lead to a reduction of 80–90 % of the saturated infiltration rate. Although rapid inputs of precipitation should mean this effect is significant, initial assessments suggested that it might be less important on the hillslope scale because of the lateral disconnection of wetted areas (Baird and Wainwright, unpublished modelling results), more recent literature suggests that the effect is important in irrigation settings in drylands (Hammecker *et al.*, 2003; Navarro *et al.*, 2008). It is thus also likely to be important in transmission losses in rills and channels, and certainly flood bores in the latter show the rapid mixing of air and water.

11.4 Runoff generation

11.4.1 Ponding and surface storage

As a precipitation event proceeds, the first stage of the runoff process is the ponding of water at the surface. Where local irregularities produce pits in the surface, these pits will form depression storage. Continuing rainfall will cause any such storage to fill, and thus produce interconnected overland flow. The nature of this flow is then controlled by the surface hydraulics. The time taken for ponding is a function of initial conditions, precipitation rate and soil conditions. In the simplest case it can be estimated for constant-intensity rainfall by rearrangement of the Philip, Green and Ampt or similar equations.

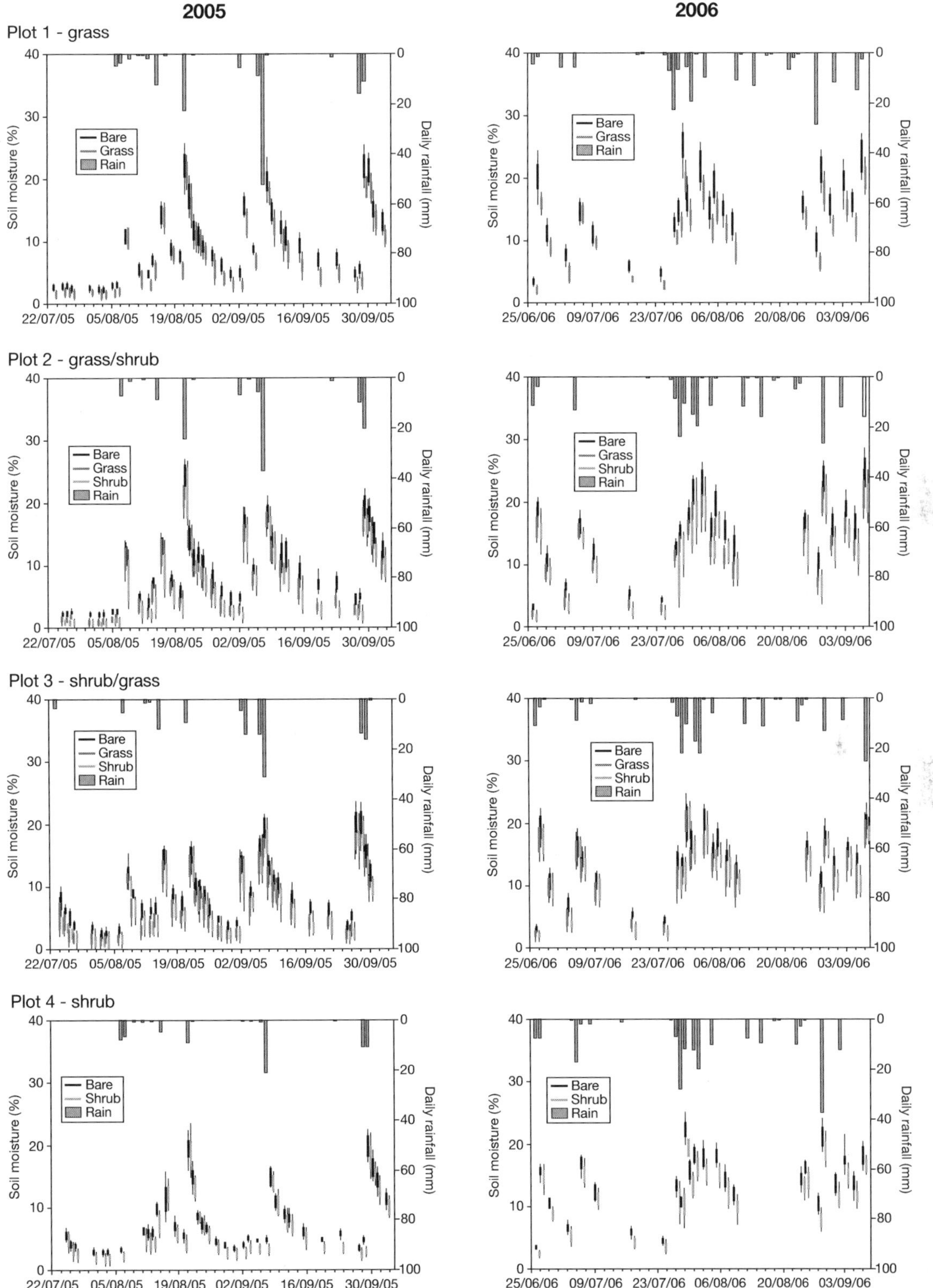

Figure 11.8 Variations in soil moisture within and between rainfall seasons (2005–2006) as a function of storm size, spacing and vegetation type from measurements by Turnbull, Wainwright and Brazier (2010) at the Sevilleta LTER site, New Mexico. Plot 1 is dominated by grass (*Bouteloua erioipoda*), Plot 2 is grass with some creosotebush (*Larrea tridentata*), Plot 3 is creosotebush with some grass and plot 4 is dominated by creosotebush.

For example, if the Philip approximation to the Richards equation is employed (with the numerous caveats noted above), the time to ponding (t_p [T]) can be estimated as

$$t_p = \left(\frac{S}{2[r - i_f]} \right)^2 \qquad (11.4)$$

where S is a parameter relating to the unsaturated infiltration rate [L/T$^{0.5}$], r is the rainfall intensity [L/T] and i_f is the final infiltration rate at saturation [L/T]. A further constraint of this equation is that it is only valid for constant rainfall intensities, which, as noted above, are rare in dryland storms. In reality, such calculations may be useful for evaluating relative spatial differences in the onset of runoff, but the complexities of the infiltration process mean that absolute estimates derived in this way need to be treated with caution.

At the plot scale, microtopography influences surface depression storage (Onstad, 1984), infiltration and its variability (Fox, Le Bissonnais and Bruand, 1998), evaporation (Allmaras *et al.*, 1977), solar radiation reflection (Allmaras, Nelson and Hallauer, 1972), overland flow hydraulics (Gilley and Finkner, 1991; Takken and Govers, 2000), soil erosion and sediment transport (Helming, Römkens and Prasad, 1998; Gómez and Nearing, 2005) and water routing (Dunne, Moore and Taylor, 1975; Darboux *et al.*, 2002). The detention of water at the soil surface is particularly important where the infiltration rate is slightly lower than the rainfall intensity and plays a regulatory role in the generation of surface runoff (Abedini, Dickinson and Rudra, 2006). This situation of precipitation excess is often found in semi-arid areas where high-intensity storms fall on soils that may exhibit a relatively low infiltration capacity. At larger scales surface roughness also affects flowpaths of subsequent runoff, the organisation of drainage patterns and the connectivity of the landscape (Govers, Takken and Helming, 2000; Bracken and Croke, 2007). Somewhere within these complex relations, feedbacks and interrelationships between vegetation including type, structure, age and pattern (Thornes, 1976; Domingo *et al.*, 1998), surface gradient (Govers, 1991) and soil (including type, distribution and crusting) (Poesen, Ingelmo Sanchez and Mucher, 1990; Le Bissonnais, 1996) also influence runoff and flowpath generation.

11.4.2 Flow hydraulics

Overland flow may be dominated by flow dynamics (where infiltration can be ignored) or, alternatively, flow velocity may be dominated by infiltration rate (e.g. Lei *et al.*, 2006). The reality will inevitably lie somewhere between these two extreme possibilities, depending on surface conditions. The Darcy–Weisbach, Chézy and Manning equations are the most widely used empirical resistance equations employed for velocity calculation in hydrological and erosion models (Jetten, de Roo and Guerif, 1998; Takken *et al.*, 2005), but these equations were developed from either experiments in pipes or observations from open-channel flows. Smith, Cox and Bracken (2007) chart the historical development of these equations and underline the difficulty of developing resistance equations for steep channels and overland flows where the assumptions underlying the more conventional approaches limit their range of applicability. Equations to predict resistance coefficients from measures of surface roughness typically relate the velocity-profile 'roughness height' of the Keulegan (1938) equation to grain-size measurements (Robert, 1990; Clifford, Robert and Richards, 1992; Lawless and Robert, 2001), and so should not be applied uncritically to overland flows where the velocity profile is likely to be distinct from that predicted by the 'law of the wall' (Prandtl, 1935). The Keulegan equation relates flow resistance to the ratio of flow depth d and a roughness height ε (defined as the inundation ratio Λ) (see also Ferguson, 2007). Similarly, the Manning–Strickler approach (Manning, 1891; Strickler, 1923) calculates resistance as a function of Λ with the added approximation that conveyance (the inverse of resistance) increases with the sixth-root of the hydraulic radius (see Smith, Cox and Bracken, 2007).

Where resistance formulae are applied to open channel flows, the focus is often simply on grain resistance. However, studies of overland flows have shown that the other components (e.g. form and wave resistance) make a substantial contribution to total resistance (Abrahams, Parsons and Hirsch, 1992; Parsons, Abrahams and Wainwright, 1994). The hydraulic equations for the friction factor and laminar/turbulent flow are used to determine the influence of roughness on surface flow. When roughness elements are fully submerged, hydraulic roughness theoretically declines with flow depth (an inverse relationship between the Darcy–Weisbach friction factor and the Reynolds number; see Savat, 1980; Govers, 1992b; Bryan, 2000). However, when roughness elements equal or exceed flow depth, the Darcy–Weisbach friction factor–Reynolds number relationship becomes positive (Abrahams, Parsons and Luk, 1986; Abrahams and Parsons, 1991b; Nearing *et al.*, 1998). This results in the decreasing importance of grain resistance (Govers and Rauws, 1986; Abrahams, Parsons and Luk, 1986), and sediment movement is controlled by form resistance. The transition from dominance by grain or form resistance

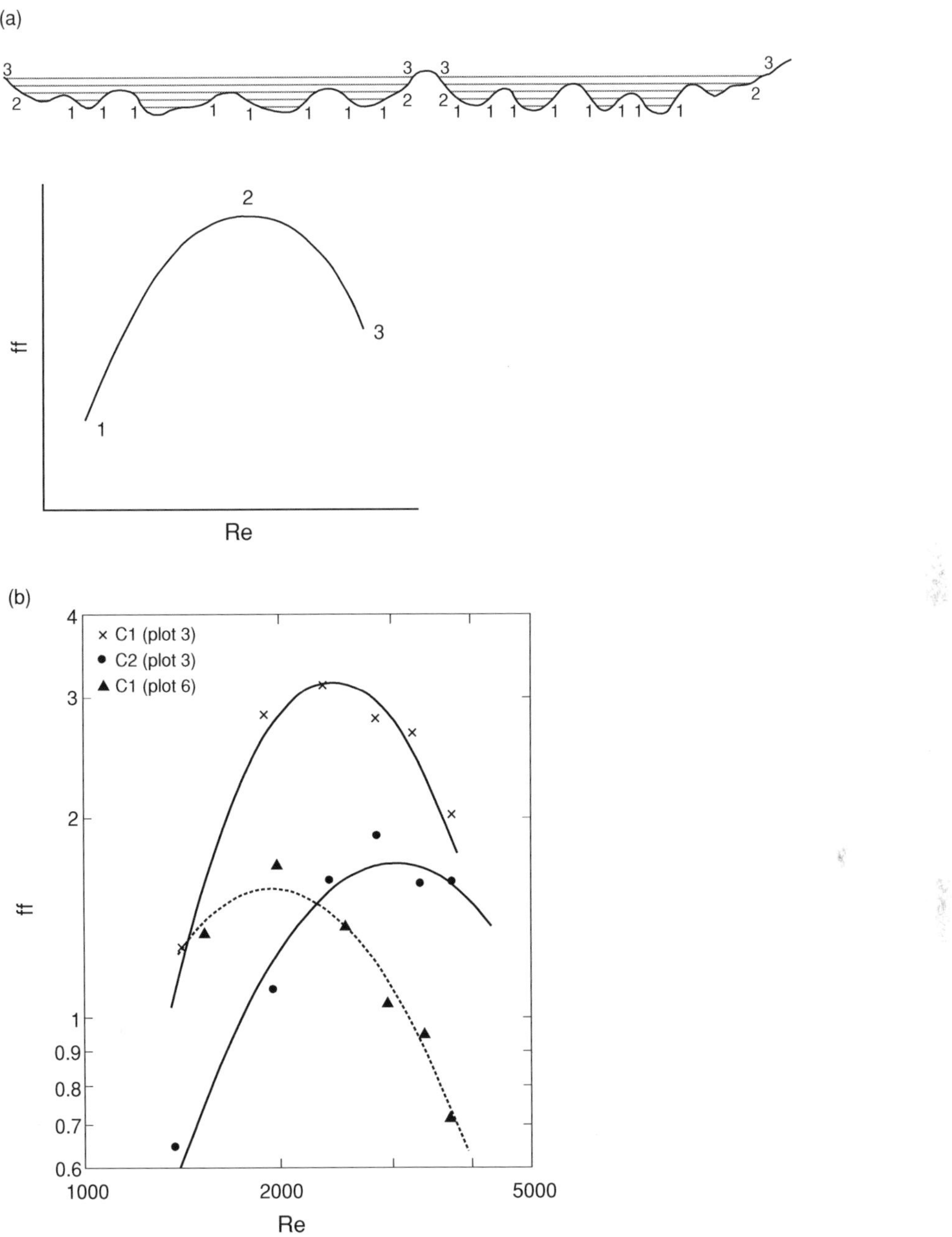

Figure 11.9 Hypothetical relationships between the Darcy–Weisbach friction factor and flow Reynolds number and measured relationships on surfaces from Walnut Gulch, Arizona (Abrahams, Parsons and Luk, 1990).

varies with soil characteristics (Bryan, 2000). It is thus necessary to relate current flow conditions to the observed roughness, with important feedbacks in the dynamics of runoff and erosion processes (Figure 11.9).

This apparent complexity raises a wider conceptual issue of what surface roughness actually *is* and, consequently, how should it be measured. Parsons, Abrahams and Wainwright (1994) demonstrated that different techniques of measurement are underpinned by different assumptions about flow roughness, and can produce very different estimates of roughness for the same surfaces. Lane (2005) defines roughness as that component of topography that must be dealt with implicitly at the scale of enquiry. Yet there is no single property that can be uniquely defined as 'surface roughness'. It can be characterised in a variety of ways, as, for example, deviations from a plane, tortuosity or semi-variance (Kuipers, 1957; Boiffin, 1984; Linden and Van Doren, 1986). Even studies using the term 'random roughness' to characterise deviations from a plane employ it in a range of different ways (Onstad, 1984). Moreover, it seems necessary to distinguish between roughness as applied in modelling studies, often as a calibration factor and a property of the flow itself implicitly representing those processes not directly included in any model including some fluid boundary (Lane and Ferguson, 2005), and a more field-based definition of roughness describing the topographic form of a surface. More generally, all processes operating at the Earth's surface interact with surface roughness (at whatever scale) in a multitude of ways. These processes often exhibit a complex relationship with roughness, and so any attempt to represent surface roughness must be sensitive to and informed by such behaviour.

Field exploration has shown that flow resistance is highly variable, even within a single plot (Smith, Cox and Bracken, 2010). Simple relationships, as predicted by conventional equations based on pipe-flow experiments, were found to provide an inadequate description of overland flow processes. The complexity of the surface forms demonstrated by natural soil surfaces and progressive inundation of roughness elements may explain the poor performance of these equations, which represent surface roughness using a single roughness measure (Smith, Cox and Bracken, 2010).

11.4.3 Pipes and macropore flow

Soil pipes form under a range of conditions, but are most common on clay-rich soils. More generally, macropores develop as a result of soil moisture or temperature cycling, plant-root decay and animal burrowing. The importance of this type of flow is that it occurs in a non-Darcian manner and thus is hydraulically very efficient. Farifteh and Soeters (1999) demonstrated in southern Italy that pipes can be extensive parts of the drainage pattern (Figure 11.10) and generally produced the first flows to reach the channel from the slope system.

11.4.4 Scales of overland flow

An overarching framework for understanding runoff and runon is the concept of hydrological connectivity. This is a relatively new term but is a subject where research is developing at an exponential rate (Bracken and Croke, 2007). Hydrological connectivity refers to 'the transfer of water and related matter from one point in a catchment to another' (Pringle, 2003; Tetzlaff *et al.*, 2007). The development of hydrological connections via overland flows is a function of both water volume (supplied by rainfall and runon, depleted by infiltration, evaporation and transpiration) and rate of transfer (a function of flow resistance). These processes interact with flow resistance varying as a function of flow depth. This interaction establishes a feedback between rainfall, infiltration and flow routing, which produces the nonlinearity seen in river hydrographs and scale dependence of runoff coefficients. Existing research has determined two elements to hydrological connectivity: static/structural and dynamic/functional connectivity (Bracken and Croke, 2007; Turnbull, Wainwright and Brazier, 2008). Structural connectivity refers to spatial patterns in the landscape, such as the spatial distribution of landscape units that influence water transfer patterns and flow paths. Functional aspects of connectivity refer to how these spatial patterns interact with catchment processes to produce runoff, connected flow and hence water transfer in catchments (Turnbull, Wainwright and Brazier, 2008). Research to date has been good at describing the elements defining structural connectivity (Lexartza-Artza and Wainwright, 2009; Kirkby, Bracken and Reaney, 2002; Bull *et al.*, 2003). However, the elements defining functional aspects of hydrological connectivity are more important in understanding the concept, but are more difficult to measure and quantify due to their dynamic nature, complexity and variability (Bracken and Croke, 2007; Lexartza-Artza and Wainwright, 2009).

Previous studies attempting to understand functional aspects of hydrological connectivity have been conducted in a range of environments selected from a continuum of catchment types and hydrological behaviours. For example, research in rangeland catchments in southeast Australia and New Zealand demonstrated that patterns in shallow soil moisture can be used as an indication of saturated excess processes that control the fluxes of water in their catchments (Western *et al.*, 2004). However, results from studies conducted in bedrock-controlled catchments in the USA disagree and demonstrate that soil depth and bedrock topography control the pattern of active flow generated during storm events (Tromp van Meerveld and McDonnell, 2006). At an intermediate point on the continuum between these two environments, work conducted in temperate forest watersheds suggested a nonlinear response in runoff for small variations in antecedent moisture, but did not observe a significant change in geostatistical hydrologic connectivity with variations in antecedent

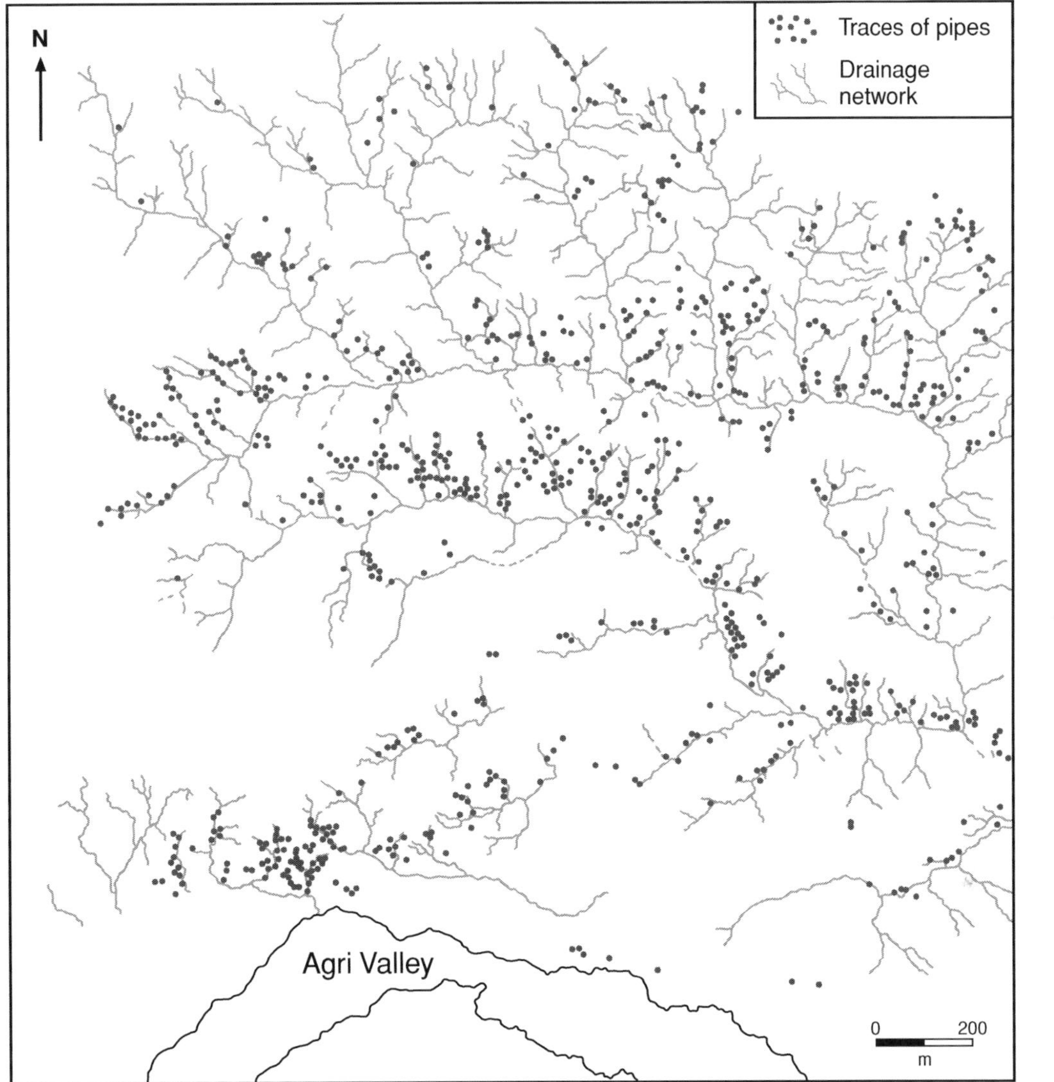

Figure 11.10 Patterns of pipes mapped in the Agri basin, southern Italy by Farifteh and Soeters (1999).

conditions (James and Roulet, 2007). Thus, a spectrum of approaches and understandings of hydrological connectivity has resulted from research conducted in different environments.

Current debates around hydrogeomorphic connectivity have centred on quantifying indices of hydrological connectivity (Troch *et al.*, 2009; Antoine, Javaux and Bielders, 2009) and investigating how these vary between catchments. Studies can be divided into those deriving pathways from topography (e.g. Lane, Reaney and Heathwaite, 2009; Lesschen, Schoorl and Cammeraat, 2009; Tetzlaff *et al.*, 2009) (in a similar vein to using topographic wetness indices to predict variable source area runoff; see Beven and Kirkby 1979), those developing understandings informed by water infiltration and transfer at the plot or catchment scale (Buda *et al.*, 2009) and occasionally by bringing these two approaches together (Jensco *et al.*, 2009; Meerkerk, van Wesemael and Bellin, 2009).

A key soil variable that modifies the connectivity of overland flow is the surface roughness. Surface roughness is complex because it operates at a number of spatial scales and can be highly dynamic. For instance, plough furrows are relatively large and influence flow pathways (*form roughness*; see Kirkby, Bracken and Reaney, 2002) and individual stones and soil aggregates can be larger than the flow depth (*grain roughness*), yet on the other hand disintegration of aggregates, differential swelling and

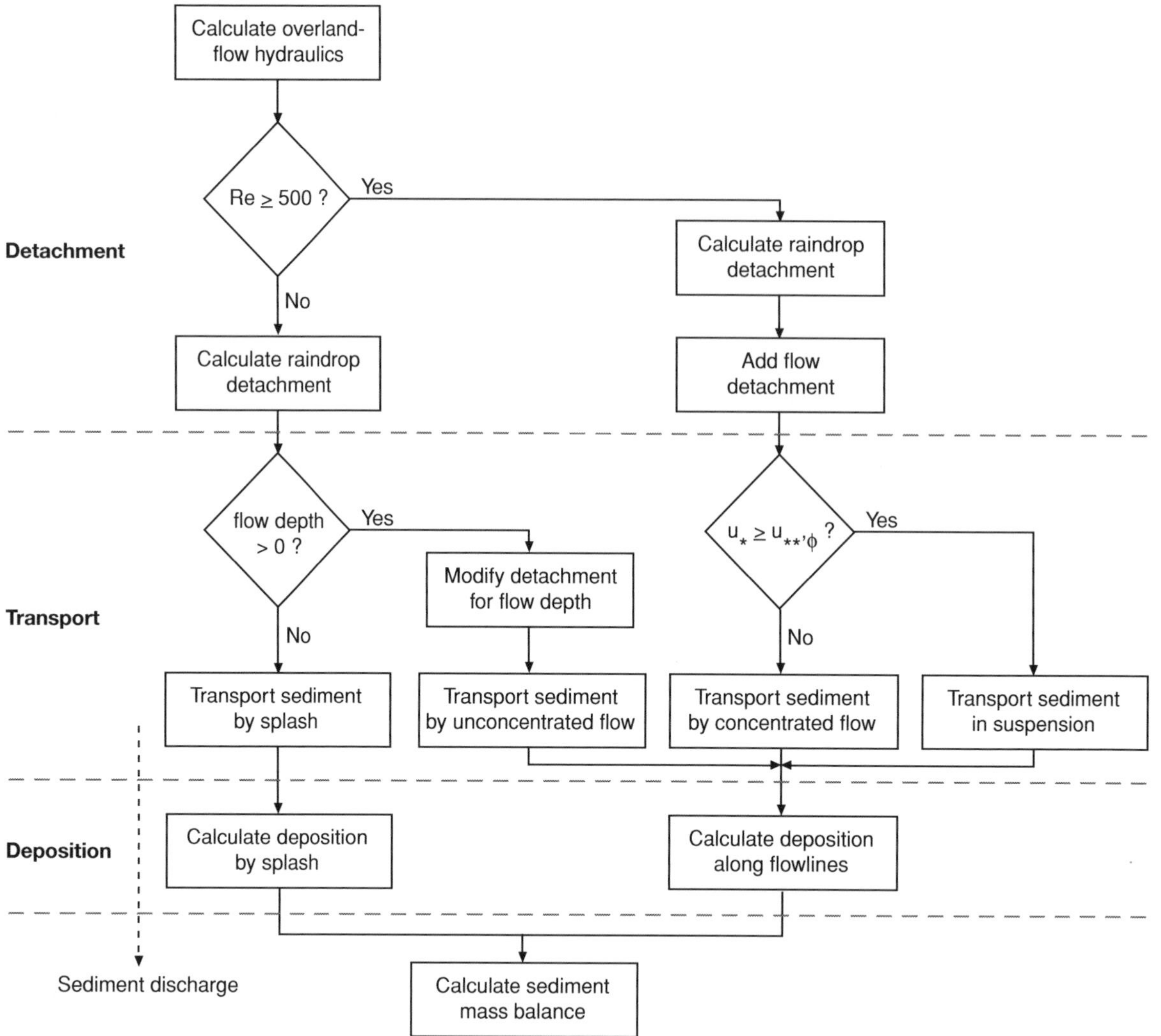

Figure 11.11 Key thresholds in detachment and transport processes in overland flows (after Wainwright *et al.*, 2008).

surface sealing are also dynamic over the duration of a storm as small-scale roughness elements (Slattery and Bryan, 1992). On noncultivated slopes, rills may form a dynamic feedback between roughness and infiltration characteristics, enhancing connectivity of the surface (Müller, Wainwright and Parsons, 2008). This observation emphasises the rôle of geomorphic feedbacks in controlling the functional connectivity of the system, requiring a recognition of the discontinuity or continuity of the various erosion processes along a slope. In many dryland systems, the interactions of continuity of water and sediment movement, and consequently the distribution of nutrients across the landscape, produce patterns of vegetation that further emphasise the connectivity of the landscape (Thornes, 1985, 1990; Wainwright, 2009a).

11.5 Erosion processes on hillslopes

During a storm event, different processes of erosion will be dominant at different points in time and at different locations on the slope. To understand these patterns, it is important to recognise the different processes of particle detachment, entrainment, transport and deposition (Figure 11.11). Detachment refers to the loosening of materials at the surface as a precursor to actual movement. It occurs as the result of energy applied to a sediment particle by raindrop impact or by flow, enhanced by the effects of gravity on slopes. Entrainment is the initial movement of the sediment at the surface and may under certain circumstances be continuous with the detachment process. It, too, is controlled by the energy of raindrop impact and/or water flow. Transport relates to the actual movement of a sediment particle and may be by direct movement through the air following the ejections of particles by raindrop impact or carried in rebounding water droplets from impacting raindrops. Alternatively, it may occur as a form of bedload, with particles rolling or sliding along the surface, by saltation within a water flow or by suspension in the water flow. Deposition will then be controlled by the mode of transport. Splashed particles return following parabolic trajectories to the surface and may rebound depending on conditions at the point of return. Bedload-type transport leads to deposition in relation to the local interaction of flow and sediment conditions, while deposition from suspension is as a consequence of settling of the sediment through the water column. On any slope, erosion and deposition are likely to relate to a continuum of several of these processes, with specific thresholds controlled by storm, flow and surface dynamics. This continuum is best described in terms of variations in splash, unconcentrated and concentrated erosion.

11.5.1 Splash

While all splash requires raindrop detachment, not all raindrop detachment leads to splash erosion. This distinction is critical and leads to fundamental problems with the characterisation of erosion processes, in that while splash is measurable, detachment is not (Kinnell, 1993; Wainwright *et al.*, 2009). Poesen and Savat (1981) suggest that the minimum energies required for detachment fall in the same range as the maximum splash rate, which assumes that detachment and splash are directly correlated. Van Dijk, Meesters and Bruijnzeel (2002) suggest that this assumption is reasonable as long as the splash distance is known, which is consistent with the results of Savat and Posen (1981) using radioactive tracers. Raindrop detachment is controlled by particle size and density. The dominant particle size detached by typical raindrops, the diameter of which is generally <5 mm, is fine sand (Ellison, 1944; Poesen and Savat, 1981; Parsons, Abrahams and Simanton, 1992). Finer particles require more energy, due largely to cohesion developed by electrostatic forces between particles; coarser particles require greater energies. Practical thresholds of maximum movement are of the order of 10–20 mm, but will depend on the density of the particle and the exact configuration of the particle on the surface (Wainwright, 1992). This maximum threshold means that fine sediment can be preferentially removed from the surface, leaving stones sitting atop columns of sediment, known as splash pillars. Control of raindrop energy is usually characterised as a function of rainfall energy (Morgan *et al.*, 1998) or the square of rainfall energy (e.g. Ghadiri and Payne, 1977; Meyer, 1981), although some authors have questioned the use of intensity as the underlying control and have suggested raindrop momentum as a better way of characterising the process (Salles and Poesen, 2000). Increasing slope angle is generally considered as enhancing the effect of detachment. For example, the results of Quansah (1981) suggest a power-law dependency on slope angle, with the coefficient dependent on the particle size. However, Torri and Poesen (1992) suggest that slope effects at practical scales of measurement are more problematic, and certainly on rough surfaces the local microtopography may counterbalance average slope effects. Particle size and slope controls on detachment are further affected by the interaction with wind intensity and direction (Wainwright, 1992; Erpul, Gabriels and Norton, 2005), so that the direction of travel of a storm event in relation to slope and slope aspect may be a significant control on the spatial pattern of the process.

Soil cohesion is a further control on detachment, typically with an exponential decrease in the energy required for detachment as cohesion decreases (Al Durrah and Bradford, 1982; Nearing and Bradford, 1985; Bradford, Ferris and Remley, 1987b). These decreases will occur dynamically through an event as the surface becomes saturated. Surface sealing (Bradford, Ferris and Remley, 1987a), aggregate stability (Farres, 1987) and organic matter content (Tisdale and Oades, 1982) all have the opposite effect, but again can change dynamically during a storm, as noted above. Soil compaction leads to a significant reduction in detachment (Drewry, Paton and Monaghan, 2004).

Splash erosion occurs as the combination of raindrop detachment and transport, either by direct ejection of particles or by entrainment in droplets that rebound from the surface. The exact combination of these transport mechanisms is poorly understood but relates to the effect of elastic versus plastic deformation of the surface on impact (Huang, Bradford and Cushman, 1982, 1983). Splashed sediment follows parabolic trajectories away from source and can travel distances of a few cm to tens of cm (Savat and Poesen, 1981; Moeyersons, 1983; Poesen and Torri, 1988; Van Dijk, Meesters and Bruijnzeel, 2002; Legout *et al.*, 2005). Splash is essentially a diffusive process, in that it can move sediment in all directions. Movement can occur in both and upslope and downslope directions, with a progressively smaller proportion of upslope movement as the slope angle increases (Savat, 1981). This diffusion of sediment by splash has been interpreted as the mechanism explaining convexities in the upper profile of slopes in drylands.

As the surface becomes progressively wetter in an event, rates of detachment will change, particularly for finer particles as the cohesion of the surface decreases. Al-Durrah and Bradford (1982) suggest an exponential relationship of reduced energy required for detachment as the water content of the surface increases. Splash can continue after the onset of ponding and in the initial overland flows. Exact thresholds of cessation of splash are debated in the literature, but relate to the changes in raindrop detachment as flow depth increases, which will be considered further in the next section.

11.5.2 Unconcentrated overland-flow erosion

Once overland flow starts, it occurs in an unconcentrated form. Flow in the laminar régime ($Re < 500$) is unable to entrain sediment in shallow flows (Ellison, 1945; Young and Wiersma, 1973), so that initially overland flows require detachment by raindrop action to produce erosion. As flow depths increase, there is a negative feedback whereby the energy of the raindrop is reduced so that less detachment and entrainment is caused. This feedback has been described as a negative exponential function of flow depth (Torri, Sfalanga and del Sette, 1987), so that flow depths of more than about 12 mm effectively reduce detachment to zero, depending on the particle size of the surface. Although some studies have suggested that in thin flows entrainment efficiency increased with rainfall rate (Kilinc and Richardson, 1973; Kinnell, 1991), presumably due to acceleration of the drop through the flow, this observation may be an artefact of the experimental methods used (see the discussion in Wainwright *et al.*, 2009). Once flows enter the transitional régime ($500 < Re \leq 2000$), some entrainment seems to occur through flow detachment, although the relative amounts seem to be of the same order of magnitude as to compensate for the corresponding decline in raindrop detachment (see the discussion in Wainwright *et al.*, 2008). The shallow depth of flow typically means that such entrainment is by flow shear alone, as there is insufficient development of a pressure gradient to produce a lift force. Where exfiltration forms an important fraction of the overland flow, the shear force is augmented by a small lift force associated with the emergence of water from the soil (Kochel, Howard and MacLane, 1982). Controls on detachment are thus very similar in unconcentrated overland flows as in splash, with the addition of the feedback with flow depth.

Transport in unconcentrated overland flows is by bedload-related processes or by flow-aided saltation. The latter may be common where particles are initially lifted up into the flow by raindrop impact. Most unconcentrated flows are too shallow to produce conditions for transport in suspension, even for the finest particle sizes (see the discussion in Wainwright *et al.*, 2008). Although unconcentrated overland flows are more competent at transporting sediment than splash, the overall rate of transport is a function of the interrelationship between flow energy and flow depth, which thus is affected by the surface roughness. Distances of travel have been described as a function that is the product of the inverse of the particle mass, the flow energy and the raindrop energy (Parsons, Stromberg and Greener, 1998). Deposition is thus a function of the ability of the continued energy put into the flow to overcome surface roughness, and particles are often deposited in local hollows in the bed (Tatard *et al.*, 2008) or behind dams formed by vegetation (usually grasses; see Parsons *et al.*, 1997; Cerdà, 1997; Boer and Puigdefábregas, 2005). The effect of vegetation on unconcentrated flows is generally one of deceleration and diffusion or deposition of transported sediment.

11.5.3 Concentrated overland-flow erosion

Concentration of overland flow is caused by incision into the surface by turbulent flows ($Re > 2000$). As noted above, there is likely to be a transitional régime where threads of flow develop due to the combination of small amounts of local flow detachment with existing surface irregularities. Parsons and Wainwright (2006) discussed the conditions required for the development of permanent concentrations as rills as a result of this combination, which overcomes limitations with existing approaches based on a critical shear stress (Figure 11.12). For example, Rauws and Govers (1988) described a function where the critical shear stress for incision is significantly lower than the apparent shear strength of the surface, and thus theoretically should be insufficient to produce the erosion causing incision. Incision occurs where the total shear stress of the flow is (a) high enough to disrupt the root mat, (b) high enough to incise bare patches of soil between vegetation, (c) where initially unconcentrated flows through vegetation are able to evacuate sediment loosened (detached) by biogenic disruption or (d) where seepage erosion triggers the initial incision through the vegetated surface so the bare soil is exploited by saturation overland flow. Grazing of large herbivores has apparently triggered concentrated erosion in a number of environments because it not only contributes to the first three of these mechanisms but also reduces the shear strength of the soil (Parsons and Wainwright, 2006). Once incision occurs,

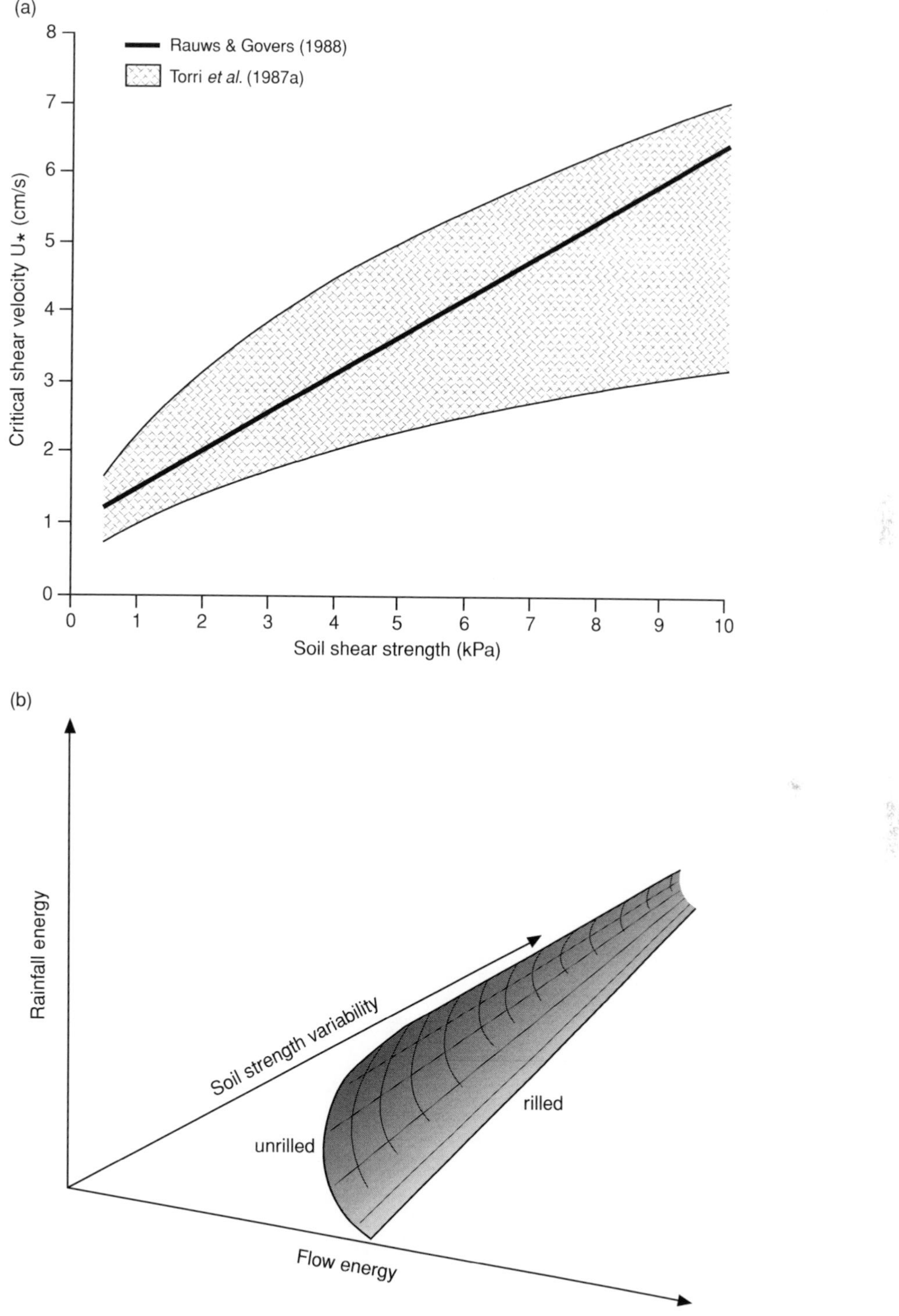

Figure 11.12 Comparison of approaches to understanding thresholds of rill initiation: (a) Rauws and Govers (1988) and (b) Parsons and Wainwright (2006).

channel form is strongly influenced by vertical differences in soil strength, caused by a tougher surface or a plough pan or other resistant layer at depth.

Transport by concentrated overland flow may also be by bedload-type processes, especially in its early stages, or where stones are being transported. However, as flow energy continues to increase, transport of the finer particles becomes increasingly likely to be in the form of suspension (Wainwright *et al.*, 2008, 2009, 2010). This distinction is important as transport is much more efficient in suspension, and as particles travel further before deposition from the flow, overall rates of transport increase. Concentrated flows in rills have been observed to transport stones over distances of tens of metres (Poesen, 1987). Although the distance of movement has been described as a function of excess flow shear stress (Hassan, Church and Ashworth, 1992), there is some disagreement between observations in unconcentrated and concentrated flows, suggesting that investigations that consider the problem in an integrated fashion are required (Wainwright *et al.*, 2008). Combined with the greater efficiency of flows at detachment, the overall increase in sediment flux can be in the region of an order of magnitude. The threshold for flow concentration is thus a critical one for accurate estimates of erosion from a slope overall.

While many studies consider it useful to define a transport capacity for overland flows (Govers, 1985, 1992a, 1992b; Govers and Rauws, 1986; Abrahams and Atkinson, 1993; Li and Abrahams, 1997), this concept can be considered problematic on a number of levels (Wainwright *et al.*, 2008). First, as sediment transport increases, the nature of the flow changes so that it is no longer a water flow supporting sediment movement but a non-Newtonian fluid flow of a water–sediment mix. Second, theoretical definitions (e.g. Yalin, 1971) have been based on entrainment by the flow shear force, but their extension to unconcentrated overland flows (e.g. Ferro, 1998) fail

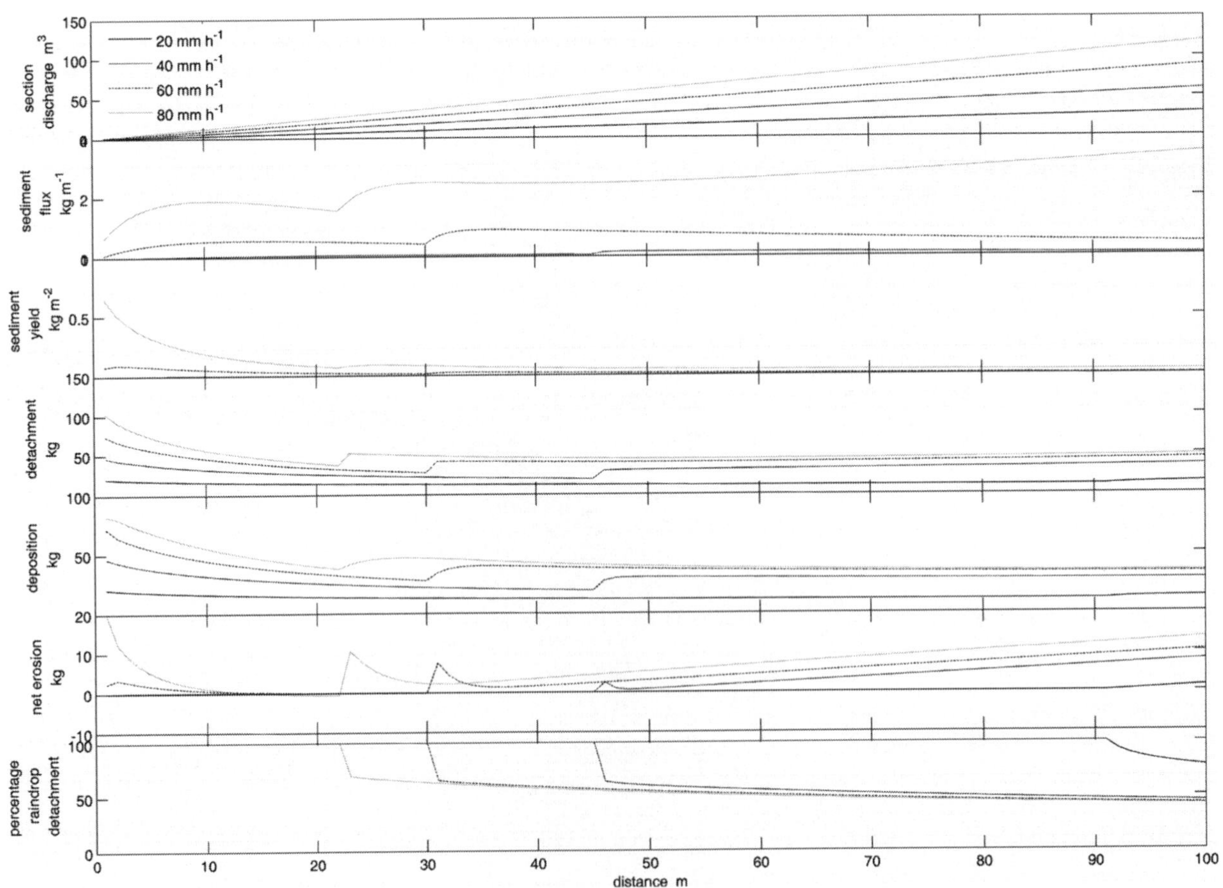

Figure 11.13 Patterns of erosion scaling as a response to storms of different intensities along a hillslope as simulated by the Mahleran model of Wainwright *et al.* (2008). Note the thresholds along the slope for the higher-intensity storms as erosion passes from being dominated by splash erosion, then interrill erosion and finally rill erosion.

to recognise that this is not the detachment mechanism. Third, even making these assumptions, the theoretical capacity is size-selective in that it is inversely related to the particle diameter and thus there is no simple definition of a single value for real soils, which are typically made up of a wide range of particle sizes.

11.5.4 Patterns and scales of sediment transport

The relative effectiveness of splash, unconcentrated and concentrated erosion on hillslopes is a product of the detachment rates and distance moved in each case. In broad terms, the rates of unconcentrated erosion are typically an order of magnitude bigger than rates of splash, but in turn are an order of magnitude lower than those in concentrated erosion. These differences can produce marked changes in the rate of sediment movement at different locations along a hillslope (Figure 11.13). An important consequence of these differences is that erosion rates do not scale linearly and that unconcentrated erosion scales in a different way (increases then decreases along a slope due to the balance in the increase in distance moved in relation to reduced detachment as the flow depth increases) from concentrated erosion (non-linear increase with slope length). This observation has significant consequences for the understanding of sediment fluxes at different spatial scales and in the way sediment delivery at catchment scale is represented (Parsons *et al.*, 2006a, 2006b).

11.6 Conclusions

Significant advances have been made over the last decades in the understanding of the dynamics of runoff and erosion processes in drylands. In particular, in recent years there have been conceptual developments that have allowed the complexity of the processes to be evaluated in a way that accounts for their highly nonlinear characteristics at point scale, but also the major nonlinearities that occur when moving from point to slope and to landscape scales. Infiltration is a complex process and there are numerous reasons why the standard models do not work well in dryland contexts. There are many processes at play, which means that simple approximations based on grain size are not appropriate and many surface and subsurface characteristics that need to be accounted for in predicting infiltration rates and thus runoff. Not least, over the last decade or so has been the importance of the development of concepts of landscape connectivity in understanding larger-scale runoff. The effects of vegetation in producing ecogeomorphic feedbacks between infiltration, runoff and erosion are becoming increasingly recognised. Similarly, in erosion studies, new models are being developed that can account for the variability of erosion rates at different scales and pick out key continua and thresholds. Methodologically, the integration of field observation, numerical modelling and theoretical conceptualisation in an iterative process is fundamental in overcoming limitations from just looking at small-scale processes or from the landscape perspective. In this way, we can hope that the coming decades will produce equally impressive strides forward in the understanding of dryland surface processes *and* landforms as a result of overland flows and related erosion.

References

Abedini, M.J., Dickinson, W.T. and Rudra, R.P. (2006) On depressional storages: the effect of DEM spatial resolution. *Journal of Hydrology*, **318**, 138–150.

Abrahams, A.D. and Atkinson, J.F. (1993) Relation between grain velocity and sediment concentration in overland-flow. *Water Resources Research*, **29**, 3021–3028.

Abrahams, A.D. and Parsons, A.J. (1991a) Relation between infiltration and stone cover on a semiarid hillslope, southern Arizona. *Journal of Hydrology*, **122**, 49–59.

Abrahams, A.D. and Parsons, A.J. (1991b) Resistance to overland flow on desert pavement and its implications for sediment transport modelling. *Water Resources Research*, **27**, 1827–1836.

Abrahams, A.D., Parsons, A.J. and Hirsch, P. (1992) Field and laboratory studies of resistance to overland flow on semiarid hillslopes, southern Arizona, in *Overland Flow: Hydraulics and Erosion Mechanics* (eds A.J. Parsons and A.D. Abrahams), UCL Press, London, pp. 1–24.

Abrahams, A.D., Parsons, A.J. and Luk, S.-H. (1986) Resistance to overland flow on desert hillslopes. *Journal of Hydrology*, **88**, 343–363.

Abrahams, A.D., Parsons, A.J. and Luk, S.-H. (1989) Distribution of depth of overland flow on desert hillslopes and its implications for modeling soil erosion. *Journal of Hydrology*, **106**, 177–184.

Abrahams, A.D., Parsons, A.J. and Luk, S.-H. (1990) Field experiments on the resistance to interrill overland flow on desert hillslopes, in *Erosion, Transport and Deposition Processes* (eds D.E. Walling, A. Yair and S. Berkowicz), International Association of Hydrological Sciences Publication 189, Wallingford, pp. 1–18.

Abrahams, A.D., Parsons, A.J. and Wainwright, J. (2003) Disposition of stemflow under creosotebush. *Hydrological Processes*, **17**, 2555–2566.

Abu-Awwad, A.M. (1997) Water infiltration and redistribution within soils affected by a surface crust. *Journal of Arid Environment*, **37**, 231–242.

Abu-Sharar, T.M. and Salameh, A.S. (1995) Reductions in hydraulic conductivity and infiltration rate in relation to aggregate stability and irrigation water turbidity. *Agricultural Water Management*, **29**, 53–62.

Al Durrah, M.M. and Bradford, J.M. (1982) Parameters for describing soil detachment due to single raindrop detachment. *Soil Science Society of America Journal*, **46**, 836–840.

Allmaras, R.R., Nelson, W.W. and Hallauer, E.A. (1972) Fall versus spring ploughing and related soil heat balance in the Western Belt, Minnesota Agricultural Experiment Station Technical Bulletin 283, 22 pp.

Allmaras, R.R., Rickman, R.W., Ekin, L.G. and Kimball, B.A. (1977) Chiseling influences on soil hydraulic-properties. *Soil Science Society of America Journal*, **41**, 796–803.

Ambroise, B. (2004) Variable 'active' versus 'contributing' areas or periods: a necessary distinction. *Hydrological Processes*, **18**, 1149–1155.

Antoine, M., Javaux, M. and Bielders, C. (2009) What indicators can capture runoff-relevant connectivity properties of the micro-topography at the plot scale? *Advances in Water Resources*, **32**, 1297–1310.

Archer, N.A.L., Quinton, J.N. and Hess, T.M. (2002) Below-ground relationships of soil texture, roots and hydraulic conductivity in two-phase mosaic vegetation in south-east Spain. *Journal of Arid Environments*, **52**, 535–553.

Baird, A.J. (1997) Overland flow generation and sediment mobilization by water, in *Arid Zone Geomorphology, Form and Change in Drylands*, 2nd edn (ed. D.S.G. Thomas), John Wiley & Sons, Ltd, Chichester, pp. 165–184.

Baird, A.J. (2003) Soil and hillslope hydrology, in *Environmental Modelling: Finding Simplicity in Complexity* (eds J. Wainwright and M. Mulligan), John Wiley & Sons, Ltd, Chichester, pp. 93–106.

Baird, A.J., Thornes, J.B. and Watts, G.P. (1992) Extending overland flow models to problems of slope evolution and the representation of complex slope-surface topographies, in *Overland Flow: Hydraulics and Erosion Mechanics* (eds A.J. Parsons and A.D. Abrahams), UCL Press, London, pp. 199–223.

Belnap, J. (2006) The potential roles of biological soil crusts in dryland hydrologic cycles. *Hydrological Processes*, **20**, 3159–3178.

Belnap, J., Welter, J.R., Grimm, N.B. *et al.* (2005) Linkages between microbial and hydrologic processes in arid and semi-arid watersheds. *Ecology*, **86**, 298–307.

Betson, R.P. (1964) What is watershed runoff? *Journal of Geophysical Research*, **69**, 1541–1552.

Beven, K.J. (2001) *Rainfall-Runoff Modelling. The Primer*, John Wiley & Sons, Ltd, Chichester.

Beven, K.J. and Kirkby, M.J. (1979) A physically based variable contributing area model of basin hydrology. *Hydrologic Science Bulletin*, **24**, 43–69.

Boer, M.M. and Puigdefábregas, J. (2005) Effects of spatially-structured vegetation patterns on hillslope erosion rates in semiarid Mediterranean environments: a simulation study. *Earth Surface Processes and Landforms*, **30**, 149–167.

Boiffin, J. (1984) Structural degradation of the soil surface by the action of rainfall, PhD Dissertation, Paris Institut National d'Agronomie Paris-Grignon.

Boix-Fayos, C., Calvo-Cases, A., Imeson, A.C. *et al.* (1998) Spatial and short-term temporal variations in runoff, soil aggregation and other soil properties along a Mediterranean climatological gradient. *Catena*, **33**, 123–138.

Bracken, L.J., Cox, N.J. and Shannon, J. (2008) The relationship between rainfall inputs and flood generation in south-east Spain. *Hydrological Processes*, **22**, 683–696.

Bracken, L.J. and Croke, J. (2007) The concept of hydrological connectivity and its contribution to understanding runoff dominated systems. *Hydrological Processes*, **21**, 1749–1763.

Bradford, J.M., Ferris, J.E. and Remley, P.A. (1987a) Interrill soil erosion processes: I. Effect of surface sealing on infiltration, runoff and soil splash detachment. *Soil Science Society of America Journal*, **51**, 1566–1571.

Bradford, J.M., Ferris, J.E. and Remley, P.A. (1987b) Interrill soil erosion processes: II. Relationship of splash detachment to soil properties. *Soil Science Society of America Journal*, **51**, 1571–1575.

Brooks, R.H. and Corey, A.T. (1964) Hydraulic properties of porous media, Colorado State University Hydrology Paper 3, Fort Collins, CO.

Bryan, R.B. (2000) Soil erodibility and processes of water erosion on hillslope. *Geomorphology*, **32**, 385–415.

Bryan, R.B., Imeson, A.C. and Campbell, I.A. (1984) Solute release and sediment entrainment on micro-catchments in the Dinosaur Park badlands, Alberta, Canada. *Journal of Hydrology*, **71**, 79–106.

Buckingham, E. (1907) *Studies on the Movement of Soil Moisture*, Bulletin 38, USDA Bureau of Soils, Washington, DC.

Buda, A.R., Kleinman, P.J.A., Srinivasan, M.S. *et al.* (2009) Factors influencing surface runoff generation from two agricultural hillslopes in central Pennsylvania. *Hydrological Processes*, **23**, 1295–1312.

Bull, L.J. and Kirkby, M.J. (1997) Gully processes and modelling. *Progress in Physical Geography*, **21**, 354–374.

Bull, L.J., Kirkby, M.J., Shannon, J. and Hooke, J.M. (2000) The impact of rainstorms on floods in ephemeral channels in southeast Spain. *Catena*, **38**, 191–209.

Bull, L.J., Kirkby, M.J., Shannon, J. and Dunsford, H. (2003) Predicting hydrologically similar surfaces (HYSS) in semi-arid environments. *Advances in Environmental Monitoring and Modelling*, **2**, 1–26.

Bunte, K. and Poesen, J. (1994) Effects of rock fragment size and cover on overland-flow hydraulics, local turbulence and sediment yield on an erodible soil surface. *Earth Surface Processes and Landforms*, **19**, 115–135.

Cable, J.M. and Huxman, T.E. (2004) Precipitation pulse size effects on Sonoran Desert soil microbial crusts. *Oecologia*, **141**, 317–324.

Cammeraat, L.H. and Imeson, A.C. (1999) The significance of soil-vegetation patterns following land abandonment and fire in Spain. *Catena*, **37**, 107–127.

Cammeraat, L.H., Willott, S.J., Compton, S.G. and Incoll, L.D. (2002) The effects of ants' nests on the physical, chemical and hydrological properties of a rangeland soil in semi-arid Spain. *Geoderma*, **105**, 1–20.

Campbell, G.S. (1985) *Soil Physics with BASIC*, Elsevier, Amsterdam.

Casenave, A. and Valentin, C. (1992) A runoff capability classification-system based on surface-features criteria in semiarid areas of West Africa. *Journal of Hydrology*, **130**, 231–249.

Ceballos, A., Martínez-Fernández, J., Santos, F. and Alonso, P. (2002) Soil-water behaviour of sandy soils under semi-arid conditions in the Duero Basin (Spain). *Journal of Arid Environments*, **51**, 501–519.

Cerdà, A. (1997) The effect of patchy distribution of *Stipa tenacissima* L. on runoff and erosion. *Journal of Arid Environments*, **36**, 37–51.

Cerdà, A. and Doerr, S.H. (2007) Soil wettability, runoff and erodibility of major dry-Mediterranean land use types on calcareous soils. *Hydrological Processes*, **21**, 2325–2336.

Clifford, N.J., Robert, A. and Richards, K.S. (1992) Estimation of flow resistance in gravel-bedded rivers; a physical explanation of the multiplier of roughness length. *Earth Surface Processes and Landforms*, **17**, 111–126.

Constantz, J., Herkelrath, W.N. and Murphy, F. (1988) Air encapsulation during infiltration. *Soil Science Society of America Journal*, **52**, 10–16.

Cooke, R.U., Warren, A. and Goudie, A. (1993) *Desert Geomorphology*, UCL Press, London.

Cunningham, G.L. and Burk, J.H. (1973) The effect of carbonate deposition layers ('caliche') on the water status of Larrea divaricata. *American Midland Naturalist*, **90**, 474–480.

Darboux, F., Davy, P. Gascuel-Odoux, C. and Huang, C. (2002) Evolution of soil surface roughness and flowpath connectivity in overland flow experiments. *Catena*, **46**, 125–139.

De Rooij, G.H. (2000) Modeling fingered flow of water in soils owing to wetting front instability: a review. *Journal of Hydrology*, **231**, 277–294.

Devitt, D.A. and Smith, S.D. (2002) Root channel macropores enhance downward movement of water in a Mojave Desert ecosystem. *Journal of Arid Environments*, **50**, 99–108.

Domingo, F., Sanchez, G., Moro, M.J. *et al.* (1998) Measurement and modelling of rainfall interception by three semi-arid canopies. *Agricultural and Forest Meteorology*, **91**, 275–292.

Drewry, J.J., Paton, R.J. and Monaghan, R.M. (2004) Soil compaction and recovery cycle on a Southland dairy farm: implications for soil monitoring. *Australian Journal of Soil Research*, **42**, 851–856.

Dunkerley, D.L. (2002) Infiltration rates and soil moisture in a groved mulga community near Alice Springs, arid central Australia: evidence for complex internal rainwater redistribution in a runoff–runon landscape. *Journal of Arid Environments*, **51**, 199–219.

Dunkerley, D.L. (2004) Flow threads in surface run-off: implications for the assessment of flow properties and friction coefficients in soil erosion and hydraulics investigations. *Earth Surface Processes and Landforms*, **29**, 1011–1026.

Dunne, T. and Black, R.D. (1970) Partial area contributions to storm runoff in a small New England watershed. *Water Resources Research*, **6**, 1296–1311.

Dunne, T., Moore, T.R. and Taylor, C.H. (1975) Recognition and prediction of runoff-producing zones in humid regions. *Hydrological Sciences Bulletin*, **20**, 305–327.

Dunne, T., Zhang, W. and Aubry, B.F. (1991) Effects of rainfall, vegetation and microtopography on infiltration and runoff. *Water Resources Research*, **27**, 2271–2285.

Elkins, N.Z., Sabol, G.V., Ward, T.J. and Whitford, W.G. (1986) The influence of subterranean termites on the hydrological characteristics of a Chihuahuan Desert ecosystem. *Oecologia*, **68**, 521–528.

Ellison, W.D. (1944) Studies of raindrop erosion. *Agricultural Engineering*, **25**, 131–136, 181–182.

Ellison, W.D. (1945) Some effects of raindrops and surface flow on soil erosion and infiltration. *Transactions of the American Geophysical Union*, **26**, 415–429.

Emmett, W.W. (1970) The hydraulics of overland flow on hillslopes, US Geological Survey Professional Paper 662-A, Washington, DC.

Erpul, G., Gabriels, D. and Norton, L.D. (2005) Sand detachment by wind-driven raindrops. *Earth Surface Processes and Landforms*, **30**, 241–250.

Escoto-Rodríguez, M. and Bullock, S.H. (2002) Long-term growth rates of cirio (*Fouquieria columnaris*), a giant succulent of the Sonoran Desert in Baja, California. *Journal of Arid Environments*, **50**, 593–611.

Farifteh, J. and Soeters, R. (1999) Factors underlying piping in the Basilicata region, southern Italy. *Geomorphology*, **26**, 239–251.

Farres, P.J. (1987) The dynamics of rainsplash erosion and the role of soil aggregate stability. *Catena*, **14**, 119–130.

Ferguson, R.I. (2007) Flow resistance equations for gravel- and boulder-bed streams. *Water Resources Research*, **43**, W05427.

Ferro, V. (1998) Evaluating overland flow sediment transport capacity. *Hydrological Processes*, **12** (12), 1895–1910.

Fitter, A.H. and Hay, R.K.M. (1987) *Environmental Physiology of Plants*, 2nd edn, Academic Press, London.

Fitzjohn, C., Ternan, J.L. and Williams, A.G. (1998) Soil moisture variability in a semi-arid gully catchment: implications for runoff and erosion control. *Catena*, **32**, 55–70.

Fox, D.M., Le Bissonnais, Y. and Bruand, A. (1998) The effect of ponding depth on infiltration in a crusted surface depression. *Catena*, **32**, 87–100.

Ghadiri, H. and Payne, D. (1977) Raindrop impact stress and the breakdown of soil crumbs. *Journal of Soil Science*, **28**, 247–258.

Gibbens, R.P. and Lenz, J.M. (2001) Root systems of some Chihuahuan Desert plants. *Journal of Arid Environments*, **49**, 221–263.

Gilley, J.E. and Finkner, S.C. (1991) Hydraulic roughness coefficients as affected by random roughness. *Transactions of the ASAE*, **34**, 897–903.

Giovannini, G., Lucchesi, S. and Giachetti, M. (1988) Effect of heating on some physical and chemical parameters related to soil aggregation and erodibility. *Soil Science*, **146**, 255–261.

Giovannini, G., Lucchesi, S. and Giachetti, M. (1990) Beneficial and detrimental effects of heating on soil quality, in *Fires in Ecosystem Dynamics. Proceedings of the Third International Symposium on Fire Ecology, Freiburg, FRG, May 1989* (eds J.G. Goldammer and M.J. Jenkins), SPB Academic Publishing, The Hague, pp. 95–102.

Gómez, J.A. and Nearing, M.A. (2005) Runoff and sediment losses from rough and smooth soil surfaces in a laboratory experiment. *Catena*, **59**, 253–266.

Govers, G. (1985) Selectivity and transport capacity of thin flows in relation to rill erosion. *Catena*, **12**, 35–49.

Govers, G. (1991) A field study on topographical and topsoil effects on runoff generation. *Catena*, **18**, 91–111.

Govers, G. (1992a) Evaluation of transporting capacity formulae for overland flow, in *Overland Flow: Hydraulics and Erosion Mechanics* (eds A.J. Parsons and A.D. Abrahams), UCL Press, London, pp. 140–273.

Govers, G. (1992b) Relationship between discharge, velocity and flow area for rills eroding loose, non-layered materials. *Earth Surface Processes and Landforms*, **17**, 515–528.

Govers, G. and Rauws, G. (1986) Transporting capacity of overland flow on plane and on irregular beds. *Earth Surface Processes and Landforms*, **11**, 515–524.

Govers, G., Takken, I. and Helming, K. (2000) Soil roughness and overland flow. *Agronomie*, **20**, 131–146.

Graef, F. and Stahr, K. (2000) Incidence of soil surface crust types in semi-arid Niger. *Soil and Tillage Research*, **55**, 213–218.

Gutierrez, J. and Hernandez, I.I. (1996) Runoff and interrill erosion as affected by grass cover in a semi-arid rangeland of northern Mexico. *Journal of Arid Environments*, **34**, 287–295.

Hamerlynck, E.P., McAuliffe, J.R. and Smith, S.D. (2000) Effects of surface and sub-surface soil horizons on the seasonal performance of Larrea tridentate (creosotebush). *Functional Ecology*, **14**, 596–606.

Hamerlynck, E.P., McAuliffe, J.R., McDonald, E.V. and Smith, S.D. (2002) Ecological responses of two Mojave Desert shrubs to soil horizon development and soil water dynamics. *Ecology*, **83**, 768–779.

Hammecker, C., Antonino, A.C.D., Maeght, J.L. and Boivin, P. (2003) Experimental and numerical study of water flow in soil under irrigation in northern Senegal: evidence of air entrapment. *European Journal of Soil Science*, **54**, 491–503.

Hassan, M.A., Church, M. and Ashworth, P.J. (1992) Virtual rate and mean distance of travel of individual clasts in gravel-bed channels. *Earth Surface Processes and Landforms*, **17**, 617–627.

Hassanizadeh, S.M. (1986) Derivation of basic equations of mass-transport in porous-media. 2. Generalized Darcy and Fick laws. *Advances in Water Resources*, **9**, 207–222.

Hassanizadeh, S.M. and Gray, W.G. (1987) High velocity flow in porous media. *Transport in Porous Media*, **2**, 521–531.

Helming, K., Römkens, M.J.M. and Prasad, S.N. (1998) Surface roughness related processes of runoff and soil loss: a flume study. *Soil Science Society of America Journal*, **62**, 243–250.

Hewlett, J.D. and Hibbert, A.R. (1967) Factors affecting the response of small watersheds to precipitation in humid areas, in *Forest Hydrology* (eds W.E. Sopper and H.W. Lull), Pergamon Press, New York, pp. 345–360.

Hiernaux, P., Bielders, C.L., Valentin, C. *et al.* (1999) Effects of livestock grazing on physical and chemical properties of sandy soils in Sahelian rangelands. *Journal of Arid Environments*, **41**, 231–245.

Horton, R.E. (1945) Erosional development of streams and their drainage basins; hydrophysical approach to quantitative morphology. *Bulletin of the American Geological Society*, **56**, 275–370.

Huang, C. (1990) Depressional storage for Markov–Gaussian surfaces. *Water Resources Research*, **26**, 2235–2242.

Huang, C., Bradford, J.M. and Cushman, J.H. (1982) A numerical study of raindrop impact phenomena – the rigid case. *Soil Science Society of America Journal*, **46**, 14–19.

Huang, C., Bradford, J.M. and Cushman, J.H. (1983) A numerical study of raindrop impact phenomena – the elastic-deformation case. *Soil Science Society of America Journal*, **47**, 855–861.

Hyder, P.W., Fredrickson, E.L., Estell, R.E. *et al.* (2002) Distribution and concentration of total phenolics, condensed tannins, and nordihydroguaiaretic acid (NDGA) in creosotebush *(*Larrea tridentata*)*. Biochemical Systematics and Ecology, **30**, 905–912.

Imeson, A.C., Verstraten, J.M., van Mulligen, E.J. and Sevink, J. (1992) The effects of fire and water repellency on infiltration and runoff under Mediterranean type forest. *Catena*, **19**, 345–361.

James, A.L. and Roulet, N.T. (2007) Investigating hydrologic connectivity and its association with threshold change in runoff response in a temperate forested watershed. *Hydrological Processes*, **21**, 3391–3408.

Janeau, J.L., Mauchamp, A. and Tarin, G. (1999) The soil surface characteristics of vegetation stripes in Northern Mexico and their influences on the system hydrodynamics – an experimental approach. *Catena*, **37**, 165–173.

Jensco, K.G., McGlynn, B.L., Gooseff, M.N. *et al.* (2009) Hydrologic connectivity between landscapes and streams: transferring reach and plot scale understanding to the catchment scale. *Water Resources Research*, **45**, article W04428, DOI: 10.1029/2008WR007225.

Jetten, V., de Roo, A. and Guerif, J. (1998) Sensitivity of the model LISEM to variables related to agriculture, in *Modelling Soil Erosion by Water* (eds J. Boardman and D. Favis-Mortlock), Springer-Verlag, Berlin, pp. 339–349.

Keulegan, G.H. (1938) Laws of turbulent flow in open channels. *Journal of Research of the National Bureau of Standards*, **21**, 707–741.

Kidron, G.J. (1999) Differential water distribution over dune slopes as affected by slope position and microbiotic crust, Negev Desert, Israel. *Hydrological Processes*, **13**, 1665–1682.

Kidron, G.J. (2005) Angle and aspect dependent dew and fog precipitation in the Negev Desert. *Journal of Hydrology*, **301**, 66–74.

Kilinc, M. and Richardson, E.V. (1973) Mechanics of soil erosion from overland flow generated by simulated rainfall, Colorado State University Hydrology Paper 63, Fort Collins, CO.

Kinnell, P.I.A. (1991) The effect of flow depth on sediment transport induced by raindrops impacting shallow flows. *Transactions of the American Society of Agricultural Engineers*, **34**, 161–168.

Kinnell, P.I.A. (1993) Sediment concentrations resulting from flow depth/drop size interactions in shallow overland flow. *Transactions of the American Society of Agricultural Engineers*, **36**, 1099–1013.

Kirkby, M.J. and Bracken, L.J. (2009) Gully formation and gully processes. *Earth Surface Processes and Landforms*, **34**, 1841–1851.

Kirkby, M.J., Bracken, L.J. and Reaney, S. (2002) The influence of landuse, soils and topography on the delivery of hillslope runoff to channels in SE Spain. *Earth Surface Landforms and Processes*, **27**, 1459–1473.

Kochel, R.C., Howard, A.D. and MacLane, C.F. (1982) Channel networks developed by groundwater sapping in fine-grained sediments: analogs in some Martian valleys, in *Models in Geomorphology* (ed. M.J. Woldenberg), Allen and Unwin, Boston, pp. 313–341.

Kuipers, H. (1957) A relief meter for soil cultivation studies. *Netherlands Journal of Agricultural Science*, **5**, 255–262.

Kumar Mandal, U., Rao, K.V., Mishra, P.K. *et al.* (2005) Soil infiltration, runoff and sediment yield from a shallow soil with varied stone cover and intensity of rain. *European Journal of Soil Science*, **56**, 435–443.

Lane, S.N. (2005) Roughness – time for a re-evaluation? *Earth Surface Processes and Landforms*, **30**, 251–253.

Lane, S.N. and Ferguson, R.I. (2005) Modelling reach-scale fluvial flows, in *Computational Fluid Dynamics Applications in Environmental Hydraulics* (eds P.D. Bates, S.N. Lane and R.I. Ferguson), John Wiley & Sons, Ltd, Chichester, pp. 217–269.

Lane, S.N., Reaney, S.M. and Heathwaite, A.L. (2009) Representation of landscape hydrological connectivity using a topographically driven surface flow index. *Water Resources Research*, **45**, article W08423.

Lange, J., Greenbaum, N., Husary, S. *et al.* (2003) Runoff generation from successive simulated rainfalls on a rocky, semi-arid, Mediterranean hillslope. *Hydrological Processes*, **17**, 279–296.

Lange, O.L., Green, T.G.A., Meyer, A. and Zellner, H. (2007) Water relations and carbon dioxide exchange of epiphytic lichens in the Namib Fog Desert. *Flora – Morphology, Distribution, Functional Ecology of Plants*, **202**, 479–487.

Lavee, H., Kutiel, P., Segev, M. and Benyamini, Y. (1995) Effect of surface roughness on runoff and erosion in a Mediterranean ecosystem: the role of fire. *Geomorphology*, **11**, 227–234.

Lawless, M. and Robert, A. (2001) Scales of boundary resistance in coarse-grained channels: turbulent velocity profiles and implications. *Geomorphology*, **39**, 221–238.

Le Bissonnais, Y. (1996) Aggregate stability and assessment of soil crustability and erodibility. 1. Theory and methodology. *European Journal of Soil Science*, **47**, 425–437.

Leffler, A.J., Peek, M.S., Ryel, R.J., Ivans, C.Y. and Caldwell, M.M. (2005) Hydraulic redistribution through the root systems of senesced plants. *Ecology*, **86**, 633–642.

Legout, C., Leguédois, S., Le Bissonnais, Y. and Mallam Issa, O. (2005) Splash distance and size distributions for various soils. *Geoderma*, **124**, 279–292.

Lei, T.W., Pan, Y.H., Liu, H., Zhan, W.H. and Yuan, J.P. (2006) A runoff–on-ponding method and models for the transient infiltration capability process of sloped soil surface under rainfall and erosion impacts. *Journal of Hydrology*, **319**, 216–226.

Lesschen, J.P., Schoorl, J.M. and Cammeraat, L.H. (2009) Modelling runoff and erosion for a semi-arid catchment using a multi-scale approach based on hydrological connectivity. *Geomorphology*, **109**, 174–183.

Levy, G.J., Goldstein, D. and Mamedov, A.I. (2005) Saturated hydraulic conductivity of semiarid soils: combined effects of salinity, sodicity, and rate of wetting. *Soil Science Society of America Journal*, **69**, 653–662.

Lexartza-Artza, I. and Wainwright, J. (2009) Hydrological connectivity: linking concepts with practical implications. *Catena*, **79**, 146–152.

Li, G. and Abrahams, A.D. (1997) Effect of saltating sediment load on the determination of the mean velocity of overland flow. *Water Resources Research*, **33** (2), 341–347.

Linden, D.R. and Van Doren, D.M. (1986) Parameters for characterising tillage-induced soil surface roughness. *Soil Science Society of America Journal*, **50**, 1560–1565.

Ludwig, J.A., Wiens, J. and Tongway, D.J. (2000) A scaling rule for landscape patches and how it applies to conserving soil resources in savannas. *Ecosystems*, **3**, 82–97.

Ludwig, J.A., Wilcox, B.P., Breshears, D.D. *et al.* (2005) Vegetation patches and runoff–erosion as interacting ecohydrological processes in semiarid landscapes. *Ecology*, **86**, 288–297.

Luk, S.-H. and Cai, Q.G. (1990) Laboratory experiments on crust development and rainsplash erosion of loess soils, China. *Catena*, **17**, 261–276.

McAuliffe, J.R. (1994) Landscape evolution, soil formation, and ecological patterns and processes in Sonoran Desert bajadas. *Ecological Monographs*, **64**, 111–148.

McDonnell, J.J. (2003) Where does water go when it rains? Moving beyond the variable source area concept of rainfall-runoff response. *Hydrological Processes*, **17**, 1869–1875.

Maestre, F.T., Huesca, M., Zaady, E. *et al.* (2002) Infiltration, penetration resistance and microphytic crust composition in contrasted microsites within a Mediterranean semi-arid steppe. *Soil Biology and Biochemistry*, **34**, 895–898.

Manning, R. (1891) On the flow of water in open channels and pipes. *Transactions of the Institution of Civil Engineers of Ireland*, **20**, 161–207.

Martinez-Mena, M., Castillo, V. and Albaladejo, J. (2002) Relations between interrill erosion processes and sediment particle size distribution in a semiarid Mediterranean area of SE of Spain. *Geomorphology*, **45**, 261–275.

Martinez-Meza, E. and Whitford, W.G. (1996) Stemflow, throughfall and channelization of stemflow by roots in three Chihuahuan desert shrubs. *Journal of Arid Environments*, **32**, 271–287.

Meerkerk, A.L., van Wesemael, B. and Bellin, N. (2009) Application of connectivity theory to model the impact of terrace failure on runoff in semi-arid catchments. *Hydrological Processes*, **23**, 2792–2803.

Meyer, L.D. (1981) How rain intensity affects interrill erosion. *Transactions of the American Society of Agricultural Engineers*, **24**, 1472–2475.

Moeyersons, J. (1983) Measurements of splash-saltation fluxes under oblique rain. *Catena Supplement*, **4**, 19–31.

Moorhead, D.L., Fisher, F.M. and Whitford, W.G. (1988) Cover of spring annuals on nitrogen-rich kangaroo rat mounds in a Chihuahuan Desert grassland. *American Midland Naturalist*, **120**, 443–447.

Morgan, R.P.C., Quinton, J.N., Smith, R.E. *et al.* (1998) The European soil erosion model (EUROSEM): a dynamic approach for predicting sediment transport from fields and small catchments. *Earth Surface Processes and Landforms*, **23**, 527–544.

Mualem, Y., Assouline, S. and Rohdenburg, H. (1990) Rainfall induced soil seal. A: A critical review of observations and models. *Catena*, **17**, 185–203.

Müller, E.N., Wainwright, J. and Parsons, A.J. (2008) Spatial variability of soil and nutrient parameters within grasslands and shrublands of a semi-arid environment, Jornada Basin, New Mexico. *Ecohydrology*, **1**, 3–12.

Mun, H.T. and Whitford, W.G. (1990) Factors affecting annual plants assemblages on banner-tailed kangaroo rat mounds. *Journal of Arid Environments*, **18**, 165–173.

Musick, H.B. (1975) Barrenness of desert pavement in Yuma County. *Arizona Academy of Science Journal*, **10**, 24–28.

Navarro, V., Yustres, A., Candel, M. and García, B. (2008) Soil air compression in clays during flood irrigation. *European Journal of Soil Science*, **59**, 799–806.

Naveh, Z. (1975) The evolutionary significance of fire in the Mediterranean region. *Vegetation*, **29**, 199–209.

Nearing, M.A. and Bradford, J.M. (1985) Single waterdrop splash detachment and mechanical properties of soils. *Science Society of America Journal*, **49**, 547–552.

Nearing, M.A., Norton, L.D., Bulgakov, D.A. *et al.* (1998) Hydraulics and erosion in eroding rills. *Water Resources Research*, **33**, 865–876.

Neary, D.G., Ryan, K.C. and DeBano, L.F. (eds) (2008) Wildland fire in ecosystems: effects of fire on soils and water, General Technical Report RMRS-GTR-42-vol.4, US Department of Agriculture, Forest Service, Rocky Mountain Research Station Ogden, UT.

Neave, M. and Abrahams, A.D. (2001) Impact of small mammal disturbances on sediment yield from grassland and shrubland ecosystems in the Chihuahuan Desert. *Catena*, **44**, 285–303.

Neuman, S.P. (1977) Theoretical derivation of Darcy's law. *Acta Mechanica*, **25**, 153–170.

Nielson, R.P. (1986) High-resolution climatic analysis and southwest biogeography. *Science*, **232**, 27–34.

Nulsen, R.A., Bligh, K.J., Baxter, I.N. *et al.* (1986) The fate of rainfall in a malle and heath vegetated catchment in southern Western Australia. *Australian Journal of Ecology*, **11**, 361–371.

Onstad, C.A. (1984) Depressional storage on tilled soil surfaces. *Transactions of the ASAE*, **27**, 729–732.

Osborn, H.B. and Renard, K.G. (1969) Analysis of two major runoff producing southwest thunderstorms. *Journal of Hydrology*, **8**, 282–302.

Owens, M.K., Lyons, R.K. and Alejandro, C.L. (2006) Rainfall partitioning within semiarid juniper communities: effects of event size and canopy cover. *Hydrological Processes*, **20**, 3179–3189.

Pariente, S. (2002) Spatial patterns of soil moisture as affected by shrubs, in different climatic conditions. *Environmental Monitoring and Assessment*, **73**, 237–251.

Parsons, A.J., Abrahams, A.D. and Simanton, J.R. (1992) Microtopography and soil-surface materials on semi-arid piedmont hillslopes, southern Arizona. *Journal of Arid Environments*, **22**, 107–115.

Parsons, A.J., Abrahams, A.D. and Wainwright, J. (1994) On determining resistance to interrill overland flow. *Water Resources Research*, **30**, 3515–3521.

Parsons, A.J., Stromberg, S.G.L. and Greener, M. (1998) Sediment-transport competence of rain-impacted interrill overland flow. *Earth Surface Processes and Landforms*, **23**, 365–375.

Parsons, A.J. and Wainwright, J. (2006) Depth distribution of interrill overland flow and the formation of rills. *Hydrological Processes*, **20**, 1511–1523.

Parsons, A.J., Wainwright, J., Abrahams, A.D. and Simanton, J.R. (1997) Distributed dynamic modelling of interrill overland flow. *Hydrological Processes*, **11**, 1833–1859.

Parsons, A.J., Wainwright, J., Brazier, R.E. and Powell, D.M. (2006a) Is sediment delivery a fallacy? *Earth Surface Processes and Landforms*, **31**, 1325–1328.

Parsons, A.J., Brazier, R.E., Wainwright, J. and Powell, D.M. (2006b) Scale relationships in hillslope runoff and erosion. *Earth Surface Processes and Landforms*, **31**, 1384–1393.

Poesen, J. (1987) Transport of rock fragments by rillflow: a field study. *Rill Erosion: Processes and Significance, Catena Supplement*, **8**, 35–54, Catena, Cremlingen.

Poesen, J.W.A. (1992) Mechanisms of overland flow generation and sediment production on loamy and sandy soils with and without rock fragments, in *Overland Flow Hydraulics and*

Erosion Mechanics (eds A.J. Parsons and A.D. Abrahams), UCL Press, London, pp. 275–305.

Poesen, J., Ingelmo Sanchez, F. and Mucher, H. (1990) The hydrological response of soil surfaces to rainfall as affected by cover and position of rock fragments in the top layer. *Earth Surface Processes and Landforms*, **15**, 653–671.

Poesen, J. and Lavee, H. (1994) Rock fragments in top soils: significance and processes. *Catena*, **23**, 1–28.

Poesen, J. and Savat, J. (1981) Detachment and transportation of sediments by raindrop splash. Part II: Detachability and transport. *Catena*, **8**, 19–41.

Poesen, J. and Torri, D. (1988) The effect of cup size on splash detachment and transport measurements: Part II. Field measurements. *Catena Supplement*, **12**, 113–126.

Prandtl, L. (1935) The mechanics of viscous fluids, in *Aerodynamic Theory: A General Review of Progress*, vol. **3** (ed. W.F. Durand), Springer, Berlin, pp. 34–208.

Pressland, A.J. (1973) Rainfall partitioning by an arid woodland (*Acacia aneura* F. Meull.) in south-western Queensland. *Australian Journal of Botany*, **21**, 235–245.

Pringle, C. (2003) What is hydrologic connectivity and why is it ecologically important? *Hydrological Processes*, **17**, 2685–2689.

Puigdefábregas, J., Sole, A., Gutierrez, L. *et al.* (1999) Scales and processes of water and sediment redistribution in drylands: results from the Rambla Honda field site in southeast Spain. *Earth-Science Reviews*, **48**, 39–70.

Quansah, C. (1981) The effect of soil type, slope, rain intensity and their interactions on splash detachment and transport. *Journal of Soil Science*, **32**, 215–224.

Rauws, G. and Govers, G. (1988) Hydraulic and soil mechanical aspects of rill generation on agricultural soils. *Journal of Soil Science*, **39**, 111–124.

Reggiani, P., Hassanizadeh, S.M., Sivapalan, M. and Gray, W.G. (1999) A unifying framework for watershed thermodynamics: constitutive relationships. *Advances in Water Resources*, **23**, 15–39.

Richards, L.A. (1931) Capillary conduction of liquids through porous materials. *Physics*, **1**, 318–333.

Richards, J.H. and Caldwell, M.M. (1987) Hydraulic lift: substantial nocturnal water transport between soil layers by *Artemisia tridentata* roots. *Oecologia*, **73**, 486–489.

Ritsema, C.J. and Dekker, L.W. (2000) Preferential flow in water repellent sandy soils: principles and modeling implications. *Journal of Hydrology*, **231**, 308–319.

Robert, A. (1990) Boundary roughness in coarse-grained channels. *Progress in Physical Geography*, **14**, 42–70.

Rockström, J., Jansson, P.E. and Barron, J. (1998) Seasonal rainfall partitioning under runon and runoff conditions on sandy soil in Niger. On-farm measurements and water balance modelling. *Journal of Hydrology*, **210**, 68–92.

Römkens, M.J.M., Prasad, S.N. and Whisler, F.D. (1990) Surface sealing and infiltration, in *Process Studies in Hillslope Hydrology* (eds M.G. Anderson and T.P. Burt), John Wiley & Sons, Ltd, Chichester, pp. 127–172.

Ryel, R.J., Caldwell, M.M., Leffler, A.J. and Yoder, C.K. (2003) Rapid soil moisture recharge to depth by roots in a stand of *Artemisia tridentata*. *Ecology*, **84**, 757–764.

Ryel, R.J., Leffler, A.J., Peek, M.S., *et al.* (2004) Water conservation in *Artemisia tridentata* through redistribution of precipitation. *Oecologia*, **141**, 335–345.

Salles, C. and Poesen, J. (2000) Rain properties controlling soil splash detachment. *Hydrological Processes*, **14**, 271–282.

Savat, J. (1979) Laboratory experiments on erosion and deposition of loess by laminar sheetflow and turbulent rill flow. *Proceedings of the Seminar Agricultural Soil Erosion in Temperate Non-Mediterranean Climate*, 20–23, 139–143.

Savat, J. (1980) Resistance to flow in rough, supercritical sheetflow. *Earth Surface Processes*, **5**, 103–122.

Savat, J. (1981) Work done by splash – laboratory experiments. *Earth Surface Processes and Landforms*, **6**, 275–283.

Savat, J. and Poesen, J. (1981) Detachment and transportation of loose sediments by raindrop splash. Part I: The calculation of absolute data on detachability and transportability. *Catena*, **8**, 1–18.

Shanyengana, E.S., Henschel, J.R., Seely, M.K. and Sanderson, R.D. (2002) Exploring fog as a supplementary water source in Namibia. *Atmospheric Research*, **64**, 251–259.

Slattery, M.C. and Bryan, R.B. (1992) Hydraulic conditions for rill incision under simulated rainfall: a laboratory experiment. *Earth Surface Processes and Landforms*, **17**, 127–146.

Smit, G.N. and Rethman, N.F.G. (2000) The influence of tree thinning on the soil water in a semi-arid savanna of southern Africa. *Journal of Arid Environments*, **44**, 41–59.

Smith, M.W., Cox, N.J. and Bracken, L.J. (2007) Applying flow resistance equations to overland flows. *Progress in Physical Geography*, **31**, 363–387.

Smith, M.W., Cox, N.J. and Bracken, L.J. (2010) Terrestrial laser scanning soil surfaces: a field methodology to examine soil surface roughness and overland flow hydraulics. *Hydrological Processes*, DOI:10.1002/hyp.7871.

Snyman, H.A. (2003) Short-term response of rangeland following an unplanned fire in terms of soil characteristics in a semi-arid climate of South Africa. *Journal of Arid Environments*, **55**, 160–180.

Sposito, G. (1972) Thermodynamics of swelling clay-water systems. *Soil Science*, **114**, 243–249.

Strickler, A. (1923) Beitraege zur Frage der Geschwindigheitsformel und der Rauhikeitszahlen für Strome Kanale und Geschlossene Leitungen, in *Mitteilungen des Eidgenössischer Amtes für Wasserwirtschaft*, vol. **16**, Bern, Switzerland.

Swartzendruber, D. (1962) Modification of Darcy's law for the flow of water in soils. *Soil Science*, **93**, 22–29.

Takken, I. and Govers, G. (2000) Hydraulics of interrill overland flow on rough, bare soil surfaces. *Earth Surface Processes and Landforms*, **25**, 1387–1402.

Takken, I., Govers, G., Jetten, V. *et al.* (2005) The influence of both process descriptions and runoff patterns on predictions from a spatially distributed soil erosion model. *Earth Surface Processes and Landforms*, **30**, 213–229.

Tatard, L., Planchon, O., Wainwright, J. *et al.* (2008) Measurement and modelling of high-resolution flow-velocity data under simulated rainfall on a low-slope sandy soil. *Journal of Hydrology*, **348**, 1–12.

Tetzlaff, D., Soulsby, C., Bacon, P.J. *et al.* (2007) Connectivity between landscapes and riverscapes – a unifying theme in integrating hydrology and ecology in catchment science? *Hydrological Processes*, **21**, 1385–1389.

Tetzlaff, D., Seibert, J., McGuire, K.J. *et al.* (2009) How does landscape structure influence catchment transit time across different geomorphic provinces? *Hydrological Processes*, **23**, 945–953.

Thornes, J.B. (1976) Semi-arid erosional systems, Geographical Papers 7, London School of Economics, London.

Thornes, J.B. (1985) The ecology of erosion. *Geography*, **70**, 222–235.

Thornes, J.B. (1990) The interaction of erosional and vegetational dynamics in land degradation: spatial outcomes, in *Vegetation and Erosion. Processes and Environments* (ed. J.B. Thornes), John Wiley and Sons, Ltd, Chichester, pp. 41–55.

Tisdale, J.M. and Oades, J.M. (1982) Organic matter and water-stable aggregates in soils. *Journal of Soil Science*, **33**, 141–163.

Torri, D. and Bryan, R.B. (1997) Micropiping processes and biancana evolution in southeast Tuscany, Italy. *Geomorphology*, **20**, 219–235.

Torri, D. and Poesen, J. (1992) The effect of soil surface slope on raindrop detachment. *Catena*, **19**, 561–578.

Torri, D., Sfalanga, M. and del Sette, M. (1987) Splash detachment: runoff depth and soil cohesion. *Catena*, **14**, 149–155.

Troch, P.A., Carrillo, G.A., Heidbüchel, I. *et al.* (2009) Dealing with landscape heterogeneity in watershed hydrology: a review of recent progress toward new hydrological theory. *Geography Compass*, **3**, 375–392.

Tromp van Meerveld, H.J. and McDonnell, J.J. (2006) On the interrelations between topography, soil depth, soil moisture, transpiration rates and species distribution at the hillslope scale. *Advances in Water Resources*, **29**, 293–310.

Turnbull, L., Wainwright, J. and Brazier, R.E. (2008) A conceptual framework for understanding semi-arid land degradation: ecohydrological interactions across multiple-space and time scales. *Ecohydrology*, **1**, 23–34.

Turnbull, L., Wainwright, J. and Brazier, R.E. (2010) Changes in hydrology and erosion over a transition from grassland to shrubland. *Hydrological Processes*, **24**, 393–414.

Valentin, C. (1993) Soil crusting and sealing in West Africa and possible approaches to improved management, in *Soil Tillage in Africa: Needs and Challenges*, FAO Soils Bulletin 69, Chapter 9.

Valentin, C. and Bresson, L.M. (1992) Soil crust morphology and forming processes in loamy and sandy soils. *Geoderma*, **55**, 225–245.

Van Der Kamp, G., Hayashi, M. and Gallen, D. (2003) Comparing the hydrology of grassed and cultivated catchments in the semi-arid Canadian prairies. *Hydrological Processes*, **17**, 559–575.

Van Dijk, A.I.J.M., Meesters, A.G.C.A. and Bruijnzeel, L.A. (2002) Exponential distribution theory and the interpretation of splash detachment and transport experiments. *Soil Science Society of America Journal*, **66**, 1466–1474.

Van Genuchten, M.T. (1980) A closed-form equation for predicting the hydraulic conductivity of unsaturated soils. *Soil Science Society of America Journal*, **44**, 892–898.

Wainwright, J. (1992) Assessing the impact of erosion on semi arid archæological sites, in *Past and Present Soil Erosion* (eds M. Bell and J. Boardman), Oxbow Books, Oxford, pp. 228–241.

Wainwright, J. (2005) Climate and climatological variations in the Jornada Experimental Range and neighbouring areas of the US Southwest. *Advances in Environmental Monitoring and Modelling*, **2**, 39–110.

Wainwright, J. (2009a) Desert ecogeomorphology, in *Geomorphology of Desert Environments*, 2nd edn (eds A.J. Parsons and A.D. Abrahams), Springer, Berlin, pp. 21–66.

Wainwright, J. (2009b) Weathering, soils and slope processes, in *The Physical Geography of the Mediterranean* (ed. J.C. Woodward), Oxford University Press, Oxford, pp. 167–200.

Wainwright, J. and Parsons, A.J. (2002) The effect of temporal variations in rainfall on scale dependency in runoff coefficients. *Water Resources Research*, **38** (12), (1271), DOI: 10.1029/2000WR000188.

Wainwright, J., Parsons, A.J. and Abrahams, A.D. (1999) Rainfall energy under creosotebush. *Journal of Arid Environments*, **43**, 111–120.

Wainwright, J., Parsons, A.J. and Abrahams, A.D. (2000) Plot-scale studies of vegetation, overland flow and erosion interactions: case studies from Arizona and New Mexico. *Hydrological Processes*, **14**, 2921–2943.

Wainwright, J. and Thornes, J.B. (2003) *Environmental Issues in the Mediterranean: Processes and Perspectives from the Past and Present*, Routledge, London.

Wainwright, J., Parsons, A.J., Müller, E.N. *et al.* (2008) A transport-distance approach to scaling erosion rates: 1. Background and model development. *Earth Surface Processes and Landforms*, **33**, 813–826.

Wainwright, J., Parsons, A.J., Müller, E.N. *et al.* (2009) Response to Kinnell's 'Comment on "A transport-distance approach to scaling erosion rates: 3. Evaluating scaling characteristics of MAHLERAN"'. *Earth Surface Processes and Landforms*, **34**, 1320–1321.

Wainwright, J., Parsons, A.J., Müller, E.N. *et al.* (2010) Standing proud: a response to 'Soil-erosion models: where do we really stand?' by Smith et al. *Earth Surface Processes and Landforms*.

Wang, X.P., Wang, X.R., Xiao, H.L., *et al.* (2007) Effects of surface characteristics on infiltration patterns in an arid shrub desert. *Hydrological Processes*, 72–79.

Warren-Rhodes, K., Weinstein, S., Piatek, J.L. *et al.* (2007) Robotic ecological mapping: habitats and the search for life in the Atacama Desert. *Journal of Geophysical Research – Biogeosciences*, **112** (G4), article G04S06.

Western, A.W., Zhou, S.L., Grayson, R.B. *et al.* (2004) Spatial correlation of soil moisture in small catchments and its relationship to dominant spatial hydrological processes. *Journal of Hydrology*, **286**, 113–134.

Whitford, W.G., Stinnett, K. and Anderson, J. (1988) Decomposition of roots in a Chihuahuan desert ecosystem. *Oecologia*, **75**, 8–11.

Yalin, M.S. (1971) *Mechanics of Sediment Transport*, Pergamon Press.

Yoder, C.K. and Nowak, R.S. (1999) Hydraulic lift among native plant species in the Mojave Desert. *Plant and Soil*, **215**, 93–102.

Young, R.A. and Wiersma, J.L. (1973) The role of rainfall impact on soil detachment and transport. *Water Resources Research*, **9**, 1629–1636.

Young, M.H., McDonald, E.V., Caldwell, T.G. *et al.* (2004) Hydraulic properties of a desert soil chronosequence in the Mojave Desert, USA. *Vadose Zone Journal*, **3**, 956–963.

12

Distinctiveness and diversity of arid zone river systems

Stephen Tooth and Gerald C. Nanson

12.1 Introduction

To those unfamiliar with arid zones, synonymous terms such as 'arid zone river', 'desert river' or 'dryland river' almost sound illogical, for aridity would not seem to be conducive to sustaining river systems. In reality, most drylands support numerous river systems, many of which play a central role in landscape change and exert a strong influence on human use of these marginal environments. Some rivers may be sourced largely from outside dryland settings (exogenous or allogenic rivers), having headwaters in more humid uplands but with sections of their lower courses located in much drier settings (Figure 12.1(a)). Moisture is derived from snowmelt and rainfall in the headwaters, possibly along with small contributions from groundwater, with relatively little runoff being received from within the dryland. These rivers typically have perennial (albeit variable) flow and tend to be major landscape features, many being hundreds or thousands of kilometres long and commonly traversing the dryland as part of exoreic systems (draining to the ocean). Exogenous dryland rivers include some of the largest and best-known rivers in the world, including the Colorado (Figure 12.1(a)) in the American southwest (sourced in the Rocky Mountains), the Nile in north Africa (sourced in the Ethiopian Highlands and East African plateau), the Orange River in South Africa (sourced in the mountains in Lesotho) and the Tigris/Euphrates system in Iraq (sourced in the mountains of eastern Turkey).

The majority of rivers, however, are sourced entirely from within dryland settings (endogenous or endogenic rivers) (Figure 12.1(b)). Moisture sources can be variable in composition, scale and seasonality, but include precipitation derived from local convective thunderstorms, frontal systems or tropical storms, possibly in combination with smaller contributions from snowmelt and groundwater.

Except in areas of groundwater resurgence, these endogenous rivers typically have intermittent flows (seasonal floods followed by little or no flow) or ephemeral flows (occasional floods being interspersed with longer periods of no flow). Many endogenous rivers fail to reach the ocean, instead terminating within drylands on lowland alluvial plains, in playa basins or among aeolian dunefields as part of endoreic systems (interior draining). Endogenous rivers tend to be much shorter than exogenous rivers but in certain physiographic settings nonetheless can still be major landscape features. Large areas of dryland Australia, Africa, South America and Asia, for example, have many endogenous river systems that arise in arid and semi-arid ranges and traverse extensive piedmont and lowland settings over hundreds of kilometres.

Most research into dryland rivers has focused on the smaller endogenous rivers, and it is commonly claimed that the dryland climate sets these rivers apart from their counterparts in wetter and/or cooler climatic settings. Much empirical evidence demonstrates, for example, how some endogenous dryland rivers are characterised by flow and sediment transport processes, channel forms or spatial and temporal patterns of channel change that appear highly distinctive in comparison with humid region rivers (e.g. Graf, 1988; Bull and Kirkby, 2002). At the same time, however, there is growing recognition that even endogenous dryland rivers are inherently diverse and difficult to typify and that in some instances there is considerable overlap with the characteristics of rivers in

Arid Zone Geomorphology: Process, Form and Change in Drylands, Third Edition. Edited by David S. G. Thomas

Figure 12.1 Photographs showing different types of dryland rivers: (a) the exogenous perennial Colorado River in the Grand Canyon, Arizona, southwest USA (flow direction from upper right to middle left); (b) an endogenous ephemeral river in the Eastern Desert, Egypt, which emerges from rugged uplands into a piedmont setting (flow direction towards camera). These photographs illustrate many of the 'textbook' characteristics of dryland fluvial landscapes (see Table 12.1).

other climatic settings, particularly (but not exclusively) the seasonal tropics (e.g. Knighton and Nanson, 1997; Tooth, 2000a; Nanson, Tooth and Knighton, 2002; Powell, 2009).

In the previous edition of this book, Knighton and Nanson (1997) provided an assessment of the distinctive, diverse and unique characteristics of dryland rivers. They recognised that dryland rivers exist along a continuum from ephemeral to perennial flow, but their proposition that dryland rivers as a group can be both distinctive and internally diverse may be seen by some as a possible inconsistency. Drawing on research advances over the past 10–15 years, this chapter has three main aims: (1) to outline why claims for the distinctiveness of dryland rivers have arisen, while also drawing attention to their still underappreciated diversity; (2) to expand the global assessment of dryland river diversity initiated by Knighton and Nanson (1997), Tooth (2000a) and Nanson, Tooth and Knighton (2002) by focusing on three contrasting dryland settings (the Mediterranean region, southern Africa and Australia); and (3) in the light of this revised assessment, to ask whether we can make any sound generalisations regarding the distinctiveness of dryland rivers. The emphasis in this chapter is mainly on the catchment- and reach-scale attributes of dryland rivers, particularly river patterns and floodplains. The short reach, cross-sectional or bed-scale hydrological, hydraulic, geometric and sedimentological attributes of dryland rivers, particularly those associated with endogenous ephemeral rivers, are the subject of Chapter 13.

12.2 Distinctiveness of dryland rivers

Until the mid 1980s, the bulk of dryland fluvial research was conducted in a limited range of environmental contexts, principally parts of the American southwest, the Mediterranean and adjacent areas of Israel and Kenya (e.g. Leopold, Wolman and Miller, 1964; Reid and Frostick, 1987; Graf, 1988; Lewin, Macklin and Woodward, 1995).

Table 12.1 Summary of the typical 'textbook' characteristics of dryland fluvial landscapes, emphasising valley and river characteristics.

Fluvial landscape feature	*Implications for the fluvial system*	*Valley/river characteristics*
High local relief, high drainage densities	Well-developed slope–channel coupling, moderate to high channel gradients	Large areas of bare rock outcrop, locally with active scree slopes, landslides, rockfalls and tributary fans
Sparse or nonexistent hillslope vegetation, rapid runoff generation	Abundant coarse-grained sediment supply, dominance of short-lived flash floods, high rates of sediment transport, dominance of bedload transport	Tendency towards development of wide, shallow, braided channels. Terraces may be present but floodplain development is typically limited. In confined valleys and gorges, finer grained slackwater flood sediments may be deposited and preserved locally
Sparse or nonexistent riparian vegetation	Limited channel bank cohesion or resistance	Tendency for major channel changes during large floods, including alluvial channel widening and deepening, and floodplain and bedrock erosion
Highly variable flow regimes	Major floods separated by long periods with little or no significant flows, erratic catchment sediment delivery	Channel changes initiated during large floods commonly persist for long periods of time owing to limited number of restorative low to moderate flows. Channels thus remain in nonequilibrium, with channel form commonly reflecting the impact of the last large flood

The physiography of these regions inevitably meant that research was focused on a restricted range of river settings, mainly small, high-gradient, endogenous rivers draining catchments that commonly have some degree of tectonic activity (Figure 12.1(b)). Almost as inevitably, this led to a particular view of dryland fluvial landscapes. Many older 'textbook' descriptions (e.g. Mabbutt, 1977; Graf, 1988; Cooke, Warren and Goudie, 1993; Thornes, 1994a, 1994b; Miall, 1996) acknowledged that there was fluvial diversity within drylands but taken together tended to emphasise the valley and river characteristics common to such settings, with Table 12.1 summarising the main generalisations. Limited hillslope vegetation and well-developed slope-channel coupling result in an abundant supply of coarse-grained (gravel, sand) sediment to rivers in confined valleys or to wide, braided channels in piedmont settings. Long periods with little or no channel activity are interrupted by intense rainfall-runoff events that lead to short-lived but high-energy flash floods. Rapid flow velocities and high sediment transport rates commonly result in a dominance of horizontal lamination or low-angle cross-bedding in fluvial deposits. Older deposits may be preserved as terrace sequences, but active floodplains may be nonexistent or limited in extent. Limited sediment cohesion and sparse or nonexistent vegetation mean that banks tend to be unstable, so that large floods result in major (even catastrophic) channel changes, including pronounced widening and/or deepening. Given the evidence for such discontinuous system behaviour, there arose a tendency to view many dryland rivers as being in a state of almost permanent nonequilibrium (e.g. Thornes, 1980; Rhoads, 1988; Graf, 1988), with flows, sediment transport and channel morphology rarely in balance. By this view, dryland rivers were commonly seen as the antithesis of many humid zone perennial rivers that possess well-defined feedback mechanisms between flow, sediment transport and channel morphology, which commonly operate to keep the channel in equilibrium.

To some extent, this early focus on rivers in small, steep, commonly tectonically active catchments was justified. These types of upland landscapes are key physiographic elements in many drylands, host a large proportion of the dryland human population and provide the source of much of the water and sediment for moderate to lower relief piedmont and lowland settings. At the same time, it left the far larger areas of drylands with tectonically stable uplands and vast, lower relief plains poorly known in fluvial terms. Since the mid 1980s, however, dryland fluvial research has expanded into such settings and has provided a broader range of examples from which to draw

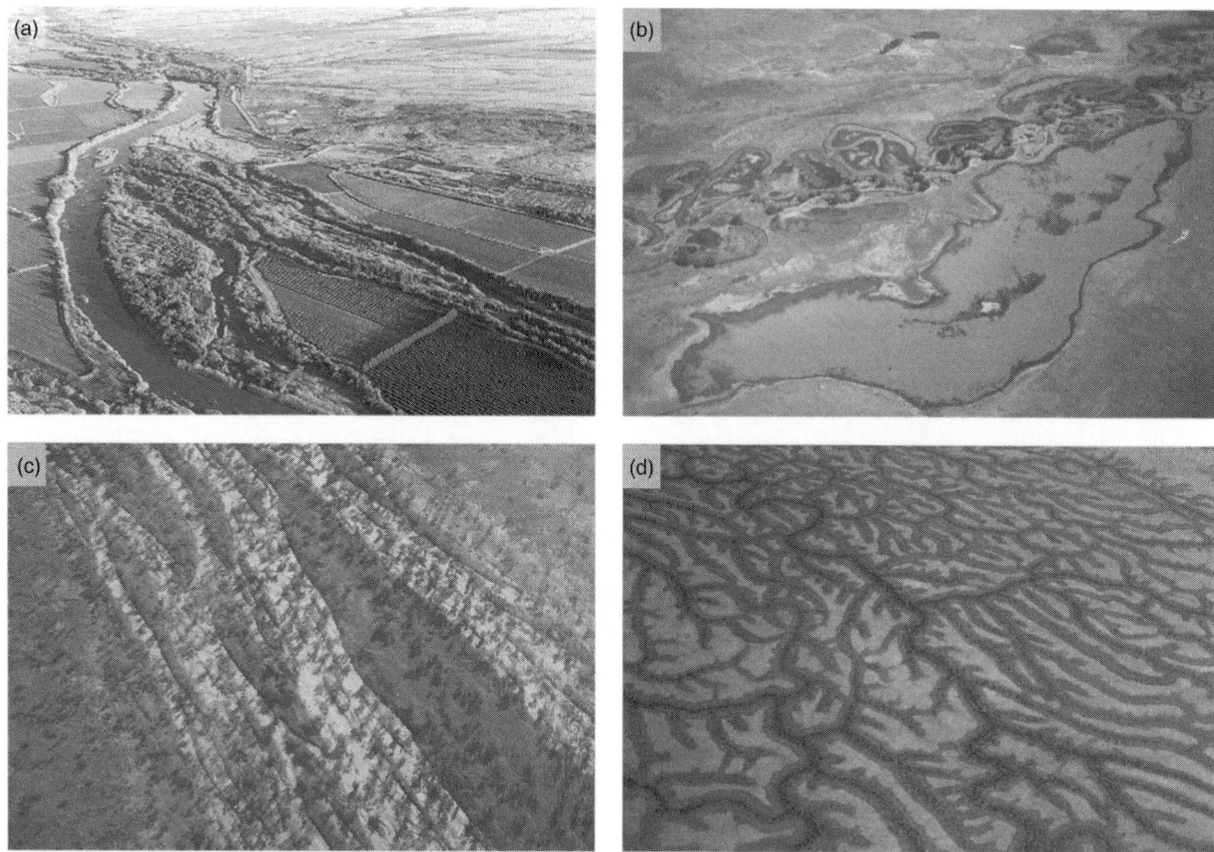

Figure 12.2 Oblique aerial views illustrating dryland rivers that contrast with many 'textbook' characteristics: (a) mixed bedrock–alluvial anabranching on the Orange River, South Africa, illustrating a shallow valley with multiple channels that divide and rejoin around large islands composed of alluvium and/or bedrock (flow direction from upper left to lower right); (b) meandering on the Klip River, eastern South Africa, showing a sinuous channel (marked by invasive willow trees) flanked by flooded oxbows and backswamps (flow direction from middle left to upper right); (c) alluvial anabranching on the Marshall River, central Australia, illustrating multiple channels dividing and rejoining around vegetated, narrow ridges and broader islands (flow direction towards lower right); (d) reticulate channels on the floodplains along Cooper Creek, eastern central Australia (general flow direction from upper left to lower right).

generalisations about dryland river characteristics. In particular, studies of larger, lower gradient rivers draining tectonically quiescent or tectonically inactive catchments in Australia and southern Africa have served to emphasise how many dryland rivers exhibit processes, forms and behaviours that contrast sharply with the 'textbook' characteristics of dryland rivers (Figure 12.2). The extensive anabranching, anastomosing and meandering rivers that are characteristic of some parts of these drylands, for example, challenge the common assumptions that braided rivers are most common in drylands and that meandering rivers are rarely developed, and also demonstrate how some dryland rivers may exhibit forms, processes and behaviours that are similar to humid rivers, including aspects of equilibrium behaviour (e.g. Nanson, Rust and Taylor, 1986; Nanson *et al.*, 1988; McCarthy and Ellery, 1998; Knighton and Nanson, 1994a, 1997, 2000; Tooth, 1999; Tooth and Nanson, 1999, 2000a, 2000b, 2004). Recent overviews have drawn attention to the global diversity of dryland rivers and have highlighted the greater overlap with rivers in other climatic settings (Nanson, Tooth and Knighton, 2002; Powell, 2009), but these messages have tended to become lost among the far more numerous studies of rivers in small, steep, commonly tectonically active catchments. For instance, in the most recent edited volume on dryland rivers (Bull and Kirkby, 2002), the thrust and balance of the chapters is focused largely on catchments in the Mediterranean region (including Israel) and the American southwest, which tend to conform to the 'textbook' characteristics (Table 12.1). Many other recent studies or overviews have also tended to focus on the river types in these drylands (e.g. Cohen and Laronne, 2005; Hassan, 2005; Thomas, 2005; Powell *et al.*, 2007; Pollen-Bankhead *et al.*, 2009), in many cases unwittingly reinforcing the impression that these characteristics apply to most or all dryland rivers. This poses some important, interrelated questions. To what extent does the climatic setting actually impart a distinctive character to dryland rivers? Do exogenous or endogenous rivers with perennial flows (e.g. as fed by snowmelt or groundwater resurgence) exhibit characteristics more akin to humid zone perennial rivers? Why are many rivers in Australia so different from many rivers in other drylands? To what extent can we actually generalise about dryland river characteristics? Such questions highlight the need for a global assessment of dryland rivers that explains their diversity and identifies their distinctiveness.

12.3 Diversity of dryland rivers

A useful starting point for assessing dryland river diversity is to highlight the fact that river systems are the integrated product of their environmental settings, including climatic, tectonic, structural, lithological and vegetative factors. Through their influence on gross physiography, runoff and sediment supply, these environmental factors impact directly and indirectly on river process, form and behaviour. In addition, none of these factors are constant but instead are subject to various degrees of change on different spatial and temporal scales: climate can vary through time, tectonic activity can wax and wane, different lithologies can be uncovered during landscape denudation, vegetation cover can fluctuate in density, composition or health, and so on. River characteristics are in part a reflection of their environmental histories (possibly including inheritance from more humid conditions) but they can also be impacted more-or-less continually by ongoing environmental changes, with geomorphic setting and catchment scale being important determinants of how different rivers will respond to changing conditions (e.g. Nanson and Tooth, 1999).

Drylands are characterised by varied degrees of aridity and exist across a wide range of tectonic, structural, lithological and vegetative settings (Chapters 1 and 2). It has been suggested, for example, that drylands encompass a higher diversity of hydrological conditions than more humid zones (e.g. Pilgrim, Chapman and Doran, 1988; Knighton and Nanson, 1997; Nanson, Tooth and Knighton, 2002). This is exemplified by Figure 12.3, which highlights the fact that dryland rivers exist across a spectrum of hydrological conditions, as expressed in terms of a linear flow occupancy scale. The location of any given dryland river on that scale influences the nature of hydrologic inputs, throughputs and outputs, as well as fluvial network and channel characteristics. Similarly, other natural environmental factors – tectonic and structural setting, lithology, vegetation – can influence dryland rivers in diverse ways (Table 12.2). In addition, different drylands have experienced different environmental (climatic, tectonic, vegetative) histories over Cenozoic, historical and instrumental timescales (Chapters 3 and 4). Although the details of these past and present changes are still being deciphered, across the global extent of drylands, these different combinations of environmental factors, environmental changes and catchment scales might be expected to lead to great diversity in dryland river characteristics.

'River style' (more broadly, 'fluvial style') is a term that is generally used to refer to the overall geomorphological and sedimentary character of a river, including cross-sectional, planform/pattern and longitudinal profile features. A first-order global assessment of dryland rivers should therefore focus on the spatial extent and frequency of occurrence of different styles, but this cannot yet be done comprehensively. As mentioned above, large parts of many drylands remain relatively poorly known in fluvial terms, including large areas of Australia and southern Africa, but even less information is known about river characteristics in the extensive South American and Asian drylands. Within the last few years, the release of virtual globes such as Google Earth (Tooth, 2006) have opened up possibilities of undertaking preliminary planform-based assessments of dryland river characteristics, but this has yet to be done systematically for the world's drylands as a whole, and certainly has not been matched by a corresponding increase in field-based investigations.

Until such comprehensive global assessments are undertaken, a comparative regional assessment is the best approach, as this will at least demonstrate how different combinations of factors can give rise to different dryland river styles. Consequently, the following sections examine river styles in three different drylands that present a spectrum of environmental conditions (Table 12.3). One region is the Mediterranean, which is taken to be broadly representative of the characteristics of relatively high energy rivers draining predominantly small, steep catchments with varying degrees of tectonic activity (Figure 12.4(a)). The other regions offer significantly different perspectives. Southern Africa is taken to be more representative of moderate to low energy rivers draining dominantly moderate-size, moderate-gradient, tectonically quiescent catchments (Figure 12.2(a) and (b)). Australia is examined as representative of relatively low-energy rivers draining

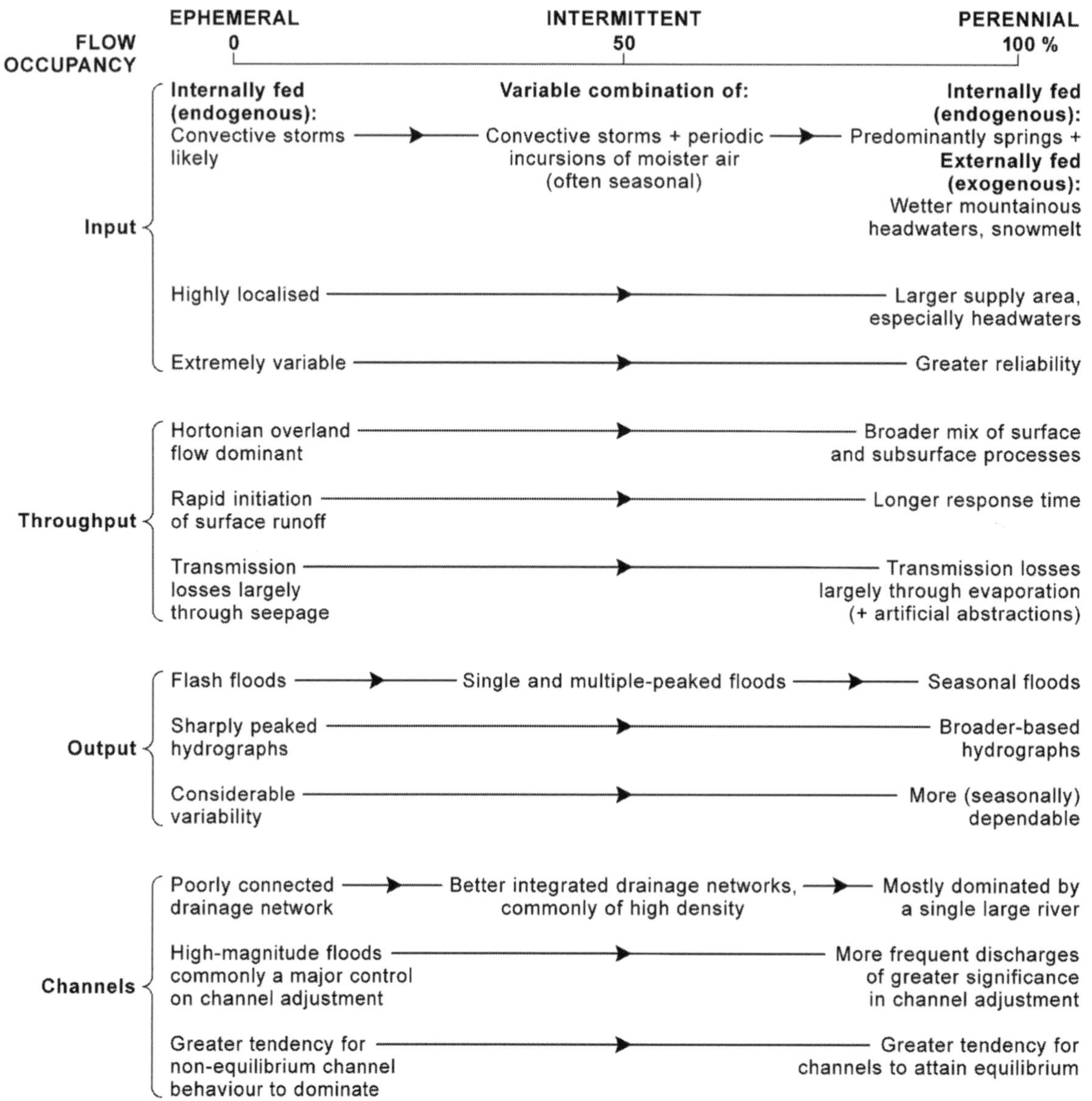

Figure 12.3 Summary of the diversity in hydrology (inputs, throughputs, outputs) and drainage network and channel characteristics within drylands, as expressed in terms of a linear flow occupancy scale (from Knighton and Nanson, 1997, and Nanson, Tooth and Knighton, 2002). Along exogenous rivers or rivers in dry subhumid regions, flow occupancy may be 100 % and the characteristics of these dryland rivers may be more akin to their counterparts in humid regions.

dominantly larger, lower gradient, essentially tectonically inactive catchments (Figure 12.2(c) and (d)).

12.3.1 Higher energy dryland rivers: the Mediterranean region

The Mediterranean region mainly incorporates those rivers that drain to the Mediterranean Sea, but is defined here loosely to also include adjacent parts of southern Europe, north Africa and the Middle East. Large parts of the region are drylands, with the regional climate being characterised by hot, dry summers and cooler, wetter winters, but rainfall decreases from west to east and from north to south so the degree of aridity varies correspondingly from subhumid through to hyper-arid. Much of the region is dominated by high relief catchments with short rivers that descend to typically rugged coastlines

Table 12.2 Examples of natural factors besides climate and hydrology that can influence dryland river characteristics (for more details see Chapters 1 to 4).

Environmental factor	*Primary subdivision*	*Examples of related factors*	*Examples of implications for river character*	*Examples of contrasting drylands*
Tectonic and structural setting	• Active continental margins (1) • Older orogenic belts (2) • Interorogenic, intercratonic areas (3) • Passive continental margins (4) • Cratons (5)	• Rates and amounts of uplift or subsidence • Degree of neotectonic warping • Nature and degree of fault and fold activity • Degree of continentality	Tectonic and structural factors influence: • Absolute and relative local relief • Catchment hypsometry • Baselevel stability/instability • Nature and distribution of sediment sources and sinks (e.g. patterns, rates, volumes and calibre of sediment supply, and the potential for sediment accommodation) • Drainage patterns (e.g. channel orientations) • Valley and channel gradients	Mediterranean drylands: dominated by a variety of tectonically active settings (including 1, 2 and 3), with many areas characterised by seismicity, uplift, subsidence or faulting. Most catchments have coastal outlets with baselevel instability commonplace Australian drylands: dominated by essentially tectonically inactive settings (including 2, 4 and 5), which are characterised by negligible or slow long-term uplift or subsidence. Most catchments are in continental settings with relatively stable baselevels
Lithology	• Igneous • Metamorphic • Sedimentary	Igneous and metamorphic rocks: • Mineralogy and grain size • Nature of foliation, folding and faulting • Joint density and orientation Sedimentary rocks: • Mineralogy and nature of cementation • Grain size • Nature of bedding (e.g. dip/strike), folding and faulting • Joint density and orientation • Nature of contacts between sedimentary units	Lithological factors influence: • Patterns, rates, volumes and calibre of sediment supply • Valley width and drainage patterns (e.g. channel orientations, channel planforms) • Valley and channel gradients • Channel boundary (bed, bank) strength, either directly for bedrock rivers or indirectly through the calibre of alluvial bank material	Southwestern Africa (western South Africa, Namibia): dominantly igneous and metamorphic rocks with varying degrees of deformation. Most areas characterised by low to moderate sediment supply, with abrupt changes in valley widths, channel planforms and gradients commonly corresponding with changes in lithology American southwest (Colorado Plateau): dominantly sedimentary rocks, albeit with exposed igneous/metamorphic basement and intrusions. Sedimentary strata dominantly horizontal to gently dipping but with some deformation (folding and faulting). Many areas are characterised by moderate to high sediment supply, with abrupt changes in valley widths and gradients commonly corresponding with changes in lithology

Table 12.2 (*Continued*)

Vegetation	• Trees • Shrubs • Grasses • Lower order plant life (e.g. algae, fungi, lichens)	• Longevity (e.g. perennial grasses versus ephemeral forbs) • Above-ground characteristics (e.g. plant density and/or distribution) • Above-ground stucture (e.g. height, width and flexibility in channel flows) • Root characteristics (e.g. density and/or distribution) • Vegetation health • Nature of litter layer	Vegetative factors influence: • Rates and volumes of hillslope runoff • Rates, volumes and calibre of hillslope sediment supply • Channel boundary (especially bank) strength • Channel flow hydraulics and sediment transport rates	Australian drylands: typically well-developed riparian vegetation assemblages, including long-lived, deep rooted trees and various shrubs and grasses, many of which establish on lower channel banks or on channel beds. Large vegetative contribution to channel boundary strength and flow resistance Peruvian/Chilean Atacama desert (coastal setting): large areas virtually devoid of vegetation cover but, where present, vegetation consists mainly of patchy dwarf shrubs and some lower order fog-dependent plants. Limited vegetative contribution to channel boundary strength and flow resistance

Table 12.3 Generalised characteristics of Mediterranean, southern African and Australian dryland fluvial environments. While considerable fluvial diversity can be found within each of the three regions (see text for details), these generalisations serve to highlight some of the main contrasts.

Mediterranean region	*Southern Africa*	*Australia*
Subhumid to hyper-arid climates:	Subhumid to hyper-arid climates	Subhumid to arid climates
Dominance of small, steep catchments	Dominance of moderate-size, moderate-gradient catchments	Dominance of large, low-gradient catchments
Mainly endogenous rivers, but some exogenous rivers	Exogenous and endogenous rivers	Mainly partly exogenous and endogenous rivers
Ongoing tectonism in many areas, albeit to varying degrees	Limited large-scale tectonism during the Quaternary	Limited large-scale tectonism during the Cenozoic
Variable geological structures and lithologies	Variable geological structures and lithologies	Variable geological structures and lithologies
Cenozoic palaeoclimatic changes with sea-level changes also impacting on lower reaches of rivers with a coastal outlet	Cenozoic palaeoclimatic changes but with limited influence of sea-level changes on most dryland rivers	Cenozoic palaeoclimatic changes but with most dryland rivers isolated from Quaternary sea-level changes
Variable development of native riparian vegetation communities, many invasive species	Variable development of native riparian vegetation communities, many invasive species	Very well-developed native riparian vegetation communities, some invasive species
Long history (1000s of years) of significant human impacts	Recent history (last 100–150 years) of significant human impacts	Recent (last 100–150 years) or negligible history of significant human impacts

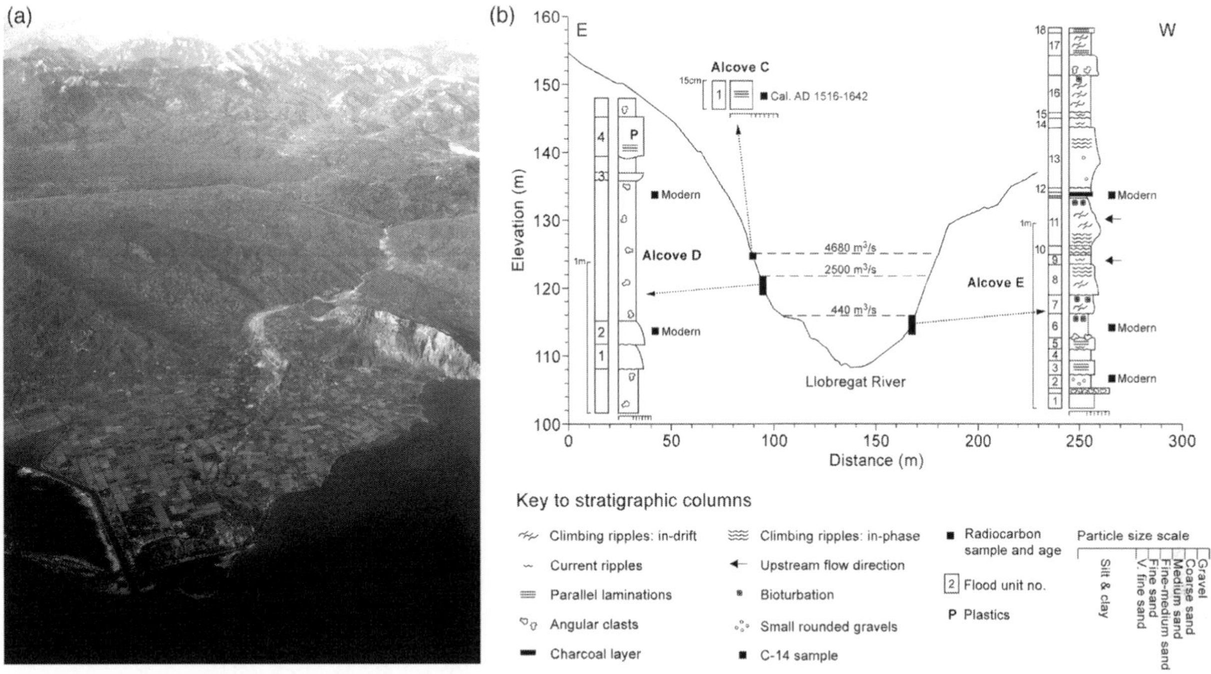

Figure 12.4 (a) Oblique aerial view of the Evinos River, Greece, showing features common to many Mediterranean catchments. In this instance, a coarse-grained, braided river emerges from mountainous headwaters and crosses a narrow piedmont and heavily cultivated coastal plain before entering the sea. (b) Almalgamated cross-section illustrating the stratigraphy and chronology of slackwater sediments preserved in three alcoves (C, D, E) in a bedrock-confined reach of the Llobregat River, northeast Spain. Alcove E represents a relatively complete record of low to high magnitude flood events over the last ~100 years, while the higher elevation alcove C only preserves a record of a single extreme flood event that occurred in the Little Ice Age. The minimum discharge estimates for the upper flood units in each alcove are indicated (from Thorndycraft *et al.*, 2005).

(e.g. Figure 12.4(a)) but endoreic basins can also be found, most notably in the Dead Sea rift valley in the eastern Mediterranean. A variety of dryland river types can be found across the region (Macklin, Lewin and Woodward, 1995), including: (1) high-gradient rivers that are commonly deeply incised into bedrock, colluvium or older alluvial fills; (2) high- to moderate-gradient rivers that traverse basin- and range-type environments; and (3) rivers in coastal alluvial plains and deltas. Endogenous intermittent or ephemeral rivers are commonplace, but the delta of the exogenous perennial Nile River is located in the far southeast of the region.

Various tectonic, structural and lithological factors have interacted with eustatic and palaeoclimatic changes to influence the characteristics of Mediterranean dryland fluvial landscapes. The region forms an active boundary zone between the Eurasian, African and Arabian plates, which has resulted in formation of extensive fold belts, horst and graben structures and local faulting and tilting. Across large parts of the region, many relatively erodible Late Cenozoic sedimentary successions (e.g. molasse, flysch) have been uplifted and eroded, promoting hillslope instability and high rates of sediment supply, commonly from badland-type environments to high-energy braided or compound channels (e.g. Hooke and Mant, 2002). During the Late Miocene (5–6 Ma) 'Messinian Salinity Crisis', the Mediterranean Sea desiccated (Krijgsman *et al.*, 1999) and the dramatic kilometre-scale baselevel fall promoted deep landscape incision around the basin margins, initiating the formation of many present-day gorges. In the western Mediterranean, diapirism and solution-induced subsidence has also influenced the long-term development of many dryland rivers (e.g. Harvey and Wells, 1987; Benito *et al.*, 2000; Mather, 2000; Stokes and Mather, 2003; Candy, Black and Sellwood, 2004; Guerrero, Gutiérrez and Lucha, 2008; Maher and Harvey, 2008; Stokes, 2008). Quaternary environmental changes have affected the region, with fluctuations in palaeoprecipitation and palaeotemperature having combined to create conditions both drier and wetter than present. Glaciation and periglaciation occurred in some catchment headwaters (e.g. Woodward *et al.*, 2008), eustatic changes impacted on the lower river reaches of rivers draining to the Mediterranean Sea (e.g. Amorosi *et al.*,

2008), while in the Dead Sea rift valley major fluctuations in the level of the Dead Sea and its precursors have also influenced river behaviour (e.g. Niemi, Ben-Avraham and Gat, 1997; Enzel, Agnon and Stein, 2006). Across much of the region, these environmental changes have combined to result in phases of enhanced hillslope sediment supply and channel aggradation, which typically have been superimposed on underlying longer-term incisional trends, leading to the production of complex but decipherable terrace records along many dryland rivers (Lewin, Macklin and Woodward, 1995; Macklin and Woodward, 2009; see also the example in Box 12.1).

Box 12.1 Gorge-bound rivers in Crete

The arid to semi-arid island of Crete contains many ephemeral and intermittent, high-gradient bedrock, boulder-bed or cobble-bed rivers. A reach of the Anapodaris River, located in a gorge in the lower part of a 500 km^2 catchment in south central Crete (Figure 12.5(a)), exemplifies how strong slope–channel coupling makes many

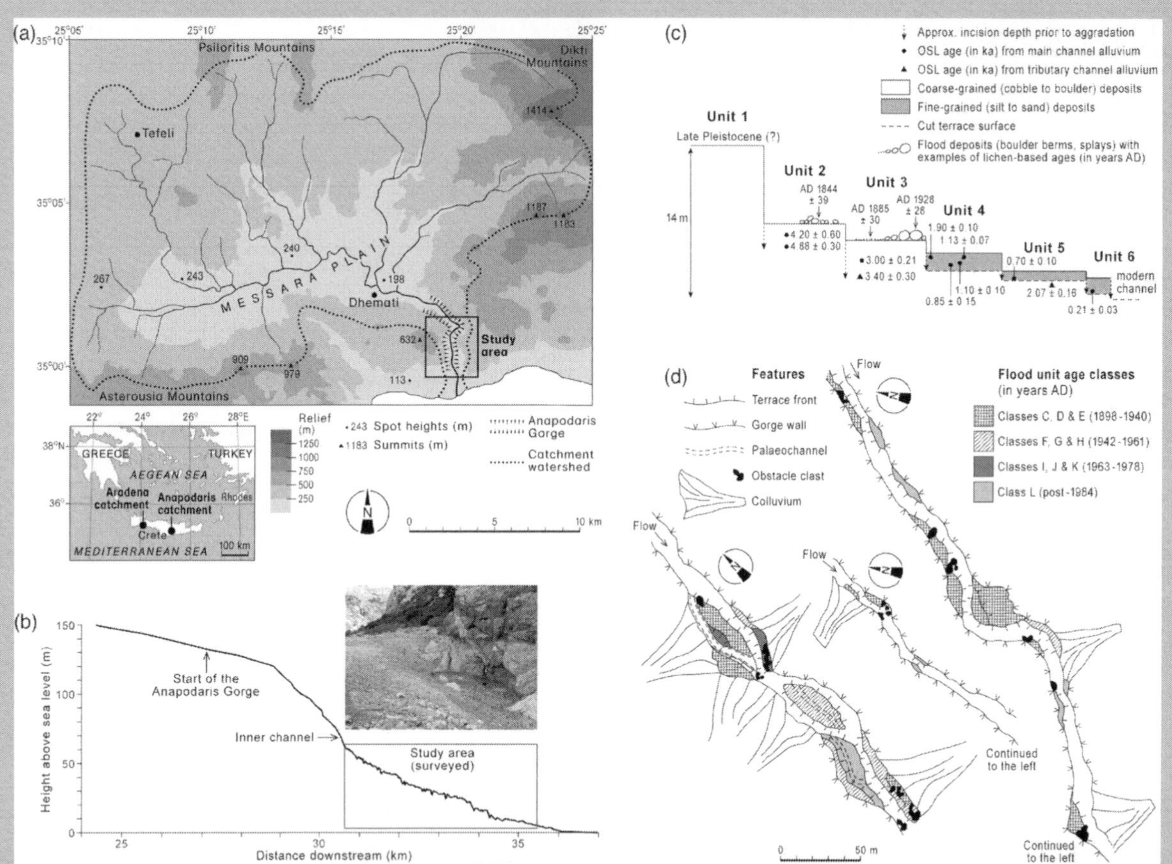

Figure 12.5 (a) Location and topography of the Anapodaris catchment, south central Crete (spot heights and contours in metres above sea level), showing the location of the study area in the gorge (modified, after Macklin *et al.*, 2010). (b) longitudinal profile of the Anapodaris River through the gorge, as derived from a combination of 1 : 5000 topographical maps and field survey data (modified, after Macklin *et al.*, 2010). The inset shows a typical scour pool developed within one of the constrictions in the surveyed part of the gorge. (c) Schematic cross-section (to scale in the vertical) to illustrate the morphological and stratigraphical relationships, sedimentology and optically stimulated luminescence (OSL) ages of the alluvial units in the Anapodaris Gorge. Examples of lichen-based age estimates for recent flood deposits are also indicated (modified after Macklin *et al.*, 2010). (d) Geomorphological map of part of the Aradena Gorge (see Figure 12.5(a) for location) showing the position of colluvial and alluvial deposits and the lichen- and tree ring-based ages of coarse-grained valley floor surfaces (modified after Maas and Macklin, 2002).

of these rivers potentially responsive to changes in catchment runoff and sediment supply driven by fluctuations in the dryland climate, but also illustrates that tectonic, structural, lithological and/or anthropogenic factors exert influences on river character (Macklin *et al.*, 2010). Upstream of the gorge, the river flows through the Messara Plain, a graben that has been infilled with Tertiary sedimentary successions (marls, flysch, sandstone) and more recent alluvium, but within the gorge, hillslope and river gradients steepen dramatically (Figure 12.5(b)). Abrupt changes in gorge orientation correspond with regional faults or fractures, with narrower sections and scour pools up to several metres deep corresponding with resistant limestone outcrop (Figure 12.5(b), inset) and wider sections with coarse gravel bars and local islands corresponding with weaker flysch outcrop. The gorge margins are characterised by a succession of coarse-grained (predominantly cobble to boulder) and fine-grained (predominantly silt to sand) alluvial terraces that locally interfinger with, or are overlain by, coarse colluvial and tributary river deposits. Coarse-grained sediment is derived locally from the steep, sparsely vegetated hillsides and tributaries surrounding the gorge, while fine material is derived principally from the Messara Plain. In this dryland climate, the lower parts of many terraces have become strongly cemented by secondary calcite that has formed in pore spaces within the gravels or as a replacement for the finer sediment matrix, and the terraces are not easily eroded by the present-day, short-lived, ephemeral flows. Geochronological investigations demonstrate that these terraces date to the Mid to Late Holocene, and provide evidence for widespread, coarse-grained aggradational episodes that have been punctuated by incisional episodes and coarse sediment export (Figure 12.5(c)). Comparison with other Mediterranean environmental change records, particularly high-resolution marine isotope records, suggests that these aggradation/incisional episodes were primarily driven by fluctuations between arid and semi-arid conditions, as reflected in a changing balance between high-energy flood events and hillslope/tributary sediment supply. By contrast, several phases of widespread fine-grained deposition within the last 2000 years have locally capped the lower terrace surfaces (Figure 12.5(c)). This provides evidence for decreases in flood competence, possibly coupled with land cover and land-use changes in the Messara Plain, which represents one of the largest and most fertile lowlands in Crete. Since the middle of the nineteenth century, several large floods have formed localised boulder berms (Figure 12.5(c)) and have contributed to stripping of the fine-grained deposits from many parts of the gorge, but even these floods typically have been unable to cause significant erosion of the cemented coarser-grained units and lateral channel activity has thus been constrained (Macklin *et al.*, 2010).

Studies of dryland rivers in southwest Crete have also revealed various Late Quaternary alluvial deposits. In many small limestone catchments, bedrock channels or coarse-grained, braided channels are widespread. Deposits include boulder berm deposits that have been related to clusters of flood events occurring over the last 100–150 years or so (Figure 12.5(d)), but finer-grained deposits tend to be less common owing to the restricted extent of suitable source material (Maas and Macklin, 2002). These findings demonstrate how river process, form and behaviour in such small, steep, dryland catchments depends on a delicate and dynamic balance between discharge and the volume and calibre of supplied sediment. In Crete and the Mediterranean more generally, many rivers have been similarly sensitive to rapid Holocene climate change (Lewin, Macklin and Woodward, 1995; Macklin and Woodward, 2009), but in most cases these climatic influences on river character and response have also been strongly conditioned by tectonic, structural, lithological and anthropogenic factors.

The Mediterranean has also been subject to a long history of human impact, including extensive land-cover and land-use changes, and various forms of direct flow regulation and channel modification (e.g. diversion, damming, realignment). The relative roles of climate and human activity in determining Late Holocene histories of deposition and erosion in Mediterranean valleys is a focus of continuing research (e.g. Benito *et al.*, 2008, 2010; Macklin *et al.*, 2010), but it is clear that the rate and extent of human impacts have clearly intensified over the last 100 years, so that relatively few rivers remain in a pristine condition. Despite the pervasive human impacts, however, climate generally has been the primary driver of 'very Late Holocene' river activity, with most activity occurring during long-duration winter flood events or more extreme summer/autumn floods (e.g. Poesen and Hooke, 1997; Maas and Macklin, 2002; Greenbaum, Schwartz and Bergman, 2010). Flood records have been reconstructed from the sedimentary deposits left by such events, such as coarse-grained (cobble and boulder gravel) deposits (e.g. Maas and Macklin, 2002; Laronne and Shlomi, 2007; see also the example in Box 12.1). In some catchments, fine-grained slackwater sediments preserved in protected embayments or alcoves in bedrock-confined reaches have been used for more detailed palaeoflood reconstructions, including palaeoflow hydraulics, changes in flood magnitude and frequency, and the relationship with atmospheric and hydrological controls (e.g. Thorndycraft *et al.*, 2005; Greenbaum *et al.*, 2006; Benito *et al.*, 2008) (Figure 12.4(b)). Many studies have documented the impacts of historical moderate-to-high energy floods on channels in the region (e.g. Harvey, 1984; Poesen and Hooke, 1997; López-Bermúdez, Conesa-García and Alonso-Sarría, 2002; Greenbaum and Bergman, 2006), with some channels having undergone dramatic widening and/or deepening, impacts that in many cases last for years or decades.

Overall, Mediterranean rivers exemplify many of the 'textbook' generalisations regarding dryland river characteristics. Across much of the region, a legacy of tectonic activity, lithological variations, eustatic fluctuations and palaeoclimatic changes has created a rugged physiography consisting of short, steep catchments or basin-and range-type settings, with coastal lowlands relatively restricted in extent. Strong slope–channel coupling promotes abundant sediment supply to high-energy rivers. River styles thus tend to be dominated by bedrock channels or coarse-grained alluvial channels (Macklin, Lewin and Woodward, 1995). Many valleys preserve terrace sequences, and although floodplain development tends to be limited, subtle changes in the nature of boundary resistance and the calibre of sediment supply can be important determinants of channel process, form and change (see Box 12.1). Even with flow regulation and various forms of channel modification, many rivers are subject to major, long-lasting change during erratic, short-lived flash floods, commonly exhibiting nonequilibrium behaviour (e.g. Poesen and Hooke, 1997).

12.3.2 Moderate and lower energy dryland rivers: southern Africa

Southern Africa can be defined as a region encompassing all or part of the countries of South Africa, Lesotho, Swaziland, Namibia, Botswana, Zimbabwe and Mozambique. With the exception of a typically narrow (<150 km) humid zone around the southern and eastern margins of the subcontinent, drylands are extensive. The interior is characterised by warm, wet summers and cool, dry winters and the western margins by hot, dry summers and cooler, wetter winters, and the degree of aridity tends to increase along an east–west gradient, from subhumid and semi-arid in the eastern interior to arid and hyper-arid in the west. The Great Escarpment roughly parallels the coastal margin and separates a narrow (<200 km wide) dissected piedmont and coastal lowland from the low relief, elevated interior, much of which lies above 1000 m a.s.l. (Summerfield, 1991). Several types of river system can be identified, including: (1) moderate-size, moderate-gradient rivers that drain seaward from the Great Escarpment or coast-parallel ranges to the ocean; (2) longer, moderate-gradient rivers associated with the extensive (~2350 km long) Orange River system; and (3) moderate-to low-gradient rivers of the endoreic Kalahari basin. Endogenous intermittent and ephemeral rivers are commonplace, but there are also large exogenous perennial rivers (e.g. Orange and Okavango Rivers).

Various tectonic, structural, lithological and palaeoclimatic factors have influenced the characteristics of southern African dryland fluvial landscapes. Following Late Mesozoic break-up of the Gondwana supercontinent, large-scale uplift has driven long-term net river incision and landscape denudation. Substantial thicknesses (~1–2 km) of basalts and sedimentary cover rocks have been removed, leading to widespread river superimposition onto underlying, typically more resistant lithologies. Along with temporal and spatial variations in hillslope sediment supply, many rivers have become closely adjusted to the variable lithologies that they traverse, leading to a variety of mixed bedrock-alluvial and bedrock styles, including anabranching rivers (Figure 12.2(a)), meandering rivers flanked by seasonally inundated floodplain wetlands (see Figure 12.2(a) and Box 12.2) and ingrown meanders (Figure 12.6(a)) (Heritage, van Niekerk and Moon, 1999; van Niekerk *et al.*, 1999; Tooth *et al.*, 2002a; Tooth and McCarthy, 2004a; McCarthy and Tooth, 2004). Some dryland river reaches are bedrock confined, most notably along the lower Orange River (Wellington, 1958; Jacob, Bluck and Ward, 1999) and in the headwaters of rivers draining the Great Escarpment in Namibia (Jacobson, Jacobson and Seeley, 1995; Codilean *et al.*, 2008). During the Cenozoic, there has been a long-term trend towards drier and more variable climates, and while the direct effects of Quaternary glaciations and eustasy on most dryland rivers has been minimal, palaeoclimatic fluctuations involving conditions both drier and wetter than present have impacted on the region. This has led to the contraction of formerly more extensive drainage networks (e.g. Goudie and Wells, 1995; Tooth and McCarthy, 2007), particularly in western Namibia, where many former exoreic river systems now terminate in aeolian dunefields far from the coastline, forming saline pans and playas (e.g. Seeley and Sandelowsky, 1974; Teller, Rutter and Lancaster, 1990; Jacobson, Jacobson and Seeley, 1995). Sedimentary records along many of these dryland rivers typically reveal phases of channel aggradation that have been superimposed on an underlying longer-term

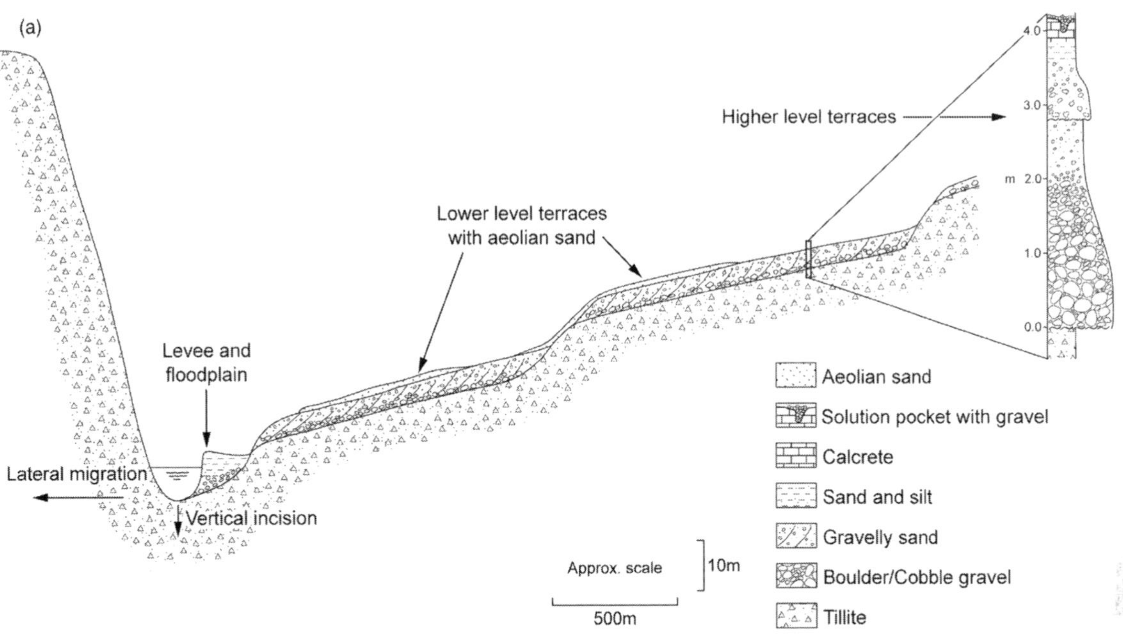

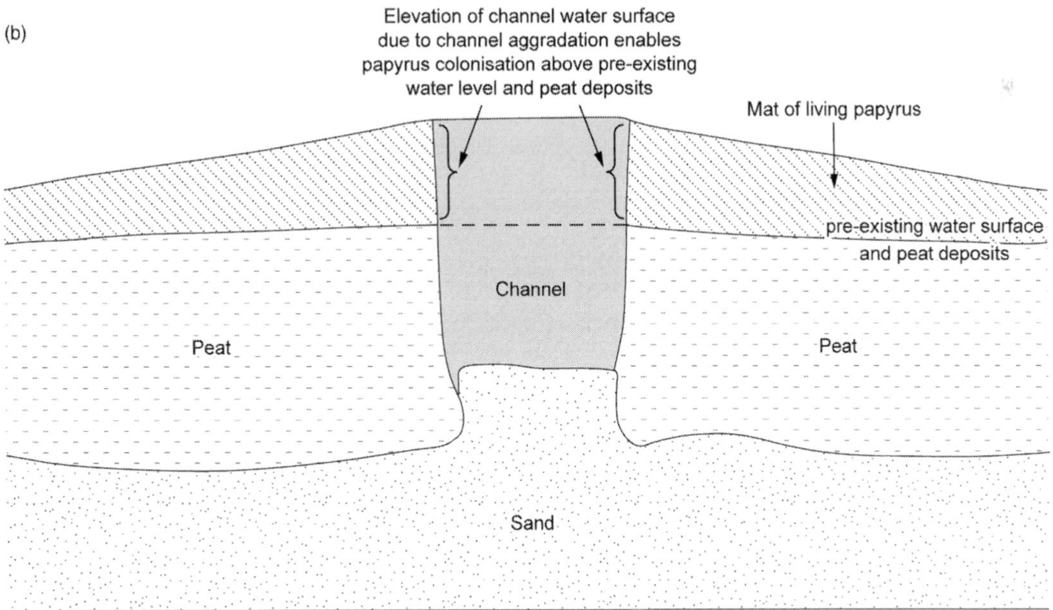

Figure 12.6 Schematic diagrams to highlight the contrasting behavioural tendencies of rivers in southern Africa. (a) Ingrown meander bend along the mixed bedrock–alluvial middle Orange River, South Africa. On the inner bend, gently sloping alluvial sheets form as a result of point bar deposition during lateral migration and minor vertical incision of the channel bed, while distinct scarps separating these sheets result from phases of more rapid vertical incision during the headward migration of knickpoints. On the outer bend, steep cutbanks are formed in tillite (from McCarthy and Tooth, 2004, reproduced by permission of Schweizerbart and Borntraeger Science Publishers, http://www.borntraeger-cramer.de). (b) Failing channel in the wetlands of the Okavango Delta, Botswana. Channel aggradation by bedload sand leads to a gradual rise of local water levels in and around the channel, which promotes invasion by papyrus (*Cyperus papyrus*). In turn, papyrus invasion promotes further aggradation and water losses and ultimately leads to channel avulsion (from Ellery, McCarthy and Smith, 2003, with kind permssion of Springer Science and Business Media).

incisional trend (e.g. de Wit, Marshall and Partridge, 2000) (Figure 12.6(a)). By contrast, in the more northerly part of the interior, the seismically active Kalahari Basin formed in a broad crustal downwarp and during the Cenozoic has been gradually infilling with colluvial, lacustrine, fluvial and aeolian sediments (Thomas and Shaw, 1991). In this largely endoreic basin, dry valley networks ('mekgacha') and ephemeral single-thread, sandy alluvial rivers are the most widespread fluvial landforms (e.g. Shaw, 1989; Nash, Thomas and Shaw, 1994). The basin also hosts the extensive (~40 000 km^2) Okavango Delta, which has formed within a depression related to the south-westerly extension of the East African Rift Systems and incorporates prime examples of permanent and seasonal wetlands in a dryland setting (McCarthy and Ellery, 1998; Tooth and McCarthy, 2007). The wetlands are sustained by seasonal floods arising in the exogenous headwater tributaries of the Okavango River and by local rainfall, and are traversed by unusual, peat- and vegetation-lined meandering, stable sinuous and straight sand-bed channels that occur within local anastomosing and more extensive distributary networks (McCarthy and Ellery, 1998; Tooth and McCarthy, 2004b). Along these channels, a combination of Late Quaternary flow regime changes, subtle tectonic activity (minor faulting and tilting) and autogenic changes (channel sedimentation and vegetation encroachment; see Figure 12.6(b)) has resulted in lateral activity (migration) and frequent avulsion, leading to large-scale shifts in the distribution of water and sediment across the delta (McCarthy and Ellery, 1998).

Box 12.2 Mixed bedrock-alluvial rivers on the South African Highveld

In subhumid to semi-arid northeastern South Africa, many perennial and intermittent rivers commonly meander within floodplain wetlands up to 2 km wide. The upper Klip River, eastern Free State (Figure 12.7(a)), exemplifies how the formation, distribution and character of these meanders and floodplain wetlands are strongly influenced by local structural and lithological factors that tend to override the dryland climatic influences (Tooth *et al.*, 2002a). Catchment geology consists of sedimentary rocks (sandstones, mudstones) of the Karoo Supergroup, which have been extensively intruded by resistant dolerite sills and dykes ('barriers'). Where these barriers crop out in the river bed, they form local baselevels for the river upstream, with the corollary that vertical erosion in the weaker sedimentary rocks cannot proceed faster than vertical erosion of the channel bed at the downstream dolerite barrier. Over short to medium timescales (decades to tens of thousands of years), dolerite erosion rates can be presumed slow, and erosion in the upstream sandstone/shale valleys is restricted to lateral erosion down to the level of the dolerite. This takes place as the river migrates across the valley floor, simultaneously planing the sedimentary rocks underlying the channel bed and reworking floodplain alluvium (Figure 12.7(b)). Over time, this has widened the valleys and created additional accommodation space for meanders, floodplain alluvium and the associated wetlands. Floodplains are inundated during summer rainfall and flooding (Figure 12.2(b)), with topographic lows (e.g. abandoned channels, oxbows, backswamps) retaining water the year round and forming significant areas of wetlands in an otherwise seasonally dry environment. Luminescence dating of alluvial deposits in abandoned channels indicates that the Klip River wetlands have formed, at a minimum, over the last 30 000 years, with regional climatic fluctuations having had remarkably little influence on channel morphology or dynamics (Tooth *et al.*, 2007).

Similarly pronounced transitions from alluvial meandering reaches with floodplain wetlands in sandstone/shale valleys to bedrock-influenced straighter reaches in dolerite valleys occur along many other dryland rivers in the region (Tooth *et al.*, 2002a). Over long timescales (>10^5 years), meandering and bedrock planing in the sandstone/shale valleys has resulted in lowering of the channel gradient in relation to gradient in the dolerite valleys so that many rivers have strongly stepped long profiles (Figure 12.7(c)). Meanders and floodplain wetlands persist in the low-gradient reaches as long as the downstream dolerites continue to form stable barriers, but ongoing river incision means that the dolerites are eventually breached, either partially or fully, which leads to a fall in local baselevel. When this happens, headward-retreating knickpoints migrate upstream into the sandstone/shale valleys, leading to river incision, river straightening (Figure 12.7(d)) and channel-floodplain decoupling (Tooth *et al.*, 2004). With the cessation of regular flooding, the former floodplain wetlands desiccate and become susceptible to erosion by dongas (gullies), as is demonstrated along a number of rivers in the region (Figure 12.7(d)). These findings show that in this part of South Africa, dryland river process, form and behaviour depends largely on the interaction

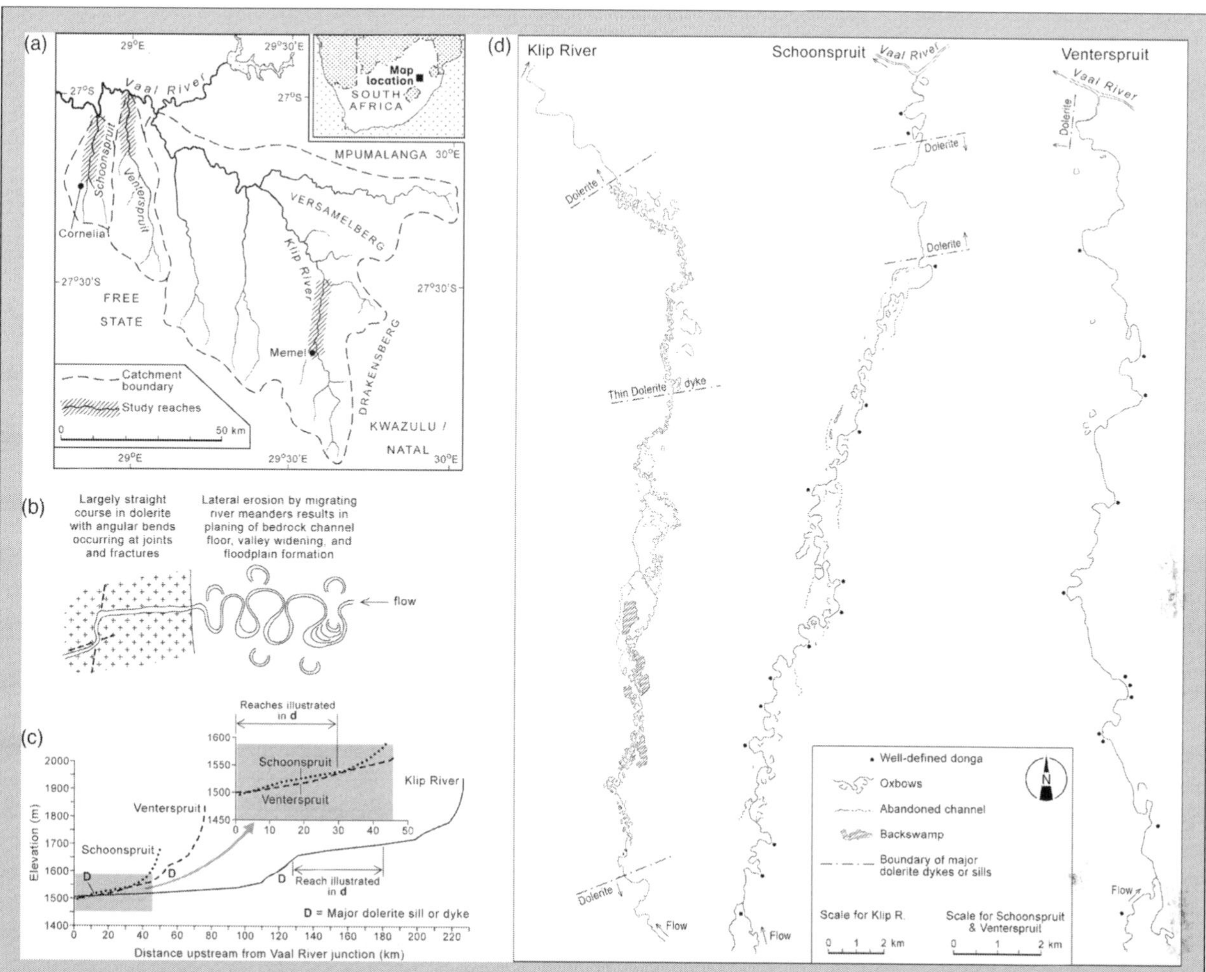

Figure 12.7 (a) Location of the Klip River catchment in eastern South Africa, also showing the location of the neighbouring Ventersspruit and Schoonspruit catchments (from Tooth *et al.*, 2004). (b) Schematic illustration of the factors promoting meandering and floodplain development in sandstone/shale valleys upstream of resistant dolerite sills or dykes (after Tooth *et al.*, 2002a). (c) Longitudinal profiles of the Klip River, Ventersspruit and Schoonspruit (constructed from contour crossings on 1 : 50 000 topographic maps), showing how channel-bed gradients are lower upstream and downstream of major dolerite sills or dykes (from Tooth *et al.*, 2004). (d) Planforms of three neighbouring rivers with different degrees of downstream dolerite control, illustrating the contrasting channel sinuosities and the differences in the number of oxbows, abandoned channels and dongas: Klip River (intact dolerite sill at lower end of study reach), Schoonspruit (partially breached dolerite sill at lower end of study reach), Ventersspruit (fully breached dolerite sill at lower end of study reach). Note the different scale for the Klip (from Tooth *et al.*, 2004). Parts (a), (c) and (d) reproduced with kind permission of Elsevier.

between alluvial and bedrock processes and that the dryland climatic influences are relatively muted. Geologically controlled, long-term cycles of wetland formation and destruction are a key natural element of these landscapes, but in some areas poor land management and artificial drainage schemes have also contributed to wetland desiccation and donga erosion.

Prior to the start of European colonisation in the seventeenth and eighteenth centuries, human impact on dryland rivers across southern Africa was minimal, but has greatly intensified since, particularly over the last 100–150 years. Land-cover and land-use changes have included reduction or removal of native fauna (especially grazing ungulates) and replacement with cattle and sheep, introduction of exotic invasive riparian vegetation (e.g. willows) and direct channel–floodplain modifications (e.g. drainage, diversion, channelisation). Unmodified flow regimes tend to be highly variable on a quasi-decadal basis, alternating between wet periods with occasional very large, destructive floods and periods with below-average flows. Late Holocene flood records have been reconstructed from flood deposits (including slackwater sediment sequences) along a number of rivers (e.g. Zawada, 1997; Srivastava *et al.*, 2006; Eitel *et al.*, 2006), and the geomorphic impacts of large historical floods have also been documented from various locations (e.g. Zawada, 1994). Along mixed bedrock–alluvial anabranching rivers in the Kruger National Park, northeastern South Africa, several floods over the past 100 years have periodically stripped alluvium down to the underlying bedrock in certain reaches, but island and floodplain rebuilding occurs in association with vegetation establishment and growth during subsequent lower flow periods (Heritage, van Niekerk and Moon, 1999; Rountree, Rogers and Heritage, 2000; Rountree, Heritage and Rogers, 2001). Other rivers, however, have been less sensitive even to large historical floods; for instance, the Orange River near Augrabies Falls, western South Africa, experienced only modest erosion and deposition even during a large (8300 m^3/s) flood in 1988 (Zawada and Smith, 1991) and the channels in the Okavango Delta at present tend to undergo only slow progressive change even during the large seasonal floods (McCarthy and Ellery, 1998).

Overall, many southern African rivers only partly match the 'textbook' generalisations regarding dryland rivers. In western Namibia, moderate-gradient, poorly vegetated catchments with relatively strong slope–channel coupling and moderately abundant sediment supply have promoted a variety of confined bedrock and alluvial braided rivers (Jacobson, Jacobson and Seeley, 1995) with morphological and sedimentological similarities to rivers in the Mediterranean and the American southwest. Many single-thread, planar-bed, sandy rivers in the Kalahari basin and adjacent areas (e.g. Shaw, 1989; Hassan, Schick and Shaw, 1999) exhibit morphological characteristics common to many broad, wide sandy channels described from the American southwest, Middle East and East Africa (e.g. Wolman and Gerson, 1978; see also Chapter 13). The rest of the region, however, is characterised by moderate-gradient catchments with variable slope–channel coupling and riparian vegetation assemblages, where a variety of river styles not commonly associated with drylands has developed, including various mixed bedrock–alluvial styles (e.g. structurally and lithologically controlled meandering and anabranching) as well as extensive floodplain wetlands (Figures 12.2(a), (b) and 12.6(a); see also Box 12.2). Although some rivers are subject to major (even catastrophic) change during occasional large floods, channel recovery tends to take place on a timescale of years to decades, so that rivers may readjust towards and possibly attain equilibrium before the next large flood (e.g. Rountree, Rogers and Heritage, 2000). Other rivers appear to be relatively insensitive to large flood events and can be viewed as being in short-term equilibrium, even on landscapes that they are slowly eroding or aggrading over longer timescales (e.g. McCarthy and Tooth, 2004).

12.3.3 Lower energy dryland rivers: Australia

With the exception of Antarctica, Australia is regarded as the driest continent, both in terms of the relative extent of its arid and semi-arid areas, which cover between two-thirds and three-quarters of the continent, and its low mean continental precipitation and runoff (Mabbutt, 1986; Warner, 1986; Finlayson and McMahon, 1988). Extensive subhumid areas surround the arid/semi-arid continental interior, and even the relatively well-watered southern and eastern margins of the continent are subject to highly variable climates that include long periods of drought. In the seasonal (semi-arid) tropics of the north, summer wet seasons are followed by a 7 to 9 month long, warm, dry season so that conditions for part of the year are also similar to many dryland areas. The easterly Great Dividing Range passes westward into extensive, low-relief, low-gradient plains, topographic depressions and extensive aeolian dunefields that are broken only by isolated, low elevation (mostly <1000 m a.s.l.) ranges. Different types of moderate- to low-gradient river system can be found across the Australian drylands, including: (1) rivers associated with the exoreic Murray–Darling system, southeastern Australia; (2) rivers associated with the endoreic drainage in the Lake Eyre basin, central Australia; and (3) rivers associated with isolated ranges and extensive shield areas such as characterise parts of South Australia and Western Australia. Endogenous intermittent and ephemeral rivers are commonplace but Australia's largest dryland rivers are partly exogenous (e.g. Cooper Creek, Murray–Darling), receiving regular (seasonal) rainfall in humid to subhumid headwaters.

Australia's low relief and elevation results from a lack of significant uplift since Gondwana break-up, with Cenozoic tectonic activity having been limited to broad, crustal warping and local, minor faulting. This tectonic and physiographic framework has facilitated the development of many large, low-gradient, dryland river systems that arise in ancient, stable headwater ranges and follow predominantly alluvial courses across extensive plains towards the centre of topographic depressions, with the Finke River in central Australia providing a prime example (Pickup, Allan and Baker, 1988). Palaeoclimatic changes have also been important; many Australian dryland rivers originated under wetter climates in the Late Mesozoic or Early Cenozoic and have since undergone many changes in erosional and depositional behaviour in response to a long-term trend to drier and more variable climates, including the breakdown of many formerly better integrated drainage networks (e.g. van de Graaff *et al.*, 1977; Salama *et al.*, 1993; Morgan, 1993). Quaternary palaeoclimatic fluctuations involving conditions both drier and wetter than present have also impacted on Australia's dryland rivers, with the continent progressively becoming drier with each glacial cycle during at least the Mid- to Late Quaternary (Nanson *et al.*, 2008), but direct glacial, periglacial and eustatic influences have been minimal. Relative tectonic stability, typically low denudation rates and limited slope–channel coupling have tended to restrict sediment supply to many Australian dryland rivers, so that in comparison with many Mediterranean and some southern African rivers, long-term vertical channel activity (pronounced aggradation and incision) tends to have been muted, except in localised areas (e.g. Croke, Magee and Price, 1996). Lateral channel activity (migration and avulsion) has been more pronounced, however, leading to the formation of many extensive low-gradient, 'fan-shaped' plains. In combination with unusually well-developed riparian vegetation assemblages and variations in sediment calibre, diverse river styles have developed across the Australian drylands, including numerous sandy, single-thread, planar-bed, straight rivers and sandy anabranching rivers (Figure 12.2(c) and see Box 12.3), muddy anastomosing rivers with extensive floodplains marked by waterholes, 'braided' channel and reticulate channel networks (Figure 12.2(d)), distributary rivers (Figure 12.8(a)) and rivers that decrease in size downstream and disappear in floodouts on alluvial plains, in playas or among aeolian dunefields (Figure 12.8(a) and (b)) (Woodyer, Taylor and Crook, 1979; Nanson, Rust and Taylor, 1986; Nanson *et al.*, 1988; Nanson, Tooth and Knighton, 2002; Dunkerley, 1992, 2008a; Knighton and Nanson, 1994a, 1997, 2000; Bourke and Pickup, 1999; Tooth, 1999, 2005; Tooth and Nanson, 1999, 2000a, 2000b, 2004; Fagan and Nanson, 2004; Wakelin-King and Webb, 2007; Fisher *et al.*, 2008). In some instances, 'reforming channels' occur farther downvalley and either join a larger river system or disappear in another floodout (Tooth, 1999). Australia has an extensive Quaternary record of widespread fluvial–aeolian interactions (e.g. Nanson, Chen and Price, 1995; Nanson *et al.*, 2008; Page *et al.*, 2001; Maroulis *et al.*, 2007; Cohen *et al.*, 2010), and across parts of the interior, river systems have interacted with aeolian dunefields to create complex landform assemblages of anabranching/anastomosing and distributary channel networks, floodplains and floodplain wetlands, floodouts and reforming channels, waterholes and pan and playa complexes (Figure 12.8(b)).

Box 12.3 Alluvial rivers on the Northern Plains of central Australia

On the Northern Plains in arid central Australia, ephemeral rivers are commonly low sinuosity ('straight'), but vary from single thread to anabranching, both along individual rivers and between neighbouring rivers (Figure 12.9(a)). The Marshall River exemplifies how the dryland climate exerts a strong control on the degree of anabranching through its direct and indirect influence on flow regime, sediment supply and riparian vegetation growth strategies (Tooth and Nanson, 2004). The Marshall follows a roughly west–east course across the Plains and is joined at various points along its length by large tributaries arising in ranges to the north (Figure 12.9(a)). Localized rainfall in these tributary catchments sometimes results in the Marshall flowing for short distances downstream of the tributary junctions, while the reaches upstream remain dry, and these inflows also provide additional gravelly sand. Over time, the more frequent provision of water to sections of the Marshall downstream of the tributary junctions encourages greater numbers and/or denser growth of river red gums (*Eucalyptus camaldulensis*) on the channel bed. These trees are deep-rooted, long-lived species and, once established, are able to withstand high flood flow velocities. By acting as obstacles to flow, these trees commonly initiate ridges as leeside accumulations of gravelly sand with minor cohesive fines. Subsequent vegetation colonisation in the intervals between floods, together with further deposition of sediment, helps these incipient ridges to stabilise and to grow longitudinally, laterally and

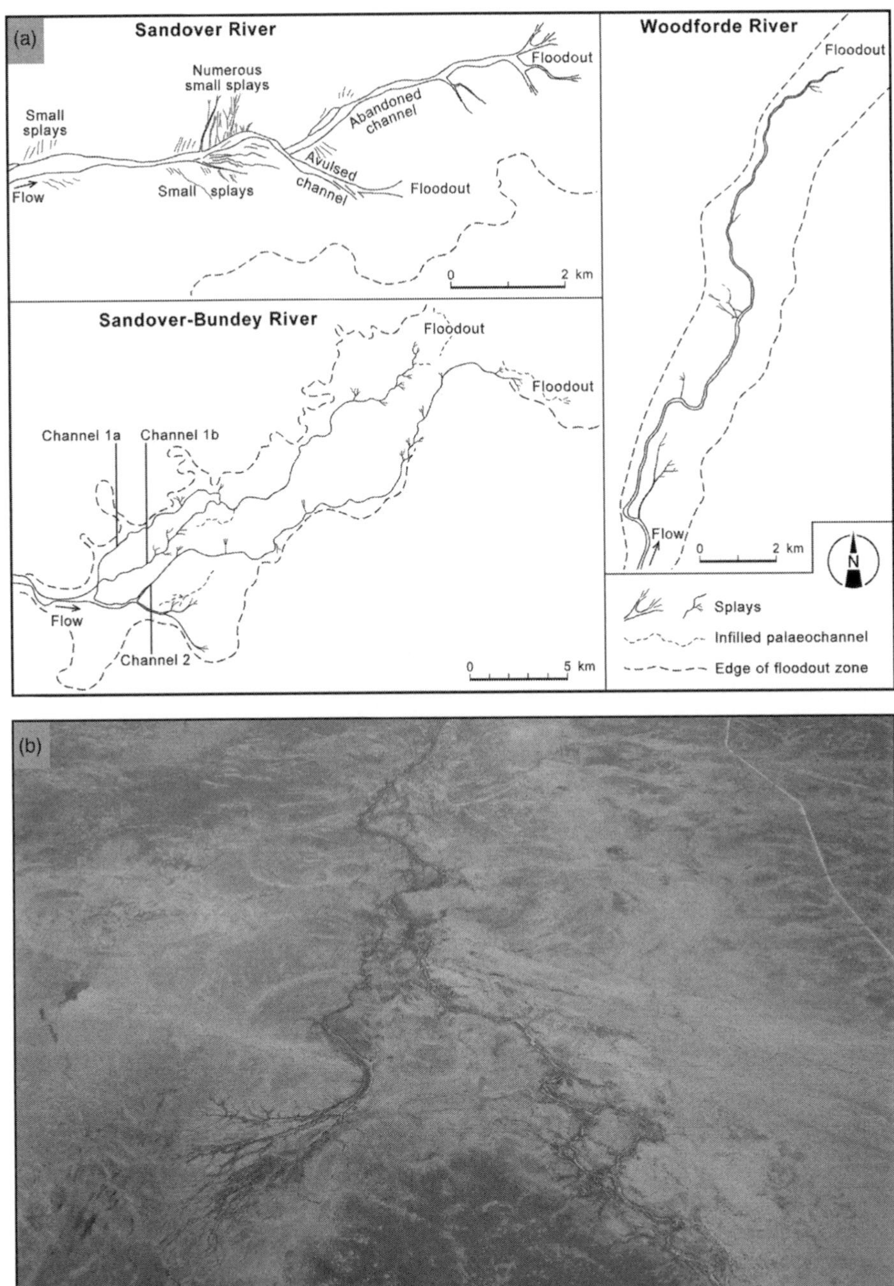

Figure 12.8 (a) Planform sketches of the lower reaches of three ephemeral rivers in central Australia, illustrating the different downstream changes in channel morphology and the floodouts at the channel termini (from Tooth, 2005). (b) Warburton Creek in the Channel Country, illustrating the complex landform assemblages characteristic of many Australian dryland plains (flow direction towards the camera). Muddy anastomosing channels divide and rejoin around large islands, with large palaeomeanders visible on the surrounding floodplains. Aeolian linear dunes are prominent in the middle right and a single dune is also developed in mid centre, alongside which a waterhole has developed. The waterhole is hosted in a wider, deeper section of channel, which divides downvalley into a complex of distributaries and splays, so that defined channels disappear. Bedrock crops out in the near foreground.

vertically. These depositional mechanisms eventually form extensive subparallel ridges or large islands that separate anabranches (Figure 12.2(c)). Alternative erosional mechanisms for anabranch formation can occur where high flows in the trunk channel cause local ponding of tributary inflows, ultimately resulting in scour of the floodplain surfaces adjacent to the vegetated banklines along the trunk channel and, over time, to the excision of ridges or large islands from the floodplain (Tooth and Nanson, 1999). The Marshall clearly operates close to a transitional condition, changing downstream of tributary junctions from single-thread or weakly anabranching reaches with relatively wide (60–150 m), straight channels to strongly anabranching reaches with numerous narrower (typically <60 m wide), straight channels (Figure 12.9(b)). Field observations and theoretical model results suggest that the formation of these narrow anabranches minimises boundary roughness and helps to maintain sediment throughput across these low-gradient plains (Tooth and Nanson, 2004).

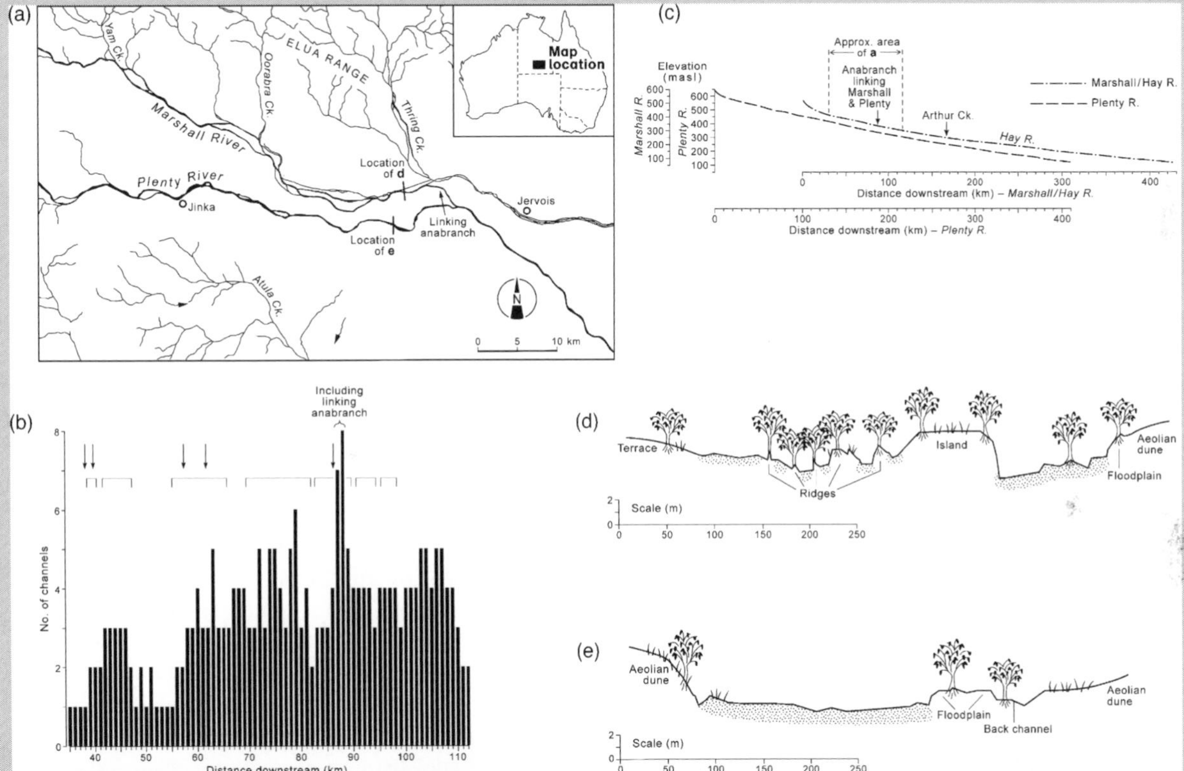

Figure 12.9 (a) Location of the Marshall and Plenty Rivers, central Australia (inset). The two rivers follow closely adjacent, subparallel courses for ~70 km and an anabranch of the Marshall River crosses the low-relief divide near Thring Bore to join the Plenty River (from Tooth and Nanson, 2004). (b) Relationship between tributary inflows (downward-pointing arrows) and the degree of anabranching along the Marshall River, based on assessment of the number of channels at 1 km intervals downstream (determined from aerial photographs and field surveys). Horizontal bars indicate where large islands are present in addition to narrower ridges, with the number of anabranches typically increasing for a short distance downstream of the tributary junctions (from Tooth and Nanson, 2004). (c) Longitudinal profiles of the Plenty and Marshall Rivers, demonstrating the similarity in channel-bed gradient through the study reach (constructed from contour crossings and spot heights on 1 : 100 000 topographic maps) (from Tooth and Nanson, 2004). (d) Examples of surveyed cross-sections from the Marshall and Plenty Rivers, illustrating the contrasting morphological characteristics, and (e) the Plenty River, illustrating the contrasting morphological characteristics. The distribution of trees is schematic only (after Tooth and Nanson, 2004). Reproduced by permission of the Geological Society of America.

The closely adjacent middle reaches of the Plenty River (Figure 12.9(a)) provides a further illustration of the importance of these tributary inputs and the riparian vegetation to anabranching development. The Plenty has a similar channel-bed gradient (Figure 12.9(c)), discharge and bank strength to the Marshall, but because it is joined by few tributaries in its middle reaches (Figure 12.9(a)), bed material is composed of medium to coarse sand and the channel remains relatively free of in-channel trees. Consequently, and in strong contrast to the anabranching Marshall (Figure 12.9(d)), the Plenty remains predominantly single thread (Figure 12.9(e)) but is variably wide (~100–1200 m), and in places appears transitional to braiding (Tooth and Nanson, 2004). These findings show that dryland river process, form and behaviour in central Australia is finely tuned to local environmental factors, particularly those related to the dryland climate. Variations along and between dryland rivers do not result from any significant changes in discharge, gradient, bed material, bank strength or human land use but instead seem to depend on complex and subtle sets of processes involving adjustments between flow hydraulics, sediment transport, vegetation density, number and geometry of channels, and boundary roughness.

As in southern Africa, human impact on dryland rivers was probably minimal prior to the start of European colonisation of the continental interior in the early to mid nineteenth century, but has intensified over the last 100–150 years. Vegetation clearance and the introduction of cattle and sheep have led to river degradation in some dryland catchments (Pickup, 1991; Fanning, 1999), but outside of the heavily regulated Murray–Darling system, there are very few dryland rivers that have been dammed or subjected to other forms of extensive flow regulation or direct channel modification, so Australia retains an unusual number of near-pristine, large dryland river systems. Flow regimes are highly variable, with long periods of below-average flows alternating with wetter periods with occasional very large floods. Late Pleistocene and Holocene palaeoflood records from headwater gorges and piedmont settings provide evidence for significant geomorphic impacts during irregular events, including bedrock erosion, aeolian dune reworking and gravel bedform deposition (Pickup, 1991; Patton, Pickup and Price, 1993; Bourke, 1998; Hollands *et al.*, 2006; Jansen and Brierley, 2004; Jansen, 2006). By contrast, the impacts of even large historical floods have tended to be much more muted, with low-energy, slow-moving floodwaves typically accomplishing little widespread or substantial geomorphic work, particularly where boundary resistance is high as a result of indurated alluvial terraces, cohesive muds or riparian vegetation. Along Cooper Creek, eastern central Australia, very few substantial morphological changes can be detected on aerial photographs dating from the last 50–60 years despite regular (seasonal) floods that commonly inundate floodplains up to 60 km wide (Knighton and Nanson, 1997; Tooth and Nanson, 2000b; Fagan and Nanson, 2004). On the Northern Plains, central Australia, channels have remained highly stable over the last few decades despite several large floods, although some changes have occurred in the lower reaches, including splay formation (Figure 12.8(a)) and channel abandonment (Tooth and Nanson, 2000b; Tooth, 2005).

Overall, many Australian rivers do not readily fit the 'textbook' generalisations regarding dryland rivers. Across the interior, a variety of upland, bedrock-confined dryland rivers can be found with some morphological and sedimentological similarities to upland rivers in other drylands, but rates of fluvial geomorphic activity tend to slow in comparison to dryland rivers draining more tectonically active dryland catchments (Jansen, 2006). Australia's numerous sandy, single-thread, planar-bed, straight rivers have morphological similarities to those described from other drylands (e.g. Wolman and Gerson, 1978; see Chapter 13) but occur in unusually large numbers across the interior, while the sandy anabranching (Figure 12.2(c)) and muddy anastomosing rivers are not commonly associated with other drylands. In particular, the Channel Country's extensive muddy floodplains with abundant waterholes (Knighton and Nanson, 1994a, 2000) and 'braided' and reticulate (Figure 12.2(d)) channel networks (Nanson, Rust and Taylor, 1986; Fagan and Nanson, 2004) are highly unusual features, and to date have not been described from other drylands. Characteristics of many other Australian dryland rivers, such as the pronounced downstream decreases in channel size, floodouts, reforming channels, floodplain wetlands and extensive aeolian–fluvial interactions, have been described from other continents, but Australia exhibits these characteristics on a frequency and over spatial scales unknown elsewhere. Although many upland gorges and piedmont settings are subject to periodic, high-magnitude flood events, Australia is better known for its extensive, relatively low-energy, lowland rivers, which are characterised by slow-moving, long-duration floods that represent hydraulic conditions very different from the flash floods typical of many other dryland rivers (Knighton and Nanson, 2001, 2002). Many of these lowland rivers appear to be insensitive even to recent large flood events, having undergone little substantial or long-lasting change, and in many instances are characterised by long-term equilibrium conditions (Tooth and Nanson, 2000b).

12.4 Reassessing distinctiveness and diversity

The three regions highlighted above cover a range of different local climatic, tectonic, structural, lithological and vegetative settings, and illustrate how dryland river characteristics vary accordingly. A full global assessment that includes the drylands of other parts of Europe, Africa, North and South America, the Middle East and Asia undoubtedly would reveal a variety of dryland river styles, owing to the different local climates, degrees of tectonic activity, geological controls and riparian vegetation assemblages, as well as their different histories of environmental change (Chapters 1 to 4). Although information is sparse for some of these regions, descriptions of bedrock gorges and coarse-grained alluvial rivers in the American southwest, Middle East and South America (e.g. Graf, 1988; Wells, 1990; McLaren *et al.*, 2004; Hoke *et al.*, 2004; Unkel *et al.*, 2007; Magilligan *et al.*, 2008) reveal morphological and sedimentological similarities with some Mediterranean and southern African rivers. Inland deltas in north Africa (e.g. McCarthy, 1993; Makaske, 1998) have some similarities with the Okavango Delta in southern Africa. Sand-bed rivers that decrease in size downstream and terminate on broad alluvial plains or among aeolian dunefields have been described from drylands in the American southwest (e.g. Langford, 1989; Clarke and Rendell, 1998), east and north Africa (e.g. Vanney, 1960; Mabbutt, 1977; Billi, 2007) and China (e.g. Yang *et al.*, 2002; Yang, 2006), many of which have similarities with some Namibian or central Australian rivers. It is notable, however, that many parts of the Australian drylands possess river styles (e.g. alluvial anabranching/anastomosing rivers, reticulate channels, waterholes) that are either unusual in global terms or that occur on a frequency and across spatial scales unknown in other drylands.

The foregoing global assessment – albeit partial – reveals that not only is there great diversity in river characteristics *between* different drylands but that there is also great diversity *within* drylands, including between neighbouring rivers and along individual rivers (see Boxes 12.1 to 12.3). Schumm's (1977) concept of the production, transfer and deposition zones in an idealised fluvial system can be adapted by using physiographic terms to subdivide catchments, which highlights the fact that the characteristics of an individual dryland river reach vary according to whether it is located in an upland, piedmont or lowland setting and whether it ends in an endoreic basin (floodout, playa or aeolian dunefield setting) or is through-flowing to the coast (Figure 12.10). This approach provides a firmer geomorphological basis for contextualising findings and thus for more rigorously comparing rivers between and within different drylands. For instance, while broad generalisations about dryland rivers are essential for making sense of highly variable natural phenomena, statements that Mediterranean dryland rivers tend to have different characteristics to Australian dryland rivers (see above) downplay or ignore the fact that some small, steep Mediterranean rivers have similarities with the characteristics of the upland or piedmont reaches of *some* Australian rivers (e.g. low-order, bedrock or coarse-grained tributaries; see Jansen and Brierley, 2004; Jansen, 2006). The key point is that while there are upland and piedmont river styles in both regions, because of the different gross physiographies in each (a function of different tectonic and geological frameworks), the spatial extent and relative proportions of the upland, piedmont, lowland and floodout zones varies dramatically. In many Mediterranean catchments with a coastal outlet, the upland and piedmont zones commonly possess high relative relief, are tectonically active and are relatively extensive, whereas the lowland zone is more restricted in extent and the floodout zone is typically nonexistent, with most deposition occurring in deltas, estuaries or offshore (Figure 12.4(a)). Consequently, river styles are dominated by those most commonly associated with the relatively high-energy upland and piedmont settings (e.g. gorge-bound rivers, braided rivers) and the styles associated with moderate or lower energy lowland and floodout settings are more limited or nonexistent. By contrast, in many Australian catchments, the upland and piedmont zones typically have limited relative relief, are tectonically stable and are relatively restricted in spatial extent by comparison with the lowland and floodout zones (Figure 12.8(a) and (b)). Hence, river styles associated with lower energy settings (e.g. single-thread straight, anabranching, anastomosing and distributary rivers, and floodouts) are most prominent. Southern Africa possesses more of a balance between the spatial extent of tectonically-quiescent, upland, piedmont, lowland and floodout zones, so river styles tend to be associated with a wide variety of energy settings, including various bedrock, alluvial and mixed bedrock–alluvial styles (e.g. braided, single-thread straight, meandering, anabranching/anastomosing and distributary rivers, and floodouts) (Figures 12.2(a) and (b), 12.6 and 12.7).

Against this backdrop of global diversity, one can choose either to emphasise the differences between river characteristics across drylands or to search for commonalities between dryland rivers. Similarly, one can choose to highlight the differences between the characteristics of dryland rivers and rivers in more humid regions or to stress greater overlap. Overlap between the

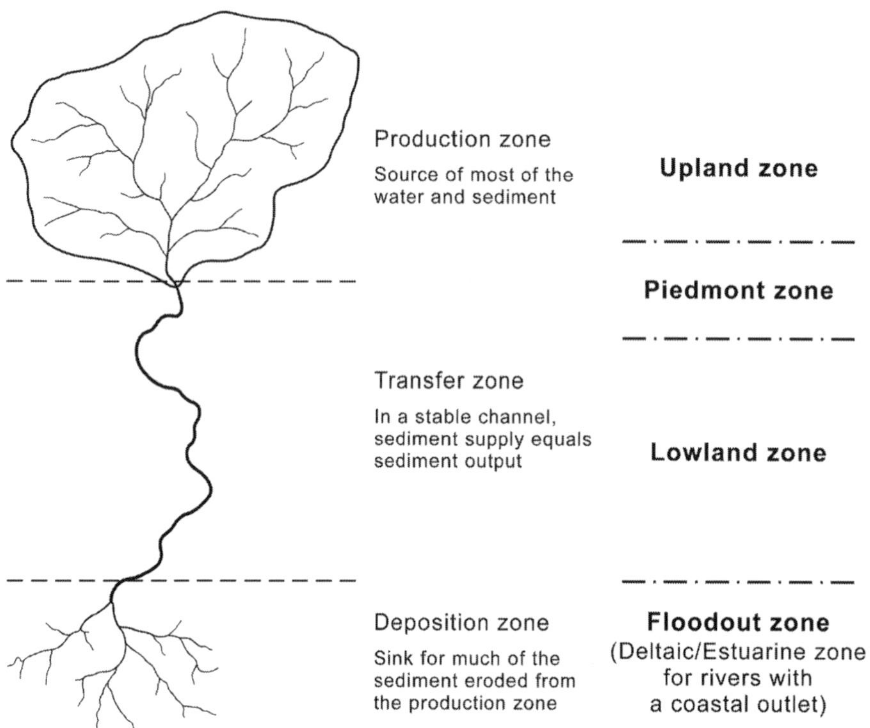

Figure 12.10 Zones of an 'idealised' fluvial system (after Schumm, 1977, and Gordon, McMahon and Finlayson, 1992), also illustrating the subdivision into upland, piedmont, lowland, and floodout or deltaic/estuarine zones. These zones are not specific to dryland rivers, but help to provide the geomorphological context for comparing findings from rivers across different drylands.

characteristics of dryland rivers and rivers in more humid regions is evidenced by exogenous rivers, many of which possess perennial flow, commonly exhibit well-developed meanders and in many cases display a greater tendency towards equilibrium behaviour, precisely the sorts of characteristics that are commonly ascribed to many humid region rivers. Examples of such rivers include the Okavango River, Botswana (McCarthy and Ellery, 1998), and the Rio Grande, southern Colorado, USA (Jones and Harper, 1998). Even discounting exogenous rivers, however, overlap with humid region river characteristics is also demonstrated by the fact that no endogenous river styles or behaviours are necessarily unique to drylands, as they may also be found in more humid settings. Gorge-bound rivers in Crete (Box 12.1), for example, have many similarities with humid rivers in similarly confined settings (e.g. Wohl, 1992). Meandering rivers in South Africa (Box 12.2) are morphometrically similar to meandering rivers in humid regions (e.g. Hooke, 2004). Anabranching rivers in central Australia (Box 12.3) are initiated and develop in response to particular sets of processes that in large part are related to the dryland climate, but nevertheless also have morphometric similarities with anabranching rivers in humid regions (Nanson and Knighton, 1996). Even floodout formation – once thought to be a key expression of dryland climatic influence on river behaviour owing to downstream discharge losses (Sullivan, 1976; Mabbutt, 1977) – has been described from fluvial settings in more humid regions (e.g. Melville and Erskine, 1986; Fryirs and Brierley, 1998; Grenfell, Ellery and Grenfell, 2009), where instead it may be promoted by lithologically controlled downstream decreases in gradient and concomitant declines in sediment transport capacity (Tooth, 2004).

If this is the case, can we therefore make any generalisations regarding dryland rivers? In particular, are there any characteristics that distinguish dryland rivers from rivers in other climatic settings? While there appear to be few (if any) river styles that could only occur in drylands and, similarly, few (if any) river styles that could not occur in drylands, some processes, forms and behaviours are nevertheless more common to dryland rivers as compared to rivers in other climatic settings. In this general sense, dryland rivers *as a group* do exhibit some distinctive characteristics, as outlined in the following sections.

12.4.1 Downstream flow decreases and localised flood patterns

One distinctive aspect of dryland rivers – both exogenous and endogenous – is a widespread tendency for downstream decreases in flow volume. In most instances, decreasing flow volumes are caused principally by transmission losses that result from floodwater infiltration into the unconsolidated alluvium forming channel boundaries, with further losses resulting from overbank flooding and from evaporative and transpirative losses (Babcock and Cushing, 1942; Keppel and Renard, 1962; Sharp and Saxton, 1962; Thornes, 1977; Walters, 1990; McCarthy and Ellery, 1998). Combined with hydrograph attenuation and a common absence of appreciable tributary inflows in the lower parts of many dryland catchments, significant downstream decreases in total flow volumes are associated with decreases in flood peaks and flow frequencies, and in larger ephemeral dryland rivers many flows will fail to travel along the full length of the channels (Vanney, 1960; Keppel and Renard, 1962; Mabbutt, 1977; Walters, 1989; Kotwicki and Isdale, 1991; Hughes and Sami, 1992; Lekach, Schick and Schlesinger, 1992; Knighton and Nanson, 1994b, 2001; Sharma, Murthy and Dhir, 1994; Sharma and Murthy, 1996; Enzel and Wells, 1997). Along some rivers in central Australia and southern Africa, the reduced flow volumes and flow frequencies lead to downstream decreases in channel size, and ultimately defined channels cannot be maintained (McCarthy and Ellery, 1998; Tooth, 1999; Tooth *et al.*, 2002b). In these instances, irregular large floods spill beyond the channel termini and spread slowly across floodouts or through floodplain wetlands (Figures 12.8(a) and (b) and 12.11(b)).

As addressed in further detail in Chapter 13, even in a given ephemeral river, transmission losses can be highly variable as they depend on many interrelated factors, including the characteristics of the storm (e.g. size, position of the storm track, location in relation to the drainage net), the hydrograph (e.g. flow volume and duration) and the channel (e.g. size of the wetted perimeter, porosity and initial moisture content of the perimeter sediments, stratigraphy of the channel fill) (e.g. Sorman and Abdulrazzak, 1993; Knighton and Nanson, 1994b, 1997, 2001; Sharma, Murthy and Dhir, 1994; Dick, Anderson and Sampson, 1997; Greenbaum *et al.*, 1998; Lekach *et al.*, 1998; Dunkerley and Brown, 1999; Lange, 2005; Dunkerley, 2008b). Storm characteristics and resulting hydrograph events are important in very large catchments or in catchments subject to discrete convective storms, for individual storms are unlikely to wet the entire catchment while successive storms are likely to wet different parts of the catchment and may produce compound rather than single floodwaves (Cooke, Warren and Goudie, 1993; Knighton and Nanson, 2001). Hence, some tributaries may be activated while neighbouring tributaries remain dormant, and these active tributaries may flow before the ephemeral trunk channel. Where such localised flood patterns occur, transmission losses along the trunk channel (and thus the flow survival length) will vary from flood to flood according to the variable pre-wetting of the channel bed by direct rainfall and tributary inflows prior to passage of the flood hydrograph.

Such localised flood patterns also have various implications for sediment transport and channel morphology in ephemeral rivers. At the catchment scale, Schumm and Hadley (1957) have described how tributaries in semi-arid catchments in eastern Wyoming and northern New Mexico can be found in all stages of integration with the trunk channel, with each tributary having its own history of alluviation and dissection. At the reach scale, in some upland and piedmont settings, asynchronous tributary flows can lead to aggradation, gradient steepening or increases in width or capacity of the trunk channel, and in more extreme cases to partial obstruction or damming (e.g. Woolley, 1946; Schick and Lekach, 1987; Patton, Pickup and Price, 1993; Macklin *et al.*, 2010), thus disrupting the continuity of downstream sediment transfer. By contrast, in the lower energy environments of the Northern Plains, central Australia, asynchronous tributary inflows and sediment supply promote the formation of anabranches that contribute to maintaining downstream sediment transfer (see Box 12.3). Hence, while the influences are varied, the localised flood patterns common to many drylands are quite different to many catchments in other climatic settings, which tend to be characterised by better integration between tributary and trunk river behaviour.

12.4.2 Induration of alluvial sediments

Along many dryland rivers, the downstream flow decreases and localised flood patterns also have implications for the nature of channel and floodplain alluvium. Stratigraphic studies of alluvial deposits in Israeli ephemeral catchments have revealed the common existence of a calcic 'soil' developed 50–100 cm beneath the surface of the channel alluvial fill and in Late Quaternary terraces (Figure 12.11(a)). In active channels, this 'fluvial pedogenic unit' occurs at the lower limit of scour and fill processes and is associated with the cumulative influence of persistent differences in floodwater availability to various parts of the channel bed. During floods, the infiltration rate is lower than the floodwave propagation rate

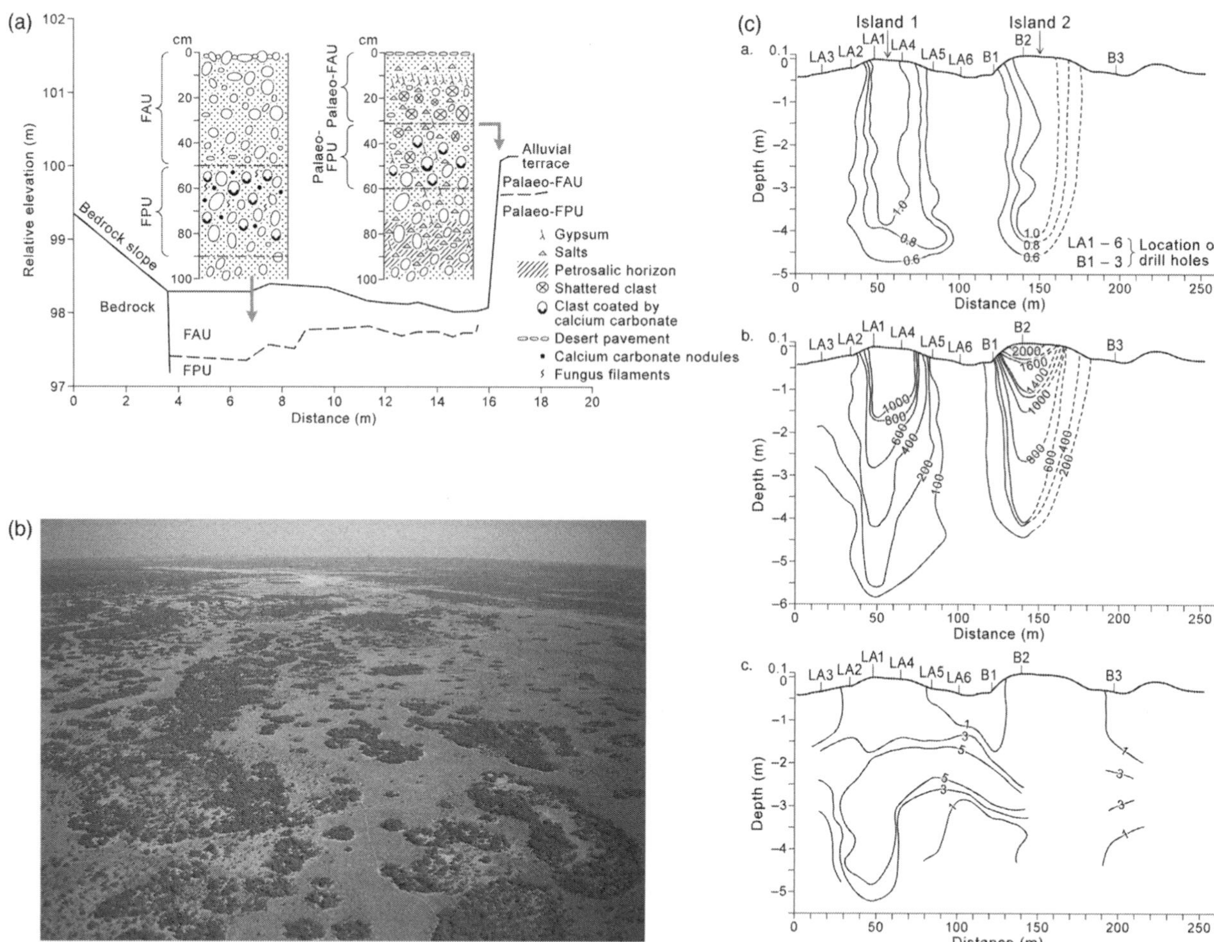

Figure 12.11 Examples of the pedogenic features and landforms associated with chemical sedimentation along dryland rivers and on floodplains. (a) Schematic cross-section across a typical ephemeral river in the Negev Desert, Israel, showing the location and relationship of the 'fluvial active unit' (FAU) and 'fluvial pedogenic unit' (FPU) in the channel and the palaeo-FAU and palaeo-FPU in late Pleistocene and Holocene alluvial terraces. The presence of the palaeo-FPU indicates that the hydrologic regime of past floods was similar to the present, with arid conditions prevailing during their formation. On the alluvial terrace, a gypsic-saline reg soil has developed on the calcium carbonate-coated clasts that represent the palaeo-FPU (from Amit *et al.*, 2007). (b) Oblique aerial view illustrating the numerous elevated, near-circular islands with woody fringes and sparsely vegetated or barren interiors on a broad unchannelled floodplain (floodout) in the Nyl River valley, northern South Africa. During extensive flooding, sheetflow occurs between the islands with the flow direction towards the camera. (c) Diagrams illustrating the location of drill holes and distribution of ions beneath two islands on the unchannelled floodplain in the Nyl River valley: sodium (Na_2O, wt %), chloride (Cl, ppm), calcium (CaO, wt %). Elevated levels of sodium and chloride are found beneath the islands owing to transpirative losses induced by the woody vegetation, but there is no difference in the abundance of calcium (from Tooth *et al.*, 2002b). Reproduced by permission Taylor & Francis Ltd., http://www.informaworld.com.

and the limited moisture left in the sediment is sufficient for calcium carbonate deposition (Lekach *et al.*, 1998; Amit *et al.*, 2007). Extreme cases of alluvial induration can occur where dryland rivers experience more seasonal flow regimes (Nanson, Tooth and Knighton, 2002), as sufficient warmth and moisture can lead to pronounced chemical alteration of bedrock and alluvium, particularly where the bedrock or alluvium is composed of relatively reactive lithologies, such as igneous rocks or carbonates. Chemical weathering releases certain elements in solution that are transported along moisture gradients and may be reprecipitated locally to form indurated or cemented horizons of ferricrete, silcrete, calcrete, gypcrete or halite, depending on local conditions of aridity and pH. Such induration, and in extreme cases lithification, of alluvium can produce hardened channel banks, islands and benches that can greatly influence river and floodplain forms and processes, including the formation of small gorges and waterfalls in the indurated or lithified alluvium (Shaw and Nash, 1998; Makaske, 1998; Nanson, Tooth and Knighton, 2002; Nash and Smith, 2003). In limestone gorges in Crete, for example, recent floods have been unable to erode the calcite-cemented lower parts of coarse-grained terraces (see Box 12.1) and lateral channel activity has been constrained. Within lower energy floodplain wetlands in southern Africa, vegetation transpiration promotes subsurface chemical precipitation, and the subsequent soil volumetric increases can lead to the initiation and growth of distinctive, near-circular vegetated islands that become raised above the general level of flooding (e.g. McCarthy, Ellery and Ellery, 1993; McCarthy and Ellery, 1995; Tooth *et al.*, 2002b) (Figure 12.11(b) and (c)). While induration and lithification of alluvium can occur in other (particularly tropical) environments, it appears to be common in some dryland fluvial settings because the characteristically strong net moisture deficits readily promote reprecipitation of minerals from solution.

12.4.3 Channel–vegetation interactions

Another distinctive aspect of some dryland channels is the fact that a variety of native and invasive vegetation types tends to grow on the lower banks and channel beds. Along ephemeral channels, vegetation types may include large tree and shrub species (Figure 12.2(c)), while wetland channels may be characterised by various grasses and sedges rooted in peat (Figure 12.6(b)). As Chapter 13 also highlights, such riparian vegetation has many diverse influences on channel process, form and behaviour, including increasing flow resistance, promoting localised bed scour and/or aggradation, contributing to bank stabilisation/destabilisation, and enabling channel recovery following destructive floods (e.g. Osterkamp and Costa, 1987; Graeme and Dunkerley, 1993; Dunkerley, 2008a; Tooth, 2000b; Rountree, Rogers and Heritage, 2000; Knighton and Nanson, 2002; Hooke and Mant, 2002; Sandercock, Hooke and Mant, 2007). Some aspects of dryland river channel–vegetation interactions are similar to rivers in more humid environments, but in other cases the interactions differ between rivers with ephemeral or intermittent flow and those with perennial flow (Tooth and Nanson, 2000a; Nanson, Tooth and Knighton, 2002). In channels that are dry year-round or seasonally, for instance, vegetation can establish on channel beds and can directly initiate the formation of depositional features such as bars, benches, ridges or islands (e.g. Hadley, 1961; Graf, 1978; Woodyer, Taylor and Crook, 1979; Pickup, Allan and Baker, 1988), and in some cases this promotes the formation of extensive anabranching (Figure 12.2(c) and Box 12.3). In contrast, as Fielding, Alexander and Newman-Sutherland (1997) have noted, vegetation in perennial rivers tends to respond to sedimentation, typically only colonising bars or islands that are sufficiently elevated to be protected from regular inundation (Nanson and Beach, 1977; Hupp, 1990; Hupp and Osterkamp, 1996).

12.4.4 Fluvial–aeolian interactions

A final distinctive aspect of some dryland rivers is the tendency for widespread interaction with surrounding aeolian dunefields (Figure 12.8(b)). Over the last few decades, numerous studies have documented a diverse range of fluvial–aeolian interactions over various temporal and spatial scales from many drylands worldwide (for overviews, see Bullard and Livingstone, 2002, and Bullard and McTainsh, 2003). In some settings, channel activity controls aeolian processes and landforms, for instance by eroding dune flanks, by breaching dune cordons or by supplying the material for winds to rework into dunes (e.g. Nanson, Chen and Price, 1995; Clarke and Rendell, 1998; Page *et al.*, 2001; Maroulis *et al.*, 2007; Cohen *et al.*, 2010). In other settings, aeolian activity influences channel processes and landforms by controlling channel alignment or by ponding floodwaters (e.g. Lancaster and Teller, 1988; Teller, Rutter and Lancaster, 1990; Yang *et al.*, 2002). Clearly, while the strength and direction of these fluvial–aeolian interactions can vary greatly depending on local circumstances, the scale and spatial extent of these interactions sets many drylands rivers apart from rivers in more humid environments, where a greater degree of soil development and a higher percentage of vegetation cover tend to limit aeolian activity.

12.5 Conclusions

Much of the older literature contains statements regarding the distinctiveness of dryland rivers, as expressed by particular flow and sediment processes, channel morphologies or behaviours, and in some respects these perceptions still persist. Supporting empirical evidence can be found, but has been derived mainly from research conducted in a limited range of dryland settings, namely small, steep rivers that commonly drain tectonically active catchments. Consequently, such statements need to be tempered by taking account of the findings from more recent research conducted in a broader range of drylands. Globally, drylands encompass a wide variety of climatic, physiographic, geological and vegetative settings. Rivers are a product of their environments; therefore, as environmental settings change within and between drylands, so do river characteristics. At the most basic level, the diversity in dryland river characteristics is reflected by subdivisions into exogenous and endogenous river types or into exoreic and endoreic drainage. Diversity can also be demonstrated by highlighting how dryland rivers operate across a spectrum of conditions, e.g. relating to flow regime (ephemeral, intermittent, perennial), sediment calibre (coarse-grained, fine-grained) or boundary characteristics (bedrock, alluvial, indurated or lithified alluvial), long-term behavioural tendencies (incisional, aggradational, migratory, avulsive) and riparian vegetation associations (nonvegetated or vegetated with trees, shrubs and/or grasses). Different combinations of natural environmental factors give rise to different river styles, and river patterns in particular can vary across a wide spectrum (braiding, single-thread straight, meandering, anabranching/anastomosing, distributary), with pattern changes commonly occurring in response to subtle variations in the controlling factors. Knighton and Nanson (1997) highlighted that dryland rivers form part of a continuum of hydrological conditions from hyperarid to very humid, and so many dryland river styles have morphological and sedimentological overlap with the characteristics of rivers in wetter regions. Climate is just one factor influencing dryland river characteristics and it is commonly mediated or overridden by other influencing factors. The key message is that when assessing dryland river characteristics, rather than focusing on only limited ranges of drylands and interpreting the findings as representing a much larger reality, there is a need to step back and look at the 'big picture' across drylands as a whole. Many previous statements regarding dryland river characteristics either can no longer be sustained as generalisations or, at the very least, the geographical and geomorphological context for those generalisations needs to be clarified. Indeed, today's understanding of dryland rivers undoubtedly will change as more examples are investigated.

Arguments for the distinctiveness and diversity of dryland rivers are not mutually exclusive. *As a group*, dryland rivers do display some common forms, processes and behaviours, some (but not all) of which are distinctive in comparison with most humid region rivers (e.g. transmission losses, aeolian–fluvial interactions). Equally, however, recognition must be given to the fact that not all dryland rivers conform to these generalisations. In any case, it is a moot point whether it is more important to stress the distinctiveness or to acknowledge the diversity of dryland rivers. From a scientific perspective, both are important for developing a comprehensive understanding of complex natural phenomena. From a practical (applied) perspective, both have implications for developing environmentally sensitive, sustainable management practices for dryland rivers, particularly as there is now greater awareness of the importance of fluvial geomorphological processes for issues such as assessing the stability of engineering structures, estimating reservoir lifespans, predicting bank erosion and water quality, and maintaining riverine ecological health. The history of river management in drylands is riddled with examples where mismanagement has occurred because of the application of inappropriate practices that commonly have been based on experience with the operation of very different types of humid rivers (e.g. Graf, 1985; Beaumont, 1989; Powell, 2008). Of equal danger, however, is the wholesale replacement of these inappropriate practices with other inappropriate management practices derived from a limited and unrepresentative subset of dryland river types. Recognition of both distinctiveness and diversity are important if fluvial geomorphologists are to facilitate the design of management practices that are tailored to the characteristics of individual dryland rivers.

References

Amit, R., Lekach, J., Ayalon, A. *et al.* (2007) New insight into pedogenic processes in extremely arid environments and their paleoclimatic implications – the Negev Desert, Israel. *Quaternary International*, **162–163**, 61–75.

Amorosi, A., Dinelli, E., Rossi, V. *et al.* (2008) Late Quaternary palaeoenvironmental evolution of the Adriatic coastal plain and the onset of Po River delta. *Palaeogeography, Palaeoclimatology, Palaeoecology*, **268**, 80–90.

Babcock, H.M. and Cushing, E.M. (1942) Recharge to groundwater from floods in a typical desert wash, Pinal County, Arizona. *Transactions of the American Geophysical Union*, **23**, 49–55.

Beaumont, P. (1989) *Drylands: Environmental Management and Development*, Routledge, London and New York.

Benito, G., Gutiérrez, F., Peréz-González, A. and Machado, M.J. (2000) Geomorphological and sedimentological features in Quaternary fluvial systems affected by solution-induced subsidence (Ebro Basin, NE-Spain). *Geomorphology*, **33**, 209–224.

Benito, G., Thorndycraft, V.R., Rico, M. *et al.* (2008) Palaeoflood and floodplain records from Spain: evidence for long-term climate variability and environmental changes. *Geomorphology*, **101**, 68–77.

Benito, G., Rico, M., Sánchez-Moya, Y. *et al.* (2010) The impact of late Holocene climatic variability and land use change on the flood hydrology of the Guadalentín River, southeast Spain. *Global and Planetary Change*, **70**, 53–63.

Billi, P. (2007) Morphology and sediment dynamics of ephemeral stream terminal distributary systems in the Kobo Basin (northern Welo, Ethiopia). *Geomorphology*, **85**, 98–113.

Bourke, M.C. (1998) *Fluvial geomorphology and palaeofloods in central Australia*, PhD thesis (unpublished), Australian National University, Canberra.

Bourke, M.C. and Pickup, G. (1999) Fluvial form variability in arid central Australia, in *Varieties of Fluvial Form* (eds A.J. Miller and A. Gupta), John Wiley & Sons, Ltd, Chichester, pp. 249–271.

Bull, L.J. and Kirkby, M.J. (eds) (2002) *Dryland Rivers: Hydrology and Geomorphology of Semi-Arid Channels*, John Wiley & Sons, Ltd, Chichester.

Bullard, J.E. and Livingstone, I. (2002) Interactions between aeolian and fluvial systems in dryland environments. *Area*, **34**, 8–16.

Bullard, J.E. and McTainsh, G.H. (2003) Aeolian–fluvial interactions in dryland environments: examples, concepts and Australia case study. *Progress in Physical Geography*, **27**, 471–501.

Candy, I., Black, S. and Sellwood, B.W. (2004) Interpreting the response of a dryland river system to Late Quaternary climate change. *Quaternary Science Reviews*, **23**, 2513–2523.

Clarke, M.L. and Rendell, H.M. (1998) Climate change impacts on sand supply and the formation of desert sand dunes in the southwest USA. *Journal of Arid Environments*, **39**, 517–531.

Codilean, A.T., Bishop, P., Stuart, F.M. *et al.* (2008) Single-grain cosmogenic ^{21}Ne concentrations in fluvial sediments reveal spatially variable erosion rates. *Geology*, **36**, 159–162.

Cohen, H. and Laronne, J.B. (2005) High rates of sediment transport by flash floods in the Southern Judean Desert, Israel. *Hydrological Processes*, **19**, 1687–1702.

Cohen, T.J., Nanson, G.C., Larsen, J.R. *et al.* (2010) Late Quaternary aeolian and fluvial interactions on the Cooper Creek Fan and the association between linear and source-bordering dunes, Strzelecki Desert, Australia. *Quaternary Science Reviews*, **29**, 455–471.

Cooke, R.U., Warren, A. and Goudie, A.S. (1993) *Desert Geomorphology*, University College London Press, London.

Croke, J.C., Magee, J.W. and Price, D.M. (1996) Major episodes of Quaternary activity in the lower Neales River, northwest of Lake Eyre, central Australia. *Palaeogeography, Palaeoclimatology, Palaeoecology*, **124**, 1–15.

de Wit, M.C.J., Marshall, T.R. and Partridge, T.C. (2000) Fluvial deposits and drainage evolution, in *The Cenozoic of Southern Africa* (eds T.C. Partridge and R.R. Maud), Oxford Monographs on Geology and Geophysics 40, Oxford University Press, Oxford, pp. 55–72.

Dick, G.S., Anderson, R.S. and Sampson, D.E. (1997) Controls on flash flood magnitude and hydrograph shape, Upper Blue Hills badlands, Utah. *Geology*, **25**, 45–48.

Dunkerley, D.L. (1992) Channel geometry, bed material and inferred flow conditions in ephemeral stream systems, Barrier Range, western N.S.W, Australia. *Hydrological Processes*, **6**, 417–433.

Dunkerley, D. (2008a) Flow chutes in Fowlers Creek, arid western New South Wales, Australia: evidence for diversity in the influence of trees on ephemeral channel form and process. *Geomorphology*, **102**, 232–241.

Dunkerley, D.L. (2008b) Bank permeability in an Australian ephemeral dry-land stream: variation with stage resulting from mud deposition and sediment clogging. *Earth Surface Processes and Landforms*, **33**, 226–243.

Dunkerley, D. and Brown, K. (1999) Flow behaviour, suspended sediment transport and transmission losses in a small (sub-bank-full) flow event in an Australian desert stream. *Hydrological Processes*, **13**, 1577–1588.

Eitel, B., Kadereit, A., Blümel, W.-D. *et al.* (2006) Environmental changes at the eastern Namib Desert margin before and after the Last Glacial Maximum: new evidence from fluvial deposits in the upper Hoanib River catchment, northwestern Namibia. *Palaeogeography, Palaeoclimatology, Palaeoecology*, **234**, 201–222.

Ellery, W.N., McCarthy, T.S. and Smith, N.D. (2003) Vegetation, hydrology and sedimentation patterns on the major distributary system of the Okavango Fan, Botswana. *Wetlands*, **23**, 357–375.

Enzel, Y., Agnon, A. and Stein, M. (eds) (2006) *New frontiers in Dead Sea research*, Geological Society of America Special Paper 401, Geological Society of America, Boulder.

Enzel, Y. and Wells, S.G. (1997) Extracting Holocene paleohydrology and paleoclimatology information from modern extreme flood events: an example from southern California. *Geomorphology*, **19**, 203–226.

Fagan, S.D. and Nanson, G.C. (2004) The morphology and formation of floodplain-surface channels, Cooper Creek, Australia. *Geomorphology*, **60**, 107–126.

Fanning, P.C. (1999) Recent landscape history in arid western New South Wales, Australia: a model for regional change. *Geomorphology*, **29**, 191–209.

Fielding, C.R., Alexander, J. and Newman-Sutherland, E. (1997) Preservation of in situ, aborescent vegetation and fluvial bar construction in the Burdekin River of north Queensland, Australia. *Palaeogeography, Palaeoclimatology, Palaeoecology*, **135**, 123–144.

Finlayson, B.L. and McMahon, T.A. (1988) Australia v. the world: a comparative analysis of streamflow characteristics, in *Fluvial Geomorphology of Australia* (ed. R.F. Warner), Academic Press, Sydney, pp. 17–40.

Fisher, J.A., Krapf, C.B.E., Lang, S.C. *et al.* (2008) Sedimentology and architecture of the Douglas Creek terminal splay, Lake Eyre, central Australia. *Sedimentology*, **55**, 1915–1930.

Fryirs, K. and Brierley, G.J. (1998) The character and age structure of valley fills in upper Wolumla Creek catchment, South Coast, New South Wales, Australia. *Earth Surface Processes and Landforms*, **23**, 271–287.

Gordon, N.D., McMahon, T.A. and Finlayson, B.L. (1992) *Stream Hydrology: An Introduction for Ecologists*, John Wiley & Sons, Ltd, Chichester.

Goudie, A.S. and Wells, G.L. (1995) The nature, distribution and formation of pans in arid zones. *Earth-Science Reviews*, **38**, 1–69.

Graeme, D. and Dunkerley, D.L. (1993) Hydraulic resistance by the river red gum, *Eucalyptus camaldulensis*, in ephemeral desert streams. *Australian Geographical Studies*, **31**, 141–154.

Graf, W.L. (1978) Fluvial adjustments to the spread of tamarisk in the Colorado Plateau region. *Geological Society of America Bulletin*, **89**, 1491–1501.

Graf, W.L. (1985) *The Colorado River: Instability and Basin Management*, Association of American Geographers, Washington, DC.

Graf, W.L. (1988) *Fluvial Processes in Dryland Rivers*, Springer-Verlag, Berlin.

Greenbaum, N. and Bergman, N. (2006) Formation and evacuation of a large gravel-bar deposited during a major flood in a Mediterranean ephemeral stream, Nahal Me'arot, NW Israel. *Geomorphology*, **77**, 169–186.

Greenbaum, N., Schwartz, U. and Bergman, N. (2010) Extreme floods and short-term hydroclimatological fluctuations in the hyper-arid Dead Sea region, Israel. *Global and Planetary Change*, **70**, 125–137.

Greenbaum, N., Margalit, A., Schick, A.P. *et al.* (1998) A high magnitude storm and flood in a hyperarid catchment, Nahal Zin, Negev Desert, Israel. *Hydrological Processes*, **12**, 1–23.

Greenbaum, N., Porat, N., Rhodes, E. and Enzel, Y. (2006) Large floods during late Oxygen Isotope Stage 3, southern Negev Desert, Israel. *Quaternary Science Reviews*, **25**, 704–719.

Grenfell, M.C., Ellery, W.N. and Grenfell, S.E. (2009) Valley morphology and sediment cascades within a wetland system in the KwaZulu-Natal Drakensberg foothills, eastern South Africa. *Catena*, **78**, 20–35.

Guerrero, J., Gutiérrez, F. and Lucha, P. (2008) Impact of halite dissolution subsidence on Quaternary fluvial terrace development: case study of the Huerva River, Ebro Basin, NE Spain. *Geomorphology*, **100**, 164–179.

Hadley, R.F. (1961) Influence of riparian vegetation on channel shape, northeastern Arizona. *United States Geological Survey Professional Paper*, **424-C**, 30–31.

Harvey, A.M. (1984) Geomorphological response to an extreme flood: a case from southeast Spain. *Earth Surface Processes and Landforms*, **9**, 267–279.

Harvey, A.M. and Wells, S.G. (1987) Response of Quaternary fluvial systems to differential epeirogenic uplift: Aguas and Feos River systems, southeast Spain. *Geology*, **15**, 689–693.

Hassan, M.A. (2005) Characteristics of gravel bars in ephemeral streams. *Journal of Sedimentary Research*, **75**, 29–42.

Hassan, M.A., Schick, A.P. and Shaw, P.A. (1999) The transport of gravel in an ephemeral sandbed river. *Earth Surface Processes and Landforms*, **24**, 623–640.

Heritage, G.L., van Niekerk, A.W. and Moon, B.P. (1999) Geomorphology of the Sabie River, South Africa: an incised bedrock-influenced channel, in *Varieties of Fluvial Form* (eds A.J. Miller and A. Gupta), John Wiley & Sons, Ltd, Chichester, pp. 53–79.

Hoke, G.D., Isacks, B.L., Jordan, T.E. and Yu, J.S. (2004) Groundwater-sapping origin for the giant quebradas of northern Chile. *Geological Society of America Bulletin*, **32**, 605–608.

Hollands, C.B., Nanson, G.C., Jones, B.G. *et al.* (2006) Aeolian–fluvial interaction: evidence for Late Quaternary channel change and wind-rift linear dune formation in the northwestern Simpson Desert, Australia. *Quaternary Science Reviews*, **25**, 142–162.

Hooke, J.M. (2004) Cutoffs galore!: occurrence and causes of multiple cutoffs on a meandering river. *Geomorphology*, **61**, 225–238.

Hooke, J.M. and Mant, J.M. (2002) Morpho-dynamics of ephemeral streams, in *Dryland Rivers: Hydrology and Geomorphology of Semi-Arid Channels* (eds L.J. Bull and M.J. Kirkby), John Wiley & Sons, Ltd, Chichester, pp. 173–204.

Hughes, D.A. and Sami, K. (1992) Transmission losses to alluvium and associated moisture dynamics in a semi-arid ephemeral channel system in southern Africa. *Hydrological Processes*, **6**, 45–53.

Hupp, C.R. (1990) Vegetation patterns in relation to basin hydrogeomorphology, in *Vegetation and Erosion: Processes and Environments* (ed. J.B. Thornes), John Wiley & Sons, Ltd, Chichester, pp. 217–237.

Hupp, C.R. and Osterkamp, W.R. (1996) Riparian vegetation and fluvial geomorphic processes. *Geomorphology*, **14**, 277–295.

Jacob, R.J., Bluck, B.J. and Ward, J.D. (1999) Tertiary-age diamondiferous fluvial deposits of the lower Orange River valley, south-western Africa. *Economic Geology*, **94**, 749–758.

Jacobson, P.J., Jacobson, K.M. and Seeley, M.K. (1995) *Ephemeral rivers and their catchments: sustaining people and development in western Namibia*, Desert Research Foundation of Namibia, Windhoek.

Jansen, J. (2006) Flood magnitude-frequency and lithologic control on bedrock river incision in post-orogenic terrain. *Geomorphology*, **82**, 39–57.

Jansen, J. and Brierley, G.J. (2004) Pool-fills: a window to palaeoflood history and response in bedrock-confined rivers. *Sedimentology*, **51**, 1–25.

Jones, L.S. and Harper, J.T. (1998) Channel avulsions and related processes, and large-scale sedimentation patterns since 1875, Rio Grande, San Luis Valley, Colorado. *Geological Society of America Bulletin*, **110**, 411–421.

Keppel, R.V. and Renard, K.G. (1962) Transmission losses in ephemeral stream beds. *Journal of the Hydraulics Division, Proceedings of the American Society of Civil Engineers*, **8** (HY3), 59–68.

Knighton, A.D. and Nanson, G.C. (1994a) Waterholes and their significance in the anastomosing channel system of Cooper Creek, Australia. *Geomorphology*, **9**, 311–324.

Knighton, A.D. and Nanson, G.C. (1994b) Flow transmission along an arid zone anastomosing river, Cooper Creek, Australia. *Hydrological Processes*, **8**, 137–154.

Knighton, A.D. and Nanson, G.C. (1997) Distinctiveness, diversity and uniqueness in arid zone river systems, in *Arid Zone Geomorphology: Process, Form and Change in*

Drylands, 2nd edn (ed. D.S.G. Thomas), John Wiley & Sons, Ltd, Chichester, pp. 185–203.

Knighton, A.D. and Nanson, G.C. (2000) Waterholes form and process in the anastomosing channel system of Cooper Creek, Australia. *Geomorphology*, **35**, 101–117.

Knighton, A.D. and Nanson, G.C. (2001) An event-based approach to the hydrology of arid zone rivers in the Channel Country of Australia. *Journal of Hydrology*, **254**, 102–123.

Knighton, A.D. and Nanson, G.C. (2002) Inbank and overbank velocity conditions in an arid zone anastomosing river. *Hydrological Processes*, **16**, 1771–1791.

Kotwicki, V. and Isdale, P. (1991) Hydrology of Lake Eyre, Australia: El Niño link. *Palaeogeography, Palaeoclimatology, Palaeoecology*, **84**, 87–98.

Krijgsman, W., Hilgen, F.J., Raffi, I. *et al.* (1999) Chronology, causes and progression of the Messinian salinity crisis. *Nature*, **400**, 652–655.

Lancaster, N. and Teller, J.T. (1988) Interdune deposits of the Namib Sand Sea. *Sedimentary Geology*, **55**, 91–107.

Lange, J. (2005) Dynamics of transmission losses in a large arid stream channel. *Journal of Hydrology*, **306**, 112–126.

Langford, R.P. (1989) Fluvial–aeolian interactions: Part 1. Modern systems. *Sedimentology*, **36**, 1023–35.

Laronne, J.B. and Shlomi, Y. (2007) Depositional character and preservation potential of coarse-grained sediments deposited by flood events in hyper-arid braided channels in the Rift Valley, Arava, Israel. *Sedimentary Geology*, **195**, 21–37.

Lekach, J., Schick, A.P. and Schlesinger, A. (1992) Bedload yield and in-channel provenance in a flash-flood fluvial system, in *Dynamics of Gravel-Bed Rivers* (eds P. Billi, R.D. Hey, C.R. Thorne and P. Tacconi), John Wiley & Sons, Ltd, Chichester, pp. 537–551.

Lekach, J., Amit, R., Grodek, T. and Schick, A.P. (1998) Fluvio-pedogenic processes in an ephemeral stream channel, Nahal Yael, Southern Negev, Israel. *Geomorphology*, **23**, 353–369.

Leopold, L.B., Wolman, M.G. and Miller, J.P. (1964) *Fluvial Processes in Geomorphology*, W.H. Freeman and Company, San Francisco.

Lewin, J., Macklin, M.G. and Woodward, J.C. (eds) (1995) *Mediterranean Quaternary River Environments*, Balkema, Rotterdam.

López-Bermúdez, F., Conesa-García, C. and Alonso-Sarría, F. (2002) Floods: magnitude and frequency in ephemeral streams of the Spanish Mediterranean region, in *Dryland Rivers: Hydrology and Geomorphology of Semi-Arid Channels* (eds L.J. Bull and M.J. Kirkby), John Wiley & Sons, Ltd, Chichester, pp. 329–350.

Maas, G.S. and Macklin, M.G. (2002) The impact of recent climate change on flooding and sediment supply within a Mediterranean mountain catchment, southwestern Crete, Greece. *Earth Surface Processes and Landforms*, **27**, 1087–1105.

Mabbutt, J.A. (1977) *Desert Landforms*, Australian National University Press, Canberra.

Mabbutt, J.A. (1986) Desert lands, in *Australia – A Geography. Volume One: The Natural Environment* (ed. D.N. Jeans), Sydney University Press, Sydney, pp. 180–202.

McCarthy, T.S. (1993) The great inland deltas of Africa. *Journal of African Earth Sciences*, **17**, 275–291.

McCarthy, T.S. and Ellery, W.N. (1995) Sedimentation on the distal reaches of the Okavango fan, Botswana, and its bearing on calcrete and silcrete (ganister) formation. *Journal of Sedimentary Research*, **A65**, 77–90.

McCarthy, T.S. and Ellery, W.N. (1998) The Okavango Delta. *Transactions of the Royal Society of South Africa*, **53**, 157–182.

McCarthy, T.S., Ellery, W.N. and Ellery, K. (1993) Vegetation-induced, subsurface precipitation of carbonate as an aggradational process in the permanent swamps of the Okavango (Delta) Fan, Botswana. *Chemical Geology*, **107**, 111–131.

McCarthy, T.S. and Tooth, S. (2004) Incised meanders along the mixed bedrock-alluvial Orange River, Northern Cape Province, South Africa. *Zeitschrift für Geomorphologie*, **48**, 273–292.

Macklin, M.G., Lewin, J. and Woodward, J.C. (1995) Quaternary fluvial systems in the Mediterranean Basin, in *Mediterranean Quaternary River Environments* (eds J. Lewin, M.G. Macklin and J.C. Woodward), Balkema, Rotterdam, pp. 1–28.

Macklin, M.G. and Woodward, J.C. (2009) River systems and environmental change, in *The Physical Geography of the Mediterranean* (ed. J.C. Woodward), Oxford University Press, Oxford, pp. 319–335.

Macklin, M.G., Tooth, S., Brewer, P.A. *et al.* (2010) Holocene flooding and river development in a Mediterranean steepland catchment: the Anapodaris Gorge, south-central Crete, Greece. *Global and Planetary Change*, **70**, 35–52.

McLaren, S.J., Gilbertson, D.D., Grattan, J.P. *et al.* (2004) Quaternary palaeogeomorphologic evolution of the Wadi Faynan area, southern Jordan. *Palaeogeography, Palaeoclimatology, Palaeoecology*, **205**, 131–154.

Magilligan, F.J., Goldstein, P.S., Fisher, G.B. *et al.* (2008) Late Quaternary hydroclimatology of a hyper-arid Andean watershed: climate change, floods, and hydrologic responses to the El Niño–Southern Oscillation in the Atacama Desert. *Geomorphology*, **101**, 14–32.

Maher, E. and Harvey, A.M. (2008) Fluvial system response to tectonically induced base-level change during the late-Quaternary: the Rio Alias, southeast Spain. *Geomorphology*, **100**, 180–192.

Makaske, B. (1998) *Anastomosing Rivers: Forms, Processes and Sediments*, Faculteit Ruimtelijke Wetenschappen, Universiteit Utrecht, Utrecht.

Maroulis, J.C., Nanson, G.C., Price, D.M. and Pietsch, T. (2007) Aeolian–fluvial interaction and climate change: source-bordering dune development over the past ~100 ka on Cooper Creek, central Australia. *Quaternary Science Reviews*, **26**, 386–404.

Mather, A.E. (2000) Adjustment of a drainage network to capture induced base-level change: an example from the Sorbas Basin, SE Spain. *Geomorphology*, **34**, 271–289.

Melville, M.D. and Erskine, W.D. (1986) Sediment remobilization and storage by discontinuous gullying in humid southeastern Australia, in *Drainage Basin Sediment Delivery* (ed. R.F. Hadley), International Association of Hydrological Sciences Publication 159, IAHS Press, Wallingford, pp. 277–286.

Miall, A.D. (1996) *The Geology of Fluvial Deposits: Sedimentary Facies, Basin Analysis, and Petroleum Geology*, Springer-Verlag, Berlin.

Morgan, K.H. (1993) Development, sedimentation and economic potential of palaeoriver systems of the Yilgarn Craton of Western Australia. *Sedimentary Geology*, **85**, 637–656.

Nanson, G.C. and Beach, H.F. (1977) Forest succession and sedimentation on a meandering river floodplain, northeast British Columbia, Canada. *Journal of Biogeography*, **4**, 229–251.

Nanson, G.C., Chen, X.Y. and Price, D.M. (1995) Aeolian and fluvial evidence of changing climate and wind patterns during the past 100 ka in the western Simpson Desert, Australia. *Palaeogeography, Palaeoclimatology, Palaeoecology*, **113**, 87–102.

Nanson, G.C. and Knighton, A.D. (1996) Anabranching rivers: their cause, character and classification. *Earth Surface Processes and Landforms*, **21**, 217–239.

Nanson, G.C., Rust, B.R. and Taylor, G. (1986) Coexistent mud braids and anastomosing channels in an arid-zone river: Cooper Creek, central Australia. *Geology*, **14**, 175–178.

Nanson, G.C. and Tooth, S. (1999) Arid-zone rivers as indicators of climate change, in *Paleoenvironmental Reconstruction in Arid Lands* (eds A.K. Singhvi and E. Derbyshire), Oxford and IBH Press Company, New Delhi, pp. 175–216.

Nanson, G.C., Tooth, S. and Knighton, A.D. (2002) A global perspective on dryland rivers: perceptions, misconceptions and distinctions, in *Dryland Rivers: Hydrology and Geomorphology of Semi-Arid Channels* (eds L.J. Bull and M.J. Kirkby), John Wiley & Sons, Ltd, Chichester, pp. 17–54.

Nanson, G.C., Young, R.W., Price, D.M. and Rust, B.R. (1988) Stratigraphy, sedimentology and late-Quaternary chronology of the Channel Country of western Queensland, in *Fluvial Geomorphology of Australia* (ed. R.F. Warner), Academic Press, Sydney, pp. 151–175.

Nanson, G.C., Price, D.M., Jones, B.G. *et al.* (2008) Alluvial evidence for major climate and flow-regime changes during the middle and late Quaternary in eastern central Australia. *Geomorphology*, **101**, 109–129.

Nash, D.J. and Smith, R.F. (2003) Properties and development of channel calcretes in a mountain catchment, Tabernas Basin, southeast Spain. *Geomorphology*, **50**, 227–250.

Nash, D.J., Thomas, D.S.G. and Shaw, P.A. (1994) Timescales, environmental change and valley development, in *Environmental Change in Drylands* (eds A.C. Millington and K. Pye), John Wiley & Sons, Ltd, Chichester, pp. 25–41.

Niemi, T.M., Ben-Avraham, Z. and Gat, J.R. (eds) (1997) *The Dead Sea: The Lake and Its Setting*, Oxford University Press, New York.

Osterkamp, W.R. and Costa, J.E. (1987) Changes accompanying an extraordinary flood on a sand-bed stream, in *Catastrophic Flooding*, vol. 18 (eds L. Mayer and D. Nash), Binghampton Symposium in Geomorphology, Allen and Unwin, Boston, pp. 201–224.

Page, K.J., Dare-Edwards, A.J., Owens, J.W. *et al.* (2001) TL chronology and stratigraphy of riverine source bordering sand dunes near Wagga Wagga, New South Wales, Australia. *Quaternary International*, **83–85**, 187–193.

Patton, P.C., Pickup, G. and Price, D.M. (1993) Holocene paleofloods of the Ross River, central Australia. *Quaternary Research*, **40**, 201–212.

Pickup, G. (1991) Event frequency and landscape stability on the floodplain systems of arid central Australia. *Quaternary Science Reviews*, **10**, 463–473.

Pickup, G., Allan, G. and Baker, V.R. (1988) History, palaeochannels and palaeofloods of the Finke River, central Australia, in *Fluvial Geomorphology of Australia* (ed. R.F. Warner), Academic Press, Sydney, pp. 105–127.

Pilgrim, D.H., Chapman, T.G. and Doran, D.G. (1988) Problems of rainfall-runoff modelling in arid and semiarid regions. *Hydrological Sciences Journal*, **33**, 379–400.

Poesen, J.W.A. and Hooke, J.M. (1997) Erosion, flooding and channel management in Mediterranean environments of southern Europe. *Progress in Physical Geography*, **21**, 157–199.

Pollen-Bankhead, N., Simon, A., Jaeger, K. and Wohl, E.E. (2009) Destabilization of streambanks by removal of invasive species in Canyon de Chelly National Monument, Arizona. *Geomorphology*, **103**, 363–374.

Powell, J.L. (2008) *Dead Pool: Lake Powell, Global Warming and the Future of the Water in the West*, University of California Press, Berkeley.

Powell, D.M. (2009) Dryland rivers: processes and forms, in *Geomorphology of Desert Environments*, 2nd edn (eds A.J. Parsons and A.D. Abrahams), Springer, London, pp. 333–373.

Powell, D.M., Brazier, R., Parsons, A., Wainwright, J. and Nichols, M. (2007) Sediment transfer and storage in dryland headwater streams. *Geomorphology*, **88**, 152–166.

Reid, I. and Frostick, L.E. (1987) Flow dynamics and suspended sediment properties in arid zone flash floods. *Hydrological Processes*, **1**, 239–253.

Rhoads, B.L. (1988) Mutual adjustments between process and form in a desert mountain fluvial system. *Annals of the Association of American Geographers*, **78**, 271–287.

Rountree, M.W., Heritage, G.L. and Rogers, K.H. (2001) In-channel metamorphosis in a semiarid, mixed bedrock/alluvial river system: implications for instream flow requirements, in *Hydro-Ecology: Linking Hydrology and Aquatic Ecology*, International Association of Hydrological Sciences Publication 266, IAHS Press, Wallingford, pp. 113–123.

Rountree, M.W., Rogers, K.H. and Heritage, G.L. (2000) Landscape state change in the semi-arid Sabie River, Kruger National Park, in response to flood and drought. *South African Geographical Journal*, **82**, 173–181.

Salama, R.B., Farrington, P., Bartle, G.A. and Watson, G.D. (1993) The role of geological structures and relict channels in the development of dryland salinity in the wheatbelt of Western Australia. *Australian Journal of Earth Sciences*, **40**, 45–56.

Sandercock, P.J., Hooke, J.M. and Mant, J.M. (2007) Vegetation in dryland river channels and its interaction with fluvial processes. *Progress in Physical Geography*, **31**, 107–129.

Schick, A.P. and Lekach, J. (1987) A high magnitude flood in the Sinai Desert, in *Catastrophic Flooding*, vol. 18 (eds L. Mayer and D. Nash), Binghampton Symposium in Geomorphology, Allen and Unwin, Boston, pp. 381–410.

Schumm, S.A. (1977) *The Fluvial System*, John Wiley & Sons, Ltd, New York.

Schumm, S.A. and Hadley, R.F. (1957) Arroyos and the semiarid cycle of erosion. *American Journal of Science*, **255**, 161–174.

Seeley, M.K. and Sandelowsky, B.H. (1974) Dating the regression of a river's end point. *South African Archaeological Bulletin*, Goodwin Series **II**, 61–64.

Sharma, K.D. and Murthy, J.S.R. (1996) Ephemeral flow modeling in arid regions. *Journal of Arid Environments*, **33**, 161–178.

Sharma, K.D., Murthy, J.S.R. and Dhir, R.P. (1994) Streamflow routing in the Indian arid zone. *Hydrological Processes*, **8**, 27–43.

Sharp, A.L. and Saxton, K.E. (1962) Transmission losses in mature stream valleys. *Journal of the Hydraulics Division, Proceedings of the American Society of Civil Engineers*, **88**, 121–142.

Shaw, P.A. (1989) Fluvial systems of the Kalahari – a review, in *Arid and Semi-Arid Environments – Geomorphological and Pedological Aspects*, vol. 14 (eds A. Yair and S. Berkowicz), Catena Supplement, pp. 119–126.

Shaw, P.A. and Nash, D.J. (1998) Dual mechanisms for the formation of fluvial silcretes in the distal reaches of the Okavango Delta Fan, Botswana. *Earth Surface Processes and Landforms*, **23**, 705–714.

Sorman, A.U. and Abdulrazzak, M.J. (1993) Infiltration-recharge through wadi beds in arid regions. *Hydrological Sciences Journal*, **38**, 173–186.

Srivastava, P., Brook, G.A., Marais, E. *et al.* (2006) Depositional environment and OSL chronology of the Homeb silt deposits, Kuiseb River, Namibia. *Quaternary Research*, **65**, 478–491.

Stokes, M. (2008) Plio-Pleistocene drainage development in an inverted sedimentary basin: Vera basin, Betic Cordillera, SE Spain. *Geomorphology*, **100**, 193–211.

Stokes, M. and Mather, A.E. (2003) Tectonic origin and evolution of a transverse drainage: the Río Almanzora, Betic Cordillera, southeast Spain. *Geomorphology*, **50**, 59–81.

Sullivan, M.E. (1976) Drainage disorganisation in arid Australia and its measurement, MSc thesis (unpublished), University of New South Wales.

Summerfield, M.A. (1991) Sub-aerial denudation of passive margins: regional elevation versus local relief models. *Earth and Planetary Science Letters*, **102**, 460–469.

Teller, J.T., Rutter, N. and Lancaster, N. (1990) Sedimentology and paleohydrology of Late Quaternary lake deposits in the northern Namib Sand Sea, Namibia. *Quaternary Science Reviews*, **9**, 343–364.

Thomas, D.S.G. (2005) Dryland processes and environments, in *An Introduction to Physical Geography and the Environment* (ed. J. Holden), Pearson Education Ltd., Harlow, pp. 383–404.

Thomas, D.S.G. and Shaw, P.A. (1991) *The Kalahari Environment*, Cambridge University Press, Cambridge.

Thorndycraft, V.R., Benito, G., Rico, M. *et al.* (2005) A long-term flood discharge record derived from slackwater flood deposits of the Llobregat River, NE Spain. *Journal of Hydrology*, **313**, 16–31.

Thornes, J.B. (1977) Channel changes in ephemeral streams: observations, problems and models, in *River Channel Changes* (ed. K.G. Gregory), John Wiley & Sons, Ltd, Chichester, pp. 317–335.

Thornes, J.B. (1980) Structural instability and ephemeral channel behaviour. *Zeitschrift für Geomorphologie*, Supplementband **36**, 233–244.

Thornes, J.B. (1994a) Catchment and channel hydrology, in *Geomorphology of Desert Environments* (eds A.D. Abrahams and A.J. Parsons), Chapman & Hall, London, pp. 257–287.

Thornes, J.B. (1994b) Channel processes, evolution and history, in *Geomorphology of Desert Environments* (eds A.D. Abrahams and A.J. Parsons), Chapman & Hall, London, pp. 288–317.

Tooth, S. (1999) Floodouts in central Australia, in *Varieties of Fluvial Form* (eds A.J. Miller and A. Gupta), John Wiley & Sons, Ltd, Chichester, pp. 219–247.

Tooth, S. (2000a) Process, form and change in dryland rivers: a review of recent research. *Earth-Science Reviews*, **51**, 67–107.

Tooth, S. (2000b) Downstream changes in dryland river channels: the Northern Plains of arid central Australia. *Geomorphology*, **34**, 33–54.

Tooth, S. (2004) Floodout, in *Encyclopedia of Geomorphology*, vol. **1** (ed. A.S. Goudie), Routledge, London, pp. 380–381.

Tooth, S. (2005) Splay formation along the lower reaches of ephemeral rivers on the Northern Plains of arid central Australia. *Journal of Sedimentary Research*, **75**, 634–647.

Tooth, S. (2006) Virtual globes: a catalyst for the re-enchantment of geomorphology? *Earth Surface Processes and Landforms*, **31**, 1192–1194.

Tooth, S. and McCarthy, T.S. (2004a) Anabranching in mixed bedrock-alluvial rivers: the example of the Orange River above Augrabies Falls, Northern Cape Province, South Africa. *Geomorphology*, **57**, 235–262.

Tooth, S. and McCarthy, T.S. (2004b) Controls on the transition from meandering to straight channels in the wetlands of the Okavango Delta, Botswana. *Earth Surface Processes and Landforms*, **29**, 1627–1649.

Tooth, S. and McCarthy, T.S. (2007) Wetlands in drylands: key geomorphological and sedimentological characteristics, with emphasis on examples from southern Africa. *Progress in Physical Geography*, **31**, 3–41.

Tooth, S. and Nanson, G.C. (1999) Anabranching rivers on the Northern Plains of arid central Australia. *Geomorphology*, **29**, 211–233.

Tooth, S. and Nanson, G.C. (2000a) The role of vegetation in the formation of anabranching channels in an ephemeral river, Northern Plains, arid central Australia. *Hydrological Processes*, **14**, 3099–3117.

Tooth, S. and Nanson, G.C. (2000b) Equilibrium and nonequilibrium conditions in dryland rivers. *Physical Geography*, **21**, 183–211.

Tooth, S. and Nanson, G.C. (2004) Forms and processes of two highly contrasting rivers in arid central Australia, and the implications for channel-pattern discrimination and prediction. *Geological Society of America Bulletin*, **116**, 802–816.

Tooth, S., McCarthy, T.S., Brandt, D. *et al.* (2002a) Geological controls on the formation of alluvial meanders and floodplain wetlands: the example of the Klip River, eastern Free State, South Africa. *Earth Surface Processes and Landforms*, **27**, 797–815.

Tooth, S., McCarthy, T.S., Hancox, P.J. *et al.* (2002b) The geomorphology of the Nyl River and floodplain in the semi-arid Northern Province, South Africa. *South African Geographical Journal*, **84**, 226–237.

Tooth, S., Brandt, D., Hancox, P.J. and McCarthy, T.S. (2004) Geological controls on alluvial river behaviour: a comparative study of three rivers on the South African Highveld. *Journal of African Earth Sciences*, **38**, 79–97.

Tooth, S., Rodnight, H., Duller, G.A.T. *et al.* (2007) Chronology and controls of avulsion along a mixed bedrock-alluvial river. *Geological Society of America Bulletin*, **119**, 452–461.

Unkel, I., Kaderiet, A., Mächtle, B. *et al.* (2007) Dating methods and geomorphic evidence of palaeoenvironmental changes at the eastern margin of the South Peruvian coastal desert (14°30′S) before and during the Little Ice Age. *Quaternary International*, **175**, 3–28.

van de Graaff, W.J.E., Crowe, R.W.A., Bunting, J.A. and Jackson, M.J. (1977) Relict Early Cainozoic drainages in arid Western Australia. *Zeitschrift für Geomorphologie*, **21**, 379–400.

Vanney, J.-R. (1960) Pluie et crue dans le Sahara Nord-Occidental, Monographies Régionales 4, Institut de Recherches Sahariennes de l'Université d'Alger.

van Niekerk, A.W., Heritage, G.L., Broadhurst, L.W. and Moon, B.P. (1999) Bedrock anastomosing channel systems: morphology and dynamics of the Sabie River, Mpumulanga Province, South Africa, in *Varieties of Fluvial Form* (eds A.J. Miller and A. Gupta), John Wiley & Sons, Ltd, Chichester, pp. 33–51.

Wakelin-King, G.A. and Webb, J.A. (2007) Threshold-dominated fluvial styles in an arid-zone mud-aggregate river: the uplands of Fowlers Creek, Australia. *Geomorphology*, **85**, 114–127.

Walters, M.O. (1989) A unique flood event in an arid zone. *Hydrological Processes*, **3**, 15–24.

Walters, M.O. (1990) Transmission losses in arid region. *Journal of Hydraulic Engineering, Proceedings of the American Society of Civil Engineers*, **116**, 129–138.

Warner, R.F. (1986) Hydrology, in *Australia – A Geography. Volume One: The Natural Environment* (ed. D.N. Jeans), Sydney University Press, Sydney, pp. 49–79.

Wellington, J.H. (1958) The evolution of the Orange River: some outstanding problems. *South African Geographical Journal*, **40**, 3–30.

Wells, L.E. (1990) Holocene history of the El Niño phenomenon as recorded in flood sediments of northern coastal Peru. *Geology*, **18**, 1134–1137.

Wohl, E.E. (1992) Gradient irregularity in the Herbert Gorge of northeastern Australia. *Earth Surface Processes and Landforms*, **17**, 69–84.

Wohl, E.E. and Grodek, T. (1994) Channel bed-steps along Nahal Yael, Negev Desert, Israel. *Geomorphology*, **9**, 117–126.

Wolman, M.G. and Gerson, R. (1978) Relative scales of time and effectiveness of climate in watershed geomorphology. *Earth Surface Processes and Landforms*, **3**, 189–208.

Woodward, J.C., Hamlin, R.H.B., Macklin, M.G. *et al.* (2008) Glacial activity and catchment dynamics in northwest Greece: long-term river behaviour and the slackwater sediment record for the last glacial to interglacial transition. *Geomorphology*, **101**, 44–67.

Woodyer, K.D., Taylor, G. and Crook, K.A.W. (1979) Depositional processes along a very low-gradient, suspended-load stream: the Barwon River, New South Wales. *Sedimentary Geology*, **22**, 97–120.

Woolley, R.R. (1946) Cloudburst floods in Utah, 1850–1938. *United States Geological Survey Water Supply Paper*, **994**.

Yang, X. (2006) Desert research in northwestern China – a brief review. *Géomorphologie: Relief, Processus, Environnement*, **4**, 275–284.

Yang, X., Zhu, Z., Jaekel, D. *et al.* (2002) Late Quaternary palaeoenvironment change and landscape evolution along the Keriya River, Xinjiang, China: the relationship between high mountain glaciation and landscape evolution in foreland desert regions. *Quaternary International*, **97–98**, 155–166.

Zawada, P.K. (1994) Palaeoflood hydrology of the Buffels River, Laingsburg, South Africa: was the 1981 flood the largest? *South African Journal of Geology*, **97**, 21–32.

Zawada, P.K. (1997) Palaeoflood hydrology: method and application in flood-prone southern Africa. *South African Journal of Science*, **93**, 111–132.

Zawada, P.K. and Smith, A.M. (1991) The 1988 Orange River flood, Upington region, Northwestern Cape Province, RSA. *Terra Nova*, **3**, 317–324.

13

Channel form, flows and sediments of endogenous ephemeral rivers in deserts

Ian Reid and Lynne E. Frostick

13.1 Introduction

Water is a significant agent of erosion in arid lands. Upon entering a desert or semi-desert area, it soon becomes obvious that access is often facilitated or hindered by the river system. Yet, until recently, the role of rivers in shaping desert landscapes has generally been underestimated by geomorphologists. The reasons for this are several. First, there has been a tendency to concentrate attention on processes and forms that are thought to be more peculiar to drylands. Seminal texts such as that of Bagnold (1941) and collections of papers such as those of McKee (1979) and Brookfield and Ahlbrandt (1983) have highlighted the action of wind and given prominence to windblown dunes. This has been reinforced by popular portrayals of the desert in novels and on film. Second, the infrequence of rainfall and runoff in drylands has made data acquisition an expensive proposition for the fluvial geomorphologist. Few have had either the financial resources or the patience to collect long-term records that consist largely of non-information and that are punctuated only spasmodically with a frenzy of relevant data. Notable exceptions are Renard and Keppel (1966) and Stone *et al*. (2008) in Arizona and Schick (1970), Reid, Laronne and Powell (1995) and Alexandrov *et al*., 2009 in Israel. Indeed, many of those who have examined ephemeral streams have done so over shorter periods of time. However, to do this, they have moved towards semi-arid areas where rivers are no less dry and where higher rainfall gives a greater likelihood of flash floods (see, for example, Leopold and Miller, 1956; Thornes, 1976; Frostick and Reid, 1977).

There is, as a consequence, a dearth of information about processes in ephemeral rivers. It would be an understatement to declare that the explosive increase in both rainfall-runoff and sediment-transport data for perennial systems of temperate latitudes has not been matched by that collected in drylands. As a result, there is little basis for stochastic or other forms of modelling of flows in ephemeral rivers. Each piece of information, each measured flood, is fairly unique and has to be treasured as a gem that gives insight into the workings of these mysterious elements of these vast landscapes. For the geomorphologist, hydrologist and sedimentologist, this remains pioneering territory, even at the start of the twenty-first century.

Notwithstanding the small total information base for ephemeral river systems, they have figured greatly in the development of process-orientated geomorphology through the incorporation of the results of work done in the southwestern United States and collated in the timely text of Leopold, Wolman and Miller (1964). However, the impetus that this text might have given to work on desert streams per se was lost in its adoption as a general statement about river processes by a generation whose imagination it had caught. There was also the fact that the rash of process studies that broke out during the 1960s and 1970s inevitably focused on the backyards of those concerned. These backyards were (and still are) predominantly in humid environments where water catchments are (or could be) fully clothed with vegetation, where the soil fauna ensure a perforated medium that accommodates infiltering rainfall and where the stream at the bottom of the hill has a perennial flow regime.

More recently, attempts have been made to redress the continuing imbalance in emphasis on windblown sediments in desert environments by giving equal attention

Arid Zone Geomorphology: Process, Form and Change in Drylands, Third Edition. Edited by David S. G. Thomas

to hydrological and fluvial processes (Frostick and Reid, 1987; Cooke, Warren and Goudie, 1993; Tooth, 2000; Bull and Kirkby, 2002; Parsons and Abrahams, 2009; and also in this volume). This is important not only because fluvial processes are a cause of so many problems in desert areas, but also because the peculiarities of dryland environments cause sufficient differences in river behaviour that the lessons learnt in humid areas are not reliably translated (Pilgrim, Chapman and Doran, 1988).

13.2 Rainfall and river discharge

13.2.1 Storm characteristics

One way in which desert streams differ from their perennial counterparts is the generation and propagation of the flood wave. However, the peculiarities of flash floods are not entirely a function of processes on the ground. Of considerable significance for runoff is the fact that rain is more often than not associated with discrete convective cells (Sharon, 1972, 1974). Data are few because the dense networks of gauges that would be required to measure such spotty rainfalls are extremely rare in drylands. Nevertheless, there is good evidence to suggest that these cells have a diameter that is generally less than 10–14 km (Diskin and Lane, 1972; Renard and Keppel, 1966) (Figure 13.1(a)). This means that rainfall measured at one gauge cannot be used to predict rainfall even a few kilometres away, in contrast with temperate regions affected by ubiquitous frontal storms (Wheater *et al.*, 1991a) (Figures 13.1(b) and 13.2). Because of the discrete nature of each convective cell, an individual storm may be unlikely to affect the entire drainage net, while successive storms are more than likely to wet different parts of a river catchment (Schick and Lekach, 1987) (Figure 13.1(a)).

In catchments of modest dimensions (10^2 to 10^3 km^2), this has implications for the flood hydrograph, because a different part of the drainage basin will contribute water during each event. Indeed, some tributaries have been noted first to run and then to remain dry during a period of successive floods in a trunk stream (Frostick, Reid and Layman, 1983). As a result of this and other variables, the shape of the flood hydrograph must change considerably.

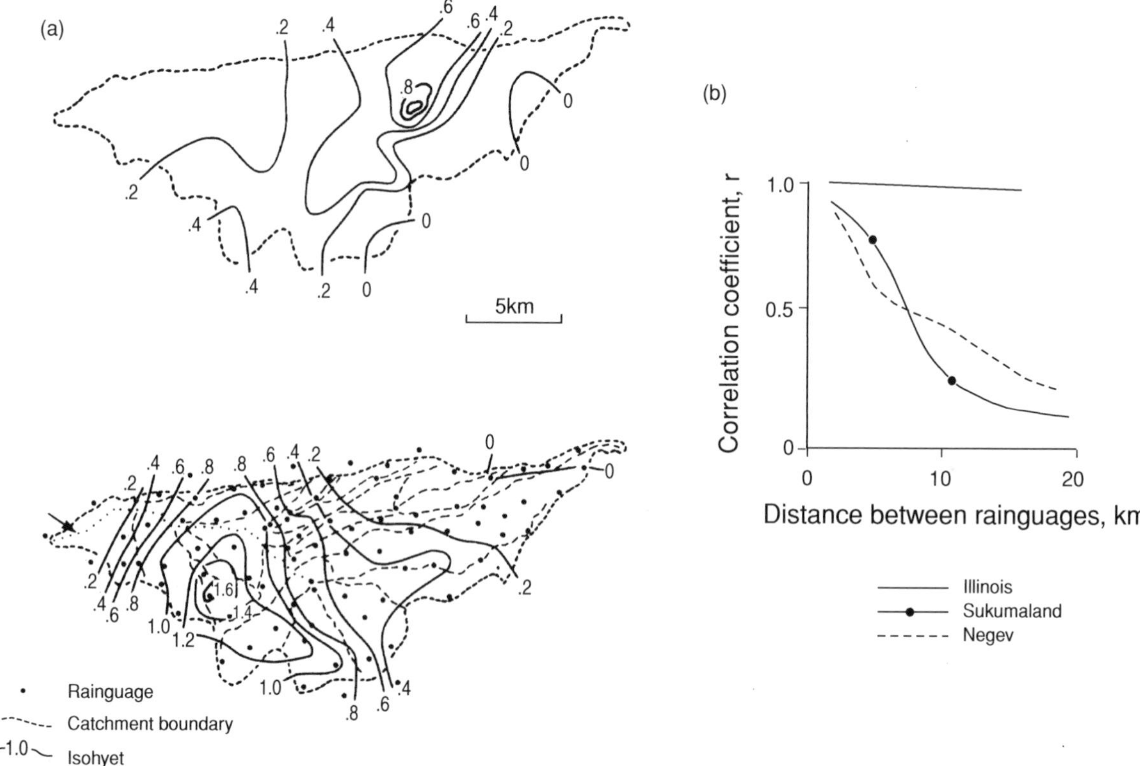

Figure 13.1 (a) Raingauge net, catchment boundary and the isohyets (in inches) of two storms over Walnut Gulch, Arizona (modified and redrawn after Renard and Keppel, 1966). (b) Correlogram of rainfall caught by point gauges spaced at distances up to 20 km for winter frontal rainfall (Illinois) and cellular convective storms (Negev Desert and Sukumaland, western Tanzania) (after Sharon, 1974).

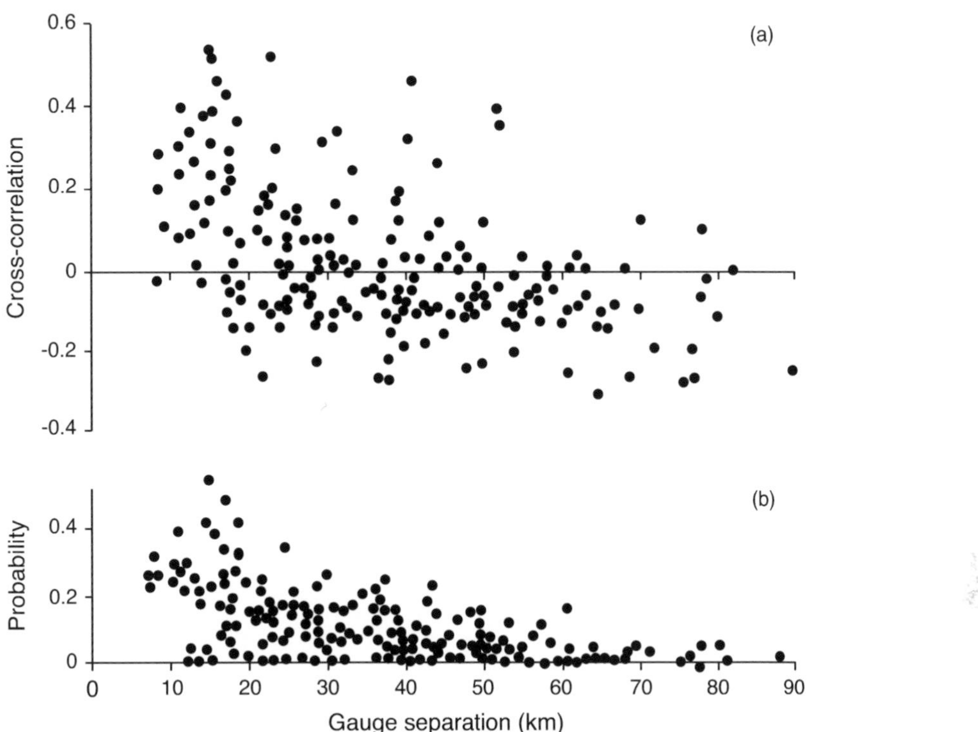

Figure 13.2 (a) Cross-correlation of hourly rainfall as a function of gauge separation, Wadi Yiba, southwest Saudi Arabia. (b) Probability of simultaneous occurrence of hourly rainfall as a function of gauge separation, Wadi Yiba, southwest Saudi Arabia. Both reflect the spotty nature of rainfall in desert regions (after Wheater *et al.*, 1991b).

There is therefore less likelihood of identifying a *characteristic* outfall hydrograph for an ephemeral system, in contrast with perennial streams affected by widespread frontal rain (Wheater and Brown, 1989). Besides this, the seemingly fixed maximum storm-cell size means that the fraction of a river basin that will be affected by rain will fall as the basin size increases. Although it is conceivable that several storms will wet a large basin more or less simultaneously, atmospheric dynamics will dictate that they are widely separated geographically. Sharon (1974) has suggested that the distance between storm centres may average 40–60 km.

One other factor that has considerable bearing on the nature of river discharge is that storm cells migrate as they deliver their rain (Sharon, 1972; Frostick and Reid, 1977; Frostick, Reid and Layman, 1983). Again, there are implications for the shape of the flood hydrograph, since a storm moving upstream across a drainage basin will cause runoff from lower tributaries to occur earlier than those of headwater regions. Indeed, the fact that overland flow is generated within minutes in arid environments (Yair and Lavee, 1976; Reid and Frostick, 1986) encourages this temporal separation of contributions from each tributary source. Schick (1988) has speculated that the multipeak nature of flash floods reflects, among other things, the piggybacked contributions of individual tributary water catchments as each receives and disposes of rainfall. Countering this in part, Ben-Zvi, Massoth and Schick (1991) have demonstrated for ephemeral streams in Israel that the downstream travel time of flood crests is governed most significantly by crest height, as would be expected from hydraulic considerations of wave celerity. This would suggest that the contributions from individual tributaries would overrun each other in a manner dictated by the spatial inhomogeneity of storm rainfall, as much as by other factors. Because spatiotemporal rainfall patterns are highly unpredictable, this has to add to the complexity and changeability of the river outfall hydrograph (Reid, Laronne and Powell, 1998).

Alexandrov, Laronne and Reid (2007) have used the climatically transitional nature of the Levant to highlight seasonal differences in catchment runoff response that arise from different types of storm. Here, atmospheric conditions dictate that winter months are characterised by frontal rainstorms, which have low rain intensities and deliver water fairly ubiquitously throughout a catchment. In contrast, autumn and spring are characterised by convectively enhanced, cellular storms that have high rain intensities and wander over the landscape delivering their water along discrete tracks. In consequence, autumn and spring storms are flashier, the flood timebase is shorter, but, *ceteris paribus*, peak discharge is greater.

In areas of pronounced topography, orographic heightening of convective storms can have a significant effect on the spatial pattern of rainfall, a factor that has to be taken into account when assessing probable stream discharge in basins that are not greatly separated geographically. Wheater *et al.* (1991b) demonstrate reasonably well-defined direct relations between both number of raindays and annual rainfall and altitude in the Asir escarpment of southwest Saudi Arabia. Martínez-Goytre, House and Baker (1994) suggest a rain-shadow effect in the Santa Catalina Mountain range of southeastern Arizona after analysing the magnitude of palaeofloods in basins disposed at different locations around the massif.

13.2.2 Flash flood hydrograph

The peculiarities of storms in desert areas mean that the relations between rainfall and runoff are extremely complex, almost precluding hydrological modelling (Wooding, 1966; Renard and Lane, 1975; Srikanthan and McMahon, 1980). Wheater and Brown (1989) (Figure 13.3) have illustrated poor relations between flood peak discharge and maximum rainfall intensity and between flood runoff and storm rainfall volume in Wadi Ghat, a 597 km^2 basin draining towards the Red Sea in southwestern Saudi Arabia. The most convincing relation they show is, ironically, between flood runoff volume and flood peak discharge, suggesting some consistency between basic hydrograph shape and flood magnitude. In contrast, Reid, Powell and Laronne (1998) show a fairly convincing relation between flood runoff and rainfall volumes for the Nahal Eshtemoa, a 112 km^2 catchment in the semi-arid northern Negev. In fact, although the information base is small for modelling purposes, despite the 15-year record, Alexandrov *et al.* (2009) provide a fairly clear-cut flood magnitude probability relation for the same catchment (Figure 13.4). The emergence of this relation is surprising. Despite reports of nonstationary relations for runoff elsewhere (specifically, the Wadi Wahrane basin of northern Algeria; see Benkhaled and Remini, 2003), it might be used to give water resource managers at least some hope that stochastic flood prediction is feasible, even in semi-arid regions. It might encourage greater investment in flow gauging, despite the infrequence and uncertainty of river floods.

For catchments of modest size, there are three characteristics apparent in the flood hydrographs of streams in widely dispersed deserts. First, every one exhibits a steep rising limb that incorporates a bore (Figures 13.5 and 13.6). McGee (1897) gave a remarkable first description of a bore, though for an unconfined sheetflood on a bajada fronting a mountain range in the Sonoran Desert. Since then the phenomenon has been noted in ephemeral channels in different deserts by various workers, including Hubbell and Gardner (1944), Leopold and Miller (1956), Renard and Keppel (1966), Gavrilovic (1969) and Frostick and Reid (1979). The bore is rarely the 'wall of water' often referred to in popular literature about flash floods. Hassan (1990a) (Figure 13.7) has related the speed of advance and the depth of some bores observed in Israel,

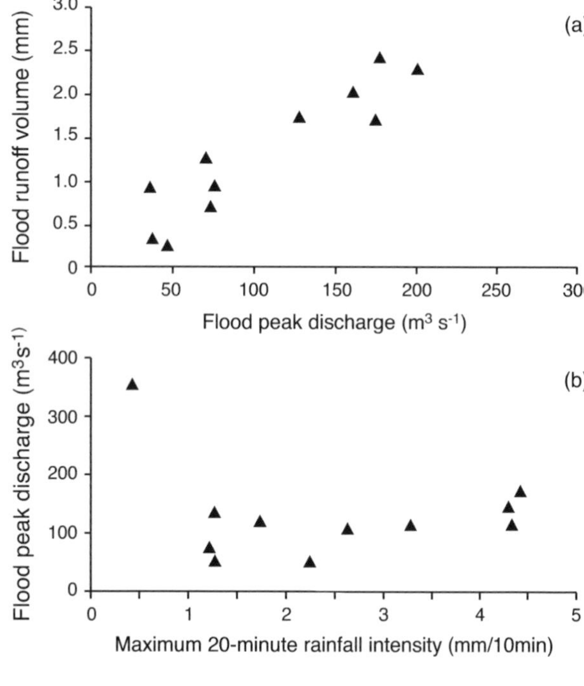

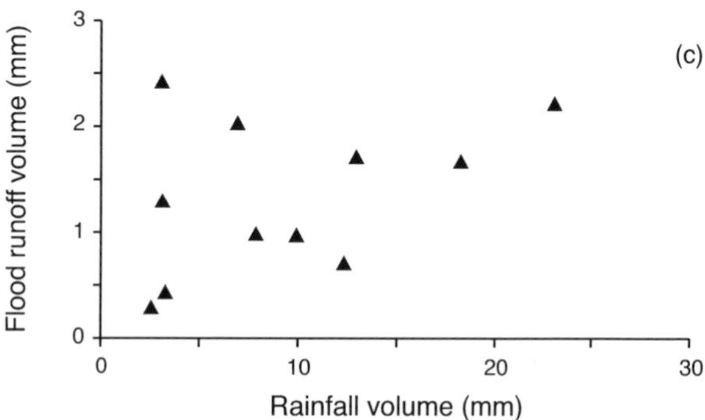

Figure 13.3 (a) Surprisingly good relation between flash-flood volume and peak discharge, Wadi Ghat, southwest Saudi Arabia. (b and c) Lack of relations between flash flood parameters and storm rainfall, Wadi Ghat, southwest Saudi Arabia (after Wheater and Brown, 1989).

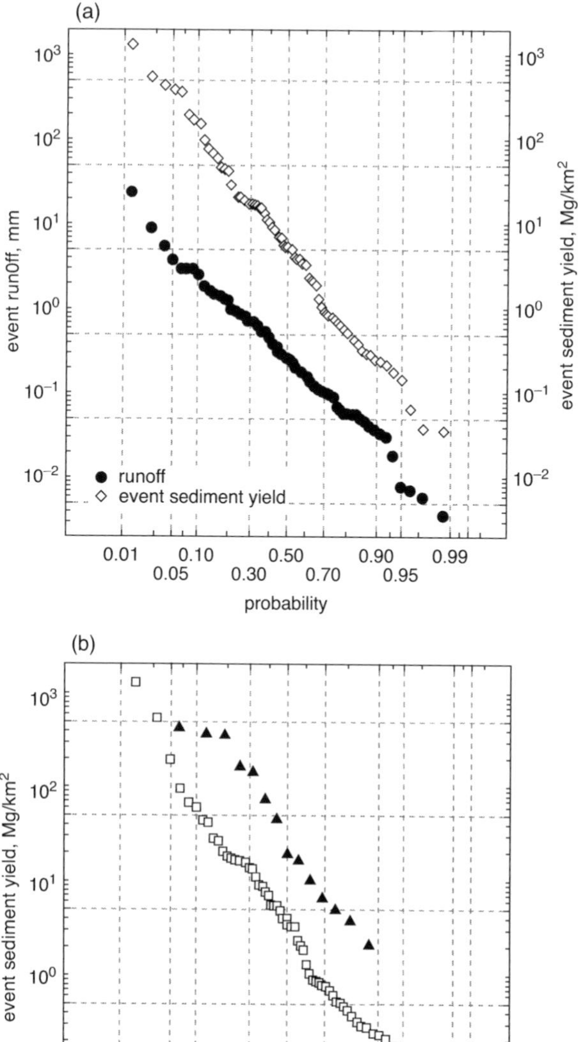

Figure 13.4 (a) Probability ($p = r/(n-1)$, where r is rank order and n is the total number of events) of exceedance for runoff volume and total (suspended sediment load plus bedload) sediment yield of flash floods in the Nahal Eshtemoa, 1991–2006. (b) Probability of exceedance for total sediment yield of flood events arising from convective (autumn–spring) and frontal (winter) storms (after Alexandrov *et al.*, 2009).

distinguishing between those travelling over antecedently dry or wet channel beds, and indicates remarkably modest velocities ≤1 m/s. These are corroborated by other observations made in Israel by Reid, Laronne and Powell (1998). Nevertheless, the arrival of a bore has often been fatal for people caught unawares (Hjalmarson, 1984). This is undoubtedly because the time of rise of the flood hydrograph is short. Reid *et al.* (1994) have measured rates of rise in water-stage as high as 0.25 m/min and average water velocities >3 m/s within a few minutes of the passage of the bore. A person caught mid-channel in these circumstances can easily lose footing and be swept away. Reports of the interval between the arrival of the bore and peak discharge for catchments of small or moderate size range from 18 to 23 minutes (Renard and Keppel, 1966), 10 minutes (Schick, 1970) and 14 to 16 minutes (Reid and Frostick, 1987).

The second characteristic of ephemeral stream hydrographs is the steepness of the flood recession (Figure 13.5). This may reflect the overriding importance of Hortonian overland flow in arid environments. Observations of the limited penetration of the wetting front in the soil (e.g. ≤210 mm, Reid and Frostick, 1986) suggest no substantial subsurface routing of water that might sustain stream discharge (Pilgrim, Chapman and Doran, 1988). Besides this, the extreme dryness of the soil always means a high gradient of matric potential, which would draw water downward into the soil profile and reduce the chance of achieving pressures greater than atmospheric that would permit soil interflow or pipeflow.

This leads to the third characteristic that is peculiar to ephemeral systems in modest-sized catchments. The flood is often extremely short-lived. The whole event from initial to final dry bed might have taken no more than a few hours (Figures 13.5 and 13.6) and rarely more than a day. Because the number of floods that might be expected ranges from around six (Reid and Frostick, 1987) to much less than one per year (Schick and Lekach, 1987), as one moves from semi-arid to arid environments, this means that the river system is active for much less than 1 or 2 % of the time. The probability of being on-site to make observations and to take measurements during an event is extremely small. This and the low rate of data acquisition are the chief reasons for the paucity of information that we have regarding desert flash floods.

13.2.3 Transmission losses

Another factor that sets ephemeral systems aside from those that flow perennially or even intermittently is the loss of water to the stream bed. The process has been quantified by Renard and Keppel (1966), Burkham (1970) and Butcher and Thornes (1978) for streams in Arizona and Spain, and noted in many other instances (such as Murphey, Lane and Diskin, 1972; Schick and Lekach, 1987; Reid, Best and Frostick, 1989). The rate of

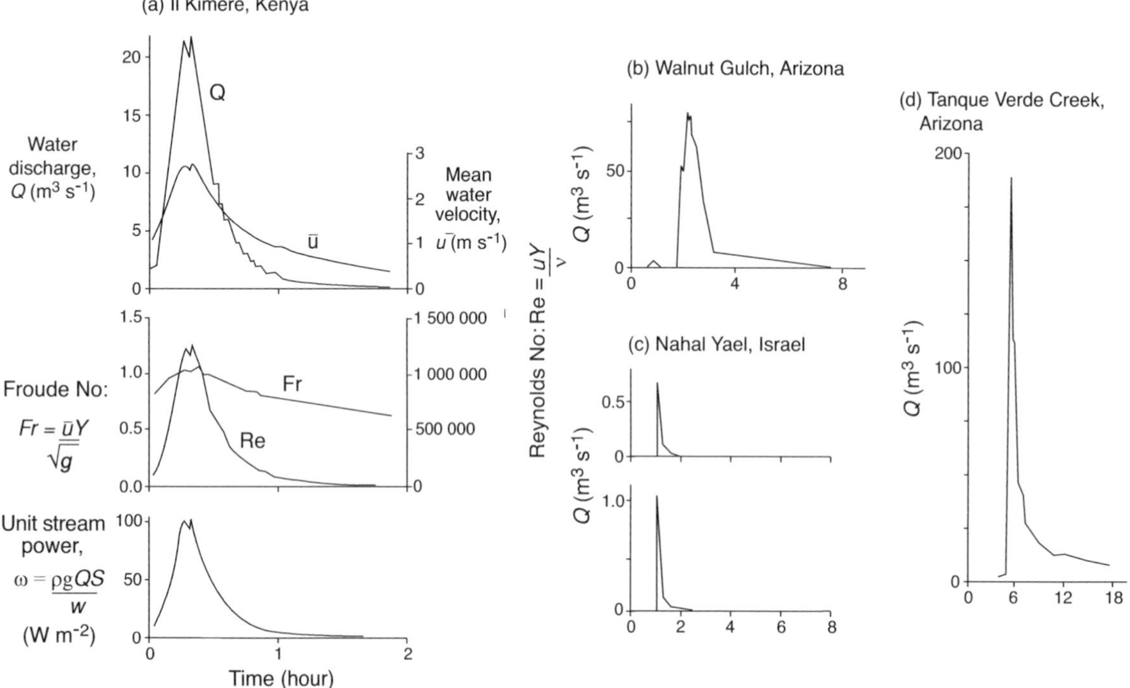

Figure 13.5 Flood hydrographs and hydraulic parameters of ephemeral rivers in desert regions: (a) Il Kimere, Kenya (after Reid and Frostick, 1987); (b) Walnut Gulch, Arizona (after Renard and Laursen, 1975); (c) Nahal Yael, Israel (after Lekach and Schick, 1982); (d) Tanque Verde Creek, Arizona (after Hjalmarson, 1984). Note that the unit of the time axis is an hour.

Figure 13.6 (a) Shallow-angle flood bore in the sand-bed ephemeral Il Kimere, northern Kenya. (b) Steep-angle flood bore in the sand-bed Il Eriet, Kenya–Ethiopia border region. (c) Steep-angle flood bore over-running Birkbeck-type bedload samplers in the gravel-bed Nahal Eshtemoa, Israel; note the trash – a typical feature of flows sweeping a channel after a period of dormancy.

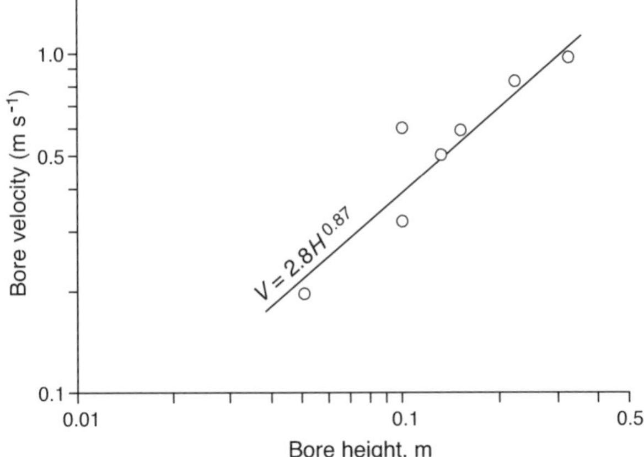

Figure 13.7 Relation between flash-flood bore velocity and height observed in Nahal Og, Judean Desert, and Nahal Hebron, Negev Desert, Israel (after Hassan, 1990a).

transmission losses will depend upon the porosity and depth of the channel fill as well as upon the hydraulic conductivity of its least permeable layer, in relation to the length of the flood period (Parissopoulos and Wheater, 1992). The fixed flow-measuring flumes installed in Walnut Gulch in Arizona (Renard and Keppel, 1966) also indicate that an increase in the wetted perimeter with increasing peak discharge is an important variable in reckoning transmission losses (Figure 13.8). Where flow exceeds channel capacity and spills on to the floodplain, as in the peculiar anastomosing channel system of Cooper Creek of the Lake Eyre basin of Australia, transmission losses are shown to increase from c. 60 % to c. 90 % over a channel length of 420 km (Knighton and Nanson, 1994).

Thornes (1977) (Figure 13.9) offers a conceptual model of the interaction of transmission losses and tributary discharge contributions for low- and high-magnitude floods that alerts the geomorphologist to the complicated downstream changes in flow that can be expected in desert floods. Almost all statistical and mathematical models of transmission losses (e.g. Lane, Diskin and Renard, 1971; Walters, 1990; Sharma and Murthy, 1995) have necessarily used channel reaches in which the outflow hydrograph is unaffected (as best as can be judged) by complications such as tributary inflow. Interestingly, they give tolerable first approximations, in some cases over channel lengths of up to 25 km. However, Knighton and Nanson (1994) draw attention to the fact that the pattern of behaviour evident in one reach of a river may not be repeated downstream, reinforcing the impression of unpredictability in ephemeral systems.

Transmission losses have two effects on the flood wave. They may steepen the bore (Butcher and Thornes, 1978), though the effect will be highly variable depending upon the location of the storm in relation to the drainage net and, therefore, how dry the bed is before being overrun (Reid and Frostick, 1987). They also ensure that flood discharge *decreases* downstream in the absence of significant tributary inflows (Figure 13.8). This is in complete contrast with perennial streams and has considerable implications for downstream changes in channel geometry and sediment transport.

13.2.4 Drainage basin size and water discharge

The low annual rainfall of arid and semi-arid areas inevitably means that the annual discharge of ephemeral streams is low compared with perennial counterparts having the same size drainage basin. In fact, Wolman and

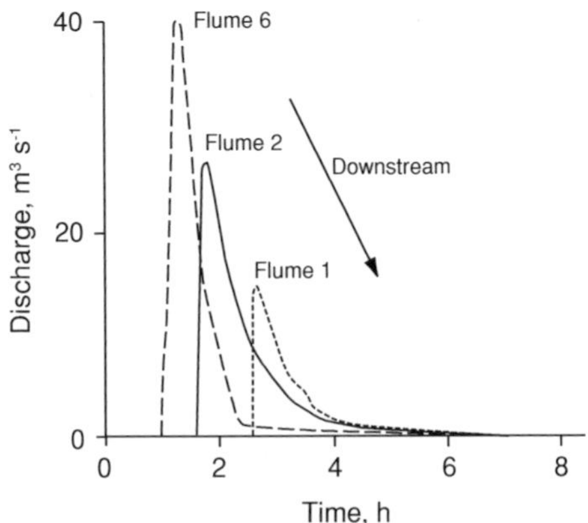

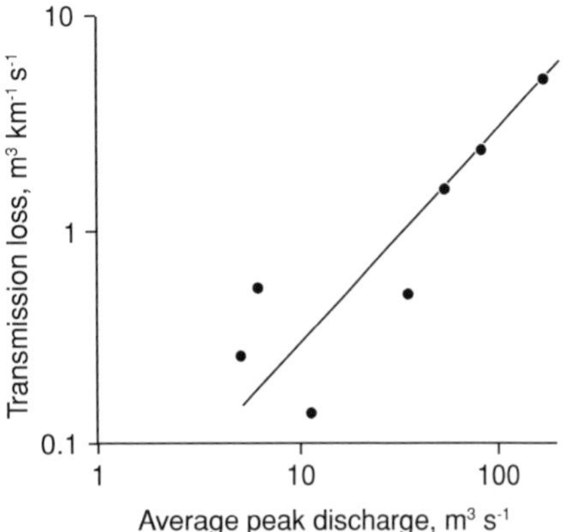

Figure 13.8 (a) Transmission losses reduce the size of the flood wave on Walnut Gulch by c. 60 % over the 10 km channel length between flumes 6 and 1. (b) Transmission losses, expressed in cubic metres per second per kilometre of channel, as a function of average peak discharge at flumes 6 and 2 on Walnut Gulch, Arizona (after Renard and Keppel, 1966).

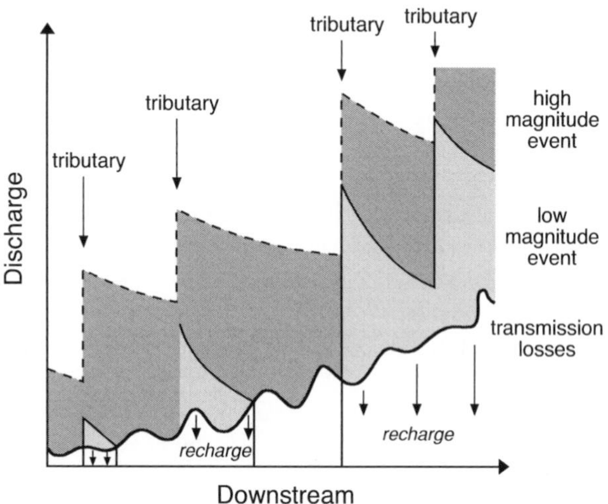

Figure 13.9 Conceptual model of the interplay of tributary inflows and transmission losses during large and small flash floods. The hachured areas represent net surface streamflow (after Thornes, 1977).

Gerson (1978) demonstrated that the exponent in the log-log relation between mean annual runoff and drainage area of ephemeral streams in California is about half that of the same relation for perennial streams of Maryland (Figure 13.10). However, perhaps just as noteworthy in the relation is the fact that there is far greater range in runoff values among ephemeral streams for any particular drainage area. It is tempting to speculate that this contrast with the perennial stream case is a function of the spatially discrete and highly variable rainfall pattern, together with wide-ranging transmission losses and a host of other factors, including lithologically and pedologically controlled runoff-rainfall ratios. Whatever the exact reasons, it highlights again the relative unpredictability of the ephemeral drainage system.

Despite the expected differences in gross annual discharge, Wolman and Gerson (1978) point to the unexpected similarities in flood peak discharge between ephemeral and perennial rivers having the same size drainage basin (Figure 13.10). This must be fortuitous from a process point of view given the now well-established role of soil interflow, the diminutive role of Hortonian overland flow and the general concept of partial-area contribution to streamflow in humid environments (Hewlett and Hibbert, 1967) and the undoubted importance of Hortonian overland flow in deserts (see Chapter 11 and Bonell and Williams, 1986). Nevertheless, it is interesting that desert streams can match perennial counterparts at least in this one flood parameter, despite the spottiness of the rainfall pattern, the transmission losses and so on.

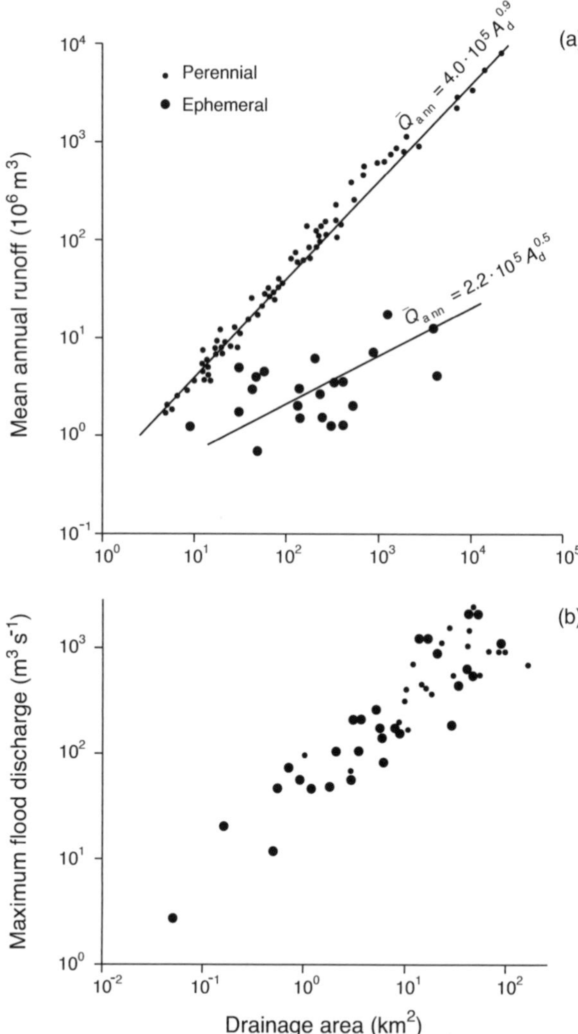

Figure 13.10 (a) Annual runoff as a function of drainage basin size for perennial rivers in Maryland and ephemeral streams in California. (b) Peak flood discharge as a function of drainage area for ephemeral and perennial streams (after Wolman and Gerson, 1978; data compiled from several sources).

Of potential value from an engineering standpoint is the fact that the relations between mean annual flood peak discharge and drainage area in different arid zones throughout a number of continents have been shown by Farquharson, Meigh and Sutcliffe (1992) to follow a reasonably common trend. They also show a degree of similarity in dimensionless flood frequency curves, although some regions yield data that does not conform as well. This holds the promise of being able to offer a best estimate of flood flow in ungauged water catchments.

Figure 13.11 (a) Sand-bed ephemeral stream showing great width and planar bed, Laga Tulu Bor, northern Kenya. (b) Gravel-bed ephemeral showing great width and planar bed, unnamed stream, Sibilot National Park, northern Kenya.

13.3 Ephemeral river channel geometry

13.3.1 Channel width

Desert streams are notable for their great width (Leopold, Emmett and Myrick, 1966; Baker, 1977; Frostick and Reid, 1979; Graf, 1983) (Figure 13.11). When visiting a desert for the first time, those whose familiarity is rather with perennial streams of humid regions might think that the ephemeral channels they come across represent much larger basins than is the case. Wolman and Gerson's (1978) compilation of data for different arid and semi-arid regions confirms this rapid increase in width with drainage area, at least for small- to moderate-sized basins. However, their diagram also reveals a fascinating and fairly universal asymptote value of between 100 and 200 m for channel width once the drainage basin exceeds about 50 km^2. (This value is only achieved in perennial rivers when the drainage basin is about 10 000 km^2 in size!)

The large rate of increase of stream width in small- to moderate-sized basins might be attributable, in part, to the high drainage densities that are sustained by rapid runoff in drylands. For example, Frostick, Reid and Layman (1983) report densities of 90 km/km^2 for a catchment in northern Kenya. The change to a fairly steady stream width once the water catchment exceeds c. 50 km^2 might result, in part, from the fact that transmission losses from flows with such a high wetted perimeter compensate for any addition of tributary water. Indeed, in a situation where there are no significant tributary inputs, Dunkerley (1992) (Figure 13.12) shows that severe transmission losses produce a commensurate decline in channel width and capacity over distances as small as a few kilometres. On the other hand, it could indicate that the finite size of convective rain cells imposes an upper limit on the discharge of an ephemeral drainage system regardless of its size. In other words, there would be no need for any further increase in channel capacity once the runoff of one rain cell is catered for. This is clearly an area where there is a need for greater information about cellular rainfall and both flows and channel geometry.

Superimposed on regional trends, Schumm (1961) provides us with an insight into the variability of channel width at a more local scale by examining a surrogate measure of bed and bank material shear strength. He shows that the lower the silt-clay content of the wettable perimeter, the higher the ratio between bankfull width and depth (Figure 13.13). Working in another part of arid North America, Murphey, Lane and Diskin (1972) have also noted downstream changes in channel width that are related to the nature of local bank materials. In their case, sandstones are more erodible than conglomerates, thus producing a less well-defined and less confined channel.

However, some reservations may be needed in assessing the geometry of the arroyos of the southwestern United States, if only because land-use changes and changes in storm size and frequency during the late nineteenth century are thought to have been responsible for subsequent channel incision and consequent confinement of the flow (Cooke and Reeves, 1976; Reid, 2009).

13.3.2 Channel bed morphology

Not only are ephemeral streams typically wide, they are noted for having a remarkably subdued bed topography.

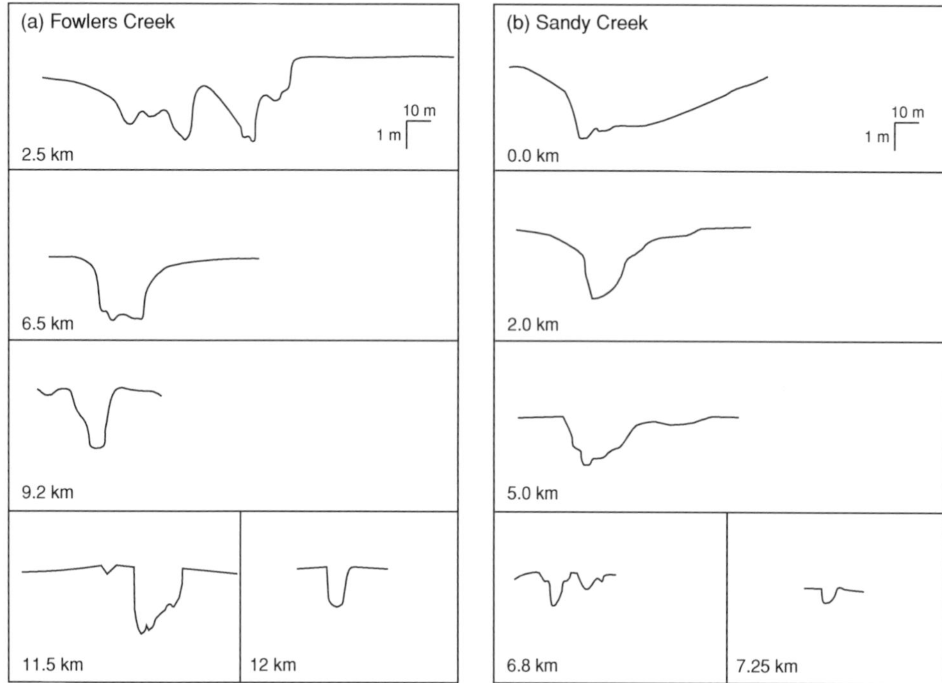

Figure 13.12 Cross-sections of two ephemeral channels at specified distances beyond the reentrant they make in the scarp of the Barrier Range of New South Wales and as they flow towards a flood basin. Transmission losses ensure rapid reduction in channel capacity and width (after Dunkerley, 1992).

In fact, the beds of single-thread streams are often near-horizontal and planar; any bar forms are also often flat-topped and rise only 10–20 cm above the thalweg (Leopold, Emmett and Myrick, 1966; Frostick and Reid, 1977, 1979) (Figure 13.11). It is likely that channel width and bed flatness are related through flow depth, since great width spreads the flow, so ensuring shallow depths. This in turn suppresses the secondary current cells that might otherwise encourage the building of bars, but it also maintains sediment transport efficiency by maintaining relative grain exposure (water depth and bed grain diameter) throughout flood events. As a result, temporary 'highs' are planed off and 'lows' are filled in.

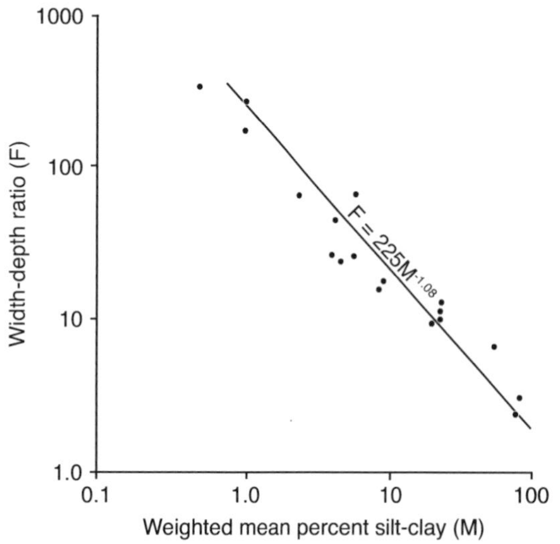

Figure 13.13 Width-to-depth ratio of stable cross-sections of four ephemeral streams in four western states of the USA as a function of weighted mean percentage silt-clay content of the stream bed and bank material (after Schumm, 1961).

This does not mean that all ephemeral stream channels are featureless. Where the bed material is sand and the last bed-forming flow was of appropriate strength, dunes and megadunes may be found (Williams, 1970). However, exaggerated bar forms tend to occur only at pronounced bends, as might be expected from analogy with perennial rivers, or they occur where the channel is diffused in a braid plan (e.g. Cooper Creek in the Lake Eyre basin of Australia; see Rust and Nanson, 1986). There is a suggestion that single-thread ephemeral streams tend to low sinuosity (Schumm, 1961). This straightness may be the reason why flat channel beds are so characteristic. However, the number of examples that have been documented and that can be categorised is too small to make definitive comments.

13.4 Fluvial sediment transport

Ephemeral streams move vast quantities of sediment during each flood event, both as bedload and as suspended load. In fact, they have produced the highest recorded values of suspended sediment concentration, reaching 68 % solids (by weight) in the Rio Puerco, New Mexico (Bondurant, 1951). Material is readily available for transport not only from the thinly vegetated or bare slopes of a catchment but also from the bed of the stream. Even in hyper-arid regions, where rainfall may be <50 mm per year, annual specific sediment yield has been shown to be high (Schick and Lekach, 1993). However, it is in semi-arid environments, where annual rainfall is 250–350 mm, that specific sediment yield reaches its highest levels in the absence of significant human degradation of soil and vegetation (Langbein and Schumm, 1958). From a 15-year study of a 112 km^2 catchment in the northern Negev, Alexandrov *et al.* (2009) provided an average annual specific yield of both suspended sediment and bedload of 290 Mg/km^2, to which bedload contributes 15 Mg/km^2, or 5 %. However, weather is capricious in all drylands and annual total yield ranged wildly from 0 to 1720 Mg/km^2 during the period of monitoring.

Before considering the entrainment and transport of sediment, attention should be drawn to a characteristic of some ephemeral streams that distinguishes them from their perennial counterparts and has a considerable effect on the flow. The perched groundwater that is held in the channel-fill makes the channel bank a favoured, and sometimes only, location for the growth of large trees (Frostick and Reid, 1979). Because flows of magnitude sufficient to sweep the channel in perennial streams are infrequent in ephemerals, trees are also able to establish themselves on the channel bed, or trees that were once bankside may find themselves within-channel as the bank is eroded. Graeme and Dunkerley (1993) (Figure 13.14) indicate that flow resistance can be greatly increased by the presence of the trunks. Indeed, because the flow may encounter more and more vegetative 'roughness' in the form of branches as stage discharge increases, they show that, instead of the 'usual' inverse relation between the roughness parameter and water depth, roughness values may increase with increasing depth. Where this is significant, it will obviously have an impact on the amount of energy available for sediment transport. The importance of riparian and in-channel tree growth for flow resistance and channel geometry has also been examined by Graf (1979) for ephemeral streams of the American southwest. Here, where valley bottom biomass is low, arroyos are entrenched, but where it is high – implying an increase in flow resistance – valley bottoms lack entrenchment.

Tooth and Nanson, in a series of papers (1999, 2000, 2004), have explored some remarkable differences in channel geometry and pattern that arise from differences of in-channel tree growth in two ephemeral rivers that follow subparallel paths less than 2.5 km apart on the Northern Plains of central Australia. The Plenty River is single thread, its channel is largely devoid of trees and it has a bed of sand, while the Marshall River, running the same longitudinal gradient, has multiple, anastomosed, narrow channels separated by tree-induced sediment ridges and islands and has a bed of coarse sand and granules (Figure 13.15). Tooth and Nanson rationalise that, in contrast with the neighbouring Plenty, ephemeral tributaries of the Marshall provide both a source of local perched groundwater in the channel-fill that sustains in-channel tree growth and coarser bedload sediment, which accumulates as shadow deposits in the lee of the trees. The growth and consolidation of these deposits has split the channel into multiple, subparallel threads, in stark contrast to the single-thread Plenty.

Positive feedback between vegetation growth and channel sedimentation, with consequences for channel geometry, also arises where land-use change brings a reduction in runoff and sediment yield. Rozin and Schick (1996) show that soil and water conservation measures implemented in the catchment of the Nahal Hoga, Israel, have produced a dramatic shift in river character over a period of 50 years (Figure 13.16). What was, at one time, a wide, single-thread channel devoid of vegetation became, first, multithreaded and, subsequently, has become an entrenched single-thread channel as vegetation progressively invaded and sediment accumulated.

13.4.1 Scour and fill

The fact that ephemeral streams allow complete access to their bed sediments between floods without the need for pumping equipment, coffer dams and so on means that they were the first river type to reveal the process of scour and fill.

By emplacing scour chains in the small but elongate Arroyo de Los Frijoles in New Mexico, Leopold, Emmett and Myrick (1966) were able to show that a single flood might incise the river bed by as much as 0.3 m. However, just as significant was the fact that this scour would be matched by a more or less equal amount of deposition during flood recession (Figure 13.17). Over and above this single flood pattern, it was also revealed that the entire channel was in grade, that is to say, in equilibrium, since the bed was shown to be restored to more or less the same altitude over four successive flood seasons. Scour

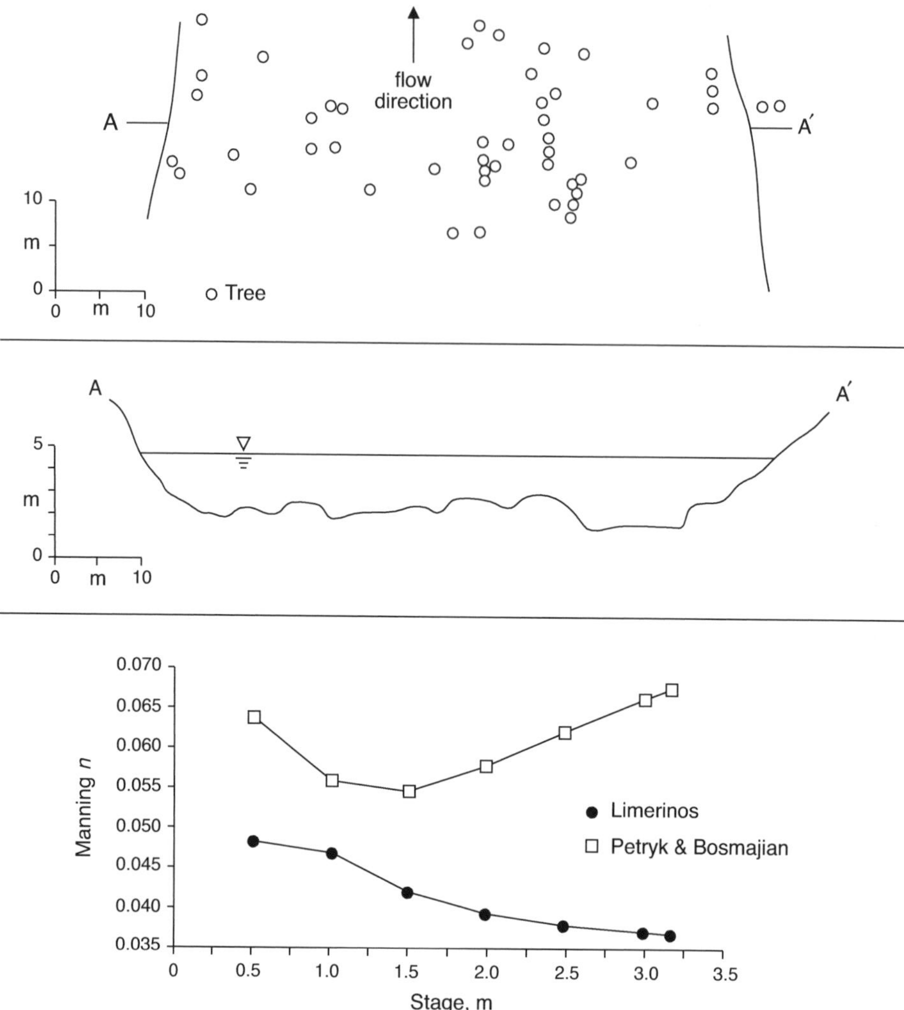

Figure 13.14 (a and b) Plan and cross-section of a reach of ephemeral Fowlers Creek, New South Wales, showing the distribution of River Red Gum trunks in the channel. (c) Manning's roughness coefficient, n, as a function of water stage taking into account the bed material size characteristics only (following Limerinos) and, in addition, the effect of vegetation (following Petryk and Bosmajian) (after Graeme and Dunkerley, 1993).

and fill has been shown subsequently to operate in other predominantly sandy (Foley, 1978) (Figure 13.18) and gravelly (Schick, Lekach and Hassan, 1987) streams.

Leopold *et al.* were able to provide a loosely deterministic relation between a hydraulic parameter – unit discharge – and the depth of scour (Figure 13.17). This has been corroborated from observations made in a small sand-bed tributary of Walnut Gulch, southern Arizona, though, here, the relation establishes itself only at flows that have unit stream power >10–15 W/m^2 (Powell *et al.*, 2007) (Figure 13.19(b)). Foley (1978) has gone further in his explanation by suggesting that the depth of scour is determined by the amplitude of migrating antidunes that form under upper flow regime conditions. Actual observations by Reid and Frostick (1987) indicate the short-lived nature of supercritical flow during a flash flood. This suggests that, if scour were indeed due to antidune migration, it occurs in a short time interval of perhaps no more than several tens of minutes, and only near peak flow. In fact, in some way corroborating the restricted nature of this interval, the values of the Manning roughness coefficient that have been derived for the major part of flash-flood duration are consistent with plane bed conditions rather than with more complex bedforms (Nordin, 1963; Reid and Frostick, 1987). It seems likely, therefore, that if antidunes were to be implicated in the process, scour is achieved much quicker than subsequent fill. However, observations of standing waves suggest that the underlying

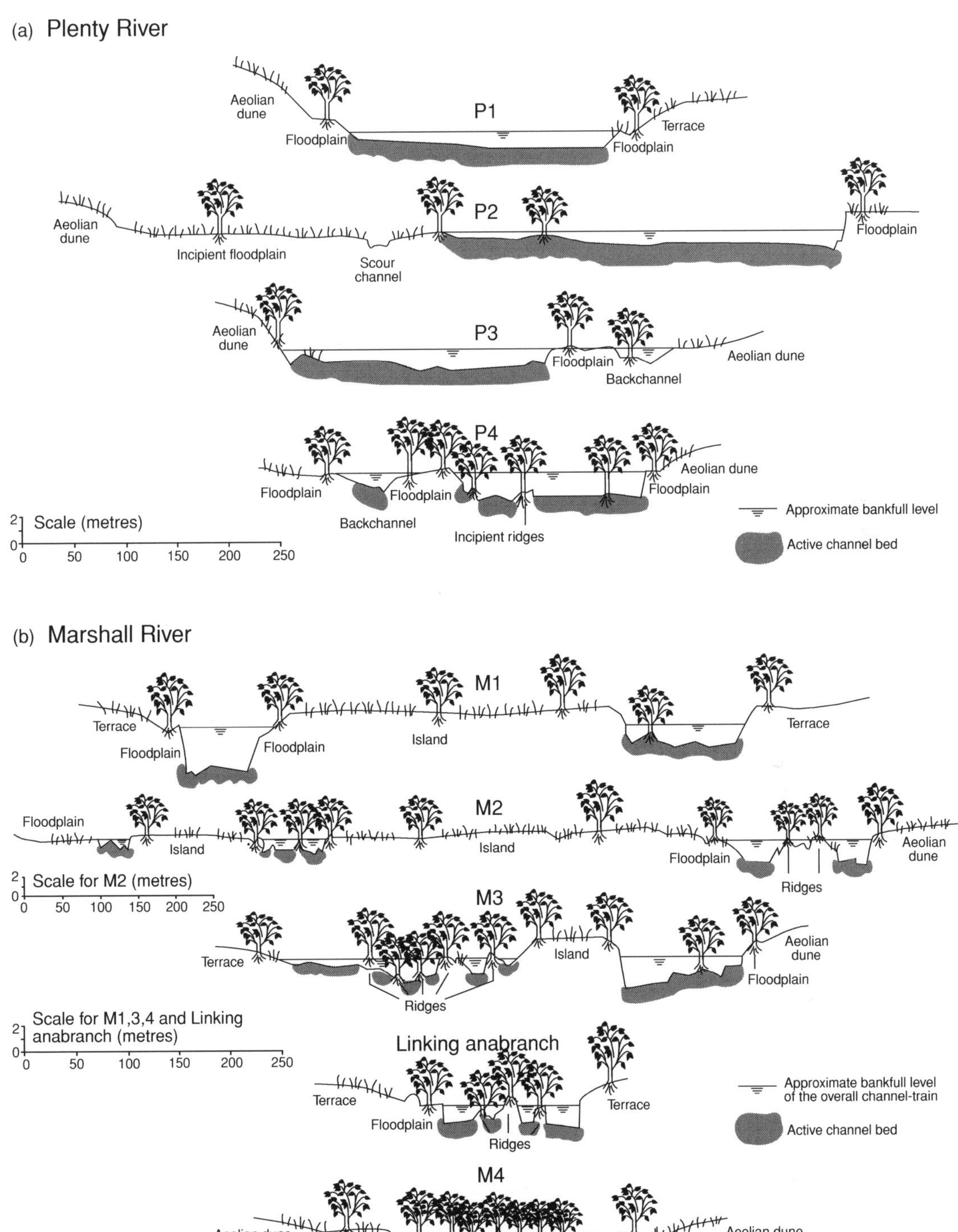

Figure 13.15 Cross-sections of two ephemeral channels that run subparallel courses less than 2.5 km apart on the Northern Plains of central Australia. (a) The Plenty River, which is single-thread, largely devoid of in-channel trees and has a bed material of sand. (b) The Marshall River, which has multiple, narrow anabranches separated by tree-induced ridges and islands and a bed material of coarse sand and granules (after Tooth and Nanson, 2004).

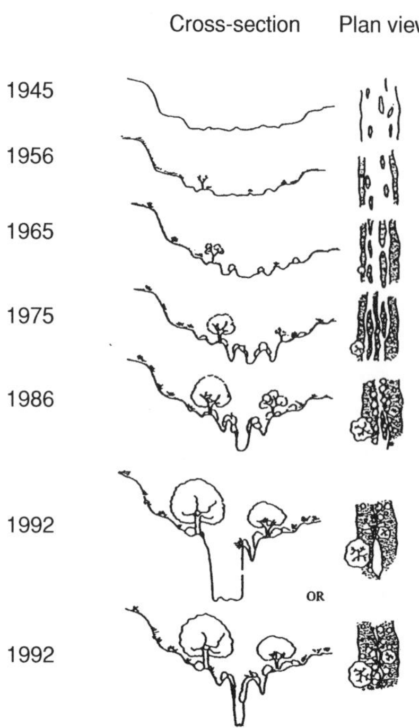

Figure 13.16 Dated stages in the development of the channel of Nahal Hoga, Israel coastal plain, following the imposition of soil and water conservation measures in the water catchment. The cross-sections are vertically exaggerated (after Rozin and Schick, 1996).

antidunes are spatially discrete and extremely transient. The channel-wide scour and fill patterns indicated by the scour chains of Leopold, Emmett and Myrick (1966) may require a different explanation, perhaps involving plane beds in the upper flow regime.

Powell *et al.* (2006) explore the possibility that depth of scour (and subsequent fill) in a sand-bed tributary of Walnut Gulch varies downstream in a fashion that has pseudo-regularity (Figure 13.19(a)). For the larger floods of their record, lozenge-shaped zones of maximum scour appear to have a longitudinal spacing that is about seven times the channel width. This is tentatively attributed to either secondary currents associated with helical flow cells or large-scale, turbulence-induced, roller eddies that generate successive zones of flow acceleration and deceleration. A spatial variation of scour and fill is also noted by Laronne and Shlomi (2007) in distal, braided, gravel-bed ephemeral channels of the Dead Sea trough. Here, the disturbance of the bed and subsequent deposition are shown to scale with flood magnitude. However, in the alluvial stratigraphy of channels in the Arava, to the south, and for longer intervals (10^3 years), Lekach, Amit and Grodek (2009) detect a fluvio-pedogentic unit that they suggest reflects long-term sediment preservation and, hence, for sediments above, the local depth of maximum scour that has occurred since genesis of the unit. From a number of channels, they develop a regional scour envelope curve (analogous to a regional flood envelope curve) to show that maximum scour increases with catchment size up to 10 km^2, beyond which it maintains an asymptotic value of c. 0.9 m, implying that this represents the limit of the scour process, regardless of palaeoflood magnitude.

13.4.2 Sediment transport in suspension

Irrespective of the point in a flood when bed scour takes place, the process provides an easily erodible source of sediment, especially in sand-bed ephemerals. Added to this is the material that is brought to the stream channel by overland flow. Together, these produce sediment concentrations that have inspired local farmers to describe the Colorado River and some of its tributaries as 'too thin to plough and too thick to drink!' (Beverage and Culbertson, 1964).

From the few measurements of suspended sediment that have been taken for flash floods (e.g. Nordin, 1963; Lekach and Schick, 1982; Reid and Frostick, 1987; Sutherland and Bryan, 1990; Alexandrov *et al.*, 2009), it can be shown that the concentration rises along with water discharge, as with perennial streams for which there is, of course, vastly greater information. However, this is the only similarity. The constant a in the relationship $C = aQ^b$, where C is sediment concentration (measured in mg/L) and Q is water discharge (measured in m^3/s), is anywhere between 6 and 4500 times higher in ephemeral streams. This reflects the fact that, even at low flows, sediment concentrations are often more than five times as high as at times of *high* flow in perennial systems. On the other hand, the exponent b is usually less than unity for ephemeral streams, but always greater than unity for perennial rivers. This suggests that perennial systems are more responsive to changes in discharge, but only in relative terms, since the range of concentrations that might be expected in an ephemeral stream will be anywhere from 35 to 1700 times higher than that expected in perennial systems (Frostick, Reid and Layman, 1983).

In a 15-year study of the Nahal Eshtemoa, of the northern Negev, Alexandrov, Laronne and Reid (2007) clearly demonstrate not only the high concentrations reached by suspended sediment but also that these are differentiated according to the storm type that generates runoff (Figure 13.20). Winter frontal storms with low 5-minute

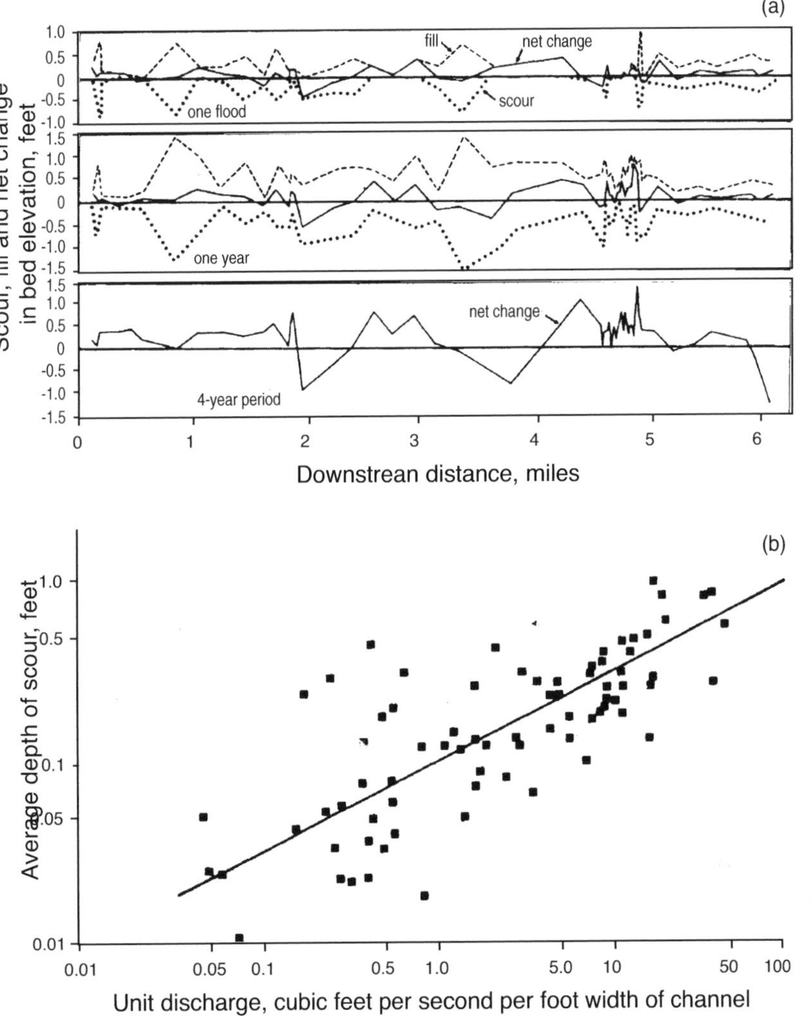

Figure 13.17 (a) Downstream pattern of bed scour and fill for different periods in the Arroyo de Los Frijoles, New Mexico. (b) Average depth of scour as a function of flood peak discharge in the Arroyo de Los Frijoles (after Leopold, Emmett and Myrick, 1966).

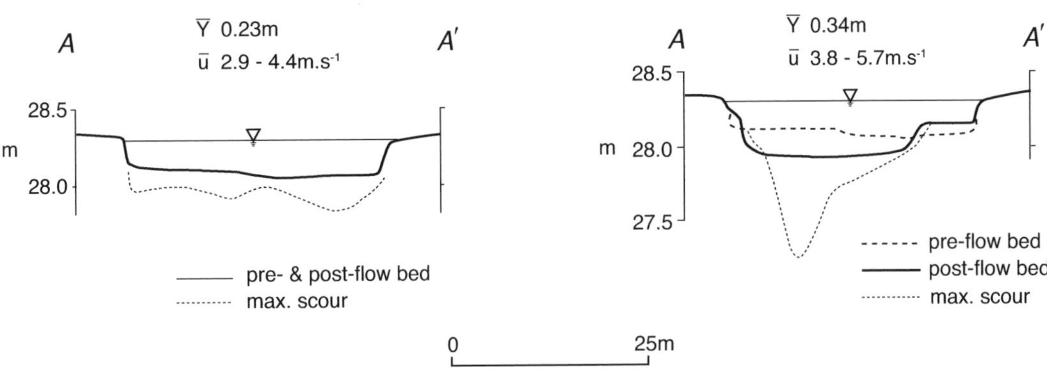

Figure 13.18 The patterns of scour and fill at one cross-section of Quatal Creek, California, during two flash floods of different magnitude. Y is the calculated average flow depth; u is the calculated average flow velocity (after Foley, 1978).

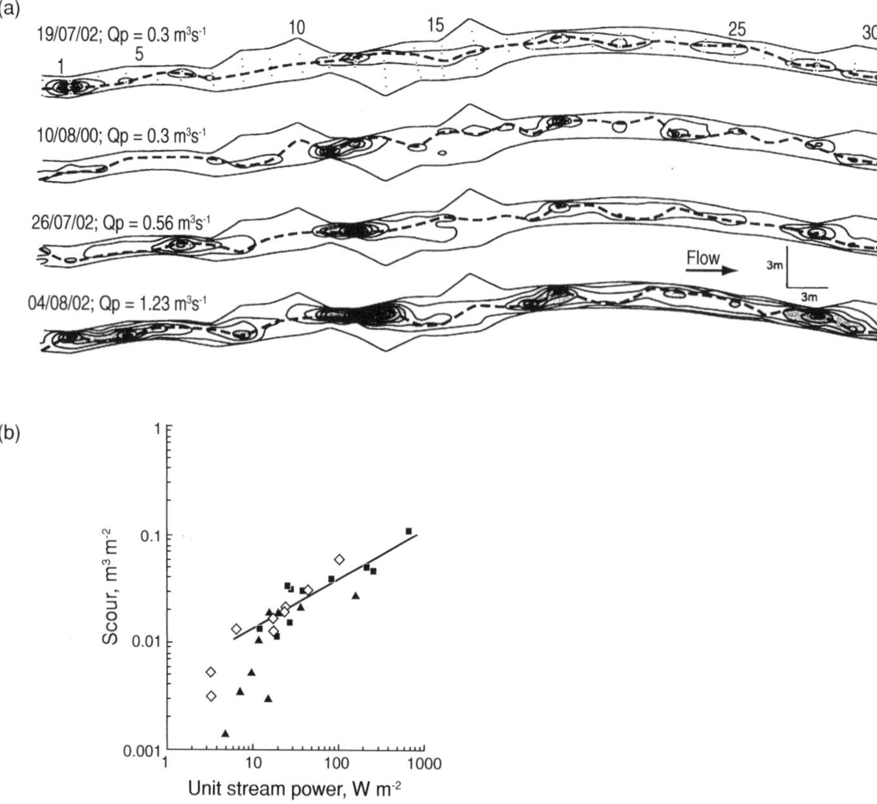

Figure 13.19 (a) Isopleths of scour along a reach of a sand-bed, headwater tributary of Walnut Gulch, southern Arizona, during four flash floods of varying magnitude showing pseudo-regularity of maximum scour with a downstream spacing of about seven times the channel width (after Powell *et al.*, 2006). (b) Specific scour volume as a function of unit stream power for reaches of three headwater channels of Walnut Gulch. The least-squares curve refers to bivariates where specific scour volume $\geq 0.1\ m^3/m^2$ (after Powell *et al.*, 2007).

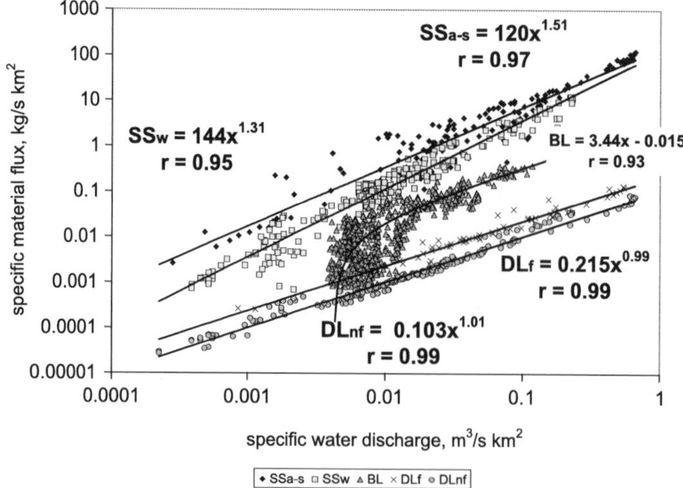

Figure 13.20 Specific flux of suspended sediment (SS) load, dissolved load (DL) and bedload (BL) as a function of contemporary specific water discharge in the Nahal Eshtemoa, northern Negev, Israel. Subscripts: a-s, autumn–spring; w, winter; f, first event of rain season; nf, nonfirst events of rain season (after Alexandrov *et al.*, 2009).

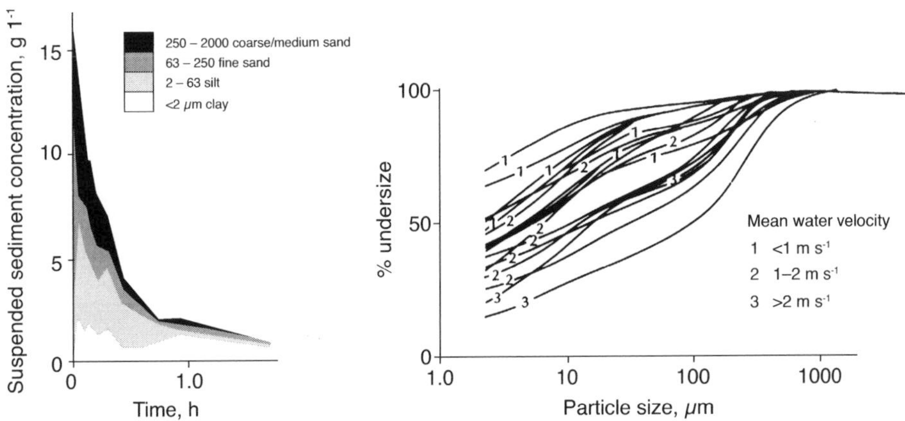

Figure 13.21 (a) Sedigraph of suspended sediment by grain size class during one flood in the Il Kimere, northern Kenya. (b) Size distribution of suspended sediment sampled at various flow velocities during one flood in the Il Kimere (after Frostick, Reid and Layman, 1983; Reid and Frostick, 1987).

rainfall intensities, commonly of 2 mm/h, produce concentrations that range, typically, up to 25 000 mg/L. This is in stark contrast to convectively enhanced, cellular storms of autumn and spring, which have 5-minute intensities typically of 20–40 mm/h, with maxima approaching 160 mm/h. In these cases, the average concentration of suspended sediment is 67 300 mg/L and maxima reach nearly 300 000 mg/L.

There is hysteresis in the relation between suspended sediment concentration and flow, as in perennial rivers. However, from the limited number of detailed studies that yield sufficient information to identify the intraflood pattern, the phenomenon appears to vary, depending on the availability of sediment by size. For a sand-bed ephemeral, the Il Kimere of northern Kenya, it appears to be almost nonexistent. This might be because particles of all sizes are available for transport and transport is dictated by changes in the hydraulic environment rather than by either limitations on sediment supply or the superfluity of finer material of clay and silt size. In this case, it has been shown that the size distribution of suspended sediment varies systematically with water velocity or discharge (Reid and Frostick, 1987) (Figure 13.21). In contrast, the Nahal Eshtemoa, a gravel-bed ephemeral set in a landscape clothed with Holocene loess, shows strong hysteresis between suspended sediment concentration and water discharge. Again, as with levels of concentration, the sense in the direction of hysteresis often depends on storm type. Convectively enhanced storms show a clockwise relation, reflecting the flushing of sediment from proximal hillslopes and the channel bed, while low-intensity frontal storms show an anticlockwise relation, reflecting gentler processes of splash and runoff generation and delayed arrival in the stream of further-travelled material originating on distal hillslopes (Alexandrov, Laronne and Reid, 2003) (Figure 13.22).

13.4.3 Sediment transport along the stream bed

In keeping with river sediment studies in all environments, much less is known about bedload transport in desert streams than about transport in suspension. Ironically, in relative terms there is probably more information about bedload in desert streams because of recent field monitoring programmes, mainly in Israel. However, this 'anomaly' only reflects the paucity of studies of any sort of fluvial sediment transport in the world's drylands!

Leopold, Emmett and Myrick (1966) were able to show that large clasts of pebble to cobble size can be moved by as much as 3 km during a single flood event by overpassing a predominantly sand-bed channel. Schick, Lekach and Hassan (1987) and Hassan (1990b) have extended this work to cover gravel-bed streams. They have traced tagged clasts as they move downstream in the Nahel Hebron in the northern Negev. The use of a two-coil metal detector allowed them to locate buried clasts as well as those exposed on the surface of the stream bed. As a result, they have been able to highlight two important processes: for the first time, scour and fill has been shown to operate in a gravel-bed channel; besides this, there is a clear pattern of exchange between buried and exposed clasts, flood by flood, which Schick *et al.* relate to the depth of scour in each event (Figure 13.23). In addition, using some of the same data, Church and Hassan (1992) have indicated the effects of clast interlock and exposure on transport distance as conditioned by clast size and flood magnitude, suggesting that bed structure appears more

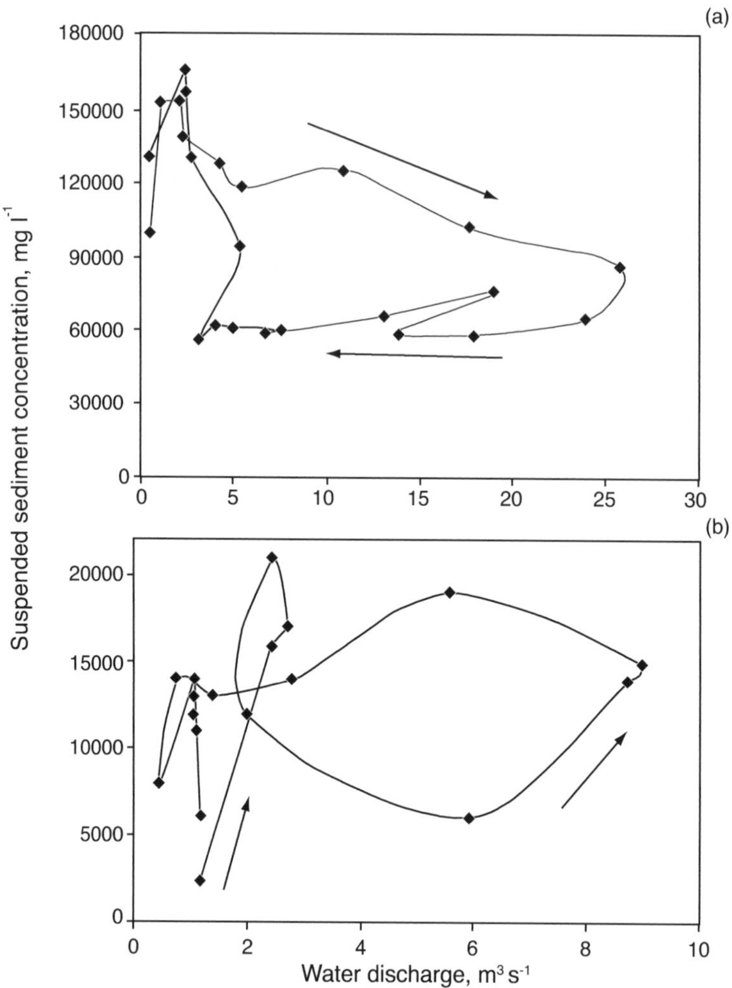

Figure 13.22 Suspended sediment concentration versus water discharge for a sequence of samples taken during two flash floods in the Nahal Eshtemoa, northern Negev: (a) 18 October 1997 event arising from high-intensity, convectively enhanced, cellular storm, showing clockwise hysteresis as a result of sediment flushing; (b) 5 November 1994 event arising from low-intensity, frontal storm, showing counterclockwise hysteresis in both waves of the flood hydrograph (after Alexandrov, Laronne and Reid, 2003).

important in controlling downstream displacement during moderate floods in which values of excess shear stress are comparatively small. In this context, of interest is a comparative study of perennial and ephemeral channels (Wittenberg *et al.*, 2007), which finds the spatial density of pebble/cobble clusters (implicated in Church and Hassan's 'locked' category of surface clasts) to be lower in ephemerals, where abundant sediment supply and lack of low-flow winnowing ensures a smaller grain sorting index and, therefore, less tendency to form grain clusters. The implication is that bed structure is less of a restraint on entrainment, a factor contributing to high bedload flux in ephemerals.

These studies provide valuable information about bed material translocation, but the data are of a 'black box' nature in that they report displacement of clasts after the event. Indeed, in some cases (e.g. Schick and Lekach, 1987), even the flood hydrograph has been reconstructed from slackwater deposit evidence of maximum water-stage rather than recorded information. Live-bed information in ephemeral streams is inevitably rare given the dangers of wading in flash floods with portable bedload samplers, let alone the fact that the chance of being on-site during an event is extremely small (Lekach and Schick, 1983).

This absence of information has been remedied in recent years by the installation of Birkbeck-type bedload samplers in the Nahal Yatir and Nahal Eshtemoa, two gravel-bed ephemeral streams in the northern Negev Desert, Israel (Laronne *et al.*, 1992, 2003). These have

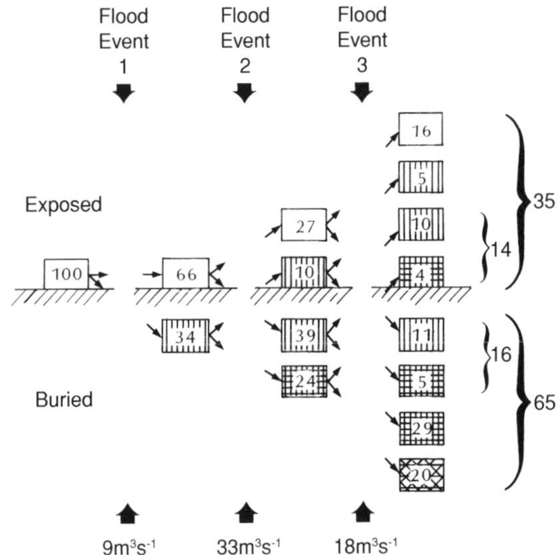

Figure 13.23 Pattern of burial and exhumation of tagged gravel clasts during downstream transport by three successive floods of increasing magnitude in the Nahal Hebron, northern Negev, Israel. The tagged clasts were seeded initially on the surface of the gravel bed (after Schick, Lekach and Hassan, 1987).

revealed patterns of behaviour that, as with other characteristics, set desert streams apart from those of humid environments. First, bedload transport rates are much higher over the same range of shear stress and when allowance is made for differences in bed material grain size. The differences might be as high as a millionfold at the threshold of entrainment and tenfold at moderate levels of shear stress (Figure 13.24) (Laronne and Reid, 1993; Reid and Laronne, 1995). Second, the relation between bedload transport rate and flow appears to be simpler and, therefore, more predictable (Figure 13.25) (Reid, Powell and Laronne, 1996). These differences have been attributed to a lack of armour development, which is thought to arise because the sparsely vegetated slopes of the desert are capable of supplying large quantities of sediment. Indeed, Pickup (1991) draws attention to the effect of reducing vegetation cover through drought or grazing in the Todd River basin, central Australia, pointing to a tenfold increase in sediment transport over those rates measured during marginally wetter periods when the sward is better developed. The ready supply of sediment to the channel system counteracts the tendency for selective entrainment of finer clasts on the channel

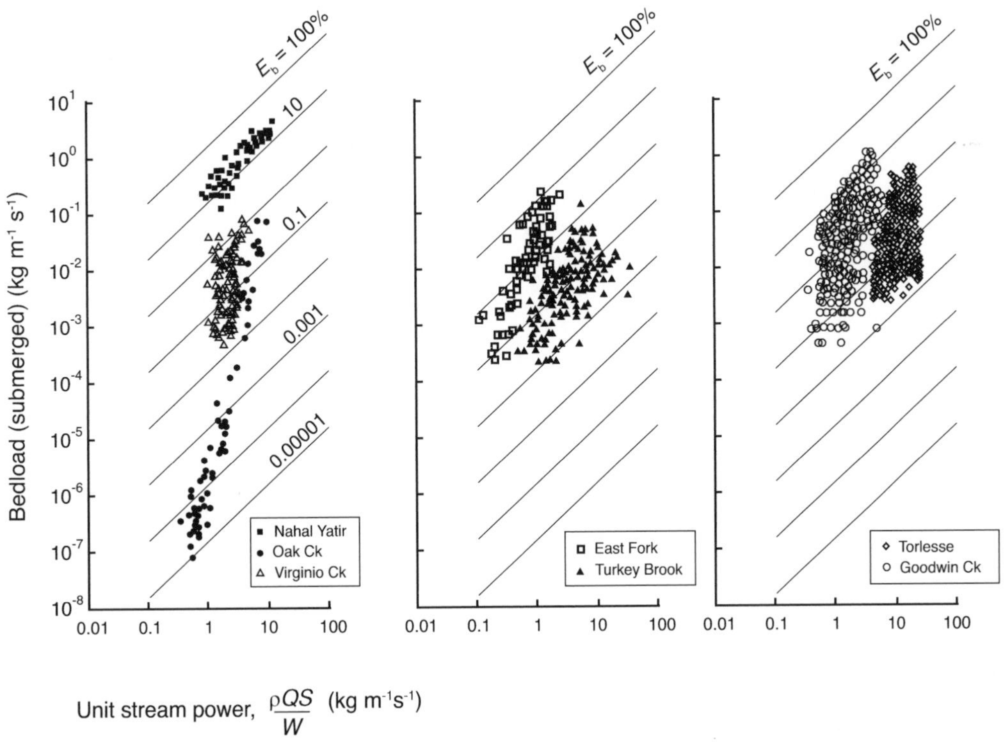

Figure 13.24 Comparison of bedload transport rates measured by slot samplers in the ephemeral Nahal Yatir, Israel, the seasonal Goodwin Creek, Mississippi, the perennial East Fork River, Wyoming, Oak Creek, Oregon, Torlesse Stream, New Zealand, Turkey Brook, England, and Virginio Creek, Tuscany. East Fork River has a gravelly sand bed. The other streams have gravel beds. E_b is Bagnold's efficiency parameter (after Reid and Laronne, 1995).

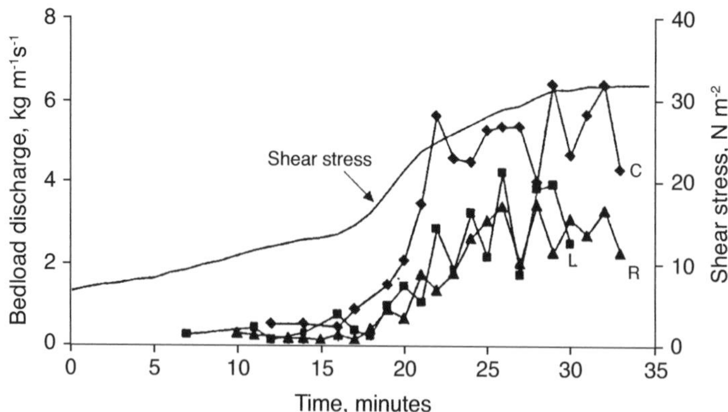

Figure 13.25 Unit bedload flux and channel average boundary shear stress during the rising limb of a flash flood of 8 February 1991 in the gravel bed at Nahal Yatir, northern Negev, Israel, showing the sympathetic response of bedload to increasing hydraulic stress and the impact of sidewall drag on sediment flux. Birkbeck-type samplers were located at: C, channel centreline; L, leftward half-channel; R, rightward half-channel (after Powell *et al.*, 1998).

bed, so reducing armour development and permitting high bedload transport rates through an absence of protection (Dietrich *et al.*, 1989; Laronne *et al.*, 1994).

The virtually unique bedload data sets of the Nahal Yatir and Nahal Eshtemoa have also provided conclusive evidence that settles a controversy that has exercised river scientists for several decades. It has long been known that sands and granules may be moved at a comparatively low transport stage that just exceeds the shear stress associated with initial entrainment of bedload. There have been flume experiments that show that increasing shear stress brings progressively larger grains into motion until all grain sizes of the bed material are mobile. One data set, for a perennial stream, has provided a hint that this condition of 'equal mobility' is approached as discharge increases, but, unfortunately, the maximum measured transport stage fell short of achieving this condition. Using the Eshtemoa bedload flux data, Powell, Reid and Laronne (2001) have shown that equal mobility of all grain sizes is achieved when boundary shear stresses rise above four-and-a-half times that associated with initial entrainment. The coarseness of bedload sediment transported by each flash flood is directly related to the duration of flood flow that exceeds this threshold. This has implications for the coarseness of sediment transported into water reservoirs, all of which provide a vital resource in drylands and all of which have comparatively small half-lives because of sedimentation (Syvitski, 2003). The spectrum of flash-flood magnitude therefore controls the growth and nature of reservoir deltas through its impact on the calibre of the bedload. It also carries implications for the grain-size distribution of ancient desert fluvial sediments. Deposits laid down predominantly by small- and medium-sized floods of modest duration will be finer than those laid down by a spectrum of floods that includes larger flow durations and flow magnitudes that ensure some periods when there is equal mobility of all the grain sizes available for transport.

Because of its relative importance, bedload cannot be ignored as a component of sediment yield in arid environments in the same way that it often and conveniently is in more humid environments.

13.5 Desert river deposits

It is perhaps curious that much of what has been written about the *dynamic* behaviour of desert streams is speculative and comes from deductions about *static* fluvial deposits. Perhaps even more curious is that there is more written about the dynamics of desert sediments that may be as much as 400 million years old (see, for example, Allen, 1964; Schumm, 1968a; Tunbridge, 1984; Olsen, 1987; Frostick *et al.*, 1988; Frostick, Linsey and Reid, 1992; North and Taylor, 1996) than about those of modern streams, for which there is more information about the likely character of flows that were responsible, the nature of the channel and so on. This might reflect Schumm's (1968b) interesting observation that, prior to development of higher land plants during the Devonian, all fluvial sediments were derived in landscapes akin to deserts, regardless of how much rainfall they received and where they were in the world. There are, however, a few studies of modern flash-flood deposits. They are often associated with high-magnitude events that have achieved local notoriety (see, for example, McKee, Crosby and Berryhill,

Table 13.1 Attributes of endogenous desert ephemeral streams and rivers.

Attribute	*Upland, headwater channels*[a]	*Lowland, distal channels*[b]
System size	Small to medium, regional; $<1^0$–10^3 km^2	Large to subcontinental, 10^5 km^2
Terrain	Mountainous, rifted or block-faulted	Peneplaned and cratonic
Channel slope	Moderate to high; 10^{-3}–10^{-2}	Low; 10^{-4}
Channel type	Single-thread (often straight), braided	Anastomosed, braided, single-thread
Flow regime	Ephemeral; predominantly upper flow regime	Intermittent and ephemeral; predominantly lower flow regime
Flood timebase	10^0–10^2 h	10^2–10^3 h
Sediment concentration:		
(a) Suspended load	High to very high; 15–285 g/L	Low to moderate; 0.5–12 g/L
(b) Bedload	High; maximum recorded: 5 g/L (7 kg/m s at a boundary shear stress of 36 Pa)	Unknown
Bedforms in sand	Plane bed (predominant); dunes (occasional)	Ripples (20–30 %); dunes (25–45 %); plane bed (7–15 %)

[a]Alexandrov *et al.* (2009); Bull (1979); Dunkerley (1992); Frostick, Reid and Layman (1983); Graf (1983); Hassan (1990b); Karcz (1972); Laronne and Reid (1993); Leopold, Emmett and Myrick (1966); Meerovitch, Laronne and Reid (1998); Powell, Reid and Laronne (2001); Renard and Keppel (1966); Schick, Lekach and Hassan (1987); Schumm (1961); Sharma and Murthy (1994); Sneh (1983); Thornes (1977).
[b]Bourke and Pickup (1999); Maroulis and Nanson (1996); Nanson and Knighton (1996); Tooth (2000); Tooth and Nanson (2004); Williams (1971).

1967; Williams, 1971; Sneh, 1983), though some are less specific (such as Picard and High, 1973).

Before assessing what lessons might be gleaned from the present in order to interpret ancient desert fluvial sediments, it is essential to remind ourselves that research on modern arid-zone rivers has been conducted largely in two distinct morphogenetic provinces. The first (and probably most visited) might be characterised as headwaters, involving small- to medium-sized catchments. Much of this research has been conducted in or near orogens or rifts of Cenozoic age and comes from the American southwest, the Levant and Maghreb, Iberia, East Africa and northwest India. The second emanates from cratonic lowlands, often involving distal reaches of systems of regional to subcontinental scale. Most of this research has been carried out in Australia, though some is located in southern Africa. Table 13.1 attempts to summarise key attributes of the drainage systems in these two provinces. Nanson, Tooth and Knighton (2002) have issued a strong caution that the riverine processes and landforms of montane and rifted terrains might only be of marginal relevance when characterising and understanding systems in cratonic terrain. It is axiomatic that the reverse is true. It behoves the sedimentologist, therefore, to contextualise the geomorphological arena within which sediments were accumulating, where possible keeping in mind that knowledge about desert stream sediments is, as yet, in its infancy.

Since the hydraulic processes that entrain a clast are universal regardless of the environmental setting, the question arises as to whether or not there are sedimentary characteristics that can be used to distinguish ephemeral from perennial stream deposits. Some doubt has been cast (e.g. North and Taylor, 1996; Martin, 2000). However, we contend that there appear to be at least three attributes that might especially lead a sedimentologist to conclude that an ancient sedimentary sequence was laid down by an ephemeral stream.

13.5.1 Thin beds

There is no quantification that can be used to justify a claim that desert river deposits are composed characteristically of thin beds, i.e. 0.1–0.3 m thick. However, deposits of widely differing age – Devonian (Tunbridge, 1984) (Figure 13.26), Triassic (Frostick *et al.*, 1988; Reid, Linsey and Frostick, 1989) and Holocene (Frostick and Reid, 1986) – undoubtedly have an easily recognisable affinity because of the nature of their bedding (Figure 13.27). There is plenty of evidence of the minor incision that is associated with scour in the nested fill-sets of ancient deposits and the range in bed thickness is consistent with the depth of scour and fill as defined by field experiments in modern sand-bed ephemeral streams (Leopold, Emmett and Myrick, 1966; Powell *et al.*, 2006) (Figures 13.17 and 13.27). Modern gravel-bed ephemerals also exhibit similar thicknesses of scour and fill (Laronne and Shlomi, 2007; Lekach, Amit and Grodek, 2009). Hassan, Marren and Schwartz (2009) (Figure 13.28) describe the tabular morphology and alluvial architecture of

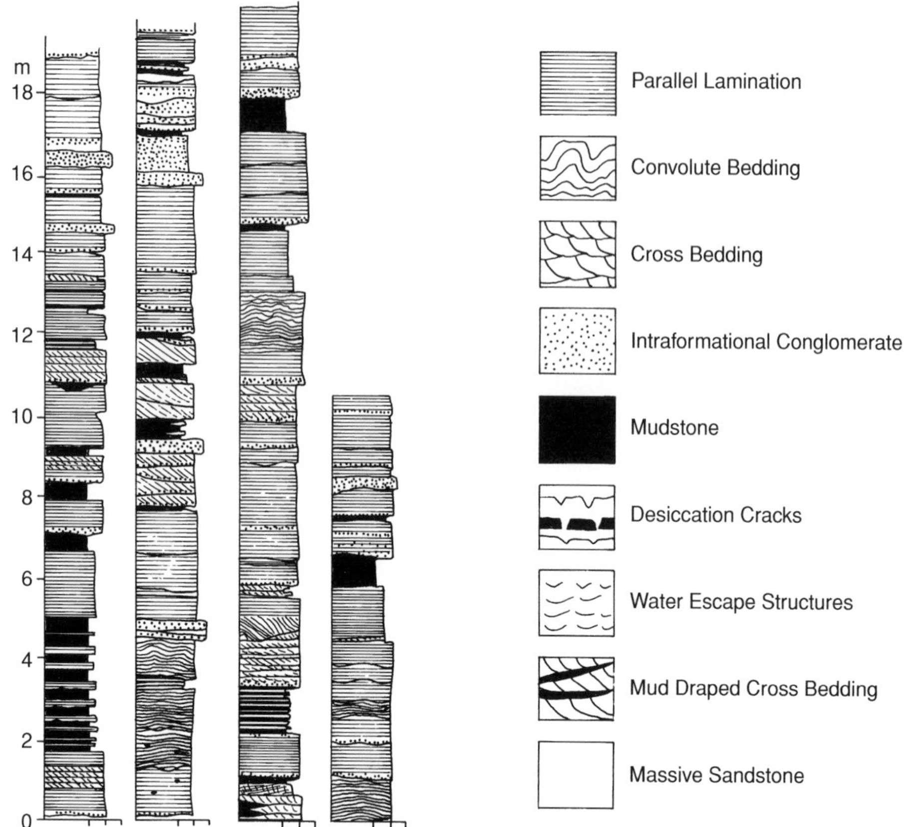

Figure 13.26 Measured sections of flash flood deposits of the Middle Devonian Trentishoe Formation, exposed in the sea cliffs of North Devon, England (after Tunbridge, 1984).

mid-channel bars in the gravel-bed Nahal Zin, which drains to the Dead Sea trough, suggesting a sediment turnover of c. 20 % every four or five years on the basis of flood magnitude frequency and erosional trimming and regrowth. For the same bars, Hassan (2005) gives an average scour depth of 0.1 m for the mean annual flood, 0.3 m for the 1-in-10 year event and 0.4 m for larger floods with a recurrence interval of 1-in-20 years.

13.5.2 Predominance of horizontal lamination in sand beds

The planar surface of many single-thread ephemeral streams has already been remarked on. This surface reflects the apparent ubiquity (Picard and High, 1973) and prevalence (McKee, Crosby and Berryhill, 1967) (Figure 13.29) of horizontal primary structures in ephemeral stream deposits. There is, however, some confusion over the conditions of flow responsible for their formation. This stems largely from the singular paucity of observations of floods in ephemeral channels, but it is more than probable that they represent upper flow regime plane beds. They usually comprise a set of couplets, however, in which laminae of coarser and finer particles alternate (Figure 13.27). An examination of the deposits of floods associated with one storm but in four contiguous river basins of northern Kenya suggests that the separation of particles by size is related to small pulses of sediment-laden water that are superimposed on the main flood wave by staggered contributions from tributary channels. From the very strong correlations in four separate basins, it is conjectured that each pair of laminae of the event channel-fill is associated with a single pulse (Frostick and Reid, 1977, 1979) (Figure 13.30).

Of interest is that gravel-bed ephemerals are similarly characterised by a predominance of subhorizontal bedding (Laronne and Shlomi, 2007). The absence of cross-bedding is attributed by Hassan, Marren and Schwartz (2009) to the shallowness of competent flows (a part-product of wide channels) and the low relief of the bars, which precludes leeside flow separation and gives

Figure 13.27 (a) Flash flood deposits of the Triassic Burghead Beds on the Moray coast, Scotland. (b) Section in a modern ephemeral stream channel-fill, northern Kenya, showing couplets of horizontal laminae. (c) Part-desiccated, thick, mud drape in a modern ephemeral sand bed stream channel, northern Kenya, laid down in the last stages of recent flood flow.

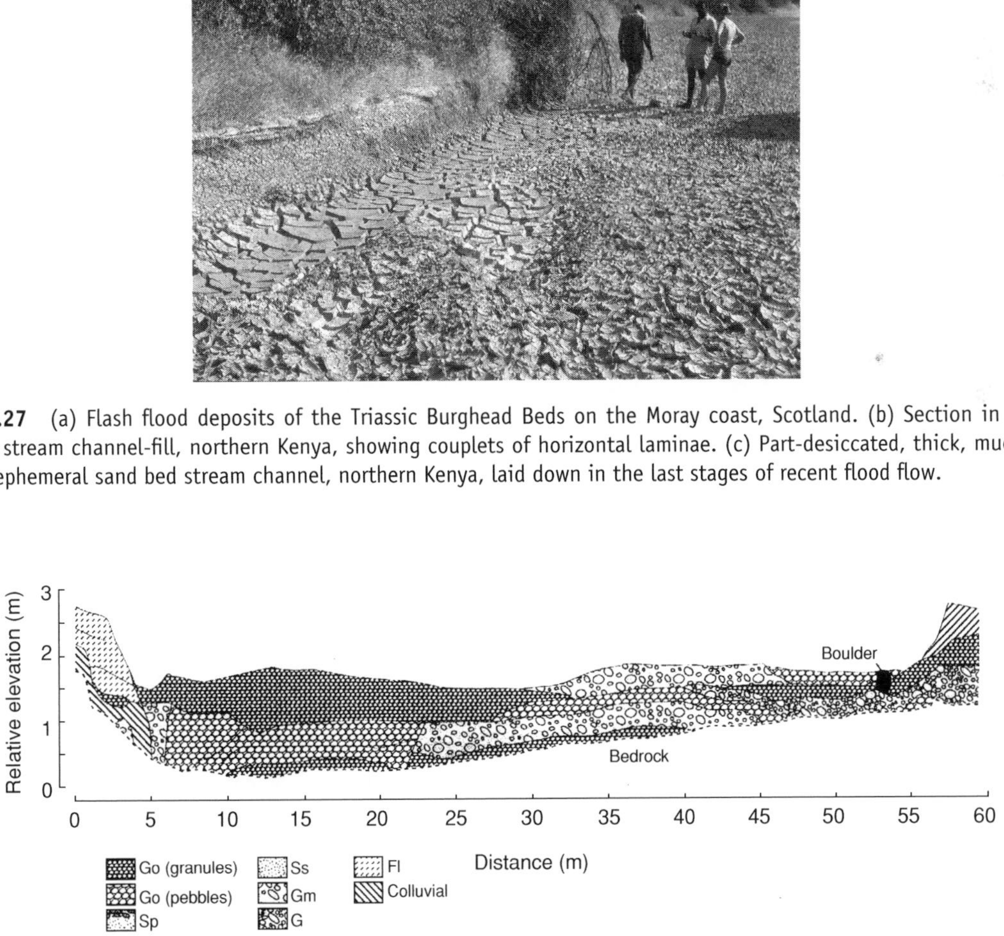

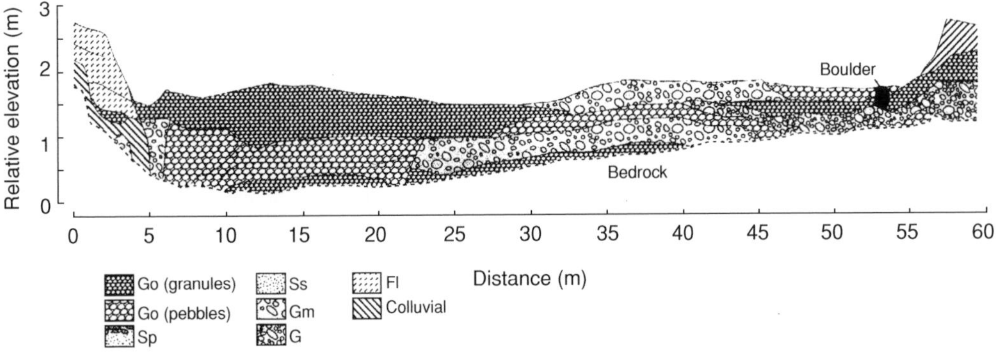

Figure 13.28 Alluvial architecture of a mid-channel bar in the lower reaches of the gravel bed Nahal Zin, Arava, Israel, revealed by cross-channel trenching. Go, openwork gravel; Gm, matrix-supported gravel; G, framework-clast supported gravel; Sp, pebbly medium to coarse sand; Ss, medium to coarse sand; Fl, interbedded fine sand and mud (after Hassan, Marren and Schwartz, 2009).

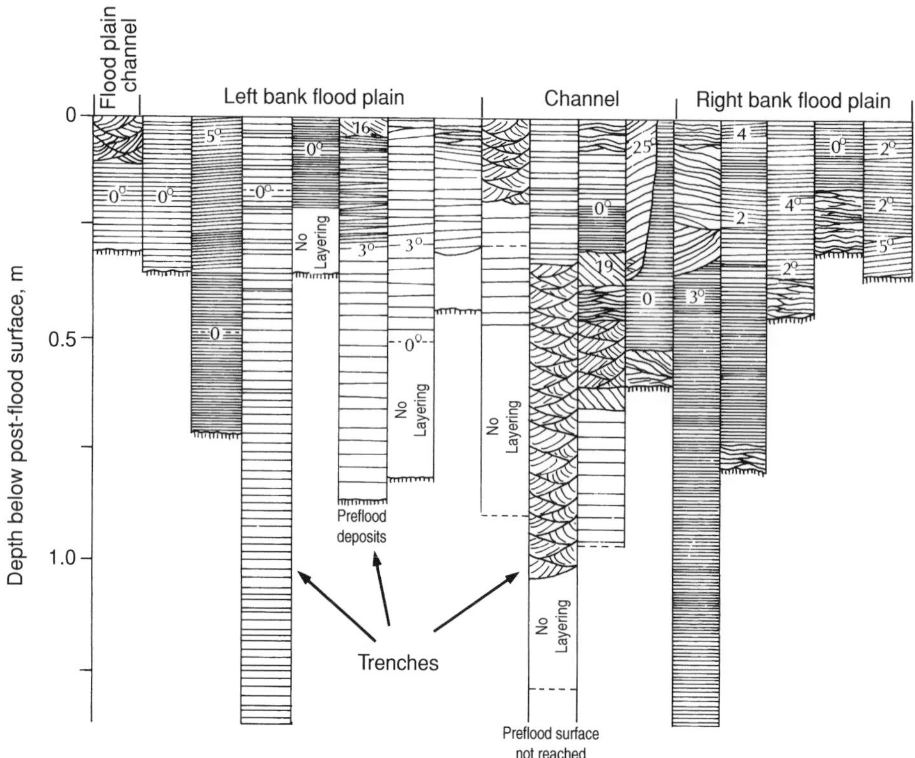

Figure 13.29 Flash-flood deposits of Bijou Creek, Colorado, exposed in trenches cut in the flood plain and channel. Note the predominance of horizontal primary structures (after McKee, Crosby and Berryhill, 1967).

no opportunity for the development of avalanche faces and their associated cross-bedding. As a consequence, deposits are predominantly clast-supported but, because of high bedload flux over the subhorizontal planar surfaces involving a wide range of grain sizes, interstices are matrix-filled.

13.5.3 Mud drapes and mud intraclasts

The fact that the concentration of suspended sediment is often in excess of several thousand milligrams per litre, even in the last stages of flow, that much of this is of clay size and that the channel dries out after each flood so that fines are not winnowed as they might be in a perennial system, all mean that ephemeral stream sequences are often punctuated by comparatively thick (say up to 0.1 m) drapes of clay that have settled from the flow or from stagnant pools in the channel (Tunbridge, 1984; Reid, Lindsey and Frostick, 1989) (Figure 13.27). However, not only are there intact clay drapes but also an abundance of clay intraclasts. These have started as mudcurls lying on the stream bed, have been entrained by a subsequent flood and have ended up incorporated in a superposed bed downstream.

13.6 Conclusions

Rivers play an important role in shaping the Earth's deserts despite the fact that they run for only a vanishingly small fraction of time as one moves towards the hyper-arid core regions. The thick sequences of fluviatile desert sediments that range in age from Precambrian (Williams, 1969), through Devonian (Carruthers, 1987) and Triassic (Steel and Thompson, 1983), to the Plio-Pleistocene (Vondra and Bowen, 1978) are more than adequate testimony.

When they run, desert rivers are more efficient erosional agents than their perennial counterparts (Alnedeij and Diplas, 2005; Reid, Laronne and Powell, 1999). This is due to the fact that material is readily available for

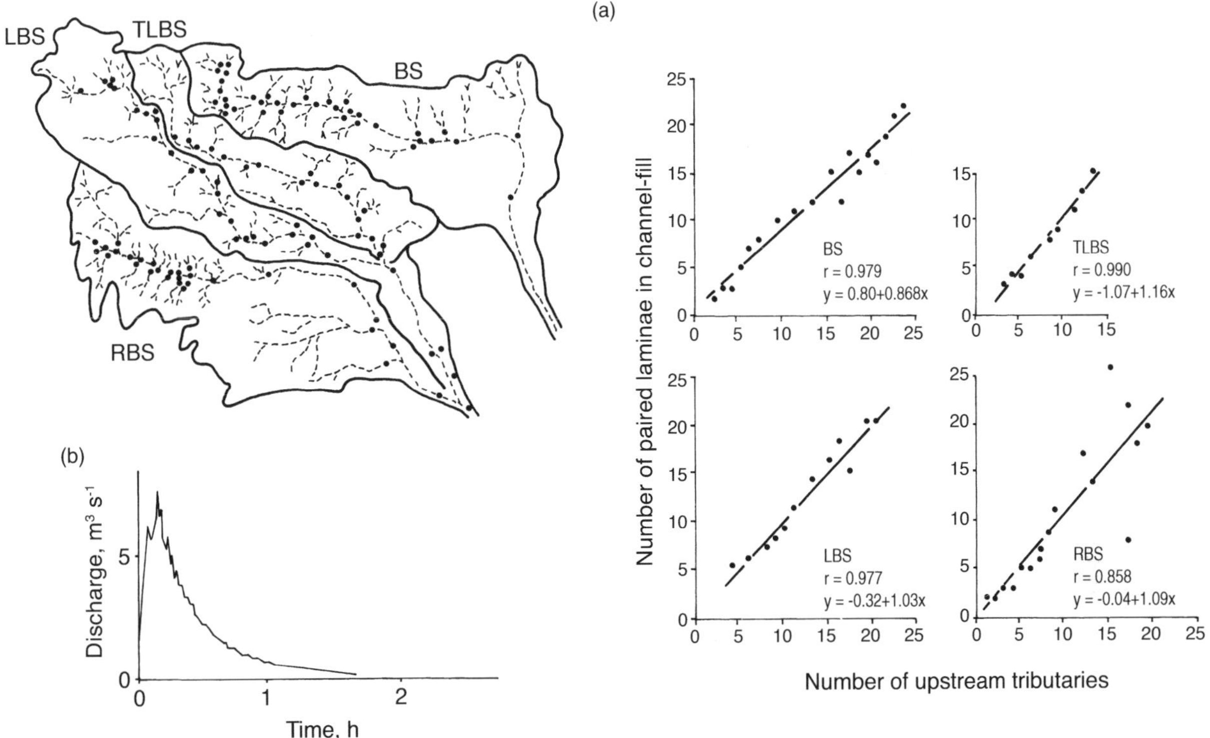

Figure 13.30 (a) Drainage nets and catchments of four unnamed contiguous sand-bed ephemeral streams, northern Kenya, showing bed material sample locations, and the relations between the number of coarse–fine couplets of horizontal laminae and the number of influent tributaries upstream of any sampling trench in the channel-fill of the most recent flash flood (after Frostick and Reid, 1979). (b) Flood hydrograph of the Il Kimere, northern Kenya, showing the discharge pulses that arise from tributary contributions to the trunk stream (after Frostick, Reid and Layman, 1983).

erosion on the poorly vegetated slopes of a water catchment, as well as from the channel bed. Sediment transport is not supply-limited. As a result, suspended sediment concentrations reach record levels, as do bedload transport rates. However, the consequences of such efficiency are far from beneficial. For instance, the life-expectancy of water impoundment structures is shortened considerably as deltas prograde into dam lakes and fine-grained material settles right up to dam walls. Even where the dammed trunk river is perennial, it may be being supplied with sediment by its highly efficient ephemeral tributaries as it passes through a desert region. As a result, the half-life of the reservoirs – so essential to human existence in such regions – will be much reduced. This is the case with the Elephant Butte Reservoir on the Rio Grande (Vanoni, 1975) and with the Tarbella Reservoir on the Indus, where 6 % of the capacity had been lost within five years of its construction (Ackers and Thompson, 1987). In fact, loss of capacity can be so rapid as to call into question the sustainability of this means of eking out water supply in drylands. Thus, for example, the capacity of the Sefid Rud reservoir in northern Iran was reduced from 1.8 to 1.2 billion m^3 in only 16 years after closure of the dam, a loss of 2 % per year (Tolouie, West and Billam, 1993).

However, the spottiness of rainfall in desert regions, together with the highly variable effects of transmission losses, make ephemeral rivers highly unpredictable in comparison with perennial equivalents. As a result, there is less chance of achieving a stochastic model of flows and, therefore, less chance of predicting their nuisance – though recent research in semi-arid environments gives encouragement that predictive tools might eventually emerge. There is a great need for an increase in information about channel form, flood hydrology and sediment transport through an increase in fixed gauging stations, however unrewarding it may seem to document floods that often occur as infrequently as less than once per year.

Box 13.1 A century of channel change in the Gila River, southern Arizona

The ephemeral streams and rivers of the American southwest have been a focus of geomorphological attention for more than a century and research into their changing channel forms has been encapsulated in two seminal texts (Leopold, Wolman and Miller, 1964; Graf, 1988), each of which has been influential in the general development of fluvial geomorphology. Towards the end of the nineteenth century, landowners, government agencies and academics from a number of disciplines became aware that significant changes were occurring in once flat-floored valleys. Arroyos – deep, steep-sided river incisions – were rapidly destroying grazing lands (Cooke and Reeves, 1976).

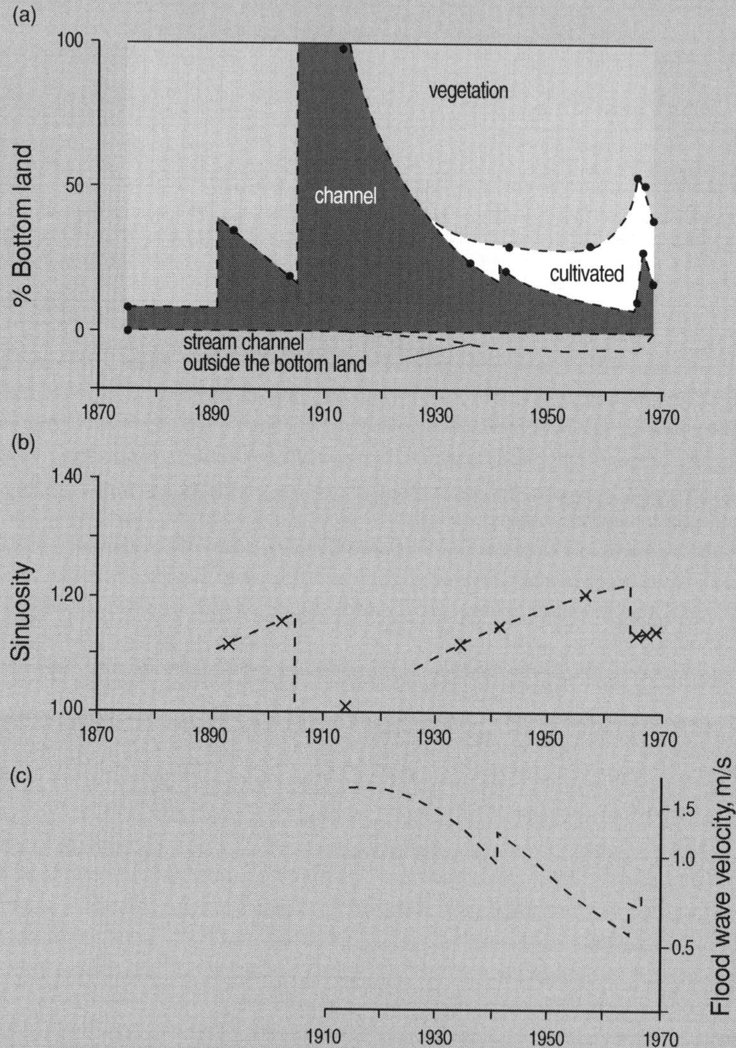

Figure 13.31 Changes in channel morphology and land use along a subreach of the Gila River in Safford Valley, southeast Arizona during the period 1870 to 1970. (a) Dramatic increases in channel width arose principally from floods of 1891, 1905–1917 (eight separate events), 1941, 1965 and 1967. (b) Channel sinuosity showing successive episodes of straightening by damaging floods and the slow subsequent recovery sympathetic with gradual reduction in channel width due to the reestablishment of in-channel bars. (c) Estimates of flood wave celerity showing the impact of increasing flow resistance that arose from progressive reduction in channel width and increasing sinuosity following the damaging flood events of the 1905–1917 period (after Burkham, 1970, 1972).

Another threat to this semi-arid landscape emerged in the early part of the twentieth century. Unlike the arroyo 'problem', around which controversy raged as to its cause – overgrazing or climate change or both – this other threat involved the spread of invasive species such as tamarisk, introduced from Europe in the late nineteenth century (Graf, 1978). The dense woodlands of phreatophytes that developed on bottomlands and river terraces drew heavily on shallow groundwater and, hence, were a threat to the water resources of the increasing human population of the region.

Burkham (1972), reporting as part of the US Geological Survey Gila River Pheatophyte Project, reviewed channel changes along a 70 km reach that runs through Safford Valley, southeast Arizona, examining archives of photographs, documents and surveys that dated back as far as 1846 (Figure 13.31). The Gila catchment above Safford Valley is c. 20 000 km^2, but the valley also receives water from immediately adjacent ephemeral channels with a combined catchment area of c. 9000 km^2. Large floods are associated either with frontal rain on snow in winter or convectively enhanced frontal storms in summer. Burkham notes that, through the end nineteenth century, channel width of the Gila in Safford Valley averaged c. 45 m, the banks being well defined and clothed with cottonwoods, willow and mesquite. Flood-induced widening occurred in 1891 but this was followed by natural restoration. However, a series of eight winter floods during the period 1905–1917 led to dramatic widening of the channel to an average of 488 m, with considerable areas of riparian vegetation uprooted and farmland destroyed. The highest flood peak of this wet period was estimated to have been c. 4255 m^3/s and flows such as this caused dramatic straightening of the channel. Sinuosities of 1.2, typical of the nineteenth century channel, were reduced to unity. Because the flood occurred in winter, Burkham suggests that the sediment load entering Safford Valley was low and speculates that this was a factor in maximising local erosion. Cottonwood trees, uprooted from the river's banks, contributed because their root bowls 'took large chunks of alluvium with them'.

The straightening of the Gila channel meant an increase in gradient of c. 20 % and a commensurate increase in stream power. The floods of 1905–1917 are thought to have set in train sediment waves that progressed downstream within tributaries and, subsequently, within the trunk channel over the ensuing decades. This provided the material that, in the absence of major floods in the period through to 1967, accreted as bars, which grew into islands that were colonised by trees and were eventually attached to the remnant floodplain. Recovery of channel width to nineteenth century levels took c. 50 years (Figure 13.31), though this was accelerated in places by artificial means of channel containment and tree planting. Restoration through sedimentation led to increasing channel sinuosity, values eventually reaching those prior to the 1905 flood. Burkham uses gauge data for the latter part of his study period to indicate the impact of increasing flow resistance on the celerity of the flood wave, showing this to fall to about one-third of the value of the flood-straightened channel. The benefits of groundwater recharge that ensued from greater transmission losses from the slower passage of floods would have been offset in part by the increased probability of flood-plain inundation.

The history of the Gila River serves to highlight both the fragility of ephemeral channels subject to high magnitude floods and the long recovery period characteristic of dryland environments, where flood magnitude frequency is highly skewed (Wolman and Gerson, 1978; Knighton and Nanson, 1997).

References

Ackers, P. and Thompson, G. (1987) Reservoir sedimentation and influence of flushing, in *Sediment Transport in Gravel-Bed Rivers* (eds C.R. Thorne, J.C. Bathurst and R.D. Hey), John Wiley & Sons, Ltd, Chichester, pp. 845–868.

Alexandrov, Y., Laronne, J.B. and Reid, I. (2003) Suspended sediment concentration and its variation with water discharge in a dryland ephemeral channel, northern Negev, Israel. *Journal of Arid Environments*, **53**, 73–84, DOI: 10.1006/jare.2002.1020.

Alexandrov, Y., Laronne, J.B. and Reid, I. (2007) Intra-event and inter-seasonal behaviour of suspended sediment in flash floods of the semi-arid northern Negev, Israel. *Geomorphology*, **85**, 85–97.

Alexandrov, Y., Cohen, H., Laronne, J.B. and Reid, I. (2009) Suspended sediment load, bed load, and dissolved load yields from a semiarid drainage basin: a 15-year study. *Water Resources Research*, **45**, W08408, DOI: 10.1029/2008WR007314.

Allen, J.R.L. (1964) Studies in fluviatile sedimentation: 6. Cyclothems from the lower Old Red Sandstone, Anglo-Welsh Basin. *Sedimentology*, **3**, 89–108.

Alnedeij, J. and Diplas, P. (2005) Bed load sediment transport in ephemeral and perennial gravel bed streams. *EOS Transactions of the American Geophysical Union*, **86**, 429–434.

Bagnold, R.A. (1941) *The Physics of Blown Sand and Desert Dunes*, Methuen, London.

Baker, V.R. (1977) Stream channel response to floods, with examples from central Texas. *Geological Society of America Bulletin*, **88**, 1057–1071.

Benkhaled, A. and Remini, B. (2003) Temporal variability of sediment concentration and hysteresis phenomena in the Wadi Wahrane Basin, Algeria. *Hydrological Sciences Journal*, **48**, 243–255, DOI: 10.1623/hysj.48.2.243.44698.

Ben-Zvi, A., Massoth, S. and Schick, A.P. (1991) Travel time of runoff crests in Israel. *Journal of Hydrology*, **122**, 309–320.

Beverage, J.P. and Culbertson, J.K. (1964) Hyper-concentrations of suspended sediment. *Proceedings of the American Society of Civil Engineers, Journal of Hydraulics Division*, **90**, HY6, 117–128.

Bondurant, D.C. (1951) Sedimentation studies at Conchas Reservoir in New Mexico. *Transactions, American Society of Civil Engineers*, **116**, 1292–1295.

Bonell, M. and Williams, J. (1986) The generation and redistribution of overland flow on a massive oxic soil in a eucalypt woodland within the semi-arid Tropics of north Australia. *Hydrological Processes*, **1**, 31–46.

Bourke, M.C. and Pickup, G. (1999) Fluvial form variability in arid central Australia, in *Varieties of Fluvial Form* (eds A.J. Miller and A. Gupta), John Wiley & Sons, Ltd, Chichester, pp. 249–271.

Brookfield, M.E. and Ahlbrandt, T.S. (eds) (1983) *Eolian Sediments and Processes*, Developments in Sedimentology 38, Elsevier, Amsterdam.

Bull, W.B. (1979) Threshold of critical power in streams. *Geological Society of America Bulletin*, **90**, 453–464.

Bull, L.J. and Kirkby, M.J. (eds) (2002) *Dryland Rivers: Hydrology and Geomorphology of Semi-arid Channels*, John Wiley & Sons, Ltd, Chichester.

Burkham, D.E. (1970) Depletion of streamflow by infiltration in the main channels of the Tucson Basin, southeastern Arizona, United States Geological Survey Water Supply Paper 1939.

Burkham, D.E. (1972) Channel changes of the Gila River in Safford Valley, Arizona 1846–1970, US Geological Survey Professional Paper 655-G.

Butcher, G.C. and Thornes, J.B. (1978) Spatial variability in runoff processes in an ephemeral channel. *Zeitschrift für Geomorphologie, Supplementbund*, **29**, 83–92.

Carruthers, R.A. (1987) Aeolian sedimentation from the Galtymore Formation (Devonian), Ireland, in *Desert Sediments: Ancient and Modern* (eds L.E. Frostick and I. Reid), Geological Society of London Special Publication 35, Blackwell Scientific, Oxford, pp. 251–268.

Church, M. and Hassan, M.A. (1992) Size and distance of travel of unconstrained clasts on a streambed. *Water Resources Research*, **28**, 299–303.

Cooke, R.U. and Reeves, R.W. (1976) *Arroyos and Environmental Change in the American Southwest*, Clarendon Press, Oxford.

Cooke, R.U., Warren, A. and Goudie, A.S. (1993) *Desert Geomorphology*, UCL Press, London.

Dietrich, W.E., Kirchner, J.W., Ikeda, H. and Iseya, F. (1989) Sediment supply and the development of the coarse surface layer in gravel-bedded rivers. *Nature*, **340**, 215–217.

Diskin, M.H. and Lane, L.J. (1972) A basinwide stochastic model of ephemeral stream runoff in southeastern Arizona. *International Association of Scientific Hydrologists Bulletin*, **17**, 61–76.

Dunkerley, D.L. (1992) Channel geometry, bed material, and inferred flow conditions in ephemeral stream systems, Barrier Range, Western N.S.W., Australia. *Hydrological Processes*, **6**, 417–433.

Farquharson, F.A.K., Meigh, J.R. and Sutcliffe, J.V. (1992) Regional flood frequency analysis in arid and semi-arid areas. *Journal of Hydrology*, **138**, 487–501.

Foley, M.G. (1978) Scour and fill in steep, sand-bed ephemeral streams. *Geological Society of America Bulletin*, **89**, 559–570.

Frostick, L.E., Linsey, T.K. and Reid, I. (1992) Tectonic and climatic control of Triassic sedimentation in the Beryl Basin, northern North Sea. *Journal of the Geological Society of London*, **149**, 13–26.

Frostick, L.E. and Reid, I. (1977) The origin of horizontal laminae in ephemeral stream channel-fill. *Sedimentology*, **24**, 1–9.

Frostick, L.E. and Reid, I. (1979) Drainage-net control of sedimentary parameters in sand-bed ephemeral streams, in *Geographical Approaches to Fluvial Processes* (eds A.F. Pitty), Geoabstracts, Norwich, pp. 173–201.

Frostick, L.E. and Reid, I. (1986) Evolution and sedimentary character of lake deltas fed by ephemeral rivers in the Lake Turkana Basin, in *Sedimentation in the African Rifts*, vol. **25** (eds L.E. Frostick, R.W. Renaut, I. Reid and J.J. Tiercelin), Special Publication Geological Society, London, pp. 113–125.

Frostick, L.E. and Reid, I. (eds) (1987) *Desert Sediments: Ancient and Modern*, Geological Society of London Special Publication 35, Blackwell Scientific, Oxford.

Frostick, L.E., Reid, I. and Layman, J.T. (1983) Changing size distribution of suspended sediment in arid-zone flash floods. *Special Publication International Association of Sedimentologists*, **6**, 97–106.

Frostick, L.E., Reid, I., Jarvis, J. and Eardley, H. (1988) Triassic sediments of the Inner Moray Firth, Scotland: early rift deposits. *Journal of the Geological Society of London*, **145**, 235–248.

Gavrilovic, D. (1969) Die Überschwemmungen im Wadi Bardague im Jahr 1968 (Tibesti, Rep. du Tchad). *Zeitschrift für Geomorphologie*, **NF, 14**, 202–18.

Graeme, D. and Dunkerley, D.L. (1993) Hydraulic resistance by the River Red Gum, *Eucalyptus camaldulensis*, in ephemeral desert streams. *Australian Geographical Studies*, **31**, 141–154.

Graf, W.L. (1978) Fluvial adjustments to the spread of tamarisk in the Colorado Plateau region. *Geological Society of America Bulletin*, **89**, 1491–1501.

Graf, W.L. (1979) The development of montane arroyos and gullies. *Earth Surface Processes*, **4**, 1–14.

Graf, W.L. (1983) Flood-related channel change in an arid-region river. *Earth Surface Processes and Landforms*, **8**, 125–139.

Graf, W.L. (1988) *Fluvial Processes in Dryland Rivers*, Springer-Verlag, Berlin.

Hassan, M.A. (1990a) Observations of desert flood bores. *Earth Surface Processes and Landforms*, **15**, 481–485.

Hassan, M.A. (1990b) Scour, fill and burial depth of coarse material in gravel bed streams. *Earth Surface Processes and Landforms*, **15**, 341–356.

Hassan, M.A. (2005) Characteristics of gravel bars in ephemeral streams. *Journal of Sedimentary Research*, **75**, 29–42.

Hassan, M.A., Marren, P.M. and Schwartz, U. (2009) Bar structure in an ephemeral stream. *Sedimentary Geology*, **221**, 57–70, DOI: 10.1016/j.sedgeo.2009.07.012.

Hewlett, J.D. and Hibbert, A.R. (1967) Factors affecting the response of small watersheds to precipitation in humid areas, in *Forest Hydrology* (eds W.E. Sopper and H.W. Lull), Pergamon, Oxford, pp. 267–290.

Hjalmarson, H.W. (1984) Flash flood in Tanque Verde Creek, Tucson, Arizona. *Journal of Hydraulic Engineering, ASCE*, **110**, 1841–1852.

Hubbell, D.S. and Gardner, J.L. (1944) Some edaphic and ecological effects of water spreading on rangelands. *Ecology*, **55**, 27–44.

Knighton, A.D. and Nanson, G.C. (1994) Flow transmission along an arid zone anastomosing river, Cooper Creek, Australia. *Hydrological Processes*, **8**, 137–154.

Knighton, A.D. and Nanson, G.C. (1997) Distinctiveness, diversity and uniqueness in arid zone river systems, in *Arid Zone Geomorphology*, 2nd edn (ed. D.S.G. Thomas), John Wiley & Sons, Ltd, Chichester, pp. 185–203.

Lane, L.J., Diskin, M.H. and Renard, K.G. (1971) Input–output relationships for an ephemeral stream channel system. *Journal of Hydrology*, **13**, 22–40.

Langbein, W.B. and Schumm, S.A. (1958) Yield of sediment in relation to mean annual precipitation. *EOS Transactions of the American Geophysical Union*, **39**, 1076–1084.

Laronne, J.B. and Reid, I. (1993) Very high bedload sediment transport in desert ephemeral rivers. *Nature*, **366**, 148–150.

Laronne, J.B. and Shlomi, Y. (2007) Depositional character and preservation potential of coarse-grained sediments deposited by flood events in hyper-arid braided channels in the Rift Valley, Arava, Israel. *Sedimentary Geology*, **195**, 21–37, DOI: 10.1016/j.sedgeo.2006.07.008.

Laronne, J.B., Reid, I., Yitshak, Y. and Frostick, L.E. (1992) Recording bedload discharge in a semiarid channel, Nahal Yatir, Israel, in *Erosion and Sediment Monitoring Programmes in River Basins* (eds J. Bogen, D.E. Walling and T.J. Day), International Association of Hydrological Sciences Publication 210, pp. 79–86.

Laronne, J.B., Reid, I., Yitshak, Y. and Frostick, L.E. (1994) The non-layering of gravel streambeds under ephemeral flow regimes. *Journal of Hydrology*, **159**, 353–363.

Laronne, J.B., Alexandrov, Y., Bergman, N. *et al.* (2003) The continuous monitoring of bed load flux in various fluvial environments. Erosion and sediment transport measurement in rivers: technological and methodological advances. *Proceedings of the Oslo Workshop, June 2002. International Association of Hydrological Sciences Publications*, **283**, 134–145.

Lekach, J., Amit, R. and Grodek, T. (2009) Scour envelope curve (SEC), Negev Desert, Israel. *Israel Journal of Earth Sciences*, **57**, 189–197, DOI: 10.1560/IJES.57.3-4.189.

Lekach, J. and Schick, A.P. (1982) Suspended sediment in desert floods in small catchments. *Israel Journal of Earth Sciences*, **31**, 144–56.

Lekach, J. and Schick, A.P. (1983) Evidence for transport of bedload in waves: analysis of fluvial sediment samples in a small upland stream channel. *Catena*, **10**, 267–279.

Leopold, L.B., Emmett, W.W. and Myrick, R.M. (1966) Channel and hillslope processes in a semiarid area of New Mexico, United States Geological Survey Professional Paper 352G.

Leopold, L.B. and Miller, J.P. (1956) Ephemeral streams – hydraulic factors and their relation to the drainage net, United States Geological Survey Professional Paper 282A.

Leopold, L.B., Wolman, M.G. and Miller, J.P. (1964) *Fluvial Processes in Geomorphology*, Freeman & Company, San Francisco.

McGee, W.J. (1897) Sheetflood erosion. *Geological Society of America Bulletin*, **8**, 87–112.

McKee, E.D. (ed.) (1979) A study of global sand seas, United States Geological Survey Professional Paper 1052.

McKee, E.D., Crosby, E.J. and Berryhill, H.L. (1967) Flood deposits of Bijou Creek, Colorado, June (1965). *Journal of Sedimentary Petrology*, **37**, 829–851.

Maroulis, J.C. and Nanson, G.C. (1996) Bedload transport of aggregated muddy alluvium from Cooper Creek, central Australia: a flume study. *Sedimentology*, **43**, 771–790.

Martin, A.J. (2000) Flaser and wavy bedding in ephemeral streams: a modern and an ancient example. *Sedimentary Geology*, **136**, 1–5.

Martínez-Goytre, J., House, P.K. and Baker, V.R. (1994) Spatial variability of small-basin paleoflood magnitudes for a southeastern Arizona mountain range. *Water Resources Research*, **30**, 1491–1501.

Meerovitch, L., Laronne, J.B. and Reid, I. (1998) The variation of water-surface slope and its significance for bedload transport during floods in gravel-bed streams. *Journal of Hydraulic Research*, **36**, 147–157.

Murphey, J.B., Lane, L.J. and Diskin, M.H. (1972) Bed material characteristics and transmission losses in an ephemeral stream. Hydrology and water resources in Arizona and the Southwest, in *Proceedings of 1972 Meeting Arizona Section, American Water Association and the Hydrology*

Section, vol. **2**, Arizona Academy Science, Prescott, Arizona, pp. 455–472.

Nanson, G.C. and Knighton, A.D. (1996) Anabranching rivers: their cause, character and classification. *Earth Surface Processes and Landforms*, **21**, 217–239.

Nanson, G.C., Tooth, S. and Knighton, A.D. (2002) A global perspective on dryland rivers: perceptions, misconceptions and distinctions, in *Dryland Rivers: Hydrology and Geomorphology of Semi-Arid Channels* (eds L.J. Bull and M.J. Kirkby), John Wiley & Sons, Ltd, Chichester, pp. 17–54.

Nordin, C.F. (1963) A preliminary study of sediment transport parameters, Rio Puerco, near Bernardo, New Mexico, United States Geological Survey Professional Paper 462C.

North, C.P. and Taylor, K.S. (1996) Ephemeral-fluvial deposits: integrated outcrop and simulation studies reveal complexity. *Association of American Petroleum Geologists Bulletin*, **80**, 811–830.

Olsen, H. (1987) Ancient ephemeral stream deposits: a local terminal fan model from the Bunter Sandstone Formation (L. Triassic) in the Tønder-3, -4 and -5 wells, Denmark, in *Desert Sediments: Ancient and Modern* (eds L.E. Frostick and I. Reid), Geological Society of London Special Publication 35, Blackwell Scientific, Oxford, pp. 69–86.

Parissopoulos, G.A. and Wheater, H.S. (1992) Experimental and numerical infiltration studies in a wadi stream bed. *Hydrological Sciences Journal*, **37**, 27–37.

Parsons, A.J. and Abrahams, A.D. (eds.) (2009) *Geomorphology of Desert Environments*, 2nd edn, Springer Science+Business Media B.V., DOI: 10.1007/978-1-4020-5719-9.

Picard, M.D. and High, L.R. (1973) *Sedimentary Structures of Ephemeral Streams*, Developments in Sedimentology 17, Elsevier, Amsterdam.

Pickup, G. (1991) Event frequency and landscape stability on the floodplain systems of arid central Australia. *Quaternary Science Reviews*, **10**, 463–473.

Pilgrim, D.H., Chapman, T.G. and Doran, D.G. (1988) Problems of rainfall-runoff modelling in arid and semiarid regions. *Hydrological Sciences Journal*, **33**, 379–400.

Powell, D.M., Reid, I. and Laronne, J.B. (2001) Evolution of bed load grain size distribution with increasing flow strength and the effect of flow duration on the caliber of bed load sediment yield in ephemeral gravel bed rivers. *Water Resources Research*, **37**, 1463–1474.

Powell, D.M., Reid, I., Laronne, J.B. and Frostick, L.E. (1998) Cross stream variability of bedload flux in narrow and wide ephemeral channels during desert flash floods, in *Gravel-Bed Rivers in the Environment* (eds P. Klingemann, R.L. Beschta, P.D. Komar and J.B. Bradley), Water Resources Publications, Highland Ranch, Colorado, pp. 177–196.

Powell, D.M., Brazier, R., Wainright, J. *et al.* (2006) Spatial patterns of scour and fill in dryland sand bed streams. *Water Resources Research*, **42**, W08412, DOI: 10.1029/2005WR004516.

Powell, D.M., Brazier, R., Parsons, A. *et al.* (2007) Sediment transfer and storage in dryland headwater streams. *Geomorphology*, **88**, 152–166, DOI: 10.1016/j.geomorph.2006.11.001.

Reid, I. (2009) River landforms and sediments: evidence of climatic change, in *Geomorphology of Desert Environments*, 2nd edn (eds A.J. Parsons and A.D. Abrahams), Springer Science+Business Media B.V, pp. 695–721, DOI: 10.1007/978-1-4020-5719-9_23.

Reid, I., Best, J.L. and Frostick, L.E. (1989) Floods and flood sediments at river confluences, in *Floods* (eds K. Beven and P.A. Carling), British Geomorphological Research Group Symposium Series, John Wiley & Sons, Ltd, Chichester, pp. 135–150.

Reid, I. and Frostick, L.E. (1986) Slope process, sediment derivation and landform evolution in a rift valley basin, northern Kenya, in *Sedimentation in the African Rifts* (eds L.E. Frostick, R.R. Renaut, I. Reid and J.-J. Tiercelin), Geological Society of London Special Publication 25, Blackwell, Oxford, pp. 99–111.

Reid, I. and Frostick, L.E. (1987) Flow dynamics and suspended sediment properties in arid zone flash floods. *Hydrological Processes*, **1**, 239–253.

Reid, I. and Laronne, J.B. (1995) Bedload sediment transport in an ephemeral stream and a comparison with seasonal and perennial counterparts. *Water Resources Research*, **31**, 773–781.

Reid, I., Laronne, J.B. and Powell, D.M. (1995) The Nahal Yatir bedload database: sediment dynamics in a gravel-bed ephemeral stream. *Earth Surface Processes and Landforms*, **20**, 845–857.

Reid, I., Laronne, J.L. and Powell, D.M. (1998) Flash-flood and bedload dynamics of desert gravel-bed streams. *Hydrological Processes*, **12**, 543–557.

Reid, I., Laronne, J.B. and Powell, D.M. (1999) Impact of major climate change on coarse-grained river sedimentation – a speculative assessment based on measured flux, in *Fluvial Processes and Environmental Change* (eds A.G. Brown and T.A. Quine), John Wiley & Sons, Ltd, Chichester, pp. 105–115.

Reid, I., Linsey, T. and Frostick, L.E. (1989) An automatic bedding descriminator for use with digital wireline logs. *Marine and Petroleum Geology*, **6**, 364–369.

Reid, I., Powell, D.M. and Laronne, J.B. (1996) Prediction of bedload transport by desert flash-floods. *Journal of Hydraulic Engineering, American Society of Civil Engineers*, **122**, 170–173.

Reid, I., Powell, D.M. and Laronne, J.B. (1998) Flood flows, sediment fluxes and reservoir sedimentation in upland desert rivers, in *Hydrology in a Changing Environment, Volume II* (eds H. Wheater and C. Kirby), John Wiley & Sons, Ltd, Chichester, pp. 377–386.

Reid, I., Laronne, J.B., Powell, D.M. and Garcia, C. (1994) Flash floods in desert rivers: studying the unexpected. *EOS, Transactions of the American Geophysical Union*, **75**, 452.

Renard, K.G. and Keppel, R.V. (1966) Hydrographs of ephemeral streams in the Southwest. *Proceedings of the*

American Society of Civil Engineers, Journal of Hydraulics Division, **92**, **HY2**, 33–52.

Renard, K.G. and Lane, L.J. (1975) Sediment yield as related to a stochastic model of ephemeral runoff. Present and prospective technology for predicting sediment yields and sources, in *Proceedings of Sediment-Yield Workshop*, USDA Sedimentation Laboratory, Oxford, MS, 1972, pp. 253–263.

Renard, K.G. and Laursen, E.M. (1975) Dynamic behaviour model of ephemeral streams. *Proceedings of the American Society of Civil Engineers, Journal of Hydraulics Division*, **101**, **HY5**, 511–28.

Rozin, U. and Schick, A.P. (1996) Land use change, conservation measures and stream channel response in Mediterranean/semiarid transition zone: Nahal Hoga, southern Coastal Plain, Israel, in *Erosion and Sediment Yield: Global and Regional Perspectives* (eds D.E. Walling and B. Webb), Proceedings of the Exeter Symposium, July 1996, International Association of Hydrological Sciences Publication 236, pp. 427–444.

Rust, B.R. and Nanson, G.C. (1986) Contemporary and palaeochannel patterns and the late Quaternary stratigraphy of Cooper Creek, southwest Queensland, Australia. *Earth Surface Processes and Landforms*, **11**, 581–590.

Schick, A.P. (1970) Desert floods, in *Symposium on the Results of Research on Representative Experimental Basins*, International Association Scientific Hydrologists/Unesco, pp. 478–493.

Schick, A.P. (1988) Hydrologic aspects of floods in extreme arid environments, in *Flood Geomorphology* (eds V.R. Baker, R.C. Kochel and P.C. Patton), John Wiley & Sons, Inc., New York, pp. 189–203.

Schick, A.P. and Lekach, J. (1987) A high magnitude flood in the Sinai Desert, in *Catastrophic Flooding* (eds L. Mayer and D. Nash), Allen and Unwin, Boston, pp. 381–410.

Schick, A.P. and Lekach, J. (1993) An evaluation of two ten-year sediment budgets, Nahal Yael, Israel. *Physical Geography*, **14**, 225–238.

Schick, A.P., Lekach, J. and Hassan, M.A. (1987) Vertical exchange of coarse bedload in desert streams, in *Desert Sediments: Ancient and Modern* (eds L.E. Frostick and I. Reid), Geological Society of London Special Publication 35, Blackwell Scientific, Oxford, pp. 7–16.

Schumm, S.A. (1961) Effect of sediment characteristics on erosion and deposition in ephemeral stream channels, United States Geological Survey Professional Paper 352C.

Schumm, S.A. (1968a) River adjustment to altered hydrologic regimen – Murrumbidgee River and paleochannels, Australia, United States Geological Survey Professional Paper 598.

Schumm, S.A. (1968b) Speculations concerning palaeohydrologic controls of terrestrial sedimentation. *Geological Society of America Bulletin*, **79**, 1573–1588.

Sharma, K.D. and Murthy, J.S.R. (1994) Modelling sediment transport in stream channels in the arid zone of India. *Hydrological Processes*, **8**, 567–572.

Sharma, K.D. and Murthy, J.S.R. (1995) Hydrological routing of flow in arid ephemeral channels. *Journal of Hydraulic Engineering, ASCE*, **121**, 466–471.

Sharon, D. (1972) The spottiness of rainfall in a desert area. *Journal of Hydrology*, **17**, 161–75.

Sharon, D. (1974) The spatial pattern of convective rainfall in Sukumaland, Tanzania – a statistical analysis. *Archives of Metrological Geophysical Booklet Series B*, **22**, 201–218.

Sneh, A. (1983) Desert stream sequences in the Sinai peninsula. *Journal of Sedimentary Petrology*, **53**, 1271–1279.

Srikanthan, R. and McMahon, T.A. (1980) Stochastic generation of monthly flows for ephemeral streams. *Journal of Hydrology*, **47**, 19–40.

Steel, R.J. and Thompson, D.B. (1983) Structures and textures in Triassic braided stream conglomerates ('Bunter' Pebble Beds) in the Sherwood Sandstone Group, north Staffordshire, England. *Sedimentology*, **30**, 341–67.

Stone, J.J., Nichols, M.H., Goodrich, D.C. and Buono, J. (2008) Long-term runoff database, Walnut Gulch Experimental Watershed, Arizona, United States. *Water Resources Research*, **44**, W05S05, DOI: 10.1029/2006WR005733.

Sutherland, R.A. and Bryan, R.B. (1990) Runoff and erosion from a small semiarid catchment, Baringo District, Kenya. *Applied Geography*, **10**, 91–109.

Syvitski, J.P.M. (2003) Supply and flux of sediment along hydrological pathways: research for the 21st century. *Global Planetary Change*, **39**, 1–11, DOI: 10.1016/S0921-8181(03)00008-0.

Thornes, J.B. (1976) Semi-arid erosional systems: case studies from Spain, London School of Economics Geographical Paper 7.

Thornes, J.B. (1977) Channel changes in ephemeral streams: observations, problems and models, in *River Channel Changes* (ed. K.G. Gregory), John Wiley & Sons, Ltd, Chichester, pp. 317–335.

Tolouie, E., West, J.R. and Billam, J. (1993) Sedimentation and desiltation in the Sefid-Rud reservoir, Iran, in *Geomorphology and Sedimentology of Lakes and Reservoirs* (eds J. McManus and R.W. Duck), John Wiley & Sons, Ltd, Chichester, pp. 125–138.

Tooth, S. (2000) Process, form and change in dryland rivers: a review of recent research. *Earth-Science Reviews*, **51**, 67–107.

Tooth, S. and Nanson, G.C. (1999) Anabranching rivers on the Northern Plains of arid central Australia. *Geomorphology*, **29**, 211–233.

Tooth, S. and Nanson, G.C. (2000) The role of vegetation in the formation of anabranching channels in an ephemeral river, Northern Plains, arid central Australia. *Hydrological Processes*, **14**, 3099–3177.

Tooth, S. and Nanson, G.C. (2004) Forms and processes of two highly contrasting rivers in arid central Australia, and the implications for channel-pattern discrimination and prediction. *Geological Society of America Bulletin*, **116**, 802–816.

Tunbridge, I.P. (1984) Facies model for a sandy ephemeral stream and clay playa complex: the Middle Devonian

Trentishoe Formation of north Devon, UK. *Sedimentology*, **31**, 697–715.

Vanoni, V.A. (ed.) (1975) *Sedimentation Engineering. American Society of Civil Engineers Manual on Sedimentation*, American Society Civil Engineers, New York.

Vondra, C.F. and Bowen, B.E. (1978) Stratigraphy, sedimentary facies and palaeoenvironments, East Lake Turkana, Kenya, in *Geological Background to Fossil Man* (ed. W.W. Bishop), Scottish Academic Press, Edinburgh, pp. 395–414.

Walters, M.O. (1990) Transmission losses in arid region. *Journal of Hydraulic Engineering, ASCE*, **116**, 129–138.

Wheater, H.S. and Brown, R.P.C. (1989) Limitations of design hydrographs in arid areas – an illustration from south west Saudi Arabia, in Proceedings of the British Hydrological Society National Symposium, Sheffield, September 1989, pp. 3.49–3.56.

Wheater, H.S., Butler, A.P., Stewart, E.J. and Hamilton, G.S. (1991a) A multivariate spatial-temporal model of rainfall in southwest Saudi Arabia. I. Spatial rainfall characteristics and model formulation. *Journal of Hydrology*, **125**, 175–199.

Wheater, H.S., Onof, C., Butler, A.P. and Hamilton, G.S. (1991b) A multivariate spatial-temporal model of rainfall in southwest Saudi Arabia. II. Regional analysis and long-term performance. *Journal of Hydrology*, **125**, 201–220.

Williams, G.E. (1969) Characteristics and origins of a Precambrian pediment. *Journal of Geology*, **77**, 183–207.

Williams, G.E. (1970) The central Australian stream floods of February–March 1967. *Journal of Hydrology*, **11**, 185–200.

Williams, G.E. (1971) Flood deposits of the sand-bed ephemeral streams of central Australia. *Sedimentology*, **17**, 1–40.

Wittenberg, L., Laronne, J.B. and Newson, M.D. (2007) Bed clusters in humid perennial and Mediterranean ephemeral gravel-bed streams: the effect of clast size and bed material sorting. *Journal of Hydrology*, **334**, 312–318, DOI: 10.1016/j.jhydrol.2006.09.028.

Wolman, M.G. and Gerson, R. (1978) Relative scales of time and effectiveness of climate in watershed geomorphology. *Earth Surface Processes*, **3**, 189–208.

Wooding, R.A. (1966) A hydraulic model for the catchment-stream problem. *Journal of Hydrology*, **4**, 21–37.

Yair, A. and Lavee, H. (1976) Runoff generative process and runoff yield from arid talus mantled slopes. *Earth Surface Processes*, **1**, 235–47.

14

Dryland alluvial fans

Adrian Harvey

14.1 Introduction: dryland alluvial fans – an overview

14.1.1 Definitions, local occurrence, general morphology

Alluvial fans are depositional landforms that occur where confined stream channels emerge from mountain catchments into zones of reduced stream power. The abrupt reduction of stream power results in the deposition especially of the coarse fraction of the sediment load. The result is progressive deposition to form a fan-like body of sediment. A common situation where this occurs is at mountain fronts (Bull, 1977), either at erosional, pediment-controlled mountain fronts or at fault-bounded mountain fronts (Figures 14.1(a) and 2(a)), often at the margins of a subsiding sedimentary basin. Alternatively, such situations occur within intermontane basins or at tributary junctions, where steep confined tributary stream channels join a more open, lower-gradient main valley (Figures 14.1(a) and 2(b)). In these cases the fans may be confined by the valley walls, whereas mountain-front fans may only be confined by adjacent fans. Fans at faulted mountain fronts and at the margins of sedimentary basins may involve a thick sequence of deposits, whereas those in erosionally controlled situations will be thinner.

Alluvial fans generally have a conical surface form, modified by whatever confinement may be present, with slopes radiating away from an apex at the point where the channel issues from the mountain catchment (Figure 14.1(b)). Some fans, especially along active faults, may have a single stream source at the apex; others, especially where pedimentation has occurred prior to burial by fan deposits or where the fan sediments have backfilled into the mountain catchments, may have multiple sources (Figure 14.2(c)). Basin-margin fans may coalesce down-fan to form a depositional apron, known as a bajada (Figure 14.2(d)). Distally, fan deposits may interdigitate with river, lake or windblown deposits and may form part of a continuous suite of deposits extending from mountain-front, basin margin locations to basin centre. Fans that terminate in standing water, lakes or the sea are known as fan deltas. They may include a substantial subaqueous portion and their distal facies may interact with lacustrine or marine coastal sediments. The long profiles of alluvial fans tend to be more or less planar or exhibit a slight upwards concavity and the cross-profiles tend to be upwardly convex (Hooke, 1967; Bull, 1977).

Within the mountain catchment the stream channels would have a conventional tributary pattern, but on the fan the feeder channel may become a multithread distributary system. On fully aggrading, nondissected fans this transition may take place at the fan apex, but on many fans the feeder channel may be incised into the proximal fan surfaces as a fanhead trench (Figure 14.2(e)), emerging on to the fan surface mid-fan at an intersection point (Hooke, 1967). On some fans the channel may by trenched below the fan surface throughout the fan, in which case the fan surface is essentially a fossil surface receiving no further sedimentation. The main channel may be an axial fan channel running more or less down the centre of the fan, but may switch to run down the steeper flanks to become or join an interfan channel draining the depression between two adjacent fans (Figure 14.1(b)). As the result of such switching the focus of deposition will switch from one part of the fan to another (Denny, 1967)(Figure 14.1(c)). Other small channels may form on the fan surface as fan surface washes, draining parts of the fan surface, but having no mountain source areas.

Arid Zone Geomorphology: Process, Form and Change in Drylands, Third Edition. Edited by David S. G. Thomas

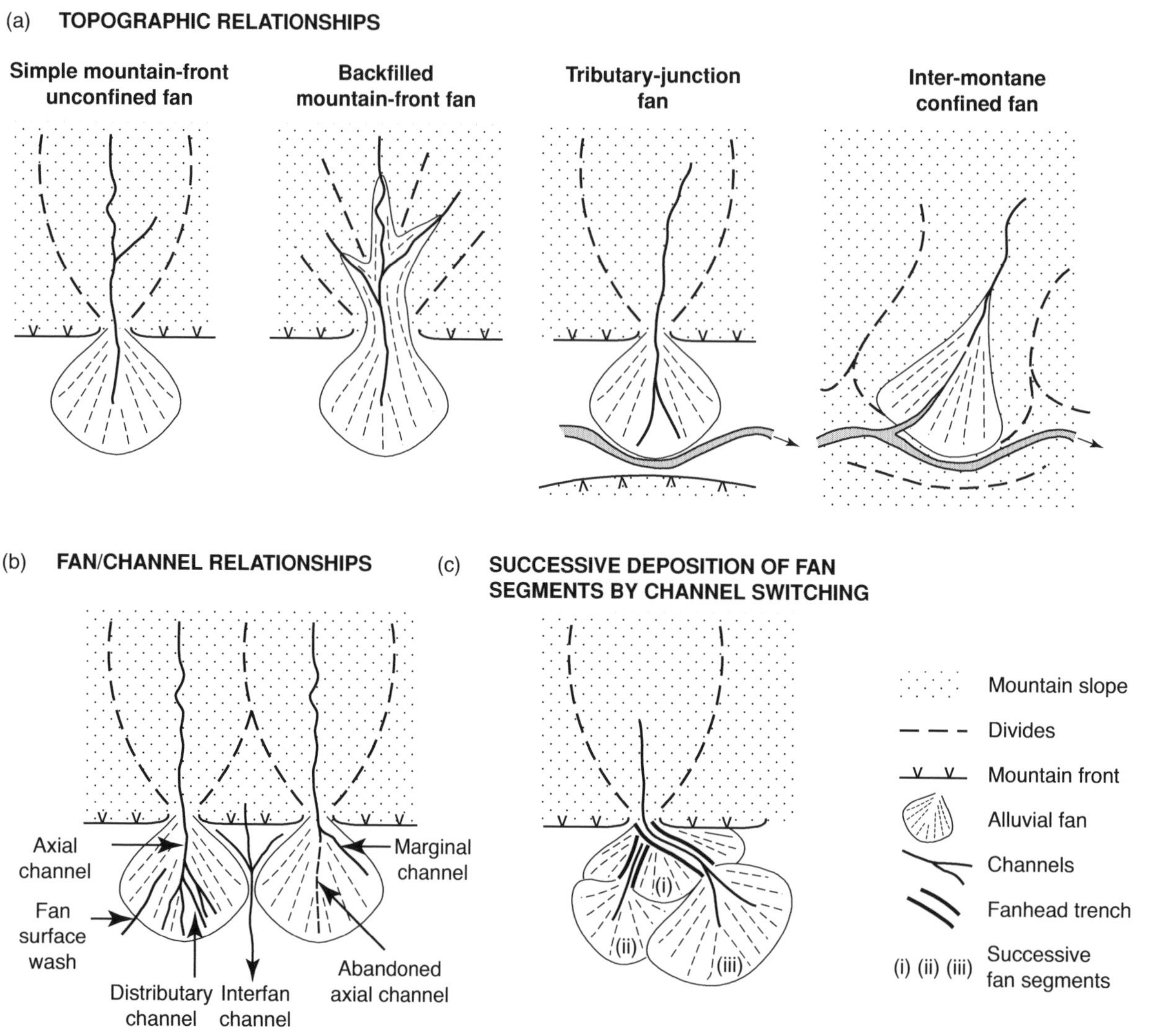

Figure 14.1 Topographic and locational characteristics of alluvial fans: (a) topographic relationships; (b) fan/channel relationships; (c) channel switching and surface age of fan segments (modified from Denny, 1967).

14.1.2 Global occurrence and distribution of dryland alluvial fans

Alluvial fans are not restricted to dry regions. Indeed they can occur in any mountain region where there is juxtaposition of a high rate of sediment supply from mountain catchments and a sudden drop in unit stream power at mountain-front or tributary-junction locations. They have been described in many environments (for examples see Rachocki and Church, 1990; Harvey, Mather and Stokes, 2005), including arctic environments (e.g. Leggett, Brown and Johnston, 1966; Ritter and Ten Brink, 1986), alpine environments (e.g. Kostaschuk, Macdonald and Putnam, 1986; Derbyshire and Owen, 1990), humid temperate regions (e.g. Harvey *et al.*, 1981; Kochel, 1990; Oguchi and Ohmori, 1994; Oguchi, 1997; Saito and Oguchi, 2005; Chiverrell, Harvey and Foster, 2007) and even in the humid tropics (e.g. Kesel and Spicer, 1985).

The conditions that favour alluvial fan deposition are, however, particularly well developed in arid and semi-arid mountain regions. Several conditions seem to be important. As the result of sparse vegetation cover, intense storm rainfall and the dominance of overland flow processes, desert mountains have high rates of sediment production and steep desert streams have high sediment concentrations (Laronne and Reid, 1993). The domination of fluvial environments by flash floods, with associated high unit stream power (Baker, 1977: Wolman and Gerson, 1978), is also important. The tendency for discharge to diminish downstream by transmission losses through the channel bed and by evaporation may accentuate the drop in unit stream power. Finally, within the context of the spatial

Figure 14.2 Fan topography. (a) Air view of mountain-front fans: Zagros Mountains, Iran. Note the multiple-age fan surfaces: older surfaces are darker and eroded by fan-surface washes, younger surfaces are paler and smooth. (b) Tributary-junction fan: Wadi Al-Bih, Oman. Wadi Al-Bih flows from left to right, a large fluvially dominant tributary-fan with secondary debris cones at right. (c) Backfilled mountain-front fan: Northern mountain front, Sierra de Carrascoy, Murcia, Spain. (d) Coalescent alluvial fans forming a *bayada*: Death Valley, east-side fans. (e) Fanhead trench: Nijar fan, Almeria, southeast Spain. Note the calcrete-crusted fan surfaces; fanhead trench has cut through the Quaternary fan sediments into the underlying Miocene rocks.

discontinuity (Brunsden and Thornes, 1979) of arid environments, in tributary junction situations there is a high probability of a lack of synchroneity in the timing of flood peaks between the tributary and the main stream. Whether these factors result in dry-region fans that are distinct from those in other regions is a theme that will be discussed later in this chapter.

Whether or not dry-region fans are distinct, until recently they have dominated the research literature on alluvial fans. This is particularly true of the fans of the American West. Indeed, until 30 years ago almost all published research on alluvial fans related to that area, particularly to the mountain-front, basin-margin fans of the Basin and Range region, so that they became the prototype against which all fans would be viewed and the factors controlling them were assumed to be universal. The first papers that really established descriptions of fan morphology and processes were published in the 1920s (Eckis, 1928; Blackwelder, 1928), but it was really the 1950s and 1960s that saw the emergence of a substantial body of alluvial fan research. Several important overview papers were published (Blissenbach, 1954; Hooke, 1967; Denny, 1967). There was work on fan profiles (Blissenbach, 1952; Bull, 1961), fan sediments (Bull, 1962a, 1963; Bluck, 1964; Lustig, 1965) and on fan morphometry as a major theme in alluvial fan research (Bull, 1962b; Hooke, 1968). Bull's work (synthesised by Bull, 1964) was based on the California Coast Ranges, but most of the other work, including a number of other studies (Beaty, 1963, 1970, 1974, 1990), focused on the Basin and Range region of Nevada and eastern California and particularly on the Death Valley Region (Denny, 1965; Hunt and Mabey, 1966; Hunt, 1975) (see Box 14.1). The main emphases within this body of work were fan morphology and morphometry in relation to geology and tectonics.

Box 14.1 Death Valley alluvial fans

The Death Valley fans are the archetype fans in much of the literature, which deals with three main themes: tectonic control, climatic control and modern fan processes. Tectonic control includes both passive control of fan geometry and style, and active tectonics (minor faults across the west-side fans; major faults bounding the mountain front of east-side fans; tectonic subsidence creating accommodation space, the result being burial of older fan sediments by modern aggrading east-side fans). Quaternary climatic control expressed in alternating fan

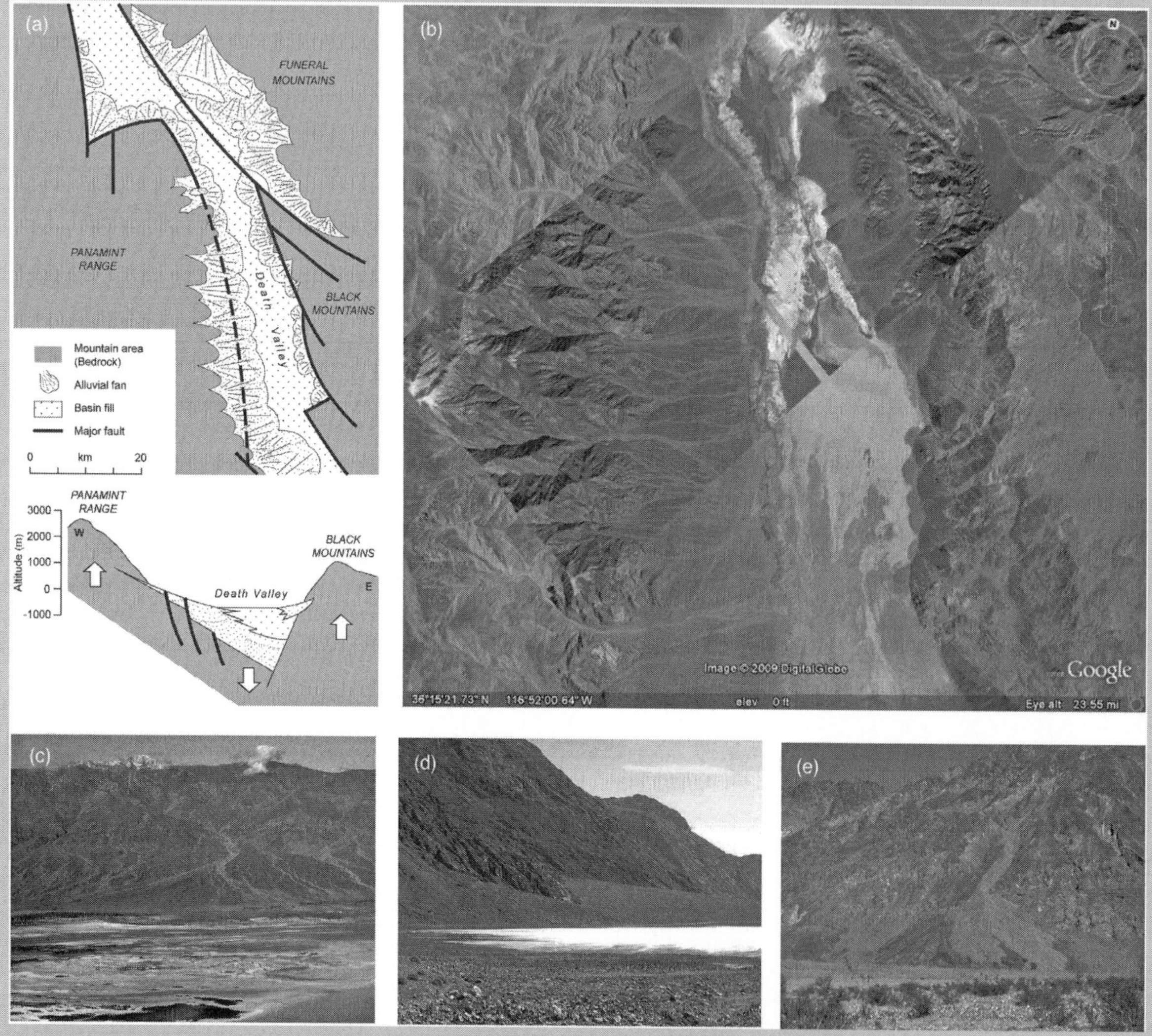

(a) Summary map showing the main tectonic units and schematic cross-section across Death Valley; (b) composite satellite image of the central part of Death Valley (taken from Google Earth); (c) west-side fans: Hanaupah fan seen from Dante's View above Death Valley. Note the large backfilled fans on flanks of the uplifted Panamint Mountains, with Telescope Peak on the skyline, and composite telescopic fan surfaces with well-developed fanhead trenches; different surface ages are picked out by surface texture and varnish colour, with the basal zone trimmed by the Lake Manley Late Pleistocene pluvial lake shoreline. (d) East-side fans, Badwater fan, on the downthrown side of Badwater fault: small sheetflood dominated aggrading fan; little or no fanhead trench; near-uniform modern fan surface age. (e) Small (frequently photographed) east-side debris cone, near Mormon Point. Note the multiple ages of debris flows picked out by varnish development.

aggradation and dissection/progradation phases provided fan aggradation, under excess sediment supply transitions probably during cooler wetter Pleistocene glacial conditions or during glacial-to-interglacial conditions. During peak glacials the floor of the valley was occupied by a terminal pluvial lake (Lake Manley); Badwater playa is its modern (partly spring-fed) shrunken remnant. Today's hyper-arid environment is subject to occasional storms, flash floods and debris flows, adding sediment by debris flows from small steep catchments, channelised streamflows from larger catchment feeder channels and through braided channels within fanhead trenches, and sheetfloods on aggrading fans and distal fan surfaces.

Although fans are largely not coupled with adjacent environments, interactions between fan processes and other environments are important for identifying past phases of fan evolution, interaction with the Amargosa fluvial system in the southern section of the valley, the Badwater playa lake in the central section and dunefields to the north. For a selection of Death Valley alluvial fan references see Denny (1965), Hunt and Mabey (1966), Hunt (1975), Hooke and Dorn (1992), Dorn (1988), Blair and McPherson (1994a) and Blair (1999). For other illustrations of Death Valley fans see Figures 14.2(d), 14.5(c,ii), 14.14(d) and 14.17.

In the years that followed there has been consolidation of these themes (Hooke and Rohrer, 1977, 1979; Lecce, 1991; Blair and McPherson, 1994a, 1998; Blair, 1999), but the response of fans to Quaternary climatic change has emerged as a major theme (Dorn *et al.*, 1989; Bull, 1991; Hooke and Dorn, 1992; Weissmann, Mount and Fogg, 2002; Weissmann, Bennett and Lansdale, 2005). This has been aided by development of understanding of soil/geomorphic relationships (Gile, Peterson and Grossman, 1966; Harden, 1982; Harden and Taylor, 1983; Machette, 1985), desert varnish (Dorn and Oberlander, 1981, 1982; Dorn, 1988, 1994) and desert pavement development, especially aiding the relative dating of fan surfaces (Lattman, 1973; Wells, McFadden and Dohrenwend, 1987; McFadden, Wells and Jercinavich, 1987; McFadden, Ritter and Wells, 1989). Furthermore, increasing knowledge of Quaternary environments in the American deserts (Grayson, 1993; Enzel, Wells and Lancaster, 2003), particularly in relation to the basin-centre pluvial lakes (e.g. Adams and Wesnowsky, 1998, 1999), has enabled research into fan-lake relationships to focus on Quaternary fan dynamics (Harvey and Wells, 1994, 2003; Ritter, Miller and Husek-Wulforst, 2000; Harvey, Wigand and Wells, 1999; Harvey, 2005; Garcia and Stokes, 2006).

Dry-region alluvial fans occur in all the major dryland mountain regions of the world. Since the 1970s, building on the American work, alluvial fan research has developed in other dry regions. Only recently has much attention been given to the dryland fans in South America, in the Atacama Desert (e.g. Mather and Hartley, 2005; Hartley *et al.*, 2005) and in the semi-arid lands of the Argentinian Andes (Colombo, 2005; Robinson *et al.*, 2005). In Australia, Williams (1973) and Wasson (1974, 1979) described fan morphology and sediments in the Flinders Ranges and elsewhere. In Asia, although there has been little primary research other than in Chinese on the large and impressive fans in the Taklimakan Desert in northwest China and other deserts in central Asia, in India there has been some research. Of particular significance is the recognition of enormous 'megafans', first described for the Kosi fan (Gohain and Parkash, 1990), although not really a dryland fan. A similar megafan, the Okavango fan in southern Africa (Stanistreet and McCarthy, 1993), has been described as a terminal fan, where the Okavango River peters out at the margins of semi-arid northern Botswana.

In the Middle East, there has been research in Israel focusing on the interplay between tectonic, climatic and base-level controls (Bowman, 1978, 1988; Frostick and Reid, 1989; Gerson *et al.*, 1993; Amit, Harrison and Enzel, 1995). Elsewhere, there have been studies in Kuwait (Al Sarawi, 1988), Turkey (Roberts, 1995), Iran, (Arzani, 2005) and in the UAE and Oman (Al Farraj and Harvey, 2000, 2004, 2005).

Apart from the ongoing work in the American West, it is the drier parts of the Mediterranean region that have provided the basis for many alluvial fan studies since the late1970s. In North Africa, White (1991), White and Walden (1994, 1997) and White *et al.* (1996) have pioneered two developments, the application of remote sensing to mapping fan surfaces and the analysis of pedogenic iron oxide geochemistry, as an aid to correlating fan surfaces. In the drier parts of Europe other studies include those by Sorriso-Valvo, Antronico and Le Pera (1998) in southern Italy, stressing topographic and tectonic controls, and in Greece by Pope and Van Andel (1984), emphasising geoarchaeological implications, Nemec and Postma (1993) and Pope and Wilkinson (2005), the latter stressing tectonic and climatic controls and using OSL dating among a range of sophisticated geotechnical methods.

There has been a range of studies in semi-arid southeast Spain, initially dealing with fan sediments, morphological sequences and morphometry (Harvey, 1978, 1984a, 1984b, 1987, 1988, 1992a, 1992b, 2002a; Somosa *et al.*, 1989; Silva *et al.*, 1992), and more recently with fan evolution and dynamics in relation to tectonic, climatic and base-level controls (Harvey, 1990, 1996, 2002b; Calvache, Viseras and Fernandez, 1997; Harvey *et al.*, 1999, 2003; Viseras *et al.*, 2003; Silva *et al.*, 2008). In addition there have been important studies of calcretes on fan surfaces, a characteristic of Pleistocene fan surfaces in this region (e.g. Alonso-Zarza *et al.*, 1998; Stokes, Nash and Harvey, 2007).

14.1.3 The role of alluvial fans within dryland fluvial systems

With their location between sediment-source areas and either enclosed sedimentary basins or arterial river systems, alluvial fans have an important role in either coupling or buffering dry-region fluvial systems. In the long term alluvial fans act as major sediment stores, trapping the coarse fraction of the incoming sediment. Locally, deposition will occur if the sediment supply rate is greater than the transporting capacity, in other words, if the actual stream power through the fan is less than that needed to carry the sediment through the system, a threshold defined by Bull (1979, 1991) as the threshold of critical stream power. Spatial variations in stream competence will cause variations in deposition on the fan in relation to stream power behaviour (Figure 14.3(a)). Temporal variations will cause variations in sedimentation rate or style or even switches between net aggradation on, or dissection of, the fan surface (Figure 14.3(b)).

Fans undergoing aggradation, whether untrenched and proximally aggrading or fanhead trenched and distally aggrading (telescopic or prograding fans; see later in Section 14.2.4), will act as buffers within the fluvial system, breaking the continuity of sediment movement between the sediment source area and either the sedimentary basin or the main channel system downstream (Harvey, 2002c, 2010). The implications are important not only in the context of sediment movement through the system but also in the system's response to environmental change. Alluvial fans may absorb such changes by varying their rates or patterns of sedimentation (Harvey, 1987) and inhibit the downstream transmission of the effects of such changes. On the other hand, where fans become dissected throughout their length there may be continuity of coarse sediment movement from the source area to arterial drainage. For these reasons the coupling/buffering role of alluvial fans is fundamental to the dynamics of sediment movement through the fluvial system.

Because alluvial fans are sensitive environments (Brunsden and Thornes, 1979), responding to changes in flood power and sediment supply by changing the style or rate of sedimentation or by switching between erosional and depositional regimes, they may preserve in their sediments and morphology a sensitive record of environmental change in their source areas, particularly as affecting sediment supply. Indeed, this may be a much more sensitive and complete record than that preserved in fluvial sediments further downstream, which tend to respond primarily to variations in flood power (Macklin and Lewin, 2003) and would be more affected by changes in coupling relationships.

There are, however, two problems. One is dating. In contrast with fan deposits in humid areas, which may preserve organic materials suitable for radiocarbon dating, such material is rare in dry-region fan sediments. On the other hand, advances in luminescence and cosmogenic nuclide dating have potential (see Section 14.5.2.3).

The other problem relates to scale and the preservation potential of fan sediments. For modern or Quaternary fans there is little problem, but for fan sediments preserved in the rock record, larger bodies of sediment, especially those at basin margins, are more likely to be preserved than those within valley systems. This may lead to a mismatch between interpretations or models based on the rock record and those based on modern fans. One of the challenges facing alluvial fan research has been to relate observations on modern or Quaternary fans to those made on ancient fan sediments. One approach that shows some potential has been to apply the sequence stratigraphy concept, normally applied to marine sediments in the rock record, to the Quaternary alluvial fan sediments of the central valley of California (Weissmann, Mount and Fogg, 2002; Weissmann, Bennett and Lansdale, 2005). Instead of sea-level change controlling sequence boundaries, interstratified palaeosol horizons are seen as the key and Quaternary climates the primary control of the sequence. Glacials caused fan aggradation and interglacials caused fan trenching and soil formation on the abandoned fan surfaces. In this way, from a three-dimensional interpretation of the changing fan morphology and sediment sequence a meaningful environmental reconstruction has been achieved.

14.2 Process and form on dryland alluvial fans

14.2.1 Sediment supply, transport and depositional processes

The processes delivering sediment from the feeder channel to the fan range from debris flow to fluvial processes.

(a)

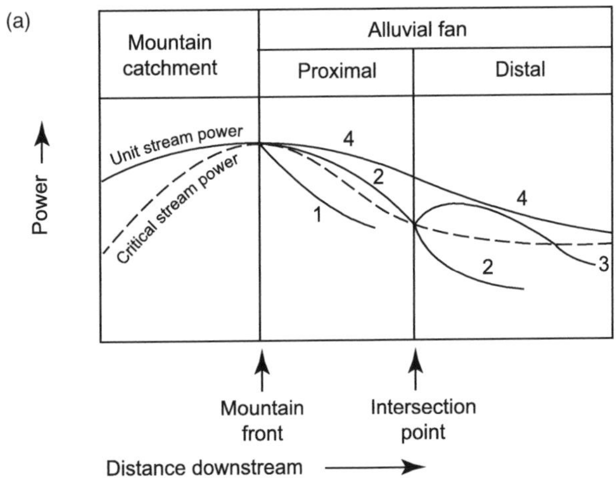

(b)

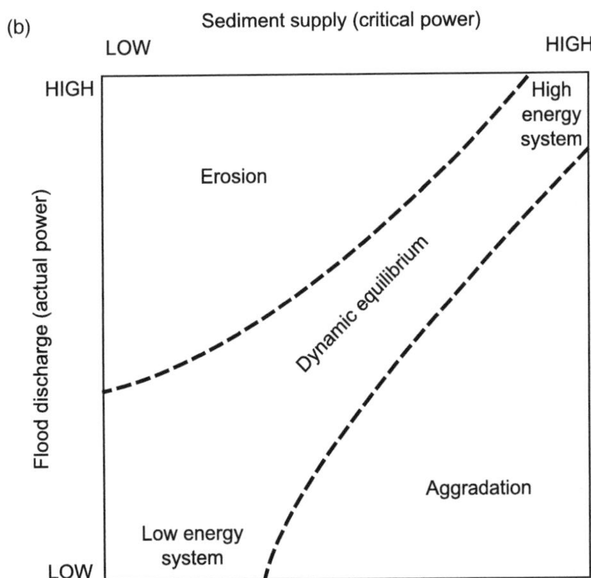

Figure 14.3 Critical power relationships. (a) Theoretical stream power relationships through alluvial fans (amended from Harvey, 1992a, based on the concept of the threshold of critical stream power of Bull, 1979). Solid lines show actual (unit) stream power for 1, aggrading fans; 2, proximally trenched, distally aggrading fans; 3, midfan trenched fans; 4, through-trenched fans (for context see Section 14.3.2). (b) Theoretical relationships between critical power and erosion/deposition behaviour.

Sediment transport on the fan can be as debris flow or as fluvial tractional processes by sheetflows or by channelised flows. Sediments within dry-region fans are often composite, ranging from fluvial or sheetflood to debris-flow dominance.

14.2.1.1 Debris flows

Debris flows on arid region fans have been described by Blackwelder (1928), Beaty (1974, 1990), Blair (1999), Blair and McPherson (1994a, 1994b, 1998) and Whipple and Dunne (1992), but much of the research on debris flows is based on nonarid areas (Pierson, 1980, 1981; Suwa and Okuda, 1983; Johnson and Rodine, 1984; Pierson and Scott, 1985; Wells and Harvey, 1987; McArthur, 1987; Rickenmann and Zimmerman, 1993; DeGraff, 1994).

Debris flows consist of an unsorted mixture of water, sediment matrix and coarse clasts. They vary in behaviour depending on the water-to-sediment ratio, the particle size of the matrix, especially the clay content, the matrix-to-clast ratio and the clast size.

They range from viscous flows (the 'classic' debris flows), within which deformation is by flowage and by shearing. They have some internal strength by which the larger clasts may be supported in the upper parts of the flow. Deposition (i.e. the cessation of movement) is brought about by friction with the underlying surface, reduction of gradient and by the reduction of the volume of the flow partly by marginal deposition during transport and the formation of debris-flow levees. The deposits generally are massive and show matrix-supported clasts (facies Gms of Rust, 1978, 1979, and Pierson, 1981), but there is some interaction between the larger clasts, resulting in a clast fabric indicative of compression and internal shearing (Wells and Harvey, 1987) (Figure 14.4(a)). This may result in clast alignment parallel to the base of the flow, but near-vertical clasts aligned across the flow at the top and towards the front of a debris-flow lobe. The depositional topography is of the 'lobe and levee' type (Blair and McPherson, 1998).

With higher water content and increased fluidity, possibly also smaller clast sizes, deformation is largely by plastic flowage, which on deposition may produce a tendency for clast alignment with the flow direction. Such deposits would be common in fan-delta environments. With even higher water content and increased fluidity, the mixture behaves as a hyperconcentrated flow (Pierson and Scott, 1985) near the threshold between transport by plastic and turbulent flow. On deposition the water may drain through the sediment, taking much of the sediment matrix vertically down through the flow, leaving the clasts to exhibit an open, matrix-poor, clast-supported, collapsed fabric (Wells and Harvey, 1987).

14.2.1.2 Fluvial processes

Beyond a threshold water content the flow becomes fully turbulent, transporting coarse sediment by traction and finer sediment by saltation and in suspension. This is the realm of fluvial sediment transport, described in many fluvial geomorphology and sedimentology texts (e.g. Reinech and Singh, 1983; Schumm, 1977; Nilsen, 1982). Transport takes place either in relatively confined channels (open-channel flow) either single- or multithread channels, or as wide shallow unconfined sheetflow. Under fluvial sediment transport, bedload transport capacity depends on unit stream power (Richards, 1982), dependent on flow depth and gradient. Deposition of bedload clasts occurs as stream power falls below the transport threshold, through inceased bed friction, reduced depth or reduced gradient. The threshold depositional gradient is much lower for tractional flows than for debris flows and lower for channelised flows than for sheetflows.

On deposition the sediments exhibit much more organisation than do debris -flow sediments (Figure 14.4(b)). There is often a channelised or erosional base; then the sediments themselves may exhibit distinctive sedimentary structures (e.g. bedding, cross-bedding, interbedding of gravels, cobbles and finer sediments, grading, reverse grading, clast imbrecation; facies Gm, Gp, Gt of Miall, 1977, 1978). Sheetflood deposits may comprise thin sheets of gravels and sands, often forming couplets (Blair and McPherson, 1994a). The depositional topography on alluvial fan surfaces will be much smoother than that on debris-flow depositional surfaces It may resemble featureless sheets of imbricated gravels or may show an intricate 'bar and swale' topography (Figure 14.4(c)). One bar form that has been described is the 'sieve bar' (Hooke, 1967; Wasson, 1974), a lobate bar showing nose-to-tail and vertical fining of the coarsest clasts, which sieve out the finer sediments (Figure 14.4(d)).

On most fans, especially dry-region fans, whether debris-flow or fluvially- dominant, sediment supply is episodic, related to major storm and flood events (Beaumont and Oberlander, 1971; Whipple *et al.*, 1998; Arzani, 2005; Mather and Hartley, 2005). On any one fan the water-to-sediment ratio may vary during a storm (Wells and Harvey, 1987) and may vary downfan (Pierson and Scott, 1985), resulting in a complex of depositional facies, but with the overall pattern, facies-dominance and the differences between fans reflecting the gross catchment controls over water and sediment production. As indicated earlier, fans issuing from small steep catchments tend towards debris-flow dominance, especially if the geology is conducive to the production of fine sediments, and those issuing from larger or less steep catchments tend to be fluvially dominant (Harvey, 1984a, 1992a; Kostaschuk, Macdonald and Putnam, 1986; Wells and Harvey, 1987). In this way the style of sedimentation could be said to be sediment-led. Ironically, the processes are event-based and therefore could be said to be flood-led.

14.2.2 Post-depositional modification of dry-region fan surfaces

In addition to purely depositional facies, arid area fans are often characterised by other sedimentary features, related to weathering and pedogenic processes, as well as by surface modification through erosion (see Bull, 1991). Surface modifications are time-dependent and have proved invaluable in relative dating and correlating fan surfaces (McFadden, Ritter and Wells, 1989; Bull, 1991).

Older fan surfaces in deserts are usually characterised by a closely interlocked pavement of angular clasts (Figure 14.4(e)), below which is a fine silty soil that may fill the spaces between the subsurface clasts or be entirely clast-free. Such surfaces are described as desert pavement or rock pavement in the USA, gibber in Australia and reg

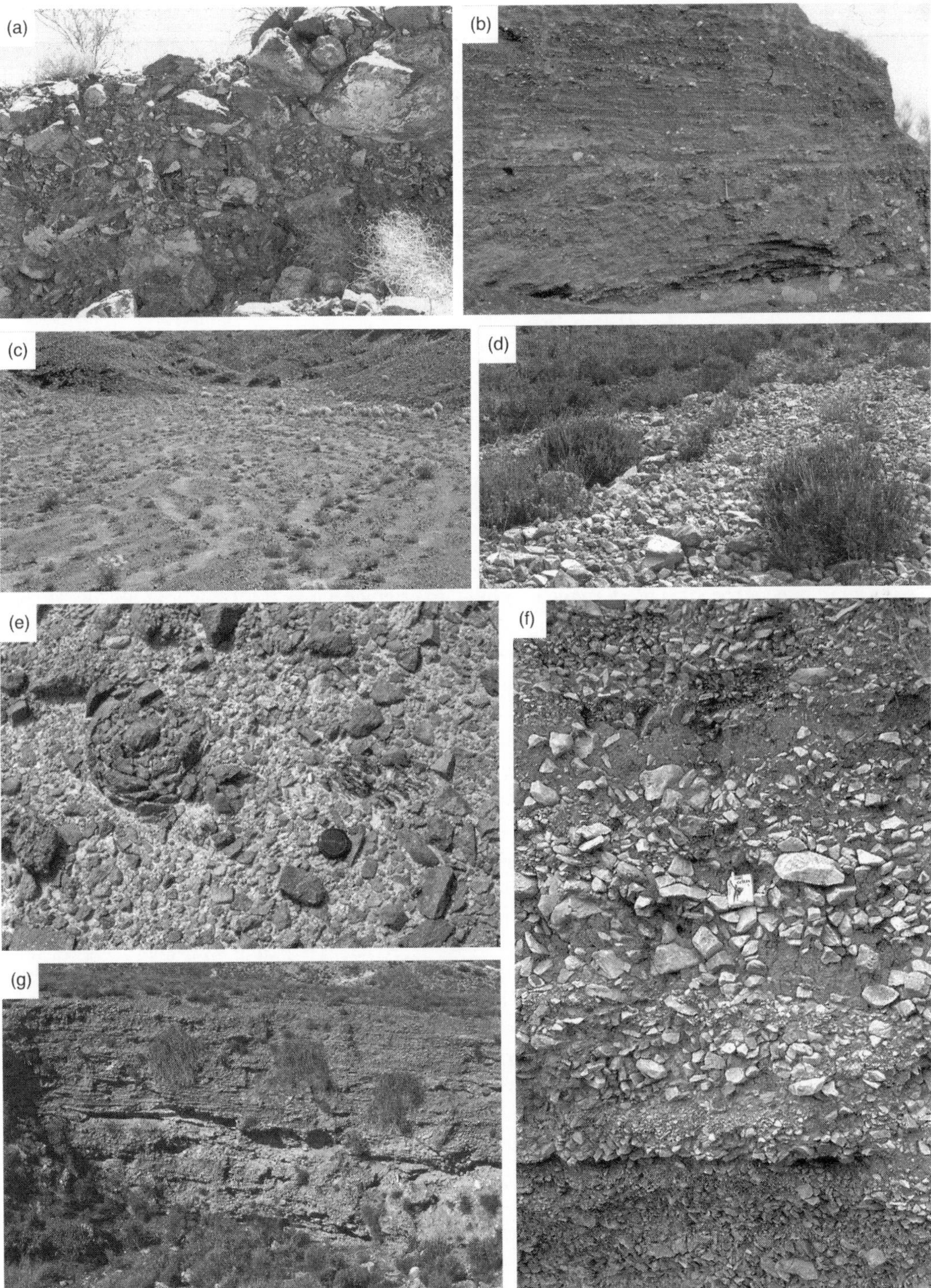

Figure 14.4 Fan sediments. (a) Bouldery debris-flow deposits, Zzyzx fans, California. Note the clast fabric – flow was from right to left. Hammer in centre of photo is for scale. (b) Stratified channel gravels (below) and sheetflood gravels (above) with sand lenses: Nudos fan, Tabernas, southeast Spain. Hammer for scale. Flow direction was away from the camera. Note that this sequence approximates to 'Scott'-type sequences of Miall (1978); see also Figure 14.5(a). (c) Bar and swale fan topography: Osbourne Wash fan, Arizona. (d) Intersection-point seive lobe deposits on a small aggrading fan near Vera, southeast Spain. Note the nose-to-tail fining and transverse clast alignment; flow direction was towards the camera from right to left. (e) Desert pavement on fan surfaces at Sinai, Egypt. Note the 'ghost' clasts disintigrating by fracture. Lens cap for scale. (f) Stratified fan deposits, approximating to the Trollheim type (Miall, 1978; see also Figure 14.5(a)): Cayola fan, Alicante, Spain (d = debris flows, c = channel gravels). Flow was from right to left. Cigarette pack for scale. (g) Ceporro fan, Almeria, southeast Spain. Fanhead trench section showing early debris-flow deposits, followed by erosional scour then sheet and channel gravels, capped by a calcrete crust. Height of section c. 10 m. Flow was from left to right (see also Figure 14.5(b,ii)).

in the Middle East (see Chapter 9). Two concurrent processes operate to produce desert pavements: mechanical weathering of the clasts exposed on the surface and the accumulation of windblown dust. Several theories have been advanced for the origin of desert pavements (Cooke, 1970), aeolian winnowing of fine material and wetting and drying of the fines, causing swelling and contraction, which in turn causes the clasts to rise through the fine material, but the general consensus now is that mechanical weathering of the clasts occurs concurrently with the aeolian accumulation of dust to form the pavement and the underlying Av soil horizon (McFadden, Wells and Jercinavich, 1987; McFadden *et al.*, 1999). Pavement development is time-dependent (Yaalon, 1970; Dan *et al.*, 1982; Amit and Gerson, 1986; Amit, Gerson and Yaalon, 1993). Any initial depositional morphology, e.g. bar forms, is obliterated as the pavement develops. The clasts themselves are reduced in size and become more angular with progressive fracturing through mechanical weathering (Al Farraj and Harvey, 2000).

The clasts on desert fan surfaces commonly carry a rock varnish (desert varnish) of iron and manganese salts (Chapter 8), which increases in thickness and darkens with age to a deep red on the undersides of the clasts and a dark brown to black above. Again, there is debate concerning the formation of desert varnishes (see Dorn and Oberlander, 1981, 1982), but their age-related characteristics have allowed the relative dating and correlation of fan surfaces (e.g. Hunt and Mabey, 1966; Harvey and Wells, 2003). It has been argued that trace carbon content in varnish may allow the radiocarbon dating of desert surfaces (Dorn *et al.*, 1989) and that other detailed analytical techniques may not only reveal age information (Dorn, 1983) but also give other palaeoenvironmental signals (Dorn, 1984; Dorn, De Niro and Ajie, 1987). There is, however, some controversy about this methodology.

Many studies in arid and semi-arid regions have used soil characteristics (Chapter 7) as indicators of surface age (see Birkeland, 1985). Some have used characteristics of the soil profile as a whole (Harden, 1982; Harden and Taylor, 1983; McFadden, Ritter and Wells, 1989; Harvey and Wells, 2003), while others have focused on the B-horizon, dealing with soil colour (see Hurst, 1977), pedogenic iron oxides (see Alexander, 1974; White and Walden, 1997) or magnetic mineral behaviour (White and Walden, 1994; Harvey *et al.*, 1999, 2003; Pope, 2000; Pope and Millington, 2000).

Perhaps the best-known aspect of arid and semi-arid area soils is the accumulation of pedogenic carbonate. The stages of carbonate accumulation defined by Gile *et al.* (1966) and elaborated by Machette (1985) have been used in many studies as aids to correlation and relative dating of geomorphic surfaces, including alluvial fan surfaces. On exposure pedogenic carbonate indurates to form calcrete (caliche) (Chapter 8) and undergoes a complex sequence of brecciation and recementation. Many geomorphic surfaces in arid and semi-arid regions, including but not exclusively alluvial fan surfaces, are crusted by various forms of calcrete (Butzer, 1964; Lattman, 1973; Goudie, 1983; Wright and Alonso-Zarza, 1990; Alonso-Zarza *et al.*, 1998; Nash and Smith, 1998; Stokes, Nash and Harvey, 2007). As well as providing the basis for correlation and relative dating of geomorphic surfaces (Dumas, 1969; Harvey, 1978), absolute dating is also possible using the uranium/thorium methodology (Ku *et al.*, 1979; Candy *et al.*, 2003; Candy, Black and Sellwood, 2004) (see below, Section 14.5.2.3). Calcrete is not only important for its correlation or dating potential but it also modifies geomorphic processes on fan surfaces by modifying infiltration behaviour (McDonald *et al.*, 1997) and erosional resistance (Harvey, 1978, 1987; Van Arsdale, 1982).

A final point to make in relation to alluvial fan sediments relates to interactions with nonfan sediments, particularly aeolian and lacustrine sediments. Interactions between alluvial fan and aeolian sediments may have palaeoclimatic implications (e.g. Nanson, Chen and Price, 1995; Enzel, Wells and Lancaster, 2003; Al Farraj and Harvey, 2004). Interaction with lacustrine sediments, and particularly with dated lacustrine shorelines of pluvial lakes, does, especially in the Basin and Range area of the USA, provide the basis for calibrating relatively dated fan surfaces with a known Quaternary chronology (see Harvey and Wells, 1994, 2003; Adams and Wesnowsky, 1998, 1999; Harvey, Wigand and Wells, 1999; Ritter, Miller and Husek-Wulforst, 2000; Harvey, 2005; Garcia and Stokes, 2006).

14.2.3 Alluvial fan sediment sequences and spatial variations

Assemblages of depositional facies produce characteristic vertical and spatial variations within alluvial fan deposits. A common vertical sedimentary sequence is that described by Miall (1978) as the 'Trollheim' type, named after the Trollheim fan in California, northwest of Death Valley (Hooke, 1967). It comprises a sequence dominated by alternating debris-flow deposits, massive sheet and channel gravels (Figures 14.4(f) and 14.5(a), showing facies Gms, Gm, Gt, after Miall, 1978). Further downfan or on fluvially dominant fans the sequence may be of the 'Scott' type (Figures 14.4(b) and 14.5(a)), dominated by sheet and channel gravels (facies Gm, Gp, Gt, after Miall, 1978) and various sand facies.

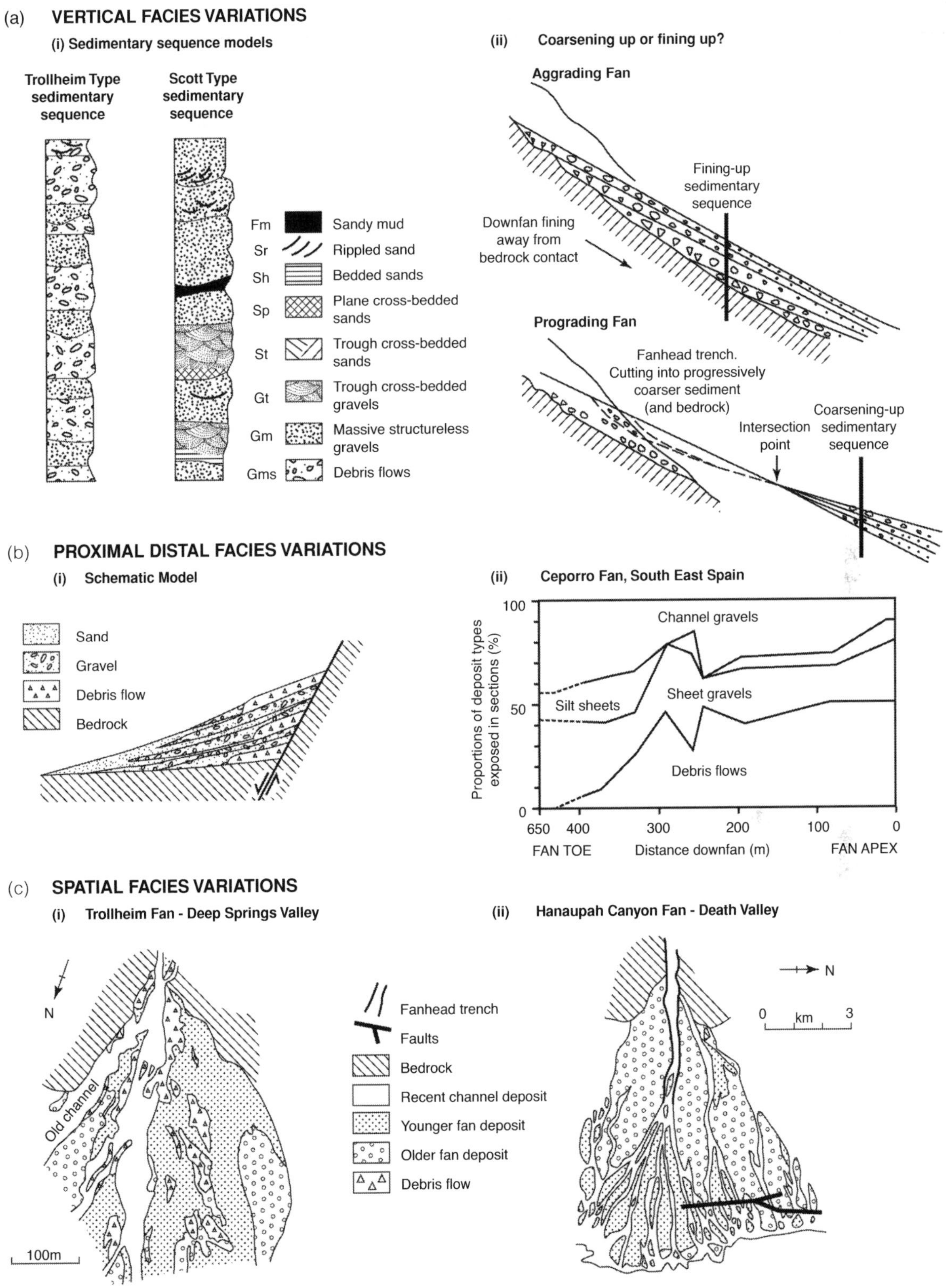

Figure 14.5 Alluvial fan sedimentary sequences. (a) Vertical facies variations: (i) sedimentary sequence models of Trollheim type and Scott type, after Miall (1978); (ii) coarsening-up or fining-up? Schematic model for proximal and distal locations. (b) Proximal-distal variations: (i) schematic model, after Rust (1979); (ii) downfan variations in proportion of facies exposed in fan sections on Ceporro fan, Almeria, southeast Spain (after Harvey, 1984b). (c) Spatial facies variations: (i) Trollheim fan, California (after Hooke, 1967): apex area of a proximally aggrading fan; (ii) Hanaupah Canyon fan, Death Valley, California (after Hunt and Mabey, 1966): a proximally trenched, distally aggrading fan.

Short-term or intrinsic variations in sedimentary processes produce facies assemblages of these simple sequences, but major climatic or tectonic changes or the long-term progressive erosion of the source area ('ageing') may cause a progressive vertical trend in sedimentary style. Such changes have been identified in Quaternary fan sequences in Australia (Williams, 1973; Wasson, 1979), Nevada (Bluck, 1964) and Spain (Harvey, 1978, 1984b, 1990). In the Spanish case many fans show early dominantly debris-flow deposition, followed later by sheet and channel gravels and finally by trenching or dissection becoming dominant over aggradation (Figure 14.4(g)). A similar sequence has been observed at Zzyzx, California, but with a strong climatic signal (Harvey and Wells, 1994; Harvey, Wigand and Wells, 1999). Late Pleistocene fan aggradation, dominantly by debris flows, was followed by Holocene fanhead trenching and fan progradation under fluvial processes (Figure 14.6). The timing of these phases is constrained by the interaction of the fan system with two dated shorelines of pluvial Lake Mojave (Harvey and Wells, 2003).

Ancient alluvial fan sequences often show progressive changes in sedimentary environment, often involving upwards coarsening of megasequences (Steel, 1974; Steel *et al.*, 1977; Rust, 1979; Nilsen, 1982; Mack and Rasmussen, 1984). In most cases the progressive changes are seen largely as tectonically or 'ageing' controlled. The coarsening-up model may be appropriate for distal locations on fans that are undergoing fanhead trenching, but the proximal parts of an aggrading fan would be subject to a fining-up tendency as backfilling progressively buries the bedrock topography and any one site becomes progressively further from bedrock outcrop and thus effectively more distal (Figure 14.5(a)). It may be that on extant Quaternary fans, where sections in proximal sediments are commonly exposed in fanhead trenches, fining-up trends are common (Harvey, 1990), but from the ancient record where preservation potential may favour more distal environments a coarsening-upward trend may be more common.

Over geological timescales, phases of fan sedimentation may only be temporary stages in the evolution of sedimentary basins associated with basin uplift and inversion (Mather, 1993). In the Neogene basins of southeast Spain a phase of Plio-Pleistocene alluvial fan sedimentation followed earlier marine and then low-energy terrestrial environments. Basinwide deposition ceased following uplift and dissection of the basins (Mather, 1993; Mather and Harvey, 1995) and the conversion of the fans to through drainage (Harvey, 2006; Maher, 2006; Stokes,

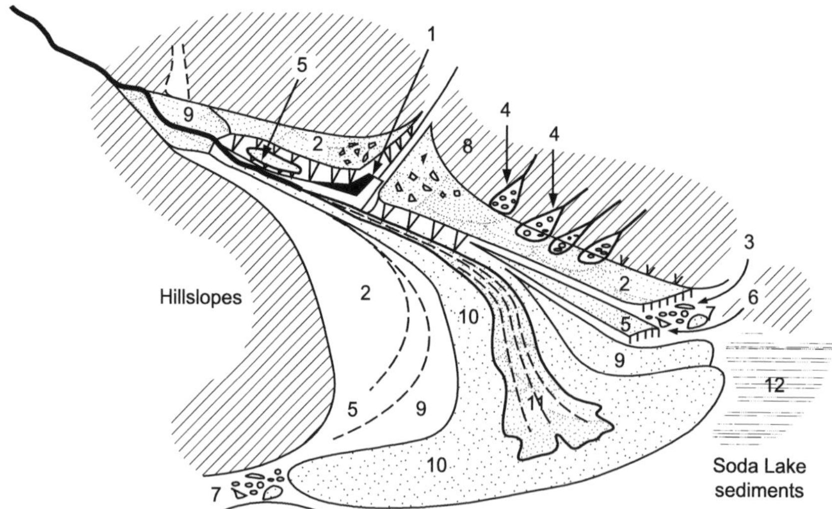

Figure 14.6 Summary sequence of late Pleistocene and Holocene alluvial fan erosion and deposition in Zzyzx, California: 1, Mid Pleistocene heavily calcreted fan deposits; 2, main phase of Late Pleistocene aggradation: debris flow in proximal areas and from small tributary fans, fluvial deposits in main distal fan; 3, Late Pleistocene shoreline of pluvial Lake Mojave; 4, hillslope debris flows, contemporaneous with phases 2 and 5; 5, latest Pleistocene fan deposits resulting from limited fanhead trenching and distal progradation; 6, Early Holocene shoreline of pluvial Lake Mojave; 7, lake shoreline sediments relating to phases 3 and 6; 8, hillslopes now stabilised during the Holocene; 9, Mid Holocene fan sedimentation, limited deposition at fan apex, fanhead trenching in midfan and distal progradation by fluvial deposition; 10, further fanhead trenching and distal progradation by processes during the Late Holocene; 11, active fan sediments; 12, modern salt flats of Soda Lake (modified after Harvey and Wells, 1994, 2003).

2008; Silva *et al.*, 2008). A similar sequence, but operating over longer timescales and over a larger spatial scale, has been demonstrated for the Tertiary fans of the Ebro basin, northern Spain (Hirst and Nichols, 1986; Nichols, 2005). In the southeast Spain case, smaller fans developed around the margins of the basins later in the Pleistocene, largely in response to climatic change (Harvey, 1987, 1990).

Spatial variations in fan deposition partly reflect proximal-to-distal variations in depositional processes and sorting mechanisms. Debris-flow dominance in proximal zones may give way downfan to sheetflood deposition (Figure 14.5(b)) as debris-flow runout distances (D'Agostino, Cesca and Marchi, 2010) may be limited to the upper parts of a fan. Channel deposits may also give way distally to sheetflood deposits as flow spreads over the distal fan surfaces. Where this transition takes place, often in midfan below the intersection point, sieve deposits may be common (Hooke, 1967; Wasson, 1974). In distal environments sand and silt sheets may interdigitate with thin gravel sheets. Within the sediments as a whole there is normally a downfan decrease in clast size and an increase in sorting (Bull, 1962a, 1963; Bluck, 1964, 1987; Lustig, 1975).

Channel switching and fanhead trenching will produce detailed variations in the distribution of sedimentary units over the fan surface, resulting in a mosaic of fan deposits and of depositional surfaces of different ages. There is a contrast in spatial pattern between the random pattern produced by channel switching near the apex of an undissected aggrading fan and the more ordered age-progression of deposits younging downfan and telescoped (Bowman, 1978) into the fanhead trench of trenched prograding fans (Figure 14.5(c)).

14.2.4 *Alluvial fan morphology and style*

The morphological style of a fan reflects its recent history and its current process regime, controlled by the critical power relationships discussed earlier (Harvey, 2002b). Four basic styles can be recognised (Figure 14.7(a)). These are:

(i) A passive/inactive fan is a relict feature inherited from a more active environment in the past. Under present processes little or no sediment is yielded from the feeder catchment and flood power is insufficient to cause any erosion on the fan, floodwater merely infiltrating into the fan sediments.

(ii) Under conditions of excess sediment supply a fan will undergo aggradation, the style of sedimentation ranging from debris-flow to fluvial dominance, dependent on the processes feeding sediment to the fan.

(iii) Progradational fans are common, exhibiting proximal erosion within a fanhead trench and distal deposition downfan from an intersection point. Such fans may be mildly progradational, transferring sediment from proximal to distal zones, or may be telescopic (Bowman, 1978) and strongly progradational, extending the fan boundaries distally.

(iv) Under conditions of excess flood power, fans may undergo dissection. This may be focused on one of three zones of the fan, proximally to form a fanhead trench, in midfan, downfan from an intersection point (see Section 14.3.2.2), or distally, in which case it may reflect either erosion of the fan toe (Leeder and Mack, 2001) or base-level change (see Section 14.3.2.3).

Given the topographic location and available accommodation space (Viseras *et al.*, 2003), apart from the influence of fan toe processes, the fan regime responds to water and sediment supply conditions fed from the fan catchment. Should these conditions change, then the fan regime will respond, effectively by a change in plotting position on Figure 14.7(b), resulting in either a change in the intensity of processes or a change in the process regime of the fan.

14.3 Factors controlling alluvial fan dynamics

Alluvial fan dynamics, the response of alluvial fan systems to external controls, depend on two sets of factors: (a) those that are essentially passive over the timescale of fan development and (b) those that are dynamic, changing within that timescale, and to which fan processes and morphology respond. Passive factors include the fan setting, dependent on the tectonic and geomorphic history and influencing accommodation space and confinement (see Section 14.1.1), catchment geology, affecting rock resistance to erosion, and catchment size and relief, influencing water and sediment supply. Of course, over long timescales these factors can change through, for example, tectonic uplift of the source area, progressive erosion, reducing catchment relief or through river capture affecting drainage area (Mather, 2000; Mather, Harvey and Stokes, 2000), but in the context of, say, Late Quaternary alluvial

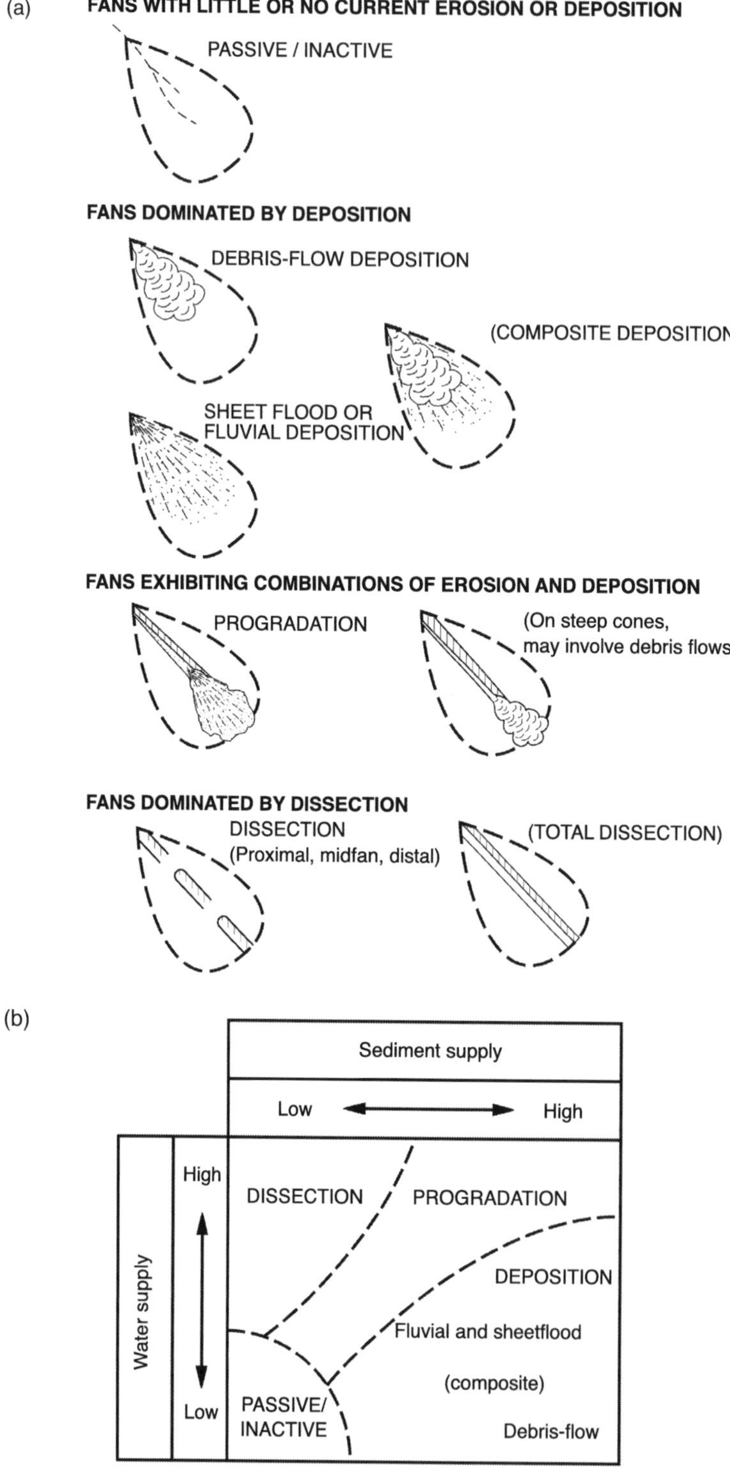

Figure 14.7 Fan style: (a) schematic representation of fan style (modified from Harvey, 2002b); (b) schematic model of influence of water and sediment supply (hence critical power relationships) on fan style, based on Figure 14.3(b) (modified from Harvey, 2002b).

fans, except in exceptional circumstances, they can be regarded as passive factors. The dynamic factors include ongoing tectonic activity, climate and base-level conditions (Harvey, Mather and Stokes, 2005).

14.3.1 Passive factors: influence on fan morphology

14.3.1.1 Setting

The setting of the fan (e.g. mountain front, tributary junction, etc., see Section 14.1.1) is controlled by long-term tectonics (see also below for a discussion of relations between passive and active tectonic control), geological and geomorphological history, which, in turn control the available accommodation space (Viseras *et al.*, 2003) and the degree of confinement offered by the surrounding bedrock topography (Sorriso-Valvo, Antronico and Le Pera, 1998; Al Farraj and Harvey, 2005).

14.3.1.2 Catchment controls

The geology of the catchment area, especially rock resistance to erosion, may affect the volume and nature of the sediment delivered to the fan, which may in turn influence the processes of sediment delivery (large amounts of fine sediment would be conducive to debris flows) and the volume of sediment delivered. This, in turn, affects the fan aggradation rate and the fan size (Hooke and Rohrer, 1977; Lecce, 1991). Most important are the topographic characteristics of the catchment, the drainage network and particularly the catchment size and relief characteristics. Larger catchments produce higher flood discharges, and with the greater potential for within-catchment sediment storage (Harvey, 2010) are likely to have higher water-to-sediment ratios during floods than smaller catchments. Smaller and especially steeper catchments are likely to produce higher sediment concentrations, coarser sediments and more debris flows. The combined effects of catchment size and relief on sediment delivery to the fan system are well known (Harvey, 1984a, 1992a; Kostaschuk, Macdonald and Putnam, 1986; Wells and Harvey, 1987), whereby larger and less steep catchments generate fluvially dominated fans and smaller steeper catchments generate debris cones and debris-flow dominated fans. These relationships are expressed in fan morphology, where debris-flow fans are in general steeper than fluvially dominated fans (Figure 14.8(a)) and there is a general positive relationship between drainage area and fan area (Figure 14.8(b)).

14.3.1.3 Fan morphometry

A common approach to the analysis of fan morphology has been to consider the morphometric relationships between catchment controls, especially drainage area, and fan properties, especially fan area and fan gradient, using regression analysis (Bull, 1962b; Denny, 1965; Hooke, 1968; Harvey, 1987, 1990, 1992b, 2002a). The results have usually been expressed in the form below (after Harvey, 1987):

$$F = pA^q \tag{14.1}$$

$$G = aA^b \tag{14.2}$$

where A is the drainage area (km^2), F is the fan area (km^2) and G is the fan gradient (dimensionless). Figure 14.9 summarises these regression relationships from a selection of earlier studies.

For the fan area relationship, from a whole series of studies of dry-region fans there is a clear positive correlation between the drainage area and fan size. The exponent q shows very little variation between c. 0.7 and c. 1.1, but values for constant p show a wide range of between c. 0.1 and c. 2.1, reflecting regional differences in fan age and history as well as in catchment geology (Hooke, 1968; Hooke and Rohrer, 1977; Lecce, 1991) and fan setting, especially confinement (see below).

For the fan gradient relationship, again from a whole series of studies (Blissenbach, 1952; Bull, 1961, 1962b; Harvey, 1987, 1990, 2002a, 2002b), values for the exponent b show a limited range between −0.15 and −0.35, but values for the constant a show a much greater range from c. 0.03 to c. 0.17, probably reflecting different sedimentary processes, with debris-flow fans being much steeper than fluvially dominant fans.

The fan gradient relationship is important because different depositional processes have different threshold depositional gradients (see above, Section 14.2.1). This can be illustrated by data from fans from a number of regions (Figure 14.10), where the fans have been classified on the basis of either surface morphology or deposits exposed in fanhead trenches into debris-flow or fluvially dominant (Harvey, 1992b). For these areas debris-flow fans are generated only from catchments smaller than 5 km^2. For every group, fan gradients for debris-flow fans are almost an order of magnitude steeper than those for fluvial fans draining similar drainage areas.

This difference in depositional gradient between debris flows and fluvial sediments has been used to account for the differences in gradient between fans and

Figure 14.8 Size and gradient of alluvial fans. (a) Musandan mountain front, Wadi Al-Bih, Oman. Note the steep angle debris-flow cone behind the lower-angle fluvially dominant fan in the centre of the photo. The main wadi cuts the toe of the fluvially dominant fan in the foreground. (b) Mountain-front fans, Panamint Valley, California. Note the triangular slope facets indicating a faulted mountain front. Note also the general positive relationship beween catchment size and fan area and the inverse relationship with the fan gradient.

rivers (Blair and McPherson, 1994b), defined as the 'slope gap'. While there is undoubtedly a difference in threshold depositional gradient, the concept of a specific slope gap is flawed. Blair and McPherson compared the moderate-sized Death Valley fans with larger rivers. Saito and Oguchi (2005) have demonstrated that fan gradients do, in fact, span the slope gap and Harvey (2002a) has demonstrated that gradients on small rivers also span the gap.

In an attempt to improve on the general drainage area to fan gradient regression, Harvey (1987), on a sample of 77 Spanish fans, used multiple regression, also taking into account drainage basin relief characteristics. The overall simple regression was

$$G = 0.066A^{-0.20} \tag{14.3}$$

(correlation coefficient $= -0.64$ and standard error of the estimate $= 0.170$ log units).

Subdividing the data into three groups (shown as three separate lines in Figure 14.9: lower), in two of the three cases correlations improve and standard errors reduce marginally. When multiple regression is used the correlations improve markedly to between 0.74 and 0.89, and standard errors are reduced further (see Harvey, 1987, 1990, for details).

Morphometric analysis has also be used to examine other fan properties, particularly those of fan channels (Harvey, 1987, 2002a). For channels within fanhead trenches, the channel gradient relationship to drainage area can be expressed by the equation

$$S = fA^{g} \tag{14.4}$$

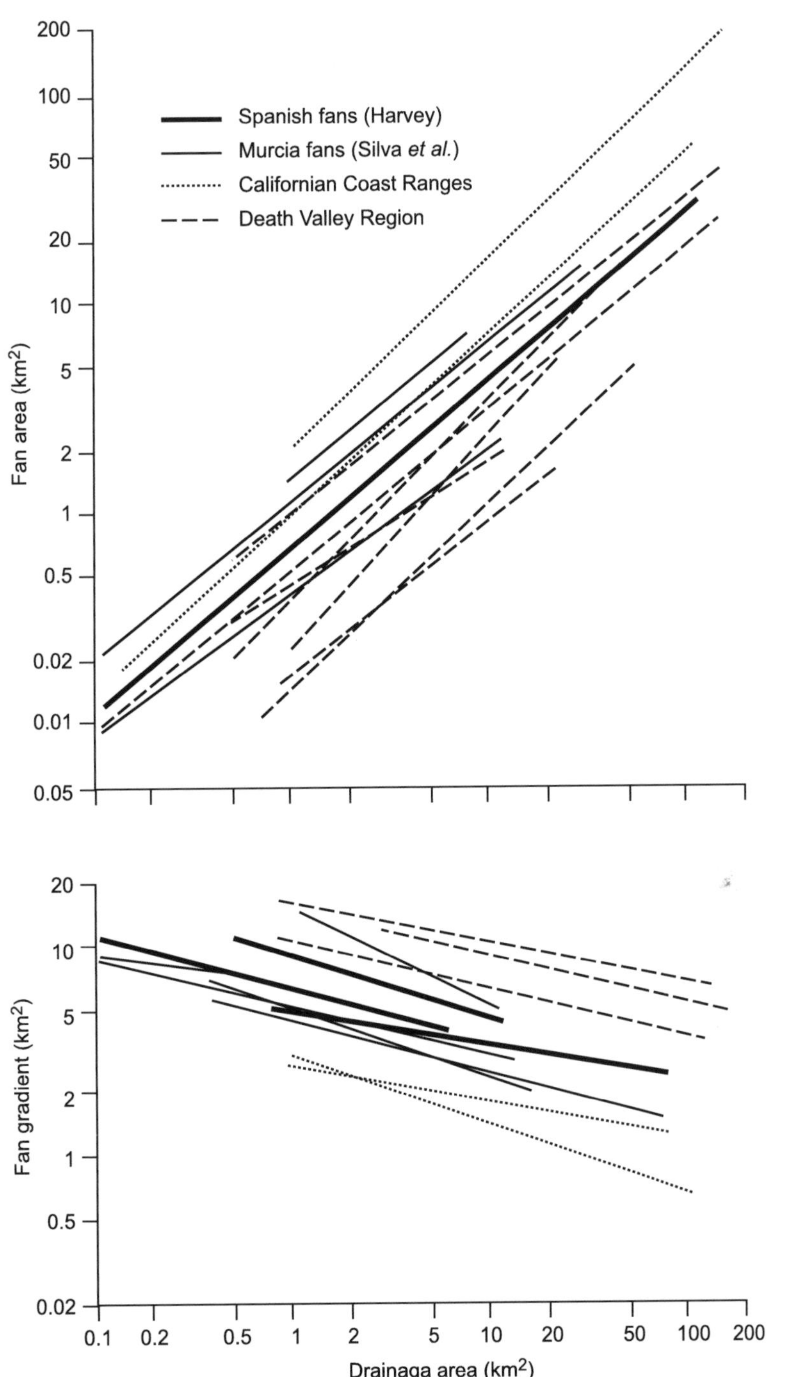

Figure 14.9 Summary of regression relationships of drainage area to fan area and fan gradient (data for Spanish fans from Harvey, 1987, 1997; for Murcia (Spain) from Silva *et al.*, 1992; for California Coast Ranges from Bull, 1962b, 1977; for Death Valley region from Denny, 1965, Hooke,1968, and Hooke and Rohrer, 1977).

(where S is the dimensionless channel slope). For the three Spanish fan groups, the exponent g ranges from -0.2 to -0.4 and the constant f is between c. 0.02 and 0.06. The actual gradients are of course less than the fan surface gradients. There are insufficient data for a comparison with American fan channels, but a comparison can be made with small Spanish nonfan stream channels (Harvey, 2002a). For drainage areas less than c. 5 km^2 there

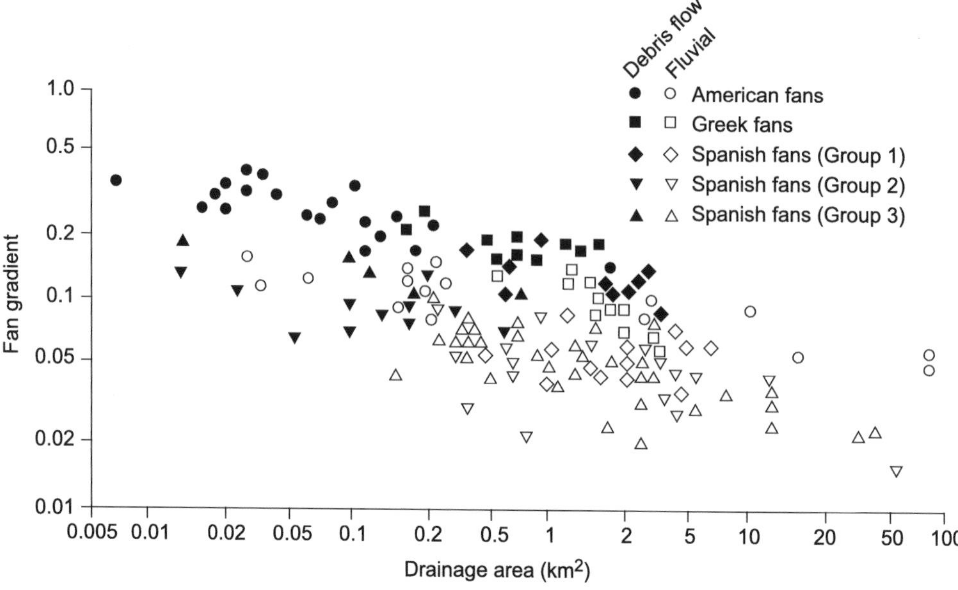

Figure 14.10 Drainage area to fan gradient relationships for selected Spanish, Greek and American fans, subdivided into debris-flow and dominantly fluvial fans (data from Harvey, 1992b).

is no significant difference in gradient between fan and nonfan channels. Only for larger drainage areas are the fan channels steeper than the nonfan channels.

For fanhead trench channels the width (W, metres) relationship to drainage area can be expressed by the equation

$$W = hA^k \tag{14.5}$$

For the three Spanish fan groups, the exponent k ranges from 0.17 and 0.34 and the constant h between 6 and 13. Again there are few comparable American studies, but Denny (1965) quotes values of 0.5 for exponent k and 23.3 for constant h. It is clear that American fan channels, at least in the Death Valley region, are much wider than the Spanish channels and tend to increase in width down the fanhead trench whereas the Spanish ones maintain a near-constant width through the fanhead trenches. One characteristic that is common on some American fans and many Spanish fans is the presence of calcrete on the channel floors (Van Arsdale, 1982; Harvey, 1987), restricting the widths of these fan channels.

More recent research than much of that quoted above has used morphometric analysis to focus on how fan morphology reflects interactions between the controlling factors (e.g. Calvache, Viseras and Fernandez, 1997). For example, from a modified sample from the Spanish fans, using 67 fans with a single apex, Harvey (2002a) calculated the two basic morphometric regression relationships, yielding the following results:

$$F = 0.807A^{0.675} \tag{14.6}$$

(correlation coefficient = 0.908 and standard error of the estimate = 0.208 log units) and

$$G = 0.068A^{0.249} \tag{14.7}$$

(correlation coefficient = −0.693 and standard error of the estimate = 0.173 log units).

In Figure 14.11 plotting positions are identified by fan style. Different groups of fans tend to cluster either above or below the regression lines, suggesting the need for an analysis of the two groups of residuals. This is the approach first used by Silva *et al.* (1992b) and helped to identify the influence of tectonic setting on the spatial geometry of fans in Murcia, Spain (Figure 14.12(a)). It has subsequently been used by Harvey *et al.* (1999a) on the Cabo de Gata fans, identifying the effects of confinement, and of coastal erosion of the fan toes (Figure 14.12(b)) by Harvey (2005) in Nevada, identifying the effects of base-level change (see below), and by Al Farraj and Harvey (2005) in the UAE and Oman, identifying the effects of source-area geology (Figure 14.12(c)) as well as those of fan setting and confinement.

Another application of fan morphometry has been to infer past drainage basin properties from ancient fan

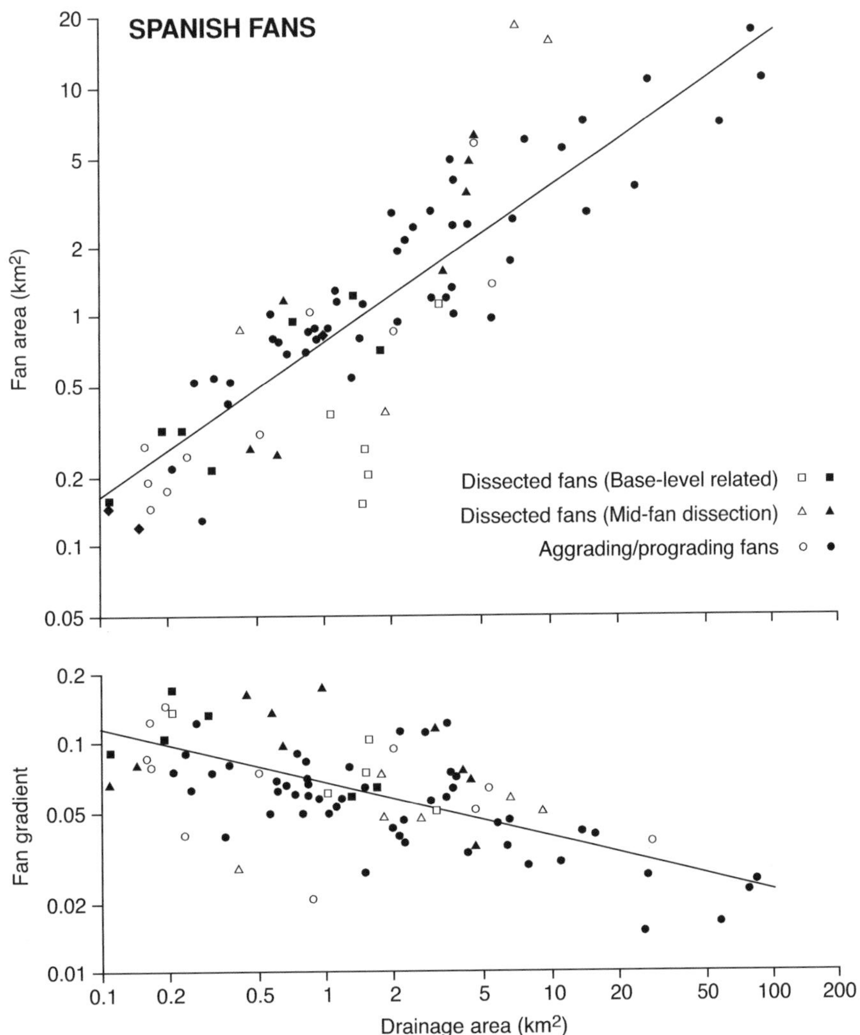

Figure 14.11 Morphometric regression relationships between drainage area and fan area and gradient for 67 Spanish Quaternary fans (modified from Harvey, 2002a), differentiating between dissected and aggading/prograding fans. Open symbols are for multiapex fans, not used in the regression analysis (see text for details).

sediments, using the morphometric properties of modern fans (Mather, Harvey and Stokes, 2000).

The morphometric approach has proven to be an important tool in the analysis of the controls of fan morphology, but it has its limitations in that it tends to generalise rather than focus on the depositional processes themselves (Whipple and Dunne, 1992; Whipple *et al.*, 1998).

14.3.2 Dynamic controls

14.3.2.1 Ongoing tectonics

Tectonics can be both a passive and a dynamic control on alluvial fan processes and morphology. As a passive control, past tectonics may determine the position and overall relief of mountain-front environments and hence the settings of alluvial fans. Cessation of tectonic activity tends to result in the backfilling of fans into the mountain catchments and to translate a former linear mountain front into an irregular and recessed mountain front (Bull, 1978), on which fan sediments lap unconformably onto the underlying bedrock.

As an active control, ongoing tectonic activity during the existence of the fan directly affects the fan itself, significantly altering the processes or morphology, or acts by modifying regional base levels, which in turn affect the fan environment as base-level change propagates through the system. The distinction between passive and active

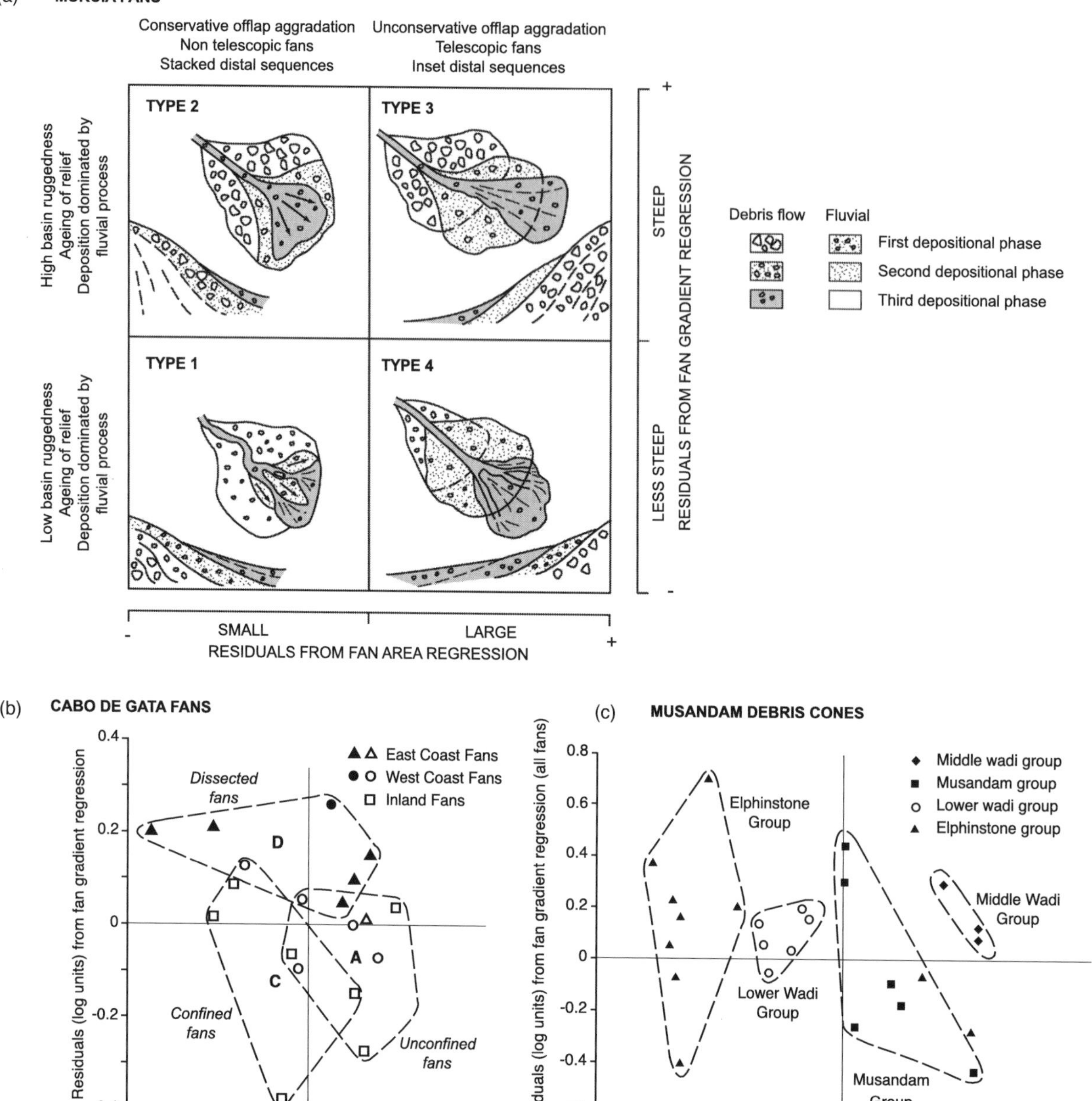

Figure 14.12 Examples of analyses of regression residuals: (a) Murcia fan groups: influence of tectonic context (after Silva *et al.*, 1992); (b) Cabo de Gata fans: influence of fan setting and base-level conditions (after Harvey *et al.*, 1999). Solid symbols relate to fans subject to base-level induced dissection; A,C,D relate to the mean plotting positions for the three groups of fans. (c) Musandam debris cones (after Al Farraj and Harvey, 2005): residuals from the 'all-fans' regressions of fan area and gradient on drainage area. The various groups differ in geology and basin relief characteristics.

influence is not always obvious and sometimes can be seen as a matter of scale. Three examples can be used to illustrate this. Large beheaded fans exist on the Jordanian side of the Dead Sea rift (Enzel, personal communication). Their position reflects essentially passive structural control, but their beheading reflects active tectonic movement along the main strike-slip fault.

The Sierra de Carrascoy, Murcia, southeast Spain, is flanked on both northern and southern margins by alluvial fans (Harvey, 1988). The mountain range is bounded to the north by a major thrust fault system and to the south by basically an unfaulted but formerly pedimented margin (Figure 14.13(a)). The structure is essentially a passive control over alluvial fan location and style, with simple mountain-front fans to the north backfilling slightly into the mountain catchments (Figure 14.2(c)), but with a complex of pediment-burying irregular mountain-front and intramontane fans to the south. However, the faults are not totally inactive; mid-Pleistocene fan sediments on the northern mountain front are faulted and the older fan segments have probably been tilted, steepening the upper fan gradients (Silva *et al.*, 1992). On the south there are minor faults crossing the piedmont area (Somoza *et al.*, 1989), but these have little or no discernible effect on the fan geomorphology.

Death Valley, California (see Box 14.1), is a half-graben bounded to the west by the uplifting Panamint Range and to the east by the normal-fault-bounded margin of the Black Mountains (Hunt and Mabey, 1966; Hunt, 1975). This passive structural setting results in the contrast in fan style between the large, telescopic west-side fans and the smaller stacked east-side fans (Denny, 1965). The Panamint mountain front is essentially a nonfaulted uplifted block, affected only by small antithetic faults. The fans backfill into the mountain catchments. The normal fault along the Black Mountains mountain front is highly active, resulting in a straight fault-bounded mountain front, simple mountain-front fans, aggrading from the apices down, essentially because accommodation space is increased as the floor of the valley subsides. A similar complex pattern of different styles, dependent on local active tectonics, including sustained aggradation, has been identified on downfaulted fans in the Dixie Valley, Nevada (Bell and Katzer, 1987; Harvey, 2005).

The response of fans to tectonically induced base-level change can be illustrated by the fans in the Tabernas basin, southeast Spain (Figure 14.13(b)). The basin is bounded to the north by the tilted and uplifting basement schists of the Sierra de los Filabres and to the south by more schists of the the Sierra de Alhamilla thrusting forward along a frontal reverse fault system. To the north of the Sierra de Alhamilla is the Serrata del Marchante, an anticline in Neogene conglomerates forming a growth fold over a blind basement fault also thrusting northwards (Mather *et al.*, 2003). Quaternary alluvial fans formed along these three mountain fronts (Harvey *et al.*, 2003), as huge backfilled fans on the Filabres mountain front, simple mountain-front fans along the Marchante front and a confined fan complex where the Rambla de la Sierra issues from the Sierra de Alhamilla (Harvey, 1978, 1984b; Delgardo-Castilla, 1993). During the Late Pleistocene, uplift along a growth fold to the west of the basin ponded the drainage, causing a lacustrine or palustrine environment to form to the west of the fans (Harvey *et al.*, 2003). During the Late Pleistocene the barrier caused by the growth fold was breached and the drainage rapidly incised, creating a spectacular badland landscape in the western part of the basin, but also cutting back into the alluvial fan zone in the eastern part of the basin. At present the incision wave has cut back into the Sierra de Alhamilla, dissecting the Sierra fan; dissection of the Marchante fans is, at least in geomorphological timescales, imminent, but is nowhere near dissecting the Filabres fans. This illustrates an important point related to coupling through drainage systems (Harvey, 2002c). The tectonic signal, which may be synchronous only in the vicinity of the structure themselves, is otherwise variable and time transgressive, as the effects are propagated through the system. A similar incision wave, but related to Mid Pleistocene deformation has been identified by Garcia *et al.* (2004) in the Alpujarras, west of the Tabernas basin, whereby tectonically induced dissection has propagated through the system, eventually causing fan entrenchment at the head of the system.

14.3.2.2 Climatic change

Although alluvial fans are not exclusively dry-region forms, they are particularly well developed in arid mountain areas. What is important is not climate itself, but climatic change (Bull, 1991). Fan surface form adjusts to the prevailing flood and sediment regime (see above, Section 14.2.4, Figure 14.7(b)), and if a climatic change causes a change in this regime, the fan will adjust accordingly. Indeed, in a whole wealth of studies of Quaternary dry-region fans, the climatic signal seems to dominate, and even in areas of considerable tectonic activity or base-level change the climatic signal is only modified by the tectonic or base-level signal (e.g. Frostick and Reid, 1989; Roberts, 1995; Ritter *et al.*, 1995; Bowman, 1988; Harvey, 2002a, 2003; Pope and Wilkinson, 2005).

Three examples illustrate the alluvial fan response to Quaternary climatic change. On many Spanish fans, although the fan catchments themselves were not glaciated during Pleistocene cold phases, the cold dry climates generated large volumes of sediment, which resulted in sustained periods of regional aggradation affecting fans and river valleys (Harvey, 1987, 2003). During the intervening interglacials the fans were largely sediment starved and underwent fanhead trenching and limited progradation. During the current interglacial period, the Holocene, this has resulted in many fans undergoing incision and headcut formation at or immediately downfan from the intersection point (Figure 14.14; compare Figure 14.15(a) and (b)). The reasons for this appear to be, first, that many of

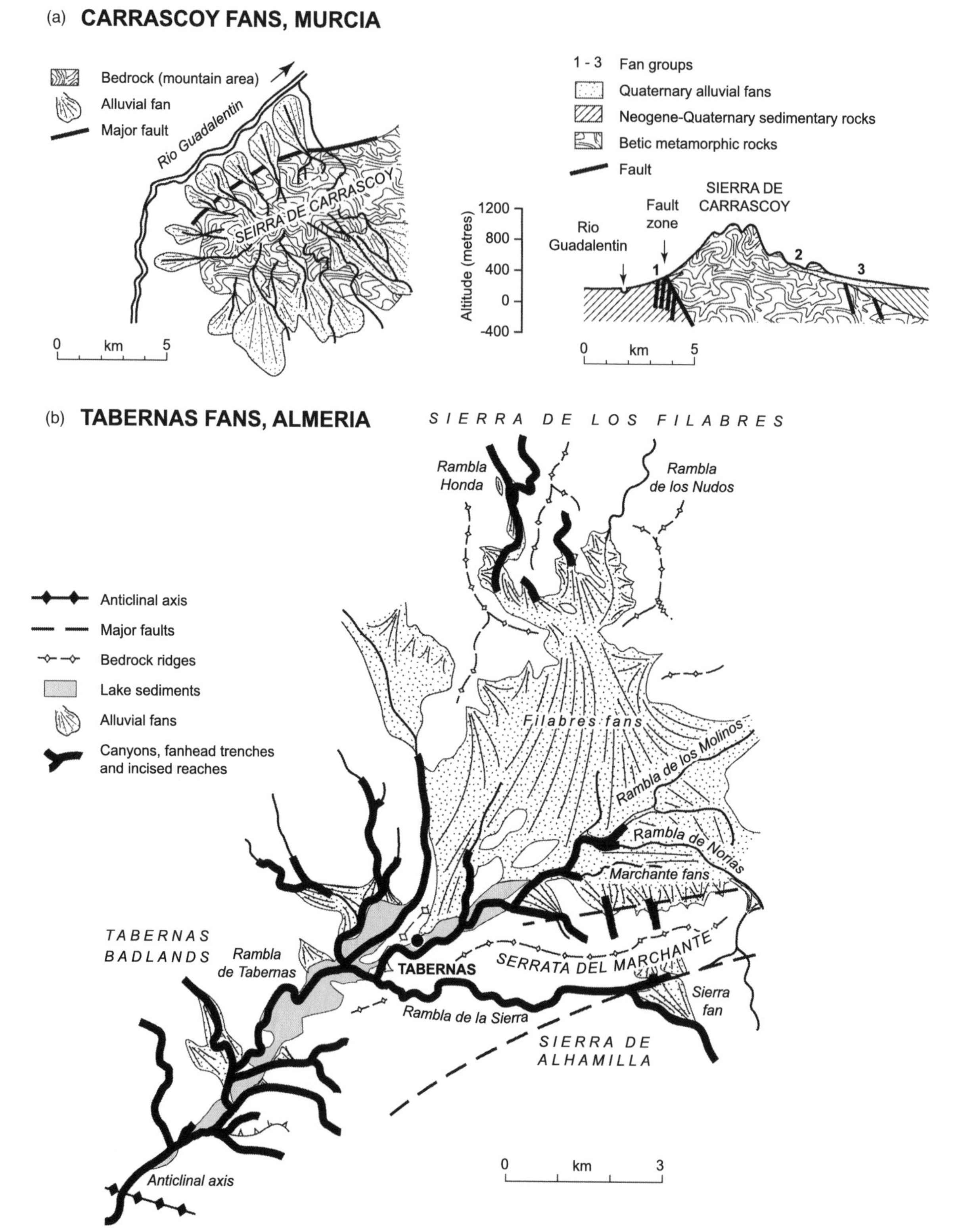

Figure 14.13 Examples of tectonic influences on fan style: (a) contrasts between fans on the northern and southern margins of the Sierra de Carrascoy, Murcia, Spain (modified from Harvey, 1988); (b) contrasting tectonic setting of the Tabernas fans, illustrating the different coupling relationships following the tectonically induced headwards erosion of the Rambla de Tabernas into the upper part of the basin (after Harvey *et al.*, 2003).

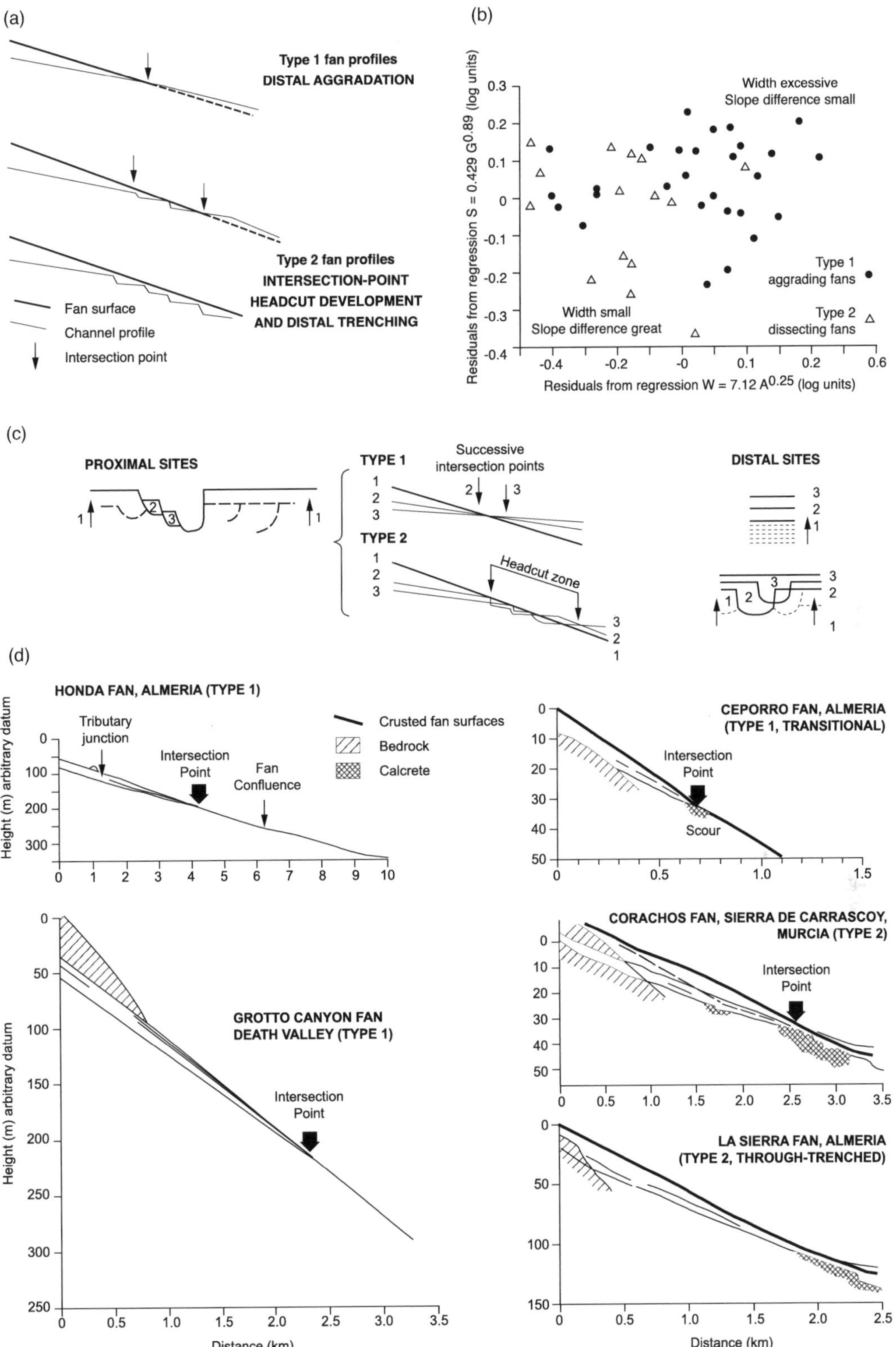

Figure 14.14 Mid-fan (intersection point) dissection. (a) Schematic representation of mid-fan (intersection-point) profiles: Type 1, 'normal' intersection-point aggradation; Type 2, intersection-point headcut development and distal trenching. (b) Residuals from channel gradient on a gradient regression plotted against those from channel width on drainage area regression (from Harvey, 1987). Note how dissected fans of Type 2 are crusted fans associated with large gradient differences and small widths. (c) Implications for distal fan stratigraphy. Note how cut-and-fill stratigraphy is characteristic of sequences prone to trenching whereas stacked stratigraphy is characteristic of aggrading fans. (d) Example fan profiles for selected fans illustrating intersection-point dissection types (modified after Harvey, 1987); morphometric data for these fans are given in Table 14.1. Note that La Sierra fan (Tabernas, southeast Spain) has been affected by base-level induced dissection.

the older fan surfaces are indurated by a calcrete crust, reducing the erodibility of the older sediments and restricting the width of the fan channels (see above). This factor when combined with the increase in gradient at the intersection point, where the channel gradient gives way to an inherited fan surface gradient, resulted in an increase in unit stream power sufficient to allow erosion, rather than the more normal decrease in unit stream power associated with a reduction in confinement and associated intersection point deposition (Wasson, 1974). Intersection-point headcut development appears to be favoured by calcrete-crusted fan surfaces, on which there is a large increase in channel gradient at the intersection point and where channels are restricted in width (Figure 14.14). Successive headcuts may form downfan to create throughfan dissection (e.g. Figure 14.15(c)), although this type of dissection can also result from tectonically or base-level induced distal dissection that propagates headwards. These relationships can be illustrated by the fan profiles for selected individual fans (Figure 14.14, with summary gradient and channel width data given in Table 14.1). For the example fans, note that on all fans except La Sierra (subjected to tectonically induced dissection; see Section 14.3.2.1 above) distal fan gradients are greater than channel gradients within the fanhead trench. Note also that Corachos is a strongly crusted fan and has the greatest *G/S* ratio and a relatively narrow channel, all properties conducive to the formation of intersection-point headcuts.

A second, simpler, response to climatic change is illustrated by the response of the Nevada and Zzyzx, California, fans to the Pleistocene to Holocene climatic change (Harvey, Wigand and Wells, 1999). In both cases large volumes of alluvial fan sediments pre-date the Late Pleistocene pluvial Lakes Lahontan and Mojave. The Zzyzx fans were active during the Pleistocene–Holocene climatic transition (Harvey and Wells, 1994, 2003; see also Figure 14.6), undergoing some hillslope debris-flow activity, limited fanhead trenching and distal progradation. On the other hand, the Stillwater fans in northern Nevada were inactive at that time (Harvey, Wigand and Wells, 1999). Lake levels fell and there was some base-level induced distal incision, but it was not until climatic aridification in the Mid Holocene that significant fan sedimentation resumed. The reasons for these differences are partly due to vegetation differences (themselves climatically controlled) and partly due to climate itself. Even during the Late Pleistocene the lower elevation, more southerly Zzyzx area was under desert scrub vegetation, whereas at the same time the Stillwater fans were under juniper woodland and the mountan catchments were under coniferous forest. Only in the Mid Holocene did desert scrub vegetation become established in northern Nevada (Harvey, Wigand and Wells, 1999), allowing a reduction of erosional thresholds. Climatically the Early Holocene saw the penetration of unstable 'monsoonal' air masses, with their associated intense storm rainfall, into southeastern California. It was only in the Mid Holocene that there appeared to be penetration of these air masses further north (Harvey, Wigand and Wells, 1999).

A final example deals with the alluvial fans of the Musandam peninsular in northern UAE and Oman (Al Farraj and Harvey, 2000, 2005). Here three main periods of Late Quaternary alluvial fan sedimentation are identified. The first, of an uncertain date, involved substantial sedimentation of relatively small gravels throughout the region, but especially in the high mountains, suggesting a colder climatic regime with mechanical weathering important in the mountains. The second phase, in the Late Pleistocene, involved much coarser sediments and in many places can be directly linked to slope failures, suggesting a wetter climate. The third phase, in the Early Holocene, saw reworking of older deposits and was probably associated with a wetter climate than today.

14.3.2.3 Base-level change

Most fans toe out to stable base levels, so over time fans would extend or prograde distally (Figure 14.16). Base-level change is irrelevant for fan dynamics on these fans. Even for fans on the margins of basins of interior drainage where base levels may slowly rise as sedimentation takes place in the basin centre, the rate and amount of change is likely to have little or no effect on fan dynamics. However, for fans that toe out into main river valleys or on coastlines, both marine and lacustrine, base-level change can be a major control over alluvial fan dynamics.

Conventionally a base-level fall would trigger incision of fan toe zones and a base-level rise would promote sedimentation in fan toe zones. In reality, the situation is more complex. Even fans toeing out on to a stable valley-floor base level may be subject to 'toe trimming' by lateral migration of the main river (Leeder and Mack, 2001) or in coastal areas by wave erosion of the distal fan areas. Such erosion could effectively shorten and steepen the distal fan profile, causing a wave of incision to work headwards up the fan (Figure 14.16).

Distal incision can, of course, be produced by a fall in base level, in the fluvial case by a fall in local base level caused by the incision of a main stream, a situation documented on Wadi Al Bih, UAE (Al Farraj and Harvey, 2000, 2005; Harvey, 2002b). Where fans toe out at a marine or lacustrine margin, such incision will *only* occur if the newly exposed sea or lake floor has a steeper gradient than the threshold erosional gradient for the fan

Figure 14.15 Intersection-point contrasts. (a) The 'normal' pattern of intersection point aggradation, looking downfan from the intersection point on Villaescusa fan, Murcia, Spain. (b) Intersection-point headcut development, Torres fan, Murcia, Spain. (c) Throughfan dissection at Virgin Mountain fans, Nevada.

channels (Figure 14.16). That was the case for the Stillwater fans, Nevada, after Lake Lahontan fell, exposing the steep former lake margins (Harvey, 2002d, 2005), but was not the case for the Cabo de Gata fans, Spain, as the Mediterranean Sea fell after the last interglacial highstand and exposed only a low gradient sea floor (Harvey *et al.*, 1999; Harvey, 2002d). In that case sediments generated during the subsequent glacial would have simply prograded over the exposed former sea floor, a situation similar to that described for recent falling levels of the Dead Sea, Israel (Bowman, 1988).

A rise in lake or sea levels would transform a fan into a fan delta if the distal zones of the fan are inundated, causing interaction between fan and coastal sediments at the shoreline and creating a new subaqueous sediment wedge offshore.

Table 14.1 Morphometric data for the example fans illustrated on Figure 14.14(d).

	1	*2*	*3*	*4*	*5*	*6*
Honda *Tabernas, SE Spain*	27.0	0.032	0.020	0.029	22–30	1.6
Grotto Canyon *Death Valley, California*	10.6	0.085	0.064	0.075	60–120	1.3
Ceporro[a] *Tabernas, SE Spain*	0.4	0.081	0.043	0.063	2.0–2.5	1.9
Corrachos[a] *Carrascoy, SE Spain*	4.6	0.033	0.014	0.020	7–10	2.4
La Sierra[b] *Tabernas, SE Spain*	3.3	0.055	0.045	0.045	20–25	1.2

1. Drainage area (km^2).
2. Fan gradient (dimensionless) G.
3. Channel gradient (dimensionless) S.
4. Distal fan gradient (dimensionless).
5. G/S.

[a]Fan surfaces crusted by calcrete.
[b]Fan surfaces crusted by calcrete; fan subjected to tectonically induced distal dissection (see Section 14.2.4, Figure 14.13(b) and Figure 14.14(d)).

Alternatively, if exposed to sufficient wave energy, the distal zones may be subject to erosion, and hence profile shortening and steepening (Figure 14.16). This was the case for the Cabo de Gata east coast fans during the last interglacial marine highstand and again during the Holocene (Harvey *et al.*, 1999), when distally induced dissection cut back through the fans, one of the few documented cases of a base-level rise-triggering incision (see also Scheepers and Rust, 1999).

14.4 Alluvial fan dynamics

14.4.1 Expressions of fan dynamics

Combinations of the passive and dynamic factors controlling fan evolution are expressed in the fan sediment sequences and particularly by the fan morphology, both in plan view (Figure 14.17) and in profile. Within the context of the passive controls (see above Section 14.3.1), alluvial fans respond to changes in the dynamic controls (see above, Section 14.3.2), by changes in erosional or sedimentational regime and/or changes in morphological style following changes in critical power relationships (see above, Section 14.3.2; Figures 14.3(b) and 14.7(b)). Some such changes, though probably relatively minor, may be brought about through intrinsic thresholds operating within the fan morphosedimentological system (Schumm, Mosley and Weaver, 1987), but major 'regime changes' are likely to be the result of extrinsically controlled thresholds related to major changes in the dynamic factors: tectonics, climate or base level.

How these factors interact, and how their signals are preserved in the sediment or morphological record, has been the subject of much research (Ritter *et al.*, 1995). Interestingly, most geological work dealing with the ancient rock record and necessarily larger bodies of sediment, often in the context of the evolution of sedimentary basins over relatively long timescales, stresses tectonic controls (Harvey, Mather and Stokes, 2005). Research on Quaternary fans, on the other hand, operating over the generally shorter timescales related to Quaternary climatic cycles, stresses climatic controls.

The controls themselves are not totally independent. One of the responses to tectonic activity would be a change in base level; base-level changes related to main valley incision or to changes in sea level or pluvial-lake level are either indirectly or directly related to the global climatic changes of the Quaternary.

For Quaternary fans the overriding control appears to be climatic change and its influence on sediment generation and supply to fan environments (Frostick and Reid, 1989; Roberts, 1995; Harvey, 2002b, 2005; Pope and Wilkinson 2005). In cases of local tectonic stability, and where fans toe out to stable base levels, the sedimentological and morphological sequences are wholly proximally and climatically controlled. In cases where there is local tectonic activity or base-level change the primary climatic signal is modified by tectonics or base-level change (see below).

14.4.2 Interactions between the dynamic controls: case studies of alluvial fan response to Late Quaternary environmental change

Three examples of interactions between the dynamic controls are developed from some of the examples given in the text in the previous sections (Figure 14.18).

14.4.2.1 Southeast Spain

Throughout southeast Spain a common sequence of Quaternary alluvial fan aggradation and dissection can be identified (Harvey, 1978, 1987), which can be ascribed

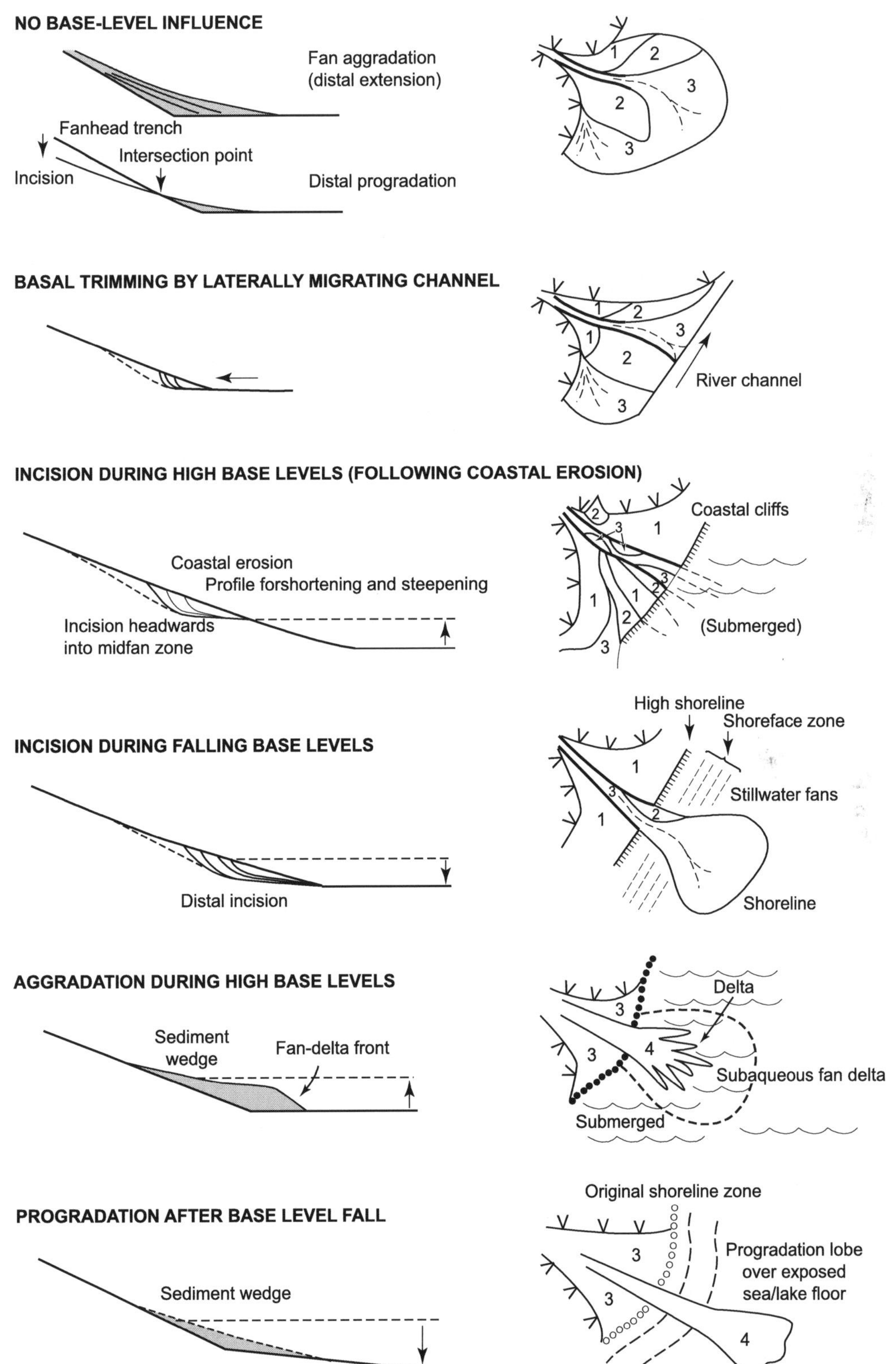

Figure 14.16 Influence of base-level conditions on alluvial fan profiles and plan geometry (modified from Harvey, 2002b, 2002d, 2003).

Figure 14.17 Segmented fan surface: Grotto Canyon fan, Death Valley, California. Note the older dissected basin-fill sediments (left foreground), three phases of fan aggradation, represented by the three fan surfaces, differentiated by the degree of dissection and colour as influenced by varnish development. Separating each aggradational phase was a dissectional phase. As a whole the fan is a telescopic, prograding fan.

to the effects of Quaternary climatic change, modified locally by fan setting and catchment characteristics (Harvey, 1990). Aggradation dominates the Quaternary cold stages, which in this area were cold and dry (Harvey, 2002b, 2003), and some form of fan dissection, fanhead trenching, midfan trenching or throughfan dissection, characterised the milder semi-arid Mediterranean interglacials. In zones of local tectonic and base-level stability the sequence is a simple climatically induced one. In some areas the climatic signal is modified by a tectonic signal (e.g. Carrascoy fans: Harvey, 1988; Silva *et al.*, 1992) or by a differential, tectonically induced base-level effect (e.g. Tabernas fans: Harvey *et al.*, 2003).

On the Cabo de Gata fans (Harvey *et al.*, 1999) the same climatic signal can be identified, but on those fans facing the coast it is complicated by the effects of base-level change following the Quaternary sea-level changes of the Mediterranean Sea. High interglacial sea levels caused coastal erosion of the fan toes, which foreshortened the fan profiles, triggering incision, which cut back through the fans (see above, Section 14.3.2.3). During interglacials reduced sediment supply had resulted in fanhead trenching at the same time as the basal-induced distal incision occurred (Figure 14.18(a)), resulting in throughfan trenching (Harvey, 2002b, 2002d, 2003). During the intervening glacials proximal fan aggradation resumed and as sea level fell, exposing only a low-gradient seafloor sediment simply prograded on to the exposed sea floor.

14.4.2.2 Northern Nevada

On the Stillwater fans in northern Nevada, major periods of sedimentation appear to have coincided with periods of aridity, and during the colder periods of the Late Pleistocene the fans appeared to have been receiving very little sediment (Harvey, Wigand and Wells, 1999). At those times lake levels of pluvial Lake Lahontan lapped on to the fan toes. As lake levels fell at the end of the Pleistocene the steep slope of the foreshore allowed distal fan incision to take place, but it was only during the increasing aridity of the Mid Holocene that the fans prograded substantially on to the exposed lake shore (Figure 14.18(b)), resulting in extensive lobes of sediment extending on to the former lake floor (Harvey, 2002b, 2003, 2005).

14.4.2.3 Musandam, UAE and Oman

The Quaternary fans of the Musandam peninsula can be grouped into two, mountain front and tributary junction, fans (Al Farraj and Harvey, 2000, 2005). Base-level change has had little or no influence on the dynamics of the mountain front fans, except in the very northern part of the UAE where they are lapped by the Arabian Gulf coastline (Al Farraj and Harvey, 2004). On the other hand, the tributary junction fans in Wadi Al Bih, the major west-coast drainage system, have been subjected to major local base-level changes, related to wadi-floor aggradation and dissection, to form a series of wadi terraces. Three main

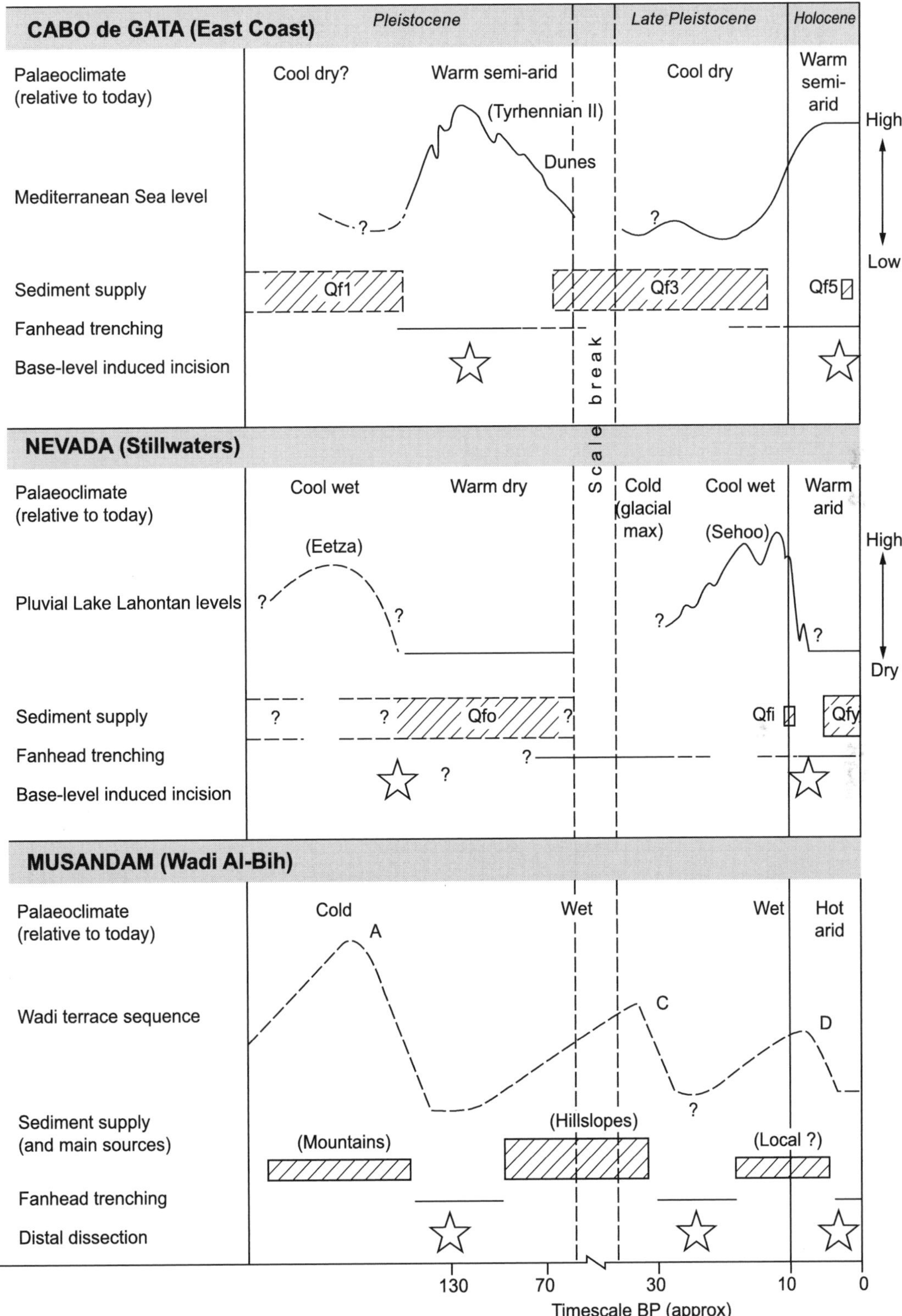

Figure 14.18 Summary of Late Quaternary sequences of aggradation and dissection on (a) the Cabo de Gata fans, Almeria, Spain, (b) the Stillwater fans, Nevada, (c) the Wadi Al-Bih fans, UAE and Oman (modified from Harvey, 2003).

phases of aggradation affecting both the wadi floor and the fans can be identified (see Figure 14.18(c)), which appear to be primarily responses to climatic changes. The first aggradation phase (see above, Section 14.3.2.2; see also Al Farraj and Harvey, 2000, 2004) appears to have been caused by cold climates in the mountain source area, though a tectonic influence cannot be entirely ruled out. The second and third phases appear to be related to climates wetter than today. Under the arid climate of today, there appears to be no excess sediment supply to the fans, simply the transport of wadi sediments through the channel system. Because of the synchroneity of excess sediment supply to both the fans and the wadi, fan aggradation coincided with rises in local base level and fan dissection with wadi incision (Harvey, 2002b, 2003).

These three examples illustrate how alluvial fan dynamcs respond to interaction of the dynamic controls, but the nature of the response depends on the precise timing of sediment supply in relation to the timing of either tectonic activity or base-level change.

14.5 Discussion: significance of dry-region alluvial fans

14.5.1 Commonly held myths and outdated concepts

Before discussing the significance of dry-region alluvial fans, it would be appropriate to consider some of the myths and misconceptions that persist in the literature. The first relates to the common association of wet fans/dry fans (Schumm, Mosley and Weaver, 1987) with wet climates/dry climates. Debris-flow or fluvial dominance is primarily controlled by sediment supply and transport mechanisms (see above Section 14.2.1). Arid and semi-arid environments are more prone to flash floods than humid areas (Baker, 1977). Erosion in humid areas is more likely to involve soils, providing the basis for debris-flow activity. If anything, these two factors would be likely to favour the development of fluvially (or at least sheetflood) dominant fans in arid areas (Harvey, 1992a). Some corroboration is afforded by the tendency for a climatic switch to aridity to involve a reduction in debris-flow and other mass movement processes and an increase in runoff-related processes (Gerson, 1982; Gerson and Grossman, 1987).

A second 'myth' relates to the concept of the 'slope gap' put forward by Blair and McPherson (1994b). As demonstrated above (Section 14.3.1.3), Blair and McPherson were correct in identifying different depositional slopes for various depositional processes, but the concept of a specific 'slope gap' differentiating fluvial from alluvial fan deposits is flawed.

A third myth, though not really a myth, relates to misconceptions in sedimentary sequence models, particularly in relation to coarsening-up sedimentary sequences. Many descriptions of alluvial fan sedimentary sequences in the rock record identify coarsening-up sequences as characteristic. This may relate to the preservation potential of proximal and distal environments As shown above (Section 14.2.3; Figure 14.5(a)), coarsening-up may be characteristic of distal fan environments on prograding fans, but fining-up is more characteristic of proximal environments on aggrading fans.

A fourth misconception is similar and relates to the different emphasis on tectonic control in relation to the rock record but climatic control of Quaternary alluvial fan sequences. This may again be related to preservation potential and to the different temporal and spatial scales involved (see above, Section 14.4.1).

14.5.2 Significance to science

14.5.2.1 The distinctiveness of dry-region alluvial fans?

Alluvial fans occur in a variety of climatic regions, but the conditions leading to the formation of alluvial fans are particularly well developed in dryland mountain regions (see Section 14.1.2). Furthermore, it is open to question whether sediment transport and depositional processes differ significantly between arid and humid regions. However, where there are important differences lie in the processes of post-depositional modification of fan surfaces. Interactions between fans and adjacent environments (e.g. pluvial lakes, dunefields) often contain important palaeoclimatic information (see Section 14.2.2). Soils developed on dry-region fans not only allow correlation and relative dating of fan surfaces but on induration the carbonate horizons create indurated calcreted fan surfaces, modifying infiltration rates and increasing erosional resistance (see Section 14.2.2).

One important proviso is that the modern alluvial fans in today's drylands are largely Pleistocene landforms, developed under climates that may have been different from today's. While this may be an aid to palaeoclimate reconstruction, caution is needed when linking modern process observations to fan morphology.

14.5.2.2 Significance of alluvial fan research to geomorphology and sedimentology

Alluvial fan research has traditionally involved two disciplines, geomorphology and sedimentology. In the first case the focus has been to understand extant (i.e. Quaternary) alluvial fan morphology; in the second to understand sediment sequences in the ancient rock record. For this reason, sometimes sedimentologically and geomorphologically based explanations have been at odds with one another (see above, Section 14.5.1). It is only through the more recent study of Quaternary and modern fan sediments that the two approaches have been integrated. Alluvial fans therefore have significance for developments in both geomorphology and sedimentology, both individually and in bringing the two disciplines closer together.

Within geomorphology, because fans have an important buffering/coupling role within dry-region drainage basins, understanding alluvial fan dynamics is essential for understanding the sediment flux and dynamics of dry-region mountain fluvial systems (see Section 14.1.3). Within Quaternary science, understanding alluvial fan sequences has made important contributions to reconstructing environmental sequences in dry regions. Within sedimentology attention has been focused on depositional mechanisms, and this emphasis, together with input from geomorphology and Quaternary science, has enhanced our understanding of sedimentary basins within sedimentary geology.

Over the last 20 years there has been an explosion of alluvial fan research, involving not only a much wider geographical cover than hitherto (see Section 14.1.3) but significantly greater integration between studies dealing with depositional processes, fan morphology and fan sequences, applicable to understanding the Quaternary and to interpretations of the ancient rock record.

14.5.2.3 Future research challenges

Within the range of modern research on dryland alluvial fans, several major current research challenges can be identified. The first relates to the causes and behaviour of 'regime change' (see Section 14.4.1). This is fundamental to the buffering/coupling role of fans and also represents a threshold fan response to a radical change in the controlling dynamic factors. Understanding this area would be important for predicting the response of dryland geomorphic systems to climatic change. It also has implications for understanding the ancient rock record in that a fan 'regime change' appears to be one of the main causes of unconformities in the rock record.

A second challenge lies in the application of modern dating methods to Quaternary alluvial fan sequences. Until recently, dryland fan surfaces were best correlated by relative dating methods (see Section 14.2.2) and then dated through their relationships with other 'datable' sediments. Several modern methods are now available for numerically dating alluvial fans These range from dating the sediments themselves, e.g. luminescence dating (Wintle, 1991), recently applied by Pope and Wilkinson (2005) and Robinson *et al.* (2005), to dating pedogenic calcretes, using the U/Th method (Ku *et al.*, 1979; Candy *et al.*, 2003; Candy, Black and Sellwood, 2004), to dating surface exposure using cosmogenic nucliides (e.g. Harbor, 1999; Anderson, Repka and Dick, 1996). Application of these methods will greatly add precision to environmental reconstructions based on alluvial fan sequences, as well as increasing the potential for modelling alluvial fan sediment budgets.

A third challenge needs to address questions of scale, both temporal scale, which will benefit from the application of modern dating methods, and spatial scale. Crucial here is the need to consolidate the linkages between geomorphological and sedimentological approaches to understanding Quaternary fans and stratigraphic approaches to understanding the rock record. Knowledge of the three-dimensional geometry of fan sediment bodies is important and there is scope for applying sequence-stratigraphy concepts to alluvial fan environments, using climatic change as the fundamental sequence control, perhaps following the lines suggested by Weissman, Mount and Fogg (2002) and Weissman, Bennet and Lansdale (2005). Most work so far on fan morphometry has dealt with populations of individual fans within clearly defined groups. Perhaps more important in relation to quantifying the links between fan geometry and the geology of sedimentary basins would be to develop a methodology for considering fan complexes or mountain-front fan groups as units, rather than individual fans (A.J. Hartley, personal communication).

14.5.3 Significance of dry-region alluvial fans for society

In simple terms dry-region alluvial fans comprise gently sloping surfaces within or adjacent to mountain areas. As such they are favourable sites for settlement, whether for small settlements in the mountains of the Middle East or for large desert cities; Las Vegas, Nevada, is built largely on the distal surfaces of coalescent Quaternary alluvial fans. In coastal areas in desert mountains, resort development may take place on the distal surfaces of coastal alluvial fans. For example, Elat in Israel, Sharm ash Shaykh in Sinai and the resorts on the East coast of the UAE are all built in part on distal alluvial fan surfaces.

Alluvial fans are composed largely of permeable materials; hence in arid areas they have important groundwater potential, for potable water and particularly for irrigation water. In the UAE, apart from at a few desert oases, what agriculture there is relates to irrigation from springs at the toes of alluvial fans. In southeast Spain large-scale irrigation from groundwater resources within alluvial fans supports modern agriculture that has been developed largely on alluvial fan surfaces, agriculture that produces a wide range of crops, including citrus fruit, salad crops, melons and peppers. In the long term, whether this agriculture is sustainable is open to question. Water does have to be imported from outside the region and there have been recent developments in desalination to combat local groundwater shortages.

Groundwater occurrence within alluvial fans is not necessarily simple. The finer horizons, debris flows and palaeosols, act as partial aquicludes and groundwater flow tends to follow buried channel systems. Reconstructing the three-dimensional geometry is important for modelling groundwater potential of alluvial fans (Weissmann *et al.*, 2004). The reservoir potential of bodies of alluvial fan sediments for oil and gas has long been recognised. Clearly, increased understanding of the hydrology of alluvial fan sediments has potential in exploration for hydrocarbons.

Dry-region alluvial fans therefore have importance for settlement, land use and or their water resource potential. They are, however, hazardous environments, prone to flash floods and related sediment problems. It is significant that many of the documented flash flood disasters in the USA have involved settlement on alluvial fans. Similarly, in Alpine Europe, though not in any sense a dry region, there have been a number of flood disasters on alluvial fans. In planning the safe development of such zones, it is important to understand the flash-flood risk. Modelling the flash-flood hazard (Zarn and Davies, 1994) should take account of the geomorphology of alluvial fans. For example, such models should take into account the likelihood of channel switching to the flanks of a fan rather than assuming that the zone of greatest risk is 'straight down the middle'.

Dry-region alluvial fans are important features in their own right. Understanding their processes and morphology is fundamental to understanding the function of fluvial systems in dry-region mountain environments. They are also of considerable importance in unravelling Quaternary environmental change in dry regions. Understanding their sedimentary processes has implications for understanding alluvial fan sediment sequences in the ancient rock record. Finally, alluvial fan science has its place in the safe and sustainable use of the world's arid lands.

Acknowledgements

The author thanks the cartographics section of the Department of Geography, University of Liverpool, and especially Sandra Mather, for producing the illustrations.

References

Adams, K.D. and Wesnowsky, S.G. (1998) Shoreline processes and the age of the Lake Lahontan highstand in the Jessup embayment, Nevada. *Geological Society of America Bulletin*, **110**, 1318–1332.

Adams, K.D. and Wesnowsky, S.D. (1999) The Lake Lahontan highstand: age, surficial characteristics, soil development, and regional shoreline correlation. *Geomorphology*, **30**, 257–292.

Alexander, E.B. (1974) Extractable iron in relation to soil age in terraces along the Truckee River, Nevada. *Soil Science Society of America, Proceedings*, **38**, 121–124.

Al Farraj, A. and Harvey, A.M. (2000) Desert pavement characteristics on wadi terrace and alluvial fan surfaces: Wadi Al-Bih, UAE and Oman. *Geomorphology*, **35**, 279–297.

Al Farraj, A. and Harvey, A.M. (2004) Late Quaternary interactions between aeolian and fluvial processes: a case study in the northern UAE. *Journal of Arid Environments*, **56**, 235–248.

Al Farraj, A. and Harvey, A.M. (2005) Morphometry and depositional style of Late Pleistocene alluvial fans: Wadi Al-Bih, northern UAE and Oman, in *Alluvial Fans: Geomorphology, Sedimentology, Dynamics* (eds A.M. Harvey, A.E. Mather and M. Stokes), Geological Society, London, Special Publication 251, pp. 85–94.

Alonso-Zarza, A.M., Silva, P.G., Goy, J.L. and Zazo, C. (1998) Fan-surface dynamics and biogenic calcrete development: interactions during ultimate phases of fan evolution in the semiarid SE Spain (Murcia). *Geomorphology*, **24**, 147–167.

Al Sarawi, A.M. (1988) Morphology and facies of alluvial fans in Kadhmah Bay, Kuwait. *Journal of Sedimentary Petrology*, **58**, 902–907.

Amit, R. and Gerson, R. (1986) The evolution of Holocene reg (gravelly) soils in deserts – an example from the Dead Sea region. *Catena*, **13**, 59–79.

Amit, R., Gerson, R. and Yaalon, D.H. (1993) Stages and rate of gravel shattering process by salts in desert reg soil. *Geoderma*, **57**, 295–324.

Amit, R., Harrison, J.B.J. and Enzel, Y. (1995) Use of soils and colluvial deposits in analyzing tectonic events – the southern Arava Rift, Israel. *Geomorphology*, **12**, 91–107.

Anderson, R.S., Repka, J.L. and Dick, G.S. (1996) Explicit treatment of inheritance in dating depositional surfaces using in situ 10Be and 26Al. *Geology*, **24**, 47–51.

Arzani, N. (2005) The fluvial megafan of the Abarfoh Basin (central Iran): an example of flash-flood sedimentation in arid lands, in *Alluvial Fans: Geomorphology, Sedimentology,*

Dynamics (eds A.M. Harvey, A.E. Mather and M. Stokes), Geological Society, London, Special Publication 251, pp. 41–59.

Baker, V.R. (1977) Stream channel response to floods, with examples from central Texas. *Geological Society of America Bulletin*, **88**, 1057–1071.

Beaty, C.B. (1963) Origin of alluvial fans, White Mountains, California and Nevada. *Annals of the Association of American Geographers*, **53**, 516–535.

Beaty, C.B. (1970) Age and estimated rate of accumulation of an alluvial fan, White Mountains, California, U.S.A. *American Journal of Science*, **268**, 50–77.

Beaty, C.B. (1974) Debris flows, alluvial fans and a revitalized catastrophism. *Zeitschrift fur Geomorphologie*, Supplementbund, **21**, 39–51.

Beaty, C.B. (1990) Anatomy of a White Mountains debris flow – the making of an alluvial fan, in *Alluvial Fans: A Field Approach* (eds A.H. Rachocki and M. Church), John Wiley & Sons, Ltd, Chichester, pp. 69–89.

Beaumont, P. and Oberlander, T.M. (1971) Observations on stream discharge and competence at Mosaic Canyon, Death Valley, California. *Geological Society of America Bulletin*, **82**, 1695–1698.

Bell, J.W. and Katzer, T. (1987) Surficial geology, hydrology, and the Quaternary tectonics of the IXL Canyon area, Nevada, as related to the 1954 Dixie Valley earthquake. *Nevada Bureau of Mines Bulletin*, **102**, 52pp.

Birkeland, P.W. (1985) Quaternary soils of the western United States, in *Soils and Quaternary Landscape Evolution* (ed. J. Boardman), John Wiley & Sons, Ltd, Chichester, pp. 303–324.

Blackwelder, E. (1928) Mudflow as a geologic agent in semi-arid mountains. *Geological Society of America Bulletin*, **39**, 465–484.

Blair, T.C. (1999) Sedimentology of the debris-flow-dominated Warm Springs Canyon alluvial fan, Death Valley, California. *Sedimentology*, **46**, 941–965.

Blair, T.C. and McPherson, J.G. (1994a) Alluvial fan processes and forms, in *Geomorphology of Desert Environments* (eds A.D. Abrahams and A.J. Parsons), Chapman & Hall, London, pp. 354–402.

Blair, T.C. and McPherson, J.G. (1994b) Alluvial fans and their natural distinction from rivers based on morphology, hydraulic processes, sedimentary processes and facies assemblages. *Journal of Sedimentary Research*, **A64**, 450–489.

Blair, T.C. and Mc Pherson, J.G. (1998) Recent debris-flow processes and resultant forms and facies of the Dolomite alluvial fan, Owens Valley, California. *Journal of Sedimentary Research*, **68**, 800–818.

Blissenbach, E. (1952) Relation of surface angle distribution to particle size distribution on alluvial fans. *Journal of Sedimentary Petrology*, **22**, 25–28.

Blissenbach, E. (1954) Geology of alluvial fans in semi-arid regions. *Geological Society of America Bulletin*, **65**, 175–190.

Bluck, B.J. (1964) Sedimentation of an alluvial fan in Southern Nevada. *Journal of Sedimentary Petrology*, **34**, 395–400.

Bluck, B.J. (1987) Bed forms and clast size changes in gravel-bed rivers, in *River Channels: Environment and Process* (ed. K. Richards), Blackwell, Oxford, pp. 159–178.

Bowman, D. (1978) Determination of intersection points within a telescopic alluvial fan complex. *Earth Surface Processes*, **3**, 265–276.

Bowman, D. (1988) The declining but non-rejuvenating base-level – the Lisan Lake, the Dead Sea, Israel. *Earth Surface Processes and Landforms*, **13**, 239–249.

Brunsden, D. and Thornes, J.B. (1979) Landscape sensitivity and change. *Institute of British Geographers, Transactions*, New Series **4**, 463–484.

Bull, W.B. (1961) Tectonic significance of radial profiles of alluvial fans in western Fresno County, California, United States Geological Survey, Professional Paper 424B, pp. 182–184.

Bull, W.B. (1962a) Relation of textural (CM) patterns to depositional environment of alluvial fan deposits. *Journal of Sedimentary Petrology*, **32**, 211–216.

Bull, W.B. (1962b) Relations of alluvial fan size and slope to drainage basin size and lithology, in western Fresno County, California, United States Geological Survey, Professional Paper 430B, pp. 51–53.

Bull, W.B. (1963) Alluvial fan deposits in western Fresno County, California. *Journal of Geology*, **71**, 243–251.

Bull, W.B. (1964) Geomorphology of segmented alluvial fans in western Fresno County, California, United States Geological Survey, Professional Paper 352E, pp. 89–129.

Bull, W.B. (1977) The alluvial fan environment. *Progress in Physical Geography*, **1**, 222–270.

Bull, W.B. (1978) Geomorphic tectonic activity classes of the south front of the San Gabriel Mountains, California, United States Geological Survey, Contract Report 14-08-001-G-394, Office of Earthquakes, Volcanoes and Engineering, Menlo Park, California, 59 pp.

Bull, W.B. (1979) The threshold of critical power in streams. *Geological Society of America Bulletin*, **90**, 453–464.

Bull, W.B. (1991) *Geomorphic Responses to Climatic Change*, Oxford University Press, Oxford, 326pp.

Butzer, K.W. (1964) Climatic-geomorphologic interpretation of Pleistocene sediments in the Eurafrican sub-tropics, in *African Ecoloqy and Human Evolution* (eds F.C. Howell and F. Bouliere), Methuen, London, pp. 1–25.

Calvache, M.L., Viseras, C. and Fernandez, J. (1997) Controls on fan development – evidence from fan morphometry and sedimentology: Sierra Nevada, Spain. *Geomorphology*, **21**, 69–84.

Candy, I., Black, S. and Sellwood, B.W. (2004) Quantifying time series of pedogenic calcrete formation using U-series disequilibria. *Sedimentary Geology*, **170**, 177–187.

Candy, I., Black, S., Sellwood, B.W. and Roan, J.S. (2003) Calcrete profile development in Quaternary alluvial sequences, southeast Spain: implications for using calcretes as a basis for landform chronologies. *Earth Surface Processes and Landforms*, **28**, 169–185.

Chiverrell, R.C., Harvey, A.M. and Foster, G.C. (2007) Hillslope gullying in the Solway Firth–Morecambe Bay region,

Britain: responses to human impact and/or climatic deterioration?, *Geomorphology*, **84**, 317–343.

Colombo, F. (2005) Quaternary telescopic-like alluvial fans, Andean Ranges, Argentina, in *Alluvial Fans: Geomorphology, Sedimentology, Dynamics* (eds A.M. Harvey, A.E. Mather and M. Stokes), Geological Society, London, Special Publication 251, pp. 69–84.

Cooke, R.U. (1970) Stone pavements in deserts. *Annals of the Association of American Geographers*, **60**, 560–577.

D'Agostino, V., Casca, M. and Marchi, L. (2010) Field and laboratory investigations on the runout distances of debris flows in the Dolomites (Eastern Italian Alps). *Geomorphology*, **115**, 294–304.

Dan, J., Yaalon, D.H., Moshe, R. and Nissim, S. (1982) Evolution of Reg soils in southern Israel and Sinai. *Geoderma*, **28**, 173–202.

DeGraf, J.V. (1994) The geomorphology of some debris flows in the southern Sierra Nevada, California. *Geomorphology*, **10**, 231–252.

Delgardo-Castilla, L. (1993) Estudio sedimentologico de los cuerpos sedimentarios Pleistocenos en la Rambla Honda, al N. de Tabernas, provincia de Almeria (SE de Espana). *Cuaternario y Geomorfologia*, **7**, 91–100.

Denny, C.S. (1965) Alluvial fans in the Death Valley region, California and Nevada. United States Geological Survey, Professional Paper 466: 59 pp.

Denny, C.S. (1967) Fans and pediments. *American Journal of Science*, **265**, 81–105.

Derbyshire, E. and Owen, L.A. (1990) Quaternary alluvial fans in the Karakoram Mountains, in *Alluvial Fans: A Field Approach* (eds A.H. Rachocki and M. Church), John Wiley & Sons, Ltd, Chichester, pp. 27–53.

Dorn, R.I. (1983) Cation-ratio dating: a new rock varnish age-determination technique. *Quaternary Research*, **20**, 49–73.

Dorn, R.I. (1984) Cause and implications of rock varnish microchemical laminations. *Nature*, **310**, 767–770.

Dorn, R.I. (1988) A rock varnish interpretation of alluvial-fan development in Death Valley, California. *National Geographic Research*, **4**, 56–73.

Dorn, R.I. (1994) The role of climatic change in alluvial fan development, in *Geomorphology of Desert Environments* (eds A.D. Abrahams and A.J. Parsons), Chapman & Hall, London, pp. 593–615.

Dorn, R.I., De Niro, M.J. and Ajie, H.O. (1987) Isotopic evidence for climatic influence on alluvial fan development in Death Valley, California. *Geology*, **15**, 108–110.

Dorn, R.I. and Oberlander, T.M. (1981) Rock varnish origin, characteristics and usage. *Zeitschrift fur Geomorphologie*, **NF**, **25**, 420–436.

Dorn, R.I. and Oberlander, T.M. (1982) Rock varnish. *Progress in Physical Geography*, **6**, 317–367.

Dorn, R.I., Jull, A.J.T., Donahue, D.J. *et al.* (1989) Accelerator mass spectrometry radiocarbon dating of rock varnish. *Geological Society of America Bulletin*, **101**, 1363–1372.

Dumas, M.B. (1969) Glacis et croutes calcaires dans le levant espanol. *Association des Geographes Bulletin*, **375**, 553–561.

Eckis, R. (1928) Alluvial fans of the Cucamunga district, southern California. *Journal of Geology*, **36**, 225–247.

Enzel, Y., Wells, S.G. and Lancaster, N. (2003) Palaeoenviroments and palaeohydrology of the Mojave and southern Great Basin Deserts, Geological Society of America, Special Paper 368, 249 pp.

Frostick, L.E., and Reid, I. (1989) Climatic versus tectonic controls of fan sequences: lessons from the Dead Sea, Israel. *Journal of the Geological Society, London*, **146**, 527–538.

Garcia, A.F. and Stokes, M. (2006) Late Plestocene highstand and recession of a small, high altutude pluvial lake, Jakes Valley, central Great Basin, USA. *Quaternary Research*, **65**, 179–186.

Garcia, A.F., Zhu, Z., Ku, T.L. *et al.* (2004) An incision wave in the geologic record, Alpujarran Corridor, southern Spain (Almeria). *Geomorphology*, **60**, 37–72.

Gerson, R. (1982) Talus relics in deserts: a key to major climatic fluctuations. *Israel Journal of Earth Science*, **31**, 123–132.

Gerson, R. and Grossman, S. (1987) Geomorphic activity on escarpments and associated fluvial systems in hot deserts as an indicator of environmental regimes and cyclic climatic changes, in *Climate: History, Periodicity, Predictability* (eds M.R. Rampino, J.E. Sanders and W.S. Newman), Van Nostrand Reinhold, Stroudsburg, PA, pp. 300–322.

Gerson, R., Grossman, S., Amit, R. and Greenbaum, N. (1993) Indicators of faulting events and periods of quiescence in desert alluvial fans. *Earth Surface Processes and Landforms*, **18**, 181–202.

Gile, L.H., Peterson, F.F. and Grossman, R.B. (1966) Morphological and genetic sequence of carbonate accumulation in desert soils. *Soil Science*, **101**, 347–360.

Gohain, K. and Parkash, B. (1990) Morphology of the Kosi megafan, in *Alluvial Fans: A Field Approach* (eds A.H. Rachocki and M. Church), John Wiley & Sons, Ltd, Chichester, pp. 151–178.

Grayson, D.K. (1993) *The Desert's Past: A Natural History of the Great Basin*, Smithsonian Institution, Washington, DC, 356 pp.

Harbor, J. (ed.) (1999) Cosmogenic isotopes in geomorphology. *Geomorphology* (Special Issue), **27**, 1–172.

Harden, J.W. (1982) A quantitative index of soil development from field descriptions: examples from a chronosequence in central California. *Geoderma*, **28**, 1–26.

Harden, J.W. and Taylor, E.M. (1983) A quantitative comparison of soil development in four climatic regimes. *Quaternary Research*, **30**, 342–359.

Hartley, A.J., Mather, A.E., Jolley, E. and Turner, P. (2005) Climatic controls on alluvial fan activity, Coastal Cordillera, northern Chile, in *Alluvial Fans: Geomorphology, Sedimentology, Dynamics* (eds A.M. Harvey, A.E. Mather and M. Stokes), Geological Society, London, Special Publication 251, pp. 95–116.

Harvey, A.M. (1978) Dissected alluvial fans in southeast Spain. *Catena*, **5**, 177–211.

Harvey, A.M. (1984a) Debris flows and fluvial deposits in Spanish Quaternary alluvial fans: implications for fan morphology, in *Sedimentoloqy of Gravels and Conglomerates* (eds E.H. Koster and R. Steel), Canadian Society of Petroleum Geologists, Memoir 10, pp. 23–132.

Harvey, A.M. (1984b) Aggradation and dissection sequences on Spanish alluvial fans: influence on morphological development. *Catena*, **11**, 289–304.

Harvey, A.M. (1987) Alluvial fan dissection: relationships between morphology and sedimentation, in *Desert Sediments, Ancient and Modern* (eds L. Frostick and I. Reid), Geological Society of London, Special Publication 35, Blackwell, Oxford, pp. 87–103.

Harvey, A.M. (1988) Controls of alluvial fan development: the alluvial fans of the Sierra de Carrascoy, Murcia, Spain, in *Geomorphic Processes in Environments with Strong Seasonal Contrasts – Volume II: Geomorphic Systems* (eds A.M. Harvey and M. Sala), Catena, Supplement 13, pp. 123–137.

Harvey, A.M. (1990) Factors influencing Quaternary alluvial fan development in southeast Spain, in *Alluvial Fans: A Field Approach* (eds A.H. Rachocki and M. Church), John Wiley & Sons, Ltd, Chichester, pp. 247–269.

Harvey, A.M. (1992a) Controls on sedimentary style on alluvial fans, in *Dynamics of Gravel-Bed Rivers* (eds P. Billi, R.D. Hey, C.R. Thorne and P. Tacconi), John Wiley & Sons, Ltd, Chichester, pp. 519–535.

Harvey, A.M. (1992b) The influence of sedimentary style on the morphology and development of alluvial fans. *Israel Journal of Earth Sciences*, **41**, 123–137.

Harvey, A.M. (1996) The role of alluvial fans in the mountain fluvial systems of southeast Spain: implications of climatic change. *Earth Surface Processes and Landforms*, **21**, 543–553.

Harvey, A.M. (1997) The role of alluvial fans in arid zone fluvial systems, in *Arid Zone Geomorphology: Process, Form and Change in Drylands*, 2nd edn (ed. D.S.G. Thomas), John Wiley & Sons, Ltd, Chichester, pp. 231–259.

Harvey, A.M. (2002a) The relationships between alluvial fans and fan channels within Mediterranean mountain fluvial systems, in *Dryland Rivers: Hydrology and Geomorphology of Semi-arid Channels* (eds L. Bull and M.J. Kirkby), John Wiley & Sons, Ltd, Chichester, pp. 205–226.

Harvey, A.M. (2002b) Factors influencing the geomorphology of alluvial fans: a review, in *Apertaciones a la Geomorfologia de Espana en el Inicio de Tercer Mileno* (eds A. Perez-Gonzalez, J. Vegas and M.J. Machado), Instituto Geologico y Minero de Espana, Madrid, pp. 59–75.

Harvey, A.M. (2002c) Effective timescales of coupling within fluvial systems. *Geomorphology*, **44**, 175–201.

Harvey, A.M. (2002d) The role of base-level change on the dissection of alluvial fans: case studies from southeast Spain and Nevada. *Geomorphology*, **45**, 67–87.

Harvey, A.M. (2003) The response of dry region alluvial fans to Quaternary climatic change, in *Desertification in the Third Millennium* (eds A.S. Alsharhan, W.W. Wood, A.S. Goudie *et al.*), Swete & Zeitlinger, Lisse, The Netherlands, pp. 75–90.

Harvey, A.M. (2005) Differential effects of base-level, tectonic setting and climatic change on Quaternary alluvial fans in the northern Great Basin, Nevada, USA, in *Alluvial Fans: Geomorphology, Sedimentology, Dynamics* (eds A.M. Harvey, A.E. Mather and M. Stokes), Geological Society, London, Special Publication 251, pp. 117–131.

Harvey, A.M. (2006) Interactions between tectonics, climate and base level: Quaternary fluvial systems of Almeria, southeast Spain, in *Geomorfologia y Territorio* (eds A.P. Alberti and J.P. Bedoya), Actas de la IX Reunion Nacional de Geomorfologia, Universidade de Santiago de Compostella, Santiago de Compostella, Spain, pp. 25–48.

Harvey, A.M. (2010) Local buffers to the sediment cascade: debris cones and alluvial fans, in *Sediment Cascades: An Integrated Approach* (eds T.P. Burt and R. Allison), John Wiley & Sons, Ltd, Chichester, pp. 153–180.

Harvey, A.M., Mather, A.E. and Stokes, M. (2005) Alluvial fans: geomorphology, sedimentology, dynamics – introduction. A review of alluvial fan research, in *Alluvial Fans: Geomorphology, Sedimentology, Dynamics* (eds A.M. Harvey, A.E. Mather and M. Stokes), Geological Society, London, Special Publication 251, pp. 1–7.

Harvey, A.M. and Wells, S.G. (1994) Late Pleistocene and Holocene changes in hillslope sediment supply to alluvial fan systems: Zzyzx, California, in *Environmental Change in Drylands: Biogeographical and Geomorphological Perspectives* (eds A.C. Millington and K. Pye), John Wiley & Sons, Ltd, Chichester, pp. 66–84.

Harvey, A.M. and Wells, S.G. (2003) Late Quaternary variations in alluvial fan sedimentologic and geomorphic processes, Soda Lake basin, eastern Mojave Desert, California, in *Palaeoenviroments and Palaeohydrology of the Mojave and Southern Great Basin Deserts* (eds Y. Enzel, S.G. Wells and N. Lancaster), Geological Society of America, Special Paper 368, pp. 207–230.

Harvey, A.M., Wigand, P.E. and Wells, S.G. (1999) Response of alluvial fan systems to the Late Pleistocene to Holocene climatic transition: contrasts between the margins of pluvial Lakes Lahontan ad Mojave, Nevada and California, USA. *Catena*, **36**, 255–281.

Harvey, A M., Oldfield, F., Baron, A.F. and Pearson, G. (1981) Dating of post-glacial landforms in the central Howgills. *Earth Surface Processes and Landforms*, **6**, 401–412.

Harvey, A.M., Silva, P., Mather, A.E. *et al.* (1999) The impact of Quaternary sea-level and climatic change on coastal alluvial fans in the Cabo de Gata ranges, southeast Spain. *Geomorphology*, **28**, 1–22.

Harvey, A.M., Foster, G.C., Hannam, J. and Mather, A.E. (2003) The Tabernas alluvial fan and lake system, southeast Spain: applications of mineral magnetic and pedogenic iron oxide analyses towards clarifying the Quaternary sediment sequences. *Geomorphology*, **50**, 151–171.

Hirst, J.P.P. and Nichols, G.J. (1986) Thrust tectonic controls on Miocene alluvial distribution patterns, southern Pyrennees, International Association of Sedimentologists, Special Publication 8, pp. 247–258.

Hooke, R.L. (1967) Processes on arid-region alluvial fans. *Journal of Geology*, **75**, 438–460.

Hooke, R. le B. (1968) Steady state relationships on arid-region alluvial fans in closed basins. *American Journal of Science*, **266**, 609–629.

Hooke, R. le B. and Dorn R.I. (1992) Segmentation of alluvial fans in Death Valley, California: new insights from surface-exposure dating and laboratory modelling. *Earth Surface Processes and Landforms*, **17**, 557–574.

Hooke, R. le B. and Rohrer, W.L. (1977) Relative erodibility of source area rock types from second order variations in alluvial fan size. *Geological Society of America Bulletin*, **88**, 1177–1182.

Hooke, R. le B. and Rohrer, W.L. (1979) Geometry of alluvial fans: effect of discharge and sediment size. *Earth Surface Processes*, **4**, 147–166.

Hunt, C.B. (1975) *Death Valley: Geology, Archaeology, Ecology*, University of California Press, Berkeley, 234 pp.

Hunt, C.B. and Mabey, D.R. (1966) Stratigraphy and structure, Death Valley, California, United States Geological Survey, Professional Paper 494A, 162 pp.

Hurst, V.J. (1977) Visual estimates of iron in saprolite. *Geological Society of America Bulletin*, **88**, 174–176.

Johnson, A.M. and Rodine, J.R. (1984) Debris flows, in *Slope Instability* (eds D. Brunsden and D.B. Prior), John Wiley & Sons, Inc., New York, pp. 257–361.

Kesel, R.H. and Spicer, B.E. (1985) Geomorphic relationships and ages of soils on alluvial fans in the Rio General Valley, Costa Rica. *Catena*, **12**, 149–166.

Kochel, R.C. (1990) Humid fans of the Appalachian Mountains, in *Alluvial Fans: A Field Approach* (eds A.H. Rachocki and M. Church), John Wiley & Sons, Ltd, Chichester, pp. 109–129.

Kostaschuk, R.A., Macdonald, G.M. and Putnam, P.E. (1986) Depositional processes and alluvial fan – drainage basin morphometric relationships near Banff, Alberta, Canada. *Earth Surface Processes and Landforms*, **11**, 471–484.

Ku, T-L., Bull, W.B., Freeman, S.T. and Knauss, K.G. (1979) Th230–U234 dating of pedogenic carbonates in gravelly desert soils of Vidal Valley, southeastern California. *Geological Society of America Bulletin*, **90**, 1063–1073.

Laronne, J.B. and Reid, I. (1993) Very high rates of bedload sediment transport by ephemeral desert rivers. *Nature*, **366**, 148–150.

Lattman, L.H. (1973) Calcium carbonate cementation of alluvial fans in southern Nevada. *Geological Society of America Bulletin*, **84**, 3013–3028.

Lecce, S.A. (1991) Influence of lithologic erodibility on alluvial fan area, western White Mountains, California and Nevada. *Earth Surface Processes and Landforms*, **16**, 11–18.

Leeder, M.R. and Mack, G.H. (2001) Lateral erosion ('toe cutting') of alluvial fans by axial rivers: implications for basin analysis and architecture. *Journal of the Geological Society, London*, **158**, 885–893.

Leggett, R.F., Brown, R.J.E. and Johnston, G.H. (1966) Alluvial fan formation near Aklavik, Northwest Territories, Canada. *Geological Society of America Bulletin*, **77**, 15–30.

Lustig, L.K. (1965) Clastic sedimentation in Deep Springs Valley, California, United States Geological Survey, Professional Paper 352F, pp. 131–192.

McArthur, J.L. (1987) The characteristics, classification and origin of Late Pleistocene fan deposits in the Cass Basin, Canterbury, New Zealand. *Sedimentology*, **34**, 459–471.

McDonald, E.V., Pierson, F.B., Flerchinger, G.N. and McFadden, L.D. (1997) Application of a process-based soil-water balance model to evaluate the influence of Late Quaternary climate change on soil-water movement. *Geoderma*, **74**, 167–192.

McFadden, L.D., Ritter, J.B. and Wells, S.G. (1989) Use of multiparameter relative-age methods for estimation and correlation of alluvial fan surfaces on a desert piedmont, eastern Mojave Desert, California. *Quaternary Research*, **32**, 276–290.

McFadden, L.D., Wells, S.G. and Jercinavich, M.J. (1987) Influences of eolian and pedogenic processes on the origin and evolution of desert pavements. *Geology*, **15**, 504–508.

McFadden, L.D., McDonald, E.V., Wells, S.G. *et al.* (1999) The vesicular layer and carbonate collars in desert soils and pavements: Formation, Age and relation to climate change. *Geomorphology*, **24**, 101–145.

Machette, M.N. (1985) Calcic soils of the southwestern United States, in *Soils and Quaternary Geology of the Southwestern United States* (ed. D.L. Weide), Geological Society of America, Special Paper 203, pp. 1–21.

Mack, G.M. and Rasmussen, K.A. (1984) Alluvial fan sedimentation of the Cutler formation (Permo-Pennsylvanian), near Gateway, Colorado. *Geological Society of America Bulletin*, **95**, 109–116.

Macklin, M.G. and Lewin, J. (2003) River sediments, great floods and centennial-scale Holocene climatic change. *Journal of Quaternary Science*, **18**, 102–105.

Maher, E. (2006) The Quaternary evolution of the Rio Alias, Southeast Spain, with emphasis on sediment provenance, PhD thesis, University of Liverpool.

Mather, A.E. (1993) Basin inversion: some consequences for drainage evolution and alluvial architecture. *Sedimentology*, **40**, 1069–1089.

Mather, A.E. (2000) Impact of headwater river capture on alluvial system development: an example from SE Spain. *Journal of the Geological Society, London*, **157**, 957–966.

Mather, A.E. and Hartley, A.J. (2005) Flow events on a hyper-arid alluvial fan: Quebrada Tambores, Salar de Atacama, northern Chile, in *Alluvial Fans: Geomorphology, Sedimentology, Dynamics* (eds A.M. Harvey, A.E. Mather and M. Stokes), Geological Society, London, Special Publication 251, pp. 9–24.

Mather, A.E. and Harvey, A.M. (1995) Controls on drainage evolution in the Sorbas basin, southeast Spain, in *Mediterranean Quaternary River Environments* (eds J. Lewin, M.G. Macklin and J. Woodward), Balkema, Rotterdam, pp. 65–76.

Mather, A.E., Harvey, A.M. and Stokes, M. (2000) Quantifying long term catchment changes of alluvial fan systems. *Geological Society of America Bulletin*, **112**, 1825–1833.

Mather, A.E., Martin, J.M., Harvey, A.M. and Braga, J.C. (2003) *A Field Guide to the Neogene Sedimentary Basins of the Almeria Province, South-east Spain*, Blackwell Science, Oxford, 350pp.

Miall, A.D. (1977) A review of the braided river depositional environment. *Earth Science Reviews*, **13**, 1–62.

Miall, A.D. (1978) Lithofacies types and vertical profile models in braided river deposits: a summary, in *Fluvial Sedimentology. Canadian Society of Petroleum Geologists* (ed. A.D. Miall), Memoir 5, pp. 597–604.

Nanson, G.C., Chen, X.Y. and Price, D.M. (1995) Aeolian and fluvial evidence of changing climate and wind patterns during the past 100 ka in the western Simpson Desert, Australia. *Palaeogeography, Palaeoclimatology, Palaeoecology*, **113**, 87–102.

Nash, D.J. and Smith, R.F. (1998) Multiple calcrete profilesin the Tabernas basin, southeast Spain: their origins and geomorphic implications. *Earth Surface Processes and Landforms*, **23**, 1009–1029.

Nemec, W. and Postma, G. (1993) Quaternary alluvial fans in southwest Crete: sedimentation processes and geomorphic evolution, in *Alluvial Sedimentation: International Association of Sedimentologists* (eds M. Marzo and C. Puigdefabrigas), Special Publication 17, pp. 235–276.

Nichols, G. (2005) Tertiary alluvial fans at the northern margin of the Ebro basin: a review, in *Alluvial Fans: Geomorphology, Sedimentology, Dynamics* (eds A.M. Harvey, A.E. Mather and M. Stokes), Geological Society, London, Special Publication 251, pp. 187–206.

Nilsen, T.H. (1982) Alluvial fan deposits, in *Sandstone Depositional Environments* (eds P.A. Scholle and P. Spearing), American Association of Petroleum Geologists, Memoir 31, pp. 49–86.

Oguchi, T. (1997) Late Quaternary sediment budjet in alluvial-fan–source-basin systems in Japan. *Journal of Quaternary Science*, **12**, 381–390.

Oguchi, T. and Ohmori, H. (1994) Analysis of relationships among alluvial fan area, source basin area, basin slope and sediment yield. *Zeitschrift fur Geomorphologie*, **NF**, **38**, 405–420.

Pierson, T.C. (1980) Erosion and deposition by debris flows at Mt Thomas, North Canterbury, New Zealand. *Earth Surface Processes and Landforms*, **5**, 226–247.

Pierson, T.C. (1981) Dominant particle support mechanisms in debris flows at Mt Thomas, New Zealand, and implications for flow mobility. Sedimentology, **28**, 49–60.

Pierson, T.C. and Scott, K.M. (1985) Downstream dilution of a lahar: transition from debris flows to hyperconcentrated streamflow. *Water Resources Research*, **21**, 1511–1524.

Pope, R.J.J. (2000) The application of mineral magnetic and extractable iron (Fed) analysis for differentiating and relatively dating alluvial fan surfaces in central Greece. *Geomorphology*, **32**, 57–67.

Pope, R.J.J. and Millington, A.C. (2000) Unravelling the patterns of alluvial fan development using mineral magnetic analysis: examples from the Sparta basin, Kakonia, southern Greece. *Earth Surface Processes and Landforms*, **25**, 601–615.

Pope, K.O. and Van Andel, T.H. (1984) Late Quaternary alluviation and soil formation in the southern Argolid: its history, causes and archaeological implications. *Journal of Archaeological Science*, **11**, 281–306.

Pope, R.J.J. and Wilkinson, K.N. (2005) Reconciling the roles of climate and tectonics in Late Quaternary fan development on the Sparta piedmont, Greece, in *Alluvial Fans: Geomorphology, Sedimentology, Dynamics* (eds A.M. Harvey, A.E. Mather and M. Stokes), Geological Society, London, Special Publication 251, pp. 133–152.

Rachocki, A.H. and Church, M. (eds) (1990) *Alluvial Fans: A Field Approach*, John Wiley & Sons, Ltd, Chichester, 391 pp.

Reinech, H.E. and Singh, I.B. (1983) *Depositional Sedimentary Environments with Reference to Terrigenous Classics*, 3rd edn, Springer Verlag, Berlin, 549 pp.

Richards, K.S. (1982) *Rivers, Form and Process in Alluvial Channels*, Methuen, London, 358 pp.

Rickenmann, D. and Zimmerman, M. (1993) The 1987 debris flows in Switzerland: documentation and analysis. *Geomorphology*, **8**, 175–189.

Ritter, J.B., Miller, J.R. and Husek-Wulforst, J. (2000) Environmental controls on the evolution of alluvial fans in the Buena Vista Valley, north central Nevada, during late Quaternary time. *Geomorphology*, **36**, 63–87.

Ritter, D.F. and Ten Brink, N.W. (1986) Alluvial fan development and the glacial–glaciofluvial cycle, Nenana Valley, Alaska. *Journal of Geology*, **94**, 613–625.

Ritter, J.B., Miller, J.R., Enzel, Y. and Wells, S.G. (1995) Reconciling the roles of tectonism and climate in Quaternary alluvial fan evolution. *Geology*, **23**, 245–248.

Roberts, N. (1995) Climatic forcing of alluvial fan regimes during the Late Quaternary in Konya basin, south central Turkey, in *Mediterranean Quaternary River Environments* (eds J. Lewin, M.G. Macklin and J. Woodward), Balkema, Rotterdam, pp. 205–217.

Robinson, R.A.J., Spencer, J.Q.G., Strecker, M.R. *et al.* (2005) Luminescence dating of alluvial fans in intramontane basins of NW Argentina, in *Alluvial Fans: Geomorphology, Sedimentology, Dynamics* (eds A.M. Harvey, A.E. Mather and M. Stokes), Geological Society, London, Special Publication 251, pp. 153–168.

Rust, B.R. (1978) Depositional models for braided alluvium, in *Fluvial Sedimentology* (ed. A.D. Miall), Canadian Society of Petroleum Geologists, Memoir 5, pp. 605–625.

Rust, B.R. (1979) Facies models 2: coarse alluvial deposits, in *Facies Models* (ed. R.G. Walker), Geoscience Reprint Series 1, Kitchener, Ontario, Canada, pp. 9–21.

Saito, K. and Oguchi, T. (2005) Slope of alluvial fans in humid regions of Japan, Taiwan and The Phillipines. *Geomorphology*, **20**, 147–162.

Scheepers, A.C.T. and Rust, I.C. (1999) The Uniab River fan: an unusual alluvial fan on the hyper-arid Skeletal Coast,

Namibia, in *Varieties of Fluvial Form* (eds A.J. Miller and A. Gupta), John Wiley & Sons, Ltd, Chichester pp. 273–294.

Schumm, S.A. (1977) *The Fluvial System*, John Wiley & Sons, Inc., New York, 338 pp.

Schumm, S.A., Mosley, M.P. and Weaver, W.E. (1987) *Experimental Fluvial Geomorphology*, John Wiley & Sons, Inc., New York, 413 pp.

Silva, P.G., Harvey, A.M., Zazo, C. and Goy, J.L. (1992) Geomorphology, depositional style and morphometric relationships of Quaternary alluvial fans in the Guadalentin depression (Murcia, southeast Spain). *Zeitschrift fur Geomorphologie*, **NF**, **36**, 325–341.

Silva, P.G., Bardaji, T., Calmel-Avila, M. *et al.* (2008) Transition from alluvial to fluvial systems in the Guadalentin depression (SE Spain): Lorca fan versus Guadalentin River. *Geomorphology*, **100**, 140–153.

Somoza, L., Zazo, C., Goy, J.L. and Morner, N.A. (1989) Estudio geomorfologico de seguencias de abanicos aluviales cuaternarios (Alicante-Murcia, Espana). *Cuaternario y Geomorfolgia*, **3**, 73–82.

Sorriso-Valvo, M., Antronico, L. and Le Pera, E. (1998) Controls on modern fan morphology in Calabria, southern Italy. *Geomorphology*, **24**, 169–187.

Stanistreet, I.G. and McCarthy, T.S. (1993) The Okavango fan and the classification of subaerial fan systems. *Sedimentary Geology*, **65**, 115–133.

Steel, R.J. (1974) New red sandstone floodplain and piedmont sedimentation in the Hebridean Province, Scotland. *Journal of Sedimentary Petrology*, **44**, 336–357.

Steel, R.J., Moehle, S., Nilsen, H. *et al.* (1977) Coarsening upwards cycles in the alluvium of Homelen Basin (Devonian), Norway: sedimentary response to tectonic events. *Geological Society of America Bulletin*, **88**, 1124–1134.

Stokes, M. (2008) Plio-Pleistocene drainage development in an inverted sedimentary basin: Vera basin, Betic Cordillers, SE Spain. *Geomorphology*, **100**, 193–211.

Stokes, M., Nash, D.J. and Harvey, A.M. (2007) Calcrete fossilisation of alluvial fans in SE Spain: groundwater vs pedogenic processes and fan dynamics in calcrete development. *Geomorphology*, **85**, 63–84.

Suwa, H. and Okuda, S. (1983) Deposition of debris flows on a fan surface, Mt Yakedale, Japan. *Zeitschrift fur Geomorphologie*, Supplementband, **46**, 79–101.

Van Arsdale, R. (1982) Influence of calcrete on the geometry of arroyos near Buckeye, Arizona. *Geological Society of America Bulletin*, **93**, 20–26.

Viseras, C., Calvache, M.L., Soria, J.M. and Fernandez, J. (2003) Differential features of alluvial fans controlled by tectonic or eustatic accommodation space. *Geomorphology*, **50**, 181–202.

Wasson, R.J. (1974) Intersection point deposition on alluvial fans: an Australian example. *Geografiska Annaler*, **56A**, 83–92.

Wasson, R.J. (1979) Sedimentation history of the Mundi alluvial fans, western New South Wales. *Sedimentary Geology*, **22**, 21–51.

Weissmann, G.S., Bennett, G.L. and Lansdale, A.L. (2005) Factors controlling sequence development on Quaternary fluvial fans, San Joaquin basin, California, USA, in *Alluvial Fans: Geomorphology, Sedimentology, Dynamics* (eds A.M. Harvey, A.E. Mather and M. Stokes), Geological Society, London, Special Publication 251, pp. 169–186.

Weissmann G.S., Mount, J.F. and Fogg, G.E. (2002) Glacially driven cycles in accommodation space and sequence stratigraphy of a stream-dominated alluvial fan, San Joachim Valley, California, USA. *Journal of Sedimentary Research*, **72**, 270–281.

Weissmann, G.S., Zhang, Y., Fogg, G.E. and Mount, J.F. (2004) Influence of incised-valley-fill deposits on hydrogeology of a glacially-influenced, stream-dominated alluvial fan, in *Aquifer Characterisation* (eds J. Bridge and D.W. Hyndman), SEPM (Society for Sedimentary Geology), Special Publication 80, pp. 15–28.

Wells, S.G. and Harvey, A.M. (1987) Sedimentologic and geomorphic variations in storm generated alluvial fans, Howgill Fells, northwest England. *Geological Society of America Bulletin*, **98**, 182–198.

Wells S.G., McFadden, L.D. and Dohrenwend, J.C. (1987) Influence of late Quaternary climatic changes on geomorphic and pedogenic processes on a desert piedmont, Eastern Mojave Desert, California. *Quaternary Research*, **27**, 130–146.

Whipple, K.X. and Dunne, T. (1992) The influence of debris-flow rheology on fan morphology, Owens Valley, California. *Geological Society of America Bulletin*, **104**, 887–900.

Whipple, K.X., Parkes, G., Paola, C. and Mohrig, D. (1998) Channel dynamics, sediment transport and the shape of alluvial fans: experimental study. *Journal of Geology*, **106**, 677–693.

White, K.H. (1991) Geomorphological analysis of piedmont landforms in the Tunisian Southern Atlas using ground data and satellite imagery. *The Geographical Journal*, **157**, 279–294.

White, K. and Walden, J. (1994) Mineral magnetic analysis of iron oxides in arid zone soils, Tunisian Southern Atlas, in *Environmental Change in Arid and Semi-arid Environments* (eds A. Millington and K. Pye), John Wiley & Sons, Ltd, Chichester, pp. 43–65.

White, K. and Walden, J. (1997) The rate of iron oxide enrichment in arid zone alluvial fan soils, Tunisian Southern Atlas, measured by mineral magnetic techniques. *Catena*, **30**, 215–227.

White, K., Drake, N., Millington, A. and Stokes, S. (1996) Constraining the timing of alluvial fan response to Late Quaternary climatic changes, southern Tunisia. *Geomorphology*, **17**, 295–304.

Williams, G.E. (1973) Late Quaternary piedmont sedimentation, soil formation and palaeoclimates in arid South Australia. *Zeitschrift fur Geomorphologie*, **17**, 102–125.

Wintle, A. (1991) Luminescence dating, in *Quaternary Dating Methods – A Users Guide* (eds P.L. Smart and P.D. Frances), Technical Guide 4, Quaternary Research Association, Cambridge, pp. 108–127.

Wolman, M.G. and Gerson, R. (1978) Relative scales of time and effectiveness of climate in watershed geomorphology. *Earth Surface Processes*, **3**, 189–208.

Wright, V.P. and Alonso-Zarza, A.M. (1990) Pedostratigraphic models for alluvial fan deposits: a tool for interpreting ancient sequences. *Journal of the Geological Society, London*, **147**, 8–10.

Yaalon, D.H. (1970) Parallel stone cracking, a weathering process on desert surfaces. *Bucharest Geological Institute of Technical Economics Bulletin*, **18**, 107–110.

Zarn, B. and Davies, T.R. (1994) The significance of processes on alluvial fans to hazard assessment. *Zeitschrift fur Geomorphologie*, **38**, 487–500.

15

Pans, playas and salt lakes

Paul A. Shaw and Rob G. Bryant

15.1 The nature and occurrence of pans, playas and salt lakes

The arid zone contains few perennial lakes; examples such as the Caspian Sea, the Dead Sea and Lake Aral invariably have inflowing rivers rising in fringing uplands or distant humid areas. Instead, the majority of arid and semi-arid regions are characterised by endoreic (internal) drainage or, in extreme cases, may lack integrated surface drainage altogether. Under these circumstances surface depressions become important local and regional foci for the accumulation of water in episodic (termed here *ephemeral*) lakes. Due to the negative balance between evaporation and rainfall (often exceeding 10 : 1), these water bodies are often highly saline and, in some cases, supersaturated with salts. Such salt lakes, which have a minimum salinity of 5000 mg/L, in turn lie at one end of a spectrum of otherwise ephemeral and often relict closed basins of varying scales and origins, frequently termed playas or pans.

Pans and playas have been described in most hot dryland environments, particularly Africa, Australia, Arabia and in western USA (see Shaw and Thomas, 1997), but also occur in cold drylands such as Antarctica (Lyons *et al.*, 1998). Although mostly associated with aridity – the majority of southern Kalahari and peri-southern Kalahari pans, for example, occur on the arid side of the 500 mm mean annual isohyet and the 1000 mm free evaporation isoline (Goudie and Thomas, 1985) – some comparable features are found beyond the limits of modern aridity, e.g. the Plains of Zambia (Goudie and Thomas, 1985; Williams, 1987) and the Darwin region of Australia (McFarlane *et al.*, 1995).

Playas and pans vary in size from the frequently very small depressions of a few tens of square metres in the Kalahari, western Australia and Texas (Goudie and Thomas, 1985; Killigrew and Gilkes, 1974; Osterkamp and Wood, 1987) to massive tectonic basins, which may exceed 10 000 km^2 in area, such as Lake Eyre, south central Australia, and Lake Uyuni, Bolivia (Lowenstein and Hardie, 1985). Though pans and playas occupy as little as 1 % of the total landscape, they are important and often numerous features. In parts of southern Africa pans attain densities of up to 1.14 pans per km^2 (Goudie and Thomas, 1985) and occupy 20 % of the surface area (Goudie and Wells, 1995), while there are an estimated 30 000 to 37 000 basins on the southern High Plains of northwest Texas and adjoining New Mexico (Reddell, 1965; Osterkamp and Wood, 1987).

Playas and pans have been important to human populations from prehistoric times to the present day as sources of water and minerals. In modern times they have been used for urban development (Cooke *et al.*, 1982), for airfields and racetracks (e.g. the Blackrock Desert, Nevada, USA) and in the case of Lop Nor, China, and China Lake, USA, for testing nuclear weapons. Regrettably these uses conflict with the inherent value of pans and playas as extreme, unusual and often valuable habitats (e.g. Haukos and Smith, 1994; McCulloch *et al.*, 2008), while development itself is not without difficulties, as the flooding of Salt Lake City in the 1980s shows (Atwood, 1994). Scientifically they have become increasingly important for the elucidation of palaeoenvironmental conditions from their sediments and landforms (e.g. Telfer and Thomas, 2007), while on the negative side, they are now recognised as major sources of atmospheric dust (Gillette, 1981; Prospero *et al.*, 2002), and are monitored accordingly.

Arid Zone Geomorphology: Process, Form and Change in Drylands, Third Edition. Edited by David S. G. Thomas

15.1.1 Playa and pan terminology

Arid and semi-arid lacustrine basins have a rich terminology, particularly in Asia and the Middle East (Shaw and Thomas, 1997). The most commonly used terms, pan and playa, are interchangeable, with an increasing tendency to use pan for small basins formed by arid zone geomorphological, rather than geological, processes (Goudie and Wells, 1995) and playa for a depression with a saline surface (Rosen, 1994). However, the many descriptors in regional usage and with approximately equivalent meaning has led to confusion in terminology. For example, attempts to restrict the application of terms such as sebkha to the coastal environment have not been widely adhered to, while recent attempts to introduce broad terms such as hemi-arid basin (Currey, 1994) have done little to clarify the issue. A major concern has been the permanence and provenance of the lacustrine system.

In a review of terminology, Briere (2000) attempted to group terms into the categories of playa, playa lake and sabkha (Figure 15.1) and offers the following revised definitions:

Playa. An intercontinental arid zone basin with a negative water balance for over half of each year, dry for over 75 % of the time, with a capillary fringe close enough to the surface such that evaporation will cause water to discharge, usually resulting in evaporates (Briere, 2000, p. 3).

Playa lake. An arid zone feature, transitional between playa and lake, neither dry more than 75 % of the time nor wet more than 75 % of the time. When dry, the basin qualifies as a playa (Briere, 2000, p. 3).

Sabkha. A marginal marine mudflat where displacive and replacive evaporate minerals form in the capilliary zone above a saline water table (Briere, 2000, p. 4).

These definitions have some merit, particularly as they eliminate the need for the definition of a continental or inland sabkha, a step not met with approval by some (e.g. Barth, 2001). They fail, however, to cover clay-floored pans with little hydrological input such as the features of Texas and the Kalahari.

A broader definition of *pans* or *playas* can be arid zone basins of widely varying size and origin, which, although generally above the present groundwater table, are subject to ephemeral surface water inundation of variable periodicity and extent. Their basal and marginal sediments often display evidence of evaporite accumulation, aeolian deflation and accumulation and/or lacustrine activity.

The term *coastal sabkha* or *sebkha* (Glennie, 1970) originally applied to saline flats in arid areas that occur above the level of high tide, but nevertheless receive periodic marine incursions and associated sediments. They have many of the features and processes of inland playas (Yechieli and Wood, 2002); indeed there is a recognised continuum between coastal flats and inland playas which receive sea water (Bye and Harbison, 1991). However, coastal sebkhas are not considered in this chapter.

15.1.2 General characteristics

Despite the variability of pans and playas, a number of common characteristics emerge. Most obvious is that pans occupy topographic lows, though not necessarily the lowest areas in enclosed drainage basins, because small pans can develop almost anywhere in relatively flat arid landscapes, e.g. along deranged drainage lines (Osterkamp and Wood, 1987). With larger playas there is inevitably a strong geological control on form (e.g. Salama, 1994). Topographic position and geological framework may both influence a groundwater regime and hence the formative processes. This is nowhere more apparent than in the *boinkas* (groundwater discharge areas) of southeast Australia (Jacobson, Ferguson and Evans, 1994), where contemporary playas are nested in the topographic lows of larger groundwater-controlled landscapes and geologically older lake basins.

In terms of surface hydrology they are essentially 'closed' systems, having no surface outflow. The dominance of potential evaporation (PE) over precipitation (P) and other inputs is the essential contributory factor to closed status. Hydrological inputs may be direct precipitation, surface or subsurface inflow, or any combination of the three. Standing surface water is ephemeral, accounting for the distinct morphological and sedimentological differences between arid basins and those of more humid areas. The extent, frequency and length of surface water occupancy depends on climatological and hydrological regimes, and is a major source of pan and playa variability. Overall, arid and humid closed basins can be viewed as part of a climate-based spectrum ranging from 0 to 100 % surface water occupancy at its extremes, each with a distinct morphology and processes (Bowler, 1986).

Much research has focused on the role of groundwater in pan formation and function, particularly in terms of interaction with surface processes (Fryberger, Schenk and Krystinik, 1988; Osterkamp and Wood, 1987; Torgersen, 1984; Torgersen *et al.*, 1986; Rosen, 1994, Yechieli and Wood, 2002; Reynolds *et al.*, 2007) and of the origins of the brines themselves (Bye and Harbison, 1991; Herczeg

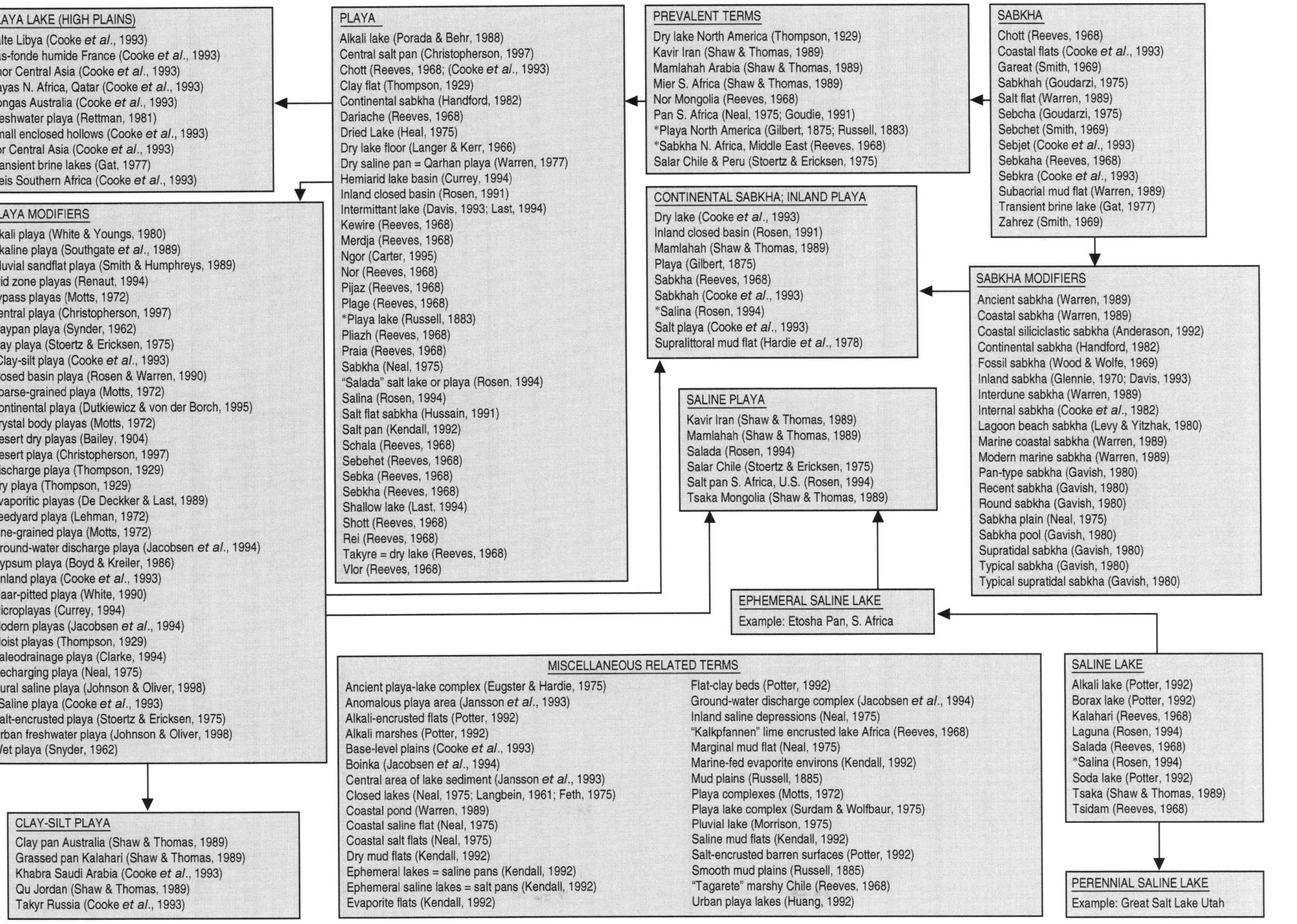

Figure 15.1 The nomenclature of playas, playa lakes and sabkhas (after Briere, 2000). In this scheme, each term in a given set has been used synonymously or interchangeably with the header of that set. Arrows connect sets that are joined by at least one common term. Asterisks denote terms that may help the reader follow the chart's flow. For original references see Briere (2000).

Figure 15.2 Playas and pans with contrasting groundwater depths: (a) Chott el Djerid, Tunisia, a discharge playa with groundwater at the surface (foreground) and evaporate mineral accumulation (background); (b) Silver Lake, USA, a clay-floored recharge playa, with little or no evaporate mineral accumulation at the surface.

and Lyons, 1991; Jankowski and Jacobson, 1989; Bryant *et al.*, 1994a, 1994b). Not only is the role of groundwater important in pan and playa formation, but there is again a spectrum of conditions and subsequent effects present. These range from pans and playas where the groundwater table intersects the basin surface (Figure 15.2(a)), accompanied by surface evaporite accumulation and evaporative effects, as in the Chotts of Tunisia (e.g. Roberts and Mitchell, 1987; Bryant *et al.*, 1994) , to those where the water table lies at depth (Figure 15.2(b)). These latter features are usually clay-floored and percolating groundwater plays a major role in deep weathering and eluviation. Such variations are thus a function of topography and geology rather than climate.

Surfaces are usually vegetation-free, particularly at their lowest elevations. Episodic flooding, vertisol or solonchak formation and salt accumulation discourage vegetation growth, although halophytic plants and shallow rooting grasses may be established. Grassed pans exist alongside bare clay surfaces, as in the Kalahari (Boocock and Van Straten, 1962), suggesting small variations in soil alkalinity and wind action. Butterworth (1982) has proposed a cycle of development linking grassed and clay pans.

Table 15.1 Origins of closed arid zone basins.

(a) *Structural controls*
Faulting and rifting
Downwarping
Fracture lines
Intrusions
Differential weathering of adjacent rock types
(b) *Erosional controls*
Deflation
Subsurface solution and karstic development
Animal scouring
(c) *Ponding*
Ephemeral or abandoned drainage lines
Floodplain
Interdune troughs
Interstandline troughs
Coastal sedimentation
Drainage disrupted by tilting
(d) *Dramatic*
Meteor impact
Volcanic crater development

15.1.3 Origins and development of pans and playas

Pans and playas have a variety of origins, which can be classified into structural, erosional and ponding controls (Table 15.1 and Figure 15.3). A few have more dramatic origins: Pretoria Saltpan (South Africa), Zuni Salt Lake (New Mexico) and Meteor Crater (Arizona) have been

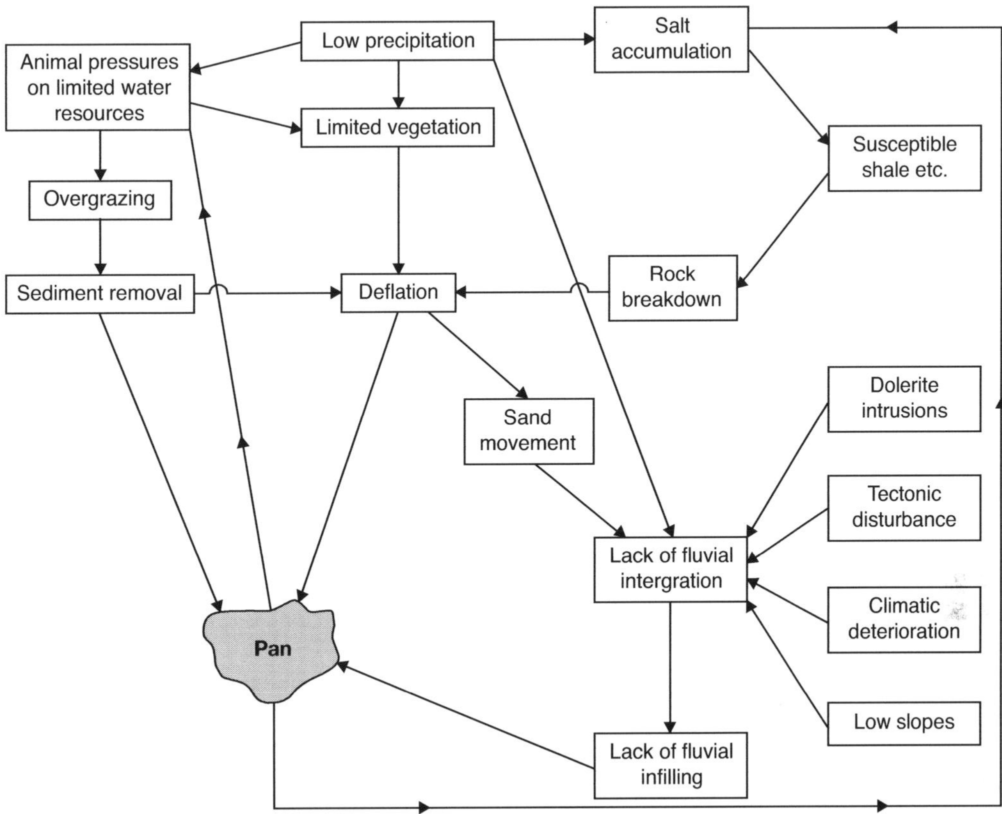

Figure 15.3 Flow diagram showing complex linkages of factors favouring pan and playa initiation and development by erosion (in this case, predominantly by deflation) (after Goudie and Thomas, 1985).

attributed to the impact of meteors or volcanic crater development (Wellington, 1955; Mabbutt, 1977; Goudie and Thomas, 1985).

Faulting and downwarping have led to the development of major regional basins of interior drainage in some arid environments. For example, Cainozoic tectonic activity, including block faulting, has been responsible for the concentration of the intermontane basins in the 'Great Basin' and the 'Basin and Range' desert regions of the southwestern United States (Smith and Street-Perrott, 1983, and see Chapter 4). Gentler Cainozoic tectonism has contributed to the development of the Etosha and Makgadikgadi basins in southern Africa (Wellington, 1955; Thomas and Shaw, 1988) and the Eyre Basin in Australia (Johns, 1989). Many of these large basins have responded to major local and regional hydrological inputs by developing massive lakes responding to Quaternary climatic fluctuations. These changes in hydrologic status are preserved in the sedimentary record or in former shorelines (strandlines), and have been termed playa-lake complexes by some authors (e.g. Eugster and Hardie, 1978; Eugster and Kelts, 1983). In some cases the lakes have lost their closed status and overflowed at times, as in the Bonneville and Lahontan Lakes of the Great Basin (Benson *et al.*, 1990).

On a smaller scale, lineaments, by acting as conduits for groundwater movement, are the preferential sites for the development of smaller pans, as suggested for some of the features of the Texas High Plain (Osterkamp and Wood, 1987) and the south and southeast Kalahari (Arad, 1984; Shaw and De Vries, 1988). Intruded bodies at depth found in association with lineaments may also influence pan location, as indicated by the geophysical studies of Lokgware and Mogatse pans in the Kalahari (Farr *et al.*, 1982) and the influence of dolerite sills on the pans of the Lake Chrissie complex in the eastern Transvaal (Wellington, 1955). Ponding may also occur in the linear depressions (*straats*) between longitudinal dunes, as in parts of the Kalahari (Mallick, Habgood and Skinner, 1981), between strandlines of palaeolakes, as in the Dautsa Ridge sequence of Lake Ngami (Shaw, 1985), or, by obstruction of ephemeral channels, the extension of dunes as cited with reference to western Australia (Gregory, 1914) and the Namib Sand Sea (Rust and Wieneke, 1974).

Erosional processes, such as deflation, contribute to the formation of the larger structural basins, as in the Qattara and Siwa Depressions, Egypt (Gindy, 1991), but are especially important in the genesis of smaller local or subregional-scale features. Both aeolian deflation and removal of material by solution during deep weathering have been proposed as erosional mechanisms, but, as the debate on small depressions in, for example, Texas and New Mexico (Reeves, 1966; Carlisle and Marrs, 1982; Osterkamp and Wood, 1987; Wood and Osterkamp, 1987) shows, there is a strong case for a polygenetic origin for many small pans.

Deflation has often been cited as an originator or contributor in pan development, e.g. Egypt (Haynes, 1980), the Kalahari (Lancaster, 1978a), Australia (Hills, 1940; Bowler, 1973), Texas (Reeves, 1960), the Argentine Pampas (Tricart, 1969) and Zaire (de Ploey, 1965). For deflation to be effective the criteria necessary for aeolian entrainment must be satisfied, while a near-surface groundwater table in a playa can act as a base-level control on the depth of deflation. Of special importance is the susceptibility of surfaces to deflation (Goudie and Thomas, 1985), both in terms of material susceptibility and in the absence of a protective vegetation cover. The latter may be effected by concentration of salts (Le Roux, 1978) or seasonal surface inundation (Bowler, 1986). In this respect Osterkamp and Wood's (1987) observation that any slight depression in an otherwise flat surface has the potential to develop into a pan or playa should be noted.

The role of deflation in playa and pan development may be indicated by the presence of fringing transverse or lunette (Hills, 1940) dunes on the downwind margin of the depression, or, indeed, by orientation of the pan transverse to prevailing winds (see Le Roux, 1978; Goudie and Thomas, 1985; Bowler, 1986). Some authors (e.g. Wood and Osterkamp, 1987) have opposed the deflationary hypothesis on the grounds that the volume of sand in the fringing dune does not represent the volume removed from the pan. However, deflated sediment can be transported beyond the margins of depressions and into the atmospheric circulation (Reheis, 2006). A more serious objection is the observation that, in many Kalahari pans, the material comprising the lunette have different sediment characteristics to the pan surface (Goudie and Thomas, 1986).

Solution, piping and subsurface karstic collapse may be locally important mechanisms in areas underlain by carbonate and other sedimentary lithologies. Wood and Osterkamp (1987) propose a model for the formation of small clay-floored pans based on studies of the Texas High Plains, where pan development has taken place in post-Pliocene times under arid to semi-arid conditions in a variety of sedimentary strata in response to lowering of the regional water table. During initial development, depressions originate by various means, including deflation, drainage ponding and along structural lineaments. Protobasins act as sites of seasonal runoff concentration on the relatively flat plains surface and through which groundwater recharge occurs. This results in sub-basin locations in the unsaturated zone becoming foci for oxidation and carbonate dissolution, leading to piping development and the disintegration of the calcrete, thus contributing to basin enlargement.

This model accords well with other regions of small pans unaffected by groundwater inputs, such as the Kalahari, although here percolating water rarely reaches the water table under present climatic conditions, leading to the precipitation of fresh calcrete at depth. The contention of Wood and Osterkamp (1987) that pans are capable of enlargement by peripheral weathering is backed by the observation of Farr *et al.* (1982) that some pans in the Kalahari are capable of migration over a long period of time.

The excavations and trampling of animals were seen as important factors in forming depressions by early investigators in Texas (Gilbert, 1895) and the Kalahari (Allison, 1899; Passarge, 1904). While clearly inapplicable to the evolution of larger basins, animal activity has been observed to contribute to depression development in areas of seasonally limited water supplies (Weir, 1969; Ayeni, 1977; see Thomas, 1988, for a review). Termites have also been implicated in the formation of small, highly saline pans on islands in swamp ecosystems, such as the Okavango Delta (McCarthy, McIver and Cairncross, 1986).

The mechanisms proposed require suitably susceptible surfaces. In southern Africa, pans are preferentially found on lithologies that readily break down to fine-grained sediments or which are generally poorly consolidated (Goudie and Wells, 1995). Susceptibility may be enhanced in lithologies that contain significant amounts of sodium sulfate, which enhances salt weathering and retards plant growth, or clays such as bentonite, which have high coefficients of expansion on hydration. Extensive low-relief terrain also seems to favour pan development, as in Texas and southern Africa. Such surfaces limit the potential to develop integrated drainage and promote the concentration of both moisture and fine-grained clastic material into surface depressions.

Playas are also aggradational features, deriving sediment through episodic inflows or aeolian inputs. The sediments that are received are almost exclusively fine grained, which can be explained in three ways. First, where playas represent drainage terminals, only fine

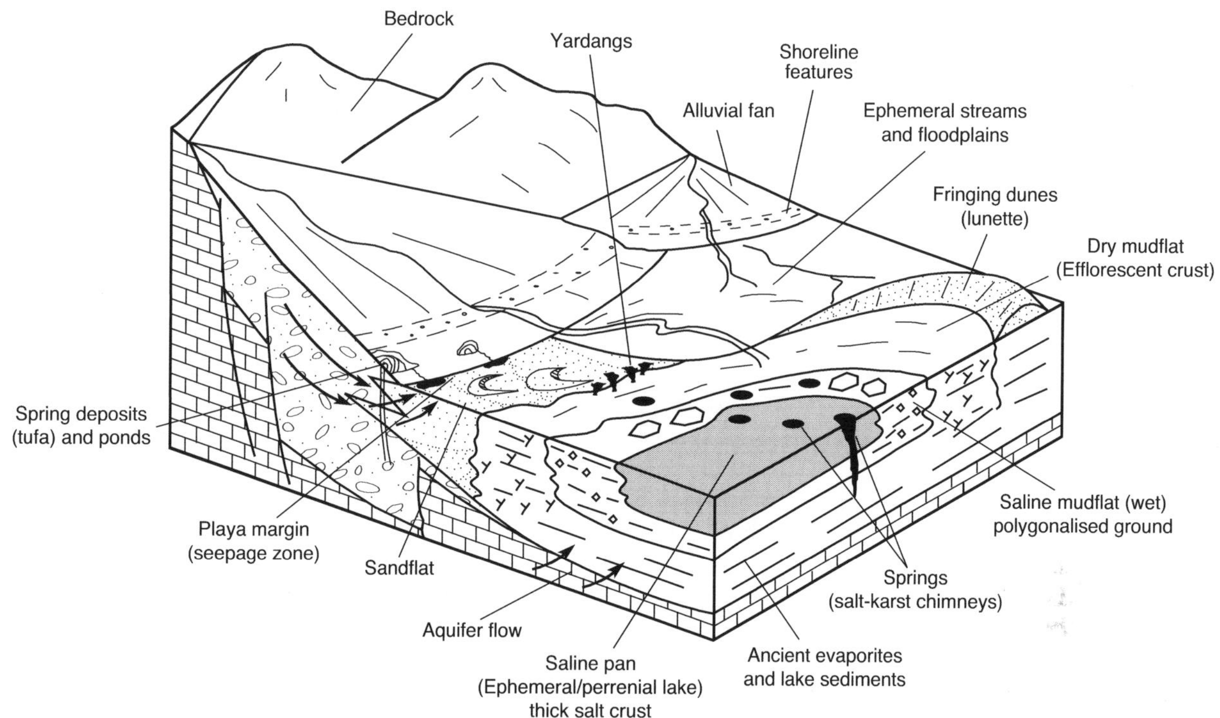

Figure 15.4 Idealised diagram of the depositional subenvironments that can occur in closed arid zone playa basins (after Hardie, Smoot and Eugster, 1978; Chivas, 2007).

sediments are likely to be transported to the basin as discharge generally decreases downstream in arid channels. Second, where playas occur in a tectonic basin closely bordered by uplands, alluvial fans at the basin–mountain interface usually act as buffers in the fluvial system, trapping sediment. Third, aeolian sediments are, by their very nature, no coarser than sand.

Surface evaporation plays a major role in pan evolution, together with the complex processes involved in salt and water transfer at the groundwater–surface water interface, leading to the accumulation of evaporite deposits on and within near-surface deposits. Where bedrock is close to, or outcrops at, the playa surface, high rates of evaporation may favour rapid breakdown by salt weathering (Goudie and Thomas, 1985). The various geomorphic and hydrologic depositional processes that operate in the basins are neither mutually exclusive in space nor time (Bowler, 1986). Consequently, a series of depositional subenvironments may be present (Hardie, Smoot and Eugster, 1978) (Figure 15.4), including concentric zonation of salts and sediments (e.g. Rosen, 1991) that may be identifiable as facies (Magee, 1991). Any individual basin may possess only some of these environments at any given time (Eugster and Kelts, 1983).

The aggradational attributes of playas contributes to their usually flat, horizontal surfaces, especially in the subenvironments subject to inundation. Given the fine-grained nature of the sediments, any irregularities, including those derived from evaporite growth (see Lowenstein and Hardie, 1985), are smoothed out by water movement and dissolution when surface water occupies the basin. Playas with highly infrequent (possibly not recorded in historical times) surface water inundation may develop uneven surfaces through evaporite growth or sand dune development. The extension of dunes (Bowler, 1986) and fluvial distributaries (Townshend *et al.*, 1989) on to playa surfaces from surrounding areas may also lead to uneven margins.

15.2 Pan hydrology and hydrochemistry

Several sedimentary subenvironments exist in playas and pans. The processes involved can be grouped into those resulting in deposition on the basin floor, the basin subsurface and the basin margins. Deposition in basin margin locations is not necessarily directly related to the processes operative in the basin itself. Depending on

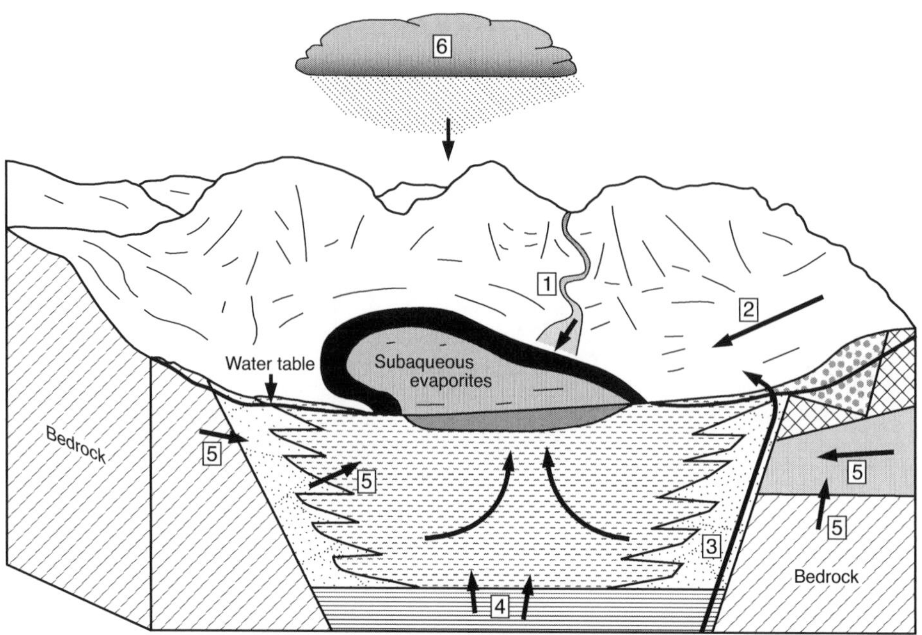

Figure 15.5 Sources of groundwater to a playa basin: 1, channelled flow in permanent or ephemeral streams; 2, unrestricted overland flow (sheet wash); 3, hydrothermal fluids from a deep source (may occur as a seep or spring or directly mix into the groundwater); 4, connate or formation water derived from when the formation was deposited; 5, meteoric groundwater derived within (hydrologically closed) or outside (throughflow) the immediate basin; 6, direct precipitation on to the playa surface or surrounding catchment (after Rosen, 1994).

individual basin settings, marginal sedimentation can be achieved by the activity of ephemeral rivers, alluvial fans, sand seas or, more rarely, by mass movement processes.

The dominant sediment types encountered within the basin are fine sediments brought in by surface flow or aeolian action, organic materials and evaporite minerals. The main ions encountered are SiO_2, Ca^{2+}, Mg^{2+}, K^+, Na^+,Cl^-, HCO_3^-,CO_3^- and SO_4^-, which are derived from both the surface and groundwater catchments. Hardie, Smoot and Eugster (1978) note that weathering reactions and catchment lithology are the first determinants of the types of salts precipitated within the basin, although airborne salts may be important in coastal locations (Jack, 1921; Eckardt and Spiro, 1999) and have, in the long term, contributed to inland playas as well (Chivas *et al.*, 1991; Jones, Hanor and Evans, 1994). Precipitation, in turn, is controlled by salt composition and concentration and the relative influence of the surface and groundwater regimes (Rosen, 1994) (Figure 15.5). Given the range of solute sources, transport mechanisms and evaporative regimes in playa basins, it is important to note the geochemical diversity of the evaporite deposits that can accumulate within any one basin over time.

15.2.1 *Inflow and water balance modelling*

The hydrological regime affects pan development and morphology in two ways: first, in relation to external factors such as climate and catchment; second, in the relationship between the surface water and groundwater inputs within the basin. Bowler (1986) addresses this first control in his model of a six-stage hydrological sequence for closed basins in Australia. The stages range from a lake with permanent surface water at one end of the continuum to an ephemeral terminal sink totally controlled by groundwater at the other (Figure 15.6). A disequilibrium index, Δ_{eL}, calculated from hydrological and climatic data, was also used to relate present conditions in a basin to those necessary to maintain a steady-state water cover, with values ranging from 0 in presently perennial basins to -1000 for those currently in the driest locales. While the results, in ignoring many of the other basin variables, are not universally applicable, they serve to emphasise the difference between surface water and groundwater processes, both in terms of the nature of the waters and sedimentation and in their interrelationships, particularly within the flooding–desiccation cycle.

Rosen (1994) addresses the second control and summarises the importance of groundwater depth in

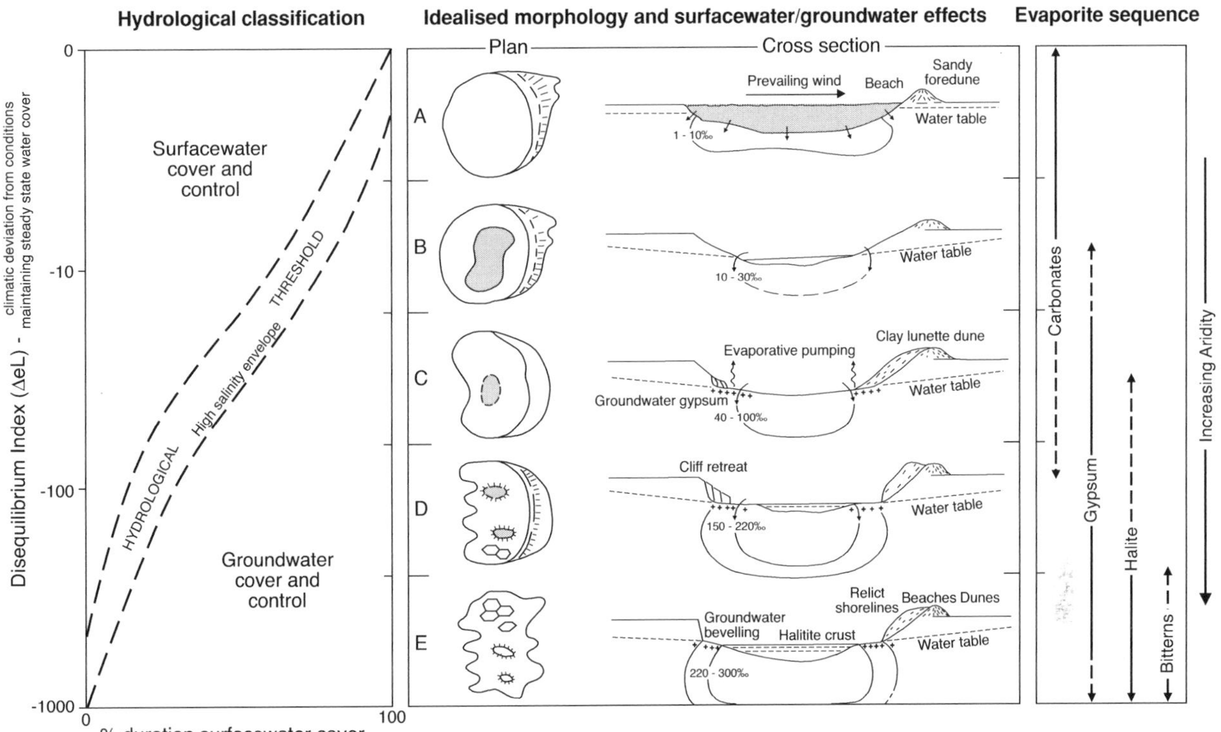

Figure 15.6 Hydrological classification, idealised morphology and groundwater–surface water interaction of evaporative basins (after Bowler, 1986).

influencing pan surface processes and development (Figure 15.7). In the first instance, groundwater may be highly saline, whereas surface waters generally have low solute loads but appreciable suspended sediment. Thereafter, the depth to groundwater controls surface evaporite accumulation, as evaporation from open water is 1–2 orders of magnitude higher than from capillary rise (Tyler *et al.*, 1997). In addition, groundwater fluxes when the water table is very close to the land surface can be dramatic, with rapid inundation resulting from little observed change in water recharge or potential evaporation (Tyler, Munoz and Wood, 2006). Conversely, when water tables are deep (>2 m below the surface), water table responses to interannual changes in inflow can lag significantly behind such changes (Tyler, Munoz and Wood, 2006). At the larger scale evaporate mineral accumulation can be a slow process, for even if basins do exhibit a large surface and subsurface inflow component, they often only develop significant evaporite deposits in the long term. Hardie, Smoot and Eugster (1978) note that the flood in Lake Eyre in 1950, which covered 8000 km^2, evaporated two years later to leave only a thin layer of halite over an area of about only one-tenth of the original flood, while Holser (1979) estimated that the evaporation of a 200 metre depth of surface inflow would be necessary to produce a 3 metre thickness of the same mineral. Smoot and Lowenstein (1991) point to the importance of repeated inflow into playa basins over long time periods in order to allow the development of stable surface salt deposits in the intervening dry periods through a net increase in the salinity of shallow groundwater, which eventually inhibits complete salt pan dissolution. The interval between episodes of surface inundation is therefore important for sedimentation, as surface water halts evaporation from subsurface water and leads to resolution of salts. Eugster and Kelts (1983) point out that the Great Salt Lake, Utah, has only deposited major halite beds in historic times on two occasions, 1930–1935 and 1960–1964, coincident with periods of drought.

Mass budget modelling (e.g. Yechieli and Wood, 2002; Tyler, Munoz and Wood, 2006) suggests that water and aeolian budgets in playas are often in dynamic equilibrium, while the salt budget may display a lag of thousands of years and reflect palaeohydrological conditions (e.g. Bristol (dry) Lake; see Rosen, 1991) or solute leakage (Wood and Sanford, 1990). This, in turn, impacts on model parameters. In an open-system playa with constant water volume both the concentration and mass of solutes will

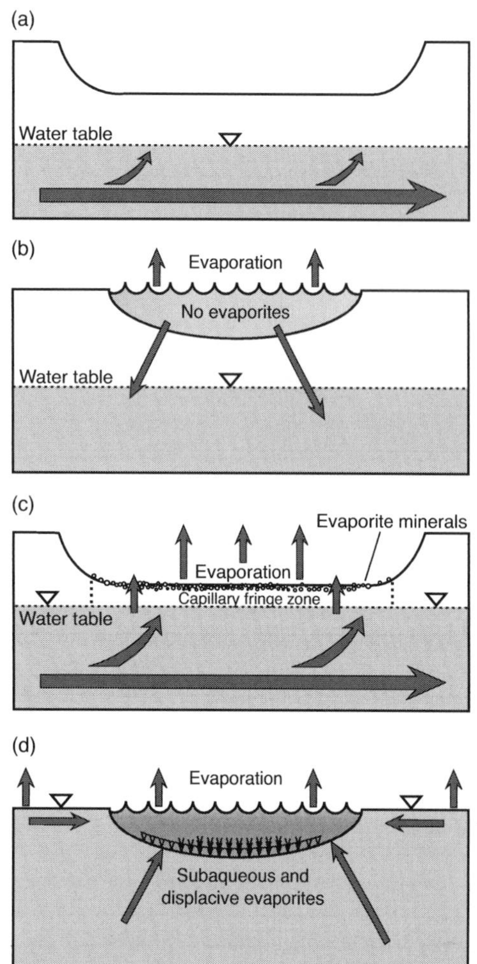

Figure 15.7 Relationship of sustained groundwater flow to evaporite mineral accumulation to all possible groundwater configurations (modified after Rosen, 1994 and Reynolds *et al.*, 2007). (a) When the water table is far below the ground surface only flow through conditions can occur and no evaporites will accumulate. (b) When water is transported or precipitated in this type of situation the depression acts as a recharge zone and leaks water to the subsurface. No significant evaporites will accumulate, although a thin crust may develop when the final solution evaporates to dryness. This crust would likely be deflated. (c) When the water table intersects the ground surface in a hydrologically closed basin (a playa situation), displacive evaporites may form, but significant accumulations of subaqueous evaporites cannot. Those that can accumulate after a rain event will likely be deflated when the lake dries out. However, significant accumulations of displaced evaporites may occur. (d) Only when the groundwater table is above the surface of the deepest part of a closed basin playa, so that groundwater input is constant, can subaqueous evaporates accumulate in a hydrologically closed basin. Although slightly more complicated for throughflow basins, this model also applies to these types of basins.

increase with time, while a closed-system playa with declining water volumes, as implied by a drying climate, will increase the concentration of solutes but leave the mass constant (Yechieli and Wood, 2002). Other factors influencing precipitation include the catchment area, pan surface area and the presence of aquatic vegetation (Russell, 2008).

15.2.2 *Geochemical processes and mineral precipitation*

Increasing near-surface salinity, resulting from either climatic or hydrological factors, can result in evolution from clay-floored pans to salt pans containing evaporites. As already noted, the chemistry of inflowing water is largely dependent on solute output from the catchment; the subsequent evolution of evaporites will be dependent on the ratios of solutes present and the precipitation gradient of the salts involved. From an understanding of fractional crystallisation of mineral phases resulting from the evaporation of seawater (e.g. a sequence with increased evaporation of calcite, anhydrite/gypsum and halite followed by epsomite, sylvite, kainite, carnellite and borates/celestite; see Valyashko, 1972), we can gain some idea of the relative type and proportion of evaporite phases that may be present in playa basins. However, given their extreme chemical variability, most nonmarine saline waters do not follow this template. Hardie, Smoot and Eugster (1978) identified four main brine types resulting from a series of evaporative concentration steps on undersaturated inflow for nonmarine brines, a model that has been subsequently modified by others (see Jankowski and Jacobson, 1989; Rosen, 1994). These represent an accepted set of geochemical pathways along which most nonmarine evaporites develop within playa basins (Figure 15.8(a)).

The three pathways outlined by Eugster and Hardie (1978) use the relationship between the molar content of bicarbonate (HCO_3) relative to molar Mg + Ca for all inflow waters to determine the ultimate brine type and mineral assemblage that may result from evaporation. In all, these authors identify five end-member brine types for nonmarine waters and a number of key mineral phases that are associated with their evaporation (Figure 15.8(b)). In this scheme, initial evaporation and degassing (Eugster and Kelts, 1983) leads to the progressive precipitation of low-Mg calcite, aragonite and high-Mg calcite (protodolomite), followed by gypsum ($CaSO_4{\cdot}2H_2O$) at concentrations of 40–100 g/L (Bowler, 1986), dependent on the type and duration of processes in the evaporation zone. Gypsum precipitation is also dependent on the degree of carbonate depletion. Halite (NaCl)

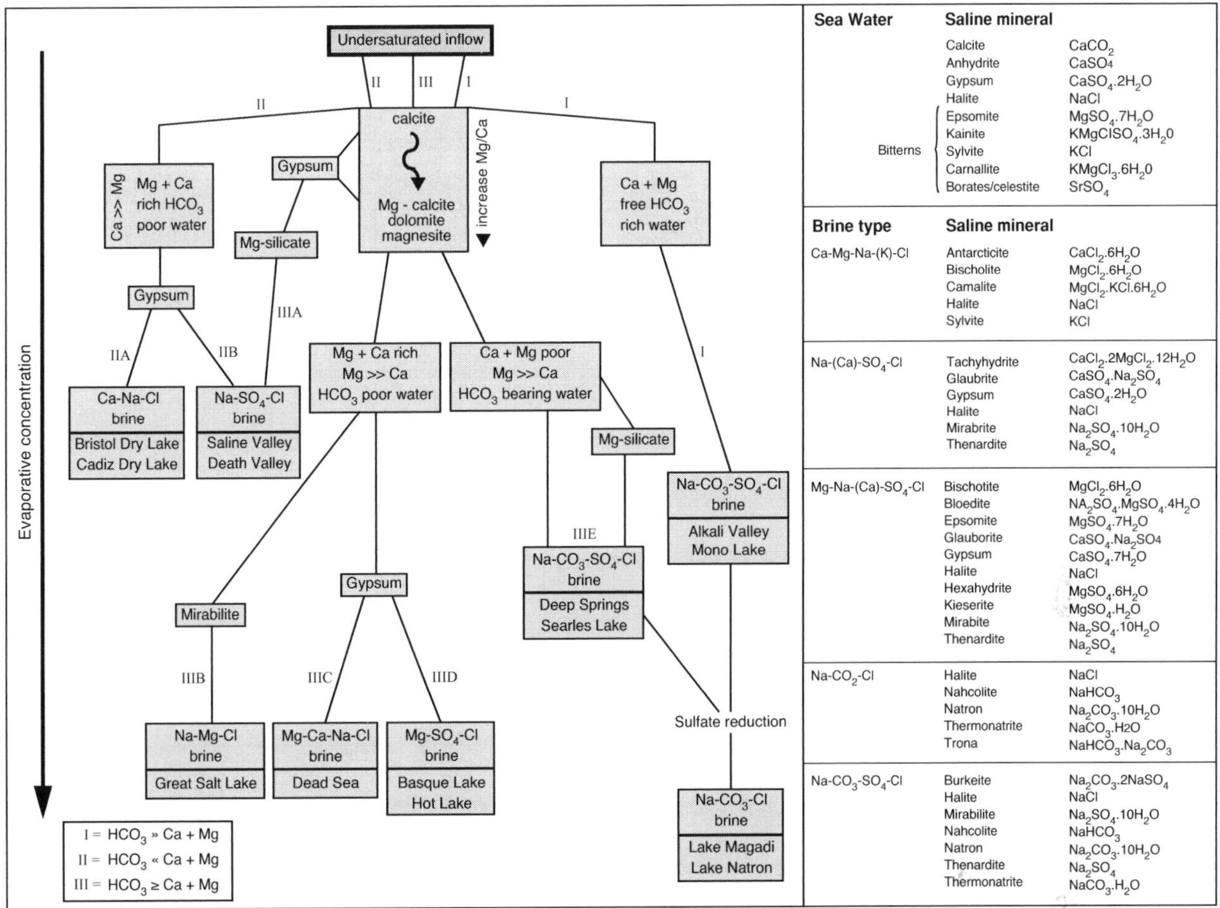

Sea Water	Saline mineral	
	Calcite	$CaCO_2$
	Anhydrite	$CaSO_4$
	Gypsum	$CaSO_4.2H_2O$
	Halite	NaCl
	Epsomite	$MgSO_4.7H_2O$
	Kainite	$KMgClSO_4.3H_20$
Bitterns	Sylvite	KCl
	Carnallite	$KMgCl_3.6H_20$
	Borates/celestite	$SrSO_4$
Brine type	**Saline mineral**	
Ca-Mg-Na-(K)-Cl	Antarcticite	$CaCl_2.6H_2O$
	Bischolite	$MgCl_2.6H_2O$
	Camalite	$MgCl_2.KCl.6H_2O$
	Halite	NaCl
	Sylvite	KCl
$Na-(Ca)-SO_4-Cl$	Tachyhydrite	$CaCl_2.2MgCl_2.12H_2O$
	Glaubrite	$CaSO_4.Na_2SO_4$
	Gypsum	$CaSO_4.2H_2O$
	Halite	NaCl
	Mirabrite	$Na_2SO_4.10H_2O$
	Thenardite	Na_2SO_4
$Mg-Na-(Ca)-SO_4-Cl$	Bischotite	$MgCl_2.6H_2O$
	Bloedite	$NA_2SO_4.MgSO_4.4H_2O$
	Epsomite	$MgSO_4.7H_2O$
	Glauborite	$CaSO_4.Na_2SO4$
	Gypsum	$CaSO_4.7H_2O$
	Halite	NaCl
	Hexahydrite	$MgSO_4.6H_2O$
	Kieserite	$MgSO_4.H_2O$
	Mirabite	$Na_2SO_4.10H_2O$
	Thenardite	Na_2SO_4
$Na-CO_2-Cl$	Halite	NaCl
	Nahcolite	$NaHCO_3$
	Natron	$Na_2CO_3.10H_2O$
	Thermonatrite	$NaCO_3.H2O$
	Trona	$NaHCO_3.Na_2CO_3$
$Na-CO_3-SO_4-Cl$	Burkeite	$Na_2CO_3.2NaSO_4$
	Halite	NaCl
	Mirabilite	$Na_2SO_4.10H_2O$
	Nahcolite	$NaHCO_3$
	Natron	$Na_2CO_3.10H_2O$
	Thenardite	Na_2SO_4
	Thermonatrite	$NaCO_3.H_2O$

Figure 15.8 (a) Geochemical pathways and classification for brine evolution with progressive evaporation (after Hardie, Smoot and Eugster, 1978, and Rosen, 1994). Several lakes are given as examples for each pathway. (b) Common evaporative minerals associated with sea water and the five principal nonmarine brine types (after Eugster and Hardie, 1978 and Chivas, 2007).

saturation occurs at around 200–350 g/L, while other common salts include trona ($Na_2CO_3{\cdot}NaHCO_3{\cdot}2H_2O$), thenardite ($Na_2SO_4$), epsomite ($MgSO_4{\cdot}7H_2O$) and burkeite ($Na_2CO_3{\cdot}2Na_2SO_4$). Where sodium-rich brines come into contact with deposits of gypsum or calcite, double salts, such as glauberite ($CaSO_4{\cdot}Na_2SO_4$) or gaylussite ($Na_2CO_3{\cdot}CaCO_3.5H_2O$), may be formed. Less common evaporites are potassium and magnesium chlorides (e.g. carnallite), which are found in the Qaidam Basin of China (Yuan, Chengyu and Keqin, 1983; Chen and Bowler, 1986; Bryant *et al.*, 1994a), and nitrates, as in the saltpetre-rich Matsap Pan of South Africa (Wellington, 1955). The three pathways (I, II and II in Figure 15.8) are generally found to be characteristic of volcanic terrains (path I), seawater (path II) and the recycling of ancient carbonate or evaporates (path III).

Evaporation and salt precipitation rates represented in the Eugster and Hardie (1978) scheme are controlled by the thermodynamic activity of the water (Langmuir, 1997); for any given set of pressure and temperature conditions an increase in dissolved ions reduces of the ability of the water to evaporate. This relationship, expressed by the Pitzer equations (Pitzer, 1987), suggests that the evaporative potential of concentrated brine is about half that of pure water. There are also variations in the precipitation of individual salts, with those minerals of retrograde solubility, such as gypsum, anhydrite and calcite, precipitating at high temperatures, while sodium and magnesium chloride become supersaturated as temperatures fall. This leads to variations in salt populations on diurnal to seasonal scales (Yechieli and Wood, 2002).

In broad terms the increasing concentration of brine leads to a zonation of the evaporites by solubility. In individual salt pans this leads to a 'bulls eye' effect of lateral zonation of facies from carbonates at the edge, through sulfates to chlorides in the sump (Jones, 1965; Hunt *et al.*,

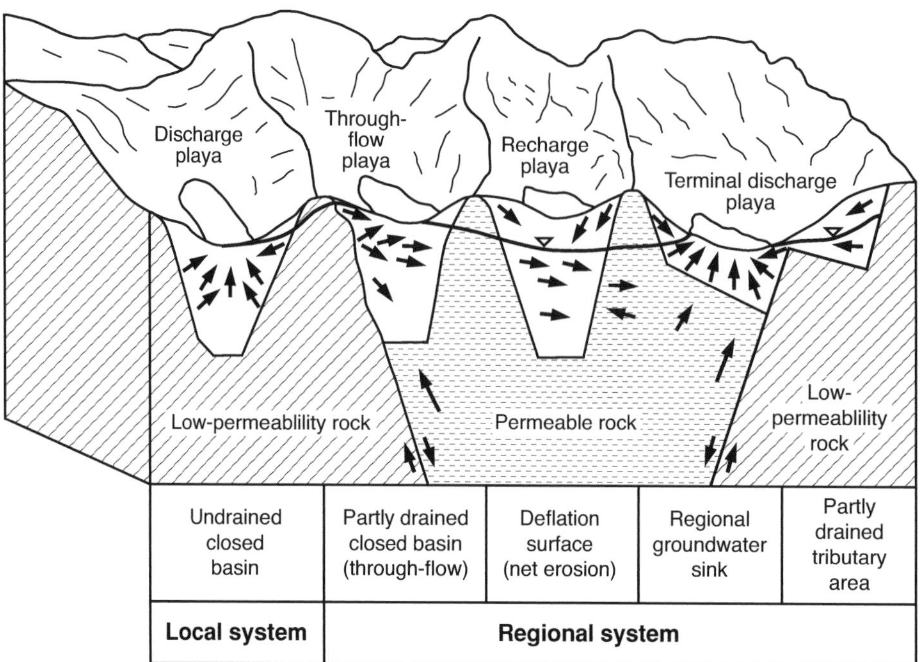

Figure 15.9 Summary diagram of hydrologic classifications based upon groundwater hydrology of playa basins (after Rosen, 1994).

1966). This zonation will also be apparent in the texture of the pan surface, with a transition from the peripheral clay floor to a soft mud with surface efflorescence, described as 'self-raising ground' (Mabbutt, 1977), which represents the capillary fringe of the groundwater. This, in turn, gives way to a salt crust whose thickness is dependent on the frequency of surface flooding and groundwater characteristics, and to a brine layer if present. The concentric surface zonation of salts may be mirrored by a vertical zonation as a result of variations in solubility, with the most soluble minerals at the surface, or as a response to subsurface processes, particularly reduction (Neev and Emery, 1967; Rosen, 1991; Bryant *et al.*, 1994a).

15.2.3 The importance of groundwater: classification of playa and pan types

The position of the water table has been used as the basis of classification of playas by Mabbutt (1977), Wood and Sanford (1990) and Rosen (1994) (Figure 15.9). In pans where the water table lies at depth, features defined as recharge playas by Rosen (1994), there will be little interaction between the surface and groundwater, and sedimentation will occur in a shallow surface and subsurface layer, with overall transfer of water towards the water table. These pans usually have clay or vegetated surfaces, with little sign of evaporite accumulation. Conversely, saline basins with near-surface water tables have complex transfers of water and salts along physical and chemical pathways in three dimensions, leading to the accumulation of surface crusts and displacive evaporites.

Clay-floored pans are characteristic of regimes with low groundwater input or where the surface lies above the influence of capillary rise from the water table, a depth of usually about 3 metres (Rosen, 1994). Usually they are composed of a flat clay or sandy clay surface, either as the base of a pan or as a higher surround to a more saline basin (Mabbutt, 1977). The dominant sediment is clastic material deposited from suspension during inundation, although lenses of sand may be deposited under higher energy conditions. The clay surface, in turn, forms an impervious layer to groundwater recharge at the pan centre. Reynolds *et al.* (2007) describe clay-floored pans as 'dry' playas, which are unlikely to be susceptible to deflation or dust production (Figure 15.7(a) or (b)).

Salt input is generally low to basins of this type. Precipitation of calcium and magnesium is common, producing a range of carbonates from calcite to dolomite, dependent upon the Ca/Mg ratio (Müller, Irion and Forstner, 1972), as cements, laminates, crusts or other structures (Eugster and Kelts, 1983). Gypsum efflorescence may follow the drying of the clay surface, while silica mobility is also apparent; silcretes as well as calcretes are found in both the bed and periphery of pans in semi-arid environments (Summerfield, 1982).

Where the water table lies close to, but does not intersect, the pan surface (Figure 15.7(c)), three zones may be identified: (a) a saturated zone; (b) a porewater zone in which capillary rise, enhanced by surface evaporation, occurs; and (c) the surface crust (Tyler, Munoz and Wood, 2006). These zones will change in extent, laterally and vertically, with water table changes, leading to corresponding changes in the surface sediments. Evaporative concentration through this system, whether the groundwater intersects the surface or lies below it, is controlled not only by rates of evaporation but also by groundwater salinity and density, hydraulic conductivity of the aquifer and the depth of the porewater zone (Bowler, 1986). Changes within the porewater zone, termed 'shallow interstitial waters' by Bowler, have profound effects on the ultimate character of the playa. For example, evaporation within this zone will cause variations in water density, often to great depth, enhancing vertical and horizontal transfers of groundwater to balance the salinity gradient. Reynolds *et al.* (2007) refer to playas of this type as 'wet' playas and highlight the importance of groundwater depth to surface characteristics and deflation rates.

Salts precipitate within this system as surface crusts, or by interstitial crystallisation within existing sediments, or as subaqueous evaporites in brine pools. Salt emplacement can arise by direct crystallisation from the brine or by reaction between the brine and surrounding sediments and organisms.

15.2.4 Implications of climate change and human impacts on playa hydrology

It is important to understand how changes in water balance, driven either by climate forcing or human intervention, might affect any equilibrium that may exist between these components. Human impacts on the hydrology of playa basins can often be both rapid and quite dramatic. In recent years, a number of notable closed basin lakes or playas (e.g. Figure 15.7(d)) that were initially fed by perennial rivers have undergone dramatic changes in water balance due to upstream water diversions, e.g. Owens (dry) Lake, USA, and the Dead Sea in the Middle East. Subsequently, human interventions have resulted in either partial or complete desiccation of the lakes and significant associated falls in regional groundwater levels (e.g. Figure 15.7(c)). In each case, desiccation has led to an accumulation of evaporite minerals at the surface (e.g. Benson *et al.*, 1996; Tyler *et al.*, 1997) and aeolian deflation at the lake margins (Gill, 1996). In the case of the Salton Sea, USA, human intervention initially led to an accidental diversion of the Colorado River's flow into the formerly dry Salton Sink in 1905. After this period, the lake, which sits 70 m below sea level, has been maintained through agricultural return flows from the Imperial, Coachella and Mexicali Valleys (90 % of total inflow), and the lake has gradually become an important biodiversity and recreational resource. However, fluctuations in water balance and salinity have led to dramatic lowering in lake level and peripheral desiccation in recent years, with associated increases in salt deposition, aeolian deflation and environmental concern (Gill, 1996).

Tyler, Munoz and Wood (2006) use a coupled (soil physics, climate data, geochemical processes) model to understand water table responses to climate change. They found that for playas with a shallow water table (<0.5 m) (Figure 15.7(c)), relatively modest changes in water table depth would result from an increase or decrease in water balance (inflow/evaporation). They also note that the surface sediments of these playas often have a saturated vadose zone and therefore, due to a lack of storage capacity, can respond rapidly (e.g. by flooding) to changes in inflow (Bryant *et al.*, 1994a; Drake and Bryant, 1994; Bryant and Rainey, 2002) or atmospheric pressure (Turk, 1975) without any necessity for climate forcing. However, where water tables are naturally deeper (>0.5 m) they found that water table changes may be much greater when accommodating similar changes in water balance, but the changes may also significantly lag behind climate forcing. When additional processes were factored in (e.g. the precipitation of minerals in the sediment column, influx of aeolian material) it was evident that changes in base level can occur without direct climate forcing due to changes in sediment pore space and associated hydraulic conductivity within the undersaturated zone above the water table.

In each case we can see that climate forcing and human intervention can lead to significant changes in groundwater levels, which can in turn impact on the status and equilibrium of playas. Playas have long been recognised as having recorded important information relating to past changes in climate (Torgersen *et al.*, 1996) and in most cases the surface hydrological regime responds in a predictable but lagged nature to climate forcing. However, it is also possible that, in some circumstances, changes in hydrological balance can occur in terminal discharge playas without the need for climate forcing. Human interventions in playa basins can at the very least allow us to study the impact of changes in hydrological regime on playa processes. Nevertheless, the rapid and short-term nature of changes in any hydrological/sedimentological equilibrium that can occur often lead to a range of significant and far-reaching and often negative environmental impacts (Gill, 1996).

15.3 Influences of pan hydrology and hydrochemistry on surface morphology

Pan surfaces change over time in the course of the pan cycle (Lowenstein and Hardie, 1985; Bryant *et al.*, 1994a) (Figure 15.10). While Stage 1 (flooding) is normally very rapid, it is worth noting that Stages 2 to 4 can take a variable period of time (0.5–100 years) depending on the nature and level of groundwater interaction with inflow. In all cases, however, influx of surface water halts evaporative loss from groundwater, reverses many of the reactions in the interstitial zone and sets up new gradients between the surface and groundwater bodies. It also introduces 'fresh' clastic material from inflow and from the atmosphere, which settles on the lake bed and provides an environment for organisms that play a part in overall sedimentation (Bryant *et al.*, 1994a). Diatoms, which store SiO_2 (Neev and Emery, 1967), and algal mats, involved in the precipitation of carbonates and the formation of kerogen-rich organic layers (Brock, 1979; Grant and Tindall, 1985), are particularly important in this respect. Larger organisms, such as the brine shrimp *Artemia*, cause bioturbation and sediment reworking (Eardley, 1938).

As surface waters evaporate (Stage 2) they become increasingly brackish, leading to precipitation of salts at the periphery of the water body and on the surface of the brine as precipitation thresholds are reached. These crystals, initially held by surface tension, sink to the lake floor and become nuclei for further, distinctive patterns of crystal growth (Lowenstein and Hardie, 1985). They may also be concentrated by wind action (Stage 3) into arcuate bands known as salt ramps (Millington *et al.*, 1995), which persist as minor landforms after evaporation of the brine. The desiccation of the pan surface (Stage 4) will lead to further interstitial crystal growth and dissolution, and an eventual return to the groundwater dominated regime. In instances where these stages (1–4) represent the long-term drawdown of the groundwater table in response to climate changes or anthropic intervention, an additional stage (Stage 5) representing degradation and reworking of the pan surface through deflation of evaporate minerals and silt-clay sediment (e.g. lunette formation) and fluvial reworking can occur (Bowler, 1986).

Given the continual reworking of surface crusts and sediments within the pan cycle, it is not surprising that evaporite deposits do not persist as sedimentary strata in many playas and, when they do so, may take thousands of years to accumulate. Many of the larger salt lakes owe their extensive evaporite deposits, usually in the form of a series of mud-salt (Hardie, Smoot and Eugster, 1978) or protodolomite–gypsarenite couplets (Dutkiewicz and von der Borch, 1995), to the gradual or repeated desiccation of larger water bodies by climatic change or tectonic activity, as in the case of Lake Magadi (Kenya), Lake Bonneville (USA), the Makgadkgadi Pans (Botswana) and the Dead Sea (Israel–Jordan). In a number of instances (e.g. playas in Tunisia and Australia) evaporate preservation is relatively poor, with salt crusts (often 0.1–1 m) often being partially or completely dissolved by groundwater or surface water inflow (e.g. Bryant *et al.*, 1994a). This can lead to the net accumulation of clay-silt sediment, which is only capped by the thin salt crust during desiccation.

15.3.1 Pan topography

Pan surface morphology is the product of periodic flooding and desiccation, including: (a) rainfall effects, (b) groundwater depth, brine concentration and associated crystal growth and dissolution at or near the sediment surface and (c) aeolian deflation. Given the potential dynamism of these three factors, surface features themselves are among the most ephemeral of geomorphological phenomena, some lasting no longer than the interval between one rainfall event and the next.

Haloturbation is an important process, usually involving gypsum or halite. In the saline mudflat and saline pan environments surface cracking is apparent, leading to polygon formation. Thin hard crusts of carbonates or puffy crusts of more soluble minerals may appear on drying (Hardie, Smoot and Eugster, 1978). Surface flaking, with gypsum precipitation, is also common.

Salt crusts have smooth surfaces only while above the level of capillary action, as at Bonneville Flats (Eugster and Kelts, 1983), or when wet; on drying, crystal expansion leads to the formation of salt blisters and salt polygons, the latter up to 10 m in diameter (Krinsley, 1970) (Figure 15.11). Plate boundaries become foci for evaporation and precipitation, producing thrust surfaces up to 50 cm above the pan floor and capable of lifting gravel size material. Extrusion at the plate edges may lead to the formation of mud and salt pinnacles (Figure 15.12(a)). Subsequent inundation and desiccation leads to a fresh cycle of polygon development.

Under artesian conditions groundwater effluents may be marked by the growth of spring mounds or dissolution of salt karst chimneys (Last, 1993), known collectively as *aioun* in North Africa (Roberts and Mitchell, 1987)

STAGE 1: FLOODING

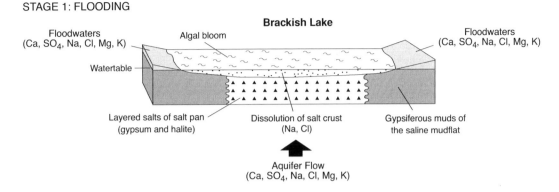

Brackish Lake

Floodwaters
(Ca, SO_4, Na, Cl, Mg, K)

Algal bloom

Floodwaters
(Ca, SO_4, Na, Cl, Mg, K)

Watertable

Layered salts of salt pan
(gypsum and halite)

Dissolution of salt crust
(Na, Cl)

Gypsiferous muds of
the saline mudflat

Aquifer Flow
(Ca, SO_4, Na, Cl, Mg, K)

STAGE 2: EVAPORATIVE CONCENTRATION

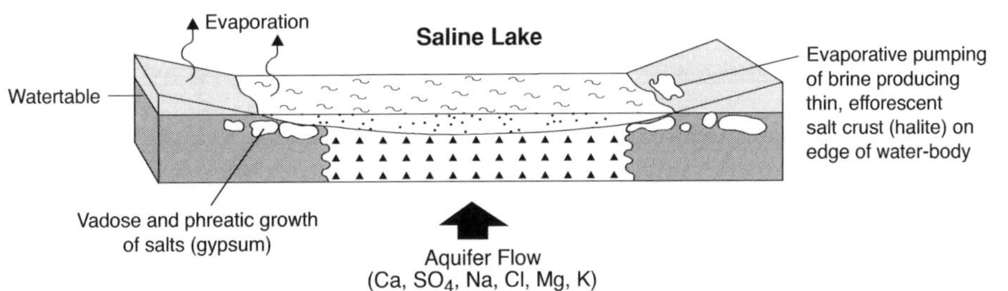

STAGE 3: BRINE POOL

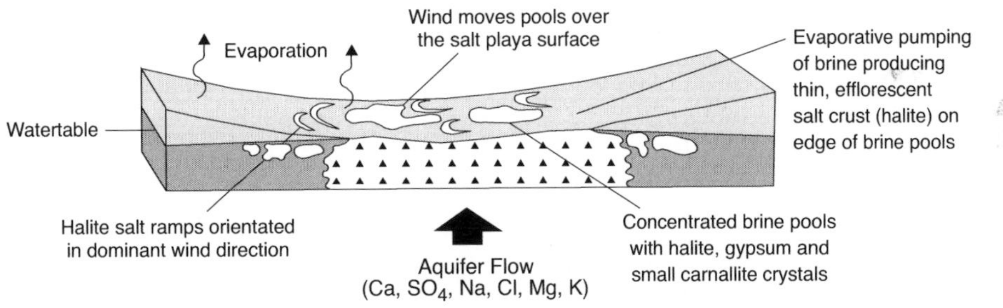

STAGE 4: DESICCATION

Dry Pan

Evaporation

Mineral assemblage:
Gypsum, Halite, Carnallite

Watertable

Concentrated
groundwater brine

Surface crust
broken into polygons

Authigenic growth of
gypsum and halite
within the mudflat

Aquifer Flow
(seasonal fluctuations in groundwater level
due to variations in evaporation)

Figure 15.10 The saline pan cycle for a typical playa; this example is taken from the Chott el Djerid, Tunisia, and charts the cycle (and associated brine geochemical evolution) as it occurred after an extreme flood event in 1990 (after Lowenstein and Hardie, 1985, and Bryant *et al.*, 1994a, 1994b).

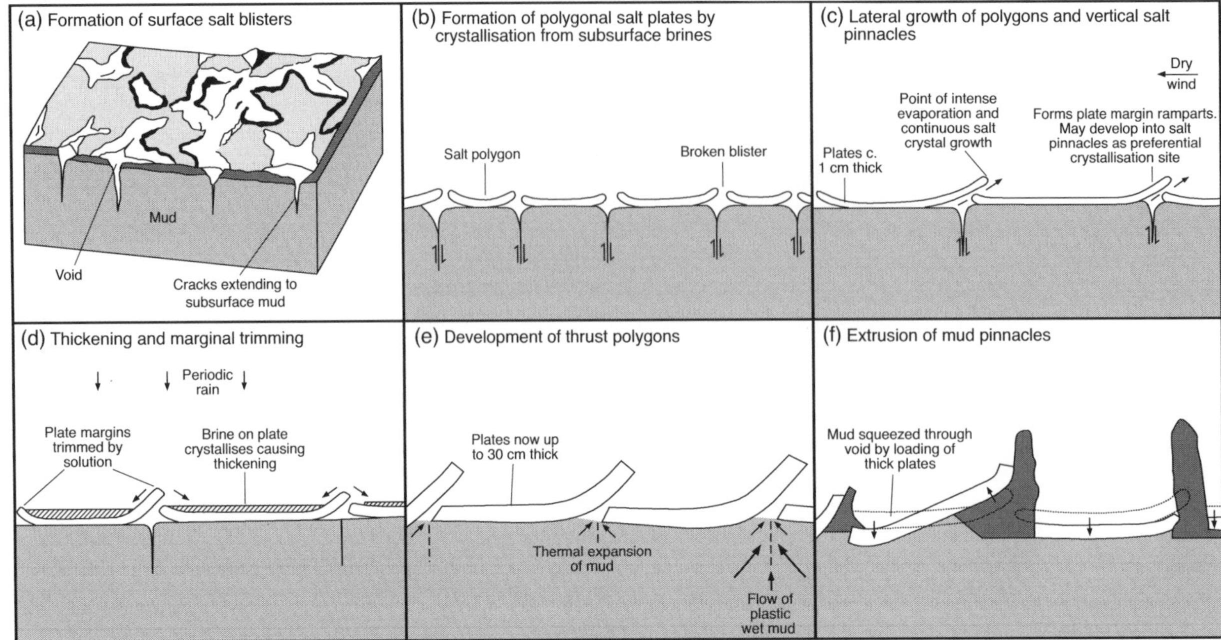

Figure 15.11 Idealised sequential development of a playa salt crust (after Krinsley, 1970).

(Figure 15.12(b)), and tufa deposits, partly organic in origin, as in many of the playas of the southwestern United States (Neal, 1969). Algae may also be preserved as calcareous or siliceous stromatolites towards pan margins, as at Urwi Pan, southern Kalahari (Lancaster, 1977) and the East African salt lakes (Casanova and Hillaire-Marcel, 1992).

15.3.2 Surface dynamics: mapping pan surface morphologies using remote sensing

Because of the difficulties of field investigation of pans and playas (Millington *et al.*, 1989), remote sensing data have been applied to the study of playa basins in a number of instances in order to overcome these issues and gain

Figure 15.12 Playa surfaces: (a) salt thrust polygons, Soda Lake, USA; (b) salt karst chimney (locally termed *aioun*), Chott el Djerid, Tunisia.

important initial information or to augment ongoing research into: (a) surface geochemistry and geomorphologic mapping, (b) hydrology and hydrological balance and (c) studies of dust emissions. Examples of each are outlined below.

As an understanding of surface mineral distributions can elucidate playa basin geochemical processes (Hardie, Smoot and Eugster, 1978) and groundwater regime (Rosen, 1994), remote sensing data have been used to map evaporite mineral assemblages on playa surfaces in a number of instances. Millington *et al.* (1989) use remote sensing data to undertake simple interpretation of playa surfaces. However, subsequent work by Crowley (1991) and Drake (1995) have allowed a link to be made between the mineralogy and surface reflectivity (0.4–2.5 μm) of a wide range of salt phases and geochemical contexts. These data have facilitated the application of a range of remote sensing approaches to both evaporite mineral mapping and facies mapping to a limited number of playas (e.g. Bryant, 1996; Castañeda, Herrero and Casterad, 2005a; White and Eckardt, 2006). Crowley and Hook (1996) and Katra and Lancaster (2008) have also utilised multispectral thermal infrared data to map mineral phases and surface types (Figure 15.13). However, despite the potential outlined in these studies, and the often routine use of playas as calibration sites for remote sensing systems, the use of remote sensing in this manner has perhaps been limited due to: (a) the often complex (heterogeneous) nature of many playa surfaces in relation to pixel size, (b) the changeable and intermittent nature of playa processes, (c) the often non-pristine nature of many mineral assemblages relative to reference data, (d) the presence of surface moisture (e.g. Bryant, 1996) and (e) the need for ground validation data (Bryant, 1996).

Changes in playa water balance can occur over short (e.g. day–month) or long (decade–century) timescales. Mapping presence or absence of surface water on playas is relatively straightforward (e.g. Prata, 1990; Verdin, 1996). In order to monitor and study the hydrological balance of playas, a number of workers have used long archives to generate monthly timeseries of high (e.g. Landsat) and moderate resolution (e.g. AVHRR, MODIS) remote sensing data spanning the last 30 years to monitor the presence, absence and magnitude of surface water bodies on playas. Bryant and Rainey (2002), Bryant (2003) and Bryant *et al.* (2007) use long times series AVHRR and MODIS data to study the hydrological balance of playas in Africa (e.g. Zone of Chotts, North Africa, and Etosha Pan and Magkadigkadi, southern Africa). These workers show that these data can be used to determine: (a) the groundwater regime operating within playas basins (as per Rosen, 1994), (b) the flooding/ponding frequency of playa basins in response to both climatic and nonclimatic factors (as per Bowler, 1986) and (c) the evaporation rate operating during lake desiccation following playa inundation. Similar approaches using a mix of high and moderate resolution data have also been used in the USA (e.g. Lichvar, Gustina and Bolus, 2004; French *et al.*, 2006), Spain (e.g. Castañeda, Herrero and Casterad, 2005b; Castañeda and Garcia-Vera, 2008) and Egypt (e.g. Bastawesy, Khalaf and Arafat, 2008) to a similar end, and confirm the potential of the approach for the wider assessment, hydrological analysis and classification of playa basins. Wadge, Archer and Millington (1994), Archer and Wadge (2001) and Wadge and Archer (2003) investigate the potential of sequential synthetic aperture radar (ERS-1) data for mapping sedimentation and evaporation rates on the Chott el Djerid, Tunisia.

Given the importance of the dust cycle of many playa basins, some workers have used remote sensing to map potential sites of dust emission on playas or observe dust emission. Katra and Lancaster (2008) use a time series of ASTER data to generate maps of surface mineral composition on Soda Lake, USA, that may be used to characterise preferential dust emission sites on the playa surface. Work by Chappell *et al.* (2007) in Australia also suggest that additional approaches utilising soil BRDF may be used to infer the erodibility of dust-producing playa surfaces. Bryant *et al.* (2007) evaluate the dust cycle of playas within the Magkadigkadi basin using a number of remote sensing data types and approaches, and are able to confirm links between the emissive nature of playas and climate feedbacks within southern Africa. Work by Bullard *et al.* (2008) in the Lake Eyre basin also demonstrate the ability of time series of remote sensing (e.g. MODIS) data to monitor in detail the dust cycle of playa basins to enable a better understanding of the spatial and temporal dynamics of dust sources.

Overall, it is apparent that the increased use of remote sensing data (particularly in the last 5–10 years) to monitor playa hydrology, mineralogy and dust emissions has enabled increased understanding of feedbacks that exist between these characteristics of playa basins and the possible wider impacts and feedbacks that exist between climate forcing and human intervention in drylands.

15.4 Aeolian processes in pan environments

Unvegetated and unconsolidated surfaces provide ideal conditions for aeolian activity. Pan surfaces experience deflation during dry periods, with transport in the dominant wind direction to form dunes on the pan and its margins.

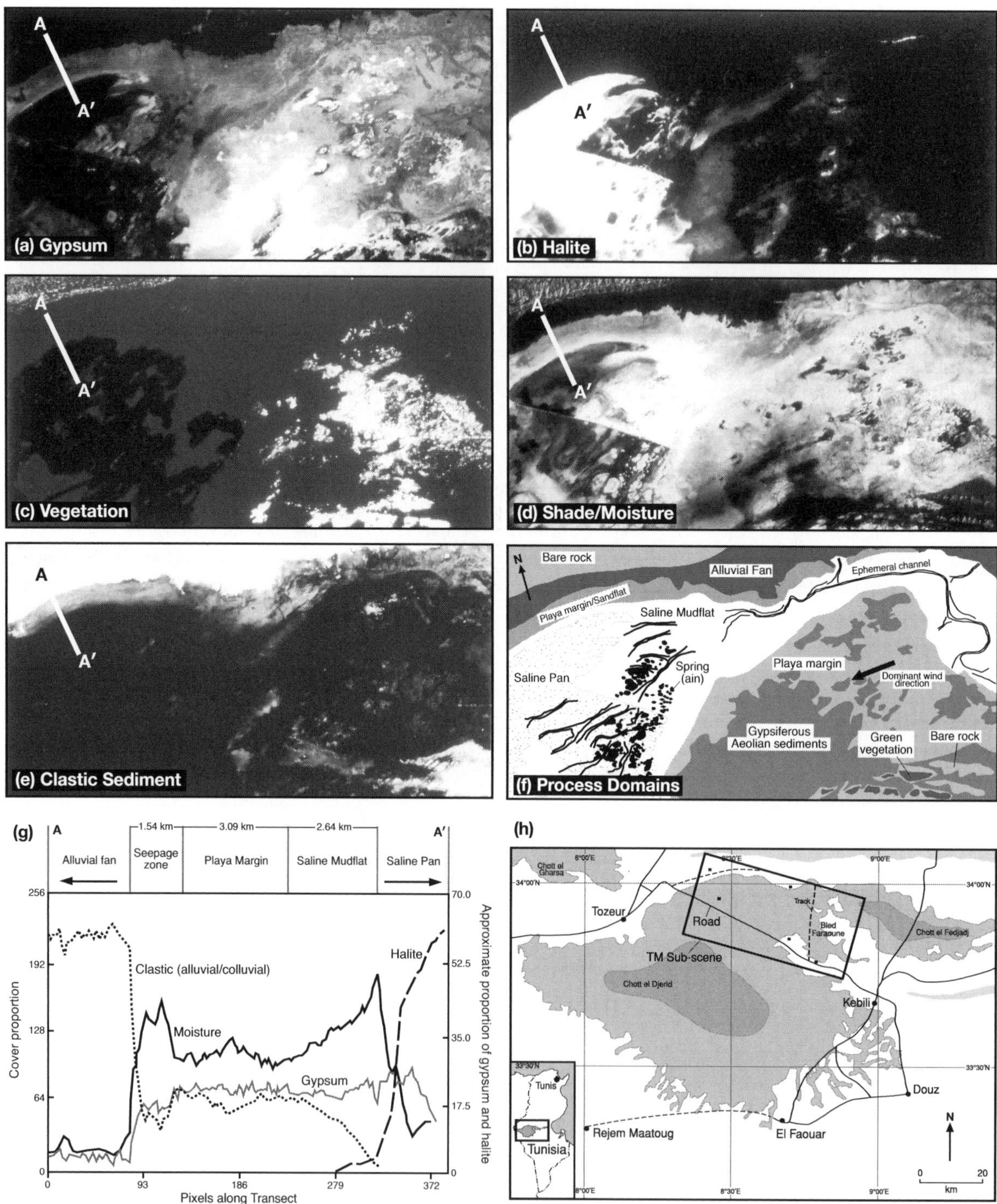

Figure 15.13 An example of the use of remote sensing for mapping and monitoring playa surfaces. In this instance, Landsat TM data of the Chott el Djerid have been processed to produce maps of the surface concentrations of: (a) gypsum, (b) halite, (c) vegetation, (d) moisture/shade and (e) clastic sediments. These data are summarised in a map of process domains (f). A graph cross-section (A–A′ c. 20 km) through these data (g) show how data for these surfaces can be used to delineate and better understand the relationships that exist between the playa depositional subenvironments outlined in Figure 15.4. A location diagram is provided (h).

15.4.1 Wind action on the pan surface

Pan surfaces, composed of dry sands, clays and salts (e.g. playa margin or dry mudflat surfaces), are often vulnerable to wind erosion, although crusting (e.g. Rice and McEwan, 2001) and residual surface moisture (e.g. Reynolds *et al.*, 2007) can reduce its effect. Wind scour can remove material to the level of the near-surface water table (capillary fringe), creating an unconformity known as a Stokes surface (Stokes, 1968; Tyler, Munoz and Wood, 2006), present in a number of depositional subenvironments (Fryberger, Schenk and Krystinik, 1988). On the pan surface wind action initially entrains surface materials, mainly fine sands and small pellets of clay of equivalent dimension, the latter produced by salt efflorescence or desiccation (Bowler, 1973). Where permanent or stable salt crusts are apparent, sand blasting and fluting of polygons and other surface forms is common. At the same time, winnowing and ejection of fine material from weak crusts and fractures separating surface plates also occurs. Removal of fines may lead to the formation of lag deposits composed of gravels, silcrete fragments or remnant crusts.

Depositional forms will include the formation of sand ripples on salt crusts, which may become accentuated by incorporation into the edges of polygon structures. Sediments will accumulate around plant stems to form phreatophytic mounds, which, in turn, may lead to the formation of *nabkha* dunes. On a larger scale parabolic or lunette dunes (Hills, 1940) form on the downwind side of the pan, particularly where wind direction is strongly controlled by basin structure (Hardie, Smoot and Eugster, 1978). These dunes may be modified by later inundation, as in the Makgadikgadi basin of Botswana (Cooke, 1980). In large basins with ample sediment supply, such as Lake Eyre, Australia, a range of dune forms may occur.

The wind may also be implicated in the movement of larger rocks, called sliding stones or playa scrapers, across the pan surface under low frictional conditions (Sharp and Carey, 1976). However, the hydraulic energy of surface runoff has also been proposed as a cause of this phenomenon (Wehmeier, 1986).

15.4.2 The emission of fine particles (dust): process and controls

Playas have for some time been recognised as a source of fine particles (e.g. Gill, 1996; Goudie and Middleton, 2001; Prospero *et al.*, 2002; Washington *et al.*, 2006; Reynolds *et al.*, 2007), and aeolian processes associated with these environments are shown to be significant in determining solute concentration in groundwater in arid and semi-arid areas (e.g. Wood and Sanford, 1995; Eckardt and Spiro, 1999; Eckardt *et al.*, 2008). Given the wider importance of dust for global climate, air quality with respect to visibility and human health, the fertilization of marine and terrestrial ecosystems, transportation and indications of desertification, an understanding of the processes and controls leading to dust emission from playas is essential.

Many studies have sought to characterise processes leading to particle entrainment and wind erosion on playas (e.g. at Owens (dry) Lake; see Box 15.1) using portable wind tunnels or combined aeolian sediment collection and detailed meteorological measurements (e.g. Cahill *et al.*, 1996; Gillette, Ono and Richmond, 2003). However, although wind strength and the availability of sand-sized sediment within a playa basin are the most important drivers of the dust emission process, a number of hydrological and geochemical factors also exist that can significantly moderate or modify the process of dust emission. Indeed, areas on playas that are found to be nonemissive are often covered by a durable salt/silt crust (Rice, Mullins and McEwan, 1997) or are wet (Reynolds *et al.*, 2007, 2009). Given the dynamism of some playa surfaces in response to natural or human-induced changes in groundwater position and surface water ponding, the locations from which dust are emitted and the magnitude of emission from playa surfaces can vary substantially in space and time (Elmore *et al.*, 2008). Some workers (e.g. Mahowald *et al.*, 2003; Bryant, 2003; Bryant *et al.*, 2007; Reynolds *et al.*, 2007) have shown conclusively that a reduction or cessation of dust emissions from playa basins accompanies significant inflow, ponding of water or increased groundwater levels. However, once the inflow waters have receded, Bryant *et al.* (2007) also note an enhanced emission of dust due to the increased availability of fine particles on the playa surface delivered through the flooding process itself. With regard to the inundation process, it is worth reiterating that these events are themselves often either difficult to predict (or are spatially/temporally discrete), as changes in hydrological balance of this nature can occur either in response to climate forcing, surface inflow or human intervention, or a combination of the three. Kotwicki and Isdale (1991), Kotwicki and Allen (1998) and Bryant *et al.* (2007) do, however, show that flooding (and hence aspects of the dust cycle) within large playa basins can be closely linked to regional climate. To assess the gross impacts of hydrological changes, Reynolds *et al.* (2007) presented a conceptual model outlining the relationship between groundwater position and dust emission magnitudes based upon detailed observations from discharge (wet) and recharge (dry) playas in the southwestern

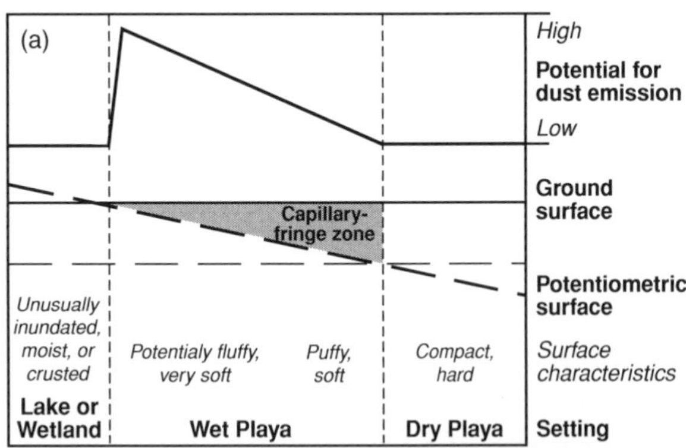

Figure 15.14 Dust emissions from playas. (a) A model, drawing on the classification system of Rosen (1994), which outlines the importance of groundwater regime and depth on the morphology of surface crusts type and dust emission potential from playas from the southwestern USA (after Reynolds *et al.*, 2007). (b) A dust storm observed emanating from the margins of Sua Pan, Botswana (photo courtesy of Frank Eckardt).

USA (Figure 15.14(a)). Pelletier (2006) has also been able to model the relationships that may exist between climate forcing, groundwater levels and dust emission magnitudes for one of these playas (Soda Lake, USA), and outlined similar groundwater-controlled influences on the emission of dust from these sources. It is apparent that the success with which playa hydrologic dynamics of this nature are integrated into models that simulate the regional or global dust cycle will impact significantly on their ability/inability to predict future dust emission magnitudes (e.g. Thomas and Wiggs, 2008). In summary, playa type, size and setting directly influence playa hydrology and surface sediment characteristics, which, in turn, have been identified as being important controls on the type and amount of atmospheric dust emitted from playas. Figure 15.14(b) documents a typical dust storm emanating from a playa surface (Sua Pan, Botswana).

15.4.3 Lunette dunes

Pan-margin dunes (Figure 15.4) are important landscape features in many regions, including southern Australia (Bowler, 1973), the southern Kalahari (Goudie and Thomas, 1986), Tunisia (Coque, 1979) and Texas (Huffman and Price, 1979). Although commonly between 10 and 50 m high, one example in Tunisia rises almost 150 m above the basin floor (Goudie and Thomas, 1986). The surfaces of lunettes are frequently vegetated, which contributes to their development through encouraging sediment accretion. Dune size is a function of a range of factors, but basin size, morphology and sediment supply are important. Cyclical episodes of lunette formation have been identified (Thomas *et al.*, 1993; Dutkiewicz and von der Borch, 1995; Telfer and Thomas, 2007) on the basis of depositional hiatuses and palaeosol formation.

Individual pans and playas can possess more than one fringing dune, with as many as three identified on the margins of some southern Kalahari pans. Differences in morphology, orientation and sedimentology between dunes on the margins of individual basins have been interpreted as indicators of changing palaeoenvironmental (wind regime and hydrological) conditions (e.g. Lancaster, 1978b). In the southern Kalahari some pans possess an outer quartz sand lunette and an inner form that has a higher silt and clay content of between 12 and 20 % by weight (Lancaster, 1978b). The former are interpreted as the outcome of deflation in the initial stages of pan development from the sandy Kalahari floor, under relatively dry conditions in an arid environment, in a manner comparable to parabolic dune development from partially vegetated surfaces. Conversely, in Australia, the orientation of quartz-rich fringing dunes reflects wet-season winds. They have therefore been seen as the outcome of deflation from pan and playa beach sediments during periods of high or seasonal lakes, in a manner comparable to coastal dune development (see, for example, Twidale, 1972).

Importance has been attached to the deflation of clay pellets from seasonally dry pan surfaces in the development of the clay lunettes of Australia (e.g. Bowler, 1973, 1986) and this is also the mechanism that Lancaster (1978b) ascribes to the development of the inner sandy-clay dunes found on pan margins in the southern Kalahari. As pellet formation is dependent on the breakdown of basin-floor clays by salt efflorescence (Australian lunettes also contain high percentages of gypsum as well as clay), clay lunette development is unlikely in extremely dry or surface water-dominated environments (Bowler, 1986).

Box 15.1 Owens (dry) Lake, USA

Owens (dry) Lake, USA, was a perennial lake at the terminus of the Owens River for most of the last 800 000 years (Smith, Bischoff and Bradbury, 1997). During the late 1800s and early 1900s the lake fluctuated between about 7 and 15 m deep and had an area of about 280 km^2, depending on drought conditions and irrigation diversions (see Figure 15.15). As a result of the diversion of inflow waters beginning in 1913 the pre-existing perennial saline lake (e.g. Figure 15.7(d)) developed into a wet groundwater discharge playa by 1928 (e.g. Figure 15.7(b)),

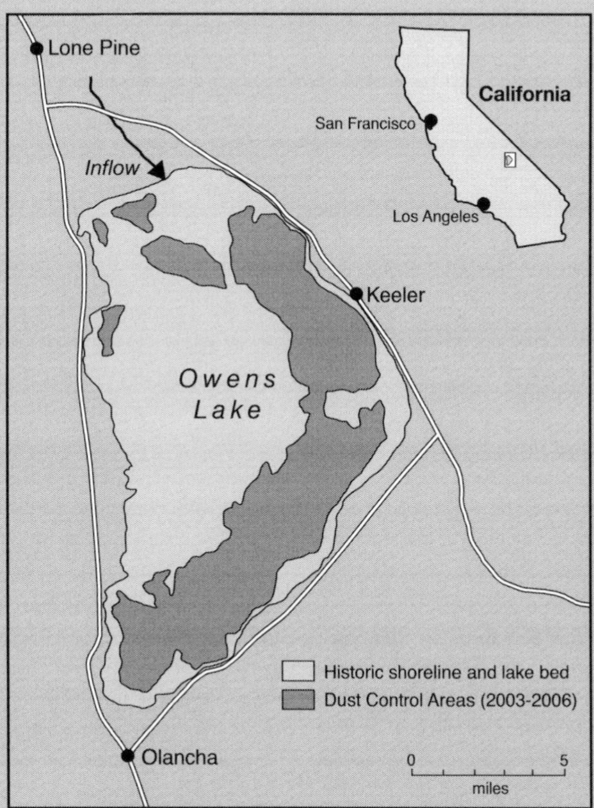

Figure 15.15 Case study: map of Owens Lake, showing dust control areas.

which has subsequently been the largest single source of particulate-matter emission (by one estimate, 900 000–8000 000 tonnes/year) in the United States (e.g. Gill and Gillette, 1991; Cahill *et al.*, 1996; Gillette, Ono and Richmond, 2003). Research on this playa has led to a greater understanding of the processes involved in the emission of dust from playas. Gillette, Ono and Richmond (2003) outline the process whereby emission of dust is initiated through creep (reptation) and saltation of sand-sized particles across the playa surface, which loosen other particles and sandblast the surface, causing finer particles, including dust, to be ejected and to mix vertically in the turbulent air stream. The amount of dust emitted was therefore suggested to be proportional to this horizontal saltation flux on any playa surface. During the period 2000 to 2006 a wide range of dust-control measures (including the use of shallow flooding, controlled vegetation and gravel) were implemented on a large part of the former lake bed.

Up to 80 % of the material in Australian lunettes is in the form of clay pellets (Bowler, 1973), though overall dune sediments range from sandy clay, comparable to that forming the inner Kalahari lunettes, to almost pure gypsum (Bowler, 1976). Analyses of the Australian clays show that the material persists in pelletal form after deposition (Bowler and Wasson, 1984), whereas in Nevada Young and Evans (1986) report that clay pellets break down after deposition on the dunes when the binding salts are broken down in subsequent rainfall events. The resultant landform is termed a *mud dune*.

15.4.4 Yardangs

On a large scale differential erosion of horizontal sediments may lead to the development of yardang topography (Figure 15.16), with a relief of up to 20 m (Mabbutt, 1977). The kalut landscape of the Kerman basin in Iran, a series of parallel ridges and troughs with a relief of 60 m (Dresch, 1968; Krinsley, 1970), is an extreme example of this landform suite. On the regional scale, playas with falling water tables may be lowered by deflation and groundwater weathering to form pan and escarpment landscapes, as in the Oasis Depressions of North Africa, including the Qattara Depression at 134 m below sea level.

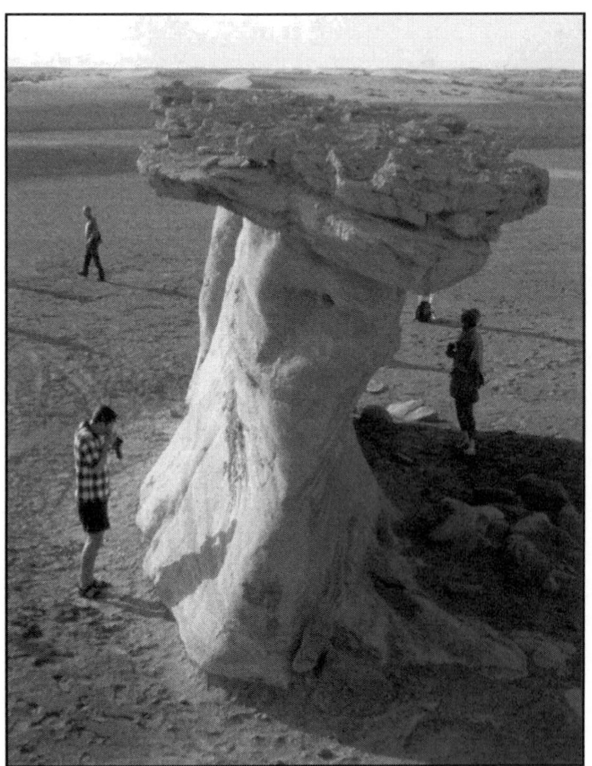

Figure 15.16 Yardang in aeolianite on a pan surface, United Arab Emirates.

15.5 Pans and playas as palaeoenvironmental indicators

Pans and playas have long been important sources of palaeoenvironmental reconstruction in arid environments, particularly in Australia (e.g. Harrison, 1993), Africa (Street-Perrott and Roberts, 1983) and the southwest USA (Benson *et al.*, 1990), even though the evidence they preserve is often discontinuous and absolute dating has been problematic. Within the past two decades absolute dating has moved on from radiocarbon (e.g. Bowler *et al.*, 1986) to the application of Th/U isotopes (e.g. Herczeg and Chapman, 1991), amino acid racemisation (Miller *et al.*, 1991) and luminescence techniques (e.g. Chen, 1995), bringing better temporal resolution over longer time spans. Alongside this, the understanding of pans and playas as dynamic, three-dimensional landforms involving the interface of aeolian, groundwater and surface water processes has reduced the degree of ambiguity in palaeoenvironmental interpretation. The evidence is based upon three aspects of playa research.

15.5.1 Identification and dating of pan shorelines

The mapping and dating of shorelines makes it possible to recreate past water budgets on the assumption that most playas are ‘amplifier lakes’ (Street, 1980) in that, lacking surface outflow, they tend to emphasise the influence of precipitation on the water budget. The evidence usually takes the form of a lake of successively decreasing volumes represented by a suite of strandlines, each controlled by a threshold within the hydrological system, such as lake morphology, or sometimes an overflow. Shorelines have been dated by thermoluminescence (TL; see Magee *et al.*, 1995) and optically stimulated luminescence (OSL; see Burrough *et al.*, 2007) dating techniques, a key factor being improved sampling to depths of c. 15 m using lightproof augers.

15.5.2 Dating and stratigraphy of lunette dunes

The study of landforms associated with pans, in particular the stratigraphy of lunette dunes (Chen, 1995; Dutkiewicz and von der Borch, 1995), supports the contention that many playas have cycles of erosion and sedimentation with a distinct local signature. In turn this can be interpreted in terms of decreased precipitation, or changes in wind strength or circulation, and can be dated using an OSL assay of sedimentary cores (e.g. Telfer and Thomas, 2007; Telfer *et al.*, 2008).

15.5.3 Stable isotope studies and pan hydrochemical evolution

Recent advances in the multidisciplinary study of playas as three-dimensional features, with distinct hydrological, sedimentary, chemical and organic budgets, has allowed the identification of distinct facies associated with the saline pan cycle, and thus allowed the interpretation of their sedimentary record, even where it is discontinuous (Chivas, 2007; Yechieli and Wood, 2002). Geochemical studies have contributed greatly to this, in particular the recognition that common minerals such as gypsum may take on different crystal forms in lacustrine, groundwater, aeolian and pedogenic environments (Magee, 1991; Magee *et al.*, 1995). Building on the work of Eugster and Jones (1979), a number of workers (e.g. Risacher and Fritz, 1991; Bryant *et al.*, 1994a; Yan, Hinderer and Einsele, 2002; Eckardt *et al.*, 2008) have undertaken systematic analyses of surface waters in order to understand and model sources of solutes and geochemical brine evolution processes associated with playa systems (Figure 15.8). In order to look more closely at the geochemical pathways associated with brine evolution over long time periods, trace element and stable isotope approaches have been used to trace the often complex geochemical provenance of evaporite deposits (e.g. groundwater, surface water or atmospheric sources; see Figure 15.5). Typical trace element analyses include determining the Br/Cl content of halite (e.g. Hardie, 1984) and the strontium content of gypsum (e.g. Rosell *et al.*, 1998). More general bromine geochemistry of playa halite has also been undertaken (e.g. Bryant *et al.*, 1994a; Risacher and Fritz, 2000) to attempt to understand their source, preservation and diagenetic history. However, given that modern playa evaporites can display a marine-like geochemical signature, these approaches have fundamental limitations (Bryant *et al.*, 1994a). As a consequence, Chivas *et al.* (1991), Vengosh *et al.* (1992), Ramesh, Jani and Bhushan (1993) and Eckardt *et al.* (2008) successfully use a combination of strontium, oxygen, boron and sulfur isotope compositions of waters and mineral phases (e.g. calcium ($^{87}Sr/^{86}Sr$), gypsum ($^{34}S/^{32}S$), borates ($^{10}B/^{11}B$)) to determine the sources of the geochemical evolution of terrestrial evaporites deposited in a range of playas from Australian, USA and Africa.

In order to elucidate geomorphic processes and dating of recent surficial deposits on playas, Reynolds *et al.* (2007) have utilised radionuclide analyses (^{137}Cs) to determine recent erosion and sediment accumulation rates for playas in the southwestern USA. At the same time, the geochemistry and regional impact of dust emanating from playas has also received much attention (e.g. Gill *et al.*, 2002; Reheis, 2006). Although organic materials do not display the same degree of preservation that occurs in temperate environments, studies involving pollen, diatoms (Burrough *et al.*, 2007), stromatolites (Casanova and Hillaire-Marcel, 1992), ostracods (Lister *et al.*, 1991) and even ostrich shells (Miller *et al.*, 1991) have added to the debate.

References

Allison, M.S. (1899) On the origin and formation of pans. *Transactions, Geological Society of South Africa*, **4**, 159–161.

Arad, A. (1984) Relationship of salinity of groundwater to recharge in the southern Kalahari Desert. *Journal of Hydrology*, **71**, 225–238.

Archer, D.J. and Wadge, G. (2001) Modelling the backscatter response due to salt crust development. *IEEE Transactions on Geoscience and Remote Sensing*, **39**, 2307–2310.

Atwood, G. (1994) Geomorphology applied to flooding problems of closed-basin lakes – specifically Great Salt Lake, Utah, in *Geomorphology and Natural Hazards* (ed. M. Morisawa), *Proceedings of the 25th Binghampton Symposium*, pp. 197–220.

Ayeni, J.S.O. (1977) Waterholes of the Tsavo National Park. *Journal of Applied Ecology*, **14**, 369–378.

Barth, H.-J. (2001) Comment on 'Playa, playa lake, sabkha: proposed definition of old terms'. *Journal of Arid Environments*, **47**, 513–514.

Bastawesy, M.A., Khalaf, F.I. and Arafat, S.M. (2008) The use of remote sensing and GIS for the estimation of water loss from Tushka lakes, southwestern desert, Egypt. *Journal of African Earth Sciences*, **52** (3), 73–78.

Benson, L.V., Currey, D.R., Dorn, R.I. *et al.* (1990) Chronology of expansion and contraction of four Great Basin lake systems during the past 35,000 years. *Palaeogeography, Palaeoclimatology, Palaeoecology*, **78**, 241–286.

Benson, L.V., Burdett, J.W., Kashgarian, M. *et al.* (1996) Climatic and hydrologic oscillations in the Owens Lake Basin and adjacent Sierra Nevada, California. *Science*, **274** (5288), 746–749.

Boocock, C. and Van Straten, J.J. (1962) Notes on the geology and hydrology of the central Kalahari region,

Bechuanaland Protectorate. *Transactions, Geological Society of South Africa*, **65**, 130–132.

Bowler, J.M. (1973) Clay dunes, their occurence, formation and environmental significance. *Earth Science Reviews*, **9**, 315–338.

Bowler, J.M. (1976) Aridity in Australia: age, origins and expression in aeolian landforms and sediments. *Earth Science Reviews*, **12**, 279–310.

Bowler, J.M. (1986) Spatial variability and hydrological evolution of Australian lake basins: analogue for Pleistocene hydrological change and evaporite formation. *Palaeogeography, Palaeoclimatology, Palaeoecology*, **54**, 21–41.

Bowler, J.M. and Wasson, R.J. (1984) Glacial age environments of inland Australia, in *Late Cenozoic Palaeoclimates of the Southern Hemisphere* (ed. J. Vogel), Balkema, Rotterdam, pp. 183–208.

Bowler, J.M., Huang Qi, Chen Kezao, Head, M.J. and Yuan Baoyin (1986) Radiocarbon dating of playa-like hydrologic changes: examples from northwest China and central Australia. *Palaeogeography, Palaeoclimatology, Palaeoecology*, **54**, 241–260.

Brock, T.D. (1979) Environmental biology of living stromatolites, in *Stromatolites* (ed. M.R. Walter), Developments in Sedimentology 20, Elsevier, Amsterdam.

Briere, P.R. (2000) Playa, playa lake, sabkha: proposed definitions for old terms. *Journal of Arid Environments*, **45**, 1–7.

Bryant, R.G. (1996) Validated linear mixture modelling of LANDSAT TM data for mapping evaporite minerals on a playa surface. *International Journal of Remote Sensing*, **17**, 315–330.

Bryant, R.G. (2003) Monitoring hydrologic controls on dust emissions: preliminary observations from Etosha Pan, Namibia. *The Geographical Journal*, **169**, 131–141.

Bryant, R.G. and Rainey, M.P. (2002) Investigation of flood inundation on playas within the Zone of Chotts using a time series of AVHRR. *Remote Sensing of Environment*, **82**, 360–375.

Bryant, R.G., Sellwood, B.W., Millington, A.C. and Drake, N.A. (1994a) Marine-like potash evaporite formation on a continental playa: case study from Chott el Djerid, southern Tunisia. *Sedimentary Geology*, **90**, 269–291.

Bryant, R.G., Drake, N.A., Millington, A.C. and Sellwood, B.W. (1994b) The chemical evolution of the brines of Chott el Djerid, Southern Tunisia, after an exceptional rainfall event in January (1990), in *The Sedimentology and Geochemistry of Modern and Ancient Saline Lakes* (eds R.W. Renaut and W.M. Last), SEPM (Society for Sedimentary Geology) Special Publication 50, Tulsa, Oklahoma, USA, pp. 3–12.

Bryant, R.G., Bigg, G.R., Mahowald, N.M., Eckhardt, F.D. and Ross, S.G. (2007) Dust emission response to climate in southern Africa. *Journal of Geophysical Research – Atmospheres*, **112**, D09207, DOI: 10.1029/2005JD007025.

Bullard, J., Baddock, M., McTainsh, G. and Leys, J. (2008) Sub-basin scale dust source geomorphology detected using MODIS. *Geophysical Research Letters*, **35** (15), article L15404.

Burrough, S.L., Thomas, D.S.G., Shaw, P.A. and Bailey, R.M. (2007) Multiple Quaternary highstands at Lake Ngami, Kalahari, northern Botswana. *Palaeogeography, Palaeoecology, Palaeoclimatology*, **253**, 280–299.

Butterworth, J.S. (1982) *The chemistry of Mogatse Pan, Kgalagadi District*, Botswana Geological Survey Department, Report JSB/14/82.

Bye, J.A.T. and Harbison, I.P. (1991) Transfer of inland salts to the marine environment at the head of the Spencer Gulf, South Australia. *Palaeogeography, Palaeoclimatology, Palaeoecology*, **84**, 357–368.

Cahill, T.A., Gill, T.E., Reid, J.S., *et al.* (1996) Saltating particles, playa crusts, and dust aerosols at Owens (dry) Lake, California. *Earth Surface Processes and Landforms*, **21**, 621–639.

Carlisle, W.J. and Marrs, R.W. (1982) Eolian features of the Southern High Plains and their relationship to windflow, Geological Society of America, Special Paper 192, pp. 89–105.

Casanova, J. and Hillaire-Marcel, C. (1992) Chronology and palaeohydrology of Late Quaternary high lake levels in the Manyara basin (Tanzania) from isotopic data (18O,13C,14C, Th/U) on fossil stromatolites. *Quaternary Research*, **38**, 205–226.

Castañeda, C. and Garćia-Vera, M.Á. (2008) Water balance in the playa-lakes of an arid environment, Monegros, NE Spain. *Hydrogeology Journal*, **16**, 87–102, DOI: 10.1007/s10040-007-0230-9.

Castañeda, C., Herrero, J. and Casterad, M.A. (2005a) Landsat monitoring of playa-lakes in the Spanish Monegros Desert. *Journal of Arid Environments*, **63**, 497–516.

Castañeda, C., Herrero, J. and Casterad, M.A. (2005b) Facies identification within the playa-lakes of the Monegros Desert, Spain, from field and satellite data. *Catena*, **63**, 39–63.

Chappell, A., Strong, C., McTainsh, G. and Leys, J. (2007) Detecting induced in situ erodibility of a dust-producing playa in Australia using a bi-directional soil spectral reflectance model. *Remote Sensing of Environment*, **106**, 508–552.

Chen, X.Y. (1995) Geomorphology, stratigraphy and thermoluminescence dating of the lunette dune at Lake Victoria, western New South Wales. *Palaeogeography, Paleoeclimatology,Palaeoecology*, **113**, 69–86.

Chen, K. and Bowler, J.M. (1986) Late Pleistocene evolution of salt lakes in Qaidam Basin, Quinghai Province, China. *Palaeogeography, Palaeoclimatology, Palaeoecology*, **54**, 87–104.

Chivas, A.R. (2007) Terrestrial evaporites, in *Geochemical Sediments and Landscapes* (eds D.J. Nash and S.J. McLaren), Blackwell Science, pp. 330–364.

Chivas, A.R., Andrew, A.S., Lyons, W.B. *et al.* (1991) Isotopic constraints on the origin of salts in Australian playas. 1. Sulphur. *Palaeogeography, Palaeoclimatology, Palaeoecology*, **84**, 309–332.

Cooke, H.J. (1980) Landform evolution in the context of climatic change and neo-tectonism in the Middle Kalahari of north central Botswana. *Transactions of the Institute of British Geographers*, **5**, 80–99.

Cooke, R.U., Brunsden, D., Doornkamp, J.C. and Jones, D.K.C. (1982) *Urban Geomorphology in Drylands*, Oxford University Press, Oxford.

Coque, R. (1979) Sur la place du vent dans l'érosion en milieu aride. L'exemple des lunettes (bourrelets éoliens) de la Tunisie. *Mediterranée*, **1/2**, 15–21.

Crowley, J.K. (1991) Visible and near-infrared (0.4–2.5 μm) reflectance spectra of playa evaporite minerals. *Journal of Geophysical Research B*, **96** (B10), 16, 231–16, 240.

Crowley, J.K. and Hook, S.J. (1996) Mapping playa evaporite minerals and associated sediments in Death Valley, California, with multispectral thermal infrared images. *Journal of Geophysical Research B: Solid Earth*, **101**, (B1), 643–660.

Currey, D.R. (1994) Hemiarid lake basins: hydrographic patterns, in *Geomorphology of Desert Environments* (eds A.D. Abrahams and A.J. Parsons), Chapman and Hall, London, pp. 405–421.

de Ploey, J. (1965) Position géomorphologique, genèse et chronologie de certains dépôts superficiels au Congo occidental. *Quaternaria*, **7**, 131–154.

Drake, N.A. (1995) Reflectance spectra of evaporite minerals (400–2500 nm): applications for remote sensing. *International Journal of Remote Sensing*, **16**, 2555–2571.

Drake, N.A. and Bryant, R.G. (1994) Monitoring the flooding ratios of Tunisian Playas using AVHRR data, in *Climatic Change in Drylands* (eds A.C. Millington and K. Pye), John Wiley & Sons, Ltd, London, pp. 87–96.

Dresch, J. (1968) Reconnaissance dans le Lut (Iran). *Bulletin, Association Géographie Français*, **362/3**, 143–153.

Dutkiewicz, A. and von der Borch, C.C. (1995) Lake Greenly, Eyre Peninsula, South Australia; sedimentology, paleoclimatic and paleohydrologic cycles. *Palaeogeography, Palaeoclimatology, Palaeoecology*, **113**, 43–56.

Eardley, A.J. (1938) Sediments of the Great Salt Lake, Utah. *Bulletin of the American Association of Petroleum Geology*, **22**, 1305–1411.

Eckardt, F.D. and Spiro, B. (1999) The origin of sulphur in gypsum and dissolved sulphate in the Central Namib desert, Namibia. *Sedimentary Geology*, **123**, 255–273.

Eckardt, F.D., Bryant, R.G., McCulloch, G. et al. (2008) The hydrochemistry of a semi-arid pan basin, case study: Sua Pan, Makgadikgadi, Botswana. *Applied Geochemistry*, **23**, 1563–1580, DOI: 10.1016/j.apgeochem.2007.12.033.

Elmore, A.J., Kaste, J.M., Okin, G.S. and Fantle, M.S. (2008) Groundwater influences on atmospheric dust generation in deserts. *Journal of Arid Environments*, **72**, 1753–1765.

Eugster, H.P. and Hardie, L.A. (1978) Saline lakes, in *Chemistry, Geology and Physics of Lakes* (ed. A. Lerman), Springer Verlag, New York, pp. 237–239.

Eugster, H.P. and Jones, B.F. (1979) Behavior of major solutes during closed-basin brine evolution. *American Journal of Science*, **279**, 609–631.

Eugster, H.P. and Kelts, K. (1983) Lacustrine chemical sediments, in *Chemical Sediments and Geomorphology* (eds. A.S. Goudie and K. Pye), Academic Press, London, pp. 321–368.

Farr, J.J., Peart, R.J., Nellisse, C. and Butterworth, J.S. (1982) Two Kalahari pans: a study of their morphometry and evolution, Botswana Geological Survey Department, Report GS10/10.

French, R.H., Miller, J.J., Dettling, C. and Carr, J.R. (2006) Use of remotely sensed data to estimate the flow of water to a playa lake. *Journal of Hydrology*, **325**, 67–81, DOI: 10.1016/j.jhydrol.2005.09.034.

Fryberger, S.G., Schenk, C.J. and Krystinik, L.F. (1988) Stokes surfaces and the effects of near-surface groundwater-table on aeolian deposition. *Sedimentology*, **35**, 21–41.

Gilbert, C.K. (1895) Lake basins created by wind erosion. *Journal of Geology*, **3**, 47–49.

Gill, T.E. (1996) Eolian sediments generated by anthropogenic disturbance of playas: human impacts on the geomorphic system and geomorphic impacts on the human system. *Geomorphology*, **17**, 207–228.

Gill, T. E. and Gillette, D.A. (1991) Owens Lake: a natural laboratory for aridification, playa desiccation and desert dust. *Geological Society of America, Abstract Programs*, **23** (5), 462.

Gill, T. E., Gillette, D. A., Niemeyer, T. and Winn, R. T. (2002) Elemental geochemistry of wind-erodible playa sediments, Owens Lake, California. *Nuclear Instruments and Methods in Physics Research, B*, **189**, 209–213.

Gillette, D.A. (1981) Production of dust that may be carried great distances, in *Desert Dust* (ed. T.L. Péwé), Geological Society of America Special Paper 186, pp. 11–26.

Gillette, D.A., Ono, D. and Richmond, K. (2003) A combined modeling and measurement technique for estimating windblown dust emissions at Owens (dry) Lake, California. *Journal of Geophysical Research*, **109**, DOI: 10.1029/2003JF000025.

Gindy, A.R. (1991) Origin of the Qattara Depression, Egypt – a discussion. *Geological Society of America Bulletin*, **103**, 1374–1376.

Glennie, K.W. (1970) *Desert Sedimentary Environments*, Elsevier, Amsterdam.

Goudie, A.S. and Middleton, N.J. (2001) Saharan dust storms: nature and consequences. *Earth-Science Reviews*, **56**, 179–204.

Goudie, A.S. and Thomas, D.S.G. (1985) Pans in southern Africa with particular reference to South Africa and Zimbabwe. *Zeitschrift für Geomorphologie*, NF, **29**, 1–19.

Goudie, A.S. and Thomas, D.S.G. (1986) Lunette dunes in southern Africa. *Journal of Arid Environments*, **10**, 1–12.

Goudie, A.S. and Wells, G.L. (1995) The nature, distribution and formation of pans in arid zones. *Earth-Science Reviews*, **38**, 1–69.

Grant, W.D. and Tindall, B.J. (1985) The alkaline environment, in *Microbes in Extreme Environments* (eds. R.A. Herbert and G.A. Codd), Society of General Microbiology, Special Publication 17.

Gregory, J.W. (1914) The lake system of Westralia. *Geographical Journal*, **43**, 656–664.

Hardie, L.A., Smoot, J.P. and Eugster, H.P. (1978) Saline lakes and their deposits: a sedimentological approach, in *Modern and Ancient Lake Sediments* (eds. A. Matter and W. Tucker), International Association of Sedimentologists, Special Publication 2, pp. 7–41.

Harrison, S.P. (1993) Late Quaternary lake-level changes and climates of Australia. *Quaternary Science Reviews*, **12**, 211–231.

Haukos, D.A. and Smith, L.M. (1994) The importance of playa wetlands to biodiversity of the Southern High Plains. *Landscape and Urban Planning*, **28**, 83–98.

Haynes, C.V. (1980) Geological evidence of pluvial climates in the Nabta area of the Western Desert, Egypt, in *Prehistory of the Eastern Sahara* (eds F. Wendorf and R. Schild), Academic Press, New York, pp. 353–371.

Herczeg, A.L. and Chapman, A. (1991) Uranium-series dating of lake and dune deposits in southeastern Australia: a reconnaissance. *Palaeogeography, Palaeoclimatology, Palaeoecology*, **84**, 285–298.

Herczeg, A.L. and Lyons, W.B. (1991) A chemical model for the evolution of Australian sodium chloride lake brines. *Palaeogeography, Palaeoclimatology, Palaeoecology*, **84**, 43–53.

Hills, E.S. (1940) The lunette: a new landform of aeolian origin. *The Australian Geographer*, **3**, 1–7.

Holser, W.T. (1979) Mineralogy of evaporites, in *Marine Minerals: Reviews in Mineralogy*, vol. 6 (ed. R.G. Burns), Mineralogical Society of America, pp. 211–294.

Huffman, G.W. and Price, W.A. (1979) Clay dune formation near Corpus Christi, Texas. *Journal of Sedimentary Petrology*, **19**, 118–127.

Hunt, C.B., Robinson, T.W., Bowles, W.A. and Washburn, A.I. (1966) Hydrologic basin, Death Valley, California, US Geological Survey, Professional Paper 494B.

Jack, R.L. (1921) The salt and gypsum resources of South Australia, Geological Survey of South Australia, Bulletin 8.

Jacobson, G., Ferguson, J. and Evans, W.R. (1994) Groundwater-discharge playas of the Mallee Region, Murray Basin, southeast Australia, in *Paleoclimate and Basin Evolution of Playa Systems* (ed. M.R. Rosen), Geological Society of America Special Paper 289, pp. 81–96.

Jankowski, J. and Jacobson, G. (1989) Hydrochemical evolution of regional groundwaters to playa brines in central Australia. *Journal of Hydrology*, **108**, 123–173.

Johns, R.K. (1989) The geological setting of Lake Eyre, in *The Great Filling of Lake Eyre in 1974* (eds C.W. Bonython and A.S. Fraser), Royal Geographical Society of Australasia, Adelaide, pp. 60–66.

Jones, B.F. (1965) The hydrology and mineralogy of Deep Springs Lake, Inyo County, California, US Geological Survey, Professional Paper 502A.

Jones, B.F., Hanor, J.S. and Evans, W.R. (1994) Sources of dissolved salts in the central Murray Basin, Australia. *Chemical Geology*, **111**, 135–154.

Katra, I. and Lancaster, N. (2008) Surface-sediment dynamics in a dust source from spaceborne multispectral thermal infrared data. *Remote Sensing of Environment*, **112**, 3212–3221, DOI: 10.1016/j.rse.2008.03.016.

Killigrew, L.P. and Gilkes, R.J. (1974) Development of playa lakes in south western Australia. *Nature*, **247**, 454–455.

Kotwicki, V. and Allan, R. (1998) La Niña de Australia – contemporary and palaeohydrology of Lake Eyre. *Palaeogeography, Palaeoclimatology, Palaeoclimatology*, **144**, 265–280.

Kotwicki, V. and Isdale, P. (1991) Hydrology of Lake Eyre, Australia: El Niño link. *Palaeogeography, Palaeoclimatology, Palaeoecology*, **84**, 87–98.

Krinsley, D.P. (1970) A geomorphological and palaeoclimatological study of the playas of Iran, US Geological Survey, Final Scientific Report CP 70-800.

Lancaster, I.N. (1977) Pleistocene lacustrine stromatolites from Urwi Pan, Botswana. *Transactions, Geological Society of South Africa*, **80**, 283–285.

Lancaster, I.N. (1978a) The pans of the southern Kalahari. *Geographical Journal*, **144**, 80–98.

Lancaster, I.N. (1978b) Composition and formation of southern Kalahari pan margin dunes. *Zeitschrift für Geomorphologie*, NF, **22**, 148–169.

Langmuir, D. (1997) *Aqueous Environmental Chemistry*, Prentice-Hall, New Jersey, 600 pp.

Last, W.M. (1993) Salt dissolution features in saline lakes of the northern Great Plains, Western Canada. *Geomorphology*, **8**, 321–334.

Le Roux, J.S. (1978) The origin and distribution of pans in the Orange Free State. *South African Geographer*, **6**, 167–176.

Lichvar, R., Gustina, G. and Bolus, R. (2004) Ponding duration, ponding frequency, and field indicators: A case study on three California, USA, playas. *Wetlands*, **24**, 406–413.

Lister, G.S., Kelts, K., Zao, C.K. et al. (1991) Lake Qinghai, China: closed-basin lake levels and the oxygen isotope record for ostracoda since the latest Pleistocene. *Palaeogeography, Palaeoclimatology, Palaeoecology*, **84**, 141–162.

Lowenstein, T.K. and Hardie, L.A. (1985) Criteria for the recognition of salt pan evaporites. *Sedimentology*, **32**, 627–644.

Lyons, W.B., Welch, K.A., Neumann, K. et al. (1998) Geochemical linkages among glaciers, streams and lakes within the Taylor Valley, Antarctica, American Geophysical Union, pp. 77–92.

Mabbutt, J.A. (1977) *Desert Landforms*, MIT Press, Cambridge, USA.

McCarthy, T.S., McIver, J. and Cairncross, B. (1986) Carbonate accumulation on islands in the Okavango Delta. *South African Journal of Science*, **82**, 588–591.

McCulloch, G.P., Irvine, K., Eckardt, F.D. and Bryant, R.G. (2008) Hydrochemical fluctuations and crustacean community composition in an ephemeral saline lake (Sua Pan, Makgadikgadi Botswana). *Hydrobiologia*, **596**, 31–46, DOI: 10.1007/s10750-007-9055-8.

McFarlane, M.J., Ringrose, S.M., Giusti, L. and Shaw, P.A. (1995) The origin and age of karstic depressions in the Darwin–Koolpinyah area of the Northern Territory of Australia, in *Geomorphology and Groundwater* (ed. A. Brown), John Wiley & Sons, Ltd, Chichester, pp. 93–120.

Magee, J.W. (1991) Late Quaternary lacustrine, groundwater, aeolian and pedogenic gypsum in the Prungle Lakes, southeastern Australia. *Palaeogeography, Palaeoclimatology, Palaeoecology*, **84**, 3–42.

Magee, J.W., Bowler, J.M., Miller, G.H. and Williams, D.L.G. (1995) Stratigraphy, sedimentology, chronology and palaeohydrology of Quaternary lacustrine deposits at Madigan Gulf, Lake Eyre, South Australia. *Palaeogeography, Palaeoclimatology, Palaeoecology*, **113**, 3–42.

Mahowald, N.M., Bryant, R.G., del Corral, J. and Steinberger, L. (2003) Ephemeral lakes and desert dust sources. *Geophysics Research Letters*, **30**, 1074, DOI: 10.1029/2002GL016041.

Mallick, D.I.J., Habgood, F. and Skinner, A.C. (1981) A geological interpretation of LANDSAT imagery and air photography of Botswana, in *Overseas Geological and Mining Research*, vol. 56, HMSO, London.

Miller, G.H., Wendorf, F., Ernst, E. et al. (1991) Dating lacustrine episodes in the eastern Sahara by the epimerization of isoleucine in ostrich eggshells. *Palaeogeography, Palaeoclimatology, Palaeoecology*, **84**, 175–189.

Millington, A.C., Drake, N.A., Townshend, J.R.G. et al. (1989) Monitoring salt playa dynamics using Thematic Mapper data. *IEEE Transactions on Geoscience and Remote Sensing*, **27**, 754–761.

Millington, A.C., Drake, N.A., White, K. and Bryant, R.G. (1995) Salt ramps: wind-induced depositional features on Tunisian playas. *Earth Surface Processes and Landforms*, **20**, 105–113.

Müller, G., Irion, G. and Forstner, U. (1972) Formation and diagenesis of inorganic Ca–Mg carbonates in the lacustrine environment. *Naturwissenschaften*, **59**, 158–164.

Neal, J.T. (1969) Playa variation, in *Arid Lands in Perspective* (eds. W.G. McGinnies and B.J. Goldman), University of Arizona Press, Tucson, USA, pp. 14–44.

Neev, D. and Emery, K.O. (1967) The Dead Sea – depositional processes and environments of evaporites. *Israel Geological Survey, Bulletin*, **41**.

Osterkamp, W.R. and Wood, W.W. (1987) Playa-lake basins on the southern High Plains of Texas and New Mexico: Part 1 – Hydrologic, geomorphic, and geologic evidence for their development. *Geological Society of America Bulletin*, **99**, 215–223.

Passarge, S. (1904) *Die Kalahari*, Dietrich Riemer, Berlin.

Pelletier, J.D. (2006) Sensitivity of playa windblown-dust emissions to climatic and anthropogenic change. *Journal of Arid Environments*, **66**, 62–75.

Pitzer, K.S. (1987) Thermodynamic model for aqueous solutions of liquid-like density, in *Thermodynamic Modeling of Geological Materials* (eds I.S.E. Carmichael and H.P. Eugster), Reviews in Mineralogy, Mineral Society of America, vol. 17, pp. 97–142.

Prata, A.J. (1990) Satellite-derived evaporation from Lake Eyre, South Australia. *International Journal of Remote Sensing*, **11**, 2051–2068.

Prospero, J.M., Ginoux, P., Torres, O. et al. (2002) Environmental characterization of global sources of atmospheric soil dust derived from Nimbus-7 TOMS absorbing aerosol product. *Review of Geophysics*, **40**, 2–32.

Ramesh, R., Jani, R.A. and Bhushan, R. (1993) Stable isotopic evidence for the origin of water in salt lakes of Rajasthan ad Gujarat. *Journal of Arid Environments*, **25**, 117–123.

Reddell, D.L. (1965) Water resources of playa lakes, *Cross Section*, **12**, 1.

Reeves, C.C. (1966) Pluvial lake basins of west Texas. *Journal of Geology*, **74**, 269–291.

Reheis, M.C. (2006) 16-year record of eolian dust in Southern Nevada and California, U.S.A.: controls on dust generation and accumulation. *Journal of Arid Environments*, **67**, 487–520.

Reynolds, R.L., Young, J.C., Reheis, M. et al. (2007) Dust emission from wet and dry playas in the Mojave Desert, USA. *Earth Surface Processes and Landforms*, **32**, 1811–1182.

Reynolds, R.L., Bogle, R., Vogel, J. et al. (2009) Dust emission at Franklin Lake playa, Mojave Desert (USA): response to meteorological and hydrologic changes 2005–2008, in *Saline Lakes Around the World: Unique Systems with Unique Values*, vol. 15 (eds A. Oren, D.L. Naftz and W.A. Wurtsbaugh), Natural Resources and Environmental Issues, pp. 105–116 (ISSN 1069-5370).

Rice, M.A. and McEwan, I.K. (2001) Crust strength: a wind tunnel study of the effect of impact by saltating particles on cohesive soil surfaces. *Earth Surface Processes and Landforms*, **26**, 721–733.

Rice, M.A., Mullins, C.E. and McEwan, I.K. (1997) An analysis of soil crust strength in relation to potential abrasion by saltating particles. *Earth Surface Processes and Landforms*, **22**, 869–883.

Risacher, F. and Fritz, B. (1991) Geochemistry of Bolivian salars, Lipez, southern Altiplano: origin of solutes and brine evolution. *Geochimica et Cosmochimica Acta*, **55**, 687–705.

Risacher, F. and Fritz, B. (2000) Bromine geochemistry of salar de Uyuni and deeper salt crusts, Central Altiplano, Bolivia. *Chemical Geology*, **167**, 373–392, DOI: 10.1016/S0009-2541(99)00251-X.

Roberts, C.R. and Mitchell, C.W. (1987) Spring mounds in southern Tunisia, in *Desert Sediments: Ancient and Modern* (eds. L.E. Frostick and I. Reid), Geological Society Special Publication 35, pp. 321–336.

Rosen, M.R. (1991) Sedimentologic and geochemical constraints on the evolution of Bristol Dry Lake Basin, California, USA. *Palaeogeography, Palaeoclimatology, Palaeoecology*, **84**, 229–257.

Rosen, M.R. (1994) The importance of groundwater in playas: a review of playa classifications and the sedimentology and hydrology of playas, in *Paleoclimate and Basin Evolution of Playa Systems* (ed. M.R. Rosen), Geological Society of America Special Paper 289, pp. 1–18.

Russell, J.L. (2008) The inorganic chemistry and geochemical evolution of pans in the Mpumalanga Lakes District, South

Africa, Unpublished MSc Thesis, University of Johannesburg.

Rust, U. and Wieneke, F. (1974) Studies on the Grammadulla Formation in the middle part of the Kuiseb River, South West Africa. *Madoqua*, **3**, 5–15.

Salama, R.B. (1994) The Sudanese buried saline lakes, in *Paleoclimate and Basin Evolution of Playa Systems* (ed. M.R. Rosen), Geological Society of America Special Paper 289, pp. 33–47.

Sharp, R.P. and Carey, D.L. (1976) Sliding stones, Racetrack Playa, California. *Geological Society of America Bulletin*, **87**, 1704–1717.

Shaw, P.A. (1985) The desiccation of Lake Ngami: an historical perspective. *Geographical Journal*, **151**, 318–326.

Shaw, P.A. and De Vries, J.J. (1988) Duricrust, groundwater and valley development in the Kalahari of southeast Botswana. *Journal of Arid Environments*, **14**, 245–254.

Shaw, P.A. and Thomas, D.S. G. (1997) Playas, pans and salt lakes, Chapter 14, in *Arid Zone Geomorphology: Process, Form and Change in Drylands* (ed. D.S.G. Thomas), John Wiley & Sons, Ltd, Chichester, pp. 293–318.

Smith, G.I., Bischoff, J.L. and Bradbury, J.P. (1997) Synthesis of the paleoclimatic record from Owens Lake core OL-92, in *An 800,000-Year Paleoclimatic Record from Core 0L-92, Owens Lake, Southeast California* (eds G.I. Smith and J.L. Bischoff), Geological Society of America Special Paper 317, pp. 143–160.

Smith, G.I. and Street-Perrott, F.A. (1983) Pluvial lakes of the western United States, in *Late Quaternary Environments of the United States*, vol. 1 (ed. H.E. Wright), Longman, London, pp. 63–87.

Smoot, J.P. and Lowenstein, T.K. (1991) Depositional environments of non-marine evaporates, in *Evapories, Petroleum and Mineral Resources*, Developments in Sedimentology vol. 50, Elsevier, Amsterdam, pp. 189–347.

Stokes, W.L. (1968) Multiple parallel truncation bedding planes – a feature of wind-deposited sandstone formations. *Journal of Sedimentary Petrology*, **38**, 510–515.

Street, F.A. (1980) The relative importance of climate and local hydrogeological factors in influencing lake-level fluctuations. *Palaeoecology of Africa*, **12**, 137–158.

Street-Perrott, F.A. and Roberts, N. (1983) Fluctuations in closed-basin lakes as an indicator of past atmospheric circulation patterns, in *Variations in the Global Water Budget* (ed. F.A. Street-Perrott), Reidel, Dortmund, pp. 331–345.

Summerfield, M.A. (1982) Distribution, nature and probable genesis of silcrete in arid and semi-arid southern Africa, in *Aridic Soils and Geomorphic Processes*, vol. 1 (ed. D.H. Yaalon), Catena Supplement, pp. 37–56.

Telfer, M.W. and Thomas, D.S.G. (2007) Late Quaternary linear dune accumulation and chronostratigraphy of the southwestern Kalahari: implications for aeolian palaeoclimatic reconstructions and predictions of future dynamics. *Quaternary Science Reviews*, **26**, 2617–2630.

Telfer, M.W., Thomas, D.S.G., Parker, A.G., Walkington, H. and Finch, A.A. (2008) Optically stimulated luminescence (OSL) dating and palaeoenvironmental studies of pan (playa) sediment from Witpan, South Africa. *Palaeogeography, Palaeoclimatology, Palaeoecology*, **273**, 50–60.

Thomas, D.S.G. (1988) The biogeomorphology of arid and semiarid environments, in *Biogeomorphology* (ed. H.A. Viles), Blackwell, Oxford, pp. 193–221.

Thomas, D.S.G. and Shaw, P.A. (1988) Late Cainozoic drainage evolution in the Zambezi Basin: evidence from the Kalahari Rim. *Journal of African Earth Sciences*, **7**, 611–618.

Thomas, D.S.G. and Wiggs, G.F.S. (2008) Aeolian systems response to global change: challenges of scale, process and temporal integration. *Earth Surface Processes and Landforms*, **33**, 1396–1418.

Thomas, D.S.G., Nash, D.J., Shaw, P.A. and Van Der Post, C. (1993) Present day lunette sediment cycling at Witpan in the arid southwestern Kalahari Desert. *Catena*, **20**, 515–527.

Torgersen, T. (1984) Wind effects on water and salt loss in playa lakes. *Journal of Hydrology*, **74**, 137–149.

Torgersen, T., De Dekker, P., Chivas, A.R. and Bowler, J.M. (1986) Salt lakes: a discussion of the processes influencing palaeoenviromental interpretations and recommendations for future study. *Palaeogeography, Palaeoclimatology, Palaeoecology*, **54**, 7–19.

Townshend, J.R.G., Quarmby, N.A., Millington, A.C. et al. (1989) Monitoring playa sediment transport systems using Thematic Mapper data. *Advances in Space Research*, **9**, 177–183.

Tricart, J. (1969) Actions éoliennes dans la Pampa Deprimada (Republique Argentine). *Revue de Géomorphologie Dynamique*, **19**, 178–189.

Turk, L.J. (1975) Diurnal fluctuations of water tables induced from atmospheric pressure changes. *Journal of Hydrology*, **26**, 1–16.

Twidale, C.R. (1972) Evolution of sand dunes in the Simpson Desert, central Australia. *Transactions of the Institute of British Geographers*, **56**, 77–110.

Tyler, S.W., Munoz, J. and Wood, W. (2006) The response of playa and sabkha hydraulics and mineralogy to climate forcing. *Groundwater*, **44**, 329–338.

Tyler, S.W., Kranz, S., Parlange, M.B. et al. (1997) Estimation of groundwater evaporation and salt flux from Owens Lake, California, USA. *Journal of Hydrology*, **200**, 110–135.

Valyashko, M.G. (1972) Scientific works in the field of geochemistry and the genesis of salt deposits in the USSR, in *Geology of Saline Deposits* (ed. G. Richter-Bernburg), UNESCO, Paris, pp. 289–311.

Vengosh, A., Starinsky, A., Kolodney, Y. et al. (1992) Boron isotope variations during fractional evaporation of sea water: new constraints on the marine vs. nonmarine debate. *Geology*, **20**, 799–802.

Verdin, J.P. (1996) Remote sensing of ephemeral water bodies in western Niger. *International Journal of Remote Sensing*, **17**, 733–748.

Wadge, G. and Archer, D.J. (2003) Evaporation of groundwater from arid playas measured by C-band SAR. *IEEE Transactions on Geoscience and Remote Sensing*, **41**, 1641–1650.

Wadge, G., Archer, D.J. and Millington A.C. (1994) Monitoring playa sedimentation using sequential radar images. *Terra Nova*, **6**, 391–396.

Washington R, Todd, M.C., Lizcano, G. et al. (2006) Links between topography, wind, deflation, lakes, and dust: the case of the Bodélé Depression, Chad. *Geophysical Research Letters*, **33**, L09401, DOI: 10.1929/2006GL025827.

Wehmeier, E. (1986) Water induced sliding of rocks on playas: Alkali Flat in Big Smoky Valley, Nevada. *Catena*, **13**, 197–209.

Weir, J.S. (1969) Chemical properties and occurences on Kalahari Sands of salt licks created by elephants. *Journal of Zoology*, **138**, 292–310.

Wellington, J.H. (1955) *Southern Africa*, vol. 1, Cambridge University Press, Cambridge.

White, K. and Eckardt, F.D. (2006) Geochemical mapping of carbonates sediments in the Makgadikgadi Basin using moderate resolution remote sensing data. *Earth Surface Processes and Landforms*, **31**, 665–681.

Williams, G.J. (1987) A preliminary LANDSAT interpretation of the relict landforms of western Zambia, in *Geographical Perspectives on Development in Southern Africa* (eds. G.J. Williams and A.P. Wood), Commonwealth Geographical Bureau, James Cook University, Queensland, pp. 23–33.

Wood, W.W. and Osterkamp, W.R. (1987) Playa-lake basins on the souther High Plains of Texas and New Mexico: Part 2 – A hydrologic model and mass-balance arguements for development. *Geological Society of America Bulletin*, **99**, 224–230.

Wood, W.W. and Sanford, W.E. (1990) Ground-water control of evaporite deposition. *Economic Geology*, **85**, 1226–1235.

Wood, W.W. and Sanford, W.E. (1995) Eolian transport, saline lake basins, and groundwater solutes. *Water Resources Research*, **31**, 3121–3129.

Yan, J.P., Hinderer, M. and Einsele, G. (2002) Geochemical evolution of closed-basin lakes: general model and application to Lakes Qinghai and Turkana. *Sedimentary Geology*, **148**, 105–122.

Yechieli Y and Wood, W.W. (2002) Hydrogeologic processes in saline systems: playas, sabkhas and saline lakes. *Earth Science Reviews*, **58**, 343–365.

Young, J.A. and Evans, R.A. (1986) Erosion and deposition of fine sediments from playas. *Journal of Arid Environments*, **10**, 103–116.

Yuan, J., Chengyu, H. and Keqin, C. (1983). Characteristics of salt deposits in the Dry Salt Lake, In *Abstracts of the 6th International Symposium on Salt*, Northern Ohio Geological Society, Cleveland.

16

Groundwater controls and processes

David J. Nash

16.1 Introduction

Beyond the arid zone, the role of groundwater as a geomorphological agent has received considerable attention in the literature. Subsurface water has long been recognised as an important factor in processes such as weathering (particularly carbonate dissolution), soil development and hillslope stability, and as a component of river discharge. However, the influence of groundwater in sculpting arid landscapes has been either underestimated or ignored (Higgins, 1984). Dryland landscapes are often viewed as an end-product of the long-term interaction of wind and surface water operating under different structural and tectonic settings, with little reference to subsurface activity. That is not to say that the importance of groundwater as a geomorphological agent in arid areas has been completely overlooked – many of the pioneering studies of piping and tunnel scour (e.g. Bryan and Yair, 1982; Parker and Higgins, 1990, and see Chapter 11), salt weathering (e.g. Cooke *et al.*, 1982, and see Chapter 6) and groundwater seepage erosion (e.g. Peel, 1941) were based upon observations made in desert environments. However, unless the geomorphological impacts of subsurface water are manifest at relatively short timescales and have an impact upon engineering structures, they are largely overlooked. This can, in part, be explained by the difficulties in identifying the long-term role of groundwater in drylands, particularly where groundwater processes contributed to early landscape development and have been overwritten by more easily observed surface-water and aeolian processes (Nash, Thomas and Shaw, 1994).

Groundwater plays an important role in dryland landscapes, which is reflected in other chapters: in the formation of nonpedogenic calcrete, silcrete, sodium nitrate, halite and gypsum crusts (see Chapter 8), in the geomorphology of dryland slopes (Chapter 11), drainage networks (Chapter 12), badland gullies (Chapter 11) and in pans and playas (Chapter 15). In addition to these various direct roles as an agent of weathering and erosion, groundwater may also act as an important control on the operation of specific processes (e.g. where the maximum extent of aeolian deflation is limited by the depth to the regional water table). This chapter will identify three areas of groundwater influence in drylands, complimenting discussion in other chapters where reference is made to specific landform suites. The water resource implications arising from the influence of geomorphology upon groundwater availability in drylands are not discussed as they have been considered elsewhere (e.g. Berger, 1992; Carter, 1994; Carter *et al.*, 1994). Karst processes and landscapes in arid environments are also not considered as, while dissolution processes operate on limestone surfaces even under conditions of limited available moisture (Smith, 1988, 1994), many dryland karst landscapes are largely relict at a macroscale (Smith, 1987; Palmer, 1990). Within the phreatic zone (the zone beneath the water table in which all voids are completely filled with water), groundwater processes and morphologies in karst terrain are virtually independent of climate (Palmer, 1990). There may be reduced dissolution in dryland regions owing to the higher salinity, low levels of recharge and subsequent antiquity of many desert groundwater sources (Lowry and Jennings, 1974). For a wider discussion of the role of groundwater in geomorphology in general, readers are referred to La Fleur (1999) and the excellent summaries provided in the volumes edited by La Fleur (1984), Higgins and Coates (1990) and Brown (1995).

Arid Zone Geomorphology: Process, Form and Change in Drylands, Third Edition. Edited by David S. G. Thomas

16.2 Groundwater processes in valley and scarp development

There are two main ways in which groundwater can act as a factor in dryland valley and scarp development. First, the processes of tunnel scour and seepage erosion associated with subsurface water emerging at a free face or along a hillside or scarp or valley floor can generate surface channels. Second, the operation of in situ deep-weathering processes (principally chemical and biochemical corrosion) associated with the lateral and vertical movement of groundwater along preferential subsurface flowpaths can progressively lower land surfaces. Neither of these processes are unique to arid environments, but some of the best-documented resultant landforms occur within drylands.

16.2.1 Erosion by exfiltrating water: definitions and mechanisms

In general, subsurface flow will discharge from the ground surface either (a) where a water table in an unconfined aquifer intersects the landscape or (b) where the land lies below the piezometric surface of a confined aquifer that is linked to the surface by a fracture or fault in the aquiclude (an impermeable rock stratum that prevents the upward passage of groundwater). Erosion by exfiltrating (i.e. emerging) subsurface water can operate in three ways. First, near-surface groundwater flow may apply stress to the walls of a pre-existing macropore, commonly within a partially or fully consolidated material, which may have originated by a variety of means (e.g. as a result of subsurface flowing water, as a shrinkage crack or from burrowing animals or plant roots). Second, sufficient drag force may be generated as water seeps through and exfiltrates from a porous, usually semi- or unconsolidated material, to entrain particles, cause failure or liquify the material (Dunne, 1990). Third, groundwater outflow may, through the operation of biological, chemical and physical weathering processes, exert stress on the walls of pores, weaken the material and ultimately lead to mass wasting (e.g. by providing a moist microenvironment for algal growth or through the precipitation of salts in pore spaces; see Laity, 1983).

The first of these three processes is termed *tunnel scour* while the second and third both contribute to the process of *seepage erosion* (Dunne, 1990). Of the two, seepage erosion appears to be the most significant in terms of scarp and valley formation (Uchupi and Oldale, 1994), with tunnel scour being most effective at smaller scales (Dunne, 1980). There has been considerable confusion over the terminology used to describe these processes, with terms such as piping, pipe formation, tunnel erosion (Bennett, 1939), sapping, spring sapping (Bates and Jackson, 1980), spring erosion, artesian sapping (Milton, 1973), basal sapping and seepage erosion (Hutchinson, 1968) often used imprecisely and interchangeably. In the following discussion the terminology suggested by Dunne (1990) is adopted.

The formation of subsurface pipes may result from both tunnel scour or seepage erosion and, if such pipes collapse, may lead to channel initiation and ultimately valley development (Dunne, 1980). Both tunnel scour and seepage erosion may also lead to sapping, in its simplest sense 'the undermining of the base of a cliff, with the subsequent failure of the cliff face' (Bates and Jackson, 1980, p. 556). If groundwater flow is sufficiently focused to emerge as a spring then spring sapping may occur, while seepage erosion may lead to weakening and collapse along a more diffuse seepage zone. The processes of tunnel scour in dryland drainage development have been discussed in Chapter 11 in the context of badland development. As such the following section focuses predominantly upon the role of seepage erosion in scarp and valley evolution.

16.2.2 Seepage erosion and valley formation

The earliest reference to the role of seepage erosion in dryland valley formation can be traced to Peel (1941), arising from observations made as part of Major R.A. Bagnold's expedition to the Gilf Kebir plateau of Libya in 1938. In this region, Peel identified wadis with flat floors and steep sides that terminated in a headward cliff, with little or no evidence of fluvial activity in the plateau region surrounding the valley head. This led him to suggest that the wadis appeared to have been 'cut out from *below* rather than 'let down from above' (Peel, 1941, p. 13, italics author's original). This description neatly summarises the main difference between scarp and valley development by groundwater seepage erosion processes as opposed to surface incision by rivers – seepage erosion and sapping effectively undermine valley heads and sides due to enhanced weathering and erosion within a zone of groundwater emergence, while erosion by flowing water operates from the surface downwards (Laity and Malin, 1985). For further details of the role of seepage erosion in the formation of various landforms, see Higgins (1984, 1990), Howard, Kochel and Holt, (1988), Baker (1990) and Howard and Selby (2009).

Since Peel's observations, seepage erosion has been identified as an important factor in the formation of scarps,

Figure 16.1 Groundwater seepage erosion valleys cut into a calcrete plateau south of the Gaub River, central Namib Desert: (a) view of an amphitheatre valley head containing well-developed alcoves, (b) close-up of the undercut base of an amphitheatre valley head wall with evidence of salt weathering along a seepage line, (c) detail of the zone of seepage with extensive salt weathering damage.

canyons and drainage systems in a variety of terrestrial and extraterrestrial settings (Figure 16.1 and Table 16.1). These include submarine canyons (e.g. Robb *et al.*, 1982; Robb, 1990), erosion cirques (Issar, 1983) and valleys in environments ranging from some of the Earth's driest (e.g. southwestern Egypt; see Maxwell, 1979) to its wettest (the Hawaiian Islands; see Kochel and Piper, 1986; Baker, 1990). It was, however, the identification of vast valley systems on images of Mars from the Mariner 9 mission in the mid-1970s that generated most interest in groundwater as a factor in valley formation (Baker, 1982; Baker *et al.*, 1992). Despite the difficulties in ground-truthing these images and identifying evidence for the operation of seepage erosion processes, not to mention the dangers of circular argument, many Martian valleys were suggested to have formed by groundwater 'sapping' by analogy with terrestrial valley networks (e.g. Pieri, Malin and Laity, 1980; Higgins, 1982; Tanaka *et al.*, 1998; Goldspiel and Squyres, 2000; Gulick, 2001; Aharonson *et al.*, 2002; Grant and Parker, 2002; Luo, 2002; Harrison and Grimm, 2005; Stepinski and Stepinski, 2005; Luo and Howard, 2005, 2008).

The use of terrestrial analogues to explain the origin of Martian valleys highlights one of the major problems of many studies of the role of exfiltrating water in valley development, namely that seepage erosion is often invoked purely on the basis of morphological and morphometric properties rather than by direct observation of processes (cf. Lamb *et al.*, 2008). This arises, in part, from the difficulties of making direct field observations of groundwater seepage erosion processes in operation, primarily due to the lack of accessibility at headwalls of active gullies and streams. With the notable exceptions of Laity (1983) and Onda (1994), most quantitative assessments of the role of seepage erosion in valley development have come from experimental work using stream table

Table 16.1 Valleys, scarps and canyons suggested to have formed by groundwater seepage erosion processes.

Landform	*Source*
(a) Drylands	
Australia	Baker (1980), Jennings (1979), Young (1986)
Botswana	Nash (1995), Nash, Shaw and Thomas (1994), Nash, Thomas and Shaw (1994), Shaw and de Vries (1988)
Chile	Hoke *et al.* (2004), Stepinski and Stepinski (2005)
Colorado Plateau (USA)	Ahnert (1960), Baker (1990), Howard and Kochel (1988), Howard, Kochel and Holt (1988), Laity (1980, 1983), Laity and Malin (1985), Laity, Pieri and Malin (1980), Pieri, Malin and Laity (1980)
Egypt	El-Baz *et al.* (1980), El-Baz (1982), Maxwell (1979)
Italy	Mastronuzzi and Sanso (2002)
Jordan	Rech *et al.* (2007)
Libya	Peel (1941)
Morocco	Smith (1987)
(b) Nondrylands	
Florida (USA)	Schumm *et al.* (1995)
Hawaii (USA)	Baker (1980, 1990), Hinds (1925), Kochel and Piper (1986)
Japan	Onda (1994)
Massachusetts (USA)	Uchupi and Oldale (1994)
New Zealand	Schumm and Phillips (1986)
UK	Nash (1996), Small (1964), Sparks and Lewis (1957–1958)
(c) Beach microdrainage networks	Higgins (1982, 1984)
(d) Sub-marine canyons	Robb (1990), Robb *et al.* (1982)
(e) Experimental drainage networks	Baker (1990), Gomez and Mullen (1992), Howard (1988), Howard and McLane (1988), Kochel, Howard and McLane (1985), Kochel and Piper (1986), Sakura, Mochizuki and Kawasaki, (1987)
(f) Extra-terrestrial valley networks	
Mars	Baker and Kochel (1979), Carr (1980), Baker (1982, 1983, 1985, 1990), Baker *et al.* (1992), Belderson (1983), Craddock and Maxwell (1993), El-Baz (1982), Gulick and Baker (1990), Higgins (1982, 1983, 1984), Howard, Kochel and Holt (1988), Kochel, Howard and McLane (1985), Kochel and Piper (1986), Laity, Pieri and Malin, (1980), Mars Channel Working Group (1983), Milton (1973), Pieri (1980), Pieri, Malin and Laity (1980), Sharp (1973), Sharp and Malin (1985), Stiller (1983), Tanaka and Chapman (1992), Tanaka *et al.* (1998), Carr and Malin (2000), Goldspiel and Squyres (2000), Gulick (2001), Aharonson *et al.* (2002), Grant and Parker (2002), Luo (2002), Harrison and Grimm (2005), Stepinski and Stepinski (2005) and Luo and Howard (2005, 2008)

simulations (e.g. Kochel and Piper, 1986; Howard and McLane, 1988; Gomez and Mullen, 1992; Baker, 1990) and computer modelling (Howard and Selby, 1994; Luo and Howard, 2008; Howard and Selby, 2009). Most stream table approaches use unconsolidated or semi-consolidated sediments to allow the rapid development of drainage features and are, as such, relatively limited in terms of their applicability to valley development in bedrock settings. Differences in experimental technique, particularly in the variety of initial conditions used in stream table experiments, limit the conclusions of these investigations. Field studies must also be treated with caution as much work has concentrated upon 'model' landscapes such as the Colorado Plateau (see Box 16.1) where the observation of groundwater erosion processes is relatively free from other influences that would mask the effect of seepage erosion (such as fluvial activity and extensive mass movement; see Dunne, 1990). Nonetheless, some generalisations can be made on the operation of seepage erosion processes. These observations, taken from studies in a number of different environments, appear to hold for a wide range of drainage network scale from beach microdrainage networks (Higgins, 1982) up to megascale Martian valleys (Baker, 1982, 1990).

Box 16.1 Geological controls on the development of groundwater sapping valleys in the Colorado Plateau, USA

Parts of the Colorado Plateau exhibit all of the factors required for the development of valleys and scarps via groundwater seepage erosion processes (see Table 16.3 later). The Plateau has a low overall annual rainfall (130–380 mm) but seasonal recharge. The most intensively studied valleys attributed to seepage erosion are

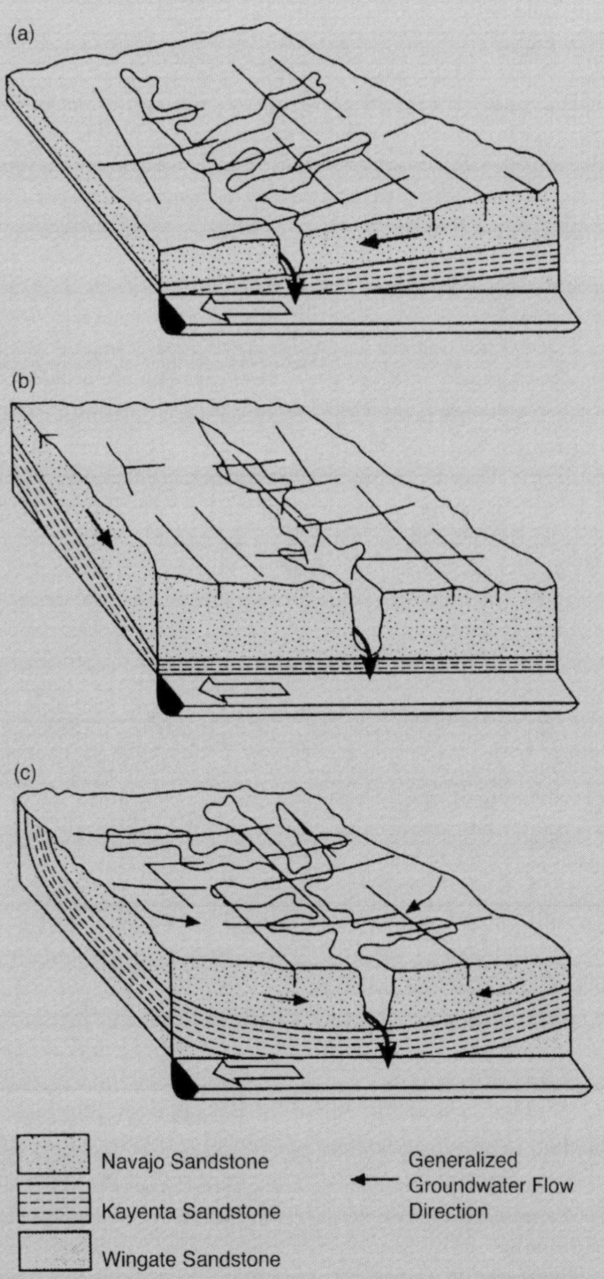

Figure 16.2 Diagrammatic representations of valleys developed by seepage erosion in the Colorado Plateau (after Laity and Malin, 1985).

amphitheatre-headed canyon tributaries to the Escalante, San Juan and Colorado Rivers. These developed at the lithologic contact between the permeable aeolian Navajo Sandstone and the impermeable mudstones and sandstones of the fluvial Kayenta Formation (Laity, in Baker, 1990), both part of the Triassic–Jurassic Glen Canyon Group. Numerous minor seeps with associated alcove development also occur in parts of the Navajo Sandstone above the main seepage zone, where thin shales interbedded with the aeolian sandstone act as aquicludes (Laity, 1988). The regional geological structure of the Colorado Plateau exerts an important control upon valley morphology and drainage patterns, with both pattern and form determined by the direction of groundwater flow (see Figure 16.2). Network length, tributary length and tributary asymmetry all vary in response to geological structure, with symmetrical systems developed in association with synclines and more asymmetric patterns found in areas of laterally dipping strata. The orientation of zones of secondary permeability, such as faults and fractures, also control the spacing and alignment of valleys by acting as both zones of preferential groundwater flow and foci for groundwater emergence where they intersect trunk valley walls. The fine-grained Navajo Sandstone is prone to granular disintegration driven by dissolution of the iron oxide, clay and carbonate cements surrounding aeolian quartz grains, as well as intergranular pressures exerted by salt and biological weathering. The fine-grained sediments released by weathering can be readily transported by infrequent surface water flows.

The process of seepage erosion in bedrock involves at least some intergranular flow, with weathering proceeding by the slow release of grains within the zone of groundwater emergence, leading to spalling and mass-wasting in the form of rockfalls around the seepage zone. Seepage erosion is usually focused into a narrow zone where the discharge of groundwater is concentrated (Howard and Selby, 2009). Experimental studies show that the main method of drainage network development is by headward erosion, which proceeds rapidly by headwall collapse during the early stages of valley formation (Kochel, Howard and McLane, 1985; Baker, 1990; Gomez and Mullen, 1992). Theoretical studies and field observations indicate that the velocity at which valley heads advance is proportional to the flux of groundwater towards the heads (Abrams *et al.*, 2009). Tributary growth occurs as a result of permeability variations (Howard, Kochel and Holt, 1988) and disturbances in subsurface flow (Dunne, 1980), and may also be influenced by joints and geological structures (Laity, Pieri and Malin, 1980; Pieri, Malin and Laity, 1980; Laity in Baker, 1990). Headward erosion has been found to occur most effectively in gently dipping lithologies with an overall regional dip of 1° to 4°, with erosion of the valley head proceeding in an up-dip direction (Howard, Kochel and Holt, 1988). In cohesionless sediment, seepage forces at the site of emergence of subsurface flow are the most important controls on headward erosion (Howard and McLane, 1988), whereas in cohesive bedrock, mechanical and chemical weathering are likely to be the dominant displacive processes (Laity, Pieri and Malin, 1980). Mechanical weathering processes are significant factors in the development of dryland canyons in the Colorado Plateau. SEM analyses of sandstone within the 20 to 25 m thick zone of groundwater emergence identify macropore development (Figure 16.3(a)), together with algal growth and the precipitation of calcite and salt efflorescences within pore spaces, as agents of rock weakening (Figures 16.1(c) and 16.3(b)) (Laity, 1983; Laity in Baker, 1990). In dryland environments, the presence of salts within pores may contribute to spalling through pressures exerted by expansion and contraction of minerals due to thermal expansion and rehydration (Cooke and Smalley, 1968). Other processes leading to rock wasting within Colorado Plateau sapping valleys are the dissolution of cements by exfiltrating groundwater and the weathering of fine shale layers within the sandstone formations.

16.2.3 Characteristics of drainage networks developed by groundwater seepage erosion

Valleys developed predominantly by seepage erosion have a number of distinctive morphological features that, to a certain extent, may be diagnostic of the operation of groundwater processes in their formation (Howard, Kochel and Holt, 1988; Baker, 1990; Luo, 2000). These are best illustrated by consideration of the characteristics of the most intensively investigated of the dryland seepage erosion valley systems, those of the Colorado Plateau (Figure 16.4). The most prominent morphometric properties of these (and other nondryland) systems are summarised in Table 16.2. They include the abrupt initiation of valleys at amphitheatre headwalls, little evidence of surface flow

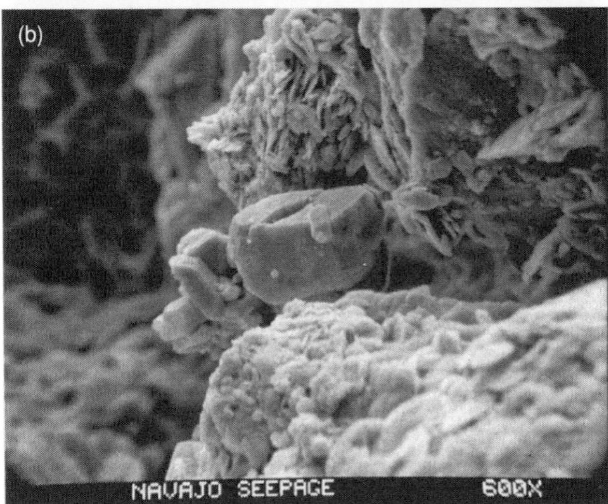

Figure 16.3 Scanning electron micrographs of weathering features associated with groundwater seepage within the Navajo Sandstone, Colorado Plateau: (a) macropore development within sandstone, (b) small calcite crystal adhering to a gypsum crystal precipitated within a sandstone pore (micrographs courtesy of Julie Laity).

above the valley head, the presence of alcoves and springs in the headward region (Figure 16.5), steep valley flanks with an abrupt change in slope angle to a flat valley floor, a long valley with a constant valley width (Figure 16.4), short first-order tributaries with possible hanging valleys and a paucity of downstream tributaries. Analyses of the hypsometric curve (the area–altitude relationship which permits the comparison of drainage area and relief for different systems) for typical terrestrial seepage erosion networks further indicates that such networks have high hypsometric interval, low hypsometric skewness, negative density skewness and high density kurtosis compared to their fluvial counterparts (Luo, 2000).

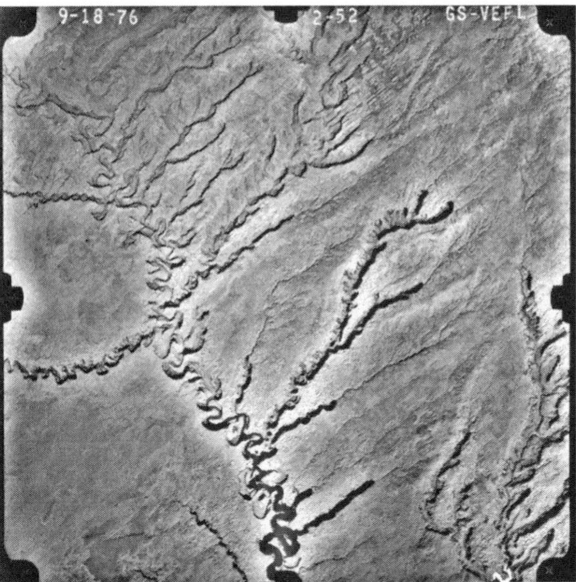

Figure 16.4 Aerial photograph of the Escalante River (Colorado Plateau) together with Long, Bowns and Explorer Canyons, which developed by groundwater seepage erosion (photograph courtesy of Julie Laity).

Not all valleys influenced by seepage erosion will exactly match all of these morphologic criteria (Dunne, 1990). In particular, the amphitheatre valley head and near-vertical valley sides often considered an essential

Table 16.2 Diagnostic morphometric features of terrestrial valley networks developed by groundwater seepage erosion processes (Howard, Kochel and Holt, 1988).

Morphological characteristics
• Abrupt channel initiation, possibly with amphitheatre valley headwalls
• Alcove development with springs or seepage zones in headwater regions
• Steep valley walls with an abrupt angle to a flat valley floor
• Small basin area-to-canyon area ratio
• Low drainage density
• Long main valley with constant valley width
• Short stubby first-order tributaries with a paucity of downstream tributaries
• Possible parallelism of tributaries
• Structurally controlled tributary asymmetry
• Flat or stepped longitudinal profile
• Hypsometric curve with high hypsometric interval, low hypsometric skewness, negative density skewness and high density kurtosis

Figure 16.5 The morphology of valleys within the Colorado Plateau developed by groundwater seepage erosion: (a) the amphitheatre valley head of Bowns Canyon with well-developed alcoves, (b) the amphitheatre valley head of Explorer Canyon showing the zone of seepage emergence (photographs courtesy of Julie Laity).

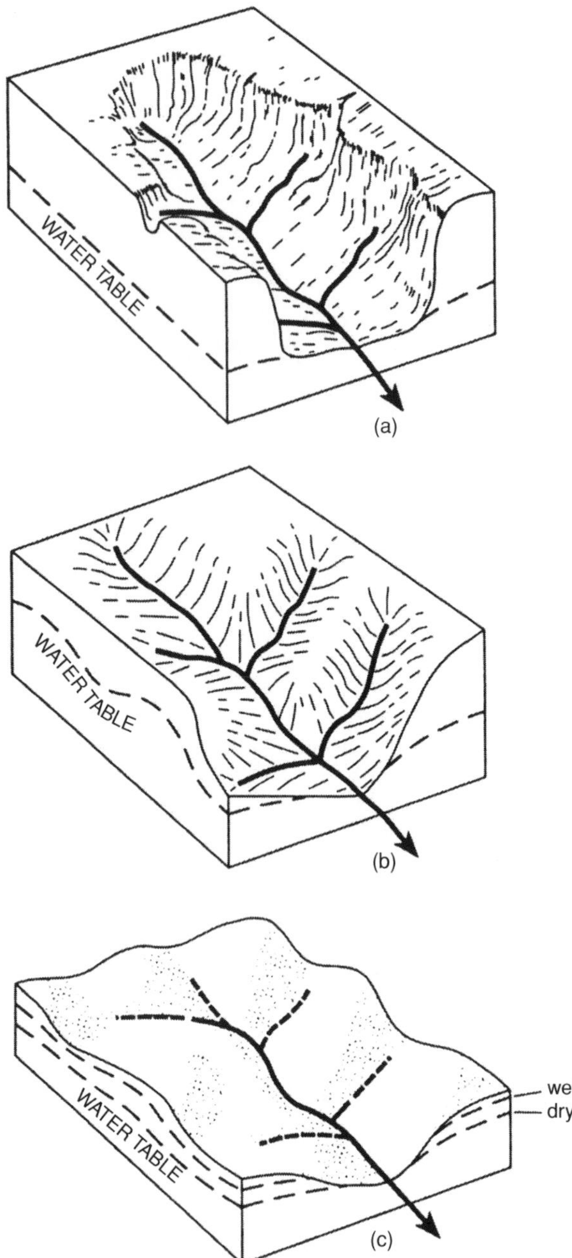

Figure 16.6 Variability in the morphology of valley heads produced by groundwater seepage erosion: (a) amphitheatre-headed valley produced by seepage erosion in rocks with a high angle of internal friction, (b) valley head formed by seepage erosion in rocks that either have a lower angle of internal friction or are more susceptible to mass-wasting, (c) valley heads where seepage erosion makes only a minor contribution to valley morphology (after Dunne, 1990).

feature of groundwater outflow networks may not be present (Figure 16.6) if erosion by surface flow and hillslope processes exceeds erosion by groundwater seepage (Sakura, Mochizuki and Kawasaki, 1987). This will partly depend upon local climatic characteristics but will also be greatly affected by the angle of internal resistance of the material within which the valley has developed. Valleys such as those with near-vertical canyon walls in the Colorado Plateau (developed predominantly by seepage erosion of well-cemented sandstones) must be viewed as forming one end of a process spectrum, with systems produced entirely by surface incision at the other end (Nash, 1996). Seepage erosion networks in drylands are often more readily identified than their temperate counterparts as there is less likelihood of extensive contemporary surface flow to mask the influence of groundwater processes. It is, however, possible in such systems that surface flows have been more prevalent during former periods of wetter climate with the balance between groundwater and surface erosion processes fluctuating with time (Nash, Thomas and Shaw, 1994).

As a cautionary note, some canyons that meet all the morphologic criteria for formation by seepage erosion may have developed by other mechanisms. Box Canyon, Idaho, USA, for example, is incised into a basalt plain, has no drainage network upstream of its valley head and has groundwater seepage of around 10 m^3/s from its headwall. However, sediment transport constraints, ^4He and ^{14}C dates, and the presence of plunge pools and scoured rocks in the valley floor suggest that a megaflood around 45 000 years ago carved the canyon (Lamb *et al.*, 2008). Field observations and topographic analyses of the amphitheatre-headed Kohala valleys in Hawaii suggest that these systems developed by vertical headward waterfall erosion rather than seepage erosion (Lamb *et al.*, 2007). These studies may imply that similar features on Mars formed as a result of surface runoff processes rather than purely seepage erosion. Recent computer modelling work (Luo and Howard, 2008) suggests that Martian valley networks may have developed through the action of *seepage weathering* combined with fluvial runoff, since seepage erosion alone would be only marginally effective at generating integrated networks under realistic rates of aquifer recharge. Overall, while seepage erosion is an important process in loose or moderately well-cemented sediments, the extent to which it can operate in more resistant rock types is less certain (Lamb *et al.*, 2007).

16.2.4 Parameters promoting the operation of groundwater seepage erosion processes

In addition to many common morphometric characteristics, there also appear to be a number of parameters common to valley systems at a variety of scales that influence the effectiveness of groundwater seepage processes. Howard, Kochel and Holt (1988) suggest five factors necessary for the operation of seepage erosion (Table 16.3). These include the need for a permeable aquifer of a transmissive rock type, a rechargeable groundwater system (ideally of a large areal extent), a free face at which water can emerge, a structural or lithological inhomogeneity to increase local hydrologic conductivity and a means of transporting material from the free face.

Table 16.3 Prerequisites for the operation of groundwater seepage erosion in valley development (Howard, Kochel and Holt, 1988; Baker, 1990).

Hydrogeological and geomorphological prerequisites
A permeable aquifer of a transmissive rock type
A rechargeable groundwater system, preferably of a large areal extent
A free face at which water can emerge
A structural or lithological inhomogeneity to increase local hydraulic conductivity
A means of transporting material from the free face

The operation of seepage erosion processes is affected by a variety of factors, which vary in significance according to scale and time (Figure 16.7). These include megascale characteristics, such as climate and regional geology, which may determine whether seepage is perennial, ephemeral or unlikely to occur. Regional water tables will also affect process operation, as will the gradual development of any valley system that will progressively change the distribution and foci of groundwater flowpaths by feedback processes. At meso- and microscales, the scales at which the actual processes of seepage erosion occur, there are a variety of complex feedback mechanisms, with, for example, the amount of surface water flow influencing the slope morphology and hence the operation of seepage erosion processes. Even in the 'textbook' seepage erosion valleys of the Colorado Plateau, Lamb *et al.* (2006) note the significant role played by flash-flood discharges upon canyon morphology. Amphitheatre heads such as Horseshoe Canyon, Utah, that drain moderate to large surface areas typically have plunge pools below their headwalls created by waterfalls.

16.2.5 Groundwater seepage erosion and environmental change

The environmental significance of seepage erosion in dryland valley development should not be overlooked (Nash, Thomas and Shaw, 1994). While theatre-headed valley forms in some arid regions are relict features, the fact that groundwater erosion is an ongoing process in many valleys within the Colorado Plateau suggests that, given ideal geological conditions (Table 16.3), seepage erosion may be an extremely important landforming process under semi-arid conditions. The process of valley development

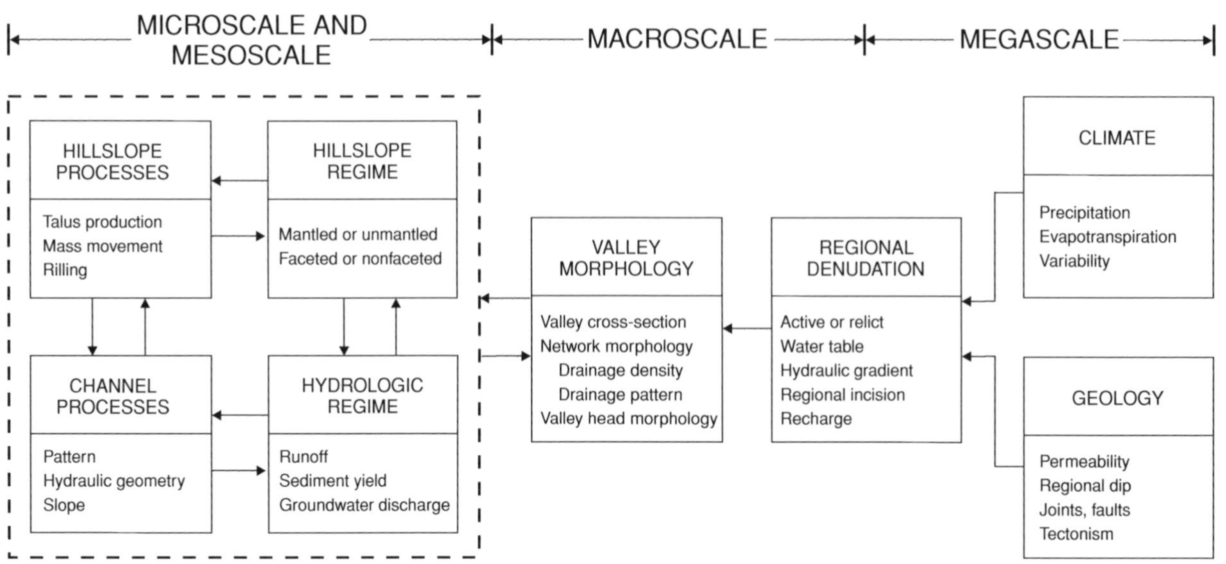

Figure 16.7 Factors influencing valley development by groundwater seepage erosion at a variety of scales (after Baker, 1990).

by seepage erosion does, however, appear to have inherent thresholds, with different modes of operation under different lithological and climatic settings. Laity (in Baker, 1990) suggests that present-day groundwater discharge and rates of cliff retreat in the Colorado Plateau may be less than during previous wetter periods, implying that a wetter climate will promote greater seepage erosion. Conversely, Howard, Kochel and Holt (1988) hypothesise that greater rainfall and groundwater outflow in this region would not increase erosion rates as it would hinder the accumulation of minerals and hence salt weathering. High rates of groundwater outflow do not, however, appear to be a hindrance to the development of seepage erosion valleys in parts of the Hawaiian Islands (Kochel and Piper, 1986) where chemical erosion and basal sapping of basalts occurs around spring sites with considerably higher discharges than those of the Colorado Plateau (Laity and Malin, 1985; Kochel and Baker in Baker, 1990). Clearly, further investigation of the variation in weathering processes associated with ephemeral and perennial groundwater seepage acting upon varying lithologies under different climatic regimes is required.

16.2.6 *In situ* deep-weathering and valley development

A second method by which groundwater can contribute to dryland valley formation is through the operation of deep-weathering processes as groundwater moves vertically and laterally along preferential subsurface flowpaths. Geological faults and fractures may, for example, act as linear aquifers due to their enhanced permeability (Buckley and Zeil, 1984), while igneous dykes may form barriers to groundwater movement and compartmentalise aquifers (Bromley *et al.*, 1994). In bedrock, subsurface flow will be greatest along fractures and joint planes which are normal to the force equipotential surface of the regional water table. Any leaching concentrated along these zones will gradually lower valley bases (Newell, 1970; Buckley and Zeil, 1984).

This process has been identified as a possible factor in the formation of a number of valleys in dryland and subhumid regions, most notably in the development of some African dambos (also termed *fadamas*, *vleis*, *bas-fonds* and *bolis*). Dambos are broad, shallow, seasonally waterlogged, grassed depressions without a marked stream channel, occupying valley floors, which commonly occur at the headwaters of stream networks (Mäckel, 1974; Acres *et al.*, 1985; Boast, 1990; Bullock, 1992). They are found in areas with strongly seasonal rainfall regimes, with present-day total annual precipitation in the range 600–1500 mm, and, like valleys attributed to formation by seepage erosion processes, are not restricted to semi-arid or dry subhumid environments. Climate does not appear to be the overriding control upon the distribution of contemporary dambos, otherwise many hydrologically active dambos would be relict features (Whitlow, 1985). Geology and the influence of lithology upon soil characteristics are important in determining drainage, while a flat, gentle relief appears to be a prerequisite for formation (Mäckel, 1974).

There are two schools of thought regarding the origin of dambos; one sees fluvial erosion and slope transport processes dominating development (e.g. Mäckel, 1974) while the other implies that dambos have developed by pseudo-karstic in situ deep-weathering during drier climates, largely independent of the fluvial network (e.g. McFarlane, 1989, 1990; McFarlane and Whitlow, 1990). Under the latter theory of formation it is envisaged that Zimbabwean dambos evolved without the action of rivers. Chemical and biochemical corrosion due to vertically and laterally moving water operating along lines of geological weakness such as concentrations of fractures, joints and faults is suggested to have led to gradual surface lowering by solute leaching and ultimately to the formation of a valley (McFarlane, 1990). This theory is based upon the identification of a number of dambos displaying strong structural control and features untypical of a fluvial valley, most notably where dambos cross drainage divides.

The coincidence of surface lowering associated with dissolution of noncarbonate bedrock along lines of fractures has also been noted in semi-arid eastern Botswana, where fractured aquifers are characterised by shallow surface depressions (Gieske and Selaolo, 1988). The alignment of many structurally controlled Kalahari valley systems (Figure 16.8) has been attributed to the same process, with similar evidence for bedrock dissolution identified from boreholes drilled within valley floors (Von Hoyer, Keller and Rehder, 1985; Nash, Thomas and Shaw, 1994; Nash, 1995). The significance of this process for dryland valley development is that surface lowering is suggested to occur in the absence of fluvial activity and does not necessarily require seasonal recharge. However, it should be noted that McFarlane's (1989, 1990) model does not totally preclude river action in dambo formation, rather it suggests an alternation between leaching and fluvial incision. Leaching would dominate drier phases but be interrupted by periods of river activity due to rejuvenation of the fluvial system, with associated incision, removal of dambo floor sediments and a concomitant lowering of regional water tables. There still remains, however, the as yet unresolved 'chicken and egg' question of which came first, the valley or the deep-weathering.

16.3 Groundwater and pan/playa development

The balance between surface- and groundwater inputs, as reflected by the position of the water table beneath the playa surface, is one of the most important determinants influencing the development, morphology and sedimentology of playas within arid zones (see Chapter 15 and Yechieli and Wood, 2002). Playas in which the water table lies at depth beneath the surface are termed *recharge playas* (Rosen, 1994; Briere, 2000). Such basins experience little interaction between the surface and groundwater, since the playa floor is frequently above the zone of capillary rise, have minimal accumulation of salts at their surface and tend to be clay-floored. In contrast, more saline *discharge playas*, where the water table is close to or outcrops at the surface, experience appreciable seasonal

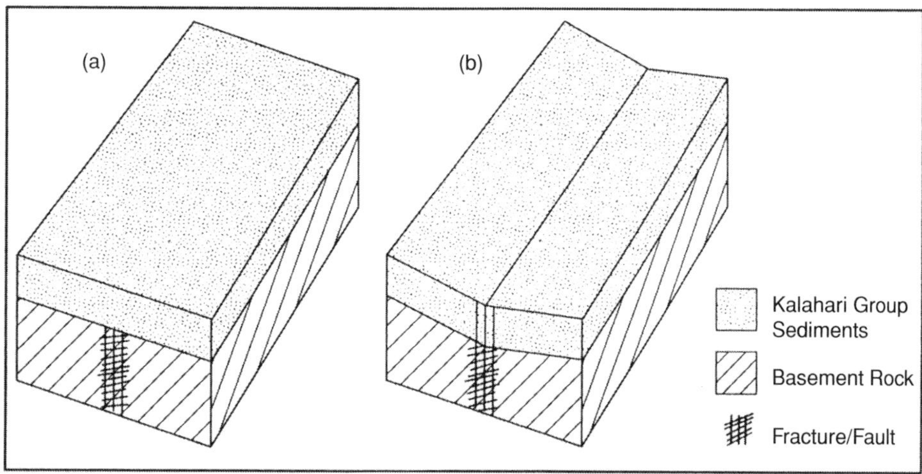

Figure 16.8 A conceptual diagram of Kalahari valley development by in situ deep-weathering processes: (a) following the deposition of the Jurassic to recent Kalahari Group sediments but prior to groundwater movement along subsurface faults and fractures and (b) once chemical and biochemical corrosion has lowered the ground surface by bedrock dissolution (modified after Nash, 1995).

groundwater outflow and are commonly associated with accumulations of evaporite deposits (e.g. Clarke, 1994).

The role of groundwater as a factor in pan formation and development has been most closely considered through studies in Australia (e.g. Bowler, 1986; Torgersen *et al.*, 1986; Jacobson, Ferguson and Evans, 1994; Dutkiewicz and von der Borch, 1995; Boggs *et al.*, 2006; Cupper, 2006; Harrington, Herczeg and La Salle, 2008), the Americas (e.g. Osterkamp and Wood, 1987; Paine, 1994; Rosen, 1994; Risacher, Alonso and Salazar, 2003; French *et al.*, 2006; Risacher and Fritz, 2009; Scuderi, Laudadio and Fawcett, 2010) and in the Kalahari (e.g. Butterworth, 1982; Farr *et al.*, 1982; Thomas *et al.*, 1993; Eckardt *et al.*, 2008), with other basins fed by groundwater seepage across North Africa described by Boulaine (1954), Coque (1962), Glennie (1970), Bryant (1999), Bryant and Rainey (2002) and Hamdi-Aissa *et al.* (2004).

Groundwater can operate as a factor in playa development in three main ways. First, percolating groundwater can lead to pan-floor subsidence by direct dissolution processes (Baker, 1915; Judson, 1950; Osterkamp and Wood, 1987), as discussed in more detail in Chapter 15. Second, a long-term reduction in groundwater head levels can lead to a change in the status of a playa, e.g. from discharge to recharge as groundwater flow is focused towards other lower-lying playas in a region. Such a change may be associated with a concomitant decrease in the surface salinity, in turn promoting an increase in the vegetation cover and ultimately leading to a change in the distribution of discharge playas by a process of playa capture (Jacobson and Jankowski, 1989). Third, the subsurface watertable can also act as a base level for wind deflation (Bowler, 1986; Thomas *et al.*, 1993). Groundwater dissolution and playa capture will now be considered, with the links between groundwater and aeolian deflation discussed in the final section of this chapter.

The processes of deep-weathering and bedrock dissolution outlined in the previous section have also been proposed as possible mechanisms (along with deflation, biogenic activity, volcanism, tectonism and meteorite impact; see Goudie and Thomas, 1985, 1986; Goudie and Wells, 1995; Sanchez *et al.*, 1998) in the formation of playas. Osterkamp and Wood (1987) and Wood and Osterkamp (1987) have proposed a lithologically specific groundwater solution model for the development of clay-floored basins based upon observations in the Southern High Plains of Texas and New Mexico, substantiated by mass-balance calculations (see Chapter 15). These authors suggest that deepening and expansion of a playa-floor area occurs essentially by dissolution and removal of material beneath the playa surface. The infiltration, weathering and downward transport of solutes by percolating groundwater (Zartman, Evans and Ramsey, 1994), along with the removal to the subsurface of clastic material along solutional pipes, is suggested to lead to the gradual subsidence of the playa surface. Subsidence is a particularly important mechanism in the Southern High Plains in areas where many playas are underlain by evaporite-bearing Permian bedrock (Paine, 1994). The dissolution of the Permian strata is suggested to have been continuous throughout the deposition of later formations during the Neogene and may be occurring today.

In addition to operating directly as a factor in playa-floor dissolution, groundwater may also be an important control on the hydrological and sedimentological characteristics of a playa through time. The depth of the water table beneath a playa surface will vary in response to seasonal and longer-term drought and also due to regional climatic change. A long-term reduction in the level of groundwater head has been suggested to cause the migration of playas by a process termed *playa capture* (Jacobson and Jankowski, 1989). This process, illustrated by Figure 16.9, is broadly analogous to river capture and results from a shift in subsurface groundwater flow arising from the combination of a fall in regional groundwater tables and the preferential deepening of one playa floor relative to adjacent basins. The model is based upon studies of discharge playas near Curtin Springs, central Australia, an area where groundwater head is known to have decayed over several thousand years (Jacobson, Arakel and Chen, 1988). In this region, Samphire Lake contains thick deposits of groundwater-derived gypsum, indicating that it was a previously active discharge playa. Lowering of the water table has reduced the levels of saline groundwater outflow on to the playa surface, allowed the encroachment of vegetation on to the playa and focused groundwater discharge into the neighbouring Spring and Glauberite Lakes. The dry playa surfaces are thus rendered susceptible to both aeolian sediment deflation as well as alluviation by sediments from around the playa periphery. This may, in part, explain the occurrence of groundwater discharge-derived playa sediments now buried by aeolian material. The decay in groundwater head therefore results in a variation in the spatial distribution of active and abandoned groundwater discharge playas through time (Jacobson and Jankowski, 1989). Glauberite Lake will become the primary focus of groundwater activity in this region, ultimately forming a regional groundwater sump.

16.4 Groundwater and aeolian processes

Groundwater has long been recognised as an important factor in the control of aeolian sediment deflation and

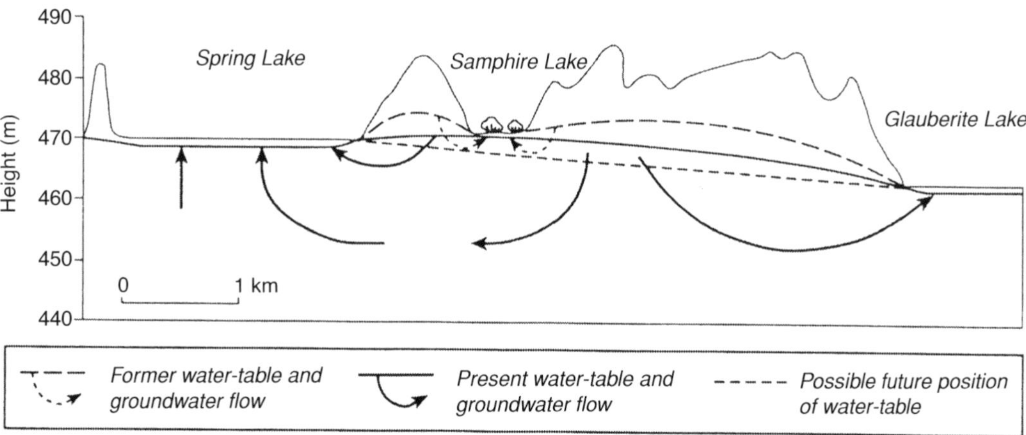

Figure 16.9 A schematic representation of the process of 'playa capture' or playa abandonment due to groundwater head decay. Samphire Lake formerly had a strong groundwater discharge but is now vegetated, with Glauberite Lake progressively capturing the groundwater flow system in this region (after Jacobson and Jankowski, 1989).

deposition within dunefields and ergs (see Chapters 18 and 19), particularly where the water table is close to the surface, as is the case in many coastal sand seas, playas and sabkha environments. Aeolian entrainment and deposition in dunefields is a function of sediment availability and transportability, both of which are controlled by factors such as sediment size, sediment dryness, the degree of surface cementation, the wind energy and the degree of vegetation or lag cover (Kocurek and Nielson, 1986; Kocurek, 1988, 1991). The presence of a high water table can substantially modify ground surface conditions and limit the extent to which the wind can act as an erosional agent (Kocurek, Robinson and Sharp, 2001).

The concept of the water table acting as a base level for wind scour was first suggested by Stokes (1968) and has been described in detail by Fryberger, Schenk and Krystinik (1988). A near-surface water table, sometimes called a *Stokes surface*, can influence aeolian processes in four ways. The first, as originally described by Stokes (1968), arises from the higher cohesivity of damp sand in proximity to the water table, primarily as a result of increased intergranular surface tension due to the presence of pore water. Damp or wet sand is therefore less easily entrained by the wind and the boundary between dry and wet sediment may act as an erosional disconformity. Stokes surfaces in modern arid environments (Table 16.4) vary in extent from extremely localised zones within interdune areas up to extensive planar surfaces of at least 25 km^2 (Fryberger, Schenk and Krystinik, 1988). Modern Stokes surfaces have a variety of features and associated sedimentary structures, the most typical of which (in aeolian systems) is the extremely sharp truncation of underlying

Table 16.4 Examples of modern Stokes surfaces.

Location	*Setting*	*Source*
Sabkha Matti, Arabian Gulf	Continental sabkha	Kinsman (1969)
White Sands dunefield, New Mexico, USA	Alkali flats in continental sand sea	Loope (1984), Fryberger, Schenk and Krystinik, (1988), Kocurek *et al.* (2007)
Jafurah Sand Sea, Saudi Arabia	Coastal offshore prograding sand sea	Fryberger, Schenk and Krystinik, (1988)
Guerrero Negro, New Mexico, USA	Coastal onshore prograding sand sea	Fryberger, Schenk and Krystinik (1988), Fryberger, Krystinik and Schenk (1990)
Great Salt Lake Desert, USA	Continental sand sea and salt flats	Stokes (1968)
Skeidararsandur, southern Iceland	Sandur plain with small dunefields separated by flooded interdune flats	Mountney and Russell (2009)

cross-beds with only a thin veneer of overlying sediments (see, for example, Reading, 1996). Although erosional in nature, Stokes surfaces also act as important controls upon deposition and are commonly associated with thin layers of sediment, indicating the role of the water table in aeolian sedimentation. The erosional topography of Stokes surfaces is not always completely planar, partly due to the water table mimicking the dune field topography, and may also exhibit surface irregularities (Reading, 1996). The capillary tension within damp sand can assist in the formation of a variety of sedimentary adhesion structures due to wind sculpturing, including adhesion ripples, laminations, warts and other steep-sided, irregular bedforms (Kocurek, 1981a).

The second influence of near-surface groundwater as a limit to wind scour is through the cementation of sediment in the vicinity of the water table. High levels of evaporation combined with a high water table may promote the precipitation of evaporite cements in the phreatic zone or of salcretes nearer the surface at the top of the capillary fringe zone (see Chapter 8). Schenk and Fryberger (1988) and Fryberger, Schenk and Krystinik (1988), for example, record massive phreatic gypsum cementation in the White Sands dune field, New Mexico, which currently acts as a basal limit to aeolian deflation and has created extensive planar surfaces in interdune areas. Wet–damp surface interdune deposits consist predominantly of gypsum salt ridges and mats of microorganisms and algae, which form within the capillary fringe. Sediment accumulation in the interdune areas at White Sands follows a distinctive seasonal cycle. Typically, interdune floors are at their dampest during the late autumn to early spring, with aeolian sediment trapping occurring due to the rise of the capillary fringe and the associated development of salt ridges and biogenic mats. During the late spring to early autumn when interdune surfaces are at their driest, salt ridges and biogenic mats are typically dry and brittle and consist of an admixture of wind-blown sand (Kocurek *et al.*, 2007). Cementation in proximity to the water table may also produce an irregular surface topography, with Fryberger, Schenk and Krystinik (1988) recording 'mini-yardangs', nonstreamlined eroded bumps, salt ridges and scour pits in halite-cemented sediments in the Jadurah Sand Sea, Saudi Arabia. In addition to cementing the sand surface, the presence of salts also raises the threshold velocity of sand (Pye, 1980).

The third link between groundwater and aeolian processes in dryland regions occurs where shallow water tables promote the establishment of a vegetation cover as opposed to acting as an erosional base level. Fixed dunes in central Niger have had their surfaces stabilised by the packing of fines due to increased rainfall and the action of cyanophytes and fungi acting as binding agents (Talbot, 1980). Former water tables are suggested by Talbot (1985) to have influenced vegetation colonisation of the dunefield and hence stabilised the dune surface. Residual dune ridges may develop as a result of the establishment of dune vegetation along a line partway up the lower stoss slope of dunes during periods of higher water table (Levin *et al.*, 2009). The causes of such water table fluctuations could include the inundation of interdune areas during fluvial flood events (Langford, 1989; Purvis, 1991) or by seawater from lagoons at high tide (Inman, Ewing and Corliss, 1966). Surface vegetation may act locally to protect sediments from wind erosion, while plant roots and intergranular surface tension due to pore water may act to bind deeper sediments, resulting in the development of an arcuate ridge once adjacent dune deposits are deflated. The impacts of variations in groundwater salinity upon vegetation cover may also influence dune field morphometry. Studies in White Sands dune field have shown that barchans dunes are

Figure 16.10 Example of a contemporary wet aeolian system: (a) linear dunes crossing a sabkha surface in the southern Rub Al Khali close to the Oman/Saudi Arabia border, (b) close-up of the interface between linear dune and sabkha sediments showing a zone of salt efflorescences.

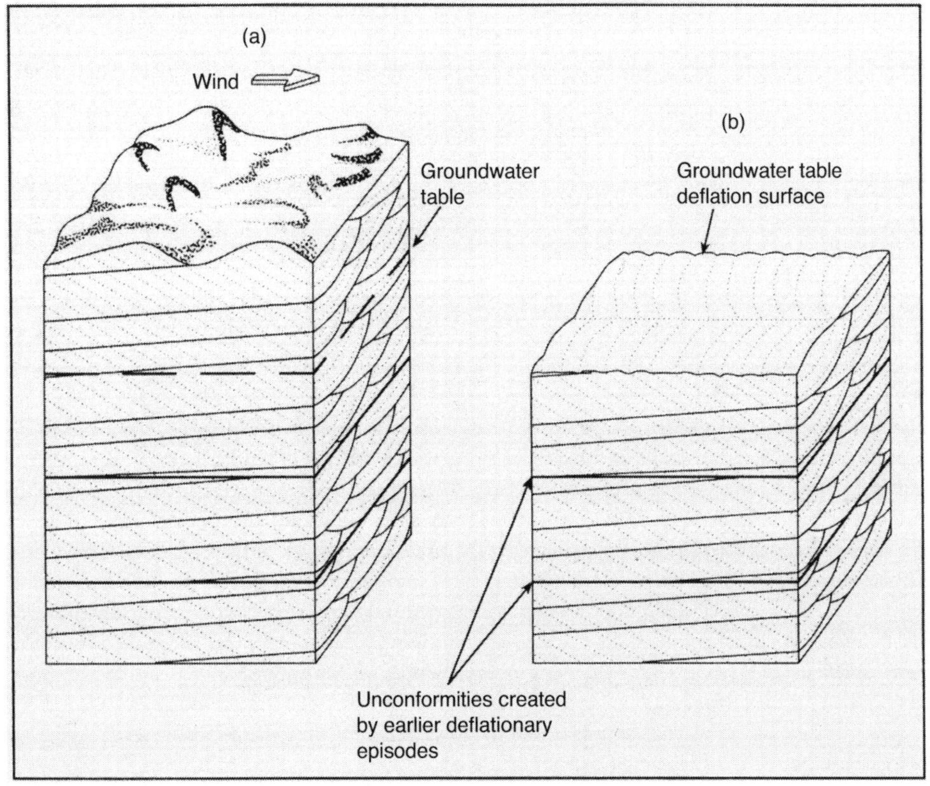

Figure 16.11 The formation of flat bedding planes by aeolian deflation to the level of the groundwater table. (a) Bedform climbing leads to the accumulation of sand in trough cross-beds. (b) Subsequent deflation to the water table occurs due to either a change in wind regime or a decrease in sand supply, with eroded sand deposited far downwind (after Loope, 1984).

most prevalent above areas of saline groundwater, while parabolic dunes typify topographic highs that accumulate a lens of fresher precipitation-derived water and hence develop a partial vegetation cover (Langford, Rose and White, 2009).

In addition to impacting upon contemporary sediment erosion and deposition, the presence of a high water table may also exert a major influence upon the longer-term accumulation and preservation of aeolian sediments. In so-called *wet aeolian systems* (e.g. Figure 16.10), the effect of a high water table is to reduce erosion in interdune areas and near the base of dunes, where they lie within the capillary fringe. Long periods of water table stability may lead to the formation of a first-order bounding surface (see Figure 16.11). However, if the water table rises over time, e.g. due to increased precipitation, reduced evaporation, land/basin subsidence or sea-level rise, both dune and interdune sediments will progressively accumulate, with interdune flats growing at the expense of dunes. In general, the vertical sediment accumulation rate will equal the rate of water table rise (Reading, 1996), although periods of more rapid water table rise may lead to deposition being dominated by water- rather than wind-laid deposits (Paim and Scherer (2007). There is some debate over whether extensive planar first-order bounding surfaces (Brookfield, 1977) forming bedding planes in ancient aeolian sandstones were all generated in this way (e.g. Rubin and Hunter, 1982, 1984; Loope, 1984; Kocurek, 1981a, 1981b, 1984, 1988, 1991). The original concept of a water table acting as a control of deflation as put forward by Stokes (1968) is now viewed as one possible scenario for the creation of bounding surfaces, with interdune migration and bedform climbing (e.g. Rubin and Hunter, 1982) viewed as less site-specific alternatives (Mader and Yardley, 1985).

Despite these reservations, there is widespread evidence for sediment cohesion, cementation and salt encrustation associated with former water tables preserved within the sedimentary architecture of ancient aeolian sandstones (e.g. Bromley, 1992; Chan and Kocurek, 1988; Gaylord, 1990; Crabaugh and Kocurek, 1993; Loope, 1985, 1988). Units of the Permian age Cedar Mesa Sandstone in southeast Utah, USA, for example, represent an ancient wet aeolian system (Mountney and Jagger, 2004),

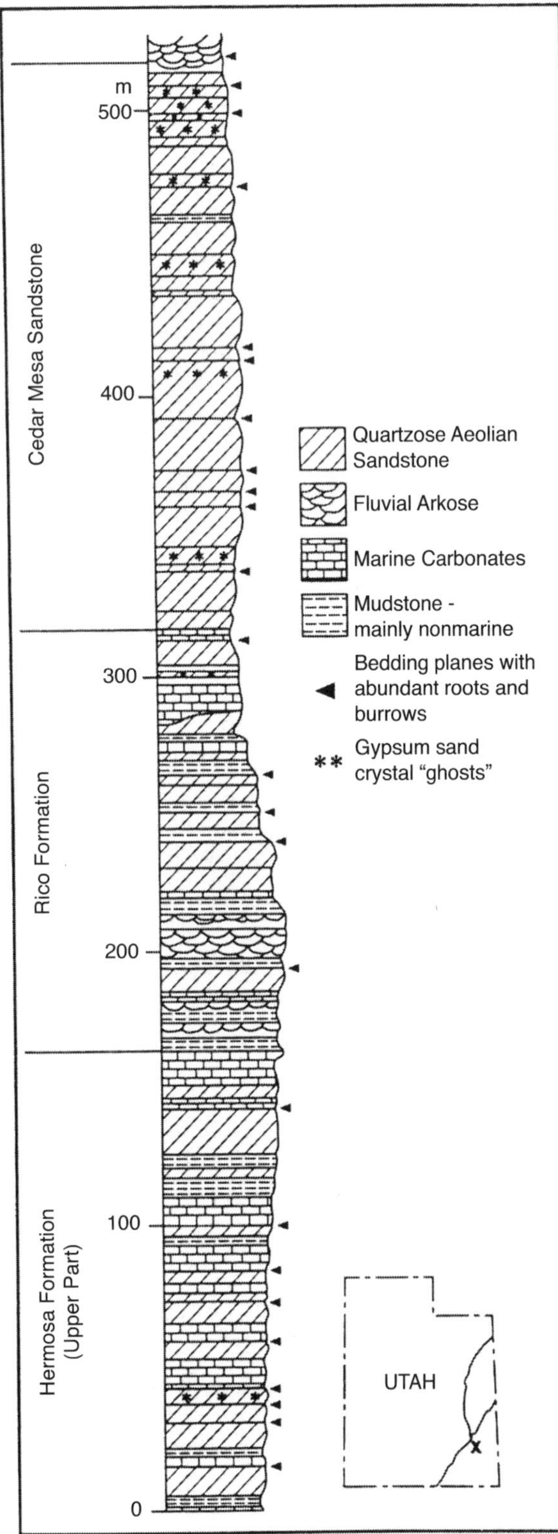

Figure 16.12 A stratigraphic section of Upper Pennsylvanian (Carboniferous) and Lower Permian rocks exposed in Canyonlands National Park, Utah (after Loope, 1985).

and comprise thin sand bodies separated by extensive planar erosion surfaces attributed to deflation to a former water table (Loope, 1985, 1988) (Figure 16.12). Rhizoliths are abundant within the sandstone beds directly beneath these deflation surfaces, suggesting a close proximity to the water table (Loope, 1984). Sedimentary evidence also suggests that topographic lows in former draa surfaces contained ponds of water (Langford *et al.*, 2008). The presence of replaced gypsum sand crystals in a number of horizons provides further supporting evidence that the water table was sufficiently close to the surface to allow evaporite formation (Loope, 1988). The Jurassic Entrada Sandstone, studied along a detailed 2.7 km traverse in northeast Utah by Crabaugh and Kocurek (1993), is also interpreted as having formed in a wet aeolian system. The sandstone architecture exhibits subaqueous ripple deposits, contorted strata, breccias, collapse features, wavy bedding, foundered sets, loaded set bases, corrugated surfaces, polygonal fractures, ball-and-pillow structures and trains of small dunes 'frozen' in place, all indicating a near-surface water table with occasional shallow flooding of interdune areas.

References

Abrams, D.M., Lobkovsky, A.E., Petroff, A.P. *et al.* (2009) Growth laws for channel networks incised by groundwater flow. *Nature Geoscience*, **2**, 193–196.

Acres, B., Blair Rains, A., King, R. *et al.* (1985) African dambos: their distribution, characteristics and use. *Zeitschrift für Geomorphologie, Supplementband*, **52**, 63–86.

Aharonson, O., Zuber, M.T., Rothman, D.H. *et al.* (2002) Drainage basins and channel incision on Mars. *Proceedings of the National Academy of Sciences of the United States of America*, **99**, 1780–1783.

Ahnert, F. (1960) The influence of Pleistocene climates upon the morphology of cuesta scarps on the Colorado Plateau. *Association of American Geographers, Annals*, **50**, 139–156.

Baker, C.L. (1915) Geology and underground waters of the Northern Llano Estacado, University of Texas Bulletin 57.

Baker, V.R. (1980) Some terrestrial analogs to dry valley systems on Mars, NASA Technical Memo TM 81776, pp. 286–288.

Baker, V.R. (1982) *The Channels of Mars*, University of Texas Press, Austin.

Baker, V.R. (1983) Large scale fluvial paleohydrology, in *Background to Palaeohydrology* (ed. K.J. Gregory), John Wiley & Sons, Ltd, Chichester, pp. 453–478.

Baker, V.R. (1985) Models of fluvial activity, in *Models in Geomorphology* (ed. M. Woldenberg), Allen and Unwin, London, pp. 287–312.

Baker, V.R. (1990) Spring-sapping and valley network development, with case studies by R.C. Kochel, V.R. Baker, J.E.

Laity and A.D. Howard, in *Groundwater Geomorphology; The Role of Subsurface Water in Earth-Surface Processes and Landforms* (eds C.G. Higgins and D.R. Coates), Geological Society of America Special Paper 252, Boulder, Colorado, pp. 235–265.

Baker, V.R. and Kochel, R.C. (1979) Martian channel morphology; Maja and Kasei Valleys. *Journal of Geophysical Research*, **84**, 7961–7983.

Baker, V.R., Carr, M.H., Gulick, V.C. *et al.* (1992) Channels and valley networks. in *Mars* (eds H.H. Kiefer, B.M. Jakosky, C.W. Snyder and M.S. Mathews), University of Arizona Press, Tucson, pp. 493–522.

Bates, R.L. and Jackson, J.A. (eds) (1980) *Glossary of Geology*, 2nd edn, American Geological Institute, Falls Church, Virginia.

Belderson, R.H. (1983) Comment: drainage systems developed by sapping on Earth and Mars. *Geology*, **11**, 55.

Bennett, H.H. (1939) *Soil Conservation*, McGraw-Hill, New York.

Berger, D.L. (1992) Ground-water recharge through active sand dunes in northwestern Nevada. *Water Resources Bulletin*, **28**, 959–965.

Boast R. (1990) Dambos: a review. *Progress in Physical Geography*, **14**, 153–177.

Boggs, D.A., Boggs, G.S., Ellot, I. and Knott, B. (2006) Regional patterns of salt lake morphology in the lower Yarra drainage system of Western Australia. *Journal of Arid Environments*, **64**, 97–115.

Boulaine J. (1954) La sebkha de Ben Ziane et sa 'lunette' ou Bourellet exemple de complex morphologique formé par la dégradation éolienne des sols salés. *Revue de Géomorphologique Dynamique*, **4**, 102–123.

Bowler, J.M. (1986) Spatial variability and hydrological evolution of Australian lake basins: an analogue for Pleistocene hydrologic change and evaporite formation. *Palaeogeography, Palaeoclimatology, Palaeoecology*, **54**, 21–41.

Briere, P.R. (2000) Playa, playa lake, sabkha: proposed definitions for old terms. *Journal of Arid Environments*, **45**, 1–7.

Bromley, M. (1992) Topographic inversion of early interdune deposits, Navajo Sandstone (Lower Jurassic), Colorado Plateau, USA. *Sedimentary Geology*, **80**, 1–25.

Bromley, J., Mannstrom, B., Nisca, D. and Jamtlid, A. (1994) Airborne geophysics – application to a groundwater study in Botswana. *Ground Water*, **32**, 79–90.

Brookfield, M.E. (1977) The origin of bounding surfaces in ancient aeolian sandstones. *Sedimentology*, **24**, 303–332.

Brown, A.G. (ed.) (1995) *Geomorphology and Groundwater*, John Wiley & Sons, Ltd, Chichester.

Bryan, R.B. and Yair, A. (eds) (1982) *Badland Geomorphology and Piping*, GeoBooks, Norwich.

Bryant, R.G. (1999) Application of AVHRR to monitoring a climatically sensitive playa. Case study: Chott el Djerid, southern Tunisia. *Earth Surface Processes and Landforms*, **24**, 283–302.

Bryant, R.G. and Rainey, M.P. (2002) Investigation of flood inundation on playas within the Zone of Chotts, using a time-series of AVHRR. *Remote Sensing of Environment*, **82**, 360–375.

Buckley, D.K. and Zeil, P. (1984) The character of fractured rock aquifers in eastern Botswana, in *Challenges in African Hydrology and Water Resources (Proceedings of the Harare Symposium July 1984)*, International Association of Hydrological Sciences Publication 144, pp. 25–36.

Bullock, A. (1992) Dambo hydrology in southern Africa – review and reassessment. *Journal of Hydrology*, **134**, 373–396.

Butterworth, J.S. (1982) The chemistry of Mogatse Pan – Kgalagadi District, Botswana Department of Geological Survey, Unpublished Report JSNB/14/82.

Carr, M.H. (1980) Survey of Martian fluvial features, NASA Technical Memo TM 81776, pp. 265–267.

Carr, M.H. and Malin, M.C. (2000) Meter-scale characteristics of Martian channels and valleys. *Icarus*, **146**, 366–386.

Carter, R.C. (1994) The groundwater hydrology of the Manga Grasslands, northeast Nigeria: importance to agricultural development strategy for the area. *Quarterly Journal of Engineering Geology*, **27**, S73–S83.

Carter, R.C., Morgulis, E.D., Dottridge, J. and Agbo, J.U. (1994) Groundwater modelling with limited data: a case study in a semi-arid dunefield of northeast Nigeria. *Quarterly Journal of Engineering Geology*, **27**, S85-S94.

Chan, M.A. and Kocurek, G. (1988) Complexities in eolian and marine interactions: processes and eustatic controls on erg development. *Sedimentary Geology*, **56**, 283–300.

Clarke, J.D.A. (1994) Lake Lefroy, a palaeodrainage playa in Western Australia. *Australian Journal of Earth Sciences*, **41**, 417–427.

Cooke, R.U. and Smalley, I.J. (1968) Salt weathering in deserts. *Nature*, **220**, 1226–1227.

Cooke, R.U., Brunsden, D., Doornkamp, J.C. and Jones, D.K.C. (1982) *Urban Geomorphology in Drylands*, Oxford University Press, Oxford.

Coque, R. (1962) *La Tunisie Pré-Saharienne, étude Géomorphologique*, Colin, Paris.

Crabaugh, M.C. and Kocurek, G. (1993) Entrada Sandstone: an example of a wet aeolian system, in *The Dynamics and Environmental Context of Aeolian Sedimentary Systems* (ed. K. Pye), Geological Society of London, Special Publication 72, pp. 103–126.

Craddock, R.A. and Maxwell, T.A. (1993) Geomorphic evolution of the Martian Highlands through ancient fluvial processes. *Journal of Geophysical Research*, **98** (E2), 3453–3468.

Cupper, M.L. (2006) Luminescence and radiocarbon chronologies of playa sedimentation in the Murray Basin, southeastern Australia. *Quaternary Science Reviews*, **25**, 2594–2607.

Dunne, T. (1980) Function and control of channel networks. *Progress in Physical Geography*, **4**, 211–239.

Dunne, T. (1990) Hydrology, mechanics, and geomorphic implications of erosion by subsurface flow, in *Groundwater Geomorphology; The Role of Subsurface Water in Earth-Surface*

Processes and Landforms (eds C.G. Higgins and D.R. Coates), Geological Society of America Special Paper 252, Boulder, Colorado, pp. 1–28.

Dutkiewicz, A. and von der Borch, C.C. (1995) Lake Greenly, Eyre Peninsula, South Australia – sedimentology, paleoclimatic and palaeohydrologic cycles. *Palaeogeography, Palaeoclimatology, Palaeoecology*, **113**, 43–56.

Eckardt, F.D., Bryant, R.G., McCulloch, G. *et al.* (2008) The hydrochemistry of a semi-arid pan basin case study: Sua Pan, Makgadikgadi, Botswana. *Applied Geochemistry*, **23**, 1563–1580.

El-Baz F., Boulos L., Breed C. *et al.* (1980) Journey to the Gilf Kebir and Uweinat, southwest Egypt. *Geographical Journal*, **146**, 51–93.

El-Baz, F. (1982) Desert landforms of southwest Egypt: a basis for comparison with Mars, National Air and Space Museum, Report NASA-CR-3611.

Farr, J.L., Peart, R.J., Nelisse, G. and Butterworth, J.S. (1982) Two Kalahari pans: a study of their morphology and evolution, Botswana Department of Geological Survey, Unpublished Report GS 10/10.

French, R.H., Miller, J.J., Dettling, C. and Carr, J.R. (2006) Use of remotely sensed data to estimate the flow of water to a playa lake. *Journal of Hydrology*, **325**, 67–81.

Fryberger, S.G., Krystinik, L.F. and Schenk, C.J. (1990) Tidally flooded back-barrier dunefield, Guerrero Negro area, Baja California, Mexico. *Sedimentology*, **37**, 23–43.

Fryberger, S.G., Schenk, C.J. and Krystinik, L.F. (1988) Stokes surfaces and the effects of near-surface groundwater-table on aeolian deposition. *Sedimentology*, **35**, 21–41.

Gaylord, D.R. (1990) Holocene paleoclimatic fluctuations revealed from dune and interdune strata in Wyoming. *Journal of Arid Environments*, **18**, 123–138.

Gieske, A. and Selaola, E. (1988) A proposed study of recharge processes in fractured aquifers of semi-arid Botswana, in *Estimation of Natural Groundwater Recharge* (ed. I. Simmers), NATO ISI Series C, vol. 222, Reidel, Dordrecht, pp. 117–124.

Glennie, K.W. (1970) *Desert Sedimentary Environments*, Developments in Sedimentology 14, Elsevier, Amsterdam.

Goldspiel, J.M. and Squyres, S.W. (2000) Groundwater sapping and valley formation on Mars. *Icarus*, **148**, 176–192.

Gomez, B. and Mullen, V.T. (1992) An experimental study of sapped drainage network development. *Earth Surface Processes and Landforms*, **17**, 465–476.

Goudie, A.S. and Thomas, D.S.G. (1985) Pans in southern Africa with particular reference to South Africa and Zimbabwe. *Zeitschrift für Geomorphologie*, **29**, 1–19.

Goudie, A.S. and Thomas, D.S.G. (1986) Lunette dunes in southern Africa. *Journal of Arid Environments*, **10**, 1–12.

Goudie, A.S. and Wells, G.L. (1995) The nature, distribution and formation of pans in arid zones. *Earth-Science Reviews*, **38**, 1–69.

Grant, J.A. and Parker, T.J. (2002) Drainage evolution in the Margaritifer Sinus region, Mars. *Journal of Geophysical Research – Planets*, **107** (E9), article 5066.

Gulick, V.C. (2001) Origin of the valley networks on Mars: a hydrological perspective. *Geomorphology*, **37**, 241–268.

Gulick, V.C. and Baker, V.R. (1990) Origin and evolution of valleys on Martian volcanoes. *Journal of Geophysical Research*, **95**, 14,325–14,344.

Hamdi-Aissa, B., Valles, V., Aventurier, A. and Ribolzi, O. (2004) Soils and brine geochemistry and mineralogy of hyperarid desert playa, Ouargla basin, Algerian Sahara. *Arid Land Research and Management*, **18**, 103–126.

Harrington, N.M., Herczeg, A.L. and La Salle, C.L. (2008) Hydrological and geochemical processes controlling variations in Na^{+}–Mg^{2+}–$Cl_2SO_4{}^{2-}$ groundwater brines, south-eastern Australia. *Chemical Geology*, **251**, 8–19.

Harrison, K.P. and Grimm, R.E. (2005) Groundwater-controlled valley networks and the decline of surface runoff on early Mars. *Journal of Geophysical Research – Planets*, **110** (E12).

Higgins, C.G. (1982) Drainage systems developed by sapping on Earth and Mars. *Geology*, **10**, 147–152.

Higgins, C.G. (1983) Reply: drainage systems developed by sapping on Earth and Mars. *Geology*, **11**, 55–56.

Higgins, C.G. (1984) Piping and sapping: development of landforms by groundwater outflow, in *Groundwater as a Geomorphic Agent* (ed. R.G. La Fleur), Allen and Unwin, London, pp. 18–58.

Higgins, C.G. (1990) Seepage-induced cliff recession and regional denudation, with case studies by W.R. Osterkamp and C.G. Higgins, in *Groundwater Geomorphology; The Role of Subsurface Water in Earth-Surface Processes and Landforms* (eds C.G. Higgins and D.R. Coates), Geological Society of America Special Paper 252, Boulder, Colorado, pp. 291–318.

Higgins, C.G. and Coates, D.R. (eds) (1990) *Groundwater Geomorphology; The Role of Subsurface Water in Earth-Surface Processes and Landforms*, Geological Society of America Special Paper 252, Boulder, Colorado.

Hinds, N.E.A. (1925) Amphitheatre valley heads. *Journal of Geology*, **33**, 816–818.

Hoke, G.D., Isacks, B.L., Jordan, T.E. and Yu, J.S. (2004) Groundwater-sapping origin for the giant quebradas of northern Chile. *Geology*, **32**, 605–608.

Howard, A.D. (1988) Groundwater sapping experiments and modelling at the University of Virginia, in *Sapping Features of the Colorado Plateau – A Comparative Planetary Geology Fieldguide* (eds A.D. Howard, R.C. Kochel and H. Holt), NASA Publication SP-491, pp. 71–83.

Howard, A.D. and Kochel, A.D. (1988) Introduction to cuesta landforms and sapping processes on the Colorado Plateau, in *Sapping Features of the Colorado Plateau – A Comparative Planetary Geology Fieldguide* (eds A.D. Howard, R.C. Kochel and H. Holt), NASA Publication SP-491, pp. 6–56.

Howard, A.D., Kochel, R.C. and Holt, H.E. (1988) *Sapping features of the Colorado Plateau - a Comparative Planetary Geology Fieldguide*, NASA Publication SP-491.

Howard, A.D. and McLane, C.F. (1988) Erosion of cohesionles sediment by groundwater seepage. *Water Resources Research*, **24**, 1659–1674.

Howard, A.D. and Selby, M.J. (1994) Rock slopes, in *Geomorphology of Desert Environments* (eds A.D. Abrahams and A.J. Parsons), Chapman & Hall, London, pp. 123–172.

Howard, A.D. and Selby, M.J. (2009) Rock slopes, in *Geomorphology of Desert Environments*, 2nd edn (eds A.J. Parsons and A.D. Abrahams), Chapman & Hall, London, pp. 189–232.

Hutchinson, J.N. (1968) Mass movement, in *Encyclopedia of Geomorphology* (ed. R.W. Fairbridge), Reinhold, New York.

Inman, D.L., Ewing, G.C. and Corliss, J.B. (1966) Coastal sand dunes of Guerrero Negro, Baja California, Mexico. *Geological Society of America Bulletin*, **77**, 787–802.

Issar, A. (1983) Emerging groundwater, a triggering factor in the formation of the Makhteshim (erosion cirques) in the Negev and Sinai. *Israel Journal of Earth-Sciences*, **32**, 53–61.

Jacobson, G., Arakel, A.V. and Chen, Y.J. (1988) The central Australian groundwater discharge zone – evolution of associated calcrete and gypcrete deposits. *Australian Journal of Earth Sciences*, **35**, 549–565.

Jacobson, G., Ferguson, J. and Evans, W.R. (1994) Groundwater-discharge playas of the Mallee Region, Murray Basin, southeast Australia, in *Paleoclimate and Basin Evolution of Playa Systems* (ed. M.R. Rosen), Geological Society of America Special Paper 289, pp. 81–96.

Jacobson, G. and Jankowski, J. (1989) Groundwater-discharge processes at a central Australian playa. *Journal of Hydrology*, **105**, 275–295.

Jennings, J.N. (1979) Arnhem Land, city that never was, *Geographical Magazine*, **51**, 822–827.

Judson, S. (1950) Depressions of the northern portion of the Southern High Plains of eastern New Mexico. *Geological Society of America Bulletin*, **61**, 253–274.

Kinsman, D.J.J. (1969) Modes of formation, sedimentary associations, and diagnostic features of shallow-water and supratidal evaporites. *Bulletin of the American Association of Petroleum Geologists*, **53**, 830–840.

Kochel, R.C., Howard, A.D. and McLane, C.F. (1985) Channel networks developed by groundwater sapping in fine-grained sediments: analogs to some Martian valleys, in *Models in Geomorphology* (ed. M.J. Woldenberg), Allen and Unwin, London, pp. 313–341.

Kochel, R.C. and Piper, J.F. (1986) Morphology of large valleys on Hawaii: evidence for groundwater sapping and comparisons with Martian valleys. *Journal of Geophysical Research*, **91** (b13), e175–e192.

Kocurek, G. (1981a) Significance of interdune deposits and bounding surfaces in aeolian dune sands. *Sedimentology*, **28**, 753–780.

Kocurek, G. (1981b) Erg reconstruction: the Entrada Sandstone (Jurassic) of Northern Utah and Colorado. *Palaeogeography, Palaeoclimatology, Palaeoecology*, **36**, 125–153.

Kocurek, G. (1984) Reply – origin of first-order bounding surfaces in aeolian sandstones. *Sedimentology*, **31**, 125–127.

Kocurek, G. (1988) First-order and super bounding surfaces in eolian sequences – bounding surfaces revisited. *Sedimentary Geology*, **56**, 193–206.

Kocurek, G. (1991) Interpretation of ancient aeolian sand dunes. *Annual Review of Earth and Planetary Science*, **19**, 43–75.

Kocurek, G. and Nielson, J. (1986) Conditions favourable for the formation of warm climate aeolian sand sheets. *Sedimentology*, **33**, 795–816.

Kocurek, G., Robinson, N.I. and Sharp, J.M. (2001) The response of the water table in coastal aeolian systems to changes in sea level. *Sedimentary Geology*, **139**, 1–13.

Kocurek, G., Carr, M., Ewing, R. *et al.* (2007) White Sands Dune Field, New Mexico: age, dune dynamics and recent accumulations. *Sedimentary Geology*, **197**, 313–331.

La Fleur, R.G. (ed.) (1984) *Groundwater as a Geomorphic Agent*, Allen and Unwin, London.

La Fleur, R.G. (1999) Geomorphic aspects of groundwater flow. *Hydrogeology Journal*, **7**, 78–93.

Laity, J.E. (1980) Groundwater sapping on the Colorado Plateau, in *Reports of Planetary Geology Programme 1980*, NASA Technical Memo TM 82385, pp. 358–360.

Laity, J.E. (1983) Diagenetic controls on groundwater sapping and valley formation, Colorado Plateau, as revealed by optical and electron microscope. *Physical Geography*, **4**, 103–125.

Laity, J.E. (1988) The role of groundwater sapping in valley evolution on the Colorado Plateau, in *Sapping Features of the Colorado Plateau – A Comparative Planetary Geology Fieldguide* (eds A.D. Howard, R.C. Kochel and H.E. Holt), NASA Publication SP-491, pp. 63–70.

Laity, J.E. and Malin, M.C. (1985) Sapping processes and the development of theater-headed valley networks in the Colorado Plateau. *Geological Society of America Bulletin*, **96**, 203–217.

Laity, J.E., Pieri, D.C. and Malin, M.C. (1980) Sapping processes in tributary valley systems, NASA Technical Memo TM 81776, pp. 295–297.

Lamb, M.P., Howard, A.D., Johnson, J. *et al.* (2006) Can springs cut canyons into rock? *Journal of Geophysical Research – Planets*, **111** (E7), article E07002.

Lamb, M.P., Howard, A.D., Dietrich, W.E. and Perron, J.T. (2007) Formation of amphitheater-headed valleys by waterfall erosion after large-scale slumping on Hawaii. *Geological Society of America Bulletin*, **119**, 805–822.

Lamb, M.P., Dietrich, W.E., Aciego, S.M. *et al.* (2008) Formation of Box Canyon, Idaho, by megaflood: implications for seepage erosion on Earth and Mars. *Science*, **320**, 1067–1070.

Langford, R.P. (1989) Fluvial-aeolian interactions: Part I, modern systems. *Sedimentology*, **36**, 1023–1035.

Langford, R.P., Rose, J.M. and White, D.E. (2009) Groundwater salinity as a control on development of eolian landscape: an example from the White Sands of New Mexico. *Geomorphology*, **105**, 39–49.

Langford, R.P., Pearson, K.M., Duncan, K.A. *et al.* (2008) Eolian topography as a control on deposition incorporating lessons from modern dune seas: Permian Cedar Mesa Sandstone, SE Utah, USA. *Journal of Sedimentary Research*, **78**, 410–422.

Levin, N., Tsoar, H., Herrmann, H.J. *et al.* (2009) Modelling the formation of residual dune ridges behind barchan dunes in north-east Brazil. *Sedimentology*, **56**, 1623–1641.

Loope, D.B. (1984) Origin of extensive bedding planes in aeolian sandstones: a defence of Stokes' hypothesis. *Sedimentology*, **31**, 123–125.

Loope, D.B. (1985) Episodic deposition and preservation of eolian sands: a late Paleozoic example from southeastern Utah. *Geology*, **13**, 73–76.

Loope, D.B. (1988) Rhizoliths in ancient eolianites. *Sedimentary Geology*, **56**, 315–339.

Lowry, D.C. and Jennings, J.N. (1974) The Nullarbor karst Australia. *Zeitschrift für Geomorphologie*, **18**, 35–81.

Luo, W. (2000) Quantifying groundwater-sapping landforms with a hypsometric technique. *Journal of Geophysical Research – Planets*, **105** (E1), 1685–1694.

Luo, W. (2002) Hypsometric analysis of Margaritifer Sinus and origin of valley networks. *Journal of Geophysical Research – Planets*, **107** (E10).

Luo, W. and Howard, A.D. (2005) Morphometric analysis of Martian valley network basins using a circularity function. *Journal of Geophysical Research – Planets*, **110** (E12).

Luo, W. and Howard, A.D. (2008) Computer simulation of the role of groundwater seepage in forming Martian valley networks. *Journal of Geophysical Research – Planets*, **113** (E5), article E05002.

McFarlane M.J. (1989) Dambos - – their cCharacteristics and gGeomorphological eEvolution in pParts of Malawi and Zimbabwe, with pParticular rReference to their rRole in the hHydrogeological rRegime of sSurviving aAreas of African sSurface, in. Proceedings of the Groundwater Exploration and Development in Crystalline Basement Aquifers Workshop, *Harare, Zimbabwe, 15–24 June 1987*, Commonwealth Science Council 1 (Session 3), pp. 254–308.

McFarlane, M.J. (1990) Aa review of the development of tropical weathering profiles with particular reference to leaching history and with examples from Mmalawi and Zzimbabwe, in. Proceedings of the Groundwater Exploration and Development in Crystalline Basement Aquifers Workshop, *Harare, Zimbabwe, 15-24 June 1987*, Commonwealth Science Council 1 (Session 8), pp. 93–145.

McFarlane, M.J. and Whitlow, R. (1990) Key factors affecting the initiation and progress of gullying in dambos in parts of Zimbabwe and Malawi. *Land Degradation and Rehabilitation*, **2**, 215–235.

Mäckel R. (1974) Dambos: a study in morphodynamic activity on the plateau regions of Zambia. *Catena*, **1**, 327–365.

Mader, D. and Yardley, M.J. (1985) Migration, modification and merging in aeolian systems and the significance of the depositional mechanisms in Permian and Triassic dune sands of Europe and North America. *Sedimentary Geology*, **43**, 85–218.

Mars Channel Working Group (1983) Channels and valleys on Mars. *Geological Society of America Bulletin*, **94**, 1035–1054.

Mastronuzzi, G. and Sanso, P. (2002) Pleistocene sea-level changes, sapping processes and development of valley networks in the Apulia region (southern Italy). *Geomorphology*, **46**, 19–34.

Maxwell, J.A. (1979) Field investigation of Martian canyonlands in southwestern Egypt, NASA Conference publication 2072.

Milton, D.J. (1973) Water and processes of degradation in the Martian landscape. *Journal of Geophysical Research*, **78**, 4037–4047.

Mountney, N.P. and Jagger, A. (2004) Stratigraphic evolution of an aeolian erg margin system: the Permian Cedar Mesa Sandstone, SE Utah, USA. *Sedimentology*, **51**, 713–743.

Mountney, N.P. and Russell, A.J. (2009) Aeolian dune-field development in a water table-controlled system: Skeidararsandur, Southern Iceland. *Sedimentology*, **56**, 2107–2131.

Nash, D.J. (1995) Structural control and deep-weathering in the evolution of the dry valley systems of the Kalahari, central southern Africa. *Africa Geoscience Review*, **2**, 9–23.

Nash, D.J. (1996) Groundwater sapping and valley development in the Hackness Hills, North Yorkshire, England. *Earth Surface Processes and Landforms*, **21**, 781–795.

Nash, D.J., Shaw, P.A. and Thomas, D.S.G. (1994) Duricrust development and valley evolution: process-landform links in the Kalahari. *Earth Surface Processes and Landforms*, **19**, 299–317.

Nash, D.J., Thomas, D.S.G. and Shaw, P.A. (1994) Timescales, environmental change and dryland valley development, in *Environmental Change in Drylands* (eds A.C. Millington and K. Pye), John Wiley & Sons, Ltd, Chichester, pp. 25–41.

Newell, W.L. (1970) Factors influencing the grain of the topography along the Willoughby Arch in northeastern Vermont. *Geografisca Annaler*, **52A**, 103–112.

Onda, Y. (1994) Seepage erosion and its implication to the formation of amphitheatre valley heads: a case study at Obara, Japan. *Earth Surface Processes and Landforms*, **19**, 627–640.

Osterkamp, W.R. and Wood, W.W. (1987) Playa lake basins on the Southern High Plains of Texas and New Mexico: Part 1. Hydrologic, geomorphic and geologic evidence for their development. *Geological Society of America Bulletin*, **99**, 215–223.

Paim, P.S.G. and Scherer, C.M.S. (2007) High-resolution stratigraphy and depositional model of wind- and water-laid deposits in the ordovician Guaritas rift (Southernmost Brazil). *Sedimentary Geology*, **202**, 776–795.

Paine, J.G. (1994) Subsidence beneath a playa basin on the Southern High Plains, U.S.A.: evidence from shallow seismic data. *Geological Society of America Bulletin*, **106**, 233–242.

Palmer, A.N. (1990) Groundwater processes in karst terranes, in *Groundwater Geomorphology; The Role of Subsurface Water in Earth-Surface Processes and Landforms* (eds C.G. Higgins and D.R. Coates), Geological Society of America Special Paper 252, Boulder, Colorado, pp. 177–210.

Parker Sr, G.G. and Higgins, C.G. (1990) Piping and pseudokarst in drylands, with case studies by G.G. Parker, Sr. and W.W. Wood, in *Groundwater Geomorphology; The Role of Subsurface Water in Earth-Surface Processes and Landforms* (eds C.G. Higgins and D.R. Coates), Geological Society of America Special Paper 252, Boulder, Colorado, pp. 77–110.

Peel, R.F. (1941) Denudational landforms of the central Libyan desert. *Journal of Geomorphology*, **4**, 3–23.

Pieri, D.C. (1980) Martian valleys: morphology, distribution, age and origin. *Science*, **210**, 895–897.

Pieri, D.C., Malin, M.C. and Laity, J.E. (1980) Sapping: network structure in terrestrial and Martian valleys, NASA Technical Memo TM-81979, pp. 292–293.

Purvis, K. (1991) Stoss-side mud drapes: deposits of interdune pond margins. *Sedimentology*, **38**, 153–156.

Pye, K. (1980) Beach salcrete and aeolian sand transport, evidence from North Queensland. *Journal of Sedimentary Petrology*, **50**, 257–261.

Reading, H.G. (1996) *Sedimentary Environments: Processes, Facies, and Stratigraphy*, Blackwell, Oxford.

Rech, J.A., Quintero, L.A., Wilke, P.J. and Winer, E.R. (2007) The lower paleolithic landscape of 'Ayoun Qedim, al-Jafr Basin, Jordan. *Geoarchaeology – An International Journal*, **22**, 261–275.

Risacher, F., Alonso, H. and Salazar, C. (2003) The origin of brines and salts in Chilean salars: a hydrochemical review. *Earth-Science Reviews*, **63**, 249–293.

Risacher, F. and Fritz, B. (2009) Origin of salts and brine evolution of Bolivian and Chilean salars. *Aquatic Geochemistry*, **15**, 123–157.

Robb, J.M. (1990) Groundwater processes in the submarine environment, in *Groundwater Geomorphology; The Role of Subsurface Water in Earth-Surface Processes and Landforms* (eds C.G. Higgins and D.R. Coates), Geological Society of America Special Paper 252, Boulder, Colorado, pp. 267–281.

Robb, J.M., O'Leary, D.W., Booth, J.S. and Kohout, F.A. (1982) Submarine spring sapping as a geomorphic agent on the East Coast Continental Slope. *Geological Society of America Abstracts and Programs*, **14**, 600.

Rosen, M.R. (1994) The importance of groundwater in playas: a review of playa clasifications and the sedimentology and hydrology of playas, in *Paleoclimate and Basin Evolution of Playa Systems* (ed. M.R. Rosen), Geological Society of America Special Paper 289, pp. 1–18.

Rubin, D.M. and Hunter, R.E. (1982) Bedform climbing in theory and nature. *Sedimentology*, **29**, 121–138.

Rubin, D.M. and Hunter, R.E. (1984) Reply. *Sedimentology*, **31**, 128–132.

Sakura, Y., Mochizuki, M. and Kawasaki, I. (1987) Experimental studies on valley headwater erosion due to groundwater flow. *Geophysical Bulletin of Hokkaido University*, **49**, 229–239.

Sanchez, J.A., Perez, A., Coloma, P. and Martinez-Gil, J. (1998) Combined effects of groundwater and aeolian processes in the formation of the northernmost closed saline depressions of Europe: north-east Spain. *Hydrological Processes*, **12**, 813–820.

Schenk, C.J. and Fryberger, S.G. (1988) Early diagenesis in eolian dune and interdune sands at White Sands, New Mexico. *Sedimentary Geology*, **55**, 109–120.

Schumm, S.A. and Phillips, L. (1986) Composite channels of the Canterbury Plains, New Zealand: a Martian analogue. *Geology*, **14**, 326–330.

Schumm, S.A., Boyd, K.F., Wolff, C.G. and Spitz, W.J. (1995) A ground-water sapping landscape in the Florida Panhandle. *Geomorphology*, **12**, 281–297.

Scuderi, L.A., Laudadio, C.K. and Fawcett, P.J. (2010) Monitoring playa lake inundation in the western United States: modern analogues to late-Holocene lake level change. *Quaternary Research*, **73**, 48–58.

Sharp, R.P. (1973) Mars: fretted and chaotic terrains. *Journal of Geophysical Research*, **78**, 4063–4072.

Sharp, R.P. and Malin, M.C. (1985) Channels on Mars. *Geological Society of America Bulletin*, **86**, 593–609.

Shaw, P.A. and de Vries, J.J. (1988) Duricrust, groundwater and valley development in the Kalahari of south-east Botswana. *Journal of Arid Environments*, **14**, 245–254.

Small, R.J. (1964) The escarpment dry valleys of the Wiltshire Chalk. *Transactions of the Institute of British Geographers*, **34**, 33–52.

Smith, B.J. (1987) An integrated approach to the weathering of limestone in an arid area and its role in landscape evolution: a case study from southeast Morocco, in *International Geomorphology 1986 Part 2* (ed. V. Gardiner), John Wiley & Sons, Ltd, Chichester, pp. 637–657.

Smith, B.J. (1988) Weathering of superficial limestone debris in a hot desert environment. *Geomorphology*, **1**, 355–367.

Smith, B.J. (1994) Weathering processes and forms, in *Geomorphology of Desert Environments* (eds A.D. Abrahams and A.J. Parsons), Chapman and Hall, London, pp. 39–63.

Sparks, B.W. and Lewis, W.V. (1957–1958) Escarpment dry valleys near Pegsdon, Hertfordshire. *Proceedings of the Geologists Association*, **68**, 26–38.

Stepinski, T.F. and Stepinski, A.P. (2005) Morphology of drainage basins as an indicator of climate on early Mars. *Journal of Geophysical Research – Planets*, **110** (E12).

Stiller, D. (1983) Comment: drainage systems developed by sapping on Earth and Mars. *Geology*, **11**, 54–55.

Stokes, W.L. (1968) Multiple parallel truncation bedding planes – a feature of wind-deposited sandstone formations. *Journal of Sedimentary Petrology*, **38**, 510–515.

Talbot, M.R. (1980) Environmental responses to climatic change in the West African Sahel over the past 20,000 years, in *The Sahara and the Nile* (eds M.A.J. Williams and H. Faure), A.A. Balkema, pp. 37–62.

Talbot, M.R. (1985) Major bounding surfaces in aeolian sandstones – a climatic model. *Sedimentology*, **32**, 257–265.

Tanaka, K.L. and Chapman, M.G. (1992) Kasei Valles, Mars: interpretation of canyon materials and flood sources. *Proceedings of Lunar and Planetary Science*, **22**, 73–83.

Tanaka, K.L., Dohm, J.M., Lias, J.H. and Hare, T.M. (1998) Erosional valleys in the Thaumasia region of Mars: hydrothermal and seismic origins. *Journal of Geophysical Research-Planets*, **103** (E13), 31407–31419.

Thomas, D.S.G., Nash, D.J., Shaw, P.A. and Van Der Post, C. (1993) Present day lunette sediment cycling at Witpan in the arid southwestern Kalahari Desert. *Catena*, **20**, 515–527.

Torgersen, T., De Deckker, P., Chivas, A.R. and Bowler, J.M. (1986) Salt lakes: a discussion of processes influencing palaeoenvironmental interpretations and recommendations for future study. *Palaeogeography, Palaeoclimatology, Palaeoecology*, **54**, 7–19.

Uchupi, E. and Oldale, R.N. (1994) Spring sapping origin of the enigmatic relict valleys of Cape Cod and Martha's Vineyard and Nantucket Islands, Massachusetts. *Geomorphology*, **9**, 83–95.

Von Hoyer, M., Keller, S. and Rehder, S. (1985) Core borehole Lethlakeng 1, Botswana Department of Geological Survey, Unpublished Report MVH/4/85, Lobatse.

Whitlow J.R. (1985) Dambos in Zimbabwe. A review. *Zeitschrift für Geomorphologie, Supplementband*, **52**, 115–146.

Wood, W.W. and Osterkamp, W.R. (1987) Playa lake basins on the Southern High Plains of Texas and New Mexico: Part 2. A hydrologic model and mass-balance arguments for their development. *Geological Society of America Bulletin*, **99**, 224–230.

Yechieli, Y. and Wood, W.W. (2002) Hydrogeologic processes in saline systems: playas, sabkhas, and saline lakes. *Earth-Science Reviews*, **58**, 343–365.

Young R.W. (1986) Sandstone terrain in a semi-arid littoral environment: the lower Murchison Valley, Western Australia. *Australian Geographer*, **17**, 143–153.

Zartman, R.E., Evans, P.W. and Ramsey, R.H. (1994) Playa lakes on the Southern High Plains in Texas: reevaluating infiltration. *Journal of Soil and Water Conservation*, **49**, 299–301.

IV

The work of the wind

17

Aeolian landscapes and bedforms

David S.G. Thomas

17.1 Introduction

Arid zone aeolian deposits have long attracted the attention of geographers, geomorphologists and allied scientists. They represent for many people the seminal landscape of deserts, though the actual geographical distribution and extent of, for example, dune fields is no more than 20 % of the total area of drylands today, and in some deserts, such as the southwestern USA, they cover less than 1 % of the total area. Scientifically, Bagnold's (1941) seminal work on sand transport processes and dune formation established a depth of understanding and rigour that set the framework for research in the ensuing half-century. Subsequently, at markedly different scales, the availability of remotely sensed data coving otherwise difficult to access desert areas (e.g. Breed and Grow, 1979; Paisley *et al.*, 1991) and the growing number of reductionist studies involving the detailed monitoring of wind and sand flow over dunes (e.g. Weng *et al.*, 1991; Wiggs, 1993) led to advances in aeolian research at very contrasting spatial scales (Thomas and Wiggs, 2008) (Figure 17.1).

More recently, four scientific developments have provided the opportunity for across-scale enhancements in understanding the nature and behaviour of aeolian systems (Thomas and Wiggs, 2008): (1) further advances, largely driven by technological developments (e.g. Arens, 1996; van Boxel, Sterk and Arens, 2004), in empirical measurement that allow the relationships between sediment movement and wind power to be elucidated; (2) mathematical modelling of dune (e.g. Baas and Nield, 2007) and atmospheric dust (e.g. Zender, Bian and Newman, 2003) dynamics and their interactions with key controls; (3) technological developments in remote monitoring systems that allow (a) high-resolution observations of sediment movement (via the plethora of remote sensing technologies now available from space; e.g. Washington *et al.*, 2003) and (b) the internal structures of aeolian deposits to be understood (i.e. via ground penetrating radar, e.g. Bristow *et al.*, 2000); and (4) chronometric advances, such as optically stimulated luminescence dating (e.g. Aitken, 1994), which are allowing the timing and frequency of aeolian deposit accumulation to be established, as well as the relationships to drivers of system dynamics.

Given these advances, we are now better placed than ever to understand the spatial and the temporal behaviour of aeolian systems, the drivers of dynamic behaviour and the relationships in behaviour across scales from the movement of individual sand and silt grains to the development of dunefields and sand seas. To develop an understanding of how aeolian systems behave and accumulate, it is nonetheless necessary to make separate considerations of processes, landforms and landscapes. In this chapter the larger spatial and temporal scales are examined, by looking at the distribution of aeolian deposits at the global scale and what determines that distribution and, by examining the long-term Quaternary timescale, the development of aeolian systems and their relationships to climatic changes. The chapter considers both sandy deposits – sand seas and their constituent parts – and the landscape features comprised of wind-lain silt, termed loess.

17.2 Aeolian bedforms: scales and relationships

Because of their fine-grained nature, loessic deposits (primarily silt-sized material with 3.9-63 μm particle diameters) tend to be rather amorphous, landscape-blanketing

Arid Zone Geomorphology: Process, Form and Change in Drylands, Third Edition. Edited by David S. G. Thomas

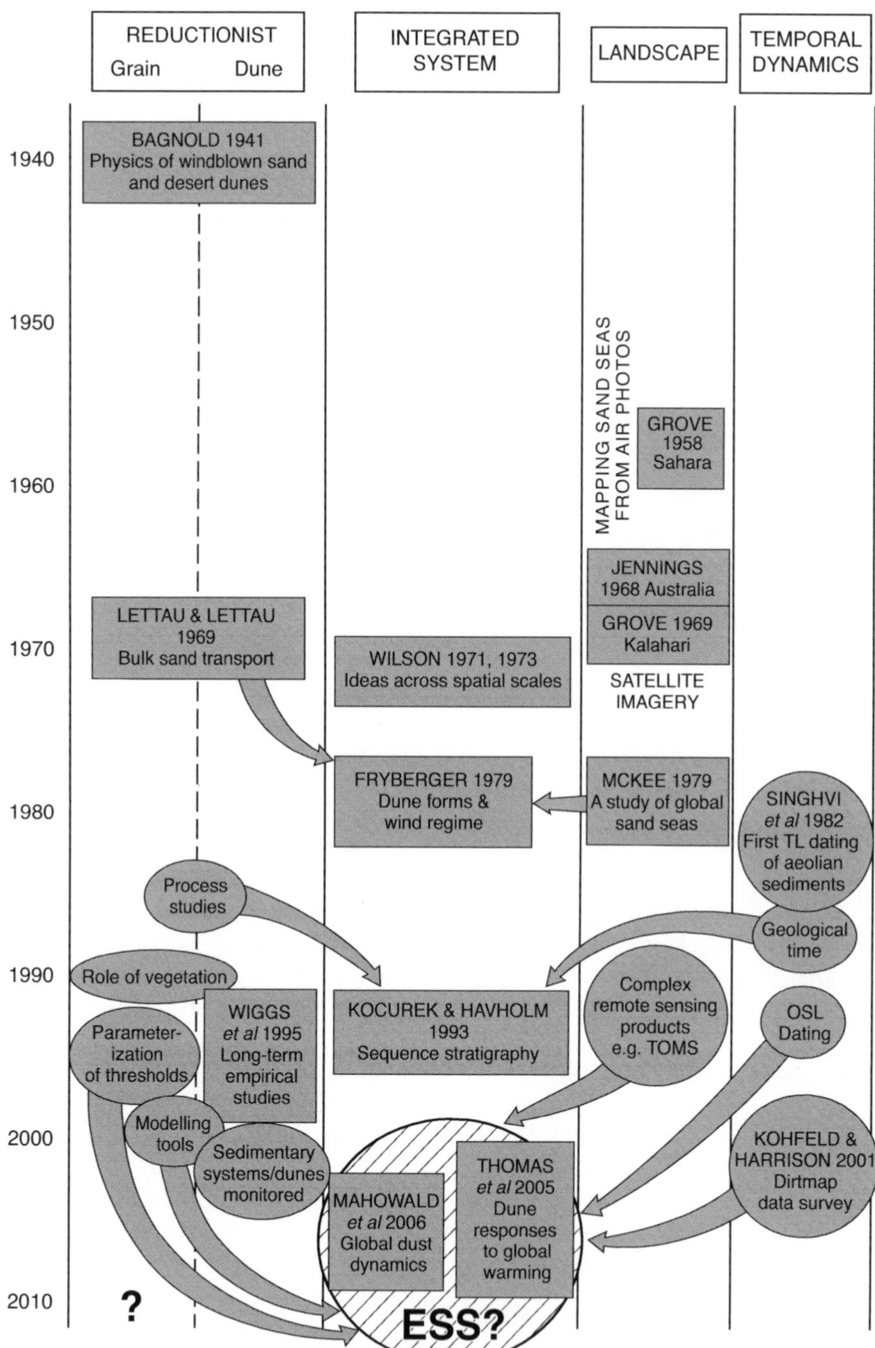

Figure 17.1 The routes and scales of aeolian research since Bagnold (1941). The themes focus on reductionist process research, research that treats aeolian systems in an integrated (time/space or scale) manner, research focused at the landscape scale and research with a strong focus on understanding the temporal history of aeolian systems. Since 2000 there has perhaps been a growing emphasis on research that attempts to integrate all these elements and which might be termed 'earth system science' (after Thomas and Wiggs, 2008).

features (Pye and Sherwin, 1999). Loess is largely devoid of landforms that are a direct result of shaping by the wind; being comprised of fine-grained sediment, it is generally unable to hold a morphology once deposition from suspension transport by the wind has occurred, other than that determined by the underlying topography over which it lies. Thus the great loess plateau of China and the loessic deposits of eastern Europe and America are rather flat and featureless. There are exceptions to this, and low and high amplitude linear structures ('loess hills' and 'loess dunes') have been reported from the loess deposits of eastern Germany and Bulgaria. These are parallel to the wind direction that was responsible for deposition (Leger, 1990). In Bulgaria, where the features are locally known as 'breda', they are reported as having a wavelength of 300–1200 m and a trough–ridge amplitude of up to 40 m (Rozsycki, 1967). However, loess is very susceptible to erosion by running water, especially, paradoxically, in drier areas where vegetation cover is sparse, and most landforms associated with the material are erosional in nature, such as gullies, piping and features resulting from landsliding and slumping (Derbyshire *et al.*, 1993).

The wider grain-size range of sand particles transported by the wind (particle diameters from 2000 to 63 μm) means that constructional aeolian forms – dunes and ripples – can develop due to the packing that can occur between particles of different sizes. Over 99 % of arid zone aeolian sand deposits are found in extensive sand seas or *ergs* (Wilson, 1973), the largest landform unit of aeolian deposition (Table 17.1), within which smaller-scale depositonal landforms – predominantly dunes – occur. Independent, isolated, dunes can occur outside sand seas, e.g. where aeolian sand supply is limited and a topographic obstacle causes sand accumulation, but usually dunes occur in association with other dunes. This is significant when dune-forming processes are considered because it is increasingly being recognised that the interactions that can occur between adjacent dunes, due to self-organisation (Werner, 1995), are a key component of explaining how dunes form. As such, the beautiful patterns that exist within dunefields (Figure 17.2 and see Chapter 19), and which remote sensing from space has made so visible, are not merely aesthetically pleasing but result from the processes of organisation that occur in complex environmental systems (Ewing, Kocurek and Lake, 2006; Baas, 2007).

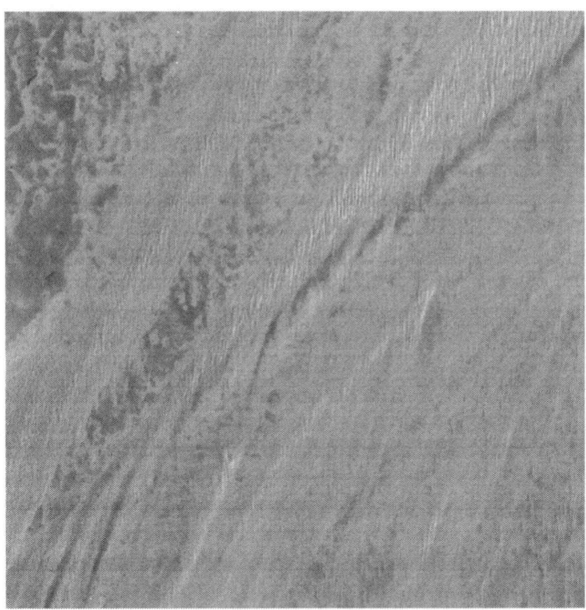

Figure 17.2 Bedform patterns revealed in a sand sea by satellite imagery. Unravelling the complex evolution of such a system requires scientific data at a range of temporal and spatial scales.

Table 17.1 Scales of aeolian depositional landforms.

Scale	*Description*	*Name*[a]
I	Deposits of regional extent	Sand seas or ergs
II	Deposits commonly devoid of scale III bedforms	Sand sheets, streaks and stringers
IIIa	Bedforms superimposed on scale I deposits	Complex, compound and megadunes
IIIb	As scale IIIa or independently	Simple dunes
IV	Bedforms superimposed on scale I, II or III deposits	Normal, fluid drag and megaripples

[a]Many deposits have alternative names. These are given in the table of Breed and Grow (1979).

The form and size of dunes can vary within and between sand seas, through the interaction of and variations in sand supply, vegetation cover and wind regime. While there may be spatial variability through a dunefield in dune size, statistical analyses tend to show that patterns are consistent over quite large areas of 10^1–10^2 km^2 (e.g. Thomas, 1986; Bullard *et al.*, 1996), resulting from self-organisational principles (Ewing, Kocurek and Lake, 2006; Zheng, Bo and Zhu, 2009). Over time, patterns may mature and multiple patterns may become superimposed (Ewing, Kocurek and Lake, 2006), possibly resulting from dunefield development during different arid phases (Kocurek, 1998). Sand seas or parts of sand seas that are largely devoid of dunes are called *sand sheets, stringers* or *streaks*. *Ripples* are the smallest bedform unit and can occur on any larger-scale aeolian bedform.

Scale III and IV features (Table 17.1) are *bedforms*, a term that applies to any regularly repeated erosional or depositional surface feature, whether aeolian or subaqueous (Wilson, 1972a). This section considers general aspects of aeolian bedform development, while the formation of ripples and the controls on development of different dune types are discussed in greater detail in Chapters 20 and 21.

17.2.1 Scale effects in aeolian bedform development

Contrary to early opinion (Cornish, 1914), ripples do not develop over time into dunes. Ripples and dunes develop independently, at different scales, in response to different factors. Wilson (1972a, 1972b) developed the concept of bedform scale into one of a bedform hierarchy with four components: ripples (two orders), dunes and draas (Table 17.2), draas being large whaleback sand bodies, or megadunes.

The spacing of the three highest-order bedforms was attributed to secondary atmospheric flows of differing sizes. As bigger sand grains require higher shear velocities for their mobilisation, particle size was deemed to set the scale of lower-atmospheric eddy currents (Wilson, 1972a, 1972b). Grain size would therefore be coarser in the higher-order, more widely spaced, bedforms.

The universal applicability of this 'granulometric control' hypothesis has not survived the scrutiny of subsequent investigations. Draas cannot be separated from dunes in terms of grain size (Wasson and Hyde, 1983), probably being compound or complex dune forms that have developed in areas of high net sand accumulation, complex wind regimes and long-term aridity (see, for example, McKee, 1979; Lancaster, 1985) or through repetitive periods of accumulation, as in the case of the megadunes of the Badain Jaran Desert in China (Yang, Liu and Xiao, 2003; Dong, Wang and Wang, 2004). On the basis of height–wavelength relationships, Dong *et al.* (2009) confirm that megadunes are simply a higher hierachical aeolian form controlled by wind conditions. More controversially, by using field data and numerical modelling, Andreotti *et al.* (2009) argue that megadune formation is controlled by the depth of the atmospheric boundary layer, while Bubenzer and Bolten (2008) suggest distinct atmospheric conditions have been responsible for the formation of draa in the eastern Sahara and the linear dunes that overlie them. These debates demonstrate that there is still much that is unclear regarding the nature of aeolian system development.

Aeolian deposition and the initiation of landforms takes place when mobile particles come to rest in a sheltered position on the ground surface, through *accretion* (Bagnold, 1941) or *tractional deposition* (Kocurek and Dott, 1981). It also occurs when wind velocities decrease, causing a reduction in shear stress and a fall in transportational potential, termed *grain-fall deposition*, and when particles come to rest after rolling down a slope under the effects of gravity, called *avalanching or grainflow deposition*, which is important for particle movement down topographically induced or bedform slopes (see Tsoar, 1982). In many regards, the processes of deposition and accumulation are scale-independent; i.e. the same general conditions leading to sand deposition must apply whether a sand sea, dune or ripple is being initiated or a nuclei or focal point for accumulation is required (Table 17.3). The processes of dune and ripple formation and initiation are discussed in subsequent chapters; those for sand seas are discussed below in the context of understanding their global distribution.

17.3 The global distribution of sand seas

Sand seas are neither clearly nor consistently defined in the literature. Following Wilson (1971), Fryberger and

Table 17.2 Wilson's hierarchy of aeolian bedforms.

Order	*Name*	*Wavelength (m)*	*Height (m)*	*Origin*
1	Draas or mega dunes	300–5500	20–450	Aerodynamic instability
2	Dunes	3–600	0.1–100	Aerodynamic instability
3	Aerodynamic ripples	0.015–0.25	0.002–0.05	Aerodynamic instability
4	Impact ripples	0.05–2.0	0.0005–0.1	Impact mechanism

Source: Wilson (1972a).

Table 17.3 Possible nuclei for sand dune development.

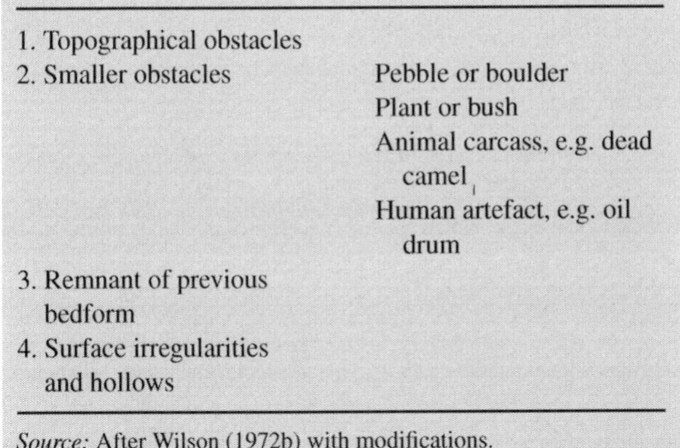

1. Topographical obstacles	
2. Smaller obstacles	Pebble or boulder
	Plant or bush
	Animal carcass, e.g. dead camel
	Human artefact, e.g. oil drum
3. Remnant of previous bedform	
4. Surface irregularities and hollows	

Source: After Wilson (1972b) with modifications.

Ahlbrandt (1979) and others, an *active sand sea* may be defined as an area of at least 125 km^2 where no less than 20 % of the ground surface is covered by windblown sand. Lancaster (1995) has defined sand seas as 'dynamic sedimentary bodies that form part of local- and regional-scale sand transport systems in which sand is moved from source zones to depositional sinks'. Dune forms are commonly present, but in areas where sand throughflow dominates over sand accumulation, sand sheets occur and dune forms are normally absent.

Figure 17.3 shows the global distribution of major active and fixed sand seas. The term 'fixed' is applied here to sand seas where dune forms are identifiable, but are presently inactive or have limited surface activity, because current environmental factors do not favour the significant operation of aeolian processes. This is largely evidenced by the presence of a significant vegetation cover on dunes, e.g. in large parts of Australia and the northern Kalahari sand sea in southern Africa, but also through other evidence such as the occurrence of soils on dune surfaces, and degradation of dune forms. Although it is the interaction between atmospheric (*erosivity*, by wind energy) and surface (*erodibility*, linked in particular to vegetation cover) conditions that determines the potential for sand transport (Lancaster, 1988; Thomas, Knight and Wiggs, 2005), most of the major sand seas that are obviously active today lie within the 150 mm mean annual isohyet (Wilson, 1973) (Figure 17.4), where given a supply of sediment and suitable wind conditions, aeolian transport is uninhibited by plants. The implication of fixed sand seas is that climatic and environmental conditions in the past have been more conducive to aeolian sand transport than today, particularly during the Late Quaternary Period.

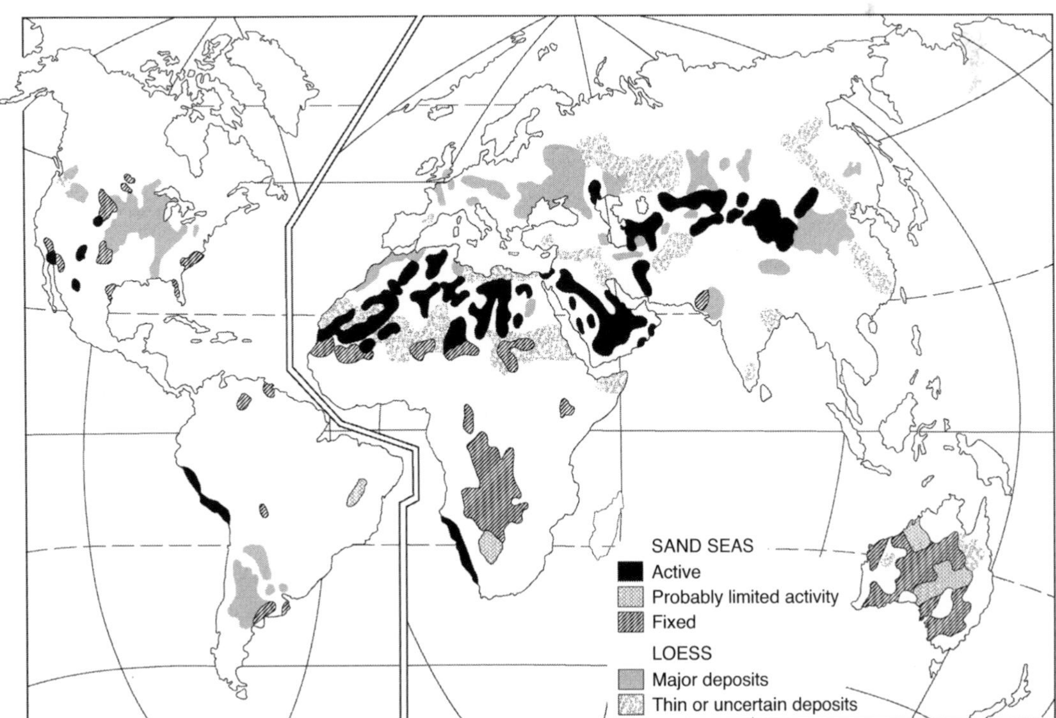

Figure 17.3 The global distribution of aeolian deposits (after Snead, 1972, modified and supplemented with data from numerous sources).

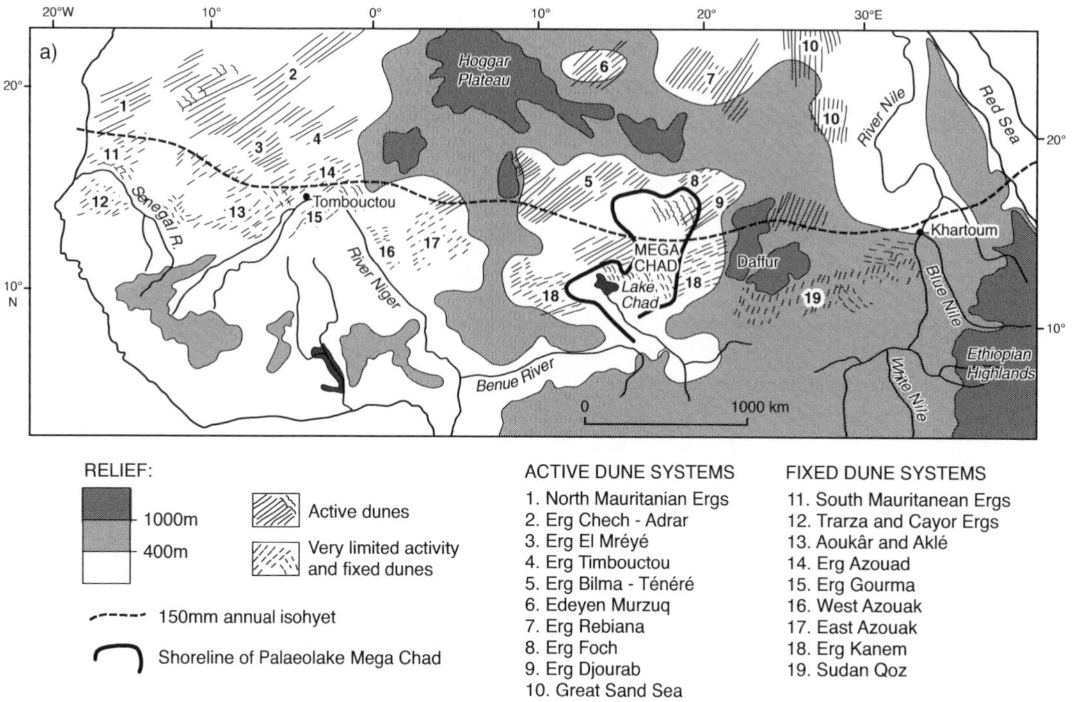

Figure 17.4 The distribution of active and fixed sand seas in the Sahelian-Sahel belt of Africa. Note that the degree of activity/inactivity is not considered (after Grove, 1958, and Grove and Warren, 1968).

17.3.1 Sand sea development

The formation of a sand sea is contingent on sediment availability and aeolian transport capacity. Kocurek (1999) breaks these conditions down further such that the sediment issue is considered both in terms of its sourcing (which he confusingly terms 'supply') and its availability to the wind – i.e. just because suitable sediment exists, it may not be available to the wind until, for example, vegetation cover declines or sediment in a lake is exposed by declining water levels. Overtime, it is the interaction between supply (*senso* Kucurek), sediment availability to the wind and the transport capacity of the wind that determines whether a sand sea is stable, accumulating or degrading. Simply put, sand sea construction occurs when a large supply of sediment is available to wind of sufficient energy to transport it to a place where transport energy is reduced and deposition occurs. These conditions may vary over time (Figure 17.5) at a range of scales, such that climate cycles may affect both wind energy and the availability of sediment on the ground.

17.3.2 Sediment supply in sand seas

The initial sourcing of sediment for sand sea development is normally non-aeolian. Locally, aeolian abrasion of bedrock may contribute material for sand sea development (e.g. in the eastern Namib Sand Sea; see Besler, 1980), but fluvial, lacustrine and coastal sources are the principal locations from which sediment is derived by the wind and moved towards areas where sand seas can accumulate (Lancaster, 1999).

Aeolian transportation sorts sediment effectively, so that usually over 90 % of particles in active sand seas are sand-sized. Finer silt, possibly generated by the grain-on-grain impact of sand in transport is commonly transported out of the desert aeolian system in suspension, where it may contribute to loess deposits (Crouvi *et al.*, 2008). Clay is normally only transported by the wind in pellets, which then saltate and creep like the more familiar quartzose sands (Bowler and Wasson, 1984; Al Janabi *et al.*, 1988). Dare-Edwards (1984) has called such deposits *sand loess*.

Sand may move between source areas and sand seas along preferred transport pathways. In some areas, such as the Mojave Desert in California, these pathways may cross drainage basins and divides prior to accumulation in sand seas (Zimbelman, Williams and Tchskerian, 1994). The highest rates of sand transport occur across the duneless sand sheets and stringers, which often lead into areas of sand seas with dunes (Sarnthein and Walger, 1974, give a range of 62.5–162.5 m^3/m width per year for Mauritania). Rates of sand throughflow in the Selima Sand Sheet, Egypt, for example, are as high as 1000 m/yr (Maxwell

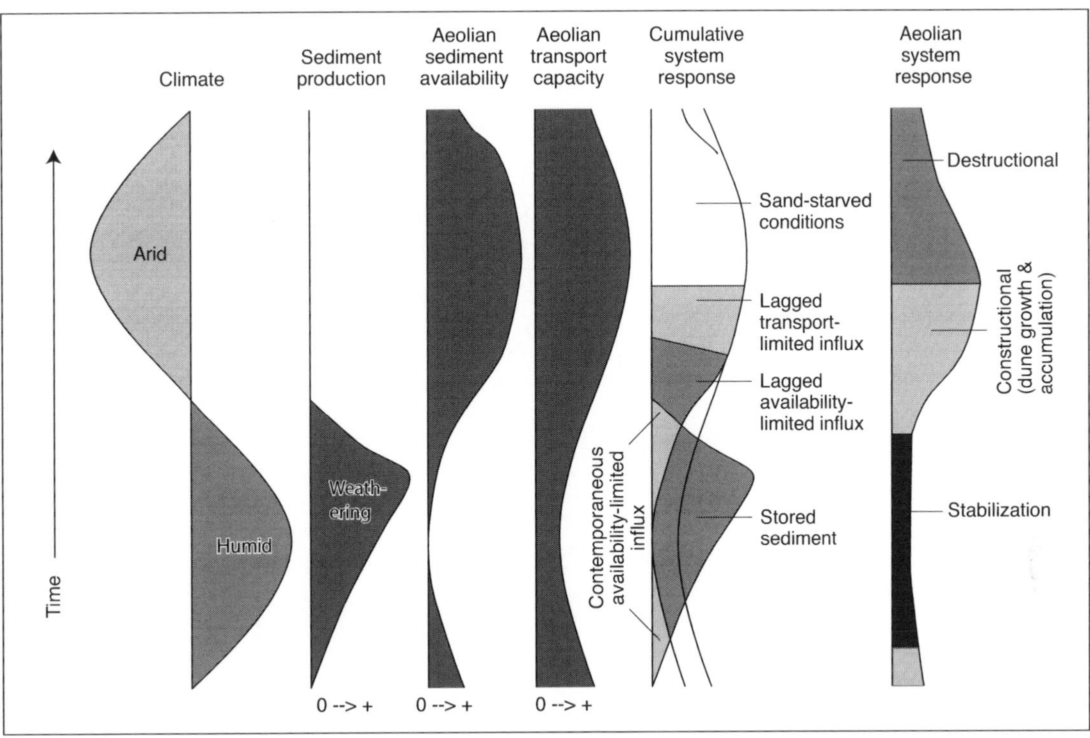

Figure 17.5 Theoretical model of aeolian system response to climatic changes (after Kocurek, 1998). The model aims to show how sediment production, availability and aeolian transport interact to leave (or not) a sedimentary record of past conditions in sand seas.

and Haynes, 2001) and occurs through the movement of extremely low-amplitude giant ripples (500–1000 m wavelength; see Breed, McCauley and Davis, 1987) or even higher amplitude 'chevrons' (Maxwell and Haynes, 1989). This may appear counterintuitive to earlier statements that sand sheets do not possess bedforms; however, these transport features are not perceptible on the ground and are only visible via remote sensing (Maxwell and Haynes, 2001). When vegetation is present, it is not so dense as to significantly inhibit aeolian sand transport.

Mobile barchan dunes represent bulk transport rates of up to 3.49 m^3/m width per year. Substantial quantities of sand may be transported huge distances across sand sheets before accumulating into dunes, as in the case of the Makteir erg in Mauritania (Fryberger and Ahlbrandt, 1979). Sand transport pathways may be sinuous, showing sensitivity to local topography and being analogous to 'rivers of sand' (Zimbelman, Williams and Tchakerian, 1994).

Dune sediments can reach sand seas along dominant non-aeolian pathways. In the Taklamakan Desert, China (Zhu, Zou and Yang, 1987), central Australia (Twidale, 1972; Nanson *et al.*, 2008) and other areas, so-called 'source bordering dunes' have been deflated from dry river valleys or seasonally inundated channel systems. Wadi sediments may be particularly important local aeolian forms as they dry out through drainage rather than evaporation and are therefore uncemented (Glennie, 1970). It is in fact likely that alluvial sources have made direct or indirect contributions to the sediments in parts of many sand seas, especially in low relief situations (Cooke, Warren and Goudie, 1993).

Namib Desert dune sands have arrived at their destination through the combined effects of river, sea and wind transport (Lancaster and Ollier, 1983). In the Wahiba Sands of Oman, hyperaridity has precluded fluvial contributions to Holocene sand sea accumulation (Warren, 1988). Instead, sediment supply is attributed to coastal erosion of soft sandy materials (Figure 17.6). The extensive dune systems of the Kalahari represent periods of aeolian reworking of pre-existing continental basin sediments, which have accumulated by various means since the Mid-Jurassic (Thomas, 1987). These have been supplemented in the southwestern Kalahari by aeolian inputs operating along thin sand sheets (Fryberger and Ahlbrandt, 1979) (Figure 17.7). Such *intraergal* and *extraergal* sands can sometimes be distinguished by grain size and mineralogical differences (see Bowler and

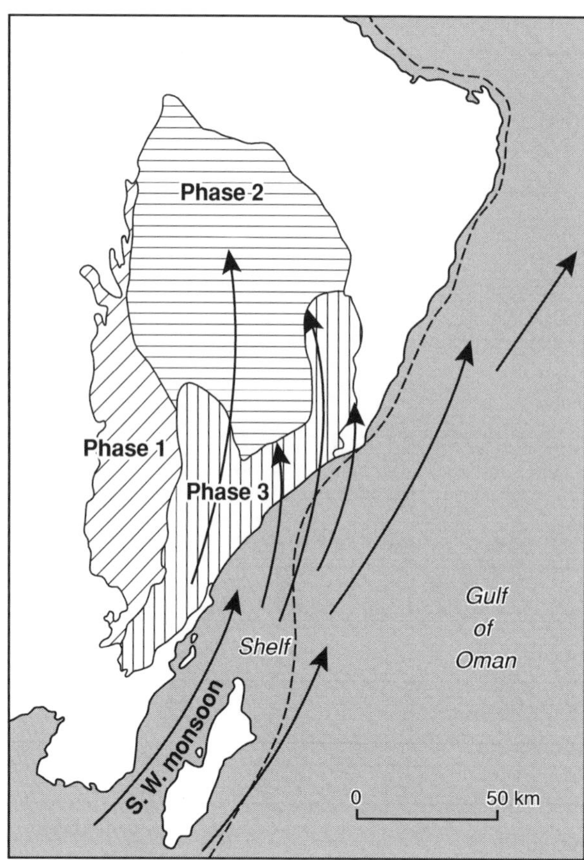

Figure 17.6 Phases of deposition in the Wahiba Sands, Oman. Sands have been derived by the inland transport of coastal carbonate sands in at least three phases; in the third, Holocene, phase, cemented cliff sands have been deflated and transported inland (redrawn after Cooke, Warren and Goudie, 1993).

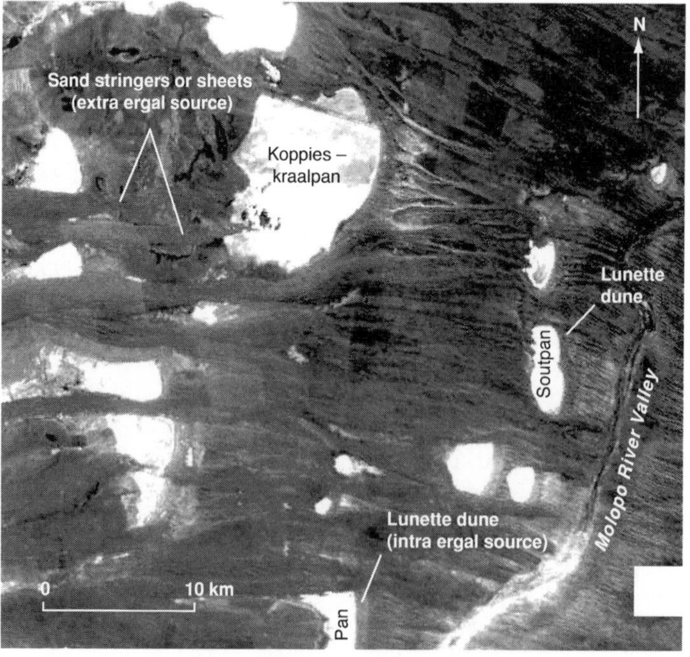

Figure 17.7 Space shuttle hand-held camera photograph of part of the southwestern Kalahari dunefield, southern Africa. Linear dunes dominate this sand sea, but larger, more amorphous sand sheets or stringers, also trending west–east, can also be identified. These may represent paths of extraergal sand inputs. Linear dunes also extend downwind from local sand sources, notably pan-margin lunette dunes and dry river valleys.

Wasson, 1984). Finally, it can be noted that for many sand seas there is evidence for multiple phases of accumulation (e.g. Figure 17.5). One reason for hiatuses in accumulation is climate change. It has been suggested that during the last 30 000 years active accumulation of sediment in sand seas may have been restricted to less than a third of the time (Lancaster, 1990), with episodicity of accumulation even affecting sand seas in the driest regions (Haney and Grolier, 1991), because even with aridity present, sufficient wind energy is required for sand transport and deposition. Other reasons for disruptions in accumulation are tectonic activity and, in coastal situations, base level changes. The overall duration of sand sea accumulation may possibly be established by examining the mineralogical maturity of the materials they comprise (Muhs, 2004), since a dominance of quartz suggests long-establishment or sand seas that they have experienced long periods of stability, as other weaker minerals have been removed by weathering (Muhs, 2004).

17.3.3 Sandflow conditions and sand sea development

It has been noted that sand sea development requires both favourable sediment budgets and wind conditions. Where net sand transport rates are high, throughflow exceeds deposition and bedform development is limited to sheets or stringers with ripples or highly mobile dune types such as barchans (Wilson, 1971). In such throughflow-dominated locations, sandflow is usually unsaturated and ground sand cover incomplete, because unsaturated throughflow has the potential to erode pre-existing deposits (Lancaster, 1999). If surface conditions are favourable, however, isolated bedforms can develop so long as the sand received by the bedform is equal to that which is lost downwind. Wilson (1971) termed this situation *metasaturated flow*. Sand seas proper accrue where sandflow is saturated and accumulation exceeds net transport, with consequential bedform development and the vertical or lateral accumulation of sand.

Wilson (1971) provided theoretical models of erg development in terms of wind regime and sandflow variations, demonstrating how sand seas should grow in locations of convergent windflow and wind speed deceleration, e.g. in intermontane basins (Figure 17.8). An empirical study (Fryberger and Ahlbrandt, 1979) has supported Wilson's major assertions and identified the synoptic and topographical conditions favouring sand sea development (Figure 17.8(b)).

Fryberger (1979) classified sand seas into low-, intermediate- and high-energy environments. Most sand seas experience less than 27 m^3/m width per year potential sandflow (Table 17.4); however, estimating sandflow rates over large areas using limited meteorological data can be problematic. Many low-energy sand seas occur near the centre of semi-permanent high- and low-pressure cells, while the high-energy sand seas fall in the Trade Wind zones near the margins of these pressure systems. The transference of sand from high- to low-energy locations within sand seas has been demonstrated by studies in the Jafurah erg, Saudi Arabia (Fryberger *et al.*, 1984), the Namib Desert (Lancaster, 1985) and the Gran Desierto, Mexico (Lancaster, Greeley and Christensen, 1987). Clearly, therefore, the general classification of wind environments by Fryberger and Ahlbrandt (1979) masks considerable intraregional variations in sandflow potential. This may account for the difference between their classification of the Simpson Desert (Table 17.3) and Ash and Wasson's (1983) assertion that the major limitation on sand movement there today is the low wind speeds.

17.3.4 Sand sheets

Sand sheets can develop in aeolian environments where conditions do not favour dune development, though they may exhibit low-relief aeolian features such as ripples and *zibar* (see Chapter 18). They can be small, local, features of a few square kilometres, often on the margins of dunefields (Kocurek, 1986) and associated with the previously referred-to movement of sand from high- to lower-energy locations. Sand sheets can, however, also represent major regional landscape components (Breed, McCauley, and Davis, 1987). For example, the Selima Sand Sheet of southern Egypt is a 'single utterly flat sheet of firm sand, known to cover an area of nearly 100 000 square miles' (Maxwell and Haynes, 2001, p.1623) (Figure 17.9).

Kocurek and Nielson (1986) recognised five major controls on aeolian sand sheet development. *Vegetation*, especially grasses, may encourage the accretion of low-angle laminae while limiting the construction of dunes. *Coarse sand*, which is not readily formed into dunes, can characterise sand sheets, sometimes upwind of a dunefield, where it remains as a surface lag deposit, as at the Algodone dunefield, California (Kocurek and Nielson, 1986). A coarse-sand sheet can therefore sometimes be regarded as a deflational remnant, rather like some desert pavement or *regs*. A *near-surface groundwater table* (Stokes, 1968) or 'Stokes surface' (Fryberger, Schenk and Krystinik, 1988) can be an important control on sand sheet development, acting as a base level to the action of wind scour. This can occur in locations deprived of upwind

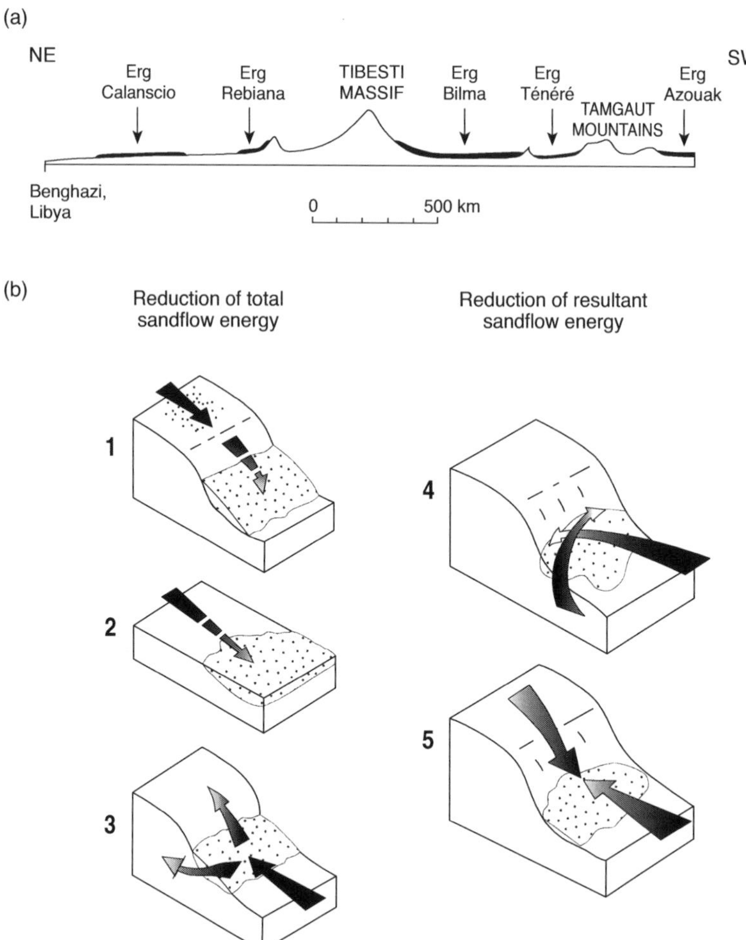

Figure 17.8 (a) The relationships of ergs in part of North Africa to topography (after Wilson, 1973). (b) Models of topographic influences on sand sea development (after Fryberger and Ahlbrandt, 1979). (1) In the shadow of a topographic barrier, (2) in shallow desert depressions and (3) by the direct blocking of wind, all leading to a total reduction in sand transporting energy. Resultant energy may be reduced (4) when surface winds are deflected, leading to sites of favourable accumulation, and (5) when katabatic winds off a highland oppose a dominant wind. Sand seas can also develop where water bodies interrupt regional sand flow patterns.

aeolian sand supply, or playa, sebkha or coastal sand sea situations (Fryberger, Schenk and Krystinik, 1988), where *periodic* or *seasonal* flooding can also prevent dune development. Langford (1989) identified a number of ways in which interactions between aeolian and fluvial processes could lead to different sedimentary deposits in areas with fluctuating groundwater tables. The formation of *surface crusts*, e.g. salcretes (see Chapter 6), or the growth of algal mats (Fryberger, Schenk and Krystinik,1988) can also inhibit deflation and contribute to the formation of sand sheets, as in the White Sands, New Mexico.

With the exception of the presence of coarse sand, the controls identified by Kocurek and Nielson (1986) are probably inapplicable to the development of the hyper-arid Selima Sand Sheet, because their effectiveness requires either the presence of some moisture or proximity to a dunefield. Rather, Breed, McCauley and Davis (1987) ascribe its development to the blanketing of an ancient fluvial landscape by aeolian redistribution of the sand fraction contained within widespread alluvium deposits. The absence of major topographic barriers able to inhibit long-distance wind transport accounts for the great size of this sand sheet, while the coarseness of the sand has probably caused the general lack of dune development and the formation of gently undulating subhorizontal tabular deposits 1–10 cm thick.

Table 17.4 Sand sea wind environments.

Sand sea (m^3/m width per year)	*Rate of sand drift*
1. *High-energy environments*	25–40
Northern Arabian sand seas	
Northwestern Libya	
2. *Intermediate-energy environments*	15–25
Simpson Desert, Australia	
Western Mauritania	
Peski Karakumy, Kazakhstan	
Peski Kyzylkum, Kazakhstan	
Erg Oriental, Algeria	
Erg Occidental, Algeria	
Namib Sand Sea, Namibia	
Rub'al Khali, Saudi Arabia	
3. *Low-energy environments*	<15
SW Kalahari, southern Africa	
Sahelian zone sand seas	
Gobi Desert, China	
Thar Desert, India	
Taklimakan Desert, China	

17.4 The global distribution of loess

Figure 17.2 shows the global distribution of loess deposits, which cover up to 10 % of the Earth's land surface (Pecsi, 1968). Despite a complex definitional history, where the role of water in loess deposition and the role of post-depositional processes has been debated, loess can principally be defined as a terrestrial accumulation of aeolian silt (dust) (e.g. Pye, 1987) or as a deposit, primarily aeolian in origin, that is principally comprised of silt-size sediment that is dominated by quartz. Fine sand and clay can also be present. Because of its fine-grained nature, silt is readily transported in suspension and is prone to long-distance transport (see Chapter 18). Much dust therefore ends up being deposited in the oceans (e.g. Hesse and McTainsh, 1999) and not on the land surface.

17.4.1 Loess production and distribution

In the mid-latitudes, loess deposits are strongly associated with the generation, transport and deposition of silt from glacial erosion systems during cold phases in the Quaternary (Smalley and Vita-Finzi, 1968; Smalley and Krinsley, 1978). This is not to say that conditions during deposition were not rather arid, since the continental location of many of these deposits, in central Europe, interior North and South America and Asia is in contexts that even today tend towards arid–semi-arid conditions or at least seasonal dryness – parameters that in many cases would have been enhanced at glacial times by enhanced continentality.

The extensive blanket loess deposits of Central Asia (particularly in the Karakum, in Tajikistan and in Uzbekistan) extend to over 200 m and in the Huangtu Gaoyuan, or Loess Plateau, of China deposits attain a thickness of up to 500 m (Huang, Pang and Zhao, 2000), although 150 m is more representative of consistent maximum depths (Derbyshire and Goudie, 1998; Goudie, 2002). In both regions these deposits date back to the Pliocene, and the loess sequences in fact comprise alternating loess and palaeosol units, demonstrating that periods of deposition (during cold phases) alternated with warmer periods of greater landscape stability. While these mid-latiutude loess deposits are commonly regarded as 'cold' loessic deposits, their distribution actually shows close affinity to the aeolian processes that have affected the desert areas that they border. In both regions, the loess deposits occur at the downwind end of a sequence of aeolian depositional units. Thus in China (Figure 17.10) the sandy desert of the Mu Us grades southeastwards to sandy loess and then to the extensive loess plateau. This in effect represents changing aeolian transport efficiency southeastwards away from sediment source areas.

17.4.2 Peridesert loess

In lower latitudes, dust deposits are less widespread but are clearly associated with desert conditions. It is to these deposits that the term 'peridesert loess' is more usually applied. These are the loess deposits comprised of material attributed to generation by abrasion during aeolian sand saltation (e.g. Smith, Wright and Whalley, 2002). Interestingly, most of the major desert dust sources today that contribute material to long-distance atmospheric transport are associated with deflation from dry lake basins and not directly from aeolian abrasion. However, the distinction between glacial and peridesert loess is not as clear-cut as sometimes presented, as implied above. Derbyshire (1983) demonstrated, both mineralogically and through mapping the limits of Quaternary glaciations in China, that the extensive Chinese loess deposits were not solely comprised of glacial erosion material, but of sediments derived from desert contexts. Indeed, several authors have now identified a range of regularly occurring dryland processes that can produce ample silt that is available for dust generation and ultimately the deposition of loess on desert margins. These include a range of weathering processes, aeolian abrasion, and abrasion and fracture during

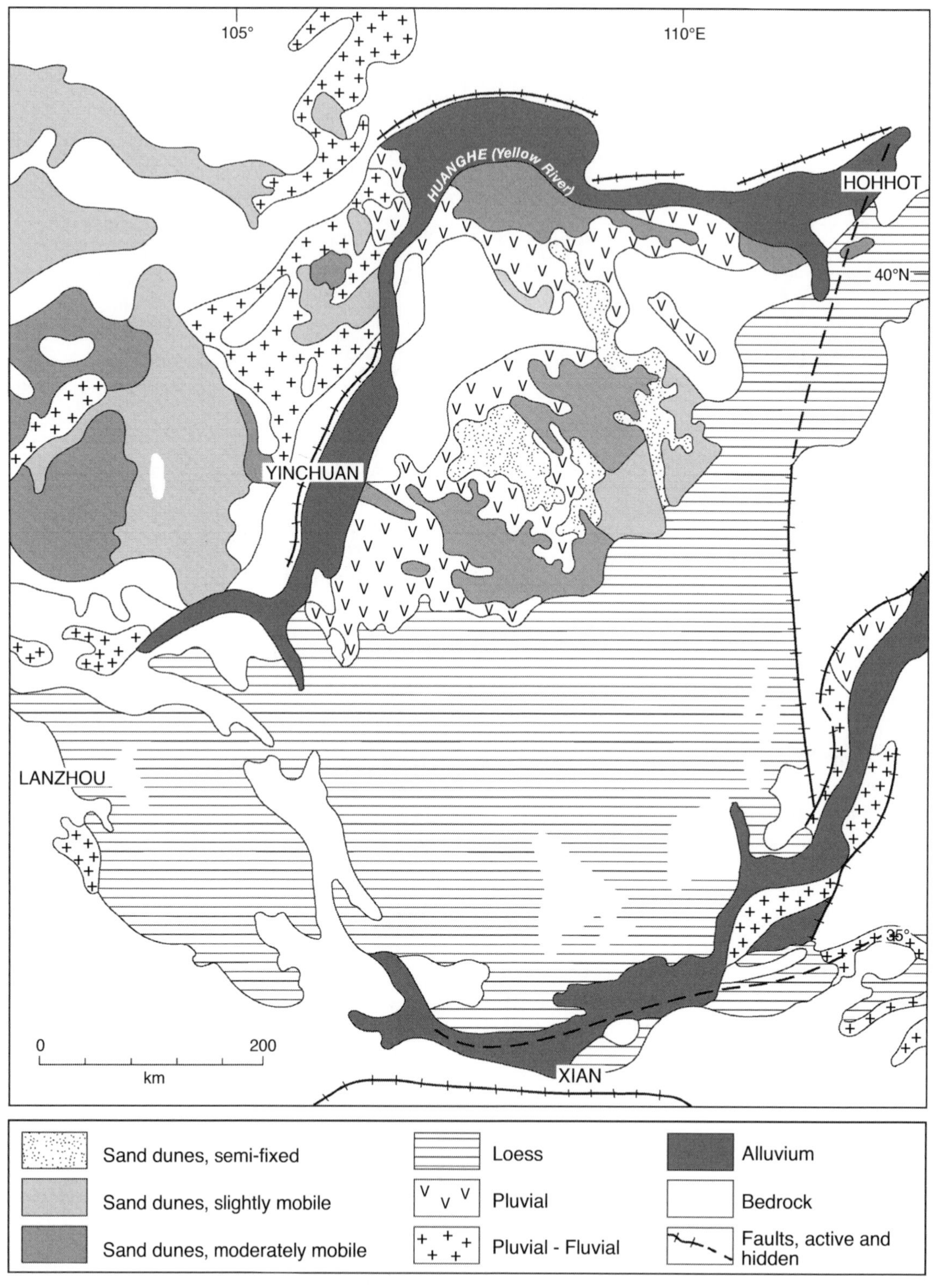

Figure 17.9 Quaternary sedimentary facies of the Ordos Platform and the Loess Plateau, China, showing the spatial relationship between sand and silt units (simplified from *Geomorphological Map of China and Its Adjacent Area* (scale 1:4M), attached to Chen, 1993).

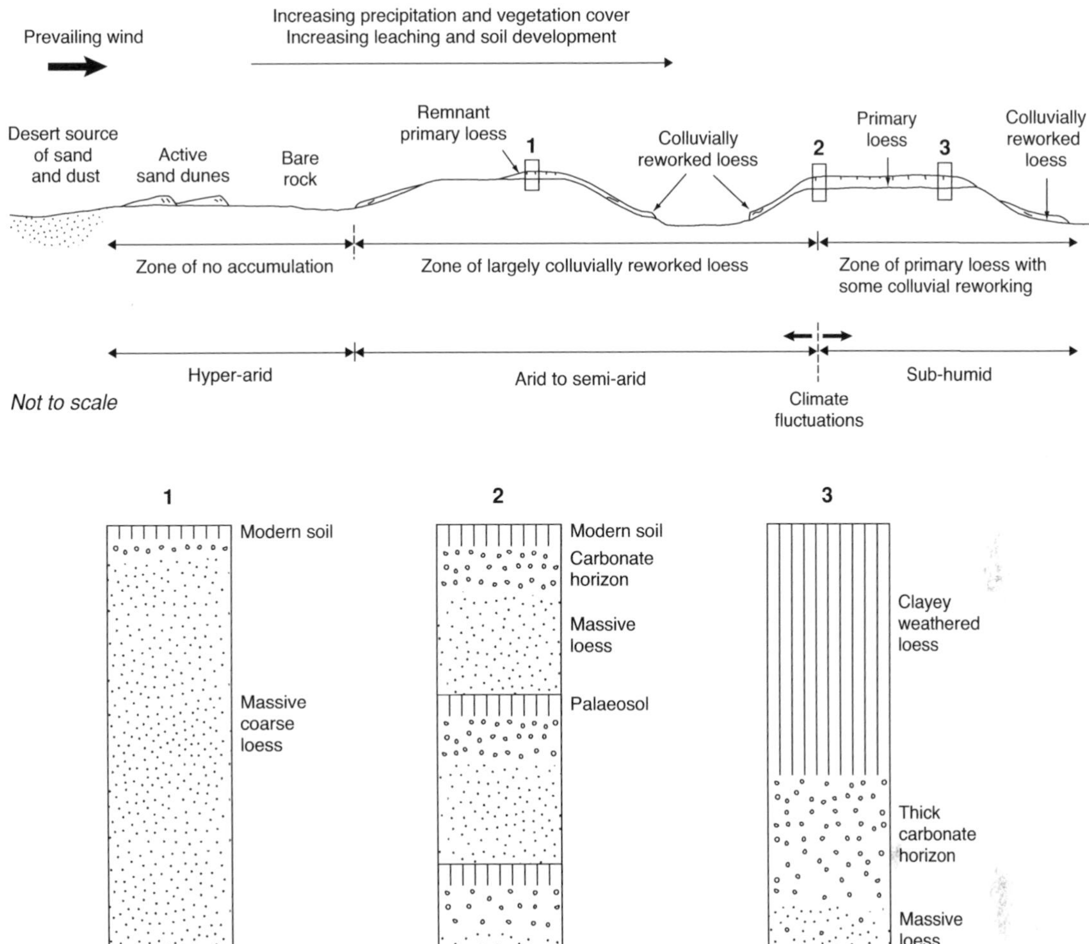

Figure 17.10 Schematic model of the distribution of peridesert loess in the landscape (modified after Pye and Sherwin, 1999).

episodic fluvial transport (McTainsh, 1987; Assallay *et al.*, 1998; Smith, Wright and Whalley, 2002).

The limited extent of low-latitude peridesert loess is now sometimes attributed to the lack of suitably vegetated dust-trapping surfaces in these areas; thus most of the dust escapes into the atmosphere and is available for wider global dispersal, including into oceans (Tsoar and Pye, 1987). It may also be due to a more prosaic reason: it has not been properly identified in the field (McTainsh, 1987; Hesse and McTainsh, 2003), partially because it is incorporated into soils as dust production and weathering tends to occur at similar rates, limiting opportunities for blanket deposits to develop (Vine, 1987).

Pye and Sherwin (1999) have produced a schematic model of the characteristics of peridesert loess (Figure 17.11). They distinguish nondepositional zones on desert margins where there is insufficient vegetation cover to trap dust: primary dust accumulation settings in vegetated zones of higher relief and, due to its susceptibility to reworking, areas of colluvially reworked loess on downslope locations.

17.5 Dynamic aeolian landscapes in the Quaternary period

Environmental change during the Quaternary period has contributed greatly to the nature of today's deserts and drylands (Chapter 3). Sarnthein's (1978) hypothesis of expanded aridity during glacial times (Figure 17.11) was based both on an analysis of aeolian sediments in marine cores and on identifying the extent of currently 'fixed' sand dunes on the margins of today's active sand seas. This built very much on identifying an association between

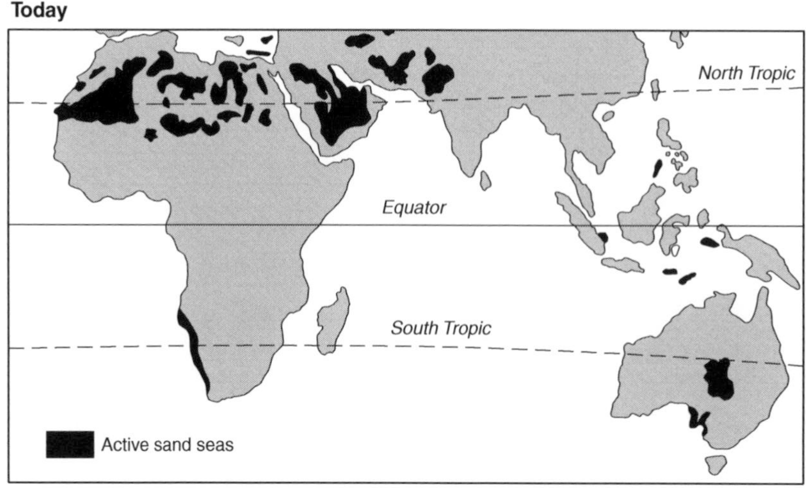

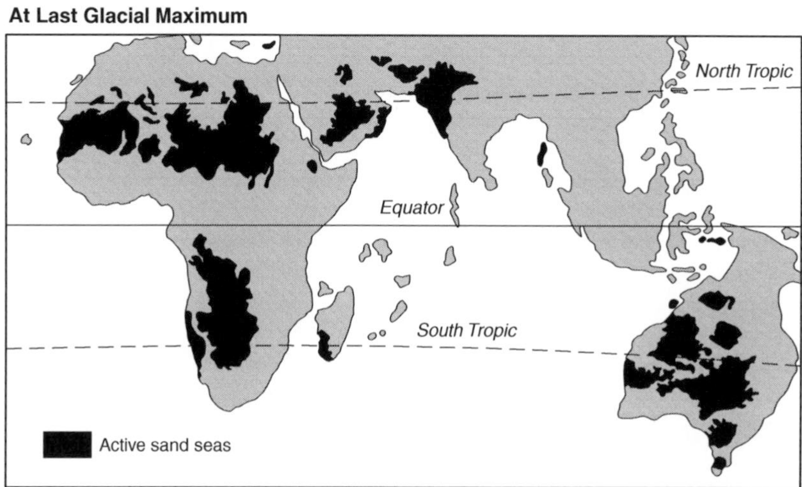

Figure 17.11 The distribution of active sand seas today and as suggested for the Last Glacial Maximum (after Sarnthein, 1978).

dune activity, rainfall and vegetation (e.g. Grove, 1958), with the role of wind energy now recognised as critically important too (e.g. Lancaster, 1987).

The identification of inactive or degraded sand dunes and palaeo-sand seas has been facilitated by the examination initially of aerial photography (Grove, 1958, 1969) and subsequently satellite imagery. In linear dunefields distinct vegetation zonation aids identification (Figure 17.13). Compared to interdune areas, the dune crests may be relatively sparsely vegetated, as in much of Australia (Madigan, 1936), or densely covered by trees because the soft sands favour the development of deep root systems, as in the northern Kalahari (Thomas, 1984). In areas where smaller dune types occur and their morphological patterns are less obvious, identification is still feasible using imagery because quartz-rich sand has a high reflectivity (Muhs, 1985).

The extent of ancient dunefields is well established in Africa India, Australia, North America and to a lesser extent in South America. Reviews of the fixed sand seas of the Kalahari have been provided by Thomas and Shaw (2002), for Australia by Wasson (1986) and for North America by Wells (1983). Goudie's (2002) review of desert environments includes summaries of sand sea histories from around the globe.

A range of criteria has been used to designate fixed or relict dune status (Table 17.5). In the Thar Desert (India), Makgadikgadi basin (Botswana) and Chad basin (Nigeria), dunes have been flooded and overlain by lacustrine deposits (Singh, 1971; Cooke, 1984; Grove, 1958). Other

Figure 17.12 Landsat image (5 August 1973) of part of southwestern Zambia, southeastern Angola and the Caprivi Strip, Namibia. Vegetation contrasts between the crests of fixed dune ridges and interdune areas allow the identification of ancient dune systems in the area covered by the left half of the image.

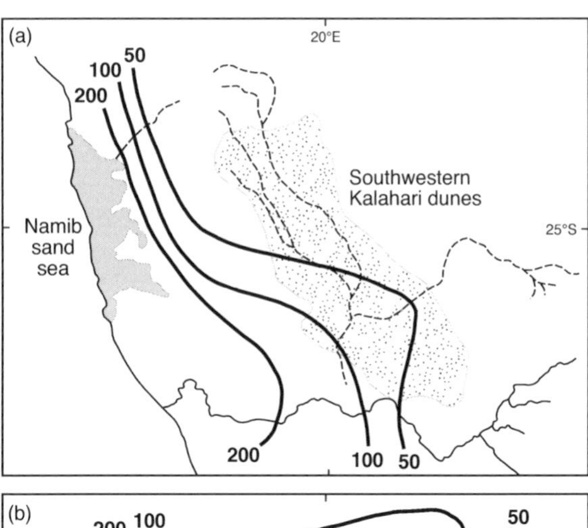

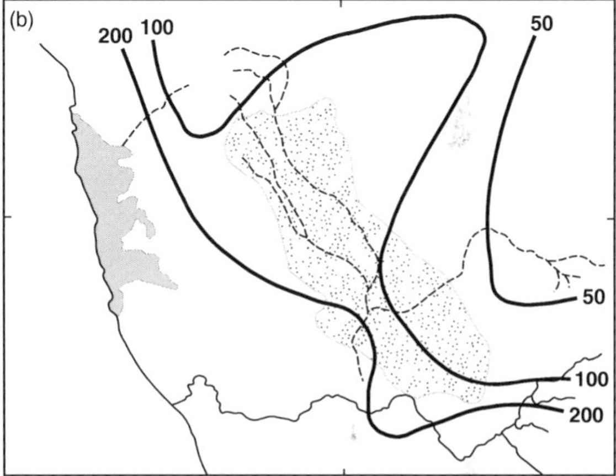

Figure 17.13 Suggested spatial variations in the distribution of dune mobility index M values between today and the last Glacial Maximum, in the southwestern Kalahari, according to Lancaster (1988). $M = W/(P/PE)$ where W is the % time wind exceeds the threshold velocity for sand transport and P/PE is the ratio of annual precipitation to annual potential evapotranspiration. Index values are interpreted in terms of $M > 200$ = dunefield fully active; 100–200 = dune plinths and interdunes vegetated; 50–100 = dune crests only active and <50 = dunes inactive.

dunes have been subjected to eluviation (Smith, 1963), pedogenesis and calcification, resulting from an increase in available moisture and reduction in aeolian activity (Allchin, Goudie, and Hegde, 1978). The nature of post-depositional dune degradation may, however, complicate the subsequent interpretation of the resultant forms, as well as the dating of phases of accumulation (Munyikwa, 2005; McFarlane *et al.*, 2005).

The presence of dune surface vegetation is perhaps the most widely used indicator of fixed dune status, but needs careful interpretation as an indicator of inactivity. This is especially so for sand seas composed of linear dunes, which are in any case relatively immobile forms compared to transverse forms (Thomas and Shaw, 1991a) and which may widely support an element of plant cover even under arid conditions (Tsoar and Moller, 1985). While there can be little dispute that dunefields covered by dense woodland are inactive, it is not so clear in the case of sparsely or partially vegetated dunes (Hesse and Simpson, 2006). In present arid environments, there is not a distinct threshold between aeolian activity and inactivity, but rather different levels of activity (Livingstone and Thomas, 1993; Mason *et al.*, 2008) and a gradual decline in sand movement in the direction of increasing vegetation density (Ash and Wasson, 1983; Wiggs *et al.*, 1995). Depending on wind velocities, significant sand transport may occur where up to 35 % of the ground surface is plant-covered – a situation that occurs in several dunefields, which have been designated fixed or inactive, e.g. the southwestern Kalahari and parts of the Australian and Indian sand seas. Some authors have concluded that present-day dune inactivity is not necessarily a function of increased effective precipitation but may be due to a decline in windiness (Ash and

Table 17.5 Criteria used in the literature to designate dune 'relict' status.

1. A cover of vegetation sufficient to inhibit sand transport
2. Dune form degraded (slope angles) compared to active dunes
3. Subject to pedogenesis
4. Presence of surface crusts/lithification
5. Dune slopes gullied
6. Internal structures destroyed by burrowing animals
7. Sediment size distribution bimodal due to post-formation dust inputs
8. Dunes overlain by other depositional landforms (including second generation dunes)
9. Drowned in valleys/lakes/off-shore

Wasson, 1983; Wasson, 1984; Bullard *et al.*, 1996). For example, as Williams (1985, p. 226) has noted:

> In the case of the source bordering dunes of Australia, Sudan and, presumably, of Zaire and Amazonia, the three prerequisities for dune formation seem to be a seasonally abundant supply of river sands, strong unidirectional seasonal winds, and a less dense tree cover than today. Such palaeo dunes do not connote extreme aridity.

After analysis of a range of variables, Hesse and Simpson (2006) concluded that over millennial timescales, it is vegetation cover that is the primary control on linear dunefield activity and on dune accumulation, rather than wind strength. A recent study by Mason *et al.* (2008) also suggested that wind strength is not a primary control of activity, though some authors attach primary importance to this parameter (e.g. Tsoar, 2005). In an attempt to resolve the issue of the relative roles of erodibility (vegetation cover) and erosivity (wind energy) in explaining dunefield inactivity and the level of climatic changes that have occurred since dunes were active, Talbot (1994), Wasson (1984), Lancaster (1988) and Chase and Brewer (2009) have all proposed the use of dune mobility indices (Figure 17.14). Similar indices may also be used with modelled climate data to predict sand sea responses to climatic changes in the future (Thomas, Knight and Wiggs, 2005; Wang *et al.*, 2009).

Relict dune systems have also been used to reconstruct palaeocirculation patterns as well as the former extent of aridity (Lancaster, 1981; Wells, 1983; Wasson, 1984; Thomas and Shaw, 1991b). The orientation of the relict dunes is compared to the resultant direction of modern sand-moving winds, calculated from wind data. Results have frequently suggested that circulation patterns differed from those of today when dune development occurred. Because dunes form in a wide variety of wind regimes today, often with considerable seasonal or diurnal directional variability (see Chapter 19), the conclusions that have been drawn are frequently speculative.

17.5.1 *Dating aeolian landscape change*

Major advances have occurred in understanding Quaternary period sand sea dynamics through the advent and application of thermoluminescence (TL) (Singhvi, Sharma and Agrawal, 1982) and subsequently optically stimulated luminecscence (OSL) (e.g. Stokes, 1992; Stokes, Thomas and Washington, 1997) dating to aeolian sediments (see Box 17.1). Prior to the advance of these dating methods, quartz-rich aeolian sands and silts were datable only if bracketed by other sediments to which other dating methods could be applied. For example, the alternating loess–palaeosol sequences in the Loess plateau of China have been dated by radiocarbon or uranium-series dating of the palaeosol units, of which over 30 occur, by palaeomagnetic studies (Liu and Ding, 1998) or by 'curve fitting' of sedimentological properties of the loess with isotope records from polar ice cores (Balsam, Elwood and Ji, 2005).

Box 17.1 Luminescence dating

The origins of luminescence dating lie more than four decades ago within the discipline of archaeological science and a need to determine the age of pottery. Since then it has undergone a rapid series of methodological and technological developments and has emerged in the last decade as an important and effective Quaternary environmental dating method. It is particular useful in arid zones where it is used to date directly the depositional age of sediments within landforms such as sand dunes, loess deposits and lake shorelines. The application of luminescence dating has enabled a much greater understanding of environmental change in situations where the absence of organic material (e.g. in former desert contexts) prevents the use of radiocarbon dating and also where sediments are older than the age limits of radiocarbon dating.

What is luminescence dating?

Within all sediments the presence of naturally occurring radioisotopes of uranium-238, thorium-232 and potassium-40 undergoing radioactive decay create a low-level background radiation field (alpha, beta and gamma radiation). Cosmic rays from the Sun also provide a significant contribution of radiation to near-surface sediments. The specific nature of quartz and feldspar minerals allows them to behave as dosimeters: recording the total sum of ionizing radiation that they have been exposed to since they were buried (see Preusser *et al.*, 2009, for a review). This is possible because structural defects and impurities within the minerals behave as traps in which unbound electrons (freed from their parent atom by energy derived from the radiation field of the sediment) will accumulate over time. The rate at which electrons are displaced depends on the strength of the level of radiation in the sediment during its burial history and is typically known as the 'dose rate'. These displaced 'trapped' electrons act as a store of potential energy that can be naturally, or artificially, dissipated in the form of photons (luminescence). Each time the minerals are exposed to sunlight during transport (e.g. the sediment is blown by the wind or eroded and washed downstream), the trapped electrons are able to obtain enough energy to be freed from this state and the dosimetric clock is reset ('bleached'). The grains may at some point be deposited within an indicative landform such as a river terrace or sand dune and again become shielded from sunlight. The accumulation of electrons in defect sites is time-dependent and it is this property that can be utilised to obtain burial ages. A simple analogy that is frequently used to explain this process is to imagine the quartz or feldspar grains as a bucket underneath a tap into which water is dripping at a constant rate: If we know the rate of dripping (analogous to the dose rate) and we can measure the total amount of water in the bucket (analogous to the total amount of trapped electrons), then it is possible to calculate how long the bucket has been located under the dripping tap. Figure 17.14 summarises the main principles in the accumulation of a measurable luminescence signal in a sediment grain.

How are luminescence ages produced?

To avoid mineral grains being exposed to light prior to measurement, sediment samples collected in the field must be extracted and stored in lightproof containers. In the laboratory, under controlled wavelength low-light conditions, the sediment undergoes a series of chemical treatments to isolate the quartz or feldspar component. The total population of trapped charge within the minerals can then be measured by stimulating the material with external energy in the form of light (optically stimulated luminescence: OSL) or heat (thermoluminescence: TL). As the trapped electrons are released, energy is dissipated in the form of luminescence. Specific measurement conditions are used that include pre-heating the sample and monitoring any change of sensitivity to laboratory stimulation during measurement in order to isolate the luminescence signal from only those traps that are thermally stable over the burial period. The natural luminescence signal of the quartz or feldspar can be measured using a photomultiplier tube (cf. Bøtter-Jensen *et al.*, 2003) and calibrated against the luminescence emitted when the same minerals are exposed to known laboratory doses of radiation. From this it is possible to establish the equivalent laboratory radiation dose (D_e) to that received during the burial period (the Palaeodose) (see Wintle and Murray, 2006, for a review of dating protocols).

To complete the age equation the rate of electron displacement, determined by the ionizing radiation of the burial environment (D'), must also be known. This can be estimated by either directly measuring the gamma radiation (in the field or laboratory) or by laboratory determination of the radioisotope concentration within the sediment (see Aitken, 1998).

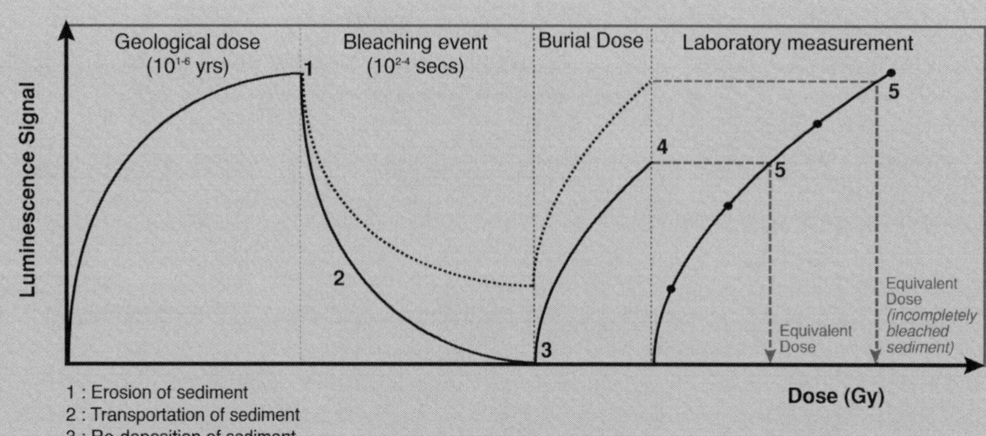

Figure 17.14 Principles of the accumulation of a luminescence signal in sediment and the effect of incomplete bleaching on determining an accurate estimate of the burial age (adapted from Bailey and Arnold, 2006). Quartz or feldspar grains deposited by the wind or another geomorphic process, once buried, begin to accumulate electrons at imperfections within the crystal structure. Over time, an accumulated dose (trapped charge population) builds up. If exposed to sunlight, these electrons are removed and the dosimetric clock is reset. Laboratory measurements of the trapped electron population allow an estimate of the radiation dose received during burial to be made. If the resetting process is incomplete (dashed line), a fraction of the geological dose will remain and the measured trapped charge population will equate to part of the geological dose plus the burial dose. Without recognition, this results in an overestimate of the sediment burial age (refer to text and Table 17.6 for details).

An age is therefore determined by

$$\text{Age (ka)} = \frac{D_e\,(\text{Gy})\,(\text{total absorbed radiation dose})}{D\,(\text{Gyka})\,(\text{radiation dose rate})}$$

Luminescence dating procedures are mineral- and site-specific and may be affected by a number of potential difficulties, some of which are detailed in Table 17.6.

Sallie L. Burrough

Luminescence dating of sand seas has shown that many have histories of multiple phases of accumulation. Where multiple generations of dune forms overlap within a sand sea, e.g. in Mauritania (Lancaster *et al.*, 2002, see Chapter 19, Box 19.1, for a detailed consideration of this dune system) and the eastern Sahara (Bubenzer and Bolten, 2008), ages can appear to suggest that each dune pattern represents a single phase of accumulation, when subtle differences in atmospheric circulation led to different pattern orientations. In other contexts, more complex relationships are emerging, where phases of accumulation are not linked to specific morphological expressions; i.e down-profile age changes occur within dunes, sometimes separated by weak palaeosols (e.g. Fitzsimmons *et al.*, 2007), which identify periods of dune accumulation at 73–66, 35–32, 22–18 and 14–10 ka in the Strzlecki Desert, Central Australia.

In other sand seas, as the number of direct age determinations from dune sand has increased, so the interpretation of sand sea development has become more complex. Initial accumulation phase histories for the Kalahari were based on a relatively limited number of OSL ages

Table 17.6 Causes and consequences of luminescence dating challenges.

Potential difficulty	*Physical explanation*	*Environmental context*	*Consequence*	*Possible solution*
Signal saturation	Geologically stable electron traps fill up	Sample is buried for a very long period of time or the dose rate is particularly high	Possible to obtain a minimum age	Application of new techniques such as isothermal TL may extend the datable burial age limit of some sediments
Incomplete bleaching	Some electrons remain trapped even after a bleaching event	Sediment is insufficiently exposed to light during transport	Residual dose remains and D_e is overestimated. Differential exposure of grains results in a wide distribution of D_e values	Signal distribution analysis allows identification of the problem. In most cases single grain dating and the use of age models enables the correct burial age to be determined
Post-depositional mixing		Biotic activity within the sediment mixes grains from chronologically distinct depositional events	Wide (and often skewed) D_e distributions	The use of single grain dating, mixture modelling and Bayesian statistics may enable the identification of depositional events, even when mixed
Signal instability	Electrons 'leak' from traps. Some traps (responsible for a part of the luminescence signal) are more prone to this happening than others	Dependent on specific properties of the quartz or feldspar grains	Incorrect age	A stable part of the signal can be isolated and measured using a number of techniques
Dose rate change over time	The level of ionising radiation is not constant over time	Disequilibrium may occur between the parent and daughter isotopes where one or the other is removed from the system. Ingrowth of duricrete material during burial will also change the dose rate	Incorrect age	Disequilibrium can be measured and corrected for in young sediments. In older sediments, equilibrium may have been reestablished and past disequilibrium will remain undetected
Beta heterogeneity	The level of ionising radiation is not spatially constant within the sample	Hot spots and buffer zones may occur within the sediment matrix, causing some grains to receive disproportionately more radiative dose during the burial history than others	Wide (and often skewed) D_e distributions. Difficult to differentiate from bioturbation	Using particular age models, a good estimate of burial age can be obtained. Errors will be larger

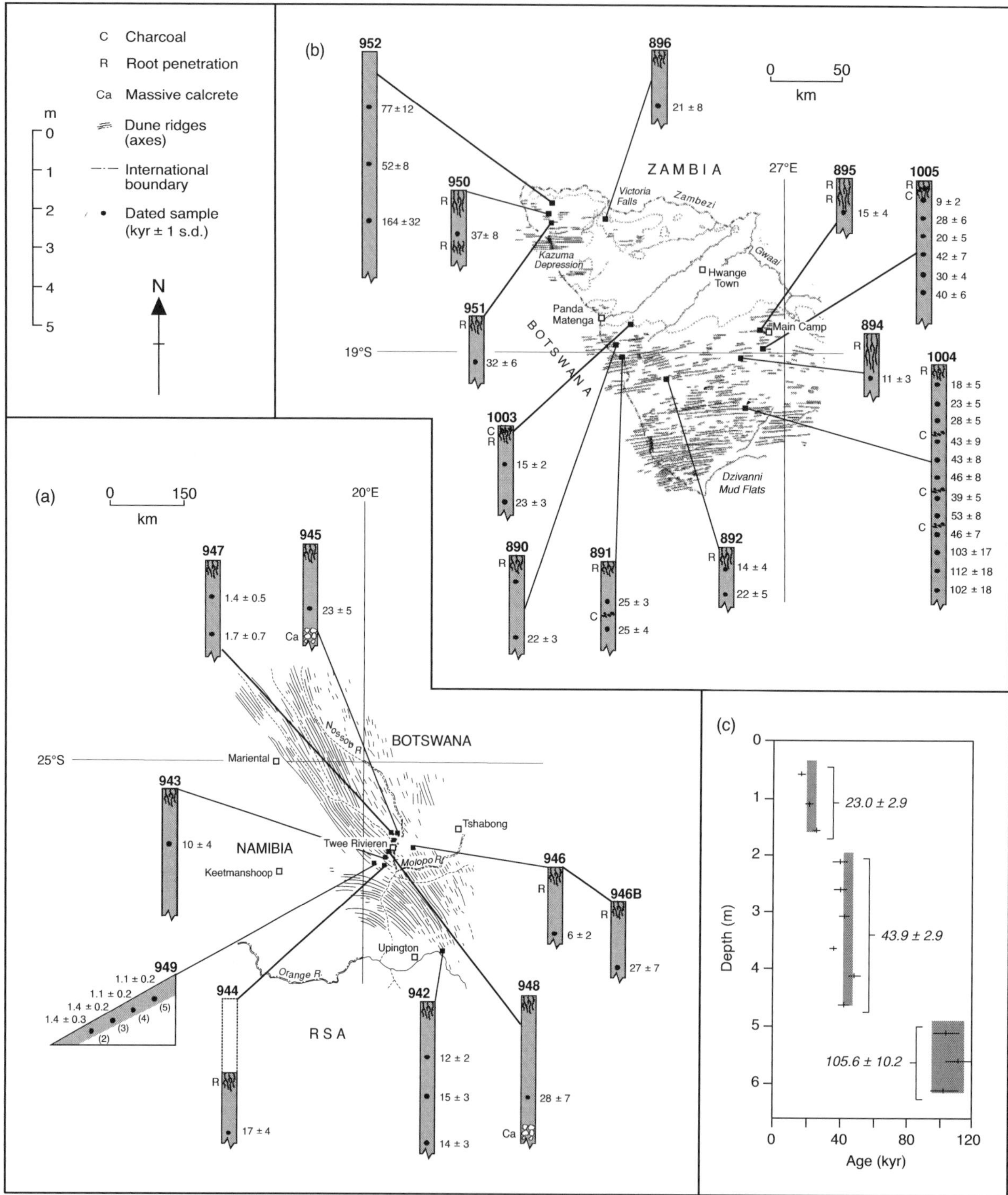

Figure 17.15 Chronology of linear dune deposition in the Kalahari Desert, based on OSL dating of sediment samples. Ages were interpreted as demonstrating discrete periods of dune deposition during the Late Quaternary (after Stokes, Thomas and Washington, 1997).

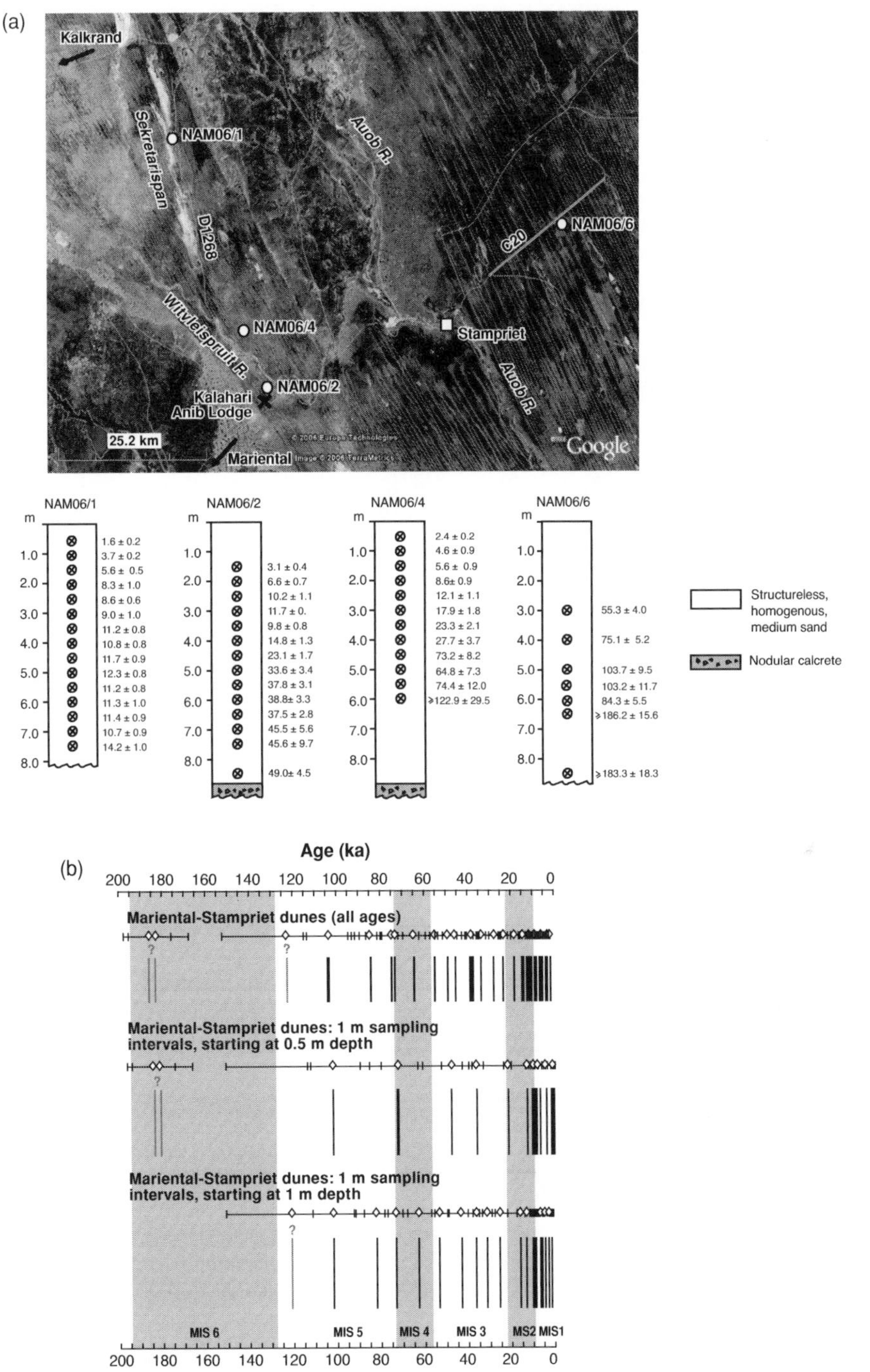

Figure 17.16 (a) High vertical intensity OSL dating of samples from sediment cores from linear dunes in the western Kalahari suggest a record of almost continuous deposition in the Late Quaternary, in contrast to Figure 17.15, which was based on fewer samples. (b) Demonstrates how the number of samples dated can potentially impact on the interpretation of the periodicity of dune depositional history (after Stone and Thomas, 2008).

from samples collected from shallow pits in dunes, which gave a relatively simple picture of sand sea history, dune accumulation and inferred aridity (Stokes, Thomas and Washington, 1997) (Figure 17.16).

As more age determinations have been achieved, pictures that are complex to interpret emerge. These include spatially variable accumulation histories along short distances of individual dunes (Telfer and Thomas, 2006), but also that, as more ages are added to data sets and sampling strategies become more intense, apparent discrete phases of accumulation can disappear and also apparent continuous deposition can be represented by full sets of age determinations (Stone and Thomas, 2008) (Figure 17.16). While a number of factors could contribute to such a situation (e.g. the error range on ages, possible post-depositional sediment mixing, sampling issues, the complex suite of factors that affect dune activity and sediment preservation), this demonstrates a number of important factors in elucidating sand sea histories. First, simple relationships between dunefield accumulation and broad global climate trends *senso* Sarnthein (1978) have not occurred. Second, variability in depositional ages can be marked between dunes within an individual sand sea, because net accumulation and preservation does not appear to occur simultaneously at all points within a dunefield. Third, care needs to be exerted in terms of how chronologies are produced and how many ages are used to produce conclusions for individual sand seas: the dunefield, not the individual dune, is the natural unit from which to gain an appropriate picture of a system history (Stone and Thomas, 2008).

Luminescence dating is also shedding new light on the records of past climate change preserved within loess bodies. Independent direct dating of loess has challenged previous chronologies based on magnetic susceptibly or record correlation (e.g. Singhvi *et al.*, 2001; Stevens *et al.*, 2007). Contrary to views that loess deposits provide a spatially uniform record of deposition over large areas, these studies show that accumulation histories can be as complex and variable as those derived from within sand seas.

17.6 Conclusions

Aeolian deposits are a major – though not as extensive as widely assumed – component of dryland environments. Determining where they are, and why they are there, requires the application of scientific approaches that elucidate the timing of deposition, the controls on deposition and the processes responsible for sediment mobilisation, transport and settlement. Gaining this knowledge requires investigations as a range of spatial and temporal scales (Thomas and Wiggs, 2008), and in this chapter we have primarily considered what controls the nature and character of the landscape features of aeolian systems: sand seas and loess deposits. These spatially extensive features provide the context in which the operation of aeolian processes and the generation of bedforms play out, and these process- and landform-focused dimensions of aeolian systems are examined in detail in the next chapters.

References

Aitken, M.J. (1994) Optical dating: a non-specialist review. *Quaternary Geochronology*, **13**, 503–508.

Aitken, M.J. (1998) *An Introduction to Optical Dating: The Dating of Quaternary Sediments by the Use of Photo-Stimulated Luminescence*, Oxford University Press, Oxford, 267 pp.

Al Janabi, K.Z., Jaward, A. Al-Taie, F.H. and Jack, F.J. (1988) Origin and nature of sand dunes in the alluvial plain of southern Iraq. *Journal of Arid Environments*, **14**, 27–34.

Allchin, B., Goudie, A.S. and Hegde, K.T.M. (1978) *The Prehistory and Palaeogeography of the Great Indian Desert*, Academic Press, London.

Andreotti, B., Fourriere, A., Ould-Kaddopur, F. *et al.* (2009) Giant aeolian dune size determined by the average depth of the atmospheric boundary layer. *Nature*, **457**, 1120–1123.

Arens, S.M. (1996) Rates of aeolian transport on a beach in a temperate humid climate. *Geomorphology*, **17**, 3–18.

Ash, J.E. and Wasson, R.J. (1983) Vegetation and sand mobility in the Australian desert dunefield. *Zeitschrift für Geomorphologie*, Supplementband **45**, 7–25.

Assallay, A.M., Roigers, C.D.F., Smalley, I.J. and Jefferson, I. (1998) Silt: 2–64μm, 9–4ø. *Earth Science Review*, **4**, 61–88.

Baas, A.C.W. (2007) Complex systems in aeolian geomorphology. *Geomorphology*, **91**, 311–331.

Baas, A.C.W. and Nield, J.M. (2007) Modelling vegetated dune landscapes. *Geophysical Research Letters*, **34**, L06405. DOI:10.1029/2006GL029152.

Bagnold, R.A. (1941) *The Physics of Blown Sand and Desert Dunes*, Methuen, London.

Bailey, R.M. and Arnold, L.J. (2006) Statistical modelling of single grain quartz of distributions and an assessment of procedures for estimating burial dose. *Quaternary Science Reviews*, **25**, 2475–2502.

Balsam, W., Ellwood, B. and Ji, J. (2005) Direct correlation of the marine oxygen isotope record with the Chinese Loess Plateau iron oxide and magnetic susceptibility records. *Palaeogeography, Palaeoclimatology, Palaeoecology*, **221**, 141–152.

Besler, H. (1980) Die Dunen-Namib: Entstehung und Dynamik eines Ergs. *Stuttgarter Geographische Studien*, **96**.

Bøtter-Jensen, L., Andersen, C.E., Duller, G.A.T. and Murray, A.S. (2003) Developments in radiation, stimulation and observation facilities in luminescence measurements. *Radiation Measurements*, **37**, 535–541.

Bowler, J.M. and Wasson, R.J. (1984) Glacial age environments of inland Australia, in *Late Cainozoic Palaeoclimates of the Southern Hemisphere* (ed. J.C. Vogel), Balkema, Rotterdam, pp. 183–208.

Bowler, J.M., Hope, G.S., Jennings, J.N. *et al.* (1976) Late Quaternary climates of Australia and New Guinea. *Quaternary Research*, **6**, 359–394.

Brackenrige, G.R. (1978) Evidence for a cold, dry full-glacial climate in the American southwest. *Quaternary Research*, **9**, 22–40.

Bradley, R.S. (1985) *Quaternary Palaeoclimatology*, Allen & Unwin, Boston.

Breed, C.S. and Grow, T. (1979) *Morphology and distribution of dunes in sand seas observed by remote sensing*, US Geological Survey, Professional Paper 1052, pp. 253–302.

Breed, C.S., McCauley, J.F. and Davis, P.A. (1987) Sand sheets of eastern Sahara and ripple blankets on Mars, in *Desert Sediments Ancient and Modern* (eds L. Frostick and I. Reid), Geological Society Special Publication 35, London, pp. 337–359.

Breed, C.S., Fryberger, S.G., Andrews, S. *et al.* (1979) *Regional studies of sand seas using Landsat (ERTS) imagery*, US Geological Survey, Professional Paper 1052, pp. 305–397.

Bubenzer, O. and Bolten, A. (2008) The use of new elevation data (SRTM/ASTER) for the detection and morphometric quantification of Pleistocene megadunes (draa) in the eastern Sahara and the southern Namib. *Geomorphology*, **102**, 221–231.

Bullard, J.E., Thomas, D.S.G., Livingstone, I.P. and Wiggs, G.F.S. (1996) Wind energy variations in the southwestern Kalahari Desert and implications for linear dunefield activity. *Geomorphology*, **11**, 189–203.

Chase, B.M. and Brewer, S. (2009) Last Glacial Maximum dune activity in the Kalahari Desert of southern Africa: observations and simulations. *Quaternary Science Reviews*, **28**, 301–307.

Chen, Z. (ed.) (1993) *An Outline of China's Geomorphology*, China Cartographic Publishing House, Beijing, 133 pp.

Cooke, H.J. (1984) The evidence from northern Botswana of Late Quaternary climatic change, in *Late Cainozoic Palaeoclimates of the Southern Hemisphere* (ed. J.C. Vogel), Balkema, Rotterdam, pp. 265–278.

Cooke, R.U., Warren, A. and Goudie, A.S. (1993) *Desert Geomorphology*, UCL Press, London.

Cornish, V. (1914) *Waves of Sand and Snow*, Fisher-Unwin, London.

Crouvi, O., Amit, R., Enzel, Y. *et al.* (2008) Sand dunes as a major proximal dust source for Late Pleistocene loess in the Negev Desert, Israel. *Quaternary Research*, **70**, 275–282.

Dare-Edwards, A.J. (1984) Aeolian clay deposits of south-eastern Australia: parna or loessic clay? *Institute of British Geographers, Transactions*, **9**, 337–344.

Derbyshire, E.D. (1983) On the morphology, sediments and origin of the loess plateau of China, in *Megageomorphology* (eds R. Gardner and H. Scoging), Oxford University Press, Oxford, pp. 172–194.

Derbyshire, E.D. and Goudie, A.S. (1998) Asia, in *Arid Zone Geomorphology*, 2nd edn (ed. D.S.G. Thomas), John Wiley & Sons, Ltd, Chichester, pp. 487–506.

Derbyshire, E.D., Dijkstra, T.A., Billard, A. *et al.* (1993) Thresholds in a sensitive landscape: the loess region of central China, in *Landscape Sensitivity and Change* (eds D.S.G. Thomas and R.J. Allison), John Wiley & Sons, Ltd, Chichester, pp. 97–127.

Dong, Z., Wang, T and Wang, X. (2004) Geomorphology of the megadunes in the Badain Jaran Desert. *Geomorphology*, **60**, 191–203.

Dong, Z., Qjan, G., Luo, W. *et al.* (2009) Geomorphological hierarchies for complex mega-dunes and their implications for mega-dune evolution in the Badain Jaran Desert. *Geomorphology*, **106**, 180–185.

Ewing, R.C., Kocurek, G. and Lake, L.W. (2006) Pattern analysis of dune-field parameters. *Earth Surface Processes and Landforms*, **31**, 1176–1191.

Fitzsimmons, K.E., Rhodes, E.J., Magee, J.W. and Barrows, T.T. (2007) The timing of linear dune activity in the Strzelecki and Tirari Deserts, Australia. *Quaternary Science Reviews*, **26** (19–21), 2598–2616.

Fryberger, S.G. (1979) *Dune form and wind regime*, US Geological Survey, Professional Paper 1052, pp. 137–169.

Fryberger, S.G. and Ahlbrandt, T.S. (1979) Mechanism for the formation of aeolian sand seas. *Zeitschrift Für Geomorphologie*, **NF**, **23**, 440–460.

Fryberger, S.G., Ahlbrandt, T.S. and Andrews, S. (1979) Origin, sedimentary features and significance of low-angle eolian 'sand sheet' deposits, Great Sand Dunes, National Monument and vicinity, Colorado. *Journal of Sedimentary Petrology*, **49**, 733–746.

Fryberger, S.G., Schenk, C.J. and Krystinik, L.F. (1988) Stokes surfaces and the effects of near surface groundwater-table on aeolian deposition. *Sedimentology*, **35**, 21–41.

Fryberger, S.G., Al-Sari, A.M., Clisham, T.J. *et al.* (1984) Wind sedimentation in the Jafurah Sand Sea, Saudi Arabia. *Sedimentology*, **31**, 413–431.

Glennie, K.W. (1970) *Desert Sedimentary Environments*, Elsevier, Amsterdam.

Goudie, A.S. (2002) *Great Warm Deserts of the World*, Oxford University Press, Oxford.

Grove, A.T. (1958) The ancient ergs of Hausaland, and similar formations on the south side of the Sahara. *Geographical Journal*, **124**, 528–533.

Grove, A.T. (1969) Landforms and climatic change in the Kalahari and Ngamiland. *Geographical Journal*, **135**, 191–212.

Grove, A.T. and Warren, A. (1968) Quaternary landforms and climate on the south side of the Sahara. *Geographical Journal*, **134**, 194–208.

Haney, E.M. and Grolier, M.J. (1991) Geologic map of major Quaternary aeolian features, northern and central coastal Peru, US Geological Survey Miscellaneous Investigation I-2162.

Hesse, P.P. and McTainsh, G.H. (1999) Last glacial maximum to early Holocene wind strength in the mid-latitudes of the Southern Hemisphere from aeolian dust in the Tasman Sea. *Quaternary Research*, **52**, 343–334.

Hesse, P.P. and McTainsh, G.H. (2003) Australian dust deposits: modern processes and the Quaternary record. *Quaternary Science Reviews*, **22**, 2007–2035.

Hesse, P.P. and Simpson, R.L. (2006) Variable vegetation cover and episodic sand movement on longitudinal desert sand dunes. *Geomorphology*, **81**, 276–291.

Huang, C.C., Pang, J. and Zhao, J. (2000) Chinese loess and the evolution of the East Asian monsoon. *Progress in Physical Geography*, **2**, 75–96.

Jennings, J.N. (1968) A revised map of the desert dunes of Australia. *Australian Geographer*, **10**, 408–409.

Kocurek, G. (1986) Origins of low-angle stratification in aeolian deposits, in *Aeolian Geomorphology. The Binghampton Symposia in Geomorphology, International Series 17* (ed. W.G. Nickling), Allen and Unwin, Boston, pp. 177–193.

Kocurek, G. (1998) Aeolian system response to external forcing factors – a sequence stratigraphic view of the Saharan region, in *Quaternary Deserts and Climatic Change* (eds A. Alsharan, K. Glennie, G. Whittle and C. Kendall), Balkema, Rotterdam, pp. 327–337.

Kocurek, G. (1999) The aeolian rock record, in *Aeolian Environments, Landforms and Sediments* (eds A.S. Goudie, I. Livingstone and S. Stokes), John Wiley & Sons, Ltd, Chichester, pp. 239–259.

Kocurek, G. and Dott, R.H. (1981) Distinctions and uses of stratification types in the interpretation of aeolian sand. *Journal of Sedimentary Petrology*, **51**, 579–595.

Kocurek, G. and Havholm, K.G. (1993) Eolian sequence stratigraphy – a conceptual framework, in *Siliclastic Sequence Stratigraphy* (eds P. Weiner and H. Posamentier), American Association of Petroleum Geologists, Tulsa, pp. 393–409.

Kocurek, G. and Nielson, J. (1986) Conditions favourable for the formation of warm-climate eolian sand sheets. *Sedimentology*, **33**, 795–816.

Kohfeld, K.E. and Harrison, S.P. (2001) DIRTMAP: the geologic record of dust. *Earth Science Reviews*, **54**, 81–114.

Lancaster N. (1987) Formation and reactivation of dunes in the south-western Kalahari: palaeoclimatic implications. *Palaeoecology of Africa*, **18**, 103–110.

Lancaster, N. (1981) Palaeoenvironmental implications of fixed dune systems in southern Africa. *Palaeogeography, Palaeoclimatology, Palaeoecology*, **33**, 327–346.

Lancaster, N. (1985) Wind and sand movements in the Namib Sand Sea. *Earth Surface Processes and Landforms*, **10**, 607–619.

Lancaster, N. (1988) Development of linear dunes in the southwestern Kalahari. *Journal of Arid Environments*, **14**, 233–244.

Lancaster, N. (1990) Palaeoclimatic evidence from sand seas. *Palaeogeography, Palaeoecology, Palaeoclimatology*, **76**, 279–290.

Lancaster, N. (1995) *The Geomorphology of Desert Dunes*, Routledge, London.

Lancaster, N. (1999) Geomorphology of desert sand seas, in *Aeolian Environments, Landforms and Sediments* (eds A.S. Goudie, I. Livingstone and S. Stokes), John Wiley & Sons, Ltd, Chichester, pp. 49–69.

Lancaster, N., Greeley, R. and Christensen, P.R. (1987) Dunes of the Gran Desierto Sand Sea, Sonora, Mexico. *Earth Surface Processes and Landforms*, **12**, 277–288.

Lancaster, N. and Ollier, C.D. (1983) Sources of sand for the Namib Sand Sea. *Zeitschrift für Geomorphologie*, Supplementband **45**, 71–83.

Lancaster, N., Kocurek, G., Singhvi, A. *et al.* (2002) Late Pleistocene and Holocene dune activity and wind regimes in the western Sahara Desert of Mauritania. *Geology*, **30**, 991–994

Langford, I. (1989) Fluvial–aeolian interactions. Part 1. Modern systems. *Sedimentology*, **3**, 273–289.

Leger, M. (1990) Loess landforms. *Quaternary International*, **7–8**, 53–61.

Lettau, K. and Lettau, H. (1969) Bul transport of sand by the barchans of the Pampa La Toya in Southern Peru. *Zeitschrift für Geomorphologie*, NF, **13**, 182–195.

Liu, T. and Ding, Z. (1998) Chinese loess and the palaeomonsoon. *Annual Review of Earth and Planetary Sciences*, **26**, 111–145.

Livingstone, I. and Thomas, D.S.G. (1993) Modes of linear dune activity and their palaeoenvironmental significance: an evaluation with reference to southern African examples, in *The Dynamic and Context of Aeolian Sedimentary Systems* (ed. K. Pye), Geological Society Special Publication 72, London, pp. 91–101.

McFarlane, M.J., Eckardt, F.D., Ringrose, S. *et al.* (2005) Degradation of linear dunes in Northwest Ngamiland, Botswana, and the implications for luminescence dating of periods of aridity. *Quaternary International*, **135**, 83–90.

McKee, E.D. (1979) *Introduction to a study of global sand seas*, US Geological Survey, Professional Paper 1052, pp. 1–19.

McTainsh, G.H. (1987) Desert loess in northern Nigeria. *Zeitschrift fur Geomorphologie*, **31**, 145–165.

Madigan, C.T. (1936) The Australian sand ridge deserts. *Geographical Review*, **26**, 205–227.

Mahowald, N.M., Muhs, D.R., Levis, S. *et al.* (2006) Change in atmospheric mineral aerosols in response to climate: last glacial period, preindustrial, modern, and doubled carbon dioxide climates. *Journal of Geophysical Research*, **111**, D10202, DOI: 10.1029/2005JD006653.

Mason, J.A., Swinehart, J.B., Lu, H. *et al.* (2008) Limited change in dune mobility in response to a large decrease in wind power in semi-arid northern China since the 1970s. *Geomorphology*, **102**, 351–363.

Maxwell, T.A. and Haynes Jr., C.V. (1989) Large-scale, low-angle bedforms (chevrons) in the Selima Sand Sheet. *Science*, **243**, 1179–1182.

Maxwell, T.A. and Haynes Jr., C.V. (2001) Sand sheet dynamics and Quaternary landscape evolution of the Selima Sand Sheet, southern Egypt. *Quaternary Science Reviews*, **20**, 1623–1647.

Muhs, D.R. (1985) Age and paleoclimatic significance of Holocene dune sand in northeastern Colorado. *Annals of the Association of American Geographers*, **75**, 566–582.

Muhs, D.R. (2004) Mineralogical maturity in dunefields of North America, Africa and Australia. *Geomorphology*, **59**, 247–269.

Munyikwa, K. (2005) The role of dune morphogenetic history in the interpretation of linear dune luminescence chronologies: a review of linear dune dynamics. *Progress in Physical Geography*, **29**, 317–336.

Nanson, G.C., Price, D.M., Jones, B.G. *et al.* (2008) Alluvial evidence for major climate and flow regime changes during the middle and late Quaternary in eastern central Australia. *Geomorphology*, **101**, 109–129.

Paisley, E.C.I., Lancaster, N., Gaddis, L.R. and Greeley, R. (1991) Discrimination of active and inactive sand from remote sensing: Kelso Dunes, Mojave Desert. *California Remote Sensing of the Environment*, **37**, 153–166.

Pecsi, M. (1968) Loess, in *Encyclopedia of Geomorphology* (ed. R.W. Fairbridge), Reinhold, New York, pp. 674–678.

Preusser, F., Chithambo, M.L., Götte, T. *et al.* (2009) Quartz as a natural luminescence dosimeter. *Earth-Science Reviews*, **97**, 196–226.

Pye, K. (1987) *Aeolian Dust and Dust Deposits*, Academic Press, London, 334 pp.

Pye, K. and Sherwin, D. (1999) Loess, in *Aeolian Environments, Landforms and Sediments* (eds A.S. Goudie, I. Livingstone and S. Stokes), John Wiley & Sons, Ltd, Chichester, pp. 213–238.

Rozsycki, S.Z. (1967) Le sens des vents portant la poussière de loess à la lumière de l'analyse des formes d'accumulation du loess en Bulgarie et en Europe Centrale. *Revue de Géomorphologie Dynamique*, **1**, 1–9.

Sarnthein, M. (1978) Sand deserts during Glacial Maximum and climatic optimum. *Nature*, **272**, 43–46.

Sarnthein, M. and Walger, E. (1974) Der äolische Sandstrom aus der W-Sahara zur Atlantikküste. *Geologische Rundschau*, **63**, 1065–1087.

Singh, G. (1971) The Indus valley culture seen in context of post-glacial climate and ecological studies in northwest India. *Archaeology and Anthropology in Oceania*, **6**, 177–189.

Singhvi, A.K., Sharma, Y.P. and Agrawal, D.P. (1982) Thermoluminescence dating of sand dunes in Rajesthan, India. *Nature*, **295**, 313–315.

Singhvi, A.K., Bluszcz, A., Bateman, M.D. and Someshwar Rao, M. (2001) Luminescence dating of loess–palaeosol sequences and coversands: methodological aspects and palaeoclimatic implications. *Earth-Science Reviews*, **54**, 193–211.

Smalley, I.J. and Krinsley, D.H. (1978) Eolian sedimentation on Earth and Mars: some comparisons. *Icarus*, **40**, 276–288.

Smalley, I.J. and Vita-Finzi, C. (1968) The formation of fine particles in sandy deserts and the nature of 'desert' loess. *Journal of Sedimentary Petrology*, **38**, 766–774.

Smith, H.T.U. (1963) *Eolian geomorphology, wind direction and climatic change in North Africa*, US Air Force, Cambridge Research Laboratories, Report AF19 (628).

Smith, B.J., Wright, J.S. and Whalley, W.B. (2002) Sources of non-glacial, loess-size quartz silt and the origins of desert loess. *Earth Science Reviews*, **59**, 1–26.

Snead, R.E. (1972) *Atlas of World Physical Features*, John Wiley & Sons, Inc., New York.

Stevens, T., Thomas, D.S.G., Armitage, S.J. *et al.* (2007) Reinterpreting climate proxy records from Late Quaternary Chinese loess: a detailed OSL investigation. *Earth-Science Reviews*, **80**, 111–136.

Stokes, W.L. (1968) Multiple parallel truncation bedding planes – a feature of wind-deposited sand-stone formations. *Journal of Sedimentary Petrology*, **38**, 510–515.

Stokes, S. (1992) Optical dating of independently-dated Late Quaternary eolian deposits from the Southern High Plains. *Current Research in the Pleistocene*, **9**, 125–129.

Stokes, S., Thomas, D.S.G. and Washington, R. (1997) Multiple episodes of aridity in southern Africa since the last interglacial period. *Nature*, **388**, 154–158.

Stone, A.E.C. and Thomas D.S.G. (2008) Linear dune accumulation chronologies from the southwest Kalahari, Namibia: challenges of reconstructing Late Quaternary palaeoenvironments from aeolian landforms. *Quaternary Science Reviews*, **27**, 1667–1681.

Talbot, M.R. (1994) Late Pleistocene rainfall and dune building in the Sahel. *Paleoecology of Africa*, **16**, 203–214.

Telfer, M.W. and Thomas, D.S.G. (2006) Complex Holocene lunette dune development, South Africa: implications for paleoclimate and models of pan development in arid regions. *Geology*, **34**, 853–856.

Thomas, D.S.G. (1984) Ancient ergs of the former arid zones of Zimbabwe, Zambia and Angola. *Transactions, Institute of British Geographers*, NS, **9**, 75–88.

Thomas, D.S.G. (1986) Dune pattern statistics applied to the Kalahari Dune Desert, southern Africa. *Zeitschrift für Geomorphologie*, NF, **30**, 231–242.

Thomas, D.S.G. (1987) Discrimination of depositional environments, using sedimentary characteristics, in the Mega

Kalahari, central southern Africa, in *Desert Sediments Ancient and Modern* (eds L. Frostick and I. Reid), Geological Society of London Special Publication 35, Blackwell, Oxford, pp. 293–306.

Thomas, D.S.G., Knight, M. and Wiggs, G.F.S. (2005) Remobilization of southern African desert dune systems by twenty-first century global warming. *Nature*, **435**, 1218–1221.

Thomas, D.S.G. and Shaw, P.A. (1991a) 'Relict' desert dune systems: interpretations and problems. *Journal of Arid Environments*, **20**, 1–14.

Thomas, D.S.G. and Shaw, P.A. (1991b) *The Kalahari Environment*, CUP, Cambridge.

Thomas, D.S.G. and Shaw, P.A. (2002) Late Quaternary environmental change in central southern Africa: new data, synthesis, issues and prospects. *Quaternary Science Reviews*, **21**, 783–798.

Thomas, D.S.G. and Wiggs, G.F.S. (2008) Aeolian system response to global change: challenges of scale, process and temporal integration. *Earth Surface Processes and Landforms*, **33**, 1396–1418.

Tsoar, H. (1982) Internal structure and surface geometry of longitudinal (seif) dunes. *Journal of Sedimentary Petrology*, **52**, 823–831.

Tsoar, H. (2005) Sand dunes mobility and stability in relation to climate. *Physica A: Statistical Mechanics and Its Applications*, **357**, 50–56.

Tsoar, H. and Møller, J.T. (1986) The role of vegetation in the formation of linear dunes, in *Aeolian Geomorphology* (ed. W.G. Nickling), Binghampton Symposia in Geomorphology, International Series 17, Allen & Unwin, Boston, pp. 75–95.

Tsoar, H. and Pye, K. (1987) Dust transport and the question of desert loess formation. *Sedimentology*, **34**, 139–153.

Twidale, C.R. (1972) Evolution of sand dunes in the Simpson Desert, central Australia. *Transactions, Institute of British Geographers*, **56**, 77–109.

van Boxel, J.H., Sterk, G. and Arens, S.M. (2004) Sonic anemometers in aeolian sediment transport research. *Geomorphology*, **59**, 131–147.

Vine, H. (1987) Wind-blown material and West African soils: an explanation of the 'ferallitic soil over loose sandy sediments' profile, in *Desert Sediments Ancient and Modern* (eds L. Frostick and I. Reid), Geological Society Special Publication 35, London, pp. 171–186.

Wang, X., Yang, Y., Dong, Z. and Zhang, C. (2009) Responses of dune activity and desertification in China to global warming in the twenty-first century. *Global and Planetary Change*, **67**, 167–185.

Warren, A. (1988) The dunes of the Wahiba Sands. *Journal of Oman Studies*, Special Report **3**, 131–160.

Washington, R., Todd, M., Middleton, N.J. and Goudie, A.S. (2003) Dust-storm source areas determined by the total ozone monitoring spectrometer and surface observations. *Annals, Association of American Geographers*, **93**, 297–313.

Wasson, R.J. (1984) Late Quaternary palaeoenvironments in the desert dunefields of Australia, in *Late Cainozoic Palaeoclimates of the Southern Hemisphere* (ed. J.C. Vogel), Balkema, Rotterdam, pp. 419–432.

Wasson, R.J. (1986) Geomorphology and Quaternary history of the Australian continental dunefields. *Geographical Review of Japan*, **59B**, 55–67.

Wasson, R.J. and Hyde, R. (1983) A test of granulometric control of desert dune geometry. *Earth Surface Processes and Landforms*, **8**, 301–312.

Wells, G.L. (1983) Late-glacial circulation over central north America revealed by aeolian features, in *Variations in the Global Water Budget* (eds F.A. Street-Perrott, M. Beran and R. Ratcliffe), Reidel, Dordrecht, pp. 317–330.

Weng, W.S., Hunt, J.C.R., Carruthers, D.J. *et al.* (1991) Air flow and sand transport over dunes. *Acta Mechanica*, Supplement **2**, 1–22.

Werner, B.T. (1995) Eolian dunes: computer simulations and attractor interpretation. *Geology*, **23**, 1107–1110.

Wiggs, G.F.S. (1993) Desert dune dynamics and the evaluation of shear velocity: an integrated approach, in *The Dynamics and Environmental Context of Aeolian Sedimentary Systems* (ed. K. Pye), Geological Society Special Publication 72, London, pp. 32–46.

Wiggs, G.F.S., Livingstone, I., Thomas, D.S.G. and Bullard, J.E. (1995) Dune mobility and vegetation cover in the southwest Kalahari Desert. *Earth Surface Processes and Landforms*, **20**, 515–529.

Williams, M.A.J. (1985) Pleistocene aridity in tropical Africa, Australia and Asia, in *Environmental Change and Tropical Geomorphology* (eds I. Douglas and T. Spencer), George Allen and Unwin, London, pp. 219–233.

Wilson, I.G. (1971) Desert sand flow basins and a model for the development of ergs. *Geographical Journal*, **137**, 180–197.

Wilson, I.G. (1972a) Aeolian bedforms – their development and origins. *Sedimentology*, **19**, 173–210.

Wilson, I.G. (1972b) Universal discontinuities in bedforms produced by the wind. *Journal of Sedimentary Petrology*, **42**, 667–669.

Wilson, I.G. (1973) Ergs. *Sedimentary Geology*, **10**, 77–106.

Wintle, A.G. and Murray, A.S. (2006) A review of quartz optically stimulated luminescence characteristics and their relevance in single-aliquot regeneration dating protocols. *Radiation Measurements*, **41**, 369–391.

Yang, X., Liu, T. and Xiao, H. (2003) Evolution of megadunes and lakes in the Badain Jaran Desert, Inner Mongolia, China, during the last 31,000 years. *Quaternary International*, **104**, 99–112.

Zender, C.S., Bian, H. and Newman, D. (2003) Mineral dust entrainment and deposition (DEAD) model: description and 1990s dust climatology. *Journal of Geophysical Research*, **108** (D14), 4416, DOI: 10.1029/2002JD002775.

Zheng, X.-J., Bo, T.-L. and Zhu, W. (2009) A scale-coupled method for simulation of the formation and evolution of aeolian dune field. *International Journal of Nonlinear Sciences and Numerical Simulation*, **10**, 387–395.

Zhu, Z., Zou, B. and Yang, Y. (1987) The characteristics of aeolian landforms and the control of mobile dunes in China, in *International Geomorphology* (ed. V. Gardiner), John Wiley & Sons, Ltd, Chichester, pp. 1211–1215.

Zimbelman, J.R., Williams, S.H. and Tchakerian, V.P. (1994) Sand transport paths in the Mojave Desert, southwestern United States, in *Desert Aeolian Processes* (ed. V.P. Tchakarian), Chapman & Hall, New York.

18

Sediment mobilisation by the wind

Giles F. S. Wiggs

18.1 Introduction

Wind can be a particularly effective medium for sediment movement in drylands where a relatively sparse vegetation cover and thin soils combine to create highly erosive conditions on highly erodible surfaces. There is little evidence to suggest that dryland winds are any stronger than their counterparts in humid regions, but the sparse vegetation cover in deserts allows the winds to contact the surface more effectively, and the lack of root systems and moisture to bind sediment together renders it far more susceptible to erosion (Pye and Tsoar, 1990). The efficiency of dryland winds is underlined by the continuous advance and development of desert dunes in the Earth's sand seas (e.g. Bristow, Duller and Lancaster, 2007), the long-range transport of Saharan dust across the Atlantic to the Amazon basin (Swap *et al.*, 1992), the deposition of >400 m thick loess deposits in the Lanzhou region of China (Porter, 2001) and the human and environmental distress caused by agricultural wind erosion in the 1930s Dust Bowl in mid-west USA (Worster, 1979).

To interpret the evolution of landscapes such as those mentioned above, the processes of aeolian erosion, transportation and deposition of sediment in drylands have to be understood. Understanding of aeolian processes has been inspired by the work of Bagnold (1941, 1953, 1956). Indeed, no discussion of aeolian processes is complete without reference to his north African research and much can still be learned from these comprehensive works. However, there has been a great deal of progression in our comprehension in the intervening 60–70 years and an understanding of aeolian processes has been used to clarify a wide range of natural and environmental questions and problems.

Improved understanding and measurement of aeolian processes has allowed the development and calibration of models of contemporary dune dynamics and distribution (Andreotti, Claudin and Douady, 2002; Baddock, Livingstone and Wiggs, 2007; Hersen, 2005; Livingstone, Wiggs and Weaver, 2007; Parsons *et al.*, 2004; Tsoar, Blumberg and Stoler, 2004; Walker and Nickling, 2002; Weaver and Wiggs, 2010; Wiggs, Livingstone and Warren, 1996), which has proved essential to our interpretation of palaeodune sequences (Mason *et al.*, 2004; Stone and Thomas, 2008; Thomas *et al.*, 2000; Telfer and Thomas, 2006), while our improved understanding of the controls on aeolian processes has also allowed us to consider the potential impact of global warming scenarios on future dune activity (Knight, Thomas and Wiggs, 2004; Thomas, Knight and Wiggs, 2005; Wang *et al.*, 2009).

Our understanding of aeolian processes has also been enhanced through investigations of wind erosion from susceptible soils and agricultural fields where models of potential erosion, including the effects of implementing management strategies, have been developed and calibrated (Buschiazzo and Zobeck, 2008; Webb *et al.*, 2009). Previously, such models provided a fundamental framework from which dust emission schemes have been developed for application to regional and global scale dust dynamics models (Zender, Bian and Newman, 2003; Marticorena and Bergametti, 1995; Raupach, Gillette and Leys, 1993). Finally, our application of aeolian process understanding has progressed to extra-terrestrial surfaces with a particular focus on the dunes and wind-streaks visible on Mars (Laity and Bridges, 2009; Kok and Renno, 2009a; Bourke, 2010; Gillies *et al.*, 2010).

Sediment movement is essentially a function of both the power of the wind (*erosivity*) and grain characteristics

Arid Zone Geomorphology: Process, Form and Change in Drylands, Third Edition. Edited by David S. G. Thomas

holding particles in place (*erodibility*). In general terms, where erosivity exceeds the resistance due to erodibility parameters, particle dislodgement and erosion will take place. This simple equation is complicated by a variety of surface characteristics (such as roughness, slope and moisture content), which can affect both erosivity and erodibility. This chapter investigates the key factors influencing erosivity and erodibility and examines some of the methods with which we can measure and characterise them. In particular, there is a focus on our burgeoning understanding of how both the presence of vegetation and wind turbulence may affect sediment entrainment – both of which have seen considerable advances in understanding since the previous edition of *Arid Zone Geomorphology*.

18.2 The nature of windflow in deserts

As air moves over the surface of the Earth it is retarded by friction at its base and a *velocity profile* develops. The zone of flow where air is affected by surface friction is called the *boundary layer* and within this layer there is a gradation from zero velocity in a very thin layer at the surface to free stream velocity at a height beyond the effects of surface friction. The atmospheric boundary layer (ABL) is approximately 1–2 kilometres thick.

The structure of the velocity profile within the boundary layer is highly dependent on the type of airflow: laminar or turbulent. In *laminar* flow, there is little mixing between the different layers of fluid and faster layers slip over slower layers, while at the surface itself the air is stationary. With this type of flow, momentum transfer between layers of fluid is accomplished by means of molecular transfer. Slower moving molecules of air drift into faster moving layers, thus causing a drag on the faster overlying airflow. Such a transfer of momentum produces shearing forces between layers of air.

Turbulent flow also comprises zero flow velocity at the surface, but the exchange of momentum in the boundary layer is achieved through the action of gusts and turbulent eddies mixing between layers. Such momentum exchange is far more efficient than the molecular exchange seen in laminar flow and this is represented by a differing velocity profile. The greater mixing in turbulent flow results in a steeper velocity gradient at the surface and hence higher shearing stresses (Figure 18.1).

Laminar and turbulent flows can be distinguished from each other by the Reynolds number (Re):

$$\mathrm{Re} = \frac{\rho h V}{\nu} \tag{18.1}$$

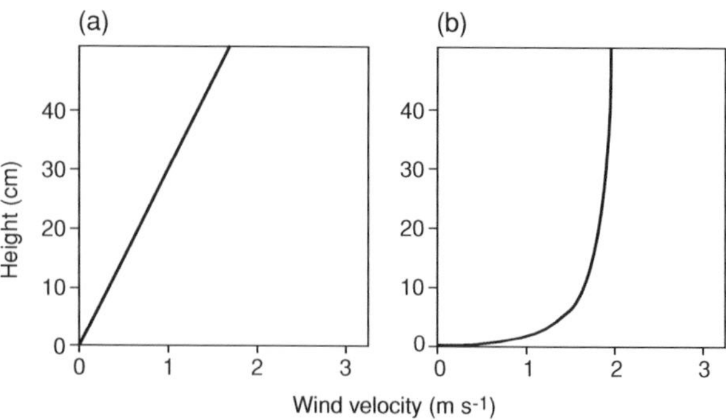

Figure 18.1 Near-surface vertical velocity profiles showing (a) the smaller surface shear stresses in laminar flow (b) when compared to turbulent flow (after Bagnold, 1941).

where

ρ = fluid density

h = flow depth

V = flow velocity

ν = viscosity

An increasing Reynolds number signifies an increasing turbulence intensity and a shift towards inertial forces dominating over viscous forces, such that where Re < 500–2000 laminar flow exists and where Re $\geq$ 2000 turbulent flow predominates.

In the atmosphere, airflow is nearly always turbulent because air has a low viscosity and boundary layer depths are normally quite high. Only in very viscous, slow or thin flows do laminar characteristics develop.

18.2.1 The turbulent velocity profile

Under normal atmospheric conditions on flat, unvegetated surfaces and in the absence of intense solar heating, the turbulent velocity profile plots as a straight line on a semi-logarithmic chart (Figure 18.2). The gradient of the semi-logarithmic profile is a result of the *surface roughness* producing a drag on the overlying airflow. Hence, if the gradient of the velocity profile is known, the *shear stress* (drag) at the surface can be determined. In sedimentological research, a common method for describing the gradient of the velocity profile is in terms of the *shear velocity* (u_*). It is the value of u_* that has been traditionally used in calculations for determining thresholds of erosion and fluxes of sediment transport. The shear velocity (u_*) is

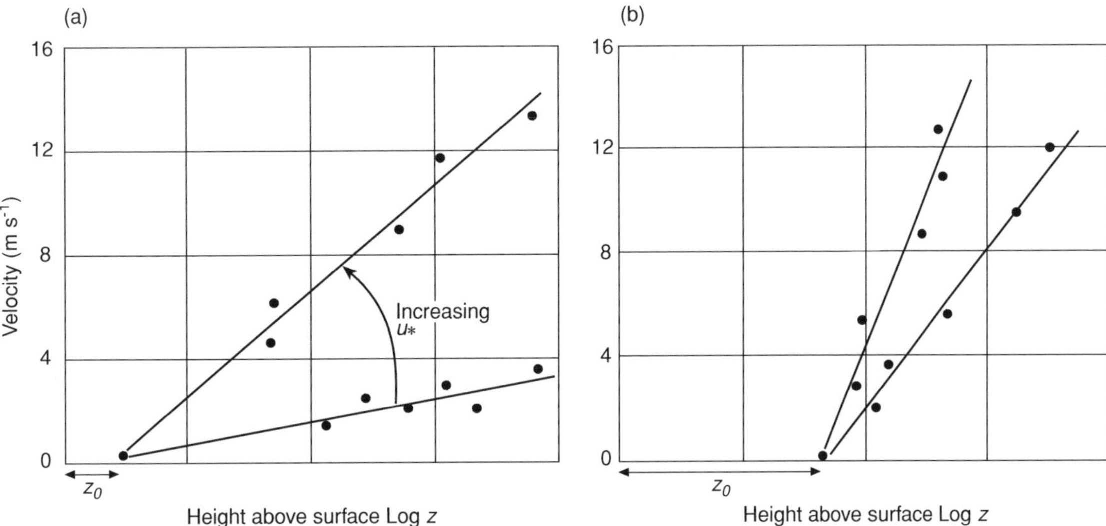

Figure 18.2 Semi-logarithmic velocity profiles showing a focus at height z_0, the aerodynamic roughness. The aerodynamic roughness in (a) is less than that in (b); hence shear velocities (u_*) in (b) are greater. See text for details.

proportional to the velocity profile gradient (Figure 18.2) and is related to the actual surface shear stress (τ_0) by the following expression:

$$u_* = \sqrt{\tau_0/\rho} \tag{18.2}$$

In Figure 18.2(a) the semi-logarithmic turbulent velocity profile does not reach the surface. This is because very close to the surface the wind velocity is zero. The height of this zero-velocity region is termed the *aerodynamic roughness length* (z_0) and it is an important parameter for it is a function of the surface roughness, it has an impact on surface erodibility and it partly controls the gradient of the velocity profile and hence u_*. For example, over a rougher surface (Figure 18.2(b)), the aerodynamic roughnesss (z_0) is larger, and with all other parameters remaining constant, the velocity profile gradient becomes steeper. Hence, shear velocities (u_*) increase and, consequent upon this, sediment transport would be likely to rise. Another way in which shear velocity (u_*) might increase is by a rise in the overall environmental wind velocity, as shown in Figure 18.2(a). Note in Figure 18.2 that height is the independent variable and is therefore plotted on the x axis of the graph. This is important when calculating shear velocities using a least squares regression technique (as described below).

The relationship between aerodynamic roughness (z_0), shear velocity (u_*) and wind velocity (u) at a height (z) are described by the Karman–Prandtl velocity distribution, otherwise known as the 'law-of-the-wall':

$$\frac{u}{u_*} = \frac{1}{\kappa} \ln \frac{(z-d)}{z_0} \tag{18.3}$$

where

κ = von Karman's constant ($\approx$0.4)

d = zero-plane displacement

The *zero-plane displacement* (d) is a measure of the vertical displacement of z_0 on surfaces with large surface roughnesses (e.g. long grass or bushes). It can largely be ignored on smooth and unvegetated desert surfaces.

18.2.2 Measuring shear velocity (u_*) and wind stress

Shear velocity (u_*) is commonly determined from least squares regression analysis of time-averaged velocity measurements at several known heights. Wilkinson (1983/1984) reports that at least five velocity data points are required to determine u_* and z_0 with any statistical significance and so equipment arrays such as that shown in Figure 18.3 are typically erected. When plotted in the manner of Figure 18.2, u_* can be determined from the slope component of the regression equation by

$$u_* = \kappa m \tag{18.4}$$

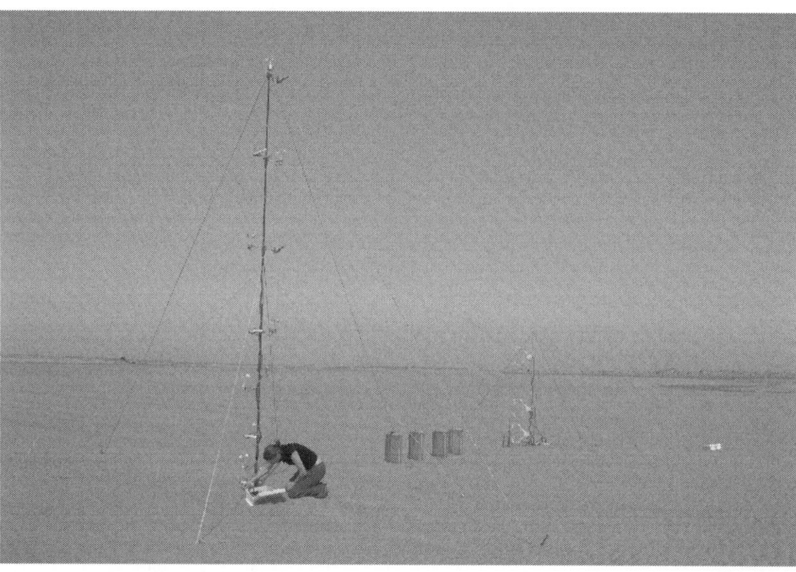

Figure 18.3 Photograph of a 6 m high anemometer tower incorporating eight logarithmically spaced cup anemometers to measure changing shear stress (u_*) and aerodynamic roughness (z_0) on the Skeleton Coast in Namibia. The array also consists of four wedge-shaped sand traps, two saltation impact responders and a vertical array of three sonic anemometers (photo: author).

where

$\kappa =$ von Karman's constant ($\approx$0.4)

$m =$ gradient statistic of the regression equation

The averaging time required to measure values of u_* meaningful for sediment transport calculations using this least squares approach is a matter of debate. While for regional scale wind erosion modelling a u_* calculated over several hours may be sufficient, many geomorphologists investigating relationships between changing u_* and sediment transport flux have calculated u_* on the order of seconds or minutes. However, at high measurement frequencies ($\sim$1 second) it is clear that the log-linear nature of the turbulent wind velocity profile collapses and it is not possible to calculate a time-averaged u_* (Bauer, Sherman and Wolcott, 1992; Wiggs, Atherton and Baird, 2004). The highest frequencies for successful u_* calculation are of the order of 30–40 seconds (Wiggs, Atherton and Baird, 2004; Weaver, 2008), although Namikas, Bauer and Sherman (2003) report success with measurement frequencies as high as 10–20 seconds.

Fluctuating topography also hinders the successful calculation of u_*. Process research on sand dunes has been particularly hampered by the recognition that acceleration of wind flow up the stoss slopes of dunes results in the disruption of the log-linear velocity profile (thus negating the calculation of u_*), with maximum acceleration occurring in an inner-layer depth close to the dune surface (Frank and Kocurek, 1996; Lancaster *et al.*, 1996; Wiggs, Livingstone and Warren, 1996; also see Chapter 19). In such cases, calculated values of u_* measured above a dune's surface become meaningless.

Corrections to measured values of u_* also have to be made for the effect of thermal stability or instability, which can significantly alter the shape of velocity profiles. If strong thermal heating or atmospheric stratification occurs, both of which are common in arid environments when wind velocities are low, then wind turbulence may be driven by buoyant forces rather than mechanical forces and the velocity profile may again become non-log-linear (Frank and Kocurek, 1994). Rasmussen, Sørensen and Willetts (1985) found that omitting stability corrections could lead to errors in the measurement of u_* of the order of 15–25 %, with subsequent estimates of sand transport in error by as much as 15 times (Frank and Kocurek, 1994). Correction factors involve the calculation of the Richardson number (RI), a measure of the relative strength of buoyancy to shear generated turbulence (Oke, 1987), and require knowledge of the local atmospheric temperature gradient. However, Rasmussen, Sørensen and Willetts (1985) suggested that errors in measured velocity due to such instability was a function of height and that at elevations below $\sim$0.5 m the correction required was unimportant. From their measurements in White Sands, New Mexico, Frank and Kocurek (1994) also found that correction factors were not necessary at wind speeds higher than $\sim$10 m/s.

An additional measure of stress within the airflow is the time-averaged Reynolds shear stress ($-\overline{u'w'}$). This involves the calculation of the average of the instantaneous fluctuating components of the wind (in comparison to the mean value) in the horizontal (u) and vertical (w) directions. Such data can be measured at-a-point in the airflow by a sonic anemometer (see Box 18.1) and so the requirement for a log-linear velocity profile over a significant depth of boundary layer is dismissed. Measurements of the Reynolds shear stress are becoming more common in aeolian process research (van Boxel, Sterk and Arens, 2004; Weaver and Wiggs, 2010; Weaver, 2008), with fluctuating components measured at 10 Hz and shear stress averaged over 10 minutes showing a good correlation with sediment transport rates (Weaver, 2008).

Box 18.1 From bagnold to sonics: measuring boundary layer airflow

Over the last 60 years one of the greatest contributions to improving our understanding of the processes of sediment mobilisation by the wind has been the extraordinary technological development in the measurement of airflow. From the 1940s to 1960s aeolian researchers often coped with bulky mechanical weather stations where data were recorded on paper rolls controlled by clockwork and hourly average windspeeds were considered high resolution. The bulk of these instruments meant that velocity could only reasonably be measured at a single height and often several metres away from the surface. It is a sensation then that so much of our understanding of the physics of blowing sand comes from this time period and that reference to the classic works of the time is still so prevalent.

In the 1970s to 1990s small, fast response cup anemometers were developed for use with electronic logging systems. These allowed much higher resolution measurements in both space and time. The small diameter of the anemometer bodies allowed several to be placed in vertical arrays without obstructing the airflow, thus effectively measuring the vertical velocity profile and allowing a detailed investigation of shear velocity (u_*). The small plastic cups on these instruments responded quickly to accelerations and decelerations in airflow and these short response and lag times, combined with electronic logging, enabled data to be recorded at minute intervals or better. However, these instruments were prone to failure and required heavy maintenance if used in sandy environments for extended periods. This was particularly the case as investigators attempted to measure the enigmatic relationships between sand flux and wind velocity right at the top of the saltation layer. At this time it also became clear from wind tunnel studies, where near-surface wind speeds could be measured over stabilised (nonsaltating) sand beds with fast-response hot-wire anemometers, that measurement of turbulent frequencies in the airflow was going to be required in the field.

As the twenty-first century began, aeolian geomorphologists had started using sonic anemometers in field situations (Figure 18.4). These instruments measure wind velocity by calculating the delay in travel time of an ultrasonic acoustic signal over a set path between paired sonic transducer heads. The difference in time delay between both directions along the same path is proportional to the wind speed. Sonic anemometers offer huge advantages in measurement techniques to desert geomorphologists. First, they calculate wind velocity within an empty volume of air between the transducer heads, and hence there is no obstruction to the flow. Second, they

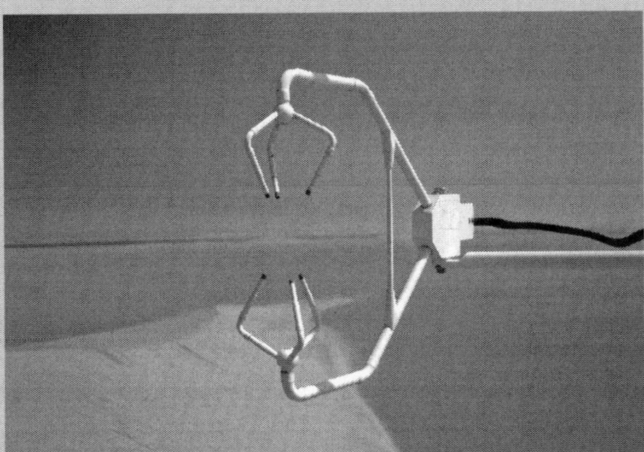

Figure 18.4 Modern equipment: a sonic anemometer in the field at Skeleton Coast, Namibia.

have no moving parts prone to attack from saltating particles. Third, and most importantly, they can simultaneously measure the three high-frequency (>10 Hz) components of wind velocity in the horizontal streamwise component (u), the lateral or spanwise component (v) and the vertical component (w). From these instantaneous velocity measurements it is possible to determine turbulence statistics including the uw covariance, the Reynolds stress and coherent flow structures, all of which are now thought to be relevant in saltation dynamics. It has taken some time for investigators to devise protocols for using the anemometers in complex field situations and also to develop analytical schemes to make best use of the huge volumes of data that can be collected. However, the field application of sonic anemometers now offers a new paradigm in aeolian geomorphology.

Given some of the difficulties in assessing the erosivity of the wind by measuring various windflow characteristics (as decribed above) an alternative approach is to measure the surface shear stress (τ_0) directly. This can be achieved by using calibrated Irwin sensors or rugged drag balances placed directly on the sediment surface. Irwin sensors are small, omnidirectional skin friction meters that measure the near-surface vertical pressure gradient (Irwin, 1980), which, once calibrated, can be used to measure high-frequency surface shear stresses (Wu and Stathopoulos, 1994). A drag balance, buried with the measuring element flush to the surface or attached to a roughness element, directly measures the force applied by the wind and records it on a sensitive load cell. Both technologies have been successfully tested in wind tunnels and dryland field locations to measure wind forces on bare sediment and vegetated surfaces in a variety of flow conditions and surface roughness configurations (Gillies *et al.*, 2000; King, Nickling and Gillies, 2005; Gillies, Nickling and King, 2007; Walker and Nickling, 2003).

18.2.3 Measuring aerodynamic roughness (z_0)

The aerodynamic roughness (z_0) of desert surfaces varies widely both temporally and spatially. Typical values for z_0 may be 0.0007 m for stationary sand surfaces (Stull, 1988), 0.003 m for surfaces with moving sand (Rasmussen, Sørensen and Willetts, 1985) to 0.2 m and greater for vegetated or semi-vegetated desert surfaces (Wiggs *et al.*, 1994). Owing to its importance in determining wind shear velocity, it is a vital parameter to calculate correctly in studies of aeolian processes (Levin *et al.*, 2008). However, on many desert surfaces it is a parameter that is difficult to resolve effectively because of its sensitivity to the frequently changing scales of nonerodible elements and surface roughness.

Bagnold (1941) found a relationship between z_0 and surface roughness. He noted that $z_0 = d/30$, where d is the mean surface particle diameter. However, this relationship assumes a homogeneous surface and well-sorted sediment. Furthermore, it ignores the effect of roughness element spacing and scale where the aerodynamic roughness might be determined by larger particles on a stone pavement. For example, Greeley and Iversen (1985) noted that z_0 may reach a maximum value of $d/8$ for widely spaced elements, before returning to a $d/30$ ratio as element spacing increased further (see Figure 18.5). Further, it is clear that z_0 also varies in relation to changes in microtopography and this suggests that mean particle size may not be a good determinant of z_0 for surfaces with a large range in particle sizes (Lancaster, Greeley and Rasmussen, 1991) or changing surface patterning (e.g. sand ripples).

For this reason aerodynamic roughness has often been determined from the intercept of velocity profiles with the height axis, as shown in Figure 18.2. This intercept clearly describes z_0 as defined as the depth of air at the surface with an effective zero velocity. When plotted in the manner of Figure 18.2, z_0 can be determined from the slope and intercept components of the regression equation by

$$z_0 = \exp(-n/m) \tag{18.5}$$

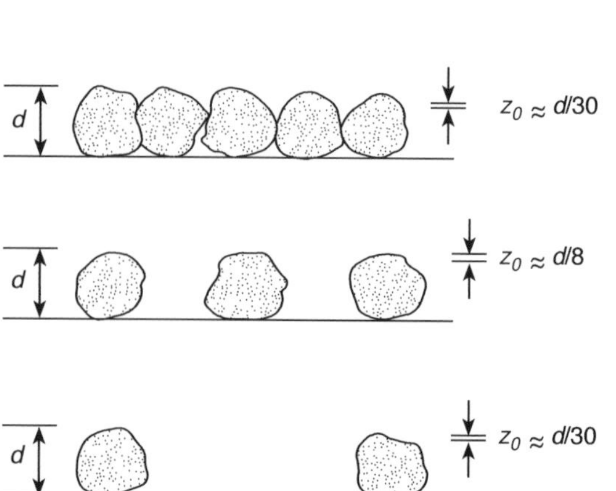

Figure 18.5 Aerodynamic roughness height (z_0) as a function of roughness element spacing (after Greeley and Iversen, 1985).

where

n = intercept statistic of the regression equation

m = gradient statistic of the regression equation

Such an approach is common practice and works well for flat and noncomplex surfaces (Bauer *et al.*, 2009; Gillies, Nickling and King, 2007; Gillies *et al.*, 2000; King, Nickling and Gillies, 2005; Sherman *et al.*, 1998; Wiggs *et al.*, 1994; Weaver, 2008). However, defining a clear intercept on complex or patterned (e.g. ploughed) surfaces is problematic where small variations in wind direction during measurement can have a dramatic effect on calculated values of z_0 as the wind profile responds to the changing effective configuration of surface properties that provide the surface drag (Gillette, Herrick and Herbert, 2006; Wiggs and Holmes, 2010). Furthermore, on sloping surfaces the regression equation approach to calculating z_0 suffers from the same problems as when calculating u_*, the disruption of the log-linear velocity profile due to flow acceleration.

An additional issue arises when a sand bed responds to increasing wind shear by allowing grain entrainment and saltation to take place. In such cases the saltating sand extracts momentum from the wind, carried in the form of a grain-borne shear stress, and this acts as an additional drag on the airflow and so increases the value of z_0 (McKenna-Neuman and Nickling, 1994; Sherman and Farrell, 2008; Bauer, Houser and Nickling, 2004, 2009). In order to satisfy the concept of an increasing z_0 with increasing u_* when saltation is occurring, several workers (Rasmussen, Sørensen and Willetts 1985; Anderson and Haff, 1991; McEwan, 1993) have turned their attention to the relationship applied by Owen (1964):

$$z_0' = C_0 u_*^2 / 2g \tag{18.6}$$

where

z_0' = aerodynamic roughness height during saltation

C_0 = constant ($\approx$0.02)

In this case, the depth of saltation (associated with the aerodynamic roughness height, z_0') is related to the lift-off velocity of the individual sand grains, which in turn is governed by the shear velocity (u_*). While this relationship has been widely used to determine u_* in wind tunnel studies, it has yet to be fully utilised in field situations where the constant C_0 is difficult to define (although see Rasmussen, Sørensen and Willetts, 1985; Sherman, 1992; Sherman and Farrell, 2008).

The Owen (1964) equation has been used as the basis for constructing self-regulatory models of saltation (Anderson and Haff, 1988, 1991; Werner, 1990; McEwan and Willetts, 1991, 1993). These steady-state models work on the premise that as u_* (and hence saltation) increases, so too does the effective aerodynamic roughness (z_0'). This increased aerodynamic roughness exerts an extra drag on the airflow and this effect propagates upwards through the velocity profile as an internal boundary layer. This results in near-surface winds being reduced, eventually to reach a steady-state value where as many grains are leaving the surface as are falling onto it (Anderson, Sørensen and Willetts, 1991). An equilibrium is therefore established between u_*, z_0' and saltation load (Sherman, 1992).

The difficulties in determining a meaningful z_0, as described above, are a function of the complexity of the parameter and its response to changing surface characteristics in time and space (King, Nickling and Gillies, 2006, 2008). Blumberg and Greeley (1993) regard one of the most difficult hurdles as being able to take account of the effects of a surface roughness change into the development of a velocity profile, for as there is a transition from a smooth to rough surface (or rough to smooth), a boundary layer grows downwind in response to that transition (Figure 18.6). Hence, different parts of the velocity profile are likely to be responding to different surface roughnesses with many subsidiary boundary layers responding to roughnesses provided at different scales by such elements as individual grains, surface ripples or isolated vegetation.

In the example shown in Figure 18.6, the wind passes from a smooth surface (perhaps a flat sand sheet) to a

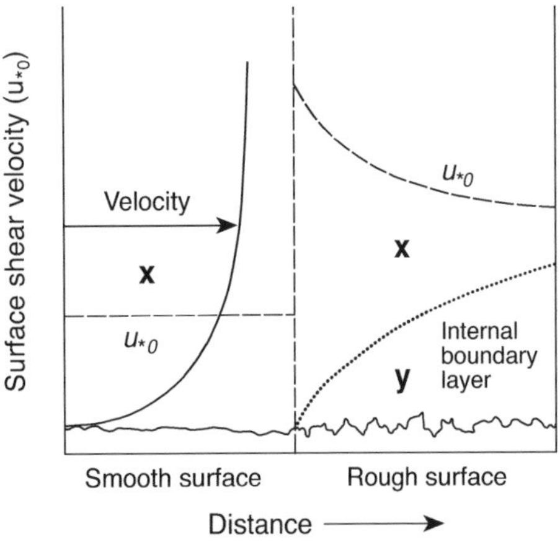

Figure 18.6 The growth of an internal boundary in response to a changing surface roughness (after Greeley and Iversen, 1985). See text for details.

rougher surface (such as coarse-grained ripples or stone pavement). At the smooth–rough transition, airflow close to the surface is quickly decelerated by the additional drag, and the aerodynamic roughness (z_0) and surface shear velocity (u_{*0}) increase rapidly in response. As the internal boundary layer responding to the rougher surface grows downwind, shear velocity steadily decreases to reach an equilibrium level. It should be noted that such a rise in u_* may not necessarily lead to an increase in sediment transport or erosion because the increase in surface roughness may result in more grains lying below the height of z_0 and hence in a zone of zero wind velocity, although velocity may be enhanced around individual, large, nonerodible roughness elements and result in additional local erosion (Ash and Wasson, 1983).

From Figure 18.6, it can be seen that at points marked X the wind is still in equilibrium with the smooth upstream surface, while at point Y the wind is responding to the rougher downstream surface. Hence, z_0 and u_* measured at any particular point in a velocity profile are functions not only of the size and spacing of surface roughness elements in the immediate locality but also of changes in surface roughness within the fetch of the wind (Blumberg and Greeley, 1993).

Clearly, although z_0 is a critical value to determine in terms of assessing aeolian sediment transport, there are many difficulties in its successful and effective calculation. There has been some success in interpreting values and spatial variations in z_0 from remote sensing. Greeley *et al.* (1997) found good correlations between subregional variations in z_0 and radar backscatter in the Mojave and Namib Deserts, demonstrating the potential to map z_0 for large vegetation-free areas from orbit using radar systems. Such an approach may make it easier to determine the temporal variation in large-scale z_0 values with changing climatic and environmental conditions, but often the scale of interest in aeolian processses is small enough that field and wind tunnel studies are the only means of providing the relevant data.

18.2.4 The effect of nonerodible roughness elements on velocity profiles

The distribution of cobbles, rocks and vegetation on otherwise homogeneous desert surfaces act as a surface roughness, providing significant drag on overlying airflow and considerably altering velocity profile parameters (King, Nickling and Gillies, 2008; Gillies *et al.*, 2000; Wiggs *et al.*, 1994, 1996). By altering values of u_* and z_0 such nonerodible elements can have a considerable role in controlling wind erosion (Ash and Wasson, 1983; Wasson and Nanninga, 1986; Wolfe and Nickling, 1993; Wiggs *et al.*, 1994, 1995). However, the varying geometry, spatial organisation and density of such roughness elements are difficult to account for in models of aeolian sediment transport (King, Nickling and Gillies, 2005).

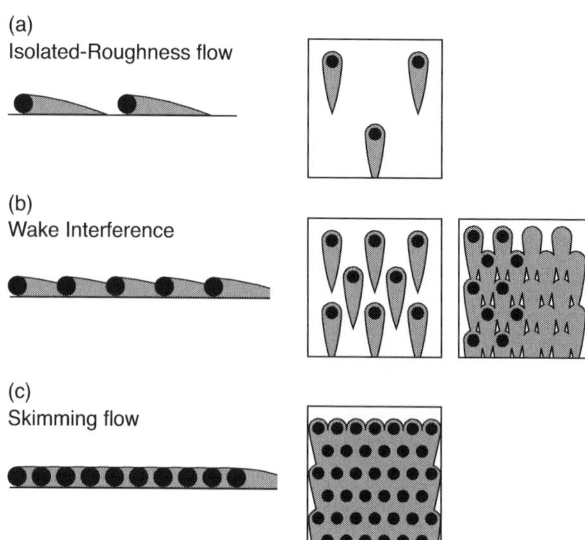

Figure 18.7 Flow regimes associated with different roughness element spacings and geometries. The shaded areas represent zones of reduced wind speed (after Wolfe and Nickling, 1993).

The way in which nonerodible roughness elements might interact with the wind flow was demonstrated theoretically by Wolfe and Nickling (1993) and is shown diagrammatically in Figure 18.7. With an increasing density of roughness elements the flow regime changes from one where the elements act individually on the flow (isolated roughness flow) to a regime where the interaction between the wakes downwind of the elements is such that the flow skims across the top of the elements (skimming flow). In semi-arid areas the distribution of plants is often such that the airflow is responding to the isolated roughness or wake interference regimes (Wolfe and Nickling, 1993).

It is clear from Figure 18.7 that an increasing roughness element density reduces near-surface flow velocities and this is reflected in the corresponding velocity profiles. Using vegetation as an example, Figure 18.8 demonstrates that the additional drag on the airflow provided by a vegetated surface not only increases the aerodynamic roughness (z_0) but also displaces it upwards by a value d, resulting in a greater depth of flow with zero velocity at the surface. The greater z_0 of the vegetated surface also

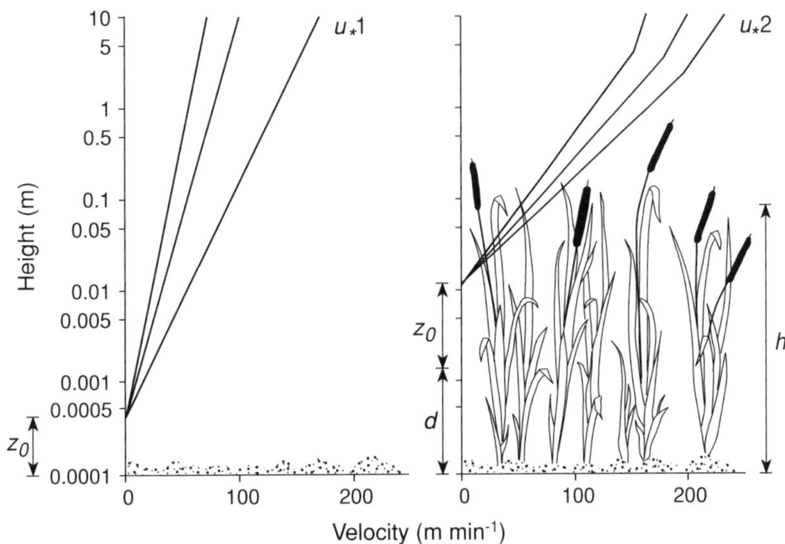

Figure 18.8 Vegetation effect on velocity profile structure: d = zero-plane displacement, h = mean vegetation canopy height, $u_*^2 > u_*^1$.

results in an increased u_* above the vegetation canopy. From measurements on vegetated and bare dunes in the Kalahari Desert, Wiggs *et al.* (1994) found threefold increases in above-canopy u_* and 200 % reductions in near-surface wind velocity on vegetated dune surfaces.

The value d in Figure 18.8 (and Equation (18.3)) is the *zero-plane displacement* height and is the level of the mean momentum sink and the elevation at which the mean drag appears to act (Thom, 1971; Jackson, 1981). A practical explanation of the physical meaning of d can be given within the context of Figure 18.7. In an isolated-flow roughness regime where roughness elements (e.g. vegetation) are widely spaced, each individual roughness element provides a drag on the airflow and so z_0 is determined by the geometry of the roughness elements and d would be small enough to ignore. At the opposite extreme, with dense roughness elements, the wakes of each element interact, resulting in wake interference or skimming flow (Figure 18.7). In this case it is clear that the aerodynamic roughness (z_0) would be controlled only by the very tops of the vegetation canopy and so the effective drag provided by each individual roughness element is considerably reduced. The value of z_0 therefore decreases markedly (in comparison to isolated roughness flow) but the height at which that aerodynamic roughness operates is displaced upwards by the zero-plane displacement height (d), a value that, in this example, may approach the scale of the roughness elements themselves.

Vegetated surfaces are therefore seen to have conflicting influences on aeolian erosion. When compared to a bare surface, a vegetated surface induces a lower near-surface wind velocity (which would tend to diminish erosion) but also a higher u_* in the airflow (which would tend to enhance erosion). A question therefore arises as to the height above the surface at which the additional energy from an enhanced u_* is effective. In the case of skimming flow (Figure 18.7) with a large value of d, that energy is functional near the top of the vegetation canopy and therefore has little influence on surface erosion. However, with more isolated roughness elements the vertical displacement of the velocity profile may not be sufficient to counteract the increased erosional effects of an enhanced u_*, and the extra wind stresses may be applied on the surface sediment, thus increasing erosion potential.

A real difficulty from a sediment transport perspective wth regard to predicting both dust emission and sand entrainment is in determining the proportion of u_* that acts on the surface, compared to the proportion that acts on any distributed nonerodible roughness elements. Much research has been undertaken to define a parameter that would successfully take account of the impact on the airflow of the geometry and spacing of nonerodible roughness elements. Methods for partitioning the shear stress between that acting on a vegetation canopy and that affecting the surface often make reference to the roughness density (λ) (Gillette and Stockton, 1989; Musick and Gillette, 1990; Stockton and Gillette, 1990):

$$\lambda = nbh/s \tag{18.7}$$

where

n = number of roughness elements

b = width of roughness elements

h = height of roughness elements

s = surface area

However, such approaches have been seen to be unsatisfactory when describing complex roughness distributions and three-dimensional objects (Minvielle *et al.*, 2003; Musick, Trujillo and Truman, 1996); problems may arise in the case of vegetation where effective height may vary in response to changing wind speeds and plant pliability, and the drag provided by vegetation may vary with stem porosity.

However, there are two models for determining the impact of shear stress partitioning in the presence of nonerodible roughness elements that have become widely recognised in the research literature. While both models compare the ratio of stress required for erosion on a bare surface to the stress required for erosion on a surface including roughness elements, they each have a distinct approach to the problem (see King, Nickling and Gillies, 2005, for a full comparison). The first is the model of Marticorena and Bergametti (1995), which defines the ratio between the aerodynamic roughness of the surface between roughness elements (z_{0s}) and the total aerodynamic roughness (z_0). This model often forms the basis for the emission schemes in global-scale dust models (Zender, Bian and Newman, 2003; see Chapter 20). The other approach is that of Raupach (1992) and Raupach, Gillette and Leys (1993), in which the wake development behind individual roughness elements is modelled to generate a ratio between the erosion threshold on a bare surface and that incorporating nonerodible roughness elements. The input parameters to the Raupach, Gillette and Leys (1993) model are more complex than for the Marticorena and Bergametti (1995) model and include the vegetation roughness density (λ) combined with terms describing the vegetation aspect ratio and erosion threshold delineated as a function of the maximum shear stress.

The Raupach, Gillette and Leys (1993) model shows good general agreement with field and wind tunnel data (Figure 18.9) and experimental testing of the procedure has confirmed its robustness (Brown, Nickling and Gillies, 2008; King, Nickling and Gillies, 2005, 2006; Gillette, Herrick and Herbert, 2006; Gillies *et al.*, 2000, 2010; Gillies, Nickling and King, 2007). However, very few applications of the theory have been undertaken and those that have been carried out involve a very limited range of vegetation types (e.g. Lancaster and Baas, 1998; Crawley and Nickling, 2003); far more experimental data are required to develop the model further. Furthermore, application of the Raupach, Gillette and Leys (1993) model at a regional or global scale may be hampered by the relatively complex input parameters required.

Okin (2008) has recently presented a new model for determining the effect of vegetation on wind erosion that partly resolves this problem. In recognition of the observation that vegetation in drylands is rarely regularly patterned, Okin has devised a model that focuses on the size distribution of *gaps* between nonerodible roughness elements as being the driving force controlling the amount of potential erosion (rather than the structural parameters of the roughness elements themselves). Initial testing of the model has shown good results but a significant advantage is that it requires relatively straightforward input data that could be acquired from fieldwork or large-scale image analysis.

18.3 Sediment in air

18.3.1 Grain entrainment

Sediment is entrained into the airflow when forces acting to move a stationary particle overcome the forces resisting sediment movement. The relevant forces for the entrainment of dry, bare sand are shown diagrammatically in Figure 18.10.

Particles are subjected to three forces of movement: lift, surface drag and form drag. Lift is a result of the air flowing directly over the particle forming a region of low pressure (in contrast to relatively high pressure beneath the particle); hence there is a tendency for the particle to be 'sucked' into the airflow. Surface drag is the shear stress on the particle provided by the velocity profile and the form drag is also related to upwind and downwind pressure differences around the particle. When these forces overcome the forces of particle cohesion, packing and weight, the particle tends to shake in place and then lift off, spinning into the airstream.

It has been shown that aerodynamic entrainment is primarily a function of the mean grain size of the particles involved combined with the erosivity of the wind, with u_* the preferred measure (Williams, Butterfield and Clark, 1990). Bagnold (1941) studied these relationships and derived values of *critical threshold shear velocity* (u_{*ct})

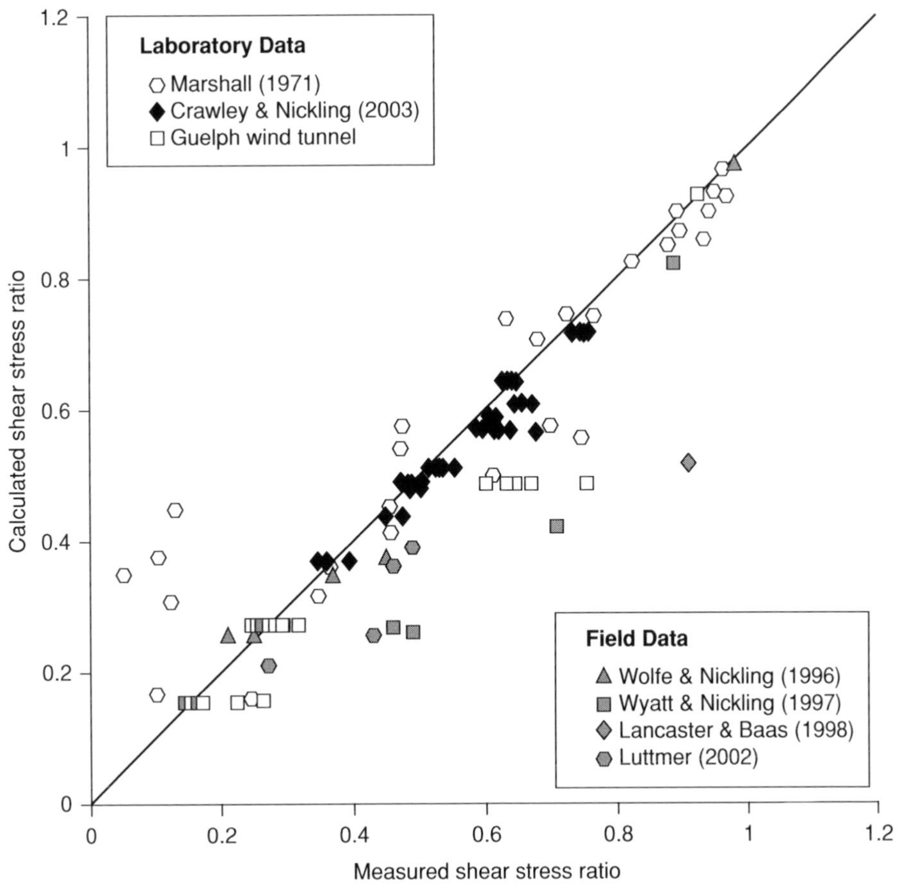

Figure 18.9 The relationship between measured shear stress ratios in the presence of nonerodible roughness elements and calculations using the model of Raupach, Gillette and Leys (1993) (from King, Nickling and Gillies, 2005).

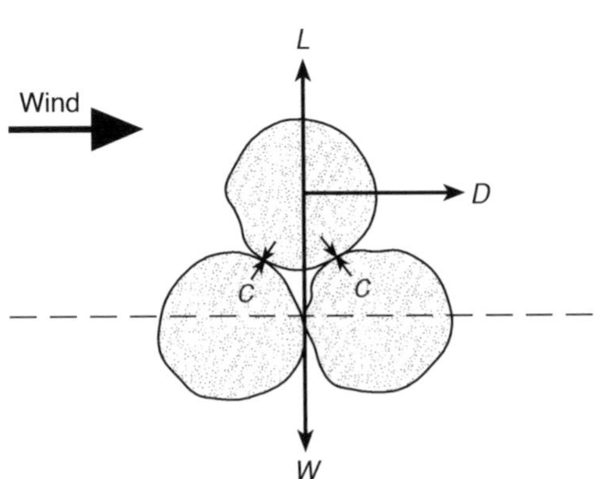

Figure 18.10 Forces exerted on a grain by the wind: l = lift, d = surface and form drag, D = form drag, w = weight, c = interparticle cohesion.

for a wide range of particle sizes, using the square root of the grain diameter as the principal determinant:

$$u_{*ct} = A\sqrt{\left(\frac{\sigma - \rho}{\rho}\right)gd} \tag{18.8}$$

where

ρ = particle density

ρ = air density

g = acceleration due to gravity

d = grain diameter

A = constant dependent upon the grain Reynolds number (≈ 0.1)

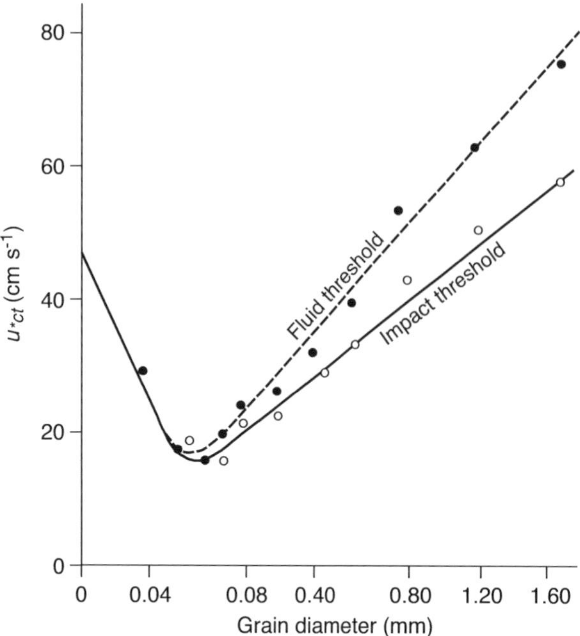

Figure 18.11 The idealised relationship between grain diameter (d) and threshold shear velocity (u_{*ct}) showing the fluid and impact thresholds (after Chepil, 1945). See text for details.

Figure 18.11 shows the idealised relationship between fluid threshold shear velocity and grain size. From this figure, it can be seen that, in general, larger particles have a higher threshold of entrainment. However, smaller particles (with diameters less than about 0.07 mm) also require higher shear velocities to entrain them. This is because particles in this size range and smaller tend to have additional molecular and electrostatic forces of cohesion. They also have an affinity for the retention of moisture, often become protected from erosion by larger particles and frequently rest in the zero-velocity layer (beneath the height of z_0). The most susceptible grain size for entrainment is seen to be between about 0.07 and 0.08 mm, i.e. fine textured sand. The heightened mobility of sediment in this size range in drylands enables large volumes of sand to be moved and accumulate, resulting in the extensive dunefields present in many drylands (see Chapter 17).

The relationships shown in Figure 18.11 tend to result in a downwind fining of aeolian sediment both at the scale of the individual ripple and across whole sand seas. Such a downwind sorting can be used to identify process–form relationships; sand surfaces subject to aeolian erosion tend to consist of a coarse lag material (Nickling and McKenna-Neuman, 1995), while those subject to aeolian deposition are commonly composed of fine material, such as loess.

The discussion above refers to the value of the *static* or *fluid* critical threshold of entrainment where the grains are entrained only by the drag and lift forces of the moving air. However, one of the most significant influences on the entrainment of grains is the existence of grains already in motion. In particular, the impact of saltating grains on the bed surface imparts momentum to previously stationary grains, hence 'splashing' them into saltation (see the later discussion). A second threshold of grain entrainment can therefore be identified: the *impact* (or *dynamic*) threshold, which refers to the threshold of entrainment where other grains are already in motion. Anderson and Haff (1988) note that the impact threshold is about 0.8 of the static threshold (see Figure 18.11) because saltating grains bring extra momentum gained from higher in the velocity profile towards the surface. It is the action of impacting grains that drives the saltation process in this case, rather than the lift and drag forces of the wind.

This impact mechanism of entrainment is seen to be particularly important in the emission of dust-sized particles (<62.5 μm; see Goudie and Middleton, 2006). Many highly emissive sources of dust are often found where the entrainment of surface sand at relatively low wind velocities results in the emission of dust via the impact of saltating sand grains on the surface, rather than by direct fluid entrainment (Rice, Willetts and McEwan, 1996; Lu and Shao, 1999; Goudie and Middleton, 2006).

Once entrainment of grains has been accomplished, however, the relative importance of the two processes of grain entrainment is still not entirely clear (Ungar and Haff, 1987). Rice (1991) has shown that when impact dislodgement becomes established, direct fluid entrainment still accounts for a significant proportion of erosion. However, the consensus from empirical observations (Bagnold, 1941; Willetts and Rice, 1985a) and theoretical examinations (notably by Anderson and Haff, 1988; Werner, 1990; Haff and Anderson, 1993; McEwan and Willetts, 1994) is that grain impact is the principal mechanism by which saltation is maintained. Hence, once grains are entrained at the fluid threshold, a reduction in wind velocity will not necessarily result in a reduction in sand transport so long as the velocity remains above the impact threshold, as shown in Figure 18.11.

18.4 Determining the threshold of grain entrainment

The determination of the wind velocity (u_t) or shear velocity (u_{*t}) threshold for grain entrainment is a fundamental necessity in the understanding and prediction of the aeolian sediment transport system. There are many means by which threshold values can be ascertained, although the majority of techniques have been developed in the controlled conditions of the wind tunnel and involve the observation of the onset of sediment motion with increasing wind velocity (see, for example, Bagnold, 1941; Kawamura, 1951; Zingg, 1953). While such studies have helped in the development of relationships such as that shown in Figure 18.11 they have not been able to greatly improve our knowledge of thresholds in natural field environments where the identification of the onset of sediment ransport is complicated by a constantly varying wind velocity.

However, the development of electronic grain impact sensors such as the Sensit (Stockton and Gillette, 1990) and the Safire (Baas, 2004) have enabled us to investigate entrainment thresholds in the field more fully. A typical field data series of wind velocity and sediment transport occurrence is shown in Figure 18.12 and demonstrates the difficulties inherent in identifying one single entrainment threshold. While a threshold value could be considered as the wind velocity at which the first grains are seen to be entrained into transport (A in Figure 18.12), it is also clear that there are many instances where no sediment transport is detected despite winds being far greater than this. In order to provide a quantitative means by which a single threshold can be determined from such data series, Davidson-Arnott, MacQuarrie and Aagaard (2005) recommend calculating the mean of the five minimum windspeeds at which transport is observed and the mean of the five maximum windspeeds at which no sand transport is observed, thus providing a range in the value of possible entrainment thresholds.

Stout and Zobeck (1996, 1997) and Stout (2004) tackled the problem by developing the time fraction equivalence method, which has begun to be applied in a variety of field situations (Wiggs, Atherton and Baird, 2004; Wiggs, Baird and Atherton, 2004; Davidson-Arnott, MacQuarrie and Aagaard, 2005). The technique is based on the observation that sand transport is intermittent (when measured at a frequency of ~1 Hz) and the assumption that the fraction of time in which sand transport events occur (described by the intermittency factor, γ) should be equivalent to the fraction of time that the wind velocity is equal to or greater than the threshold value for a particular surface and at a particular time. By converting measurements of sand transport and wind speed into two binary series, the threshold value that gives rise to this equivalence can be determined by iteration (Stout and Zobeck, 1996). A modification of the procedure employing a cumulative frequency plot of wind velocity has also been presented by Wiggs *et al.* (2004a) and is demonstrated in Figure 18.13.

An entrainment threshold calculated using the time fraction equivalence method is shown as B in Figure 18.11. While the technique allows a good objective method by which to calculate a threshold, it is clear that the inherent variability in sand transport data again results in a threshold whereby there are instances of both sand transport events below threshold and no sand

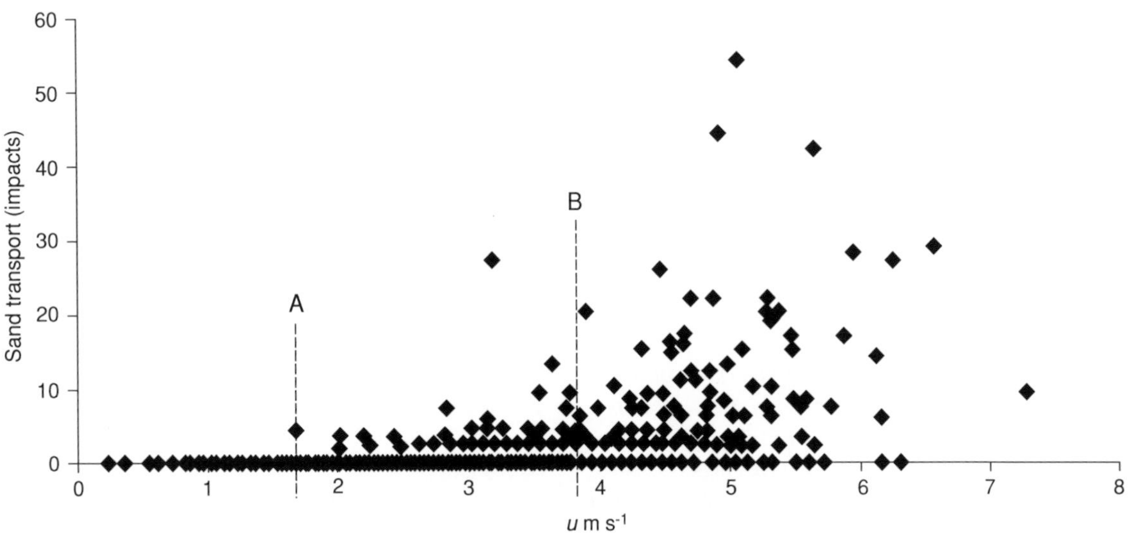

Figure 18.12 Time series data of wind speed and detected sand transport (at 1 Hz sampling frequency). Entrainment thresholds defined by (A) the onset of sediment transport and (B) the time fraction equivalence method are shown (from Wiggs, Atherton and Baird, 2004).

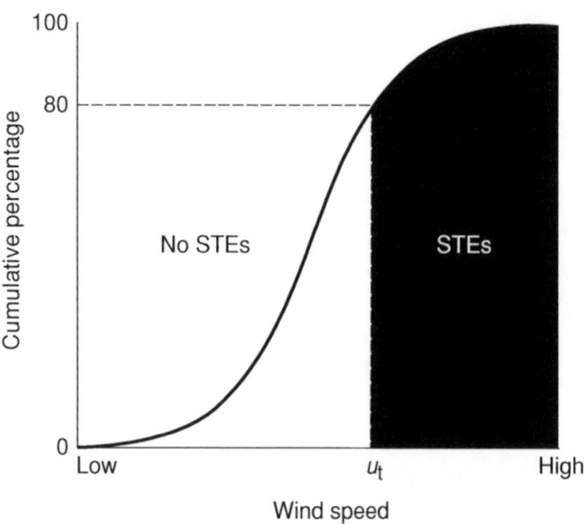

Figure 18.13 An idealised example of the modified time fraction equivalence method for determining the threshold of sediment entrainment using a cumulative frequency plot of wind velocity and a calculated saltation intermittency value (γ) of 20 %. The threshold corresponds to the 80th percentile value of wind velocity (from Wiggs, Atherton and Baird, 2004).

transport events above threshold. Wiggs, Atherton and Baird (2004) noted that in their data set such problems resulted in <60 % of their sand transport events accounted for by the calculated threshold.

Improvements to the explanatory power of thresholds calculated in this way can be achieved if a lag time is assumed between the changing wind velocity and the response of the sand transport mechanism. Many investigators have found that the response time of the sand transport system to wind variability is of the order of ~1 second (Wiggs, Atherton and Baird, 2004; Davidson-Arnott and Bauer, 2009; Weaver, 2008) and so lagging sand transport behind velocity by this amount results in a much closer association between these two data series.

While the identification of a single entrainment threshold remains an appealing prospect, it is clear that variability inherent in the natural system (in terms of paraemeters such as grain size, microtopography, packing, moisture) makes it an unachievable goal. Such complexity in the system means that entrainment thresholds for a particular surface are perhaps more appropriately represented by a range of possible thresholds (Wiggs, Atherton and Baird, 2004; Davidson-Arnott, MacQuarrie and Aagaard, 2005; Davidson-Arnott *et al.*, 2008) whereby the probability of erosion may be modelled by investigation of the cumulative probability density functions of time series data of wind velocity and sand transport (Davidson-Arnott and Bauer, 2009).

An alternative to comparing field-measured time series of wind and sand transport data involves the direct determination of entrainment thresholds from small-scale field wind tunnels. Such experiments reduce wind and surface variability and so tend to give better correlations between sand transport and wind velocity. While traditional portable wind tunnels have provided useful data on entrainment thresholds (Belnap and Gillette, 1998) they are bulky to manoeuvre and so there are limitations on the surfaces that can be sampled. A recent development has been that of the Portable In-Situ Wind Erosion Laboratory (PI-SWERL) (Etyemezian *et al.*, 2007), which is a relatively small circular wind tunnel. The portability of the PI-SWERL offers the ability to obtain numerous replicates of the critical wind speed required for erosion over a broad range of surfaces that have previously been impractical to measure in the field.

18.5 Surface modifications to entrainment thresholds and transport flux

While the relationships shown in Figure 18.11 have been found to be satisfactory for loose, dry, flat and homogeneous surfaces (Williams, Butterfield and Clark, 1994), the critical thresholds of motion and grain transport fluxes on natural sediment beds are also influenced by variations in factors such as sediment mixtures, surface crusting, surface slope, moisture and vegetation. The relationships between these parameters are frequently complex and are not yet fully understood. The effects of vegetation are dealt with above. Three other especially important modifications are provided by surface crusting, surface slope and moisture content.

18.5.1 Surface crusting

A surface crust lying above loose, erodible sediment significantly inhibits the entrainment of sediment into the wind. In wind tunnel studies Zobeck (1991) found that soils without surface crusts were 40–70 times more erodible than crusted soils and similar impacts on erodibility have been found by investigations in wind tunnels (Rice, Mullins and McEwan, 1997; Rice and McEwan, 2001; McKenna-Neuman and Maxwell, 2002) and the field (Rajot *et al.*, 2003; Houser and Nickling, 2001a, 2001b; Hupy, 2004). However, the effect of crust characteristics on aeolian processes is still poorly understood and so its impact is often disregarded in aeolian transport models (Zobeck *et al.*, 2003).

Physical crusts form where limited rainfall impacts on a soil with at least 5 % clay content (Rajot *et al.*, 2003), although Rice and McEwan (2001) state that the greater the proportion of fine (cementing) material, the greater the strength of the crust with surfaces containing <12 % fines being easily eroded. Hupy's (2004) field investigation found that thick, silty, physical crusts could be as protective of a surface as a cover of gravel or grass. In contrast, McKenna-Neuman and Maxwell (2002) undertook a wind tunnel test of the breakdown of microphytic (moss) crusts and found that they were weak compared to physical crusts and could deteriorate under the impact of saltating grains even at low wind velocities. However, they also found that in isolated cases the elasticity of such microphytic crusts could protect the soil beneath under heavy saltation bombardment if the integrity of the filament net in the crusts was strong.

On silt/clay surfaces a crust may significantly protect the underlying fine, dust-sized particulates from erosion. However, where such a sediment bed is of mixed size and includes sand-sized particles that are prone to saltation, the impact of saltating particles at relatively low wind velocities can break down the crust and induce intense dust entrainment. Such bombardment is thought to be the primary process in the entrainment of dust-sized material (Houser and Nickling, 2001a), with the most severe sources of dust emission from playa surfaces being those associated with a surface cover of erodible sand (Cahill *et al.*, 1996).

18.5.2 Bedslope

Despite the potential importance of surface slope on the threshold of entrainment and saltation trajectories of particles (some windward dune slopes may reach angles aproaching 30°), there has been relatively little empirical research in this area. Theoretical analyses have been presented by Allen (1982) and Dyer (1986) while Hardisty and Whitehouse (1988), Iversen and Rasmussen (1994) and Rasmussen, Iversen and Rautahemio (1996) have used portable field and tilting laboratory wind tunnels. Hardisty and Whitehouse (1988) found good agreement between theory and practice, with results indicating a relative increase in critical threshold on positive slopes (upslope) and a decrease on negative slopes (downslope) (Figure 18.14). Computational fluid dynamics (CFD) modelling by Huang, Shi and van Pelt (2008) confirmed this influence of dune slope angle on particle entrainment while the field work, modelling and wind tunnel testing of Tsoar, White and Berman (1996) and White and Tsoar (1998) suggest that only particles < 230 μm are capable

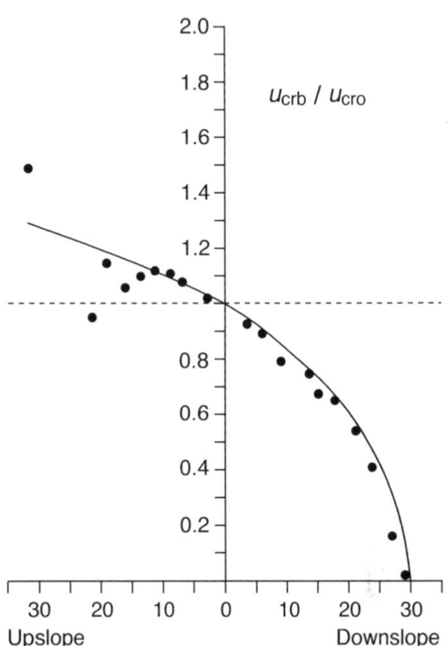

Figure 18.14 The effect of bedslope on threshold of grain entrainment: solid line = theoretical models of Allen (1982) and Dyer (1986), circles = experimental data (from Hardisty and Whitehouse, 1988).

of climbing a 20° slope under normal wind conditions, explaining the accumulation of large particles at the base of dune windward slopes.

Surface slope may not only have an important effect on the *threshold* of sediment movement but also on the *rate* (or flux) of sediment transport. Despite much discussion of the potential effect (Hunt and Nalpanis, 1985; Nalpanis, 1985; Hardisty and Whitehouse, 1988; Whitehouse and Hardisty, 1988; Iversen and Rasmussen, 1994), conclusions are far from complete. Bagnold (1941) derived a simple geometric relationship to describe the effect of bedslope on the sand transport rate, although Howard *et al.* (1977) found that the Bagnold formula had only a minor influence on transport rate predictions and characterised the actual sand transport rate no better than if bedslope was not taken into account. In contrast, Hardisty and Whitehouse (1988) found that the sand transport rate was much more dependent upon surface slope than predicted by the Bagnold relationship (Figure 18.15).

18.5.3 Moisture content

It is clear that a high moisture content in surface sediment can substantially increase entrainment thresholds and reduce transport potential, except where the erosive power provided by wind speed is particularly forceful

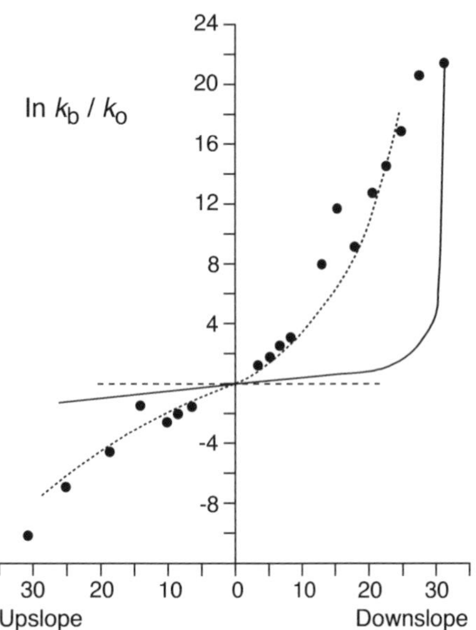

Figure 18.15 The effect of bedslope on sand transport rate: solid line = geometric relationship (Bagnold, 1941), circles = experimental data (from Hardisty and Whitehouse, 1988).

(Wiggs, Baird and Atherton, 2004; Bauer *et al.*, 2009). However, its influence is complex, variable and not well understood. The majority of our knowledge comes from wind tunnel investigations (e.g. Cornelis and Gabriels, 2003; Han *et al.*, 2009; McKenna-Neuman and Nickling, 1989), although recent field studies on beaches have developed our comprehension of the complexities involved (Bauer *et al.*, 2009; Davidson-Arnott and Bauer, 2009; Wiggs, Baird and Atherton, 2004). The theoretical basis for available predictive models is that the critical shear velocity required for entrainment increases as a function of the increased surface tension associated with pore moisture. However, as shown in Figure 18.16, there is much disagreement as to the precise mechanism by which this should be applied. In this figure the Hotta *et al.* (1984) and Belly (1964) data are based on empirical wind tunnel measurements while those of Kawata and Tsuchiya (1976) are theoretical.

The data shown in Figure 18.16 suggest that the entrainment system is perhaps very sensitive to changes in moisture status in the range 0–4 %. However, in field experiments conducted by Wiggs, Baird and Atherton (2004) significant limitation on sediment entrainment was found only at higher moisture values (4–6 %). They argued that this was due to spatial inhomogeneity in the wetness of surface sediments such that drier sediment on topographic highs (such as crests of ripples) could be entrained and transported at high values of average surface moisture content. Furthermore, Wiggs, Baird and Atherton (2004) noted that moisture that was adhered to saltating particles could reach 2 % (gravimetric) before any influence on the transport rate could be detected. Saltation of dry sediment is therefore possible across wet surfaces and Sarre (1988, 1990) found that surface moisture contents of up to 14 % had little effect on sand already in transport. Such natural inhomogeneity in surface moisture is not taken account of in the static conditions of wind tunnel tests and such data tend to show a shutdown of the saltation system at lower average surface moisture values.

Complexities in the influence of soil moisture on the saltation system are now being investigated on beaches. Studies by Bauer *et al.* (2009), Davidson-Arnott and Bauer (2009), Jackson and Nordstrom (1997, 1998), Wiggs, Atherton and Baird (2004), Wiggs, Baird and Atherton (2004) and Davidson-Arnott *et al.* (2008) have all noted the spatial and temporal variability in surface moisture content on beaches and have investigated the impacts of such variability on saltation dynamics. In particular, the sensitivity of sediment entrainment to the moisture status of a very thin layer of surface sediment has been explored. Once these surface grains have dried to a sufficient extent to allow entrainment the damper grains below are revealed and critical entrainment thresholds rise once again. This high temporal dynamism in entrainment thresholds, coupled with variation in wind speeds, produces a highly intermittent saltation system where erosion of surface sediment at a specific location can occur at a range of wind speeds within a very short space of time. Davidson-Arnott and Bauer (2009) therefore suggest that there exists a range of entrainment thresholds, rather than a single fixed value based on an average moisture content and grain size. Such sensitivity of the saltation system to the moisture status of the top few grains results in the dynamics of the system being readily influenced by air humidity. McKenna-Neuman and Sanderson (2008) and McKenna-Neuman (2003, 2004) note that colder airflows support much higher sand transport rates than warmer air. They partly explain this by the decreased adsorption of moisture from the air to surface grains and hence reduced interparticle cohesion and lower entrainment thresholds in the cold air case.

With the requirement for higher spatial and temporal resolution measurements of the moisture status of surface sediment new methods of measurement are being explored that do not rely on gravimetric analysis of grab samples. Such grab samples may include substantial amounts of subsurface (and wetter) sediment. One new method receiving attention is that of measurement of surface brightness using calibrated digital photography, which has been shown to be a good indicator of the amount of pore water

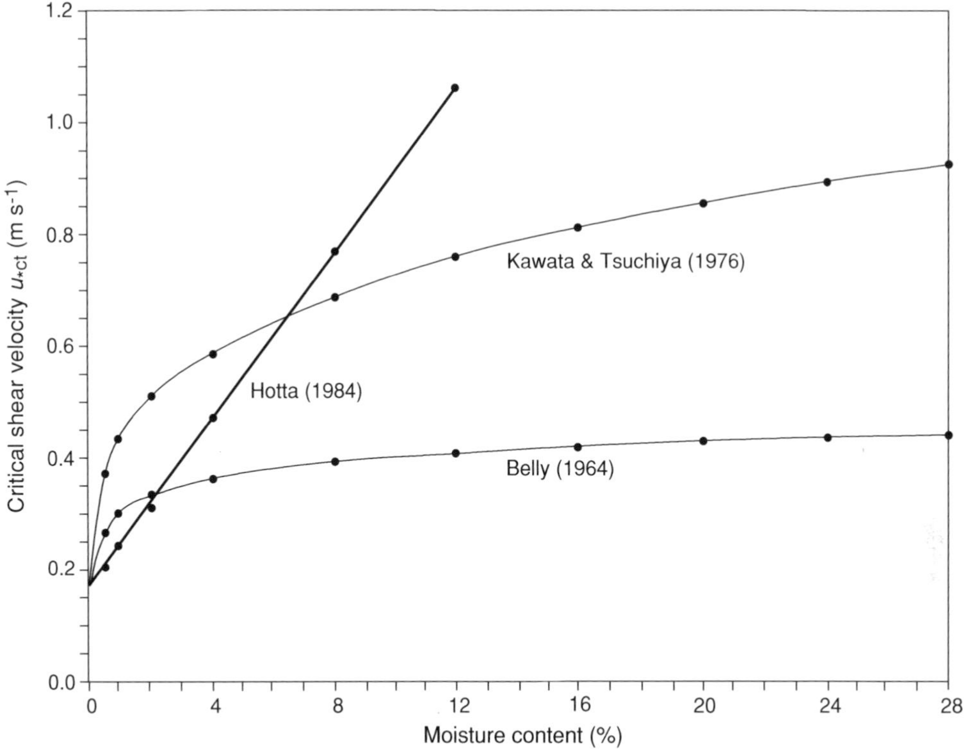

Figure 18.16 Critical shear velocity for sediment entrainment as a function of sand moisture content (from Sherman, 1990).

present (McKenna-Neuman and Langston, 2006; Darke and McKenna-Neuman, 2008; Darke, Robin and Ollerhead, 2009).

18.6 Modes of sediment transport

Once entrained into the airflow, sediment may be transported by any one of four mechanisms, principally dependent upon the sediment grain size (Bagnold, 1941, and Figure 18.17). The following modes of transport are not discreet classes and the transition from one to another may not be well defined.

18.6.1 Suspension

Small particles of less than 62.5 μm (Goudie and Middleton, 2006) whose settling velocity may be very small in comparison to the combined effects of wind lift and drag might be transported in *suspension,* with the vertical profile of flux decreasing with height and described by a power function (Nickling, McTainsh and Leys, 1999; Wang *et al.*, 2008). The turbulent motion of airflow can keep very fine sediment suspended for many days, high in the atmosphere, which may ultimately be deposited as loess or dust (see Chapter 20). At the coarse end of this spectrum, some material may be transported in modified saltation (Hunt and Nalpanis, 1985), where saltation particle trajectories are affected by wind turbulence.

18.6.2 Creep

Larger particles tend to be transported by one of the three modes of bedload (or contact) transport: creep, reptation or saltation. Surface *creep* describes the rolling action of coarse particles (0.5–2.0 mm) as a result both of wind drag on the grain surfaces and the impact of high-velocity saltating grains. In the case of the finer particles in this size range, a creep movement may become apparent immediately prior to the onset of saltation (Nickling, 1983; Willetts and Rice, 1985a). The difficulty in isolating creep in experimental observations has made the task of defining the relative importance of creep in terms of other transport mechanisms problematic. Willetts and Rice (1985b) estimate that creep accounts for approximately one quarter of the bedload transport rate while Dong *et al.* (2002) found in their wind tunnel tests that the creep transport rate increased with wind speed but decreased with grain size, while the creep fraction varied widely, ranging from about 4 to 29 %, averaging at about 9 %.

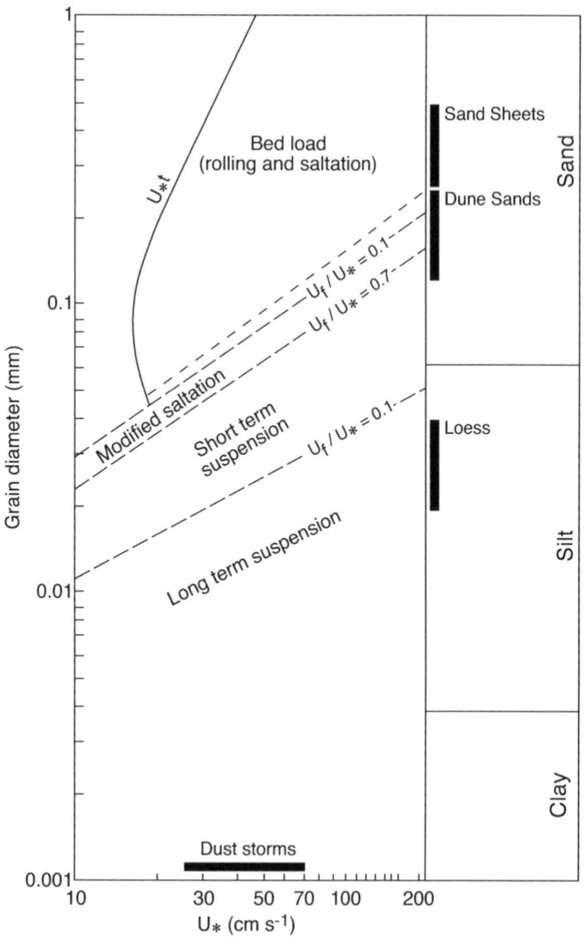

Figure 18.17 The relationships between grain diameter, shear velocity and mode of sediment transport showing the distinction between the suspension of dust-sized sediment and the bedload transport of sand-sized sediment; U_f is the particle fall velocity (after Tsoar and Pye, 1987).

18.6.3 Reptation

The low hopping of several grains consequent upon the high-velocity impact of a single saltating grain has been termed *reptation* (Anderson and Haff, 1988) and is an important transitional state between the modes of creep and saltation. On impact and subsequent rebound, a saltating grain may lose 40 % of its velocity (Willetts and Rice, 1989; Anderson and Haff, 1991; Haff and Anderson, 1993). This energy imparted to the grain bed results in the ejection (or 'splashing') of perhaps 10 other grains (Werner and Haff, 1988) with velocities at approximately 10 % of the impact velocity, often too low to enter into saltation (Willetts and Rice, 1989; Anderson and Haff, 1991). Hence, each grain takes a single hop (Figure 18.18), the majority in a downwind direction (Haff and Anderson, 1993). Much more research is required to understand this process fully, but Anderson, Sørensen and Willetts (1991) suggest that it could be extremely important in near-surface aeolian transport and more recent modelling work confirms that transport flux models produce more realistic simulations with the inclusion of a reptation component (Andreotti, 2004; Namikas, 2003).

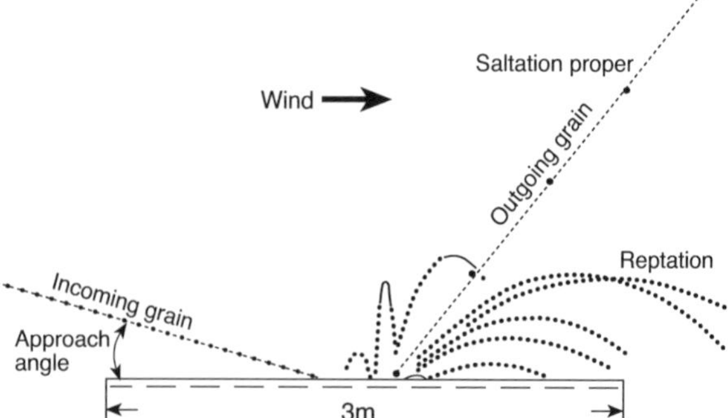

Figure 18.18 The process of reptation where the impact of a high-velocity saltating grain ejects other grains into the airflow (after Anderson, 1987).

18.6.4 Saltation

The most intensively researched mode of aeolian transport is that of *saltation*. This is the characteristic ballistic trajectory of grains (≈0.06–0.5 mm in diameter) as they are ejected from the grain bed, are given horizontal momentum by the airflow, descend to impact the grain bed and then continue 'leaping' downwind (see Figure 18.19). The impact of saltating grains on the surface often leads to the splash of other grains into the airflow, which may then undergo reptation or, if sufficient momentum is transferred, saltation. Hence, saltation is a very efficient transporting mechanism whereby a few saltating grains can rapidly

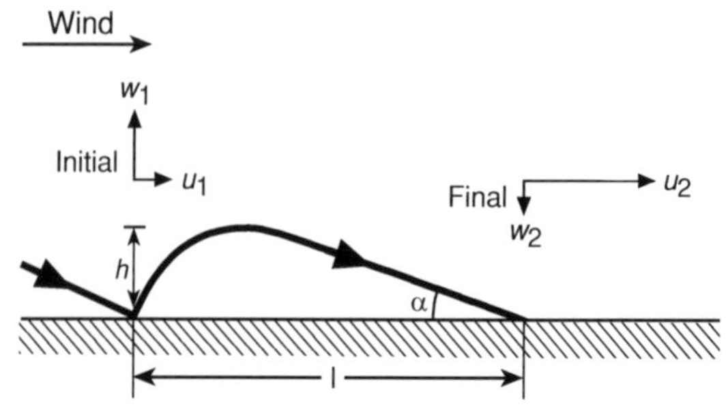

Figure 18.19 The ballistic trajectory of a saltating sand grain; w and u represent vertical and horizontal velocities respectively (after Bagnold, 1941).

Figure 18.20 Saltating sand being swept from the crest of an 80 m high linear dune in the Namib Desert. Increasing wind speeds and saltation impacts result in progressively more intense saltation from dune foot to dune crest (photo: author).

induce mass transport of sediment in a cascading system (Figure 18.20), often termed the 'fetch effect' (Gillette *et al.*, 1996). A limit to the amount of sediment in saltation is set when an equilibrium is established between the fluxes of sediment and air as a result of the momentum extracted from the airflow by the saltating grains. This momentum extraction causes a considerable reduction in near-surface wind speeds during saltation (Owen, 1964) and results in the majority of grain ejections being intiated by impact from other saltating grains rather than by direct fluid entrainment.

Bagnold (1936, 1941) was the first to appreciate the ballistic trajectory of saltating grains and there are numerous studies that have investigated the subprocesses of grain entrainment, trajectory, bed collision and velocity profile modification, principally relying on wind tunnel and numerical modelling approaches (e.g. Willetts and Rice, 1986a, Anderson and Hallett, 1986; Werner and Haff, 1988; Anderson and Haff, 1988; Rasmussen and Mikkelsen, 1991; Werner, 1990; McEwan, Willetts and Rice, 1992; McEwan, 1993; Haff and Anderson, 1993; Dong *et al.*, 2002; Kok and Renno, 2009b; Namikas, 2003; Rasmussen and Sørensen, 2008; Bauer, Houser and Nickling, 2004).

Bagnold (1941) and Chepil (1945) both identified steep initial take-off angles of grains at angles approaching 90°. However, other studies have recognised a range of lift-off angles of between 15 and 70° (White and Schultz, 1977; Nalpanis, 1985; Willetts and Rice, 1985a; Anderson, 1989; Namikas, 2003; Kok and Renno, 2009b). The actual grain trajectory depends very much on the height of bounce of the grain into the boundary layer. Wind speeds increase at a logarithmic rate away from the surface (see Figure 18.2); hence, the higher a grain leaps, the more momentum it can extract from the airflow and the faster and longer is its saltating jump. The manner of momentum extraction from the airflow is demonstrated by the wind tunnel experiments of Rasmussen and Sørensen (2008), who recorded 1:1 ratios between air and grain speed within 80 mm of the surface, increasing to a ratio of 2.0 at 5 mm height. Grain size also has an influence on trajectory paths, with larger grains rebounding from the surface at lower speeds and following a shorter and lower trajectory (Namikas, 2003). Similarly, Willetts (1983) noted that grain shape influences the trajectory with platy grains tending to have lower and longer paths than spherical grains. The length of jump is thought to be of the order of 12–15 times the height of bounce (Livingstone and Warren, 1996) and the trajectory is also influenced if the grain starts to spin after a glancing impact. Reports of spin rates reaching over 400 r.p.s (White and Schultz, 1977; White, 1982) can induce a lift force (termed the Magnus effect) that may extend saltation trajectories. Particle collisions with the surface effectively convert near-horizontal momentum (with angles of descent up to 10°) to a vertical momentum via ejection or rebound. Ungar and Haff (1987) noted that the number of grains splashed up into the airstream by an impacting grain was proportional to the square of the impacting grain velocity, which may be up to 5 times its initial velocity (Anderson and Hallett, 1986).

The height of the saltation layer is dependent on the wind speed (Dong *et al.*, 2002), the grain size in transport and surface characteristics. Bagnold (1941) recognised that saltation leaps were higher on a pebbly or hard rock surface than on a loose sand surface because hard surfaces are less absorbent of the grain momentum on each bounce. Pye and Tsoar (1990) quote a maximum saltation height on such surfaces of 3 m, although heights of ~0.2 m are more commonly found. However, the mass of saltating particles decreases very rapidly with height and has been described by a declining exponential function (Rasmussen, Sørensen and Willetts, 1985; Dong *et al.*, 2002; Rasmussen and Sørensen, 2008), with up to 80 % of all transport taking place within 2 cm of the surface (Butterfield, 1991).

18.7 Ripples

The sediment transport mechanisms of creep, reptation and saltation combine to create ripples, a mobile bedform that contributes to bulk sediment transport. Despite decades of research these features still remain enigmatic and our understanding of them is incomplete. Wind tunnel and numerical modelling investigations on the controls on aeolian ripple dynamics and wavelengths was common up to the 1990s. More recent modelling work has treated the problem from a nonlinear perspective (Anderson, 1990; Yizhaq, Balmorth and Provenzale, 2004) or by using a complex systems approach employing aspects of emergent behaviour and self-organisation in cellular automaton models (Werner and Gillespie, 1993; Anderson and Bunas, 1993; Baas, 2002, 2007; Pelletier, 2009).

Bagnold (1941) recognised three categories of ripples: *normal* or *ballistic*; *granule*, *sand ridge* or *mega*; and *fluid drag* or *aerodynamic*. They generally form transverse to the wind direction in repeated patterns that continually adjust in response to variability in windflow. Ripples commonly have wavelengths of 1–25 cm and heights of 0.5–1.0 cm (Sharp, 1963) with a asymmetrical profile consisting of windward slopes of about 10° (Mabbutt, 1977). However, megaripples can achieve crest-to-crest wavelengths exceeding 20 m, are characterised by a bimodal distribution of coarse and fine particle sizes (Yizhaq *et al.*, 2008) and tend to be more symmetrical in profile (Greeley and Iversen, 1985). This symmetry may be related to shifts in wind direction, suggesting that the form and spacing of larger ripples are less dynamic. However, cellular automaton models of ripple formation also suggest that grain size has a fundamental control on ripple geometry (Anderson and Bunas, 1993; Baas, 2007). Theoretical and empirical studies by Ellwood, Evans and Wilson (1975) led to the conclusion that mega- and normal ripples do not form distinct populations, but are the upper and lower bounds of a continuum in wavelengths as a result of differences in grain size (Ellwood, Evans and Wilson, 1975) and wind speed, with wavelength responding positively to increases in both controls (Sharp, 1963; Claudin and Andreotti, 2006; Andreotti, Claudin and Pouliquen, 2006).

18.7.1 Ballistic ripples

Bagnold's (1941) impact mechanism theory has been widely used to explain ripple formation. In this theory ripple wavelength is related to a characteristic or mean saltation path length. Surface irregularities act as erosional and depositional nuclei for moving particles (Figure 18.21). The windward side of an irregularity is bombarded by more saltating grains per unit area than the sheltered lee

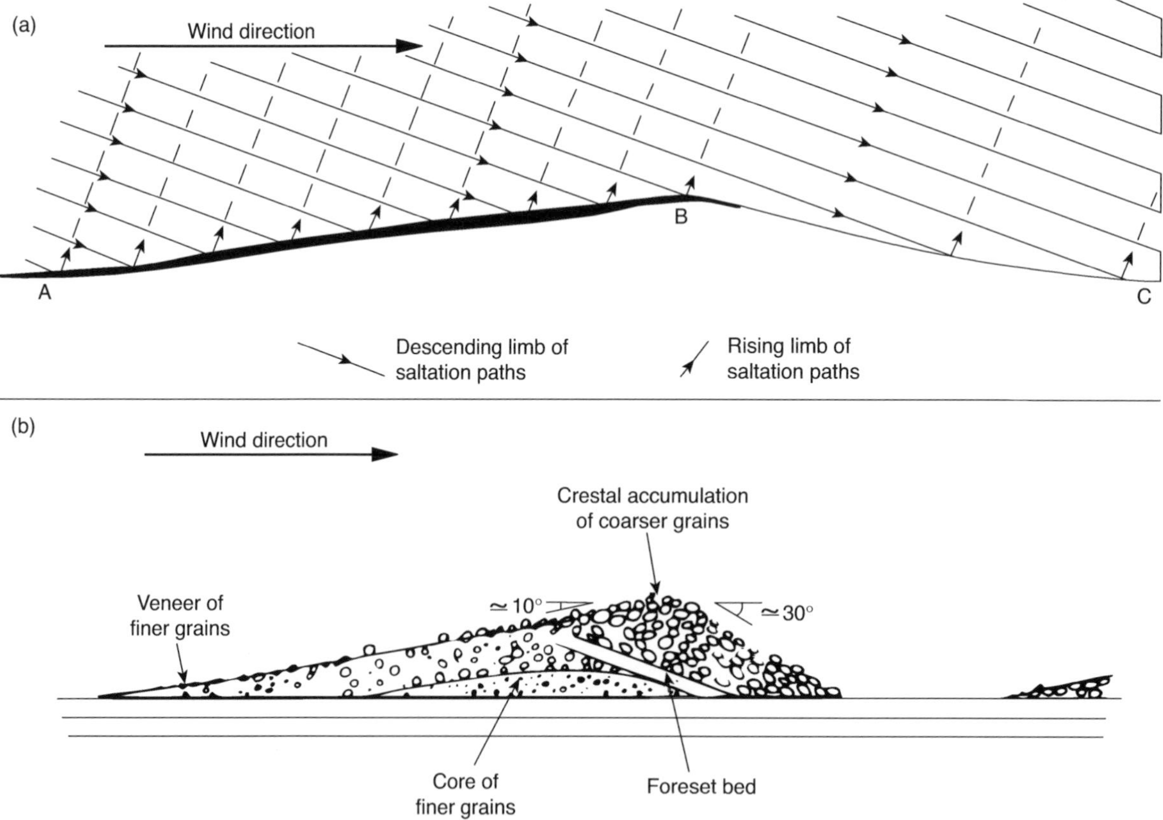

Figure 18.21 (a) Ripple development (after Bagnold, 1941). *AB* is preferentially bombarded by descending grains compared with *BC* (see text for details). (b) Laminae and grain-size distribution within a ripple (from diagrams in Sharp 1963).

side and so is an area of net erosion. The preferential loss of particles on the windward side of the emergent ripple creates a new zone of bombardment downwind, at a distance equivalent to the mean saltation path length. Hence, the ripple pattern migrates downwind as a series of alternating zones of erosion and protection. Coarser grains, which are not set into saltation by the impact of bombarding grains, creep forward to accumulate in the less bombarded crestal zone.

However, Sharp (1963) notes that ripple spacing increases with time and so it is unlikely to be related to the mean saltation path length under steady wind conditions. In contrast, he suggested that the angle of incidence of descending particles and ripple height determined ripple spacing. Height can perhaps be seen as crucial; given the narrow range of ripple slope angles that have been recorded, it must geometrically affect the length of the windward and lee slopes. Thus, the minimum wavelength must increase with height (Brugmans, 1983) and ripple wavelength must increase correspondingly as ripple growth proceeds. As the ripple protrudes further into the boundary layer and coarser grains on the crest are more readily moved forward, the height of the ripple becomes limited (Greeley and Iversen, 1985).

Wind tunnel and theoretical studies have indicated how saltation dynamics might influence ripple wavelengths. Saltating particles require vertical lift-off momentum while creeping grains require forward momentum. Willetts and Rice (1986a, 1986b) have calculated that bombarding grains transfer relatively more vertical energy on steeper slopes and more horizontal energy on shallower slopes. Thus, saltation is favoured from the middle of the windward side of ripples and creep activity increases towards the crest, favouring the accumulation of coarse grains (Figure 18.21).

Anderson (1987) has suggested that while ripple wavelength is affected by grain trajectory lengths, it is not the same distance as the mean saltation path length. A criticism of Bagnold's (1941) theory was that it relied on a very narrow range in saltation path lengths, although experimental studies showed that saltation trajectories were often widely distributed and variable (Mitha *et al.*, 1986). Anderson's (1987) model therefore includes a wide range of saltation trajectory lengths and speeds that drive a reptating population. In this model ripples are formed through the accumulation of these reptating grains, with a wavelength that is a function of the probability distribution of the total trajectory population and the ejection rate of the reptating grains. This reptation control of ripple development has support from the wind tunnel experiments of Willets and Rice (1986b) and simulations of Werner and Haff (1988).

18.8 Prediction and measurement of sediment flux

In order to quantify surface erosion or deposition, geomorphologists are often interested in the changing rate of sediment flux in space and/or time. There are many equations to calculate mass sand flux from wind velocity data, nearly all derived from theoretical or wind tunnel experimental work (see Greeley and Iversen, 1985, for a review). All the expressions tend to the form of

$$q = \mathrm{A}u_*^3 \tag{18.9}$$

The two types of relationship frequently used to calculate sand flux are typified by the expressions of Bagnold (1941) and Kawamura (1951), the latter incorporating a specific term for the threshold shear velocity for entrainment:

$$q = C(d/D)^{0.5}u_*^3\rho/g \quad \text{(Bagnold, 1941)} \tag{18.10}$$

$$q = K_k(u_* - u_{*\mathrm{ct}})(u_* - u_{*\mathrm{ct}})^2\rho/g \quad \text{(Kawamura, 1951)} \tag{18.11}$$

where

q = sand transport rate ($\mathrm{gm}^{-1}\ \mathrm{s}^{-1}$)

C = constant (1.8 for naturally graded dune sand)

d = grain diameter

D = standard grain diameter (0.25 mm)

ρ = air density

K_k = constant (2.78)

u_{*ct} = threshold of grain entrainment

Well-known problems with the Bagnold (1941) expression include the fact that it predicts sand movement below the threshold of entrainment and it commonly predicts rates that are considered too low at high values of shear velocity (Sarre, 1987). Owing to the inclusion of a threshold term in the Kawamura (1951) calculation, it is more accurate at lower levels of shear velocity. However, this expression only incorporates the effect of grain size in the threshold term (u_{*ct}), despite the fact that it is also likely to have an important effect on the transfer of momentum on grain impact.

Zingg (1953) followed a similar argument to Bagnold (1941) but used a $^3/_4$ power function:

$$q = C_2(d/D)^{3/4} u_*^3 \rho / g \qquad (18.12)$$

where

$C_2 =$ constant (0.83)

Owen (1964) found that this expression was more accurate over a wider range of particle sizes than the Bagnold (1941) formula, but it still omits a term for the threshold of entrainment. Such a term was incorporated into a later analysis of Bagnold (1956), which was subsequently refined by Lettau and Lettau (1978) into a very commonly used expression:

$$q = C_3(d/D)^{0.5}(u_* - u_{*ct})u_*^2 \rho / g \qquad (18.13)$$

where

$C_3 =$ constant (4.2)

Despite this array of transport equations, no specific model has proven to be broadly applicable under a variety of natural environmental conditions. This is partly because the majority of published models have been derived from theory or experimentally from wind tunnel observations with homogeneous sediment beds (Spies, McEwan and Butterfield, 1995; Butterfield, 1999; Bauer, Houser and Nickling, 2004), without sufficient testing or development in the field. Field testing that has been accomplished (Sarre, 1987; Sherman, 1990; Wiggs, 1992) has shown a great deal of variation between the observed rates and those predicted as a function of u_* (Figure 18.22). Given the natural variability and complexity over short spatial and temporal scales in grain size, moisture status, bedslope and wind conditions, as previously described, such weak relationships are not surprising.

Furthermore, field calibration of sediment flux equations is severely hampered by the errors inherent in measuring sand flux in the field. In order to measure the sand transport rate, a collecting device has to be inserted into the airflow for a known period of time. This inevitably causes airflow disturbance and hence error in flux measurement. Two major problems with sand traps are those of back pressure and scouring of sand around the base of traps. To reduce these errors, the majority of traps are of a

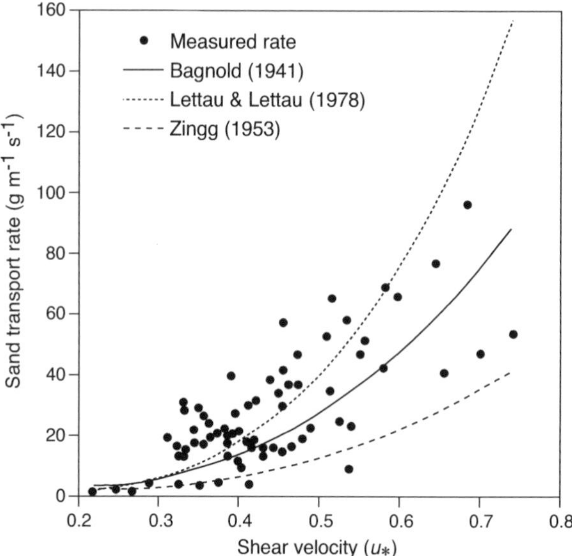

Figure 18.22 Relationship between shear velocity, measured sand transport rate and that calculated from three sand transport rate formulae (after Wiggs, 1992).

thin vertical design (hence presenting a minimum disturbance to the wind flow) and incorporate the bleeding of air from within the trap to reduce backpressure. The sampling efficiencies of the various types of trap vary considerably and have been reported to be as low as 20 % (Knott and Warren, 1981) and as high as 70 % (Gillette and Goodwin, 1974; Marston, 1986). However, the determination of efficiencies is fraught with problems and Jones and Willetts (1979) have shown that small differences in the operation of traps in the field may give rise to large contrasts in efficiency.

Substantial improvements in sampling efficiency have been provided by the incorporation of wings extending downwind from trap openings, giving the trap a wedge-shaped appearance (Figure 18.23). This further decreases backpressure and allows sand flux collectors to become almost isokinetic in terms of their flow characteristics (Nickling and McKenna-Neuman, 1997). While these designs have achieved near 90 % sampling efficiencies in wind tunnel tests their performance in the field is severely restricted by their sensitivity to the incident angle of the wind. Nickling and McKenna-Neuman (1997) report that incident angles as low as 5°, which may occur frequently during periods of measurement, severely reduce the sampling efficiency of such traps (McKenna-Neuman, Lancaster and Nickling, 2000; Namikas, 2003).

More recent attempts to characterise saltation activity in the field have involved the use of saltation impact responders. A variety of designs exist but all operate on a

Figure 18.23 A series of isokinetic wedge-shaped sand traps measuring sand flux in the field. For comparison two saltation impact responders can also be seen at the base of the sonic anemometers towards the left of the photograph (photo: author).

similar basis of responding to the individual impacts of saltating grains on a sensitive surface. In the case of the Saltiphone, this surface is a highly responsive microphone (Arens, 1996; Schönfeldt and von Löwis, 2003; Sterk, Jacobs and van Boxel, 1998; Zobeck *et al.*, 2003), while in the case of the Sensit and Safire, the surface is a piezoelectric transducer (Gillette and Stockton, 1989; Stout and Zobeck, 1997; Baas, 2004). The advantage of such instruments over traditional sand traps is that they are small, offer little resistance to the flow and are omnidirectional in response. They also allow high-frequency and instantaneous assessment of saltation activity, something that can only be achieved with a sand trap if a dynamic load cell is incorporated into the design, a technique that is generally restricted to wind tunnel applications (although see McKenna-Neuman, Lancaster and Nickling, 2000). However, while such sensors have been shown to be very capable of identifying the onset of saltation and intermittency in saltation dynamics, they are less reliable for quantifying the actual saltation flux (van Pelt, Peters and Visser, 2009).

A considerable difficulty in characterising saltation activity using sand flux collectors or impact responders in the field arises from the very nature of saltation. Baas and Sherman (2005) and Baas (2008) recognise that saltation does not occur as a homogeneous 'curtain' across an eroding surface; rather it most often occurs in the form of temporally and spatially discontinuous sand streamers that 'snake' across the surface in an ever-changing weaving pattern (Figure 18.24). This poses significant measurement problems, for within a continuously saltating environment sand traps and saltation impact responders are sometimes in contact with these streamers but, equally,

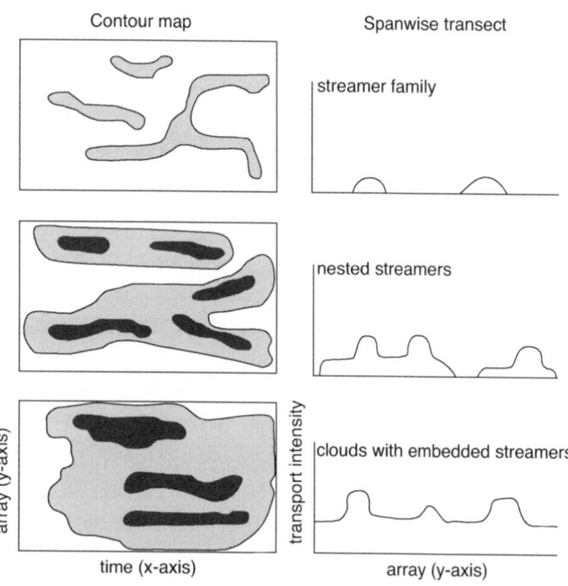

Figure 18.24 Characteristics of observed sand streamer patterns in plan view (left) and cross-section (right) (from Baas and Sherman, 2005).

often they are not. The spatiotemporal dynamics of these sand streamers in differing wind conditions is not known and so the impact they have on measurements of saltation activity and flux provided by sand traps and impact responders is also uncertain. The controls on the formation and development of sand streamers is unclear, but it seems likely that they are governed by near-surface turbulent structures in the wind (Baas, 2008).

18.9 The role of turbulence in aeolian sediment transport

As discussed above, there is an increasing understanding that sediment entrainment and transport can be highly intermittent under many environmental conditions. While variability in surface conditions can account for some irregularity in entrainment, it is now clear that, in a similar manner to river flows (Bennett and Best, 1996; Kostaschuk and Villard, 1996; Venditti and Bauer, 2005), turbulence in the boundary layer is also a major driving force behind aeolian sediment entrainment and transport (Livingstone, Wiggs and Weaver, 2007; Weaver, 2008; Sterk, Jacobs and van Boxel, 1998; Schönfeldt and von Löwis, 2003). Indeed, it has also been shown that the influence of peak instantaneous turbulent stresses on sand transport dynamics is required to explain the observed development of mobile sand dunes (Castro and Wiggs, 1994; Wiggs, Livingstone and Warren, 1996; Walker and Nickling, 2002, 2003; Weaver and Wiggs, 2010) and sand streamers (Baas, 2008).

The practical difficulties inherent in measuring high-frequency turbulence in sand-laden airflows has resulted in a moderate rate of advance in research. In his wind tunnel experiments, Butterfield (1991, 1993, 1999) investigated the impact of temporally varying winds (of the order of seconds) on saltation behaviour. He discovered that mass flux and aerodynamic roughness responded within about 1 second to an acceleration in flow, but that the response was several seconds longer in a decelerating flow due to the momentum of the grains. These findings led Butterfield (1993) to conclude that with naturally fluctuating winds over a sand bed, the grain-laden boundary layer is in constant adjustment and may rarely achieve an equilibrium state. With the increasing application of sonic anemometers in the field (Walker, 2005; van Boxel, Sterk and Arens, 2004; Weaver, 2008) and the use of high-frequency grain impact sensors (Schönfeldt and von Löwis, 2003; Baas, 2004), investigations have now begun to observe the relationship between saltation dynamics and instantaneous turbulent peaks in wind velocities at much higher frequencies (1 to > 10 Hz). While the relationship between turbulent wind flow and sand transport is complex at such frequencies (Figure 18.25) the association between the parameters

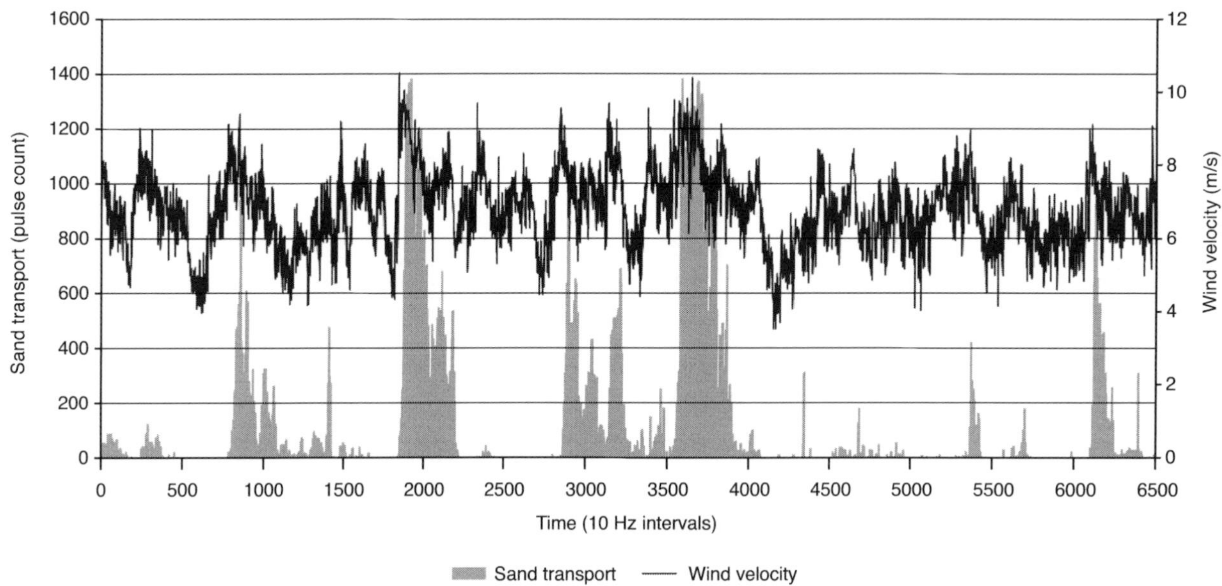

Figure 18.25 Time series of horizontal wind velocity (using a sonic anemometer) and saltation impacts (on a grain impact sensor) measured at 10 Hz. The association is complex but the turbulent frequencies in both series question the application of 'mean' measures of flow for the prediction of sediment transport (from Weaver 2008).

has led many to question the efficacy of 'mean' measures of erosivity, such as mean wind velocity or shear velocity (u_*) to predict sediment fluxes (Stout, 1998; Baas and Sherman, 2005; Weaver, 2008).

Research is now focused towards incorporating instantaneous flow velocities into assessments of shear stress and turbulence structure. By using two- or three-dimensional sonic anemometers, the instantaneous velocity deviations from mean values can be assessed in both the horizontal (u component) and vertical (w component). These data can be used to determine Reynolds shear stress (rather than mean u_*) to assess the magnitude of turbulent shear stresses or in quadrant analysis to assess the existence of structure in the turbulence signal (Lu and Willmarth, 1973). Quadrant analysis divides the flow into four discrete turbulent events depending on the relative signs of instantaneous velocity deviations away from mean values in both the horizontal and vertical flow components (Figure 18.26). Sweeps and ejections contribute to a 'bursting process' in the flow where low-speed ejections of fluid away from the bed are followed by high-speed sweeps of flow towards the bed, and this process has been found to be important in accounting for sediment transport in rivers (Best, 1993). However, field investigations by Sterk, Jacobs and van Boxel (1998), Schönfeldt and von Löwis (2003), van Boxel, Sterk and Arens (2004), Leenders, van Boxel and Sterk (2005) and Weaver (2008) have found that the majority of instances of high aeolian saltation flux are rather associated with sweeps and outward interactions, i.e. flow events that have instantaneous horizontal velocities greater than the mean (Figure 18.26). The reduced significance of the vertical flow component in air (in comparison to water) is likely to be due to its lower density.

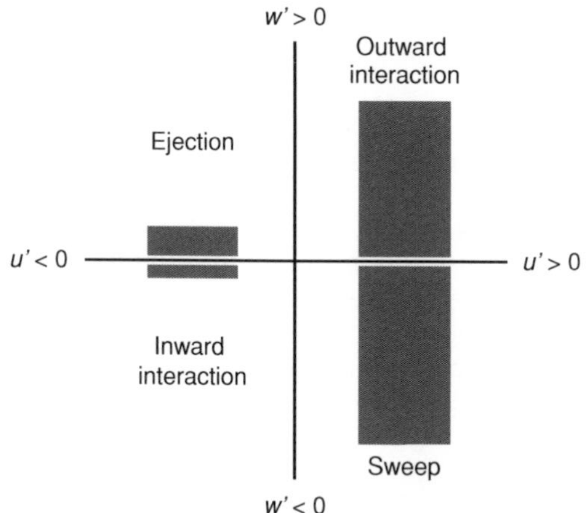

Figure 18.26 Quadrant plot of the four coherent turbulent flow structures, based on instantaneous horizontal (u') and vertical (w') velocity fluctuations from the mean (from Best, 1993). Bars represent the occurrence (% time) of saltation in each quadrant as measured by Weaver and Wiggs (2010).

Much further research is required into the relationships between turbulence structures, sediment transport mechanics and evolving bed topographies. In the aeolian case, it seems likely that initial future research directions will focus on developing sediment transport equations that incorporate some aspect of the instantaneous horizontal wind speed or establishing probabilistic analyses of potential sediment flux given wind turbulence characteristics.

18.10 Conclusions

Continued investigations over the last decade of aeolian sediment mobilisation using fieldwork, wind tunnels and mathematical modelling have significantly enhanced our understanding. Huge progress has particularly been made regarding the impacts of vegetation on erosivity and erodibility, intermittency in entrainment thresholds and the role of turbulence in sediment transport rate determination. Much of this work has necessarily involved a small-scale approach and the continuing challenge is to apply our new knowledge at this scale to research questions at landscape and global scales. With growing concerns about the potential impact of global warming on dryland landscapes, knowledge of aeolian processes, through which much landscape change will occur, will be of increasing importance.

References

Allen, J.R.L. (1982) Simple models for the shape and symmetry of tidal sand waves: (1) statically stable equilibrium forms. *Marine Geology*, **48**, 31–49.

Anderson, R.S. (1987) Eolian sediment transport as a stochastic process: the effects of a fluctuating wind on particle trajectories. *Journal of Geology*, **95**, 497–512.

Anderson, R.S. (1989) Saltation of sand: a qualitative review with biological analogy, in *Symposium: Coastal Sand Dunes* (eds C.H. Gimmingham, W. Ritchie, B.B. Willetts and A.J. Willis), Royal Society of Edinburgh, Proceedings B96.

Anderson, R.S. (1990) Eolian ripples as examples of self-organisation in geomorphological systems. *Earth Science Reviews*, **29** (1–4), 77–96.

Anderson, R.S. and Bunas, K.L. (1993) Grain size segregation and stratigraphy in aeolian ripples modelled with a cellular automaton. *Nature*, **365**, 740–743.

Anderson, R.S. and Haff, P.K. (1988) Simulation of eolian saltation. *Science*, **241**, 820–823.

Anderson, R.S. and Haff, P.K. (1991) Wind modification and bed response during saltation of sand in air. *Acta Mechanica (Suppl.)*, **1**, 21–51.

Anderson, R.S. and Hallett, B. (1986) Sediment transport by wind: toward a general model. *Geological Society of America Bulletin*, **97**, 523–535.

Anderson, R.S., Sørensen, M. and Willetts, B.B. (1991) A review of recent progress in our understanding of aeolian sediment transport. *Acta Mechanica (Suppl.)*, **1**, 1–19.

Andreotti, B. (2004) A two-species model of aeolian sand transport. *Journal of Fluid Mechanics*, **510**, 47–70.

Andreotti, B., Claudin, P. and Douady, S. (2002) Selection of dune shapes and velocities. Part 1: Dynamics of sand, wind and barchans. *European Physical Journal, B*, **28**, 321–329.

Andreotti, B., Claudin, P. and Pouliquen, O. (2006) Aeolian sand ripples: experimental study of fully developed states. *Physical Review Letters*, **96** (2), DOI: 10.1103/PhysRevLett.96.028001.

Arens, S. M. (1996) Rates of aeolian transport on a beach in a temperate humid climate. *Geomorphology*, **17**, 3–18.

Ash, J.E. and Wasson, R.H. (1983) Vegetation and sand mobility in the Australian desert dunefield. *Zeitschrift für Geomorphologie [Suppl.]* **45**, 7–25.

Baas, A.C.W. (2002) Chaos, fractals and self-organisation in coastal geomorphology: simulating dune landscapes in vegetated environments. *Geomorphology*, **48** (1–3), 309–328.

Baas, A. C. W. (2004) Evaluation of saltation flux impact responders (Safires) for measuring instantaneous aeolian sand transport intensity. *Geomorphology*, **59**, 99–118.

Baas, A.C.W. (2007) Complex systems in aeolian geomorphology. *Geomorphology*, **91**, 311–331.

Baas, A. C. W. (2008) Challenges in aeolian geomorphology: investigating aeolian streamers. *Geomorphology*, **93** (1–2), 3–16.

Baas, A. C. W. and Sherman, D. J. (2005) Formation and behaviour of aeolian streamers. *Journal of Geophysical Research*, **110**, F03011, DOI: 10.1029/2004JF000270.

Baddock, M., Livingstone, I. and Wiggs, G.F.S. (2007) The geomorphological significance of airflow patterns in transverse dune interdunes. *Geomorphology*, **87**, 322–336.

Bagnold, R.A. (1936) The movement of desert sand. *Royal Society of London, Proceedings*, **A157**, 594–620.

Bagnold, R.A. (1941) *The Physics of Blown Sand and Desert Dunes*, Methuen, London.

Bagnold, R.A. (1953) *The Surface Movement of Blown Sand in Relation to Meteorology*, Research Council of Israel, Special Publication **2**, pp. 89–93.

Bagnold, R.A. (1956) The flow of cohesionless grains in fluids. *Royal Society of London, Philosophical Transactions*, **A249**, 29–297.

Bauer, B. O., Houser, C.A. and Nickling, W.G. (2004) Analysis of velocity profile measurements from wind tunnel experiments with saltation. *Geomorphology*, **59** (1–2), 81–98.

Bauer, B.O., Sherman, D.J. and Wolcott, J.F. (1992) Sources of uncertainty in shear stress and roughness length estimates derived from velocity profiles. *Professional Geographer*, **44** (4), 453–464.

Bauer, B.O., Davidson-Arnott, R.G.D., Hesp, P.A. *et al.* (2009) Aeolian sediment transport on a beach: surface moisture, wind fetch, and mean transport. *Geomorphology*, **105** (1–2), 106–116.

Belly, P.Y. (1964) Sand movement by wind, US Army Coastal Engineering Research Centre, Technical Memo 1.

Belnap, J. and Gillette, D.A. (1998) Vulnerability of desert biological soil crusts to wind erosion: the influences of crust development, soil texture, and disturbance. *Journal of Arid Environments*, **39** (2), 133–142.

Bennett, S. J. and Best, J. L. (1996) Mean flow and turbulence structure over fixed ripples and the ripple-dune transition, in *Coherent Flow Structures in Open Channels* (eds P.J. Ashworth, S.J. Bennett, J.L. Best and S.J. McLelland), John Wiley & Sons, Ltd, Chichester, pp. 281–304.

Best, J. L. (1993) On the interactions between turbulent flow structure, sediment transport and bedform development: some considerations from recent experimental research, in *Turbulence: Perspectives on Flow and Sediment Transport* (eds N.J. Clifford, J.R. French and J. Hardisty), John Wiley & Sons, Inc., New York, pp. 61–92.

Blumberg, D.G. and Greeley, R. (1993) Field studies of aerodynamic roughness length. *Journal of Arid Environments*, **25**, 39–48.

Bourke, M.C. (2010) Barchan dune asymmetry: Observations from Mars and Earth. *Icarus* **205** (1), 183–197.

Bristow, C.S., Duller, G.A.T. and Lancaster, N. (2007) Age and dynamics of linear dunes in the Namib. *Desert Geology*, **35** (6), 555–558.

Brown, S., Nickling, W.G. and Gillies, J.A. (2008) A wind tunnel examination of shear stress partitioning for an assortment of surface roughness. *Journal of Geophysical Research*, **113**, F2, DOI: 10.1029/2007JF000790.

Brugmans, F. (1983) Wind ripples in an active drift area in the Netherlands: a preliminary report. *Earth Surface Processes and Landforms*, **8**, 527–534.

Buschiazzo, D.E. and Zobeck, T.M. (2008) Validation of WEQ, RWEQ and WEPS wind erosion for different arable land management systems in the Argentinean Pampas. *Earth Surface Processes and Landforms*, **33** (12), 1839–1850.

Butterfield, G.R. (1991) Grain transport rates in steady and unsteady turbulent airflows. *Acta Mechanica (Suppl.)*, **1**, 97–122.

Butterfield, G.R. (1993) Sand transport response to fluctuating wind velocity, in *Turbulence, Perspectives on Flow and Sediment Transport* (eds N.J. Clifford, J.R. French and J. Hardisty), pp. 305–335.

Butterfield, G.R. (1999) Near-bed mass flux profiles in aeolian sand transport: high-resolution measurements in a wind tunnel. *Earth Surface Processes and Landforms*, **24** (5), 393–412.

Cahill, T.A., Gill, T.E., Reid, J.S. *et al.* (1996) Saltating particles, playa crusts and dust aerosols at Owens (dry) Lake, California. *Earth Surface Processes and Landforms*, **21** (7), 621–639.

Castro, I. P. and Wiggs, G. F. S. (1994) Pulsed-wire anemometry on rough surfaces, with application to desert sand dunes.

Journal of Wind Engineering and Industrial Aerodynamics, **52**, 53–71.

Chepil, W.S. (1945) Dynamics of wind erosion: 1. Nature of movement of soil by wind. *Soil Science*, **60**, 305–320.

Claudin, P. and Andreotti, B. (2006) A scaling law for aeolian dunes on Mars, Venus, Earth and for sub-aqueous ripples. *Earth and Planetary Science Letters*, **252** (1–2), 30–44.

Cornelis, W.M. and Gabriels, D. (2003) The effect of surface moisture on the entrainment of dune sand by wind: an evaluation of selected models. *Sedimentology*, **50** (4), 771–790.

Crawley, D.M, and Nickling, W.G. (2003) Drag partition for regularly-arrayed rough surfaces. *Boundary-Layer Meteorology*, **107**, 445–468.

Darke, I., McKenna-Neuman, C. (2008) Field study of beach water content as a guide to wind erosion potential. *Journal of Coastal Research*, **24** (5), 1200–1208.

Darke, I., Robin, D.-A. and Ollerhead, J. (2009) Measurement of beach surface moisture using surface brightness. *Journal of Coastal Research*, **25** (1), 248–256.

Davidson-Arnott, R.G.D. and Bauer, B.O. (2009) Aeolian sediment transport on a beach: thresholds, intermittency, and high frequency variability. Geomorphology, **105** (1–2), 117–126, DOI: 10.1016/j.geomorph.2008.02.018.

Davidson-Arnott, R.G., MacQuarrie, K. and Aagaard, T. (2005) The effect of wind gusts, moisture content and fetch length on sand transport on a beach. *Geomorphology*, **68** (1–2), 115–129, DOI: 10.1016/j.geomorph.2004.04.008.

Davidson-Arnott, R.G.D., Yang, Y., Ollerhead, J., Hesp, P.A. and Walker, I.J. (2008) The effects of surface moisture on aeolian sediment transport threshold and mass flux on a beach. *Earth Surface Processes and Landforms*, **33** (1), 55–74, DOI: 10.1002/esp.1527.

Dong, Z., Liu, X., Wang, H. *et al.* (2002) The flux profile of a blowing sand cloud: a wind tunnel investigation. *Geomorphology*, **49**, 219–230.

Dyer, K. (1986) *Coastal and Estuarine Sediment Dynamics*, John Wiley & Sons, Ltd, Chichester.

Ellwood, T.M., Evans P.D. and Wilson, I.G. (1975) Small scale aeolian bedforms. *Journal of Sedimentary Petrology*, **45**, 554–561.

Etyemezian, V. *et al.* (2007) The portable in-situ wind erosion laboratory (PiSwerl): a new method to measure PM10 windblown dust properties and potential for emissions. *Atmospheric Environment*, **41** (18), 3786–3796.

Frank, A. and Kocurek, G. (1994) Effects of atmospheric conditions on wind profiles and aeolian sand transport with an example from White Sands National Monument. *Earth Surface Processes and Landforms*, **19**, 735–745.

Frank, A. and Kocurek, G. (1996) Airflow up the stoss slope of sand dunes: limitations of current understanding. *Geomorphology*, **17**, 47–54.

Gillette, D.A. and Goodwin, P.A. (1974) Microscale transport of sand-sized soil aggregates eroded by wind. *Journal of Geophysical Research*, **7927**, 4080–4084.

Gillette, D.A., Herrick, J.E. and Herbert, G.A. (2006) Wind characteristics of Mesquite Streets in the northern Chihuahuan Desert, New Mexico, USA. *Environmental Fluid Mechanics*, **6** (3), 241–275.

Gillette, D.A. and Stockton, P.A. (1989) The effects of nonerodible particles on wind erosion of erodible surfaces. *Journal of Geophysical Research*, **94** (D10), 12885–12893.

Gillette, D.A., Herbert, G., Stockton, P.A. and Owen, P.R. (1996) Causes of the fetch effect in wind erosion. *Earth Surface Processes and Landforms*, **21** (7), 641–659.

Gillies, J.A., Nickling, W.G. and King, J. (2007) Shear stress partitioning in large patches of roughness in the atmospheric inertial sublayer. *Boundary-Layer Meteorology*, **122**, 367–396. DOI: 10.1007/s10546-006-9101-5.

Gillies, JA, Lancaster, N, Nickling, WG, Crawley, D. (2000) Field determination of drag forces and shear stress partitioning effects for a desert shrub (*Sarcobatus vermiculatus*, Greasewood). *Journal of Geophysical Research: Atmosphere*, **105** (D20),24871–24880.

Gillies, J.A., Nickling, W.G., King, J. and Lancaster, N. (2010) Modeling aeolian sediment transport thresholds on physically rough Martian surfaces: A shear stress partitioning approach. *Geomorphology*, **121**, 139–154.

Greeley, R. and Iversen, J.D. (1985) *Wind as a Geomorphological Process*, Cambridge University Press, Cambridge.

Greeley, R., Blumberg, D.G., McHone, J.F. *et al.* (1997) Applications of spaceborne radar laboratory data to the study of aeolian processes. *Journal of Geophysical Research E: Planets*, **102** (E5), 10971–10983

Goudie, A.S. and Middleton, N.J. (2006) *Desert Dust in the Global System*, Springer, Heidelberg.

Haff, P.K. and Anderson, R.S. (1993) Grain scale simulations of loose sedimentary beds: the example of grain-bed impacts in aeolian saltation. *Sedimentology*, **40**, 175–198.

Han, Q., Qu, J., Zhang, K. *et al.* (2009) Wind tunnel investigation of the influence of surface moisture content on the entrainment and erosion of beach sand by wind using sands from tropical humid coastal southern China. *Geomorphology*, **104** (3–4),230–237.

Hardisty, J. and Whitehouse, R.J.S. (1988) Evidence for a new sand transport process from experiments on Saharan dunes. *Nature*, **332**, 532–534.

Hersen, P. (2005) Flow effects on the morphology and dynamics of aeolian and subaqueous barchan dunes. *Journal of Geophysical Research*, **110**, F04S07, DOI: 10.1029/2004JF000185.

Hotta, S., Kubota, S., Katori, S. and Horikawa, K. (1984) Blown sand on a wet sand surface, in 19th Coastal Engineering Conference, Proceedings, American Society of Civil Engineers, New York, pp. 1265–1281.

Houser, C.A. and Nickling, W.G. (2001a) The factors influencing the abrasion efficiency of saltating grains on a clay-crusted playa. *Earth Surface Processes and Landforms*, **26** (5), 491–505.

Houser, C.A. and Nickling, W.G. (2001b) The emission and vertical flux of particulate matter <10 μm from a disturbed clay-crusted surface. *Sedimentology*, **48** (2), 255–267.

Howard, A.D., Morton, J.B., Gad-el-Hak, M. and Pierce, D.B. (1977) *Simulation model of erosion and deposition on a barchan dune*, NASA Contractor Report CR-2838, Washington, D.C.

Huang, N., Shi, F. and van Pelt, R.S. (2008) The effects of slope and slope position on local and upstream fluid threshold friction velocities. *Earth Surface Processes and Landforms*, **33** (12), 1814–1823.

Hunt, J.C.R. and Nalpanis, P. (1985) Saltating and suspended particles over flat and sloping surfaces. i. Modelling concepts, in Proceedings of International Workshop on Physics of Blown Sand, Memoirs 8, Department of Theoretical Statistics, Aarhus University, Denmark.

Hupy, J.P. (2004) Influence of vegetation cover and crust type on wind-blown sediment in a semi-arid climate. *Journal of Arid Environments*, **58** (2), 167–179.

Irwin, H.P.A.H. (1980) A simple omnidirectional sensor for wind tunnel studies of pedestrian level winds. *Journal of Wind Engineering and Industrial Aerodynamics*, **7**, 219–239.

Iversen, J.D. and Rasmussen, K.R. (1994) The effect of surface slope on saltation threshold. *Sedimentology*, **41**, 721–728.

Jackson, P.S. (1981) On the displacement height in the logarithmic velocity profile. *Journal of Fluid Mechanics*, **111**, 15–25.

Jackson, N.L. and Nordstrom, K.F. (1997) Effects of time-dependent moisture content of surface sediments on aeolian transport rates across a beach, Wildwood, New Jersey, USA, *Earth Surface Processes and Landforms*, **22** (7), 611–621.

Jackson, N.L. and Nordstrom, K.F. (1998) Aeolian transport of sediment on a beach during and after rainfall, Wildwood, NJ, USA. *Geomorphology*, **22** (2), 151–157.

Jones, J.R. and Willetts, B.B. (1979) Errors in measuring uniform aeolian sandflow by means of an adjustable trap. *Sedimentology*, **26**, 463–468.

Kawamura, R. (1951) *Study of sand movement by wind* (in Japanese), Report of the Institute of Science and Technology, University of Tokyo, Translated to English in NASA Technical Transactions F14.

Kawata, Y. and Tsuchiya, Y. (1976) Influence of water content on the threshold of sand movement and the rate of sand transport in blown sand (in Japanese). *Proceedings of the Japanese Society of Civil Engineers*, **249**, 95–100.

King, J., Nickling, W.G. and Gillies, J.A. (2005) Representation of vegetation and other non-erodible elements in aeolian shear stress partitioning models for predicting transport threshold. *Journal of Geophysical Research – Earth Surface*, **110**, F04015, DOI: 10.1029/2004JF000281.

King, J., Nickling, W.G. and Gillies, J.A. (2006) Aeolian shear stress ratio measurements within mesquite-dominated landscapes of the Chihuahuan Desert, New Mexico, USA. *Geomorphology*, **82** (3–4), 229–244.

King, J., Nickling, W.G. and Gillies, J.A. (2008) Investigations of the law-of-the-wall over sparse roughness elements. *Journal of Geophysical Research*, **113**, F02S07, DOI: 10.1029/2007JF000804.

Knight, M., Thomas, D.S.G. and Wiggs, G.F.S. (2004) Climate change in the 21st century and the impact on dunefield mobility in the Kalahari. *Geomorphology*, **59** (1–4), 197–213.

Knott, P. and Warren, A. (1981) Aeolian Processes, in *Geomorphological Techniques* (ed. A. Goudie), Allen and Unwin, London.

Kok, J.F. and Renno, N.O. (2009a) Electrification of wind-blown sand on Mars and its implications for atmospheric chemistry. *Geophysical Research Letters*, **36** (5), article L05202.

Kok, J.F. and Renno, N.O. (2009b) A comprehensive numerical model of steady state saltation (COMSALT). *Journal of Geophysical Research*, **114**, D17204, DOI: 10.1029/2009JD011702.

Kostaschuk, R. and Villard, P. (1996) Turbulent sand suspension events: Fraser River, Canada, in *Coherent Flow Structures in Open Channels* (eds P.J. Ashworth, S.J. Bennett, J.L. Best and S.J. McLelland), John Wiley and Sons, Ltd, Chichester, pp. 305–319.

Laity, J.E. and Bridges, N.T. (2009) Ventifacts on Earth and Mars: Analytical, field, and laboratory studies supporting sand abrasion and windward feature development. *Geomorphology*, **105** (3–4), 202–217.

Lancaster, N. and Baas, A. (1998) Influence of vegetation cover on sand transport by wind: field studies at Owens Lake, California. *Earth Surface Processes and Landforms*, **23**, 69–82.

Lancaster, N., Greeley, R. and Rasmussen, K.R. (1991) Interaction between unvegetated desert surfaces and the atmospheric boundary layer: a preliminary assessment. *Acta Mechanica (Suppl.)*, **2**, 89–102.

Lancaster, N., Nickling, W.G., McKenna-Neuman, C. and Wyatt, V.E. (1996) Sediment flux and airflow on the stoss slope of a barchan dune. *Geomorphology*, **17**, 55–62.

Leenders, J. K., van Boxel, J. H. and Sterk, G. (2005) Wind forces and related saltation transport. *Geomorphology*, **71**, 357–372.

Lettau, K. and Lettau, H.H. (1978) Experimental and micrometeorological field studies on dune migration, in *Exploring the World's Driest Climate* (eds H.H. Lettau and K. Lettau), University of Wisconsin-Madison, Institute for Environmental Studies, Report **101**, 110–147.

Levin, N., Ben-Dor, E., Kidron, G.J. and Yaakov, Y. (2008) Estimation of surface roughness (z_0) cover a stabilizing coastal dune field based on vegetation and topography. *Earth Surface Processes and Landforms*, **33** (10), 1520–1541.

Livingstone, I. and Warren, A. (1996) *Aeolian Geomorphology: An Introduction*, Longman, London.

Livingstone, I, Wiggs, G.F.S. and Weaver, C.M. (2007) Geomorphology of desert sand dunes. *Earth Science Reviews*, **80**, 239–257.

Lu, H. and Shao, Y. (1999). A new model for dust emission by saltation bombardment. *Journal of Geophysical Research*, (D14), **104**, 16827–16842.

Lu, S. S. and Willmarth, W. W. (1973) Measurements of the structure of the Reynolds stress in a turbulent boundary layer. *Journal of Fluid Mechanics*, **60**, 481–511.

Mabbutt, J.A. (1977) *Desert Landforms*, ANU Press, Canberra.

McEwan, I.K. (1993) Bagnold's kink: a physical feature of a wind velocity profile modified by blowing sand. *Earth Surface Processes and Landforms*, **18**, 145–156.

McEwan, I.K. and Willetts, B.B. (1991) Numerical model of the saltation cloud. *Acta Mechanica (Suppl.)*, **1**, 53–66.

McEwan, I.K. and Willetts, B.B. (1993) Sand transport by wind: a review of the current conceptual model, in *The Dynamics and Environmental Context of Aeolian Sedimentary Systems* (ed K. Pye), Special Publication of the Geological Society of London 72, Geological Society, London, pp. 7–16.

McEwan, I.K. and Willetts, B.B. (1994) On the prediction of bed-load sand transport rate in air. *Sedimentology*, **41**, 1241–1251.

McEwan, I.K., Willetts, B.B. and Rice, M.A. (1992) The grain/bed collision in sand transport by wind. *Sedimentology*, **39**, 971–981.

McKenna-Neuman, C. (2003) Effects of temperature and humidity upon the entrainment of sedimentary particles by wind. *Boundary-Layer Meteorology*, **108** (1), 61–89.

McKenna-Neuman, C. (2004) Effects of temperature and humidity upon the transport of sedimentary particles by wind. *Sedimentology*, **51** (1), 1–17.

McKenna-Neuman, C., Lancaster, N. and Nickling, W. G. (2000) The effect of unsteady winds on sediment transport on the stoss slope of a transverse dune, Silver Peak, NV, USA. *Sedimentology*, **47**, 211–226.

McKenna-Neuman, C. and Langston, G. (2006) Measurement of water content as a control of particle entrainment by wind. *Earth Surface Processes and Landforms*, **31** (3), 303–317.

McKenna-Neuman, C. and Maxwell, C. (2002) Temporal aspects of the abrasion of microphytic crusts under grain impact. *Earth Surface Processes and Landforms*, **27** (8), 891–908.

McKenna-Neuman, C. and Nickling, W.G. (1989) A theoretical and wind tunnel investigation of the effect of capillary water on the entrainment of soil by wind. *Canadian Journal of Soil Science*, **69**, 79–96.

McKenna-Neuman, C. and Nickling, W.G. (1994) Momentum extraction with saltation: implications for experimental evaluation of wind profile parameters. *Boundary-Layer Meteorology*, **68**, 35–50.

McKenna-Neuman, C., and Sanderson, S. (2008) Humidity control of particle emissions in aeolian systems. *Journal of Geophysical Research F: Earth Surface*, **113** (2), article F02S14.

Marston, R.A. (1986) Maneuver-caused wind erosion impacts, south-central New Mexico, in *Aeolian Geomorphology* (ed. W.G. Nickling), Proceedings of the 17th Annual Binghampton Symposium, pp. 273–306.

Marticorena, B. and Bergametti, G. (1995) Modeling the atmospheric dust cycle: 1. Design of a soil-derived dust emission scheme. *Journal of Geophysical Research*, **100** (D8), 16,415–16,430

Mason, J.A., Swinehart, J.B., Goble, R.J. and Loope, D.B. (2004) Late-Holocene dune activity linked to hydrological drought, Nebraska Sand Hills, USA. *Holocene*, **14** (2), 209–217.

Minvielle, F., Marticorena, B., Gillette, D.A. *et al.* (2003) Relationship between the aerodynamic roughness length and the roughness density in cases of low roughness density. *Journal of Fluid Mechanical Engineering*, **3**, 249–267.

Mitha, S., Tran, M.Q., Werner, B.T. and Haff, P.K. (1986) The grain-bed impact process in eolian saltation. *Acta Mechanica*, **63** (1–4), 267–278.

Musick, H.B. and Gillette, D.A. (1990) Field evaluation of relationships between a vegetation structural parameter and sheltering against wind erosion. *Land Degradation and Rehabilitation*, **2**, 87–94.

Musick, H. B., Trujillo, S.M. and Truman, C.R. (1996) Wind-tunnel modelling of the influence of vegetation structure on saltation threshold. *Earth Surface Processes and Landforms*, **21** (7), 589– 605.

Nalpanis, P. (1985) Saltating and suspended particles over flat and sloping surfaces. ii. Experiments and numerical simulations, in Proceedings of International Workshop on Physics of Blown Sand, Memoirs 8, Department of Theoretical Statistics, Aarhus University, Denmark.

Namikas, S.L. (2003) Field measurement and numerical modelling of aeolian mass flux distributions on a sandy beach. *Sedimentology*, **50** (2), 303–326.

Namikas, S.L., Bauer, B.O. and Sherman, D.J. (2003) Influence of averaging interval on shear velocity estimates for aeolian transport modelling. *Geomorphology*, **53** (3–4), 235–246.

Nickling, W.G. (1983) Grain size characteristics of sediment transported during dust storms. *Journal of Sedimentary Petrology*, **53**, 1011–1024.

Nickling, W.G. and McKenna-Neuman, C. (1995) Development of deflation lag surfaces. *Sedimentology*, **42**, 403–414.

Nickling, W.G. and McKenna-Neuman, C. (1997) Wind tunnel evaluation of a wedge-shaped aeolian sediment trap. *Geomorphology*, **18** (3–4), 333–345.

Nickling, W.G., McTainsh, G.H. and Leys, J.F. (1999) Dust emissions from the Channel Country of western Queensland, Australia. *Zeitschrift fur Geomorphologie, Supplementband*, **116**, 1–17.

Oke, T.R. (1987) *Boundary Layer Climates*, Methuen, New York.

Okin, G.S. (2008) A new model of wind erosion in the presence of vegetation. *Journal of Geophysical Research*, (113), DOI: 10.1029/2007JF000758.

Owen, P.R. (1964) Saltation of uniform grains in air. *Journal of Fluid Mechanics*, **20**, 225–242.

Parsons, D.R., Wiggs, G.F.S., Walker, I.J. *et al.* (2004) Numerical modelling of airflow over an idealised transverse dune. *Environmental Modelling and Software*, **19**, 153–162.

Pelletier, J.D. (2009) Controls on the height and spacing of eolian ripples and transverse dunes: a numerical modelling investigation. *Geomorphology*, **105**, 322–333.

Porter, S.C. (2001) Chinese loess record of monsoon climate during the last glacial–interglacial cycle. *Earth-Science Reviews*, **54**, 115–128.

Pye, K. and Tsoar, H. (1990) *Aeolian Sand and Sand Dunes*, Unwin Hyman, London.

Rajot, J.L., Alfaro, S.C., Gomes, L. and Gaudichet, A. (2003) Soil crusting on sandy soils and its influence on wind erosion. *Catena*, **53** (1), 1–16.

Rasmussen, K.R., Iversen, J.D. and Rautahemio, P. (1996) Saltation and wind-flow interaction in a variable slope wind tunnel. *Geomorphology*, **17** (1–3 SPEC. ISS.), 19–28.

Rasmussen, K.R. and Mikkelsen, H.E. (1991) Wind tunnel observations of aeolian transport rates. *Acta Mechanica (Suppl.)*, **1**, 135–144.

Rasmussen, K.R. and Sørensen, M. (2008) Vertical variation of particle speed and flux density in aeolian saltation: measurement and modelling. *Journal of Geophysical Research F: Earth Surface*, **113** (2), article F02S12.

Rasmussen, K.R., Sørensen, M. and Willetts, B.B. (1985) Measurement of saltation and wind strength on beaches, in Proceedings of International Workshop on Physics of Blown Sand, Memoirs 8, Department of Theoretical Statistics, Aarhus University, Denmark.

Raupach, M. R. (1992) Drag and drag partition on rough surfaces. *Boundary Layer Meteorology*, **60**, 375–395.

Raupach, M.R., Gillette, D.A. and Leys, J.F. (1993) The effect of roughness elements on wind erosion threshold. *Journal of Geophysical Research*, **98** (D2), 3023–3029.

Rice, M.A. (1991) Grain shape effects on aeolian sediment transport. *Acta Mechanica(Suppl.)*, **1**, 159–166.

Rice, M.A. and McEwan, I.K. (2001) Crust strength: a wind tunnel study of the effect of impact by saltating particles on cohesive soil surfaces. *Earth Surface Processes and Landforms*, **26** (7), 731–733.

Rice, M.A., Mullins, C.E. and McEwan, I.K. (1997) An analysis of soil crust strength in relation to potential abrasion by saltating particles. *Earth Surface Processes and Landforms*, **22** (9), 869–883.

Rice, M.A., Willetts, B.B. and McEwan, I.K. (1996) Observations of collisions of saltating grains with a granular bed from high-speed cine-film. *Sedimentology*, **43** (1), 21–31.

Sarre, R.D. (1987) Aeolian sand transport. *Progress in Physical Geography*, **11**, 157–181.

Sarre, R.D. (1988) Evaluation of aeolian sand transport equations using intertidal-zone measurements, Saunton Sands, England. *Sedimentology*, **35**, 671–679.

Sarre, R.D. (1990) Evaluation of aeolian sand transport equations using intertidal-zone measurements, Saunton Sands, England – reply. *Sedimentology*, **37**, 389–392.

Schönfeldt, H-J. and von Löwis, S. (2003) Turbulence-driven saltation in the atmospheric surface layer. *Meteorologische Zeitschrift*, **12**, 257–268.

Sharp, R.P. (1963) Wind ripples. *Journal of Geology*, **71**, 617–636.

Sherman, D.J. (1990) Discussion: evaluation of aeolian sand transport equations using intertidal-zone measurements, Saunton Sands, England – discussion. *Sedimentology*, **37**, 385–392.

Sherman, D. J. (1992) An equilibrium relationship for shear velocity and roughness length in aeolian saltation. *Geomorphology*, **5**, 419–431.

Sherman, D.J. and Farrell, E.J. (2008) Aerodynamic roughness lengths over movable beds: comparison of wind tunnel and field data. *Journal of Geophysical Research F: Earth Surface*, **113** (2), article F02S08.

Sherman, D.J., Jackson, D.W.T., Namikas, S.L. and Wang, J. (1998) Wind-blown sand on beaches: an evaluation of models. *Geomorphology*, **22** (2), 113–133.

Spies, P.J., McEwan, I. and Butterfield, G.R. (1995) On wind velocity profile measurements taken in wind tunnels with saltating grains. *Sedimentology*, **42**, 515–521.

Sterk, G., Jacobs, A. F. G. and van Boxel, J. H. (1998) The effect of turbulent flow structures on saltation sand transport in the atmospheric boundary layer. *Earth Surface Processes and Landforms*, **23**, 877–887.

Stockton, P.H. and Gillette, D.A. (1990) Field measurement of the sheltering effect of vegetation on erodible land surfaces. *Land Degradation and Rehabilitation*, **2**, 77–85.

Stone, A.E.C. and Thomas, D.S.G. (2008) Linear dune accumulation chronologies from the southwest Kalahari, Namibia: challenges of reconstructing Late Quaternary palaeoenvironments from aeolian landforms. *Quaternary Science Reviews*, **27** (17–18), 1667–1681.

Stout, J. E. (1998) Effect of averaging time on the apparent threshold for aeolian transport. *Journal of Arid Environments*, **39**, 395–401.

Stout, J. E. (2004) A method for establishing the critical threshold for aeolian transport in the field. *Earth Surface Processes and Landforms*, **29**, 1195–1207.

Stout, J.E. and Zobeck, T.M. (1996) The Wolfforth field experiment: a wind erosion study. *Soil Science*, **161** (9), 616–632.

Stout, J. E. and Zobeck, T. M. (1997) Intermittent saltation. *Sedimentology*, **44**, 959–970.

Stull, R.B. (1988) *An Introduction to Boundary Layer Meteorology*, Kluwer, Dordrecht.

Swap, R., Garstang, M., Greco, S. *et al.* (1992) Saharan dust in the Amazon Basin. *Tellus*, Series B, **44 B** (2), 133–149.

Telfer, M.W. and Thomas, D.S.G. (2006) Complex Holocene lunette dune development, South Africa: implications for paleoclimate and models of pan development in arid regions. *Geology*, **34** (10), 853–856.

Thom, A.S. (1971) Momentum absorption by vegetation. *Quarterly Journal of the Royal Meteorological Society*, **97**, 414–428.

Thomas, D.S.G., Knight, M. and Wiggs, G.F.S. (2005) Remobilization of southern African desert dune systems by twenty first century global warming. *Nature*, **435**, 1218–1221.

Thomas, D.S.G., O'Connor, P.W., Bateman, M.D. *et al.* (2000) Dune activity as a record of late Quaternary aridity in the Northern Kalahari: new evidence from northern Namibia interpreted in the context of regional arid and humid chronologies. *Palaeogeography, Palaeoclimatology, Palaeoecology*, **156** (3–4), 243–259.

Tsoar, H., Blumberg, D.G. and Stoler, Y. (2004) Elongation and migration of sand dunes. *Geomorphology*, **57**, 293–302.

Tsoar, H. and Pye, K. (1987) Dust transport and the question of desert loess formation. *Sedimentology*, **34**, 139–154.

Tsoar, H., White, B. and Berman, E. (1996) The effect of slopes on sand transport – numerical modelling. *Landscape and Urban Planning*, **34** (3–4), 171–181.

Ungar, J.E. and Haff, P.K. (1987) Steady state saltation in air. *Sedimentology*, **34**, 289–299.

van Boxel, J.H., Sterk, G. and Arens, S.M. (2004) Sonic anemometers in aeolian sediment transport research. *Geomorphology*, **59** (1–4), 131–147.

van Pelt, R.S., Peters, P. and Visser, S. (2009) Laboratory wind tunnel testing of three commonly used saltation impact sensors. *Aeolian Research*, **1** (1–2), 55–62.

Venditti, J.G. and Bauer, B.O. (2005) Turbulent flow over a dune: Green River, Colorado. *Earth Surface Processes and Landforms*, **30**, 289–304.

Walker, I.J. (2005) Physical and logistical considerations of using ultrasonic anemometry in aeolian sediment transport research. *Geomorphology*, **68**, 57–76.

Walker, I.J. and Nickling, W.G. (2002) Dynamics of secondary airflow and sediment transport over and in the lee of transverse dunes. *Progress in Physical Geography*, **26**, 47–75.

Walker, I.J. and Nickling, W.G. (2003) Simulation and measurement of surface shear stress over isolated and closely spaced transverse dunes in a wind tunnel. *Earth Surface Processes and Landforms*, **28** (10), 1111–1124.

Wang, H.T., Zhou, Y.H., Dong, Z.B. and Ayrault, M. (2008) Vertical dispersion of dust particles in a turbulent boundary layer. *Earth Surface Processes and Landforms*, **33** (8), 1210–1221.

Wang, X., Yang, Y., Dong, Z. and Zhang, C. (2009) Responses of dune activity and desertification in China to global warming in the twenty-first century. *Global and Planetary Change*, **67** (3–4), 167–185.

Wasson, R.J. and Nanninga, P.M. (1986) Estimating wind transport of sand on vegetated surfaces. *Earth Surface Processes and Landforms*, **11**, 505–514.

Weaver, C.M. (2008) Turbulent flow and sand dune dynamics: identifying controls on aeolian sediment transport, Unpublished DPhil Thesis, University of Oxford, 228 pp.

Weaver, C.M. and Wiggs, G.F.S. (2010) Field measurements of mean and turbulent airflow over a barchan sand dune. *Geomorphology* (in press).

Webb, N.P., McGowan, H.A., Phinn, S.R. *et al.* (2009) A model to predict land susceptibility to wind erosion in western Queensland, Australia. *Environmental Modelling and Software*, **24** (2), 214–227.

Werner, B.T. (1990) A steady-state model of wind-blown sand transport. *The Journal of Geology*, **98** (1), 1–17.

Werner, B.T. and Gillespie, D.T. (1993) Fundamentally discrete stochastic model for wind ripple dynamics. *Physical Review Letters*, **71**, 3230–3233.

Werner, B.T. and Haff, P.K. (1988) The impact process in eolian saltation: two dimensional simulations. *Sedimentology*, **35**, 189–196.

White, B.R. (1982) Two-phase measurements of saltating turbulent boundary layer flow. *International Journal of Multiphase Flow*, **9**, 459–473.

White, B.R. and Schultz, J.C. (1977) Magnus effect in saltation. *Journal of Fluid Mechanics*, **81**, 497–512.

White, B.R. and Tsoar, H. (1998) Slope effect on saltation over a climbing sand dune. *Geomorphology*, **22** (2), 159–180.

Whitehouse, R.J.S. and Hardisty, J. (1988) Experimental assessment of two theories for the effect of bedslope on the threshold of bedload transport. *Marine Geology*, **79**, 135–139.

Wiggs, G.F.S. (1992) *Sand dune dynamics: field experimentation, mathematical modelling and wind tunnel testing*, Unpublished PhD Thesis, University of London.

Wiggs, G.F.S., Atherton, R.A. and Baird, A.J. (2004) Thresholds of aeolian sand transport: establishing suitable values. *Sedimentology*, **51** (1), 95–108.

Wiggs, G.F.S., Baird, A.J. and Atherton, R.A. (2004) The dynamic effects of moisture on the entrainment and transport of sand by wind. *Geomorphology*, **59** (1–4), 13–30.

Wiggs, G.F.S. and Holmes, P.J. (2010) Dynamic controls on wind erosion and dust generation on west-central Free State agricultural land, South Africa. *Earth Surface Processes and Landforms* (in press).

Wiggs, G.F.S., Livingstone, I and Warren, A. (1996) The role of streamline curvature in sand dune dynamics: evidence from field and wind tunnel measurements. *Geomorphology*, **17**, 29–46.

Wiggs, G.F.S., Livingstone, I., Thomas, D.S.G. and Bullard, J.E. (1994) The effect of vegetation removal on airflow structure and dune mobility in the southwest Kalahari. *Land Degradation and Rehabilitation*, **5**, 13–24.

Wiggs, G.F.S., Thomas, D.S.G., Bullard, J.E. and Livingstone, I. (1995) Dune mobility and vegetation cover in the southwest Kalahari Desert. *Earth Surface Processes and Landforms*, **20**, (6), 515–529.

Wiggs, G.F.S., Livingstone, I., Thomas, D.S.G. and Bullard, J. E. (1996) Airflow and roughness characteristics over partially-vegetated linear dunes in the southwest Kalahari Desert. *Earth Surface Processes and Landforms*, **21**, 19–34.

Wilkinson, R.H. (1983/1984) A method for evaluating statistical errors associated with logarithmic velocity profiles. *Geo-Marine Letters*, **3**, 49–52.

Willetts, B.B. (1983) Transport by wind of granular materials of different grain shapes and densities. *Sedimentology*, **30**, 669–679.

Willetts, B.B. and Rice, M.A. (1985a) Inter-saltation collisions, in Proceedings of the International Workshop on the Physics of Blown Sand, 1, Department of Theoretical Statistics, Aarhus University (Denmark), Memoirs 8, pp. 83–100.

Willetts, B.B. and Rice, M.A. (1985b) Wind tunnel tracer experiments using dyed sand, in Proceedings of the International Workshop on the Physics of Blown Sand, 1, Department of Theoretical Statistics, Aarhus University (Denmark), Memoirs **8**, pp. 225–242.

Willetts, B.B. and Rice, M.A. (1986a) Collisions in aeolian saltation. *Acta Mechanica*, **63**, 255–265.

Willetts, B.B. and Rice, M.A. (1986b) Collisions in aeolian transport: the saltation/creep link, in *Aeolian Geomorphology, Proceedings of the 17th Annual Binghampton Symposium* (ed. W.G. Nickling), September 1986, pp. 1–18.

Willetts, B.B. and Rice, M.A. (1989) Collisions of quartz grains with a sand bed: the influence of incidence angle. *Earth Surface Processes and Landforms*, **14**, 719–730.

Williams, J.J., Butterfield, G.R. and Clark, D. (1990) Aerodynamic entrainment thresholds and dislodgement rates on impervious and permeable beds. *Earth Surface Processes and Landforms*, **15**, 255–264.

Williams, J.J., Butterfield, G.R. and Clark, D.G. (1994) Aerodynamic entrainment thresholds: effects of boundary layer flow conditions. *Sedimentology*, **41**, 309–328.

Wolfe, S.A. and Nickling, W.G. (1993) The protective role of sparse vegetation in wind erosion. *Progress in Physical Geography*, **17** (1), 50–68.

Worster, D. (1979) *Dust Bowl*, Oxford University Press, Oxford.

Wu, H. and Stathopoulos, T. (1994) Further experiments on Irwin's surface wind sensor. *Journal of Wind Engineering and Industrial Aerodynamics*, **53**, 441–452.

Yizhaq, H., Balmorth, N.J. and Provenzale, A. (2004) Blown by wind: nonlinear dynamics of aeolian sand ripples. *Physica D: Nonlinear Phenomena*, **195** (3–4), 207–228.

Yizhaq, H., Isenberg, O., Wenkart, R. *et al.* (2008) Morphology and dynamics of aeolian mega-ripples in Nahal Kasuy, southern Israel. *Israel Journal of Earth Sciences*, **57** (3–4), 149–165.

Zender, CS., Bian, H. and Newman, D. (2003) Mineral dust entrainment and deposition (DEAD) model: description and 1990s dust climatology. *Journal of Geophysical Research D: Atmospheres*, **108** (14), AAC 8-1–AAC 8-19.

Zingg, A.W. (1953) Wind tunnel studies of movement of sedimentary material. *Proceedings of 5th Hydraulic Conference Bulletin*, **34**, 111–134.

Zobeck, T.M. (1991) Abrasion of crusted soils: influence of abrader flux and soil properties. *Soil Science Society of America Journal*, **55** (4), 1091–1097.

Zobeck, T. M., Sterk, G., Funk, R. *et al.* (2003) Measurement and data analysis methods for field-scale wind erosion studies and model validation. *Earth Surface Processes and Landforms*, **28**, 1163–1188.

19

Desert dune processes and dynamics

Nick Lancaster

19.1 Introduction

Desert sand dunes form part of a self-organised hierarchical system of aeolian bedforms, which comprises: (a) wind ripples, (b) individual simple dunes or superimposed dunes on mega dunes and (c) mega dunes (also called draa or compound and complex dunes) – characterised by superimposition of simple or elemental dunes on larger forms. The majority of dunes are composed of quartz and feldspar grains of sand size, although dunes composed of gypsum, carbonate and volcanic sand also occur. Most sand dunes occur in contiguous areas of dunes known as sand seas or ergs (with an area of > 100 km^2). Smaller areas of dunes are called dune fields (see Chapter 17).

Dunes occur in self-organised patterns that develop over time as the response of sand surfaces to the wind regime (especially its directional variability) and the supply of sand (Werner, 1995). The dune types discussed below represent the steady state attractors of the aeolian sand transport system and can evolve from a wide range of initial conditions. Key variables that determine dune morphology and dynamics are the spatial and temporal characteristics of: (1) the wind regime (especially its directional variability), (2) sand supply (the amount of sand available for dune building) and (3) vegetation cover. Sand particle size does not appear to be as important as once thought. As dunes and dune patterns evolve over time, the legacy of past climates and wind regimes influences present-day dune morphology and dune patterns in many areas.

19.2 Desert dune morphology

Desert dunes occur in a variety of morphologic types (see Box 19.1 and Figure 19.1), each of which displays a range of height, width and spacing. Aerial photographs and satellite images show the general regularity of dune patterns (manifested by close correlations between dune height, width and spacing) and also indicate that essentially similar dune forms occur in widely separated localities. Even dunes on other planetry bodies (e.g. Mars, Titan) have a simliar morphology to terrestrial dunes. These commonalities indicate that there are general physical principles that govern dune form and dynamics.

Dunes are created and modified by interactions between sand transport rates, dune topography and airflow (Figure 19.2). As the dune projects into the atmospheric boundary layer, the primary air flow is modified by interactions between the dune form and the airflow, which give rise to modifications of the local wind speed, shear stress and turbulence intensity, and therefore sand transport rates. Such interactions also create secondary flow circulations, especially in the lee of the dune, as a result of flow separation and flow diversion. In addition, superimposed dunes on megadunes or draa respond to changes in airflow and sediment transport on the megadune itself. The nature of these interactions is discussed in detail below.

19.3 Dune types and environments

19.3.1 Crescentic dunes

In wind regimes characterised by a narrow range of wind directions (unidirectional wind regimes), crescentic dunes form with their crests aligned normal to the direction of sand transport.

Where sand supply is low relative to the capacity of the wind to transport sand (undersaturated sand transport conditions), isolated crescentic dunes or barchans form

Arid Zone Geomorphology: Process, Form and Change in Drylands, Third Edition. Edited by David S. G. Thomas

Box 19.1 Dune classifications

Many different classifications of types of dunes have been proposed: (1) those based on the external morphology of the dunes (morphological classifications) and (2) those that imply some relationship of dune type to formative winds or sediment supply (dynamic classifications). Morphologic classifications of dunes facilitate mapping of dune types from satellite image data (e.g. Breed *et al.*, 1979) with no assumptions regarding formative mechanisms and relations to external variables. Dynamic classifications of dunes imply some causal relationship between external variables and dune type – so that dunes may be identified as transverse, longitudinal or oblique to the formative winds (Hunter, Richmond and Alpha, 1983). Even though understanding of the conditions in which different types of dunes form has improved significantly in recent years, this knowledge remains imperfect, so that a largely morphological classification scheme is preferred. Figure 19.1 provides a framework for classifying dunes based on external morphology, sediment volume and other key parameters.

Dunes → Mega-dunes
Relative Sediment Thickness
Wind regime complexity
CRESCENTIC
Barchans
Dome dunes
Crescentic ridges
Compound crescentic dunes
LINEAR
Simple
- Vegetated
- Sinuous
Compound Linear dunes
Complex Linear dunes
REVERSING
STAR
coarse sand
SAND SHEETS
ZIBARS
vegetation anchored
NEBKHAS
BLOW OUT DUNES
PARABOLIC DUNES
source bordering
LUNETTES
topographic influence
ECHO DUNES
FALLING DUNES
CLIMBING DUNES

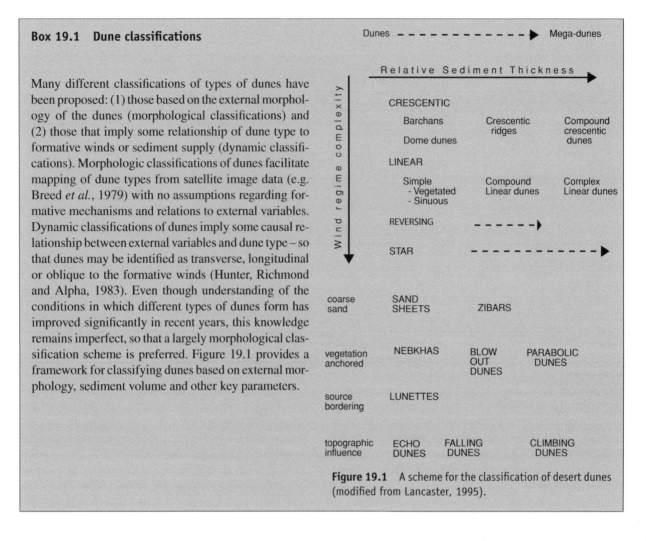

Figure 19.1 A scheme for the classification of desert dunes (modified from Lancaster, 1995).

(Figure 19.3). Barchans have been described from most desert regions and have been extensively studied in the field, using aerial and satellite images, and most recently via numerical models. Conditions favouring the formation of barchans are encountered on the margins of sand seas and dune fields as well as in the transport corridors extending from sand source areas (Figure 19.3(b)), where a clear size selection process involves merging smaller, faster moving dunes with slower moving, larger dunes so that the pattern becomes more uniform. In plan view,

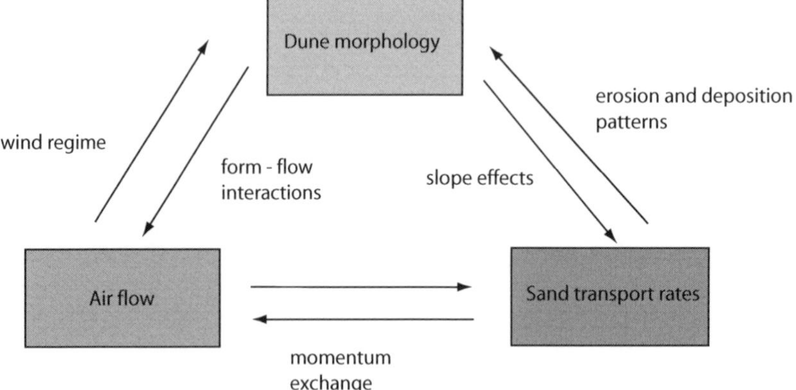

Figure 19.2 A conceptual framework for process–form interactions on aeolian dunes.

barchans are elliptical in shape with a concave lee face and horns extending downwind (Figure 19.3). Dune width is typically about ten times dune height. Good examples of barchan dunes are to be found in the Pampa La Joya, Peru (Finkel, 1959), the Tarfaya–Laayoune area of Morocco (Elbelrhiti, Andreotti and Claudin, 2008) and at several localities along the coast of the Namib Desert (Bourke and Goudie, 2009). Most barchans occur at the simple dune scale, but megabarchans are identified from a few areas (e.g. Grolier *et al.*, 1974).

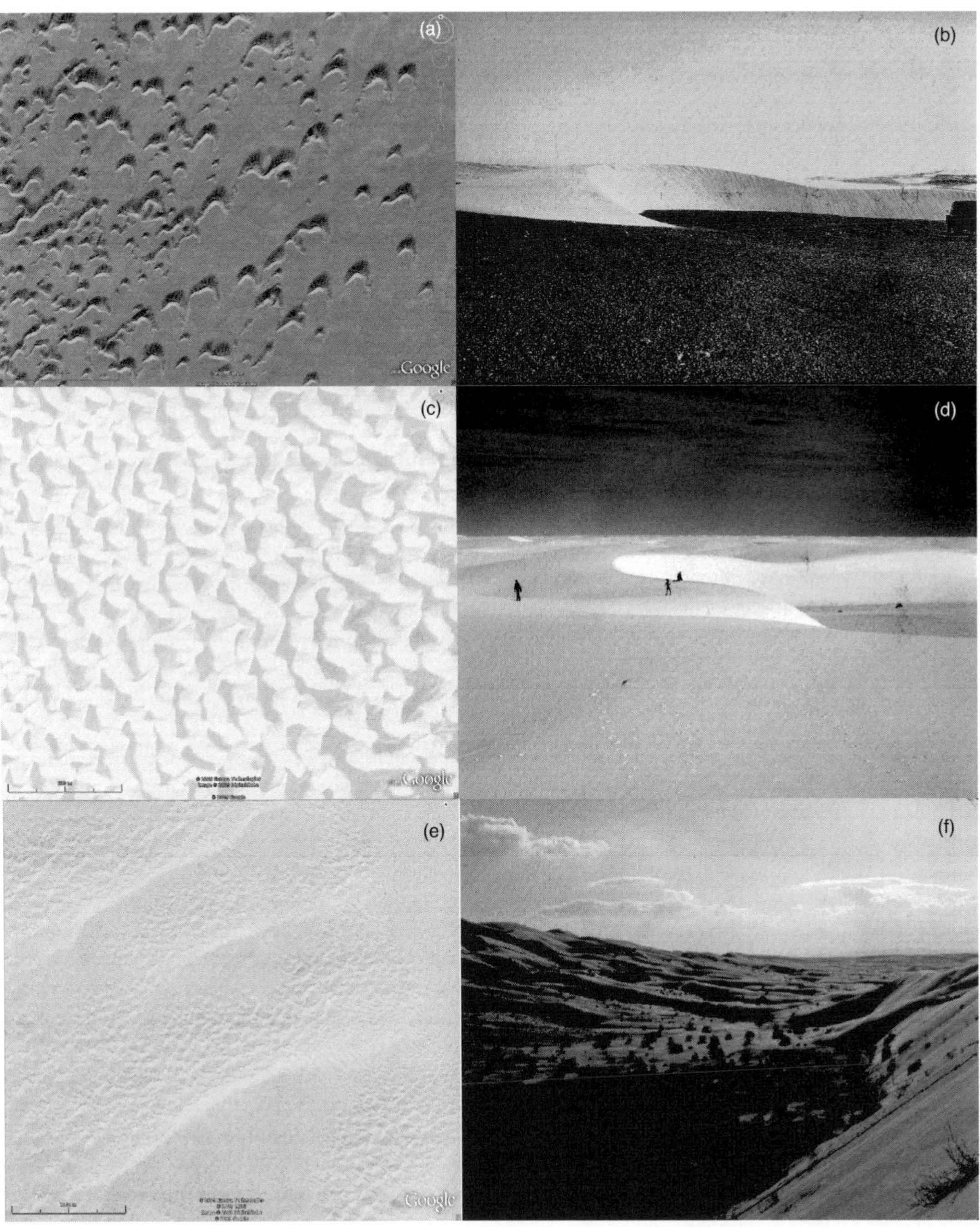

Figure 19.3 Crescentic dunes: (a) barchans in transport corridor (Laayoune, Morocco); (b) asymmetric barchan in Skeleton Coast dune field, Namibia; (c, d) simple crescentic dunes, White Sands, New Mexico; (e, f) compound crescentic dunes, Algodones Dune Field, California; (g) sand rose for an area of crescentic dunes; (h) anatomy of crescentic dune morphology.

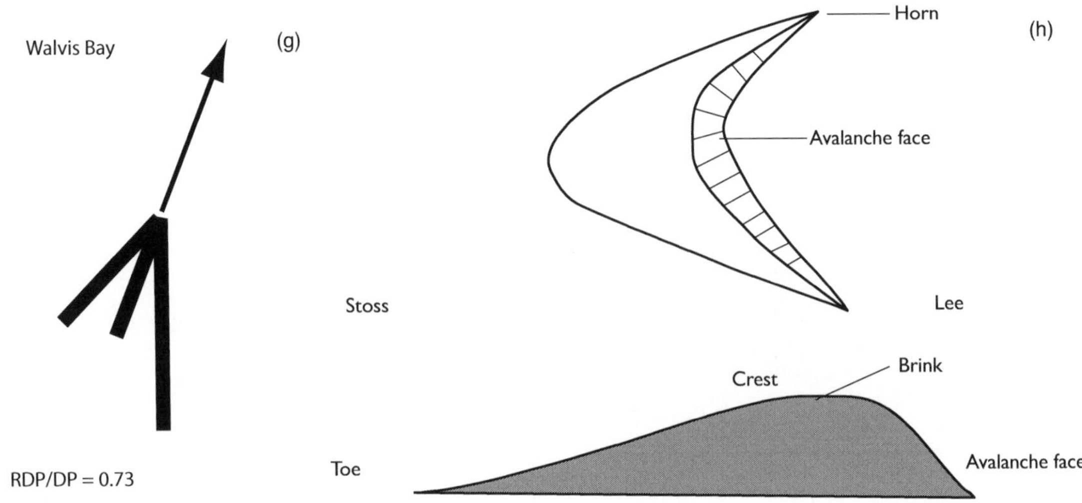

Figure 19.3 (*Continued*)

Where sand supply increases, crescentic ridges or transverse dunes occur (Figure 19.3). Worldwide, such dunes occupy about 10 % of the area of sand seas. Good examples of simple crescentic ridges can be found at White Sands, New Mexico (Kocurek *et al.*, 2007; McKee and Douglass, 1971) and in the Skeleton Coast dune field of Namibia (Lancaster, 1982a). At White Sands, the dunes are typically sinuous crescentic ridges and barchanoid dunes, 8–12 m high, with a crest length of about 250 m and an average spacing of 136 m. These dunes are migrating towards the ENE at 1–7 m/yr. The Skeleton coast dune field is characterised by straight to slightly sinuous crescentic dunes with a spacing between 150 and 300 m and a height ranging between 4 and 30 m, migrating towards the north or northeast. Dune size increases northwards (downwind) and the eastern side of the dune field is characterised by smaller crescentic dunes that are changed to barchans as the sand supply decreases.

Compound or crescentic megadunes are characterised by the superimposition of smaller crescentic dunes on their stoss (windward) slopes (Figure 19.4(d) and (e)). These smaller dunes migrate across the main form at an order of magnitude greater than that of the main dune. Classic examples of crescentic megadunes can be found in the Algodones dune field of southeast California (Derickson *et al.*, 2008; Havholm and Kocurek, 1988), the Liwa area of the UAE (El-Sayed, 2000; Stokes and Bray, 2005) and along the coastal areas of the Namib Sand Sea (Lancaster, 1989c). In the Algodones, the main dunes are 40–60 m high with a mean spacing of 1025 m, whereas the superimposed coalesced crescentic dunes are 12–20 m high with a mean spacing of 93 m (Derickson *et al.*, 2008). Superimposed dunes migrate at 0.76–1.73 m/yr, compared to 0.09 m/yr to 0.35–0.40 m/yr for the main dunes (Sharp, 1979).

19.3.2 Linear dunes

Dunes of linear form (Figure 19.4 and Figure 19.5) are widespread in deserts worldwide and dominate in Australia and the Kalahari, as well as in Namibia, Arabia and the southern and western Sahara. Linear dunes occur in simple, compound and complex varieties (the latter two at megadune scales). There are two main varieties of the simple linear dune: (1) sinuous, relatively short dunes (also known as seif dunes) and (2) longer, straight, vegetated linear dunes such as those that occur in the Simpson Desert of Australia and the Kalahari.

Sinuous simple linear dunes (seif dunes) have been documented from many desert regions (Figure 19.4). They typically consist of a sinuous ridge with alternate peaks and saddles and a more or less triangular cross-section, the sinuosity increasing with dune height. Such dunes are typically a few hundred metres long. In many areas, it appears that dunes of this type are very young, comprising dunes that have developed in modern wind regimes as a modification of older patterns, e.g. in the Azefal Sand Sea, Mauritania (Lancaster *et al.*, 2002), the Wahiba sands (Warren and Allison, 1998) or representing the extension of larger linear dunes (Bristow, Bailey and Lancaster, 2000).

Vegetated linear dunes in the southwest Kalahari and the Simpson–Strzelecki dune fields (Figure 19.4(c) and (d)) are very similar in form and provide good examples of this dune type. These dunes typically comprise straight subparallel ridges 2–35 m high and 200–450 m

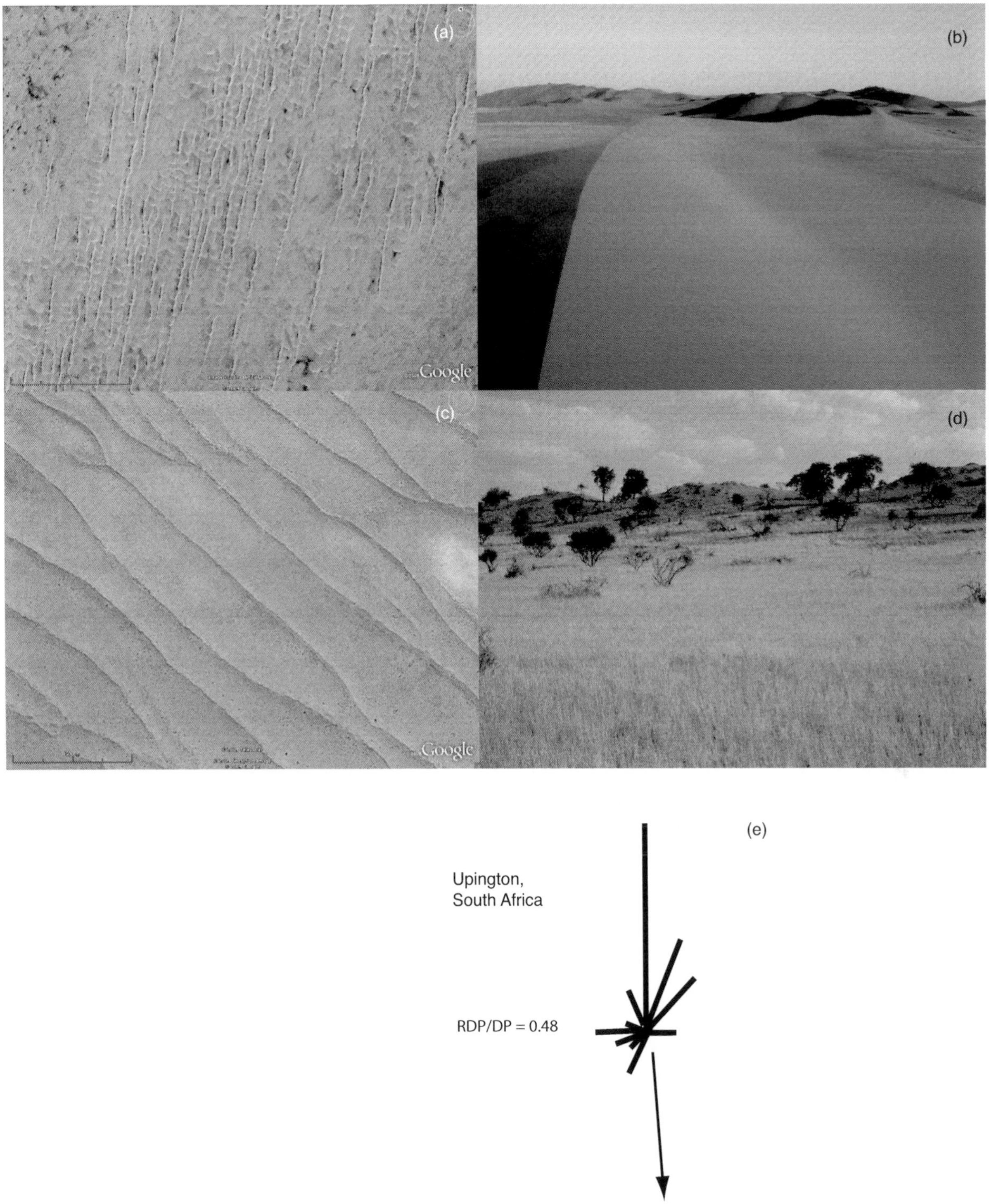

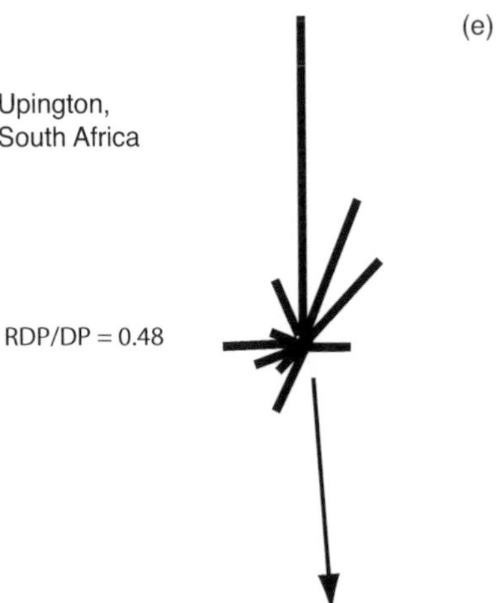

Figure 19.4 Simple linear dunes: (a) Azefal Sand Sea, Mauritania; (b) Northern Namib sand Sea; (c,d) vegetated linear dunes, southwest Kalahari.

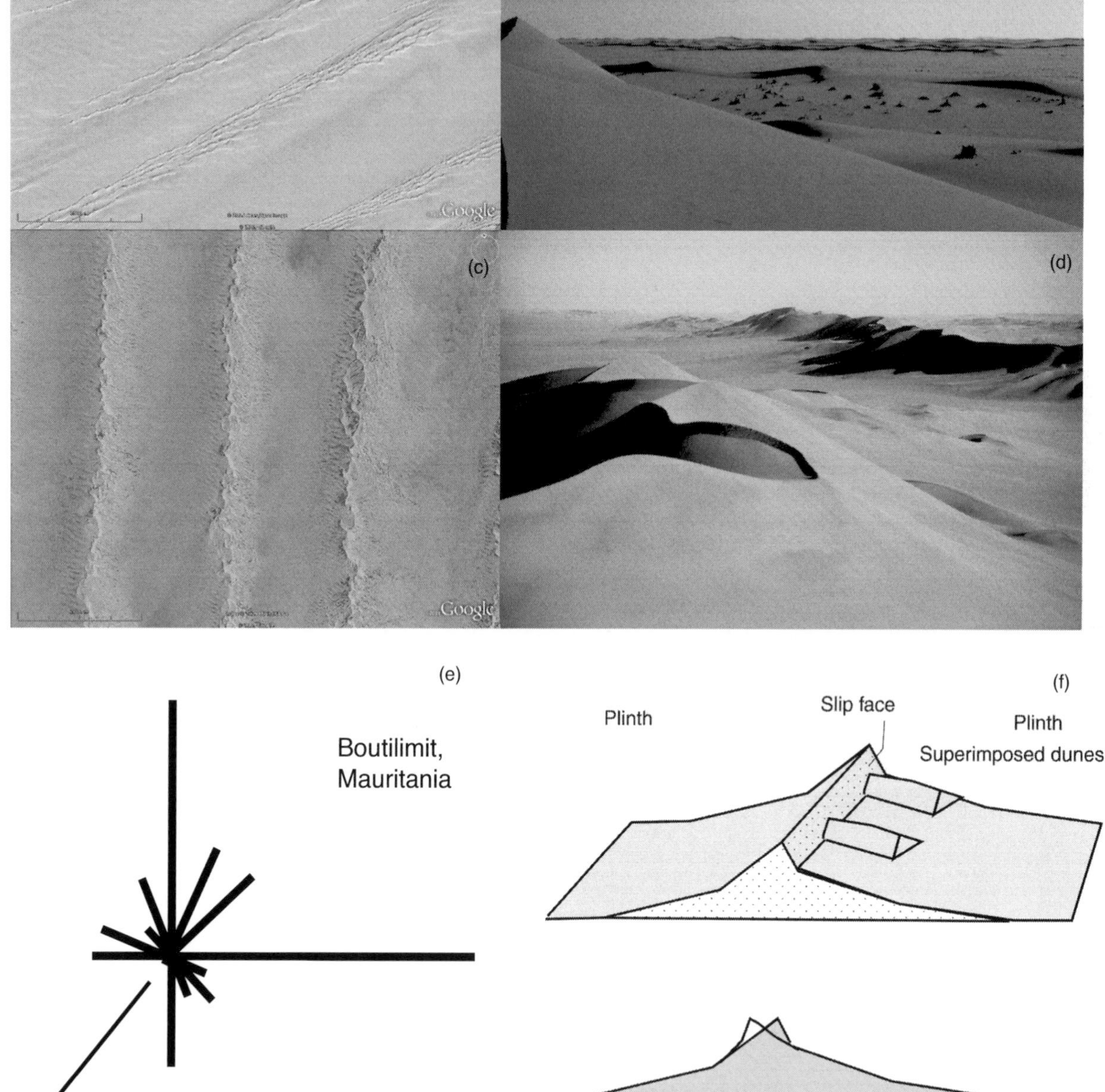

Figure 19.5 Compound and complex linear dunes: (a) compound linear dunes, Erg Fachi Bilma, Niger; (b) southern Namib Sand Sea; (c, d) complex linear dunes, Namib Sand Sea; (e) wind regimes of compound and complex linear dunes; (f) anatomy of complex linear dunes.

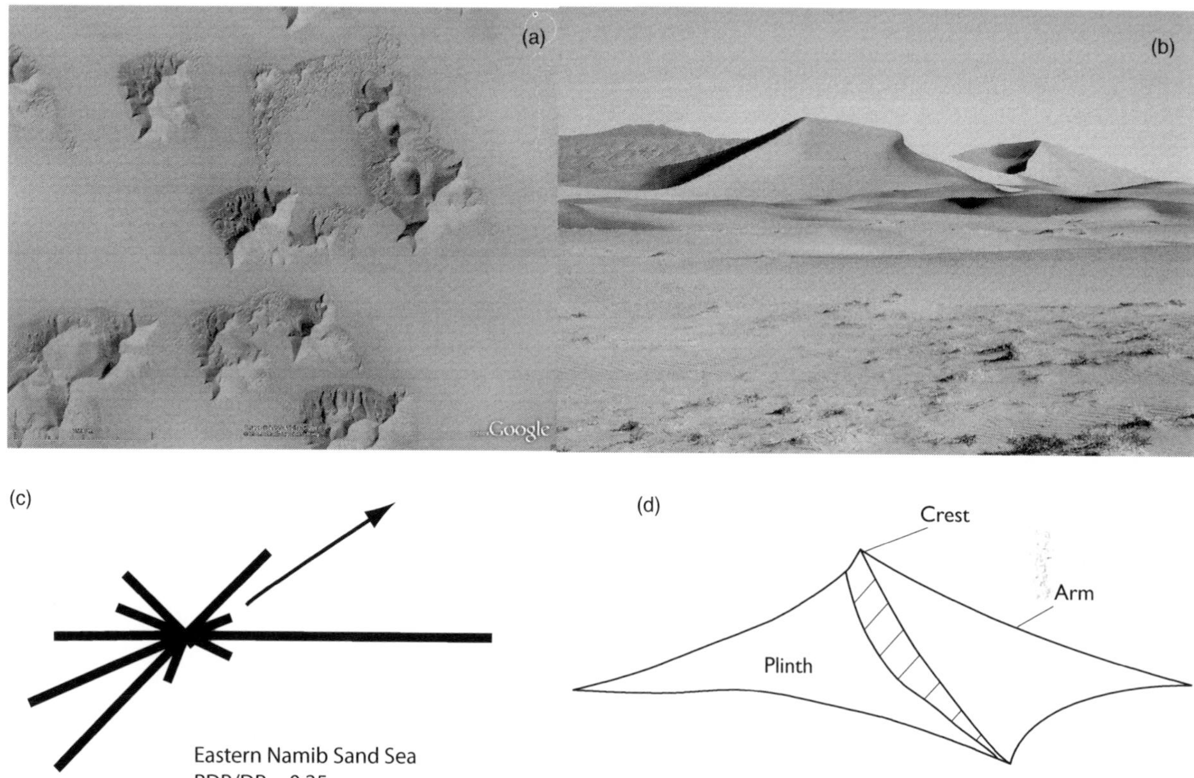

Figure 19.6 Star dunes: (a) clusters of star dunes, eastern Rub al Khali, Saudi Arabia; (b) Silver Peak, Nevada; (c) wind regime of star dunes; (d) anatomy of star dunes.

apart (Bullard *et al.*, 1995; Fitzsimmons, 2007), with a length of 20–25 km. The ridges may join in Y-junctions. Several varieties of this pattern can be identified (Figure 19.6), related to sand supply, distance from sand sources, vegetation cover and persistence of dune-forming winds (Bullard *et al.*, 1995; Wasson *et al.*, 1988).

Compound linear dunes (Figure 19.5(a) and (b)) consist of several narrow, closely spaced (100–200 m) seif-like ridges on the crest of broad linear plinths, which are spaced as much as 2.5 km apart. Good examples occur in the southern Namib Sand Sea (Lancaster, 1989c) and the Fachi Bilma Erg in Niger (Mainguet and Callot, 1978).

Complex linear dunes (Figure 19.5(c) and (d)) are comprised of relatively straight, continuous, linear ridges on which other dune types are superimposed. Such dunes are typically spaced 1–2.5 km apart and reach heights of 50–250 m. Complex linear dunes in the Namib Sand Sea and Saudi Arabia consist of a sinuous main crestline with superimposed crescentic dunes on the dune flanks (Lancaster, 1989c; Livingstone, 1993). In places, the crest line may consist of a series of peaks similar to star dunes separated by lower sinuous sharp crest lines. Such complex linear dunes are transitional to chains of star dunes in some areas (e.g. the eastern Namib Sand Sea and southeastern parts of the UAE).

The origins of linear dunes and their relationship to formative wind directions have been the subject of considerable debate (Lancaster, 1982b; Livingstone, 1988; Tsoar, 1989). A substantial body of empirical evidence now indicates that linear dunes form in bidirectional wind regimes with the two modes separated by 90° or more. This includes correlations between the occurrence of linear dune and information on the directional variability of wind regimes in these areas (e.g. Fryberger, 1979): studies of internal sedimentary structures (Bristow, Bailey and Lancaster, 2000; McKee, 1982; McKee and Tibbitts, 1964), experiments in flumes (Reffet *et al.*, 2008; Rubin and Ikeda, 1990) and numerical simulations (Werner and Kocurek, 1997), as well as detailed process studies on linear dunes (Livingstone, 1986, 1988, 1993; Tsoar, 1983). Models for linear dune formation that invoke boundary layer roller vortices in which helicoidal flow sweeps sand from interdune areas to dunes (Hanna, 1969) are not generally supported by empirical data – evidence for the occurrence of such boundary layer structures is limited.

Linear dunes primarily tend to extend downwind, but may also migrate laterally, because sand transport by the two modes of a bimodal wind regime are rarely equal (Rubin, Tsoar and Blumberg, 2008). Evidence for lateral migration has been documented from several areas, including northwest China, the Strzelecki Desert, Sinai and the Namib Sand Sea (see Tsoar, Blumberg and Stoler, 2004, and references therein). Rates of lateral variation vary considerably, ranging from 0.13 m/yr over the past 2400 years in Namibia (Bristow, Duller and Lancaster, 2007) to ~0.50 m/yr in the Sinai and up to 3 m/year in China (Rubin, Tsoar and Blumberg, 2008).

Some linear dunes do apparently form parallel to a dominant wind direction. This occurs when the sand is partially stabilised by vegetation or anchored by topographic obstacles (Hesp, 1981; Tsoar, 1989). Tsoar (1989) suggests that all vegetated linear dunes form parallel to the dominant wind direction. Recent studies suggest that partial induration of sediment may produce similar longitudinally oriented linear dunes (Rubin and Hesp, 2009).

19.3.3 *Star dunes*

Star dunes (Figure 19.6) are the largest and often the most complex type of desert sand dunes, with a pyramidal form and radiating arms (Lancaster, 1989a). Star dunes rarely occur in isolation and several patterns can be recognised, including chains and clusters of star dunes. Star dunes occur in the northern Sahara (Mainguet and Chemin, 1984), the northeastern part of the Rub al Khali, the Badain Jaran desert of China (Dong, Wang and Wang, 2004) and in parts of the Namib Sand Sea, and may reach a height of 300 m. They occur in areas of complex wind regimes, with strong seasonal changes in wind direction, involving winds from at least two directions separated by 180°. Although the mode of formation of star dunes is not well understood, evidence from several studies indicates that modification of a pre-existing form (linear or crescentic) is important (Nielson and Kocurek, 1987). Thus star dunes may form as crescentic dunes migrate into areas of seasonally reversing wind directions (Lancaster, 1989b), or by reworking of large linear or crescentic dunes as a result of climatic and wind regime changes (Beveridge *et al.*, 2004).

19.3.4 *Parabolic dunes*

Parabolic dunes (Figure 19.7) form in areas that are partially vegetated and experience unidirectional winds. They are characterised by a U or hairpin plan morphology, with an active 'nose' or dune front and trailing partly to completely vegetated arms that extend upwind. Because a sparse vegetation cover is important to the formation and maintenance of parabolic dunes, they have a restricted distribution in truly arid regions, but are common in many semi-arid dune areas, including the Great Plains of the USA (Sun and Muhs, 2007), the Thar Desert of India (Kar, 1993) and White Sands, New Mexico (Kocurek *et al.*, 2007).

19.3.5 *Zibars and sand sheets*

Sand sheets are common in the upwind areas of many sand seas and dune fields. They occur in environments that are unfavourable for the formation of dunes, often in conditions in which bypassing or net erosion of sand occurs. Sand sheets may develop in areas of coarse sands, such as in the eastern Sahara and southern Namib Sand Sea, the presence of a high water table and/or periodic flooding, as in parts of the White Sands dunefield, or a moderate vegetation cover, as in the northern Gran Desierto of Mexico (Kocurek and Nielson, 1986; Lancaster, 1992). Many sand sheets as well as interdune areas between linear and star dunes are organised into low rolling dunes, without slipfaces, known as zibars (Holm, 1960; Nielson and Kocurek, 1986; Warren, 1972), with a spacing of 50–400 m and a maximum relief of 10 m (see Figure 19.5(a)).

19.4 Airflow over dunes

The airflow around and over dunes is strongly influenced by the topography of the dune, which projects into the atmospheric boundary layer. This results in compression of streamlines on the upwind side of the dune, leading to acceleration of the wind and an increase in wind shear stress towards the dune crest. In the lee of the dune, the flow expands and/or separates from the surface, resulting in flow deceleration and secondary flows; at a point some distance downwind of the dune, the flow reattaches to the surface and gradually recovers to its upwind profile characteristics, assuming that there is no additional dune present.

Boundary layer theory provides a conceptual framework for understanding air flow over dunes (Nickling and McKenna Neuman, 1999). The flow can be divided into an outer inviscid region and an inner region, which follows the topography of the dune (Figure 19.8). The inner layer is further divided into two sublayers: (1) a very thin inner surface layer, in which the shear stress is constant,

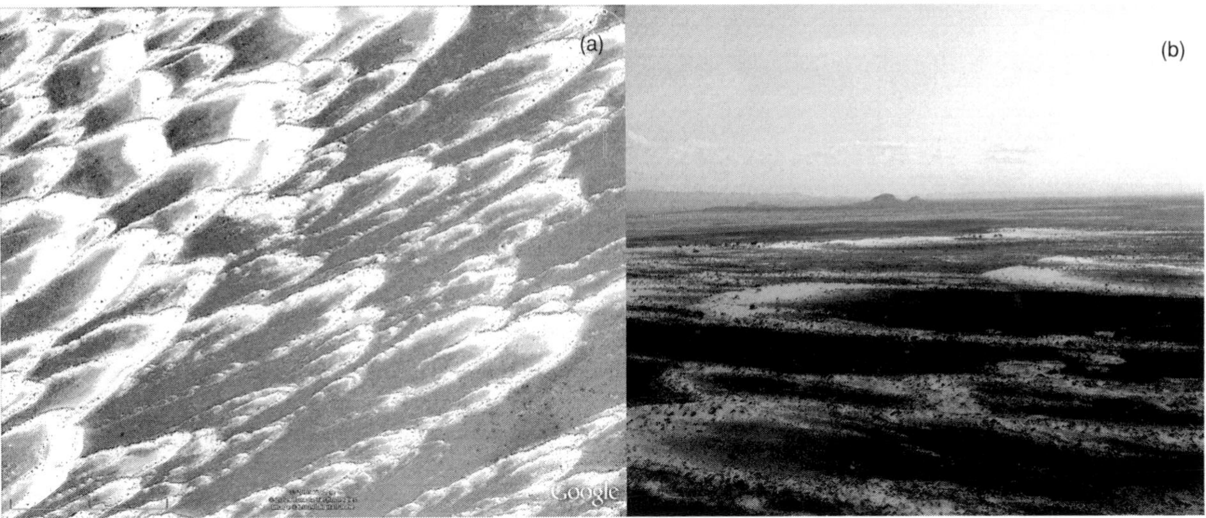

Figure 19.7 Parabolic dunes: (a, b) White Sands, New Mexico.

with a thickness equivalent to the surface roughness (1/30 of the grain diameter for a sand surface), and (2) a shear stress layer less than 1 m thick in which shear stress effects decrease with height above the surface. The shear stress developed in this part of the boundary layer generates sand transport on dunes. The basic features of this conceptual model have been confirmed by field studies of air flow over dunes (Livingstone, Wiggs and Weaver, 2007) and reproduced in numerical models (Parsons, Walker, and Wiggs, 2004; Weng *et al.*, 1991).

19.4.1 The stoss or windward slope

Winds approaching the upwind toe of a dune stagnate slightly and are reduced in velocity (Livingstone, Wiggs and Weaver, 2007; Wiggs, 1993). On the stoss, or windward slope of the dune, streamlines are compressed and winds accelerate up the slope. The magnitude of the velocity increase varies with dune height and dune steepness or aspect ratio (Figure 19.9) and is characterised by the speed-up ratio Δs or amplification factor Az (U_2/U_1,

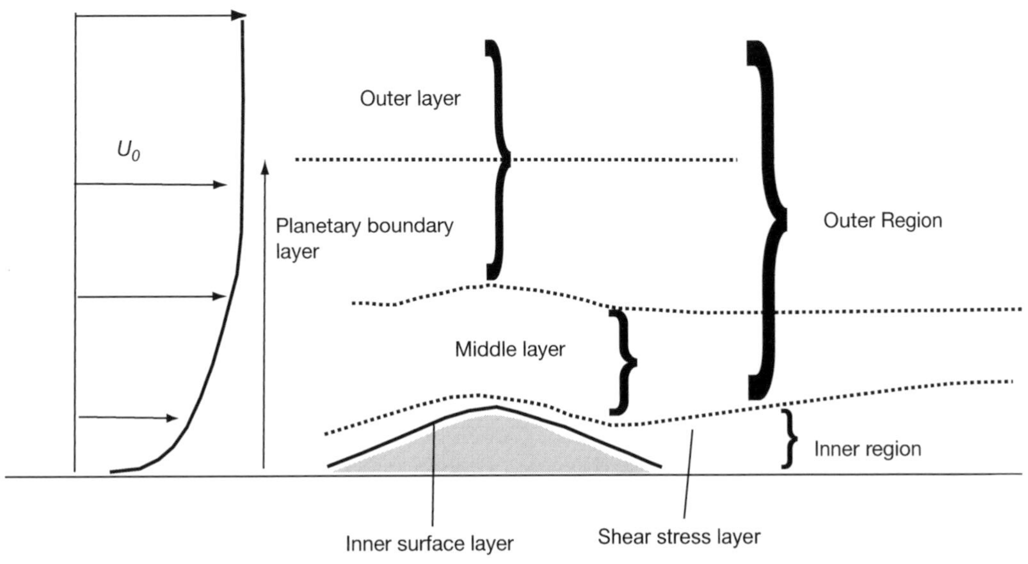

Figure 19.8 Airflow over an isolated two-dimensional hill, showing components of the boudary layer (after Nickling and McKenna Neuman, 1999).

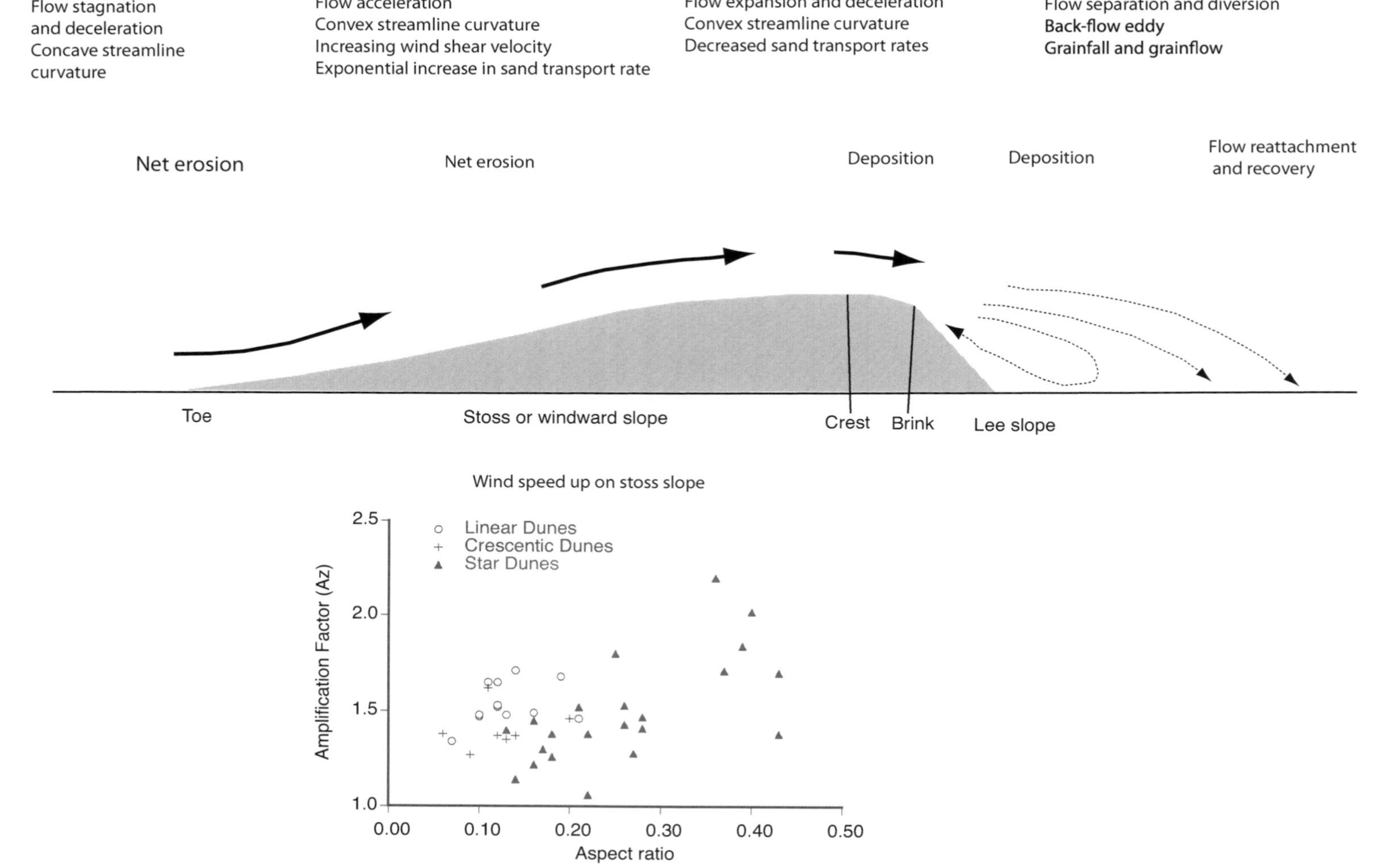

Figure 19.9 Airflow over dunes showing streamline convergence and divergence (after Livingstone, Wiggs and Weaver, 2007) and velocity speedup (after Lancaster, 1995).

where U_2 is the velocity at the dune crest and U_1 is the velocity at the same height above the surface at the upwind base of the dune). Wind tunnel simulations and numerical models show that shear stress increases on dune slopes in a manner similar to wind speed, but field measurements of wind shear velocity and shear stress on dunes are difficult to achieve where wind profiles are not log-linear as a result of flow acceleration and development of internal boundary layers (Frank and Kocurek, 1996a). In addition to the effects of velocity amplification, curvature of wind streamlines may play an important role in the pattern of wind shear stress on dunes (Figure 19.9). Concave upward curvature of streamlines at the toe of the dune enhances shear stress, whereas convex streamline curvature on the mid stoss slope and between the crest and the brink acts to decrease shear stress (Wiggs, Livingstone and Warren, 1996).

19.4.2 Lee-side flow

In the lee of the crest of dunes, wind velocities and transport rates decrease rapidly as a result of flow expansion between the crest and brink of the lee or avalanche face and flow separation on the avalanche face itself. There is a complex pattern of flow separation, diversion and reattachment on the lee slopes of dunes, which is determined by the angle between the wind and the dune crest (angle of attack), atmospheric stability and the dune aspect ratio (Walker and Nickling, 2002; Dong *et al.*, 2009; Sweet and Kocurek, 1990). Secondary flows, including lee-side flow diversion, are especially important where winds approach the dune obliquely, and are an important component of air flow on linear and many star dunes (Figure 19.10).

Conceptual models for airflow in the lee of flow-transverse dunes have been developed based on field and wind tunnel studies (Baddock, Livingstone and Wiggs, 2007; Frank and Kocurek, 1996b; Walker, 1999; Walker and Nickling, 2002). These models identify a series of distinct regions of flow that vary in wind speed, shear and turbulence intensity (Figure 19.11(a)), including a separation cell extending for 4–10 dune heights downwind, two wake regions that merge at 8–10 h and an internal boundary layer that grows downwind of the point at which the separated flow reattaches to the surface. According to Walker and Nickling (2002), the internal boundary layer is identifiable at 8–10 h and comes into equilibrium at around 25–30 h, downwind of isolated dunes. For multiple transverse dunes, Baddock, Livingstone and Wiggs (2007) identify four main zones (Figure 19.11(b)): separation, reattachment, recovery and interaction with the next dune downwind.

When flow is oblique to the dune crest a helical vortex develops in the lee (Walker and Nickling, 2002). The oblique flow is deflected along the lee slope parallel to the dune crest, with the degree of deflection being inversely proportional to the incidence angle between the crestline and the primary wind. When the angle between the dune and the wind is less than 40° the velocity of the deflected wind is greater than that at the crest and sand is transported along the lee side of the dune (Tsoar, 1983). This process is especially important on linear dunes, where it leads to dune extension. When winds are at more than 40° to the crestline the velocity of the deflected wind is reduced, giving rise to lee-side deposition. Changes in the local incidence angle between primary winds and a sinuous dune crest result in a spatially varying pattern of deposition and along-dune transport on the lee face. Deposition dominates where winds cross the crest line at angles approaching 90° and erosion or along-dune transport occurs where incidence angles are $<40°$ (Figure 19.10).

19.5 Dune dynamics

Dunes may (1) migrate downwind or laterally as a result of erosion of windward slopes and deposition in the lee, (2) extend downwind or (3) accrete vertically (Thomas, 1992). Different dune morphologic types exhibit different degrees of these three processes. Thus, the arms of barchans may extend as small linear features (Tsoar, 1984), linear dunes may migrate laterally as well as extending down wind, and also accrete vertically (Bristow, Duller and Lancaster, 2007); star dunes are dominated by vertical acretion, but their arms may migrate laterally and/or extend (Lancaster, 1989b).

19.5.1 Erosion and deposition patterns on dunes

All dunes exhibit a high degree of interaction between the shape of the dune, the amount of change in wind velocities and sand transport rates, and the rate and pattern of erosion or deposition. These interactions are manifested in several ways.

The windward slopes of dunes are characterised by an exponential increase in sediment transport rates towards the dune crest (Figure 19.12), as a result of flow acceleration, coupled with effects of stream line curvature (Lancaster *et al.*, 1996; McKenna Neuman, Lancaster and Nickling, 1997, 2000). Although net erosion tends to characterise the windward slopes of dunes, the pattern of erosion and deposition varies with overall wind speed, so that areas near dune crests may experience erosion at lower wind speeds and deposition at wind speeds

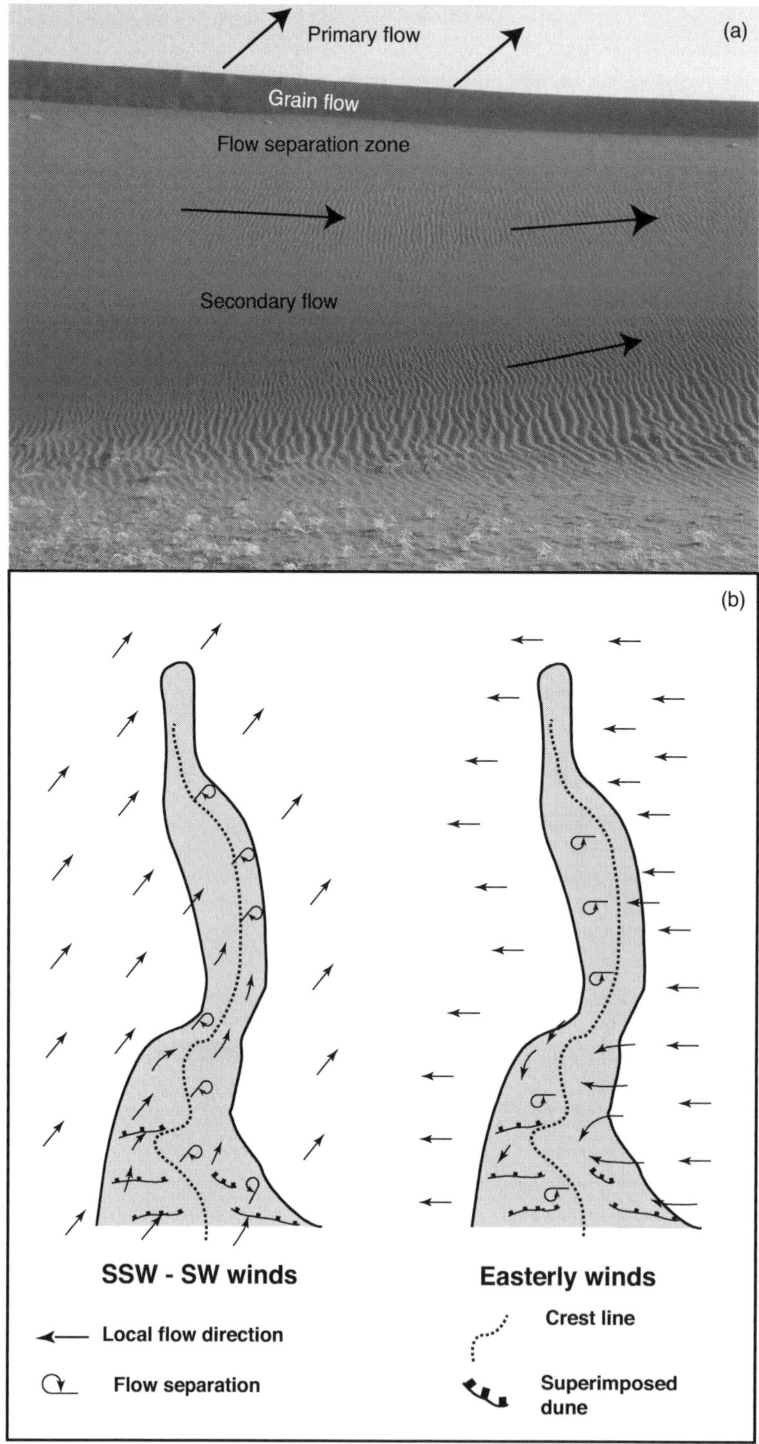

Figure 19.10 Flow separation and diversion in the lee of a linear dune: (a) field view of flow in the lee of a simple linear dune; (b) schematic of flow over the same dune in different wind regimes (after Bristow, Bailey and Lancaster, 2000).

significantly above threshold (McKenna Neuman, Lancaster and Nickling, 2000).

Separation of airflow at the brink of the lee face leads to grain fall from the overshoot of saltating grains transported up the windward slope. The rate of deposition by grainfall decreases exponentially with distance from the brink of the lee face (Figure 19.13), so that most of the sediment is deposited within 1 m of the brink (Nickling, McKenna Neuman and Lancaster, 2002). Grain flows occur when grainfall deposits build up so that the lee slope is steepened above the angle of repose and fails, initiating a grain flow or avalanche. The frequency of grain flows is dependent on the wind speed and sand transport rate for the period preceding their initiation. The interval between grain flows decreases exponentially with incident wind speed from about 60 minutes at 5 m/s to 8 minutes at 8 m/s; the magnitude of the flows is inversely proportional to the interval between flows (Breton, Lancaster and Nickling, 2008).

Seasonal patterns of erosion and deposition measured on linear and star dunes show that the greatest amount of change occurs when winds change direction seasonally

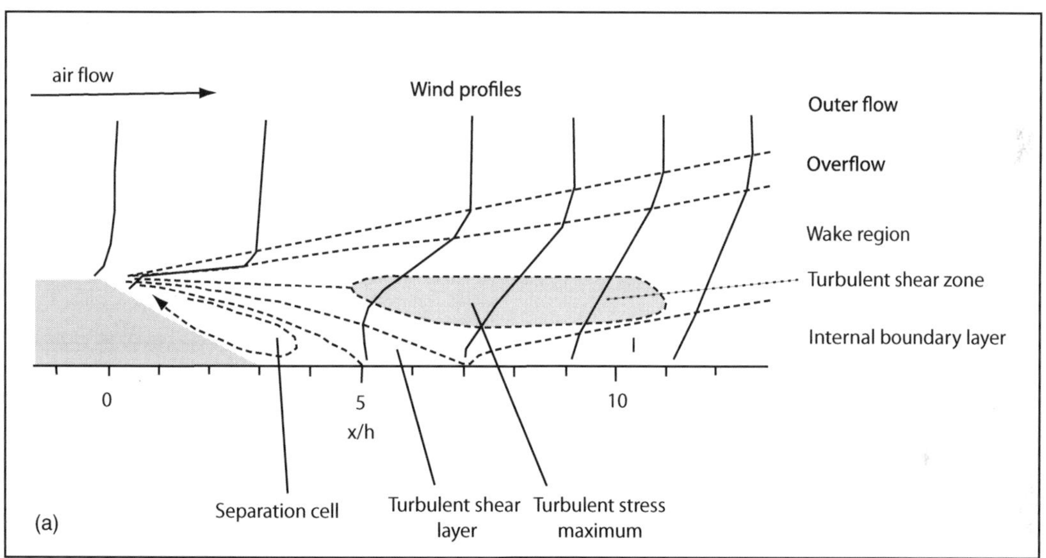

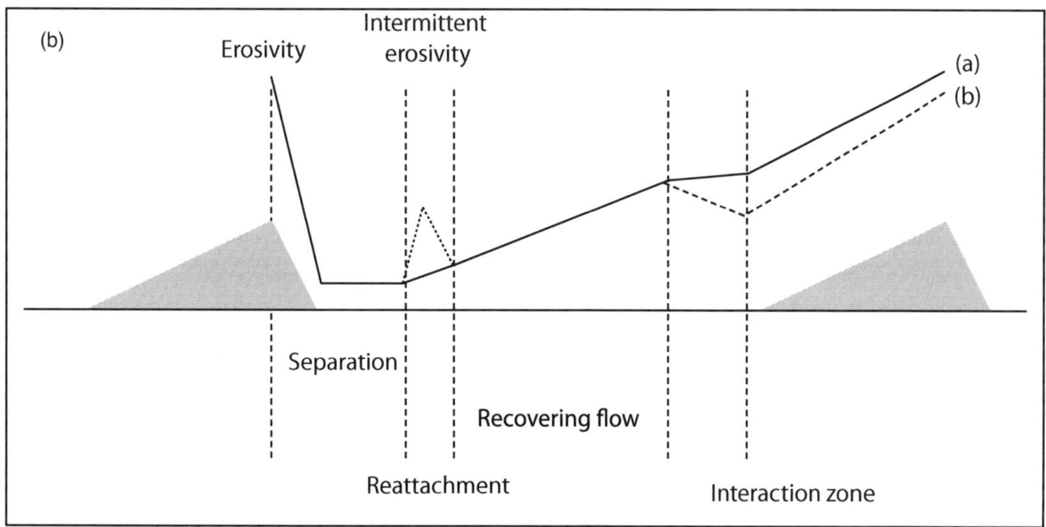

Figure 19.11 Conceptual model for flow in the lee of a transverse dune: (a) isolated dune (after Walker and Nickling, 2002); (b) (after Baddock, Livingstone and Wiggs, 2007).

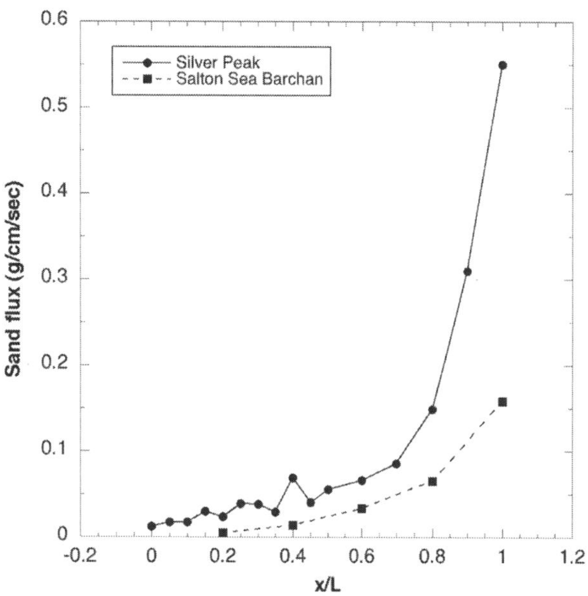

Figure 19.12 Windward slope erosion and deposition patterns (data from Lancaster *et al.*, 1996, and McKenna Neuman, Lancaster and Nickling, 1997).

(Lancaster, 1989b, 1989c). This is the result of winds encountering a dune form that is out of equilibrium with a new wind direction. Field observations and models show that the crestal profiles of linear and star dunes tend towards a convex form similar to that of transverse dunes (Tsoar *et al.*, 1985), and erosion and deposition rates near the crest decline in time as the dune comes into equilibrium with a new wind direction.

Erosion and deposition patterns on linear, reversing and star dunes show that they consist of a crestal area where erosion and deposition rates are high and a plinth zone in which there is little surface change but considerable throughput of sand (Figure 19.14). The crest lines of Namibian complex linear dunes migrate over a lateral distance of as much as 14 m over a 12-month period but show little net change over decadal timescales (Livingstone, 1989, 1993, 2003; Sharp, 1966; Wiggs *et al.*, 1995).

19.5.2 Long-term dune dynamics

Insights into dune dynamics over annual to decadal timescales can be gained from a variety of sources. A large data set on migration rates of barchans and crescentic dunes has been developed by comparing the position of dunes on time series of aerial photographs or satellite images, demonstrating that there is an inverse relationship between barchan dune height and migration rate (Figure 19.15). This relationship is important to the development of self-organised dune patterns in that more rapidly moving smaller dunes tend to catch up and merge with larger

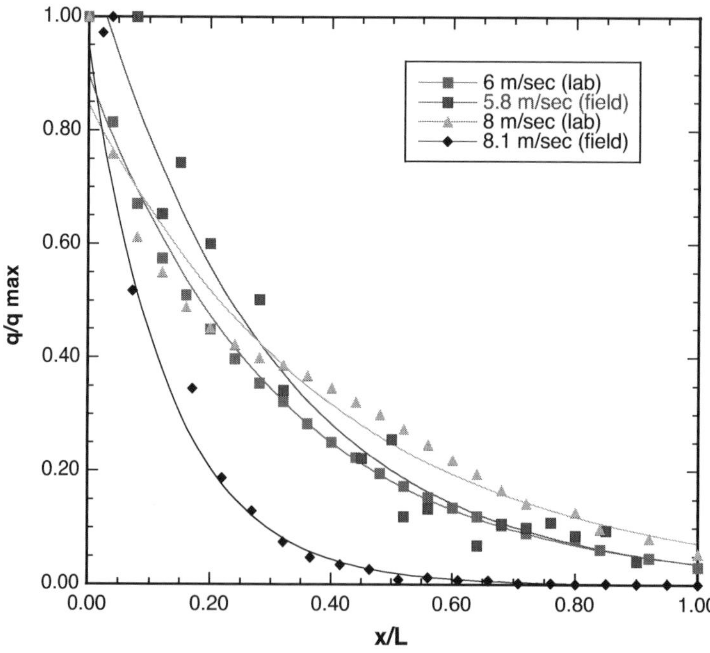

Figure 19.13 Patterns of deposition in the lee of a transverse dune (field data after Nickling, McKenna Neuman and Lancaster, 2002; laboratory wind tunnel data after Cupp *et al.*, 2005).

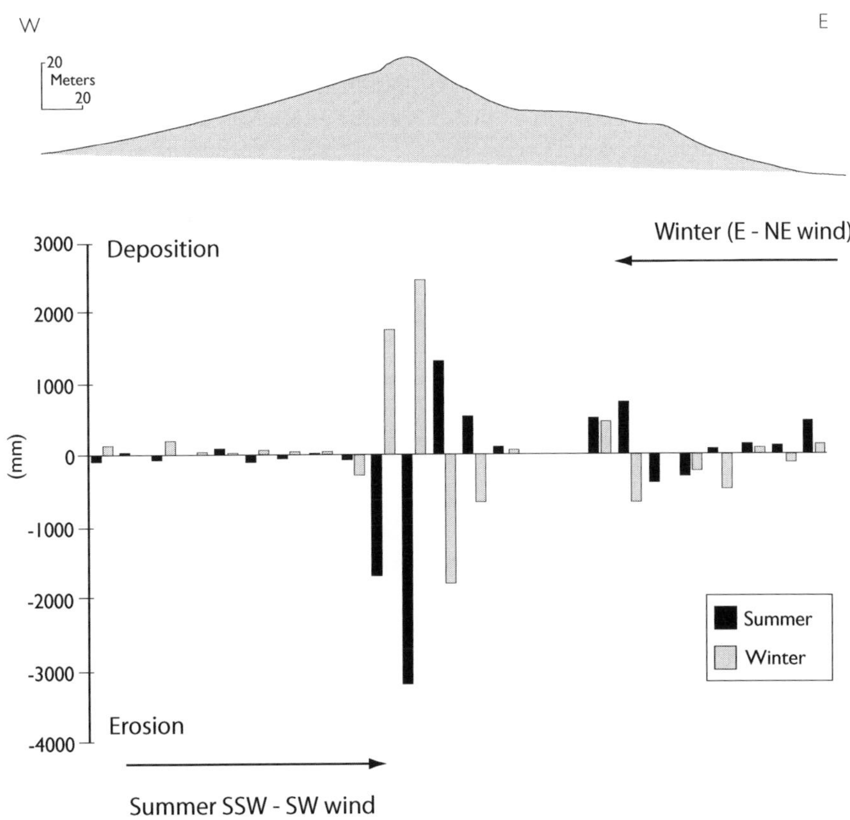

Figure 19.14 Patterns of erosion and deposition cross a Namib linear dune (after Livingstone, 1989).

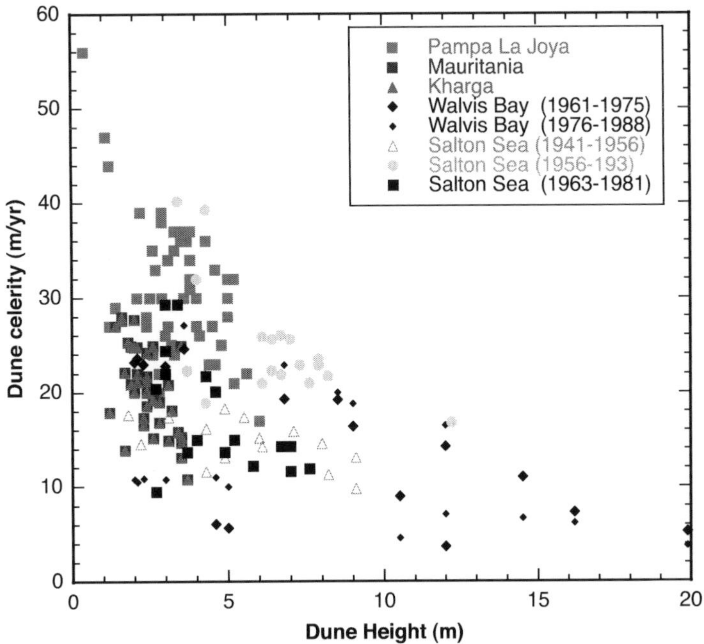

Figure 19.15 Relations between barchan migration rates and dune height (data from Finkel, 1959, Slattery, 1990, Ould Ahmedou *et al.*, 2007, Embabi, 1982, and Haff, 1984).

dunes, so promoting an overall increase in dune size and spacing downwind (pattern coarsening).

At best, these studies span the last few decades. A longer-term (centuries to millennia) perspective on dune dynamics can now be gained through the application of high-resolution optically stimulated luminescence (OSL) dating of dunes, especially when used in combination with ground penetrating radar (GPR) studies of dune sedimentary structures. Crescentic dunes superimposed on the southern edge of a north–south-oriented linear dune in the northern part of the Namib Sand Sea have migrated to the west at an average rate of 0.12 m/yr over the past 1570 years (Bristow, Duller and Lancaster, 2005), while a nearby 60 m-high linear dune shows evidence of episodic development and lateral migration at rates of up to 0.13 m/yr to the east over the past 6000 years (Bristow, Duller and Lancaster, 2007) and 20-m high crescentic dunes in the Liwa area of the UAE have migrated to the south over the past 320 years at an average rate of 0.78 m/yr, with a more rapid migration of 0.91 m/year between 220 and 110 years ago (Stokes and Bray, 2005).

19.6 Dune development

The origins and growth of individual dunes and the development of dune patterns are intimately linked, because dunes rarely occur in isolation.

Dune initiation is a poorly understood process, but is likely to involve localised reductions in sand transport rates by convergence of streamlines, changes in surface roughness (e.g. vegetation cover), surface particle size or by variations in microtopography (slope changes, relict bedforms), leading to deposition of sand, which forms a nucleation point for a protodune (Kocurek *et al.*, 1992; Lancaster, 1996). There is a minimum size for a protodune, determined by the saturation length or the distance required for actual sand mass flux to reach the capacity of the wind to transport sand. The elementary dune wavelength is $\sim$20 m on Earth and $\sim$600 m on Mars (Claudin and Andreotti, 2006). Once initiated, growth of crescentic dunes occurs by merging of smaller, faster-moving dunes with larger, slower-moving dunes, in a series of constructive interactions.

The primary response of sand surfaces to bedform development is to form a asymmetric flow-transverse dune with a convex stoss slope. Crescentic dunes dominate in unidirectional wind regimes; small dunes in multidirectional wind regimes are also of this form, because they are small enough to be completely reworked in each wind season.

Following Tsoar (1989), vegetated linear dunes are nucleated by vegetation, and extend downwind from nebkhas. Vegetated linear dunes can also develop from source bordering dunes; closely spaced dunes link by Y-junctions to form fewer, larger dunes downwind (Wasson, 1983).

The formation of sinuous linear dunes (seif dunes) by elongation of one arm of a barchan as it migrates into a bimodal wind regime was first proposed by Bagnold (1941) and supported by field observations in the Namib (Lancaster, 1980). Strong winds from an oblique direction add sand to one horn and gentler winds from the original direction extend the horn (Figure 19.16(a)). A different model is suggested by observations in the Sinai (Tsoar, 1984), where winds from both the primary and secondary directions extend the horn on the side of the dune opposite to the secondary wind direction (Figure 19.15(b)). The linear element of the dune extends more rapidly than the original barchan can migrate, and the dune evolves to a linear form.

Linear dunes develop by extension downwind and by slow accretion of sand. Bristow, Bailey and Lancaster (2000) present a new model for development of sinuous linear dunes based on GPR visualisation of sedimentary structures. In this model (Figure 19.17), secondary flows and form–flow interactions become more important as the dune increases in size, leading ultimately to the establishment of superimposed bedforms on the dune flanks.

Star dunes appear to develop mainly by modification of existing dunes (Lancaster, 1989b; Nielson and Kocurek, 1987), either as they migrate or extend into areas experiencing multidirectional winds (Figure 19.18) or as a result of changes in wind regime as a result of climate change, as in star dunes developed on linear dunes in the Gran Desierto (Beveridge *et al.*, 2006) and the southeast UAE.

Many megadunes possess superimposed dunes, which are in equilibrium with the current wind regime (e.g. on Algodones and Namib compound crescentic dunes). The slopes of these large dunes present an effectively planar surface on which sand transport takes place. Therefore, variations in sand transport rates on megadunes in time or space will lead to the formation of superimposed dunes if the major dune is sufficiently large (Andreotti *et al.*, 2009; Elbelrhiti, Claudin and Andreotti, 2005).

In some areas, however, the larger primary (mostly linear) dunes are clearly a product of past wind regime and sediment supply conditions, as demonstrated by OSL dating of the major form (Lancaster, 2007). Such dunes have considerable inertia (represented by the reconstitution time, or the time required for the dune to migrate its length in the direction of net transport) and require significant periods of time to adjust to changed conditions of

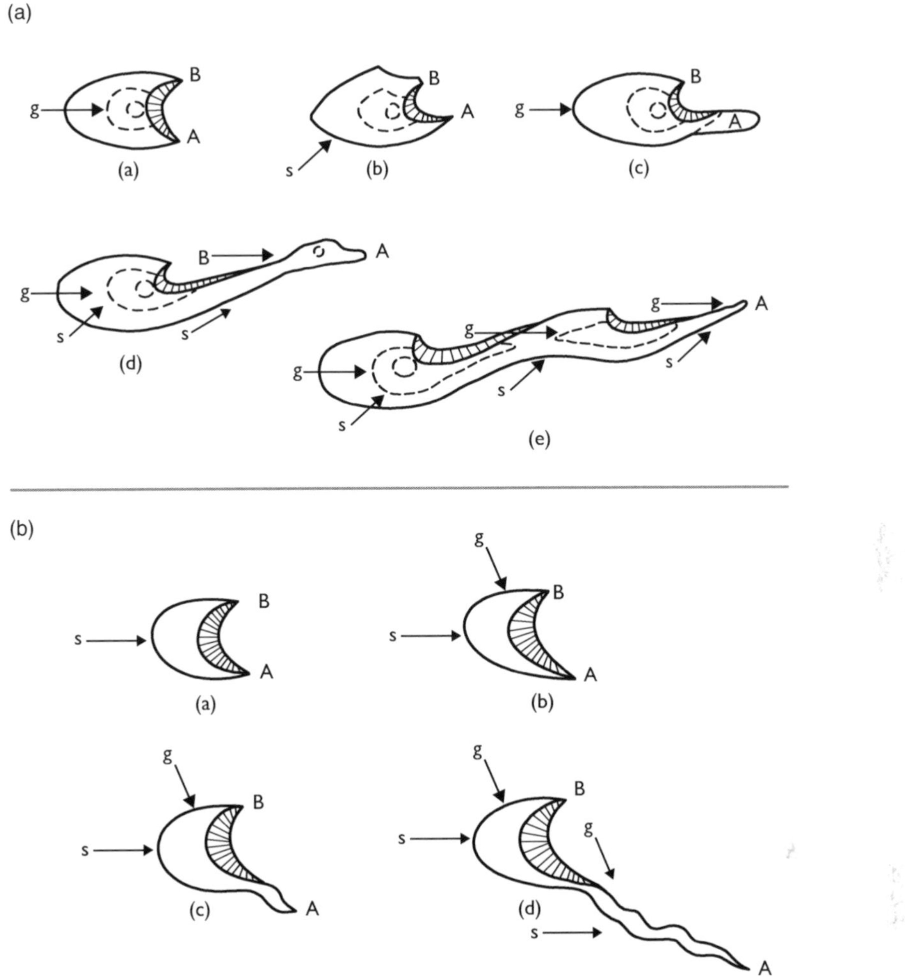

Figure 19.16 Development of linear dunes from barchans: (a) after Bagnold (1941); (b) after Tsoar (1984).

wind regime and sediment supply (Warren and Allison, 1998). Reconstitution time increases by several orders of magnitude from simple to complex dunes. The lifespan of crescentic dunes superimposed on Namib linear dunes is 30–50 years, compared to a reconstitution time of about 6000 years for the main dune, based upon estimates of migration rates derived from combined GPR and OSL studies (Bristow, Duller and Lancaster, 2007).

19.7 Controls of dune morphology

Empirical studies as well as numerical modelling have established that the directional variability of the wind regime is a major determinant of dune type. Wind speed, grain size and vegetation play subordinate roles. The effects of sand supply are uncertain. The controls of dune size and spacing are, however, less well understood.

19.7.1 Sediment characteristics

Although sand sheets and zibar are often comprised of coarse, poorly sorted sands, there is no evidence for general relationships between sand particle size and sorting and dune morphological types. Although Wilson (1972) hypothesised that the spacing of aeolian bedforms at all scales is related to sand particle size, these relationships are not generally supported by empirical data (Wasson and Hyde, 1983a).

19.7.2 Wind regimes

Global (Fryberger, 1979; Wasson and Hyde, 1983b) and regional comparisons (e.g. Lancaster, 1983) between the occurrence of major dune types and the characteristics of local wind regimes support the hypothesis that the

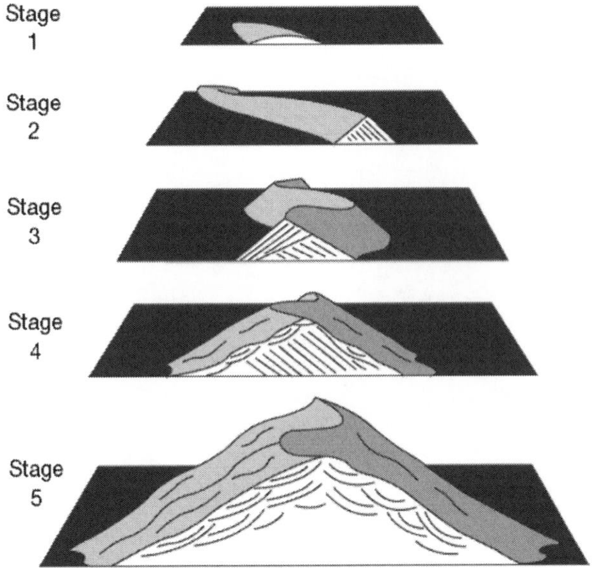

Figure 19.17 Model for linear dune development (after Bristow, Bailey and Lancaster, 2000).

major control on dune type is the directional variability of the sand-transporting winds, as characterised by the ratio between total potential sand transport (drift potential, or DP) and resultant (vector sum) potential sand transport (resultant drift potential, or RDP) (Figure 19.19). The directional variability or complexity of the wind regime increases from environments in which crescentic dunes are found to those where star dunes occur (Fryberger, 1979). Crescentic dunes occur in areas where RDP/DP ratios exceed 0.50 (mean RDP/DP ratio 0.68) and frequently occur in unimodal wind regimes, often of high or moderate energy. Linear dunes develop in wind regimes with a much greater degree of directional variability and commonly form in wide unimodal or bimodal wind regimes with mean RDP/DP ratios of 0.45. Star dunes occur in areas of complex wind regimes with RDP/DP ratios less than 0.35. Such a model is further supported by laboratory experiments and numerical simulations of bedforms in directionally varying flows (Reffet *et al.*, 2008; Rubin and Ikeda, 1990; Werner, 1995), which clearly show that crescentic dunes are not stable in directionally variable flows and that linear dunes are the product of bimodal wind regimes in which the two modes are separated by at least 90°.

19.7.3 Sand supply

The supply of sand has long been considered a factor influencing dune morphology (Hack, 1941; Wilson, 1972).

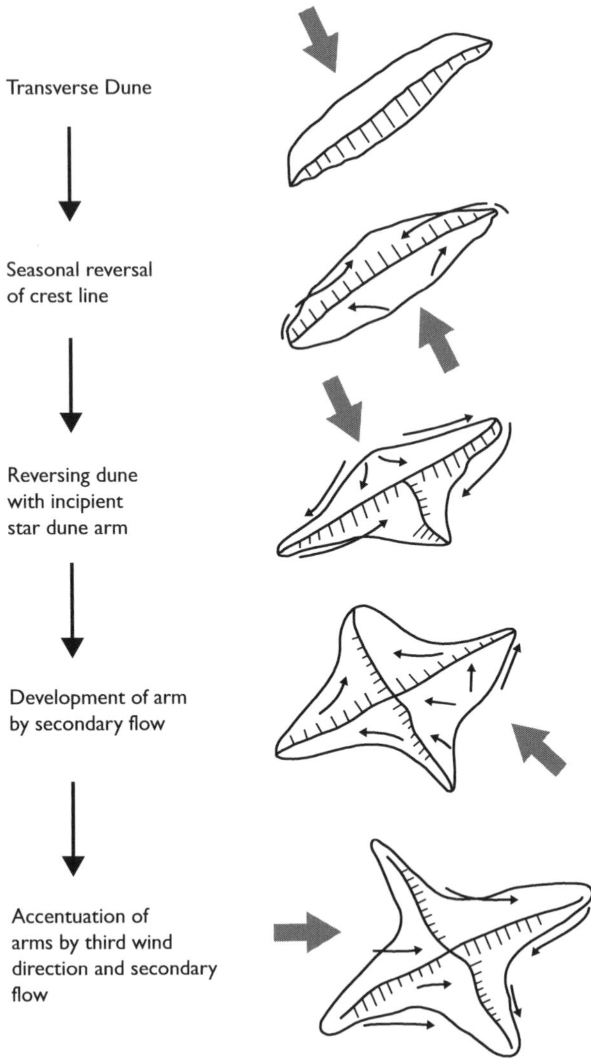

Figure 19.18 Formation of star dunes by modification of crescentic dunes (after Lancaster, 1989b).

Wasson and Hyde (1983a) confirmed that it was possible to discriminate between dune types on the basis of the complexity of the wind regime (the RDP/DP ratio) and the equivalent (or spread-out) sand. Barchans occur where sand supply is low and winds are unidirectional, crescentic dunes are located where sand is more abundant. linear dunes develop where sand supply is relateively low but winds are more variable and star dunes form in complex wind regimes with abundant sand supply. Similar relationships are evident in the Namib Sand Sea, although the range of directional variability is lower (Figure 19.19).

EST is, however, a measure of the volume of sand contained in the dunes and may be a product of dune type,

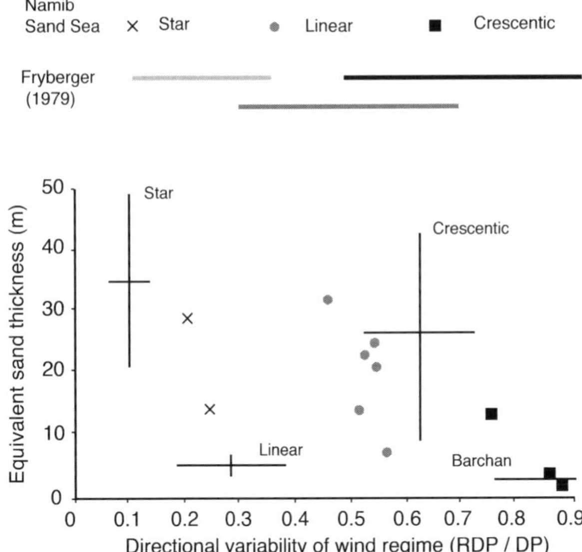

Figure 19.19 Relations between dune type, wind regime variability (RDP/DP ratio) and sand thickness (RDP/DP data from Fryberger, 1979, Wasson and Hyde, 1983a, and Lancaster, 1989).

with the dune type being influenced by the other factors, especially the wind regime (Rubin, 1984). In the Namib Sand Sea, dune types clearly occupy separate wind regime domains and the EST data suggest that there is more sand in complex linear and star dunes than in compound linear and all types of crescentic dunes.

19.7.4 Vegetation

Vegetation affects the rate of sand transport via direct protection of the surface, absorption of momentum and partitioning of shear stress between the surface and plants (Wolfe and Nickling, 1993). As a result, sand transport rates decrease exponentially with vegetation cover and sand surfaces are effectively stabilised when it exceeds about 15 % (Lancaster and Baas, 1998; Wiggs *et al.*, 1994).

The effects of vegetation on dune morphology are less well known. Hack (1941) suggested that in northeastern Arizona crescentic dunes changed to parabolic dunes with increased vegetation cover and that linear dunes occurred in areas with less sand and lower vegetation cover than parabolic dunes. In the Negev Desert, Tsoar and Møller (1986) documented changes in the morphology of linear dunes as vegetation cover decreased as a result of grazing, including development of sharp sinuous 'seif' dune crests and braided patterns from vegetated linear dunes. Recent work (Tsoar, 1989) has suggested that there is a distinct class of linear dunes that are partially vegetated. Their staight crests and orientation parallel to the dominant wind result from the anchoring of dunes by vegetation, as well as the deposition of sand downwind of vegetation.

19.7.5 Controls of dune size and spacing

Dunes exhibit a wide range of sizes, with close relationships between dune height and crest-to-crest spacing, indicating that the patterns are self-organised (Figure 19.20). In many sand seas, there is also a spatial pattern of dune size and spacing, such that small, closely spaced dunes occur on the margins, with large dunes in areas of net sand accumulation.

The controls on dune size and spacing have been debated for many years, but recent work has clarified some key issues. For example, the hypothesis of Wilson (1972) that sand particle size controls dune spacing is not supported by empirical data (Wasson and Hyde, 1983a). Likewise, there is no simple relationship between dune size and sand transport rates (Elbelrhiti, Andreotti and Claudin, 2008; Lancaster, 1988). In the Namib Sand Sea, small dunes occur in areas of high sand transport rates (as indicated by wind data), whereas large dunes occur in areas of low net sand transport, as a result of their growth in conditions of abundant sand supply and a wind regime that promotes vertical accretion, rather than migration or extension. The size of superimposed dunes on larger forms, however, scales with the local sand transport rate (Havholm and Kocurek, 1988; Lancaster, 1988). The minimum size of such simple dunes is a function of the sand transport saturation length (Claudin and Andreotti, 2006), which on Earth is around 20 m.

Studies of dune patterns and numerical models indicate that the spacing of crescentic dunes is a dynamic property of the pattern (Elbelrhiti, Andreotti and Claudin, 2008). The developed pattern results from merging and linking of dunes with a highly variable spacing, size and migration rate into fewer dunes with a much more regular pattern of size and spacing (Figure 19.21), in a similar way to the development of patterns in wind ripples and subaqueous environments. Such 'pattern coarsening' leads to a more ordered and stable pattern with fewer interactions between dunes. Similar downwind pattern development has been documented from crescentic dunes at White Sands (Ewing and Kocurek, 2010b), as well as linear dunes in the Simpson Desert (Fitzsimmons, 2007).

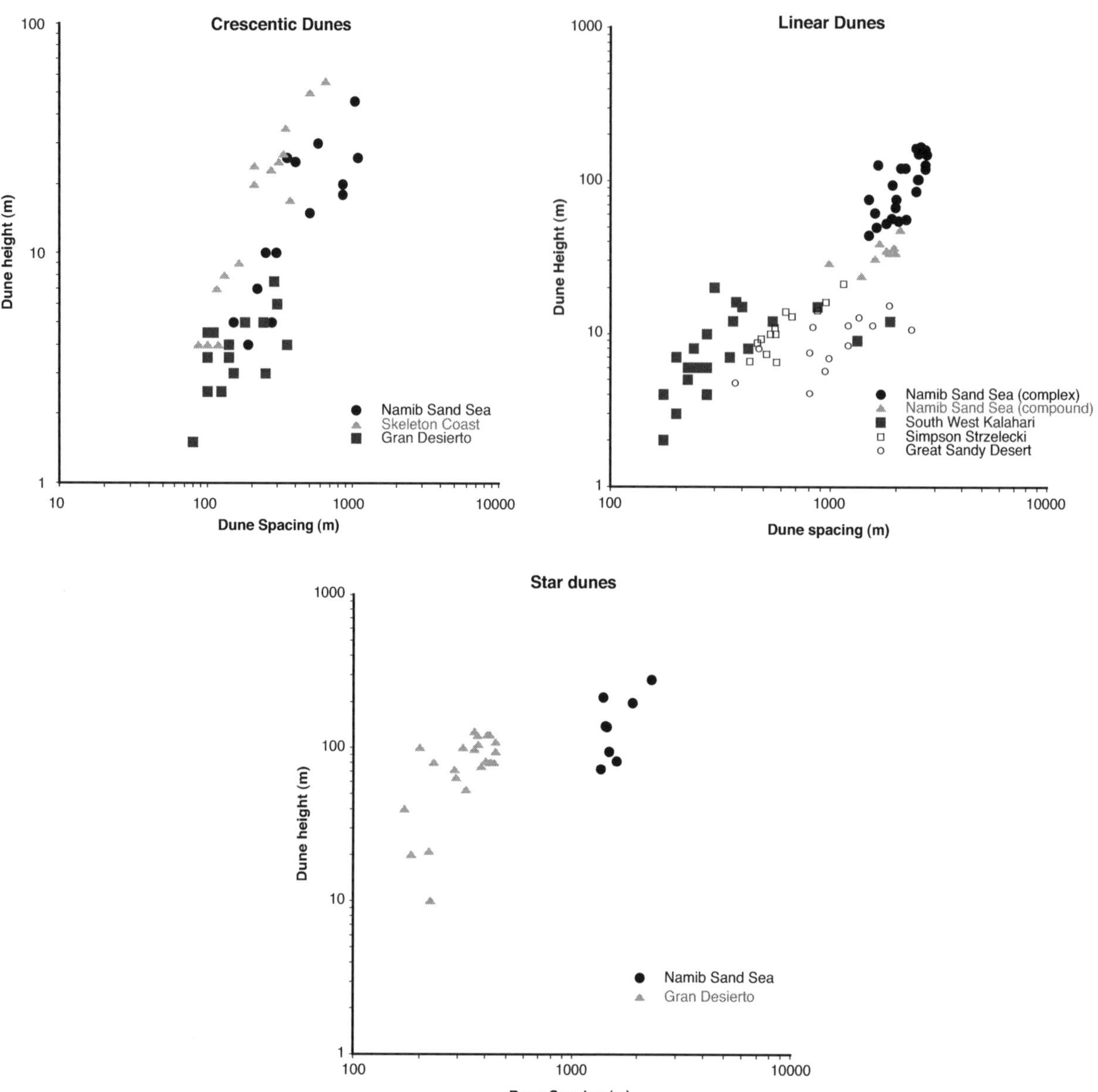

Figure 19.20 Dune height and spacing: (a) crescentic dunes, (b) linear dunes, (c) star dunes.

Fundamentally, the limits to dune size are set by the long-term supply of sand from sources external and/or internal to the sand sea or dune field. This is a function of the regional setting of the sand sea and also its size, so that large dunes almost always occur in relatively large sand seas. Ewing and Kocurek (2010a) have documented an increase in average dune spacing with dune field size, which may reflect the degree to which the dune pattern can evolve.

A further limiting factor on dune size may be the depth of the atmospheric boundary layer in a region, which is a function of the annual variation of surface temperature (Andreotti *et al.*, 2009). The depth of the boundary layer sets a limit to megadune growth by amalgamation and nonlinear interactions between simple or elemental dunes. GPR and OSL dating studies show that large dunes are the product of multiple periods of dune construction, in some cases spanning at least the last 25,000 years

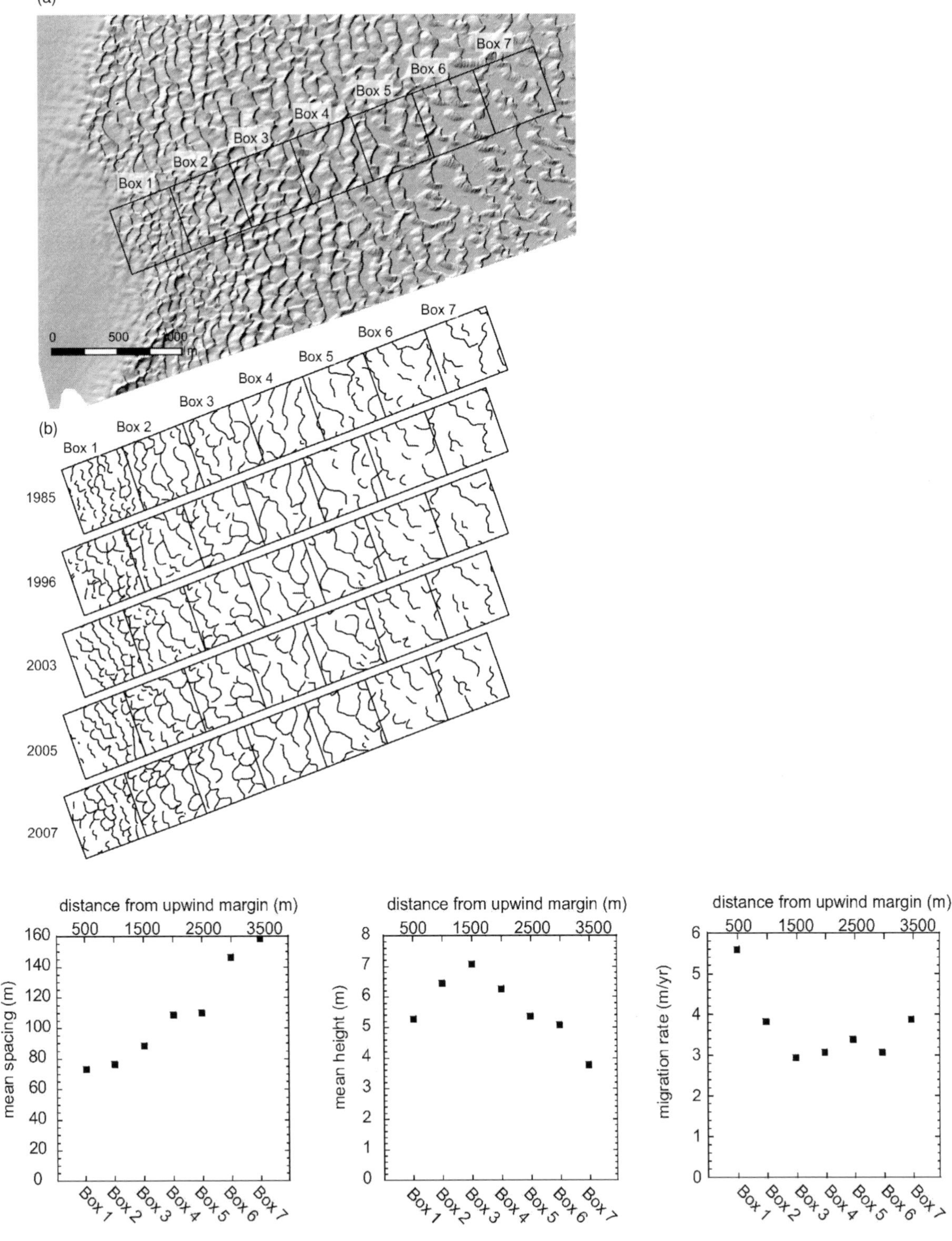

Figure 19.21 Dune pattern development in a downwind direction, White Sands, New Mexico (from Ewing and Kocurek, 2010b).

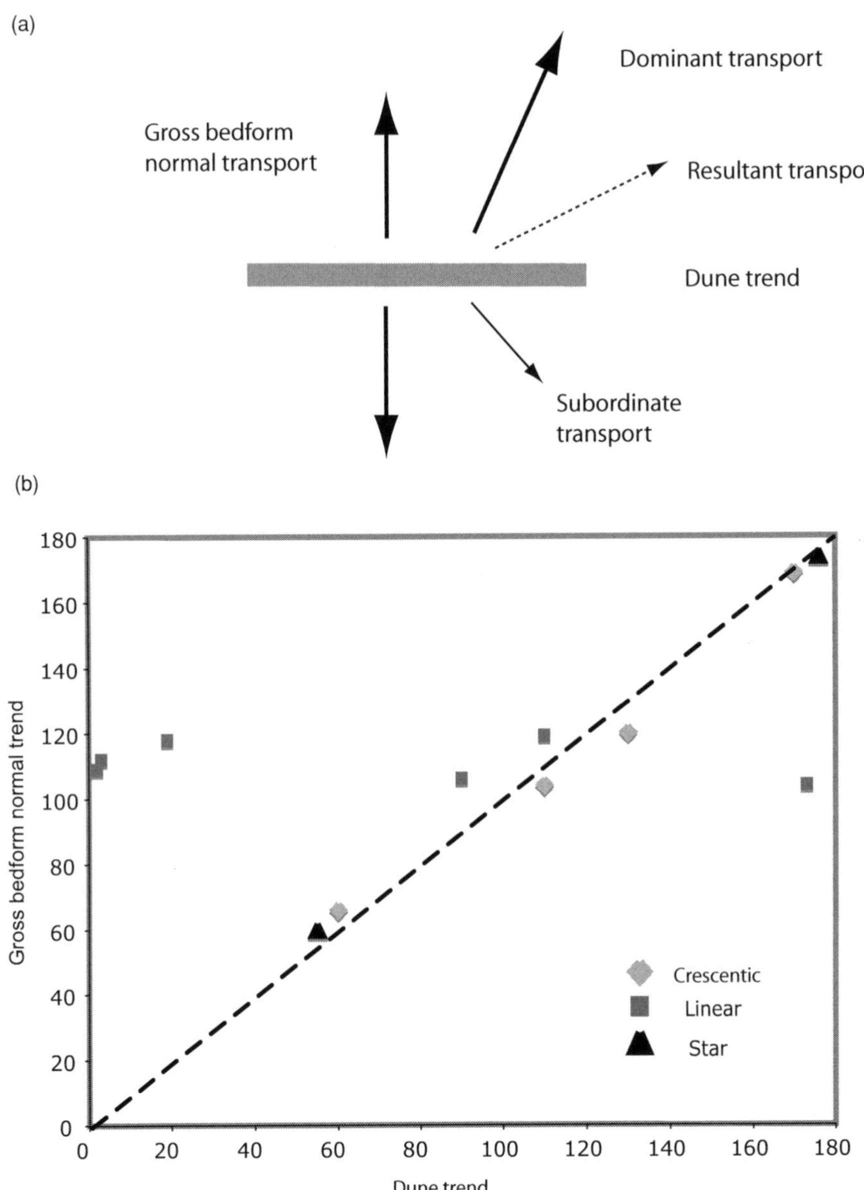

Figure 19.22 Gross bedform normal (GBN) transport direction and dune trends: (a) explanation of the gross bedform normal model (after Rubin and Ikeda, 1990); (b) relations between dune trend and GBN direction (after Lancaster, 1991, with additional data).

(Bristow, Duller and Lancaster, 2007; Kocurek *et al.*, 1991; Lancaster *et al.*, 2002), so age of the megadunes is also an important factor.

19.7.6 Dune trends

The factors that determine the alignment of dunes with respect to the wind have long been a subject for speculation. Many workers have concluded that dunes are oriented relative to the resultant or vector sum of sand transport (Fryberger, 1979). Thus dunes can be classified as transverse (strike of crestline approximately normal to resultant), longitudinal (crest parallel to resultant) or oblique (15–75° to resultant direction) (Bagnold, 1953; Glennie, 1970; Hunter, Richmond and Alpha, 1983; Mainguet and Callot, 1978).

Field and laboratory experiments with wind ripples and subaqueous dunes (Rubin and Hunter, 1987; Rubin and Ikeda, 1990) suggest that all bedforms are oriented in the direction subject to the maximum gross bedform normal transport across the crest (Figure 19.22(a)), so that sand transport from all directions contributes to bedform development. These experiments and subsequent numerical simulations show that the type of dune that forms is determined by the divergence angle between the dominant and subordinate transport vectors and the ratio between the two primary transport directions (transport ratio). A trend parallel to or normal to the resultant direction of sand transport is purely coincidental.

This model is supported by empirical data on dune trends and winds. There is a close agreement between observed and predicted gross bedform normal orientations in the case of many barchans, crescentic dunes, simple linear dunes and star dunes (Figure 19.22(b)), suggesting that all major dune types are oriented to maximise gross bedform normal sediment transport and therefore are dynamically similar (Lancaster, 1991). This approach can also be used to identify dune trends that are out of equilibrium with the modern sand transporting wind regime, as demonstrated for Mauritania, and to suggest wind regimes that could have produced the trends of dunes constructed in the past. (Lancaster *et al.*, 2002).

19.8 Dune patterns

Dune systems exhibit clear patterns of dune morphology, size, spacing and crestline alignment on several spatial scales: (1) regional, (2) within a sand sea and (3) within an area of dunes of similar morphology. The pattern of dunes is a reflection of the external and internal factors that determine the present and past dynamics of the system.

On a regional scale, mapping of dune trends shows a pattern that is determined by regional atmospheric circulations (Figure 19.23), especially the pattern of winds outblowing around anticyclonic cells, as in the Kalahari (Lancaster, 1981) and Australia (Wasson *et al.*, 1988). In the Sahara, the pattern of dune trends corresponds broadly to the of the Trade Wind circulation (Mainguet, 1984), modulated in some areas by topography and the development of low-level jets (Washington *et al.*, 2006), while in Arabia, dune trends follow the Shamal circulation (Glennie, 1998).

Within individual sand seas and dune fields, mapping of dune patterns from aerial photographs and satellite images shows distinct spatial patterns of dune type, size and spacing (Breed and Grow, 1979; Ewing, Kocurek and Lake, 2006). In part, these patterns are the product of regional changes in wind regimes – with star dunes in areas of

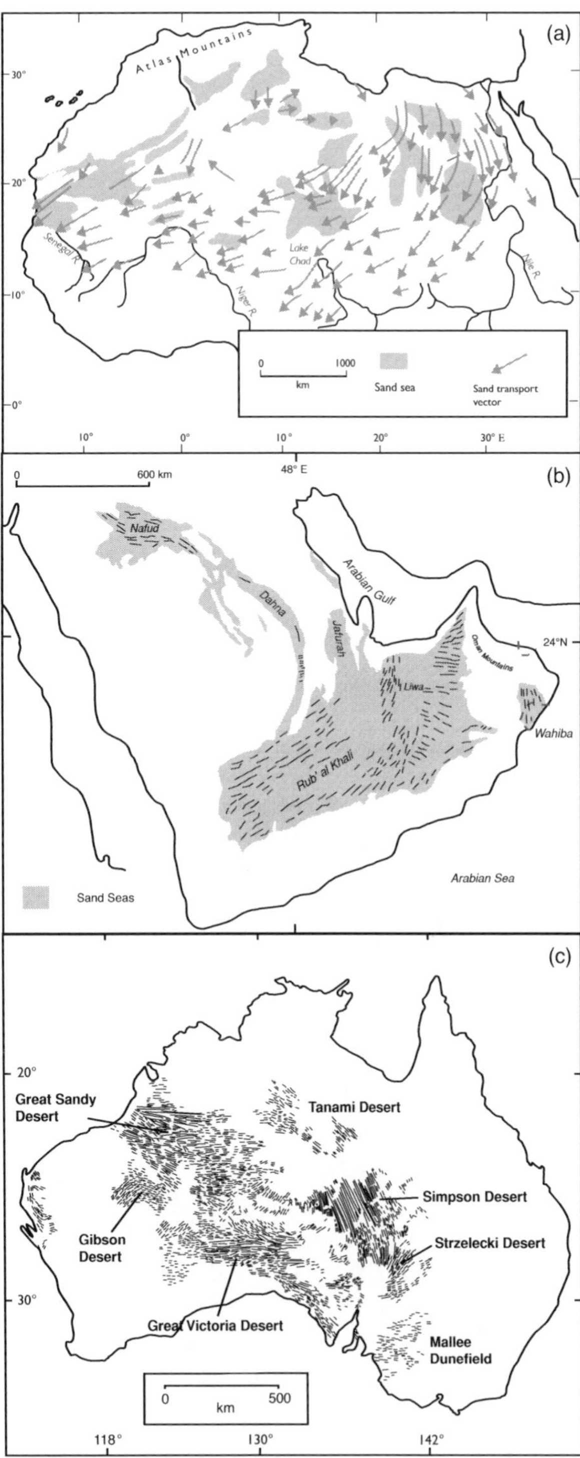

Figure 19.23 Dune morphology patterns – regional: (a) sand transport patterns in the Sahara; (b) dune trends in the Arabian Peninsula; (c) linear dune trends in Australia.

Box 19.2 Dune generations in Mauritania

Linear dunes dominate sand seas in the western Sahara. In the Azefal, Agneitir and Akchar sand seas of western Mauritania, three distinct crossing trends of linear dunes can be identified on satellite images (Figure 19.24). From oldest to youngest they are: (1) a northeast–southwest trending class of large, degraded red brown linear ridges, (2) a north–northeast trending class of moderate-sized linear dunes, commonly with active crests, that appear and (3) a north trending class of small simple linear dunes. Field studies of dune sediments and stratigraphy show that the surface of the two oldest dune generations is littered with Neolithic stone artefacts and pottery and is underlain by a bioturbated, slightly indurated red-brown sand enriched in silt and clay, which is interpreted as a paleosol. OSL ages obtained from dunes of the different generations cluster between 25–15 ka (spanning the Last Glacial Maximum), 10–13 ka (during the Younger Dryas event) and after 5 ka. Modelling of the wind regimes that produced these dunes using the gross bedform normal (GBN) approach shows that the wind regimes that occurred during each of these periods were significantly different, leading to the formation of dunes on three superimposed trends – northeast–southwest, north–northeast to south–southwest and north to south. The trend of the N–S dunes corresponds closely to the GBN trend for modern winds near the coast, whereas the NE–SW and NNE–SSW dune trends imply increased sand transport from the east and north to northeast directions, corresponding to enhanced trade wind circulations (Lancaster *et al.*, 2002).

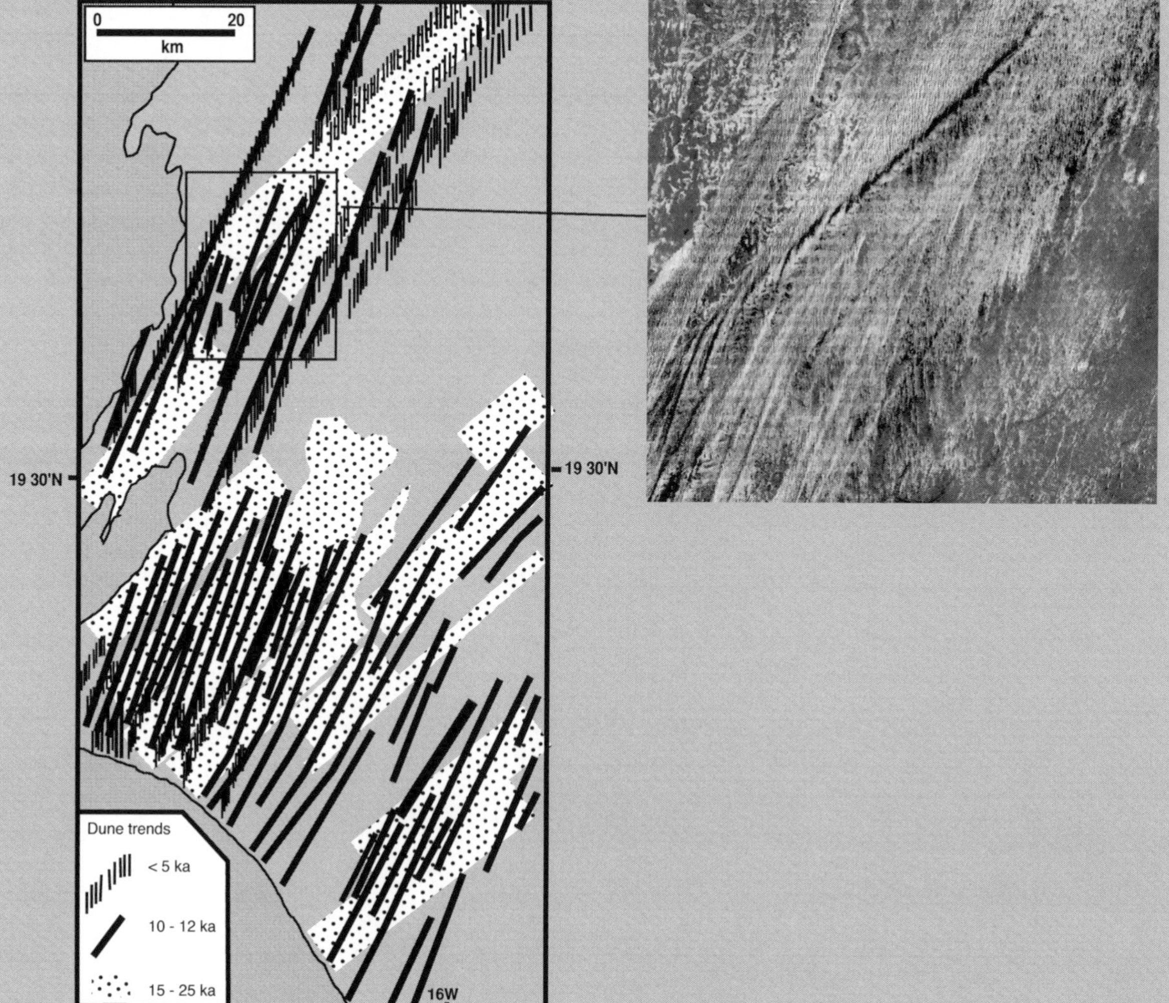

Figure 19.24 Generations of linear dunes: western Mauritania (after Lancaster *et al.*, 2002).

multidirectional winds, linear dunes in areas of bimodal winds and crescentic dunes where winds are unimodal (Figure 19.19). Such paradigms were used to explain the distribution of dune types in the Namib Sand Sea (Lancaster, 1983) (Figure 19.21) and elsewhere (Breed and Grow, 1979). Closer examination of dune patterns reveals sharp transitions in dune type, sediment composition, dune spacing and dune trend, which cannot be attributed to changes in wind regime (Lancaster, 1999). These patterns are better explained by the concept of dune generations, in which dune fields and sand seas are comprised of multiple genetically distinct groups of dunes, each formed in a different set of initial conditions of wind regime and sediment configuration (Kocurek and Ewing, 2005; Lancaster, 1999).

Dune generations can be recognised by differences in dune morphology, which are manifested by statistically distinct populations of dunes with a different crest orientation, spacing, length and defect density (Beveridge *et al.*, 2006; Derickson *et al.*, 2008), by differences in sediment composition and colour, by chronometric and relative age of different elements of a dune pattern and by geomorphic relations between dunes of different characteristics, e.g. crossing patterns (Lancaster *et al.*, 2002) (see Box 19.2).

Because only one set of dunes can form at a time in a given set of boundary conditions, it follows that dune fields with simple patterns of dune morphology most likely represent relatively young systems, whereas complex patterns are the product of multiple periods of dune construction. Such information can be used to infer the age of a dune field as well as the degree to which it has been subject to changes in the factors that control its development and current dynamics.

Within a group of dunes, patterns of dune size and spacing and the parameters of dunefield organisation vary systematically. In areas of crescentic dunes, such as at White Sands dune field, dunes become larger and more widely spaced in the direction of migration away from the linear source area (Figure 19.21). Dunes close to the source of sand are smaller, migrate more rapidly and have a high defect density. Downwind, these dunes interact with each other to produce a self-organisation of the pattern in which the dunes are larger, with longer crestlines and fewer defects (Ewing and Kocurek, 2010b). Similar patterns of dune pattern evolution have been noted in barchan dune corridors (Elbelrhiti, Andreotti and Claudin, 2008). In areas of linear dunes, such as the Simpson–Strzelecki Deserts of Australia, dune spacing is apparently related to substrate type, reflecting sediment supply from local sources (Fitzsimmons, 2007; Mabbutt and Wooding, 1983; Wasson *et al.*, 1988). Within this pattern, dune spacing increases downwind and the number of Y-junctions decreases as dunes become more widely spaced and the pattern becomes more organised away from sediment source areas (Fitzsimmons, 2007; Mabbutt and Wooding, 1983). By contrast, in the southwestern Kalahari dune field, the regional scale pattern of linear dunes becomes more complex downwind and also from southwest to northeast (Bullard *et al.*, 1995). These patterns probably reflect decreased aridity to the south and east of the dune field, as well as local sources of sediment in river valleys and pans. At a local scale, however, self-organisation of dunes downwind of river valleys and pans is expressed by joining of smaller dunes in Y-junctions to create a more organised pattern (Bullard and Nash, 1998).

19.9 Conclusions

Significant advances in understanding of the dynamics of desert dunes and the factors that determine their morphology have occurred in the past three decades. This has come about as a result of several intersecting methodological and conceptual advances, which include, but are not limited to, the application of remote sensing imagery to understand dune patterns, field studies of key dune-forming processes, OSL dating of periods of dune formation and GPR investigations of dune sedimentary structures. A rapidly evolving new development is the numerical modelling of dunes and dune systems, facilitated by vastly increased computing power, as well as better understanding of fundamental dune processes.

As a result, there is now a general understanding of the processes that form or at least maintain most major dune types. The importance of scale effects is recognised, so that the processes that form simple dunes need to be considered at different temporal and spatial scales from those that form compound and complex dunes and sand seas. The importance of past conditions of sediment supply, availability and mobility, determined by climatic and sea level changes, in the formation of mega dunes and sand seas and dune fields is now well recognised.

References

Andreotti, B., Fourriere, A., Ould-Kaddour, F. *et al.* (2009) Giant aeolian dune size determined by the average depth of the atmospheric boundary layer. *Nature*, **457** (7233), 1120–1123.

Baddock, M.C., Livingstone, I. and Wiggs, G.F.S. (2007) The geomorphological significance of airflow patterns in transverse dune interdunes. *Geomorphology*, **87**, 322–336.

Bagnold, R.A. (1941) *The Physics of Blown Sand and Desert Dunes*, Chapman & Hall, London, 265 pp.

Bagnold, R.A. (1953) The surface movement of blown sand in relation to meteorology, desert research, in *Proceedings of the International Symposium, Research Council of Israel, Jerusalem*, pp. 89–93.

Beveridge, C., Kocurek, G., Lancaster, N. *et al.* (2004) Origin and evolution of the Gran Desierto sand sea. *Geological Society of America Abstracts with Programs*, **36** (5).

Beveridge, C. *et al.* (2006) Development of spatially diverse and complex dune-field patterns: Gran Deserierto Dune Field, Sonora, Mexico. *Sedimentology*, **53**, 1391–1409.

Bourke, M.C. and Goudie, A.S. (2009) Varieties of barchan form in the Namib Desert and on Mars. *Aeolian Research*, **1**, 45–54.

Breed, C.S. and Grow, T. (1979) Morphology and distribution of dunes in sand seas observed by remote sensing, in *A Study of Global Sand Seas* (ed. E.D. McKee), United States Geological Survey, Professional Paper, pp. 253–304.

Breed, C.S. *et al.* (1979) Regional studies of sand seas using LANDSAT (ERTS) imagery, in *A Study of Global Sand Seas* (ed. E.D. McKee), United States Geological Survey, Professional Paper, pp. 305–398.

Breton, C., Lancaster, N. and Nickling, W.G. (2008) Magnitude and frequency of grain flows on a desert sand dune. *Geomorphology*, **95** (518–523).

Bristow, C.S., Bailey, S.D. and Lancaster, N. (2000) Sedimentary structure of linear sand dunes. *Nature*, **406**, 56–59.

Bristow, C.S., Duller, G.A.T. and Lancaster, N. (2005) Combining ground penetrating radar surveys and optical dating to determine dune migration in Namibia. *Journal of the Geological Society (London)*, **162** (2), 315–321.

Bristow, C.S., Duller, G.A.T. and Lancaster, N. (2007) Age and dynamics of linear dunes in the Namib Desert. *Geology*, **35** (6), 555–558.

Bullard, J.E. and Nash, D.J. (1998) Linear dune pattern variability in the vicinity of dry valleys in the southwest Kalahari. *Geomorphology*, **23**, 35–54.

Bullard, J.E., Thomas, D.S.G., Livingstone, I. and Wiggs, G.F.S. (1995) Analysis of linear sand dune morphological variability, southwestern Kalahari Desert. *Geomorphology*, **11**, 189–203.

Claudin, P. and Andreotti, B. (2006) A scaling law for aeolian dunes on Mars, Venus, Earth, and for subaqueous ripples. *Earth and Planetary Science Letters*, **252** (1–2), 30–44.

Derickson, D., Kocurek, G., Ewing, R.C. and Bristow, C.S. (2008) Origin of a complex and spatially diverse dune-field pattern, Algodones, southeastern California. *Geomorphology*, **99**, 186–204.

Dong, Z., Wang, T. and Wang, X. (2004) Geomorphology of megadunes in the Badain Jaran Desert. *Geomorphology*, **60** (1–2), 191–204.

Dong, Z., Qinan, G., Lu, P. *et al.* (2009) Turbulence fields in the lee of two-dimensional transverse dunes simulated in a wind tunnel. *Earth Surface Processes and Landforms*, **34** (2), 204–216.

El-Sayed, M.I. (2000) The nature and possible origin of megadunes in Liwa, Ar Rub' Al Kahli, UAE. *Sedimentary Geology*, **134**, 304–330.

Elbelrhiti, H., Andreotti, B. and Claudin, P. (2008) Barchan dune corridors: field characterization and investigation of control parameters. *Journal of Geophysical Research, Earth Surface*, **113**, F02S15, DOI: 10.1029/2007JF000767.

Elbelrhiti, H., Claudin, P. and Andreotti, B. (2005) Field evidence for surface-wave-induced instability of sand dunes. *Nature*, **437**, 720–723.

Ewing, R., C., Kocurek, G. and Lake, L.W. (2006) Pattern analysis of dune-field parameters. *Earth Surface Processes and Landforms*, **31** (9), 1176–1191.

Ewing, R.C. and Kocurek, G. (2010a) Aeolian dune-field pattern boundary conditions. *Geomorphology*, **114** (3), 175–187.

Ewing, R.C. and Kocurek, G. (2010b) Aeolian dune interactions and dune-field pattern formation: White Sands Dune Field, New Mexico, *Sedimentology*, **57** (5), 1199–1219.

Finkel, H.J. (1959) The barchans of Southern Peru. *Journal of Geology*, **67**, 614–647.

Fitzsimmons, K.E. (2007) Morphological variability in the linear dunefields of the Strzelecki and Tirari Deserts, Australia. *Geomorphology*, **91**, 146–160.

Frank, A. and Kocurek, G. (1996a) Airflow up the stoss slope of sand dunes: limitations of current understanding. *Geomorphology*, **17** (1–3), 47–54.

Frank, A. and Kocurek, G. (1996b) Towards a model for airflow on the lee side of aeolian dunes. *Sedimentology*, **43** (3), 451–458.

Fryberger, S.G. (1979) Dune forms and wind regimes, in *A Study of Global Sand Seas* (ed. E.D. McKee), United States Geological Survey, Professional Paper, pp. 137–140.

Glennie, K.W. (1970) *Desert Sedimentary Environments*, Developments in Sedimentology 14, Elsevier, Amsterdam, 222 pp.

Glennie, K.W. (1998) The desert of southeast Arabia: a product of Quaternary climatic change, in *Quaternary Deserts and Climatic Change* (ed. A.S. Alsharan, K.W. Glennie, G.L. Whittle and C.G.S.C. Kendall), A.A. Balkema, Rotterdam/Brookfield, pp. 279–292.

Grolier, M.J., Ericksen, G.E., McCauley, J.F. and Morris, E.C. (1974) The desert land forms of Peru: a preliminary photographic atlas, United States Geological Survey Interagency Report, Astrogeology 57, USGS, Washington, DC, 146 pp.

Hack, J.T. (1941) Dunes of the Western Navajo County. *Geographical Review*, **31** (2), 240–263.

Hanna, S.R. (1969) The formation of longitudinal sand dunes by large helical eddies in the atmosphere. *Journal of Applied Meteorology*, **8**, 874–883.

Havholm, K.G. and Kocurek, G. (1988) A preliminary study of the dynamics of a modern draa, Algodones, southeastern California, USA. *Sedimentology*, **35**, 649–669.

Hesp, P.A. (1981) The formation of shadow dunes. *Journal of Sedimentary Petrology*, **51**, 101–112.

Holm, D.A. (1960) Desert geomorphology in the Arabian Peninsula. *Science*, **123**, 1369–1379.

Hunter, R.E., Richmond, B.M. and Alpha, T.R. (1983) Storm-controlled oblique dunes of the Oregon Coast. *Geological Society of America Bulletin*, **94**, 1450–1465.

Kar, A. (1993) Aeolian processes and bedforms in the Thar Desert. *Journal of Arid Environments*, **25**, 83–96.

Kocurek, G. and Ewing, R.C. (2005) Aeolian dune field self-organization – implications for the formation of simple versus complex dune field patterns. *Geomorphology*, **72**, 94–105.

Kocurek, G. and Nielson, J. (1986) Conditions favourable for the formation of warm-climate aeolian sand sheets. *Sedimentology*, **33**, 795–816.

Kocurek, G., Havholm, K.G., Deynoux, M. and Blakey, R.C. (1991) Amalgamated accumulations resulting from climatic and eustatic changes, Akchar Erg, Mauritania. *Sedimentology*, **38** (4), 751–772.

Kocurek, G., Townsley, M., Yeh, E., Havholm, K. and Sweet, M.L. (1992) Dune and dunefield development on Padre Island, Texas, with implications for interdune deposition and water-table-controlled accumulation. *Journal of Sedimentary Petrology*, **62** (4), 622–635.

Kocurek, G. *et al.* (2007) White Sands Dune Field, New Mexico: age, dune dynamics, and recent accumulations. *Sedimentary Geology*, **197**, 313–331.

Lancaster, N. (1980) The formation of seif dunes from barchans – supporting evidence for Bagnold's hypothesis from the Namib Desert. *Zeitschrift fur Geomorphologie*, **24**, 160–167.

Lancaster, N. (1981) Palaeoenvironmental implications of fixed dune systems in southern Africa. *Palaeogeography, Palaeoclimatology, Palaeoecology*, **33**, 327–346.

Lancaster, N. (1982a) Dunes on the Skeleton Coast, SWA/Namibia: geomorphology and grain size relationships. *Earth Surface Processes and Landforms*, **7**, 575–587.

Lancaster, N. (1982b) Linear dunes. *Progress in Physical Geography*, **6**, 476–504.

Lancaster, N. (1983) Controls of dune morphology in the Namib Sand Sea, in *Eolian Sediments and Processes* (eds T.S. Ahlbrandt and M.E. Brookfield), Developments in Sedimentology, Elsevier, Amsterdam, pp. 261–289.

Lancaster, N. (1988) Controls of eolian dune size and spacing. *Geology*, **16**, 972–975.

Lancaster, N. (1989a) Star dunes. *Progress in Physical Geography*, **13** (1), 67–92.

Lancaster, N. (1989b) The dynamics of Star Dunes: an example from the Gran Desierto, Mexico. *Sedimentology*, **36**, 273–289.

Lancaster, N. (1989c) *The Namib Sand Sea: Dune Forms, Processes, and Sediments*, A.A. Balkema, Rotterdam, 200 pp.

Lancaster, N. (1991) The orientation of dunes with respect to sand-transporting winds: a test of Rubin and Hunter's gross bedform-normal rule, NATO Advanced Research Workshop on sand, dust, and soil in their relation to aeolian and littoral processes, University of Aarhus, Sandbjerg, Denmark, pp. 47–49.

Lancaster, N. (1992) Relations between dune generations in the Gran Desierto, Mexico. *Sedimentology*, **39**, 631–644.

Lancaster, N. (1996) Field studies of proto-dune initiation on the northern margin of the Namib Sand Sea. *Earth Surface Processes and Landforms*, **21**, 947–954.

Lancaster, N. (1999) Geomorphology of desert sand seas, in *Aeolian Environments, Sediments and Landforms* (eds A.S. Goudie, I. Livingstone and S. Stokes), John Wiley & Sons, Ltd, Chichester, pp. 49–70.

Lancaster, N. (2007) Low latitude dune fields, in *Encyclopedia of Quaternary Science* (ed. S.A. Elias), Elsevier, Amsterdam, pp. 626–642.

Lancaster, N. and Baas, A. (1998) Influence of vegetation cover on sand transport by wind: field studies at Owens Lake, California. *Earth Surface Processes and Landforms*, **23** (1), 69–82.

Lancaster, N., Nickling, W.G., McKenna Neuman, C.K. and Wyatt, V.E. (1996) Sediment flux and airflow on the stoss slope of a barchan dune. *Geomorphology*, **17** (1–3), 55–62.

Lancaster, N. *et al.* (2002) Late Pleistocene and Holocene dune activity and wind regimes in the western Sahara of Mauritania. *Geology*, **30** (11), 991–994.

Livingstone, I. (1986) Geomorphological significance of wind flow patterns over a Namib linear dune, in *Aeolian Geomorphology* (ed. W.G. Nickling), Allen and Unwin, Boston, pp. 97–112.

Livingstone, I. (1988) New models for the formation of linear sand dunes. *Geography*, **73**, 105–115.

Livingstone, I. (1989) Monitoring surface change on a Namib linear dune. *Earth Surface Processes and Landforms*, **14**, 317–332.

Livingstone, I. (1993) A decade of surface change on a Namib linear dune. *Earth Surface Processes and Landforms*, **18** (7), 661–664.

Livingstone, I. (2003) A twenty-one-year record of surface change on a Namib linear dune. *Earth Surface Processes and Landforms*, **28** (9), 1025–1032.

Livingstone, I., Wiggs, G.F.S. and Weaver, C.M. (2007) Geomorphology of desert sand dunes: a review of recent progress. *Earth Science Reviews*, **80** (3–4), 239–257.

Mabbutt, J.A. and Wooding, R.A. (1983) Analysis of longitudinal dune patters in the northwestern Simpson Desert, central Australia. *Zeitschrift fur Geomorphologie*, Supplement **45**, 51–69.

McKee, E. (1982) Sedimentary structures in dunes of the Namib Desert, South West Africa, Geological Society of America Special Paper 188, p. 60.

McKee, E.D. and Douglass, J.R. (1971) Growth and movement of dunes at White Sands National Monument, New Mexico, 750-D, United States Geological Survey.

McKee, E. and Tibbitts Jr, G.C. (1964) Primary structures of a seif dune and associated deposits in Libya. *Journal of Sedimentary Petrology*, **34** (1), 5–7.

McKenna Neuman, C., Lancaster, N. and Nickling, W.G. (1997) Relations between dune morphology, air flow, and sediment flux on reversing dunes, Silver Peak, Nevada. *Sedimentology*, **44** (6), 1103–1114.

McKenna Neuman, C., Lancaster, N. and Nickling, W.G. (2000) The effect of unsteady winds on sediment transport on the stoss slope of a transverse dune, Silver Peak, Nevada. *Sedimentology*, **47** (1), 211–226.

Mainguet, M. (1984) Space observations of Saharan aeolian dynamics. in *Deserts and Arid Lands* (ed. F. El Baz), Nyhoff, The Hague, pp. 59–77.

Mainguet, M. and Callot, Y. (1978) L'erg de Fachi-Bilma (Tchad-Niger). *Mémoires et Documents CNRS*, **18**, 178.

Mainguet, M. and Chemin, M.-C. (1984) Les dunes pyramidales du Grand Erg Oriental. *Travaux de l'Institut de Géographie de Reims*, **59–60**, 49–60.

Nickling, W.G. and McKenna Neuman, C. (1999) Recent investigations of airflow and sediment transport over desert dunes, in *Aeolian Environments, Sediments and Landforms* (eds A.S. Goudie, I. Livingstone and S. Stokes), John Wiley & Sons, Ltd, Chichester.

Nickling, W.G., McKenna Neuman, C. and Lancaster, N. (2002) Grainfall processes in the lee of transverse dunes, Silver Peak, Nevada. *Sedimentology*, **49** (1), 191–211.

Nielson, J. and Kocurek, G. (1986) Climbing zibars of the Algodones. *Sedimentary Geology*, **48**, 1–15.

Nielson, J. and Kocurek, G. (1987) Surface processes, deposits, and development of star dunes: Dumont dune field, California. *Geological Society of America Bulletin*, **99**, 177–186.

Parsons, D.R., Walker, I.J. and Wiggs, G.F.S. (2004) Numerical modelling of flow structures over an idealised transverse dunes of varying geometry. *Geomorphology*, **59**, 149–164.

Reffet, E., Du Pont, S.C., Hersen, P. *et al.* (2008) Longitudinal dunes on Titan: a laboratory approach, Planetary Dunes Workshop, LPSC, Alamogordo, NM.

Rubin, D.M. (1984) Factors determining desert dune type (discussion). *Nature*, **309**, 91–92.

Rubin, D.M. and Hesp, P.A. (2009) Multiple origins of linear dunes on Earth and Titan. *Nature Geoscience*, **2** (9), 653–658.

Rubin, D.M. and Hunter, R.E. (1987) Bedform alignment in directionally varying flows. *Science*, **237**, 276–278.

Rubin, D.M. and Ikeda, H. (1990) Flume experiments on the alignment of transverse, oblique and longitudinal dunes in directionally varying flows. *Sedimentology*, **37** (4), 673–684.

Rubin, D.M., Tsoar, H. and Blumberg, D.G. (2008) A second look at western Sinai seif dunes and their lateral migration. *Geomorphology*, **93**, 335–342.

Sharp, R.P. (1966) Kelso Dunes, Mohave Desert, California. *Geological Society of America Bulletin*, **77**, 1045–1074.

Sharp, R.P. (1979) Intradune flats of the Algodones chain, Imperial Valley, California. *Geological Society of America Bulletin*, **90**, 908–916.

Stokes, S. and Bray, H.E. (2005) Late Pleistocene eolian history of the Liwa region, Arabian Peninsula. *Geological Society of America Bulletin*, **117** (11/12), 1466–1480.

Sun, J. and Muhs, D.R. (2007) Mid latitude dune fields, in *Encyclopedia of Quaternary Science* (ed. S.A. Elias), Elsevier, Amsterdam, pp. 607–626.

Sweet, M.L. and Kocurek, G. (1990) An empirical model of aeolian dune lee-face airflow. *Sedimentology*, **37** (6), 1023–1038.

Thomas, D.S.G. (1992) Desert dune activity: concepts and significance. *Journal of Arid Environments*, **22**, 31–38.

Tsoar, H. (1983) Dynamic processes acting on a longitudinal (seif) dune. *Sedimentology*, **30**, 567–578.

Tsoar, H. (1984) The formation of seif dunes from barchans–a discussion. *Zeitschrift fur Geomorphologie*, **28** (1), 99–103.

Tsoar, H. (1989) Linear dunes – forms and formation. *Progress in Physical Geography*, **13** (4), 507–528.

Tsoar, H., Blumberg, D.G. and Stoler, Y. (2004) Elongation and migration of sand dunes. *Geomorphology*, **57**, 293–302.

Tsoar, H. and Møller, J.T. (1986) The role of vegetation in the formation of linear sand dunes, in *Aeolian Geomorphology* (ed. W.G. Nickling), Allen and Unwin, Boston, London, Sydney, pp. 75–95.

Tsoar, H., Rasmussen, K.R., Sørensen, M. and Willetts, B.B. (1985) Laboratory studies of flow over dunes, in Proceedings of the International Workshop on the Physics of Blown Sand (eds O.E. Barndorff, J.T. Møller, K.R. Rasmussen and B.B. Willetts), University of Aarhus, Aarhus, Denmark, pp.327–350.

Walker, I.J. (1999) Secondary airflow and sediment transport in the lee of a reversing dune. *Earth Surface Processes and Landforms*, **24**, 437–448.

Walker, I.J. and Nickling, W.G. (2002) Dynamics of secondary airflow and sediment transport over and the lee of transverse dunes. *Progress in Physical Geography*, **26** (1), 47–75.

Warren, A. (1972) Observations on dunes and bimodal sands in the Tenere desert. *Sedimentology*, **19**, 37–44.

Warren, A. and Allison, D. (1998) The palaeoenvironmental significance of dune size hierarchies. *Palaeogeography, Palaeoeclimatology, Palaeocology*, **137**, 289–303.

Washington, R. *et al.* (2006) Links between topography, wind, deflation, lakes and dust: the case of the Bodélé Depression, Chad. *Geophysical Research Letters*, **33** (L09401).

Wasson, R.J. (1983) Dune sediment types, sand colour, sediment provenance and hydrology in the Strzelecki–Simpson Dunefield, Australia, in *Eolian Sediments and Processes. Developments in Sedimentology* (eds M.E. Brookfield and T.S. Ahlbrandt), Elsevier, Amsterdam, Oxford, New York, Tokyo, pp. 165–195.

Wasson, R.J. and Hyde, R. (1983a) A test of granulometric control of desert dune geometry. *Earth Surface Processes and Landforms*, **8**, 301–312.

Wasson, R.J. and Hyde, R. (1983b) Factors determining desert dune type. *Nature*, **304**, 337–339.

Wasson, R.J., Fitchett, K., Mackey, B. and Hyde, R. (1988) Large-scale patterns of dune type, spacing, and orientation in the Australian continental dunefield. *Australian Geographer*, **19**, 89–104.

Weng, W.S. *et al.* (1991) Air flow and sand transport over sand dunes. *Acta Mechanica Supplement*, **2**, 1–22.

Werner, B.T. (1995) Eolian dunes: computer simulations and attractor interpretation. *Geology*, **23** (12), 1107–1110.

Werner, B.T. and Kocurek, G. (1997) Bed-form dynamics: Does the tail wag the dog ? *Geology*, **25** (9), 771–774.

Wiggs, G.F.S. (1993) Desert dune dynamics and the evaluation of shear velocity: an integrated approach, in *The Dynamics and Environmental Context of Aeolian Sedimentary Systems* (ed. K. Pye), Geological Society, London, pp. 37–48.

Wiggs, G.F.S., Livingstone, I. and Warren, A. (1996) The role of streamline curvature in sand dune dynamics: evidence from field and wind tunnel measurements. *Geomorphology*, **17** (1–3), 29–46.

Wiggs, G.F.S., Livingstone, I., Thomas, D.S.G. and Bullard, J.E. (1994) Effect of vegetation removal on airflow patterns and dune dynamics in the southwestern Kalahari Desert. *Land Degradation and Rehabilitation*, **5**, 13–24.

Wiggs, G.F.S., Thomas, D.S.G., Bullard, J.E. and Livingstone, I. (1995) Dune mobility and vegetation cover in the southwest Kalahari Desert. *Earth Surface Processes and Landforms*, **20** (6), 515–530.

Wilson, I.G. (1972) Aeolian bedforms – their development and origins. *Sedimentology*, **19**, 173–210.

Wolfe, S.A. and Nickling, W.G. (1993) The protective role of sparse vegetation in wind erosion. *Progress in Physical Geography*, **17**, 50–68.

20

Desert dust

Richard Washington and Giles S. F. Wiggs

20.1 Introduction

Mineral dust is the most important export from the world's arid zones to the global Earth System and is known to affect atmospheric, oceanic, biological, terrestrial and human processes and systems, including:

- Reduction in shortwave and absorption of longwave radiation (Miller and Tegen, 1998; Milton *et al.*, 2008).
- Interaction with cloud microphysics (Ansmann *et al.*, 2008).
- Suppression of rainfall, e.g. in Sahel (Rosenfeld, Rudich and Lahav, 2001; Hui *et al.*, 2008).
- Modification of tropical storm and cyclone intensities (Evan *et al.*, 2006).
- CO_2 drawdown affecting ocean fertilisation (Jickells *et al.*, 2005; Cassar *et al.*, 2007).
- Long-distance nutrient transport and vegetation fertilisation (e.g. Okin *et al.*, 2004; Koren *et al.*, 2006).
- Human health and land use (e.g. O'Hara *et al.*, 2000; Prospero *et al.*, 2008).
- Agricultural soil erosion and productivity (Worster, 1979; Zobeck and Van Pelt, 2006).

Research on dust emissions from desert regions and its distribution in the atmosphere has seen an enormous increase in attention since the 2nd edition of *Arid Zone Geomorphology* in 1997. In terms of research activity (research campaigns, number of researchers, papers published) it is perhaps a field that has advanced further than any other in the arid zone realm. Two key drivers have been responsible for this. First, the availability of long-term satellite-derived atmospheric dust data sets has allowed the first near-global perspective on dust to emerge. Second, recognition of the importance of dust in the Earth System, particularly in terms of the interaction of dust with the planetary radiation budget, has meant that representation of the dust cycle in numerical models is critical in diagnosing and predicting the climate system at timescales ranging from weather forecasting to the climates of future decades and centuries. At the same time, deficiencies in the representation of dust in the models and uncertainties in satellite retrieval have prompted aircraft and ground-based campaigns to measure in situ dust emission, transport and deposition. As a result, considerable resources and effort have been invested in investigating dust processes and dynamics from the field scale to the continental and global scale. This chapter therefore gives prominence to new methods of dust measurement since much of what we now know about dust at the global scale has emerged from these tools and initiatives. Also reviewed is the dramatic progress that has been made with numerical modelling of dust and the current status of knowledge on key dust source regions. The chapter also covers dust transport and deposition and considers the likelihood of changing global dust loads in the future.

20.1.1 *Dust in a geomorphological context*

20.1.1.1 *Physical characteristics of dust*

It is the fine texture of dust that augments its environmental impacts. Aeolian dust derived from arid zones is

Arid Zone Geomorphology: Process, Form and Change in Drylands, Third Edition. Edited by David S. G. Thomas

commonly considered to consist of particles with diameters <0.08 mm (80 μm) (Bagnold, 1941), although a more fashionable limit is now taken to be at the silt/sand boundary of 62.5 μm (Goudie and Middleton, 2006). While particles in this size range are difficult to entrain (see Chapter 18), once they are airborne they can be held aloft by wind turbulence for a considerable amount of time (hours–days) and are therefore capable of travelling very large distances (thousands of kilometres). Saharan dust can be found deposited in the Amazon basin (Swap *et al.*, 1992), the Caribbean (Colarco, Toon and Holben, 2003) and Europe (Ansmann *et al.*, 2003), Australian dust has been found in the highlands of New Zealand (McGowan *et al.*, 2005) and Chinese dust can be carried far out into the Pacific (Zhao *et al.*, 2003). Such long residence times in the atmosphere allow dust to interact with regional and global climatology and it is the distance travelled that promotes its impact with regard to redistribution of important minerals and nutrients at a global scale, unrestricted by topography.

Further, it is these fine-textured particles in agricultural soils that are capable of retaining moisture and nutrients and when wind erosion removes these particles there can be serious deleterious impacts on agricultural productivity (Figure 20.1; see Zobeck and Van Pelt, 2006) with ensuing off-site contamination of downwind soils and habitats (Plumlee and Ziegler, 2003). The fine texture of dust particles also increases its potential significant health impacts. Dust with aerodynamic diameters of <10, 5 and 2.5 μm (termed PM10, PM5 and PM2.5 respectively) are of a size that is capable of reaching into the recesses of human lungs. While the fine texture of common SiO_2 (quartz)-based dust may be deleterious to health, such dust may be particularly injurious if contaminated by trace chemicals (e.g. arsenic, cadmium, lead; see Galloway *et al.*, 1982). Such contaminants are commonly found in dust derived from anthropogenic sources such as Owens Lake in California (Reheis *et al.*, 2009) and the dried shoreline of the Aral Sea (O'Hara *et al.*, 2000).

20.1.1.2 Dust sources

There are many mechanisms in arid regions that may contribute to the production of dust-sized particles (Goudie and Middleton, 2006). These include abrasion of sands by both wind (Bullard, McTainsh and Pudmenky, 2004) and water, the physical weathering of in situ rock (Smith, Wright and Whalley, 2002, see Chapter 6) or the production of 'fluffy' evaporate minerals on the surfaces of 'wet' playas (Figure 20.2; see Reynolds *et al.*, 2007, and Chapter 15). The numerous mechanisms for dust production result in dust sources being found in a wide variety of geomorphological contexts, although the emission processes operating in each are inadequately known. Significant dust sources are commonly found where silt-sized material has been concentrated by the action of water. Important sources are therefore found in closed topographic depressions in dryland areas such as ephemeral lakes and pans (Prospero *et al.*, 2002; Engelstaedter *et al.*, 2003; Bryant *et al.*, 2007) and alluvial surfaces, floodplains and dry river valleys (Reheis and Kihl, 1995). However, stone pavements have also been shown to be a significant source in China (Wang, Zhou and Dong, 2006), while aeolian sand can also produce large quantities of dust. Recent research suggests that such sand sources (e.g. dunefields) have been largely overlooked to date, with Bullard *et al.*

Figure 20.1 A dust cloud blowing off erodible agricultural land near Yellowhouse Canyon, north Texas, in June 2009 (courtesy John Stout).

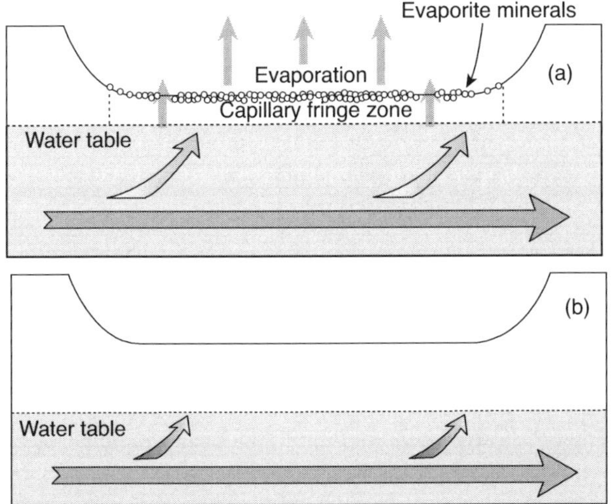

Figure 20.2 The processes by which a near-surface water table, effected by high evaporation rates and capillary rise, may result in highly erodible 'fluffy' evaporate minerals on the surface of dry lake beds (after Rosen, 1994).

(2008) noting that aeolian deposits account for 37 % of dust sources in Australia.

The spatial and temporal sensitivity of these source geomorphologies to dust emission is poorly understood. It is apparent that dust is eroded from small 'hotspots' within source areas and that susceptibility to emission is very much dependent upon complex relationships that exist between erosion potential and local surface and subsurface characteristics (see Chapter 18). Important 'hotspots' of activity have been associated with playa surfaces susceptible to saltation activity, where the impact of saltating grains can effectively break up wind-resistant crusts (Cahill *et al.*, 1996; Nickling, McTainsh and Leys, 1999; Gillette, Niemeyer and Helm, 2001). Local groundwater depth and salinity have also been shown to be a fundamental control on dust emission. Reynolds *et al.* (2007) found that the surface of 'wet' playas (where evaporation from near-surface water tables is possible) might consist of several centimetres of loose evaporite minerals highly susceptible to wind erosion. However, they also noted how dynamic such surfaces are, with erosion potential varying rapidly dependent upon changing water table depth, rainfall and rates of evaporation (Figure 20.3).

Atmospheric controls are key to dust emission, but many studies have found that it is the extreme and gusty winds (rather than the mean) that correlate best with dust emission (Engelstaedter, Tegen and Washington, 2006; Engelstaedter and Washington, 2007a; Reheis, 2006, Wiggs and Holmes, 2010). Further, some specific dust sources appear to be supply-limited, with the amount of material eroded being highly dependent upon the availability of erodible material rather than the erosivity of the wind (although see the text on the *Bodélé Depression, Chad*, later in Section 20.2.1). Surfaces may become restricted in their erodibility due to a lack of available sediment or by changing surface conditions on erodible sediment (e.g. surface crusting). In a similar manner to some Australian dust sources (McTainsh, Keys and

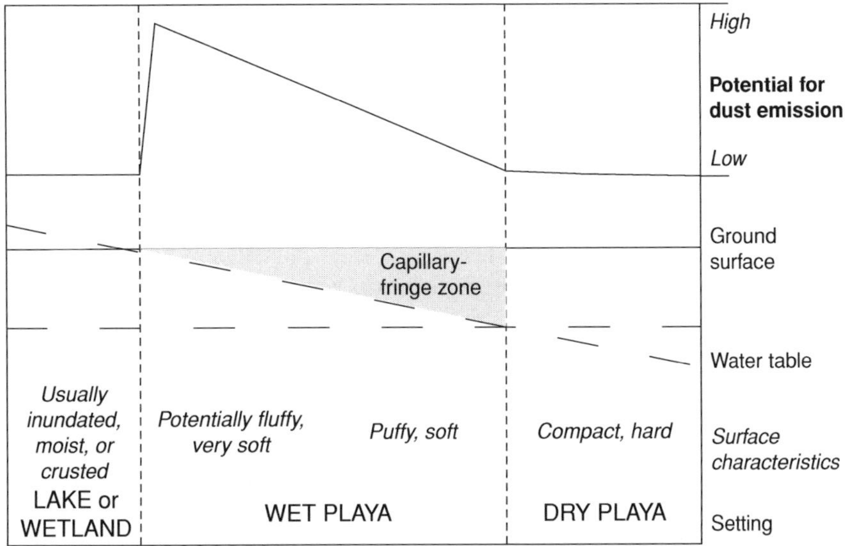

Figure 20.3 Theoretical relationship between dust emission, thickness of the capillary fringe zone and playa surface characteristics. The schematic shows the likely sensitivity of surface erodibility to minor changes in water table depth (after Reynolds *et al.*, 2007).

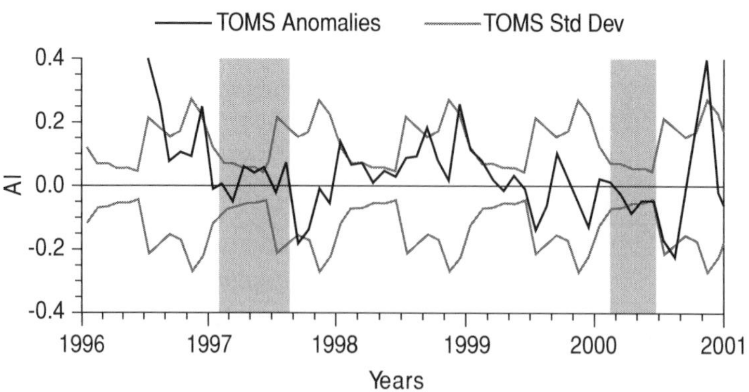

Figure 20.4 The relationship between flood inundation of Etosha Pan, Namibia, and subsequent dust events. Data show inundation of the pan due to heavy rains in 1997 and 2000 with associated lows in TOMS AI anomalies (monthly mean values minus mean of all monthly means), suggesting limited dust activity during and immediately after the flood events. However, in 1998–1999 and late 2000 there are anomalously high values in TOMS AI anomalies (larger than 1 standard deviation), possibly as a result of the replenishment of the pan with erodible material. Fluctuations in measured wind speed do not explain these dust patterns (after Mahowald *et al.*, 2003).

Nickling, 1999), investigations by Mahowald *et al.* (2003) and Bryant *et al.* (2007) have noted that the dry lake beds that constitute the most important dust source regions in southern Africa (Etosha Pan in Namibia and the Makgadikgadi Pans in Botswana) are reliant on intermittent flood inundation to provide erodible sediment for aeolian erosion in the following years (Figure 20.4). They found that a large proportion of the variability in dust emission from these sources could therefore be linked to changing rainfall patterns driven by the El Niño–Southern Oscillation (ENSO). Similar positive and lagged relationships between rainfall and dust emission have also been noted in north America (Reheis, 2006).

Human activity can have a profound influence on generating new dust sources (see Chapter 23 for a full discussion). Key human impacts that accelerate wind erosion and cause significant dust emission involve the break-up of stable dryland surfaces by both off-road vehicle (ORV) activity (Goossens and Buck, 2009) and agricultural activity on susceptible soils, as occurred during the Dust Bowl in the mid-west USA in the 1930s (Worster, 1979). Particularly significant dust sources initiated by human activity also derive from the exposure and desiccation of sediment after the draining of inland water bodies such as Owens Lake in the USA (Gill, 1996) and the Aral Sea (O'Hara *et al.*, 2000).

20.1.1.3 Dust sinks

Once entrained, dust-sized particles will be transported in the atmosphere with deposition dependent upon changing wind conditions, the roughness properties of the surface, and on the size, shape and mass of the particles themselves. Larger particles tend to have higher deposition velocities and so are deposited closer to source than fine particles (Tsoar and Pye, 1987; Pye, 1995). Thus 'dry' deposition can be a very efficient sorter of particles. In this way, Saharan dust transported across the Atlantic is seen to be much finer at distance from source, with median deposited particle sizes in Morocco measured at 22.0–37.0 μm (Khiri, Ezaidi and Kabbachi, 2004) and those deposited in the Caribbean at 4.0 μm (Petit *et al.*, 2005). In contrast 'wet' deposition, where particles are drawn out of the airflow by rain, cleanses the atmosphere and results in a much less sorted deposit (Figure 20.5). At regional scales such a fining in deposited particles with distance from source is much more difficult to ascertain because such deposits often derive from multiple dust sources (Cattle, McTainsh and Elias, 2009) and so 'fingerprinting' of sources from deposits using grain size analysis can be complicated (Wiggs *et al.*, 2003).

Rates of dust deposition can be very high close to source areas with a steep decay in deposition rates with increasing distance from source (Figure 20.6). In a 3 month study Wiggs and Holmes (2010) reported average deposition rates of 48 g/m^2 (with a maximum of nearly 30 g/m^2 in a period of 2 weeks) immediately downwind of an eroding agricultural field in South Africa, while O'Hara, Clarke and Elatrash (2006) measured rates of 276 t/km^2 yr in Libya. In Australia, rates between 31 and 44 t/km^2 yr have been measured by McTainsh and Lynch (1996) while maximum rates equivalent to 102 t/km^2 yr have been reported by Cattle, McTainsh and Elias (2009). However, given the very large seasonal and annual variation in dust

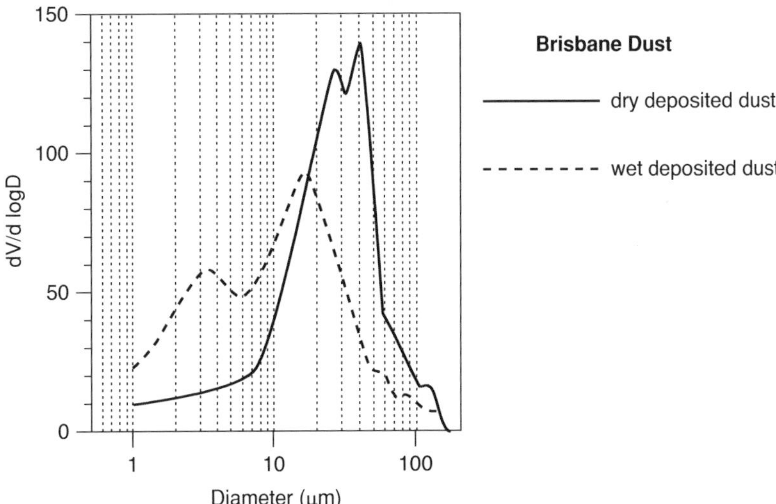

Figure 20.5 Particle size distributions of dust samples deposited by wet deposition (dashed line) and dry deposition (solid line) in Brisbane (after Hesse and McTainsh, 1999).

activity it is difficult to draw conclusions about overall dust deposition rates from relatively short-term studies (1–2 years) such as those described above, and long-term measurements of dust deposition rates are still very limited in number. Two of the best known are the 15 year study by Ta *et al.* (2004), who determined mean maximum deposition rates of 498 t/km^2 yr in the Gobi Desert region of China, and the 16 year study by Reheis (2006), who measured average rates of 20 g/m^2 yr in southern Nevada and California. These measured rates of deposition are far lower than those occurring during the Quaternary, as deduced from analysis of loess deposits (see Chapters 3 and 17 for discussion of loess). A study by Kohfeld and Harrison (2003) suggests that glacial period dust deposition rates in China were nearly 5 times higher than in interglacial periods.

Once deposited, dust can become incorporated into soils with a corresponding influence on their biogeochemistry (Okin *et al.*, 2004; Cattle, McTainsh and Elias, 2009; Reheis *et al.*, 2009). Deposited dust can also have a

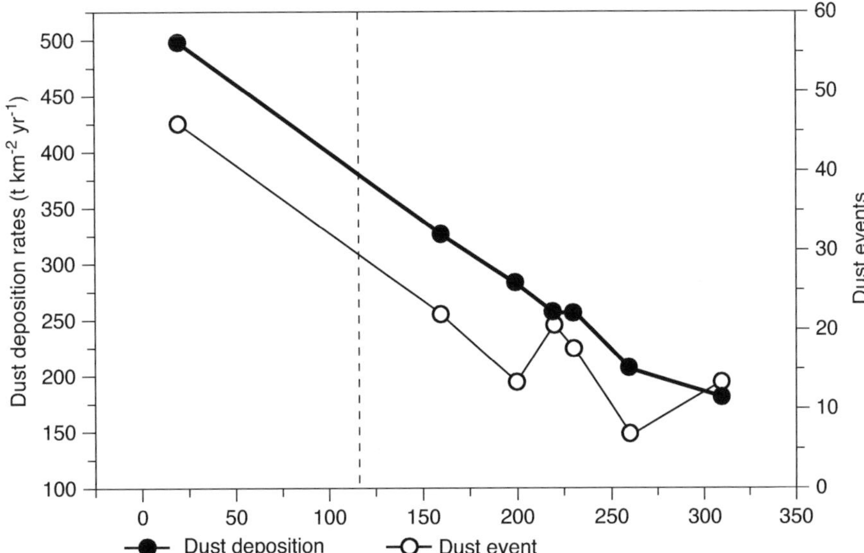

Figure 20.6 Reductions in measured dust deposition rates with increasing distance from source in the loess region of China (from Ta *et al.*, 2004).

significant impact on the formation of stone pavements (McFadden, Wells and Jercinovich, 1987) and surface crusts and duricrusts (Eckardt *et al.*, 2001). Far-travelled and fine-textured dust can become deposited in oceans and polar regions and the analysis of desert dust in ice cores and ocean sediments has proved an important source of evidence for both global palaeoclimatic reconstruction (De Deckker *et al.*, 2010) and contemporary environmental change (McConnell *et al.*, 2007).

20.1.2 Measuring dust

The major advances over the last decade in our understanding of desert dust have come from improved techniques in measuring and numerically simulating dust. Prior to the availability of remote sensing techniques suitable for the detection of dust, much of the data on global dust distribution were founded on synoptic weather reports and visibility estimates from weather stations (Middleton, 1986). Numerous problems beset these data, such as the inconsistency of reporting dust storms, lack of direct quantification of dust loadings and the subjective nature of visibility measurement. Their most important limitation is still that weather stations tend to be located at desert margins and almost all are now known, from remote sensing of dust, to be distant from key source regions. Nevertheless, they offer a useful resource, particularly in comparison with top-down satellite approaches.

20.1.2.1 Remote sensing of dust

Satellites make no discrimination against measurements from remote regions of the planet. In fact, the lack of cloud in deserts makes these regions particularly well suited to remote sensing using visible channels. Multiple satellite-derived dust data sets have been produced from many different sensors since the early 2000s (Table 20.1). Some of the most widely used are: Total Ozone Mapping Spectrometer Aersol Index (TOMS AI) (Herman *et al.*, 1997), the TOMS aerosol optical thickness (AOT) (Torres *et al.*, 2002), multiangled imaging spectroradiometer (MISR) (Meloni *et al.*, 2004) and algorithms from the Moderate Resolution Imaging Spectroradiometer (MODIS) (Kaufman *et al.*, 2005), such as the 'Deep Blue' algorithm (Hsu *et al.*, 2004). Others, such as Synergetic Aerosol Retrieval (SYNAER) from ENVISAT observations (Holzer-Popp *et al.*, 2008) are also coming on stream. All these data sources derive from polar orbiting satellites and are mostly observations in shortwave (blue to ultraviolet). In addition, there are now dust estimates from geostationary satellites such as the Infrared Difference Dust Index (IDDI) (Legrand, Plana-Fattori and N'doume, 2001) and observations from the Meteosat Second Generation (MSG) Spinning Enhanced Visible and Infrared Imager (SEVIRI) instrument (Schepanski *et al.*, 2007, 2009), which provide temporal resolution of around 15 minutes day and night, although their spatial resolution is much lower (1–4 km). The impact of satellite data on desert dust research has been profound and has been widely used in planning ground campaigns, evaluating numerical models and understanding source regions and transport. Instruments on satellites measure radiation and use algorithms to convert that radiation into measures of dust loadings, although many of these measures are nondimensional (unitless), relating only qualitatively to dust concentrations. Additionally, the algorithms rely on assumptions such as dust particle sphericity and atmospheric water vapour content for which there are no reliable data.

20.1.2.2 Aircraft campaigns

Most satellite-derived dust data sets provide a top-down, two-dimensional view of dust. Aircraft provide a means to sample the characteristics of suspended dust and its influence on the Earth System in situ. This advantage has led to numerous aircraft campaigns to measure dust, e.g. the Saharan Dust Experiment (SHADE) (Haywood *et al.*, 2003), the Mediterranean Dust Experiment (MEIDEX) (Alpert *et al.*, 2004), the Cooperative LBA Airborne Experiment (CLAIRE) (Formenti *et al.*, 2001) and the Aerosol Characterization Experiment (ACE-Asia) (Anderson *et al.*, 2003). The Dust and Biomass Experiment (DABEX) and the Dust Outflow and Deposition to the Ocean (DODO) Experiment were based out of Niamey, Niger, and Dakar, Senegal, respectively. Collectively, these campaigns have constrained and quantified the influence of dust on the radiation budget and yielded comprehensive data sets on dust characteristics, such as transport pathways, dust plume heights and dust size, shape and colour. A notable feature of these campaigns is that almost all have been distant from dust source regions.

20.1.2.3 Ground-based measurements and campaigns

In comparison with the coverage from satellite-derived data, ground measurements and campaigns remain very limited in their spatial and temporal coverage. One of the most widely used and longest running data sets on dust concentrations is the Miami Aerosol Group measurements, although measurements from the Aerosol Robotic Network (AERONET) are set to become a key data set.

Table 20.1 Examples of satellite-derived dust data sets.

Platform	*Spatial resolution*	*Temporal resolution and time of passing if daily*	*Years available*	*Features/problems*	*Reference of use in dust research*
TOMS AI (Nimbus 7)	1.25° × 1.0°	Once a day, around noon	1978–1993	Detects aerosols over land but not below 1 km, height dependency	Herman *et al.* (1997)
TOMS AI (EP)	1.25° × 1.0°	Once a day, around noon	1996–2005, has problems in some years	Detects aerosols over land but not below 1 km, height dependency	
TOMS AI (OMI)	1.0° × 1.0°	Once a day, around noon	2004–present	Detects aerosols over land but not below 1 km, height dependency	Levelt (2002)
MISR	250 × 250 m	Global coverage every 9 days, with repeat coverage between 2 and 9 days depending on latitude	2000–present		Martonchik *et al.* (1998, 2002)
SEVIRI (dust images)	3 km at nadir	Every 15 minutes, geostationary	2006–present	Detects aerosols over land, also at night, height dependency	Schepanski *et al.* (2007)
SEVIRI (DSAF)[a]	1.0° × 1.0°	8 times a day (3 hour periods), geostationary	2006–present	Day and night, height dependency	Schepanski *et al.* (2007)
MODIS (deep blue)	250–1000 m	Once a day (Aqua),	2002–present		Hsu *et al.* (2004)
SeaWiFS (deep blue)	4 km	Once a day, sun synchronous at 705 km	1998–present		Hsu *et al.* (2004)
Meteosat (IDDI)	1.0° × 1.0°	Once a day, about 11:30 UTC, product is monthly, geostationary	1984–present	Detects aerosols over land, height dependency	Brooks and Legrand (2000)
Various satellites (AOT as indicator of dust)	Varying	Varying	Varying, <daily	Based on visible WL, difficulties over bright surfaces (deserts)	

[a]Dust source activation frequency.

Ground-based subjective visibility records, which were the mainstay of inferred dust loadings prior to the mid 1990s, continue to be made at synoptic weather stations. In many stations, automatic visibility sensors are replacing observer-judged visibility. However, it remains the case that there are very few ground-based data of dust emission, flux and deposition from the vast majority of the world's key dust source regions, although the last decade has featured several field campaigns in North Africa focused on dust measurement.

20.1.2.4 *Miami Aerosol Group measurements*

A worldwide array of near-surface oceanic aerosol sampling set up by the University of Miami Aerosol Group provides the longest direct measurement of dust concentration in the atmosphere (e.g. Prospero and Lamb, 2003). Data are available from 1965 in Barbados and at about 30 stations spread across the main ocean basins from the early 1980s until late 1996. Aerosols are collected by high-volume filter samplers, some of which are controlled by a wind sensor system, which limits sampling to the open-ocean sector and minimizes the impact from local aerosol sources. Daily data are available from some stations, mainly in the North Atlantic, while elsewhere the data are weekly.

20.1.2.5 AERONET: Aerosol Robotic Network

AERONET is a network of ground-based remote sensing aerosol sensors in the form of Cimel sun photometers (Holben *et al.*, 1998). The network provides observations that include cloud-screened measures of spectral aerosol optical depth (AOD), aerosol volume size and single scattering albedo. From the perspective of measuring desert dust, the key quantification provided is that of the average total aerosol column within the atmosphere. AERONET now has more than 200 locations of quality-assured data for more than one year, 26 locations for more than 5 years and 10 locations for more than 7 years. However, most of these are not in the known dust regions of the world and very few are within even a few hundred kilometres of key dust source regions.

20.1.2.6 Field and aircraft campaigns

There are very few ground-based measurements from key dust regions of the world. Advances in dust retrieval through satellite remote sensing has led to the identification of specific source regions (see Section 20.1.4 on distribution of dust) and this, in turn, has encouraged dedicated campaigns to key dust regions. Notable to date are the Saharan Mineral Dust Experiment (SAMUM), which featured summertime observations in Morocco, the African Multidisciplinary Monsoon Analysis (AMMA) and its subprogrammes, e.g. the Atmospheric Radiation Measurement (ARM) facility, which was deployed to Niamey, Niger, and BoDEX 2005 (The Bodélé Dust Experiment) which retrieved the first ground-based data from the Bodélé Depression, Chad, in February–March 2005 (Tegen *et al.*, 2006). The Bodélé is frequently referred to as the world's most intense dust source. As described above (dust sinks, Section 20.1.1.3), there are only a very limited number of studies concerning the physical measurement of longer-term dust deposition rates using dust traps (e.g. Ta *et al.*, 2004; Reheis, 2006), although there are several examples of short-term studies in specific source areas.

20.1.3 Modelling dust

20.1.3.1 Introduction

Probably the largest effort in dust research since the late 1990s has gone into developing dust schemes for numerical models of climate. Initially these efforts were focused on the Last Glacial Maximum (Joussaume, 1993) during which dust loadings may have been an order of magnitude higher than present (Harrison *et al.*, 2001). Recognition of the important and complicated role that dust plays in climate, particularly in altering the radiation budget that drives the climate system, led to a concentration of effort in improving the dust cycle in climate (more recently Earth System) models. As a result of the progress made, knowledge of global dust loadings and transport is increasingly dependent on numerical models rather than observations.

Numerical models are representations of the workings of the atmosphere based on the laws of physics. They fall into the following categories, based on time and space distinctions: weather forecasting models (also called numerical weather prediction, NWP), global climate models, which are used for longer-term simulations, and regional models, where the numerical simulation covers a smaller domain of the Earth, typically part of a continent such as southern Africa. In all these models, space is effectively treated as discrete grid boxes, larger in the global models (e.g. 2.5 $\times$ 2.5 degrees latitude–longitude) and significantly smaller in the NWP and regional models with $^1/_2$ degree typical and better than $^1/_4$ degree possible. In many of these models, the dust cycle is now an integral component, which interacts with other important components of the climate system. The core of the dust cycle components are: dust entrainment, transport and deposition.

20.1.3.2 Dust entrainment in models

The background to physical sediment entrainment is covered in Chapter 18. In models, dust production is assessed at each grid box and is based essentially on a power function of wind friction velocity (U^*) in excess of a threshold value (U^*_t). Wind is computed by the model as part of the equations of motion and the characteristics of the wind in the boundary layer and its interaction with surface features. Height in the numerical models is dealt with in discrete layers, of which there are generally too few in the boundary layer to truly represent U^*. Near-surface wind speed is therefore sometimes used as a substitute.

In simple dust schemes, parameters important to entrainment, such as the erodible fraction, are assigned arbitrary (i.e. nonphysical) values in regions for which satellite data show large emissions and which can be defined on the basis of another surface physical attribute. One scheme, which led to a notable improvement in the match between model dust emission and qualitative satellite values of dust loadings, was the Ginoux *et al.* (2001) model. Based on TOMS AI data, Prospero *et al.* (2002) and Washington *et al.* (2003) identified topographical depressions in arid regions as key dust sources. Ginoux *et al.* (2001) also implemented these ideas in a model, thereby introducing the preferred source area concept. Similar adjustments to the emission scheme were then introduced in other models (Tegen *et al.*, 2002; Luo, Mahowald and del Corral,, 2003; Zender, Bain and Newman, 2003). There are several reasons behind the simplification in dust emissions as represented in these models. These include: (1) numerical models typically underestimate surface windspeeds, some by more than 50 % (Washington *et al.*, 2006a) and so adjustment to entrainment is more easily made to other controls, including U^*_t and/or the parameters that control the mass flux of dust as these need to be prescribed in the model rather than calculated at each time step. Any adjustment to parameters that are calculated at each time step, such as wind, would be much more complicated to implement. (2) The real values of key controls on dust emission, such as surface roughness, nonerodible fraction and particle size distribution, have never been measured across the vast majority of key source areas such as the Sahara. Therefore the easiest way to improve model dust emission is simply to alter these unknown values so that model emissions appear reasonable on the basis of either satellite estimates of dust distribution or background measurements such as those from the Miami Aerosol Group.

In contrast, more realistic, physically based and more complex dust emission schemes are increasingly being used in numerical models. Examples are the Marticorena and Bergametti (1995) and Shao, Raupach and Leys (1996) schemes (see Laurent *et al.*, 2008 and Darmenova *et al.*, 2009). In these more complex schemes, the erosion threshold is typically calculated as a function of surface roughness (both overall and that of the erodible surface component), soil moisture and particle size distribution. Saltation is a function of friction velocity, fraction of erodible to total surface and particle size distribution. Vertical flux is represented through an empirical relationship linking the ratio of the dust flux to the horizontal (sandblasting) flux to the soil clay content (Marticorena and Bergametti, 1995).

Dust emission from models has been shown to differ by an order of magnitude, even when calculated over a limited source area and for only a few days (e.g. see Todd *et al.*, 2008). Improvement to entrainment schemes hinges on direct collaboration between geomorphologists and modellers, something that has been lacking in the science. Key problems include: (a) representation of dust entrainment at the model grid box scale (tens of kilometres) when relationships between key parameters have been derived from either point source field data or idealised from wind tunnel experiments (Iversen and White, 1982); (b) lack of observed data over vast key source regions on parameters such as soil particle size; (c) inclusion of processes currently not represented in the models, e.g. supply limitations imposed by surface crusting (Lopez, 1998) and alternatives to the horizontal and vertical sediment flux distinction such as autoabrasion (Warren *et al.*, 2006).

20.1.3.3 Dust transport and deposition in models

In contrast to entrainment, transport schemes in numerical models are a relative strength since there are few other ways of determining dust pathways. Conversely, there are also few data sets with which to evaluate model performance. Tracer advection schemes in the models are typically used to determine dust transport. These include vertical motion through convection, turbulent mixing and gravitation settling. Wet deposition within and below cloud height is often determined by a precipitation rate based experimentally on a derived dependence on particle size (e.g. Woodward, 2001).

20.1.4 Distribution of dust

The release of long-term (1979–1993) data from TOMS AI (Herman *et al.*, 1997) allowed the first, albeit qualitative, complete picture of dust loadings in the atmosphere (Prospero *et al.*, 2002; Washington *et al.*, 2003) (Figure 20.7). The largest area with high values is a zone that extends from the eastern subtropical Atlantic through the Sahara Desert to Arabia and southwest Asia. In addition, there is a large zone with high AI values in central Asia, centred over the Tarim basin and the Taklimakan Desert. Central Australia has a relatively small zone, located in the Lake Eyre basin, while southern Africa has two zones, one centred on the Makgadikgadi basin in Botswana and the other on the Etosha Pan in Namibia. In Latin America, there is only one easily identifiable zone. This is in the Atacama and is in the vicinity of one of the great closed basins of the Altiplano. North America has only one very small zone with high values, located in the Great Basin (the drained Owens Lake). The importance of these different dust ‘hotspots’ can be gauged by looking not only at their areal extents but also at their relative AI values (Table 20.2). This again brings out the very clear dominance of the Sahara in particular and of the Old World deserts

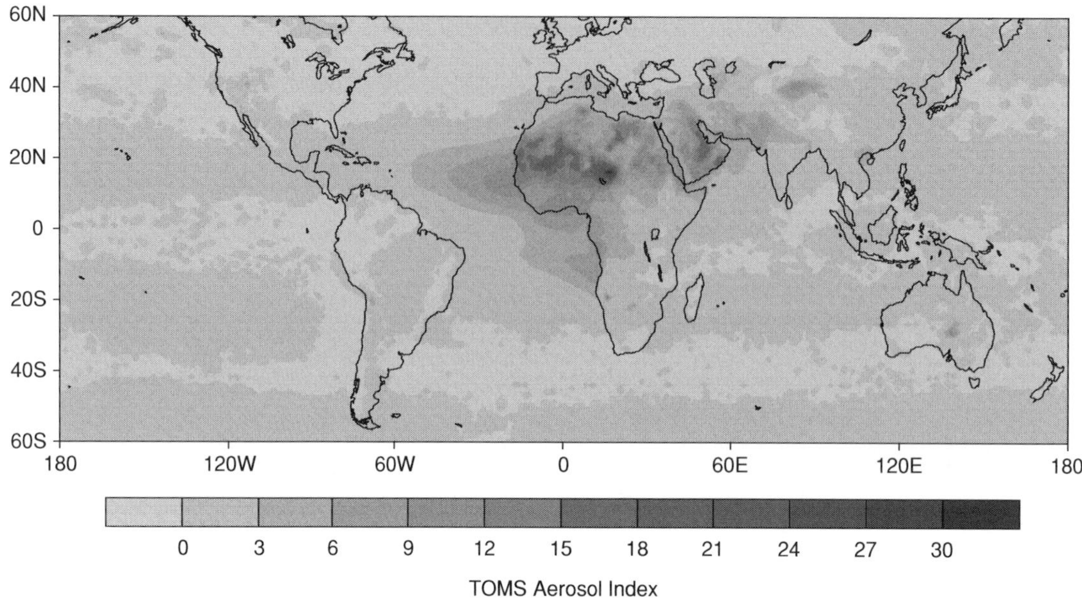

Figure 20.7 Long-term mean annual (1979–1993) TOMS aerosol index.

in general. The southern hemisphere as a whole and the Americas are notable for their relatively low AI values. Thus, for example, the AI values of the Bodélé Depression of the south-central Sahara are around four times greater than those recorded for either the Great Basin or the Salar de Uyuni in the Altiplano.

The total dust burden in the atmosphere is estimated from 15 global models to be 8–37 Tg. Simulated global dust emission is between about 1000–3000 Tg/yr (Engelstaedter, Tegen and Washington, 2006; Textor *et al.*, 2006), with the uncertainty range of about a factor of 3.

Table 20.2 Maximum mean aerosol index (AI) values for major global dust sources determined from TOMS (after Washington *et al.*, 2003).

Location	*Mean AI value*
Bodélé Depression of south central Sahara	>3.0
West Sahara in Mali and Mauritania	>2.4
Arabia (southern Oman/Saudi border)	>2.1
Eastern Sahara (Libya)	>1.5
Southwest Asia (Makran coast)	>1.2
Taklimakan/Tarim basin	>1.1
Etosha Pan (Namibia)	>1.1
Lake Eyre basin (Australia)	>1.1
Makgadikgadi basin (Botswana)	>0.8
Salar de Uyuni (Bolivia)	>0.7
Great Basin of the United States	>0.5

20.2 Key source areas

20.2.1 Bodélé Depression, Chad

Many studies have pointed to the Bodélé Depression as one of the key dust sources in the world (Kalu, 1979; Herrmann, Stahr and Jahn, 1999; Brooks and Legrand, 2000; Goudie and Middleton, 2001; Prospero *et al.*, 2002; Koren and Kaufman, 2004; Washington and Todd, 2005; Washington *et al.*, 2003, 2006a, 2006b; Schepanski *et al.*, 2007). The TOMS AI data indicate that the Bodélé is the most intense source, not only in the Sahara, but also in the world, with annual mean AI values that exceed 3.0. Dust plumes that originate in the Bodélé from a region of exposed diatomite sediment (centred near 17°N, 18°E) deposited under Lake Mega-Chad extend for hundreds of kilometres and are evident in the MODIS true colour imagery (e.g. Koren and Kaufman, 2004; Washington and Todd, 2005). The dust loading from these plumes is reflected in the AERONET data from Ilorin in Nigeria, some 1700 km downwind from the Bodélé (Pinker *et al.*, 2001; Todd *et al.*, 2008).

Bodélé dust loadings, as estimated from a variety of data sources, follow a semi-annual cycle with peaks in the boreal Spring and Autumn (Figure 20.8). There is close agreement between TOMS AOT data and the annual cycle of dust plumes over the Bodélé from MODIS imagery. TOMS AI data, on the other hand, peaks later in the year, although TOMS AI has a well-known bias for detecting

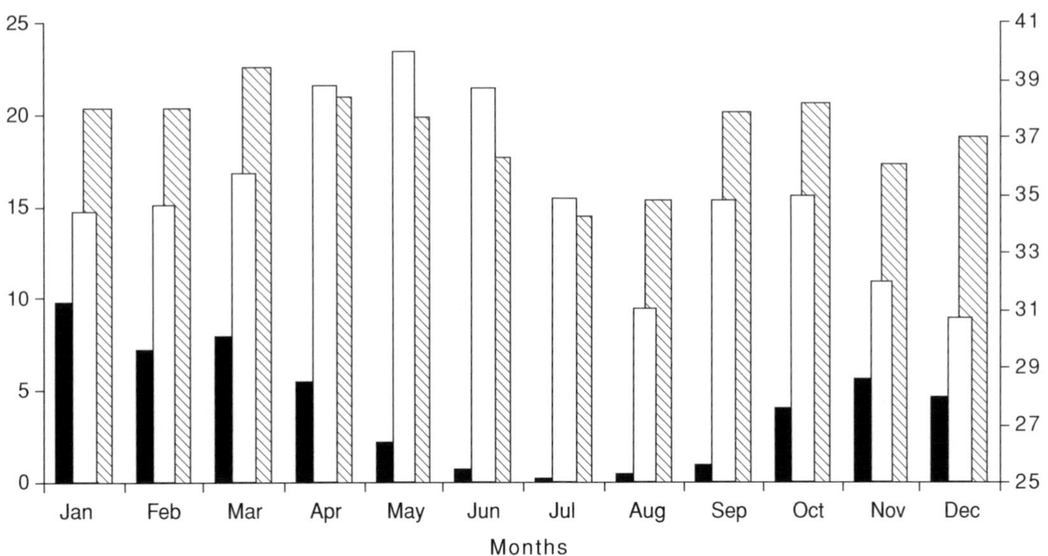

Figure 20.8 Monthly mean values for satellite-derived estimates of dust over the Bodélé Depression, Chad: TOMS aerosol optical thickness (hatched) (1979–1992), number of large dust plumes over the Bodélé (2002–2005 January to September, 2002–2004 October to December) (black), TOMS aerosol index (1979–1992) (white) using right-hand vertical axis.

dust above the surface layers (Herman *et al.*, 1997; Mahowald and Dufresne, 2004).

Dust production from the Bodélé may be explained by the colocation of the strong surface wind and a vast supply of erodible material in the Bodélé Depression. The colocation results from alternating cycles of deflation, which led to the formation of the depression during arid times and subsequent filling of the depression with erodible material during pluvials (Washington *et al.*, 2006b).

Washington and Todd (2005) point to a pronounced easterly low-level jet (LLJ) feature evident in the reanalysis wind data, peaking at close to the surface, overlying the Bodélé region near 19°E. The Bodélé LLJ is strongest in January and is present in all months of the year except August, when the northeasterlies are weak or absent. The jet is part of the northeasterly Harmattan winds of North Africa. Regional model experiments with model topography included and then excluded have shown that the northeasterly near-surface flow is likely to be accelerated between the Tibesti and Ennedi massifs, which lie 2600 m and 1000 m above the flat terrain in the Djourab Desert of Chad, respectively (Washington *et al.*, 2006b). As a result, the Bodélé LLJ is a feature that uniquely overlies the greater Bodélé region. It is absent to the west and weakens to the south, disappearing south of 10°N and north of 22°N. Sampling based on extreme dust years shows that interannual variability is associated with a strengthening of the LLJ during dusty years in the Bodélé and a retraction of the jet during years with extremely low dust loadings. Intraseasonal variability of dust over the Bodélé occurs contemporaneously with the ridging of the Libyan High and pulsing of the pressure gradient, which drives the northeasterlies in which the LLJ is embedded (Washington and Todd, 2005).

From measurements of wind speed, AOT and visual observations during BoDEx 2005, the specific threshold wind speed (15 minute average at 2 m height) for dust emission was found to be 10.0 m/s. This value is unusually high in comparison to other dust source regions. Dust emission events in the course of BoDEX 2005 were found to be associated with the pulsing of the Bodélé LLJ. Between January and March, dust from the Bodélé is transported south-southwest, crossing the coast of west Africa within 5 days. The dust is transported in a layer between 800 and 700 hPa.

The Bodélé contains large exposures of very friable diatomite (a silicious, low-density deposit) and so dust emission is transport, not supply, limited. Based on MODIS surface reflectance data, the exposed diatomite has been estimated to cover ~10 800 km^2 (Warren *et al.*, 2007). The largest extant areas of diatomite are those that were deposited in the deepest basins of Palaeolake Mega, Chad, which, some 6000 years ago, was the largest lake on earth, bigger than the present Caspian Sea (Drake and Bristow, 2006). It is diatomite, not highly weathered material, that is the source of by far the greatest proportion of the dust from the Bodélé (Warren *et al.*, 2007) (Figure 20.9).

Figure 20.9 Northern edge of the Bodélé Depression showing the diatomite surface.

20.2.2 Saharan Empty Quarter

During the boreal summer months, global dust production peaks in the Saharan Empty Quarter (SEQ) of Northern Mail, southern Algeria and Eastern Mauritania (Figure 20.10). Dust loadings are at a minimum in this region during the boreal winter. Owing to its remoteness, much of what is known about this region comes from satellite observations. The precise location of dust sources in the SEQ differs markedly between satellite data sets. On the

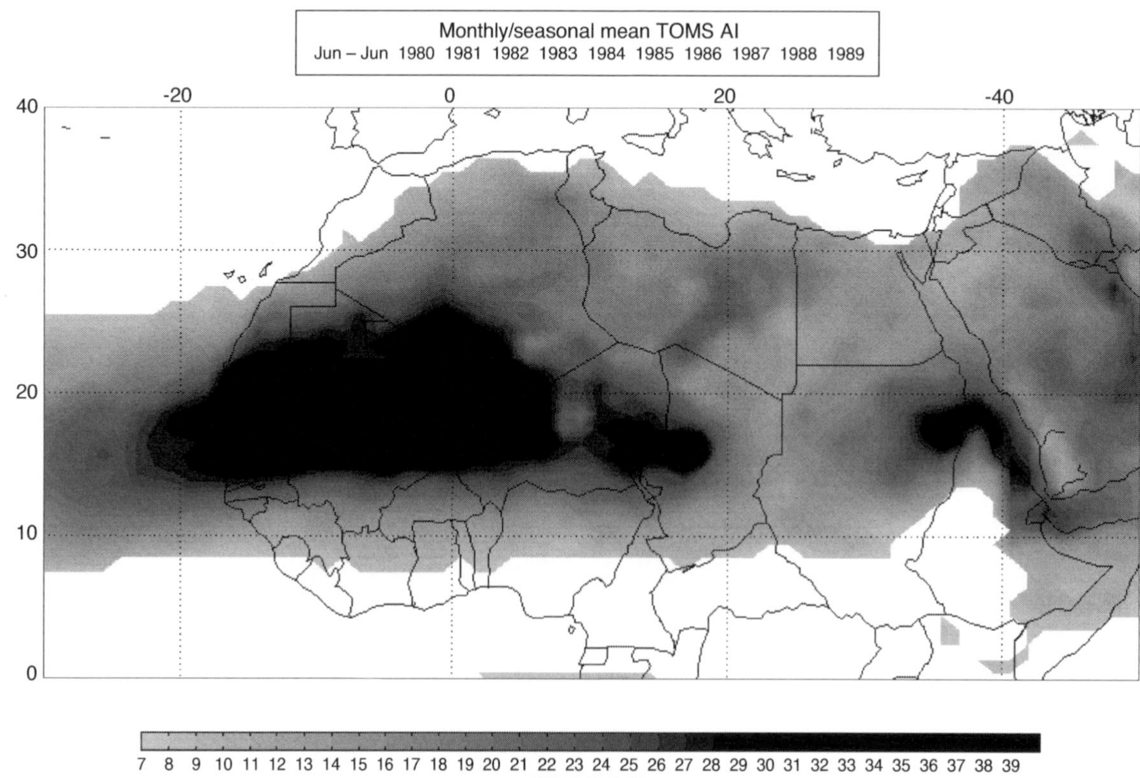

Figure 20.10 Long-term mean June TOMS aerosol index over North Africa.

basis of TOMS AI, Engelstaedter and Washington (2007a) show that the onset of dustiness at key hotspots corresponds with the northward passage of near-surface convergence and may be explained by wind gustiness rather than the mean wind (Engelstaedter and Washington, 2007b). Based on aircraft observations from a flight made during the Geostationary Earth Radiation Budget Intercomparison of Longwave and Shortwave Radiation (GERBILS) field campaign (June 2007), Marsham *et al.* (2008) show significant dust uplift into southward-propagating cold pool outflows of the monsoon flow immediately south of the intertropical discontinuity in the western Sahara. They argue that the asymmetry in the seasonal dust cycle is closely related to the downdraft convective available potential energy (DCAPE) from convective storms since there is both more dust and more DCAPE during monsoon onset than during retreat. Using 3 hourly data from the Meteosat second generation (MSG) spinning enhanced visible and infrared imager (SEVIRI), giving infrared dust index images from March 2006 to February 2008, Schepanski *et al.* (2009) have identified source regions over the SEQ that differ substantially from the TOMS AI-based source regions (Figure 20.11). On the basis of peak dust production between 0600 and 0900 hours, Schepanski *et al.* (2009) argue that the mixing of the low-level jet is a major factor in causing dust entrainment.

A notable absence in the understanding of dust production from the SEQ is any ground-based geomorphic or meteorological observations that would help to constrain the explanation of dust production in this key region. The impression from satellite data is that numerous diffuse

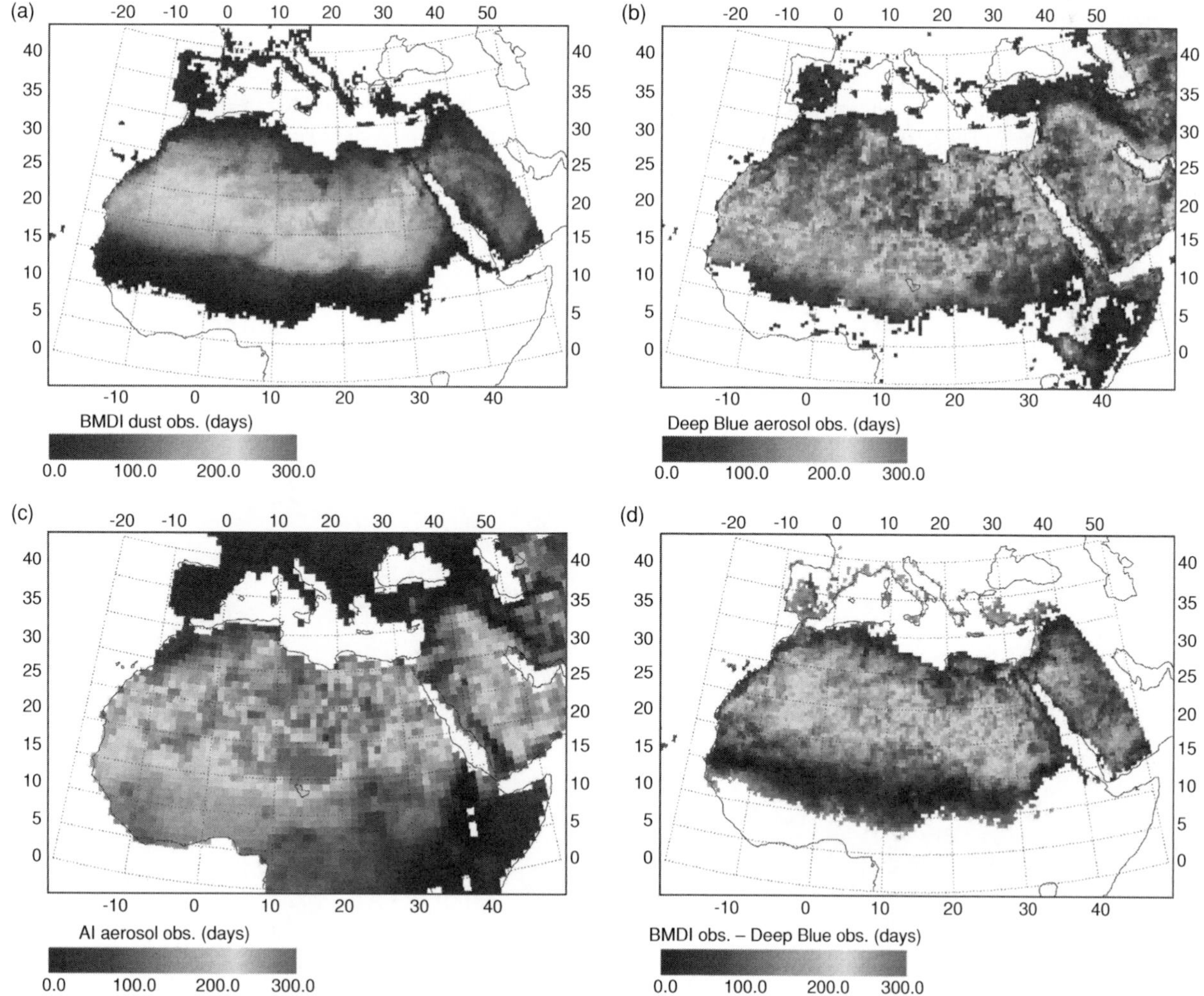

Figure 20.11 Number of days with dust detected in 2006 (a) by bitemporal mineral dust index, (b) by MODIS AOD and (c) by OMI AI. In (d) the difference between BMDI observation numbers and MODIS deep blue observation numbers is shown (after Schepanski *et al.*, 2009).

sources of erodible material are the key sources for this transport-limited region.

20.2.3 China

It has been estimated that about 800 Tg of Chinese dust is injected into the atmosphere annually, which may be as much as half of the global production of dust (Zhang, Arimoto and An, 1997), although more recent model-based dust emission estimates range between 147 and 496 Tg/yr. Studies of dust loadings and fluxes have suggested that there are two main source areas: the Taklimakan and the Badain Juran (Zhang *et al.*, 1998).

The Tarim basin experiences some of the most frequent and intense dust storm occurrences on Earth (Zhang, Arimoto and An, 1997). It is also one of the largest closed basins on the planet, being almost completely surrounded by high mountains: the Himalayan Plateau to the south, the Kunlun Mountains to the west and Tianshan Mountain to the north. Along the periphery of the Tarim basin mean annual precipitation is within a narrow range of 50–100 mm, while in the central region of the basin it is only about 10 mm (Zhao *et al.*, 2006). Most of the basin area is occupied by the Taklimakan Desert, the largest desert (338 000 km^2) in China. The sandy material in the Tarim basin is roughly 900 m thick, and mainly consists of a huge amount of weathering residues from the surrounding mountains. In contrast to other deserts, the Taklimakan Desert probably has some of the finest sands in the world (Dong, 2000; Xuan *et al.*, 2004).

Like the SEQ, there have been very few studies about hotspots in the Tarim basin owing to the sparse distribution of meteorological stations, especially in the centre of the basin, where there is only one surface station for every 10 000 km^2 (Luo *et al.*, 2005). Long-term mean TOMS AI data point to two key areas of dust production in the Tarim basin: (1) in the northeast part of the Tarim basin, near the south slope of Tianshan Mountain, centred at roughly 84°E, 40°N and (2) at the southern edge of the Tarim basin, at the north foot of Kunlun Mountain, centred at roughly 82°E, 37°N. Dust storms in this region peak in April (Gao and Washington, 2009) (Figure 20.12). Wind profiles appear to be important to differing dust generation in the two hotspots. Northeasterly winds enter the Tarim basin through the Lop Nur, the basin's opening to

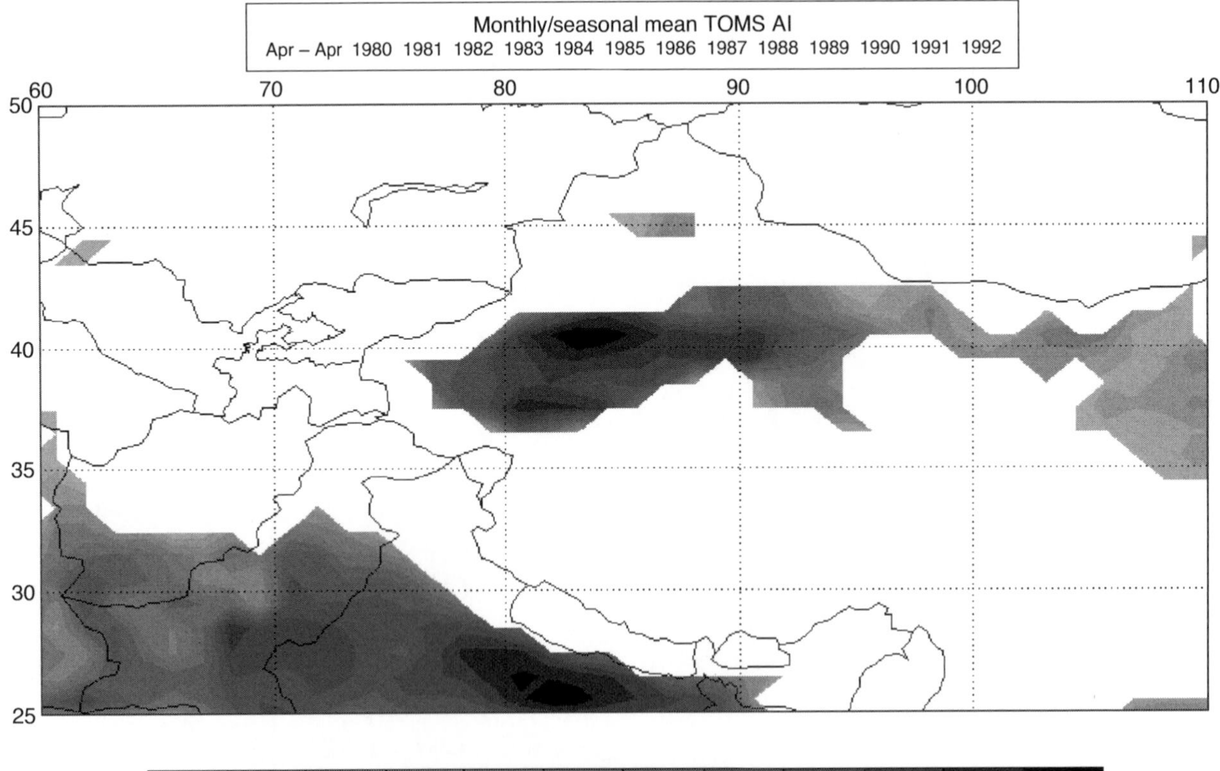

Figure 20.12 Long-term mean April TOMS aerosol index over China.

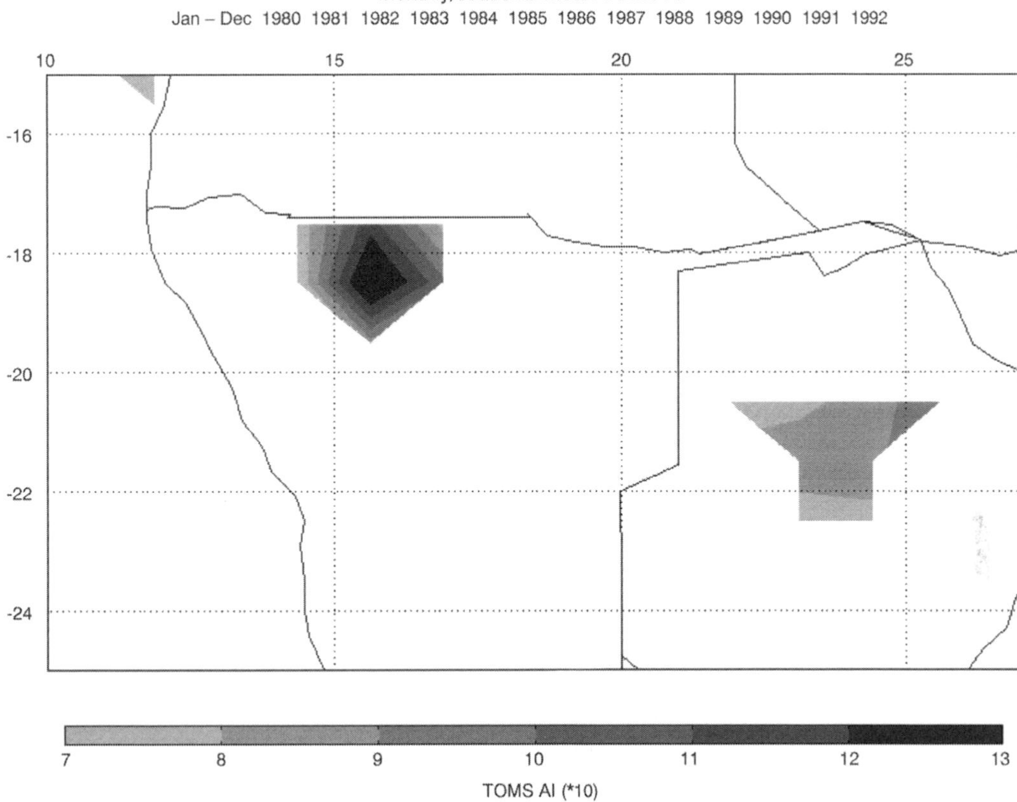

Figure 20.13 Long-term mean annual TOMS aerosol index over southern Africa.

the east. As the area to the west of the basin is blocked by high mountains, convergence forces the winds to become northwesterly at the western part of the basin. The Kunlun Mountains to the south block the flow, causing an accumulation of dust over the southern part of the basin, where the occurrence of high dust loadings is an order of magnitude higher than the northern parts of the Tarim (Xuan and Sokolik, 2002; Kurosaki and Mikami, 2005).

China has been the focus of several modelling studies (Zhao *et al.*, 2003; Laurent *et al.*, 2006; Darmenova *et al.*, 2009). The importance of dust as a health issue in urban areas has no doubt prompted efforts to simulate both the emission and transport of dust (Gong *et al.*, 2003; Shao *et al.*, 2003). Like the SEQ, specification of soil area parameters such as particle size, roughness and soil moisture remain a key limitation to better constrained modelling.

20.2.4 Southern Africa and Australia

Dust loadings in the southern hemisphere are much lower than the northern hemisphere. Two key southern hemisphere source regions are southern Africa and Australia. In contrast to China and the SEQ, these source regions are more accessible and as a result have been the feature of recent geomorphological investigation.

TOMS AI long-term mean data (Figure 20.13) indicate two relatively small but clearly developed dust sources over southern Africa. The more intense of these is centred over the Etosha Pan in northern Namibia and has an AI value of more than 1.1. The other is over the Mgkadikgadi Depression in northern Botswana and has AI values greater than 0.8.

Etosha, which covers an area of about 6000 km^2, is a salt lake that occupies the sump of a much larger basin and often floods in the summer months but is often dry enough in winter for deflation. The Mkgadikgadi is another major structural feature, which includes the Sua (Sowa) Pan. Deflation is influenced by the extent and frequency of lake inundation, sediment inflows and surface wind speed variability (Bryant *et al.*, 2007). Deflation events occur with the passage of transient eddies, in the form of west-to-east migrating anticyclones that enhance the prevailing easterly (trade) winds across southern Africa for periods of a few days at a time. On these occasions satellite images also show dust plumes blowing westwards towards the South Atlantic (Eckardt, Washington and Wilkinson, 2001) from the major river channels of the Namib coast.

While aeolian processes are of considerable importance in Australian deserts, the continent's deserts produce relatively small amounts of dust. The area of greatest dust storm frequency, as determined from meteorological station data and TOMS AI, is the huge (1.3 million km^2) internal drainage basin of Lake Eyre. Deflation operates on alluvial spreads brought by the southward flowing Eyre, Diamantina and Cooper Rivers.

Bullard *et al.* (2008) have used MODIS data to determine the spatial and temporal variability of dust emissions from different surfaces in the Lake Eyre basin. Studying data over the period 2003–2006, they classified 529 dust plumes according to their source areas. Overall 37 % of plumes originated in areas of aeolian deposits, 30 % from alluvial deposits and floodplains and 29 % from ephemeral lakes or playas. Sediment supply and availability rather than erosivity were shown to be the control on deflation.

20.3 Temporal changes in dust

Much of this chapter has been focused on understanding the long-term mean distribution and controls on dust. An important additional component concerns variability and change in dust emission, particularly given the extent of projected climate change in the twenty-first century.

20.3.1 Observational record

Most satellite and newly implemented ground-based aerosol remote sensing data sets such as Aeronet are still too short to assess changes to long-term dust loadings. Instead we rely still on ground-based visibility observations and measurements made by the Miami Aerosol Group.

Records show that the occurrence of dust storms was twice more frequent during the period 1960–1984 than during the period 1984–1997 in China. However, between 1997 and 2002, there was a significant increase in the number of dust storms occurring during spring (Zhang *et al.*, 2003). This increase is particularly noticeable in the deserts of northern Asia (Kurosaki and Mikami, 2005). Moreover, large areas of ongoing desertification due to land use have been identified in China (Xue, 1996; Zha and Gao, 1997) and Mongolia (Natsagdorj, Jugder and Chung, 2003) and may constitute additional dust sources. Part of the observed changes in the case of China relate to land use changes, including water extraction from rivers (Gao and Washington, 2009).

Using measurements from the Miami Aerosol Group's Barbados station over the period 1965–1998, Prospero and Lamb (2003) linked the great Sahel drought to increases in North African dust production, although the cause may relate to changes in atmospheric circulation, including near-surface wind and long-range transport of dust by the trade winds across the Atlantic to Barbados, rather than simply an increase in aridity in the Sahel. Mean dust concentrations in Barbados during the 1980s were about four times higher (19 $\mu g/m^3$) than the 1950s, during which rainfall in the Sahel was some 30 % higher. In addition to these multidecadal differences, dust output also increased significantly in the year following El Niño events.

20.4 Future climate change

Large and significant changes to climate are projected to take place during the Twenty-First Century as a result of increasing greenhouse gas emissions. Many of the parameters that control dust emission, such as vegetation, rainfall, soil moisture and surface wind speed, are expected to change. Some studies estimate the change in atmospheric burden of dust to decrease by 20–60 % (Mahowald and Luo, 2003) although, in a multiclimate model study, Tegen *et al.* (2004) argued that changes to dust loadings are model-dependent, with projections showing both increases and decreases globally.

There is also a research focus on both new dust sources in the twenty-first century and change to existing key source regions. The Met Office Hadley Centre coupled climate model points to major Amazon forest die-back and the emergence of this region as a key dust source late in the twenty-first century. The driver for this change is a reduction in rainfall (Betts, Sanderson and Woodward, 2008). In a multimodel study of the Bodélé Depression, Washington *et al.* (2009) show that the climate models with the most realistic present-day circulation point to a potential doubling of dust output from this key source at the end of the twenty-first century. It is important to point out that there are very large uncertainties associated with these projections, beginning with the rather poorly constrained dust emission component of the dust modules in these models.

20.5 Conclusions

Dust research has received a huge increase in attention in the decade up to 2010, driven mainly by the need to include dust in the simulation of climate change. Without doubt it has been the release of remote sensing data sets, particularly satellite-borne, that has done the most to clarify the distribution and behaviour of dust globally. Collectively, these tools have drawn attention to the key dust source regions, such as the Bodélé Depression, Chad and informed field and aircraft campaigns aimed at better constraining dust characteristics in these regions. In the next decade, many of the more recent satellite data sets will have a sufficiently long record to allow for a reassessment of this work. What is missing from this progress is the engagement of ground-based geomorphology, particularly at the interface between source region characteristics and the difficult task of representing these in models.

References

Alpert P., Kishcha P., Shtivelman, A. *et al.* (2004) Vertical distribution of Saharan dust based on 2.5-year model predictions. *Atmospheric Research*, **70** (2), 109–130.

Anderson, T., Masonis, S.J., Covert, D.S. *et al.* (2003) Variability of aerosol optical properties derived from in situ aircraft measurements during ACE-Asia. *Journal of Geophysical Research*, **108** (D23), 8647, DOI: 10.1029/2002JD003247.

Ansmann, A., Bösenberg, J., Chiakovsky, A. *et al.* (2003) Long-range transport of Saharan dust to northern Europe: the 11–16 October 2001 outbreak observed with EARLINET. *Journal of Geophysical Research D: Atmospheres*, **108** (24), AAC12-1–AAC12-15.

Ansmann, A., Tesche, M., Althausen, D. *et al.* (2008) Influence of Saharan dust on cloud glaciation in southern Morocco during the Saharan Mineral Dust Experiment. *Journal of Geophysical Research – Atmospheres*, **113** (D4), D04210.

Bagnold, R.A. (1941) *The Physics of Blown Sand and Desert Dunes*, Methuen, London.

Betts, R., Sanderson, M. and Woodward, S. (2008) Effects of large-scale Amazon forest degradation on climate and air quality through fluxes of carbon dioxide, water, energy, mineral dust and isoprene. *Philosophical Transactions of the Royal Society B – Biological Sciences*, **363** (1498), 1873–1880.

Brooks, N. and Legrand, M. (2000) Dust variability over northern Africa and rainfall in the Sahel, in *Linking Climate Change to Land Surface Change* (eds S. McLaren and D. Kniveton), Kluwer Academic Publishers, Dordrecht, pp. 1–25.

Bryant, R.G., Bigg, G.R., Mahowald, N.M. *et al.* (2007) Dust emission response to climate in southern Africa. *Journal of Geophysical Research D: Atmospheres*, **112**, article D09207.

Bullard, J.E., McTainsh, G.H. and Pudmenky, C. (2004) Aeolian abrasion and modes of fine particle production from natural red dune sands: an experimental study. *Sedimentology*, **51**, 1103–1125.

Bullard, J., Baddock, M., McTainsh, G. and Leys, J. (2008) Sub-basin scale dust source geomorphology detected using MODIS. *Geophysical Research Letters*, **35** (15), article L15404.

Cahill, T.A., Gill, T.E., Reid, J.S. *et al.* (1996) Saltating particles, playa crusts and dust aerosols at Owens (dry) Lake, California. *Earth Surface Processes and Landforms*, **21** (7), 621–639.

Cassar, N., Bender, M.L., Barnett, B.A. *et al.* (2007) The Southern Ocean biological response to Aeolian iron deposition. *Science*, **317** (5841), 1067–1070.

Cattle, S.R., McTainsh, G.H. and Elias, S. (2009) Aeolian dust deposition rates, particle sizes and contributions to soils along a transect in semi-arid New South Wales, Australia. *Sedimentology*, **56**, 765–783.

Colarco, P.R., Toon, O.B. and Holben, B.N. (2003) Saharan dust transport to the Caribbean during PRIDE: 1. Influence of dust sources and removal mechanisms on the timing and magnitude of downwind aerosol optical depth events from simulations of in situ and remote sensing observations. *Journal of Geophysical Research D: Atmospheres*, **108** (19), PRD 5-1–5-20.

Darmenova, K., Sokolik, I.N., Shao, Y. *et al.* (2009) Development of a physically based dust emission module within the Weather Research and Forecasting (WRF) model: assessment of dust emission parameterizations and input parameters for source regions in Central and East Asia. *Journal of Geophysical Research*, **114**, D14201.

De Deckker, P. A., Goodwin, I.D. *et al.* (2010) Lead isotopic evidence for an Australian source of aeolian dust to Antarctica at times over the last 170,000 years. *Palaeogeography, Palaeoclimatology, Palaeoecology*, **285** (3–4), 205–223.

Dong, Z.B. (2000) Wind erosion in arid and semiarid China: an overview. *Journal of Soil and Water Conservation*, **55**, 439.

Drake, N. and Bristow, C. (2006) Shorelines in the Sahara: geomorphological evidence for an enhanced monsoon from palaeolake Megachad. *Holocene*, **16**, 901–911.

Eckardt, F.D., Washington, R., and Wilkinson, J. (2001) The origin of dust on the West Coast of Southern Africa. *Palaeoecology of Africa*, **27**, 207–219.

Eckardt, F., Drake, N., Goudie, A.S. *et al.* (2001) The role of playas in pedogenic gypsum crust formation in the central Namib Desert: a theoretical model. *Earth Surface Processes and Landforms*, **26**, 1177–1193.

Engelstaedter, S., Tegen, I. and Washington, R. (2006) North African dust emissions and transport. *Earth-Science Reviews*, **79**, 73–100.

Engelstaedter, S. and Washington, R. (2007a) Atmospheric controls on the annual cycle of North African dust. *Journal of Geophysical Research*, **112**, D03103, DOI: 10.1029/2006JD007195.

Engelstaedter, S. and Washington, R. (2007b) Temporal controls on global dust emissions. *Geophysical Research Letters*, **34** (15), article L15805.

Engelstaedter, S., Kohfield, K.E., Tegen, I. and Harrison, S.P. (2003) Controls on dust emissions by vegetation and topographic depressions: an evaluation using dust storm frequency data. *Geophysical Research Letters*, **30**, 1294, DOI: 10.1029/2002GL016471.

Evan, A.T., Dunion, J., Foley, J.A. *et al.* (2006) New evidence for a relationship between Atlantic tropical cyclone activity and African dust outbreaks. *Geophysical Research Letters*, **33**, L19813.

Galloway, J. N., Thornton, J.D., Norton, S.A. *et al.* (1982) Trace metals in atmospheric deposition: a review and assessment. *Atmospheric Environment*, **16**, 1677–1700. DOI: 10.1016/0004-6981 (82), 90262-1.

Gao, H. and Washington, R. (2009) The spatial and temporal characteristics of TOMS AI over the Tarim Basin,China. *Atmospheric Environment*, **43**, 1106–1115.

Gill, T.E. (1996) Eolian sediments generated by anthropogenic disturbance of playas: human impacts on the geomorphic system and geomorphic impacts on the human system. *Geomorphology*, **17**, 207–228.

Gillette, D.A., Niemeyer, T.C. and Helm, P.J. (2001) Supply-limited horizontal sand drift at an ephemerally crusted, unvegetated saline playa. *Journal of Geophysical Research*, **106** (D16), 18085–18098.

Ginoux, P., Cin, M., Tegen, I. *et al.* (2001) Sources and distributions of dust aerosols simulated. *Journal of Geophysical Research – Atmospheres*, **106** (D17), 20255–20273.

Gong, S.L., Zhang, X.Y., Zhao, T.L. *et al.* (2003) Characterization of soil dust aerosol in China and its transport and distribution during 2001 ACE-Asia: 2. Model simulation and validation. *Journal of Geophysical Research – Atmospheres*, **108** (D9), 4262.

Goossens, D. and Buck, B. (2009) Dust dynamics in off-road vehicle trails: measurements on 16 arid soil types, Nevada, USA. *Journal of Environmental Management*, **90**, 3458–3469.

Goudie, A.S. and Middleton, N.J. (2001) Saharan dust storms: nature and consequences. *Earth-Science Reviews*, **56**, 179–204.

Goudie, A.S. and Middleton, N.J. (2006) *Desert Dust in the Global System*, Springer, Heidelberg.

Harrison, S.P., Kohfeld, K.E., Roelandt, C. and Claquin, T. (2001) The role of dust in climate changes today, at the last glacial maximum and in the future. *Earth-Science Reviews*, **54**, 43–80.

Haywood, J.P., Francis, S., Osborne, M. *et al.* (2003) Radiative properties and direct radiative effect of Saharan dust measured by the C-130 aircraft during SHADE: 1. Solar spectrum. *Journal of Geophysical Research*, **108** (D18), article 8577.

Herman, J.R., Bhartia, P.K., Torres, O. *et al.* (1997) Global distribution of UV-absorbing aerosols from Nimbus 7/TOMS data. *Journal of Geophysical Research*, **102**, 16911–16922.

Herrmann, K., Stahr, K. and Jahn, R. (1999) The importance of source region identification and their properties for soil-derived dust: the case of Harmattan dust sources for eastern West Africa. *Contributions to Atmospheric Physics* **72**, 141–150.

Hesse, P. and McTainsh, G.M. (1999) Last glacial maximum to early Holocene wind strength in the mid-latitudes of the Southern Hemisphere from aeolian dust in the Tasman Sea. *Quaternary Research*, **52**, 343–349.

Holben, B.N., Eck, T.F., Slutsker, I. *et al.* (1998) AERONET – a federated instrument network and data archive for aerosol characterization. *Remote Sensing of Environment*, **66**, 1–16.

Holzer-Popp, T., Schroedter-Homscheidt, M., Breitkreuz, H. *et al.* (2008) Improvements of synergetic aerosol retrieval for ENVISAT. *Atmospheric Chemistry and Physics*, **8** (24), 7651–7672.

Hsu, N.C., Tsay, S.C., King, M.D. and Herman, J.R. (2004) Aerosol properties over bright-reflecting source regions. *Transactions on Geoscience and Remote Sensing*, **42** (3), 557–569.

Hui, W.J., Cook, B.I., Ravi, S. *et al.* (2008) Dust-rainfall feedbacks in the West African sahel. *Water Resources Research*, **44** (5), W05202.

Iversen, J.D. and White, B.R. (1982) Saltation threshold on Earth, Mars and Venus. *Sedimentology*, **29** (1), 111–119.

Jickells, T.D., An, Z.S., Andersen, K.K. *et al.* (2005) Global iron connections between desert dust, ocean biogeochemistry, and climate. *Science*, **308** (5718), 67–71.

Joussaume, S. (1993) Palaeoclimatic tracers; an investigation using an atmospheric general circulation model under ice-age conditions. 1. Desert dust. *Journal of Geophysical Research*, **98** (D2), 2767–2805.

Kalu, A.E. (1979) The African dust plume: its characteristics and propagation across West Africa in winter, in *Saharan Dust: Mobilisation, Transport and Deposition* (ed. C. Morales), John Wiley & Sons, Ltd, Chichester, pp. 95–118.

Kaufman, Y.J., Koren, I., Remer, L.A. *et al.* (2005) Dust transport and deposition observed from the Terra–moderate resolution imaging pectroradiometer (MODIS) spacecraft over the Atlantic ocean. *Journal of Geophysical Research*, **110** (D10), article D10S12, 2005.

Khiri, F., Ezaidi, A. and Kabbachi, K. (2004) Dust deposits in Souss-Massa basin, south-west of Morocco: granulometirc, mineralogical and geochemical characterisation. *Journal of African Earth Science*, **39**, 459–464.

Kohfeld, K.E. and Harrison, S.P. (2003) Glacial–interglacial changes in dust deposition on the Chinese Loess Plateau. *Quaternary Science Reviews*, **22** (18–19), 1859–1878.

Koren, I. and Kaufman, Y.J. (2004) Direct wind measurements of Saharan dust events from Terra and Aqua satellites. *Geophysical Research Letters*, **31**, article L06122.

Koren, I., Kaufman, Y.J., Washington, R. *et al.* (2006) The Bodélé depression: a single spot in the Sahara that provides most of the mineral dust to the Amazon forest. *Environmental Research Letters*, **1** (1), 014005.

Kurosaki, Y. and Mikami, M. (2005) Regional difference in the characteristic of dust event in East Asia: relationship among dust outbreak, surface wind, and land surface condition. *Journal of the Meteorological Society of Japan*, **83A**, 1–18.

Laurent, B., Marticorena, B., Bergametti, G. and Mei, F. (2006) Modeling mineral dust emissions from Chinese and

Mongolian deserts. *Global and Planetary Change*, **52** (1–4), 121–141.

Laurent, B., Marticorena, B., Bergametti, G. *et al.* (2008) Modeling mineral dust emissions from the Sahara desert using new surface properties and soil database. *Journal of Geophysical Research*, **113**, D14218.

Legrand, M., Plana-Fattori, A. and N'doume, C. (2001) Satellite detection of dust using the IR imagery of Meteosat 1. Infrared difference dust index. *Journal of Geophysical Research*, **106** (D16), 18251–18274.

Levelt, R. F. (2002) OMI algorithm theoretical basis document volume 1: OMI Instrument, Level 0-1b processor, calibration and operations, Technical Report, NASA Goddard Space Flight Center, Greenbelt, Md.

Lopez, M.V. (1998) Wind erosion in agricultural soils: an example of limited supply of particles available for erosion. *Catena*, **33**, 17–28.

Luo, C., Mahowald, N.M. and del Corral, J. (2003) Sensitivity study of meteorological parameters on mineral aerosol mobilization, transport, and distribution. *Journal of Geophysical Research*, **108** (D15), article 4447.

Luo, J., Ma, S.H., Chen, S.S. *et al.* (2005) A study on the features of gale in southeast of the Taklimakan Desert. Xinjiang Meteorology 5 (in Chinese).

McConnell, J.R., Aristarain, A.J., Banta, J.R. *et al.* (2007) 20th-century doubling in dust archived in an Antarctic Peninsula ice core parallels climate change and desertification in South America. *Proceedings of the National Academy of Sciences of the United States of America*, **104**, 5743–5748.

McFadden, L.D., Wells, S.G. and Jercinovich, J. (1987) Influences of eolian and pedogenic processes on the origin and evolution of desert pavements. *Geology*, **15**, 504–508.

McGowan, H.A., Kamber, B., McTainsh, G.H. and Marx, S.K. (2005) High resolution provenancing of long travelled dust deposited on the Southern Alps, New Zealand. *Geomorphology*, **69**, 208–221.

McTainsh, G.H., Keys, J.F. and Nickling, W.G. (1999) Wind erodibility of arid lands in the Channel Country of western Queensland, Australia. *Zeitschrift fur Geomorphologie*, NF **116**, 113–130.

McTainsh, G.H. and Lynch, A.W. (1996) Quantitative estimates of the effect of climate change on dust storm activity in Australia during the last glacial maximum. *Geomorphology*, **17**, 263–271.

Mahowald, N.M. and Dufresne, J.L. (2004) Sensitivity of TOMS aerosol index to boundary layer height: implications for detection of mineral aerosol sources. *Geophysical Research Letters*, **31**, article L03103, DOI: 10.1029/2003GLO18865.

Mahowald, N.M. and Luo, C. (2003) A less dusty future? Geophysical Research Letters, **30**, 1903.

Mahowald, N.E., Bryant, R.G., del Corral, J. and Steinberger, L. (2003) Ephemeral lakes and desert dust sources. *Geophysical Research Letters*, **30**, 1074, DOI: 10.1029/2002GL016041.

Marsham, J.H., Parker, D.J., Grams, C.M. *et al.* (2008) Uplift of Saharan dust south of the intertropical discontinuity. *Journal of Geophysical Research – Atmospheres*, **113** (D21), D21102.

Marticorena B. and Bergametti G. (1995) Modelling the atmospheric dust cycle... emission scheme. *Journal of Geophysical Research*, **100**, 16415–16439.

Martonchik, J. V., Diner, D. J., Kahn, R. A. *et al.* (1998) Techniques for the retrieval of aerosol properties over land and ocean using multiangle imaging. *IEEE Transactions on Geoscience Remote Sensing*, **36**, 1212–1227.

Martonchik, J. V., Diner, D. J., Crean, K. A. and Bull, M. A. (2002) Regional aerosol retrieval results from MISR. *IEEE Transactions on Geoscience Remote Sensing*, **40**, 1520–1531.

Meloni, D., di Sarra, A., Di Iorio, T. and Fiocco, G. (2004) Direct radiative forcing of Saharan dust in the Mediterranean from measurements at Lampedusa Island and MISR space-borne observations. *Journal of Geophysical Research*, **109** (Issue D8), CiteID D08206.

Middleton, N.J. (1986) A geography of dust storms in southwest Asia. *Journal of Climatology*, **6** (2), 183–196.

Miller, R.L. and Tegen I. (1998) Climate response to soil dust aerosols. *Journal of Climate*, **11**, 3247–3267.

Milton, S.F., Grered, G., Brooks, M.E. *et al.* (2008) Modeled and observed atmospheric radiation balance during the West African dry season: role of mineral dust, biomass burning aerosol, and surface albedo. *Journal of Geophysical Research*, **113**, D00C02.

Natsagdorj, L., Jugder, D. and Chung, Y.S. (2003) Analysis of dust storms observed in Mongolia during 1937–1999. *Atmospheric Environment*, **37** (9–10), 1401–1411.

Nickling, W.G., McTainsh, G.H. and Leys, J.F. (1999) Dust emissons from the Channel Country of western Queensland, Australia. *Zeitschrift fur Geomorphologie Supplementband*, **116**, 1–17.

O'Hara, S.L., Clarke, M.L. and Elatrash, M.S. (2006) Field measurements of desert dust deposition in Libya. *Atmospheric Environment*, **40**, 3881–3897.

O'Hara, S.L., Wiggs, G.F.S., Mamedov, B. *et al.* (2000) Exposure to airborne dust contaminated with pesticide in the Aral Sea Region. Research Letter. Lancet, **355**, 627–628.

Okin, G.S., Mahowald, N., Chadwick, O.A. and Artaxo, P. (2004) Impact of desert dust on the biogeochemistry of phosphorus in terrestrial ecosystems. *Global Biogeochemical Systems*, **18** (2), 1–9.

Petit, R.H., Legrand, M., Jackowiak, I. *et al.* (2005) Transport of Sharan dust over the Caribbean islands: study of an event. *Journal of Geophysical Research*, **110** (D), DOI: 10.1029/2004JD004748.

Pinker, R.T., Pandithurai, G., Holben, B.N. *et al.* (2001) A dust outbreak episode in sub-Sahel West Africa. *Journal of Geophysical Research*, **106** (D19), 22 923–22 930.

Plumlee, G. S. and Ziegler, T.L. (2003) The medical geochemistry of dusts, soils, and other Earth materials, in *Treatise on Geochemistry*, vol. **9** (ed. B.S. Lollar), Elsevier, New York, pp. 263–310.

Prospero, J.M. and Lamb, P.J. (2003) African droughts and dust transport to the Caribbean: climate change implications. *Science*, **302** (5647), 1024–1027.

Prospero, J.M., Ginoux, P., Torres, O. and Nicholson, S.E. (2002) Environmental characterization of global sources of atmospheric soil dust derived from Niumbus-7 TOMS absorbing aerosol product. *Review of Geophysics*, **40**, 2–32.

Prospero, J.M., Blades, E., Naidu, R. *et al.* (2008) Relationship between African dust carried in the Atlantic trade winds and surges in pediatric asthma attendances in the Caribbean. *International Journal of Biometeorology*, **52** (8), 823–832.

Pye, K. (1995) The nature, origin and accumulation of loess. *Quaternary Science Reviews*, **14**, 653–667.

Reheis, M.C. (2006) A 16-year record of eolian dust in Southern Nevada and California, USA: controls on dust generation and accumulation. *Journal of Arid Environments*, **67**, 487–520.

Reheis, M.C. and Kihl, R. (1995) Dust deposition in southern Nevada and California, 1984–1989: relations to climate, source area, and source lithology. *Journal of Geophysical Research*, **100**, 8893–8918.

Reheis, M.C., Budahn, J.R., Lamothe, P.J. and Reynolds, R.L. (2009) Compositions of modern dust and surface sediments in the Desert Southwest, United States. *Journal of Geophysical Research*, **114**, F01028, DOI: 10.129/2008JF001009.

Reynolds, R.L., Yount, J.C., Reheis, M. *et al.* (2007) Dust emission from wet and dry playas in the Mojave Desert, USA. *Earth Surface Processes and Landforms*, **32**, 1811–1827. DOI: 10.1002/esp.1515.

Rosen, M.R. (1994) The importance of groundwater in playas: a review of playa classifications and the sedimentology and hydrology of playas, in *Palaeoclimate and Basin Evolution of Playa Systems* (ed. M.R. Rosen), Geological Society of America Special Paper 289, Boulder, CO, pp. 1–18.

Rosenfeld, D., Rudich, Y. and Lahav, R. (2001) Desert dust suppressing precipitation: a possible desertification feedback loop. *Proceedings of the National Academy of Sciences of the United States of America*, **98** (11), 5975–5980.

Schepanski, K., Tegen, I., Laurent, B. *et al.* (2007) A new Saharan dust source activation frequency map. *Geophysical Research Letters*, **34**, L18803.

Schepanski, K., Tegen, I., Todd, M.C. *et al.* (2009) Meteorological processes forcing Saharan dust emission inferred from MSG-SEVIRI observations of subdaily dust source activation and numerical models. *Journal of Geophysical Research*, **114**, D10201.

Shao, Y.P., Raupach, M.R. and Leys, J.F. (1996) A model for predicting aeolian sand drift and dust entrainment on scales from paddock to region. *Australian Journal of Soil Research*, **34** (3), 309–342.

Shao, Y.P., Yang, Y., Wang, J.J. *et al.* (2003) Northeast Asian dust storms: real-time numerical prediction and validation. *Journal of Geophysical Research – Atmospheres*, **108** (D22), 4691.

Smith, B.J., Wright, J.S. and Whalley, W.B. (2002) Sources of non-glacial, loess-size quartz silt and the origins of desert loess. *Earth Science Reviews*, **59**, 1–26.

Swap, R., Garstang, M., Greco, S. *et al.* (1992) Saharan dust in the Amazon Basin. *Tellus, Series B*, **44B** (2), 133–149.

Ta, W., Xiao, H., Qu, J. *et al.* (2004) Measurements of dust deposition in Gansu Province, China, 1986–2000. *Geomorphology*, **57** (1–2), 41–51.

Tegen, I., Harrison, S.P., Kohfeld, K. *et al.* (2002) Impact of vegetation and preferential source areas on global dust aerosol. *Journal of Geophysical Research D: Atmospheres*, **107** (D21), article 4576.

Tegen, I., Werner, M., Harrison, S.P. and Kohfeld, K.E. (2004) Relative importance of climate and land use in determining present and future global soil dust emission. *Geophysical Research Letters*, **31** (5), L05105.

Tegen, I., Heinold, B., Todd, M. *et al.* (2006) Modelling soil dust aerosols in the Bodélé Depression during the BoDEx campaign. *Atmospheric Chemistry and Physics*, **6**, 4345–4359.

Textor, C., Schulz, M., Guibert, S. *et al.* (2006) Analysis and quantification of the diversities of aerosol life cycles within AeroCom. *Atmospheric Chemistry and Physics*, **6**, 1777–1813.

Todd, M.C., Karam, D.B., Cavazos, C. *et al.* (2008) Quantifying uncertainty in estimates of mineral dust flux: an intercomparison of model performance over the Bodélé Depression, northern Chad. *Journal of Geophysical Research – Atmospheres*, **113**, D24107.

Torres, O., Bhartia P.K., Herman, J.R. *et al.* (2002) A long term record of aerosol optical thickness from TOMS observations and comparison to AERONET measurements. *Journal of Atmospheric Science*, **59**, 398–413.

Tsoar, H. and Pye, K. (1987) Dust transport and the question of desert loess formation. *Sedimentology*, **34**, 139–153.

Wang, X., Zhou, Z. and Dong, Z. (2006) Control of dust emissions by geomorphic conditions, wind environments and land use in northern China: an examination based on dust storm frequency from 1960 to 2003. *Geomorphology*, **81**, 292–308.

Warren, A., Chappell, A., Todd, M.C. *et al.* (2007) Dust-raising in the dustiest place on earth. *Geomorphology*, **92** (1–2), 25–37.

Washington, R. and Todd, M.C. (2005) Atmospheric controls on mineral dust emission from the Bodélé Depression, Chad: the role of the low level jet. *Geophysical Research Letters*, **32** (17), article L17701.

Washington, R., Todd, M., Middleton, N.J. and Goudie, A.S. (2003) Dust-storm source areas determined by the total ozone monitoring spectrometer and surface observations. *Annals of Association of American Geographers*, **93** (2), 297–313.

Washington, R., Todd, M.C., Engelstaedter, S. *et al.* (2006a) Dust and the low level circulation over the Bodélé Depression, Chad: observations from BoDEx 2005. *Journal of Geophysical Research – Atmospheres*, **111** (D3), D03201.

Washington, R., Todd, M.C., Lizcano, G. *et al.* (2006b) Links between topography, wind, deflation, lakes and dust: the case of the Bodélé depression, Chad. *Geophysical Research Letters*, **33**, L09401, DOI: 10.1029/2006GL025827.

Washington, R., Bouet, C., Cautenet, G. *et al.* (2009) Dust as a tipping element: the Bodélé Depression, Chad. *Proceedings of the National Academy of Sciences of the United States of America*, **106** (49), 20564–20571.

Wiggs, G.F.S. and Holmes, P.J. (2010) Dynamic controls on wind erosion and dust generation on west-central Free State agricultural land, South Africa. *Geomorphology* (in press).

Wiggs, G.F.S., O'Hara, S.L.O., Wegerdt, J. *et al.* (2003) The dynamics and characteristics of aeolian dust in dryland Central Asia: possible impacts on human exposure and respiratory health. *The Geographical Journal*, **169** (2), 142–157.

Woodward, S. (2001) Modeling the atmospheric life cycle and radiative impact of mineral dust in the Hadley Centre climate model. *Journal of Geophysical Research – Atmospheres*, **106** (D16), 18155–18166.

Worster, D. (1979) *Dust Bowl*, Oxford University Press, Oxford.

Xuan, J. and Sokolik, I.N. (2002) Characterization of sources and emission rates of mineral dust in Northern China. *Atmospheric Environment*, **36**, 4863–4876.

Xuan, J., Sokolik, I.N., Hao, J.F. *et al.* (2004) Identification and characterization of sources of atmospheric mineral dust in East Asia. *Amospheric Environment*, **38** (36), 6239–6252.

Xue, Y.K. (1996) The impact of desertification in the Mongolian and the Inner Mongolian grassland on the regional climate. *Journal of Climate*, **9** (9), 2173–2189.

Zender, C.S., Bain, H. and Newman, D. (2003) Mineral dust entrainment and deposition (DEAD) model: description dust. *Journal of Geophysical Research*, **108** (D14), article 4416.

Zha, Y. and Gao, J. (1997) Characteristics of desertification and its rehabilitation in China. *Journal of Arid Environments*, **37** (3), 19–432.

Zhang, X.Y., Arimoto, R. and An, Z.S. (1997) Dust emission from Chinese desert sources linked to variations in atmospheric circulation. *Journal of Geophysical Research*, **102** (23), 28041–28047.

Zhang, X.Y., Arimoto, R., Zhu, G.H. *et al.* (1998) Concentration, size-distribution and deposition of mineral aerosol over Chinese desert regions. *Tellus*, **50** (4), 317–330.

Zhang, X.Y., Gong, S.L., Shen, Z.X. *et al.* (2003) Characterization of soil dust aerosol in China and its transport and distribution during 2001 ACE-Asia: 1. Network observations. *Journal of Geophysical Research*, **108** (D9), 4249.

Zhao, T.L., Gong, S.L., Zhang, X.Y. and McKendry, I.G. (2003) Modeled size-segregated wet and dry deposition budgets of soil dust aerosol during ACE-Asia 2001: implications for trans-Pacific transport. *Journal of Geophysical Research D: Atmospheres*, **108** (23), ACE 33-1–ACE 33-9.

Zhao, T.L., Gong, S.L., Zhang, X.Y. *et al.* (2006) A simulated climatology of Asian dust aerosol and its trans-Pacific transport. Part I: Mean climate and validation. *Journal of Climate*, **19**, 88–103.

Zobeck, T.M. and Van Pelt, R.S. (2006) Wind-induced dust generation and transport mechanics on a bare agricultural field. *Journal of Hazardous Materials*, **132**, 26–38.

21 Wind erosion in drylands

Julie E. Laity

21.1 Introduction

Landscapes of aeolian erosion in deserts vary greatly in scale, from abraded rocks, knobs and uplands to vast regions scoured by the wind. They develop from the interaction of the wind, deflatable sediments and the underlying surface. The resultant landforms include smaller features such as pans, micro- and mesoyardangs and ventifacts, as well as enormous depressions, vast ridge-and-swale systems and exhumed landscapes. Formation times are thought to range from centuries for the smaller features, to millennia for mesoscale features, to millions of years for the largest landforms. Ventifacts are also found in coastal and periglacial environments, and yardangs and ventifacts occur on other planets, indicating the widespread nature of the fundamental erosive processes.

This chapter will examine several of the common landforms in deserts – yardangs, inverted relief and ventifacts – and consider how these features are sometimes interrelated in entire landscapes of erosion. Pans and depressions are considered separately in Chapter 15 because of their widespread significance and linkages to lake basins.

Although recent advances in access to remote imaging, most notably the widely used Google Earth, have given us a greater appreciation of the scale of these landscapes and their interrelationship with aeolian transport systems, fieldwork in remote and inhospitable corners of the Earth remains difficult. Nevertheless, progress in understanding aeolian erosion continues to be made, in part because aspects of this process, such as ventifact formation, may also be observed in other vegetation-free environments. Furthermore, yardang systems and wind-eroded basins are sometimes examined in conjunction with studies of dust generation or climatic and tectonic changes, increasing our understanding of the broader environmental context. An additional incentive for aeolian research is the growing evidence of wind erosion on Mars, manifested by extensive fields of ventifacts and yardangs.

Landforms of aeolian erosion cannot be examined in isolation, as they often form part of larger systems associated with the movement and deposition of sand. Sand flow maps of the Saharan Desert, for example, show patterns of sand movement that extend over thousands of kilometres (Wilson, 1971). At this landscape scale, aeolian environments are divided into sectors of deposition (characterised by dune fields and ergs) and of transportation (where migratory dunes move sand from areas of supply, such as littoral zones or fluvial systems, to sinks of deposition) (Corbett, 1993; Compton, 2006). The movement of saltating grains in transit abrades the landscape, eroding and lowering the surface, and providing an important mechanism by which dust is generated. Deflation removes fine, weathered particles. These transport pathways, marked by basins, ventifacts, inverted relief, yardangs or lag gravels, remain as long as they are not erased by renewed sedimentation.

The products of wind erosion may be driven by the prevailing winds across enormous expanses (Breed *et al.*, 1989; Brookes, 2003b). Where an entire landscape is involved, such as in the Borkou region of Chad (Grove, 1960; Capot-Rey, 1961; Grove and Warren, 1968; Mainguet, 1968, 1970, 1972; Wilson, 1971), the aeolian erosion basins of Namibia (Corbett, 1993) or the northeastern Arabian Peninsula (Neev and Hall, 1992), a suite of aeolian landforms may be generated, dominated by yardangs, but including smaller abraded forms (ventifacts) and zones of surface lowering (deflation hollows and depressions). These integrated landscapes develop in areas of great aridity, where wind is the principal erosional agent. In more humid deserts, such as the Mojave Desert

Arid Zone Geomorphology: Process, Form and Change in Drylands, Third Edition. Edited by David S. G. Thomas

of the southwestern United States, the scale of erosion is much less, but small yardang fields and widespread ventifaction speak of localised wind erosion, much of it relict from an earlier time period.

21.2 The physical setting: conditions for wind erosion

Wind erosion requires extreme aridity, persistent and sometimes strong winds and a supply of abrasive sediment. Aridity limits vegetation cover, thereby allowing the free sweep of the wind and the movement of sediment. The abrasive sand is usually supplied from areas where water actively provides a sediment source, either from coastal zones (Corbett, 1993), river systems that disgorge from mountainous regions, alluvial bajadas or sandy beaches (Wilson, 1971). Its transport may be regionally enhanced by topographic acceleration of the wind.

21.2.1 Processes of aeolian erosion

There are two principle processes involved in wind erosion: abrasion and deflation. *Abrasion* refers to the mechanical wear of rock or sediments by the impact of particles in saltation (Greeley *et al.*, 1984). The removal of loose, fine material from a surface, often pre-weathered by salt weathering or other processes, is referred to as *deflation*. The deflated material is transported as fine grains in atmospheric suspension (Greeley and Iversen, 1985).

Although the terms abrasion and deflation are clearly defined and signify two quite separate processes, they are not always well differentiated in the literature. The terms deflation hollow or deflation basin are commonly used, e.g. even when both abrasion and deflation are involved, with abrasion sometimes dominant. Bristow, Drake and Armitage (2009) refer to the deflation of sediments from the Bodélé Depression, while photographs and accompanying text provide evidence of abrasion on the lakebed. In this sense, deflation is often used to refer to a general lowering of the land surface by wind erosion or to the removal of fine materials produced by abrasion.

The relative significance of abrasion and deflation in landform development is rarely studied and not well understood. In the formation of ventifacts, abrasion is the key player. However, for depressions and yardangs, both processes may be important. A third process of wind erosion, *rock wedging*, has been recorded in Antarctica (Hall, 1989). This process occurs when grains moving at high velocity are packed into cracks, and then are hit by successive impacts to cause a wedging effect. Although it may potentially occur in deserts, it has not been recorded.

21.2.1.1 Abrasion

Ventifacts, abraded surfaces and the bases of yardangs develop within a curtain of saltating sand grains. The development and form of these geomorphic features is thereby dependent on particle fluxes within a zone that ranges from the surface up to a height of about 1–2 m (Hobbs, 1917). Within this general zone of abrasion, there will be a height at which the kinetic energy flux is maximised (Anderson, 1986). In general, particles travelling at greater heights have higher velocities, as wind velocity increases with height and the longer saltation paths allow for greater acceleration by the wind (Greeley *et al.*, 1984). However, the number of particles also diminishes with height and therefore the maximum kinetic energy flux must take into account both particle flux and speed.

At the elevation of the greatest kinetic energy flux, erosion profiles develop with distinct maxima of mass removal (Sharp, 1964, 1980; Wilshire, Nakata and Hallet, 1981; Anderson, 1986; Laity and Bridges, 2009), the height of which is influenced by such factors as wind speed and the degree of grain bounce. For any given surface, the height of maximum abrasion shifts upward as wind velocity increases (Jianjun *et al.*, 2001; Liu *et al.*, 2003). Winds are accelerated in constrictions, such as valleys and passes, and where they ascend over hills or escarpments, promoting erosion in these areas (Laity, 1987). Sand streams erode at greater heights when passing over hard surfaces, which promotes grain bounce (Wilshire, Nakata and Hallet, 1981). For example, the height distribution of windblown sand over Chinese gobi (desert pavement) surfaces reaches up to 2.3 m, whereas above mobile sand beds 95 % of the sand is concentrated in the lower 20 cm (Jianjun *et al.*, 2001). However, even on hard surfaces, the percentage of sand at high elevations is quite low (1–3.4 % in the Chinese example) and therefore other processes are required to explain observed abrasion at elevations greater than $\sim$1 m (Laity and Bridges, 2009). It is likely that abrasion elevations increase as the entire sand surface rises as, for example, when dune sand moves across a hill or through an interyardang passage.

The magnitude of erosion depends on the *susceptibility to abrasion* (Greeley *et al.*, 1984), S_a (a function of rock density, hardness, fracture-mechanical properties, primary texture and shape) and on the properties of the impacting particle (diameter D, density ρ_p, speed V and angle of incidence α) (Greeley *et al.*, 1984; Anderson,

1986; Bridges *et al.*, 2010). The loss of material lost per impact, A, is

$$A = S_a \rho_p (V \sin \alpha - V_0)^n (D - D_0)^m$$

where V_0 and D_0 are the threshold particle speed and diameter that will initiate erosion (Scattergood and Routbort, 1983). Anderson (1986) calculated values of $n = 2$ and $m = 3$ on the basis of abrasion experiments. The loss of material upon impact is roughly proportional to the kinetic energy of the impact. The rate of abrasion then becomes

$$\text{Rate} = s_a q f$$

where q is particle flux and f the wind frequency (Greeley *et al.*, 1984).

21.2.1.2 Deflation

Deflation plays a role in eroding softer rocks and poorly consolidated sediments and therefore influences yardang, pan and depression formation, but not the development of ventifacts. There have been few field or laboratory studies of deflation. Often, its role is suggested by field relationships, where mass loss has occurred, but features typical of sand abrasion (such as grooving) are absent.

Unlike abrasion, the process of deflation is not limited by elevation. Therefore, it can work across the entire surface of a feature, such as a yardang, particularly if other processes operate to weather and loosen material. Salt weathering and wetting and drying processes are important precursors to deflation. Even sandstone can be made friable by salt efflorescence, preparing it for deflation by the wind (Haynes, Mehringer and Zaghloul, 1979).

21.2.2 Yardangs

A yardang is an elongate ridge that shows clear signs of having been eroded by the wind (Hedin, 1903). The yardang may be formed in either hard (basalts, sandstone, granite, dolomite, etc.) or soft (lacustrine) materials, with rock type influencing form, scale and rate of formation. Yardangs vary greatly in scale, from microyardangs (centimetre-scale ridges) (Worrall, 1974), to mesoyardangs (metres in height and length), to megayardangs (tens of metres high and often more than 1000 m in length) (Cooke, Warren and Goudie, 1993; Laity, 2009). Most impressive in scale are the megayardangs (Cooke, Warren and Goudie, 1993; Goudie, 2007), whose dimensions attain heights of 60 m in the Lut Desert (Krinsley, 1970), heights of 100 m and lengths measured in kilometres in the Tablazo de Ica of the Peru–Chile Desert (McCauley, Breed and Grolier, 1977, Beresford-Jones, Lewis and Boreham, 2009) and tens of metres in height and hundreds of metres in length for the larger features in the Western Desert of Egypt (Grolier *et al.*, 1980). Megayardangs, particularly when developed in resistant rock, indicate a wind regime that has remained strong and stable over very long periods of time (Smith, 1963).

The long axes of yardangs parallel the prevailing wind or, more likely, the wind of highest velocity (Hörner, 1932; Krinsley, 1970). Yardangs seldom occur as individual landforms. They typically appear in small groupings or 'swarms', but also occur in immense'fleets', 'fields' or yardang complexes (Figure 21.1). The parallelism of the individual ridges gives a distinctive grain to the topography. Within the fields, yardang axes are largely subparallel, but the whole field may describe a gentle curve, following the direction of the prevailing winds. On the south coast of Peru, dune fields and yardang complexes curve gently through an arc of 180° to converge on the city of Ica (McCauley, Breed and Grolier, 1977). In North Africa, the yardang and ridge-and-swale systems follow, in a great arc, the deflection of the Trade Winds around the Tibesti Mountains (Laity, 2009).

The term yardang has been applied to a broad range of wind-eroded features, some highly streamlined and others much less so. Some authors limit the use of the term yardang to distinctly aerodynamic forms. 'Ridge-and-swale system' has been applied to parallel, unstreamlined forms (most notably those bordering the Tibesti Mountains of the Sahara). These are also referred to as megayardangs (Goudie, 2007). Brookes (2001) prefers the use of the term 'aeolian erosional lineation' (AEL), which incorporates both yardang and ridge systems. In this chapter, the term yardang will be used in a general sense to describe any ridge-like form modified by the wind, whether highly streamlined or not.

21.2.2.1 Yardang form and scale relationships

In their most classic form, yardangs have been described as resembling the 'hull of a racing yacht turned upside down' (Bosworth, 1922, p. 295). The windward face is typically blunt-ended, steep and high, with the leeward end declining in elevation and tapering to a point (Bosworth, 1922; McCauley, Breed and Grolier, 1977; Whitney, 1985) (Figure 21.2). The face is often undercut and sometimes has a moat.

In reality, the form of yardangs varies considerably, from 'classic' forms (Figure 21.2) to those that at first

Figure 21.1 Yardangs of the Qaidam basin, China, are formed from lacustrine sediments. The field has developed through complex cycles of sedimentation and aeolian erosion, in a deflation basin that has alternately filled and lowered (photo courtesy of R. Heermance).

glance are difficult to recognise as yardangs (Figure 21.3) (Halimov and Fezer, 1989). Within any given material, the form of the yardang is influenced by differences in resistance to weathering and erosion, determined by such factors as the degree of cementation, bedding and jointing patterns (Grolier *et al.*, 1980) and by the length of time the yardang has been eroded (Vincent and Kattan, 2006). These factors lead to spatial zonations in yardang form.

Grolier *et al.* (1980) note that one of the critical elements distinguishing yardangs from other hills or inselbergs is that they must appear to be streamlined by the wind and have a length that exceeds the width, usually by a ratio of 3 : 1 or greater. Wind tunnel experiments suggest an ideal length-to-width (L/W) ratio of 4 : 1 (Ward and Greeley, 1984), independent of scale. However, these ideal proportions are probably achieved only after very long periods of time, and actual ratios vary considerably both within and between yardang fields. Length-to-width ratios of yardangs in coastal Peru range from 3 : 1 in the northern part to 10 : 1 in the southern part of the field (McCauley, Breed and Grolier, 1977). In the Sahara, streamlining varies according to yardang size, with larger forms rather irregular in contour and smaller ones displaying a more streamlined and smoothed surface (Smith, 1963). Bosworth (1922) also observed that smaller Peruvian yardangs (<6 m) attain greater perfection of form, with corners and points, which initially offer resistance to the wind, worn off by erosion to create a ‘windform’ shape. Individual freestanding yardangs may be more streamlined than those in groupings (McCauley, Breed and Grolier, 1977).

Figure 21.2 A classic yardang form in the Qaidam basin. The windward face is steep, blunt-ended and high with the leeward end declining to a point. A sand moat lies in front of the face (photo courtesy of R. Heermance).

Figure 21.3 The form of yardangs may vary considerably within a single basin. These pyramidal forms are one of eight morphologies recognised by Halimov and Fezer (1989) in the Qaidam basin (photo courtesy of P. Kapp).

The length-to-width ratio of the yardangs is related to the time available for abrasion and the structure and lithology of the material into which it is cut. Brookes (2003b) illustrates yardangs cut into exhumed meander scrolls in a region of unidirectional winds, where elongate features develop parallel to the long axis of the scrolls and short stubby forms occur where the wind is perpendicular to the grain of the scrolls. Wind tunnel experiments by Ward and Greeley (1984) indicate that if the original form is broad, erosion will decrease the width, whereas for a more elongate form, the dominant change will be a decrease in length. Table 21.1 illustrates the large range in length-to-width ratios that have been identified in different yardang fields.

Table 21.1 Representative length-to-width ratios of yardangs.

Average L : W ratio	*Location*	*Source*
1.5 : 1	Kuwait	Al-Dousari *et al.* (2009)
1.8 : 1	Mongolia	Ritley and Odontuya (2004)
3 : 1	Western Desert, Egypt	Grolier *et al.* (1980)
4 : 1	Rogers Lake, CA, USA	Ward and Greeley (1984)
5 : 1	Central Asia	Halimov and Fezer (1989)
10 : 1	Borkou and Lut Deserts	Cooke, Warren and Goudie (1993)

Lineation and streamlining are often associated with scale-specificity (Evans, 2003). Yardangs within any given field are largely similar in scale and show great periodicity in wavelength (Mainguet, 1970, 1972), but between different areas, the scale varies according to wind activity and intensity, the time span over which erosion has acted and rock type. The crest spacing is similar within any given area: 20–40 m in yardangs formed in soft diatomites and 1.6 km in sandstone megayardangs in the Sahara Desert (Mainguet, 1970; Evans, 2003).

It has been suggested that the spacing of small meso-yardangs may be related to the regular spacing of vegetation (Cooke, Warren and Goudie, 1993). Many bushes accumulate aeolian sediments and develop into nebkhas (vegetation-anchored dunes) that became stabilised over time. They are later eroded by the wind to form yardangs, which inherit the regular spacing of the plants (Capot-Rey, 1957; Mahmoudi, 1977).

Although most yardangs are elongate or tear-dropped in form, reflecting a largely unidirectional wind regime, more unusual shapes can develop when winds come from two different directions. At a site in Farafra, Egypt, small chalk yardangs are abraded by both westerly and north–northwesterly winds to become L- or S-shaped (Donner and Embabi, 2000).

21.2.2.2 *Material properties: influence on yardang form and development*

Yardangs form in many different lithologies. They are perhaps most common and widespread in softer materials, most notably lacustrine sediments. On dry lakebeds, development can be rapid, with mature forms evolving in

Figure 21.4 Cross-section of a yardang in the Qaidam basin illustrating nearly level lacustrine beds with a thick, strong salt capping that drapes across the form. The capping protects the underlying sediments from gullying (photo courtesy of P. Kapp).

a few thousand years. Although the development of a classic aerodynamic yardang shape appears easier to attain in weakly consolidated rock, these materials are subject to other erosional processes, which modify the form, including solution (where salts are present), sheet flow, gullying and mass movement (McCauley, Breed and Grolier, 1977; Al-Dousari *et al.*, 2009). Moreover, the structure of the underlying rocks often has a significant impact on yardang form. Erosion tends to accentuate differences in the degree of cementation, jointing or bedding (Grolier *et al.*, 1980).

Some yardangs have irregular or flat tops, the consequence of more resistant cappings (Grolier *et al.*, 1980). In Kuwait, for example, calcrete caps many of the yardangs (Al-Dousari *et al.*, 2009), creating a flat top. The yardangs of the Qaidam basin commonly show a thick (40–50 cm), strong, salt capping that drapes across the form and appears to protect the underlying horizontally bedded lacustrine sediments (Figure 21.4). There is no evidence of fluvial erosion on this capping (R. Heermance, 2009, personal communication).

Given sufficient time, yardangs develop in hard rocks, including ignimbrite, basalt, dolomite, gneiss, schist and sandstone (Goudie, 1989). In such materials, the yardangs may be of great antiquity and indicate prolonged erosion (Smith, 1963). In form, harder rocks often show aeolian modification at the base and an irregular weathered top.

21.2.2.3 Megayardangs and mesoyardangs: scale variations over distance

Within a given wind regime, yardangs may vary in scale by an order of magnitude. Yardangs of the Western Desert of Egypt are developed in both hard and soft rocks, including lacustrine deposits, sandstones and limestones. 'Kharafish' topography (small-scale, sharp-ridged, intensely fluted yardangs separated by sandy corridors) extends spatially to merge with swarms of larger yardangs (hundreds of metres long and tens of metres high) in a field that is estimated to be about 150 km in length and tens of kilometres wide (Grolier *et al.*, 1980).

Yardangs of the Tibesti region and the Bodélé Depression are formed both from hard, indurated sandstones that appear as vast ridge-and-swale systems between the Tibesti and Ennedi Mountains (Grove, 1960; Capot-Rey, 1961; Grove and Warren, 1968; Mainguet, 1968, 1970, 1972; Laity, 2009) and from much softer, easily erodible lacustrine sediments derived from Lake Megachad (Bristow, Drake and Armitage, 2009). The sandstone ridges are up to 4 km in length (Mainguet, 1972) and 1 km in width, whereas the yardangs formed in diatomite on the floor of Lake Megachad have a scale of development ten times less.

The development of megayardangs requires a unique combination of features that includes hyper-aridity (rainfall <50 mm/yr), a unidirectional or narrowly bimodal wind and relatively homogeneous rocks without complex structures. They are absent in sites of active dune accumulation, in areas with large alluvial fans, in mountainous regions or where an integrated drainage system is present (Goudie, 2008). It is possible that megayardangs may take millions of years to form.

21.2.2.4 Interyardang corridors

The yardangs are separated by long, shallow depressions, variously referred to as swales, couloirs, boulevards or corridors. Although attention is often focused on the yardang form, it is likely that most of the erosion

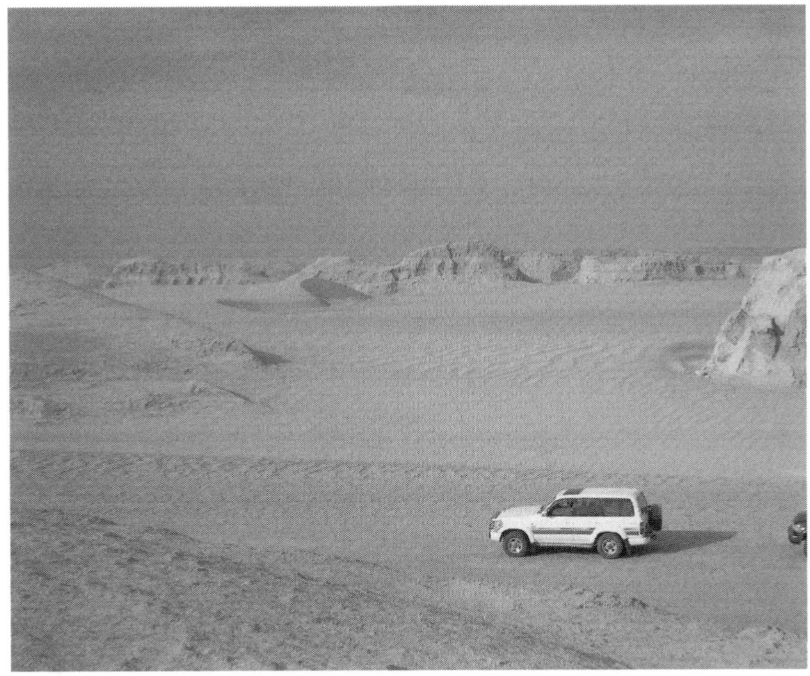

Figure 21.5 The corridors in yardang fields are cut by migrating sand, ripple trains and barchans. Location: Qaidam basin (photo courtesy of R. Heermance).

occurs in the corridors, with the yardang form adjusting to this new 'base level'. The yardang troughs act as conduits for the transport of sand (and sometimes water) from one region to another. In the Lut Desert, for example, sand is moved from the north to the south through the yardang field, trailing out into sand ridges at the southern margin of the field (Krinsley, 1970). Over time, troughs widen to become corridors (Brookes, 2001).

The troughs and corridors range in cross-sectional form from U-shaped to flat-floored (McCauley, Breed and Grolier, 1977), becoming increasingly level as the yardangs become more widely spaced (Blackwelder, 1934; Krinsley, 1970). Flattening may also be the result of erosion that has met more resistant strata. The troughs may be partly or completely buried with sand, which is present in the form of ripples, megaripples or dunes (Krinsley, 1970; Mainguet, 1970; McCauley, Breed and Grolier, 1977; Grolier *et al.*, 1980; Al-Dousari *et al.*, 2009). The ripple trains sometimes diverge at the head of the yardang and converge in the downwind direction around the flanks (McCauley, Breed and Grolier, 1977). Migrating barchans, with axes paralleling the ridges and resultant winds, also signify the movement of sand through the yardang field (Figure 21.5). The corridor floors frequently show marks of aeolian erosion, including longitudinal striations and shallow erosional basins (Mainguet, 1972; Al-Dousari *et al.*, 2009). A patina of desert varnish darkens the ridges on the southeastern flanks of the Tibesti Mountains, whereas the intervening corridors are light in tone owing to abrasion by moving sand and barchan dunes (Mainguet, 1972).

In extremely arid regions, erasure by wind action removes any morphological evidence of fluvial erosion from the corridors (Mainguet, 1970; Grolier *et al.*, 1980). In other areas, hybrid corridor forms develop that incorporate both water and wind erosion characteristics (McCauley, Breed and Grolier, 1977; Bailey *et al.*, 2007); e.g. the troughs may retain a degree of sinuosity as a relict of their former status as a watercourse (McCauley, Breed and Grolier, 1977).

21.2.3 Yardang formative processes

21.2.3.1 Environmental requirements for yardang formation

The following environmental requirements for yardang formation have been identified:

1. Strong, unidirectional or seasonally opposing winds (with one direction stronger than the other). Mean winds commonly equal or exceed the threshold velocity for sand movement, with higher gusts common, particularly in the windy season (Table 21.2).

2. Great fetch over exposed terrain (Grolier *et al.*, 1980).

Table 21.2 Climatological conditions for yardang development.

Location	*Wind velocity*	*Rainfall*	*Source*
Um Al-Rimam Depression, Kuwait	Mean at 10 m: 4.5 m/s; maximum recorded 38 m/s	Mean: 112 mm/yr	Al-Dousari *et al.* (2009)
Namibia	22–28 m/s (gusts at diurnal peak, October to March)	5–20 mm/yr	Corbett (1993)
Rio de Ica rock plateau, Peru	Unimodal from south: mean: 7.5–9 m/s; maximum gusts: 22.7–31.9 m/s	Mean: 0.3 mm/yr	Beresford-Jones, Lewis and Boreham (2009)
Lut Desert, Iran	Maximum average velocity, April: 9.35 m/s	Mean: <10 mm/yr	Ehsani and Quiel (2008)
Lop Depression		Mean: <20 mm/yr	Songqiao and Xuncheng (1984)
La Pacana crater, Atacama Desert		Mean: <10 mm/yr	Bailey *et al.* (2007)
Bodélé Depression, Chad	Mean from northeast: 6–8 m/s	17 mm at Faya Largeau	Goudie and Middleton (2001)

3. Aridity and an absence of vegetation. Annual rainfall is usually less than 100 mm and often less than 10 mm (Table 21.2.)

4. A pre-existing lineament of weakness that the wind can exploit (gullies, joints, etc.).

5. A supply of abrasive sediment: relative to most dune environments the region is 'sand poor' (McCauley, Breed and Grolier, 1977).

6. For yardangs formed on lakebeds, a low water table.

21.2.3.2 Initial conditions and yardang development

Yardangs are probably initiated during arid climatic phases when erosive winds, charged with sand, exploit the axial trends of pre-existing lineaments, most commonly relict fluvial dissection systems (Hörner, 1932; Blackwelder, 1934; Krinsley, 1970; McCauley, Breed and Grolier, 1977; Laity, 1994, 2009; Brookes, 2001; Bailey *et al.*, 2007; Al-Dousari *et al.*, 2009; Bristow, Drake and Armitage, 2009), joint systems (Mainguet, 1970; Neev and Hall, 1992; Vincent and Kattan, 2006), as for instance in sandstones, or cooling cracks in volcanic deposits (Bailey *et al.*, 2007). For yardangs of the Arabian Peninsula and the Lut Desert, Neev and Hall (1992) argue for tectonic influences, suggesting that the linear trends of yardangs, as well as their locations and patterns, are strongly influenced by pre-existing swarms of fractures.

When air enters a passageway established by a channel or fracture, it accelerates, carrying with it sand that abrades the bottom and sides of the trough, causing progressive steepening of the yardang slopes (Blackwelder, 1934). The wind (and occasionally water) erosion causes the passages to erode more rapidly than the yardangs (Halimov and Fezer, 1989). Over time, the troughs widen and eventually breach the ridges at places of weakness. As abrasion intensifies on the yardang prows, the ridges become shorter and smaller (Halimov and Fezer, 1989). Mass wasting, weathering and fluvial erosion reduce the yardang summits as the troughs deepen by erosion (Brookes, 2001).

The end product of aeolian erosion in yardang fields may be a plain. Over a distance of 75 km, Vincent and Kattan (2006) document a downwind progression from sandstone mesas and eroded canyons to megayardangs, yardangs, rock pavements and ultimately dunes. The megayardangs are up to 40 m in height and hundreds of metres in length. As sand supply and exposure increase downwind, yardangs decline in height to less than 25 m. Further downwind, they become even smaller, until the landscape transitions into a planar surface composed of hundreds of square metres of abraded bedrock, upon which can be seen the slightly raised remnants of ancient yardangs. The rock pavements are believed to be extensions of the yardang corridor floors. The downwind decline in yardang size may be related to abrasion rates, which are affected by the downwind increase in the available sand load. Finally, the pavement ends in a dune field, where the crests of barchanoid forms are oriented perpendicular to the yardangs.

It is sometimes possible to determine the original topographic surface that existed before yardang formation.

Within a given region, yardang summit elevations may be protected by cappings such as carbonates or reg derived from earlier fluvial deposits. Where summit elevations are regionally accordant, such as the 60 m elevations of the Lut Desert yardangs, earlier surfaces are suggested (Krinsley, 1970).

21.2.3.3 The role of the water table

In yardangs formed on playas, the height of the water table plays a major role in yardang formation (Haynes, 1980). As the water table rises and falls in response to climatic change, yardangs may be destroyed or formed. For instance, yardang formation in association with the excavation of the Lut Formation, Iran, was initiated by a lowering of the water table (Krinsley, 1970). In tectonically active areas, regional faulting may play a role by modifying the terrain and changing water flow and groundwater levels.

21.2.3.4 Formative processes: wind, water, mass wasting and weathering

Yardangs are formed by two fluids – wind and water. The significance of each of these processes varies both spatially and temporally. From a temporal perspective, the relative significance of wind and water vary as the yardangs form and evolve. Fluvial erosion is often deemed critical to the initial dissection of the landscape; only in later stages does the wind become dominant (Hörner, 1932; Vincent and Kattan, 2006). Furthermore, both rainfall and erosive winds have a seasonal component and one may erase the action of the other over the course of a year. Moreover, longer-scale climatic changes involving both wetter and drier climates than today profoundly shape the landscape (Haynes, 1980, 1982), changing basins from depositional to erosional systems.

Hörner (1932) details the interaction of wind (both abrasion and deflation), water and mass movement processes in his discussion of the yardangs of Lop Nor. Rainfall creates strong films of silty material on yardang surfaces, which protects them from deflation. At the foot of the yardang, the attack of drifting sand undermines the rock and unsupported blocks fall down to form talus. The silty film no longer protects the sediments revealed by block failure and loose particles of fine sand are deflated by the wind. The remains of copses on some yardang surfaces provide evidence that the area was previously better watered (Hörner, 1932).

Water erosion

Although yardangs are largely considered to be wind erosion features of extremely arid regions, both short- and long-term climatic fluctuations mean that water erosion inevitably plays a major role in yardang formation and modification. Fluvial erosion often provides the original lineations that are subsequently exploited by the wind; rain modifies the surface of the form by gullying and solution processes (Figure 21.6); interyardang floods modify and steepen the yardang sides (Figure 21.7); and fluctuations in the water table determine the depth to which yardangs are excavated in soft sediments. Thus, the yardang field

Figure 21.6 Rain modifies the surface of yardangs and gullying is commonly observed. These features may be obliterated by future aeolian corrasion. Location: Roger's Lake, Mojave Desert, California.

Figure 21.7 Yardangs of the Qaidam basin are formed from lacustrine sediments. This photograph illustrates flooding of part of the yardang complex. The field has developed through complex cycles of sedimentation and erosion, in a deflation basin that has alternately filled and lowered (photo courtesy of R. Heermance).

and the forms themselves are the result of the constant interplay between wind and water. Indeed, during climatically wetter periods, earth flows, solution and gullying may be of sufficient magnitude that aeolian features are masked (Krinsley, 1970).

Studies of contemporary yardang fields show that short-term changes to the surface of yardangs can be extensive during infrequent rains and floods. Yardangs may experience seasonal fluctuations, with gullies deepening and extending in the rainy season and aeolian processes dominating at other times of the year (Krinsley, 1970). In the Lut Desert of Iran, the upper surfaces of many yardangs are extensively gullied (see, for example, Figures 88, 89 and 90 in Krinsley, 1970, Part II) and may even be cut into buttes with vertical cliffs (Figure 21.8). The degree to which water or wind erosion affects the yardang form varies spatially within the field. The northern yardangs are more frequently inundated by floodwater and thus the hills have near-vertical cliffs, which are the result of floodwater erosion at the base, with salt stains marking the extent of these events (Krinsley, 1970). The steep lower slopes are further modified by wind abrasion. Moving southward in the field, the yardangs are increasingly modified by the wind, until they grade into sand ridges (Krinsley, 1970). The yardings of the Lut Desert are discussed in more detail in Box 21.1.

The long-term (1–2 Ma) interactions of fluvial and aeolian processes in modifying ridge and channel systems on an ignimbrite sheet in La Pacana Caldera, Atacama Desert, are reflected in the morphology of the features. Today's average rainfall is <10 mm, but the region has experienced wetter periods in the recent geologic past. Bailey *et al.* (2007) use a classification system that assumes that relatively straight ridges are associated with wind erosion and sinuous channels with fluvial erosion. A range of features were found, marking the relative significance of wind and water according to location: (1) features dominantly eroded by fluvial processes, (2) fluvial channels with some aeolian modification, (3) complex forms, resulting from fluvial and aeolian erosion, and (4) forms that were dominantly formed by aeolian processes (Bailey *et al.*, 2007).

Wind erosion: abrasion and deflation

Abrasion removes mass from the yardangs by a 'sand-blasting' process. Evidence includes polish, intense fluting and small-scale erosional lineaments on yardangs, generally within the lower one to three metres where saltation is most active, and the development of a steep, often undercut, upwind face (a re-entrant form) (Hobbs, 1917; Bosworth, 1922; Hagedorn, 1971; Grolier *et al.*, 1980; Donner and Embabi, 2000). The corridors allow the passage of sand as ripple or dune forms. Trailing ridges or streams of sand can extend from the down-wind end of yardangs (Bosworth, 1922; Smith, 1963).

Figure 21.8 Yardang in the Lut Desert, Iran. Yardangs in the northern part of the Lut Desert are modified by solution, solifluction, gullying and wind erosion and lack the streamlined form of the 'classic' yardang. The surfaces of many yardangs are extensively modified by water and may be cut into buttes with vertical cliffs (photo courtesy of Majid Karimpour Reihan).

Box 21.1 The mega yardangs of the Lut (Loot) Desert

The lowest elevation in Iran, and the most arid region of the central plateau, is the Lut Desert (Farpoor and Krouse, 2008), an area of about 80 000 km^2, which is home to yardangs, sand dunes, regs and playas. The Lut Desert is extremely dry, with an average annual rainfall of less than 10 mm, and many years may pass without precipitation. In 2004, the Lut Desert was described as the 'thermal pole of the Earth', with the hottest land surface temperature (LST) recorded (68 °C) (Mildrexler, Zhao and Running, 2006).

Yardangs occupy much of the west-central Lut in a 150 km long by 50 km wide field. The north–northwest trending yardangs cut across late Pleistocene lakebeds at altitudes that range from 100 m in the north and east to 404 m above sea level in the central and southeastern region (Ehsani and Quiel, 2008). The yardang field is bounded to the north and south by transverse faults, which, together with longitudinal bounding faults, enclose a depression in the shape of a parallelogram (Neev and Hall, 1992). The surrounding mountain ranges rise to heights of more than 3000 m. The wind is channelled into the graben, flowing northward in winter and southward in summer. At the southern end of the field, the north–northwest trending yardangs diverge to extend parallel to the east–southeast trending transverse fault, eventually merging with it. Neev and Hall (1992) suggest a genetic relationship between active tectonic processes and the yardang field. Longitudinal joints, fractures and faults offer less resistance to wind erosion owing to soil crushing processes. The wind is channelled along them, thereby reinforcing wind erosion.

The yardangs are yellowish-red in colour and sculptured into strange shapes that are sometimes referred to as *Kalut* or desert cities (Gabriel, 1938) (Figure 21.8). The ridges are up to 80 m high and are separated by troughs more than 100 m in width (McCauley, Breed and Grolier, 1977). The corridors are eroded by sand and dunes (Gabriel, 1938; Krinsley, 1970; Alavi Panah *et al.*, 2007), moving from the north to the south through the field (Krinsley, 1970).

The extensive yardang field is clearly visible on satellite imagery and represents erosional remnants of the original valley surface. The yardangs are orientated at 330° (northwest to southeast), parallel to the prevailing winds, which reach their greatest intensity in April, with an average speed of 9.35 m/s (Ehsani and Quiel, 2008). The yardangs are formed from Pleistocene lake deposits (Bobek, 1969), formed from easily erodible clays, which include illite, kaolinite, montmorillonite and chlorite (Farpoor and Krouse, 2008). The presence of palygorskite in different layers in the yardangs suggests an evaporite origin for the materials forming the huge yardangs (Farpoor and Krouse, 2008). In addition to wind erosion, cracking, piping, rilling, slumping and salt weathering modify the surface, suggesting that even in this very arid environment both wind and water are essential to the overall development of yardangs. Fluvial processes dominate during the rainy season, allowing gullies to grow, but aeolian processes prevail in other seasons. The yardangs are more affected by flooding and water erosion in the northern part of the field, with wind streamlining increasingly important in the southern section (Krinsley, 1970).

The movement of sand is episodic, and thus the corridors may be temporarily free of sand even in a wind-corrasion landscape.

The re-entrant form at the front of the yardang reflects the kinetic energy-flux profile for saltating sand. The lower part of the yardang profile mirrors that of ventifacts, indicating that similar processes are operating.

The importance of deflation as a process is not well documented, relying largely on anecdotal evidence of material loss from yardang flanks (Bosworth, 1922). Deflation removes loose surficial material, including unconsolidated sediments or grains that have weathered out. Its impact on yardang formation is greatest in poorly consolidated materials. On the basis of tree root exposures in Mongolia, Ritley and Oduntuya (2004) infer that deflation has removed up to 67 cm of material on one yardang and up to 70 cm in the interyardang corridors. Muddy yardangs in Kuwait lose mass by deflation of thin flakes of a surficial crust (Al-Dousari *et al.*, 2009). However, in some areas, strong surficial salt crusts bind poorly consolidated sediments together and limit deflation (Figure 21.4) (Halimov and Fezer, 1989). Studies of yardangs in the coastal desert of Peru by Bosworth (1922) and McCauley, Breed and Grolier (1977) suggested that abrasion is most important in the troughs and deflation on the flanks and crests. For the small and relatively low field of yardangs at Rogers Lake, California, opinion is divided: McCauley, Breed and Grolier (1977) felt that the streamlined shape and smooth ridge crests were attained largely by deflation, whereas Ward and Greeley (1984) considered abrasion the most important process, dominating trough formation and initial sculpting, with deflation helping to maintain the finished form, and Blackwelder (1934) argued that abrasion by saltating grains both over and around the yardangs shaped the forms.

21.2.3.5 *Secondary processes*

Chemical weathering, solution features and salt weathering

Chemical weathering of yardang tops, solution features and salt weathering are widely reported on yardangs, but there has been little substantive investigation of these features. The presence of salts in deserts, particularly in association with lacustrine sediments, means that salt solution processes and salt weathering play an important role in yardang formation and modification. The Lut Formation of the Lut Desert, for example, consists of horizontally bedded lacustrine sediments, 135–200 m in thickness (Krinsley, 1970), that include gypsum and other more soluble sulfate salts (Farpoor and Krouse, 2008). Elongate depressions, commonly filled with salt, are orientated at 333°, parallel to the prevailing wind and yardang axes (Krinsley, 1970).

Solution features on yardangs are not uncommon and help to elucidate the climatic history of a region and the ongoing seesaw of wetter and drier climates that affect yardang formation. They include honeycomb weathering (such as the 20–42 cm diameter alveoles in Kuwait) and deeper solution cavities (Al-Dousari *et al.*, 2009). Such features may be best developed on the leeward slopes, away from the abrasive action of the windward face. Extensive honeycomb weathering features are considered indicative of wetter conditions in the past, and their partial obliteration by aeolian abrasion on the lower yardang slopes points to a return to drier conditions (Vincent and Kattan, 2006). The strong chemical weathering of Egyptian limestone yardangs in more humid climates of the past produced solution features with interconnecting cavities (Grolier *et al.*, 1980). At the present time, relict yardangs of Precambrian dolomite in the southern Namib deflation basin are being modified by solution (Corbett, 1993).

Mass movement

Mass movement of various types modifies yardang slopes. Where soils are saline and clays swell and shrink, creep processes operate and material moves slowly downslope (Krinsley, 1970). Slump processes remove greater masses of material and may substantively change yardang form. Block failure produces talus around the base of yardangs. The blocks can be undermined by either wind or water erosion and failure may be further facilitated where joints are present (Hörner, 1932). Sandblasting of the windward slope is a common cause of undercutting and subsequent rockfalls, helping to maintain the steep inclination of the face (Al-Dousari *et al.*, 2009). Ritley and Oduntuya (2004) found that slump blocks ranged from minor forms that had little effect on the overall yardang form to much larger blocks that incorporated up to 20 % of the yardang's entire volume. South of Dakhla, Egypt, talus masks up to the lower one-third to one-half of 10-m high yardangs (Brookes, 2003b).

21.2.3.6 Yardang age, rate of formation and role as directional indicators

A wide range in yardang ages is reported, resulting from differences in material erodibility, changing climate and time available for formation. As seen in Table 21.3, the majority of yardangs formed in softer sediments probably developed in less than 2500 years. Small bedrock yardangs may develop in less than 10 000 years, but megayardangs probably take millions of years to develop fully.

The development of yardang fields over geologic time is probably more complex than suggested by many preliminary studies. Climate change influences the supply of abrasive sediment, wind strength and direction, and the infilling of yardang systems by sand and lacustrine deposits. In Namibia, subtle changes in Holocene sea level affect the supply of beach sand that is driven inland to feed an aeolian corridor (Corbett, 1993; Compton, 2006, 2007). In deflation basins associated with lakebeds, cycles of erosion and sedimentation alternately lower and fill basins (Washington *et al.*, 2006). Geologic and stratigraphic investigation of the Qaidam Basin, China, has revealed palaeoyardangs that were submerged during wetter periods by lacustrine sediments, then revealed by modern erosion as the sediments are cut into new yardang fields (Heermance, 2009, personal communication). Wind strength and frequency varies not only seasonally and spatially, but also over time frames of thousands of years. In northern Chad, for example, modern sand-laden winds blow 8 out of 12 months from the northeast, at daytime velocities of 6 -8 m s^{-1} or greater. The Bodélé low-level jet

Table 21.3 Yardang ages and rates of formation.

Location	*Age*	*Material*	*Yardang scale*	*Erosion rate (if known)*	*Source*
Northwestern Arabia	At least 400 000 years	Cambro–Ordovician sandstone			Vincent and Kattan (2006)
Um Al-Riman Depression, Kuwait	44 to 1500 years	'Muddy yardangs' formed in Quaternary playa sediments	Small (0.2–5.4 m in height)	0.5 cm/yr on flanks; 1 cm/yr at headward end	Al-Dousari *et al.* (2009)
Rogers Lake, Mojave Desert, California		Playa shoreline deposits	Small	Headward erosion: 2 cm/yr Lateral: 0. 5 cm/yr	Ward and Greeley (1984)
Qaidam basin, China	Formed in less than 1500–2000 years	Silty sediment			Halimov and Fezer (1989)
Lop-Nor region, China	Formed in less than 1500 years	Soft sediments		0.2 cm/yr	Hörner (1932), McCauley, Breed and Grolier (1977)
Western Desert, Egypt	Formed in less than 2000 years	Unconsolidated sediments	~4 m	0.2 cm/yr	Brookes (2003a)
Eastern Sahara	Several 1000 years	Soft playa muds	Small (several m in height)		Haynes (1980)
Bodélé Depression, Chad	Formed within 1200 to 2400 years	Diatomites	Small (4 m high)		Washington *et al.* (2006); Bristow, Drake and Armitage, (2009)
Payun Matru volcanic field, Argentina	Formed in less than 10 000 years	Basalt	2–3 m in height		Inbar and Risso (2001)
La Pacana Caldera	1–2 Ma	Ignimbrite sheets			Bailey *et al.* (2007)
Namibia and Sahara: bedrock yardangs	Estimate of hundreds of thousands to millions of years	Bedrock			Hagedorn (1968); Mainguet (1970); Goudie (2007)

(LLJ) increases the frequency and magnitude of episodes of wind erosion, and simulation models suggest it would have been even stronger during the Last Glacial Maximum (LGM) (Washington *et al.*, 2006).

Like ventifacts, yardangs are directional structures that are important geological indicators of modern and palaeowind directions. Donner and Embabi (2000) observed abrasional features on ventifacts and yardangs at 10 sites in the Western Desert of Egypt and concluded that east–west alignments reflected the southward displacement of the mid-latitude westerlies in a pre-Holocene cold arid phase (20 000 to 10 000 y BP). Brookes (2003a) concluded that the cross-cutting feature alignment reflected Early Holocene west and northwest flows and Late Holocene north and northeast flows.

21.2.3.7 Yardangs and desert dust

The abrasional and deflational processes associated with the formation of yardangs and their corridors bring about an attendant increase in atmospheric dust (Chapter 20), particularly when the surface material is composed of readily erodible lacustrine sediments. Studies of the main source regions for global dust indicate that many of them are large basins of internal drainage, with the Bodélé Depression of North Africa alone responsible for between 6 and 18 % of global dust emissions (Todd *et al.*, 2007). Within this depression, the palaeolake diatomite sediments of Lake Chad are moulded into yardangs as a result of atmospheric conditions conducive to wind erosion, including a strong Bodélé low-level jet (Washington and Todd, 2005; Schwanghart and Schütt, 2008), topographic channelling (Mainguet, 1996) and very gusty surface conditions (Engelstaedter and Washington, 2007; Washington *et al.*, 2006). The northeast Harmattan winds are funnelled and intensified as they pass between the Tibesti and Ennedi Mountains in northern Chad.

The deflation of material during yardang formation produces dust that is subsequently deposited elsewhere. An important geomorphic question is the relative contribution of desert dust derived from yardang fields to the loess deposits that are widespread in central Asia. Sun (2002) investigated the origin, age and provenance of loess in high mountain areas of China and concluded that the deflation of old lacustrine deposits, including the yardang-rich Junggar and Tarim Basin, provided only a minor source of silt-sized particles.

21.2.3.8 Global yardang systems

Yardangs occupy a very small percentage of the Earth's surface, as the environmental conditions required for their formation are relatively rare. Although they occur in desert regions on all continents, there is considerable variation in their scale, development and spatial extent. Comprehensive discussions of the Earth's major yardang fields are provided in McCauley, Breed and Grolier (1977) and Goudie (2007). The following section provides some additional details.

There are multiple groupings of yardangs in Asia, many of which have not been described fully in the literature. Chinese yardangs range from mesoyardangs to megayardangs. In the northwest Junggar Desert and the northeastern Taklimakan they cover an area of less than 1 % (Sun, 2002). Yardangs in the Lop Desert, or Lop Depression, located in the Tarim basin, were made famous by Hedin (1903). They are developed in old lake and alluvial sediments (Hörner, 1932; Goudie, 2007), where rainfall is generally less than 20 mm/yr, vegetation is sparse (<10–30 % cover) and there is relatively little relief (Songqiao and Xuncheng, 1984). Megayardangs, 20–25 m in height, occur in the Qaidam basin (Figure 21.9). A small (5 km^2) field of yardangs has been described in Mongolia, formed in weakly consolidated sandstones and mudstones among a group of dome dunes (Ritley and Oduntuya, 2004).

Several areas of documented yardangs are found in South America, including classically streamlined forms near the Ica Valley of Peru (up to 1 km in length) (Bosworth, 1922; McCauley, Breed and Grolier, 1977; Beresford-Jones, Lewis and Boreham, 2009), yardangs formed on ignimbrite sheets of La Pacana Caldera, Chile (Bailey *et al.*, 2007), and megayardangs up to 10 km in length in Holocene basaltic flows in the Payun Matru volcanic field in the southern Andes Mountains, Argentina (Inbar and Risso, 2001).

As discussed earlier, yardangs are extensive in the Sahara Desert. In southern Africa, they are prominent features in coastal Namibia (Corbett, 1993; Compton, 2007; Goudie, 2007). As early as 1887, Stapff described yardangs formed from bedrock in the Kuiseb Valley as 'aerodynamic landforms'.

Yardang development in North America, Australia and Europe is very limited in scale and extent. In North America, a small but well-studied group of yardangs occurs at Rogers Lake, California (Blackwelder, 1934; Ward and Greeley, 1984). Australia's deserts have extensive dune fields, but appear to lack the aridity and transport of sand required to form yardangs (Twidale, 1994; Goudie, 2007). Only minor examples have been reported (West *et al.*, 2009). Europe has a small, relict grouping of yardangs in the semi-arid Ebro Depression of Spain (Gutiérrez-Elorza, Desir and Gutiérrez-Santolalla, 2002).

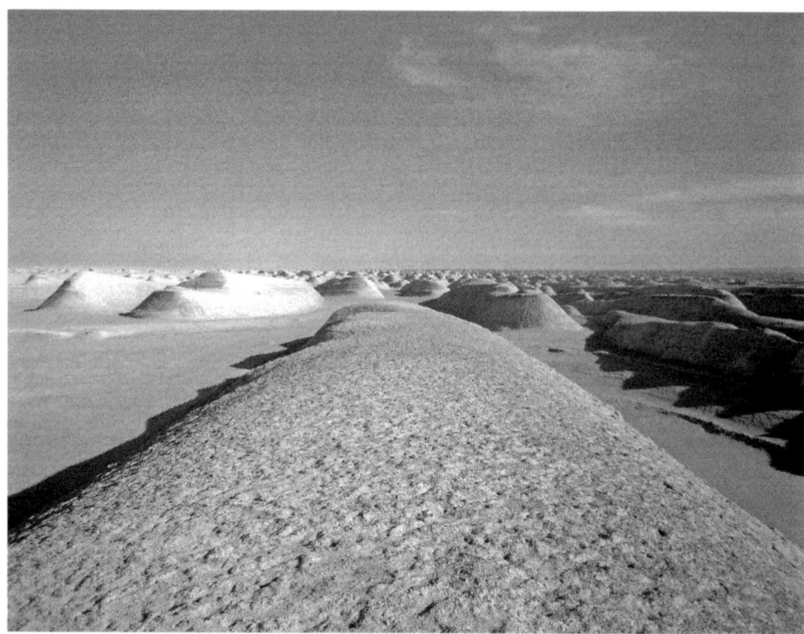

Figure 21.9 Megayardangs of the Qaidam basin, capped by a salt crust, occur in a vast swarm (photo courtesy of R. Heermance).

21.2.4 Inverted topography

A landscape develops inverted relief when previously low areas, such as river channels, are left standing in relief by wind erosion at later stages of climatic and topographic development. There is little detailed documentation of this landform and therefore this section will largely be inventory examples of these landforms. In deserts, channels are commonly more resistant than the surrounding rocks owing to induration by chemical precipitates. In the intensely arid south coast of Peru, for example, the duracrete enrichment of palaeochannels causes them to stand in relief on deflated terraces (Beresford-Jones, Lewis and Boreham, 2009). The best-studied inverted relief is the Plio-Pleistocene raised channel systems of the western Sharqiya (Wahiba), Oman. On the western edge of the Wahiba Sands, these complex high-standing palaeochannel systems cross alluvial fans that have been extensively lowered by deflation (Maizels, 1987).

Inverted relief is most commonly found within yardang fields. In the yardang landscape of the Lop-Nor basin, China, former river courses with resistant silty beds are inverted and marked by ridges and remnant hillocks (Hörner, 1932), whereas soft, young sediments have been eroded away by the wind. Similarly, in the hyper-arid region south of Dakhla, Egypt, intense aeolian erosion has brought exhumed meander scrolls into sharp relief and cut them into yardang fields (Brookes, 2003b). Whitney (1985) illustrates ancient stream systems standing in relief within yardang swarms on the limestone plateau in the Western Desert of Egypt. Inverted relief in the Bodélé Depression, Chad, is interpreted as a deltaic distributary system (Bristow, Drake and Armitage, 2009), exposed by up to 4 m of deflation.

21.2.5 Ventifacts

Ventifacts are wind-eroded rocks, characterised by their distinctive morphology and texture. They are found in periglacial, coastal and desert settings –environments with an ample supply of abrasive particles, limited vegetation cover and strong winds (Laity, 1994, 2009). Many ventifacts, particularly in periglacial and desert regions, are fossil in nature, the product of earlier surface and climatic conditions. Relict ventifacts commonly appear weathered, dulled, stained, lichen-covered or partly exfoliated (Blackwelder, 1929; Smith, 1967, 1984), darkened by rock varnish or covered by patches of grooved and fluted surfaces, with intervening areas eroded by weathering loss (Powers, 1936).

Research into ventifacts, which had been sporadic in the past, has gained momentum in recent years (Knight, 2008; Laity, 2009; Bridges *et al.*, 2010). Ventifacts are seen not as mere geological curiosities but as a link in understanding sediment movement and climate change, as part of a larger continuum of aeolian studies and as a proxy for wind direction in terrestrial and planetary studies.

21.2.5.1 Erosional forms

Unlike desert depressions or yardangs, which develop by several interacting processes, usually involving both water and wind, the sole mechanism of mass removal in ventifacts is abrasion (although weathering in between erosional episodes may alter surface characteristics). Abrasion has the potential to form one or more of several key features on a ventifact, depending on the rock type and size and duration of exposure: facets, polish (similar to glacial polish) and features (pits, grooves, flutes and so on).

A *facet* is a planar surface that evolves at right angles to the wind: ventifacts may develop more than one facet, which often join along a sharp ridge or *keel* (Figure 21.10(a)). The number of keels (kante) has been used to describe small ventifacts as einkante, zweikanter, dreikanter (one-, two-, three-ridged) and so on (Bryan, 1931). Much early research was devoted to the morphological classification of ventifacts (Bryan, 1931; King, 1936; Czajka, 1972). Multiple facets develop in response to: (1) the shifting and overturning of the rock, particularly if it is small, and (2) bidirectional or multidirectional winds (Figure 21.10(b)). If the rock is large and stable, it provides an excellent indicator of regional wind direction.

One of the more common features recorded for ventifacts is smoothing and *polishing* of the rock surface, both on the facets and within the flutes and grooves (Maxson, 1940). The sheen on ventifacts in periglacial settings may exceed that of glacial polish (Tremblay, 1961). The retention of polish over time varies according to rock type (Clark and Wilson, 1992). Although ventifact surfaces are macroscopically smooth, scanning electron micrographs (SEMs) at high magnifications reveal considerable topographic roughness and surfaces covered with microcracking or cleavage fractures formed by repeated chipping by sand grains (Laity, 1995; Laity and Bridges, 2009). After abrasion ceases, rock coatings such as silica glaze or desert varnish may cover the surface (Dorn, 1995).

While small rocks usually have both polish and facets, larger boulders are covered by additional features,

(a)

(b)

Figure 21.10 (a) Basalt ventifact with a single keel, formed by bidirectional wind flow in the Owens Valley, California. The surface with the lower facet angle shows lineations perpendicular to the keel, whereas the higher angle facet is largely pitted. (b) Multiple facets have developed on this small ventifact owing to shifting of the rock over time.

Figure 21.11 Basalts are often intensely pitted, as erosion modifies pre-existing vesicles on high-angle, windward faces. Pits merge over time and become larger. As the face angle lowers near the upper surface of the rock, pits are elongated to form flutes.

including pits, flutes, grooves, scallops and helical forms. *Pits* occur on the high-angle (55–90°) windward surfaces of boulders (Figure 21.11). The pits vary in form from the more rounded and regular shapes, characteristic of basalts, to more irregular ovoid forms, observed in tuffs (Laity, 2009). The wind erodes and modifies pre-existing indentations (such as vesicles in basalt) and softer minerals (as in the case of granites or tuffs).

As the angle of the facet decreases, *flutes* develop. In form, they appear as 'arrowheads', open at one end and closed at the other, which point in a downwind direction (Figure 21.12). The form of the flute varies with the facet angle, although these relationships have not been quantified: flutes are shorter and deeper on higher angle faces and become more elongate as the facet angle declines (Maxson, 1940; Sharp, 1949). In many cases, the scale of the flutes increases from the base of the rock to the upper face. Sometimes, smaller flutes develop within larger ones, suggesting renewed cycles of erosion, the result of a change in climate, sediment supply or wind energy. Flutes are very common forms on ventifacts: rarer are scallops and helical scores. *Scallops* are U-shaped features that are open at one end and closed at the other, and have similar length-to-width ratios. They are present on some Martian rocks. *Helical forms* begin as shallow grooves, deepen and spiral in a downwind direction and terminate in a sharp point. In theory, they can be both right- or left-handed. Observations in the Mojave Desert suggest that the handedness of the spiral is consistent across the face of the rock. Helical forms have been observed in marble, basalt and granite, often where the wind velocity is accelerated, such as within topographic saddles and near hillcrests. They form on the upper faces of boulders and, like flutes, appear to increase in scale up the rock. The most elongate element on a ventifact is the *groove*. These subparallel forms are open at both ends and best developed on surfaces gently inclined or parallel to the wind. They are sometimes found on the vertical sides of boulders, particularly in the cracks between adjacent rocks. Like flutes, they cut across mineral grains and rock structures. Grooves vary greatly in scale, from striae (fine lineations) (Figure 21.13), to grooves of intermediate scale, to channels that may be several centimetres or more in depth (Tremblay, 1961). Striae may cross an entire outcrop, but in detail each lineament is

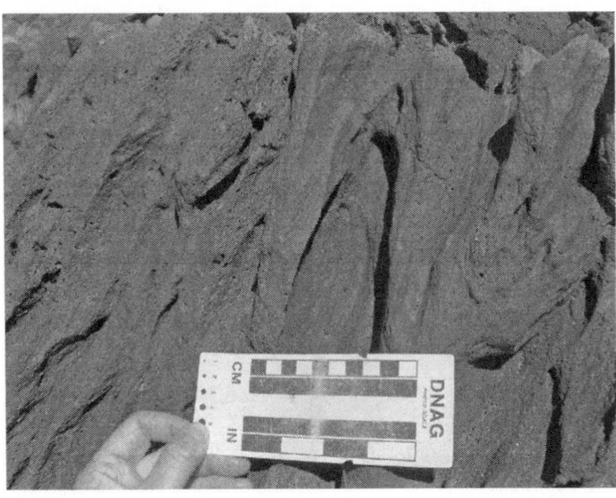

Figure 21.12 Deep flutes in basalt point in a downwind direction.

Figure 21.13 Fine lineations (grooves) on a marble ventifact parallel with the direction of the highest velocity winds. The grooves are several millimetres in width.

usually composed of a succession of short scoop-like depressions only a few centimetres in length. Groove trends are typically parallel on near-horizontal surfaces and reflect the flow direction of the highest velocity winds (Laity, 1987). Maxson (1940) suggested that grooves and flutes are initiated by vortices and then modified by saltating sand grains, whereas Schoewe (1932) proposed that the skidding action of grains on hard, smooth surfaces might be important. Whitney (1978) put forward the idea that vortex pits coalesce into flute pits and pit chains, but pit chains have not been observed in the field. Scanning electron micrographs of groove interiors (Laity and Bridges, 2009) show impact-generated cleavage fractures, suggesting that direct sand grain impact and rebounding grains are responsible for groove and flute formation.

Etching and fretting are processes of differential erosion that occur when there are strongly developed hardness differences in a rock. *Etching* develops in layered rocks such as ignimbrites, where the wind erodes away the weaker material. Etching can be replicated in layered target materials subject to sand abrasion in the field (Laity, 2009) or in the wind tunnel (Laity and Bridges, 2009; Bridges *et al.*, 2010) and subject to sand abrasion (Figure 21.14). *Fretting* occurs when there are small, hard inclusions in a rock, such that projecting points, knobs or ridges develop. The inclusions resist erosion, while the softer matrix is eroded away. The result may be finger-like projections called *dedos* or *demoiselles* (Hobbs, 1917) (Figure 21.15). When there are large, hard inclusions within the rock, a *knobby texture* may develop on high-angle windward faces (Figure 21.16). In granitic rocks, for example, large

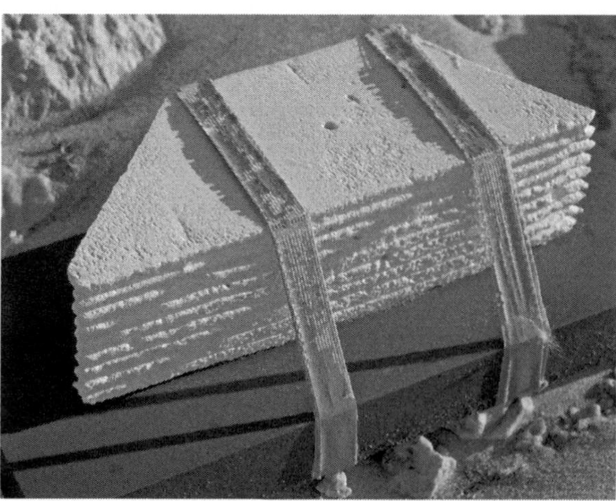

Figure 21.14 Etching, the erosion of weaker material in layered rocks, can be replicated by placing layered target materials (formed of plaster in this photograph) among a field of ventifacts being actively eroded by windblown sand.

Figure 21.15 Dedos are finger-like projections that result when small, hard inclusions occur in a rock. This dedo is about 15 cm in length.

Figure 21.16 A knobby texture can develop on high-angle windward faces when large, hard inclusions occur within the rock. This andesite boulder is located in the Mono Basin, California.

fine-grained xenolith inclusions that are resistant to erosion may stand out several centimetres in relief from the matrix.

21.2.5.2 Feature formation and feature scale

There are presently few data on the development of ventifact features. Factors that influence the type and nature of the features include rock size, homogeneity and hardness, wind velocity and face angle. It can be observed that certain features, such as flutes and helical forms, are universal in basic geometric form and occur on a range of rock types including granites, basalts, tuffs, limestones and other lithologies. In general, very small rocks (a few centimetres in diameter) are faceted and polished, but lack additional features.

The homogeneity of the rock plays an important role in feature formation. The development of features may be related to the relationship of any irregularities in the rock to the size-scale of the sand grains. This has been observed by placing manmade materials in the field. In 1993, six blocks of very fine-grained, highly homogeneous modelling foam were placed at an angle of ~45° to the horizontal adjacent to ventifacts in an active aeolian environment. Over the past 17 years, all of the blocks have lost considerable mass, with the softest almost completely abraded away (Laity and Bridges, 2009), but erosion has been essentially uniform and no flutes or lineations have formed. By contrast, formed plaster targets that were homogeneous in composition, but contained inherent air vesicles, developed flutes within a few months. Moreover, Styrofoam (a heterogeneous foam product) placed on the actively saltating bed of the Mojave River rapidly developed lineations (Laity, 1995). These observations suggest that initial irregularities large enough to be 'seen' by the sand grains will grow to macroscopic forms even in fairly homogeneous materials (P. L. Varkonyi, personal communication, 2010).

In the field, dense and texturally homogeneous rocks such as cherts, orthoquartzites or weakly vesicular basalts are faceted, with very small features that have little impact on the overall form of the ventifact. Rocks that are not homogeneous in texture develop larger and more various surface textures. In macrocrystalline or layered materials, the weaker minerals are more easily eroded and differential erosion is well developed. For example, in coarse-grained granitic rocks, near-vertical faces will be intensely pitted, with the feldspars forming the low areas, whereas at lower angles or where the wind sweeps by the sides of the rock, the more resistant quartz mineral may shield the feldspars, creating a positive relief in grooved surfaces (Fisher, 1996). Soft rocks, such as tuffs, develop large and deep features (Figure 21.17).

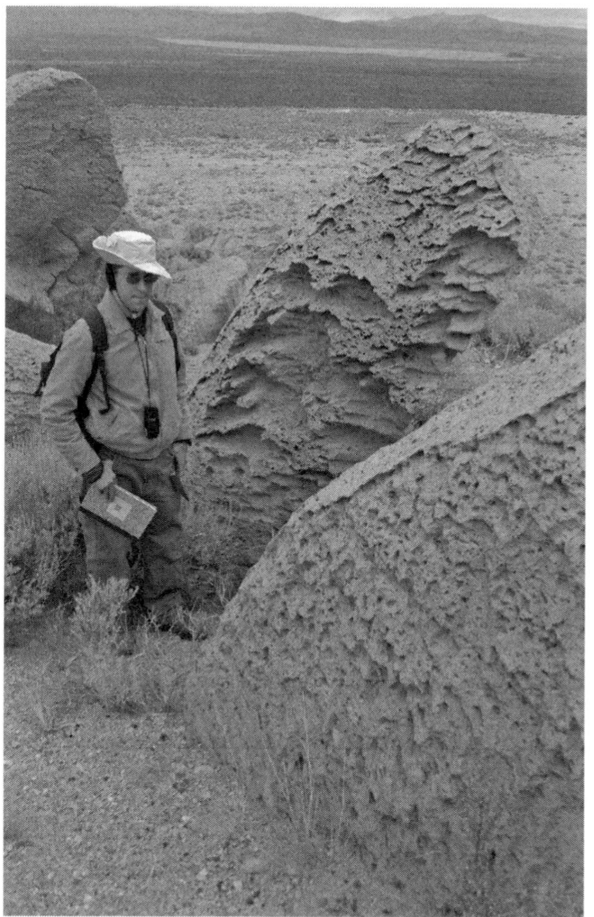

Figure 21.17 Soft rocks, such as this tuff, develop large and deep features. The surfaces of these high angle rocks are covered in deep, short flutes. A case-hardened surface on the ventifact to the right of the figure has protected much of the face of the rock, but where it has breached, significant mass has been lost. Location: Mono Basin, California.

Owing to the timescales involved in ventifact formation, it is difficult to observe feature development in the field. Therefore, laboratory studies or mathematical modelling are necessary to understand feature growth. To date, however, little work has been done in this area. The experimental work of Schoewe (1932) demonstrated that the rate of abrasion and angle of the abraded surface decrease over time. Work by Bridges *et al.* (2010) on rock simulant targets abraded by sand in a boundary layer wind tunnel indicates that the surface of a rock is initially roughened, with the surface area-to-volume ratio increasing at a rate greater than log (2/3). Initially, many small pits are produced, but over time these merge to produce larger ones. Over time, the larger pits tend to become more elongate in form, suggesting that flutes may be derived from pits as face angle lowers, an interpretation that appears to be supported by field observations.

Feature scale (length, width and depth) vary greatly on ventifacts. Factors that affect scale include rock type and size, particle velocity and the duration of abrasion. The following observations may be made based on field study:

a. Small rocks are associated with small features. For very small rocks (a few cm in diameter), macroscale features appear to be absent (Maxson, 1940) (Figure 21.10(b)). Moreover, the larger the rock, the larger the features that develop on it (Figure 21.18). During the process of ventifact development, there is both recession of the face of the rock and growth of the features. The rock must be large enough to accommodate both.

b. If there is a large rock at the base of a hill and an equally large rock at the top of a hill, then the rock at the top will have the larger features (Laity, 1987). Why is this? What accounts for scale increases in features at the hilltop – an increase in wind and particle velocity, an increase in the duration of abrasion, or a combination of these factors? The role of particle velocity and abrasion duration are difficult to disentangle. As the wind approaches a hill, the airstream compresses, resulting in an acceleration of flow and an increase in velocity and sediment transport. The largest shear-stress values and wind velocities are recorded when the near-surface flow approaches approximately normal to the topography. At the crest of dunes, McKenna Neuman, Lancaster and Nickling (1997) recorded speed-up increases (the mean velocity of the wind at height z above the hill relative to the mean velocity at the same height over a flat surface) ranging from 1.50 to 3.19, with a corresponding increase in sediment flux of one to two orders of magnitude. On a rocky hillslope, this means that sand is most active near the crest and that more sand is moved and at higher velocity (both greater particle velocity and an increased duration of time) than lower down on the slope. In the Mojave Desert, for example, many lower slopes are mantled with stable sand, with active aeolian material present only near the crest. Thus, from a historical perspective, ventifacts near the crest will be abraded for a much longer time, often extending through climatic periods where the majority of sand in a region is stabilised. Therefore, over time ventifacts at the crest are subject to more sand, at higher velocities and for longer time periods. It is difficult to separate out the velocity or duration components in considerations of feature scale.

c. Feature scale increases with height up an individual boulder, with the fundamental geometric properties remaining the same. On a single boulder, the upper part

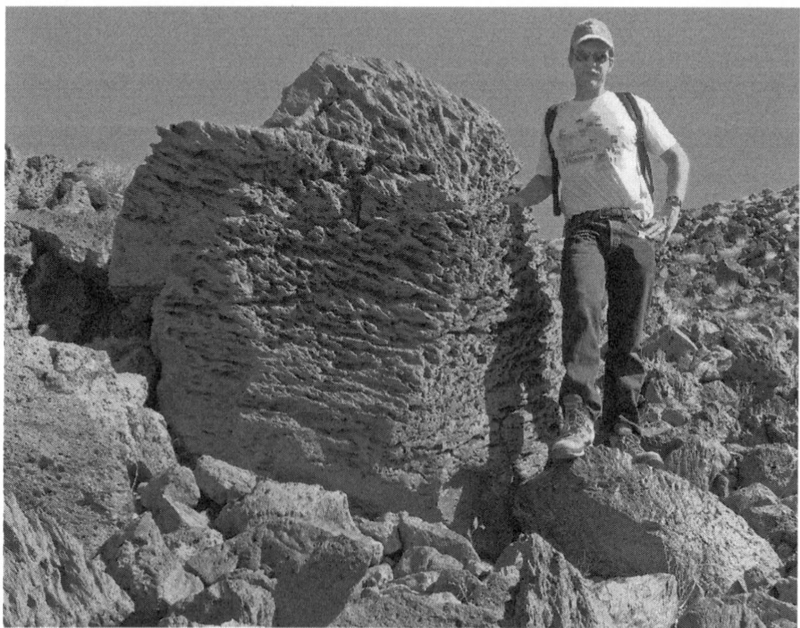

Figure 21.18 The scale of ventifact features increases with rock size. This massive basaltic ventifact is pitted on its high angle, lower windward face. On the upper face and margins of the rock, flutes occur. This rock is located on a hillcrest in the Cady Mountains, California.

protrudes into the higher velocity windstream. There are three possible explanations for the greater feature scale near the top of the rock: (1) particle impact velocity is highest here, (2) the duration of erosion is greater or (3) the features have migrated up the rock, becoming bigger with time. Quantitative studies will be necessary to unravel the answers to these questions.

21.2.5.3 *Processes of erosion*

Ventifacts form by the process of abrasion. The efficacy of this process is a function of the abradant. Although Whitney (1979) suggested that wind alone could erode rock mass, her experiments used a pressurised airblaster, which applies more force than is the case in nature. To date, no ventifacts have been recorded in an area subject to pure airflow alone. Pure air is most likely to deflate particles that have been separated from the rock mass by weathering.

Several authors have suggested that dust abrasion may have a role to play in the formation of ventifacts (Maxson, 1940; Sharp, 1949; Whitney, 1978; Lancaster, 1984; Breed, McCauley and Whitney, 1989; Schlyter, 1994). Dust particles are composed of silts and clays, less than 0.0625 mm (62.5 μm) in diameter, and transported in suspension. In dust storms in Kuwait and Saudi Arabia, for example, most particles were less than 10 μm in diameter, although 40 % fell in the 10–30 μm range (Draxler *et al.*, 2001). Dust has been suggested as an abradant because it was thought more likely to explain polish (Sharp, 1949; Lancaster, 1984) and might be able to follow vortex currents more easily and thereby cut flutes or lineations (Maxson, 1940; Whitney, 1978). However, ventifacts appear not to be found in areas subject to dust influx alone. In the Mojave Desert, a comparison may be made between the young (~18 kyr) surface of the Pisgah volcanic flow, which is traversed by sand and has abundant ventifacts, and the much older Cima volcanic field (with cones and flows formed from the Miocene to Holocene), which has accumulated up to 3 metres of aeolian dust but lacks ventifacts.

The majority of ventifact field studies have invoked sand as the abradant. Sand is a loose, granular material, 0.0635–2 mm (622.5–2000 μm) in diameter, which moves by saltation, travelling on a characteristic path (the saltation trajectory), reaching 1–2 m above the surface and extending several metres downwind. Maximum velocity is achieved at the top of the trajectory, which is approximately 0.5 to 0.66 of the wind speed. Typical abrasion heights on both ventifacts and yardangs rise to 1 m (Hobbs, 1917), although greater heights may be achieved as dunes traverse a region and raise the sand's base level.

21.2.5.4 *The nature of the abrading agent: sand versus dust*

Both sand and dust are abundant in deserts. Their relative abrasional efficacy depends upon particle velocity, mass, flux and interaction with the target (does the particle strike the target or is it deflected around it?). While dust has a high flux and velocity, particle mass is low and dust is easily deflected around an obstacle, rather than impacting it directly. By contrast, sand has a lower flux and velocity but a much higher mass and directly impacts the rock surface, often more than once owing to rebound effects. A theoretical consideration of these factors, discussed in more detail below, suggests that dust is unlikely to play a significant role in abrasion.

In contrast to dust, sand in saltation achieves sufficient momentum to be decoupled from the airstream around obstacles and impacts the target directly. Its velocity at the point of impact is ~50 % or less of the wind velocity (Bridges *et al.*, 2005) and, considering the velocity contribution to kinetic energy (v^2), the kinetic energy (KE) is much less than that of dust. However, the mass is much greater and this increases as a cube of the particle size. Therefore, a 100 μm sand grain has 1000 times the mass of a 10 μm dust particle. Upon impact, the mass loss of a rock by abrasion varies with particle diameter, D, by approximately D^3. Taking velocity and mass into consideration, the KE of sand is therefore 50–100 times that of dust (Laity and Bridges, 2009). Furthermore, dust is coupled to the airflow and deflected around the obstacle, whereas sand impacts directly. Anderson (1986) demonstrated that deflection of dust around obstacles decreased the number of impacts, such that the number of impacts from 10 μm dust is about 10 % of that of 100 μm sand, given the same number of initial particles. Laity and Bridges (2009) conclude that on a per particle basis, about 1000 times more energy is transferred to rock surfaces by sand than dust.

Given the potentially greater density of dust particles, can the additive KE contribution of dust impacts equal that of sand grains? Griffith's Theory (Griffith 1921) suggests that rocks and other solids fail from particle abrasion after a certain energy limit is exceeded (Lawn, 1995), with the criterion for failure being the size of the stress field induced by impact relative to the characteristic spacing of critical microflaws in the rock. Dust induces a stress field smaller than these microflaw spacings and is therefore much less likely to be able to abrade rock (Laity and Bridges, 2009).

Another factor that needs to be considered is the form of the ventifact relative to the kinetic energy abrasion profile. The facets on ventifacts typically slope away from the wind. On other abraded materials, such as the prows

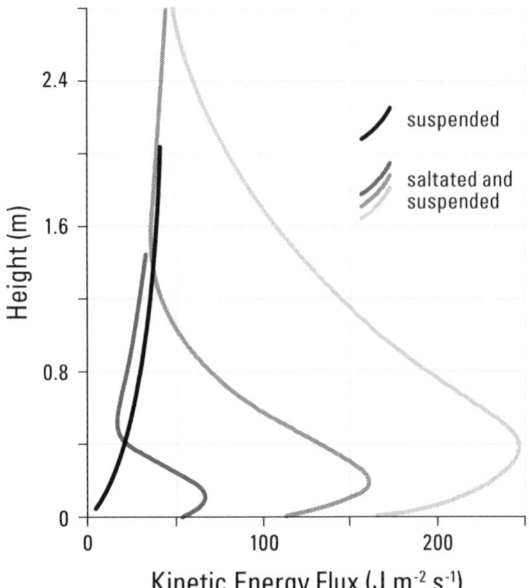

Figure 21.19 Kinetic energy flux profiles of sand and dust. The profiles of saltating grains have distinctive maxima at a height of several tens of centimetres above the surface. This profile is reflected in both ventifact and yardang forms. The height of maximum kinetic energy in sand increases with mean liftoff velocity. The kinetic energy flux of dust increases steadily with height and does not show a distinct maxima (from Laity and Bridges, 2009, after Anderson, 1986; courtesy of Elsevier Limited).

of yardangs or fence posts, a marked abrasion zone, or re-entrant, develops near the surface. This form can be explained with reference to the kinetic energy profiles of sand (Figure 21.19). Wind tunnel studies, fieldwork and analytical models indicate that the decreasing flux of sand with height combined with increasing velocity results in a kinetic energy profile peak within the lower tens of centimetres above a level surface (Sharp, 1964, 1980; Wilshire, Nakata and Hallet, 1981; Anderson, 1986; Liu *et al.*, 2003; Bridges *et al.*, 2005). The cover photograph in Greeley and Iversen (1985) shows this profile exemplified in a notched ventifact. By contrast, in the case of dust, the erosional profile should increase steadily with height, as the velocity of the dust matches closely that of the wind and the flux is fairly invariant (Anderson, 1986) (Figure 21.19). Such an erosional profile is not observed in nature.

21.2.5.5 Mass loss by sand abrasion

There has been little study of the role of particle composition on the susceptibility of the rock to abrasion. On Earth, most sand is formed of quartz grains. There appears to be little difference between the effectiveness of quartz and basalt particles. Ash appears to be less effective, but the data for both basalt and ash particles are limited (Greeley *et al.*, 1984). Aggregated grains tend to plaster against the target, creating mass gain rather than loss.

Field studies of target materials and sand-abraded rocks indicate a mass loss of between 30 and 1630 μm/yr (Sharp, 1964, 1980; Greeley *et al.*, 1984; Kuenen, 1960; Knight and Burningham, 2003). High-speed video (HSV) experiments help to elucidate the process, demonstrating that: (a) the *windward* side of rocks are subject to abrasion, (b) sand hits the targets directly and is not deflected by vortices that affect dust, (c) the outgoing particle velocity is less than the incoming, showing that kinetic energy is transferred to the target surface (Banks, Bridges and Benzit, 2005; Laity and Bridges, 2009), and (d) some grains rebound into the airstream and therefore provide a second impact to the surface, amplifying the abrasion (Bridges *et al.*, 2005).

The mechanisms of aeolian abrasion are based largely on the interpretation of scanning electron micrograph (SEM) images of disrupted surfaces. Greeley *et al.* (1984) examined brittle materials abraded under laboratory conditions and Laity (1995) and Laity and Bridges (2009) examined active ventifacts abraded in the field. According to Greeley *et al.* (1984), not all impacts will remove material; some damage the surface and prepare it for future removal. When well-rounded particles impact, a circular crack, or Hertzian fracture, develops around the indentation site. The crack diameter increases with both particle diameter and velocity. Angular particles produce a more diverse style of erosion.

Scanning electron micrographs of field specimens in an advanced stage of abrasion suggest that chipping is the most important mechanism of mass removal. A secondary mechanism, less frequently observed and perhaps only significant in relatively weak rock, involves gouging by sand grains that move essentially parallel to the rock surface (Laity, 1995). Thus, while ventifacts appear macroscopically smooth and polished, they are microscopically rough. The nature of the topographic roughness depends on material type. Basaltic ventifacts sampled from Pisgah Crater have a microcrystalline or glassy groundmass, which at high magnifications is seen to be damaged by microcracking. Exposed plagioclase phenocrysts have cleavage fractures. By contrast, marble ventifacts from the Little Cowhole Mountains, California, show a different abrasional texture (Figure 21.20(a)). At high magnifications (×2000–5000), the surface topography is shown to be formed by impact-generated cleavage fracture of the marble crystallites, with angular marble debris lying on the surface (Figure 21.20(b)). At ×20 000–50 000, the rough

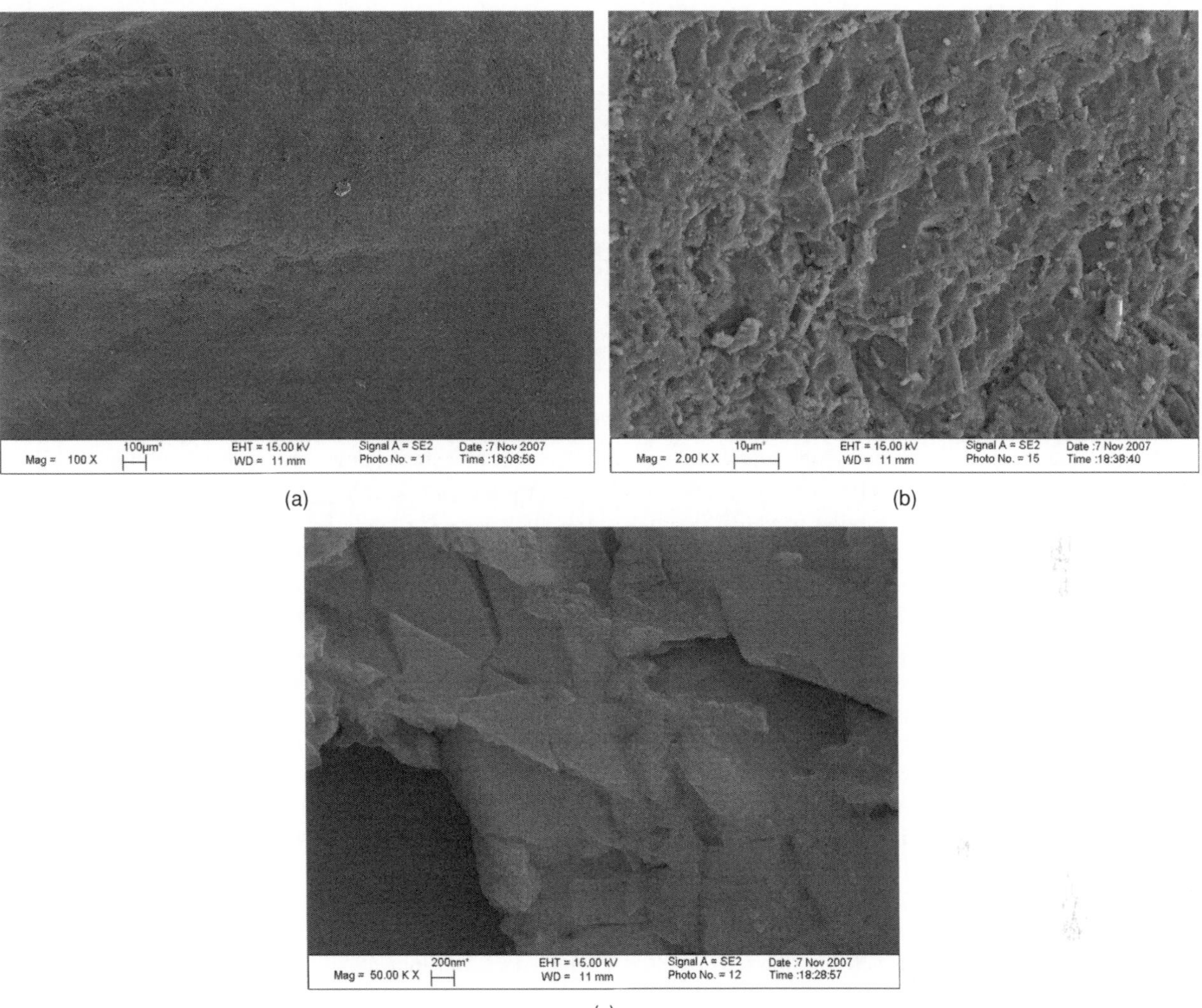

Figure 21.20 (a) The surface of a groove in an actively forming, sand-abraded marble ventifact at ×100 magnification. (b) At a scale of ×2000, cleavage fractures of marble crystals are evident. The surface is much rougher at the macroscale than suggested by field specimens. (c) At ×50 000, microcleavage of marble crystal grains and fractures in the process of propagation occur. Angular marble debris are present.

surface is clearly caused by impact damage from sand and is not reflective of the smoother surface that might be anticipated from uniform wearing by dust (Figure 21.20(c)). Such impact features are observed both within grooves and on ridges.

21.2.5.6 Rates of abrasion

Calculating rates of wind abrasion associated with ventifact formation is difficult. It is problematic to infer field rates based on laboratory studies, because field conditions are infinitely more complex. Wind abrasion in the field is aerodynamically complex and conditions of wind speed and direction, particle supply and burial/exposure of rocks vary over both short and long timespans. The exposure to abrasion is potentially very long, spanning thousands of years and several episodes of climatic change. Owing to changing environmental conditions and to changes to the rock surface itself, the erosion rate of ventifacts is nonlinear.

At the local scale, particle movement is affected by the size and spacing of rocks, resulting in areas of greater abrasion where flow is more turbulent or where grains are propelled upward from a ventifact facet into the high-velocity air flow, where they travel many metres before impacting another rock downwind. Shadow zones also occur, such as behind ventifacts or beneath sand stringers. Prolonged periods of burial by sand are common during seasonal or within-storm shifts in wind direction (Greeley *et al.*, 2002). At the scale of development of large ventifacts, climate change occurs, influencing both the character of the rock surface through weathering, the supply of sediments and the nature of wind flow, making estimates of ventifact age based on present-day meteorological conditions unrealistic.

Several studies have attempted to determine rates of ventifact formation. Sharp's (1964, 1980) experimental plot in the Coachella Valley, California, yielded a range of abrasion rates, which varied according to time and material hardness. It should be noted, however, that the materials used were not rock, but softer manmade materials. Similarly, Greeley *et al.* (1984) report a relatively wide range of inferred rates, varying from 10^{-5} to 6×10^{-1} cm/yr, determined by abrasion of a range of materials (brick, hydrocal, tuff, granite, basalt, rhyolite and obsidian) in a wind tunnel. Based on laboratory experiments, Kuenen (1960) concluded that ventifacts would develop very slowly where only fine sand is available (one to two centuries), quickly with medium sand (a few dozen years) and very rapidly on stormy beaches (a few dozen years). This latter conclusion is consistent with observations of Knight and Burningham (2003), who concluded that ventifacts along the Oregon coast, abraded by medium-grained, well-sorted sands, have a maximum age of 60–70 years. Rates of 0.24–1.63 mm/yr were calculated for this site. Clearly, the time for ventifacts to form in a desert setting is not yet well constrained. Furthermore, there is probably a considerable difference in the time required to develop initial features (pitting, polish, minor grooving) and that needed to radically change the rock form into a mature, often wedge-shaped ventifact, wherein a great deal of mass has been lost. In the central Namib Desert, Selby (1977) concluded that the faceting of boulders 10 to 50 cm in diameter would take thousands of years to complete.

21.2.5.7 Surface controls, spatial variability, feedback and topography

Within any given field of ventifacts, there is a wide array of forms and surface features, even in unidirectional winds passing over rocks of a single lithology: no two ventifacts look alike. What can account for these differences? Unlike a wind tunnel, surfaces are not flat, grain sizes are not uniform, there are not limitless supplies of dry sediment, nor are there steady and uniform winds. Natural systems are highly complicated, both from the perspective of the atmosphere (the wind system) and the topography. On a small scale, aeolian researchers recognise the influence of a range of surface roughness elements, such as patches of coarse sediment, vegetation or boulders, which determine how near-surface winds respond. In ventifact fields, these fixed elements incorporate strongly varying roughness element size (small to large boulders), density, spacing and porosity that affect sand transport and shear stress distribution. Field observations of ventifact formation by the author during strong winds show that the wind and sediment environments are constantly changing. Ventifacts are successively exposed, buried by sand and then re-exposed; sand ribbons blanket and protect ventifacts while adjacent rocks are abraded; and each ventifact affects the abrasion of downstream rocks, by shielding them or by projecting saltating grains into higher velocity layers of the atmosphere.

At the landscape scale, the influence of topography on the flow field becomes important (Gilbert, 1875; Laity, 1987, 1995). Flow is compressed and expanded based on curvature and the geometrical scales of the underlying surface. The topography affects flow acceleration and deceleration, and hence the distribution of shear stress and the pathways of sand transport. These factors have a strong influence on ventifact scale and distribution. Studies of airflow over hills and dunes (Jackson and Hunt, 1975; Bowen and Lindley, 1977; Taylor, 1977; Lancaster, 1985; Tsoar, 1985; Mulligan, 1988; McKenna Neuman, Lancaster and Nickling, 1997) have led to advances in our understanding of ventifact formation. Wind velocity increases and longer periods of abrasion near hillcrests and in topographic saddles result in scale increases in ventifact features (grooves or flutes) (Figure 21.18).

Some progress can be made in understanding field conditions by using targets to simulate rocks, but soft materials are necessary to achieve results in a reasonable period of time (Figure 21.14). Sharp (1964, 1980) placed Lucite rods in an exposed setting and was able to determine a maximum height of abrasion above the ground. Bricks showed wear after a period of 15 years, with megascopically visible effects such as polish, pitting and incipient fluting developed during an intense 10-month period of high particle flux (Sharp, 1980). Artificial gypsum targets of different internal consistencies and hardnesses have been used both in the field and in the laboratory (Figure 21.14) (Bridges *et al.*, 2004; Laity and Bridges, 2009). Field targets eroded very rapidly when placed adjacent to rock ventifacts and developed small-scale lineations and flutes within a few months, paralleling the direction of the highest velocity winds and with the same orientation as linear features on adjacent marble ventifacts.

21.2.5.8 *The role of climate change*

The role of climate change in ventifact formation is not well understood because it is very difficult to study and the ages of the rocks are poorly constrained. It has the potential to alter: (1) the surface of the rock, through case-hardening processes and varnish deposition, thereby affecting subsequent episodes of erosion; (2) wind strength, direction and the supply of particles, as well as the amount of stabilising vegetation; and (3) the surface level of the ground, through erosional or depositional processes. In the Mojave Desert, for example, much of the sand at lower elevations is stable. Active sand at hillcrests maintains modern ventifacts, whereas varnished fossil forms mantle the lower slopes and plains.

The impact of chemical and physical changes to the surface of rocks during periods between erosion episodes has not been studied. In arid environments, the formation of coatings on stabilised ventifacts (Dorn, 1995) may provide a surface that is more resistant to abrasion than the natural rock, requiring additional erosive energy to remove in subsequent aeolian episodes. Hardened surfaces are observed on ventifacts to the lee of the Sierra Nevada, California, at altitudes of ~2500–2700 m. The ventifacts were probably formed in periglacial regimes, although the environment is semi-arid today. In numerous instances, abrasion can be seen to have worked its way *under* a heavily fluted outer carapace of granite or tuff, exploiting the softer material beneath the case-hardened surface. Observations such as these support the idea that rates of erosion are not uniform over time, but fluctuate with climate and changes to the surface properties of the rock.

21.2.5.9 *The use of ventifacts to determine wind direction*

Ventifacts and abraded outcrops are an excellent proxy for wind direction. Wind direction can be determined by reference to the keel (which forms perpendicular to the wind), to pitting (perpendicular to the wind) or to grooves or flutes (parallel to the wind on low-angle facets) (Maxson, 1940; Selby, 1977; Laity, 1987; Nero, 1988; Donner and Embabi, 2000; Laity and Bridges, 2009). Where yardangs are also present, they will have the same orientation as flutes and grooves on ventifacted surfaces (Donner and Embabi, 2000).

The use of keels to map wind direction is more problematic than pits or surface lineations. Knight (2008, p. 96) questions the use of keels in the mapping of ventifacts, claiming that such features may be formed either perpendicular *or* parallel to wind flow; low elongate ventifacts with a well-defined keel and two opposing facets are thought to be formed by parallel winds, whereas wind-perpendicular keels are associated with ventifacts with only one abraded face. These statements are contradicted by other studies (Maxson, 1940), including those of contemporary ventifacts in the Mojave Desert (Laity, 1995). A weather station maintained for 17 years in the Little Cowhole Mountains, combined with a series of anemometers placed around the hillslope, clearly demonstrated that ventifacts with a sharp keel and two facets develop from opposing winds, which flow perpendicular to the rock. Nonetheless, the mapping of keel orientation to determine wind direction produces a much greater scatter of results than the use of grooves or flutes. Keels produce a good result when the ventifact is very mature, with almost no remnant of the original face or rock shape remaining. In earlier stages of development, the keel orientation tends to be rather varied. By contrast, linear features on rocks develop early in the evolution of the ventifact and are much easier to map accurately (although caveats apply). In all instances, however, care must be taken and the work must not be performed uncritically.

If a large region is mapped, palaeocirculation patterns can be reconstructed with reference to fossil ventifacts (Powers, 1936; Sharp, 1949; Tremblay, 1961; Nero, 1988; Smith, 1984; Laity, 1992, 1995; Knight, 2008). Large stable rocks provide the best medium for mapping, as smaller ventifacts may be overturned or rotated by animal movement, flooding, earthquakes or other events. In modern environments, the wind direction derived from ventifact mapping may be compared with that attained from other features, such as vegetation growth patterns, to understand the impact of topography on prevailing and erosive winds (Griffiths *et al.*, 2009). In areas of bidirectional flow, one direction may dominate in terms of erosional energy, and this is reflected in the greater development of features on that side (Corbett, 1993).

Several studies have suggested that a reverse eddy flow, caused by flow separation along the keel, may explain abrasion on the lee side of ventifacts (Breed, McCauley and Whitney, 1989; Fisher, 1996). These observations were made based on studies of relict ventifacts, for which the wind regime is unknown. The understanding of directional relationships is important in the reconstruction of wind flow fields. There are several potential problems with the concept of reverse eddy flow erosion: (1) wind velocity to the lee of rocks appears insufficient to support sand flow of this nature; (2) it is not supported by target studies either in the field or in the laboratory (Laity and Bridges, 2009); and (3) for most ventifacts, the lee of the ventifact is covered in a protective layer of sand during wind storms. Numerous authors have observed that where wind flow is strictly unidirectional, the lee side of a boulder is unabraded (Blackwelder, 1929; McKenna Neuman and Gilbert, 1986).

21.3 Conclusions

Landscapes of aeolian erosion are seldom produced by the wind alone, although it may be the dominant player. Yardangs and desert depressions (discussed in Chapter 15) evolve through a host of interacting processes, including those that are both endogenic (tectonic) and exogenic (solutional weathering, abrasion, deflation, fluvial erosion and mass movement of different types). Ventifacts are the sole form developed almost exclusively by wind erosion and, more specifically, by abrasion. However, even for ventifacts it must be recognised that weathering associated with climate change can play a role, by either softening or hardening the surface, thereby affecting the nature and rate of abrasion.

Although each aeolian landform is commonly reviewed and discussed in isolation, they are often part of a larger landscape of aeolian activity, most notably in extremely arid regions where sand flows across vast regions to form a suite of landforms. The transported sands have a strong corrasive effect on outcropping rocks, moulding them into basins, yardangs and ventifacts and, at times, creating inverted relief as they pass over ancient 'hardened' drainage systems.

Improved access to remote sensing imagery has provided a greater understanding of the landscape patterns of aeolian erosion, suggesting complicated assemblages of natural interactions within aeolian systems as a whole. Studies of tectonics, sand transport and dust emission have added new insights to our understanding of many of the large-scale processes involved. It remains to examine the spatial distribution of erosion and deposition within a broader context of the role of topography on the flow field, the distribution of shear stress, the effect of changing material type and the pathways of sediment transport.

A review of modern global wind erosion suggests that it reaches its greatest intensity and extent in hyper-arid regions. In more humid settings, such as the deserts of the American southwest or Australia, wind erosion is much more localised in its occurrence. Our understanding of contemporary processes helps us to consider the role of climate change in fashioning desert environments. Research conducted over the past few decades hints at the importance of such fluctuations to aeolian landscapes. Associated with climates that oscillate from moister to more arid states are changes in the supply of abrasive sediment, the nature of depositional processes (e.g. infilling of basins by lacustrine sediments), the height of the water table (which forms the base level for wind erosion), the biomass of vegetation and the strength and frequency of winds. During wetter phases, aeolian erosion probably ceases in all but the most favourable locales. Nonetheless, the interplay of wet and dry climates is probably essential for the nature and continuity of aeolian processes as we know them.

References

Alavi Panah, S.K., Komaki, Ch. B., Goorabi, A. and Matinfar, H.R. (2007) Characterizing land cover types and surface conditions of yardang region in Lut Desert (Iran) based upon Landsat satellite images. *World Applied Sciences Journal*, **2**, 212–228.

Al-Dousari, A.M., Al-Elaj, M., Al-Enezi, E. and Al-Shareeda, A. (2009) Origin and characteristics of yardangs in the Um Al-Riman depressions (N Kuwait). *Geomorphology*, **24**, 93–104.

Anderson, R.S. (1986) Erosion profiles due to particles entrained by wind: application of an aeolian sediment-transport model. *Bulletin of the Geological Society of America*, **97**, 1270–1278.

Bailey, J.E., Self, S., Wooller, L.K. and Mouginis-Mark, P.J. (2007) Discrimination of fluvial and eolian featues on large ignimbrite sheets around La Pacana Caldera, Chile, using Landsat and SRTM-derived DEM. *Remote Sensing of Environment*, **108**, 24–41.

Banks, M., Bridges, N.T. and Benzit, M. (2005) Measurements of the coefficient of restitution of quartz sand on basalt: implications for abrasion rates on Earth and Mars. *Lunar and Planetary Science*, **XXXVI**, 2116.

Beresford-Jones, D., Lewis, H. and Boreham, S. (2009) Linking cultural and environmental change in Peruvian prehistory: Geomorphological survey of the Samaca Basin, Lower Ica Valley, Peru. *Catena*, **78**, 234–249.

Blackwelder, E. (1929) Sandblast action in relation to the glaciers of the Sierra Nevada. *Journal of Geology*, **37**, 256–260.

Blackwelder, E. (1934) Yardangs. *Bulletin of the Geological Society of America*, **45**, 159–165.

Bobek, H. (1969) Zür Kenntnis der Südlichen Lut. *Mitteilungen der Österreicher Geographischen Gesellschaft, Wien*, **3**, 155–192.

Bosworth, T.O. (1922) *Geology of the Tertiary and Quaternary Periods in the North-west Part of Peru*, Macmillan and Company, London.

Bowen, A.J. and Lindley, D. (1977) A wind-tunnel investigation of the wind speed and turbulence characteristics close to the ground over various escarpment shapes. *Boundary-Layer Meteorology*, **12**, 259–271.

Breed, C.S., McCauley, J.F. and Whitney, M.I. (1989) Wind erosion forms, in *Arid Zone Geomorphology* (ed. D.S.G. Thomas), John Wiley & Sons, Inc., New York, pp. 284–307.

Bridges, N.T., Laity, J.E., Greeley, R. *et al.* (2004) Insights on rock abrasion and ventifact formation from laboratory and field analog studies with applications to Mars. *Planetary and Space Science*, **52**, 199–213.

Bridges, N.T., Phoreman, J., White, B.R., Greeley, R., Eddlemon, E., Wilson, G. and Meyer, C. (2005) Trajectories and energy transfer of saltating particles onto rock surfaces: application to abrasion and ventifact formation on Earth and Mars. *Journal of Geophysical Research*, **110**, E12004, DOI: 10.1029/2004JE002388.

Bridges, N. T., Razdan, A., Yin, X. *et al.* (2010) Quantification of shape and texture for wind abrasion studies: proof of concept using analog targets. *Geomorphology*, **114**, 213–226.

Bristow, C.S., Drake, N. and Armitage, S. (2009) Deflation in the dustiest place on Earth: the Bodélé Depression, Chad. *Geomorphology*, **105**, 50–58.

Brookes, I. A. (2001) Aeolian erosional lineations in the Libyan Desert, Dakhla Region, Egypt. *Geomorphology*, **39**, 189–209.

Brookes, I.A. (2003a) Geomorphic indicators of Holocene winds in Egypt's Western Desert. *Geomorphology*, **56**, 155–166.

Brookes, I.A. (2003b) Palaeofluvial estimates from exhumed meander scrolls, Taref Formation (Turonian), Dakhla Region, Western Desert, Egypt. *Cretaceous Research*, **24**, 97–104.

Bryan, K. (1931) Wind-worn stones or ventifacts – a discussion and bibliography, National Research Council Circular 98, Report of the Committee on Sedimentation, 1929–1930, Washington, DC, pp. 29–50.

Capot-Rey, R. (1957) Sur une forme d'érosion éolienne dans le Sahara Français. *Tidjschrift Konligke Nederland Aardnjsk Genootschap*, **74**, 242–247.

Capot-Rey, R. (1961) Borkou et Ounianga – Étude de Géographie Régionale, Université d'Alger, Institut de Recherches Sahariennes, Mémoire 5, Alger.

Clark, R.C. and Wilson, P. (1992) Occurrence and significance of ventifacts in the Falkland Islands, South Atlantic. *Geografiska Annaler*, **74A**, 35–46.

Cooke, R., Warren, A. and Goudie, A. (1993) *Desert Geomorphology*, UCL Press, London.

Compton, J.S. (2006) The Mid-Holocene sea-level highstand at Bogenfels Pan on the southwest coast of Namibia. *Quaternary Research*, **66**, 303–310.

Compton, J.S. (2007) Holocene evolution of the Anichab Pan on the south-west coast of Namibia. *Sedimentology*, **54**, 55–70.

Corbett, I. (1993) The modern and ancient pattern of sandflow through the southern Namib deflation basin, in *Aeolian Sediments: Ancient and Modern* (eds K. Pye and N. Lancaster), International Association of Sedimentologists Special Publication 16, Blackwell Scientific Publications, Oxford, pp. 45–60.

Czajka, W. (1972) Windschliffe als Landschaftmerkmal. *Zeitschrift für Geomorphologie*, **16**, 27–53.

Donner, J. and Embabi, N.S. (2000) The significance of yardangs and ventifacted rock outcrops in the reconstruction of changes in the Late Quaternary wind regime in the Western Desert of Egypt. *Quaternaire*, **11**, 179–185.

Dorn, R.I. (1995) Alterations of ventifact surfaces at the glacier/desert interface, in *Desert Aeolian Processes* (ed. V.P. Tchakerian), Chapman & Hall, London, pp. 199–217.

Draxler, R.R., Gillette, D.A., Kirkpatrick, J.S. and Heller, J. (2001) Estimating PM10 concentrations from dust storms in Iraq, Kuwait and Saudi Arabia. *Atmospheric Environment*, **35**, 4315–4330.

Ehsani, A.H. and Quiel, F. (2008) Application of Self Organizing Map and SRTM data to characterize yardangs in the Lut Desert, Iran. *Remote Sensing of Environment*, **112**, 3284–3294.

Engelstaedter, S. and Washington, R. (2007) Atmospheric controls on the annual cycle of North African dust. *Journal of Geophysical Research*, **112**, D03103, DOI: 10.1029/2006JD007195.

Evans, I.S. (2003) Scale-specific landforms and aspects of the land surface, in *Concepts and Modelling in Geomorphology: International Perspectives* (eds I.S. Evans, R. Dikau, E. Tokunaga *et al.*), TERRAPUB, Tokyo, pp. 61–84.

Farpoor, M.H. and Krouse, H.R. (2008) Stable isotope geochemistry of sulfur bearing minerals and clay mineralogy of some soils and sediments in Loot Desert, central Iran. *Geoderma*, **146**, 283–290.

Fisher, T.G. (1996) Sand-wedge and ventifact palaeoenvironmental indicators in north-west Saskatchewan, Canada, 11 ka to 9.9 ka BP. *Permafrost and Periglacial Processes*, **7**, 391–408.

Gabriel, A. (1938) The Southern Lut and Iranian Baluchistan. *The Geographical Journal*, **92**, 193–208.

Gilbert, G.K. (1875) Report on the geology of portions of Nevada, Utah, California, and Arizona, Geographical and Geological Surveys West of the 100th Meridian 3.

Goudie, A.S. (1989) Wind erosion in deserts. *Proceedings of the Geologists' Association*, **100**, 89–92.

Goudie, A.S. (2007) Mega-yardangs: a global analysis. *Geography Compass*, **1**, 65–81.

Goudie, A.S. (2008) The history and nature of wind erosion in deserts. *Annual Review of Earth and Planetary Sciences*, **36**, 97–119.

Goudie, A.S. and Middleton, N.J. (2001) Saharan dust storms: nature and consequences. *Earth-Science Reviews*, **56**, 179–204.

Greeley, R. and Iversen, J.D. (1985) *Wind as a Geological Process*, Cambridge University Press, Cambridge.

Greeley, R., Leach, R.N., Williams, S.H. *et al.* (1982) Rate of wind abrasion on Mars. *Journal of Geophysical Research*, **87** (B12), 10009–10014.

Greeley, R., Williams, S.H., White, B.R. *et al.* (1984) Wind abrasion on Earth and Mars, in *Models in Geomorphology* (ed. M.J. Woldenberg), Allen and Unwin, Boston, pp. 373–422.

Greeley, R., Bridges, N., Kuzmin, R.O. and Laity, J. E. (2002) Terrestrial analogs to wind-related features at the Viking and

Pathfinder landing sites on Mars. *Journal of Geophysical Research*, **107**, 5-1–5-21.

Griffith, A.A. (1921) The phenomena of rupture and flow in solids, *Philosophical Transactions of the Royal Society of London*, A **221**, 163–198.

Griffiths, P.G., Webb, R.H., Fisher, M. and Muth, A. (2009) Plants and ventifacts delineate late Holocene wind vectors in the Coachella Valley, USA. *Aeolian Research*, **1**, 63–73.

Grolier, M.J., McCauley, J.F., Breed, C.S. and Embabi, N.S. (1980) Yardangs of the Western Desert. *The Geographical Journal*, **146**, 86–87.

Grove, A.T. (1960) Geomorphology of the Tibesti region with special reference to western Tibesti. *The Geographical Journal*, **126**, 18–27.

Grove, A.T. and Warren, A. (1968) Quaternary landforms and climate on the south side of the Sahara. *The Geographical Journal*, **134**, 194–208.

Gutiérrez-Elorza, M., Desir, G. and Gutiérrez-Santolalla, F. (2002) Yardangs in the semiarid central sector of the Ebro Depression (NE Spain). *Geomorphology*, **44**, 155–170.

Hagedorn, H. (1968) Über äolische Abtragung und Formung in der Südest-Sahara. *Erdkunde*, **22**, 257–269.

Hagedorn, H. (1971) Untersuchungen über Relieftypen arider Räume an Beispielen aus dem Tibesti-Gebirge und seiner Umgebung. *Zeitschrift für Geomorphologie Supplement Band*, **11**, 1–251.

Halimov, M. and Fezer, F. (1989) Eight yardang types in central Asia. *Zeitschrift für Geomorphologie*, **33**, 205–217.

Hall, K. (1989) Wind blown particles as weathering agents? An Antarctic example. *Geomorphology*, **2**, 405–410.

Haynes Jr, C.V. (1980) Geologic evidence of pluvial climates in the El Nabta area of the Western Desert, Egypt, in *Prehistory of the Eastern Sahara* (eds. F. Wendorf and R. Schild), Academic Press, New York, pp. 353–371.

Haynes Jr, C.V. (1982) The Darb El-Arba'in desert: a product of Quaternary climatic change, in *Desert Landforms of Southwest Egypt: A Basis for Comparison with Mars* (eds F. El-Baz and T.A. Maxwell), National Aeronautics and Space Administration, Contractor Report 3611, Washington, DC, pp. 91–117.

Haynes, C.V., Mehringer Jr, P.J. and Zaghloul, E.-S.A. (1979) Pluvial lakes of North-Western Sudan. *The Geographical Journal*, **145**, 437–445.

Hedin, S. (1903) *Central Asia and Tibet*, Greenwood Press, New York.

Hobbs, W.H. (1917) The erosional and degradational processes of deserts, with especial reference to the origin of desert depressions. *Annals of the Association of American Geographers*, **7**, 25–60.

Hörner, N.G. (1932) Lop-nor. Topographical and Geological Summary. *Geografiska Annaler*, **14**, 297–321.

Inbar, M. and Risso, C. (2001) Holocene yardangs in volcanic terrains in the southern Andes, Argentina. *Earth Surface Processes and Landforms*, **26**, 657–666.

Jackson, P.S. and Hunt, J.C.R. (1975) Turbulent flow over a low hill. *Quarterly Journal of Royal Meteorological Society*, **101**, 929–955.

Jianjun, Q., Ning, H., Guangrong, D. and Weimin, Z. (2001) The role and significance of the Gobi Desert pavement in controlling sand movement on the cliff top near the Dunhuang Magao Grottoes. *Journal of Arid Environments*, **48**, 357–371.

King, L.C. (1936) Wind-faceted stones from Marlborough, New Zealand. *Journal of Geology*, **44**, 201–213.

Knight, J. (2008) The environmental significance of ventifacts: a critical review. *Earth-Science Reviews*, **86**, 89–105.

Knight, J. and Burningham, H. (2003) Recent ventifact development on the central Oregon coast, western USA. *Earth Surface Processes and Landforms*, **28**, 87–98.

Krinsley, D.B. (1970) A geomorphological and palaeoclimatological study of the playas of Iran, US Geological Survey Final Report, Contract PRO CP, pp. 70–800.

Kuenen, Ph. H. (1960) Experimental abrasion 4: eolian action. *Journal of Geology*, **68**, 427–449.

Laity, J.E. (1987) Topographic effects on ventifact development, Mojave Desert, California. *Physical Geography*, **8**, 113–132.

Laity, J.E. (1992) Ventifact evidence for Holocene wind patterns in the east-central Mojave Desert. *Zeitschrift für Geomorphologie, Supplement Band*, **84**, 1–16.

Laity, J.E. (1994) Landforms of aeolian erosion, in *Geomorphology of Desert Environments* (eds A.D. Abrahams and A.J. Parsons), Chapman & Hall, London, pp. 506–535.

Laity, J. E. (1995) Wind abrasion and ventifact formation in California, in *Desert Aeolian Processes* (ed. V.P. Tchakerian), Chapman & Hall, London, pp. 295–321.

Laity, J. E. (2009) Landforms, landscapes, and processes of aeolian erosion, in *Geomorphology of Desert Environments*, 2nd edn (eds A.J. Parsons and A.D. Abrahams), Springer Science+Business Media B.V., pp. 597–627.

Laity, J. E. and Bridges, N.T. (2009) Ventifacts on Earth and Mars: analytical, field, and laboratory studies supporting sand abrasion and windward feature development. *Geomorphology*, **105**, 202–217.

Lancaster, N. (1984) Characteristics and occurrence of wind erosion features in the Namib Desert. *Earth Surfaces Processes and Landforms*, **9**, 469–478.

Lancaster, N. (1985) Variations in wind velocity and sand transport on the windward flanks of desert sand dunes. *Sedimentology*, **32**, 581–593.

Lawn, B. (1995) *Fracture of Brittle Solids*, 2nd edn, Cambridge Solid State Science Series, Cambridge.

Liu, L.-Y., Dong, Z.-B., Gao, S.-Y. *et al.* (2003) Wind tunnel measurements of adobe abrasion by blown sand: profile characteristics in relation to wind velocity and sand flux. *Journal of Arid Environments*, **53**, 351–63.

McCauley, J.F., Breed, C.S. and Grolier, M.J. (1977) Yardangs, in *Geomorphology in Arid Regions* (ed. D.O. Doehring), Allen and Unwin, Boston, pp. 233–269.

McKenna Neuman, C. and Gilbert, R. (1986) Aeolian processes and landforms in glaciofluvial environments of southeastern Baffin Island, N.W.T., Canada, in *Aeolian Geomorphology* (ed. W.G. Nickling), Allen and Unwin, Boston, pp. 213–235.

McKenna Neuman, C., Lancaster, N. and Nickling, W.G. (1997) Relations between dune morphology, air flow, and sediment flux on reversing dunes, Silver Peak, Nevada. *Sedimentology*, **44**, 1103–1113.

Mahmoudi, F. (1977) Les nebkhas de Lut, Iran. *Annales de Géographie*, **77**, 296–322.

Mainguet, M. (1968) Le Bourkou – Aspects d'un modelé éolien. *Annales de Géographie*, **77**, 296–322.

Mainguet, M. (1970) Un étonnant paysage: les cannelures gréseuses du Bembéché (N. du Tchad). Essai d'explication géomorphologique. *Annales de Géographie*, **79**, 58–66.

Mainguet, M. (1972) *Le Modelé des Grès*, Institute Geographie National, Paris.

Mainguet, M. (1996) The Saharo–Sahelian global wind action system: one facet of wind erosion analysed at a synoptic scale, in *Wind Erosion in West Africa: The Problem and Its Control* (eds B. Buerkert, B.E. Allison and M.V. Oppen), Proceedings of the International Symposium, University of Hohenheim, Germany, 5–7 December 1994, Weikersheim, Germany, pp. 7–22.

Maizels, J.K. (1987) Plio-Pleistocene raised channel systems of the western Sharqiya (Wahiba), Oman. *Geological Society, London, Special Publications*, **35**, 31–50.

Maxson, J.H. (1940) Fluting and faceting of rock fragments. *Journal of Geology*, **48**, 717–751.

Mildrexler, D.J., Zhao, M. and Running, S.W. (2006) Where are the hottest spots on Earth? *EOS, Transactions of the American Geophysical Union*, **87**, 461–467.

Mulligan, K.R. (1988) Velocity profiles measured on the windward slope of a transverse dune. *Earth Surface Processes and Landforms*, **13**, 573–582.

Neev, D. and Hall, J.K. (1992) Counterclockwise converging basement fracturing patterns across the Arabian Peninsula and Eastern Iran, in *Basement Tectonics* (ed. R. Mason), International Basement Tectonics Association Publication 7, pp. 13–31.

Nero, R.W. (1988) The ventifacts of the Athabasca sand dunes. *The Musk Ox*, **36**, 44–50.

Powers, W.E. (1936) The evidences of wind abrasion. *Journal of Geology*, **44**, 214–219.

Ritley, K. and Odontuya, E. (2004) Yardangs and dome dunes northeast of Tavan Har, Gobi, Mongolia, Geological Society of America Abstract with Programs: Rocky Mountain and Cordilleran Joint Meeting, vol. 36, p. 33.

Scattergood, R.O. and Routbort, J.L. (1983) Velocity exponent in solid-particle erosion. *Journal of the American Ceramic Society*, **66**, C184–C186.

Schlyter, P. (1994) Paleo-periglacial ventifact formation by suspended silt or snow. Site studies in south Sweden. *Geografiska Annaler, Series A, Physical Geography*, **76A**, 187–201.

Schoewe, W.H. (1932) Experiments on the formation of wind-faceted pebbles. *American Journal of Science*, **24**, 111–134.

Schwanghart, W. and Schütt, B. (2008) Meteorological causes of Harmattan dust in West Africa. *Geomorphology*, **95**, 412–428.

Selby, M.J. (1977) Palaeowind directions in the central Namib Desert, as indicated by ventifacts. *Madoqua*, **10**, 195–198.

Sharp, R.P. (1949) Pleistocene ventifacts east of the Big Horn Mountains, Wyoming. *Journal of Geology*, **57**, 173–195.

Sharp, R.P. (1964) Wind-driven sand in Coachella Valley, California. *Bulletin of the Geological Society of America*, **75**, 785–804.

Sharp, R.P. (1980) Wind-driven sand in Coachella Valley, California: further data. *Bulletin of the Geological Society of America*, **91**, 724–730.

Smith, H.T.U. (1963) Eolian geomorphology, wind direction, and climatic change in North Africa, Air Force Cambridge Reseach Laboratories, Bedford, MA, AFCRL – 63-443, 48 pp.

Smith, H.T.U. (1967) Past versus present wind action in the Mojave Desert region, California, Air Force Cambridge Research Laboratories, 67-0683, pp. 1–26.

Smith, R.S.U. (1984) Eolian geomorphology of the Devils Playground, Kelso Dunes and Silurian Valley, California, in *Western Geological Excursions. Vol. 1: Geological Society of America 97th Annual Meeting Field Trip Guidebook, Reno, Nevada* (ed. J. Lintz), Geological Society of America, pp. 239–251.

Songqiao, Z. and Xuncheng, X. (1984) Evolution of the Lop Desert and the Lop Nor. *The Geographical Journal*, **150**, 311–321.

Stapff, F.M. (1887) Karte des unteren Khuisebtals. *Petermanns Geographische Mitteilungen*, **33**, 202–214.

Sun, J. (2002) Source regions and formation of the loess sediments on the high mountain regions of northwestern China. *Quaternary Research*, **58**, 341–351.

Taylor, P.A. (1977) Numerical studies of neutrally stratified planetary boundary-layer flow above gentle topography. *Boundary-Layer Meteorology*, **12**, 37–60.

Todd, M.C., Washington, R., Martins, J.V. *et al.* (2007) Mineral dust emission from the Bodélé Depression, northern Chad, during BoDEx 2005. *Journal of Geophysical Research*, **112**, D06207, DOI: 10.1029/2006JD007170.

Tremblay, L.P. (1961) Wind striations in northern Alberta and Saskatchewan, Canada. *Bulletin of the Geological Society of America*, **72**, 1561–1564.

Tsoar, H. (1985) Profiles analysis of sand dunes and their steady state signification. *Geografiska Annaler*, **67A**, 47–61.

Twidale, C.R. (1994) Desert landform evolution: with special reference to the Australian experience. *Cuaternario y Geomorfología*, **8**, 3–31.

Vincent, P. and Kattan, F. (2006) Yardangs on the Cambro-Ordovician Saq Sandstones, North-West Saudi Arabia. *Zeitschrift für Geomorphologie*, **50**, 305–320.

Ward, A.W. and Greeley, R. (1984) Evolution of the yardangs at Rogers Lake, California. *Bulletin of the Geological Society of America*, **95**, 829–837.

Washington, R. and Todd, M.C. (2005) Atmospheric controls on mineral dust emission from the Bodélé Depression, Chad: the role of the low level jet. *Geophysical Research Letters*, **32**, L17701.

Washington, R., Todd, M.C., Lizcano, G. *et al.* (2006) Links between topography, wind, deflation, lakes and dust: the case of the Bodélé Depression, Chad. *Geophysical Research Letters*, **33**, L09401.

West, M.D., Clarke, J.D.A., Thomas, M. *et al.* (2009) The geology of Australian Mars analogue sites. *Planetary and Space Science*, **DOI**: 10.1016/j.pss.2009.06.012.

Whitney, M.I. (1978) The role of vorticity in developing lineation by wind erosion. *Bulletin of the Geological Society of America*, **89**, 1–18.

Whitney, M.I. (1979) Electron micrography of mineral surfaces subject to wind-blast erosion. *Bulletin of the Geological Society of America*, **90**, 917–934.

Whitney, M.I. (1985) Yardangs. *Journal of Geological Education*, **33**, 93–96.

Wilshire, H.G., Nakata, J.D. and Hallet, B. (1981) Field observations of the December 1977 wind storm, San Joaquin Valley, Californisa, in *Desert Dust* (ed. T.J. Péwé), Geological Society of America Special Paper 186, pp. 233–251.

Wilson, I.G. (1971) Desert sandflow basins and a model for the development of ergs. *The Geographical Journal*, **137**, 180–199.

Worrall, G.A. (1974) Observations on some wind-formed features in the southern Sahara. *Zeitschrift für Geomorphologie*, **18**, 291–302.

V

Living with dryland geomorphology

22
The human impact

Nick Middleton

22.1 Introduction

Human use and occupation of desert and semi-desert environments has persisted over millennia and has inevitably left its mark on dryland geomorphology. The scale of impacts has grown with the rise in numbers of people living in desert areas and the efficacy of their technologies, although significant effects can also be noted by small numbers of people using rudimentary techniques. The range of impacts is wide and occurs both deliberately and inadvertently, directly and indirectly.

The number of people living in deserts was put at just over 500 million by Ezcurra (2006), where 'desert' was confined to arid and hyper-arid regions. A broader definition, adding semi-arid and dry subhumid climatic zones and more frequently called the 'drylands' (Middleton and Thomas, 1997), increases that population to about 2.1 billion people (Safriel *et al.*, 2005). These totals have been reached by rapid rates of growth in many parts of the world in recent decades. Some desert cities, in particular, have experienced very fast rates of urbanisation. They include Cairo, Nouakchott, Riyadh, Sana'a, Dubai, Tehran, Las Vegas and Phoenix. Such large agglomerations of people have made an impact on-site but also off-site via the significant ecological footprints of major urban areas. The human impact on geomorphology is not simply a numbers game, however. Many changes to landforms and geomorphological processes can also be attributed to human activities carried out at relatively low population densities, such as agricultural activities. Indeed, some of the largest human impacts on geomorphology – those attributable to desert nuclear weapons testing – have been allowed in dryland areas simply because very few people inhabit them (although such effects have actually received little attention in the geomorphological literature).

A useful distinction can be made between direct and deliberate modifications of landforms and processes, on the one hand, and indirect inadvertent changes, on the other. Generally, the former are much more easily recognised than the latter. The construction of terraces on a hillside, for example, is an ancient form of agricultural land management that is clear-cut. The acceleration of erosion in an area where the natural vegetation cover has been modified by some human activity, such as grazing or burning, is a much more difficult situation in which to attribute causation.

This chapter will assess the human impact on arid-zone geomorphology, ancient and modern, deliberate and inadvertent, direct and indirect, with a range of examples illustrating effects at a variety of spatial scales.

22.2 Human impacts on soils

Soil cover is patchy in most deserts but becomes more continuous towards desert margins. Given than human use of desert resources is generally easier in the less arid parts of drylands, most of the examples of human impact on arid-zone soil geomorphology have occurred in the wetter parts of desert regions.

22.2.1 Terracing and rainwater harvesting

A number of agricultural techniques have been developed over the generations to maximise the utility of what little surface water is available in drylands. This is achieved by modifying the land surface in some way to concentrate water into relatively small areas, which are then cultivated or used to graze livestock. Many of these ancient

Arid Zone Geomorphology: Process, Form and Change in Drylands, Third Edition. Edited by David S. G. Thomas

Table 22.1 Some examples of indigenous soil and water conservation systems from African drylands.

Country (region)	*Average annual ppt (mm)*	*Technique(s)*	*Major crops*
Burkina Faso (Mossi)	400–700	Stone lines, stone terraces, planting pits	Sorghum, millet
Chad (Ouddal)	250–650	Earth bunds	Sorghum, millet
Mali (Djenne-Sofara)	400	Pitting systems	Sorghum, millet
Niger (Ader Doutchi Maggia)	300–500	Stone lines, planting pits	Sorghum, millet
Somalia (Hiiraan)	150–300	Earth bunds	Sorghum, cowpeas
Sudan (West)	50–800	'Wadi agriculture', various water-harvesting techniques	Sorghum, vegetables

Source: after Critchley, Reij and Willcocks (1994).

techniques are well documented from deserts in Africa, the Middle East, North and South America, and are still widely used today (Gilbertson, 1986; Motsi, Chuma and Mukamuri, 2004; Oweis and Hachum, 2006). Some contemporary examples from the drier parts of Africa are shown in Table 22.1.

Terracing is a very ancient mechanical soil conservation technique that has been employed for up to 5000 years in Yemen, for example (Wilkinson, 1997). Terraces are commonly built on steep slopes and are designed to transform them by creating a series of horizontal soil strips along the slope contours. They intercept runoff, reducing its flow to a nonerosive velocity, thus also acting to conserve water. If well maintained, terraces are very effective, but they are costly to construct and their physical dimensions act as a constraint on the use of mechanised agriculture.

Check dams are constructed in a similar way to terraces and are designed to control erosion in normally dry valleys. The dam impedes the flow of water generated by the occasional rain storm, causing the deposition of sediment – which helps to build up a soil – and encouraging the ponded water to soak into this alluvial fill. A small check dam might support a single fruit tree, for example. Larger dams may allow a cereal crop to be planted in a small field. In Tunisia this type of small check dam and its associated terraced area is called a 'jessr' (plural: jessour). Jessour, which cover an area of about 4000 km^2 in the south of the country, allow crops to be grown in an environment that is otherwise too arid for agriculture. The most common crops are olive trees and drought-resistant annual grains with a short growing period, such as wheat and barley (Schiettecatte *et al.*, 2005).

More sophisticated systems for runoff farming or water harvesting involve long stone bunds or walls that snake across low-angle slopes to channel any storm-generated overland flow towards a target portion of the valley floor. A greater flow of water can be generated over a large slope area by the clearance of obstacles to overland flow such as pebbles and boulders. Removing larger obstacles means that soil pores quickly become clogged with fine material when water flows across them, effectively sealing the surface, reducing infiltration and thus generating greater runoff (Evenari, Shanan and Tadmor, 1968).

A geographical review of these techniques by Bruins, Evenari and Nessler (1986) indicates that some of the oldest examples are found in the Middle East. In the Negev, rainwater harvesting dates back at least 2000 years but a well-developed runoff farming system at Jawa in Jordan seems to be at least 5000 years old. There is also some evidence to suggest that runoff was used for agriculture at a site called Beidha in the Edom Mountains in southern Jordan nearly 9000 years ago.

A similar range of techniques has also been employed by traditionally nomadic people in the Karakum Desert in Turkmenistan to secure water supplies for consumption by their livestock and to produce some crops for themselves and their animals (Fleskens *et al.*, 2007). These methods focus on flat or slightly sloping clay surfaces – 'takyrs' – with little or no vegetation, which act as natural catchment areas.

Several types of hand-dug basin that collect rainwater runoff provide temporary storage points for watering livestock on takyr surfaces. In some cases dome-shaped brick covers have been built over these small reservoirs to reduce evaporation. Another technique is to increase the productivity of small depressions in the takyr, where water collects naturally, by digging trenches to channel a greater flow of runoff. The depressions, or 'oytaks', have a sandy topsoil and are used for hay production, but crops such as melons and gourds are also grown. Typically, a cultivated oytak of 10 m^2 requires at least 1000 m^2 of catchment to provide it with adequate moisture.

Should any of these techniques designed to maximise the agricultural utility of surface water in drylands be discontinued for whatever reason, a further set of geomorphological changes can be expected. Lesschen, Cammeraat and Nieman (2008) highlight this issue in the Mediterranean basin and note how abandoned fields in semi-arid areas are more vulnerable to gully erosion due to the formation of soil crusts with low infiltration rates and enhanced runoff. Their work in southeastern Spain demonstrates the susceptibility to greater erosion of abandoned terraces that deteriorate due to lack of maintenance.

22.2.2 Irrigated agriculture

The practice of irrigated agriculture also has a long history, particularly on some of the major rivers that flow through the deserts of North Africa, the Middle East and Southwest Asia. Irrigation can affect geomorphology in a number of ways, both directly and indirectly. The excessive use of groundwater, for example, can lead to land subsidence and declines in streamflow, as well as the reduction or complete loss of vegetation, with a range of geomorphological implications, as Zektser, Loáiciga and Wolf (2005) outline using case studies from aquifers in the southwestern USA. Soil piping may also occur on irrigated fields, resulting in the loss of much water and soil through the pipe networks (e.g. García-Ruiz and Lasanta, 1995). Perhaps the most widespread and most serious geomorphological impact of irrigated agriculture arises when previously productive soil becomes saline as a result of poor land management, so-called *secondary* salinisation.

Soil salinity problems affected farmers in Mesopotamia 4000 years ago (Jacobsen and Adams, 1958) and continue to be particularly prevalent in, although not solely confined to, drylands today. The issue receives much attention largely because major agricultural crops have a low salt tolerance. Secondary salinisation is most commonly associated with poorly managed irrigation schemes and occurs in four main ways: water leakage from supply canals, overapplication of water, poor drainage and insufficient application of water to leach salts away. One estimate suggested that nearly 50 % of all the irrigated land in arid and semi-arid regions is affected to some extent by secondary salinisation (Abrol, Yadav and Massoud, 1988) and the process is widely regarded as irrigated agriculture's most significant environmental problem (Ghassemi, Jakeman and Nix, 1995; Gardner, 1997; Middleton and van Lynden, 2000).

The impact of salinisation on crop yields acts indirectly via effects on the soil and through direct effects on the plants themselves (Rhoades, 1990; Tanji, 1990). Salt accumulation reduces soil pore space and the capability of holding soil air, moisture and nutrients, resulting in a deterioration of soil structure and a reduction in the soil's suitability as a growing medium. Plant growth is also directly impaired by salinisation through its effects on osmotic pressures and via direct toxicity. Contact between a solution containing large amounts of dissolved salts and a plant cell causes the cell's protoplasmic lining to shrink due to the osmotic movement of water from the cell to the more concentrated soil solution. As a result the cell collapses and the plant succumbs. High salt concentrations are also toxic to many plants. The toxicity effect varies according to the type of salt and the species of plant, but is especially potent during the seedling stage of all plants. Indeed, some salts, such as boron, are highly toxic to many crops when present in a soil solution at concentrations of only a few parts per million (Bingham, Rhoades and Keren, 1985).

The salt tolerance of a plant is not an exact value since it depends on factors such as soil fertility, the salt distribution in the soil profile, irrigation methods and climate, as well as biological factors including the stage of growth, plant variety and rootstock. Nonetheless, different crops are susceptible to different concentrations and forms of salinity (Maas, 1990). Any given crop has a threshold limit of tolerance to soil salinity beyond which yields will decrease linearly per unit increase in salinity. Table 22.2

Table 22.2 Salt tolerance of selected crops with respect to the electrical conductivity of the saturated-soil extract.

Crop	*Threshold (μS/cm)*	*Yield decline (%) per 1000 μS/cm increase in salinity above threshold*
Sensitive		
Apricot	1600	24.0
Carrot	1000	14.0
Grapefruit	1800	16.0
Onion	1200	16.0
Moderately sensitive		
Alfalfa	2000	7.3
Cucumber	2500	13.0
Sugarcane	1700	5.9
Tomato	2500	9.9
Moderately tolerant		
Barley (forage)	6000	7.1
Sorghum	6800	16.0
Soybean	5000	20.0
Tolerant		
Barley (grain)	8000	5.0
Cotton	7700	5.2
Sugar beet	7000	5.9

Source: after Maas (1990).

shows these thresholds and yield declines for some common crops.

Salinisation is one of the most clear-cut ways in which mismanagement can lead to desertification, an issue that is otherwise surrounded by controversy (see below). Irrigation schemes in some of the dryland countries of Asia are among the most seriously affected. In Pakistan, where irrigated land supplies more than 90 % of agricultural production, salinity problems affect about a quarter of the irrigated area or approximately 11 % of the country's land area (Middleton and van Lynden, 2000). Most of the affected soils are part of the Indus irrigation system, the largest single irrigation system in the world. Salinity and the associated problems of waterlogging are estimated to cost farmers in Pakistan about 25 % of their production potential for major crops (Qureshi *et al.*, 2008).

22.2.3 Accelerated erosion

Human-induced changes in erosion rates are well documented from drylands all over the world and can be attributed to a wide variety of activities. In Australian rangelands, for example, the introduction of both domestic and feral herbivores by Europeans in the nineteenth century is believed to have widely increased rates of soil erosion (Wasson and Galloway, 1986). The landscape of arid western New South Wales is not atypical, having experienced accelerated erosion by wind and water though sheetwash, rilling, gullying and aeolian deflation (Fanning, 1999).

Large-scale military movements have been the cause of enhanced deflation due to the disruption of desert surfaces in several cases, not least during the North African campaign in the early 1940s (Oliver, 1945). Wind erosion dramatically increased during the height of the fighting in North Africa, but deflation had returned to pre-war levels within a few years of the end of the campaign. The effects are longer lasting in desert areas used for training, where large military vehicles continually compact soil, crush and shear vegetation, and alter the structure of plant communities. The US National Training Centre in the Mojave Desert is a case in point (Caldwell, McDonald and Young, 2006).

Agricultural activities on dryland soils have often resulted in increased soil erosion. Any farmer who cultivates a crop that leaves substantial areas of soil uncovered can expect rapid soil erosion during intense rain storms or strong wind storms. Examples include woody dryland crops such as almonds, olives and grapes. In other circumstances, erosion has been exacerbated by the introduction of mechanised agriculture with its deep ploughing and large open fields, disturbing soil structure and increasing susceptibility to erosive forces. Similar outcomes have also been noted in areas where cultivation has expanded into new zones that are marginal for agricultural use because they are more prone to drought or are made up of steeper slopes that are more susceptible to erosion.

Probably the most infamous case of wind erosion came after widespread conversion of grasslands to cereal cultivation in the US Great Plains, which helped to create the notorious Dust Bowl during a period of drought in the 1930s (Worster, 1979). The most severe dust storms (so-called 'black blizzards') occurred between 1933 and 1938, with maximum wind erosion taking place during the spring of these years. At Amarillo, Texas, at the height of the period, one month had 23 days with at least ten hours of airborne dust and one in five storms had zero visibility (Choun, 1936). In 1937 the US Soil Conservation Service estimated that 43 % of a 6.5 million ha area in the heart of the Dust Bowl had been seriously damaged by wind erosion. Similar widespread deflation of soils resulted from analogous expansion of cultivation into grasslands in the Argentine Pampas during the 1930s and 1940s (Viglizzo and Frank, 2006) and after the 1950s Virgin Lands Scheme in the former USSR.

Increased wind erosion has also frequently been an off-site result of human activities, particularly those associated with drainage or water diversion that has led to the desiccation of water bodies. Dried lakebeds have, in consequence, become major new sources of airborne dust in several parts of the world (Table 22.3). A review of anthropogenically desiccated playas and their associated erosion is provided by Gill (1996). One of the best-known examples has occurred in Central Asia over 50 years or so, a period during which off-takes from two major rivers, the Amu Darya and Syr Darya, to irrigate plantations, predominantly growing cotton, have had dramatic consequences for the Aral Sea (Middleton, 2002; Micklin, 2007). Expansion of the irrigated area in the former Soviet region of Central Asia, from 2.9 million hectares in 1950 to about 7.2 million hectares by the late 1980s, resulted in the annual inflow to the Aral from the two rivers, the source of 90 % of its water, declining by an order of magnitude between the 1960s (about 55 m^3/yr) and the 1980s (about 5 m^3/yr).

In 1960, the Aral Sea was the fourth largest lake in the world, but since that time its surface area has more than halved, it has lost two-thirds of its volume and its water level has dropped by more than 20 m. The average water level in the Aral Sea in 1960 was about 53 m above sea level. By early 2003, it had receded to about 30 m a.s.l., a level last seen during the fourteenth–fifteenth centuries, but in those days for predominantly natural reasons (Boroffka *et al.*, 2006).

Table 22.3 Some examples of playas where sediments have been mobilised by wind as a result of human activity.

Site	*Country*	*Cause*	*Effects*	*Reference*
Aral Sea	Kazakhstan/ Uzbekistan	Water diversion	Severe dust storms	Stulina and Sektimenko (2004)
Sebkha Kourzia	Tunisia	Deforestation	Wind erosion, lunette formation	Perthuisot (1989)
Coastal sabkhas on Persian Gulf	Saudi Arabia	Grazing, vegetation loss	Blowing sand, dune formation	Anton (1982)
Kappakoola Swamp	Australia	Vegetation loss	Sand dune formation	Smith, Twidale and Bourne (1975)
Owens Lake	USA	Water diversion	Severe dust storms	Gill and Cahill (1992)
Lop Nor	China	Water diversion	Blowing sand and saline dust storms	Zhao (1986)

Source: after Gill (1996).

These dramatic changes have had far-reaching effects. The delta areas of the Amu Darya and Syr Darya Rivers have been transformed due to the lack of water, affecting flora, fauna and soils, while the diversion of river water has also resulted in the widespread lowering of groundwater levels. The receding sea has had local effects on climate and the exposed bed of the Aral has become a major regional dust source, from which an estimated 43 million tonnes of saline material is deposited on surrounding areas each year. The resulting hazards to local populations, which include contamination of agricultural land up to several hundred kilometres from the sea coast and suspected adverse effects on human health, are covered in detail in Chapter 23.

22.2.4 Grazing

The effects of animals on the landscape, so-called zoogeomorphology (Butler, 1995), is relatively understudied but is probably important in certain places. Among dryland environments, rangelands come particularly to mind. Herds of domesticated ungulates can affect geomorphology directly, their hooves causing soil compaction, reduced infiltration, increased runoff and sediment yield. In riparian zones, trampling can also directly degrade banks, with consequent effects on erosion (Trimble and Mendel, 1995).

A range of indirect geomorphological impacts also occur through the effects of grazing on vegetation. One widespread effect of intensive grazing is the encroachment of unpalatable or noxious shrubs into rangelands. Long-term grazing of semi-arid grasslands can typically lead to an increase in the spatial and temporal heterogeneity of water, nitrogen and other soil resources, which promotes invasion by desert shrubs, in turn leading to a further localisation of soil resources under shrub canopies in a process of positive feedback. In the barren areas between shrubs, soil fertility is decreased by erosion and gaseous emissions (Schlesinger *et al.*, 1990). Increased runoff and erosion strip the soil surface layer, promoting the formation of desert pavement in intershrub areas and the development of rills. In a similar way, grazing may contribute to the formation of banded patterns in vegetation found across desert-marginal landscapes in many parts of the world (Barbier *et al.*, 2006), which are often refered to as *brousse tigrée* after the name coined in Sahelian West Africa.

Domesticated herds have been blamed for notable transformations of landscapes in several parts of the world. One example is the series of dramatic changes in the rangeland vegetation of southern New Mexico, which have been documented since the 1880s and the beginning of intensive grazing by cattle and sheep. Typically, native grasslands have been displaced by shrublands with concomitant increases in soil erosion, stream channel cutting and the emergence of shrub coppice dunes on sandy soils (Bahre, 1991).

Other, no less significant, zoogeomorphological effects can occur through less direct human influence. A couple of examples highlighted by Butler (2006) illustrate the potential scale and magnitude of impacts, although much work remains to be done on quantifying the effects. The huge reduction through hunting of the native population of bison (*Bison bison*) in North American rangelands, and their widespread replacement with domesticated cattle, will have had significant geomorphological implications. The trampling effects of bison may be similar to those of cattle, but bison also wallow whereas cattle do not.

Pawing, rolling and trampling in roughly circular depressions significantly increases the bulk density of soil in wallows relative to the adjacent landscape and the compaction reduces infiltration, so that wallows become local ponds that can retain water for several days after a rainstorm. The number of bison wallows in North America prior to European contact has been put at more than 100 million and each displaced up to 23 m^3 of sediment.

This example can be contrasted with the impacts of the introduced European rabbit (*Oryctolagus cuniculus* L.) in Australia. Feral rabbits have had widespread effects across the semi-arid landscapes of the continent, directly through burrowing and trampling but also indirectly via profound alterations to the vegetation of Australian rangelands. The effects on microtopography, infiltration, sediment movement and hydrology are largely unmeasured.

22.3 Human impacts on sand dunes

There are numerous examples of human activities in drylands modifying the forms of sand dunes and the processes operating on them. Dunes that support some vegetation, particularly common in semi-arid desert areas, frequently attract both farmers and herders, but these systems can be easily tipped towards degradation of both vegetation and dune soils. Any reduction in vegetation cover – by factors such as grazing, cultivation and burning, but also due to natural factors such as drought – can increase the potential for sediment transport and surface change. The interaction with drought was noted by Kumar and Bhandari (1993) in their examination of semi-arid sites in northwest Rajasthan, India. This study identified a positive feedback of vegetation destruction, increased dune mobility and exacerbated surface instability as pressure from human and livestock populations has intensified.

Increasing human activities in drylands has also meant a rise in the number of instances in which mobile dunes become hazardous to economic interests. Dunes can affect transport networks, agricultural productivity, residential areas and water supplies. In the early years of the Trans-Caspian railway, each train carried a special team of workers to clear the tracks of drifting sand, a problem that led directly to the creation in 1912 of the Desert Research Station at Repetek in Turkmenistan (Babaev, 1999). Research at Repetek and elsewhere has spawned a variety of techniques to stabilise sand dunes, ranging from shelterbelts, barriers or fences to surface treatments with chemicals. The approach favoured in many Chinese deserts where dunes threaten economic activities is to plant vegetation on them using straw checkerboards initially protected from winds by barriers made of willow branches or bamboo. Li *et al.* (2003) report that this approach has been used successfully along the Baotou–Lanzhou railway in the Tengger Desert since the late 1950s. The straw checkerboards, which remain intact for 4 or 5 years, aid the establishment of xerophytic shrubs by increasing the roughness of the sand surface by 400–600 times and reducing wind velocity by 20–40 % at a height of 0.5 m above the surface. The stabilised sand surface also provides good conditions for colonisation by annual plants and the formation of biogenic soil crusts.

22.4 Human impacts on rivers

Rivers have played an important roll in moulding the geography of human activities in all biomes, but their significance in deserts has been particular, affecting not only the location of societies but also their political, economic and social development (Wittfogel, 1958; Butzer, 1976). The diversity of causes of change to the geomorphology of river systems is great and people have undoubtedly played a role in many cases. Societies have deliberately manipulated many river channels in arid areas, particularly to assist with flood control and as an aid to irrigation and navigation. Examples of such intentional changes include dams, channelisation, water extraction and river diversion, but there are also many examples of human action being responsible for the modification of river channel geometry unintentionally, and in numerous ways.

22.4.1 *Large dams*

Given the importance to society of a reliable water supply it is unsurprising that people have been building dams to manage water resources for up to 5000 years. Two of the oldest were built in desert areas to control flooding and to supply water for irrigation and domestic purposes: the Jawa Dam in Jordan and the Sadd al-Kafara in Egypt were both built on wadis some 5000 and 4600 years ago, respectively. However, the Sadd al-Kafara was destroyed by floods before it was ever used.

Flood control and water supply remain important reasons for dam construction today, along with hydroelectricity generation and the regulation of river flow. Structures above 15 m in height above their foundations are defined as large by the International Commission on Large Dams and major if they exceed 150 m. By the end of the twentieth century, there were more than 45 000 large dams in over 140 countries and approximately another 750 000 smaller dams globally (World Commission on Dams, 2000). What proportion of these structures is in deserts is unknown, but

Table 22.4 Changes in some hydrologic parameters caused by selected dams in the USA (percent difference in the means between pre-impact and post-impact conditions).

River (state)	*Dam(s)*	*1-day maximum* Q	*1-day minimum* Q	*Hydrograph rise rate*	*Hydrograph fall rate*
Bill Williams (Arizona)	Alamo	−73	19	−81	−55
Colorado (Arizona)	Glen Canyon	−64	31	22	62
Colorado (Texas)	EV Spence	−79	156	−92	−90
Coyote (California)	Coyote and Anderson	−79	0	−91	−79
Leon (Texas)	Belton	−72	−73	−78	−54

Source: after Magilligan and Nislow (2005).

the number of dams located in drylands is undoubtedly considerable. The largest of these are situated on exotic rivers (with perennial flows) that run through deserts.

Geomophologically, the effects of damming a river and creating a reservoir are numerous. Most dams reduce flow downstream and hence reduce the power and sediment-carrying capacity of the river. Peak discharges are also commonly reduced. Table 22.4 shows changes in this and some other hydrologic parameters for a number of US dams. Any resultant changes in fluvial geomorphology will depend on many different aspects of the dam, including the size and shape of the reservoir and the flow release policy. The number of dams on a river may also have an important influence.

Dam construction and reservoir impoundment raises the local base-level and can lead to sedimentation in the reservoir and channel aggradation upstream. A typology of the effects of dams on downstream geomorphology was devised by Brandt (2000) based on the balance between water discharge, sediment load, grain size and river slope. Nine cases were identified, each involving different combinations of changes in slope, cross-section, bedform and planform.

Graf (2005) notes that channel degradation and armouring were two downstream issues that received considerable early attention from geomorphologists working on dam effects in North America. Both processes result from the clear-water discharge from dams that have trapped sediments in their reservoirs.The loss of sediment due to reservoir impoundment can be considerable. On the Colorado River in the USA the amount of sediment transported declined from 1500 parts per million before the Glen Canyon Dam was closed in 1963 to just 7 ppm thereafter (Graf, 1988). In other deserts, fluvial sediment lost due to reservoir impoundment may be replaced, at least in part, by inputs of windblown sand. This is the situation on the reach of the Yellow River that flows through the Ulan Buh Desert below the Liujiaxia and Longyangxia reservoirs in northern China, where Ta, Xiao and Dong (2008) demonstrate that the river channel has experienced considerable aggradation.

Such channel changes usually begin in the initial stages of dam construction and continue throughout the life of the dam thanks to fluctuations of the water level in the reservoir. At the end of a dam's useful life it may be decommissioned and removed. The removal of large dams is a recent phenomenon and there is little geomorphological research on the effects published in formal, refereed outlets. In consequence, the development of mature scientific generalisations about fluvial processes related to dam removal is in its infancy (Graf, 2005).

22.4.2 Urbanisation

Urban development can have numerous effects on hydrological systems and fluvial geomorphology, both through deliberate interventions for hazard mitigation (e.g. from flash flooding) and resource exploitation (e.g. river sand and gravel) as well as via inadvertent effects (e.g. through large-scale introduction of impervious surfaces). However, our understanding of such effects is at a relatively early stage because the volume of research conducted in and around dryland cities lags some way behind that carried out in other environments (Cooke *et al.*, 1982; Chin, 2006). This is important because the particular dynamics of many dryland rivers may mean that they react to the effects of urbanisation in significantly different ways from those in temperate and tropical regions. The ephemeral nature of many streams (which can nevertheless typically carry high loads when water does flow), the extreme spatial and temporal variability in precipitation input and response mechanisms, along with rapid and irregular changes in channel morphology, all prompted Chin (2006, p. 475) to suggest that the impacts of urbanisation on dryland river channels are 'likely to be less predictable, more localized, and more variable than in humid–temperate streams'.

This greater variability in the morphological adjustments of desert rivers is corroborated by Laronne and Shulker (2002), who looked at several towns in the northern Negev and found cases of spectacular channel enlargement but also narrowing and a reduction in capacity. Unsurprisingly, the geomorphological effects of urbanisation are continuous, synergistic and occur concurrently with natural variability. A study of the Salt River in the Phoenix metropolitan area of central Arizona found that channel changes over more than 60 years were driven primarily by both large-scale regional flood events and local human activities (Graf, 2000). General changes in sinuosity of the low-flow channel were attributable to floods, although islands were found to have remained remarkably consistent in location and size, while channel-side bars waxed and waned. These changes occurred at the same time as extensive sand and gravel mining, which was the most important determinant of local channel form. These mining operations moved with the expanding urban fringe, which serves as a market for sand and gravel during construction.

22.4.3 Changes in vegetation

Any human activity that results in alterations to the vegetation found in an area is also likely to produce indirect changes in local geomorphology. People disturb, modify or clear the vegetation in a particular region for all sorts of reasons (e.g. fuelwood collection, cultivation, herding, construction, mining) and in many different ways (e.g. fire, selective cutting, clear-felling). Rivers can be affected by vegetation changes immediately adjacent to the channel itself, but also by more general changes in land use in a catchment.

The establishment of saltcedar (Tamarix spp.) along many rivers in the southwestern USA has caused significant aggradation of floodplains. Saltcedar stands are dense and have extensive root systems, leading to trapping of sediment and bank stabilisation. On the Brazos River in Texas, for instance, the plants encourage deposition on sandbars and at the edge of channels, where they become established. Between 1941 and 1979 the width of the channel was reduced in this way by nearly 90 metres as sediment some 3 metres thick was deposited (Blackburn, Knight and Schuster, 1982).

The effects of more general vegetation change, in this case due to long-term grazing pressure, have been studied by Allsopp *et al.* (2007) on an ephemeral river system in Namaqualand, South Africa. Both the physical and soil features of the river system were altered because of the reduced plant cover brought about by heavy grazing. Channels were found to be less braided, a higher proportion of the river width consisted of unvegetated runoff channels and the river system was generally dominated by very sandy soil. Interestingly, although plant cover increased along a section of the river system after removal of livestock for a period of nearly three years, no changes in river morphology were apparent during this time.

22.5 Cause and effect: the arroyo debate continues

The object of this chapter has been to review instances where human activities have effected geomorphological change, whether intentional or otherwise. There are many examples throughout the growing literature on environmental change in general, both contemporary and in the past, and the role played by people. The links between cause and effect are more obvious in some cases than in others. A landscape can be a complex place and the relationship between cause(s) and effect(s) is often complicated by feedbacks, timelags, thresholds and synergies.

On occasion, competing explanations for geomorphological phenomena persist for prolonged periods. Developments in our understanding of arroyo trenching in the US southwest provides a good example. Deep incision by valley-bottom gullies, or arroyos, occurred in many valleys and plains in the southwestern USA over a fairly short period between 1865 and 1915, a phenomenon that had a detrimental impact on settlement and economic activities in the area (Cooke and Reeves, 1976). Attributing causation for arroyo trenching in this part of the world has been the subject of lengthy debate and review (Elliott, Gellis and Aby, 1999).

The arrival of European settlers in the region coincides with arroyo trenching and a range of human activities that could have triggered incision has been suggested, including heavy grazing, logging, compaction along well-travelled routes and the cutting of grass for hay in valley bottoms. Conversely, a number of climatic drivers of change have also been put forward, and an essentially natural cause is supported by studies of valley fills that indicate repeated phases of aggradation and incision, some of which pre-date any possible significant human effects (Waters and Haynes, 2001). Shifts in climate towards more arid conditions, and indeed towards more humid conditions, have both been proposed. Another climatic explanation suggests that an increase in the frequency of heavy rainfall events would have enhanced the power of gullying. A third perspective (Schumm, Harvey and Watson, 1984) invokes neither climatic change nor human impact. It is also possible that arroyo incision occurred because a natural geomorphological threshold was crossed; after a period of relative stability in valley floors gullying was initiated by some triggering event.

Of course, the true explanation of arroyo incision in the US southwest during the late nineteenth and early twentieth centuries may involve any one of these competing theories – or indeed more than one, because they are not all mutually incompatible. The example serves as a reminder of the complexity of the natural world, a complexity that becomes greater still when the possible effects of people are added.

22.6 Conclusions

People have lived in deserts for a long time and have always had some influence over desert landforms and processes. These impacts have been both deliberate and inadvertent, the former being more easily assessed than the latter. The scale of human impact has undoubtedly grown with the rise in numbers of people living in desert areas, although this is not to say that impacts are predominantly recent nor that change in geomorphology cannot be induced by small numbers of people.

The human impact on desert geomorphology is certainly diverse and widespread. However, our understanding of cause and effect is not always clear and the difficulties of distinguishing between the effects of variability in purely natural geophysical drivers of change and human-induced drivers will certainly continue. Indeed, distinctions between the two are likely to become even more blurred as we grapple with changes wrought by global climate change possibly effected in part by human society (see Chapter 24). None the less, the division will continue because human actions represent the one set of driving forces for change that we can control in a predictable way. This truism is important not least because of the frequent negative feedbacks societies experience in the form of geomorphological hazards (see Chapter 23).

References

Abrol, I.P., Yadav, J.S.P. and Massoud, F.I. (1988) Salt-affected soils and their management. *FAO Soils Bulletin*, **39**.

Allsopp, N., Gaika, L., Knight, R. *et al.* (2007) The impact of heavy grazing on an ephemeral river system in the succulent karoo, South Africa. *Journal of Arid Environments*, **71**, 82–96.

Anton, D. (1982) Modern eolian deposits in the eastern province of Saudi Arabia, in *Abstracts of the Eleventh International Congress on Sedimentology*, p. 68.

Babaev, A.G. (1999) The natural conditions of central Asian deserts, in *Desert Problems and Desertification in Central Asia* (ed. A.G. Babaev), Springer-Verlag, Berlin, pp. 5–20.

Bahre, C.J. (1991) *A Legacy of Change*, The University of Arizona Press, Tucson, Arizona.

Barbier, N., Couteron, P., Lejoly, J. *et al.* (2006) Self-organized vegetation patterning as a fingerprint of climate and human impact on semi-arid ecosystems. *Journal of Ecology*, **94**, 537–547.

Bingham, F.T., Rhoades, J.D. and Keren, R. (1985) An application of the Maas–Hoffman salinity response model for boron toxicity. *Journal of the Soil Science Society of America*, **49**, 672–674.

Blackburn, W.H., Knight, R.W. and Schuster, J.L. (1982) Saltcedar influence on sedimentation in the Brazos River. *Journal of Soil and Water Conservation*, **37**, 298–301.

Boroffka, N., Oberhänsli, H., Sorrel, P. *et al.* (2006) Archaeology and climate: settlement and lake-level changes at the Aral Sea. *Geoarchaeology*, **21**, 721–734.

Brandt, S.A. (2000) Classification of geomorphological effects downstream of dams. *CATENA*, **40**, 375–401.

Bruins, H.J., Evenari, M. and Nessler, U. (1986) Rainwater-harvesting agriculture for food production in arid zones: the challenge of the African famine. *Applied Geography*, **6**, 13–32.

Butler, D.R. (1995) *Zoogeomorphology: Animals as Geomorphic Agents*, Cambridge University Press, Cambridge.

Butler, D.R. (2006) Human-induced changes in animal populations and distributions, and the subsequent effects on fluvial systems. *Geomorphology*, **79**, 448–459.

Butzer, K.W. (1976) *Early Hydraulic Civilization in Egypt: A Study in Cultural Ecology*, University of Chicago Press, Chicago.

Caldwell, T.G. McDonald, E.V. and Young, M.H. (2006) Soil disturbance and hydrologic response at the National Training Center, Ft. Irwin, California. *Journal of Arid Environments*, **67**, 456–472.

Chin, A. (2006) Urban transformation of river landscapes in a global context. *Geomorphology*, **79**, 460–487.

Choun, H.F. (1936) Dust storms in southwestern plains area. *Monthly Weather Review*, **64**, 195–199.

Cooke, R.U. and Reeves, R.W. (1976) *Arroyos and Environmental Change in the American Southwest*, Clarendon Press, Oxford.

Cooke, R.U., Brunsden, D., Doornkamp, J. and Jones, D. (1982) *Urban Geomorphology in Drylands*, Oxford University Press, Oxford.

Critchley, W.R.S., Reij, C. and Willcocks, T.J. (1994) Indigenous soil and water conservation: a review of the state of knowledge and prospects for building on traditions. *Land Degradation and Rehabilitation*, **5**, 293–314.

Elliott, J.G., Gellis, A.C. and Aby, S.B. (1999) Evolution of arroyos: incised channels of the southwestern United States, in *Incised River Channels: Processes, Forms, Engineering, and Management* (eds S.E. Darby and A. Simon), John Wiley & Sons, Inc., New York, pp. 153–185.

Evenari, M., Shanan, L. and Tadmor, N.H. (1968) 'Runoff farming' in the desert. I. Experimental layout. *Agronomy Journal*, **60**, 29–32.

Ezcurra, E. (ed.) (2006) *Global Deserts Outlook*, United Nations Environment Programme, Nairobi.

Fanning, P.C. (1999) Recent landscape history in arid western New South Wales, Australia: a model for regional change. *Geomorphology*, **29**, 191–209.

Fleskens, L., Ataev, A., Mamedov, B. and Spaan, W.P. (2007) Desert water harvesting from takyr surfaces: assessing the potential of traditional and experimental technologies in the Karakum. *Land Degradation and Development*, **18**, 17–39.

García-Ruiz J.M. and Lasanta T. (1995) The effects of irrigation on soil piping. A case study in the Ebro Depression, Spain. *Physics and Chemistry of the Earth*, **20**, 315–320.

Gardner, G. (1997) Preserving global cropland, in *State of the World 1997* (eds L. Brown, C. Flavin and H. French), Norton & Co., New York, pp. 42–59.

Ghassemi, F., Jakeman, A.J. and Nix, H.A. (1995) *Salinisation of Land and Water Resources: Human Causes, Extent, Management and Case Studies*, University of New South Wales Press, Sydney.

Gilbertson, D.D. (1986) Runoff (floodwater) farming and rural water supply in arid lands. *Applied Geography*, **6**, 5–11.

Gill, T.E. (1996) Eolian sediments generated by anthropogenic disturbance of playas: human impacts on the geomorphic system and geomorphic impacts on the human system. *Geomorphology*, **17**, 207–228.

Gill, T.E. and Cahill, T.A. (1992) Playa-generated dust storms from Owens Lake, in *The History of Water: Eastern Sierra Nevada, Owens Valley* (eds C.A. Hall Jr, V. Doyle-Jones and B. Widawski), White-Inyo Mountains White Mountain Research Station Symposium, vol. 4, University of California, Los Angeles, pp. 63–73.

Graf, W.L. (1988) *Fluvial Processes and Dryland Rivers*, Springer-Verlag, Berlin.

Graf, W.L. (2000) Locational probability for a dammed, urbanizing stream: Salt River, Arizona, USA. *Environmental Management*, **25**, 321–335.

Graf, W.L. (2005) Geomorphology and American dams: the scientific, social, and economic context. *Geomorphology*, **71**, 3–26.

Jacobsen, T. and Adams, R.M. (1958) Salt and silt in ancient Mesopotamian agriculture. *Science*, **128**, 1251–1258.

Kumar, M. and Bhandari, M.M. (1993) Human use of the sand dune ecosystem in the semi arid zone of the Rajisthan Desert, India. *Land Degradation and Rehabilitation*, **4**, 21–36.

Laronne, J.B. and Shulker, O. (2002) The effect of urbanization on the drainage system in a semiarid environment, in *Global Solutions for Urban Drainage* (eds E.W. Strecker and W.C. Huber), Proceedings of the Ninth International Conference on Urban Drainage, Portland, Oregon, 8–13 September 2002, pp. 1–10.

Lesschen, J.P., Cammeraat, L.H. and Nieman, T. (2008) Erosion and terrace failure due to agricultural land abandonment in a semi-arid environment. *Earth Surface Processes and Landforms*, **33**, 1574–1584.

Li, X.-R., Zhou, H.-Y., Wang, X.-P. and Zhu, Y.-G. and O'Conner, P.J. (2003) The effects of sand stabilization and revegetation on cryptogam species diversity and soil fertility in the Tengger Desert, Northern China. *Plant and Soil*, **251**, 237–245.

Maas, E.V. (1990) Crop salt tolerance, in *Agricultural Salinity Assessment and Management* (ed. K.K. Tanji), American Society of Civil Engineers, New York, pp. 262–304.

Magilligan, F.J. and Nislow, K.H. (2005) Changes in hydrologic regime by dams. *Geomorphology*, **71**, 61–78.

Micklin, P. (2007) The Aral Sea disaster. *Annual Review of Earth and Planetary Sciences*, **35**, 47–72.

Middleton, N.J. (2002) The Aral Sea, in *The Physical Geography of Northern Eurasia* (ed. M. Shahgedanova), Oxford University Press, pp. 497–510.

Middleton, N.J. and Thomas, D.S.G. (eds) (1997) *World Atlas of Desertification*, 2nd edn, Arnold United and Nairobi, UN Environment Programme, London.

Middleton, N.J. and van Lynden, G.W.J. (2000) Secondary salinization in South and Southeast Asia. *Progress in Environmental Science*, **2**, 1–19.

Motsi, K.E., Chuma, E. and Mukamuri, B.B. (2004) Rainwater harvesting for sustainable agriculture in communal lands of Zimbabwe. *Physics and Chemistry of the Earth*, **29**, 1069–1073.

Oliver, F.W. (1945) Dust storms in Egypt and their relation to the war period, as noted in Maryut 1939–45. *Geographical Journal*, **106**, 26–49.

Oweis, T. and Hachum, A. (2006) Water harvesting and supplemental irrigation for improved water productivity of dry farming systems in West Asia and North Africa. *Agricultural Water Management*, **80**, 57–73.

Perthuisot, J.P. (1989) Recent evaporites, in *Short Courses in Geology, Presented at the 28th International Geological Congress 3*, American Geophysical Union, Washington, DC, pp. 65–126.

Qureshi, A.S. McCornick, P.G. Qadir, M. and Aslam Z. (2008) Managing salinity and waterlogging in the Indus Basin of Pakistan. *Agricultural Water Management*, **95**, 1–10.

Rhoades, J.D. (1990) Soil salinity – causes and controls, in *Techniques for Desert Reclamation* (ed. A.S. Goudie), John Wiley & Sons, Ltd, Chichester, pp. 109–134.

Safriel, U., Adeel, Z., Niemeijer, D., Puigdefabregas, J., White, R., Lal, R., Winslow, M., Ziedler, J., Prince, S., Archer, E. and King, C. (2005) Dryland systems, Chapter 22, in *Millennium Ecosystem Assessment, vol. 1, Ecosystems and Human Well-being: Current State and Trends* (eds R. Hassan, R. Scholes and N. Ash), World Resources Institute, Washington, DC, pp. 623–662.

Schiettecatte, W., Ouessar, M., Gabriels, D., Tanghe, S., Heirman, S. and Abdelli, F. (2005) Impact of water harvesting techniques on soil and water conservation: a case study on a micro catchment in southeastern Tunisia. *Journal of Arid Environments*, **61**, 297–313.

Schlesinger, W.H., Reynolds, J.F., Cunningham, G.L. *et al.* (1990) Biological feedbacks in global desertification. *Science*, **247**, 1043–1048.

Schumm, S.A., Harvey, D.M. and Watson, C.C. (1984) *Incised Channels: Morphology, Dynamics and Control*, Water Resources Publications, Littleton, Colorado.

Smith, D.M., Twidale, C.R. and Bourne, J.A. (1975) Kappakoola dunes – aeolian landforms induced by man. *Australian Geographer*, **13**, 90–96.

Stulina, G. and Sektimenko, V. (2004) The change in soil cover on the exposed bed of the Aral Sea. *Journal of Marine Systems*, **47**, 121–125.

Ta, W., Xiao, H. and Dong, Z. (2008) Long-term morphodynamic changes of a desert reach of the Yellow River following upstream large reservoirs' operation. *Geomorphology*, **97**, 249–259.

Tanji, K.K. (ed.) (1990) *Agricultural Salinity Assessment and Management*, American Society of Civil Engineers, New York.

Trimble, S.W. and Mendel, A.C. (1995) The cow as a geomorphic agent – a critical review. *Geomorphology*, **13**, 233–253.

Viglizzo, E.F. and Frank, F.C. (2006) Ecological interactions, feedbacks, thresholds and collapses in the Argentine Pampas in response to climate and farming during the last century. *Quaternary International*, **158**, 122–126.

Wasson, R.J. and Galloway, R.W. (1986) Sediment yield in the Barrier Range before and after European settlement. *Australian Rangelands Journal*, **8**, 79–90.

Waters, M.R. and Haynes, C.V. (2001) Late Quaternary arroyo formation and climate change in the American southwest. *Geology*, **29**, 399–402.

Wilkinson, T.J. (1997) Holocene environments of the high plateau, Yemen. Recent geoarchaeological investigations. *Geoarchaeology*, **12**, 833–864.

Wittfogel, K.A. (1958) *Oriental Despotism; A Comparative Study of Total Power*, Yale University Press, New Haven.

World Commission on Dams (2000) *Dams and Development: A New Framework for Decision-Making*, Earthscan, London.

Worster, D. (1979) *Dust Bowl*, Oxford University Press, New York.

Zektser, S., Loáiciga, H.A. and Wolf, J.T. (2005) Environmental impacts of groundwater overdraft: selected case studies in the southwestern United States. *Environmental Geology*, **47**, 396–404.

Zhao, S. (1986) *Drifting Sand Hazard and Its Control in Northwest Arid China* (eds F. El-Baz, I.A. El-Tayeb and M.H.A. Hassan), Proceedings of the International Workshop on Sand Transport and Desertification in Arid Lands, Khartoum, Sudan, 17–26 November 1985, World Scientific, Singapore, pp. 253–266.

23

Geomorphological hazards in drylands

Giles F. S. Wiggs

23.1 Introduction

With more than 1 billion people estimated to be living in dryland zones (Middleton and Thomas, 1997) it is inevitable that the activities of human populations frequently overlap with the operation of geomorphological processes. While such interaction may often have negligible consequences, our increasing experience of living in drylands and utilising their resources has shown that there are specific circumstances where the interface between human activity and geomorphological dynamism can prove particularly hazardous. Such hazards can be distinguished by their two differing principle causes:

- Human activity encroaching into areas of highly active geomorphological processes (e.g. construction of urban settlements on the piedmonts of desert mountains with a significant flood risk; Rhoads, 1986).

- Human activity accelerating geomorphological processes in otherwise stable landscapes (e.g. agricultural wind erosion in the Dust Bowl of the mid-west USA; Worster, 1979).

The large range in scale of the effects of dryland geomorphological hazards is also of note. While construction of communication lines to service the oil industry in the Middle East may cultivate small and site-specific (though intense) hazards due to sand encroachment over roads and railways (Dong *et al.*, 2004; Zhang *et al.*, 2007), dust generation as a result of the draining of inland water bodies in drylands (e.g. Owens Lake in California – see Chapter 15) can result in very large off-site pollution hazards, sometimes well beyond the boundaries of drylands themselves (Gill, 1996). Such large-scale and off-site pollution does not have to be the result of human activity since geomorphological processes naturally operating in drylands can generate similar consequences (Prospero, 1999), as illustrated by the close correlation found between deposition of Saharan dust in Spain and daily mortality (Perez *et al.*, 2008). Further, environmental changes as a result of global warming may induce regional-scale remobilization of currently stable and habitable dunefields, so threatening livelihoods and food security at a national scale (Thomas, Knight and Wiggs, 2005; Wang *et al.*, 2009).

This chapter reviews the most frequently occurring hazards originating from human interaction with both aeolian and fluvial processes in drylands and, where appropriate, outlines some of the commonly employed methods by which the impacts of such hazards may be reduced. Hazards associated with these processes have been chosen for review because they have been the focus of the majority of research effort in recent years. Salt weathering of buildings and roads is not considered here as research into this hazard has declined since the 1990s (although see Goudie and Viles, 1997, for a review) and salinization of soil as a hazard is covered elsewhere in this book (see Chapter 22).

23.2 Aeolian hazards

23.2.1 Blowing sand and active dune movement

As noted in Chapter 18, where sediment is freely available and the erosivity of the wind exceeds the resistance of the sediment, sand will become entrained and transported by the windflow. While blowing sand itself can be hazardous to human activity by undermining structures, sand-blasting surfaces and depositing on roads and railways (Watson, 1985, 1990), further hazards result from

Arid Zone Geomorphology: Process, Form and Change in Drylands, Third Edition. Edited by David S. G. Thomas

Figure 23.1 Dunes migrate across a newly built road in Ras Al-Khaimah, United Arab Emirates. The most frequent dune moving hazards result from small transverse dunes that migrate quickly in a unidirectional wind (photo: author).

the movement of dunes caused by such aeolian activity where small (1–2 m high) transverse dunes may migrate >15 m/yr, therefore quickly burying roads and crops (Misak and Draz, 1997; Al-Harthi, 2002).

Such aeolian hazards tend to be associated with active dunefields and sand transport corridors in dryland regions where topographic depressions accumulate sand-sized material (see Chapter 17). Human activity within or on the edges of such regions (e.g. the Siwa Oasis in Egypt on the edge of the Great Sand Sea; Misak and Draz, 1997) involving the establishment of roads, railways, irrigation canals and farmland then become susceptible to blowing sand and dune movement (Watson, 1985, 1990) (Figure 23.1). Such hazards are therefore often predictable in both time and space and, where planning schemes to avoid such hazards cannot be utilised, there are several engineering techniques involving fences or surface treatments that can be employed to reduce the impacts of blowing sand and dune movement by enhancing upwind deposition of sediment, immobilising susceptible surfaces, enhancing sand transport through the hazardous area or deflecting sand movement away from the area to be protected (Watson, 1985, 1990). However, the maintenance of such schemes is often sporadic and so comprehensive control of sand drift can frequently be lacking (Misak and Draz, 1997).

The recent construction of roads and railways through active dunefields in China provides an excellent example of the hazards caused by blowing sand and moving dunes and mitigating measures that can be employed. The Taklimakan Desert sits in the Tarim Basin and has accumulated shifting sand across 80 % of its 338 000 km^2 area. Between 1991 and 1995 a 560 km long highway was built north–south through the desert in order to enable the exploitation of petroleum reserves (Dong *et al.*, 2004). As 447 km of the road was built through actively mobile dunefields, it has proven to be an excellent laboratory within which to test engineering measures to control blowing sand and dunes. Specific problems were found where the road dissected small (<2 m high) barchan dunes migrating between 5.0 and 7.5 m/yr (Dong *et al.*, 2000) and also where the road cut through the elongating fronts of small linear dunes and sand sheets within intermegadune flats (Dong *et al.*, 2004).

Sand control was effectively accomplished by combining four different measures, depending on the local situation. *Porous fences* about 1.1 m high and made of reed or nylon were established along a 10–20 m wide belt on the upwind edges of the highway. The fences reduce the wind speeds across a fetch of 20–25 times the fence height and enhanced sand deposition is evident both upwind and downwind of the fence. There has been much research into the design of porous fences with porosities of between 0.3 and 0.4 providing airflow characteristics most suitable for reducing wind erosion (Lee and Kim, 1999; Dong *et al.*, 2006). As fence porosity increases from 0 to 0.4, the mean wind speed reductions in the lee of the

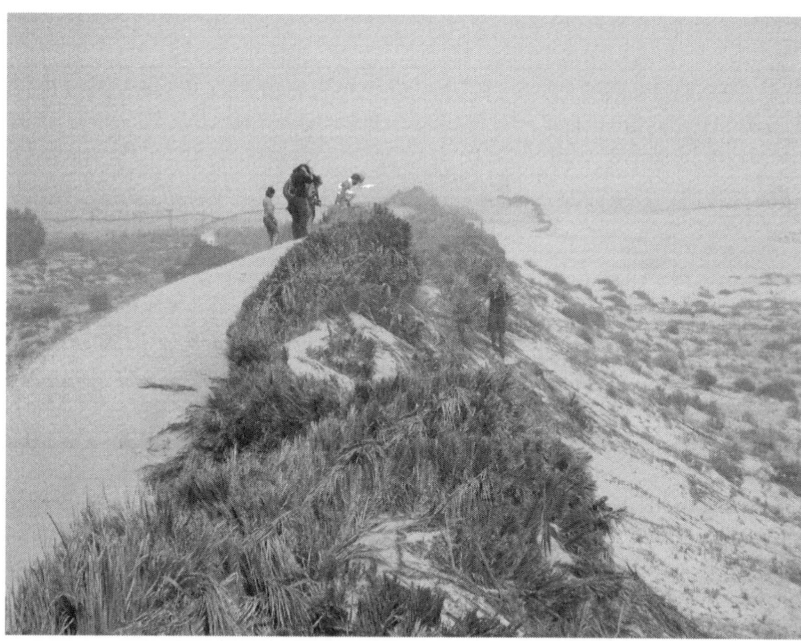

Figure 23.2 A sand fence protecting a road (on the left of the picture) in southern Tunisia. This fence is made from interwoven palm leaves that reduce wind speeds at the crest of the artificially created dune and encourage sand deposition. The fence is replaced every few years as the dune grows (photo: author).

fence decrease, but the amount of sheltered area increases as the downwind length of the reverse flow region expands (Bofah and Al-Hinai, 1986; Lee, Park and Park, 2002). A drawback with using sand fences is that as the sand builds up around the fence they require replacement so that they do not become buried and overwhelmed. Thus a maintenance plan for sand fences needs to be adhered to (Figure 23.2).

Reed checkerboards (1 m^2 and 0.15–0.2 m in height) were also established across much of the moving dune surfaces. These raised the aerodynamic roughness length (z_0) of the surface by up to 400–600 times (Dong *et al.*, 2004) and so significantly reduced surface wind speeds (Liu, 1987). The checkerboards remain intact for a period of 4–5 years during which time planted xerophytic shrubs can become established (Mitchell *et al.*, 1998). Mitchell *et al.* (1988) report that using checkerboards can transform an area with shifting sands and less than 5 % vegetation cover to areas of fixed dunes with 30–50 % cover within a few years. Where steep sand surfaces were created along road embankments erosion was reduced by using *chemical and clay fixers* to cement the sediment. These fixers included clay mixed with local saline groundwater and also emulsified crude oil and asphalt. Finally, *saline-tolerant vegetation cover* was established in areas where pumping of shallow groundwater could be achieved. Used in combination these measures have reportedly kept the highway sand free for a period of at least 8 years (Dong *et al.*, 2004).

This combination of control measures is commonly used for sand and dune control and similar engineering approaches have been taken to protect the Shapotou section of the Baotou–Lanzhou Railway (where the checkerboard control system was developed) across the Tengger Desert (Zhang *et al.*, 2007; Qiu *et al.*, 2004) and also the Qinghai–Tibet Railway (Zhang *et al.*, 2010). The checkerboard method of sand control has also been extensively used in China and elsewhere to stabilise surfaces around oases threatened by sand drift and mobile dunes (Zhang *et al.*, 2004) and also to enable reclamation of degraded agricultural land (Li *et al.*, 2009). Research has shown that after a period of decades the soil within the checkerboards shows an increase in silt and clay content and raised levels of soil carbon and nitrogen that support a diversified plant and vegetative canopy that is extremely effective at long-term control of sand drift and dune movement (Mitchell *et al.*, 1998; Li *et al.*, 2006, 2009; Su *et al.*, 2007).

23.2.2 Human disturbance of stable surfaces

Some of the most serious gemorphological hazards in drylands are created by the human disturbance of otherwise stable land and water surfaces, enabling wind erosion.

While 85 % of natural desert surfaces have been described as being not susceptible to wind erosion (Wilshire, 1980) the area at risk may be greatly enhanced by human activity (Prospero *et al.*, 1983; Tegen and Fung, 1995; Gill, 1996). Middleton (1989) describes two types of activity that may enhance wind erosion in drylands: those that break up naturally wind-resistant surfaces (such as stone pavements) and those that remove a protective vegetation cover. In the former category disturbance may be spatially distinct in the form of off-road vehicle (ORV) use (Padgett *et al.*, 2008). Research suggests that ORV activity can induce dust emissions equivalent to 564 kg/km in one year for one vehicle used on a daily basis on an unpaved road (Frazer, 2003). Compared to undisturbed surfaces, Goossens and Buck (2009a) report typical dust emission values for ORVs on dry desert soils varying between 300 and 800 g/km on mixed terrain, although they found much higher values on silty soils (Goosens and Buck, 2009b).

However, much larger scale problems can be created by dryland agricultural activity where stable vegetated surfaces may be stripped and exposed to the power of erosive winds, as occurred in the Dust Bowl era of the USA in the 1930s (Worster, 1979; see Box 23.1). It has been suggested that cultivated lands in drylands may produce 20 % more aeolian dust than uncultivated regions and construction activity may double the amount of dust in the atmosphere (Reheis and Kihl, 1995). Further, some of the most intense human-induced dust hazards globally result from the draining of inland water bodies such as Owens Lake and the Aral Sea (Gill, 1996).

23.3 The aeolian dust hazard

Aeolian erosion is hazardous on three counts. First, wind erosion can strip surfaces of soil nutrients so making them much less able to preserve a protective vegetation cover. Such erosion is particularly problematic in drylands where the lack of moisture percolation through soils results in nutrients lying very shallow beneath the surface and so they are potentially easily removed, sometimes within only one erosive event. Erosion of nutrient-rich topsoil in this manner can seriously degrade semi-arid cropland very quickly (Zobeck and Van Pelt, 2006). For example, Larney *et al.* (1998) showed that spring wheat yields on the Canadian prairies decreased linearly with increased wind erosion severity. Second, the deposition of eroded dust can bury plants, clog streams and drainage channels, contaminate food and water resources and impact on vegetation growth, soil productivity and ecosystem dynamics (Worster, 1979; Larney *et al.*, 1998; McTainsh and Strong, 2007). Gill (1996) reports that the deposition of alkaline dust from the drying Mono basin in California has the potential to alter the downwind soil chemistry and vegetation structure, leading to a decrease in biodiversity in the area and an increase in soil pH. Nearby, salt-laden depositing dust that is blown out of Owens (dry) Lake is noted to have a significant adverse environmental impact on three surrounding national parks and numerous designated areas of critical environmental concern (Gill, 1996; Reheis, 1997).

Third, the transportation of eroded sediment by high winds can abrade buildings, crops and livestock and also create visibility problems along roads and at airports. The best-known example is the 1973 aviation accident at Kano, Nigeria, when 183 people were killed in a crash-landing widely attributed to Harmattan dust haze (Adefolalu, 1984; Adedokun, Emofurieta and Adedeji, 1989). More chronically, however, the long-distance transport of dust-sized particles (<62.5 μm) by suspension can create significant health problems where fine material with aerodynamic diameters of <10 μm ($\sim$one-seventh of the width of a human hair), 5 μm and 2.5 μm (termed PM_{10}, PM_5 and $PM_{2.5}$, respectively) may be easily inhaled into the lungs of people and animals. Epidemiological studies have found that these fine particles can be drawn into the deepest portions of the human respiratory tract (Pope, Bates and Raizenne, 1996; Griffin, 2007), where they may have serious impacts on human respiratory and lung function (Bennion *et al.*, 2007). These particles may be made up of sulfates, carbonates, sodium and calcium and particular health problems are evident where these fine, respirable particles are contaminated with toxic agricultural fertilisers and pesticides or waste residue that may contain trace chemicals (e.g. arsenic, cadmium, lead, mercury; see Galloway *et al.*, 1982; Griffin, 2007) that are associated with the onset of asthma, cancers, interstitial lung disease, emphysema and lung fibrosis (O'Hara *et al.*, 2000; Bennion *et al.*, 2007; Griffin, 2007). Furthermore, windborne dust may contain high concentrations of organics composed of microorganisms, which may include fungal spores, bacteria, viruses and pollen (Prospero *et al.*, 2005; Kellog and Griffin, 2006; Griffin, 2007). Inhalation of airborne spores of the soil-dwelling *Coccidioides immitis* fungus, for example, is known to cause 'valley fever' in the southwest USA and Mexico where, during the epidemic from 1991 to 1995 the annual incidence of the infection rose from 50 to 500 per year (per 100 000 population) in Kern County, California (Zender and Talamantes, 2006).

There is a wealth of evidence that inhalation of wind-blown dryland dust has serious deleterious consequences for human health. Hefflin *et al.* (1994) recorded a 3.5 %

Box 23.1 The Dust Bowl

The 1930s Dust Bowl in the Great Plains of the United States has become synonymous with ideas of land exploitation. It was a result of the combined impacts of recurrent drought and excessive agricultural activity on dryland soils, but with the additional driving force of free enterprise and an expansionary culture (Cooke and Doornkamp, 1990). The result was 8000 ft high dust storms, 9 million hectares of damaged agricultural land, decimated wheat harvests, massacred livestock, farm abandonment and massive social hardship (Worster, 1979). Cook, Miller and Seager (2009) describe the resulting human migration as being comparable in the recent history of the USA only to the evacuation of New Orleans in 2005 as a result of hurricane Katrina.

In the 1800s the drought-resistant prairie grasslands of the Great Plains, centred at the junction between Colorado, Kansas, New Mexico, Oklahoma and Texas, were replaced with drought-sensitive wheat crops (Cook, Miller and Seager, 2009). By the late nineteenth and early twentieth century the rate of cropland expansion was boosted by an ever-growing number of steam- and gasoline-powered tractors. The term 'sodbusting' was coined as the land under wheat increased by 50 % between 1925 and 1930 (Worster, 1979). After World War I the demand for wheat was huge and the farmers responded by increasing their farm size (up to 50 square miles) and embarking on overnight ploughing (Worster, 1979). In the 1930s the region was hit by a period of severe drought with average precipitation reducing by more than 50 %. This period of additional environmental stress coincided with a severe economic depression that caused the price of wheat to collapse. With no drought plan and minimal erosion control measures many wheat farmers responded to their falling income by increasing the intensity of farming. By 1935, 33 million acres of ploughed, bare and desiccated soils were left susceptible to the wind (Worster, 1979) and the subsequent devastating dust storms reached beyond the eastern seaboard of the US. Worster (1979) argues that while the physical environment played a part in the catastrophe, social and economic considerations were equally culpable. He describes the farmers as, '... fatalistic optimists, their eyes weren't looking at the soil, they were fixed to the stockmarket... their solution was to plant more wheat'. With the dust clouds enveloping the New York Stock Exchange he suggests that '... there was no significant difference between 'Black Thursday' on Wall Street and the black days of the Dust Bowl'.

It could be argued that the Dust Bowl was the first time that western culture had come up against natural limits and it inspired the first scientific investigations of land management in dryland regions. Conservation research inspired by the Dust Bowl made it clear that agricultural activity on dryland soils required careful management to avoid accelerating aeolian erosion, and soil conservation strategies involving windbreaks, crop management and tillage operations were rolled out across the Great Plains. Nordstrom and Hotta (2004) describe how the Great Plains region was hit by an even more acute drought in the 1950s and yet problems of wind erosion were less severe. Lee, Wigner and Gregory (1993) explain this reduction in wind erosion severity in the 1950s to the successful implementation of soil conservation measures on agricultural land. However, if such conservation strategies are poorly applied or allowed to diminish then severe wind erosion problems are evident to this day.

increase in the number of daily emergency room visits for bronchitis for each 100 μg/m^3 increase in PM_{10} as a result of dust storms in Washington State. Chan *et al.* (2008) noted that long-distance transport of desert dust in Asia resulted in a 67 % increase in cardiopulmonary emergency hospital visits in Taiwan when PM_{10} concentrations were above 90 μg/m^3, with a near 5 % increase in mortality (Chen *et al.*, 2004). Similar increases in mortality as a result of Asian dust storms have also been recognised in Korea (Kwon *et al.*, 2002). Recognition of the harmful effects of PM_{10} and $PM_{2.5}$ is such that the US Environmental Protection Agency (USEPA) has, since 1986, set national standards for air quality for the 24 hour and annual average concentrations of such particles in order to protect public health and to help identify particularly hazardous dust sources.

23.4 Agricultural wind erosion

Wind erosion from agricultural land in semi-arid environments is one of the major causes of human-induced aeolian dust in the atmosphere. Much research on the problem has been carried out in the west and mid-west USA where agricultural activity combines with a semi-arid climate, highly erodible soils and seasonally active strong winds (e.g. Sharratt, Feng and Wendling, 2007; Zobeck and Van Pelt, 2006). Indeed, Nordstrom and Hotta (2004) report that 90 % of wind erosion in the USA occurs west of the Mississippi River with about 60 % in the Great Plains region (Ervin and Lee, 1994), where Lubbock, Texas, copes with the national maximum of 47.5 dust days per year (Hagen and Woodruff, 1973). The major wind erosion problems on these agricultural lands occurs during ground preparation for planting and after harvesting (Clausnitzer and Singer, 1996). At these times the cotton and wheat fields are left bare and far more susceptible to erosion, and disturbance is at its greatest through tillage (ploughing) operations, planting and weeding (Nordstrom and Hotta, 2004). Particular problems are evident during drought periods when the soil is much more erodible and protective crop residue is far sparser. As a result, Merrill *et al.* (1999) estimated that wind erosion losses in the Great Plains can be up to 6100 times greater in drought years than in wet years. The intense drought in the mid-west USA during the 1930s was a catalyst in producing the devastating dust storms of the 'Dust Bowl' (see boxed text (Box 23.1)).

The experience of the Dust Bowl promoted the development and implementation of land conservation strategies in the USA and stimulated the founding of both the US Soil Conservation Service and the Agricultural Research Service of the US Department of Agriculture. Both agencies research and encourage the application of sustainable agricultural practices and there are now many methods by which farmers and land managers can attempt to control wind erosion from agricultural fields. These include windbreaks, crop management and tillage operations.

There is a great deal of literature on the most efficient design, construction and placing of windbreaks and shelterbelts (Cleugh, 1998; Brandle, Hodges and Zhou, 2004; Cornelis and Gabriels, 2005). They usually consist of trees and shrubs placed perpendicular to the prevailing wind to reduce both upwind and downwind wind speeds, thus reducing wind erosion. Upwind velocity reductions may be experienced for a distance up to 2–5 times the height of the windbreak while the downwind protected zone can extend up to 10–30 times the windbreak height (Wang and Takle, 1995). The height, width and porosity of the windbreak all have important influences on its effectiveness. Wind tunnel experiments by Cornelis and Gabriels (2005) indicate that porosities of between 0.20 and 0.35 m^2/m are the most efficient (see Figure 23.3). While protecting the soil from erosion, windbreaks have also been shown to ameliorate the microclimate (soil temperature, evaporation) so that crop growth response is generally positive, showing increases in yields (Cleugh, 1998; Brandle, Hodges and Zhou, 2004). However, as discussed by Nordstrom and Hotta (2004), windbreaks require labour and maintenance in order to preserve their effectiveness and they also take up valuable space that could be used for arable purposes; therefore sometimes they are removed to make way for modern agricultural machinery or irrigation systems.

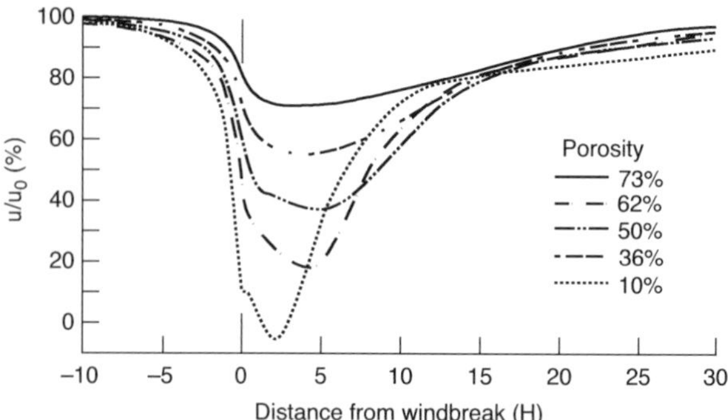

Figure 23.3 The reduction in horizontal wind speed (u) relative to an upwind reference (u_0) with varying distance and porosity (after Wang and Takle, 1997).

Crop management systems for reducing wind erosion include selecting crops that offer the maximum amount of post-harvest protective residue (Hagen, 1996). Further, crop rotation or strip cropping can be used to protect the soil surface during the most hazardous erosive periods (Nordstrom and Hotta, 2004). However, one of the most frequently used methods of wind erosion control is tillage. While tillage can enhance wind erosion if it disturbs a relatively stable soil surface, it can also reduce erosion susceptibility if practised appropriately (Merrill *et al.*, 1999). One of the key techniques, often used as an emergency wind erosion control method, is to overturn the soil such that moist clods lie on the surface in furrows perpendicular to the erosive wind (Nordstrom and Hotta, 2004). While the increased aerodynamic roughness resulting from the furrow structures has the effect of increasing wind shear stress, it also raises the depth of air with zero wind velocity (z_0), hence protecting finer particles from erosion (see Chapter 18). The excess wind shear stress is absorbed by the larger clods, which have much higher erosion thresholds (Figure 23.4). Such a tillage technique can be very successful at reducing wind erosion so long as the clods are not broken down too quickly by the action of saltation (Stout and Zobeck, 1996). Wiggs and Holmes (2010) found that the impact of employing a ridge–furrow ploughing strategy on erodible fields in South Africa was a reduction in the frequency of erosion events by almost 49 %.

Conservation tillage is also a practical method for erosion control. In this case, a strategy of minimum or no

Figure 23.4 Large soil clods brought to the surface by tillage on a field near Bloemfontein, South Africa. The clods absorb the wind stress and raise the aerodynamic roughness, thus protecting the sandy soil that is susceptible to wind erosion after harvesting (photo: author).

tillage is adopted and, where required, tillage techniques are used that do not overturn and hence disturb the soil. Such practices maximise the amount of crop residue on the soil surface, thus reducing wind erosion (Mills, Thomas and Langdale, 1991; Seta *et al.*, 1993; Tang and Zhang, 1996). Conservation tillage also improves soil conditions by increasing levels of soil organic matter content and carbon sequestration (Uri, 1998). However, while conservation tillage was adopted on 36 % of planted land in the USA in 1996 (Uri, 1998), there are also some hindrances to its adoption. These include increased herbicide and fertilizer costs, the issue of unused farm machinery and also the requirement for farmers to learn new skills. Uri (1998) argues that farmers tend to adopt conservation tillage only if they perceive a net benefit to farm finances and operations, rather than a net benefit to society (Nordstrom and Hotta, 2004).

Application of the conservation strategies described above is greatly assisted by the use of predictive wind erosion equations that enable the timing and spatial extent of the wind erosion hazard to be calculated in the context of the particular farming practices being employed. In the Great Plains region of the USA the strong scientific paradigm resulting from the Dust Bowl years of the 1930s has encouraged the continuous development of wind erosion equations such as the revised wind erosion equation (RWEQ) (Fryrear *et al.*, 1998) and the wind erosion prediction system (WEPS) (Hagen, 1991). These models are calibrated at a field or farm scale (e.g. Fryrear *et al.*, 2000) and estimate potential soil erosion based on input files often incorporating monthly weather, soil, field and management data and inclusive of farm management criteria such as cropping systems, tillage, wind barriers and irrigation. While calibration and use of these predictive equations is rare outside of the USA, similar methods for quantifying wind erosion potential have also been employed in Europe (Böhner *et al.*, 2003), Australia (Webb *et al.*, 2009) and China (Xu, Liu and Zhao, 1993).

23.5 Drainage of inland water bodies

As discussed in Chapter 20, dry or ephemeral lake beds and closed topographic depressions in dryland regions (often termed *playas*) are frequently sources of aeolian dust because they act as sinks for sediment accumulation by fluvial action (Prospero *et al.*, 2002; Engelstaedter *et al.*, 2003; Bryant *et al.*, 2007). Such accumulations of sediment are vulnerable to erosion because their saline status often inhibits the growth of stabilising vegetation and the production of surface evaporite minerals by evaporation of near-surface groundwater can offer a highly erodible surface sediment (Reynolds *et al.*, 2007). In this context, it is not unexpected that the purposeful drainage of permanent water bodies in semi-arid regions can result in the production of significant quantities of aeolian dust in the long term.

Figure 23.5 Dust storm approaching Stratford, Texas, 1935 (NOAA George E. Marsh Album).

Gill (1996) notes that human intervention in the hydrological cycle around playa systems may accelerate the desiccation of playas, lower water tables, reduce soil moisture and reduce vegetation cover, thus increasing susceptibility to wind erosion. Gill (1996) reviewed evidence of accelerated aeolian erosion from anthropogenically desiccated playa systems from across the globe, including the Great Konya Lake basin in Turkey, the Sambhar Salt Lake basin in Rajastan, Kara Bogaz Gol in Turkmenistan and Old Wives Lake in Saskatchewan. However, while the draining of the Aral Sea on the border of Kazakhstan and Uzbekistan is perhaps the best-known and largest example of human activity resulting in increased dust storm activity (Micklin, 1988), the region with the most frequently occurring problems was in the western USA. Here, Gill (1996) found numerous examples of playa-generated dust storms exacerbated by human activity, including in Antelope Valley near Los Angeles and the Salton Sea in southern California. Such disturbance of the hydrological status of these playa systems is often due to the demand for water resources for urban centres or irrigation, or because of the development of roads and communication lines that block or divert inflow or drainage waters.

The largest single source of aeolian dust in North America was Owens (dry) Lake in California (Cahill *et al.*, 1996) and much of our understanding of the impacts of human interference on hydrological and aeolian systems has come from research conducted on these lake sediments in the last 30 years. In 1913 the City of Los Angeles Department of Water and Power (LADWP) began building a 223 mile aqueduct to divert the Owens River away from the 110 square mile Owens Lake to provide water for the growing city of Los Angeles. With its main source of water diverted the lake began to dry and by 1927 there remained little standing water beyond a central brine pool. Evaporation of near-surface groundwater, with high concentrations of arsenic (Ryu *et al.*, 2002), led to the deposition of efflorescent salts that were extremely erodible (Cahill *et al.*, 1996), and from the 1930s onwards the lake and surrounding regions suffered from increasingly severe dust storms in the spring and autumn as high winds were channelled between the Sierra Nevada and the White-Inyo mountain ranges. These dust storms are heavily salt-laden and loaded with respirable PM_{10} material with concentrations measured over a 2 hour period exceeding 40 000 $\mu g/m^3$ (Cahill *et al.*, 1996) in comparison with the US federal 24 hour air quality standard of 150 $\mu g/m^3$. Estimates of the amount of eroded PM10 material vary from 540 000 tons per year to 4000 000 tons per year (GBUAPCD, 1994) and dust storms can reach as far south as Los Angeles and Orange County (Figure 23.6), covering an area >90 000 km^2 (Gill, 1996).

Since the early 1980s the Great Basin Unified Air Pollution Control District (GBUAPCD) has been coordinating efforts to reduce dust emission to within federal limits. Research has shown that, rather than the whole area of the dry lake being susceptible to erosion, dust originates from specific areas where a complex suite of conditions interact to promote the availability of erodible material. Gill (1996) lists the six conditions necessary as: saline

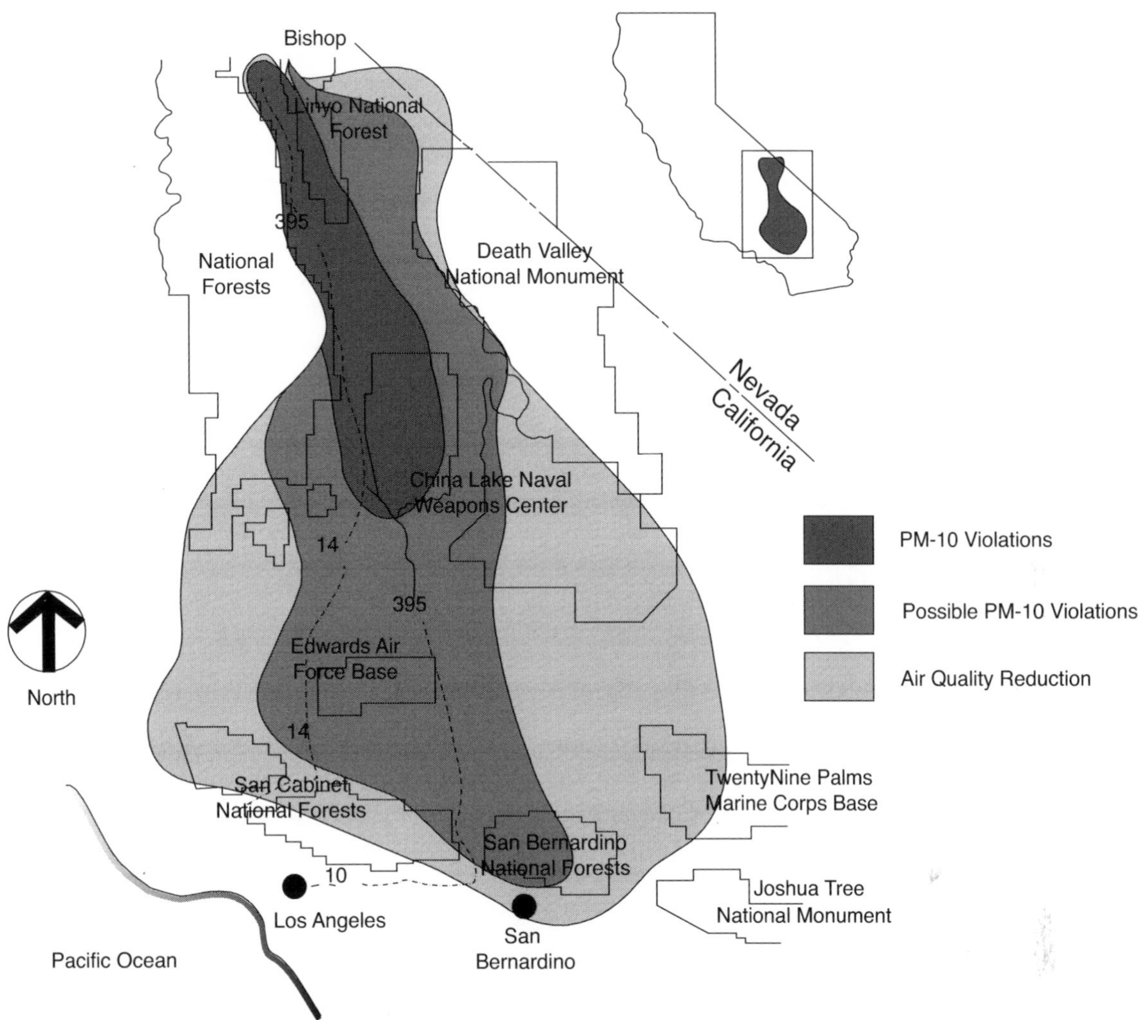

Figure 23.6 Air quality violations south of Owens (dry) Lake, California (after GBUAPCD, 1994).

groundwater near the surface, the precipitation of efflorescent salt crusts by the discharging groundwater, a sustained wind above 7 m/s, a high wind shear above a flat unobstructed surface, a 1.5 km fetch length across the playa and a supply of coarse sand that can abrade a stable salt crust as it saltates. Extensive monitoring of the lake has identified the interaction of such conditions at specific localities in the basin where the worst emissive areas are found. Such information has allowed mitigation measures to be tested and designed to reduce the likelihood of dust erosion from these surfaces (Kim, Cho and White, 2000; Tyler *et al.*, 1997; Dahlgren, Richards and Yu, 1997).

Three dust control measures have been implemented at Owens Lake. These are shallow flooding from permanently sited outlets across the lake to maintain at least 75 % surface saturated soil, the planting of native saltgrass using drip irrigation to maintain at least a 50 % ground cover and the spreading of a 4 inch thick gravel blanket (GBUAPCD, 2007). By 2006 these measures had been employed over nearly 30 square miles of the lake bed at a cost of $415 million (GBUAPCD, 2007). The success of the plan is still being determined but there are proposals to extend the mitigation measures if required.

The huge and lengthy scientific, financial and practical efforts required to reduce dust emission from the surface of Owens Lake are put into context by the disaster facing the region around the Aral Sea, where at least 42 000 km^2 of potentially erodible sediment has been exposed by the shrinking sea (Singer *et al.*, 2003). Once the fourth largest inland sea in the world, the Aral Sea has been shrinking since the early 1960s when a large-scale irrigation campaign aimed at achieving independence in cotton production was launched in Soviet Central Asia. This required a massive expansion in irrigation in what is now Uzbekistan, Kazakhstan and Turkmenistan and so water was diverted from the Aral's two feeder rivers, the Amu Darya and the Syr Darya. Between 1960 and 1980 cotton production doubled but by the 1990s river discharge into the Aral Sea was reduced to such an extent that flow ceased often for periods of 1–3 months. The volume of water reaching the sea was reduced from 56 km^3 in 1960 to 3.5 km^3 in 1987 (Saiko and Zonn, 2000). The impact of this on the Aral Sea was catastrophic. Between 1960 and 1995 there was a 17 m drop in level, a 753 km^3 reduction in volume (~70 %) and a 35 000 km^2 reduction in surface area (~52 %) (Saiko and Zonn, 2000). Furthermore, there was a tripling of the salinity of the water in the Aral and a complete decimation of the fishing industry (Figure 23.7). Between 1987 and 1989 the Aral separated into two seas (the Small and Large Aral) and up to 2006 the area of both seas had diminished by 74 % and the volume by 90 % since 1960 (Micklin, 2007).

The impact on the environment surrounding the Aral Sea has been devastating. There is evidence of widespread land degradation as the soil has become desiccated and natural vegetation has died back, biodiversity has decreased, soils have become salinised and the climate of the region has altered, with higher temperature extremes and reduced precipitation (Kotlyakov, 1991). There is also considerable evidence that the impacts of the catastrophe are extremely detrimental to human health (Smith, 1991; UNDP, 1997), with high levels of anaemia, lung disease and cancers. The exposure of the former seabed and the desiccation of surrounding soils has resulted in severe dust storms, with an estimated 43–75 million tonnes of sediment deflated annually in storms with plumes reaching 4 km in height and travelling over 500 km downwind (Micklin, 1988, 2007; Singer *et al.*, 2003; Wiggs *et al.*, 2003). This dust contains highly toxic material remnant from the agricultural operations, including presticides, insecticides and fertilizers, that may have considerable impacts on human health (O'Hara *et al.*, 2000; Bennion *et al.*, 2007). These storms are particularly severe with strong north and north-west winds, which blow the dust over the major population centres to the south.

Amelioration of the dust hazard emanating from the Aral Sea basin will be extremely problematic over such a large area where complex issues of water management policy and conservation across national boundaries have yet to materialise. Singer *et al.* (2003) suggest that, left in their current state, the exposed soils of the southern Aral Sea basin will continue to generate huge amounts of dust. They suggest that the most efficient measure may be to embark on a massive programme of conservation of natural vegetation and suitable revegetation of susceptible areas. In 2005 an $85 million World Bank project to build a 13 km long dyke to separate the Small from the Large Aral was completed and this has led to the refilling of the Small Aral (Micklin, 2007), which is now providing some inflow into the Large Aral. However, as noted by Micklin (2007), the only realistic means to increase flow substantially into the Aral is by reducing upstream demand for water by irrigation (which accounts for 92 % of withdrawals) to provide an additional 54 km^3 of annual inflow. Efforts to

Figure 23.7 A fishing boat left abandoned on the exposed bed of the drying Aral Sea (photo: author).

improve water use efficiency would be demanding technically, economically and politically. For example, Micklin (2007) calculates that to overhaul 6 million hectares of irrigation systems, and so provide an additional 12 km^3 of inflow to the Aral, would cost at least $16 billion. The likelihood of such large-scale restoration schemes is unlikely in the foreseeable future. However, partial rehabilitation schemes involving hydrological engineering in conjunction with increased irrigation efficiency, ecological conservation and improved crop management may offer a future in the short to medium term (Micklin, 2007).

23.6 Fluvial hazards

Flash floods in deserts are a common but poorly understood hazard. Indeed, the complexity of arid fluvial systems, the lack of data on rainfall–runoff relationships and the rarity of observations of flood events in channels that are dry for the majority of the time make assessments and predictions of flood hazards extremely difficult. However, the extremes of rainfall that may occur in drylands, in combination with rapid runoff from sparsely vegetated slopes, can result in major flooding events that have the capacity to bring about lasting change to both the geomorphological and human environment. For example, the 12 largest floods ever recorded in the USA in basins <1000 km^2 have all occurred in arid or semi-arid areas (Costa, 1987).

Houston (2006) reports on the Atacama flood of February 2001, where estimated maximum daily river flows (the gauging station was destroyed) for the perennial Río Loa River reached 136 m^3/s in comparison to the maximum mean daily flow of 2–8 m^3/s. The flood inundated the town of Calama, destroyed several bridges and caused major channel erosion, overbank flooding and channel reorganisation throughout the catchment. Incision of new channels reached 2 m in depth and 5 m in width, causing widespread road disruption, and silt deposition to a thickness of $\sim$1 m was evident over hundreds of square kilometres. Such floods can also come with little warning. The lag time between the centroid of rainfall and flood peak for the Atacama flood was a mere 5 hours, even though the drainage basin covers a large area of almost 33 000 km^2 (Houston, 2006). Smaller ephemeral catchments can reach peak flood in less than an hour (Ben-Zvi, Massoth and Schick, 1991).

Such floods are difficult to analyse and predict in dryland areas where rainfall is often of high intensity and, because of its convective nature, highly localised in space and time. Rainfall may therefore cover only a portion ($<$10 km^2) of any individual catchment (Sharon, 1972). The resulting partial contribution of any catchment to a flood wave makes the system extremely difficult to measure and model with complexity in both rainfall/runoff (Srikanthan and McMahon, 1980; Wainwright and Parsons, 2002) and runoff/sediment relationships (Parsons *et al.*, 2006). However, the recognition and quantification of the hazard from flooding is becoming more urgent as growing population centres begin to encroach upon mountain piedmonts and drainage zones (Rhoads, 1986; Grodek, Lekach and Schick, 2000). In such areas flash floods can be a major hazard to both human life and infrastructure (Figure 23.8), and yet our lack of knowledge concerning such floods can compound the problem as

Figure 23.8 Debris trapped beneath a bridge after a flash flood along the Gaub River, Namibia (photo: author).

settlements and communication lines are sited unknowingly in hazardous zones (Foody, Ghoneim and Arnell, 2004). Small and dry drainage channels are often ignored by urban planners who do not appreciate the flashy nature of their flooding dynamic, the long intervals between floods and the serious hazard that they present (Laronne and Shulker, 2002).

Given the unfeasibility of applying standard hydrological modelling to dryland flood events (Greenbaum *et al.*, 1998) other geomorphological techniques can be applied to determine flood hazard indirectly. High flood risks are found on the piedmont areas of dryland mountains where high-intensity floods disgorge into alluvial fans and lower slope plains. Prediction of flood washout zones here is complicated by the often unchannelised sheet floods and the sudden changes in stream course that can occur within the duration of a single flood (Rhoads, 1986; Grodek, Lekach and Schick, 2000). Rhoads (1986) investigated flood hazards in such an environment around Scottsdale in Arizona, where the landscape was formed from an assemblage of pediments, alluvial fans, alluvial slopes and alluvial plains. He used geomorphological evidence such as variations in slope, depth of channel incision, local relief, drainage texture and drainage density to identify five distinct flood hazard zones. Rhoads found the most hazardous zones to be floodplains and active alluvial fans. On floodplains, while the flow was confined by a channel, high risk was associated with overbank flows and high flow velocities leading to scour and bank failure. On active alluvial fans he found that where the channel was unentrenched flooding was possible over the entire fan surface. On such active fans the depth of flow decreases downstream as the area of flow increases. In this way the severity of flooding is greatest at the upstream end of the fan where it intersects with the mountain front, but the potential for more widespread and less severe flooding increases towards the foot of the fan where flood depths and velocities are reduced (Rhoads, 1986). Rhoads also notes the planning considerations that should be applied on such unstable, active fans, where a minor shift in the channel position at the intersection point can have dramatic consequences on the location of downfan flooding. Rhoads (1986) identified the lowest flood risks as being associated with abandoned alluvial fan segments that had undergone dissection. Here, while major floods may occur in the entrenched fan portions, the dissected interfluves suffered from only minor shallow sheetfloods.

There is general consensus that flood hazards in drylands can be most efficiently reduced by maintaining natural drainage channels and infiltration processes as far as possible (Rhoads, 1986; Schick, Grodek and Lekach, 1997). Such channels play an important role in attenuating flood peaks and distributing sediment. The zoning of urban development to avoid surfaces with a high density of such drainage channels is important (Schick, Grodek and Lekach, 1997). However, where such development cannot be avoided, Rhoads (1986) suggests that it should be kept to low densities and should be permitted first near the mountain front before proceeding downslope so that its impact on downslope drainage can be assessed.

23.7 Conclusions

As humans make growing use of dryland resources, hazards associated with aeolian and fluvial processes will intensify. While some hazards are a result of the natural processes that operate in dryland environments impinging on human activity, it is clear that the most serious issues develop where human activity has increased the erodibility of stable surfaces. In this context, the impact of aeolian hazards, such as agricultural wind erosion and dust emissions from drained inland water bodies, far outweigh the problems associated with fluvial hazards. Such wind erosion, while emanating from spatially distinct and highly localised sources, can cause serious hazards over huge downwind areas since aeolian dust is not constrained by topography and can be carried hundreds of kilometres in turbulent winds. In contrast, fluvial activity, while life-threatening, causes hazards that are more site-specific, topographically constrained and limited in downstream impact as floods are attenuated by transmission losses to groundwater.

Our appreciation of the impact of human activity on aeolian systems has grown enormously over the last 30 years as research into the issue has flourished, and it is clear that agricultural activity and water resource use in drylands cannot be undertaken without consideration of the potential acceleration in erosion that may result. However, while research has offered some measures to mitigate the hazard resulting from accelerated erosion, it has also highlighted the scientific, practical, financial and sometimes political complexities involved in identifying and controlling hazardous areas.

Typically, human activity in drylands fails to appreciate the fast-changing spatial and temporal dynamics of aeolian and fluvial processes. Droughts are a natural part of the system and can be a catalyst for accelerated wind erosion on poorly managed land; a small, dry drainage channel may be the conduit for a devastating flood. As Houston (2006) points out, in deserts '... average conditions do not exist'. Planning for the extremes is the only way forward to minimise geomorphological hazards in the future.

References

Adedokun, J.A., Emofurieta, W.O. and Adedeji, O.A. (1989) Physical, mineralogical and chemical properties of Harmattan dust at Ile-Ife, Nigeria. *Theoretical and Applied Climatology*, **40**, 161–169.

Adefolalu, D.O. (1984) On bioclimatological aspects of Harmattan dust haze in Nigeria. *Archives for Meteorology, Geophysics, and Bioclimatology*, **Series B, 33** (4), 387–404.

Al-Harthi, A.A. (2002) Geohazard assessment of sand dunes between Jeddah and Al-Lith, western Saudi Arabia. *Environmental Geology*, **42**, 360–369.

Bennion, P., Hubbard, R., O'Hara, S.L. *et al.* (2007) The impact of airborne dust on respiratory health in children living in the Aral Sea region. *International Journal of Epidemiology*, **36**, 1103–1110.

Ben-Zvi, A., Massoth, S. and Schick, A.P. (1991) Travel time of runoff crests in Israel. *Journal of Hydrology*, **122** (1–4), 309–320.

Bofah, K.K. and Al-Hinai, K.G. (1986) Field tests of porous fences in the regime of sand-laden wind. *Journal of Wind Engineering and Industrial Aerodynamics*, **23**, 309–319.

Böhner, J., Schäfer, W., Conrad, O. *et al.* (2003) The WEELS model: methods, results and limitations. *Catena*, **52** (3–4), 289–308.

Brandle, J.R., Hodges, L. and Zhou, X.H. (2004) Windbreaks in north American agricultural systems. *Agroforestry Systems*, **61**, 65–78.

Bryant, R.G., Bigg, G.R., Mahowald, N.M. *et al.* (2007) Dust emission response to climate in southern Africa. *Journal of Geophysical Research D: Atmospheres*, **112**, article D09207.

Cahill, T.A., Gill, T.E., Reid, J.S. *et al.* (1996) Saltating particles, playa crusts and dust aerosols at Owens (dry) Lake, California. *Earth Surface Processes and Landforms*, **21** (7), 621–639.

Chan, C.C., Chuang, K.J., Chen, W.J. *et al.* (2008) Increasing cardiopulmonary emergency visits by long-range transported Asian dust storms in Taiwan. *Environmental Research*, **106** (3), 393–400.

Chen, Y.S., Sheen, P.C., Chen, E.R. *et al.* (2004) Effects of Asian dust storm events on daily mortality in Taipei, Taiwan. *Environmental Research*, **95** (2), 151–155.

Clausnitzer, H. and Singer, M.J. (1996) Respirable-dust production from agricultural operations in the Sacramento Valley, California. *Journal of Environmental Quality*, **25** (4), 877–884.

Cleugh, H.A. (1998) Effects of windbreaks on airflow, microclimates and crop yields. *Agroforestry Systems*, **41**, 55–84.

Cook, B.I., Miller, R.L. and Seager, R. (2009) Amplification of the North American 'Dust Bowl' drought through human-induced land degradation. *Proceedings of the National Academy of Sciences of the United States of America*, **106** (13), 4997–5001.

Cooke, R.U. and Doornkamp, J.C. (1990) *Geomorphology in Environmental Management*, 2nd edn, Clarendon.

Cornelis, W.M. and Gabriels, D. (2005) Optimal windbreak design for wind erosion control. *Journal of Arid Environments*, **61**, 315–332.

Costa, J.E. (1987) A comparison of the largest rainfall–runoff floods in the United States with those of the People's Republic of China and the world. *Journal of Hydrology*, **96** (1–4), 101–115.

Dahlgren, R.A., Richards, J.H. and Yu, Z. (1997) Soil and groundwater chemistry and vegetation distribution in a desert playa, Owens Lake, California. *Arid Soil Research and Rehabilitation*, **11**, 221–244.

Dong, Z., Chen, G., He, X. *et al.* (2004) Controlling blown sand along the highway crossing the Taklimakan Desert. *Journal of Arid Environments*, **57**, 329–344.

Dong, Z., Qian, G., Luo, W. and Wang, H. (2006) Threshold velocity for wind erosion: the effects of porous fences. *Environmental Geology*, **51**, 471–475.

Dong, Z.B., Wang, X.M. and Chen, G.T. (2000) Monitoring sand dune advance in the Taklimakan Desert. *Geomorphology*, **35**, 219–231.

Engelstaedter, S., Kohfield, K.E., Tegen, I. and Harrison, S.P. (2003) Controls on dust emissions by vegetation and topographic depressions: an evaluation using dust storm frequency data. *Geophysical Research Letters*, **30** (6), 1294, DOI: 10.1029/2002GL016471.

Ervin, R.T. and Lee, J.A. (1994) Impact of conservation practices on airborne dust in the southern High Plains of Texas. *Journal of Soil and Water Conservation*, **49**, 430–437.

Foody, G.M., Ghoneim, E.M. and Arnell, N.W. (2004) Predicting locations sensitive to flash flooding in an arid environment. *Journal of Hydrology*, **292**, 48–58.

Frazer, L. (2003) Down with road dust. *Environmental Health Perspectives*, **111**, A892–A895.

Fryrear, D.W. Saleh, A. Bilbro J.D. *et al.* (1998) Revised wind erosion equation (RWEQ). *Technical Bulletin*, **1**, USDS-ARS Lubbock, TX.

Fryrear, D.W., Bilbro, J.D., Saleh, A. *et al.* (2000) RWEQ: improved wind erosion technology. *Journal of Soil and Water Conservation*, **55**, 183–189.

Galloway, J.N., Thornton, J.D., Norton, S.A. *et al.* (1982) Trace metals in atmospheric deposition: a review and assessment. *Atmospheric Environment*, **16**, 1677–1700. DOI: 10.1016/0004-6981(82)90262-1.

GBUAPCD (Great Basin Unified Air Pollution Control District) (1994) Owens Valley PM-10 Planning Area Best Available Control Measures State Implementation Plan, Bishop, CA.

GBUAPCD (Great Basin Unified Air Pollution Control District) (2007) 2008 Owens Valley PM10 Planning Area Demonstration of Attainment State Implementation Plan, Initial Study, Bishop, CA.

Gill, T.E. (1996) Eolian sediments generated by anthropogenic disturbance of playas: human impacts on the geomorphic system and geomorphic impacts on the human system. *Geomorphology*, **17**, 207–228.

Goossens, D. and Buck, B. (2009a) Dust emission by off-road driving: experiments on 17 arid soil types, Nevada, USA. *Geomorphology*, **107**, 118–138.

Goossens, D. and Buck, B. (2009b) Dust dynamics in off-road vehicle trails: measurements on 16 arid soil types, Nevada, USA. *Journal of Environmental Management*, **90**, 3458–3469.

Goudie, A. and Viles, H. (1997) *Salt Weathering Hazards*, John Wiley and Sons, Ltd, Chichester.

Greenbaum, N., Margalit, A., Schick, A.P. *et al.* (1998) A high magnitude storm and flood in a hyperarid catchment, Nahal Zin, Negev Desert, Israel. *Hydrological Processes*, **12** (1), 1–23.

Griffin, D.W. (2007) Atmospheric movement of micro-organisms in clouds of desert dust and implications for human health. *Clinical Microbiology Reviews*, **20** (3), 459–477.

Grodek, T., Lekach, J. and Schick, A.P. (2000) Urbanizing alluvial fans as flood-conveying and flood-reducing systems: lessons from the October 1997 Eilat flood. *IAHS-AISH Publication*, **261**, 229–249.

Hagen L.J. (1991) A wind erosion prediction system to meet user needs. *Journal of Soil Water Conservation*, **46**, 106–111.

Hagen, L.J. (1996) Crop residue effects on aerodynamic processes and wind erosion. *Theoretical Applied Climatology*, **54**, 39–46.

Hagen, L.J. and Woodruff, N.P. (1973) Air pollution from dust storms in the Great Plains. *Atmospheric Environment*, **7**, 323–332.

Hefflin, B.J., Jalaludin, B., McClure, E. *et al.* (1994) Surveillance for dust dtorms and respiratory diseases in Washington State, 1991. *Archives of Environmental Health*, **49** (3), 170–174.

Houston, J. (2006) The great Atacama flood of 2001 and its implications for Andean hydrology. *Hydrological Processes*, **20** (3), 591–610.

Kellogg, C.A. and Griffin, D.W. (2006) Aerobiology and the global transport of desert dust. *Trends in Ecology and Evolution*, **21** (11), 638–644.

Kim, D.S., Cho, G.H. and White, B.R. (2000) A wind tunnel study of atmospheric boundary-layer flow over vegetated surfaces to suppress PM10 emission on Owens (dry) Lake. *Boundary-Layer Meteorology*, **97**, 309–329.

Kotlyakov, V.M. (1991) The Aral Sea basin: a critical environmental zone. *Environment*, **33** (1), 4–9, 36–38.

Kwon, H.J., Cho, S.H., Chun, Y. *et al.* (2002) Effects of the Asian dust events on daily mortality in Seoul, Korea. *Environmental Research*, **90** (1), 1–5.

Larney, F.J., Bullock, M.S., Janzen, H.H. *et al.* (1998) Wind erosion effects on nutrient redistribution and soil productivity. *Journal of Soil and Water Conservation*, **53** (2), 133–140.

Laronne, J.B. and Shulker, O. (2002) The effect of urbanization on the drainage system in a semiarid environment, in *Global Solutions for Urban Drainage* (eds E.W. Strecker and W.C. Huber), Proceedings of the Ninth International Conference on Urban Drainage, 8–13 September 2002, pp. 1–10.

Lee, S.J. and Kim, H.B. (1999) Laboratory measurement of velocity and turbulence field behind porous fences. *Journal of Wind Engineering and Industrial Aerodynamics*, **80**, 311–326.

Lee, S., Park, K. and Park, C. (2002) Wind tunnel observations about the shelter effect of porous fences on the sand particle movements. *Atmospheric Environment*, **36**, 1453–1463.

Lee, J.A., Wigner, K.A. and Gregory, J.M. (1993) Drought, wind and blowing dust on the southern High Plains of the United States. *Physical Geography*, **14**, 56–67.

Li, X.R., Xiao, H.L., He, M.Z. and Zhang, J.G. (2006) Sand barriers of straw checkerboards for habitat restoration in extremely arid desert regions. *Ecological Engineering*, **28**, 149–157.

Li, Y., Cui, J., Zhang, T. *et al.* (2009) Effectiveness of sand-fixing measures on desert land restoration in Kerqin Sandy Land, northern China. *Ecological Engineering*, **35**, 118–127.

Liu, Y. (1987) The establishment and effect of protecting system along the Bautou–Lanzhou Railway in the Shapotou study area (Chinese with English abstract). *Journal of Desert Research*, **7** (4), 1–11.

McTainsh, G.H. and Strong, C. (2007) The role of aeolian dust in ecosystems. *Geomorphology*, **89**, 39–54.

Merrill, S.D., Black, A.L., Fryrear, D.W. *et al.* (1999) Soil wind erosion hazard of spring wheat-fallow as affected by long-term climate and tillage. *Soil Science Society of America Journal*, **63** (6), 1768–1777.

Micklin, P.P. (1988) Desiccation of the Aral Sea: a water management disaster in the Soviet Union. *Science*, **241**, 1170–1176.

Micklin, P.P. (2007) The Aral Sea disaster. *Annual Review of Earth and Planetary Sciences*, **35**, 47–72.

Middleton, N.J. (1989) Climatic controls on the frequency, magnitude and distribution of dust storms: examples from India and Pakistan, Mauritania and Mongolia, in *Palaeoclimatology and Palaeometeorology: Modern and Past Patterns of Global Atmospheric Transport* (eds M. Leinen and M. Sarnthein), NATO ASI Series C, vol. 282, pp. 97–132.

Middleton, N.J. and Thomas, D.S.G. (eds) (1997) *World Atlas of Desertification*, 2nd edn, Arnold United and Nairobi, UN Environment Programme, London.

Mills, W.C., Thomas, A.W. and Langdale, G.W. (1991) Conservation tillage and season effects on soil erosion risk. *Journal of Soil and Water Conservation*, **46** (6), 457–460.

Misak, R.F. and Draz, M.Y. (1997) Sand drift control of selected coastal and desert dunes in Egypt: case studies. *Journal of Arid Environments*, **35**, 17–28.

Mitchell, D.J., Fullen, M.A., Trueman, I.C. and Fearnehough, W. (1998) Sustainability of reclaimed desertified land in Ningxia, China. *Journal of Arid Environments*, **39**, 239–251.

Nordstrom, K.F. and Hotta, S. (2004) Wind erosion from cropland in the USA: a review of problems, solutions and prospects. *Geoderma*, **121**, 157–167.

O'Hara, S.L., Wiggs, G.F.S., Mamedov, B. *et al.* (2000) Exposure to airborne dust contaminated with pesticide in the Aral Sea region. *The Lancet*, **355**, 627–628.

Padgett, P.E., Meadows, D., Eubanks, E. and Ryan, W.E. (2008) Monitoring fugitive dust emissions from off-highway vehicles travelling on unpaved roads and trails using passive samplers. *Environmental Monitoring and Assessment*, **144**, 93–103.

Parsons, A.J., Brazier, R.E., Wainwright, J. and Powell, D.M. (2006) Scale relationships in hillslope runoff and erosion. *Earth Surface Processes and Landforms*, **31** (11), 1384–1393.

Perez, L., Tobias, A., Querol, X. *et al.* (2008) Coarse particles from Saharan dust and daily mortality. *Epidemiology*, **19** (6), 800–807.

Pope, C.A., Bates, D.V. and Raizenne, M.E. (1996) Health effects of particulate air pollution: time for reassessment? *Environment Health Perspectives*, **103**, 472–480.

Prospero, J.M. (1999) Long-range transport of mineral dust in the global atmosphere: impact of African dust on the environment of the southeastern United States. *Proceedings of the National Academy of Sciences of the United States of America*, **96**, 3396–3403.

Prospero, J.M., Charlson, R.J., Mohnen, V. *et al.* (1983) The atmospheric aerosol system: an overview. *Reviews of Geophysics and Space Physics*, **21** (7), 1607–1629.

Prospero, J.M., Ginoux, P., Torres, O. and Nicholson, S.E. (2002) Environmental characterization of global sources of atmospheric soil dust derived from Niumbus-7 TOMS absorbing aerosol product. *Review of Geophysics*, **40**, 2–32.

Prospero, J.M., Blades, E., Mathison, G. and Naidu, R. (2005) Interhemispheric transport of viable fungi and bacteria from Africa to the Caribbean with soil dust. *Aerobiologia*, **21** (1), 1–19.

Qiu, G.Y., Lee, I., Shimizu, H. *et al.* (2004) Principles of sand dune fixation with straw checkerboard technology and its effects on the environment. *Journal of Arid Environments*, **56**, 449–464.

Reheis, M.C. (1997) Dust deposition downwind of Owens (dry) Lake, 1991–1994: preliminary findings. *Journal of Geophysical Research **D***, **102** (22), 25,999–26,008.

Reheis, M.C. and Kihl, R. (1995) Dust deposition in southern Nevada and California, 1984–1989; relations to climate, source area and source lithology. *Journal of Geophysical Research*, **100**, 8893–8918.

Reynolds, R.L., Yount, J.C., Reheis, M. *et al.* (2007) Dust emission from wet and dry playas in the Mojave Desert, USA. *Earth Surface Processes and Landforms*, **32**, 1811–1827. DOI: 10.1002/esp.1515.

Rhoads, B.L. (1986) Flood hazard assessment for land-use planning near desert mountains. *Environmental Management*, **10** (1), 97–106.

Ryu, J., Gao, S., Dahlgren, R.A. and Zierenberg, R.A. (2002) Arsenic distribution, speciation and solubility in shallow groundwater of Owens Dry Lake, California. *Geochimica et Cosmochimica*, **66** (17), 2981–2994.

Saiko, T.A. and Zonn, I.S. (2000) Irrigation expansion and dynamics of desertification in the Circum-Aral region of Central Asia. *Applied Geography*, **20** (4), 349–367.

Schick, A.P., Grodek, T. and Lekach, J. (1997) Sediment management and flood protection of desert towns: effects of small catchments. *IAHS-AISH Publication*, **245**, 183–189.

Seta, K.A., Blevins, R.L., Pye, W.W. and Barfield, B.J. (1993) Reducing soil erosion and agricultural chemical losses with conservation tillage. *Journal of Environmental Quality*, **22**, 661–665.

Sharon, D. (1972) The spottiness of rainfall in a desert area. *Journal of Hydrology*, **17** (3), 161–175.

Sharratt, B., Feng, G. and Wendling, L. (2007) Loss of soil and PM10 from agricultural fields associated with high winds on the Columbia Plateau. *Earth Surface Processes and Landforms*, **32**, 621–630.

Singer, A., Zobeck, T., Poberezsky, L. and Argaman, E. (2003) The PM10 and PM2.5 dust generation potential of soils/sediments in the Southern Aral Sea Basin, Uzbekistan. *Journal of Arid Environments*, **54** (4), 705–728.

Smith, D.R. (1991) Growing pollution and health concerns in the Lower Amu Dar'ya Basin, Uzbekistan. *Soviet Geography*, **32**, 553–565.

Srikanthan, R. and McMahon, T.A. (1980) Stochastic generation of monthly flows for ephemeral streams. *Journal of Hydrology*, **47** (1–2), 19–40.

Stout, J.E. and Zobeck, T.M. (1996) The Wolfforth experiment: a wind erosion study. *Soil Science*, **161**, 616–632.

Su, Y.Z., Zhao, W.Z., Su, P.X. *et al.* (2007) Ecological effects of desertification control and desertified land reclamation in an oasis-desert ecotone in an arid region: a case study in Hexi Corridor, northwest China. *Ecological Engineering*, **29**, 117–124.

Tang, K. and Zhang, C.E. (1996) Research on minimum tillage, no tillage and mulching systems and its effects in China. *Theoretical and Applied Climatology*, **54**, 61–67.

Tegen, I. and Fung, I. (1995) Contribution to the atmospheric mineral aerosol load from land surface modification. *Journal of Geophysical Research*, **100** (D9), 18707–18726.

Thomas, D.S.G., Knight, M. and Wiggs, G.F.S. (2005) Remobilization of southern African desert dune systems by twenty-first century global warming. *Nature*, **435**, 1218–1221.

Tyler, S.W., Kranz, S., Parlange, M.B. *et al.* (1997) Estimation of groundwater evaporation and salt flux from Owens Lake, California. *Journal of Hydrology*, **200**, 110–135.

UNDP (United Nations Development Programme) (1997) Turkmenistan: human development report 1996, UNDP, Ashgabat, Turkmenistan.

Uri, N.D. (1998) Trends in the use of conservation tillage in US agriculture. *Soil Use and Management*, **14**, 111–116.

Wainwright, J. and Parsons, A.J. (2002) The effect of temporal variations in rainfall on scale dependency in runoff coefficients. *Water Resources Research*, **38** (12), 71–710.

Wang, H. and Takle, E.S. (1995) Numerical simulations of shelterbelt effects on wind direction. *Journal of Applied Meteorology*, **34**, 2206–2219.

Wang, H. and Takle, E.S. (1997) Momentum budget and shelter mechanism of boundary layer flow near a shelterbelt. *Boundary Layer Meteorology*, **82**, 417–435.

Wang, X., Yang, Y., Dong, Z. and Zhang, C. (2009) Responses of dune activity and desertification in China to global warming in the twenty-first century. *Global and Planetary Change*, **67**, 167–185.

Watson, A. (1985) The control of windblown sand and desert dunes: a review of the methods of sand control in deserts, with observations from Saudi Arabia. *Quarterly Journal of Engineering Geology*, **18** (3), 237–252.

Watson, A. (1990) The control of blowing sand and mobile desert dunes. *Techniques for Desert Reclamation*, 35–85.

Webb, N.P., McGowan, H.A., Phinn, S.R. *et al.* (2009) A model to predict land susceptibility to wind erosion in western Queensland, Australia. *Environmental Modelling and Software*, **24**, 214–227.

Wiggs, G.F.S. and Holmes, P.J. (2010) Dynamic controls on wind erosion and dust generation on west-central Free State agricultural land, South Africa. *Earth Surface Processes and Landforms* (in press).

Wiggs, G.F.S., O'Hara, S.L., Wegerdt, J. *et al.* (2003) The dynamics and characteristics of aeolian dust in dryland Central Asia: possible impacts on human exposure and respiratory health in the Aral Sea basin. *The Geographical Journal*, **169** (2), 142–157.

Wilshire, H.G. (1980) Human causes of accelerated wind erosion in California's deserts, in *Thresholds in Geomorphology* (eds D.R. Coates and J.D. Vitek), The Binghampton Symposia in Geomorphology, International Series 11, Allen and Unwin, Boston, pp. 415–433.

Worster, D. (1979) *Dust Bowl*, Oxford University Press, New York.

Xu, B., Liu, X.M. and Zhao, X.Y. (1993) Soil wind erosion of farmlands and its control in central Naiman Banner, Inner Mongolia. *Journal of Soil and Water Conservation*, **7**, 75–88.

Zender, C.S. and Talamantes, J. (2006) Climate controls on valley fever incidence in Kern County, California. *International Journal of Biometeorology*, **50**, 174–182.

Zhang, T.H., Zhao, H.L., Li, S.G. *et al.* (2004) A comparison of different measures for stabilizing moving sand dunes in the Horqin Sandy Land of Inner Mongolia, China. *Journal of Arid Environments*, **58**, 203–214.

Zhang, C.L., Zou, X.Y., Pan, X.H. *et al.* (2007) Near-surface airflow field and aerodynamic characteristics of the railway-protection system in the Shapotou region and their significance. *Journal of Arid Environments*, **71**, 169–187.

Zhang, K., Qu, J., Liao, K. *et al.* (2010) Damage by wind-blown sand and its control along Qinghai–Tibet Railway in China. *Aeolian Research*, **1**, 143–146.

Zobeck, T.M. and Van Pelt, R.S. (2006) Wind-induced dust generation and transport mechanics on a bare agricultural field. *Journal of Hazardous Materials*, **132**, 26–38.

24

Future climate change and arid zone geomorphology

Richard Washington and David S. G. Thomas

24.1 Introduction

Climate change and its impact on arid zones, particularly during the Late Quaternary, has served as an important source of knowledge of both processes and rates of change in arid zone geomorphology, highlighting the dynamism and sensitivity of arid regions. Future climates will also change as a result of a variety of mechanisms and are predicted to do so at an unprecedented rate in forthcoming decades. What is different about the twenty-first century, however, is that for the first time in the Earth's history we have predictive tools that quantify the nature and extent of future climate change. Driven largely by greenhouse gas emissions, initially emitted from fossil fuel burning and then supplemented by feedbacks in the Earth System towards the end of the century, this change can be simulated in global climate model s (GCMs), thereby allowing at least some of the changes to be anticipated.

This chapter provides an overview of the basis for uncertainties in, and nature of, climate change projections in arid zones from global climate models. This is necessary before discussing the potential geomorphological impacts of these changes, on dunes, dust and hydrological changes.

24.2 Climate change projections: basis and uncertainties

The Intergovermental Panel on Climate Change Fourth Assessment Report (IPCC AR4) featured a coordinated global climate modelling exercise involving all the major climate modelling groups in the world. As a result of this effort, monthly (and in many cases daily) data from more than 20 GCMs are available up to the year 2100 at a spatial resolution of approximately 2.5 × 2.5 degrees or roughly 250 km × 250 km, with finer resolution for higher latitudes). These models are forced with changing gaseous atmospheric composition, which results from emissions from fossil fuel consumption. Increasing greenhouse gas concentration is therefore the basis for predicting climates of future decades. Since these emissions cannot be known from first principles, several possible agreed concentrations of emissions have been established and these are referred to as the Special Report on Emission Scenarios (SRES). An often used subset of these scenarios, together with the corresponding global temperature increase, is as follows:

SRESA2 – a high future emissions scenario that results in a best estimate temperature change of ~3.4 °C by 2100.

SRESA1B – a more middle-of-the-road future emissions scenario of ~2.8 °C by 2100.

SRESB1 – a low future emissions scenario of ~1.8 °C by 2100.

These emission scenarios are often referred to as A2, A1B and B1 or high, middle and low, respectively.

There are numerous uncertainties involved in making climate change projections even before the impacts of climate change are considered. Climate models are the only approach available for making projections, but,

Arid Zone Geomorphology: Process, Form and Change in Drylands, Third Edition. Edited by David S. G. Thomas

despite continual improvements, models have key limitations. Four key sources of uncertainty are:

- It is not possible to make predictions of *future atmospheric greenhouse gas concentrations*. As a result, there will always be an envelope of climate projections corresponding to the range of emission scenarios even assuming perfect climate models.

- *Climate model limitations* impose uncertainty on the climate projections. Projected precipitation is particularly uncertain as this variable results from microscale processes, which, at the resolution of GCMs, is represented indirectly from parameters that are resolved by the models.

- *Regional climate change* is less well modelled and understood than global change. One of the limitations of current climate models is their relative inability to prescribe detail on the regional scale as compared to their abilities at the global scale; the IPCC has recognised this as a prime area for continuing research. A key feature of the Fifth Assessment will be a dedicated regional modelling exercise with resolutions at or better than 50 km. However, improvements in model resolution do not necessarily equate to improvements in the fidelity of the simulated climate.

- *Natural climate variability* will continue to exert an influence on future climate and will be relatively strong compared with greenhouse gas forcing until about 2025. Multiyear trends of warming or cooling, wetting or drying, have been observed in the past, and will continue to be a feature in the future.

24.3 Overview of global climate change projections in the context of arid zones

Projected changes in global mean surface temperature for the twenty-first century (Table 24.1) are about 0.6 °C for the period 2011–2030 (during these decades the change is largely independent of the emission scenario) to 3.1 °C for the last two decades for the high scenario. Changes are large over land areas (compared with the ocean) and larger still over some of the arid subtropics, particularly southern Africa and the Sahara (northern Mali, southern Algeria and eastern Mauritania) (Figure 24.1). Trends in observed temperature extremes (e.g. increases in the occurrence of warm nights) over the period 1951–2003 are also higher in these parts of Africa than almost anywhere else in the world (Alexander *et al.*, 2006).

Table 24.1 Global mean surface temperature warming in °C for the multimodel ensemble mean from IPCC AR4 for (four time periods relative to 1980–1999 and for three emission scenarios) (after Meehl *et al.*, 2007).

SRES	*2011–2030*	*2046–2065*	*2080–2099*	*2180–2199*
A2	0.64	1.65	3.13	
A1B	0.69	1.75	2.65	3.36
B1	0.66	1.29	1.79	2.10

Based on an ensemble mean of 14 leading IPCC AR4 GCMs, temperatures under the A2 scenario for the last two decades of the twenty-first century are projected to rise by more than 5.5 degrees in southern Africa (eastern Botswana, southern Angola and northern Namibia) during the September to November months. Over the Saharan Empty Quarter of western Algeria, northern Mali and eastern Mauritania, the warming is even greater, exceeding 5 degrees between March and December and peaking above 6 degrees between June and September inclusive, nearly double the global mean increase. These arid to hyper-arid regions are already the hottest on the continent and among the hottest regions in the world.

The interior of Australia (between approximately 22–26 degrees south and 118–130 degrees east) is projected to warm by 4.6 or more degrees by the last two decades of the twenty-first century. Peak warming of 5.2 degrees is projected for October. Similar, though slightly higher, projections (peak 5.5 degrees between June and August) are made for the interior of Saudi Arabia while eastern Iraq and western Iran are projected to warm by more than 5 degrees in between April and October, with a peak of 6.7 degrees in July and an increase above 6 degrees for May to September inclusive.

At the global scale, the hydrological response to global warming is reasonably simple (Figure 24.1, middle). There is an increase in precipitation in the tropics, a corresponding increase in subsidence, decrease in cloudiness (Figure 24.2) and therefore aridity in the subtropics. The subtropical anticyclones also expand poleward (Figure 24.1, right). Most of the reductions in precipitation in forthcoming decades results from increases in subsiding air into the subtropics. The precise nature of regional precipitation changes, while of clear importance to establishing future arid zone processes, are much harder to pinpoint because of the uncertainty with which climate models simulate precipitation and the resultant intermodel differences in the projections. Figure 24.3 provides a more integrated picture of the hydrological changes. Reduction

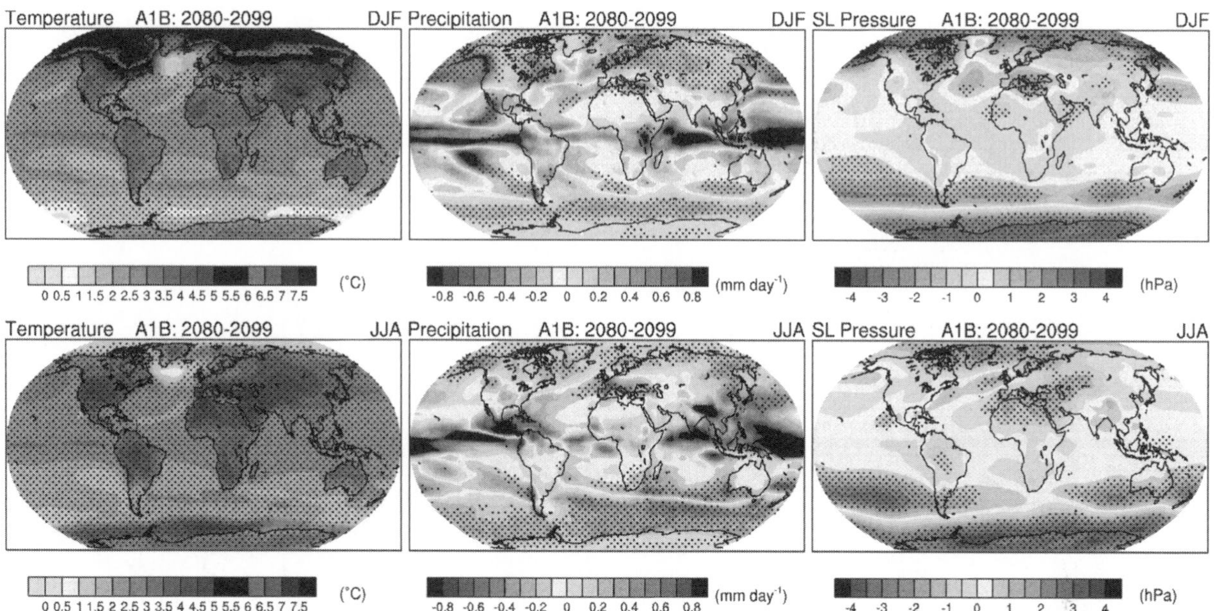

Figure 24.1 Multimodel mean changes in surface air temperature (°C, left), precipitation (mm/day, middle) and sea-level pressure (hPa, right) for December to February (DJF, top) and June to August (JJA, bottom). Changes are given for the SRES A1B scenario, for the period 2080–2099 relative to 1980–1999. Stippling denotes areas where the magnitude of the multimodel ensemble mean exceeds the intermodel standard deviation (after Meehl *et al.*, 2007).

of soil moisture and runoff is a widespread feature of arid zones, although there is consistency between the models only in some regions, notably the Mediterranean, south-western USA and parts of South America.

Sea-level rise could lead to dramatic changes to sabkhas in arid regions such as the Arabian Peninsula. Global mean sea-level rise projections from the IPCC AR4 are shown in Table 24.2. The sea-level rise estimates comprise four components: thermal expansion (calculated from the climate models, specifically the ocean component), glaciers and ice caps excluding the Greenland and Antarctic ice sheets (computed from a simple empirical formula that links the global mean temperature to mass loss based on observed data from 1963 to 2003), ice sheet surface mass balance (computed from an ice sheet surface balance model with snowfall amounts and temperature computed from a high-resolution model scaled to the coupled models) and dynamical imbalance (computed from extrapolation of observed rates from 1993 to 2003 and contributing up to 0.7 mm during this period). The range shown in Table 24.2 does not include uncertainties resulting from climate–carbon cycle feedback or changes in ice sheet flow (including dynamical processes). They do include a component due to increased ice flow from Greenland and Antarctica at the rates observed for 1993–2003.

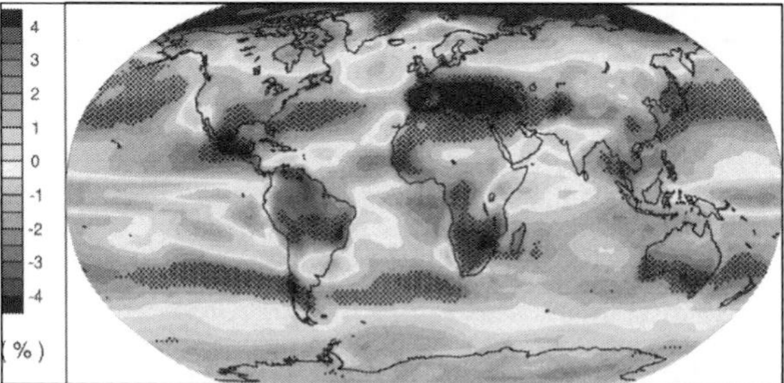

Figure 24.2 Multimodel mean changes in total cloud area fraction (percent cover from all models). Changes are given as annual means for the SRES A1B scenario for the period 2080–2099 relative to 1980–1999. Stippling denotes areas where the magnitude of the multimodel ensemble mean exceeds the intermodel standard deviation (after Meehl *et al.*, 2007).

Since IPCC AR4, there have been a number of studies that conclude that the sea-level rise by 2100 could exceed 100 cm. Rahmstorf *et al.* (2007) note that since 1990 the observed sea-level has been rising faster than the rise projected by models, as shown both by a reconstruction using primarily tide gauge data and, since 1993, by satellite altimeter data. The satellite data show a linear trend of 3.3 ± 0.4 mm/yr (1993–2006) and the tide gauge reconstruction trend is slightly less, whereas the IPCC projected a best-estimate rise of less than 2 mm/yr. Vermeer and Rahmstorf (2009) propose a simple relationship linking global sea-level variations on timescales of

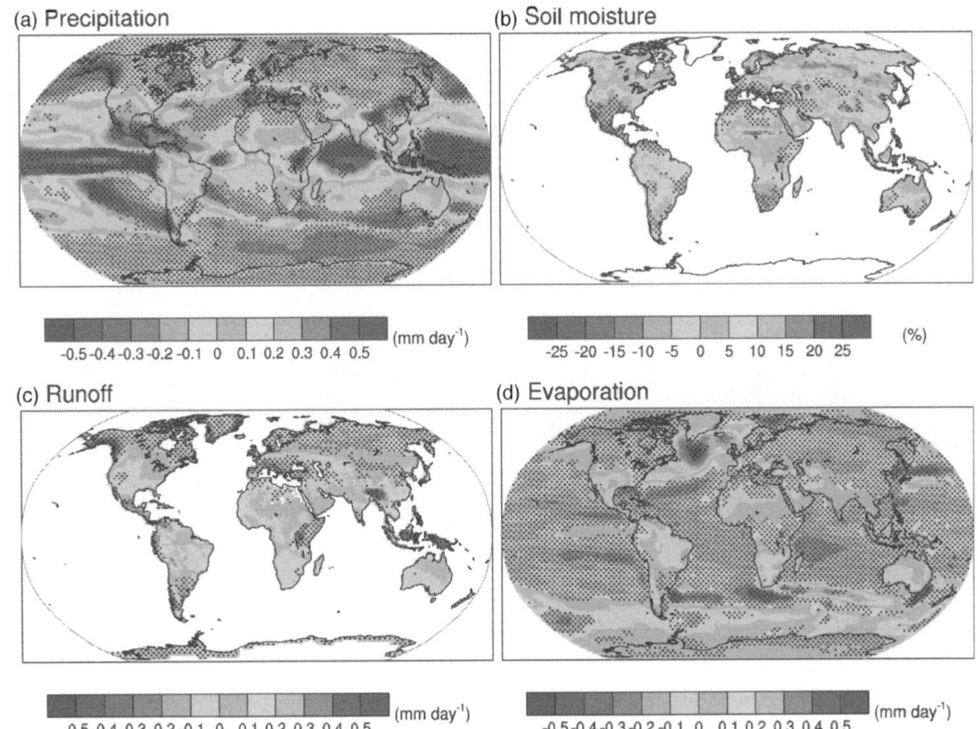

Figure 24.3 Multimodel mean changes in (a) precipitation (mm/day), (b) soil moisture content (%), (c) runoff (mm/day) and (d) evaporation (mm/day). To indicate consistency in the sign of change, regions are stippled where at least 80 % of models agree on the sign of the mean change. Changes are annual means for the SRES A1B scenario for the period 2080–2099 relative to 1980–1999 (after Meehl *et al.*, 2007).

decades to centuries to global mean temperature. For future global temperature scenarios of the Intergovernmental Panel on Climate Change's Fourth Assessment Report, the relationship projects a sea-level rise ranging from 75 to 190 cm for the period 1990–2100.

24.3.1 Methods of establishing climate change impacts in arid zones

There are at least three methods, ordered by complexity and sophistication, for establishing responses of arid zone geomorphology to future climate change. The most complex and sophisticated are not necessarily the most useful or accurate. All involve climate change projections from climate models, although a key requirement is also the thorough if not quantitative understanding of geomorphic systems and their thresholds and sensitivity to forcing factors such as precipitation, wind, temperature and evaporation.

Table 24.2 Global mean sea-level rise in cm for the end of the twenty-first century relative to the last two decades of the twentieth century from IPCC AR4. Results are shown for three emission scenarios. The range of sea level rise is therefore 18–59 cm.

Scenario	*Change in sea level in cm*
B1	0.18–0.38
A1B	0.21–0.48
A2	0.23–0.51

The simplest approach is to describe qualitatively what likely impacts will emerge from the changing climate, often on the basis of changing climate variables (precipitation, wind, temperature, wind and evaporation) over the forthcoming decades. Whereas much of arid zone geomorphology has been concerned with interpretation of past environments from the analysis of landforms, this approach is essentially the opposite – going from a set of altered climatic conditions to the inferred landscape response. While this simple approach underpins the majority of first-order assessments of climate change vulnerability across a range of disciplines, particularly at the regional and country level, the specific difficulties lie in determining the nonlinear responses of the geomorphic system to the separate climate signals and knowing how other factors, e.g. vegetation, might respond to climate change and, in turn, influence the system. Scenarios of potential response are therefore the most likely product of this approach (Goudie, 2006; Tooth, 2008).

Intermediate methods see some post-processing of the climate change projections, typically by downscaling the data to suitable spatial resolution. Empirical equations calibrated to represent the current behaviour of the system in question (e.g. dune mobility) are then used as the basis to compute the new behaviour of the system given the quantified and downscaled changed climate. Intermediate methods in the case of arid zone geomorphology are likely to be the most fruitful and have indeed made an impact on the literature (e.g. Thomas, Knight and Wiggs, 2005; De Wit and Stankiewicz, 2006).

The most sophisticated and complex approach involves driving numerical models of the system of interest with the climate change projections as the input to these models. There are few systems in geomorphology for which numerical models have been developed although the value of such models is enormous. Global climate models are one example of numerical models. Typically these models represent space in terms of grid boxes within which equations that govern the key processes in the system are solved. With the equations in prognostic form, the system can be integrated through time steps with new inputs to some of the controlling variables. Without climate models, the issue of global warming and climate change would most certainly only be at the stage of vague debate and conjecture. The models have allowed both the attribution of observed climate change to anthropogenic emissions and the projection of future climate, resulting in climate change becoming an established research priority. The background science evolved over many decades through the economic imperative of weather forecasting and was facilitated by (and indeed stimulated) the fastest available supercomputers. Geomorphology has lacked this imperative and therefore a commensurate input of resources, although hydrological models are an exception.

Of the Earth Systems for which numerical models have been developed, the cycle of mineral dust is a leading example. The need to construct models for the global simulation of dust came about largely because of the important feedbacks that dust exerts on the climate system itself. As a result, many of the leading climate models include components that simulate the deflation, advection and deposition of dust as well as its interaction with the Earth System. While there has been progress with the development numerical modelling in geomorphology, much of the capability is geared to timescales much longer than a century (Tooth, 2008; Tucker and Hancock, 2010).

24.4 Climate change and dunes

Demonstration of the dynamism and sensitivity of dune systems to changes in climate through concerted application of dating techniques has been one of the major achievements of aeolian geomorphology in recent decades (see Chapter 3). Building on work in the Kalahari, which points to episodes of punctuated aridity and dune mobilisation during the Quaternary (Stokes, Thomas and Washington, 1997), Thomas, Knight and Wiggs (2005) have demonstrated the likelihood of twenty-first century mobility of these linear dunes. Extending from northern southern Africa to Angola and Zambia, these dunes are currently stable with observed sand transport confined by low erosivity and well-developed vegetation cover. The investigation used a modified empirical index of dune mobility (an example of the intermediate approach discussed in the previous section), such that

$$A_{p,GCM} = \bar{U}_3/(P_{lag}/E_{p,lag} + \mathrm{P}_{rainy}/E_{p,rainy})$$

where:

$\bar{U}_3$ = the cube of the mean wind speed

$P_{lag}/E_{p,\,lag}$ = residual effect of recent rainfall and potential evaporation, such that $P_{lag} = (P_{-1} + P_0)/2$, where P_{-1} is precipitation in the previous month and P_0 is rainfall in the current month

$E_{p,\,lag} = (E_{p,\,-1} + E_{p,\,0})/2$, where $E_{p,\,-1}$ is potential evapotranspiration in the previous month and $E_{p,0}$ is potential evapotranspiration in the current month

$P_{rainy}/E_{p,\,rainy}$ = effect of rainy season precipitation and potential evaporation on soil moisture, such that $P_{rainy} = (P_N + P_D + P_J \cdots)/m$ and $E_{p,\,rainy} = (E_{p,\,N} + E_{p,\,D} + E_{p,\,J} \cdots)/m$, where m = N (November), D (December), J (January), and so on is the month under consideration within the rainy season

This model was therefore specifically adapted for the seasonal climate of southern Africa, and was driven with data on moisture availability and erosivity derived from the output of three global climate models, each forced by several emission scenarios. Thomas, Knight and Wiggs (2005) used model outputs, run for the twenty-first century on a monthly basis, to demonstrate that dunefields across this broad region are potentially reactivated by the end of the century, in all emission scenarios and in each of the climate models used in the study (Figure 24.4).

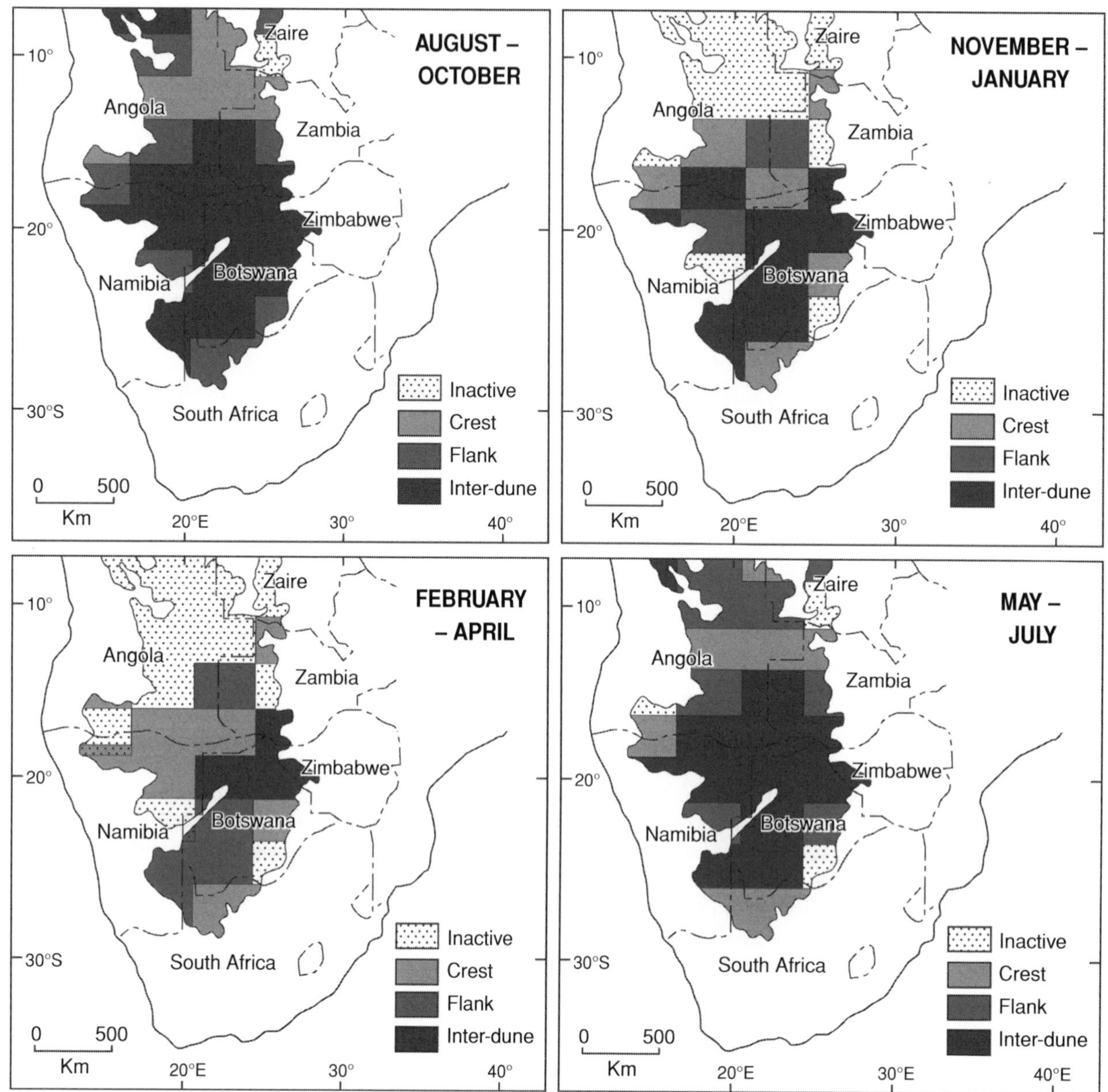

Figure 24.4 Trimonthly dune system activity status in the Kalahari, southern Africa, averaged for the period 2070–2099. The shaded area is the Kalahari Sand Sea, currently largely stable and dominated by linear dunes, with shading differences suggesting the degree of dune landscape activity through the seasons. 'Interdune' implies aeolian activity through the whole dune landscape, 'flank' refers to activity on dune slopes and 'crests' refers to activity on dune crestal areas alone (based on Thomas, Knight and Wiggs, 2005).

In a similarly posed study of the drylands of China, Wang *et al.* (2009) used Lancaster's dune mobility index (Lancaster, 1988) driven by two climate models (from an initial selection of five) and a range of emission scenarios to determine twenty-first century trends. Defining desertification to have occurred if levels of dune activity predicted by the GCMs are higher than values from 1960 to 1990, Wang *et al.* (2009) show considerable change over China. From 2040 to 2099, results from both models indicate an increase in western arid and semi-arid China. After 2070, the increases are classed as severe. They also demonstrate that anchored dunes will evolve to semi-anchored ones and semi-anchored dunes will become mobile in the dune and steppe areas of China. In contrast, the desertification is predicted to reverse in central and eastern regions of arid and semi-arid China. Overall the changes are controlled mainly by precipitation regimes and potential evaporation.

While Thomas, Knight and Wiggs (2005) and Wang *et al.* (2009) are important marker studies of the potential for dune reactivation under future climate change, they also draw attention to the importance of monitoring contemporary dune processes with a view to improving the empirical indices on which dune mobilisation was based. This is partly because of the complexity of dune processes, with a number of recent studies revealing that active and fixed dunes may coexist under similar climatic conditions (Tsoar, 2005; Yizhaq, Ashkenazy and Tsoar, 2007). A related issue is that mobility indices are calibrated against current climate conditions whereas dunes response may lag behind climatic change (Hugenholtz and Wolfe, 2005). An even more fundamental limitation is the skill with which climate models simulate precipitation and the worrying differences in this simulation, particularly at the regional scale, among different climate models. These stubborn problems cannot be resolved without concerted effort. The dramatic results of Thomas, Knight and Wiggs (2005) and Wang *et al.* (2009) show that such an effort is clearly warranted.

24.5 Climate change and dust

The development of numerical models that simulate the dust cycle has been ongoing since at least the early 1990s, making this component of aeolian geomorphology among the most sophisticated and complex methodologically. Recognition that numerical weather prediction benefits from inclusion of dust aerosols (e.g. Pérez *et al.*, 2006; Rodwell and Jung, 2008) has led to a number of experimental and operational dust forecast systems, e.g. the Navy Aerosol Analysis and Prediction System (NAAPS), the global and regional Earth-System (atmosphere) monitoring using satellite and in situ data (GEMS) at the European Centre for Medium Range Weather Forecasts (Morcrette *et al.*, 2007) and the dust regional atmospheric model (DREAM) (Nickovic *et al.*, 2001). Simultaneously, inclusion of dust aerosols has been an important step in the further development of climate and Earth System models. In the Fourth Assessment Report of the IPCC, dust was included in several of the 20 or so climate models used in climate change projections.

While operational dust forecasting schemes provide the opportunity to evaluate, constrain and improve dust model performance (not something that is possible in future climate change projections), there remains large uncertainty in the future projections of dust loadings. In the Fourth Assessment Report of the IPCC, Meehl *et al.* (2007) note the conflicting outcomes of several studies. Mahowald and Luo (2003), for example, find that the global atmospheric burden of soil dust aerosols could decrease by between 20 and 60 % in association with climate change, a result confirmed by a similar, later study (Mahowald *et al.*, 2006). Dust production in this set of studies was mainly shown to respond to changes in the source areas, which result from vegetation changes. Winds or soil moisture changes are argued to be less important. Tegen *et al.* (2004a, 2004b), using model runs by the European Centre for Medium Range Weather Forecasts/Max Planck Institute for Meteorology Atmospheric GCM (ECHAM4) and UKMO-HadCM3 forced by identical greenhouse gas scenarios, demonstrate that the future projections of dust loadings are model-dependent, with the loadings increasing in one model but decreasing in the other. These latter two simulations included changes to atmospheric conditions and vegetation cover. Tegen *et al.* (2004a, 2004b) are, however, able to conclude that dust loadings from agriculture are small (less than 10 %) of the total dust loadings, even with a maximum estimate of increased agricultural area by 2050.

Using the UKMO-HadAM3 atmospheric model, which included feedback of vegetation on climate, Woodward, Roberts and Betts (2005) point to an order of magnitude increase in atmospheric dust loadings over 100 years up to 2100 as a result of desertification and climate change. More recent work by the same group, but based on a new version of the Met Office Hadley Centre coupled climate–carbon cycle model, shows a severe drying over the Amazon, which, through carbon feedbacks on the atmosphere following forest loss, leads to the Amazon becoming an important global dust source (Betts, Sanderson and Woodward, 2008).

There are a number of key problems with numerical models of the dust cycle that contribute to the ambiguity evident in changes to projected dust loadings as reported in the Fourth Assessment Report:

- Dust models are effectively unconstrained at dust source regions since the available data on dust concentrations needed to calibrate models at source does not exist. It is widely recognised from analysis of satellite data that dust emission occurs primarily in a relatively small number of extremely remote preferential source regions (Herman *et al.*, 1997; Prospero *et al.*, 2002; Torres *et al.*, 2002; Washington *et al.*, 2003). Current estimates of global dust emission vary by over a factor of two

(Cakmur *et al.*, 2006). Models tend to be calibrated to data remote from source regions using Aeronet data or the Miami Aerosol Group measurements (see Chapter 20) (Cakmur *et al.*, 2006).

- The surface winds in most dust models are known to seriously underestimate winds over the key source regions such as the Bodélé (Koren and Kaufman, 2004; Washington *et al.*, 2006), many by 50 % or more.

- The climate models have a coarse resolution, particularly global climate models, making it difficult to represent the processes of deflation realistically. As a result, the model components that specify emission tend to be tuned to background dust loadings remote from source rather than made to satisfy observed physical processes crucial in deflation. Even when emission data from key source regions are available to constrain models and resolution is improved by running regional climate models at horizontal resolutions of 26 km or better, such as the study of five regional models for a single deflation event covering several days in the Bodélé Depression, model emission still differed by an order of magnitude (Todd *et al.*, 2008).

- The soil particle size distribution used in many model simulations are necessarily fictitious as the source regions are extremely remote. There have nevertheless been commendable efforts to recover better particle size data from remote regions using remote sensing (e.g. Christopher and Jones, 2010) as they have to map sub-basin scale dust sources (Bullard *et al.*, 2008).

- Like several other components of arid zone geomorphology, projected dust emissions under climate change suffer from the degree of spread in climate model precipitation projections. In the case of the Sahel and Sahara, this is a particular difficulty given the degree divergence of precipitation futures for this region, with some models predicting a wet future and others a dry future (Cook and Vizy, 2006; Hoerling *et al.*, 2006). The Bodélé Depression, the world's largest source of mineral aerosols (Washington *et al.*, 2003), for example, lies just a few hundred kilometres north of the northernmost fringe of the Sahel grasslands – less than a gridbox in some global climate models. The Bodélé is currently therefore critically close to being stabilised by vegetation and small changes to the current regime are unlikely to be captured with sufficient accuracy by the climate models.

A notable concern of the emission schemes of several models is that they derive ultimately from idealised wind tunnel experiments rather than more realistic assessments in the field. Solutions to some of these difficulties lie in implementing targeted research efforts such as instrumentation of key dust source regions at a scale that is sympathetic to the resolution of the climate models.

In the meantime, some research on projected change in dust emissions has resorted to less-sophisticated methods, which correspond more with the intermediate category of approaches to projecting climate change impacts. Washington *et al.* (2009), for example, analysed climate projections from the IPCC AR4 data set for the Bodélé in Chad. Of the 10 models studied, 8 show an increase in mean annual surface wind speed by the last decade of the twenty-first century compared with 1971–2000 and all 10 show an increase during January to March (maximum +0.8 m/s, minimum +0.2 m/s). Choosing only those models with a realistic simulation of current climate limits the selection to three models, two of which (MRI and GFDL) show dry conditions over the Sahel during the twenty-first century. Near-surface zonal winds over the Bodélé Depression in MRI enhance considerably during the course of the twenty-first century based on decadal means. In the decade 1991–2000, easterly winds exhibit an average peak speed of 9 m/s within the core of the low-level jet, extending from 15–17°N, 17–20°E and fragmenting over Lake Chad. By 2091–2100 the spatial coverage of the jet has increased, extending from 3–21°E across the region with a clear focus of expansion over the Bodélé.

Since dust mobilisation arises from synoptic-scale events in the Bodélé, the extremes of daily wind speed distribution and their frequency of occurrence in each month are crucial components in modelling dust output. The frequency of winds stronger than 10 m/s is markedly greater in the latter half of the twenty-first century (Figure 24.5). The percentage of January to March (JFM) days with winds exceeding 11 m/s (a rough threshold for deflation in the basin) increases from 45 % (1991–2000) to 49 % (mid-twenty-first century), exceeding 56 % by the end of the century. The Bodélé winds show a doubling in the number of February days with wind speeds exceeding 11 m/s by the end of the century ($p = 0.01$ %). Given the cubic sensitivity of dust mobilisation to wind speed, these results point to the possibility of a substantial increase in dust output from the world's largest mineral aerosol source towards the end of the twenty-first century.

In the 5 years leading up to 2010, much of the effort in dust modelling seems to have gone into refining and calibrating climate models against observed data rather than extended efforts on long-term climate change projections of dust (e.g. Yoshioka *et al.*, 2007). Much of this work has focused on regional climate models. This

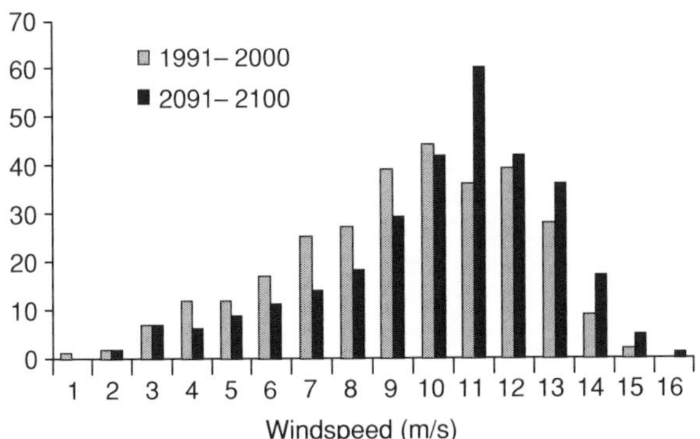

Figure 24.5 Histogram (frequency of occurrence on vertical axis versus wind speed on horizontal axis) of January to March Bodélé 925 hPa winds in m/s for 1991-2000 and 2091-2100 from the MRI model.

refinement effort is aided substantially by a number of field campaigns in desert and dryland regions such as BoDEx (Bodele Dust Experiment), AMMA (African Multidisciplinary Monsoon Analysis) and the Fennec (the Saharan Climate System) project.

24.6 Climate change and fluvial systems

Changes to hydrology, including runoff, have already been briefly discussed at the global scale earlier in this chapter. A clear result from many models forced with increased greenhouse gas concentrations is an intensification of the hydrological cycle (Meehl *et al.*, 2007), with decreases in precipitation in many parts of the already arid subtropics. Since precipitation is zero bounded, precipitation decreases in model simulations tends be accompanied by decreases in precipitation variance (e.g. Wetherald, 2009).

The dynamics of runoff and hydrology cannot be properly simulated at the resolution of the global climate models. As a result, some of the more significant research results have emerged from efforts to represent the large-scale forcing of climate change at higher resolution and to simplify complex, often nonlinear hydrological relationships. One such study, representative of the intermediate approach to assessing the impact of climate change on arid zone geomorphology, approaches the problem of assessing changes in streamflow in Africa by relating high-resolution drainage patterns to precipitation (De Wit and Stankiewicz, 2006). Using a digitised database of 2 million km^2 of river networks, cross-checked against a digital elevation model from the SRTM (Shuttle Radar Topography Mission) and statistically/empirically downscaled precipitation data from six global climate models for the twenty-first century, De Wit and Stankiewicz (2006) calculate a decrease in perennial drainage area across 25 % of Africa by the end of the century. In a series of sensitivity tests, they show that for regions receiving 500 mm/yr of precipitation, a 10 % decrease in precipitation results in a drop of 50 % of surface drainage. A significant dimension of this study is that its focus was to understand implications for population access to water, not landscape/geomorphological change per se. Its geomorphological significance is, however, considerable.

A similarly empirical approach was adopted by Ellis *et al.* (2008) for semi-arid central Arizona. Combining a statistical downscaling routine of climate change simulations from six global climate models with a climatic water budget model based on evapotranspiration, precipitation and soil moisture capacity, they show that runoff varies from 50 to 127 % of observed levels.

Where hydrological models have been applied to assess the hydrological impacts of climate change on runoff, it is similarly the case that small differences in the inputs to the hydrological model, such as temperature from different climate models, result in an amplification of the uncertainty by the hydrological model (e.g. Seguí *et al.*, 2010).

The studies of changing arid zone hydrology under climate change discussed thus far all make use of either annual or monthly precipitation. For some hydrological regimes in arid regions, flow results from one or two precipitation events during the course of the year. The capacity to simulate these systems accurately and precisely hinges on projections of precipitation that capture numerous characteristics, including amount, intensity, duration, type and timing (Goudie, 2006). Very few of these parameters have ever been extensively investigated in climate models and, for many arid regions, observed data with which to confront the models is in any case extremely scarce. The potential to take projections further to actually assess impacts on specific fluvial processes within river basins or channels is extremely limited. This is not simply due to issues associated with the outputs that climate models can generate, but because of the problems of systematically interpreting the behaviour and consequences for processes and forms of dryland rivers today.

24.7 Conclusions

Numerous components of arid zone geomorphology and the potential impact of climate change upon them have not been covered in this chapter. These include the inevitability (even with no further increase of carbon dioxide levels in the atmosphere) of sea-level rise for centuries to come and the implication for the delicate balance between deltaic aggradation and increasing marine inundation in sabkhas, the effects on groundwater and lakes in arid regions and changes to salinity.

Goudie (2006) has remarked that 'geomorphologists have yet to devote to this theme [climate change] the same amount of attention that has been expended by, for example, life scientists and hydrologists. Remarkably few scenarios for future geomorphological changes have been developed. This is a major research priority'. Given that the discipline is generally well practised in interpreting the effects of climate change on landscapes, it could be expected that geomorphologists have much to offer this important research theme. Elements of the problem of interpreting the impact of projected climate change on geomorphological processes, in arid regions and elsewhere, are likely to remain intractable for at least two decades, notably the issue of precipitation simulation in climate models. The opportunity, urgent as it is, therefore remains for geomorphologists to develop the tools to quantify the behaviour of arid zone processes. Until this occurs, projections will remain largely conjecture and qualitative, with the exception of the types of examples provided in this chapter, which at the very least provide hypotheses of the nature and rates of change that can be evaluated and enhanced as new data appear.

References

Alexander, L.V., Zhang, X., Peterson, T.C. *et al.* (2006) Global observed changes in daily climate extremes of temperature and precipitation. *Journal of Geophysical Research*, **111**, D05109, DOI: 10.1029/2005JD006290.

Betts, R., Sanderson, M. and Woodward, S. (2008) Effects of large-scale Amazon forest degradation on climate and air quality through fluxes of carbon dioxide, water, energy, mineral dust and isoprene. *Philosophical Transactions of the Royal Society – Biological Sciences*, **363**, 1873–1880.

Bullard, J., Baddock, M., McTainsh, G. and Leys, J. (2008) Sub-basin scale dust source geomorphology detected using MODIS. *Geophysical Research Letters*, **35**, L15404.

Cakmur, R.V., Miller, R.L., Perlwitz, J. *et al.* (2006) Constraining the magnitude of the global dust cycle by minimizing the difference between a model and observations. *Journal of Geophysical Research – Atmospheres*, **111** (D6), D06207.

Christopher, S.A. and Jones, T.A. (2010) Satellite and surface-based remote sensing of Saharan dust aerosols. *Remote Sensing of the Environment*, **114**, 1002–1007.

Cook, K.H. and Vizy, E.K. (2006) Coupled model simulations of the west African monsoon system: twentieth- and twenty-first-century simulations. *Journal of Climate*, **19**, 3681–3703.

De Wit, M. and Stankiewicz, J. (2006) Changes in surface water supply across Africa with predicted climate change. *Science*, **311**, 1917–1921.

Ellis, A.W. Hawkins, T.W., Balling R.C. and Gober, P. (2008) Estimating future runoff levels for a semi-arid fluvial system in central Arizona, USA. *Climate Research*, **35**, 227–239.

Goudie, A.S. (2006) Global warming and fluvial geomorphology. *Geomorphology*, **79**, 384–394.

Herman, J. R., Bhartia, P.K., Torres, O. *et al.* (1997) Global distribution of UV-absorbing aerosols from Nimbus7/TOMS data. *Journal of Geophysical Research*, **102** (D14), 16,911–16,922, DOI: 10.1029/96JD03680.

Hoerling, M., Hurrell, J., Eischeid, J. and Phillips, A. (2006) Detection and attribution of twentieth-century northern and southern African rainfall change. *Journal of Climate*, **19**, 2989–40008.

Hugenholtz, C.H. and Wolfe, S.A. (2005) Biogeomorphic model of dunefield activation and stabilization on the northern Great Plains. *Geomorphology*, **70**, 53–70.

Koren, I. and Kaufman. Y.J. (2004) Direct wind measurements of Saharan dust events from Terra and Aqua satellites. *Geophysical Research Letters*, **31**, L06122, DOI: 10.1029/2003GL019338.

Lancaster, N. (1988) Development of linear dunes in the southwestern Kalahari, southern Africa. *Journal of Arid Environments*, **14**, 233–244.

Mahowald, N.M. and Luo, C. (2003) A less dusty future? *Geophysical Research Letters*, **30**, 1903.

Mahowald, N.M., Muhs, D.R., Levis, S. *et al.* (2006) Change in atmospheric mineral aerosols in response to climate: last glacial period, preindustrial, modern, and doubled carbon dioxide climates. *Journal of Geophysical Research – Atmospheres*, **111** (D10), 10202.

Meehl, G.A., Stocker, T.F., Collins, W.D. *et al.* (2007) Global climate projections, in *Climate Change 2007: The Physical Science Basis. Contribution of Working Group I to the Fourth Assessment Report of the Intergovernmental Panel on Climate Change* (eds S. Solomon, D. Qin, M. Manning *et al.*), Cambridge University Press, Cambridge, United Kingdom, and New York, USA.

Morcrette, J.-J., Jones, L., Kaiser, J. *et al.* (2007) Toward a forecast of aerosols with the ECMWF Integrated Forecast System. *ECMWF Newsletter*, **114**, 15–17.

Nickovic, S., Kallos, G., Papadopoulos, A. and Kakaliagou, O. (2001) A model for prediction of desert dust cycle in the atmosphere. *Journal of Geophysical Research*, **106** (D16), 18,113–18,130, DOI: 10.1029/2000JD900794.

Pérez, C., Nickovic, S., Baldasano, J.M. *et al.* (2006) A long Saharan dust event over the western Mediterranean: Lidar, Sun photometer observations, and regional dust modeling. *Journal of Geophysical Research*, **111**, D15214, DOI: 10.1029/2005JD006579.

Prospero, J.M., Ginoux, P., Torres, O. *et al.* (2002) Environmental characterization of global sources of atmospheric soil dust identified with the Nimbus 7 total ozone mapping spectrometer (TOMS) absorbing aerosol product. *Reviews of Geophysics*, **40**, 1002.

Rahmstorf, S., Cazenave, A., Church, J.A. *et al.* (2007) Recent climate observations compared to projections. *Science*, **316**, 709–709.

Rodwell, M.J. and Jung, T. (2008) Understanding the local and global impacts of model physics changes: an aerosol example. *Quarterly Journal of the Royal Meteorological Society*, **134**, 1479–1497.

Segui, P.Q., Ribes, A., Martin, E. *et al.* (2010) Comparison of three downscaling methods in simulating the impact of climate change on the hydrology of Mediterranean basins. *Journal of Hydrology*, **383**, 111–124.

Stokes, S., Thomas, D.S.G. and Washington, R. (1997) Multiple episodes of aridity in southern Africa since the last interglacial period. *Nature*, **388**, 154–158.

Tegen, I., Werner, M., Harrison, S.P. and Kohfeld K.E. (2004a) Relative importance of climate and land use in determining present and future global soil dust emission. *Journal of Geophysical Research*, **31** (5), L05105.

Tegen, I. Werner, M. Harrison, S.P. and Kohfeld K.E. (2004b) Reply to comment by N.M. Mahowald *et al.*, on 'Relative importance of climate and land use in determining present and future global soil dust emission'. *Journal of Geophysical Research*, **31** (24), L24106.

Thomas, D.S.G., Knight, M. and Wiggs, G.F.S. (2005) Remobilization of southern African desert dune systems by twenty-first century global warming. *Nature*, **435**, 1218–1221.

Todd, M.C., Karam, D.B., Cavazos, C. *et al.* (2008) Quantifying uncertainty in estimates of mineral dust flux: an intercomparison of model performance over the Bodélé Depression, northern Chad. *Journal of Geophysical Research – Atmospheres*, **113**, D24107.

Tooth, S. (2008) Arid geomorphology: recent progress from an Earth System science perspective. *Progress in Physical Geography*, **32**, 81, DOI: 10.1177/0309133308089500.

Torres, O., Bhartia, P.K., Herman, J.R. *et al.* (2002) A long-term record of aerosol optical depth from TOMS observations and comparison to AERONET measurements. *Journal of Atmospheric Science*, **59**, 398–413, DOI: 10.1175/1520.

Tsoar, H. (2005) Sand dunes mobility and stability in relation to climate. *Physica A*, **357**, 50–56.

Tucker, G.E. and Hancock, G.R. (2010) Modelling landscape evolution. *Earth Surface Processes and Landforms*, **35**, 28–50.

Vermeer, M. and Rahmstorf, S. (2009) Global sea level linked to global temperature 2009. *Proceedings of the National Academy of Sciences of the United States of America*, **10**, 21527–21532.

Wang, X.M., Yang, Y., Dong, Z.B. and Zang, C. (2009) Responses of dune activity and desertification in China to global warming in the twenty-first century. *Global and Planetary Change*, **67**, 167–185.

Washington, R., Todd, M., Middleton, N.J. and Goudie A.S. (2003) Dust-storm source areas determined by the total ozone monitoring spectrometer and surface observations. *Annals of the Association of American Geographers*, **93**, 297–313.

Washington, R., Todd, Engelstaedter, S. *et al.* (2006) Dust and the low level circulation over the Bodélé depression, Chad: observations from BoDEx 2005. *Journal of Geophysical Research – Atmospheres*, **111** (D3), D03201.

Washington, R., Bouet, C., Cautenet, G. *et al.* (2009) Dust as a tipping element: the Bodele Depression, Chad. *Proceedings of the National Academy of Sciences of the United States of America*, **106**, 20564–20571.

Wetherald, R.T. (2009) Changes of variability in response to increasing greenhouse gases. Part II: Hydrology. *Journal of Climate*, **22**, 6089–6103.

Woodward, S., Roberts, D.L. and Betts R.A. (2005) A simulation of the effect of climate change-induced desertification on mineral dust aerosol. *Geophysical Research Letters*, **32** (18), L18810.

Yizhaq, H., Ashkenazy, Y. and Tsoar, H. (2007) Why do active and stabilized dunes coexist under the same climatic conditions? *Physical Review Letters*, **98**, 188001.

Yoshioka, M., Mahowald, N.M., Conley, A.J. *et al.* (2007) Impact of desert dust radiative forcing on Sahel precipitation: relative importance of dust compared to sea surface temperature variations, vegetation changes, and greenhouse gas warming. *Journal of Climate*, **20**, 1445–1467.

Index

Acasus Massif, Libya, 85, 86
Ader Doutchi Maggia, Niger, 572
aeolian
  abrasion *see* wind erosion
  accumulation, 342
  deposits (*see also* loess, sand *etc.*)
  bedforms, 427–48
  deposition, 497
  depression, 430, 432–5
  dunes *see* dunes
  dust *see* dust
  entrainment, 524–5
  erosion *see* wind erosion
  fines (*see also* dust), 191, 192
  landforms, 69–70, 75, 78–9
  landscapes, 427–448
  scales, 427–30
  scour pits, 223
  sediment flux, 475–8
  sediments, 379
  sequences, 23
aeolian processes, 391, 415–418, 455–79
  aerodynamic roughness, 457, 460–2
  entrainment (*see also* entrainment), 464–8
  hazards, 583
  roughness elements, 462–4
  shear stress, 456, 465, 464
  shear velocity, 456–60, 465–6, 471
  turbulent flow, 456, 471, 478
  vegetation effects, 462–4
  velocity profile, 456, 457
  windflow, 456
aeolian transport
  creep, 209, 471
  reptation, 472
  saltation, 466, 470, 472–3, 540, 559
  suspension, 471
aerodynamic instability, 430
aerodynamic roughness, 457, 588
Aerosol Robotic Network (AERONET), 522, 524
aerosols, 517–33
Africa, 4, 5, 8, 9, 10, 11, 12, 13, 21, 28, 33, 35, 38, 42, 43, 55, 56, 107, 212, 269, 272, 273, 274, 275, 276, 280, 284, 288, 289, 291, 293, 373, 455, 572
African Humid Period, 38
African plate, 277
Aglaonice Crater, Venus, 79
Aglaonice dunefield, Venus, 79
Agri Basin, Italy, 253
Aguas Basin, Spain, 217
aioun, 386, 388
air entrapment, 248
airflow, 494–7
  secondary, 497, 502
  separation, 497, 499
Alberta, Canada, 212, 216–7, 218, 220, 221
alcoves, 409
alcrete, 133
Alacante, Spain, 214–215, 246, 341
alfisols, 104
algae, 106, 111, 149, 386
Algeria, 139, 185, 436, 528, 600
Algodones, California, 435, 489, 502
alluvial fans, 35, 36, 150, 188, 289, 190, 199, 200, 304, 333–62, 379
  base level changes, 356
  channels, 333, 340, 344–5
  climate changes, 353–6, 358, 360–2
  debris flows, 339, 340, 362
  depositional processes, 340
  development, 339–350, 360–2
  dynamics, 351, 358
  morphology, 333, 334, 345
  morphometric data, 335, 348, 358
  occurrence, 333–4
  pediment supply to, 338
  sediments, 340
  styles, 345
  tectonic factors, 336, 345, 351–3
  water supply to, 333
Almeria, Spain, 217, 335, 343, 354
Alpujaras, Spain, 353
aluminium, 139, 140
alveoli *see* weathering features
Amarillo, Texas, 574
Amazon, 33, 35, 445, 518, 532,
Amazonia, 36, 442
Amazonian era, Mars, 64, 66, 69
American northwest, 269
American southwest, 272, 275, 284, 289
Amphitheatre head valleys, 97, 410

Arid Zone Geomorphology: Process, Form and Change in Drylands, Third Edition. Edited by David S. G. Thomas

Ampt equation, 248
Amu Darya, 574–5, 592
Anapodaris Gorge, Crete, 278
Anapodaris River, Crete, 278
Andean, 35
Andes, 186, 337, 553
anemometers, 459
Angola, 441, 603
animal
  burrowing, 196, 404
  pressures, 377, 378
Antarctica, 7, 28, 71, 139, 284, 601
Antelope Valley, California, 590
ants, 248
aquiclude, 404
aquifer, 404
Arabia, 9, 28, 38, 39, 54, 222, 373, 490, 509, 525, 526, 539, 546, 551, 601
Arabian Gulf, 135, 415
Arabian plate, 277
Arabian Sea, 33
Aradena Gorge, Crete, 278
Aral Sea, 373, 520, 574–5, 586, 590, 591
archic, 334
Arctic, 144
Ares Vallis, Mars, 67
Argentina, 37, 182, 378, 551, 574
argillic horizons, 104
arid environments
  age of, 8–9
  causes of, 8–9
  characteristics, 87
  diversity of, 53–59
  extent of, 3–16
  nature of, 3–16
arid zone
  ancient, 27–8
  contraction, 36–8
  dating fluctuations, 39
  definitions, 5–7
  distinctiveness, 4–5
  extension, 35–6
  extraterrestrial, 61–79
  terminology, 5–7
aridisols, 104
aridity indices, 7
Arizona, 21, 58, 197, 199, 239, 270, 301, 302, 304, 306, 307, 312, 314, 316, 326–7, 341, 377, 505, 594, 607
arroyos, 61, 578
Arroya de los Frijoles, New Mexico, 311, 315
artefacts, 163
Asia, 3, 4, 8, 9, 10, 28, 139, 269, 273, 289, 437, 532
Asir escarpment, Saudi Arabia, 304
Atacama Desert, 8, 9, 18, 19, 21, 42, 54, 61, 88, 95, 96, 98, 112, 131, 132–5, 137, 141, 163, 186, 195, 276, 337, 546, 548
Atacama flood, 593
ASTER, 389
Atlantic Ocean, 33, 102, 520, 532
atmospheric stability, 9
Augrabies Falls, South Africa, 284
Australasia, 9
Australia (*see also* Western Australia *etc.*), 4, 8, 9, 10, 11, 18, 19, 21, 28, 33, 36, 38, 42, 43, 53, 54, 55, 75, 120, 141, 143, 146, 150, 152, 154, 164, 193, 201, 222, 223, 252, 269, 272, 273, 275, 276, 284–8, 289, 290, 291, 373, 377, 386, 389, 392, 394, 395, 307, 311, 321, 337, 340, 406, 413, 428, 433, 441, 490, 509, 519, 520, 531, 553, 575, 589, 600
Australian boinkas, 68
AVHRR, 389
avulsion, 23
Ayers Rock, Australia, 223
Azafel Sand Sea, Mauritania, 490, 491, 510

bacteria, 147, 149, 162, 243, 244
Badain Joran Desert, China, 430, 494
badlands, 403, 212–22
  composite landscape, 213
  human activity, 219
  lithological control, 214–5
  processes, 219
  topography, 212
Badlands National Monument, South Dakota, 220
Badwater, Death Valley, 336
bajada, 333
ballard pillow structures, 417
Barbados, 524, 532
barchans *see* dune types
bare rock, 87
Barrier Ranges, Australia, 122, 310
bas-fonds, 412
basal sapping, 225
basalt, 94, 139, 186, 189, 197, 202–3, 280
base level changes, 356
Basin and Range region, USA, 335, 377
basin and range topography, 18, 21, 38, 57
basins (*see* pans, playas, salt lakes, etc.)
batholiths, 223
beach ridges, 188, 189
bedding planes, 417
bedload, 317–320
bedforms, 427–48
Benguela Current, 10
Bijou Creek, Colorado, 324
Bill Williams River, Arizona, 577
biodiversity loss, 592
biofilms, 87, 93
biogenic evidence, 29
biological crusts (*see* crusts)
bioturbation, 510
Bishop Ash, 23
bison, 575
Black Mountains, California, 23, 353
Blackrock Desert, Nevada, 373
blistering, 89
Bloemfontein, South Africa, 589
Bodélé Depression, Chad, 519, 524, 526–8, 532, 533, 540, 544, 546, 551, 552, 553, 606
BODEX, 524, 607
boinka, 68
bolis, 412
Bolivia, 135, 377, 526
bornhardts, 222
Borton, Chad, 539, 543

Botete River, Botswana, 157
Botswana, 36, 37, 40, 145, 147, 155, 156, 157, 280, 281, 290, 337, 386, 388, 391, 392, 406, 520, 526, 531
boundary
  layer, 456, 459, 461, 462, 473, 475, 478, 494, 506
  surfaces, 418
Bown's Canyon, Idaho, 410
Box Canyon, Idaho, 98
Brazil, 36
Brazilian Shield, 33
Brazos River, Texas, 578
Bristol, arid in so many ways, 239
brousse tigré, 575
bryophyte, 111
Bulgaria, 429
Bunday River, Australia, 286
bunds, 572
Burkino Faso, 242, 572

Cairo, 571
Cainozoic *see* Cenozoic
Calama, Chile, 593
calcium carbonate (*see also* calcretes, desert crusts), 29, 142, 292
calcrete, 89, 131, 132, 133, 137, 164, 182, 292, 341, 342, 356, 403
  brecciated, 142, 143, 150
  channel, 142, 150, 151
  characterisation, 141–3
  chemistry, 146–8
  classification, 142
  dating, 164
  distribution, 143–6
  groundwater, 142, 148, 150
  hardpans, 142, 143, 148
  laminar, 142, 143
  micromorphology, 146–8
  morphology, 143
  origin, 148–51
  palaeoenvironmental use, 164
  pedogenic, 142, 144, 148, 149, 163
  vadose, 142
caliche (*see* calcrete)
  alluvial, 133
California (*see also* Mojave Desert etc), 21, 23, 37, 57, 58, 196, 308, 315, 335, 338, 343, 348, 349, 353, 489, 518, 521, 550, 556, 560, 576, 586, 590
Cambrian, 28
Canada, 57, 212, 216, 586
Canadian Northwest Territories, 7
Canadian prairies, 9
Canyonlands National Park, Utah, 418
canyons, 405
Cape Coast, South Africa, 152, 153, 154
capillary rise, 87, 140, 149
Caprivi Strip, Namibia, 441
carbon
  cycling, 102
  dioxide, atmospheric content, 28, 39
  storage, 105
carbonate
  dissolution, 403
  enrichment, 104
Carboniferous, 418
Caribbean, 518, 520
Carrascoy, Spain, 354
Cascade Range, USA, 57
case hardening, 89
Caspian Sea, 137, 373
Cassini Mission, Titan, 75
Catalina Mountains, Arizona, 304
catchments, 302
cavernous weathering, 94
Cedar Mesa Sandstone, 417
Cenozoic, 18, 19, 28, 39, 273, 276, 277, 280, 282, 285, 377
Chad, 38, 40, 432, 519, 532, 533, 540, 544, 546, 553, 572, 606
chalcedony, 154
channel
  characteristics, 274
  fill, 311
  geometry, 309
  morphology, 309–10
Channel Country, Australia, 286, 288
chasmoendoliths, 113
checkdams, 572
checkerboards, 576, 585
chemical weathering, 93–4, 156,
chert, 22
Chezy equation, 250
Chihuahua, Mexico, 151
Chihuahuan Desert, 19, 57, 122, 296
Chile, 96, 130, 137, 212, 276, 406, 593
China, 9, 30, 120, 182, 183, 212, 219, 289, 357, 377, 429, 430, 455, 494, 518, 521, 530–531, 532, 541, 547, 551, 553, 575, 584, 585, 589, 604–5
Chinese deserts, 9
Chott el Djerid, Tunisia, 376, 387, 388, 390
chotts, 376, 389
Chryse Planitia, Mars, 66, 67,
Cima volcanic field, California, 190, 199, 201–202
clay
  dispersal, 109
  lunettes (*see* dunes, lunette)
  pellets, 392
  playa, 374
clays, swelling, 244
climate change, 17, 29, 40, 64, 353–6, 358, 360–2, 385–6, 432, 435, 532, 563, 599–608
  and geomorphology, 599–608
climate gradient, 101
climate variability, 10–1, 39–41, 211, 217
CLORPT framework, 89
Coachella Valley, California, 562
colluvial mantles, 36
closed basins *see* pans
colluvium, 139, 277
Colorado, 36, 101, 290, 325, 587
Colorado Plateau, 57, 223, 275, 406, 407–8, 409, 411–2
Colorado River, 22, 58, 269, 270, 314, 407, 577
Columbia Hills, Mars, 62, 67, 68, 69
conservation tillage, 588–9
continental margins, 18, 19, 20, 275
continentality, 9
Cooper Creek, Australia, 272, 284, 288, 307, 310, 532
cosmogenic nuclide dating, 17, 19, 20, 21, 22, 23, 91
Coso Range, 22

Cowhole Mountain, California, 191, 560, 563
Coyote River, California, 576
craters, 61–79
cratons, 18, 19, 20, 275
creosote bush, 200, 244, 246, 249
Cretaceous, 8, 18, 28
Crete, 278–9, 293
creep *see* aeolian transport
critical power relationships, 339
crusts (on dunes, calcrete, silcrete, ferricrete, *see separate entry*), 221, 227, 242, 468
  biological, 104, 111, 121, 226, 227, 237, 242, 244
  classification, 115
  halite, 135–137, 403
  rugose, 115
  smooth, 115
cryptoendoliths, 113
Cubango River, Angola, 41
Curtin Springs, Australia, 414
cut and fill, 213
cyanobacteria, 112, 114, 115, 120, 149, 162, 226, 243

Dakar, 522
Daklar, Egypt, 553
dambos, 412
dams, 576
Dansgaard-Oeschger event, 43
Darcy-Weisbach equation, 250, 251
Darcy's Law, 240
Darwin region, Australia, 373
datalogging, 90
Dating (*see also under specific dating methods / families: luminescence, radiocarbon, etc.*), 20–1, 34
Dautsa Ridge, Botswana, 377
dead camel, 431
Dead Sea, 23, 137, 188, 200, 277, 278, 314, 321, 352, 357, 373, 386
Death Valley, 22, 23, 36, 57, 58, 335–7, 343, 348, 349, 350, 353, 355, 358, 360
debris flows, 339, 340, 362
dedos, 556
deep sea sediments, 8
deep weathering, 412–3
deflation *see* wind erosion
deflation hollows (*see also* pans), 539
depositional
  records, 21–3
  seals, 109
demoiselles, 556
detachment, 23, 254
desert pavements, 22–23, 105–6, 162, 181, 182, 185, 228, 242, 337, 340–2, 575
  characteristics, 189–90
  clast orientation, 195
  clast reduction, 191
  pitting, 195
  regional differences, 195
  rubification, 196
  soils, 191–195
desert pavement formation, 185–9, 190–1
  aeolian aggradation, 185, 188–9
  concentration, 185, 186–7
  deflation, 185, 186
  upward migration, 185, 187–8
desert rose, 139
desert varnish (*see* rock varnish)
desiccation, 13
Devon, England, 322
Devonian, 27, 28, 320, 324
dew, 243
Diamantina River, 532
diatoms, 162, 394
diatomite, 527
Didwana, India, 30
Digital Elevation Models (DEMs), 209
Dinosaur Provincial Park, Alberta, 212, 216–7
Disequilibrium Index, 379, 380
Dixie Valley, Nevada, 353
Djenné-Sofarra, Mali, 572
dolerite, 282, 283
dolocrete, 131, 148
donga, 283
draa, 430
drainage patters, 21–22
Drakensberg Mountains, South Africa, 90
Drôme, France, 214, 220
drought, 587
Dryas cooling event, 42
drylands *see* arid environments
Dubai, UAE, 571
Dunaliella salivia 135
dune
  accretion, 497, 502
  airflow, 494–8
  alignment, 508–9
  avalanche face, 490, 497
  classification, 487–8
  crest, 490, 492, 493, 496
  dynamics, 497
  encroachment, 13
  erosion, 293
  extension, 494, 502
  formation, 493, 502–9
  initiation, 502
  migration, 490, 494, 497, 501
  mobility index, 603
  morphology, 487–8, 490, 503–9
  nuclei, 431
  patterns, 429–30, 487, 509, 510–11
  plinth, 492, 493
  process-form relationship, 488
  reactivation, 13, 583, 604–605
  relict status, 442
  sand, 139, 146
  size and spacing, 490, 505–508, 511
  slip face, 492
  stoss slope, 458, 495, 496
  vegetation, 505
  windward slope, 495–6, 497, 500
dune fields, 85, 269, 285, 293, 415, 440–2, 518, 539
dunes, 27, 36, 70
  barchan, 36, 65, 432, 435, 488–90, 497, 501, 502, 503
  complex, 429–30, 490, 493
  compound, 429–30, 490, 493, 502

crescentic, 488–90, 494, 502, 504, 505, 506
linear, 434, 490–5, 497, 498, 501, 502, 503, 504, 506, 510, 512, 584
longitudinal, 377
lunette, 38, 378, 379, 392–5, 434
mega, 429, 430
megabarchan, 489
nebkha, 391
obstacle, 494
parabolic, 36, 494, 495
seif, 490
simple, 429
source bordering, 433
star, 493, 494–497, 502, 504, 506
subaqueous, 508
transverse, 499, 500, 584
zibar, 435, 494
duricrusts (*see* silcrete, ferricrete etc)
dust, 33, 104, 119, 122, 141, 148, 159–60, 190, 193, 342, 391, 455, 469, 517–33, 553, 559–60, 583, 586–93, 605–7
and pans/playas, 519, 520
characteristics, 517
deposition rates, 520–1
dynamics, 455
entrainment, 524
hazard, 586–7
measurement, 522–4
modelling, 524–5, 605–7
sinks, 520–522
sources, 518–20, 526
storms, 13, 68, 575, 590
transport, 102
Dust and Biomass Experiment (DABEX), 522
Dust Bowl, 455, 520, 574, 586, 587–8
Dust Outflow and Deposition to the Ocean Experiment (DODO), 522
dykes, 412

Earth, 7, 13, 17, 61–4, 66, 70–3, 75, 76, 77
Earth System Science, 428
earthquakes, 197
East Africa, 42, 223, 284
East Africa plateau, 269
East African Rift, 21, 22, 131, 282
East Fork River, Wyoming, 319
Eastern Desert, Egypt, 270
Eastern Europe, 429
Eberswalde crater, Mars, 65, 72
Ebro Basin Spain, 4, 220
Ebro Depression, Spain, 552
ecohydrology, 198
ecological evidence, 38–9
ecosystems, 5, 10–12
edaphophytes, 113
Edom Mountains, Jordan, 572
Egypt, 3, 97, 139, 185, 188, 193, 270, 341, 378, 406, 432, 435, 541, 543, 544, 550, 553, 576, 584
Elat, Israel, 363
Elephant Butte, Colorado, 325
El Niño, 10, 88
endoliths, 113
Endurance Crater, Mars, 69
England, 5
Ennedi, Chad, 553
ENSO, 120, 520, 532
entisol, 101, 104
entrainment, 464–8
crusts, 468
moisture effects, 469
slope effects, 469
environmental cabinets, 91
Environmental Scanning Electron microscopy (*also see* SEM), 91
ENVISAT, 522
epedaphs, 113
ephemeral rivers, 301–27
ephemeral lakes (*see also* pans), 589
epiliths, 113
Erebus crater, Mars, 67
Erg
Aoukâr, Algeria, 432
Aklé, Algeria, 432
Azouad, Mali, 432
Azouak, Algeria, 436
Bilma, Niger, 436
Cabriscio, Libya, 436
Cayor, Mauritania, 432
Chech-Adrar, Mauritania, 432
Djourab, Chad, 432
El Mréyé, Mauritania, 432
Fachi Bilma, Niger, 492, 493
Foch, Chad, 432
Kanem, Libya/Niger, 432
Occidental, Algeria, 437
Oriental, Algeria, 432
Rebiana, Libya, 436
Ténéré Niger/Chad, 432, 436
Timbouctou, Niger, 432
Trarza, Mauritania, 432
ergs, 415, 429, 539
erodability, 455
erosion, aeolian *see* aeolian erosion
erosion, 18–20, 301
accelerated, 574
cycle of, 4
hillslope, 253, 256–8
interill, 107
erosional records, 21–3
erosivity, 455
Escalante River, Colorado, 408, 409
Ethiopia, 19, 54, 269
Ethiopian Highlands, 269
Etosha, Namibia, 377, 389, 520, 525, 526
Eucalyptus camaldulensis, 285
Euphrates River, Iraq, 269
Eurasian plate, 277
Europe, 9, 274, 289, 429, 518
European colonisation, 284, 288
evaporites (*see also* desert crust), 27, 28, 30, 152, 518, 519
evaporation, 93, 149
potential, 374
evapotranspiration, 5, 9, 149
potential, 7
Evinos River, Greece, 277
exfiltration flow, 237
exfiltrating water, 404–5
exfoliation, 223

Explorer Canyon, 409, 410
extraterrestrial arid zones, 61–79

fadamas, 412
fan development, 23
ferricrete, 131, 132, 133, 292
Finke River, Australia, 285
fire, 246
fission track dating, 17, 20, 21
flaking, 89
flash floods, 302, 304, 305, 308, 322
Flinders Range, Australia, 43, 185, 337
Florida, 406
fluvial (*see also* river)
  active unit (FAU), 292
  hazards, 593–4
  pedogenic unit (FPU), 291–2
  processes, 223, 340
  systems, 21–22, 71–2, 75–6
  terraces, 188
fylsch, 277, 279
flood patterns, 291, 326
floodplains, 518, 594
flow hydraulics, 250
fluid system, 606–7
fog, 88, 96, 135, 232, 244
Fortuna–Meshkenet dune field, 77
Fowler's Creek, New South Wales, 310, 312
Fowler's Gap, Australia, 122
Fram crater, Mars, 70
France, 153, 214, 217, 218, 220
Free State, South Africa, 282
freeze thaw, 184
frontal rain, 303
frost heave, 196
fungi, 162, 226, 586

Ganab, Namibia, 88
Gaub Valley, Namibia, 405, 593
Gauberite Lake, Australia, 414
GCMs, 599
geochemical pathways, 383
geochemistry, 382–5
geomonotony, 55–7
geomorphological hazards, 583–94
  aeolian, 583–93
  fluvial, 593–4
gerbils, 529
Germany, 429
gibber, 181, 185, 201, 340
Gila River, Arizona, 326
Gilf Kibir, Libya, 404
glacial, 338
  cycles, 18, 71, 285
  times, 41–2
glacial equals pluvial hypothesis, 39
glaciations, 277
glacis, 139
Glen Canyon Dam, USA, 577
global warming, 10, 583, 600–2
Gobi Desert, 9, 181, 182–3, 201, 521, 540
Gondwana, 28, 280, 285

Goodwin Creek, Mississippi, 318
Goudie, hotspot pioneer, 85
graben *see* rift systems
grainsize, 430
Gran Desierto, Mexico, 435, 502
Grand Canyon, Arizona, 270
granite, 21, 139
granular disintegration, 89
gravel bed, 319, 322
grazing, 540
Great Basin Desert, USA, 19, 22, 57, 58, 377, 526
Great Divide, 9
Great Dividing Range, Australia, 284
Great Escarpment, southern Africa, 280
Great Himalaya, India, 19
Great Karoo, South Africa, 19
Great Konya Lake, Turkey, 590
Great Plains, North America, 574, 587–8, 589
Great Salt Lake, USA, 58, 381
Great Salt Lake Desert, Western Australia, 415
Great Sand Sea, Egypt, 432
Great Sandy Desert, Australia, 9
Greece, 219, 277, 337, 350
Green equation, 258
ground penetrating radar, 89, 427, 502, 506, 512
groundwater, 89, 140, 141, 148, 150, 162, 225, 311, 364, 379, 381, 403–18, 590
  and aeolian processes, 415–8
  and pan and playa development, 413
  and sand sheets, 435
  and valley development, 404–15
  network characteristics, 409–10
  sapping, 97, 98
  seepage erosion, 404–12
  solution, 404–14
Gross Bedform Normal Approach (GBNA), 508–10
grus, 190
guano, 134
Gulf of Aden, 22
gullies, 212, 217
gully development, 13
gullying, 218, 219, 221
gum trees, 285
Gusev Crater, Mars, 62, 67, 68, 69, 70
gypcrete (*see also* gypsum crusts), 89, 133, 137–41, 164, 292
gypsum crusts, 133, 137–41, 164, 403, 417
  chemistry, 140
  distribution, 139
  formation, 140–1
  micromorphology, 140
  pedogenic, 140

Hadley Cell circulation, 42
Hardness tester, 90
haematite (*see also*) iron, 29
Haiwee spillway, California, 22
halite, 292
  crusts (*see* crusts), 403
halolites, 136, 164
haloturbatoin, 386
hamadas, 181, 187
Hanehai, Botswana, 145, 147

Harmattan wind, 552
Haute Alpes, France, 220
Hawaii, 11, 62
Hawaiian Islands, 405, 406, 412
headwalls, 405
headward erosion, 408
Heinrich event, 43
Henry Mountains, Utah, 212
Hesperian era, Mars, 64, 66, 68
High Plains, Texas, 414
Hiiraan, Somalia, 572
Himalayan Plateau, 528
Himalayan region, 19
Hoggar Plateau, 432
Holden North East Crater, Mars, 72
Holocene, 33, 35, 38, 42, 200, 202, 202, 279, 284, 288, 292, 321, 344, 358
Home Plane, Mars, 68
Hornby Bay Group, 7
Horseshoe Canyon, Utah, 412
Hortonian overland flow (*see* overland flow)
human
  activity, 219, 279
  populations in drylands, 5, 12
human impact
  on drylands, 571–9
  on rivers, 576–7, 578–9
  on sand dunes, 576
  on soils, 571–6
  on vegetation, 578
Humberstone, Chile, 134
humid
  environments, 86, 88, 301
  regions, 373
  tropics, 334
  zone, 102, 131, 153
Husband Hill, Mars, 68, 69, 70
Huygens probe, Titan, 72, 73, 74, 75
hydration, 93
hydrocarbons, 27
hydrograph, 303
hydrological connectivity, 239, 252–3
hydrology
  ephemeral streams, 303–4
  transmission losses, 304, 309
hydrolysis, 93
hypoliths, 113
hyrax middens, 39

Iberia, 321
iButtons, 90
Ica Valley, Peru, 553
ice caps, 28
ice cores, 17
Idaho, 98, 410
Il Eriat, Kenya, 306
Il Kimere, Kenya, 317
impact seals, 106–9
India, 19, 30, 33, 36, 42, 43, 222, 321, 337, 437, 441, 576
Indian Ocean, 43
induration, 291–3
Indus River, 23, 325
inselbergs, 222
infiltration, 102–3, 110
  factors affecting, 241–8
  ponding, 248–50
  processes, 240–241
  rates, 122, 237
intercratonic, 18, 19, 20, 275
interglacial cycles, 18
intergranular flows, 408
intermontane basins, 333
interorogenic, 18, 19, 20, 21, 275
Intertropical Convergence Zone (ITCZ), 9, 42,
Inyo Mountains, California, 22
IPCC, 599
Iraq, 139, 269
Iran, 19, 139, 335, 394, 541, 543, 545, 546, 547, 548, 549–50
iron fertilisation, 102
iron films, 162–3
ironstone concretions, 68
irrigation, 573–4
isostatic uplift, 210
Israel, 23, 111, 118, 139, 182, 185, 188, 189, 194, 195, 199, 201, 272 291, 292, 305, 306, 307, 311, 314, 316, 319, 320, 322, 324, 325, 337, 357, 386
Italy, 219, 221, 253, 337, 406

Jafurah Sand Sea, Saudi Arabia, 415, 416, 435
Japan, 212, 406
Jawa, Jordan, 572, 576
Jordan, 85, 139, 189, 192, 198, 386, 406, 572, 576
Jornado, New Mexico, 239
Judean Desert, Israel, 307
Junggar, China, 553
Jupiter, 62
Jurassic, 407, 413, 417, 433

Kalabagh fault zone, Pakistan, 23
Kalahari Desert, 4, 5, 6, 9, 10, 13, 28, 30, 36, 37, 40, 41, 54–9, 144, 151, 152, 154, 158, 282, 284, 373, 374, 376, 378, 388, 392, 413, 428, 433, 434, 437, 441, 446–8, 490, 491, 509, 512, 603–4
Kalahari Group of sediments, 6
Kalahari Sands, 41
Kalut, Iran, 549
Kang, Botswana, 155
Kansa, 587
kaolinite, 140
Kappakoola Swamp, Australia, 575
Kara Boyaz Gol, Turkmenistan, 590
Karakoram, Pakistan, 19
Karakum Desert, Turkmenistan, 437, 572
Karman-Prandtl velocity distribution, 457
Karoo, South Africa, 6, 19, 220
Karoo Supergroup, 282
karstic collapse, 378
Kazakhstan, 437, 575, 590, 591
Keetmanshoop, Namibia, 86
Kenya, 306, 308, 317, 324, 386
Kerman Basin, Iran, 394
Keulegan equation, 250
kintic energy, 106, 107
Klip River, South Africa, 272, 282, 283
Kohala Valley, Hawaii, 411

Kruger National Park, South Africa, 284
Kunlun Mountain, China, 530
Kuwait, 139, 543, 646
Kwando River, 41

Laayonne, Morocco, 489
lacustrine sediments, 342
lacustrine sequences, 22, 30
La Pacana Caldera, Atacama, 548, 551, 553
Laga Tulu Bor, Kenya, 309
lake basin studies, 40
lake basins, 57–8
lakes (*see* pans), 373
  ephemeral, 520
Lake Aral, 373
Lake Bonneville, USA, 38, 40, 58, 386
Lake Chad, 137, 432
Lake Estancia, New Mexico, 40
Lake Eyre, Australia, 37, 40, 192, 284, 307, 310, 373, 381, 391, 525, 532
Lake Lahontan, Nevada, 38, 40, 357, 377
Lake Lefroy, Western Australia, 138
Lake Lisan (Dead Sea), 200
Lake Mega Chad, 544
Lake Magadi, Kenya, 386
Lake Manly, California, 58
Lake McLeod, Western Australia, 137
Lake Ngami, Botswana, 38, 41, 377
Lake Turkana, 22
landslides, 217
Lanzhou, China, 455
Las Vegas (*where several geomorphologists have lost their wealth*), 137, 160, 363, 571
Last Glacial Maximum (LGM), 38, 39, 43, 440, 509, 524, 552
laterite, 152
lava flows, 162
leaching, 101, 194, 413
leaf drop, 238
Leon River, Texas, 576
Lesotho, 269, 280
Levant, 321
Libya, 37, 54, 63, 85, 86, 90, 182, 406, 436, 520
Libyan plateau, 201
lichen, 93, 94, 111, 112, 114, 119, 162, 243
limestone, 139
lithification, 292–3
lithological factors, 275
lithophytes, 113
Little Ice Age, 277
Llobregat River, Spain, 277
linear dunes *see* dune types
Liujiaxia, China, 577
Liwa, UAE, 490, 502
Lluta Valley, Chile, 96
loess (*see also* dust), 104, 146, 219, 427, 429, 431, 437–9, 521
  distribution, 431, 439
  dunes, 429
  glacial, 437
  peridesert, 437–8
  thickness, 437
Loess Plateau, China, 429, 437, 438
Lokwana Pan, Botswana, 378
Long Canyon, Colorado, 409
Longjangxia, China, 577
Lop Nor, China, 373, 547, 551, 552, 553, 575
Los Angeles, 590
Loot Desert *see* Lut Desert
Louth Crater, Mars, 71
low level jet, 509, 527, 553
Lubbock, Texas, 587
luminescence dating, 21, 23, 24, 30, 32, 39, 394, 442
  principles, 442–3
lunette dunes *see* dunes
Lut Desert, Iran, 541, 543, 545, 546, 547, 548, 549–50

macropores, 251, 404, 408
Mafikeng, South Africa, 56
Magellan missions, Venus, 76, 77
Maghreb, 321
Makgadikgadi Basin, Botswana, 40, 56, 377, 386, 389, 391, 520, 525, 526, 531
Makgadikgadi Pan, Botswana, 41
Makran, Iran, 19
Maktir Erg, Mauritania, 432
Mali, 528, 572, 600
man in shorts and socks, 388
Manning
  Equation, 250
  Roughness, 312
mammals
  larger, 246
  small and furry, 246
Mancos Shale Badlands, Utah, 221–2
marble, 139
marine sediments, 32–3
Mariner, 9 405
Mars, 13, 61, 62, 64–72, 77, 79, 186, 405, 406, 411, 455, 502
Marshall River, Australia, 272, 285, 287–8, 311, 313
Martian landing sites, 186
Martian valleys, 411
Maryland, 308
mass movement, 209, 217, 221
Massachusetts, 406
material hazards, 13
Matsap Pan, South Africa, 383
Mauritania, 432, 447, 490, 509, 510, 528, 600
Mediterranean, 12, 41, 161, 212, 219, 237, 246, 270, 272, 273, 274–80, 284, 285, 289, 337, 358–72, 601
Mediterranean Dust Experiment (MEDEX), 522
Mercury, 61
Meridiani Planum, Mars, 63, 67, 68, 69, 70
Meshkenet dunefield, Venus, 79
Meshkenet Tssera, Venus, 79
Mesopotamia, Egypt, 3
Mesozoic, 280, 285
Messara Plain, Crete, 278, 279
Messinian Salinity Crisis, 277
metasaturated flow, 435
Meteor Crater, Arizona, 377
meteorite impact, 414
methane cycle, Titon, 73–4, 76
methane sink, 102
Mexico, 9, 37, 57, 122, 502, 586
Miami aerosol group, 522, 523, 532

mica, 140
micrite, 146
micro-erosion meter, 91
microorganisms, 113, 135
microphytic plants, 112
microtopography, 106
Middle East, 144, 201, 274, 284, 289, 337, 342, 373, 571, 572, 573, 583
Mie Crater, Mars, 66, 67
mineral magnetics, 342
Miocene, 18, 22, 202, 277
Mississippi, 319, 587
Miyamoto crater, Mars, 65
Mobility Index, 441
modelling, 427
Modern analogue approach, 27
MODIS (*also see* remote sensing), 389, 522
moisture sources, 269
molasse, 277
Molepolole, Botswana, 56
mollisols, 104
Molopo River, Botswana, 56
moisture availability
  indices of, 7
  role of, 12
Mojave Desert, California, 9, 19, 21, 22, 57, 121, 122, 184, 185, 187, 189, 191, 194, 195, 198, 202–3, 432, 547, 555, 558, 563, 574, 586
Mojave River, California, 557 22
Molopo River, 433
Mongolia, 182, 532, 543, 550
Moon, 62, 63
Morocco, 132, 406, 489
moss, 112, 114, 119
Mossi, Burkino Faso, 572
Mozambique, 280
Mu Us Desert, China, 437
Mud drapes, 324
Mugatse Pan, Botswana, 378
Murcia, Spain, 246, 335, 349, 350, 352, 354, 357
Murray Darling system, Western Australia, 284, 288
Musandum, UAE/Oman, 360

nabkha (nebkha) *see* dunes
Nahal Eshlemoa, Israel, 306, 314, 318, 326
Nahal Hebron, Israel, 317, 319
Nahal Hoga, Israel, 311
Nahal Og, Israel, 307
Nahal Yatir, Israel, 318, 319, 320
Nahal Yeal, Israel, 305, 306
Nahal Zin, Israel, 322, 324
Namaqualand, South Africa, 578
Namib Desert, 5, 6, 10, 17, 19, 32, 42, 54, 55, 88, 94, 96, 135, 137, 138, 139, 141, 405, 432, 433, 435, 462, 473, 489, 562
Namib Sand Sea, 372, 490, 491, 494, 501, 502, 505
Namibia, 6, 8, 32, 41, 86, 91, 96, 275, 280, 284, 289, 431, 447, 458, 459, 489, 490, 520, 525, 526, 531, 539, 593
Nanga Parbat, Pakistan, 19
Navaho Sandstone, 224, 408, 409
Navier-Stokes equations, 240
nebkha *see* dunes
Negev Desert, 18, 19, 22, 88, 101, 113, 137, 181, 185, 192, 201, 219, 292, 302, 311, 314, 316, 317, 318, 505, 572
Neogene, 157, 344
Neolithic, 510
neotectonics, 20
Nevada, 23, 40, 58, 95, 160, 335, 350, 353, 356, 357, 358, 361, 373, 394, 493, 521
New Hanehai, Botswana, 145
New Mexico, 40, 148, 239, 246, 291, 311, 315, 377, 414, 415, 416, 489, 507, 575, 587
New South Wales, 104, 107, 116, 120, 122, 132, 574
New Zealand, 19, 252, 408
Niamey, Niger, 524
Niger, 36, 37, 111, 118, 242, 416, 492, 522, 572
Nigeria, 586
Nile River, 22, 35, 36, 269, 277
Nyl River, South Africa, 292
Noachian era, Mars, 66, 68
non equilibrium systems, 88
non linear systems, 89
Norfolk, UK, 144
North Africa, 139, 144, 161, 337, 386, 413, 436, 528, 541, 552, 573, 574
North America, 4, 9, 23, 55, 57, 161, 182, 189, 200, 201, 212, 289, 309, 434, 520, 525, 571
North Atlantic, 524
North Dakota, 213, 219
North Yorkshire, UK, 144
Northern Cape, 13
Northern Plains, Australia, 285, 288, 291
Nouakchott, Mauritania, 571
nutrient transport, 57

ocean
  currents, 10
  fertilisation, 517
  sediments, 17, 29
oceanic cores, 39
off loading, 225
off-road vehicles, 198, 520, 586
Okavango Delta, 6, 41, 57, 281, 282, 289, 332, 378
Okavango River, 280, 282, 290
Oklahoma, 587
Okwa River, Botswana, 56, 146
Old Wives Lake, Saskatchewan, 590
Oligocene, 8
Oman, 37, 136, 185, 200, 201, 335, 337, 348, 350, 360, 416, 433, 434
onions, 223
Orange River, South Africa, 269, 272, 281, 282, 284
Ordos Plateau, China, 438
Oregon, 58, 562
Orogenic belts, 18, 19, 20
orogeny, 8
Osborn Wash, Arizona, 340
OSL dating (*see also* luminescence), 337, 394–5, 428, 446, 502, 506, 512
Ouddal, Chad, 572
overland flow, 102, 110, 115, 117, 237, 239, 308
  Hortonian, 237
  saturation, 237
Owens Lake, California, 187, 391, 393, 518, 520, 525, 554, 575, 583, 586, 590–1
Owens River, California, 22, 58
Owens Valley, California, 22, 186

Pacific Ocean, 135, 518
Pakistan, 19, 574
palaeodune sequences, 455
Palaeogene, 157
Palaeolake Mega Chad, 528
palaeosurfaces, 21
Palomas Basin, Mexico, 151
Pampa Blanca, Chile, 133
Pampa La Joya, Peru, 489
Pampa Lina, Chile, 133
Pampas (not nappies), 378, 574
Panamint Valley, California, 188, 336, 348, 353
pans, 373–95, 403, 413, 598
  aeolian processes in, 389
  as aggradational features, 378
  as palaeoenvironmental indicators, 394–6
  characteristics, 374–6
  cycle, 386–7
  definition, 374
  density, 373
  erosional controls on, 378
  geochemistry, 382–5
  groundwater role, 379–82
  human inmpact on, 385
  hydrology, 379–381
  nature and occurrence, 373–4
  nomenclature, 374
  origins, 376
  surface features, 386–9
  terminology, 374
particle size, 427, 429
pastoralism, 101
Patagonia, 11, 19, 125
Patagonian Desert, 19
Payun Matru, Argentina, 551, 552
peat, 293
pediments, 18, 21, 22, 36, 139
pedogenic features, 292
Penman's Formula, 7
periglacial, 217
periglaciation, 71, 277
Permian, 28, 414, 417
Peru, 276, 489, 539, 553
Peski Karakumy, Kazakhstan, 437
Peski Kyzlkum, Kazakhstan, 437
petrocalcic horizons, 104, 541, 546, 550, 553
pipes, 238, 251
piping, 378, 403, 404
Phanerozoic, 18
Philip equation, 248
Phoebe Regio, Venus, 62, 77
Pheonix, Arizona, 571
Phoenix mission, 68, 71
photosynthesis, 244
Phyllocian era, Mars, 64
phytolith assemblages, 39
phreatic zone, 403
piezometric surface, 404
piping, 216, 219
Pisgah Crater, California, 560
Pisgah volcanic field, California, 201–2
plains (*see* pans), 373
plastic flow, 340
plate margins, 18
plate tectonics, 17
playas (*see also* pans), 22, 30, 75, 104, 137, 141, 150, 189, 285, 269, 373–95, 518, 519
Pleistocene, 33, 35, 38, 42, 58, 181, 212, 292, 338, 344, 353
Plenty River, Australia, 287–8, 311, 313
plinth, 492
Pliocene, 28, 437
Plio-Pleistocene, 324, 344
pluvial lakes, 337
phytoliths, 162
phytoplankton, 102
Pretoria Salt Pan, South Africa, 377
poikilohydric, 112
pollen, 162, 394
popcorn surface, 217
Prairie region, North America, 57
Precambrian, 7, 27, 324
precipitation
  snow, 8
Proterozoic, 27
Providence Mountains, California, 195, 201
puddingstone, 151
Puna Plateau, Argentina, 182

Qattara Depression, Egypt, 97, 377, 394
Qaidam Basin, China, 30, 383, 541, 542, 551, 553
Qinghai-Tibet Railway, 585
Qoz, Sudan, 432
quartz, 148
Quatal Creek, California, 315
Quaternary, 5, 6, 17, 18, 23, 24, 27, 28, 29, 30, 31, 32, 33, 34, 35, 38, 39, 44, 53, 56, 57, 58, 96, 102, 105, 277, 279, 280, 282, 285, 291, 336, 337, 338, 353, 358–67, 377, 427, 437, 439–43, 521, 603

radiocarbon dating, 20, 21, 342, 394
Rajasthan, India, 139
rain
  events, 309
  frontal, 303
  gauges, 309
  intensity, 237
raindrop
  characteristics, 106–7
  impact, 102, 106–7, 186, 255, 469
rainsplash, 186
Rajasthan, India, 576, 590
Rambla de la Sierra, Spain, 353
Rambla de Tabernas, Spain, 354
Ras Al-Khaimah, UAE, 584
recent vertical crustal movements (RVCMs), 20
Red Mountain, Arizona, 93
Red Sea, 22, 304
reg, 181, 185, 340, 435
relative dating, 200
remote sensing, 10, 388, 427,
  AERONET, 522, 524, 526
  Sea WIFS, 523
  Meteosat, 523
  MODIS, 522, 523, 526

SEVIRI, 522, 523
TOMS, 520, 522, 523, 525, 528, 529, 530
Repetek, Turkmenistan, 576
reptation *see* aeolian transport
Reynold's number, 250, 251
Reynold's sheer stress, 458–9
Rhea Mons, Venus, 77
rhizoliths, 417
Richard's equation, 240, 250
Richardson number, 458
rift
systems, 18
zones, 18, 57
rills, 211, 248
Rio Grande, Colorado, 290
Rio Loa, Chile, 593
Rio Puerco, New Mexico, 311
ripples, 224, 429–30, 435, 473–4, 509
aerodynamics, 430
ballistic, 473–4
fluid drag, 429, 473
granule, 473
impact, 430
mega, 429, 473
normal, 429
river style, 273
river systems, 269–94
allogenic, 269
characteristics, 271, 290
deposits, 320–304
distinctiveness, 269–73, 289
diversity, 273–94
endogenic, 269, 273, 276, 291, 294
exogenic, 269, 273, 274, 276, 280, 290, 291, 294
exoreic, 269, 277, 284, 294
zones of, 290
rivers
gorge bound, 278, 279
behavioural tendencies, 281
Riyadh, Saudi Arabia, 571
rock
breakdown processes, 69, 75, 77, 78
coatings (*see also* rock varnish), 158–163
fall, 89, 97
rock pavements, 546 *see also desert pavements*
rock varnish, 34–5, 159–62, 195, 336, 342, 545
Rock Mass Strength classification, 222
Rocky Mountains, 9, 57, 269
rodents, 200
Rogers Lake, USA, 543, 548, 560
roots, 147, 246, 247, 404
roughness, 251, 253, 462
Rub' Al Khali, Saudi Arabia, 416, 437, 493, 494
rugose crests, 115
runoff, 197–8, 214, 216
discontinuity, 239
generation, 237–51
runon, 110

sabkhas, 30, 136, 374, 375, 575
Sadd al-Kafara, Egypt, 576
Safford Valley, Arizona, 326–7
Sahara Desert, 4, 9, 10, 19, 27, 28, 33, 35, 36, 37, 42, 53, 54, 63, 88, 182, 185, 201, 428, 430, 448, 490, 509, 520, 525, 539, 541, 543, 583, 600
Saharan Dust Experiment (SHADE), 524
Sahel, 10, 11, 12, 42, 432, 437, 527, 532, 575, 606
Sahel Great Drought, 10
Sahelian zone, 53
Salar del Carmen, Chile, 133
Salar del Lagunas, Chile, 133
salars, 133
salcrete, 131, 436
saline pan cycle, 387
Saline Valley, California, 137
salinization, 572, 574
salitre deposits, 132
salt, 185, 191, 194
crust, 388
diapirs, 18
lakes *see* pans, 373–95
precipitation, 408
tolerance, 573
Salt Lake City, 373
salt weathering *see* weathering
saltation *see* aeolian transport
saltcedar, 578
Salton Sea, California, 590
salts, 91, 92, 93, 94, 95, 97, 104
Sambhar Salt Lake, Rajasthan, 590
Samphire Lake, Australia, 414
San Juan Mountains, Colorado, 101
San Juan River, Colorado, 408
Sana'a, Yemen, 571
sand
bed, 322
dunes *see* dunes
fence, 584–5
ramps, 23
rose, 490
size range, 429
sheets, 429, 430, 433, 435–7, 494
supply, 504–5
traps, 477
transport pathways, 433
sand seas, 22, 23, 27, 35, 429, 430–5
development, 432–5
distribution, 432, 440–1
energy, 437
sand flow conditions, 432–5
sediment sources, 432–5
topographic control, 436
Sandover River, Australia, 286
Sandy Creek, New South Wales, 310
Saskatchewan, 590
Saturn, 72
Saudi Arabia, 139, 303, 304, 415, 416, 493, 575
scale, 338
scarp development, 224
Schmidt hammer, 91
Schoonspruit catchment, South Africa, 283
Scotland, 323
scour and fill, 311–7
screes, 37

sea level rise, 601, 602
Searles Lake, California, 22
seaspray, 148
sebkhas *see* sabkhas
Sebkha Kourzia, Tunisia, 575
sediment
  deposition, 20
  gauging, 19
  size, 518
  suspended, 311, 314–7
  transport, humid, 311–20
sedimentary records, 29–32
sedimentation, 18–20
Sedna Planita, Venus, 76
seepage erosion, 404–5
seif *see* dunes
Selima sand sheet, Egypt, 432, 435, 436
SEM, 408, 554, 560, 561
Senegal, 524
serir, 181, 182, 185
SEVIRI *see* remote sensing
Shamal wind, 509
sheer stress, 456, 492, 505
Sharm ash Shaykh, Egypt, 363
sheet flow, 199
shorelines, 394
Sibilot, Kenya, 309
Siderikian era, Mars, 64
Sierra de Alhamilla, Spain, 353
Sierra de Carrascoy, Spain, 335, 352
Sierra de los Filibres, Spain, 353
Sierra Nevada, California, 22, 57, 58, 182, 563, 590
silcrete, 182, 185, 131–3, 151–8, 292, 403
  chemistry, 153–156
  classification, 151–2
  evaporate, 132
  formation, 156
  general character, 151–2
  groundwater, 152, 156
  pedogenic, 152, 156, 158
  micromorphology, 153–6
  palaeoenvironmental use, 164
  pan, 158
silica glaze, 162–163
Silver Peak, Nevada, 493
Silver Lake, California, 22, 376
Simpson Desert, Australia, 9, 435, 437, 490, 505, 512
Sinai, 19, 341, 363, 494
Siwa Depression, Egypt, 378
Siwa Oasis, Egypt, 584
Skeleton Coast, Namibia, 137, 458, 459, 489, 490
slaking, 244
slick rock slopes, 222
slip face *see* dunes
slope streaks, 70–1
slope units, 209, 210
slopes (*see also* pediments), 36, 209–29
  badlands, 212–22
  geological contexts, 209–10
  lithological variability, 210
  rates, 225–6, 227–8
  rock slopes, 222–9
  vegetation control, 217–9
smectite, 140
Snake River, Idaho, 58
snow, 101
snowmelt, 269
sod busting, 587
Soda Lake, California, 22, 344, 388, 389
sodium nitrate, 132–4
soil, 185, 191–5, 199, 211, 521
  aggregates, 106
  cohesion, 255
  crusts (*see* also crusts), 103, 104, 111–3
  development, 104
  effect of fire, 246
  erosion, 13
  formation, 238
  horizonation, 247
  horizons, 102, 103, 104
  human impact on, 571
  hydrolic properties, 103
  lack of, 87
  moisture, 247, 249, 470
  mosaics, 123–4
  profile, 342
  taxonomy, 103–4
  seals, 105–6, 108–10, 244
  vesicular, 121–2
  washin layer, 108
Sola de Uyuni, Bolivia, 525, 526
Somalia, 37, 572
Sonoran Desert, 19, 57, 161, 185, 198, 241
South Africa, 6, 13, 54, 90, 156, 157, 219, 220, 272, 275, 281, 282, 283, 284, 290, 292, 377, 588
South America, 10, 42, 54, 139, 161, 269, 273, 289, 437, 572, 601
South Australia, 105, 155, 157
South Dakota, 220
southern Africa, 11, 273, 276, 280, 282, 283, 288, 289, 600, 603–4
Spain, 119, 144, 145, 151, 213, 213–215, 217, 219, 228, 244, 246, 277, 335, 337, 341, 344, 349, 350, 357, 358, 552, 583
Special Report on Emissions Scenarios (SRES), 599–600
speleothems, 37, 38, 39, 94
splash, 255–256
spring sapping, 225, 404
Spring Lake, Australia, 414
Spring Valley, Nevada, 40
stable isotopes, 394
stem flow, 238, 245
Stokes Surface, 415–6, 435
stone mantle, 105–6, 121
stone pavements, 227
stony tableland, 181, 187
storm characteristics, 302–3
strandlines, 37
Stratford, Texas, 589
streamers, 477
stringers, 430, 432, 434
stromatolites, 395
strontium isotopes, 148
Strzelecki Desert, Australia, 31, 448, 490, 494, 512
Stuart Creek, South Australia, 155
Sturt Stony Desert, 185

subsidence, 18–20
Sua Pan, Botswana, 56, 155, 392
substrate ramps, 223
subsurface pipes *see* pipes
Sudan, 12, 36, 432, 572
Sukumaland, Tanzania, 302
surface roughness, 456
surface wash, 185
suspension *see* aeolian transport
Swakopmund, Namibia, 88
Swaziland, 36, 280
Sya Darya, Uzbekistan/Tajikistan/Kazakhstan, 574–5, 592
Syria, 139

Tabernas, Spain, 119, 145, 212, 341, 353, 358
Tablazo de Ica, Peru/Chile, 541
Tacna Desert, Chile, 134
tafoni *see* weathering features
Taiwan, 586
Taklimakan Desert, 337, 525, 526, 530, 584
takyr, 572
talus, 226
  flat ions, 227–8
Tanque Verde Creek, Arizona, 306
Tanzania, 219, 302
Taram Basin, China, 525, 526, 530, 553, 584
Tarapaca, Chile, 134
tectonic
  disruption of fluvial systems, 23
  fan development, 23
  frameworks, 17–24
  plate movements, 8
  processes, 17
  settings, 17–24, 57, 275
  tellurian, 76
tectonics, 209
Tehran, Iran, 571
temperate region, 374
Tengger Desert, China, 576, 585
tephrachronology, 165
termites, 124, 248
terraces, 572
Terres Noires, 217
Tertiary, 8, 17, 19, 28, 279, 345
Texas, 373, 374, 377, 378, 391, 587, 589
Thar Desert, 30, 43
Tharsis, Mars, 71
Theia Mons, Venus, 77
Theiikian era, Mars, 64
thermochronology, 17, 19, 21
thermoluminescence dating, 21, 394
Thin beds, 321
Thring Bore, Australia, 287
Tianshan Mountain, China, 530
Tibesti Mountains, Chad/Niger, 541, 544
Tibesti region, Chad, 544, 545, 553
Tibetan Plateau, 28
Tigris River, Iraq, 269
titanium, 153
Titon, 61, 61, 64, 72–79
Todd River, Australia, 319
TOMS *see* remote sensing
topography, 9
Torres Fan (not admirer of Spain and Liverpool football player), 357
Trade winds, 509
transmission losses *see* hydrology
Triassic, 28, 321, 323, 324
Triton, 62
Tsodilo Hills, Botswana, 56
Tshabong, Botswana, 56
tufa, 94, 386
tuff, 557
Tumas River, Namibia, 138
Tunisia, 94, 139, 141, 376, 386, 387, 388, 392, 572, 575, 585
Turkey, 269, 337, 590
Turkish Steppes, 9
Turkmenistan, 9, 572, 576, 590, 591
Tuscany, 220

UAE, 185, 490, 493, 502, 584
UK, 406
Ulan Buh Desert, China, 577
Um al-Riman, Kuwait, 546, 551
Um as Salim, Oman, 136
ungulates, 575
uniformitarian approach, 27, 30,
United Arab Emirates (UAE), 30, 31, 337, 350, 356, 358, 364
United Nations, 5
uplift, 18–20
uranium series dating, 21, 342
urbanisation, 577–8
Urwi Pan, Botswana, 388
US Soil Conservation Service, 574
USA, 4, 10, 11, 12, 21, 22, 35, 36, 38, 40, 41, 53, 57, 62, 101, 120, 220, 222, 223, 228, 237, 270, 290, 301, 310, 340, 373, 377, 393, 394, 410, 425, 427, 445, 539, 543, 578, 586, 593, 601
Utah, 58, 85, 95, 196, 221, 381, 412, 417, 418
Utopia Planitia, Mars, 66, 67
Uyuni, Bolivia, 373
Uzbekistan, 437, 575, 590

valley fever, 586
varnish coatings, 21
varnished pavements, 182, 183, 190, 200
Vasitas borealis, Mars, 67, 69
Ventespuit catchment, South Africa, 283
ventifacts, 539, 540, 553
  formation, 553–9
  wind direction, 563
vegetation, 276, 293, 311, 435, 585
  canopy, 103
  cover, 10–13, 198
  in drylands, 244, 356
  lack of cover, 102
  mosaic, 122
  non equilibrium systems, 88
  patchy, 237
  patterns, 212
ventifacts, 183, 186
Ventura Valley, California, 23
Venus, 61, 62, 64, 76–9
vertisols, 104
vesicular horizon, 191–3
Victoria Falls, 56

Viking mission, 66–8, 70
Virgin Lands Scheme, former USSR, 574
viscous flows, 339
vleis, 412
volcanic
  activity, 18
  ash, 21, 22, 149
volcanism, 413
von Karman's Constant, 458

Wadi Al Bih, Oman/UAE, 335, 348, 356, 360, 361
Wadi Rum, Jordan, 85
wadi sediments, 28, 433
Wadi Wahraine, Algeria, 304
Wadi Yiba, Saudi Arabia, 303, 304
wadis, 85
Wahibi Sands, Oman, 433, 434, 490
Walnut Gulch, Arizona, 197, 259, 241, 251, 302, 305, 306, 312, 314, 316
Warburton Creek, Australia, 286
Warrego Vallis, Mars, 65, 66, 72
Washington State, 587
water
  harvesting, 571–2
  hazards, 13
  lain deposits, 29
  not a dryland, 212
  repellency, 244
  table, 142, 148, 150, 519, 546
water cycle
  Earth, 66
  Martian, 66
weapons testing in deserts, 571
weathering, 85–98, 209, 218, 404, 518
  aeolian abrasions, 437
  deep, 412–3
  etching, 556
  effects, 89
  rates, 97
  systems, 87, 79
weathering features
  alveoli, 85, 95, 96
  cavernous, 95, 96
  landforms, 95
  tafoni, 85, 95, 96
weathering processes
  chemical, 93–4, 550
  corrosion, 412
  exfoliation, 222
  hydration, 93
  hydrology, 93
  insolation, 92, 93
  lichen, 93, 94
  mechanical, 408
  salt, 13, 91, 92, 93, 94, 95, 97, 550, 583
  solution, 550
weathering study methods, 90–2
  cabinets, 91
  ESEM, 91
  GPR, 91
  laboratory, 91
  laser scanning, 91
  micro erosion meters, 91
  non-destructive, 91
  resistivity tests, 91
  Schmidt hammer, 91
West Africa, 38
Western Australia, 125, 135, 137, 138
Western Cape, South Africa, 153
Western Desert, Egypt, 541, 543, 544, 551, 553
wet aeolian systems, 416–7
Whitesands, New Mexico, 415, 416, 436, 488, 490, 494, 507, 511
Williams Point, Australia, 37
wind erosion, 111, 120, 121, 221, 301, 497, 539–64, 574–5, 586–9
  abrasion, 437, 540, 548, 560
  agricultural, 587–9
  and vegetation, 462–4
  deflation, 415–8, 541
  equation, 589
  flutes, 555
  inverted topography, 553
  pits, 554–5, 557
  processes, 540–1
  sand and dust, 559
  rates, 561–2
  rock wedging, 540
wind
  breaks, 587
  hazards, 13
  mobilisation of sediments (*see* aeolian processes), 455–79
  regions, 503–504
  transport, 140 (*see also* aeolian)
  tunnel experiments, 184, 470, 475, 497, 562, 588
  velocity, 467
windflow characteristics, 456
Woodforde River, Australia, 286
World War I, 132
World War II, 198
Wyoming, USA, 140, 291, 319

Xanadu Regio, Titan, 62, 74

yardangs, 65, 79, 379, 394, 539
  definition, 541
  form, 541–4
  lithologies, 543–4
  processes, 545–51
  rates of formation, 559
Yellow River, China, 577
Yeman, 571
Younger Dryas, 510

Zagros, Iran, 19
Zagros Mountains, Iran, 335
Zaire, 442
Zambesi River, 41, 56
Zambia, 6, 41, 373, 441, 603
zeroplane displacement, 457, 463
zibar *see* dune types
Zimbabwe, 6, 41, 280, 412
Zin Badlands, Negev, 219
zoogeomorphology, 575
Zuni Salt Lake, New Mexico, 377
Zzyzx, California, 341, 344